CONTEMPORARY ORGANIC CHEMISTRY

CONTEMPORARY ORGANIC CHEMISTRY

STEPHEN J. WEININGER
Worcester Polytechnic Institute

HOLT, RINEHART and WINSTON, Inc.
New York Chicago San Francisco Atlanta
Dallas Montreal Toronto London Sydney

ACKNOWLEDGMENTS

Infrared and Nuclear Magnetic
Resonance Spectra courtesy of:

The Perkin-Elmer Corporation
Sadtler Research Laboratories
Varian Associates
Spectral Exercises in Structural Determination of Organic Compounds,
by Robert H. Shapiro; © 1969 by Holt, Rinehart and Winston, Inc., N.Y.

Gas Chromatogram courtesy of Applied Science Laboratories, Inc.

Library of Congress Catalog Card Number 74-171952

ISBN: 0-03-079670-9

Printed in the United States of America

2 3 4 5 071 9 8 7 6 5 4 3 2 1

Design by Lenni Schur
Illustrations by Felix Cooper

PREFACE

Contemporary Organic Chemistry was written not only in response to my own experiences in the lecture hall and the laboratory, its development was also prompted by a personal assessment of the present state and future direction of organic chemistry. I think that all teachers experience the influence of these two forces, and from that experience there arises a personal philosophy of teaching. This book has been written in order to give concrete expression to one such philosophy.

In my view, the most momentous problem confronting the instructor is the flood of new information generated by his colleagues. Every week new synthetic methods, new mechanistic information, and new techniques for structure determination present themselves to the harried reader of the chemical journals. Current indications are that this situation will worsen in the future.

Under the circumstances, I feel that it is pointless to attempt to introduce first-year students to all the "important" reactions. Our aim must be rather to elucidate, with the aid of a relatively small number of examples, those physicochemical forces which operate throughout the realm of organic chemistry, albeit in a profusion of guises. Furthermore, it is not enough to present the student with a vast number of reactions and then casually remark that these cases all exemplify the operation of the same set of underlying principles. We must actively *demonstrate* that it is possible to obtain a sound grasp of—a "feel" for—organic chemistry without being overwhelmed by the vast number of specific phenomena.

If these are the conditions that impose themselves on the teacher of organic chemistry, then they must also confront the textbook writer. It seems to me that attempts to write encyclopedic textbooks are pointless and foredoomed to failure. Our texts must be tailored to our pedagogic aims.

The tremendous upsurge of activity in the field of reaction mechanisms over the last four decades has been a great boon not only to the research chemist but to the teacher as well. The course of a very wide variety of reactions has been immensely clarified, and we have also developed insights which enable us to predict the outcome of as yet untried or unfamiliar reactions. Mechanistic studies have provided a logical framework that not only allows us to make sense of any given reaction but also suggests the underlying similarities among apparently unrelated reactions. Yet a heavy reliance on the mechanistic approach is not without its dangers for chemical pedagogy.

Perhaps I can best point out this pitfall by saying that we risk passing from a situation in which students memorize reactions to one in which they memorize mechanisms. Considering that sound scientific observations are always valid, while mechanisms are ever subject to revision or rejection, such a trend in chemical education would be a backward one. We can avoid this danger in part by referring back to those atomic and molecular properties which determine mechanistic behavior, and by making a more intensive effort to introduce thermodynamic and kinetic considerations into our discussions of organic chemistry. This has been the general approach which I have tried to adhere to in writing this text, and I welcome comments on the validity of this outlook and the degree to which the book is successful in promoting it.

Of necessity, some material has been "left out" of the book, although several topics not covered in the text are to be found in the problems. Whether this or that topic should be included or not seems to me an argument lacking definitive resolution; it is only important that the examples chosen give an adequate picture of the scope of organic chemistry. Although *Contemporary Organic Chemistry* is a relative lightweight by prevailing standards, it still contains a hefty body of material for a one-year course.

This book has not been primarily written with the idea of extending its use into a third term. My own experience has indicated that students require more detailed and sophisticated information in the third term than can be accommodated in a general textbook. Furthermore, it is my feeling that they should begin reading chemical periodicals as a matter of course. This textbook has been designed to introduce them to the periodical literature, and the supplementary readings at the end of the chapter have been selected to match the students' level. Thus, the early chapters contain references to the *Journal of Chemical Education* rather than to *Organic Reactions*.

One significant aspect in which *Contemporary Organic Chemistry* differs from many other organic chemistry textbooks is in the extent to which it incorporates biochemical topics into the general chemical discussion. Organic chemistry owes its birth as a separate discipline to man's curiosity about the chemical functioning of life processes. Subsequently, organic chemists turned their attention away from the animate world, but that dissociation is now being reversed. Biochemical systems are now viewed as chemical entities of great subtlety and complexity, but entities that are nonetheless constructed from familiar materials. As a result, organic chemists recognize that their theories and rationalizations will be put to the most comprehensive test when they are applied to biochemical phenomena.

No adequate picture of current organic chemistry can be given, then, without introducing students to an area which claims the attention and effort of so many organic chemists. In particular, I have been at pains to show that the insights gained from studies of nonenzymatic reactions may be of substantial help in understanding enzymatic processes. Hence, I felt it best to discuss the biochemical topics in concert with, rather than separately from, the discussion of the standard material.

Such virtues as are possessed by *Contemporary Organic Chemistry* cannot be ascribed to the author alone. In science, perhaps more than in any other endeavor, we are nourished by the collective effort of our colleagues and predecessors. A number of students, graduate and undergraduate, submitted to the concepts and philosophy advocated in this book. They also read and used the manuscript version and they were good enough to give me the benefit of their reactions. I must acknowledge in particular the help I have received from Ronald Greenberg and Wayne Turnblom.

Dr. Richard Balsley kindly gave me access to some of his results in advance of their publication. My former teacher, Professor Louis Sattler, gave me the benefit of his long experience as a teacher and as a carbohydrate chemist. He also helped foster in me the feeling that teaching organic chemistry was a calling worthy of one's best efforts.

Mr. John Barkham advised me on matters of style and organization, and did not stint on encouragement. The staff of Holt, Rinehart and Winston did their best to ease a difficult task. I have had the excellent editorial assistance of Dan Serebrakian, and Don Schumacher has marshalled all kinds of organizational support.

A number of my colleagues at Worcester Polytechnic Institute have aided and advised me. Professors Alfred Scala and David Todd were particularly selfless in suggesting problems, commenting on various ways of approaching the subject matter, and generally providing a thoughtful critique of the entire enterprise. Several teachers of organic chemistry outside of WPI read most or all of the manuscript and made a number of acute comments and observations, which have resulted in a substantial strengthening of the book. Professors Dann Sargent and Richard Schowen were of the greatest assistance here, while Professors Grace Borowitz and Elizabeth Thornton were also generous with their help.

I would never have even contemplated starting this book had I not been assured of the active collaboration of Professor Edward Thornton. A former teacher and continuing mentor and friend, he has been instrumental in helping me to formulate my own educational philosophy. He has made a large number of specific and useful suggestions on a variety of topics, but his greatest contribution has been in the example he has provided as a teacher and as a scientist.

This project came to fruition only because of the sacrifice and support of my entire family. It would be difficult to overvalue the assistance given me by my wife Jennifer. Aside from undertaking innumerable onerous tasks, she has sustained the entire project by her faith in its value. *Contemporary Organic Chemistry* is dedicated to her with great gratitude and affection.

Worcester, Mass. S.J.W.
December 1971

CONTENTS

CONTEMPORARY
ORGANIC
CHEMISTRY

CHAPTER 1

ORIGINS AND STRUCTURE OF ORGANIC CHEMISTRY

The division of chemistry into organic and inorganic domains has its origins in historical circumstances. Over the centuries mankind had become familiar with various substances that had been created by the direct action of living organisms. Common examples would be ethyl alcohol and acetic acid; these substances had, in fact, been obtained in a reasonably pure state. During the eighteenth and nineteenth centuries, when the codification of chemical knowledge began in earnest, compounds originating in living matter were classified as *organic compounds*.

For the preparation of inorganic compounds, various stones, minerals, earths, and ores provided the starting materials. A number of investigators noted that there were several qualitative differences between organic and inorganic compounds. All organic compounds seemed to contain carbon and hydrogen and very few other elements beyond oxygen and nitrogen. By contrast, inorganic compounds appeared to have been formed from a profusion of elements. Most inorganic compounds were also nicely crystalline substances and were quite stable to heat, while many organic compounds were difficult to crystallize as well as being relatively sensitive to thermal destruction.

Most significant, perhaps, was the fact that the elements could be mixed in almost any conceivable proportions in organic compounds, while experience had indicated that this state of affairs did not apply in inorganic chemistry. Thus, virtually all chemical investigators of the eighteenth and early nineteenth centuries were led to believe that there were qualitative differences between organic and inorganic substances. In particular, it was held that the **synthesis** (construction of a compound from its elements) of inorganic compounds could be accomplished by the chemist in his laboratory, but that the synthesis of organic compounds could only take place in living organisms that possessed a *vital force* indispensable for these syntheses.

WÖHLER'S EXPERIMENTS

The nineteenth century German chemist Friedrich Wöhler was studying the salts of cyanic acid, HOCN. In 1828 he reported on his attempts to make ammonium cyanate, $NH_4^+OCN^-$. After allowing lead cyanate to react with ammonia in aqueous solution, Wöhler filtered the lead oxide and isolated the other reaction product by boiling off the water.

(1.1) $$Pb(OCN)_2 + 2NH_3 + H_2O \longrightarrow PbO\downarrow + 2NH_4OCN$$

He was rewarded with the precipitation of a white crystalline substance that did not, however, exhibit properties expected from ammonium cyanate. For instance, its solutions did not liberate ammonia when treated with strong bases.

Wöhler was able to show, with the aid of a chance observation, that his compound was identical to urea, which had previously only been isolated from and shown to be a major constituent of human urine.

(1.2) $$NH_4^+OCN^- \xrightarrow[\text{heat}]{H_2O} \underset{\text{urea}}{H_2N{-}\overset{\overset{\displaystyle O}{\|}}{C}{-}NH_2}$$

He had, in fact, created an organic compound in the laboratory using inorganic starting materials and without the intervention of any living organism. A number of eminent chemists of the time were deeply impressed by Wöhler's results and although his paper did not by itself destroy the vitalistic theory, it did seriously undermine the tenability of the hypothesis. Evidence continued to accumulate that contradicted the predictions of vitalism, and by 1860 few prominent scientists were willing to espouse it.

Before leaving the subject of Wöhler's classic experiment, it is worth noting several interesting developments of that event. Wöhler himself had some doubts as to whether his results constituted a refutation of vitalism because of the origins of the cyanate ion that he employed. The cyanate had been produced by oxidation of cyanide that had, in turn, been obtained from a reaction in which the hooves and horns of animals constituted some of the ingredients. Wöhler remarked that a defender of vitalism could well maintain that the vital force had in fact been carried over into the cyanide and was actively responsible for the formation of urea.

Several later commentators took up this view, maintaining that Wöhler's claim to being the slayer of vitalism was unfounded. The great British chemical historian J. R. Partington searched the contemporary chemical literature, however, and discovered that predecessors of Wöhler's had synthesized both potassium cyanide and ammonia from indisputably inorganic sources. Since the inorganically derived potassium cyanide and the "horns and hooves" material were identical, as far as was known, Wöhler's experiment constituted (unbeknown to him) a true synthesis of urea from the elements — a *total synthesis.*

The construction in the laboratory of organic molecules found in nature (often quite complex) is a flourishing branch of organic chemistry (*natural product* synthesis). The final aim of workers in this field is almost always a total synthesis of the compound in question. One does not actually start with carbon, hydrogen, oxygen, etc., but rather one uses as starting materials compounds derived from simpler ones, which may be related by a string of known reactions back to the elements. This obsession with total synthesis may be seen, at least in part, as a psychological legacy of the vitalistic controversy.

ORGANIC CHEMISTRY AS A SEPARATE DISCIPLINE

As vitalism died, the term "organic chemistry" came to be synonymous with "chemistry of carbon compounds." If it was no longer necessary to believe that carbon-containing compounds were created by different forces than those responsible for the synthesis of molecules lacking carbon atoms, then what was the justification for singling out this element and elevating its study into an entire branch of chemistry?

In fact, carbon is rather unique among the elements. It forms fairly strong, unreactive bonds with itself, enabling extremely complex and varied molecular structures to exist because of their durable carbon atom skeletons. Vast numbers of organic compounds have been discovered or synthesized to date; they outnumber all the known inorganic compounds. The mere number of compounds that fall under the organic heading has served to encourage chemists to specialize in the study of these substances.

Carbon is almost always quadrivalent in its compounds, and one result of this quadrivalency is the astonishing variety of spatial arrangements of atoms found in organic compounds. The three-dimensional character of organic compounds directly affects their chemical behavior, particularly in systems of biological interest. An entire subbranch of organic chemistry is concerned with methods of uncovering the architecture of organic molecules and relating that architecture to their functional behavior.

Recent years have witnessed an intensive growth of interest on the part of organic chemists in that area that was the concern of their original forbears — the chemistry of living matter. Numerous investigations throughout the nineteenth and twentieth centuries had established the fact that most physiologically active compounds were "organic" (in the classificatory sense). The enormous chemical complexity of living systems frustrated efforts at comprehending the molecular basis of biological function. But as the data began slowly to accumulate, and in particular as the various intermediates in metabolic processes were identified, it seemed more and more plausible that there were chemical reactions taking place in cells that were closely analogous to transformations well-known in the laboratory.

This insight provided a tremendous stimulus to the study of **biochemistry** because it meant that the considerable body of data that had been produced by "pure" organic chemists could serve as a guide to those interested in physiological systems. Conversely, it also motivated organic chemists to concern themselves to an increasing extent with biochemical problems because these problems provided a stringent test of current theories and generalizations in complex and subtle systems. Although our understanding of the detailed chemistry of living systems is still in its infancy, we can clearly see the not too distant prospect of man's ability to control and alter all the vital functions of plants and animals, himself included. When this comes to pass, scientists will have placed in the hands of mankind the most awesome power yet grasped by any humans.

Finally, the economic importance of applied organic chemistry to industrialized nations would be difficult to exaggerate. Fuels, plastics, artificial fibers, pesticides, drugs, high temperature polymers, detergents, etc. — these are all composed of organic compounds. Moving organic chemistry out of the laboratory and into the plant has enabled some men to exert the most extraordinary control for material purposes yet recorded. It has also presented us with some of the most dismal prospects ever faced by the race as a whole.

HOW ORGANIC CHEMISTS SPEND THEIR TIME

The first investigations of organic compounds were confined largely to analytical experiments — interest centered on the question of which elements were present in the molecule and in what proportions. These experiments were of the utmost importance for the future development of organic chemistry, for they established the apparently qualitative differences between organic and inorganic compounds. To this day, the molecular formula is one of the prime pieces of data that one might wish to obtain about a compound of interest.

The most abundant elements in organic compounds are carbon, hydrogen, nitrogen, and oxygen. A quantitative measure of the amounts of the first three available in any compound may be obtained directly. Once the absence of other elements (such as halogen or sulfur) has been established, the amount of oxygen present is obtained by subtracting from the total weight of the sample the sum of the weights of carbon, nitrogen, and hydrogen.

The method of decomposing the organic compound was established at the start of the nineteenth century and has been used in the same essential form ever since. A weighed sample is completely oxidized with the aid of a catalyst. The carbon is converted to carbon dioxide, the hydrogen to water, and the nitrogen to molecular nitrogen. In the chemical procedure the water was adsorbed by an anhydrous salt such as calcium chloride or calcium sulfate and it was then weighed; the carbon dioxide was reacted with sodium or potassium hydroxide and the increase in weight then determined; the weight of nitrogen was determined from its volume at STP.

(1.3) $$2NaOH + CO_2 \longrightarrow Na_2CO_3 + H_2O$$

Representative calculations are given at the end of this chapter.

At one time each organic chemist had to be able to do his own analyses; it is probably fair to say that only a small percentage of chemists is able to do so today. Elemental analyses are performed by specialists who provide accurate and efficient service. Determination of the carbon, hydrogen, and nitrogen composition of an organic compound has now been automated; the quantities of carbon dioxide, water, and nitrogen in the combustion mixture can now be determined directly without the need for any chemical transformation, with the aid of gas chromatography (Chapter 5).

Elemental analysis supplies only the empirical formula of the compound being analyzed; its molecular weight must also be determined in order for the molecular formula to be known. Even when the correct molecular formula has been deduced, the organic chemist is just at the beginning of his labors, for as we shall shortly see there are usually a sizable number of different structural arrangements possible for the set of atoms specified in any single molecular formula. An immense amount of effort has been expended on organic structure determinations.

One of the tools first available for the study of organic structures was that of chemical transformation. Early workers quickly perceived that when an organic substance is treated with a given reagent, a certain atom or group of atoms is removed from the original compound while a new atom or group of atoms originating in the reagent takes its place. The remainder of the molecule seemed to remain intact while the reaction pursued its course. Furthermore, subsequent reaction could result in yet another displacement of one group of atoms by another; yet the original molecular residue would remain unperturbed.

These experiments established the existence of a number of molecular fragments that could remain intact throughout a lengthy series of chemical reactions. By gathering data on particular molecular fragments from a number of separate reactions, chemists were often enabled to deduce the bonding arrangement of atoms within that fragment. The groups of atoms being displaced or altered were known as a **functional group,** the structure of which was often illuminated by its chemistry. In fact, it became clear after awhile that most of the chemical reactivity of a molecule was concentrated in its functional groups. Furthermore, different compounds that possessed the same functional group often displayed many similar chemical characteristics.

The functional group soon emerged as a most valuable construct for organizing and classifying the burgeoning tide of organic chemical data. It also served to focus attention on the reactive sites of molecules and on the relationship between structure and reactivity. A detailed knowledge of functional group interconversions served two ends. First, it conferred great power on the synthetic organic chemist, who was now able to design and synthesize molecules with quite specific chemical properties. Second, data about the behavior of functional groups provided the raw material by which various theories of bonding and reactivity could be tested.

Bonding is an area of great concern to a large number of organic chemists. The theory of the electron pair bond (1916) and the discovery of wave mechanics (1926) provided very enlightening insights into just how molecules were held together. They also provided rationalizations for data that had already been established by experiment. At the present time theoretical organic chemists are seeking to do more than just reproduce by calculation previously established properties of molecules and of reacting systems; they are attempting rather to predict a priori the characteristics of molecules and systems not yet subjected to empirical investigation.

While the modes of reaction of a given molecule give a good deal of information about just which atoms are bonded together, they provide much less information on the three-dimensional array of these atoms. Progress in this particular area has benefited greatly from studies of the interaction of organic compounds with electromagnetic radiation (**spectroscopy**). As a result, most organic chemists are spectroscopists of one sort or another, even if only on a very part-time basis. The wavelength of the radiation used covers a wide gamut from X rays to microwave radiation.

Despite the progress made by organic chemists in sorting and codifying the tremendous number of known organic reactions, clear patterns of reactivity were by no means always apparent, and there seemed to be more diverse phenomena than could be encompassed within one large picture. Furthermore, there were relatively few clear-cut indications of why certain functional groups behaved in the ways that they did or why in some molecules deviations from the expected behavior were encountered.

Approximately 50 years ago the discipline of **physical organic chemistry** came into existence, and its prime concern was the **mechanism** of organic reactions. The mechanism of a reaction is a detailed description of all the changes in nuclear and electronic positions and energies that occur as reactants are converted into products. A stoichiometrically simple reaction may, in fact, proceed by way of complicated mechanism; the conversion of A + B to X might require the steps shown in equations 1.4 to 1.6.

(1.4) $A \longrightarrow A'$

(1.5) $A' + B \longrightarrow C$

(1.6) $C \longrightarrow X$

If one has a fairly good idea of a number of reaction mechanisms, then one can attempt to relate these to the properties of various atoms, groups of atoms, and associated electrons within the reacting molecules. The *chemistry* of the molecule is thus being correlated with the *physics* of its constituent atoms. In this way the underlying physical reasons for the observed functional group behavior may be uncovered. As a result, we are able to understand the similarity in behavior of apparently dissimilar groups, as well as the deviations from "normal" behavior of functional groups in certain molecules. One of the most valuable effects of mechanistic studies is the light they have shed on the role of catalysts in various reactions.

By focusing attention on the physical origins of molecular behavior, reaction mechanism studies helped to partially dissolve the boundaries between organic and inorganic chemistry since the behavior of all compounds could be analyzed in the light of a common set of tenets. It was found, in fact, that many generalizations that had been arrived at for compounds of one classification were actually useful in understanding the behavior of compounds in the other category. And it was the study of reaction mechanisms that put organic chemists in such a good position to help unravel biochemical puzzles.

Last, but certainly not least, mechanisms can be a tremendous aid in learning and exploring organic chemistry. By drawing attention to the basic physical processes involved in a chemical reaction, they enable one to think about modes of reaction rather than memorizing an endless string of specific cases. A mechanistic approach will often give one a handle for dealing with previously unencountered reactions, and it will also serve as a guide in planning forays into unchartered territory.

The rate of increase in the production of scientific data is staggering. It is absolutely necessary that a chemist have some sort of theoretical framework within which he can accommodate his knowledge if he has any hope of ranging widely over the field of chemistry. Mechanistic information, functional group properties, etc., all help to establish those generalizations that are necessary for such an approach.

One word of caution about reaction mechanisms is perhaps appropriate here. There are many pieces of data that one can gather about any given reaction — the dependence of its rate on the solvent, the effect of making structural changes in the reactants, the effects of potential catalysts, etc. From all these data one can *infer* one or more reaction mechanisms. As more data accumulate, it is usually possible to discard more and more of these mechanisms and to further refine those that remain.

One may eventually arrive at only one conceivable mechanism that is in accord with all the data. But the fact that one cannot *conceive* of another equally valid mechanism *does not mean that one does not exist.* It is quite common for well-accepted mechanisms to undergo significant change and even outright rejection because of their incompatibility with newly discovered facts. Therefore, one must always discuss mechanisms within a framework of *probability;* the phrase "the true mechanism" should really not be used by chemists.

A reaction mechanism is therefore potentially a very ephemeral construction; a breeze of new data could blow it away. One would therefore be foolish to blindly memorize a whole batch of mechanisms; memorizing individual reaction data is also hopeless but at least it has the advantage of allowing one to absorb "hard" information. The chief benefit to students in learning reaction mechanisms comes when the student tries to judge for himself the plausibility of a proposed mechanism in the light of the data *and* in light of his own understanding of how molecules behave.

FUNCTIONAL GROUPS

In order to provide a qualitative idea of just what constitutes a functional group, a representative sampling of such groups has been collected in Table 1-1. As seen, they range in complexity from a single atom to groups containing 20 or so atoms. Furthermore, while apparent resemblances between functional groups sometimes indicates a certain commonality of behavior, this is not invariably so.

The carboxyl group, $\mathrm{-\underset{\underset{\textstyle O}{\|}}{C}OH}$, contains as a subunit the hydroxyl group,

TABLE 1-1
Some Representative Functional Groups and Classes of Compounds

Group Structure	Functional Group Name	Class of Compounds Containing This Group
—Cl	Chlorine	Chloride
—OH	Hydroxyl	Alcohol
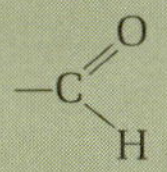	Formyl	Aldehyde
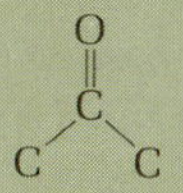	Keto	Ketone
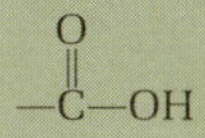	Carboxyl	Carboxylic acid
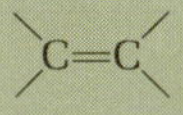	Double bond	Alkene
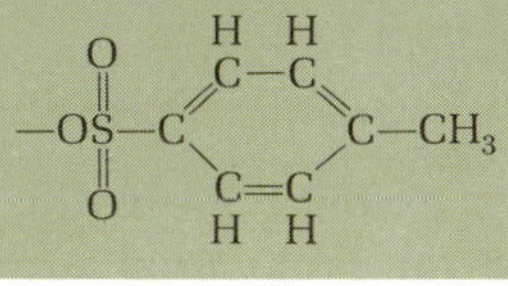	Tosyloxy	Tosylate

—OH, which is characteristic of alcohols (in alcohols, the —OH group is attached to a carbon atom that in turn is bonded only to carbon or hydrogen atoms). Yet alcohols and carboxylic acids differ in behavior in a number of important respects. On the other hand, the chemistry of chlorides and of tosylates is often closely comparable.

Functional groups not only provide a means of classifying compounds in terms of their reactive properties, they also serve as the basis for the nomenclature of organic compounds. The name of an organic compound will be built around two prime pieces of data — a description of the basic carbon skeleton and an enumeration of functional groups present in the substance. Thus the nomenclature, reactive properties, and physical structure of a compound all focus on the same area and can serve to reinforce one another for the astute learner.

CALCULATION OF EMPIRICAL FORMULAS

A few examples of empirical formula calculations are given below to illustrate the method and its limits. A quantitative determination of carbon, hydrogen, and nitrogen can be achieved with a 0.01 g sample. It is quite common, however, to get a duplicate analysis in order to increase precision so that a sample twice the size stated would have to be submitted to the analyst.

A 0.03035 g sample of an organic compound was combusted, 0.07745 g of

CO_2 and 0.03179 g of H_2O being given off in the process. From these figures we may calculate that there were $(12.01/44.01) \times 0.07745 = 0.2113$ g of carbon and $(2.016/18.02) \times 2 \times 0.03179 = 0.00356$ g of hydrogen in the original sample. A quick check reveals that there is a substantial portion of the weight of the original sample that is unaccounted for — $0.03035 - (0.02113 + 0.00356) = 0.00566$ g, to be precise. In the absence of indications to the contrary (qualitative tests for other elements, origin of the sample), we may assume that this weight is due to oxygen in the sample compound. The number of moles of carbon, hydrogen, and oxygen atoms in the sample is given below:

$$\text{Moles C: } \frac{0.02113}{12.01} = 1.759 \times 10^{-3} \qquad \text{Moles H: } \frac{0.00356}{1.008} = 3.532 \times 10^{-3}$$

$$\text{Moles O: } \frac{0.00566}{16.00} = 3.54 \times 10^{-4}$$

An empirical formula tells us only the *ratio* of moles of atoms in the compound (in order to obtain the *molecular formula,* we need to know the molecular weight of the compound). So we take the element that is present in the smallest molar quantity (in this case, oxygen) and set its relative molar quantity equal to 1.00. If we then correct the absolute values for the other elements in the same way (by multiplying them by $1.00/3.54 \times 10^{-4}$), we obtain the empirical formula $C_{4.97}H_{9.97}O_{1.00}$; within the limits of error of the analysis this is equivalent to $C_5H_{10}O$.

One rarely gets the analytical data in the form that it has been given here. The analyst merely reports the percentage by weight of each element for which an analysis has been made. When using the results of analysis as evidence in support of a proposed structure, the formula is usually considered acceptable by most chemists if the calculated and experimental weight percentages for the common elements (C, H, O, and N) agree to within $\pm 0.3\%$ (absolute value). Because of the possibility of error inherent in any determination, there will usually be several empirical formulas that are compatible with a given set of analytical data.

A typical set of analytical data could be presented as follows: C, 45.90%; H, 10.6%; N, 13.11%. There is 30.32% of the weight of the material to be accounted for, and we again assume the presence of oxygen. Since these percentages are weight values, we may divide each of them by the appropriate atomic weight to obtain relative molar proportions. If we normalize so that the relative molar quantity of the least abundant element is made equal to 1.00, then the empirical formula $C_{4.08}H_{11.31}N_{1.00}O_{2.02}$ is obtained, which reasonably corresponds to $C_4H_{11}NO_2$.

An independent measurement might also tell us that the molecular weight of the compound is 202 ± 10. Since the empirical unit $C_4H_{11}NO_2$ has a molecular weight of 105, the real molecule must contain two of these units, and the molecular formula of the compound is $C_8H_{22}N_2O_4$.

PROBLEMS

1 Elemental analysis of a compound reveals that it contains 87.31% C and 12.19% H. What is its empirical formula? If a portion of the compound is vaporized and its vapor density measured at STP, the value found is 3.70 g/l. What is the molecular formula of the compound?

2 A compound reveals the following elemental composition upon analysis: C, 76.31%; H, 10.18%; N, 13.52%. Allowing for an error of ±0.3% in each determination, suggest three or four empirical formulas that are compatible with the analytical data.

3 An organic compound is analyzed and found to have the following partial composition: C, 38.37%; H, 6.37%. Qualitative tests indicate the absence of nitrogen and sulfur and the presence of chlorine. A separate sample of the compound (0.04583 g) is decomposed chemically in such a way that all the chlorine is converted to chloride ion. Titration of the chloride ion requires 5.816×10^{-4} moles of silver ion. Calculate the empirical formula of the compound.

SUPPLEMENTARY READING

O. T. Benfey, *From Vital Force to Structural Formulas*. Boston: Houghton Mifflin Co., 1964. (P)

CHAPTER 2

ATOMIC STRUCTURE AND MOLECULAR BONDING

QUANTIZATION AND ATOMIC ORBITALS

The "planetary" picture of the atom, in which the negatively charged electrons revolve about the positive nucleus, is undoubtably a familiar one. The electrons are grouped in shells, at increasing distances from the nucleus; the shells are designated *K*, *L*, *M*, ... corresponding to the values 1, 2, 3, ... for the principal quantum number. Within each shell the electrons are further arranged into subshells, designated *s*, *p*, *d*, *f*, ... The *K*-shell contains only an *s*-subshell, the *L* shell contains *s*- and *p*-subshells, and so on. If we "create" an atom by adding just enough electrons to a bare nucleus to neutralize its charge, then the shell of lowest energy (closest to the nucleus) will be filled first, followed by the shell of next lowest energy, and so forth. Within each shell the subshells will also be filled in ascending order of energy, that is, *s* first, then *p*, etc. The resulting atom will be in its lowest possible energy state, or **ground state.**

This is essentially the Bohr-Sommerfeld picture of the atom. While it has been superseded by a more sophisticated model, it bequeathed several important postulates to later workers. The most important of these was probably the concept of **quantization** (Latin *quantum* — how much), which held that the electrons in a given atom could only assume certain predetermined values of total energy and angular momentum, and no others. The Bohr-Sommerfeld model introduced quantization in a rather arbitrary manner, however, in order to fit data obtained from atomic spectral measurements. Furthermore, the existence and nature of the subshells, etc., were not adequately explained.

It remained for Schrödinger, Heisenberg, and others to construct an atomic model that led naturally to quantization and the detailed structure of the atom. It had already been shown that in some experiments electrons masqueraded as waves; for example, they were capable of being diffracted. Schrödinger used a **wave function,** ψ, to describe the behavior of the electron, quite similar to the mathematical functions used to describe electromagnetic and hydrodynamic waves. It is not now possible to ascribe any clear physical meaning to the wave function itself. If, however, the wave function is incorporated into a **wave equation,** then it is possible to calculate the energy and angular momentum of the electron around a given nucleus under a given set of physical conditions.

The square of the wave function, ψ^2, has a precise physical meaning attached to it. It gives the probability of finding the electron at any particular point in space. According to the old Bohr theory, electrons followed precisely defined paths about their nuclei. Heisenberg proposed, however, that it is not possible to know simultaneously the exact position and exact momentum of a particle (the Uncertainty Principle). Consequently, we calculate instead the probability that the particle will be found at a given spot.

By using ψ and ψ^2, one can construct an entity that is very useful for visualizing the bonding process — the **atomic orbital.** If the wave function is plotted in three-dimensional space, a surface is obtained inside which the electron will be found a certain percentage of the time. By making this surface the proper size, it will be possible to find the electron inside with very high probability, say 90 or 95%. This then is the atomic orbital — a region of space around the nucleus, having a definite shape, within which there is a very high probability of finding an electron with a given energy and angular momentum. The shape and size of the orbital will be a function of the energy and angular momentum of the electron. The orbitals correspond to the subshells of the Bohr atom and are denoted by the same designations — 1*s*, 2*p*, 3*p*, 4*d*, etc.

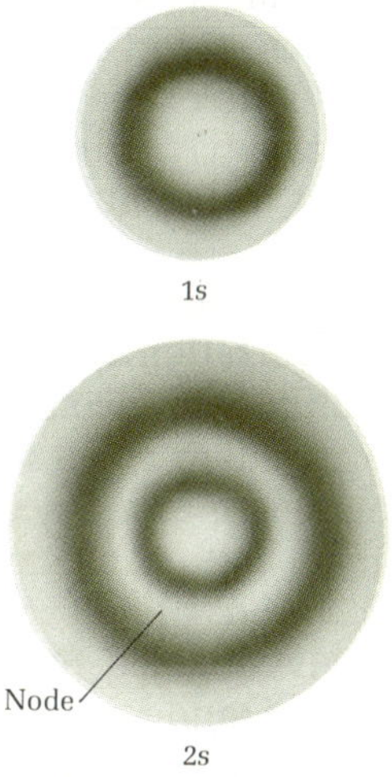

FIGURE 2-1
Hydrogen 1s- and 2s-Orbitals

Figure 2-1 shows sections through the 1s- and 2s-orbitals of atomic hydrogen. Note that these orbitals are actually spheres. The **electron density distribution** is also shown — this is simply a plot of the probability function; the regions where the dots are densest are those where the electron would probably be found. In the 2s-orbital there is a thin spherical shell in which

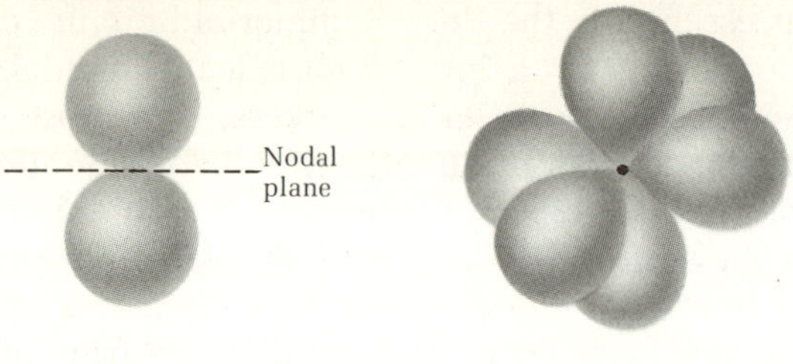

FIGURE 2-2
A Single Hydrogen 2*p*-Orbital; Arrangement of the Three 2*p*-Orbitals

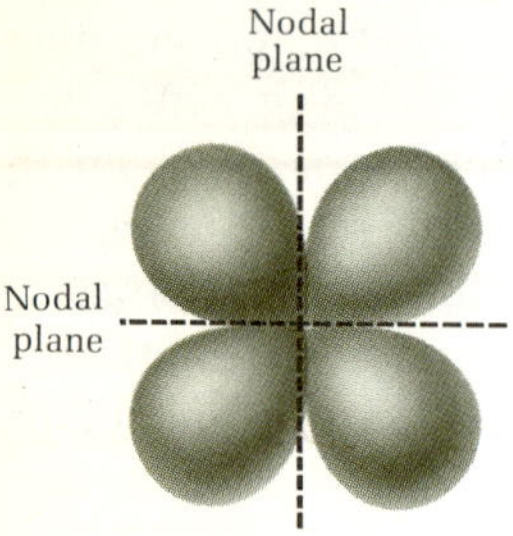

FIGURE 2-3
Hydrogen 3*d*-Orbital

the probability of finding the electron is zero. Such a region is called a **node;** every orbital that belongs to the second principal shell has one node, those in the third shell possess two nodes, etc.

The *p*-orbitals are very different in shape from the *s*-orbitals. They are not spherically symmetrical, but they are strongly directional in character. They all contain a minimum of one nodal plane, which passes through the nucleus and which separates the two lobes of the orbital. The presence of this nodal plane means that *p*-orbitals are not encountered in the first shell. All the other shells contain three *p*-orbitals; they are arranged so that their axes are mutually perpendicular.

d-Orbitals each possess at least two nodal planes and are first encountered in the third shell. The most common shape assigned to the *d*-orbitals is that shown in Figure 2-3; other configurations are possible by mathematically combining the *d*-orbitals each having the basic shape shown in Figure 2-3.

Before the process of interatomic bonding is discussed, a few other facts should be recalled. Any atomic orbital can accommodate a maximum of only two electrons; if the orbital is doubly occupied, the spins of the electrons must be paired. The orbital shapes given above have been worked out for the hydrogen atom. It is generally assumed that the atomic orbitals on other atoms have corresponding shapes; the dimensions of any particular atomic orbital will be sensitive to the nuclear charge and to the number and distribution of other electrons.

The nuclear charge and total number of electrons also determine the **ionization energy** and **electron affinity** of the atom. The ionization energy is the minimum energy required to remove the *least* tightly held electron from an atom. An electron is most easily removed from an atom when its ionization energy is low. A number of atomic ionization energies are collected in Table 2-1. Note that they decrease from top to bottom in any particular column; this is reasonable since the valence electrons are placed increasingly further from the nucleus and are increasingly shielded from it by the inner shell electrons. On the other hand the ionization energy generally increases as one traverses a row from left to right, making it progressively more difficult to remove an electron. All electrons in a given shell are approximately the same average distance from the nucleus; as the atomic number is increased, they are subjected to a more and more intense

TABLE 2-1
Ionization Energies of Gaseous Atoms (kcal/mole)

H 314							He 567
Li 124	Be 215	B 191	C 260	N 325	O 314	F 402	Ne 495
Na 118	Mg 176	Al 138	Si 188	P 253	S 239	Cl 299	Ar 362
K 100	Ca 141				Se 225	Br 273	Kr 323
						I 241	

nuclear field. The deviations that do occur may be ascribed to two factors in particular: *p*-electrons are somewhat more energetic than *s*-electrons of the same shell; and there is much more repulsion between electrons in the same orbital compared to electrons in different orbitals.

The electron affinity corresponds to the energy released when the neutral atom acquires one *additional* electron and is converted to a negative ion. As expected, those atoms having low ionization energies generally have small electron affinities and vice versa. In other words, atoms that will readily accept an additional electron are usually reluctant to part with those that they already have. The principal exceptions are the rare gases, which show little inclination to either lose or gain electrons. The trends in electron affinities are not as clear-cut as those of the ionization energies because repulsion effects are exceedingly important.

TABLE 2-2
Electron Affinities of Some Gaseous Atoms (kcal/mole)

H 17						
Li 14	Be 0	B 7	C 29	N 0	O 34	F 80
Na 19	Mg 0	Al 12	Si 32	P 17	S 47	Cl 83
K 21						Br 77

Both the ionization energy and the electron affinity are related to the **electronegativity** — the tendency of an atom to attract to itself electrons that it is sharing with another atom. The electronegativity is a *defined* property of the atom rather than a directly measurable one. Its magnitude for any given atom depends on just how it is defined. More important than the absolute magnitude of the electronegativity is the difference in electronegativity between any pair of atoms. The larger this difference is, the greater will be the tendency of a shared electron pair to spend more time near the atom of *higher* electronegativity.

If the electronegativity difference between two atoms is sufficiently large (usually 2.0 or more on the scale given), then electron transfer will be complete, and ions will result. When the electronegativity differences are smaller, polarized covalent bonds result [the lower case Greek delta (δ) indicates an unspecified fractional quantity].

$$\overset{\delta+}{A}-\overset{\delta-}{B} \qquad \text{(B is more electronegative than A)}$$

Since the elements near the center of the periodic table tend to have intermediate electronegativities, their reactions with other elements, whether to the right or to the left, will tend to result in covalent bond formation. It has been noted that, among the lighter elements particularly, stronger covalent bonds seem to result when the combining atoms are different rather than similar in electronegativity. This added bond strength may be due to an ionic contribution from the partially charged atoms.

Since the electrons in different orbitals of the same shell are not bound with equal strength, it might be expected that the apparent electronegativity of an atom in a compound depends on just which electrons are involved in bond formation. We shall see later that this is so. It would probably be more exact if we were to speak of the *orbital electronegativities* of a particular element, but for simplicity the less precise system will be retained.

TABLE 2-3

Pauling Electronegativities of Some Atoms

H 2.1						
Li 0.97	Be 1.5	B 2.0	C 2.5	N 3.1	O 3.5	F 4.1
Na 1.0	Mg 1.2	Al 1.5	Si 1.7	P 2.1	S 2.4	Cl 2.8
K 0.90						Br 2.7

FIGURE 2-4

Signs of the Wave Function in 1s-, 2s-, 2p- and 3d-Orbitals. (2p- and 3d-Orbitals not Drawn to Scale)

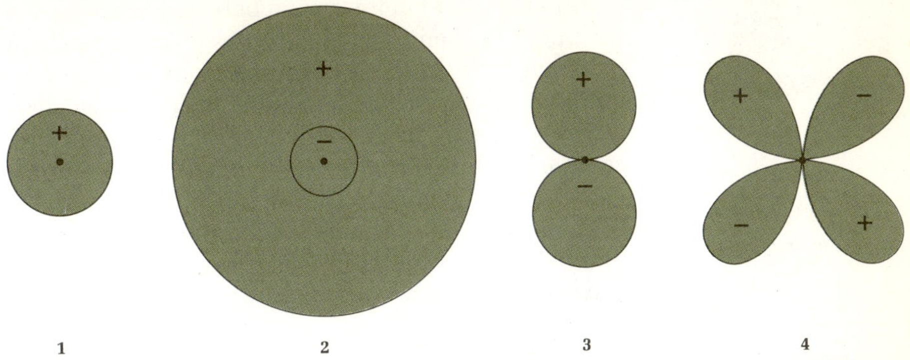

The last property of atomic orbitals to be noted at this time concerns the sign of the wave function. This sign may be either + or − (*no connection with electric charge*); the wave function is either positive or negative at every point in space, except at nodes. The particular value of the sign has no significance but it is important to note that the sign changes every time a node is traversed. Figure 2-4 contains representative examples. The signs of the wave functions have relevance to the mechanism of interatomic bonding.

MOLECULAR BONDING AND ORBITAL HYBRIDIZATION

The formation of a chemical bond between two atoms may be visualized as the coalescence of two atomic orbitals, one on each atom, to form a **molecular orbital** that encompasses within it both of the combining atoms. Those electrons originally occupying the two atomic orbitals now reside in the molecular orbital (Figure 2-5). Two electrons, with their spins paired, are the maximum number that can be accommodated in one molecular orbital. The coalescence of two half-filled atomic orbitals, or one empty and one full one, both lead to the same result.

Each electron in the molecular orbital continues to spend a sizable fraction of its time in the vicinity of one nucleus. But there is a strong possibility

FIGURE 2-5

Formation of the Hydrogen Molecular Orbital from Two Hydrogen Atoms

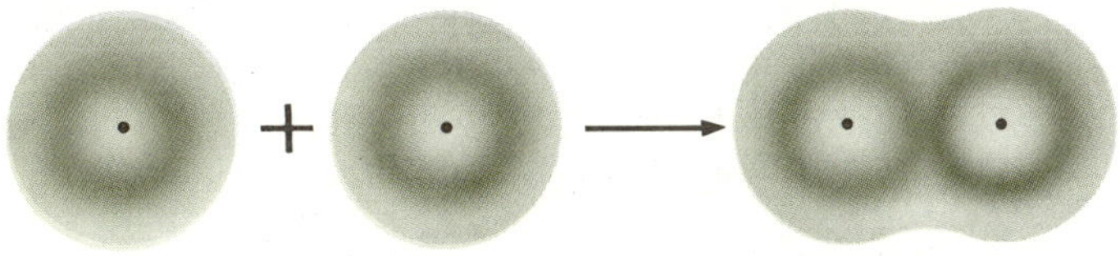

of its being found in the region between the two nuclei where it is interacting with both of them. The electron also serves to shield each nucleus from the repulsive field of the other nucleus. It is these factors that give rise to the binding energy of the molecule. That the electrons have indeed a lower potential energy in the molecule than they did in the atom can be seen by comparing the respective ionization energies of H_2 and H•: 356 kcal/mole versus 314 kcal/mole.

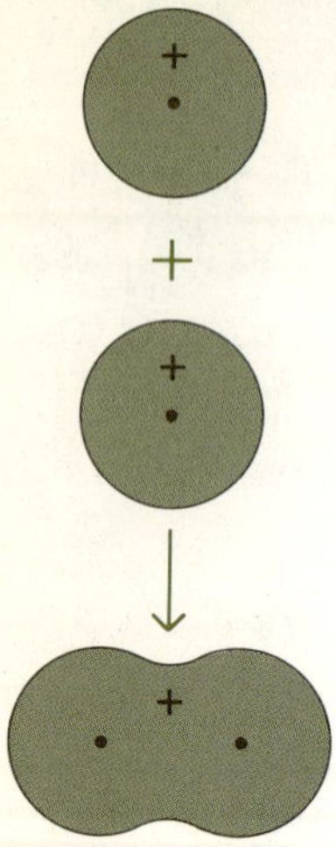

FIGURE 2-6

Combination of Two 1s-Orbitals with the Same Sign

The formation of the molecular orbital [represented by the mathematical function, Ψ (capital Greek psi)] may be symbolized as the addition of the wave functions for each atomic orbital.

$$\Psi_1^{H_2} = a(\psi_1^{H} + \psi_2^{H}) \tag{2.1}$$

Pictorially, this means that the orbitals combine with the same sign. It is completely permissible, however, to combine two atomic orbitals that have opposite signs. The molecular orbital that results has a node in its center; furthermore, the electrons occupying it are at a higher potential energy than they were in the separated atoms.

$$\Psi_2^{H_2} = b(\psi_1^{H} - \psi_2^{H}) \tag{2.2}$$

In general, the linear combination of m atomic orbitals results in the formation of m molecular orbitals. In the example cited above the molecular orbital of lowest energy, Ψ_1, is referred to as a **bonding orbital.** Transfer of electrons from an atomic orbital to this molecular orbital results in a decrease in potential energy, thus providing the binding energy for the molecule. Ψ_2 is called an **antibonding orbital;** if one or both electrons occupy this orbital, there tends to be a net repulsive force between the atoms. The bonding orbital, Ψ_1, may be further classified as a **σ-orbital** (pronounced *sigma orbital;* the term is analogous to s-orbital on an atom). There are no nodes in σ-orbitals, and they are circularly symmetrical in cross section when viewed end on. Their antibonding counterparts are **σ^*-orbitals.** In the ground state of molecular hydrogen the σ-orbital is doubly occupied and the σ^*-orbital is empty.

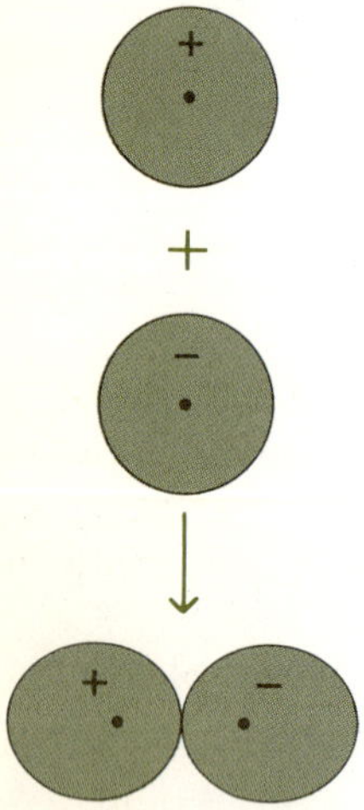

FIGURE 2-7

Combination of Two 1s-Orbitals to Produce an Antibonding Molecular Orbital

The situation for hydrogen may be summed up by means of an energy diagram (Figure 2-8). There are in actuality many more molecular orbitals than are shown in the diagram. These arise from combinations of the higher atomic orbitals and they are sufficiently high in energy that they have no effect on the ordinary chemistry of the molecule.

σ-Orbitals may also arise from **overlap** (coalescence) of s- and p-orbitals (formation of HF) and from overlap of two p-orbitals (formation of F_2). There are other ways of overlapping atomic orbitals as well. Side-by-side overlap of p-orbitals leads to molecular **π-orbitals.** This mode of overlap allows the formation of double and triple bonds between molecules; it will be explored further in this and subsequent chapters.

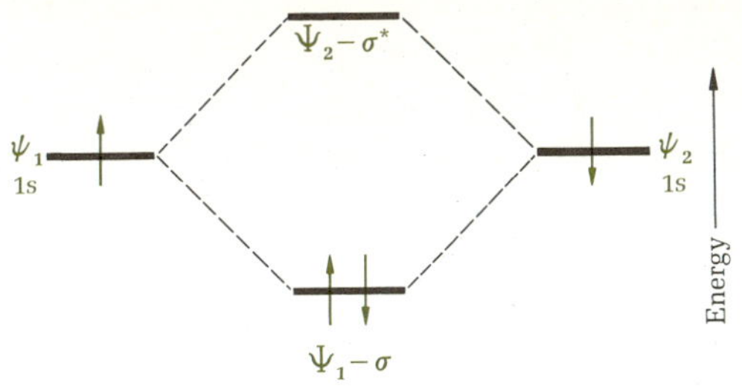

FIGURE 2-8

Energy Diagram for the Formation of Molecular Hydrogen from Hydrogen Atoms

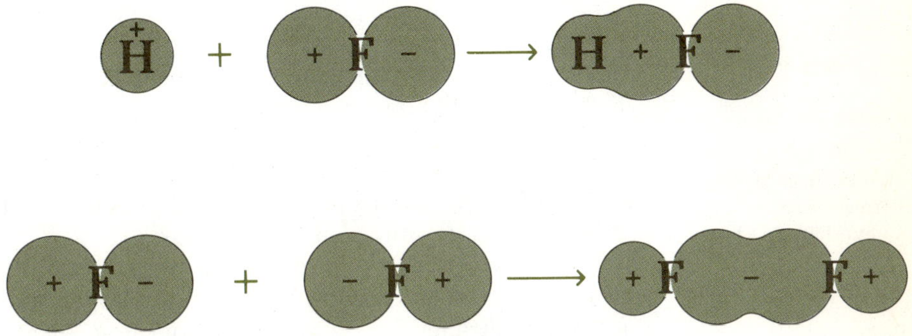

FIGURE 2-9

σ-Orbital Formation in Hydrogen Fluoride and in Molecular Fluorine

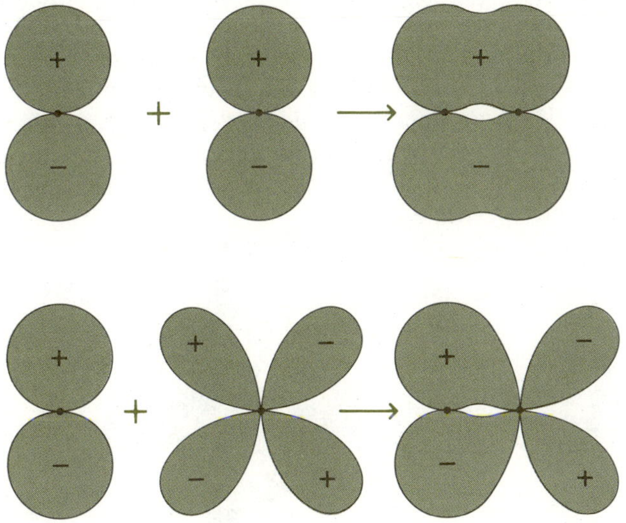

FIGURE 2-10

π-Orbital Formation by p-p and p-d Overlap

We need to take note of one more fact of molecular life at this point, namely, the phenomenon of **electron correlation.** In brief, the idea is that an electron in a molecule "feels" the presence of the other electrons and tries to stay as far away from them as possible. Thus, the Pauli principle states that if two electrons have paired spins, they may both occupy the same region of space (orbital). Within that orbital, however, there is no need for the electrons to stay close to one another; for electrostatic reasons they would probably be far apart on the average. This simple fact has important chemical consequences.

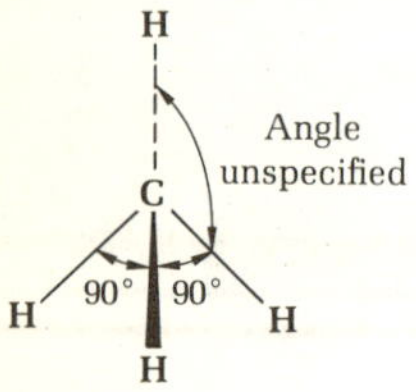

FIGURE 2-11
Hypothetical Methane Formed from Unhybridized Orbitals

The geometrical arrangement of atomic orbitals naturally inspires some predictions about the geometry of molecules. On the basis of the preceding rationalizations, one would expect that the bonding orbitals of methane, CH_4, are formed by overlap of four hydrogen 1s-orbitals with one each of the three 2p- and the one 2s-orbital of carbon. Since the 2p-orbitals are at right angles to one another, we would expect three of the H—C—H angles to be 90°; s-orbitals are spherically symmetrical so that it would not be possible to rigorously predict its position of attachment. One form of a molecule with this geometry is shown in Figure 2-11.

It is known with a high degree of certainty that this is *not* the geometry of methane. In the real molecule the hydrogen atoms lie at the apexes of a regular tetrahedron, with the carbon atom located at the center; all the H—C—H angles are the same and equal 109.5°.

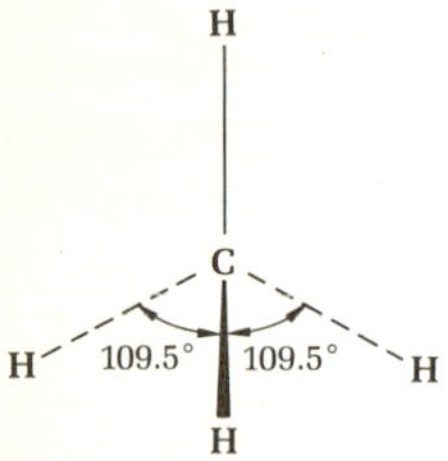

FIGURE 2-12
A Methane Molecule

How can this fact be reconciled with our atomic orbital picture of bonding? It can only be done if we assume that atomic orbitals **on the same carbon atom** can combine with another, somewhat analogously to the combination of atomic orbitals on different atoms. If we mathematically combine the 2s-orbital and the three 2p-orbitals of carbon, all equally and in all possible independent ways, four new **hybrid atomic orbitals** will result. Called sp^3-orbitals (pronounced *ess pea three*), these hybrid orbitals have their major lobes oriented toward the corners of a regular tetrahedron. The geometry of methane, and of a large number of other carbon compounds, is thus correctly predicted.

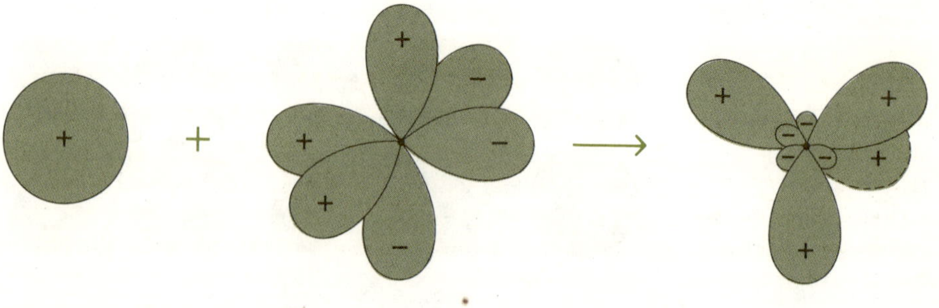

FIGURE 2-13
Combination of Carbon 2s- and 1p-Orbitals to Yield sp^3-Orbitals

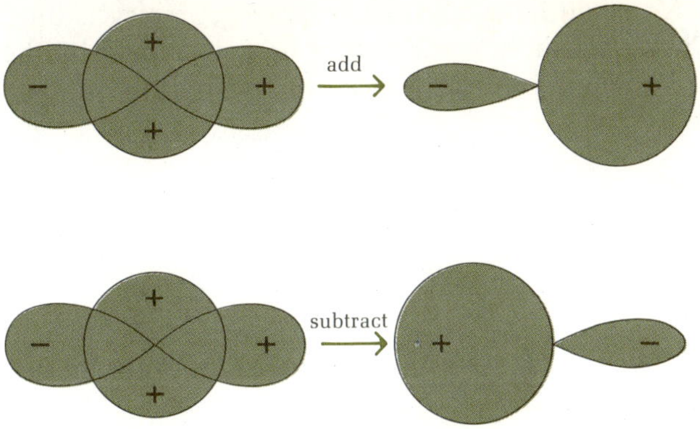

FIGURE 2-14
Formation of *sp*-Hybrid Orbitals by Addition and Subtraction of *s*- and *p*-Orbitals

Can the hybridization scheme, which may seem very contrived at first glance, be justified on other than geometric grounds? It is apparent that the lobes of the hybrid orbitals are unequal in size; use of the larger lobe for bond formation allows greater overlap than is obtainable with a corresponding *p*-orbital, resulting in a stronger bond. So if we started out by asking what combination of hydrogen and carbon atomic orbitals would lead to molecular orbitals of lowest overall energy, we would arrive at a solution equivalent to that obtained from the hybridization hypothesis. Thus the driving force for atomic rehybridization is the tendency for any system to seek the state of minimum potential energy, and the geometry of the molecule is a natural consequence of this tendency.

Once the hybridization postulate is accepted, it may be used to effectively rationalize many other bonding situations. Zinc iodide, ZnI_2, is linear. Zinc has two valence electrons; these can attain the lowest potential energy if the bonding molecular orbitals are formed from *sp*-hybrid orbitals on zinc. The hybrid orbitals, in turn, may be envisioned as arising from the addition and subtraction of atomic 4*s*- and 4*p*-orbitals. As pictured, the two atomic orbitals reinforce one another in those regions of space where their wave functions have the same sign and subtract from one another where the wave functions have opposite signs. Subtraction of the two functions is equivalent to changing the sign of one of them and then adding. Note that two of the *p*-orbitals are not used at all in the bonding; the zinc atom is capable of forming additional bonds by accepting electrons, as in the $[ZnCl_4]^{2-}$ ion.

sp-Hybrid orbitals are oriented 180° to one another; atoms that are hybridized in this way are said to be in the **digonal state.** By contrast, the carbon atom in methane is in the **tetrahedral state.** There is also a **trigonal state,** in which the σ-bonds to the central atom all lie in one plane 120° apart. The boron halides, BX_3 (X = F, Cl, Br, I), provide examples of this type of bonding. The boron atom uses three sp^2-orbitals for bonding to the halogens. The unused *p*-orbital is oriented at right angles to the plane of the hybrid orbitals.

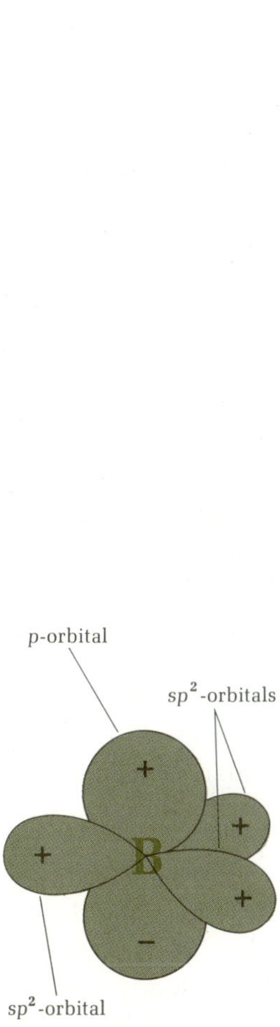

FIGURE 2-15
Hybrid Orbitals for Boron

No given hybridization state or geometry is unique to a particular atom. These parameters will rather be determined by the number of electrons in the valence shell of that atom and by the number of other atoms bonded to it.

(2.3) $ZnCl_2 + 2Cl^- \longrightarrow [ZnCl_4]^{2-}$
digonal — tetrahedral

(2.4) $BF_3 + F^- \longrightarrow BF_4^-$
trigonal — tetrahedral

(2.5) $CH_4 + H^+ \longrightarrow CH_3^+ + H_2$
tetrahedral — trigonal

ANOTHER APPROACH TO MOLECULAR GEOMETRY

Even in compounds where the bond angles deviate from the ideal values, the results are explicable by a slight extension of the ideas developed thus far. The H—N—H bond angle in NH_3 is 107° rather than the 109.5° required for tetrahedral hybridization. This discrepancy can be rationalized by assuming that the orbital containing the nonbonded electrons is larger than those with bonding electrons. Since the nonbonding electrons are subject to only one nuclear field, instead of two, this assumption is entirely reasonable; their increased repulsive interactions with the other electrons causes a contraction in the H—N—H angle. The effect should be more pronounced in H_2O, which carries two pairs of nonbonded electrons. Indeed, the H—O—H angle in this compound is 104.5°, even more distorted from tetrahedral than ammonia.

Very good quantitative guesses about molecular geometry may in fact be made by using "common sense." Electrons tend to repel one another. If an atomic nucleus is surrounded by four pairs of bonding electrons, then these pairs will try to be as mutually far apart as possible. This condition is best met if they are tetrahedrally disposed. Similarly, three electron pairs would arrange themselves 120° apart in order to minimize repulsive interactions. Deviations from the ideal angles could still be explained by the larger "size," or repulsive power, of nonbonded electrons, and so forth.

Since very adequate notions of molecular geometry may be developed without resorting to hybridization, molecular orbitals, etc., what is the justification for developing the latter? We shall see that these concepts provide a convenient and consistent method for rationalizing many molecular phenomena, including the extensive delocalization of electrons, the interaction of radiation with electrons, the stereochemistry of certain reactions, and many others. The fact that these concepts are effective in explaining many observations does not necessarily mean that they are "true"; it

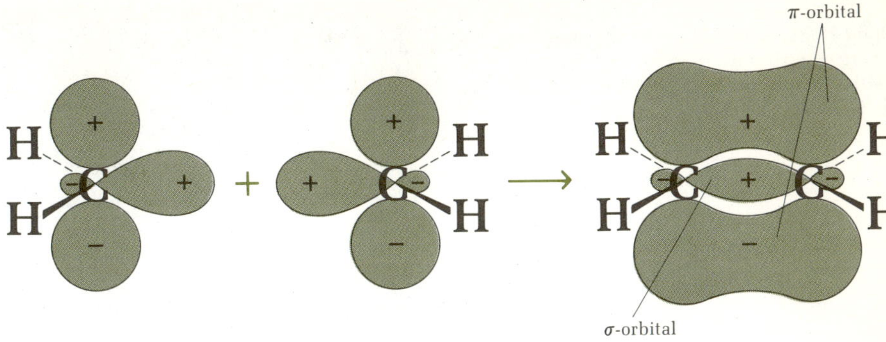

FIGURE 2-16
Combination of Two sp^2-Hybridized CH_2 Units to Produce Ethylene

is questionable, in fact, whether it is possible at all to apply labels such as true and false to hypotheses of this type.

In attempting to understand and rationalize atomic and molecular phenomena, we are in the position of the blind men touching the elephant. Each time we perform another experiment it is equivalent to grasping a different part of the elephant, and it is tempting to describe the entire elephant in terms of the characteristics of that one part. The hybridization and electron pair repulsion models are two different ways of describing the *same phenomenon* — the tendency of the molecule to find that geometry that minimizes its potential energy.

One way in which the hybrid orbital hypothesis has been very useful is in providing a clear picture of the nature of the **double bonds** in compounds such as ethylene, 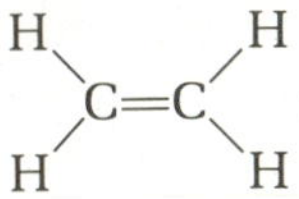 ; formaldehyde, 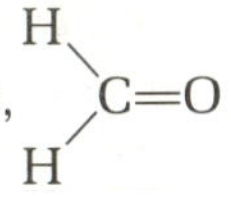 ; etc. The term double bond describes a situation in which two atoms share four electrons. Ethylene may be envisaged as arising from the combination of two sp^2-hybridized carbon atoms. Each carbon atom has two of its sp^2-orbitals involved in bonding to the hydrogen atoms; that leaves it one sp^2-orbital and one *p*-orbital for bonding to the other carbon atom. As these two CH_2 units are brought together, the sp^2-orbitals overlap to form a σ-bond and, simultaneously, the *p*-orbitals overlap to form a π-bond. The double bond should then consist of two types of molecular orbitals, σ and π; the chemistry of compounds containing these multiple bonds is in accord with such a conclusion. These and other facets of double bond chemistry are explored more fully in Chapter 8 and subsequent chapters.

BOND STRENGTHS

The unique position in chemistry occupied by organic compounds is attributable in large part to the tendency of carbon atoms to bond to one another

TABLE 2-4
Bond Dissociation Energies of Some Diatomic Molecules (kcal/mole)

H_2	104	HF	136
Li_2	26	HCl	103
F_2	38	HBr	88
Cl_2	58	HI	71
Br_2	46		
I_2	36		

in chains and to the stability of the resulting compounds. In order to appreciate the unusual features of carbon chemistry, some relevant **bond dissociation energies** and **average bond energies** must be examined.

The bond dissociation energy for the molecule A—B is defined as the energy necessary to bring about reaction 2.6 on a molar scale, in the gas phase, at 25° and 1 atm (A and B may be atoms or polyatomic fragments).

(2.6) $A{-}B \longrightarrow A + B$

The bond dissociation energy for a number of diatomic molecules of interest is given in Table 2-4; they all possess single bonds. With the striking exception of hydrogen it may be seen that **homonuclear** (Greek *homos* — the same) diatomic molecules have comparatively weak single bonds.

The average bond energy is simply an average of the bond dissociation energy for a particular type of bond, such as $-\overset{|}{\underset{|}{C}}-H$ or $-O-N\langle$, in a number of different molecules, and sometimes includes values for equivalent bonds in the same molecule. The actual bond dissociation energies that are averaged may vary over a wide range so that the values listed in Table 2-5 are themselves not of great significance. Nonetheless, several trends are apparent that have general validity. Single bonds between identical atoms seem to be of low energy, with the exception of carbon. In addition, for elements in this part of the periodic table, there seems to be a correlation between bond strength and the electronegativity difference between atoms.

TABLE 2-5
Average Bond Energies of Some Single Bonds (kcal/mole)

C—C	83	C—N	73	C—H	94
N—N	50	C—O	85	N—H	93
O—O	34	C—F	116	O—H	111

STABILITY AND REACTIVITY OF COMPOUNDS

If we inquire further into the reasons for the uniquely high C—C single bond energy, two explanations seem quite plausible. First of all, the sp^3-orbitals on each carbon atom overlap efficiently with one another. Second, there are no nonbonded electrons on the carbon atoms; the repulsive interactions among these electrons could be strong, thereby substantially weakening the bond. This seems to be the case with N—N and O—O single bonds. Each of these atoms possesses unshared electron pairs which undoubtedly repulse one another.

As a result, compounds containing carbon-carbon and carbon-hydrogen single bonds are very **stable;** i.e., they show no tendency toward spontaneous decomposition at 25°.

The compounds mentioned above also exhibit a very low **reactivity** toward other molecules, including oxygen, water, and a wide variety of acids and bases. Their stability is one factor that accounts for the unreactive nature of these compounds, but it is not the only one. We have noted the lack of nonbonded electrons around normal, quadrivalent carbon atoms. These nonbonded pairs, when present on other atoms such as oxygen or nitrogen, provide a focus for attack by electron-deficient reagents. On the other hand, carbon also lacks low energy, empty orbitals. Its valence shell is full, and it is therefore immune to the advances of electron-rich reagents.

This unique combination of properties — the high homonuclear bond strength and the resistance to both electron-deficient and electron-rich reagents — is to be found almost uniquely in the element of atomic number six. **In this collection of attributes lies the explanation for the growth of organic chemistry as a separate branch of chemical science.**

STRENGTHS OF DOUBLE BONDS

Not unexpectedly, the strengths of double bonds are considerably greater than those of the corresponding single bonds. The bond strength values rarely stand in a 2 : 1 relationship, however. For the small number of cases cited, it can be seen that the average bond energy for a double bond can be both smaller than and greater than twice the value for the corresponding

TABLE 2-6
Average Bond Energies of Some Single and Double Bonds (kcal/mole)

Single Bonds		Double Bonds	
C—C	83	C=C	146
N—N	50	N=N	100
O—O	34	O=O[a]	119
C—N	73	C=N	142
C—O	85	C=O	178

[a]For molecular O_2.

single bond. π-Bond formation by *p-p* overlap is most common among the elements from boron to fluorine. Although little thermochemical data is available, it seems clear from chemical evidence that atoms further down in the periodic table form rather weak π-bonds. This fact has been attributed to poorer *p-p* overlap with these atoms, due in turn to their longer σ-bond lengths and larger, more diffuse orbitals.

BONDS INVOLVING ISOTOPES

In many circumstances it becomes desirable to mark a certain atom or group of atoms in a molecule — one would like to be able to "paint" them, somehow. Such a procedure is often particularly useful in reaction mechanism studies. This objective can be achieved by synthesizing the compound using **isotopes** of the atom(s) in question that are not the normal ones.

Two isotopes of an element, it will be recalled, have identical numbers of protons in their nuclei but differ in the number of nuclear neutrons. They therefore have the same atomic number but different atomic weights. The predominant isotope of hydrogen, for instance, is protium, which has a nucleus consisting of only one proton. However, a small percentage of naturally occurring hydrogen atoms are deuterium (symbolized by D or ^{2}H), the nuclei of which contain one proton and one neutron. Furthermore, as a by-product of a number of nuclear reactions, tritium (T or ^{3}H) is produced; its nucleus consists of a proton and two neutrons, and it is radioactive.

A number of compounds are now available commercially that have a very high percentage of one of the rarer isotopes of a given element. For instance, one may purchase "heavy" water in which the hydrogen is virtually all deuterium or in which a large fraction of the oxygen is oxygen-18.

Since the electric field generated by a nucleus is virtually unaffected by the number of neutrons in that nucleus, two isotopes of the same element behave very similarly by chemical criteria. The dissociation energies of bonds involving separate isotopes of the same element will be different,

TABLE 2-7
Some Chemically Important Isotopes

Element	Abundant Isotope(s)	Other Useful Isotopes
Hydrogen	^{1}H	D (^{2}H), T[a] (^{3}H)
Carbon	^{12}C	^{13}C, ^{14}C[a]
Nitrogen	^{14}N	^{15}N
Oxygen	^{16}O	^{17}O, ^{18}O
Phosphorus	^{31}P	^{32}P[a]
Sulfur	^{32}S	^{34}S, ^{35}S[a]
Chlorine	^{35}Cl, ^{37}Cl	^{36}Cl[a]

[a]Denotes a radioactive isotope.

although the difference will usually be small. The exception to this generalization is provided by bonds to protium and deuterium since these isotopes differ in mass by 100%. Thus, the bond dissociation energy of H_2 is 104 kcal/mole, while that of D_2 is 106 kcal/mole. (The relationship between atomic mass and dissociation energy is discussed in Chapter 3.)

Two reactions that involve the same reactants but that differ in isotope composition may proceed at different rates. This rate difference, known as an **isotope effect,** can often provide valuable information about the reaction mechanism.

MOLECULAR FORMULAS AND ISOMERISM

It is customary when drawing molecular structures to show the bonding electron pairs as lines and nonbonding valence electrons as dots.

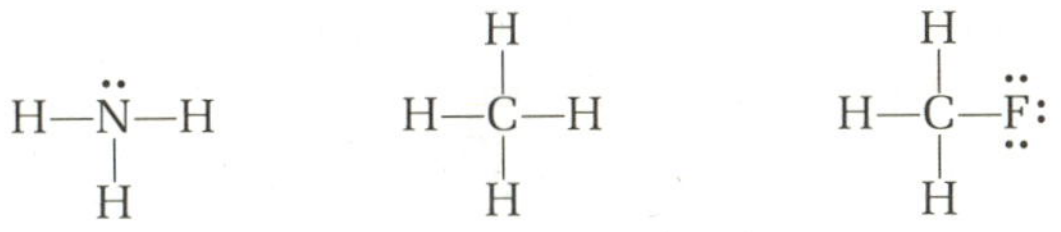

ammonia methane methyl fluoride

This system can become very cumbersome with even moderate size organic compounds, and it is not at all necessary. If we keep in mind the "normal" valences for the elements common in organic compounds — carbon, four; hydrogen, one; oxygen, two; nitrogen, three; sulfur, two; halogen, one — then compact **structural formulas** may be used. For example, ethyl chloride, C_2H_5Cl, can only correspond to

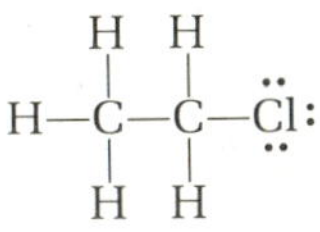

This can be compactly rendered as CH_3CH_2Cl. Similarly, propane, C_3H_8, must correspond to $CH_3CH_2CH_3$.

When we come to the molecular formula C_3H_8O, we are suddenly faced with more than one possible structure. Keeping in mind the normal valence rules, we must conclude that there are three structural formulas that correspond to the molecular formula C_3H_8O: $CH_3OCH_2CH_3$, $CH_3CH_2CH_2OH$, and $CH_3\underset{\substack{|\\ OH}}{C}HCH_3$. Each of these three is a different compound, and they are all **isomers** of one another (Greek *isomeres* — equally divided). Isomers are compounds that contain the identical number of each type of atom but have distinguishably different structures; that is, the bonding arrangements of the atoms are different in each isomer. Because of the phenomenon of isomerism, it is necessary to write the structural formulas for compounds — in most cases the molecular formula gives too little information.

CHEMICAL REACTIONS

Chemical reactions in organic systems almost invariably involve the breaking of covalent bonds in the reactants with the concomittant formation of new covalent bonds in the products. Bond rupture may take place in two ways:

(2.7) **Homolytically:** $Y{-}Z \longrightarrow Y\cdot + Z\cdot$

(2.8) **Heterolytically:** $Y{-}Z \longrightarrow Y\!: + Z$

In many cases the same bond will break in different ways, depending on the temperature and the reaction conditions. Molecular chlorine provides a convenient example. When it is heated in the gas phase, it dissociates homolytically to chlorine atoms.

(2.9) $Cl_2 \xrightleftharpoons{\text{heat}} 2Cl\cdot$

Its reaction with water, however, seems to occur heterolytically because chloride ions are produced.

(2.10) $Cl_2 + H_2O \rightleftharpoons HOCl + H^+ + Cl^-$

In attempting to grasp the logic behind the reactions of organic compounds, attention must not be focused exclusively on the covalent bonds being formed and cleaved. For instance, in the general replacement reaction symbolized by equation 2.11, CH_3I reacts most rapidly, followed by CH_3Br, then CH_3Cl, and finally CH_3F, which reacts very sluggishly.

(2.11) $Y\!: + CH_3{-}X \longrightarrow CH_3{-}Y + X\!:$ $\quad (X = F, Cl, Br, I)$

This order is not unexpected since the carbon-halogen bond is being broken and the carbon-halogen bond strengths are in the order $C{-}I < C{-}Br < C{-}Cl < C{-}F$. But the relative magnitudes of their rates of reaction will vary appreciably with the solvent, even when the same Y: is used. The solvation energies of the X: anions will be substantial, and their relative order very solvent-dependent; herein lies the main source of the variation in reaction rates.

Furthermore, it is necessary to look not only at those quantities that are characteristic of the reactants and products, such as bond strengths and solvation energies, but also at transient phenomena that arise as the reaction proceeds. As Y: approaches CH_3X, in equation 2.11, its valence electrons will be attracted by the carbon nucleus; at the same time their approach will tend to repel the electrons in the C—X bond. The resulting electron

FIGURE 2-17
Electron Polarization Upon Approach of Y: to CH_3X

polarization materially advances the progress of the reaction. The degree to which this occurs, at any given moment in the reaction sequence, depends on the **polarizability** of Y and X — that is, the susceptibility of the valence electrons in these atoms to the influence of an external electric field. Large atoms are highly polarizable because the valence electrons are far from their own nucleus; conversely, small atoms with their tightly held electrons are much less polarizable. The effect of atomic polarizability on organic reactions is considerable, although its role in any single case is often difficult to assess.

SUMMARY

1 The behavior of electrons in atoms and molecules is best described by means of a wave function. It is not possible to locate with certainty the position of a specific electron at a given instant. By using the square of the wave function, however, it is possible to calculate the **probability** that a particular electron will appear at a specific point at any given time.

If a probability distribution function is plotted in space, then an orbital is obtained — a region of space within which there is a high probability of finding an electron having a certain total energy and angular momentum. Atomic orbitals have different shapes (depending on the angular momentum of the electron) and different sizes (depending on the total energy of the electron). Atomic s-orbitals are spherical, *p*-orbitals have two lobes separated by a nodal plane, and *d*-orbitals have four lobes separated by a pair of nodal planes. The total number of nodes in any particular orbital equals $n - 1$, where n is the principal quantum number of the shell in which the orbital is found. Each time a node is crossed, the sign of the wave function changes.

2 The formation of a covalent bond may be envisaged as the coalescence, or overlap, of atomic orbitals on different atoms. Molecular orbitals are produced as a result. The molecular orbital is bonding if the atomic orbitals have the same sign in the region of overlap; it is antibonding if they overlap with opposite signs. Overlap of *s*- with *s*- or *p*-orbitals, or end-to-end overlap of *p*-orbitals, results in the formation of molecular σ-orbitals. They have no nodes and are cylindrically symmetrical. Side-by-side overlap of *p*- or *d*-orbitals leads to π-orbital formation. π-Orbitals possess a nodal plane.

3 Molecular orbital predictions concerning the geometry of molecules, and the experimental facts, can only be reconciled if hybrid atomic orbitals are allowed. They can be conceived of as arising from the mixing of atomic orbitals on the same atom. The hybrid orbitals of greatest importance in organic chemistry are the *sp* (two arranged 180° apart), sp^2 (three in a plane at 120°), and sp^3 (four disposed tetrahedrally). Deviations from the ideal bond angles can usually be rationalized by assuming that nonbonding orbitals are more space-filling than bonding ones.

4 Most homonuclear single bonds are rather weak; carbon-carbon bonds are a notable exception. With heteronuclear single bonds, there is some correlation between bond strength and the electronegativity difference between the atoms. Elements in the first atomic period are capable of forming strong double bonds; these double bonds consist of both σ- and π-orbitals. Elements further down the periodic table form much weaker double bonds.

Reaction rates are heavily influenced by the strength of the bonds being broken and formed. Other factors, such as solvation energies and polarizabilities, play a very important part as well.

PROBLEMS

1 Of the sets of pairs of atoms, which pair will form the most highly polarized bond? Indicate by δ^+ and δ^- the positive and negative ends of the bond.

a. Li—F or C—F
b. C—H or C—O
c. C—H or Cl—H
d. C—F or C—Cl
e. Br—F or I—F
f. B—C or B—O
g. B—C or Al—C
h. B—C or Be—C
i. Li—C or K—C

2 Which of the following compounds would you expect to be ionic in structure, and which covalent (in the solid state)? For those that you believe to be ionic, draw the structure of the ions.

a. CH_4
b. CO_2
c. CF_4
d. $LiCH_3$
e. Li_3N
f. LiF
g. HF
h. H_2O
i. NH_3
j. NaCl
k. NaBr
l. $MgCl_2$
m. $MgBr_2$
n. BCl_3
o. BBr_3
p. CCl_4
q. PCl_3

3 What is the most likely hybridization state of the designated atom in the following ions and molecules?

a. NH_3
b. NH_4^+
c. H_3O^+
d. BF_3
e. AlH_3
f. $(CH_3)_3P$
g. $:CH_3^-$
h. $:CH_2$
i. $\cdot CH_3$ (contains one unpaired electron)
j. CH_3^+

4 In each of the following compounds, designate the bond or bonds that would most easily undergo *homolytic* cleavage if the compound were heated in the gas phase.

a. CH_3CH_3
b. CH_3F
c. CH_3OH
d. CH_3OOCH_3
e. 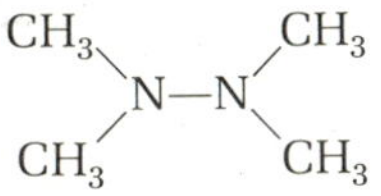
f. $CH_3—N=N—CH_3$
g. 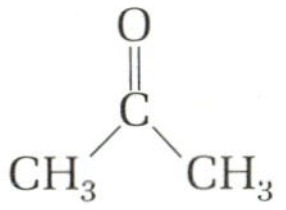
h. $HC≡CH$
i. SiH_3SiH_3
j. CH_2D_2

5 Which ion or molecule in each of the following pairs would have the most polarizable electrons?

a. Br^- versus F^-
b. Mg^{2+} versus Ba^{2+}
c. Na^+ versus Mg^{2+}
d. I_2 versus IF
e. OH^- versus F^-
f. OH^- versus SH^-
g. CH_4 versus NH_4^+
h. NH_3 versus CH_3NH_2
i. NH_3 versus PH_3

SUPPLEMENTARY READINGS

M. Orchin and H. H. Jaffé, *The Importance of Antibonding Orbitals.* Boston: Houghton Mifflin Co., 1967, Chapters 1 and 2. (P)

G. C. Pimentel and R. D. Spratley, *Chemical Bonding Clarified Through Quantum Mechanics.* San Francisco: Holden-Day, Inc., 1969. (P)

W. F. Luder, "The Electron Repulsion Theory of the Chemical Bond," *Journal of Chemical Education,* **44,** 206, 269 (1967). An interesting approach to chemical bonding, substantially different from the one given here.

CHAPTER 3

SOME PHYSICAL METHODS OF STRUCTURE IDENTIFICATION

It was noted in Chapter 1 that one of the areas of major concern to chemists is that of molecular structure — how atoms are connected together to form molecules. Virtually all chemists, even those not pursuing structural investigations per se, have need for this sort of information at one time or another. One of the best ways to obtain structural information is by means of **spectroscopy** — measuring the absorption of radiant energy by a compound, and in particular how the absorption changes as the wavelength of the radiation changes. In this chapter, the absorption of infrared (2 to 40 μ; 1 $\mu = 10^{-4}$ cm) and radio-frequency (10^2 to 10^3 cm) radiation will be examined.

The development of quantum mechanics was aided immensely by the concept that particles could exhibit wave-like properties. This idea had no doubt been prompted in part by the discovery, some years earlier, that electromagnetic waves acted at times as if they were collections of particles. At the turn of the century Max Planck showed that the energy in the waves was contained in little bundles or **quanta** (often called **photons**). He also found a very simple relationship between the amount of energy (E) in each quantum and the classical frequency (ν) of the wave (equation 3.1).

$$E = h\nu \tag{3.1}$$

Thus the higher the frequency (or shorter the wavelength, λ; $\lambda = c$ (speed of light)/ν) of the radiation, the greater the energy of each quantum.

How do these quanta interact with atoms? Consider a collection of hydrogen atoms in their ground states, $1s^1$. Suppose we wanted to excite these hydrogen atoms to the $2s^1$ state. The difference in energy between an electron in the 1s- and 2s-orbitals of hydrogen is 1.64×10^{-11} ergs, or 3.9×10^{-19} cal. Light with this quantity of energy in each photon would have a wavelength of 1215 A (A = Ångström = 10^{-8} cm) and would lie in the far ultraviolet region of the spectrum. Shining light of longer wavelength on the atoms (near ultraviolet or visible light) will produce no effect **no matter how intense the beam.** But if 1215 A light is used to irradiate the system, it will be seen that much of the light is absorbed and excited hydrogen atoms simultaneously produced.

Suppose now that the wavelength of the light is further decreased so that the photons each contain more than enough energy to bring about the electronic transition — what happens? Nothing — none of the radiation is absorbed. The energy of the photon must match very closely the energy difference between the two states of the system for absorption to occur. We have here a reaffirmation of the principle of quantization (Chapter 2) — since the system may only exist in certain energy states, there are only a limited number of possible energy changes that it can experience. Atoms (or molecules) *rarely* absorb only part of a quantum, or several quanta, in the course of undergoing an energy transition.

The principle may be clarified somewhat with the aid of a mechanical example. Consider a simplified pinball machine comprised only of a pair of flippers, a ball, and a hole part way up the board. As is well-known, excessive daintiness gets one nowhere in pinball — the ball never acquires enough "oomph" to make it. On the other hand, the ham-handed don't fare well either. The ball crashes to the top of the machine and then proceeds to roll right back to the starting point. Only when the right amount of energy has been imparted to the ball will it reach its target.

INFRARED SPECTROSCOPY

There are energy levels in molecules as well as in atoms, and transitions between them are governed by the same rules. In molecules, however, there are several different ways in which energy may be absorbed; one need not change the energy of the electrons. The entire molecule may rotate, for instance; and it may also vibrate because the atoms are held together by elastic rather than rigid forces. The C—C single bond distance (i.e., internuclear distance) is listed as 1.54 A, but this does not mean that if we were to look at a collection of molecules having C—C bonds, all of them would be that long at a given instant. Some would be longer and some shorter. The equilibrium value is 1.54 A but the bond length oscillates back and forth around this value within each molecule.

The molecular vibrations are quantized — the atoms may vibrate back and forth with only certain fixed energies. For an ideal diatomic molecule, the allowed vibrational energies are given by equation 3.2.

$$E = (n + 1/2)h\nu, \qquad n = 0, 1, 2, 3, \ldots \tag{3.2}$$

The adjacent energy levels are all $h\nu$ ergs apart, and it takes photons of energy $h\nu$ to excite the system from one vibrational level to the next. The quantity ν is the characteristic vibrational frequency of the molecule and is determined by the masses of the atoms and the strength of the bond. Note that it is impossible for the molecule to stop vibrating altogether — it has a zeroth-vibrational level, in which its vibrational energy is still $\frac{1}{2}h\nu$. If the molecules did stop vibrating, then we could determine the atomic positions (from the known bond distance) and momenta (zero). This measurement would violate Heisenberg's Uncertainty Principle.

Real molecules deviate somewhat from the theoretical model; in particular,

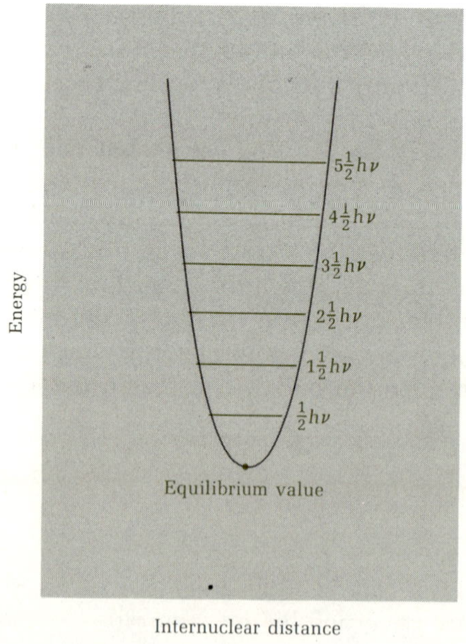

FIGURE 3-1
Vibrational Energy Levels of a Diatomic Molecule

the energy levels are not equally spaced but tend to squash together, especially for large values of n. Vibrational energy levels are much closer together than electronic ones, and so the energy needed to cause vibrational excitations is such that the radiation lies in the infrared part of the spectrum. Thus, measuring the absorption of infrared radiation by a molecule gives information about its internal vibrations, which are directly affected by its structure. Therein lies the value of infrared spectroscopy to the majority of organic chemists.

Consider the real molecule, hydrogen chloride. If one shines a beam of infrared light of continuously varying frequency on a sample of hydrogen chloride gas, nothing happens until the frequency gets close to 2885 cm^{-1},† when considerable absorption of the infrared radiation occurs. This corresponds to 8.7×10^{13} cps, which is very close to the fundamental frequency of the molecule. Since, at room temperature, most HCl molecules are in their zeroth-vibrational level ($n = 0$), those that absorb the infrared radiation will find themselves mostly in the first energy level ($n = 1$) and contain $h\nu = 5.7 \times 10^{-13}$ ergs more energy than the unexcited molecules.

For complex molecules with N atoms the number of possible independent modes of vibration is $3N - 6$. Each of these would have its own characteristic frequency, ν, as in the diatomic molecule. All the vibrational modes will not necessarily absorb infrared radiation, however, even when the frequencies match. To be infrared "active," a vibration must involve a change of **dipole moment.**

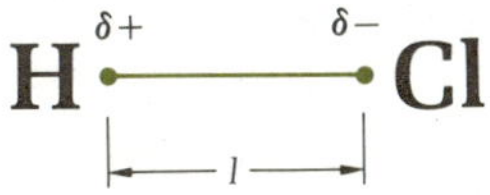

μ(dipole moment) $= \delta \times l$

FIGURE 3-2
Dipole Moment of Hydrogen Chloride

In HCl the bond is strongly polarized, and there is a center of negative charge near the chlorine atom and a positive center near the hydrogen atom. A molecule in which there are separate centers of positive and negative charge is known as a **dipole.** These dipoles have a dipole moment, μ, associated with them (Figure 3-2). As the molecule vibrates, the dipole moment changes because l changes — it is for this reason that HCl absorbs infrared radiation. A comparison with molecular chlorine, Cl_2, is instructive. Chlorine has no dipole moment, the centers of negative and positive charge exactly coinciding at the center of the bond. This situation remains unchanged as the atoms vibrate, and Cl_2 does not absorb radiation corresponding to its vibration frequency. In more complex molecules, there may also be vibrational modes that do not involve dipole moment changes.

Each vibrational mode in a complex molecule usually involves some movement by each one of the atoms. It often turns out, however, that in a given mode one group of atoms moves a good deal more than the rest do. In addition, the characteristic frequency for that motion is often found in other molecules that have the same group of atoms. So it becomes possible to speak of **group frequencies** that are characteristic of a functional group or particular arrangement of atoms in molecules. For instance, virtually all

---

†The frequency in cm^{-1} (called *wave numbers*) is obtained by dividing the frequency ν in cps by c, the speed of light. This is done because the resulting values are convenient in size.

molecules that contain an —OH group — a hydrogen bound to an oxygen — absorb infrared light that has a frequency in the range 3400 to 3600 cm^{-1}. Many more examples will follow shortly.

As the number of atoms in a molecule goes up, the number of possible isomers increases very steeply. One of the great values of infrared (ir) spectroscopy is its use in rapidly eliminating many of the possibilities from consideration. To take an example treated in Chapter 2, we know that a molecule of formula C_3H_8O can have one of three structures: $CH_3OCH_2CH_3$, $CH_3CH_2CH_2OH$, $CH_3CH(OH)CH_3$. Inspection of the ir spectrum of a C_3H_8O isomer would quickly reveal the structure of the compound in question, depending on the absence or presence in the spectrum of bands in the 3400 to 3600 cm^{-1} region (Figure 3-3). (In Figure 3-3 we have used the spectrum of $CH_3CH_2OCH_2CH_3$ rather than that of $CH_3OCH_2CH_3$, since the latter is a gas at room temperature. The two have quite similar spectra.)

FIGURE 3-3
Infrared Spectra of $(CH_3CH_2)_2O$ and $(CH_3)_2CHOH$

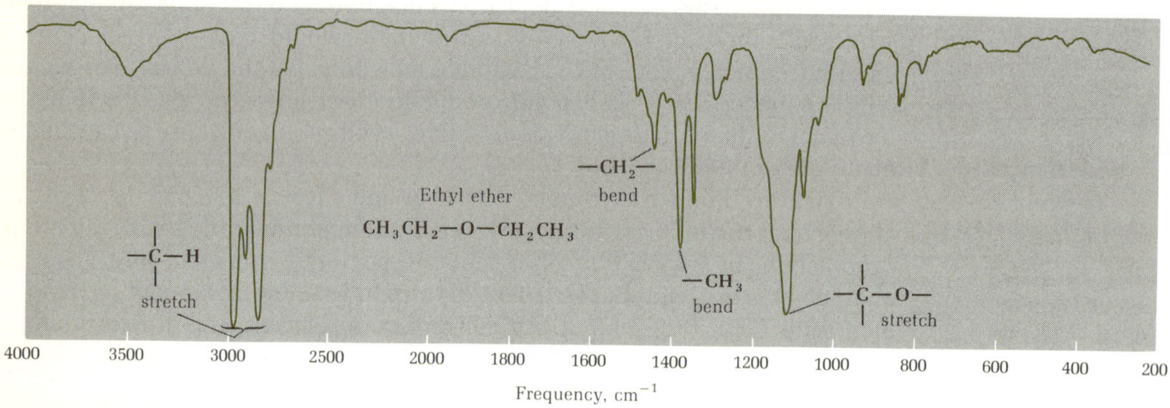

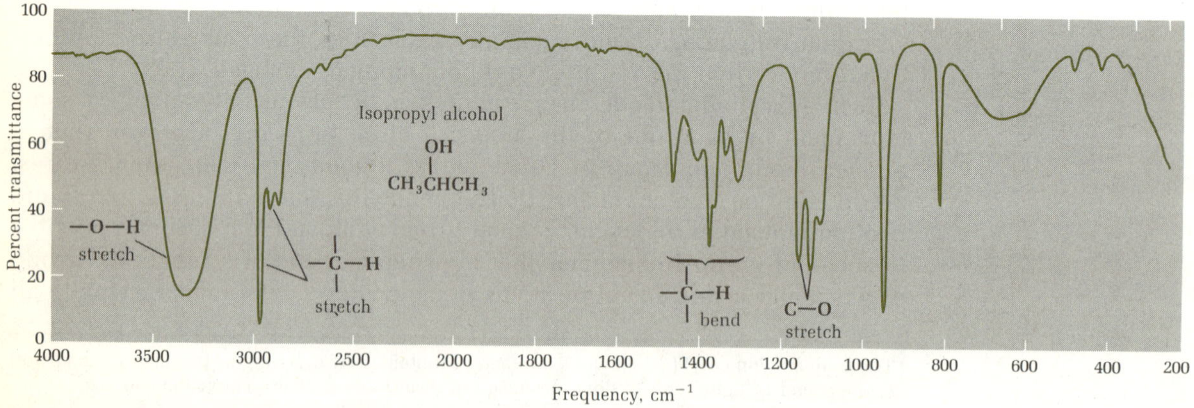

The presence or absence of an —OH group can usually be determined with ease by means of chemical tests. Ferreting out all the functional groups in a complex molecule by chemical means can be a tedious process, however. The ir spectrum gives information on a large number of these groups simultaneously, and one need consume *none* of the sample to obtain it. Finally, one should note that, in the case of the C_3H_8O isomers, the presence of an —OH group is very easily determined by ir spectroscopy, but its location within the molecule is less immediately obvious. Nuclear magnetic resonance spectroscopy is very effective in this regard, and the two types of spectroscopy in combination are a very powerful tool indeed.

A CLASSICAL MOLECULAR MODEL

We have noted that the atoms in a molecule are bound together by rather elastic forces. To make use of a classical analogy, we might imagine a set of hard spherical atoms joined by springs. Surprisingly enough, this model fits molecular behavior quite well and provides an excellent qualitative framework for understanding molecular vibrations.

Classically, the strength of a spring is directly related to a quantity called the **force constant,** *f*. The larger *f* is, the stronger the spring and the higher the frequency of vibration of two objects connected by that spring. If we take as a model a system A—B, in which A and B may be atoms or polyatomic units, then the frequency with which they will vibrate with respect to one another depends on the masses of A and B and on the force constant of the bond between them (which is the "spring"). Equations 3.3 and 3.4 give the approximate relationship; *m* is called the *reduced mass,* and m_A and m_B are the masses of A and B, respectively.

$$\nu = \frac{1}{2\pi}\sqrt{\frac{f}{m}} \qquad (3.3)$$

$$m = \frac{m_A m_B}{m_A + m_B} \qquad (3.4)$$

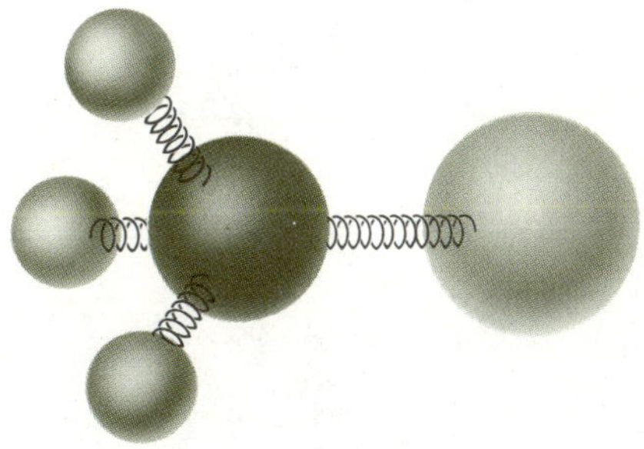

FIGURE 3-4
Ball and Spring Model of Methyl Chloride

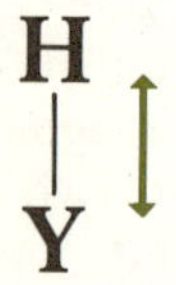

Y-H stretching mode

Two H-Y bending modes

FIGURE 3-5
Bending and Stretching Vibrations for a Hydrogen Atom

These equations can be qualitatively useful. Suppose that m_A in equation 3.4 is much smaller than m_B; then $(m_A + m_B) \simeq m_B$, and $m \simeq m_A$. A common such case is when A is a hydrogen atom. Although both halves of the system move during the vibration, the lion's share of the motion falls to the light hydrogen atom. The molecule behaves then as though the heavy atoms were motionless and the hydrogen atom(s) alone moved.

It now becomes necessary to take note of the fact that there are two different ways in which a hydrogen atom can move relative to another atom. The bond may be *stretched* so that H and the other atom move further apart or it may be *bent* so that the bond length remains the same but the bond angles change (Figure 3-5). It takes more energy to stretch the bond than to bend it; this finding accords with our own experiences of the relative ease of stretching and bending springs.

The stretching vibrations of hydrogen atoms lie at the high frequency end of a normal ir spectrum. By referring to equation 3.4 again, with $m \simeq m_H$ ($m = 1$), we see that the frequencies of hydrogen stretching vibrations will vary with the size of the force constants of the bonds between hydrogen and various other elements. We might also expect a correlation between these force constants and the bond energies. Table 3-1 gives the ranges of a number of hydrogen stretching frequencies, and it does indicate that there is a rough correlation between bond energies and force constants.

The sensitivity of the hydrogen stretching vibration to the nature of the atom bearing the hydrogen atom is very useful analytically. One can, for instance, even deduce the hybridization states of various carbon atoms within the molecule. Figure 3-6 contains the infrared spectra of several simple molecules on which the origins of some of the more prominent bands have been indicated. Note the marked differences in the spectra of allyl alcohol and acetone. They are isomers of one another (C_3H_6O) but their ir spectra serve to differentiate them immediately.

TABLE 3-1
Stretching Frequencies of Some H—Y Bonds

H—Y Bond	Frequency Range (cm^{-1})	Average Bond Energies (kcal/mole)
H—F	3958	136
H—O—	3650–3400	111
H—N<	3500–3300	~92
H—C≡ (*sp*-hybridized)	3300	~125
H—C< (*sp*^{2}-hybridized)	3100–3000	103
H—C— (*sp*^{3}-hybridized)	3000–2850	95
H—S—	2600–2550	~88

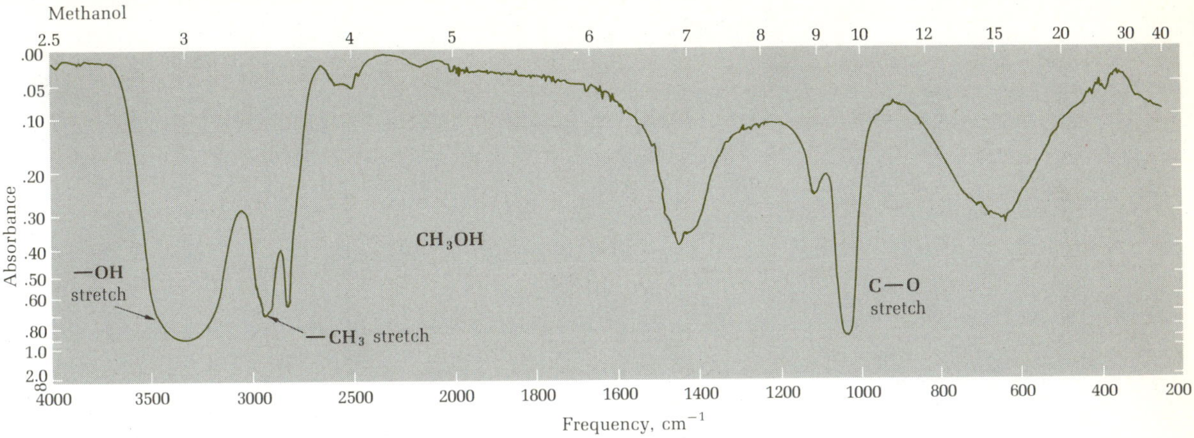

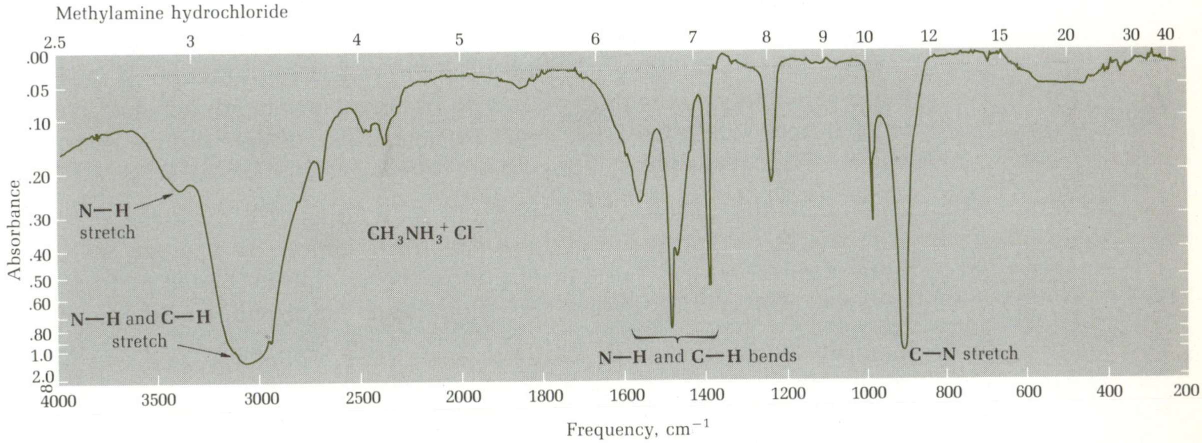

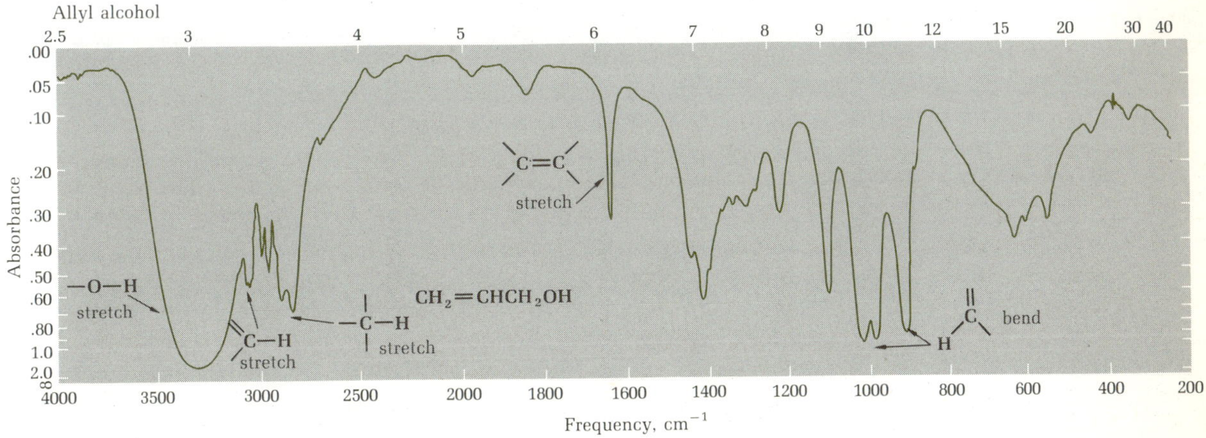

FIGURE 3-6
Infrared Spectra of Methanol, Methylamine Hydrochloride, Allyl Alcohol, and Acetone (p. 40)

Acetone

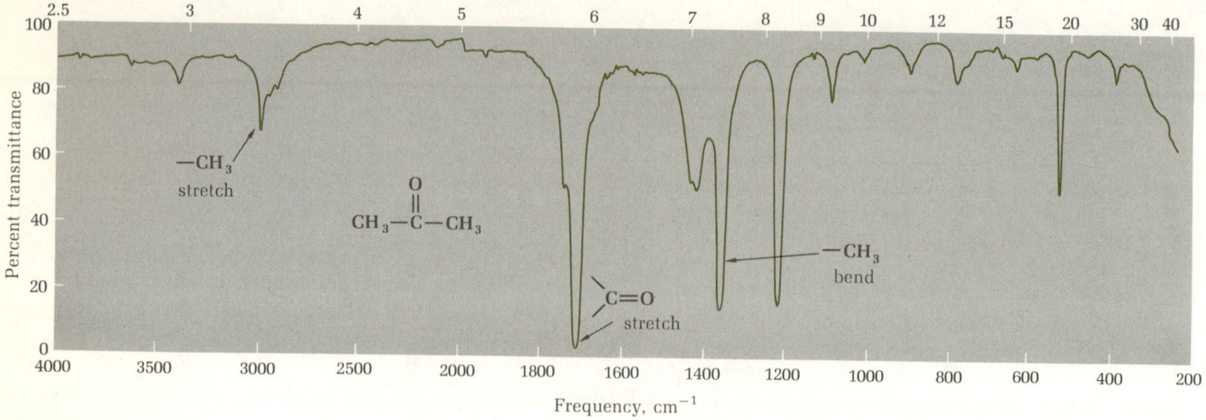

The bending vibrations of X—H bonds are not as generally useful as are the corresponding stretching vibrations for structure determination. It is sometimes valuable to know that —CH_3 groups give rise to bending bands at 1360 to 1450 cm^{-1}, while the corresponding bands for —CH_2— groups lie in the range 1460 to 1480 cm^{-1}.

Among the most useful absorptions for characterizing molecular structures are those due to the double and triple bonds. By analogy with the pattern found for the Y—H absorptions, those multiple bonds that are the most polar have the highest vibration frequencies and absorb the most intensely. Peaks due to >C=O are always strong, while those due to >C=C< are usually weak and sometimes not detectable.

TABLE 3-2
Stretching Frequencies of Multiple Bonds

Multiple Bond	Frequency (cm^{-1})
—C≡N	2260–2215
—C≡C—	2250–2100
>C=O	1900–1675
>C=N—	1690–1630
>C=C<	1680–1600

PRACTICAL CONSIDERATIONS

At this point it would be worthwhile to indicate just how an infrared spectrum is obtained. It is possible to obtain spectra of molecules in the gas phase, and there are certain advantages in having such spectra. In order to be suitable for gas-phase spectroscopy with commonly available equipment, however, a compound must have a vapor pressure of 10 to 20 mm at room temperature, a condition that serves to eliminate most organic molecules.

The usual procedure is to make up a 5 to 10% solution of the solid or liquid in a suitable solvent, preferably one that absorbs relatively little infrared radiation in the range being studied. Carbon tetrachloride, CCl_4, and chloroform, $CHCl_3$, are very widely used since between them they dissolve a variety of organic molecules. The solutions are put into cells that are then placed in the spectrometer. Ordinary glass absorbs infrared radiation and is unsuitable for making cells. It turns out that sodium chloride is transparent to infrared radiation at the frequencies ordinarily used. Cells are made from rock salt plates, and the hollow into which the liquid is injected is usually only 0.1 to 0.2 mm thick or less. Very little sample is needed to obtain a spectrum, and if necessary it can be recovered from the solvent afterward.

Many compounds, such as salts, will not dissolve to a sufficient extent in these organic solvents to give a useful spectrum. Water is clearly a forbidden solvent with sodium chloride cells. The most common method for avoiding these difficulties involves grinding a small quantity of the sample with potassium bromide and subjecting the mixture to high pressures in a special die. The potassium bromide melts and then resolidifies, forming a thin translucent disc. The bromide is transparent to infrared radiation so that only the organic compound, finely dispersed, will be responsible for absorption.

The spectrometer itself is a device for obtaining a **monochromatic** (Greek — of a single color, containing only one wavelength) beam of radiation, continuously changing its wavelength, and measuring its absorption at each wavelength by the sample. Most organic chemists use double beam spectrometers. One beam passes through a cell containing the solution of the sample while the other passes through a cell containing pure solvent. The difference in absorption in the two beams is measured and the effect of solvent absorption is thereby canceled out. The graph that the spectrometer produces, showing absorption versus frequency, is called the **infrared spectrum** of the compound.

The peaks are often broad rather than very sharp as one would expect. Molecules undergo rotation as well as vibration; the rotational energy levels lie very close together and are associated with the vibrational energy levels. Various molecules will be in different rotational energy levels when they are excited and will absorb radiation at a number of slightly different frequencies so that an absorption **band** is produced. In addition, molecules in solution interact with solvent molecules, which also results in a wide spread of differences between energy levels. For this reason, gas-phase spectra are often sharper and better defined than the corresponding solution spectra.

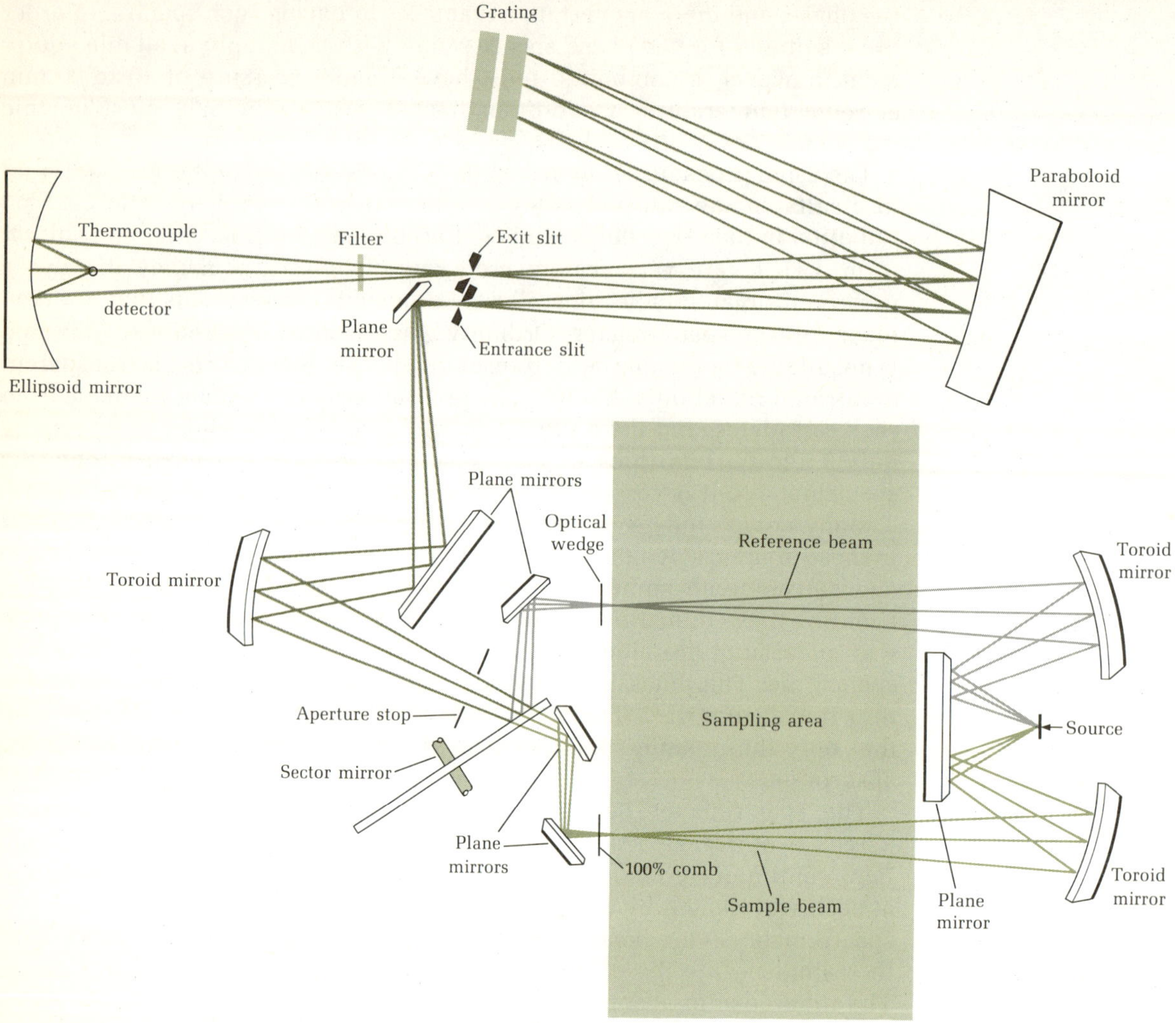

FIGURE 3-7
Optical Diagram of an Infrared Spectrophotometer

NUCLEAR MAGNETIC RESONANCE SPECTROSCOPY

In order to rationalize completely the behavior of electrons, it is necessary to assume that they possess an intrinsic spin. They are not the only particles that appear to do so; protons and neutrons also act as if they were spinning. All three have the same spin quantum number, $I = \frac{1}{2}$. These spinning particles act like bar magnets and tend to line up when in an external magnetic field. The number of ways in which the axis may orient itself in the field is equal to $(2I + 1)$ — in the case of electrons this value is 2. And indeed we know that there are only two allowable orientations for the electron spin.

Many nuclei exhibit spin properties like those associated with electrons. All nuclei except ^1H are made up of both protons and neutrons. The spins of the individual neutrons and protons may cancel one another out or they may combine so that nuclei with $I = 0$ (no spin) and nuclei with $I > \frac{1}{2}$ are also known. Any isotope that has an even number of both protons and neutrons has $I = 0$. Three are of particular interest to us, each of them constituting the major isotopic species of that element: ^{12}C, ^{16}O, and ^{32}S. Nuclei with $I = \frac{1}{2}$ include ^1H, ^{13}C, ^{19}F, and ^{31}P. Of the many with $I = 1$, the most significant for us are ^2H and ^{14}N. In the discussion that follows we shall be concerned solely with proton spectra, but the general principles apply to any nuclei with $I > 0$.

The two orientations of the proton in a magnetic field are not identical in energy. That orientation in which the north pole of the nuclear "magnet" points toward the south pole of the field is slightly lower in energy than the alternative arrangement. If a collection of bare protons were placed in a magnetic field, some would be aligned in each orientation, with a majority in the lower energy state. If the nuclei were irradiated at such a frequency that the energy in one quantum exactly equaled the energy difference between the two spin states, many nuclei would absorb those quanta and "flip their spins." This is the **nuclear magnetic resonance** condition. The correct value of that frequency is given by $\nu = \gamma B/2\pi$, where B is the strength of

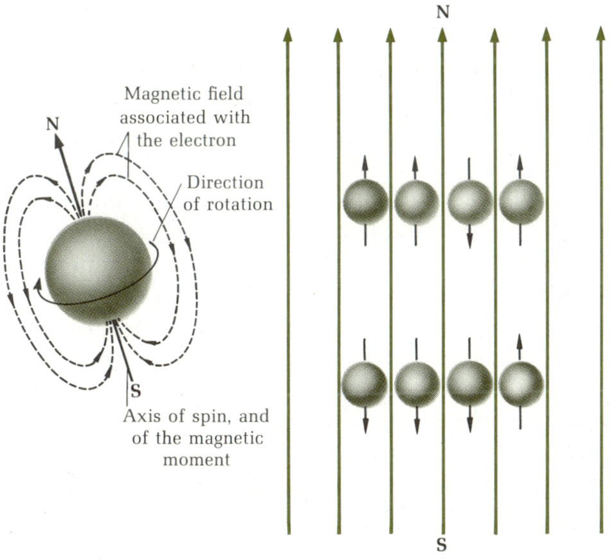

FIGURE 3-8
The Magnetic Field of a Spinning Electron. Alignment of Electron Magnetic Moments in a Magnetic Field

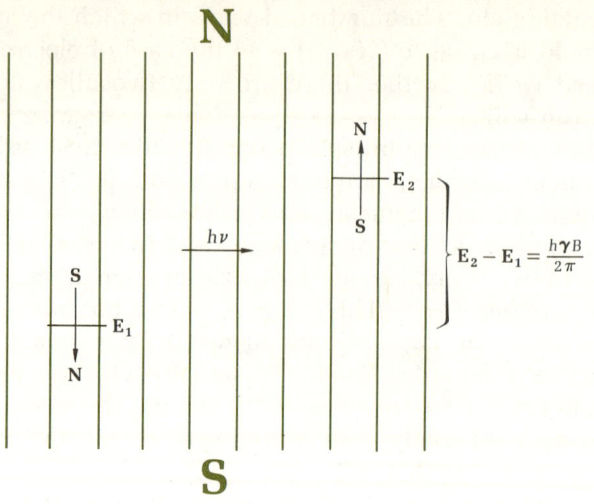

FIGURE 3-9
Flipping of a Nuclear Spin Aligned in a Magnetic Field

the magnetic field and γ is a constant characteristic of the proton (and different for each different nucleus). In most nuclear magnetic resonance (nmr) spectrometers, radiation of a fixed frequency is used, and the magnetic field strength, B, is adjusted until resonance between photons and nuclei takes place. It is also possible to hold B fixed and vary the radiation frequency.

If we were considering only bare hydrogen nuclei placed in a magnetic field, then the situation would be a simple one. All the protons would absorb energy and change their spin orientations at exactly the same magnetic field strength (for a given frequency of radiation). If a hydrogen atom is part of a molecule, however, bonded to another atom by means of a two-electron bond, the situation becomes slightly more complex. Under the influence of the magnetic field, the electrons circulate in certain preferred paths; this motion gives rise to a weak *induced* magnetic field, which opposes the applied field. The induced field tends to lessen the overall magnetic force felt by the hydrogen nuclei — the electrons are said to **shield** the nucleus. The applied field must therefore be increased in order to cancel out the shielding and allow the protons (hydrogen nuclei) to come into resonance.

If protons in different chemical environments are shielded to different extents, then each type will flip when a different value of the magnetic field strength is applied. This is the entire basis of nuclear magnetic resonance spectroscopy. Some straightforward examples follow.

The proton of a hydride ion, H^-, is surrounded by two electrons; each proton in a hydrogen molecule, H_2, has only a half-time claim on the pair of electrons. The proton in H^- is more shielded than the one in H_2, and it requires a higher field strength before it will flip. Analogously the protons in H_2^+ are less shielded than those in H_2.

To take examples somewhat more chemically interesting, the protons of methane, ammonia, and water should each give rise to peaks at a different value of the magnetic field strength. Such is indeed the case, and we shall further see that different groups of protons within the *same* molecule may be differentiated by means of nmr spectroscopy. Note that all four protons of methane are identically equivalent and that the nmr spectrum of methane therefore consists of only *one* peak (ammonia and water as well).

THE CHEMICAL SHIFT

It is experimentally difficult to measure accurately the value of the magnetic field strength at which a resonance occurs. The usual procedure is to include some reference substance along with the sample and then measure the positions of the peaks from the sample relative to that of the reference. This difference in position between a given sample peak and the reference peak is called the **chemical shift.** The usual reference compound for proton work is *tetramethylsilane*, $(CH_3)_4Si$ (commonly known as TMS). TMS is chemically unreactive, volatile, and its hydrogen atoms, which are all equivalent, are very heavily shielded and tend to resonate at a higher field strength than almost any other protons.

The difference in the field strength needed to cause a proton resonance in a sample and in the reference — methane and TMS, for example — is not a constant. If a powerful magnet is used, then these resonances will be relatively far apart; as the field strength of the magnet (*and* the frequency of the radiation) goes down, the resonances tend to lie closer together. Most nmr spectrometers use radiation sources that give radiation of 60×10^6 cps [60 megaHerz (MHz)]. The magnetic field strength necessary to cause proton resonances (they all lie within a very small range) is then about 14,000 gauss. Spectrometers are coming into use that employ 100 MHz and even 220 MHz radiation sources; their magnetic fields must be proportionately stronger.

It is convenient to use chemical shift values that are the same for a given compound and reference, regardless of the spectrometer used to obtain the data. This has been done by defining the δ unit (equation 3.5), in which the difference in field strength between sample and reference resonance peaks are divided by the actual value of the field strength for the reference peak. The chemical shift is now independent of the operating parameters of the spectrometer.

field strength for sample resonance ↘ ↙ field strength for reference resonance

(3.5) $$\delta = \frac{(B \text{ sample} - B \text{ reference})}{B \text{ reference}} \times 10^6$$

Multiplication by 10^6 is done for numerical reasons, and the "units" of δ are parts per million (ppm). Since the protons in most samples resonate at a lower field strength than the protons of TMS do, δ as defined is usually negative although the negative sign is frequently omitted when data are reported. It is important to remember that **the larger the negative value of δ, the more deshielded are the nuclei.**

In order that nmr spectroscopy be useful to the organic chemist, the chemical shifts for hydrogens in different molecular surroundings should be measurably different. Consider the methyl halides, CH_3X. All three hydrogen atoms are completely identical, and each compound gives a proton nmr spectrum consisting of only one peak. Their chemical shifts are in δ units: CH_3F, 4.26; CH_3Cl, 3.05; CH_3Br, 2.69; CH_3I, 2.19.

The hydrogens in methyl fluoride are the least shielded (most *deshielded*), while those in methyl iodide are the most shielded. Fluorine is a highly electronegative atom and would be expected to draw electrons away from carbon. The carbon atom, in turn, would tend to attract the electrons in the carbon-hydrogen bonds so that these electrons would be less able to shield the hydrogen nuclei. The shielding of the methyl hydrogens increases as the electronegativity of the attached halogen decreases. The halogen nuclei themselves can also undergo spin change, but this takes place at very different values of magnetic field strength from those at which hydrogens resonate.

There is a general correlation between the shielding of hydrogen atoms bound to carbon and the electronegativity of other atoms attached to that carbon atom. The correlation is by no means quantitative, however. It is even less reliable for hydrogens attached directly to atoms other than carbon. It must be remembered that in the molecule the electrons are circulating about *all* the nuclei and not only the hydrogen atoms. Electron circulation around a given atom (call it A) sets up a magnetic field that almost always leads to shielding of the A nucleus. But that magnetic field might have the opposite effect on neighboring nuclei, as the diagram in Figure 3-10 indicates.

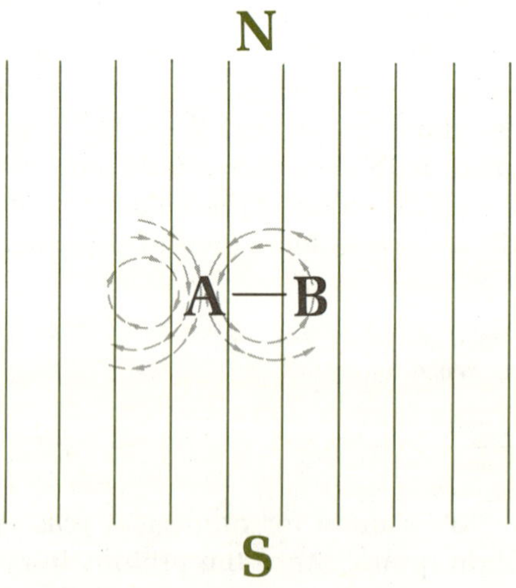

FIGURE 3-10

Induced Magnetic Fields in the Molecule A—B

The importance of this result depends on many factors, such as the susceptibility of the electrons around A to the influence of the applied magnetic field and how that susceptibility varies with the orientation of A—B. These and other phenomena serve to complicate attempts to correlate chemical shifts with simple physical parameters.

SPIN-SPIN SPLITTING

A most useful effect, known as spin-spin splitting, occurs when *chemically different* hydrogen atoms occupy adjacent nuclei. Ethane, CH_3CH_3, gives only one peak in its nmr spectrum because all six protons are identical. The spectrum of ethyl bromide, CH_3CH_2Br, shows two peaks under low resolution, the larger of which is at higher field. The peak at δ1.65 has an area 1.5 times as large as the peak at δ3.37; this would seem to indicate that the high field peak arises from the CH_3— protons while the low field peak is due to the —CH_2Br protons. This might have been anticipated on the basis of the electronegativity of bromine. The fact that the areas under the peaks stand in the same ratio to one another as the number of each type of hydrogen atom can be very useful.

When ethyl bromide is examined under high resolution, the two large peaks noted above can each be seen to be subdivided into several smaller peaks. The resonance centered at δ1.65 consists now of three peaks and is labeled a **triplet;** the δ3.37 peak has become a **quartet.** The total areas of the triplet and of the quartet still stand in a ratio of 1.5 : 1.0 (Figure 3-12).

The splitting of the methyl peak is due to the fact that the resonance of the methyl protons is sensitive to the spin arrangements of the *neighboring* protons. These neighbors are, after all, also spinning and thereby creating yet another magnetic field that contributes to the overall field experienced

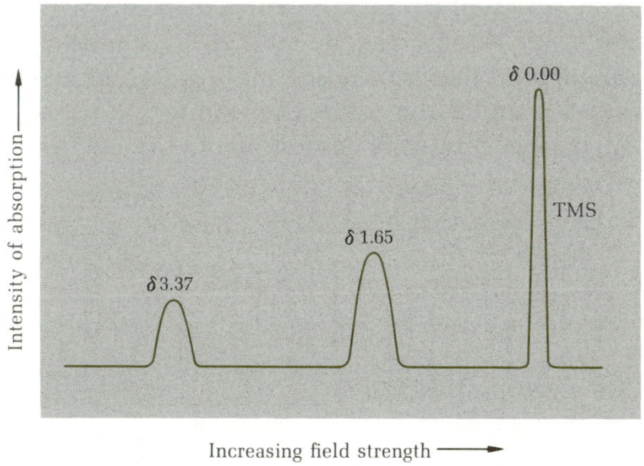

FIGURE 3-11
Low Resolution NMR Spectrum of Ethyl Bromide

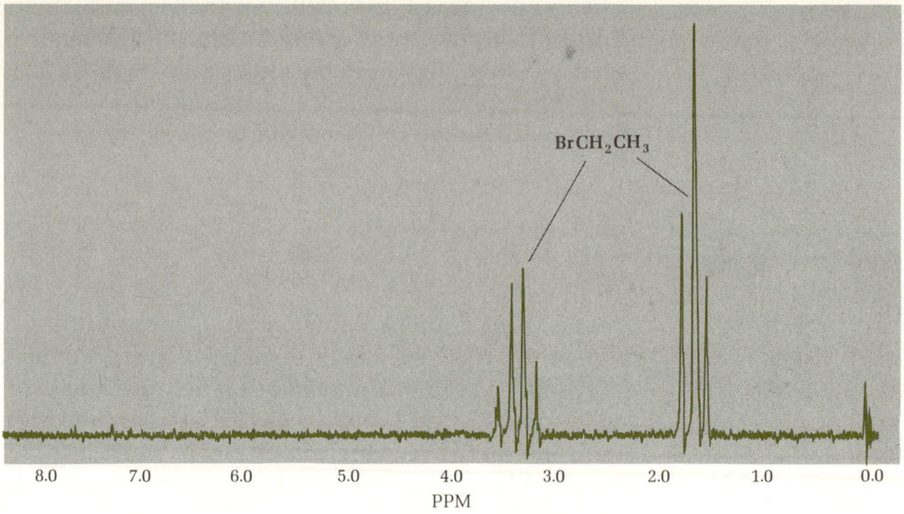

FIGURE 3-12
High Resolution NMR Spectrum of Ethyl Bromide

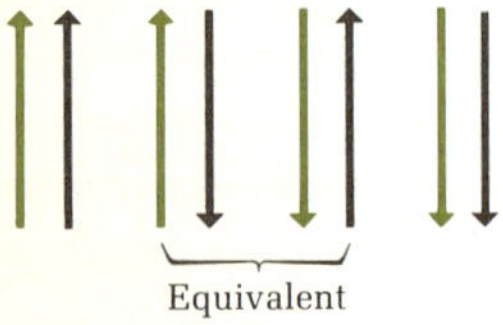

FIGURE 3-13
Possible Nuclear Spin Arrangements of Methylene (-CH_2-) Protons

by the methyl protons. The particular effect of the field generated by a given proton depends on whether its magnetic moment is aligned with or against the main magnetic field arising from the instrument. A system consisting of a single proton has only two possible spin arrangements — either "up" or "down." A two-spin system has three possible spin states, however — both up, both down, or one up and one down (Figure 3-13).

Since there are two equivalent ways of arriving at a one up, one down arrangement, this configuration is twice as probable as the other two. What, then, will a spectrometer "see" as it scans a large collection of ethyl bromide molecules in the region where the methyl protons absorb energy? In 25% of the molecules the methyl group will be next to a methylene in which both proton spins are up; 50% of the molecules will contain methyl groups whose neighboring protons have one spin up and one down; and 25% of the molecules will contain methylene groups with both proton spins down. In each of the three cases the methyl resonance comes at a slightly different field value, and that is why the methyl resonance of ethyl bromide actually appears as a triplet, the components of which have areas in the ratio of 1:2:1.

Similar reasoning tells us that the methylene proton peak will be affected

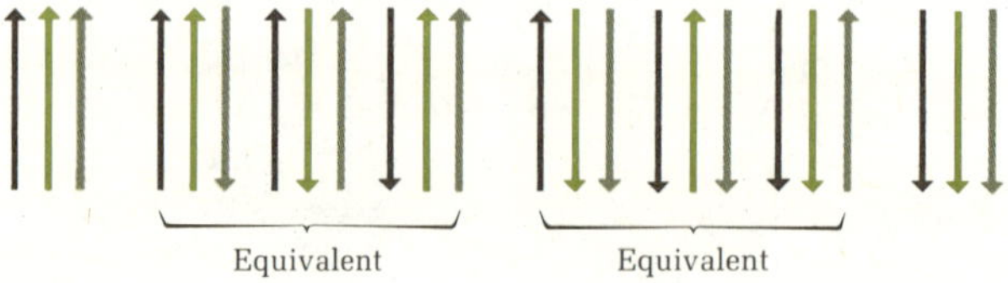

FIGURE 3-14
Possible Nuclear Spin Arrangements of Methyl (CH_3—) Protons

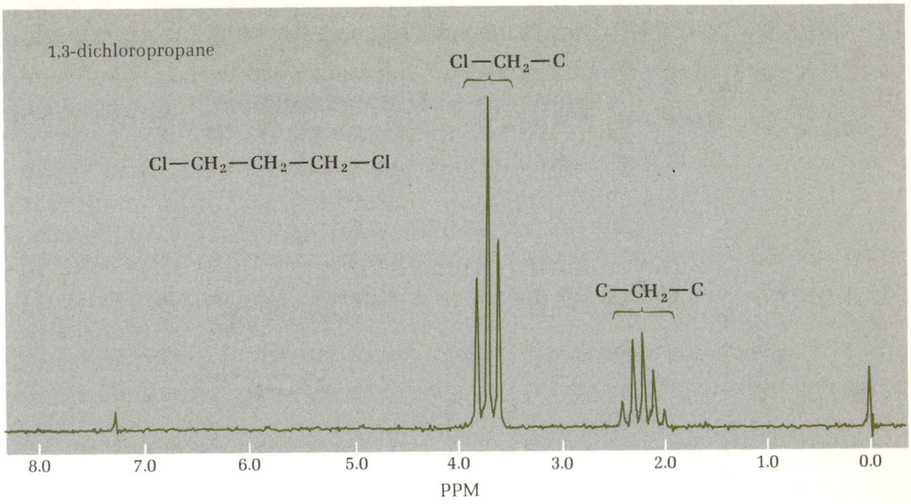

FIGURE 3-15
High Resolution NMR Spectrum of 1,3-Dichloropropane

by the spin arrangements of the methyl protons. The three methyl proton spins may be combined in four independent ways (Figure 3-14). As a result, the methylene proton peak appears as a quartet, the area of its component peaks standing in the ratio 1:3:3:1. In general, if n equivalent protons are adjacent to another set of hydrogen atoms, and the resonance of these nuclei is being observed, it will be split into $n + 1$ peaks. If the hydrogen atoms being studied have n neighbors of one kind and m of another kind, the total splitting of their peak will be $(n + 1)(m + 1)$.

To further illustrate the principles behind the splitting patterns, let us examine the spectrum of 1,3-dichloropropane, $ClCH_2CH_2CH_2Cl$. The four hydrogen atoms at the ends of the molecule are exactly equivalent and act as a single unit even though they are on different carbon atoms. The areas of the resulting triplet and quintet are in the ratio 2:1. The components of each divided peak are evenly spaced, and the distance between adjacent components is known as the **coupling constant,** J (expressed in cycles per second). The larger the value of J, the stronger is the interaction between the two types of hydrogen atoms.

Although significant coupling is normally observed only when the protons involved are attached to adjacent atoms, one occasionally observes strong coupling between protons more distantly separated. This occurs most often when the atoms in question are separated by a double or triple bond.

EQUIVALENCE OF PROTONS ATTACHED TO THE SAME CARBON ATOM

An assumption has been made throughout — namely, that each hydrogen atom in the methyl group is identical to the other two. If this were true, then these protons would not be coupled to one another and no splitting attributable to the interaction of two methyl protons should be observable. The spectrum actually obtained certainly substantiates this interpretation.

The equivalence of the methyl protons arises from the very rapid internal rotations that are taking place about the carbon-carbon bond (Chapter 5). This rotation tends to average out the methyl hydrogen atoms, as far as the neighboring nuclei are concerned. If the nmr measurement had been made at very low temperatures, such that the rotations were substantially slowed, the methyl hydrogen atoms would become nonequivalent and the spectrum would increase considerably in complexity.

INTERPRETATION OF NMR SPECTRA

It might be worthwhile at this point to list explicitly the kind of information that can be obtained from a proton nmr spectrum:

1 The position of a peak, in δ units, gives a reasonable idea of the type of carbon atom to which the protons are attached.

2 The area under the peak is a measure of the *relative* number of that type of hydrogen atom.

3 The splitting pattern indicates the number of each type of *neighboring* hydrogen atoms.

If we return to our set of C_3H_8O isomers, we recall that application of infrared spectroscopy would enable us to tell very quickly the isomers containing the —OH group from the one that did not. Differentiating between the two hydroxyl-containing isomers, *n*-propyl alcohol ($CH_3CH_2CH_2OH$) and isopropyl alcohol ($CH_3\underset{\displaystyle OH}{\underset{|}{C}}HCH_3$), was not quite so straightforward. This latter problem would be a very simple task, indeed, if one had nmr capability.

The other side of the coin is illustrated by a comparison of $CH_3\overset{\displaystyle O}{\overset{\|}{C}}CH_3$ (acetone) and CH_3CN. They each give rise to one singlet in their nmr spectra, the chemical shifts of which are very similar. Without some prior knowledge one could not distinguish them by this method. Ir spectroscopy, on the other hand, would immediately settle the question.

TABLE 3-3
Chemical Shift Ranges for Some Representative Protons

Functional Group	Chemical Shift Range (δ)
$-\overset{\vert}{\underset{\vert}{C}}-OH$[a]	0.5–0.8
$-\overset{O}{\overset{\Vert}{C}}-OH$[b]	12.1–12.3
$CH_3-\overset{\vert}{\underset{\vert}{C}}-$	0.77–1.88
$CH_3-N\langle$	2.12–3.10
CH_3-O-	3.24–4.02
$CH_3-\overset{O}{\overset{\Vert}{C}}-$	1.95–2.68
$-CH_2-\overset{\vert}{\underset{\vert}{C}}-$	0.98–2.03
$-CH_2-N\langle$	2.28–3.60
$-CH_2-O-$	3.36–4.48
$-CH_2-\overset{O}{\overset{\Vert}{C}}-$	2.07–2.42

[a] Infinitely dilute solution.
[b] Dimeric form.

A representative, although by no means complete, list of proton chemical shifts is presented in Table 3-3. The position of —OH protons is very dependent on solvent and on concentration.

OTHER NUCLEI

We have concentrated our attention on proton nmr spectra because these are the most useful to the organic chemist. Nmr spectroscopy of other nuclei is now fairly commonplace, however. The simplest nucleus to which one can extrapolate from protons is ^{19}F because it also has $I = \frac{1}{2}$. In fact, fluorine nmr spectra resemble proton nmr spectra very closely, except for the fact that the chemical shifts and coupling constants are much larger when fluorine is being scanned.

PRACTICAL DETAILS

Nmr spectra are usually obtained from solutions of the sample compound. Interactions between the solvent and the sample can have a profound effect on chemical shifts, so the choice of solvent is important. The solvent should not, of course, produce a peak on top of peaks due to the sample. Two of the best solvents are carbon tetrachloride, CCl_4, and carbon disulfide, CS_2, which contain no protons. They are not capable, however, of dissolving all compounds to a sufficient extent (3 to 15%) to allow a spectrum to be obtained. If other organic solvents must be used, one trick that is widespread is to replace their protons by deuterium (D). Deuterium nuclei possess spin, but they undergo transitions at different field strengths than protons do. The net effect of substituting deuterium for protium (light hydrogen) is to blank out the peaks that normally arise from those protons. Two fairly common nmr solvents are deuterochloroform, $CDCl_3$, and dimethyl sulfoxide-d_6.

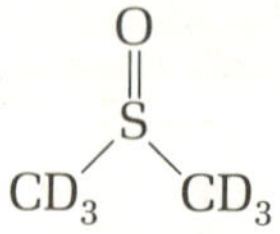

If samples are fairly insoluble in organic solvents, then heavy water (D_2O) may be tried.

When organic solvents are used, about 1% of TMS is added directly to the solution. The solution is then placed in a narrow glass tube that sits in a cavity directly between the poles of a magnet. The tube is spun rapidly while the spectrum is being recorded, in order that the sample "see" as uniform a magnetic field as possible.

Somewhat more detailed discussions of the spectroscopic properties of various functional groups will be integrated with discussions of the chemistry of these groups. The applications of spectroscopy to real chemical problems, examples of which will be brought forward from time to time, will allow the beginner to familiarize himself with the power and limitations of spectroscopic methods.

SUMMARY

1 The forces holding atoms together in a molecule are elastic; consequently the molecules vibrate. The vibrations are quantized, only certain vibrational energies being allowed; each vibration is characterized by a fundamental frequency, ν. These frequencies lie in the infrared region of the electromagnetic spectrum. The infrared spectrum of a compound is obtained by

passing a continuously changing beam of light through a solution of the sample and automatically recording the absorption at every wavelength.

2 There are peaks in the infrared spectra of various compounds that may be associated with particular atoms or groups of atoms; the various hydrogen stretching frequencies give rise to such peaks. In these cases, the vibration frequencies are related to the bond strengths. It is possible to distinguish hydrogen atoms attached to carbon in its three different hybridization states by this means. Hydrogen atoms also undergo bending motions, the frequencies of which can be connected with the molecular structure.

3 The stretching of various double and triple bonds also gives rise to characteristic infrared peaks. In order for absorption of infrared radiation to occur, even when its frequency matches that of a fundamental vibrational mode, the particular vibration involved must cause a change in the dipole moment.

4 Many nuclei arc spinning and will line up in certain allowable orientations when subjected to a magnetic field. The proton has two allowed orientations, one of which is slightly lower in energy than the other. If radiation of the correct frequency is supplied, some of these nuclei will absorb it and pass from the lower energy state to the one of higher energy.

5 The induced circulation of electrons around a nucleus gives rise to a local magnetic field opposing the applied field at that nucleus. This leads to shielding of the nucleus in question and a displacement of its resonance frequency. The electron circulation may bring about the opposite result at neighboring nuclei.

The resonance positions of protons in different chemical environments are non-identical. These chemical shifts are standardized by measuring them in relation to some reference substance, usually TMS. There is often a correlation between the chemical shifts of various protons and the electronegativity of nearby atoms. This connection is not automatic, however; many other factors have an effect on the chemical shift.

6 Protons on neighboring nuclei are magnetically coupled to one another. The different spin orientations of one set of equivalent nuclei result in slightly different chemical shifts for the neighboring nuclei. In general, n equivalent nuclei will split the resonance of neighboring protons into $n + 1$ components. The equivalence of the three protons in a methyl group, or the two protons in a methylene group, is a consequence of the very rapid internal rotations that take place in most compounds at room temperature.

7 Many nuclei other than ^1H will give rise to resonance spectra; the principal requirement is that they have $I > 0$. The number of allowable orientations

of the nucleus in a magnetic field equals $(2I + 1)$. Fluorine, for example, has $I = \frac{1}{2}$, just as hydrogen does. The spectra of fluorine-containing organic compounds resemble those of the more usual hydrogen-containing analogues. The resonance peaks show splitting due to fluorine-fluorine and fluorine-hydrogen coupling. A principal difference between fluorine and hydrogen is that chemical shift differences are much larger with fluorine.

8 Ir and nmr spectra often give complementary information about the structure of a molecule. Ir spectra can indicate the presence of functional groups containing atoms such as carbon and oxygen, which do not show up directly in nmr spectra. On the other hand, nmr spectra can indicate which atoms and groups are neighbors. Spin-spin splitting patterns are particularly valuable sources of this type of information, which is difficult or impossible to obtain from ir spectra.

PROBLEMS

1 The O—H and C—H stretching frequencies for methanol, CH_3OH, occur at about 3600 and 2900 cm^{-1}, respectively. Suppose that you had a sample of completely deuterated methanol, CD_3OD. Where would you expect the O—D and C—D stretching bands to be found? (Remember that equations 3.3 and 3.4 connect frequency with mass and that the *potential energy* of a chemical bond is unaffected by isotopic substitution.)

2 The infrared spectrum of carbon monoxide consists of a single band centered at 2143 cm^{-1}. On the basis of the data supplied in the tables, what sort of bond does there seem to be between the carbon and oxygen atoms? Draw a structure for carbon monoxide, showing all the valence electrons, which is in accord with your answer.

3 Write out structures for all the isomers that correspond to each of the molecular formulas listed below. Do you expect the isomers to be differentiable by means of infrared spectroscopy? Which ir bands in particular would aid in assigning structures to each isomer?

a. CH_3ClO b. C_2H_6O c. $C_2H_6O_2$ d. C_2H_4O e. C_4H_{10}

4 a. Phosphorus is a monoisotopic element consisting only of ^{31}P, for which $I = \frac{1}{2}$. Describe the proton nmr spectrum of phosphine, PH_3. Suppose that your spectrometer were able to operate at frequencies such that phosphorus nuclei could absorb energy. What would the phosphorus nmr spectrum of phosphine look like?

b. Borane, BH_3, can exist in solution if present in a complex. Boron consists mainly of the isotope ^{11}B, for which $I = \frac{3}{2}$. Draw both the proton and boron nmr spectra of borane.

c. Although ^{12}C, the principal isotope of carbon, has no spin, ^{13}C has $I = \frac{1}{2}$. How many peaks would you expect to find in the ^{13}C, H, and F spectra of each of the following compounds?

i. $^{13}CH_4$; ii. $^{13}CH_3F$; iii. $^{13}CH_2F_2$; iv. $^{13}C_2F_6$.

5 Describe the expected nmr spectra of the isomers in problem 3.

6 Propose structures for the compounds, the molecular formulas and ir spectra of which are given as follows:

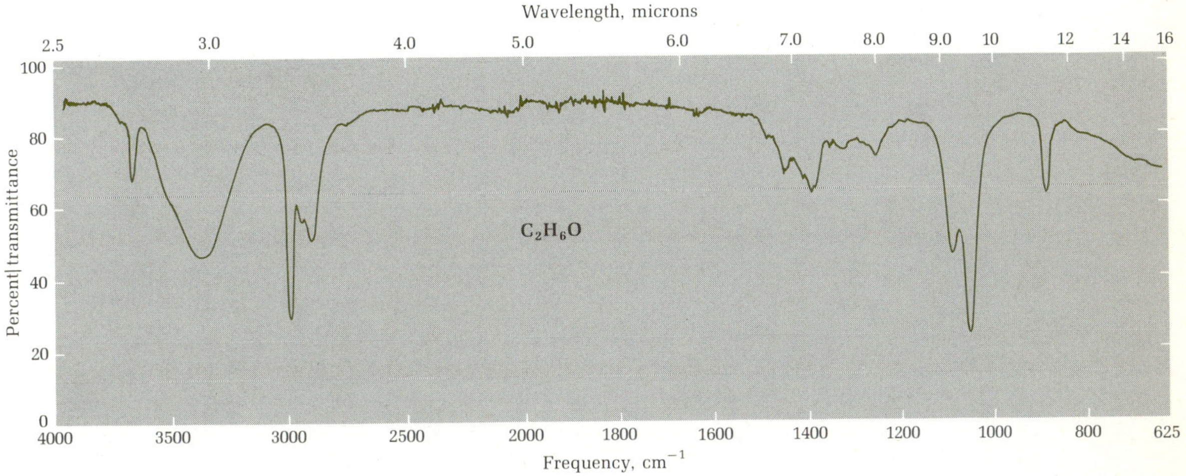

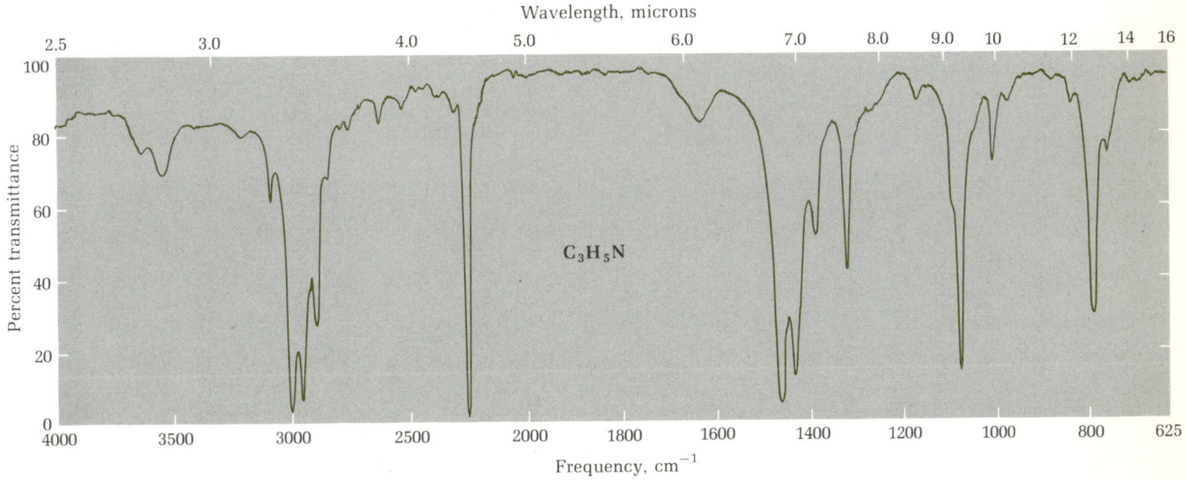

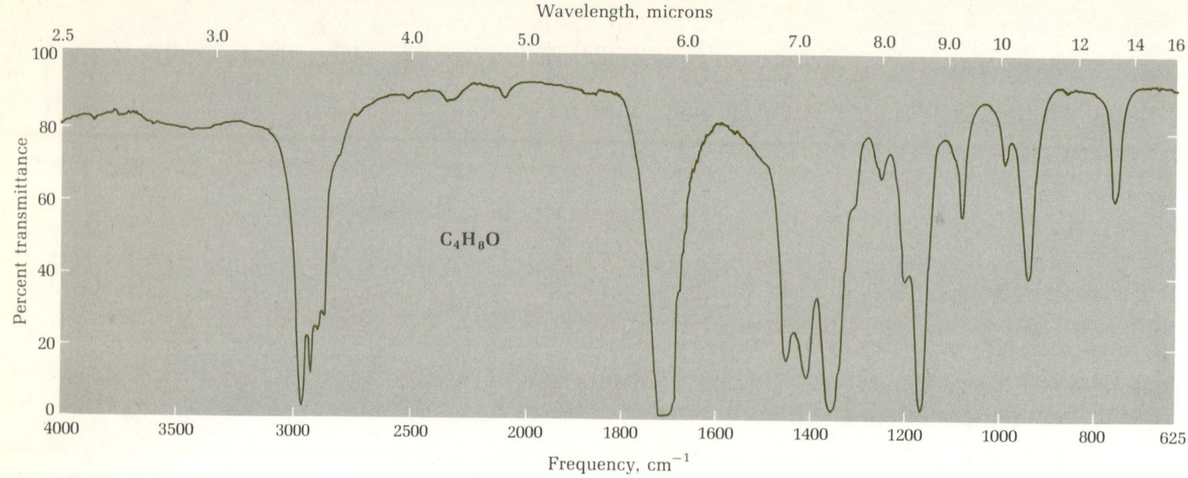

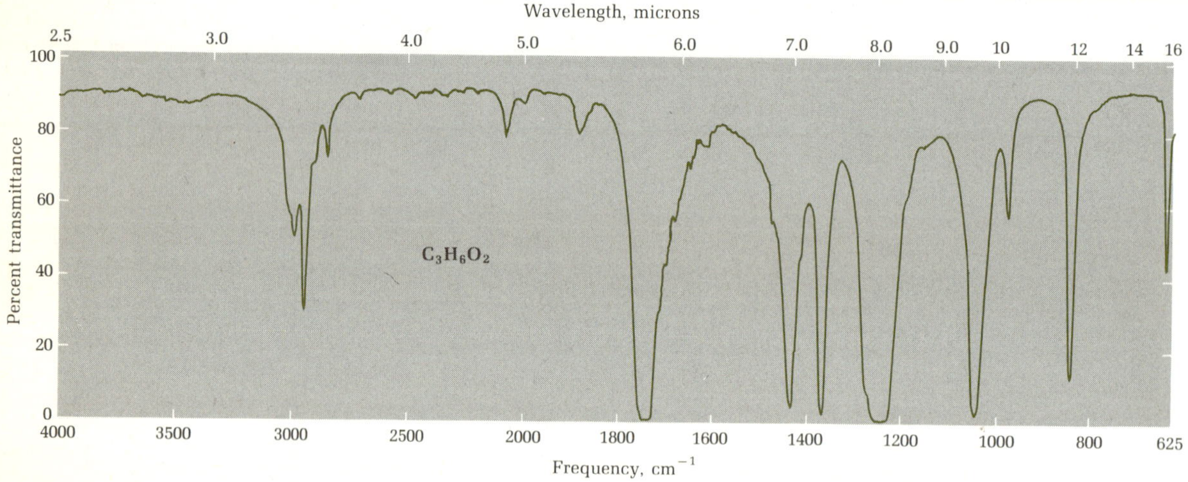

7 Propose structures for the compounds, the molecular formulas and nmr spectra of which are given as follows:

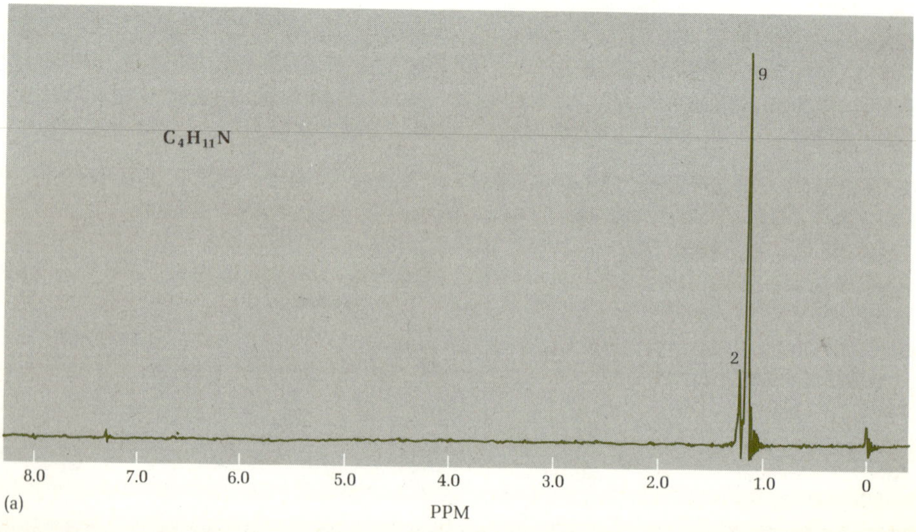

(a)

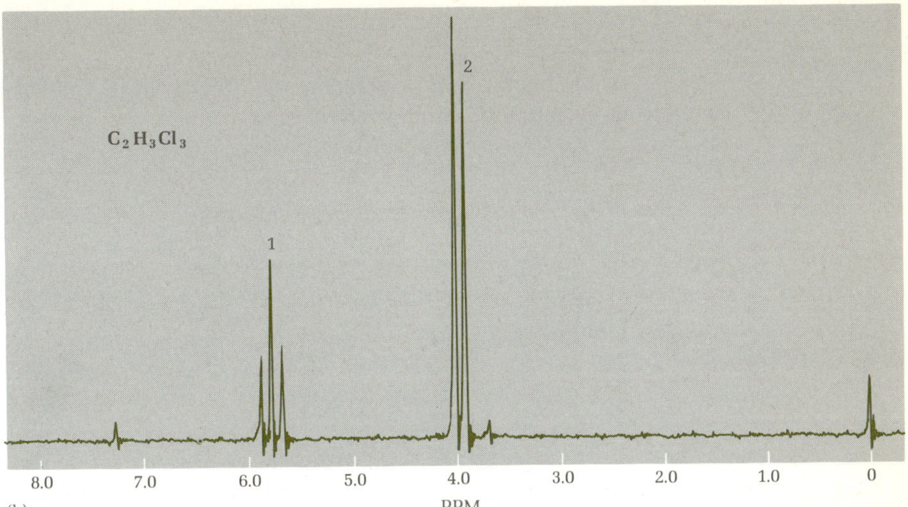

$C_2H_3Cl_3$
1
2
8.0
7.0
6.0
5.0
4.0
3.0
2.0
1.0
0
PPM

(b)

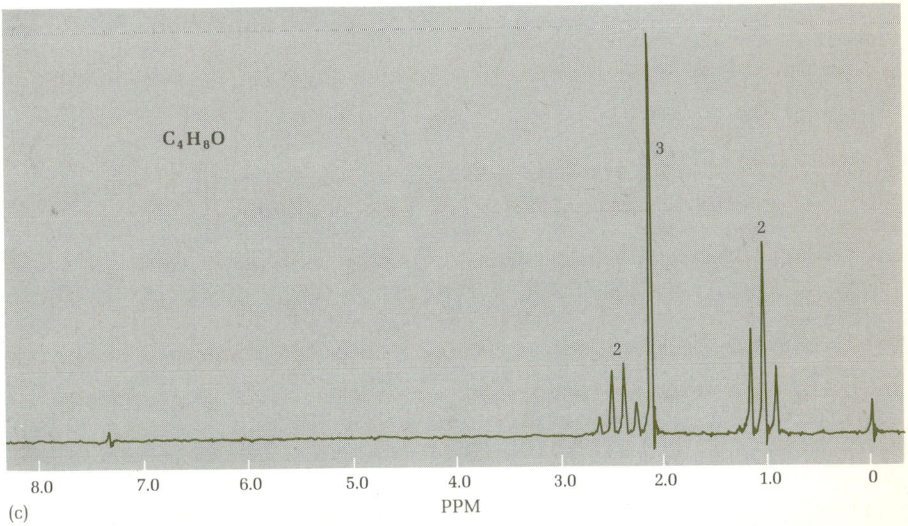

C_4H_8O
3
2
2
8.0
7.0
6.0
5.0
4.0
3.0
2.0
1.0
0
PPM

(c)

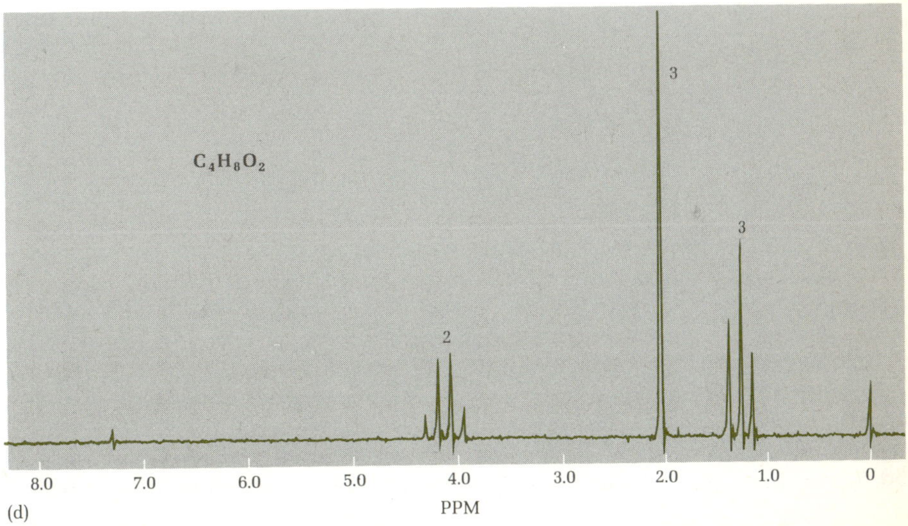

$C_4H_8O_2$
3
3
2
8.0
7.0
6.0
5.0
4.0
3.0
2.0
1.0
0
PPM

(d)

8 Here are the pairs of ir and nmr spectra for four different compounds. Deduce structures for each of the compounds that fit the molecular formulas and are also in harmony with the spectral data.

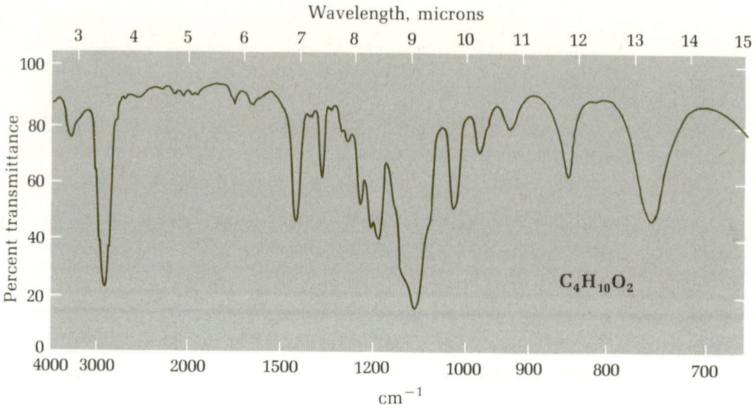

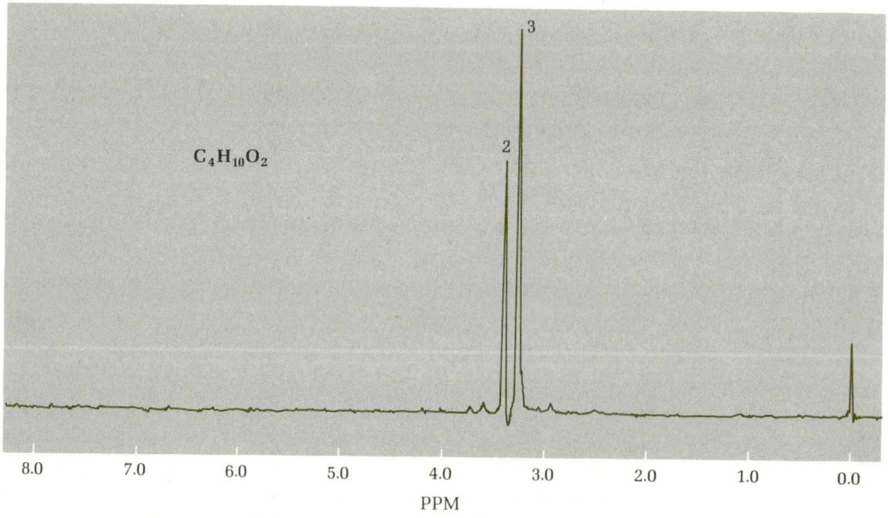

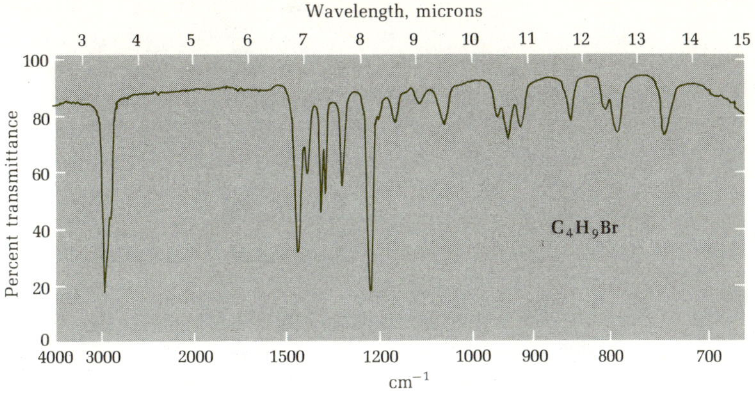
Wavelength, microns
3
4
5
6
7
8
9
10
11
12
13
14
15
100
80
60
40
20
0
Percent transmittance
C4H9Br
4000
3000
2000
1500
1200
1000
900
800
700
cm−1

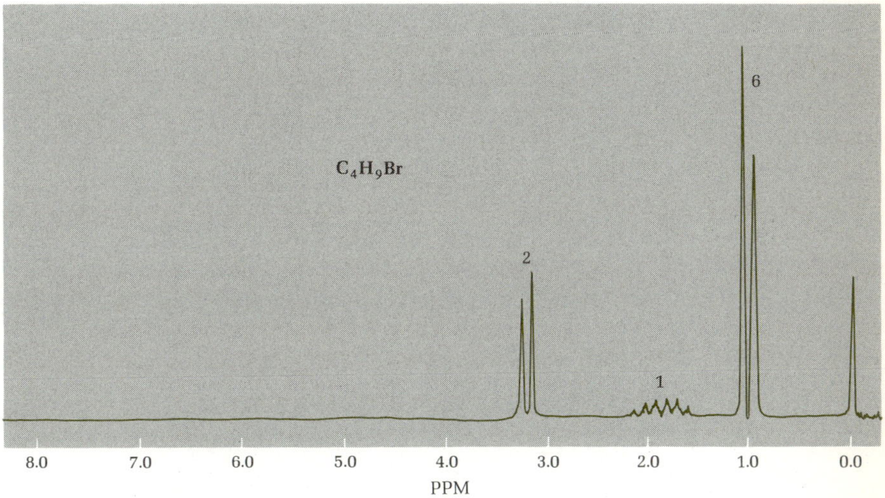
C4H9Br
6
2
1
8.0
7.0
6.0
5.0
4.0
3.0
2.0
1.0
0.0
PPM

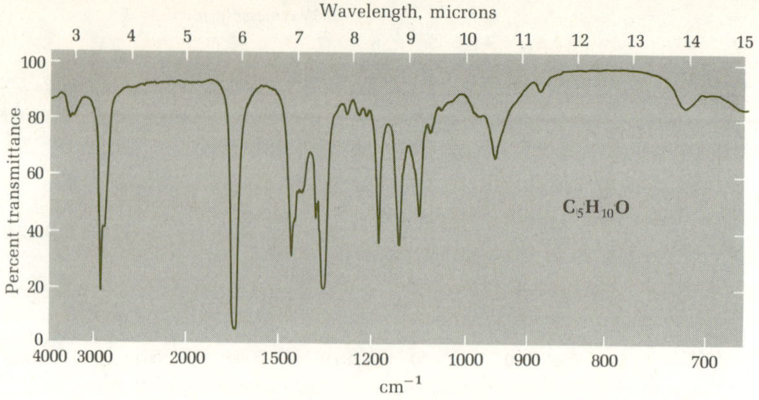
Wavelength, microns
3
4
5
6
7
8
9
10
11
12
13
14
15
100
80
60
40
20
0
Percent transmittance
$C_5H_{10}O$
4000
3000
2000
1500
1200
1000
900
800
700
cm^{-1}

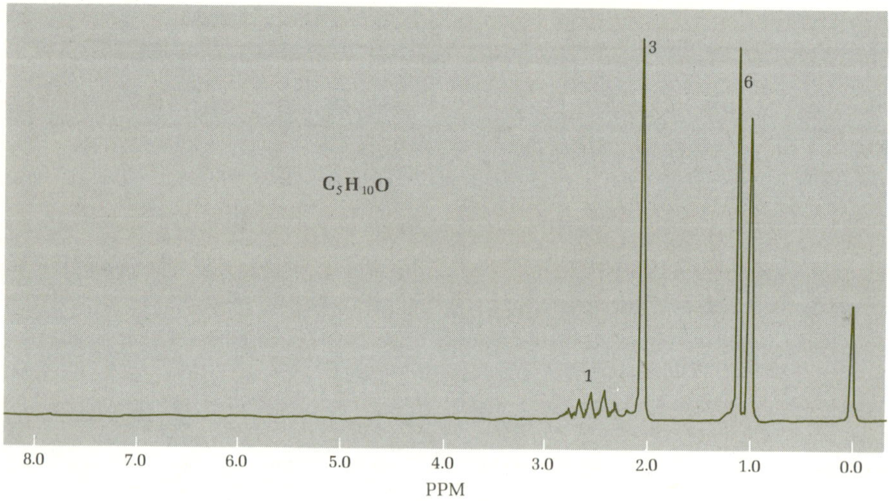
$C_5H_{10}O$
1
3
6
8.0
7.0
6.0
5.0
4.0
3.0
2.0
1.0
0.0
PPM

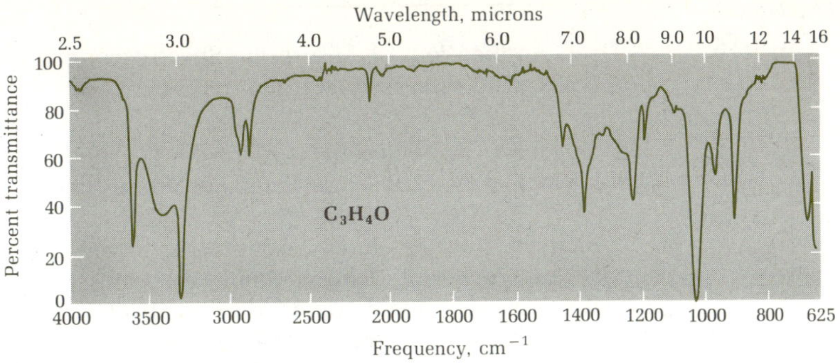
Wavelength, microns
2.5
3.0
4.0
5.0
6.0
7.0
8.0
9.0
10
12
14
16
100
80
60
40
20
0
Percent transmittance
C3H4O
4000
3500
3000
2500
2000
1800
1600
1400
1200
1000
800
625
Frequency, cm−1

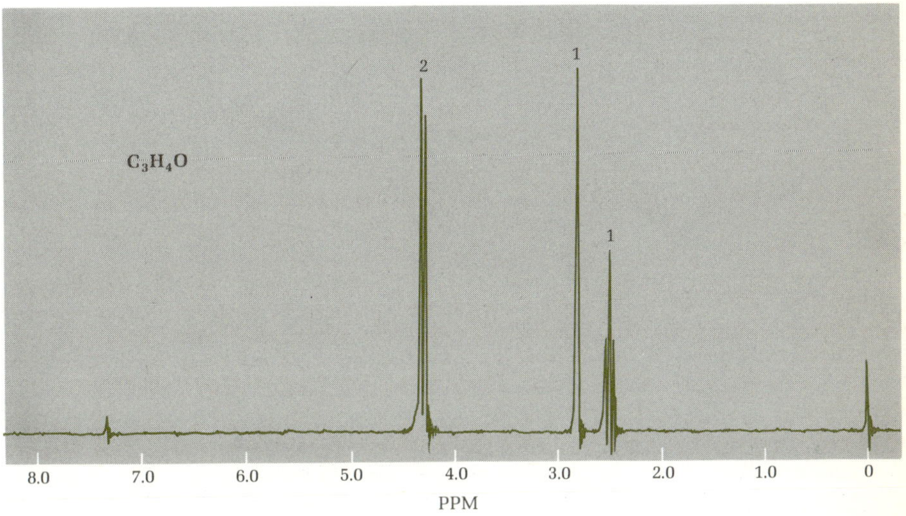
2
1
1
C3H4O
8.0
7.0
6.0
5.0
4.0
3.0
2.0
1.0
0
PPM

SUPPLEMENTARY READINGS

J. R. Dyer, *Applications of Absorption Spectroscopy of Organic Compounds.* Englewood Cliffs, N.J.: Prentice-Hall, Inc., 1965. (P)

L. M. Jackman, *Applications of Nuclear Magnetic Resonance Spectroscopy in Organic Chemistry.* New York: Pergamon Press, 1969.

R. H. Shapiro, *Spectral Exercises in Structural Determination of Organic Compounds.* New York: Holt, Rinehart and Winston, Inc., 1969. (P). This book contains spectral exercises of a wide range of difficulty. The student should be able to solve an increasing number of them as he progresses through the course, especially since chemical information is given for each unknown.

CHAPTER 4

SOLUTIONS AND CHEMICAL REACTIONS

Whether or not a particular reaction will take place, and at what rate, depends on a good deal more than the chemical nature of the reacting species. The reaction medium, the temperature, the pressure, and other factors can each exert a decisive influence on the course of the reaction. An understanding of the dynamics of chemical change yields two principal benefits: Conditions can be devised to give the best yield of product; and by noting how the reaction is affected by changes in various parameters, conclusions may be drawn about the detailed mechanism of the reaction.

Although our interest will center almost exclusively on solutions, many reaction principles may be divined by studying the somewhat simpler gas-phase reactions. We shall be concerned with the effect of various physical changes on the reaction **rate** — the number of product molecules formed within a fixed period of time or, equivalently, the number of reactant molecules consumed per unit time.

REACTIONS IN THE GAS PHASE

Theory predicts, and experiment confirms, that the greater the number of collisions between reacting molecules the faster the reaction is. If we double the number of molecules per cubic centimeter — i.e., double the pressure — we shall nearly always increase the rate (although not necessarily by a factor of 2). If we are studying the reaction of A and B, then it is logically necessary that the greater the number of AB collisions, the faster will C be produced.

(4.1) $A + B \longrightarrow C$

Raising the temperature also increases the reaction rate, even at constant pressure, because the molecules are traveling more rapidly and thus the number of collisions per unit time will increase. In addition, the molecules will bring more energy to each impact. The importance of this fact will be brought out below.

At any given instant, the rate of a reaction will depend on the concentration of at least one species (except for a few rare examples). Suppose that it was found by experiment that for reaction 4.1 the rate depended on the pressure of both A and B.

(4.2) $\text{Rate} = kP_AP_B$

Doubling the pressure of either reactant causes the initial rate of production of C to go up by a factor of 2. The rate is constantly falling as reaction proceeds, however, because A and B are being used up. If we divide the rate at any instant by the pressures of A and B at the same instant, we shall obtain k, the **rate constant.** *As long as the temperature and the reaction medium do not change, k will always be the same for a given reaction, regardless of what concentrations or pressures are used.*

The rate constant does vary with temperature, however. Arrhenius found that the rate constants for most reactions varied with temperature according to equation 4.3. A is the Arrhenius factor; R, the gas constant; T, the absolute temperature; and E_a is the **energy of activation.** The concept of energy of activation is central to our present thinking about chemical reactions. As well as indicating a direction for quantitative calculations of reaction rates, it also provides a framework for developing a useful qualitative approach to reaction rates. When molecules collide, there is no guarantee that reaction will follow.

(4.3) $k = Ae^{-E_a/RT}$

It appears that a *minimum quantity of energy* must be brought to the collision before reaction takes place; this quantity is the energy of activation. We are now in a position to understand the effect of temperature on reaction rate.

In any collection of molecules at one temperature there is a distribution of kinetic energies — one value of the kinetic energy is more probable than the others, but many molecules have energies appreciably greater or smaller than this value. As the temperature increases, the most probable energy also increases, as shown in Figure 4-1, where $T_2 > T_1$. Furthermore, the fraction of molecules having an energy greater than some arbitrary value E will always be larger at the higher temperature. In short, it is likely that at a higher temperature a molecular collision will involve energies equal to or greater than E_a for a reaction between them.

The Arrhenius factor A measures the **probability** that reaction will take place, *independent of the energy requirements.* The number of collisions per unit time is obviously of importance and affects the size of the Arrhenius constant; the collisions, in turn, depend on the size and speed of the molecules. Of equal importance with the number of collisions is the *orientation* of the reacting species upon contact. In many reactions the spatial arrangement of the atoms at the moment of contact is very critical. Such reactions are said to have a large *steric requirement* (Greek *stereos* — solid); a large steric requirement means that A will be relatively small and the probability of reaction correspondingly low.

The reaction of a sodium atom with a chlorine atom, to produce sodium chloride, is not subject to any orientation restrictions. Both atoms are spherically symmetrical, and any one collisional path is as good as another.

(4.4) $Na + Cl \longrightarrow Na^+Cl^-$

When molecules are involved, however, directional effects are very important. Consider the reaction of bromine atoms with hydrogen iodide to yield hydrogen bromide and iodine atoms (equation 4.5).

(4.5) $Br + HI \longrightarrow HBr + I$

It seems clear that in order for reaction to take place, the bromine and hydrogen atoms must be reasonably proximate to one another when the

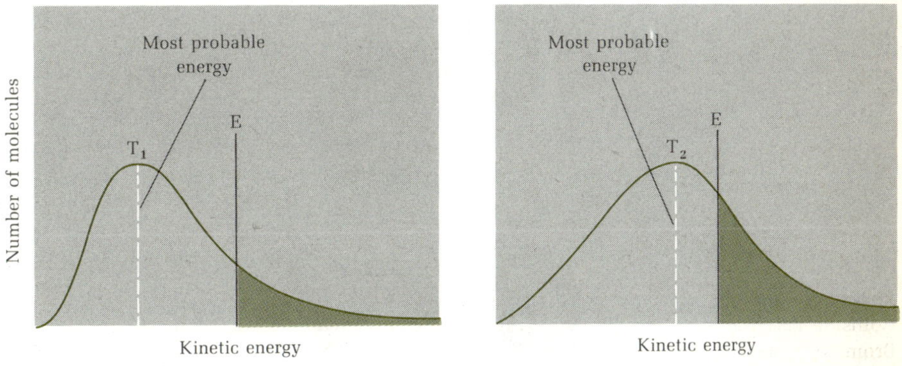

FIGURE 4-1
Distribution of Molecular Kinetic Energies as a Function of Temperature

reactants collide (Figure 4-2). A collision between the bromine and the iodine end of the molecule, no matter how energetic, cannot lead to product formation. As the reactants become increasingly complex in structure, the likelihood of fruitful collision invariably decreases.

A review of just what the Arrhenius equation tells about reactions is in order at this point.

1 The greater the number of collisions per unit time, the larger the rate constant. Raising the temperature increases the frequency of collision. The factor $e^{-E_a/RT}$ is always less than one, so that $k < A$. As $T \longrightarrow \infty$, however, this factor approaches one, so that $k = A$ and the reaction rate depends only on the steric factors.

2 In order to be effective, collisions must involve a minimum quantity of energy, the activation energy. The larger E_a is, the smaller the rate constant is. At higher temperatures more collisions will meet this energy requirement than at lower temperatures.

3 Even collisions that are energetically favorable must involve a reasonable *orientation* of the reacting species or no product will be formed.

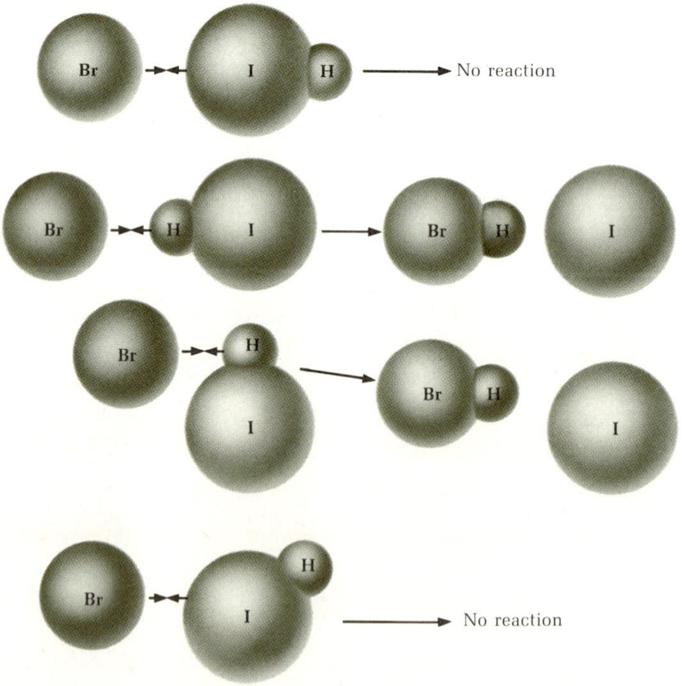

FIGURE 4-2
Possible Modes of Collision between Bromine Atoms and Hydrogen Iodide

What has been said about gas-phase reactions applies also to reactions in solution, with some additions and modifications. In order to understand reactions in solution, one must have some idea of the nature of solutions themselves, and one must therefore have at least an acquaintance with **entropy.**

The discussion of molecular properties has so far been implicitly based on one well-known postulate, namely, that all systems tend naturally to drop to the state of lowest potential energy. A corollary of this is that energy must be supplied from the outside in order to cause a system to go from a lower to a higher potential energy state. There is another factor, however, that is as important a determinant of molecular behavior as the potential energy. This factor is the *universal drive toward maximum randomness and chaos.* The entropy of any given system is a measure of the randomness of that system. **In the universe as a whole, the entropy is always increasing.**

There exists, therefore, a natural tendency, or driving force, for complex and highly organized molecules or collections of molecules to break down into simpler, less ordered units, i.e., for the entropy of the system to increase. It frequently happens, however, that a state of greater disorder is at higher potential energy than a more highly ordered state, so that it requires the input of energy before the entropic driving force can take over. The dissolution of a crystalline solid in a liquid provides a good case for studying the balance of these forces.

A crystal represents a highly ordered arrangement of its constituent units (atoms, ions, or molecules). If the crystal symmetry is known, one may know the positions of all the atoms within a small uncertainty factor. The crystal therefore represents a *low* entropy phase of matter. On the other hand, the units are generally as closely packed as possible, and the attractive forces between individual units is maximized. The crystal is thereby a low potential energy phase as well. If the crystal units are ions, the electrostatic forces holding the lattice together are quite substantial. It requires 186 kcal/mole to convert 1 mole of crystalline sodium chloride into 1 mole of gaseous sodium ions and 1 mole of gaseous chloride ions. Performing the same operation on zinc oxide costs 964 kcal/mole.

If the components of the crystal are molecules rather than ions, then the forces holding the crystal together are much weaker. When the individual molecules are dipoles, then the principal intermolecular force is that arising from the attraction of the negative end of one dipole to the positive end of another (Figure 4-3). The energy needed to vaporize a crystal made up of molecules (a *molecular crystal*), even strongly dipolar molecules, rarely exceeds 10 to 15% of the energy needed to vaporize an ionic crystal. The ease with which many organic solids dissolve in a variety of solvents is directly connected to the fact that they are molecular rather than ionic solids.

If the molecules are not even dipoles, then the energy required to vaporize

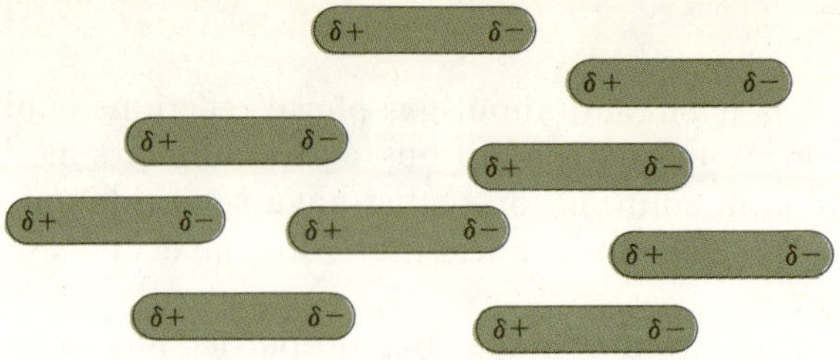

FIGURE 4-3
Alignment of Dipoles in a Molecular Crystal

1 mole of the crystal (heat of sublimation) is quite small. The melting and boiling points of nonpolar F_2 are considerably below those for strongly dipolar HCl, even though they both contain the same number of electrons and have almost the same molecular weight. Crystals formed from nonpolar molecules are held together by **van der Waals,** or **dispersion,** forces. These arise from the fact that a nucleus in one molecule exhibits a small, but real, attraction for the electrons in a neighboring molecule. This **intermolecular** attraction is much weaker than the **intramolecular** force which that nucleus exerts on its own electrons, since the electrons are much further away in the first case as compared to the second.

SOLUBILITY

On the basis of the preceding analysis we might expect that when an ionic solid dissolves in a liquid, a sizable quantity of energy must be supplied to overcome the crystal forces; a dipolar solid would require much less energy compensation, and a nonpolar solid relatively little. In virtually all cases the process of solution is **entropically** favorable since the molecules are free to move about and are much more randomly arranged in solution than they were in the crystal. The question of whether or not a particular solid will dissolve in a given solvent, and to what extent, is ultimately decided by the balance between potential energy and entropy driving forces.

When a mole of sodium chloride dissolves in a very large volume of water, only about 1 kcal of heat is absorbed from the surroundings. There must be a very strong interaction between the water and the ions to compensate for the destruction of the crystal lattice. This interaction is known as **hydration.** Water molecules are small and highly polarized; when a sodium ion dissolves in water, it is surrounded by these dipoles with their negative ends turned toward the positive ion. The chloride ions are similarly solvated by the positive end of the dipoles. These solvation interactions serve to lower the potential energy of the separated ions and compensate for the lost lattice energy of the crystal (Figure 4-4).

Sodium chloride is slightly soluble in other polar solvents containing the

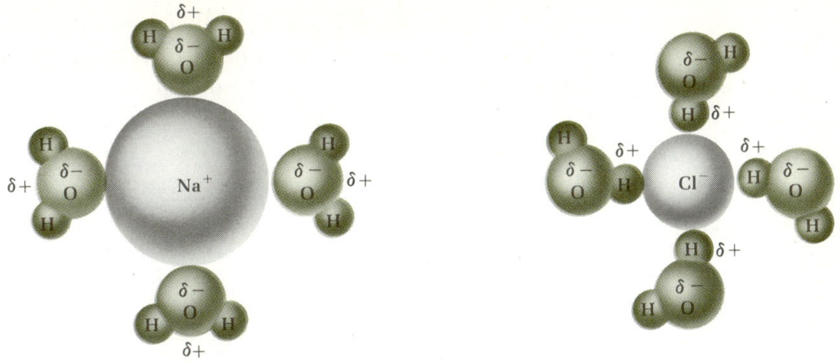

FIGURE 4-4
Hydrated Sodium and Chloride Ions

—OH— group, such as methanol, CH_3OH, and ethanol, CH_3CH_2OH. It is completely insoluble in a nonpolar solvent such as pentane, C_5H_{12}, which cannot take part in any strong interactions with the ions that would overcome the powerful lattice forces. One general measure of the solvating power of a liquid is its **dielectric constant;** this is a bulk property and is related to the ability of the solvent to neutralize the electric field emanating from an ion or a molecule. A solvent like water, which can surround a solute ion with strong dipoles, is very effective at this; a nonpolar solvent such as pentane, $CH_3CH_2CH_2CH_2CH_3$, is very ineffective in this role. It should be noted that a high dielectric constant is usually connected with the presence in the molecule of one or more highly electronegative atoms, such as oxygen. Doubly bonded oxygen atoms almost always ensure a reasonably high dielectric constant.

TABLE 4-1
Dielectric Constants of Some Representative Solvents (25°)

Name	Structure	Dielectric Constant
Formamide	$HC(=O)NH_2$	109.5
Water	H_2O	78.5
Dimethyl sulfoxide	$(CH_3)_2S{=}O$	48.9
Dimethylformamide	$HC(=O)N(CH_3)_2$	36.7
Ethanol	CH_3CH_2OH	24.3
Acetone	$(CH_3)_2C{=}O$	20.5
Ethyl ether	$CH_3CH_2OCH_2CH_3$	4.2
Hexane	$CH_3CH_2CH_2CH_2CH_2CH_3$	1.9

Ions and highly polar molecules tend to be soluble in highly polar solvents, which can interact strongly with the solute and thereby overcome the considerable attractive force between the solute molecules themselves. Nonpolar solutes, which are held together by weak forces, can dissolve in relatively nonpolar solvents, which are capable of solvating the solute to only a slight extent. These generalizations apply to liquids as well — polar liquids are miscible with one another, and nonpolar liquids tend also to be miscible.

HYDROGEN BONDING AND WATER STRUCTURE

One apparent anomaly is the immiscibility of very polar liquids, like water, with very nonpolar liquids such as hexane. It happens that liquid water is structured, although not nearly so completely as ice. The structure arises from the **hydrogen bonds** within the liquid. When a hydrogen atom is bonded to a fairly electronegative atom (principally O, N, and F), it acquires a substantial partial positive charge because of bond polarization. This electron-deficient hydrogen atom is able to coordinate electrostatically with the nonbonded electrons on other nitrogen and oxygen atoms. Halide ions are also hydrogen bond acceptors, decreasing in effectiveness in the order $F^- > Cl^- > Br^-$. Hydrogen bonds are longer and weaker than ordinary covalent bonds, their strength amounting to about 5 kcal/mole.

Any water molecule may donate two hydrogen bonds and accept two as well (since it possesses two nonbonding electron pairs); liquid water therefore consists of interchanging three-dimensional aggregates of hydrogen-bonded water molecules (Figure 4-5). When a solute dissolves in water, it partially disrupts the structure. If the solute is ionic, then the hydration energy of the ions serves to compensate for this disruption. The hydrogen bonding of water molecules to appropriate atoms of the solute may result in additional energy release. If the solute is nonpolar, however, no compensating interaction can take place.

A very useful, although hardly infallible, maxim to use when trying to predict solubilities is "like dissolves like." Polar solutes tend to be soluble in polar solvents, and nonpolar solutes in nonpolar solvents. The rationalizations for this behavior are summarized below.

1 The solution of one substance in another, whether the substances are solids or liquids, is always entropically favorable.

2 Polar solutes, particularly ionic ones, are held together by strong forces, which can only be overcome by their interaction with a polar solvent.

3 Nonpolar solutes are held together by comparatively weak forces and are therefore readily soluble in nonpolar solvents.

4 Nonpolar solutes are not easily soluble in hydrogen-bonding solvents because these solvents have relatively strong intermolecular forces that must be disrupted.

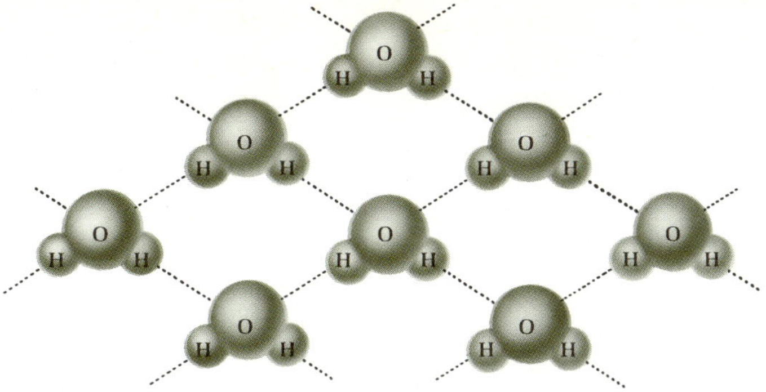

FIGURE 4-5
Hydrogen-Bonded Aggregates of Water Molecules

For illustrative purposes our discussion has centered on the extreme cases but it should be kept in mind that there is a wide spectrum between "polar" and "nonpolar," and many solutes and solvents exhibit some characteristics of each. Solvents in the middle range are often very useful when a reaction between inorganic and organic compounds is desired.

The prime purpose of the solvent is to disperse the reactants and confer the maximum mobility upon each molecule so that they can collide frequently and at all possible angles with one another. In some cases the solvent will be one of the reactants. Even when it is not, it can affect the reaction profoundly because of differential solvation of reactants and products, etc.

THERMODYNAMICS AND EQUILIBRIUM

The solubility problem was analyzed from the point of view of whether or not the overall potential energy and entropy changes would favor dissolution of the solute in the solvent. Even when they are favorable initially, however, they will change as the reaction proceeds. Eventually a point will be reached after which no more solute may be dissolved in the solvent; the system is then said to be at **equilibrium.** In order to predict in general whether or not a system is at equilibrium and how it must change to achieve equilibrium, one must know how the **free energy,** G, changes as the reaction proceeds.

The free energy of a system measures its tendency to undergo chemical transformation. If such a transformation, at a particular temperature, leads to an overall *decrease* in the free energy of the system, *then it is a favorable one.* For instance, the combined free energy of crystalline sodium chloride and pure water is higher than that of a dilute aqueous solution of sodium chloride, so the crystals dissolve. As more salt is added, the free energy difference between the crystalline salt and the dilute solution on the one hand and a more concentrated solution on the other hand starts to decrease.

As the solution becomes more and more concentrated, the free energy difference continues to get smaller until it reaches zero, at which point no more salt will dissolve.

The free energy itself is a combination of two terms, the **enthalpy,** H (also called the heat content — a measure of the total energy of the system), and a temperature-entropy term (equation 4.6).

(4.6) $$G = H - TS$$

If an infinitesimal change in the free energy occurs (dissolution of a minute amount of sodium chloride) at constant temperature, this will be accompanied by minute changes in the enthalpy and entropy (equation 4.7).

(4.7) $$dG = dH - T\,dS$$

Since a spontaneous change results in a decrease in the free energy, dG should be negative. Generally speaking, either a negative dH (release of heat) or a positive dS (increase in entropy) would be most likely to bring about a favorable free energy change. When equilibrium is reached, $dG = 0$, and $dH = T\,dS$; thus any small favorable change in enthalpy is counterbalanced by an unfavorable change in entropy and vice versa.

The overall free energy change, as reactants are transformed into products (sodium chloride and water into a saturated solution, for instance), is ΔG, which equals $\Delta H - T\,\Delta S$ at constant temperature. The more negative ΔG is, the more completely will reactants be converted to products; if ΔG is large and positive, however, there will be very little conversion of reactants to products.

The thermodynamic considerations put forward apply not only to the solution of salts in water but to the mixing of all phases of matter and to all chemical reactions in general.

(4.8) $$\mathrm{A + B \rightleftharpoons C + D}$$

Consider the reaction in equation 4.8. The overall free energy change in going from A + B to C + D is ΔG; if this number is sizably negative, then A + B will have mostly disappeared and C + D mostly formed at equilibrium. Another measure of the extent of this conversion is the **equilibrium constant,** K.

(4.9) $$\mathrm{K} = \frac{\mathrm{(C)(D)}}{\mathrm{(A)(B)}}$$

This constant and the free energy must be related; the relationship is given by equation 4.10.

(4.10) $$\Delta G^\circ = -RT \ln \mathrm{K}$$

ΔG° is the free energy change under specified standard conditions. It should be noted that equilibrium constants and free energy changes are in general both pressure- and temperature-dependent.

It is useful to form some idea of the magnitude of the quantities just under discussion for "average" chemical reactions. If the standard free energy change of a reaction such as equation 4.8 is zero, then the equilibrium constant equals unity and the reaction has proceeded 50% of the way toward completion. A negative free energy change of 1 kcal/mole (a modest amount) at 25° results in an equilibrium constant of 5.3 and a reaction that is 85% complete. When the free energy change is −2 kcal/mole, the reaction is 96% complete at equilibrium; and when the change equals −4 kcal/mole, the reaction is well over 99% complete. Chemists, therefore, often work within a rather narrow band of free energy changes and, by making relatively small changes in the reaction conditions, they can often shift the reactant:product ratio quite substantially.

TRANSITION STATE THEORY

Even if ΔG for a reaction is quite negative under the conditions given, this fact by no means assures that the reaction will proceed at a measurable rate. Sodium chloride dissolves in water fairly quickly, but a mixture of hydrogen and oxygen gases, even though much higher in free energy than water under the same conditions, might remain unchanged for eons. There is nothing in the thermodynamic equations that indicates at what *rate* equilibrium will be obtained. For this we have to turn back to a study of reaction rates, or reaction **kinetics** (Greek *kinetikos* — of motion).

Equations 4.1 and 4.2 described a hypothetical gas-phase reaction and a reasonable rate law governing that reaction. Since our concern will be almost exclusively with reactions in solution, the proposed rate law has been recast in terms of concentrations in equation 4.11. Many reactions of this type do indeed follow rate laws like equation 4.11.

(4.11) $\text{Rate} = k(\text{A})(\text{B})$

Other reactions of the same stoichiometric type are known, however, that obey different rate laws, such as those listed in equations 4.12 and 4.13.

(4.12) $\text{Rate} = k(\text{A})$

(4.13) $\text{Rate} = k(\text{A})^2(\text{B})$

In order to account for this and many other data, the **theory of absolute reaction rates,** or **transition state theory,** was put forward in the 1930's.

The central postulate of the transition state theory is that the reactants must attain a critical configuration, the **activated complex** or **transition state,** on their way to being converted to products. The transition state represents the configuration of **maximum free energy** on the reactant to product (or vice versa) path. In Figure 4-6, the reaction coordinate measures the extent

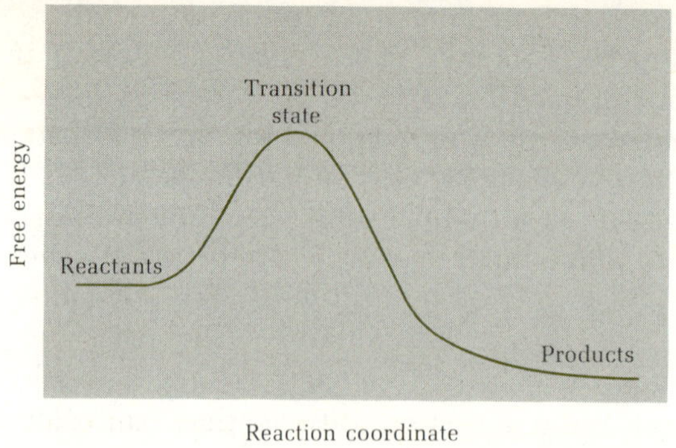

FIGURE 4-6
Plot of Free Energy vs. Reaction Coordinate Showing the Transition State

to which reactants have become converted to products. Once the transition state has been reached, then, energetically speaking, the reaction may coast downhill.

In order to use the transition state theory to calculate rate constants, it was necessary to assume that the reactants and the transition state are in equilibrium. The free energy of the transition state is always higher than that of the reactants so that the **free energy of activation,** $\Delta G^{\ddagger}$ ($G_{\text{transition state}} - G_{\text{reactants}}$), is always positive, even when the *overall* free energy change ΔG is negative. The concentration of the transition state at any given moment is thus extremely small; given the extreme instability of the transition state as well, the chances of isolating it or of even detecting it spectroscopically are vanishingly small.

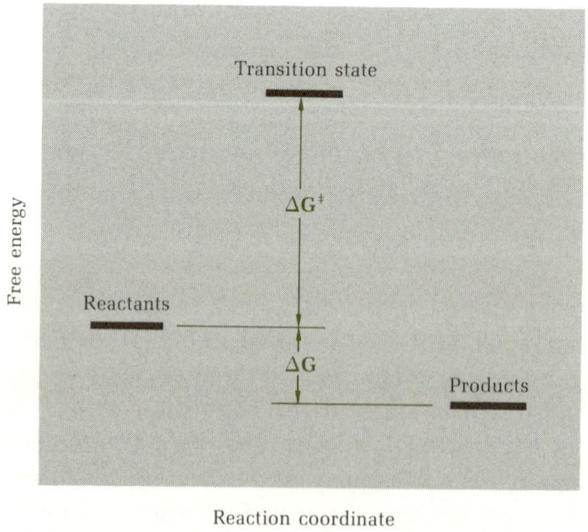

FIGURE 4-7
Free Energy of Activation and Overall Free Energy Change for a Reaction.

In any given reaction the path by which reactants are converted to products may be direct (equation 4.14) or it may be quite circumlocutious and require several steps (equations 4.15 to 4.17).

(4.14) $A + B \xrightarrow{k} C$

Each step in a multistep reaction will, in general, proceed at a different rate from the others and will have a different rate constant. One of these steps will invariably be slower than the rest, and the rate of this step will determine the *overall* rate of conversion of reactants to products. This critical portion of the reaction is called the **rate-determining** or **rate-limiting step.**

(4.15) $A \xrightarrow{k_{I}} X$

(4.16) $B + X \xrightarrow{k_{II}} Y$

(4.17) $Y \xrightarrow{k_{III}} C$

The fact that one step in this sequence is rate-limiting has important and useful consequences. Suppose that, in the conversion of A and B to C, reaction 4.15 is the slow, rate-limiting step. Then the overall reaction rate will be expressed by equation 4.18, and we see that the *rate* of formation of C depends *only* on the concentration of A despite the fact that B is also involved in the reaction.

(4.18) $\text{Rate} = k_{I}(A)$

If, on the other hand, equation 4.16 were the slow step in the reaction, then the overall rate would be given by equation 4.19.

(4.19) $\text{Rate} = k_{II}(X)(B)$

Since the concentration of X is directly dependent on the rate of the preceding step, however, we would quickly conclude that the overall rate must be a function of the concentrations of both A and B (equation 4.20).

(4.20) $\text{Rate} = k'(A)(B)$

The rate constant k' is a combination of rate constants k_{I}, k_{II}, and the rate constant for the reverse of step I.

From this analysis we can see how the *experimentally determined* rate equation can give us information about the reaction mechanism. A general rule to remember is that those reactants that are involved in the rate-limiting step or those steps preceding it will appear in the rate equation; reactants that take part in steps after the rate-limiting step will not appear in the rate equation.

INTERMEDIATES AND TRANSITION STATES

Each step in a multistep reaction mechanism must pass through its own transition state so there is an energy of activation and a free energy of activation for each individual step. Generally speaking, the slowest (rate-limiting) step will have the largest free energy of activation, and the transition state for this step will represent the point of highest free energy on the path from reactants to products.

Figure 4-8 shows a free energy diagram for a three-step reaction mechanism, in which the transition state for the second step has the highest free energy of all. Once this point has been passed, then the reaction is principally "coasting" downhill as far as free energy changes are concerned. This particular transition state may then be designated as the overall transition state for the entire process $A + B \longrightarrow C$. Note that the *identical* free energy path is followed whether one goes forward or backward in this reaction (**principle of microscopic reversibility**).

In equations 4.15 to 4.17 the substances X and Y appear; they are created and then consumed in the course of the reaction. These substances are known as **intermediates;** they may be molecules, ions, reactive species, etc. Unlike transition states, intermediates have finite lifetimes and they may, in some cases, be isolated in the course of the reaction. In a number of cases the appearance and disappearance of the intermediates may be followed spectroscopically.

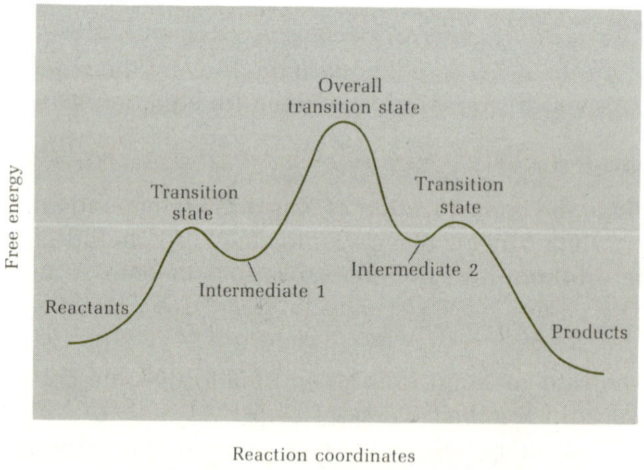

FIGURE 4-8
Energy Diagram for the Multistep Reaction

AN EXAMPLE

The mechanisms of a variety of different reactions will be discussed in the ensuing chapters and a number of transition state structures proposed. It must be stressed at the outset that one can only *infer* the structure of the transition state since direct measurements of its properties cannot be made at this time. In order to give the beginner some "feel" for what a transition state may reasonably be like, we shall analyze a simple inorganic reaction, the oxidation of iodide ion by bromine in solution at 25°.

(4.21) $2I^- + Br_2 \longrightarrow 2Br^- + I_2$

A mechanism consistent with the known facts is presented in equations 4.22 and 4.23.

(4.22) 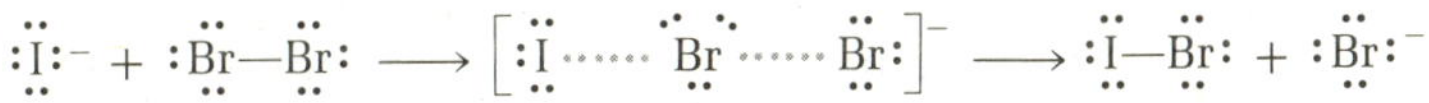

transition state 1

(4.23) 

transition state 2

In the proposed transition states, iodine and bromine atoms are joined together by "partial bonds" or "long bonds;" these are appreciably longer and weaker than ordinary single bonds and are very easily broken.

If this mechanism is the correct one overall, there still remains the question of which step is the rate-determining one. If it is equation 4.22; i.e., transition state 1 is the transition state of highest energy, then the rate equation for the reaction should be that shown in equation 4.24.

(4.24) $\text{Rate} = k(I^-)(Br_2)$

If the second step is rate-determining, however, the rate equation should have the form shown in equation 4.25.

(4.25) $\text{Rate} = k(I^-)^2(Br_2)$

The two possibilities are therefore kinetically distinguishable.

Note that iodine bromide, IBr, is postulated to be an intermediate in this mechanism. This molecule may be prepared separately, and when mixed with iodide ion it does indeed give iodine and bromide ion.

THERMODYNAMICS OF ACTIVATION

From the absolute rate theory one can derive a direct relationship between the rate constant and the free energy of activation (equation 4.26; **k** is Boltzmann's constant, R/N).

(4.26) $$k = \frac{\mathbf{k}T}{h} e^{-\Delta G\ddagger/RT}$$

This equation has a straightforward physical interpretation. The factor $e^{-\Delta G\ddagger/RT}$ represents the fraction of reactants that are present as activated complexes, and $\mathbf{k}T/h$ is the rate at which these activated complexes decompose to products. As would be expected, the more positive is $\Delta G^\ddagger$, the smaller is the fraction of activated complexes.

Working from the definition of ΔG, the rate constant equation can be rewritten (equation 4.27).

(4.27) $$k = \frac{\mathbf{k}T}{h} e^{-\Delta H^\ddagger/RT} e^{\Delta S^\ddagger/R}$$

The activation process can then be seen to depend on two factors — an **enthalpy of activation,** $\Delta H^\ddagger$, and an **entropy of activation,** $\Delta S^\ddagger$. These terms recall the ones used in the Arrhenius equation (equation 4.3) and there are in fact direct relationships between them, as there must be if both equations are valid formulations of the rate constant.

(4.28) $$\Delta H^\ddagger \simeq E_a - RT$$

(4.29) $$A \simeq \frac{\mathbf{k}T}{h} e^{\Delta S^\ddagger/R}$$

One of the ways in which chemists gain some insight into the mechanisms of chemical reactions is by changing systematically one or more of the reactants, or perhaps the solvent, and then seeing how the rate is affected by these changes. A change of this type could affect the rate of altering $\Delta H^\ddagger$, $\Delta S^\ddagger$, or both. Measuring these parameters and how they change can be very informative. It must be kept in mind, however, that if a change in the reactant alters $\Delta H^\ddagger$, then the basic cause could be a change in H_{reactant}, or a change in $H_{\text{transition state}}$, or both. Altering the reactants will usually result in a change in the overall equilibrium because ΔH and ΔS will usually be affected.

This approach will be applied to a good deal of material in the subsequent chapters; hopefully this procedure will give the student a firmer grasp of the principles and a greater feel for their utility. As an illustration, we shall close this chapter with a look at a real case.

The equilibrium shown in equation 4.30 is rather solvent-dependent, differing by a factor of 15 between acetone, $(CH_3)_2C{=}O$, and methanol, CH_3OH. Since the C—Br bond is stronger than the C—I bond, the equilibrium should lie to the right, as it does in acetone.

(4.30) $$CH_3I + Br^- \rightleftharpoons CH_3Br + I^-$$

$K^{40°}_{\text{acetone}} = 1.6$ $\quad E_a = 16.3$ kcal/mole

$K^{40°}_{\text{methanol}} = 0.11$ $\quad E_a = 21.4$ kcal/mole

The methanol results may be rationalized if one realizes that methanol, unlike acetone, is capable of donating hydrogen bonds. Since bromide is a smaller ion than iodide, and its charge more concentrated, it is a much better hydrogen-bond acceptor. The solvation energy difference between bromide ion and iodide ion is relatively large in methanol and relatively small in acetone. Thus while bond strength effects seem to determine the position of equilibrium in acetone, solvation energy effects rule in methanol.

Are these conclusions compatible with the energy of activation data? In both of the solvents the same rate equation applies (equation 4.31) and we can assume that the same mechanism is operative in both cases.

(4.31) $\text{Rate} = k(CH_3I)(Br^-)$

The mechanism currently favored involves a one-step, continuous displacement of iodide ion by bromide (equation 4.32). In order for bromide ion to do this, it must shake off at least some of the solvent molecules clinging to it.

(4.32) $Br^- + CH_3{-}I \longrightarrow [Br\cdots\cdot CH_3\cdots\cdot I]^- \longrightarrow CH_3Br + I^-$

This requires energy; if methanol molecules are attached much more strongly to bromide ion than are acetone molecules, then the activation energy should be higher in methanol. This is exactly what is found.

Analyses such as these not only help us understand the reaction much more intimately, they also allow us to plan the best conditions for making one product or the other predominate.

SUMMARY

1 The rates of gas-phase reactions generally show a dependence on both pressure and temperature. This dependence can in turn be correlated with three molecular collision phenomena that directly affect the reaction rate:

a. The frequency of collision.
b. The total energy of impact.
c. The configuration of the molecules as they collide.

2 The same principles govern reactions in solution, except that solute-solvent interactions must be taken into account. Ionic and highly polar molecular solutes tend to be soluble only in highly polar solvents, which are capable of overcoming their powerful lattice energies. These solutes are attracted to solvent molecules by dipolar and hydrogen bonds and are relatively strongly solvated. By contrast, nonpolar solutes are soluble in nonpolar solvents, in which the solvation forces are of the weak, van der Waals variety.

Nonpolar solutes are usually rather insoluble in hydrogen-bonding solvents because they cannot overcome the solvent structure.

3 Chemical reactions tend to occur spontaneously if a decrease in the free energy of the system will result. The reaction continues as long as the free

energy change is favorable; when a point is reached at which a small change in the reactants produces no change in the free energy, the system is at equilibrium.

A change in free energy involves an alteration of the enthalpy (energy) and entropy of the systems. It often happens that a reaction that exhibits a favorable enthalpy change (decrease) also creates an unfavorable entropy change (decrease); the reverse is also true. At equilibrium the energetic and entropic driving forces are in balance. The overall free energy change for the reaction is directly related to the equilibrium constant.

4 The requirement that some minimum energy must be available before reaction can occur may be understood if it is assumed that the reactants must first reach a critical stage, known as the activated complex or transition state. The transition state lies at a potential energy maximum and is highly unstable. One usually assumes that the transition state is in equilibrium with the reactants. This equilibrium is governed by the free energy of activation, which contains in turn an enthalpy of activation and an entropy of activation. Any change in the reaction conditions that alters the rate does so by affecting either, or both, of these parameters.

5 The mechanism by which reactants are transformed into products may involve several discrete steps. That step that goes by way of the transition state of highest energy overall is the rate-determining step. Only those reactants that take part in the rate-determining step, or those steps preceding it, can affect the rate through a change in their concentrations. A study of the reaction kinetics can be very useful then for narrowing the number of possible reaction mechanisms.

PROBLEMS

1 The following questions may be answered by reference to the Arrhenius equation (equation 4.3).

a. What will be the effect on the rate constant if the temperature is increased?

b. Will a reaction rate be more sensitive to temperature if the activation energy is large or small? Explain your reasoning. (You may assume that A is independent of temperature.)

c. The rate constants k_A and k_B refer to two different reactions. At 25° $k_A > k_B$. Under what circumstances might $k_B > k_A$? Provide a quantitative answer.

d. A given pair of reactants may combine in two different ways. The Arrhenius equation for one reaction path is characterized by large values of E_a and A. The second path, by contrast, has an Arrhenius equation in which both E_a and A are small. Which reaction will predominate at low temperatures and which at high temperatures? Of the two factors present in the Arrhenius equation, activation energy and probability, which one predominates at which temperature?

2 Why is the lattice energy of zinc oxide so much higher than that of sodium chloride (p. 67)?

3 Indicate which compound of the following pairs would be expected to have the higher boiling point, and state your reasons.

a. $LiCl$ or CH_3Cl
b. CCl_4 or CBr_4
c. HBr or HI
d. $CH_2{=}CH_2$ or $CH_2{=}O$
e. CH_3CH_3 or CH_3F
f. $(CH_3)_3N$ or $(CH_3)_3CH$
g. H_2O or H_2S
h. CH_3OH or $CH_2{=}O$

4 Will each of the following compounds be more markedly soluble in water or in hexane or about equally soluble in each?

a. KCN
b. $MgBr_2$
c. CH_3CH_2Br
d. $(CH_3)_2C{=}O$
e. CH_3OH
f. HCl
g. CCl_4

5 For the reaction shown in equation 1, the equilibrium constant equals 1.5 at 25° and $\Delta H = -6.0$ kcal/mole.

$$CH_3\overset{\overset{\displaystyle O}{\|}}{C}H\ (aq.) + H_2O \rightleftharpoons CH_3CH(OH)_2 \qquad (1)$$

a. Calculate ΔG and ΔS at 25°.

b. Calculate K at 0° and at 100°.

c. Which driving force, the energetic or the entropic, dominates the equilibrium at low temperature? at high temperatures?

6 Are ΔH and ΔS positive, negative, or about zero for the following reactions? Explain your answer.

a. $2I \longrightarrow I{-}I$

b. $CH_3O{-}OCH_3 \longrightarrow 2CH_3O\cdot$

c. $NaCl_{(s)} + H_2O \longrightarrow Na^+\ (hydr) + Cl^-\ (hydr)$

d. $Br^- + CH_3I \xrightarrow{acetone} CH_3Br + I^-$

e. $^*I^- + CH_3I \xrightarrow{acetone} CH_3I^* + I^-$ (*I = radioactive iodine)

f. $Cl\cdot + HI \longrightarrow HCl + I\cdot$

g. Ether and hexane $\longrightarrow$ solution

7 The reaction of methyl chloride, CH_3Cl, with hydroxide ion might reasonably occur by way of a one-step mechanism (2),

$$CH_3Cl + OH^- \longrightarrow CH_3OH + Cl^- \quad (2)$$

or equally plausibly, by a two-step mechanism (3) and (4):

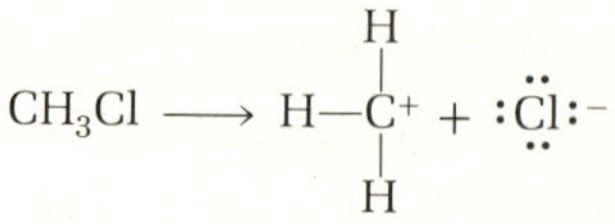

(3)

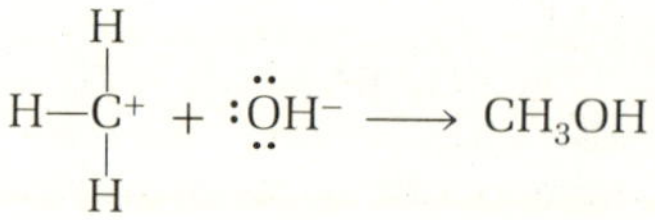

(4)

To what extent are these mechanisms distinguishable by kinetic means?

SUPPLEMENTARY READINGS

B. H. Mahan, *University Chemistry*. Reading, Mass.: Addison-Wesley Publishing Co., 1969, Chapters 8 and 9.

E. L. King, *How Chemical Reactions Occur*. New York: W. A. Benjamin, Inc., 1963. (P)

T. R. Blackburn, *Equilibrium: A Chemistry of Solutions*. New York: Holt, Rinehart and Winston, Inc., 1969, Chapters 1 and 2. (P)

CHAPTER 5

ALKANES — INTRODUCTION TO MECHANISM

PETROLEUM AND ITS FRACTIONATION

In primordial times a good part of the land surface of the earth, including many regions that are deserts today, was covered by fertile swamps. A variety of geological upheavals resulted in the burying of many swamps. The sand and earth, along with their entrapped plant and animal life, were subjected to moderate temperatures and pressures while sealed off from atmospheric oxygen. The sand was often compressed into soft rock, known as **shale,** and the organic matter was converted to **petroleum** (Latin *petra* + *oleum* — rock oil). Starting in the latter half of the last century, a vast industry has grown up around the extraction and processing of petroleum from the earth. Substances derived from petroleum have two principal uses — as fuels and as starting materials for the synthesis of other industrial staples.

Petroleum consists in the main of **hydrocarbons** — compounds containing only carbon and hydrogen. Of the various classes of hydrocarbons this chapter will concern itself only with those known as **paraffins** (Latin *parum* — having little affinity) or **alkanes.** The molecular formulas of the alkanes all fit the general formula C_nH_{2n+2} (n = 1, 2, 3, ...). Alkanes are **saturated** hydrocarbons — they contain no double or triple bonds, and each molecule contains the maximum number of hydrogen atoms possible for that number of carbons.

The lowest molecular weight alkanes are gases at ordinary temperatures, and gas pockets are often found in connection with petroleum deposits. The liquid petroleum is subjected to **distillation** — it is heated to boiling and the vapors that form are condensed and collected. This is done continuously as the temperature is steadily raised. Distillation results in the petroleum's being divided into various fractions on the basis of their boiling point ranges. These fractions each consist of a mixture of compounds of similar volatility. The proportion of each compound in the mixture is dependent on the source of the petroleum.

In a representative petroleum sample from an Oklahoma field, 50% of the hydrocarbons had boiling points of 230° or less. Table 5-1 indicates the composition of the various volatile fractions from the Oklahoma petroleum. The higher boiling material not listed in the table is the source of light and heavy lubricating oils, paraffin wax, and a variety of other materials.

It is possible, by **fractional distillation,** to resolve each fraction still further and often to obtain the individual compounds in a fairly pure state. For illustrative purposes, imagine that we have a mixture of two hydrocarbons A and B, of which A is the more volatile. Suppose that the mixture consisted of 50 mole % each of A and B. At any given temperature the vapor in equilibrium with this liquid contains *more than* 50% A because the vapor pressure of A exceeds that of B. If this vapor could be collected and condensed, it would provide a liquid mixture enriched in A. If the liquid collected were allowed to vaporize, the vapor in equilibrium with it would be even richer in A. If such a series of condensations and vaporizations could be carried out enough times, one could eventually obtain an essentially pure sample of A. This procedure is, of course, completely impractical as outlined, but it can be carried out continuously and automatically through use of a distilling column.

The column is usually made of glass for laboratory work but may be constructed of other materials for industrial applications. It is thermally insulated and sits atop the flask or other container holding the mixture to

TABLE 5-1
Volatile Constituents of Petroleum

Name of Fraction	Gas	Gasoline	Kerosine
Boiling point range (°C)	40	40–180	180–230
Number of constituents	7	107	37
Number of carbon atoms in constituents	1–5	6–10	11 and 12
Percent of total petroleum sample	4	33	13

be fractionated. The column is packed either with small rings or balls made of metal or glass or with metal gauze — the main criterion for a good column packing is that it provide a high surface area without obstructing the flow of vapor. As the mixture is heated, its vapor rises into the column and condenses on the packing, where it then establishes a new vapor-liquid equilibrium. This step repeats itself up the length of the column, the condensed liquid returning to the flask and the new vapor continuing to rise. Samples of the vapor taken at various heights indicate its gradual enrichment in the volatile components as it rises. The vapor that reaches the top of the column is condensed and a portion of it led off to a receiver, where it is collected. The efficiency with which the various components of the mixture are separated is a function of several factors — separation is more difficult if the individual compounds have similar boiling points. Columns also differ in their inherent effectiveness; the more effective a column in separating a given mixture, the greater the number of **theoretical plates** it possesses.

GAS-LIQUID CHROMATOGRAPHY

Distillation has been in use for several thousand years. A much more recent separation method is **gas-liquid chromatography,** often (incorrectly) called **vapor-phase chromatography.** A sample of a mixture is vaporized all at once by being injected into a hot chamber. The vapor is pushed through a tightly packed column several feet in length by means of an inert gas, usually nitrogen or helium. The packing consists of small particles of a solid, often

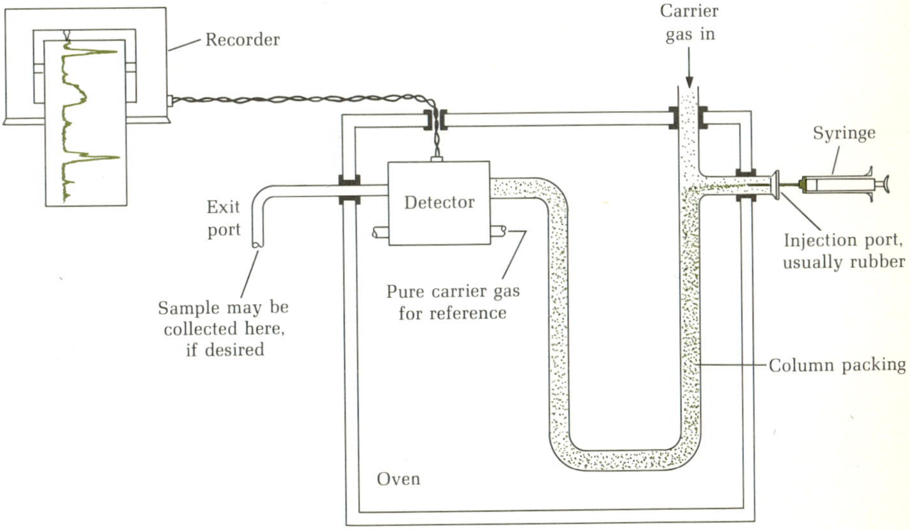

FIGURE 5-1
Schematic Diagram of Gas-Liquid Chromatography Apparatus

crushed firebrick, lightly coated with one of a variety of nonvolatile liquids. As the vapor travels through the column, its various components are continually dissolving in and revaporizing from the packing liquid. Their interaction with this liquid is not uniform, however, and depends on factors such as the boiling point of the compound, its polarity, and possibility of hydrogen bonding. The net result is that different compounds move at different rates through the column, and the mixture is thereby resolved into its components.

As the vapor of each compound emerges from the column, it activates a detector that in turn causes a peak to be traced by the recorder. The system may be calibrated so that the amount of each component in the mixture may be calculated from the area of its peak on the chart. Gas-liquid chromatography can be very sensitive, and a complete composition analysis may be performed with 1 μl (one microliter $= 10^{-3}$ ml) of sample or less. In their infancy, glc columns were designed to take only small quantities of sample — 50 to 100 μl. Later columns were built that would accommodate several milliliters and more in one injection, and the highly pure separated vapors could be condensed and collected. These columns generally have many times the number of theoretical plates of a good distilling column. Although units are now being built for commercial use, it is unlikely that glc will replace routine batch distillation. For relatively difficult separations and for routine analysis, however, it is far superior.

Complete analysis of an unknown mixture by glc is a tedious undertaking.

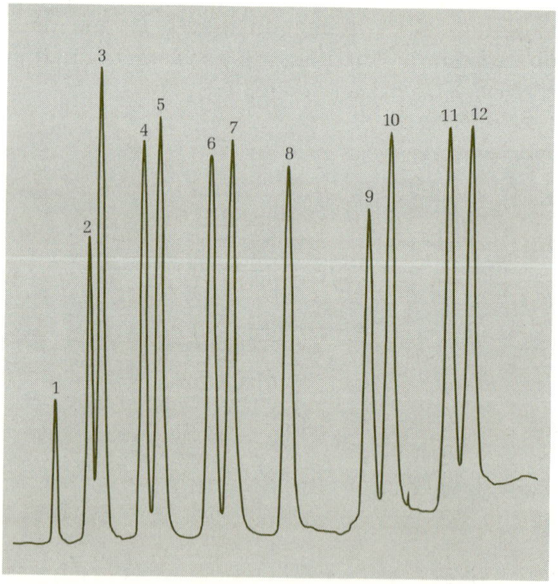

FIGURE 5-2

Gas Chromatogram of a Mixture of Hydrocarbons

1,2-C_{10} compounds
3,4-C_{11} compounds
5,6-C_{12} compounds
7,8-C_{13} compounds
9,10-C_{14} compounds
11,12-C_{15} compounds

The time that it takes for a given compound to pass through a specific column (its **retention time**) under specified conditions is definitely not unique for that compound. One could collect each fraction and analyze it chemically and spectroscopically, but it might well be a long, painstaking process. The need to analyze mixtures of hydrocarbons from petroleum refining with speed and accuracy encouraged the development of **mass spectrometers** for chemical use.

MASS SPECTROMETRY

The principle of mass spectrometry is quite simple. In a high vacuum a small quantity of the sample is vaporized, and the vapor is allowed to diffuse into a beam of electrons. These bombarding electrons have sufficient energy to knock a valence electron out of a gaseous molecule — analogous to the collision of a moving billiard ball and a stationary one. The molecule left behind is deficient by one electron and is therefore positively charged—it is known as the **molecular ion.** This molecular ion is rich in energy and may spontaneously break some bonds and decompose into fragments, one of which will retain the positive charge. These fragments may themselves undergo further decomposition. New bonds may also be formed that did not exist in the original molecule. A representative decomposition pattern of a molecule in a mass spectrometer is shown in Figure 5-3.

Some of the decomposition paths are more probable than others and some ions will therefore be more abundant than others. The entire ionization-decomposition process takes place in about 10^{-5} sec. The positive ions alone are then accelerated away from their point of origin by a negatively charged plate with a slit in it. They fly through the slit and are then separated — in most mass spectrometers this is accomplished with the aid of a magnetic field at right angles to the beam of ions (Figure 5-4). The field causes the ions to follow a curved rather than a straight path. But the curvature of the path depends on the mass of the ion, and ions of different mass will separate and follow different flight paths (Figure 5-4).

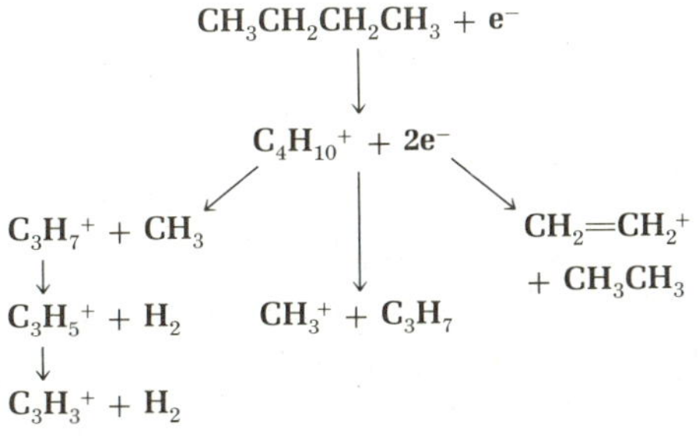

FIGURE 5-3
Decomposition of Butane after Electron Bombardment

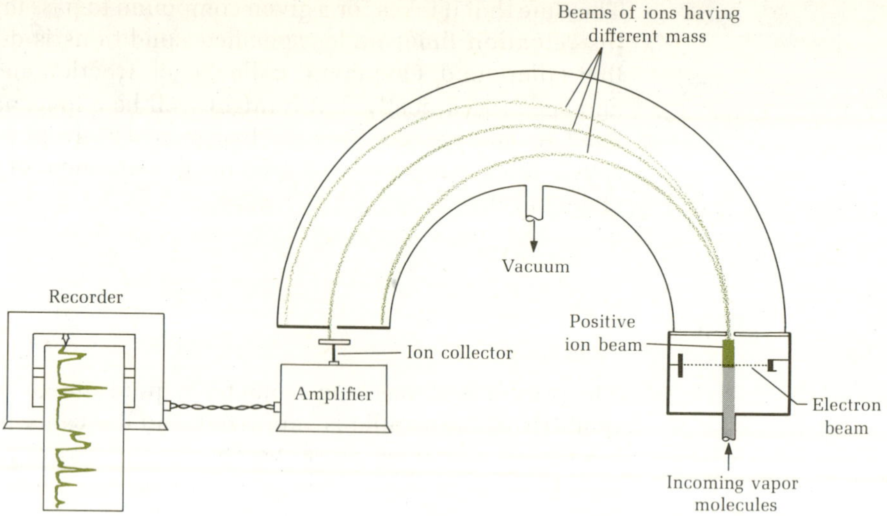

FIGURE 5-4
Schematic Diagram of a Mass Spectrometer

After a short time the ions strike another plate that also has a narrow slit in it. If one of the ion beams happens to be following a particular path, it will go through the slit and strike the collector on the other side. An electrical signal is produced that is then amplified and recorded. The strength of the signal is directly proportional to the number of ions in that beam. By changing the magnetic field strength or the voltage that accelerated the ions, it is possible to focus ions of any desired mass onto the collector. If this is done continuously and automatically and the electrical signals simultaneously recorded, the mass spectrum of the substance is obtained.

It is customary to assign to the most intense peak in the spectrum (the **base peak**) a value of 100. The intensities of the other peaks are then calculated on that basis. For any one compound the mass and intensity data together are essentially unique. The mass of the molecular ion is, of course, identical to the molecular weight of the compound. Figure 5-5 and Table 5-2 show two ways of representing a partial mass spectrum of *n*-butane, C_4H_{10}. The mass spectrum of a mixture of compounds is simply the superposition of the mass spectra of each pure compound. By finding a mass peak that is due to only one compound, it is possible to calculate the remainder of the mass spectrum of that compound. Subtraction of this spectrum from the total spectrum leaves a residual spectrum for the remaining compounds. Repetition of this procedure eventually leads to analysis of the entire mixture.

Such procedures have now been computerized. It is also possible to attach a gas chromatograph directly to a mass spectrometer and obtain a spectrum of each individual peak. This is possible, in part, because the mass spectrometers require only very small samples — a complete spectrum can be obtained from 10^{-6} g or less. Furthermore, the entire mass spectrum of one

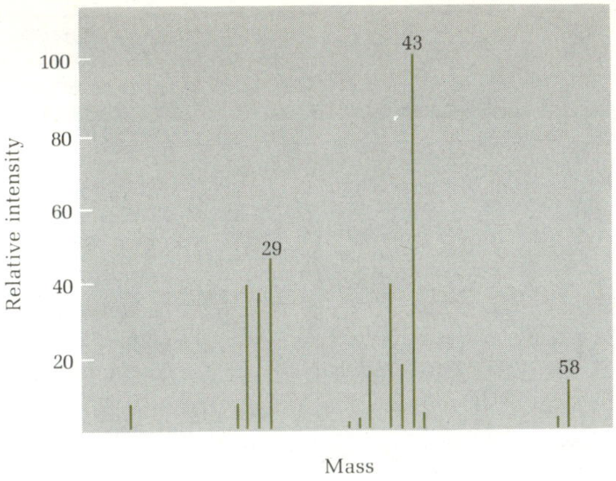

FIGURE 5-5
Partial Mass Spectrum of *n*-Butane

TABLE 5-2
Partial Mass Spectrum of n-Butane

Mass	15	26	27	28	29	37	38	39	40	41	42	43	44	50	51	57	58*
Relative Intensity	5.3	6.2	37	32	44	1.0	1.9	12.5	1.6	28	12	100	3.3	1.3	1.1	2.4	12

*Molecular ion.

compound may be scanned before another compound emerges from the gas chromatograph.

It is important to note the differences between mass spectrometry, on the one hand, and nmr and ir spectroscopy, on the other. The mass spectrometer measures only the *masses* of the various fragments produced, but it gives no information on the *structure* of these fragments. For the identification of unknown compounds it is often necessary to *assume* structures for various fragments; there are chemically based rules for making these assumptions. Ir and nmr spectroscopy, on the other hand, give structural information directly. Despite the fact that mass spectrometric analysis requires the destruction of a portion of the sample, so little is required to generate a spectrum that it is often the sole method of spectral analysis available when very small quantities of material are on hand.

ALKANES – PHYSICAL PROPERTIES AND NOMENCLATURE

Table 5-3 records the names and physical properties of a number of alkanes. The boiling points rise regularly with increasing molecular weight, although there is some variation among isomers. Melting points behave much less predictably; highly symmetrical compounds often melt at abnormally high

TABLE 5-3
Physical Properties of Some Alkanes

Number of Carbon Atoms (mol. wt.)	IUPAC Name	Semisystematic Name	Boiling Point (°C)	Melting Point (°C)
C_1 (16)	Methane	Methane	−161.5	−182.5
C_2 (30)	Ethane	Ethane	−88.6	−182.8
C_3 (44)	Propane	Propane	−44.5	−189.9
C_4 (58)	Butane	*n*-Butane	−0.5	138.3
	2-Methylpropane	Isobutane	−11.7	−159.6
C_5 (72)	Pentane	*n*-Pentane	36	−129.7
	2-Methylbutane	Isopentane	27.9	−160
	2,2-Dimethylpropane	Neopentane	9.5	−20
C_6 (86)	Hexane	*n*-Hexane	68	−95
	2-Methylpentane	Isohexane	60.3	−153.7
	3-Methylpentane		63.3	−118
	2,3-Dimethylbutane		58	−128.8
	2,2-Dimethylbutane	Neohexane	49.7	−99.8
C_7 (100)	Heptane	*n*-Heptane	98.4	−91
C_8 (114)	Octane	*n*-Octane	125	−56.5
	2,2,3,3-Tetramethylbutane		106.5	100.7
C_{10} (142)	Decane	*n*-Decane	174.1	−29.7
C_{11} (156)	Undecane	*n*-Undecane	195	−25
C_{12} (170)	Dodecane	*n*-Dodecane	216.3	−9.6
C_{15} (212)	Pentadecane	*n*-Pentadecane	270.5	10
C_{19} (268)	Nonadecane	*n*-Nonadecane	330.0	32
C_{20} (282)	Eicosane	*n*-Eicosane	343	36.8

temperatures, reflecting the fact that the molecules pack together well in the crystal lattice. All those alkanes having their carbon atoms arranged in a straight chain form a **homologous series** — their chemical properties are very similar, and their physical properties vary in a regular manner. Similarly, all alkanes with a $(CH_3)_2CH$— group at one end of the chain are homologues.

There are two principal procedures presently in use for naming alkanes and their derivatives. One was devised by the International Union of Pure and Applied Chemistry (IUPAC) and provides for the naming of any alkane and its derivatives in a systematic manner according to a straightforward set of rules. An older system employs semisystematic names and is easily applicable only to low molecular weight alkanes and their derivatives. These names are usually of historical origin.

In both systems the suffix "-ane" denotes an alkane, while the root of the name indicates the number of carbon atoms. From five on, these roots are derived from the Greek names for the appropriate numbers. In the older, semisystematic nomenclature the total number of carbon atoms in the alkane, regardless of their arrangement, was the criterion for assigning a name. Thus, all five-carbon alkanes would be pentanes. If all the carbon atoms were joined in a continuous line, then the compound would be a "normal," or straight chain, alkane, denoted by the prefix "*n*-." All others are branched-chain alkanes. If one end of the chain terminates in a $(CH_3)_2CH$— group, it is called an **isoalkane.**

$CH_3CH_2CH_2CH_2CH_3$ $(CH_3)_2CHCH_2CH_3$ 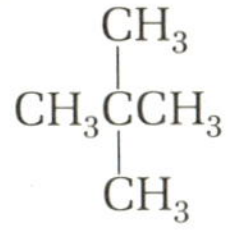

n-pentane isopentane neopentane

Finally, that isomer in which one end of the chain bears a $(CH_3)_3C$— group is the *neo*-isomer. From hexane on, there are many isomers that do not fit into any of these categories, and the use of the IUPAC system is almost mandatory.

CH_4	CH_3CH_3	$CH_3CH_2CH_3$	$CH_3CH_2CH_2CH_3$	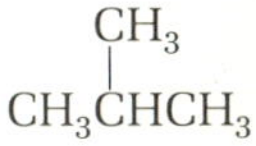
methane	ethane	propane	*n*-butane (IUPAC: butane)	isobutane (IUPAC: 2-methylpropane)

In the IUPAC system the longest **continuous** chain of carbon atoms determines the name of an alkane, while the branches are labeled as substituents attached to the principal chain. The name of the branch is derived from the name of the hydrocarbon of the same number of carbons by dropping the "-ane" and adding "-yl."

IUPAC Names

3-methylhexane
(*not* 4-methylhexane)

$CH_2CH_2CH_2CH_2CH_3$

pentane

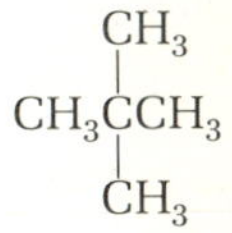

2,2-dimethylpropane

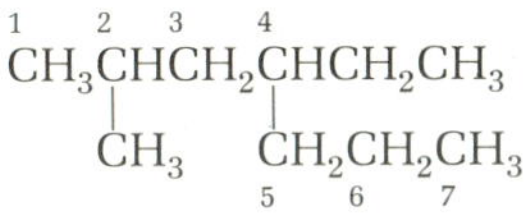

4-ethyl-2-methylheptane

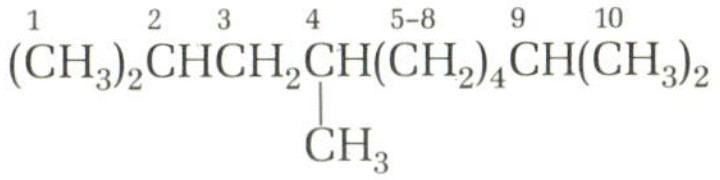

2,4,9-trimethyldecane

The atoms in the longest chain are numbered consecutively, starting from one end, in such a way that the first substituent number is the lowest one possible. If either method of numbering provides the same number for the first substituent, then that sequence is chosen that will provide the lowest number for the second substituent, and so forth. If there are two substituents on the same carbon atom, the number must be used twice, even if the substituents are identical.

The 20-carbon alkane is called eicosane, and its next few homologues are heneicosane, docosane, tricosane, etc. Starting with triacontane (30 carbons), alkanes having $n \times 10$ ($n = 3, 4, ..., 9$) carbon atoms have a numerical root attached to "-contane." Homologues are denoted by the prefixes hen-, di-, tri-, etc.

Certain alkyl substituents may be denoted by names derived from semisystematic nomenclature. Note that the prefixes *sec*- and *tert*- indicate attachment at **secondary** and **tertiary** positions, respectively.

$(CH_3)_2CH—$
isopropyl

sec-butyl
(*s*-butyl)

$(CH_3)_2CHCH_2—$
isobutyl

$(CH_3)_3C—$
tert-butyl
(*t*-butyl)

$(CH_3)_3CCH_2—$
neopentyl

$(CH_3)_2CHCH_2CH_2—$
isopentyl

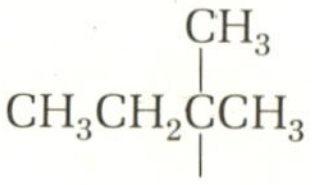
tert-pentyl
(*t*-pentyl)

A secondary carbon atom is one attached to two other carbon atoms; a tertiary carbon is attached to three other carbon atoms. When substituents are present in the side chain itself, their location is indicated numerically in the usual manner, the carbon atom by which the side chain is attached to the main chain being numbered one.

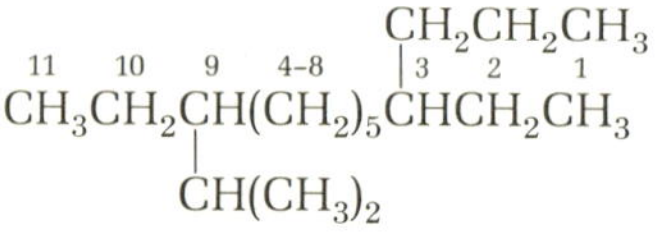
3-propyl-9-isopropylundecane

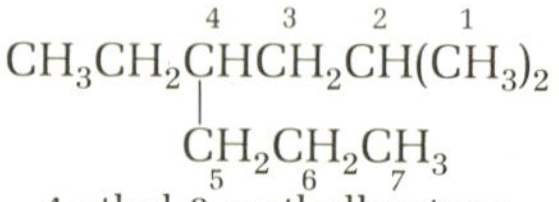
4-ethyl-2-methylheptane
(*not* 2-methyl-4-propylhexane)
(*not* 3-isobutylhexane)

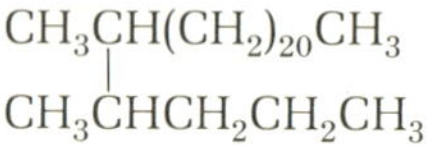
4,5-dimethylhexacosane

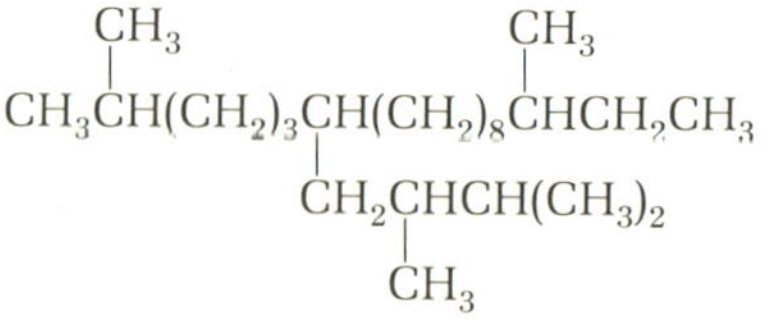
2,15-dimethyl-6-(2,3-dimethylbutyl)heptadecane

A few hydrocarbons may be named by considering them derivatives of simple alkanes such as methane and ethane. This practice is rather old-fashioned and it is going out of style.

$(CH_3)_2CHCH_2CH(CH_3)_2$
diisopropylmethane

$(CH_3)_3C—C(CH_3)_3$
hexamethylethane

HALOGEN-SUBSTITUTED ALKANES

Substituents such as halogens are treated differently by the semisystematic and IUPAC conventions. In the former, compounds are named as bromides, chlorides, etc., as in inorganic chemistry. The name of the alkyl group is used analogously to the cation name. If two halogen atoms are attached to either end of a straight chain, the number of interposed methylene (CH_2) groups is given. Amyl and pentyl are used interchangeably in semisystematic names (in the examples that follow, the semisystematic name is in *italics*).

$(CH_3)_2CHCH_2CH_2Cl$

1-chloro-3-methylbutane
isoamyl chloride

$Br(CH_2)_6Br$

1,6-dibromohexane
hexamethylene bromide

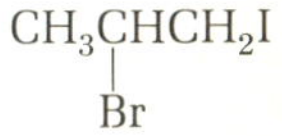

2-bromo-1-iodopropane

$(CH_3)_3CF$

2-fluoro-2-methylpropane
tert-butyl fluoride

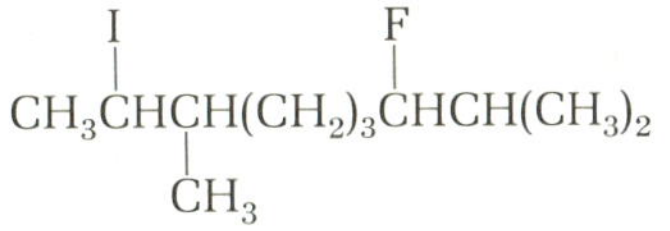

7-fluoro-2-iodo-3,8-dimethylnonane

Halogen atoms are treated as substituents on the chain, just like methyl groups, in the IUPAC system. When the substituents are listed, the halogens take precedence over alkyl groups. Within any particular class of substituents, such as halogen atoms or alkyl groups, the order of precedence is usually alphabetical. Bromo takes precedence over chloro, ethyl over methyl, methyl over propyl, etc. Prefixes such as di-, tri-, *sec*-, and *tert*- are not considered when alphabetical precedence is being established.

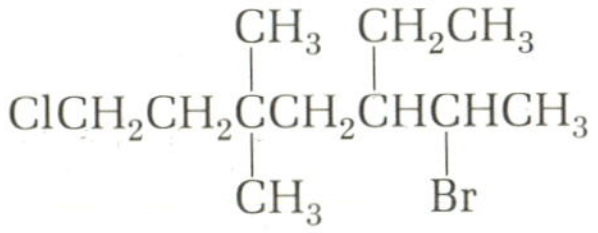

6-bromo-1-chloro-5-ethyl-3,3-dimethylheptane

STABILITY AND SYNTHESIS OF ALKANES

As indicated in Chapter 2, **carbon-carbon and carbon-hydrogen bond strengths are high, and the bonds are nonpolar. These characteristics, along with the absence of either nonbonded electrons or empty, low energy orbitals, account for the general lack of chemical reactivity of the alkanes** — thus the name paraffins. They resist the action of most strong acids and strong bases and are stable in air. Interestingly, bacteria have been found that feed on hydrocarbons, and they represent a potential source of protein from this abundant starting material. Liquid alkanes are colorless, mobile liquids considerably less dense than water. In the laboratory they find their principal use as inexpensive, easily purified, nonpolar solvents. Pentane is widely employed in this capacity because its low boiling point makes it easy to remove after it has served its purpose.

Alkanes will undergo a variety of thermal reactions at temperatures of about 400° and greater. Among the most important of these is **cracking,** in which a given alkane is converted into a mixture of saturated and unsaturated hydrocarbons of lower molecular weight (equation 5.1). This process is usually carried out over a silica-alumina catalyst and is a very important source of gasoline.

(5.1) $$n\text{-}C_4H_{10} \xrightarrow[780^\circ]{\text{gas phase}} H_2 + CH_4 + CH_2{=}CH_2 + CH_3CH_3 + CH_3CH{=}CH_2 + \text{other products}$$

Alkanes may also undergo isomerization, even at low temperatures, in the presence of Lewis acid catalysts. Low temperatures tend to favor the more branched isomer.

(5.2) $$CH_3CH_2CH_2CH_3 \underset{AlBr_3}{\overset{3\text{ atm}}{\rightleftharpoons}} (CH_3)_2CHCH_3 \qquad K^{25^\circ} \sim 4$$

Neither of these reactions is of great utility in the laboratory.

In addition to their isolation from petroleum, alkanes may be produced from other compounds by chemical means. One process that is applicable both industrially and in the laboratory is the addition of hydrogen to the double bonds of unsaturated hydrocarbons (see Chapter 9).

(5.3) $$CH_3CH_2CH_2CH{=}CH_2 + H_2 \xrightarrow[25^\circ]{Pt} CH_3CH_2CH_2CH_2CH_3$$

A catalyst is required, usually platinum, palladium, or nickel, and the reaction may be run at atmospheric pressure. Carbon-halogen bonds (Cl, Br, I) may be cleaved reductively giving an alkane and hydrogen halide or halide ion. This can be done with molecular hydrogen and a metal catalyst or by means of an active metal and a mineral acid.

(5.4) $$\underset{\text{1,2-dibromoethane}}{BrCH_2CH_2Br} + H_2 \xrightarrow[\text{ethanol, }100^\circ]{Pd/CaCO_3,\ H_2O} \underset{(100\%)}{CH_3CH_3} + 2\,HBr$$

(The figures in parentheses give the **yield** — that fraction of the maximum possible quantity of product which was actually obtained in practice.)

(5.5) $$\underset{\text{1-iodohexadecane}}{CH_3(CH_2)_{14}CH_2I} \xrightarrow[\text{acetic acid, }100^\circ]{Zn,\ HCl} \underset{(85\%)}{CH_3(CH_2)_{14}CH_3}$$

Under ordinary circumstances one never has any occasion to synthesize low or medium molecular weight alkanes, as a great variety of them are obtainable from petroleum. Indeed, the hydrocarbons from this source are the ultimate starting materials for most laboratory or plant syntheses. It occasionally becomes necessary to synthesize a rare, highly branched alkane or, more commonly, an alkane that is specifically labeled with a carbon or hydrogen isotope.

Single bonds between carbon atoms have a cylindrically symmetrical electron distribution, and rotation of one of the carbon atoms and its attached groups about this bond would not involve any change in electron distribution. It was at first assumed that this rotation was free and unhindered. All the commonly used sets of molecular models permit such free rotation. (These models can be of great value in understanding the material in this and many subsequent sections.)

Calculations of the thermodynamic properties of ethane, from first principles, gave values that were substantially at odds with the measured ones. This discrepancy could be resolved if it were assumed that rotation about the C—C bond was not entirely free but that there was instead an energy barrier to such rotation. Of all the possible instantaneous arrangements of the atoms in ethane **(conformations),** two in particular are of interest: the **eclipsed** and the **staggered** (Figure 5-7). The representations that we are using are known as **Newman projections** — the line of sight is directly down the C—C bond, and the point at the center of the circle represents the forward carbon atom. The circle is then the rearward carbon atom. Examination of a model will quickly reveal the perspective of the projection.

Experiment indicates that the internal energy of the eclipsed conformer is 2.8 kcal/mole greater than that of the staggered conformer. The internal energy of ethane is therefore at a maximum when the two methyl groups are eclipsed, and it then begins to drop as one group rotates with respect to the other; it is finally at a minimum after a 60° rotation. The origin of this energy difference has not yet been definitely identified and may in fact be quite complex. Two of the most plausible "simple" explanations are as follows:

1 It is due to repulsion between the electrons in the C—H bonds on adjacent carbon nuclei.

2 It is due to repulsion between the positive hydrogen nuclei attached to adjacent carbon atoms.

A calculation based on premise 2 led to a prediction of the strength of the rotational barrier in close agreement with the actual value. In any case, it should be noted that adjacent nuclei and the electrons associated with them are closer to one another in the eclipsed form compared to the staggered form; repulsive interactions among them will be correspondingly more intense.

The rotational barrier in propane equals 3.3 kcal/mole, reflecting the stronger methyl-hydrogen repulsion in the eclipsed conformation, compared with a hydrogen-hydrogen repulsion in the same conformation of ethane. The barrier to rotation about the central carbon-carbon bond in *n*-butane

FIGURE 5-6
Several Different Models of Ethane

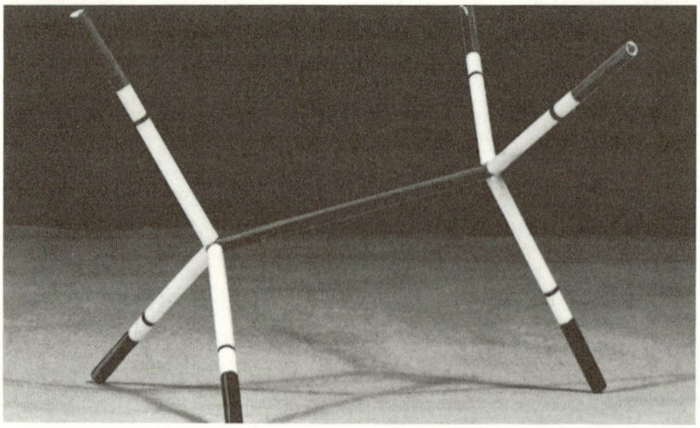

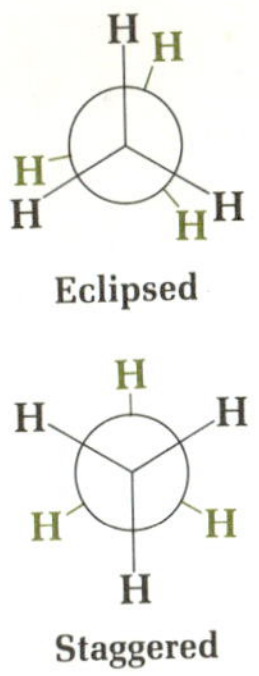

FIGURE 5-7
Eclipsed and Staggered Conformations of Ethane

is more complex than that in ethane or propane. There are several different eclipsed and staggered conformers, as indicated in Figure 5-10. The *anti* conformer is the most stable while the *skew* (*gauche*) conformer, where the methyl groups are somewhat closer together, is more energetic by 0.8 kcal/mole. The least favorable conformation is the fully eclipsed one, and this has been estimated to be 5 to 6 kcal/mole less stable than the *anti* conformation. Going from the *anti* to the partly eclipsed conformation requires about as much energy in butane as it would to go from the staggered to the eclipsed conformation in propane.

In none of the molecules so far discussed is the energy barrier anywhere near large enough to prevent rotation at room temperature. The rotation is slowed but certainly not stopped. It has, however, proved possible to isolate separate rotational conformers by operating at very low temperatures or by using molecules with very bulky groups attached to the singly bonded atoms.

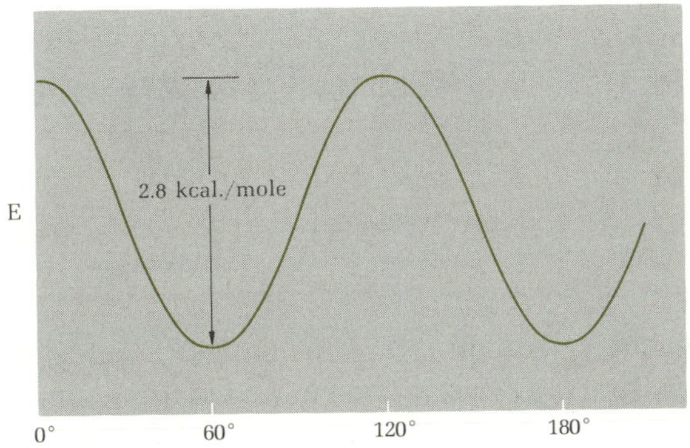

FIGURE 5-8
Energy of an Ethane Molecule as a Function of Angle of Internal Rotation

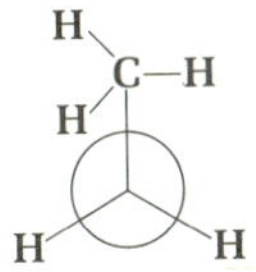

FIGURE 5-9
Eclipsed Conformation in Propane

When several substituted ethane derivatives are cooled past their freezing point, crystals are obtained that contain only one molecular conformer. In the liquid state at room temperature, however, all the possible conformers are in rapid equilibrium with one another, due to the relatively low energy barrier between conformers. The energy differences are still sufficiently large, however, that a sample of ethane at room temperature consists very largely of molecules in the staggered conformation.

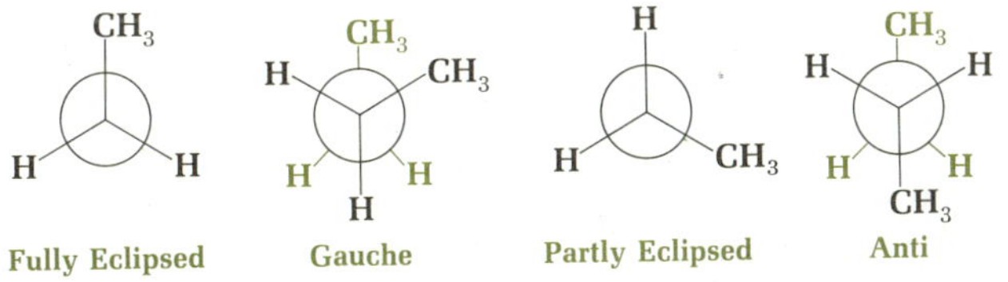

FIGURE 5-10
Eclipsed and Staggered Conformations for *n*-Butane

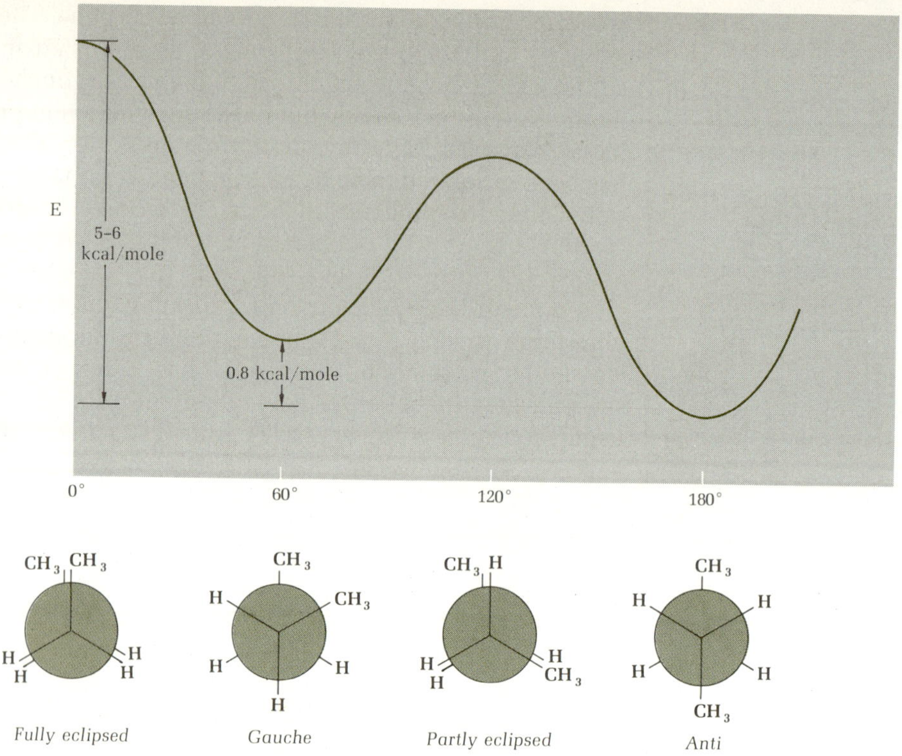

FIGURE 5-11
Energy Diagram for Internal Rotation in *n*-Butane

Long-chain alkanes are quite flexible, and although their most stable conformation is a "zigzag" one in which there is an *anti* conformation about each carbon-carbon bond, they are not confined to this conformation in solution. They may "thrash about" and bend back on themselves, and hydrogen atoms separated by quite a few carbons may momentarily come quite close together.

HALOGENATION OF ALKANES

We have noted the relative lack of chemical reactivity among the alkanes. They can be induced to undergo some reactions but certain high energy species must be produced initially to get the process underway. One of the two most important reactions of alkanes is halogenation—the replacement of one or more hydrogen atoms in the alkane by halogen atoms. For economic and other reasons, chlorine is the halogen of choice for industrial use. The alkyl chlorides so produced are useful as solvents, and also as starting materials for the synthesis of other compounds because the chlorine atom is readily replaced by a variety of different groups.

Reactions between molecular halogens and alkanes may be conducted in the gas phase or in the liquid phase. If the halogen (excepting fluorine) is mixed with the hydrocarbon and the two allowed to stand in the dark at room temperature, no detectable change occurs for some time. If the mixture is heated, or exposed to light, then a reaction will begin and may proceed very rapidly with much evolution of heat. The products are haloalkanes and hydrogen halides, along with any unreacted starting material.

It has been found that radiation of a wavelength corresponding to that of blue light is effective in initiating the chlorination of hydrocarbons. The energy in this light corresponds to about 60 kcal/mole, a good deal less than the energy needed to break carbon-carbon or carbon-hydrogen bonds. The dissociation energy of the chlorine molecule is about 58 kcal/mole, however. Furthermore, the temperatures at which the chlorination reaction begins to proceed appreciably are too low to cause any detectable dissociation of any hydrocarbon molecules. It seems very likely, therefore, that the function of the heat and light is to bring about the decomposition of chlorine molecules to chlorine atoms.

(5.6) $Cl_2 \rightleftharpoons 2Cl\cdot$

Assume that methane is the hydrocarbon being chlorinated. The chlorine atoms, with seven electrons in their valence shells, will react in such a way as to fill that shell with an eighth electron. One possibility is simple recombination with another chlorine atom to form molecular chlorine. Another is the *abstraction* of a hydrogen atom from methane to yield hydrogen chloride. In the gas phase the activation energy for this second reaction is only 3.9 kcal/mole.

(5.7) $H_3C{:}H + Cl\cdot \rightleftharpoons HCl + H_3C\cdot$

The organic molecule left behind is a **methyl radical,** 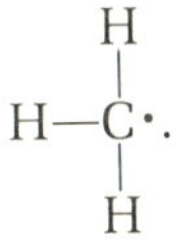.

Any species other than an atom that has one or more unpaired electrons is called a **radical,** or sometimes a **free radical.** Note that of the seven electrons in the valence shell of methyl radical, four "belong" to the carbon atom so that the radical is uncharged.

The methyl radical is itself quite reactive for the same reason that the chlorine atom is reactive.

(5.8) $CH_3\cdot + CH_3\cdot \rightleftharpoons CH_3CH_3$

(5.9) $CH_3\cdot + Cl\cdot \rightleftharpoons CH_3Cl$

It could satisfy its valence requirements by reacting with another methyl radical or with a chlorine atom, but the extremely low instantaneous concentration of each of those renders both reactions highly unlikely. Abstraction of a hydrogen atom from another methane molecule would not lead to any measurable chemical change. A chemically attractive possibility is the attack of the methyl radical on a molecule of chlorine, for which an E_a of only 2.3 kcal/mole is required (equation 5.10).

(5.10) $CH_3\cdot + Cl_2 \rightleftharpoons CH_3Cl + Cl\cdot$

The reaction shown in equation 5.10 results in the creation of a new chlorine atom that may, in its turn, attack another methane molecule and produce another methyl radical. The repeated sequential occurrence of these two steps (equations 5.7 and 5.10) results in a **radical chain reaction.** Equation 5.6, in which the entire cycle is started, is known as the **chain-initiating step.** Equations 5.7 and 5.10 are the **chain-propagating steps;** the net number of radicals is not changed in either of these reactions. Equations 5.8 and 5.9, on the other hand, represent **chain-terminating steps** because they result in a diminution in the number of available radicals. A summary of the proposed mechanism is given below.

$$Cl_2 \xrightarrow[\text{light}]{\text{heat or}} 2Cl\cdot \qquad \text{chain initiation}$$

$$\left.\begin{array}{l} CH_4 + Cl\cdot \longrightarrow CH_3\cdot + HCl \\ CH_3\cdot + Cl_2 \longrightarrow CH_3Cl + Cl\cdot \end{array}\right\} \text{chain propagation}$$

$$\left.\begin{array}{l} CH_3\cdot + CH_3\cdot \longrightarrow C_2H_6 \\ CH_3\cdot + Cl\cdot \longrightarrow CH_3Cl \\ Cl\cdot + Cl\cdot \longrightarrow Cl_2 \end{array}\right\} \text{chain termination}$$

The overall stoichiometry of the reaction is given by equation 5.11.

(5.11) $CH_4 + Cl_2 \longrightarrow CH_3Cl + HCl$

EVIDENCE FAVORING THE RADICAL MECHANISM

The radical mechanism is supported by the observation that molecular oxygen inhibits the chlorination. Oxygen is a diradical in its ground state, possessing two unpaired electrons. The interpretation of its action is that it reacts with methyl radicals to produce methylperoxy radicals that do not carry on the chain. Further support comes from the preparation of stable free radicals.

(5.12) $CH_3\cdot + O_2 \longrightarrow CH_3{-}O{-}O\cdot$

Even reactive radicals may be generated at very low temperatures in solids, where they exist long enough to be detected spectroscopically (Chapter 11).

The existence of radicals was inferred from evidence obtained around 1930, well before spectroscopic techniques became available. A stream of inert gas carrying the vapor of tetramethyllead, $(CH_3)_4Pb$, was conducted through a glass tube. When the tube was heated strongly, the compound decomposed, as evidenced by the deposition of metallic lead on the inside surface of the tube. Hydrocarbon products, mostly ethane, were swept out by the carrier gas. If the heat source were moved further upstream, then a new metallic "mirror" could be formed by decomposition of the tetramethyllead. Most important, the lead mirror *downstream* began to disappear as the new one *upstream* was being formed. Furthermore, the rate of disappearance of the downstream mirror became slower as the distance between it and the decomposition zone was increased.

These results were interpreted as indicating that the decomposition of tetramethyllead gave rise to a very reactive intermediate that could combine with lead as well as several other metals, to produce volatile products. Furthermore, this intermediate reacted very rapidly with itself to produce stable products. The most reasonable structure for this intermediate was that corresponding to methyl radical, and its most prominent reactions in this system were those corresponding to equations 5.13, 5.14, and 5.8.

(5.13) $(CH_3)_4Pb \xrightarrow{\text{heat}} 4CH_3\cdot + Pb$

(5.14) $4CH_3\cdot + Pb \xrightarrow{25°} (CH_3)_4Pb$

(5.8) $2CH_3\cdot \longrightarrow CH_3CH_3$

KINETICS AND THERMODYNAMICS OF THE HALOGENATION OF METHANE

Halogenations with fluorine, chlorine, and bromine result in the release of thermal energy; once initiated, these reactions are usually self-sustaining and go to completion. Iodination, on the other hand, is endothermic and does not usually achieve anything near complete reaction. Some more insight into this entire series of reactions may be gained by examining some of the relevant kinetic and thermodynamic data.

For a reaction in the gas phase, ΔH is simply the sum of the dissociation energies of the bonds being broken and formed. From the values given in Table 5-4 it is possible to calculate the ΔH's for the individual steps in the halogenation of methane. The E_a's that are given have been determined experimentally.

(5.15) $F\cdot + CH_4 \longrightarrow HF + CH_3\cdot$ $\Delta H = -32$ kcal/mole, $E_a = 1.2$ kcal/mole

(5.16) $CH_3\cdot + F_2 \longrightarrow CH_3F + F\cdot$ $\Delta H = -70$ kcal/mole

(5.17) $CH_4 + F_2 \longrightarrow CH_3F + HF$ $\Delta H = -102$ kcal/mole

Fluorination is highly exothermic and very difficult to control. The fluorine is usually diluted with nitrogen and the reaction run in the presence of copper packing that serves to absorb much of the heat. Even so, a multiplicity

TABLE 5-4
Bond Dissociation Energies[a]

H—H	104	CH_3—H	104
F—F	38	CH_3—F	108
Cl—Cl	58	CH_3—Cl	83.5
Br—Br	46	CH_3—Br	70
I—I	36	CH_3—I	56
H—F	136	CH_3CH_2—H	98
H—Cl	103	$(CH_3)_2CH$—H	94.5
H—Br	87.5	$(CH_3)_3C$—H	91
H—I	71	CH_2=CH—H	103
H—OH	119	CH_3—CH_3	88
H—OCH_3	102	$(CH_3)_2CH$—CH_3	84
HO—CH_3	91.5	$(CH_3)_3C$—CH_3	80
HO—OH	51	CH_2=CH—CH_3	92

[a]The data are in kilocalories per mole and refer to the dissociation of 1 mole of gaseous molecules, in the manner indicated, at 25° and 1 atm. The reader must take care when comparing these data with the results of thermochemical experiments since many of the molecules listed here are not gases at 25° and 1 atm and energy has to be supplied to vaporize them. The values given refer in all cases to **homolytic** bond cleavage. Data taken from S. W. Benson, *J. Chem. Ed.*, **42,** 502 (1965).

of products usually results. When alkanes higher than methane are fluorinated, C—C as well as C—H bonds are broken.

(5.18) $$CH_4 + F_2 \xrightarrow[25^\circ]{\text{Cu packing}} CH_3F + CH_2F_2 + CHF_3 + CF_4 + C_2F_6 + C_3F_8$$

Chlorination is much more easily controlled than fluorination, as it is much less violently exothermic.

(5.19) $$Cl\cdot + CH_4 \longrightarrow HCl + CH_3\cdot \qquad \Delta H = 1 \text{ kcal/mole}, E_a = 3.8 \text{ kcal/mole}$$
(5.20) $$CH_3\cdot + Cl_2 \longrightarrow CH_3Cl + Cl\cdot \qquad \Delta H = -25.5 \text{ kcal/mole}$$
(5.21) $$CH_4 + Cl_2 \longrightarrow CH_3Cl + HCl \qquad \Delta H = -24.5 \text{ kcal/mole}$$

Note that the first step, abstraction of a hydrogen atom by a chlorine atom, is actually slightly endothermic; but the exothermicity of the methyl radical-molecular chlorine step makes the overall process exothermic. The difference in enthalpy between the two different steps of the halogenation is even more marked when one turns to bromination, in which the first step is substantially endothermic; it is a very slow step as well, due to the high energy of activation.

Finally, we may note that direct iodination of methane is not a feasible process, because it is markedly endothermic.

(5.22) $$CH_4 + Br\cdot \longrightarrow CH_3\cdot + HBr \qquad \Delta H = 16.5 \text{ kcal/mole}, E_a = 18.6 \text{ kcal/mole}$$
(5.23) $$CH_3\cdot + Br_2 \longrightarrow CH_3Br + Br\cdot \qquad \Delta H = -24 \text{ kcal/mole}$$
(5.24) $$CH_4 + Br_2 \longrightarrow CH_3Br + HBr \qquad \Delta H = -7.5 \text{ kcal/mole}$$

Since the entropy changes in all these halogenations are nearly zero, it is the energy term which controls the position of equilibrium at 25°.

(5.25) $I\cdot + CH_4 \longrightarrow HI + CH_3\cdot$ $\Delta H = 31$ kcal/mole, $E_a = 33.5$ kcal/mole

(5.26) $CH_3\cdot + I_2 \longrightarrow CH_3I + I\cdot$ $\Delta H = -20$ kcal/mole

(5.27) $CH_4 + I_2 \longrightarrow CH_3I + HI$ $\Delta H = 11$ kcal/mole

The first step in each sequence is apparently the rate-determining one. The exothermicities of these reactions fall in the order F > Cl > Br > I; on the other hand, the energies of activation are arranged in the sequence F < Cl < Br < I. Such a correlation, in which the most exothermic of a series of related reactions also has the lowest activation energy, is found for a great many reactions; it is not by any means a universal correlation, however. Understanding the origin of this parallelism, and its limits in practice, can be very helpful for developing a qualitative "feel" for the effect of structure on reactivity.

RATE-REACTIVITY CORRELATIONS

Consider a generalized gas-phase reaction such as that given by equation 5.28.

(5.28) $AB + C \longrightarrow A + BC$

A, B, and C may be atoms, or polyatomic fragments that do not change appreciably in the course of the reaction. If we plot the potential energy of the entire system as a simultaneous function of the AB and BC distances, then the result is a three-dimensional surface. This surface rather resembles that of a golf course with small hills, valleys, sand traps, etc. It contains two prominent depressions, or basins. The bottom of one corresponds to the equilibrium configuration of the reactants and the bottom of the other, to the configuration of the products. As reaction proceeds, the reactants "climb" out of the reactant basin, go over a hump, and drop into the product basin.

Atoms and molecules, being not unlike people, will tend to climb over the lowest hump available. The configuration at the top of that hump is the transition state, and the energy needed to get from the bottom of the reactant basin to the top of the hump is the activation energy. If we take a slice through this surface, in such a manner that we cut through both basins and the transition state, the profile of this cut might look much like Figure 5-12. The portion on the left represents a section through the reactant basin, and the descending line on the right represents the "slide" into the product basin.

Suppose we compare reaction 5.28 with reaction 5.29, in which the product BC′ has a lower potential energy than BC.

(5.29) $AB + C' \longrightarrow A + BC'$

That is, reaction 5.29 is **more exothermic** than reaction 5.28. The bottom of the product basin is then lower, and so is the slide leading to it. **The net result**

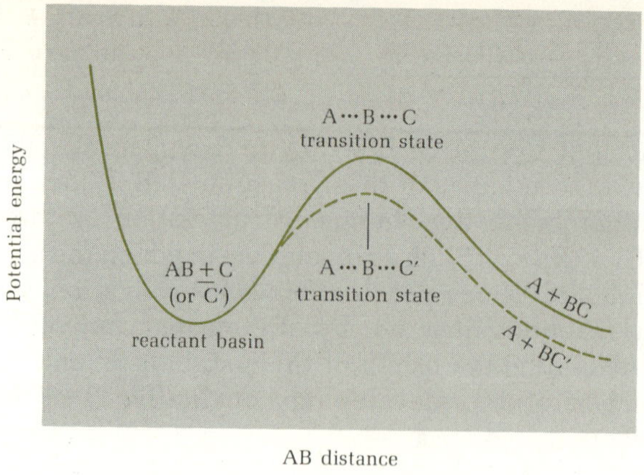

FIGURE 5-12

Slice through a Potential Surface for the Reaction of AB with C and C′

will be a lowering of the activation energy (see the dotted line in Figure 5-12). This is exactly the situation we encounter if we compare the reactions of fluorine atoms and chlorine atoms with methane. The formation of HF is much more exothermic than the formation of HCl, and therefore the activation energy for hydrogen abstraction by F· is lower than the activation energy for hydrogen abstraction by Cl·.

The conclusion to be drawn from this section is that in a series of related reactions the strength of the bond being *formed* can have a marked effect on the activation energy. The next section will suggest that the strength of the bond being *broken* also affects the activation energy. These two factors are quite important, but they are not the only ones. In particular, the repulsive interactions between the reacting species must be considered (steric effect).

In light of the foregoing analysis it is perhaps not difficult to see why an alternative mechanism for the chlorination of methane, embodied in equations 5.30 and 5.31, is not operative.

(5.30) $Cl\cdot + CH_4 \longrightarrow CH_3Cl + H\cdot \qquad \Delta H = 20.5$ kcal/mole

(5.31) $H\cdot + Cl_2 \longrightarrow HCl + Cl\cdot \qquad \Delta H = -45$ kcal/mole

Although the *overall* enthalpy change must be identical for both mechanisms, equation 5.30 is so endothermic compared with equation 5.19 that it is many, many times slower and makes no significant contribution to the overall reaction.

HALOGENATION OF HIGHER ALKANES

Ethane chlorinates somewhat more readily than methane, the hydrogen abstraction step being more exothermic and lower in activation energy. Since a C—H bond in ethane is more easily broken than one in methane (Table 5-4), the conclusion that is suggested is that making a bond easier to break (thermodynamically) results in a lower activation energy for the step in which it is broken.

(5.32) $Cl\cdot + CH_3CH_3 \longrightarrow HCl + CH_3CH_2\cdot \quad \Delta H = -5$ kcal/mole, $E_a = 1.0$ kcal/mole

In terms of the energy profile, making the bond easier to break results in a lower slope for the walls of the reactant basin (dotted line in Figure 5-13).

The monochlorination of propane now introduces a new variable since two different products are possible.

(5.33) $CH_3CH_2CH_3 + Cl_2 \longrightarrow CH_3CH_2CH_2Cl + CH_3CH(Cl)CH_3 + HCl$

n-propyl chloride; isopropyl chloride

If all the hydrogen atoms were equally reactive, a 75:25 mixture of *n*-propyl chloride and isopropyl chloride would be expected (six ways of forming *n*-propyl chloride and two ways of forming isopropyl chloride). In fact, when the reaction is run in the gas phase at 300°, a 50:50 mixture of the two chlorides is obtained. The discrepancy between statistical predictions and experimental fact is even more marked when the monochlorination of isobutane is examined.

(5.34) $(CH_3)_3CH + Cl_2 \longrightarrow (CH_3)_2CHCH_2Cl + (CH_3)_3CCl + HCl$

isobutyl chloride; *t*-butyl chloride

Two products are again possible, isobutyl chloride and *t*-butyl chloride; statistically, they should be found in a mole ratio of 9:1. The observed product ratio is 3:2.

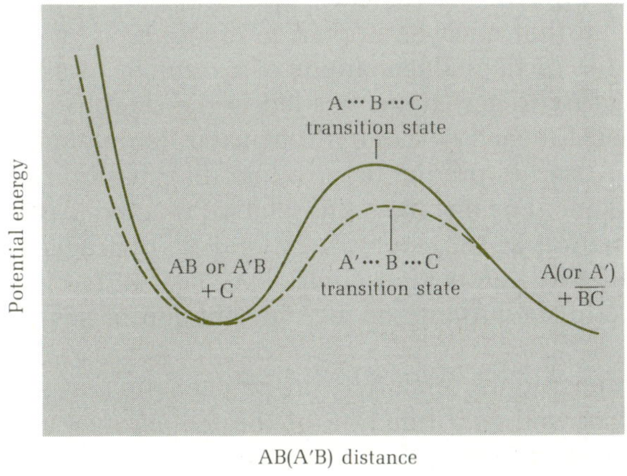

FIGURE 5-13
Section through the Potential Surface for the Reaction of AB and A′B with C

Table 5-4 reveals that the *secondary* (—CH_2—) carbon-hydrogen bonds (p. 102) of propane are more easily broken than the *primary* (—CH_3) carbon-hydrogen bonds of ethane. Furthermore, the *tertiary* carbon-

$$(-\overset{|}{\underset{|}{C}}-H)$$

hydrogen bond in isobutane is the most easily cleaved of all. The experimental results are now comprehensible since we are not warranted in assuming that all the hydrogens were of equal reactivity.

RADICAL STABILITIES

Secondary hydrogen atoms are abstracted more rapidly by halogen atoms than primary ones because of a lower activation energy. The activation energy difference is directly related to a difference in bond dissociation energies (Figure 5-13), but one might ask why this situation should exist. If a secondary hydrogen atom is abstracted from an alkane, then a secondary (2°) radical is produced. **The order of stability of alkyl radicals is 3° > 2° > 1°**; in other words, replacing a hydrogen atom attached to a radical site by an alkyl group increases the stability of the radical (explanations for this observation will be developed later).

$CH_3CH_2\cdot$	$(CH_3)_2CH\cdot$	$(CH_3)_3C\cdot$
$CH_3CH_2CH_2\cdot$	$CH_3\underset{\bullet}{C}HCH_2CH_3$	$(CH_3)_2\underset{\bullet}{C}CH_2CH_3$
$(CH_3)_2CHCH_2CH_2\cdot$	$(CH_3)_3C\underset{\bullet}{C}HCH_2CH_3$	$CH_3CH_2\underset{\bullet}{\overset{CH_3}{\overset{\vert}{C}}}CH_2CH_2CH_3$
primary radicals	secondary radicals	tertiary radicals

The stability of the radical obviously has a strong effect on the energy of the bond that must be broken to create it.

Studies of the halogenations of a number of alkanes containing different types of hydrogen atoms has led to the establishment of relative reactivity numbers for each type of hydrogen at different temperatures. These numbers may be used to predict the product distribution from monohalogenation of any alkane. For instance, the relative reactivity **per hydrogen atom** toward chlorination at 150°, for 1°, 2°, and 3° hydrogen atoms, is 1.00:3.35:3.93. At 80°, the numbers become 1.00:3.63:4.51. Figure 5-14 illustrates how product distributions for the chlorination of several alkanes may be computed.

The positional selectivity in halogenation is a function of the halogen atoms as well as a function of the temperature. Bromine atoms are much more discriminating than chlorine atoms—the relative reactivity numbers

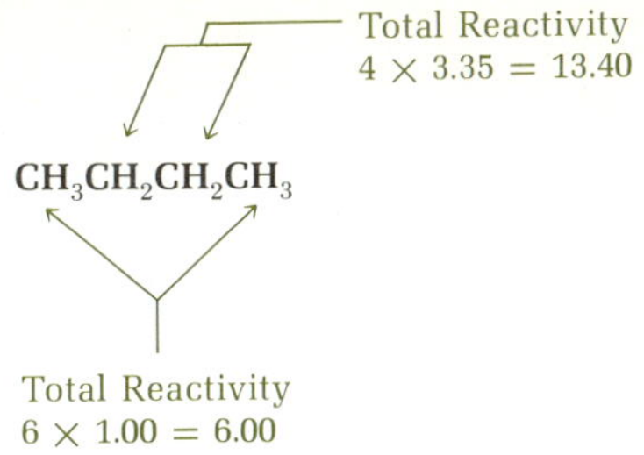

$$\% \text{ 2-chlorobutane} = \frac{13.40}{19.40} \times 100 = 69\%$$

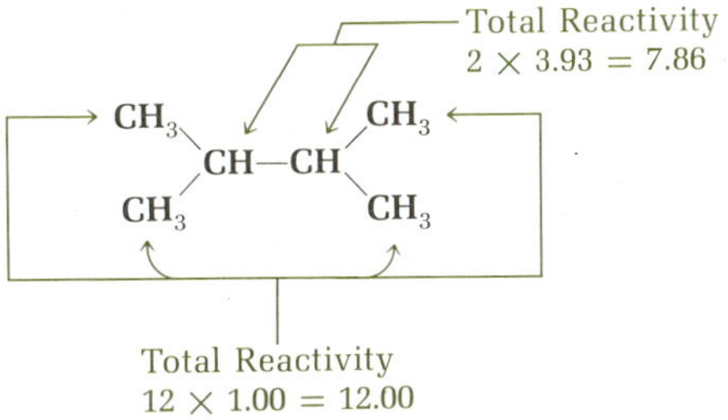

$$\% \text{ 2-chloro-2,3-dimethylbutane} = \frac{7.86}{19.86} \times 100 = 40\%$$

FIGURE 5-14
Calculation of Product Distributions from Monochlorination of *n*-Butane and 2,3-Dimethylbutane at 150°

for bromination at 150°, for 1°, 2°, and 3° hydrogen atoms, is 1.0:82:1640! In a molecule containing tertiary hydrogen atoms, primary hydrogen atoms are essentially immune to attack by bromine.

REACTIVITY AND SELECTIVITY

Hypothetical energy diagrams for strongly exothermic and endothermic reactions are given in Figure 5-15. While the generality of these diagrams cannot be proved, they do accord with the little we know about the nature of energy surfaces, and the conclusions that can be drawn with the aid of these diagrams are usually in harmony with the organic chemist's experience. A central point in this hypothesis is that for exothermic reactions the transition state resembles the reactants, while for endothermic reactions the transition state resembles the products (*Hammond's postulate*).

When we say that the transition state resembles the reactants, we mean that the bond angles and bond lengths found in the transition state do not differ markedly from the values of these parameters found in the reactants. In particular, the C—H bond being broken in a hydrogen abstraction reaction has not been stretched very much, and there is relatively little "radical character" at the carbon atom. Such would be the case if a fluorine atom

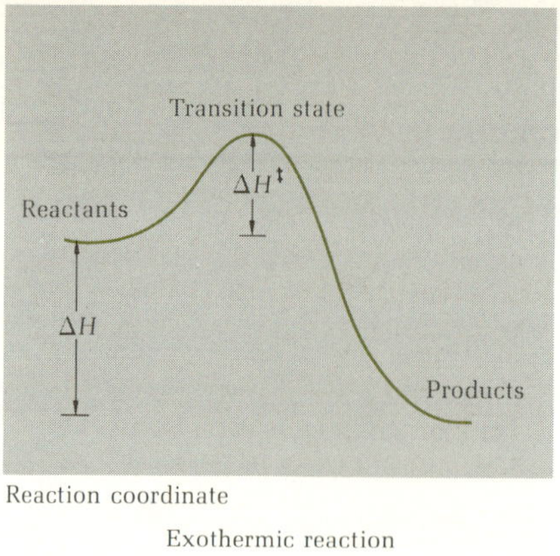

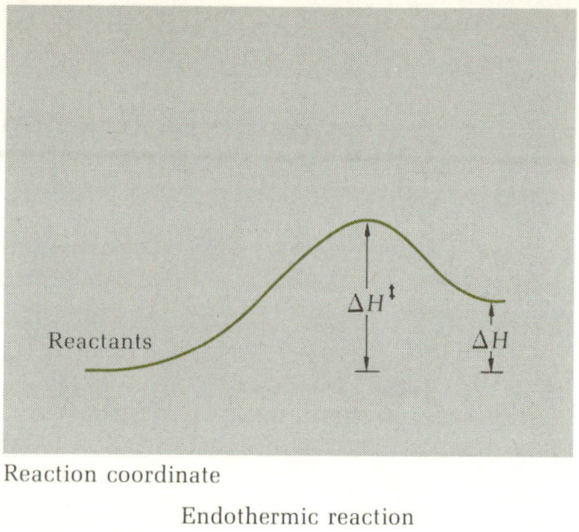

FIGURE 5-15
Energy Diagrams for Exothermic and Endothermic Reactions

were the abstracting species. Since the bond is not stretched much at the transition state, the reaction is not very sensitive to the strength of one C—H bond as compared with another. Thus, fluorine attacks all C—H bonds virtually indiscriminately.

The reverse situation arises in the endothermic case; the C—H bond is nearly broken at the transition state, and the energy of activation is very sensitive to the bond dissociation energy. This behavior would characterize the bromination of alkanes. The chlorination reaction is intermediate in thermal character, and chlorine atoms display an intermediate range of selectivities.

THE STRUCTURE OF ALKYL RADICALS

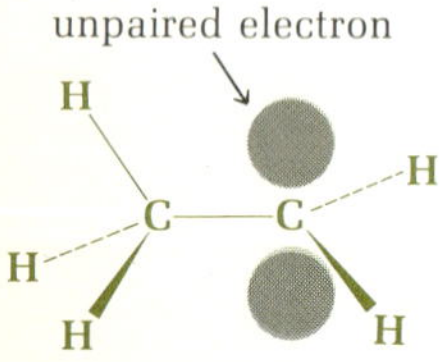

FIGURE 5-16
sp^2-Hybridized Ethyl Radical

Most of the available evidence concerning the structure of alkyl radicals is compatible with sp^2-hybridization at the carbon atom bearing the unpaired electron. This electron presumably occupies the *p*-orbital (Figure 5-16). A few radicals may prefer to have a tetrahedral configuration with the unpaired electron in an sp^3-orbital. It has been proposed that this is true of trifluoromethyl radical. It is very likely that tetrahedral radicals can undergo rapid inversion, just as ammonia does (Figure 5-17). Experimentally, it is difficult to distinguish between a planar radical and a rapidly inverting tetrahedral one.

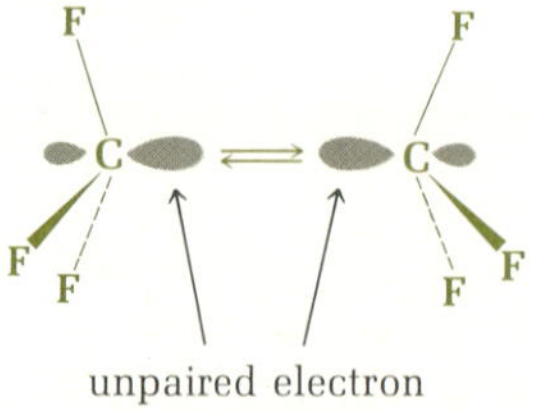

FIGURE 5-17
Equilibrating Tetrahedral Trifluoromethyl Radicals

The most prevalent commercial halogenation is chlorination, which is quite inexpensive. As a result, various alkyl chlorides are important commercial synthetic ingredients. Fluorination is also carried out on a commercial scale, despite the relative expense and the difficulty in handling fluorine, because of the unique properties of many *fluorocarbons* (compounds containing only fluorine, hydrogen, and carbon atoms). The fluorocarbons are characterized by their chemical inertness and their unexpectedly low boiling points (due, no doubt, to the very low polarizability of the electrons around fluorine).

Even partially fluorinated compounds exhibit many unusual and useful properties. The Freons (derivatives of methane and ethane containing fluorine along with other halogens) are extensively used as refrigerants because of their volatility, stability, and low toxicity. Certain simple fluorine-containing compounds show very marked anesthetic behavior; 2-bromo-2-chloro-1,1,1-trifluoroethane is an example of such a molecule.

When a laboratory halogenation is carried out, it is almost invariably a bromination or chlorination. Alkyl fluorides and iodides are usually synthesized by means of ionic replacement reactions (Chapter 7). Bromination may be conveniently carried out in the liquid phase since bromine is a volatile red-brown liquid. An inert solvent such as CCl_4 may be used, although it is not necessary. Since chlorine is a very reactive, corrosive gas at room temperature, it is rather inconvenient to handle in the laboratory. Other compounds are commonly used to generate chlorine atoms—these will generally have a chlorine atom bonded to a nonmetallic element. Two examples are sulfuryl chloride and *t*-butyl hypochlorite, $(CH_3)_3COCl$. They

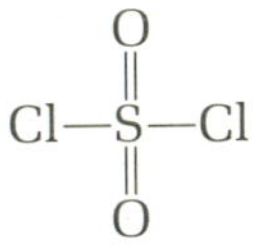

(sulfuryl chloride)

are both liquid at room temperature and not too difficult to handle. Reactions are usually run in the pure liquid hydrocarbon or in an inert solvent and initiated in most cases either by light or by a chemical **initiatior.** These latter are materials that readily decompose to give radicals that serve to start the chain reactions. They are often peroxides (RO—OR) since the O—O single bond is quite weak.

The by-products from SO_2Cl_2 reactions are SO_2 and HCl; the by-product from $(CH_3)_3COCl$ is *t*-butyl alcohol, $(CH_3)_3COH$. Sulfuryl chloride is slightly more selective toward various types of hydrogens than molecular chlorine is, suggesting that some of the abstraction is carried out by $\cdot SO_2Cl$ radicals. *t*-Butyl hypochlorite is much more discriminating than chlorine — its relative selectivity toward 1°, 2°, and 3° hydrogen atoms at 40° is 1.0:7.9:44. The chain propagating steps are

(5.35) $RH + (CH_3)_3CO\cdot \longrightarrow R\cdot + (CH_3)_3COH$

(5.36) $R\cdot + (CH_3)_3COCl \longrightarrow RCl + (CH_3)_3CO\cdot$

Attention has been confined thus far to monohalogenated products. Many polyhalogenated alkanes are of commercial and laboratory importance. Adjusting the ratio of halogen to hydrocarbon enables one to alter the balance between mono- and polyhalogenated products. When the former are desired, the hydrocarbon is usually in large excess.

Chlorination of methane past the monochloride stage leads to three additional products, each of which is an important solvent. They are shown below with their common names, which are universally employed.

(5.37) $CH_4 \xrightarrow{Cl_2} CH_3Cl \xrightarrow{Cl_2} CH_2Cl_2 \xrightarrow{Cl_2} CHCl_3 \xrightarrow{Cl_2} CCl_4$

	methyl chloride	methylene chloride	chloroform	carbon tetrachloride

COMBUSTION OF ALKANES

The other very important use of alkanes is as fuels. Their combustion leads to the release of sizable quantities of energy. Interestingly, the heat released per gram of alkane decreases as the molecular weight increases.

(5.38) $C_6H_{14}\,(\text{liq.}) + 9\tfrac{1}{2}O_2 \xrightarrow{25^\circ} 6CO_2 + 7H_2O\,(\text{liq.}) + 992\ \text{kcal/mole}$

The percent by weight of hydrogen goes down with increasing molecular weight; since formation of O—H bonds is highly exothermic, each hydrogen atom represents a sizable quantity of potential heat for very little mass.

Oxidation is a free radical chain process, like halogenation, and depends on the radical character of molecular oxygen. Activation energy must be supplied initially to induce the oxygen molecules to abstract hydrogen atoms from the alkanes. Once underway, the chain is self-sustaining. The mechanisms of the oxidations are much more complex than those of halogenations, as the example of methane will show.

(5.39) $CH_4 + O_2 \longrightarrow CH_3\cdot + HO_2\cdot$

(5.40) $CH_3\cdot + O_2 \longrightarrow [CH_3OO\cdot] \longrightarrow CH_2{=}O + \cdot OH$

(5.41) $CH_4 + \cdot OH \longrightarrow CH_3\cdot + H_2O$

(5.42) $CH_2{=}O + \cdot OH \longrightarrow H\overset{\displaystyle O}{\overset{\|}{C}}\cdot + H_2O$

(5.43) $H\overset{\displaystyle O}{\overset{\|}{C}}\cdot + O_2 \longrightarrow [H\overset{\displaystyle O}{\overset{\|}{C}}{-}O{-}O\cdot] \longrightarrow CO + HO_2\cdot$

(5.44) $HO_2\cdot + CH_2O \longrightarrow \cdot OH + H_2O + CO$

(5.45) $$CH_2O + O_2 \longrightarrow \underset{\underset{\displaystyle O}{\|}}{HC\cdot} + HO_2\cdot$$

(5.46) $$2CO + O_2 \longrightarrow 2CO_2$$

SUMMARY

1 Alkanes are obtained in very large quantities by the fractionation of petroleum. In order to get pure, or nearly pure, compounds from the mixtures, it is necessary to distill them. The mixtures may be analyzed by gas-liquid chromatography or a combination of glc and mass spectrometry. The mass spectrum will usually indicate the molecular weight of the compound, as well as showing a fragmentation pattern characteristic of the structure.

2 Alkanes are chemically unreactive, nonpolar compounds. The IUPAC system of nomenclature is based upon the longest continuous chain of carbon atoms; the number of atoms in this chain determines the name of the alkane. Carbon atoms not in the chain are named as substituents on it; the order of numbering is such as to provide the lowest possible number for the first substituent.

Halogen-containing alkanes may be named in the same way, the halogen atoms being treated as substituents on the principal hydrocarbon chain. When both halogen and alkyl substituents are present, the halogens take precedence in naming; within each group, the order is alphabetical.

3 Rotation about carbon-carbon single bonds in alkanes is not quite unhindered. There is a barrier to rotation that arises from the greater internal energy of the molecule when it is in an eclipsed conformation. Hydrogen-hydrogen repulsions are weaker than hydrogen-methyl repulsions, which in turn are less severe than methyl-methyl repulsions. Alkane molecules spend much more time in their various staggered conformations than they do in eclipsed ones, at ordinary temperatures.

4 Alkanes can react with molecular fluorine, chlorine, or bromine to yield halogenated alkanes and hydrogen halides. The reaction of iodine with alkanes is endothermic and does not proceed very far toward completion. Chlorination and bromination require initiation by light, heat, or chemical initiators.

The generally accepted mechanism for halogenation involves dissociation of a small percentage of halogen molecules into atoms; abstraction of hydrogen atoms from the alkanes by the halogen atoms then follows. The chain is maintained by the attack of the resulting alkyl radicals on molecular halogen. The hydrogen abstraction step is the rate-determining one; there is a parallel between the exothermicity of this step and its activation energy. There is also a correlation between the energy of the bond being broken and the activation energy.

The dissociation energies of various carbon-hydrogen bonds are in the order 3° < 2° < 1°. When an alkane containing different types of hydrogen atoms is halogenated, the more reactive halogens are abstracted preferentially. By using relative reactivity numbers, it is possible to calculate the product distribution from halogenation of any alkane.

For laboratory purposes, chlorinations may be carried out with the aid of sulfuryl chloride or *t*-butyl hypochlorite.

5 Heating alkanes in the presence of oxygen leads to combustion. If oxygen is absent, pyrolysis of the alkane takes place. If large alkyl radicals are formed, they tend to decompose to alkenes and smaller alkyl radicals.

PROBLEMS

1 Write out all the isomers corresponding to $C_5H_{11}Cl$, and name them.

2 Write out the structures corresponding to the following names:

a. 2-Methylpropane

b. 2,2-Dimethylbutane

c. 2,4-Dimethylpentane

d. 5-Ethyl-3-methyloctane

e. 4-Isopropyldecane

f. 3-Ethylpentane

g. 3,6-Dimethyl-5-(2-chloropropyl)nonane

h. 5-Fluoro-2,2,5-trimethylheptane

i. 2,7-Diiodo-4-*t*-butyl-8-ethyl-3,8-diisopropyldodecane

j. 3,5-Diethyl-4-isobutyl-4-(1-methylpropyl)octane

3 Name the following structures:

a. $CH_3CHCH_2CH_2CH_3$ (with CH_2CH_3 on the second carbon)

$$\begin{array}{l} CH_3\underset{\displaystyle |}{C}HCH_2CH_2CH_3 \\ \quad\; CH_2CH_3 \end{array}$$

b.

$$\begin{array}{l} CH_3CH_2\underset{\displaystyle |}{C}HCH_2CH_2CH_3 \\ \qquad\quad CH_2CH(CH_3)_2 \end{array}$$

c. 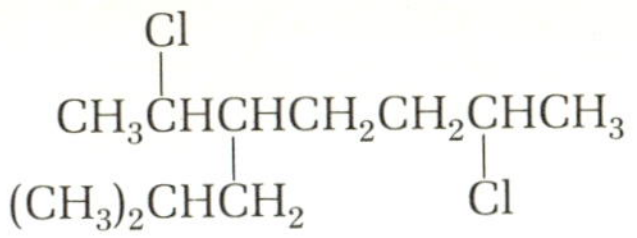

d. 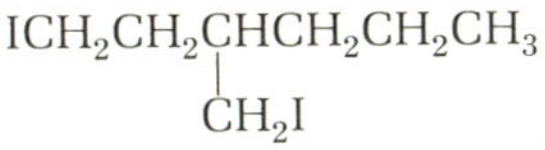

e. 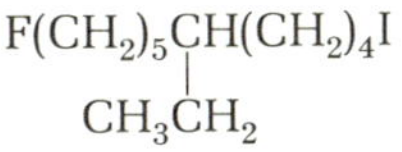

4 We have concentrated our attention on the rotation about the central C—C bond in butane. There is, however, also a barrier to rotation of the terminal methyl groups in butane. Give an estimate of the size of this rotational barrier.

5 Using the data given for propane and butane, calculate the barriers to rotation in 2-methylbutane.

6 a. Of the two possible staggered conformers of butane, the *anti* is lower in energy than the *gauche*. Which of the two is the more *probable*?

b. If the probability factor between two isomers is x (x = 1, 2, 3, ...), then the entropy difference between them contains a term $R \ln x$ (the more probable isomer has the higher entropy). Assume that the *only* source of an entropy difference between the *anti* and *gauche* conformers of butane is the probability difference. Then calculate ΔG and K for the interconversion of these two conformers at 25°.

7 The reaction of chlorine with ethane could lead to the formation of methyl chloride:

$$Cl_2 + CH_3CH_3 \longrightarrow 2CH_3Cl$$

Analyze this reaction thermodynamically. Could it compete effectively with the normal chlorination reaction? Do the same for fluorination.

8 a. Write all the products of monochlorination of *n*-pentane, 2,2-dimethylbutane, and 2-methylpentane.

b. Calculate the product distribution from each starting material at 150°.

c. Calculate the product distribution from monobromination of these alkanes.

9 a. When methane was brominated in the presence of deuterium bromide (DBr), it was found that the methyl bromide product had incorporated some deuterium. Furthermore, the unconverted methane was also found to contain deuterium. Explain these findings.

b. Chlorination of methane in the presence of deuterium chloride resulted in very much less deuterium incorporation into the methyl group. What is the difference between this experiment and the one described above?

10 Organic iodides may be synthesized by allowing compounds with certain types of functional groups to react with hydrogen iodide (equation 1).

$$ROH + HI \longrightarrow RI + H_2O \tag{1}$$

In the presence of excess hydrogen iodide, some reduction of the product often takes place (equation 2).

$$RI + HI \longrightarrow RH + I_2 \tag{2}$$

This reductive side reaction is no problem when other hydrogen halides are employed. Rationalize these findings.

11 Assume that you have, in large excess, an equimolar mixture of propane and butane. This mixture is allowed to react with 1 mole of chlorine at 80°. What are the expected proportions of the products?

12 Monochlorination of ethane-*d* (CH_3CH_2D) leads to a product that is 93% C_2H_4DCl and 7% C_2H_5Cl.

a. What conclusions would you draw from this result?

b. How would you actually *analyze* the mixture of ethyl chlorides?

13 Consider the reactions 3 and 4. Is it possible to predict which one will have the higher energy of activation? Explain your answer.

$$CH_4 + Br\cdot \longrightarrow CH_3\cdot + HBr \tag{3}$$

$$CH_3Br + Br\cdot \longrightarrow CH_3\cdot + Br_2 \tag{4}$$

SUPPLEMENTARY READINGS

W. R. Pryor, *Introduction to Free Radical Chemistry.* Englewood Cliffs, N.J.: Prentice-Hall, Inc., 1966, Chapters 1, 6 and 7. (P)

CHAPTER 6

CYCLOALKANES—CONFORMATIONAL ANALYSIS

If one hydrogen atom were clipped from either end of an *n*-alkane and the ends then joined, the resulting compound would be a **cycloalkane**—a monocycloalkane, to be precise. These compounds form a homologous series having the empirical formula $(CH_2)_n$, $n = 3, 4, 5, \ldots$. Note that this empirical formula requires two fewer hydrogens than the formula for the alkanes. In polycyclic compounds (more than one ring) the number of hydrogen atoms is two fewer *per ring* than the number of hydrogen atoms in an alkane with the same number of carbon atoms.

The alkanes are one of many classes of open-chain compounds, known collectively as **aliphatics** (Greek *aleiphar*—oily; derived from oil). Cycloalkanes are frequently known as **alicyclic** compounds—ring compounds that belong chemically to the aliphatic group. The chief function of this nomenclature is to distinguish them from **aromatic** compounds, which are chemically quite distinct.

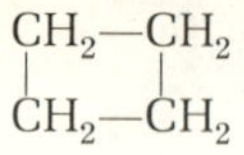

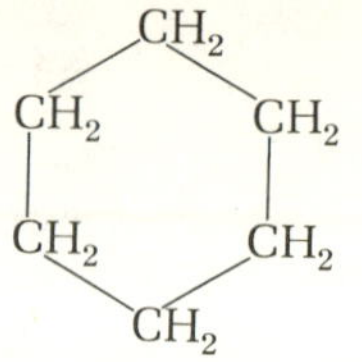

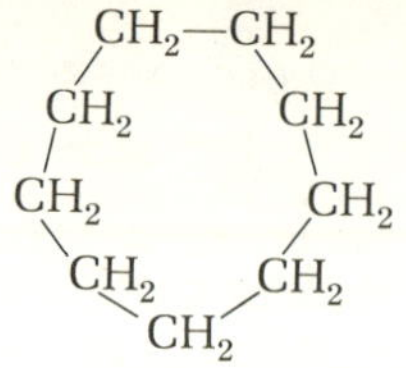

cyclobutane　　cyclohexane　　cyclononane

Naming the cycloalkanes is a refreshingly simple matter — merely tack the prefix **cyclo** onto the name of the alkane with the same number of carbon atoms. Depicting the cycloalkanes has also been made quite simple. To avoid the tedium of drawing all the CH_2 groups, the molecule is represented by a regular plane polygon, with the same number of vertices as there are carbon atoms in the ring. It is understood that each vertex represents a CH_2 group unless substituents are specifically shown. These representations are symbols, and it is important to remember that they are not meant to imply that the molecules are planar.

Substituted cycloalkanes are named according to a set of rules very similar to those used in the naming of alkanes. For a monosubstituted cycloalkane, it is unnecessary to specify the position of substitution since all carbon atoms in the ring are identical. When two or more substituents are present, the numbers are chosen for the lowest possible values as described for alkanes. Halogen substituents take precedence over alkyl substituents and are assigned the lower numbers.

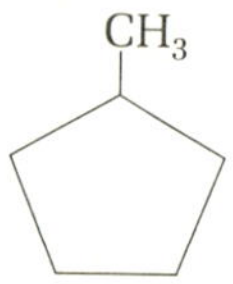

methylcyclopentane

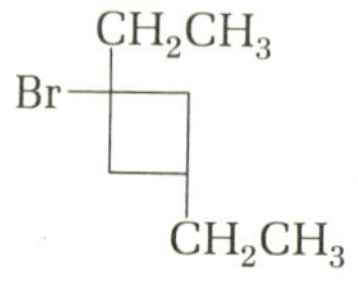

1-bromo-1,3-diethylcyclobutane

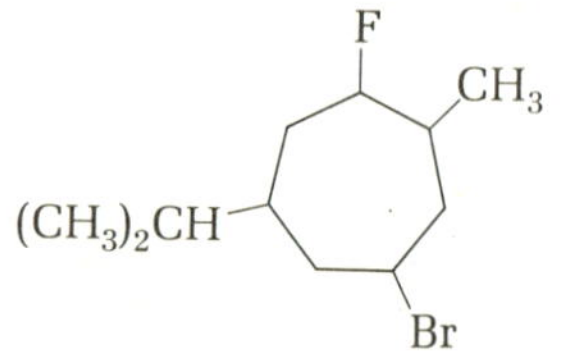

4-bromo-1-fluoro-6-isopropyl-2-methylcycloheptane

When the ring carries only a single halogen substituent, then the alkyl halide nomenclature may be used. It is quite common to refer to cyclopentyl chloride, cycloheptyl fluoride, cyclododecyl bromide, etc. These radical names may also be used when a ring is attached as a substituent to another structure, as in cyclohexylcycloheptane.

Cyclohexane often serves as a standard among cycloalkanes, and we shall discuss its structure and properties at length before turning to the other members of the class. Organic chemists have long been intrigued by cyclic molecules — at one time it was thought that certain members of this group could not be synthesized because of their inherent instability. Once these rings had in fact been synthesized and also found in naturally occurring substances, attempts were made to formulate a structural theory that would account for their properties.

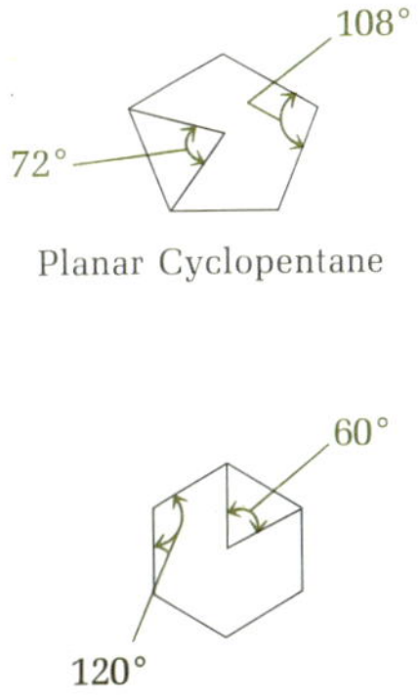

FIGURE 6-1
Bond Angles in Planar Cyclopentane and Cyclohexane

Adolph von Baeyer proposed a theory based on two assumptions:

1 The normal bond angles at carbon in saturated compounds were tetrahedral.

2 The monocycloalkanes were regular plane polygons.

If von Baeyer's hypothesis were true and cyclopentane were planar, the C—C—C bond angles would be 108°. Since the optimum value for those bond angles is 109.5°, there would not be a great deal of discrepancy between the actual situation and the ideal one. Planar cyclohexane, on the other hand, would have C—C—C bond angles of 120°, a significant deviation from the ideal 109.5°. As the ring size increased, so would the gap between the real and the ideal values of the bond angle. Von Baeyer proposed that bonds connected at abnormal angles would be "strained" and that this strain would result in a higher energy (lower stability) for the entire molecule.

We can recast von Baeyer's ideas using modern concepts, in an attempt to pinpoint the origin of the "strain energy." Since the carbon atoms of the ring are all sp^3-hybridized, the bonding orbitals originating from any single carbon atom would all be tetrahedrally disposed with respect to one another. If the positions of the carbon nuclei defined angles other than 109.5°, however, *then the axes of the orbitals would not lie on the internuclear line* (Figure 6-2). Overlap would then be less effective, the bonds weaker, and the entire molecule more energetic.

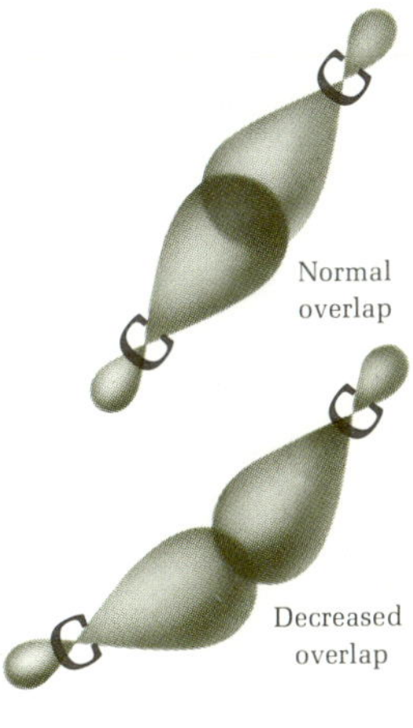

FIGURE 6-2
Decreased Overlap of Non-colinear Orbitals

The combustion of cycloalkanes, which is always an exothermic process, is described by equation 6.1.

$$\textbf{(6.1)}\quad (CH_2)_n + \frac{3n}{2}O_2 \longrightarrow nCO_2 + nH_2O$$

When the molar heat of combustion is divided by n, then the heat of combustion per —CH_2— group is obtained. If we could compare this number with the heat of combustion for a —CH_2— in a clearly unstrained molecule, then we could have a quantitative measure of the strain energy in the ring. As our model for an unstrained molecule, we take n-alkanes and determine their heat of combustion per $—CH_2—$ group. This approach is valid because of the following:

1 The strained molecule has more energy to start with and will therefore release more in the process of exothermic decomposition.

2 Both strained and unstrained compounds lead to the same products: CO_2 and H_2O.

PROBLEM

1 Linear alkanes consist of chains of $—CH_2—$ groups with one additional hydrogen atom at each end. How would you determine the heat of combustion *per methylene group* in linear alkanes?

When this is done, it is found that the heat of combustion per CH_2 group in cyclohexane (157.4 kcal/mole) and that for *n*-alkanes (157.4 kcal/mole) are identical. Von Baeyer's theory does not apply to cyclohexane, and to see why not we may turn to a postulate put forth by Sachse sometime after von Baeyer's proposal. Sachse noted that a six-membered ring could be formed in which all the C—C—C bond angles equaled 109.5°, provided only that the ring was puckered rather than flat. Two possible geometries were found for this ring, one of which was labeled the **chair** and the other the **boat** form. (It is strongly recommended that models of both forms be constructed and that the student familiarize himself thoroughly with their geometries.)

By using Sachse's theory, Mohr was able to correctly predict isomerism in certain polycyclic alkanes, and from then on the theory gained currency.

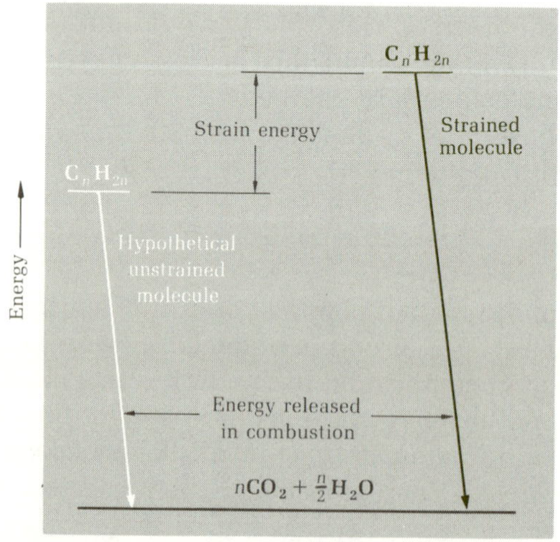

FIGURE 6-3
Energy Diagram for Combustion of CH_2 Groups in Strained and Unstrained Molecules

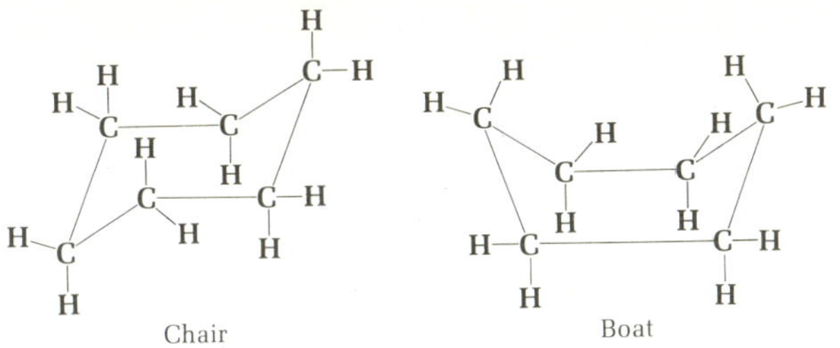

FIGURE 6-4
Chair and Boat Forms of Cyclohexane

The nonplanar nature of cyclohexane rings was confirmed by direct observation, particularly by means of **X-ray diffraction.** The atoms in a molecule tend to scatter radiation of about 1 A wavelength (X rays). If the molecules are contained in a crystal lattice, then the scattered radiation will form a characteristic pattern that can be captured on film. Mathematical analysis of hundreds of these patterns (usually with the aid of a computer) leads to a complete picture of the molecular structure including bond lengths and angles. X-Ray analysis of a number of compounds containing cyclohexane rings confirms their nonplanarity and further indicates that the chair form is preferred in the great majority of cases.

CONFORMATIONS OF CYCLOHEXANE AND THEIR INTERCONVERSION

Figure 6-5 is a Newman projection of cyclohexane in the chair conformation. Note that all the C—H and C—C bonds are staggered, very much like the *gauche* conformer of *n*-butane. Since all the CH_2 groups are equivalent, the same projection is obtained if the ring is rotated around its central axis. A Newman projection of the boat, seen "bow on," indicates that there is extensive eclipsing of hydrogen atoms and methylene groups. The resem-

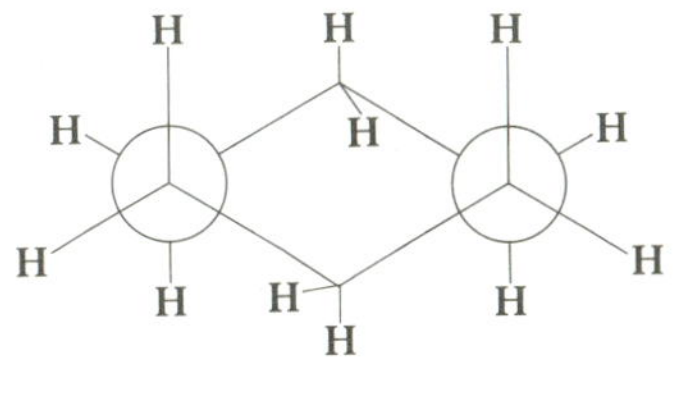

FIGURE 6-5
Newman Projection of Chair Cyclohexane

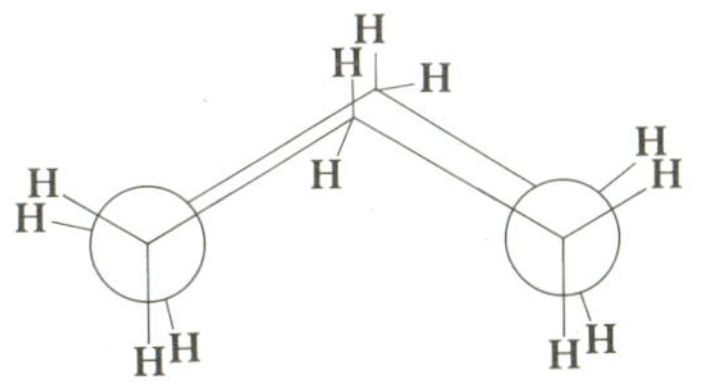

FIGURE 6-6
Newman Projection of Boat Cyclohexane

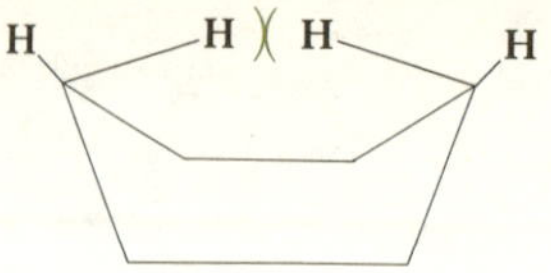

FIGURE 6-7
Side View of Boat Cyclohexane

blance here is to the fully eclipsed form of *n*-butane. A side view of the boat form reveals that there is a particularly strong repulsion between the inside hydrogens on the "bow" and "stern" atoms.

As is readily apparent from a model, the boat form is quite flexible in contrast to the rigid chair. Some of the repulsion energy in the boat may be lost if the "bow" twists to one side while the "stern" moves in the opposite direction (Figure 6-8). This **twist boat** conformation is of somewhat lower energy than the pure boat but still about 5.5 kcal/mole more energetic than the chair. Entropy favors the less rigid boat form, but this preference cannot overcome the energy difference, and the twist boat has approximately 4.4 kcal/mole more free energy than the chair conformation at 25°. As a result, the equilibrium between boat and chair is very much in favor of the latter.

(6.2) $$K^{25^\circ} = \frac{(\text{chair})}{(\text{boat})} \sim 1000$$

If a model of chair cyclohexane is held in a horizontal orientation and then viewed from the side and from above, two different types of hydrogen atom may be discerned:

1 Six hydrogen atoms form a belt around the periphery of the ring, connected to it by nearly horizontal C—H bonds; there is one of these **equatorial** hydrogen atoms attached to each carbon atom.

2 Six hydrogen atoms are connected to the ring by vertical C—H bonds; these **axial** hydrogen atoms lie alternately above and below the plane of the ring (see Figure 6-9).

The twist boat conformation is more energetic than the chair by 5.5 kcal/mole. Conversion of the chair to the boat requires an E_a of only 11.3 kcal/mole; consequently, the two conformations are in rapid equilibrium at room temperature (the *concentration* of boat conformer remains very low,

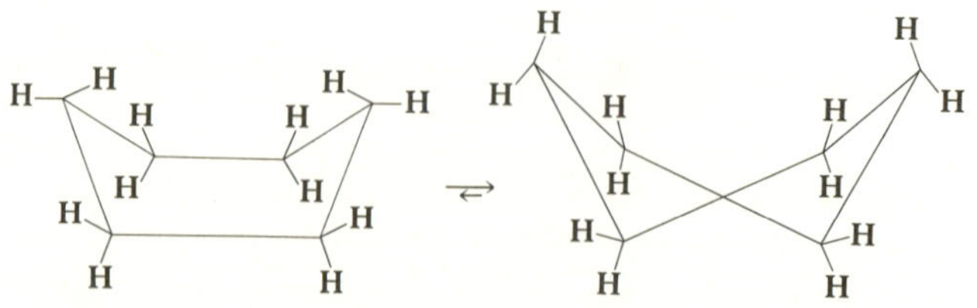

FIGURE 6-8
Conversion of Pure Boat to Twist Boat Conformation

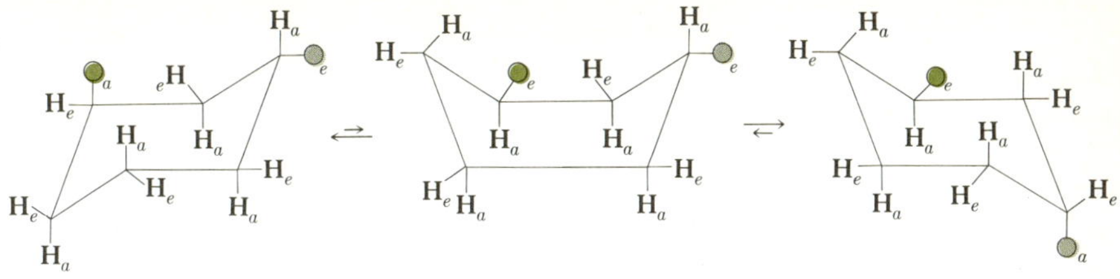

FIGURE 6-9
Interconversion of Boat and Chair Forms of Cyclohexane

of course). In the chair form one methylene group may be arbitrarily considered the "foot" and the one opposite it the "head." The "foot" then flips up to form either "bow" or "stern" (both are equivalent). When the boat reverts to the chair conformation, the methylene group that was the "foot" may now become the "head" — this process is known as **ring inversion.** In the course of ring inversion, **all axial hydrogens become equatorial and vice versa.** This is readily apparent if one follows the path of the shaded atoms in Figure 6-9.

The axial and equatorial protons of cyclohexane have different nmr chemical shifts. Ring inversion interconverts the axial and equatorial protons so rapidly at room temperature, however, that the spectrometer cannot distinguish between them. The nmr spectrum of cyclohexane at 25° consists of a single sharp line; its chemical shift value is midway between those for the axial and equatorial protons. This spectrum is called the *time-averaged* spectrum; the protons are oscillating back and forth between two different environments with such rapidity that the spectrometer only "sees" an average environment for the proton.

If cyclohexane-d_{11} (the purpose of the deuterium is to simplify the spectrum) is cooled to −100°, then ring inversion becomes quite slow.

(6.3)

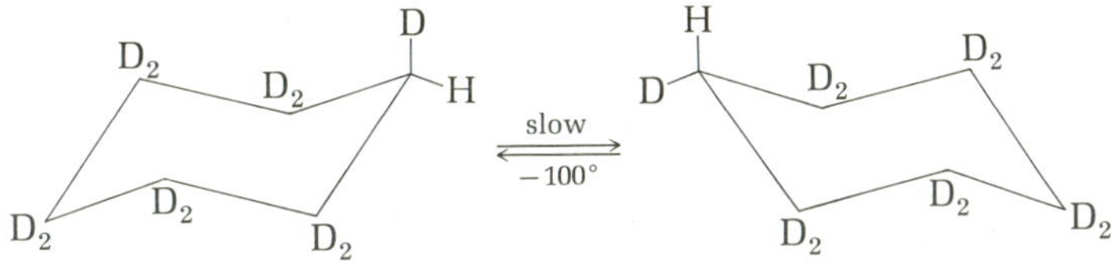

The spectrometer now "sees" half the molecules with their single proton in the equatorial position, while the other half have axial protons. A spectrum obtained at this temperature consists of *two* singlets, separated by about 0.5δ.

SUBSTITUTED CYCLOHEXANES

If a methyl group is attached to a cyclohexane ring, it may find itself in either an axial or an equatorial orientation. Due to the rapid ring inversion at 25°, however, the methyl group oscillates back and forth between these

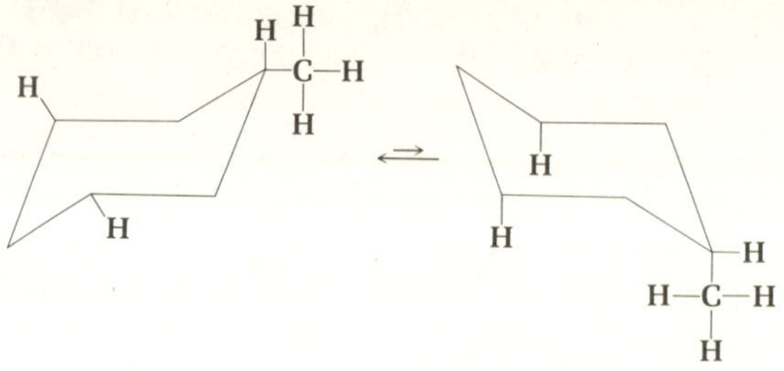

FIGURE 6-10
Equilibrium Between Equatorial and Axial Methylcyclohexane

two positions at a very high rate. (The two conformations of methylcyclohexane are not separable at room temperature, a fact that initially made chemists skeptical of Sachse's theory of nonplanar cyclohexane rings.) The two conformers are not of equal energy. Examination of a Newman projection of a cyclohexane ring with an axial methyl group (Figure 6-11) shows that the methyl is separated by a projection angle of only 60° from the CH_2 groups attached to the neighboring carbons. This is very reminiscent of the steric situation in the *gauche* butane conformer (Figure 5-8). The equatorial alkyl group, on the other hand, is *anti* to these methylenes. The *anti* conformer is more stable than the *gauche* in *n*-butane and the equatorial correspondingly more stable than the axial in methylcyclohexane. The free energy of the axial conformer is greater than that of the equatorial by 1.7 kcal/mole; this means that a sample of methylcyclohexane at room temperature consists of about 95% equatorial conformer.

The atomic origin of the repulsive interaction can perhaps be better appreciated with the aid of Figure 6-12. Here it is seen that the methyl group, as it rotates, will collide with the axial hydrogens on carbon atoms 3 and 5. These are known as 1,3-diaxial interactions. They will obviously become

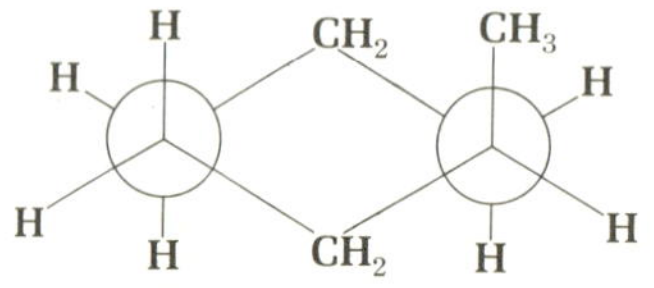

FIGURE 6-11
Newman Projection of Cyclohexane with Axial Methyl Group

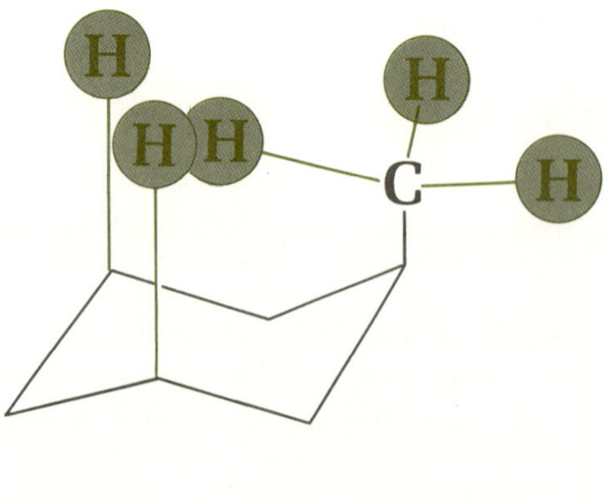

FIGURE 6-12
1,3-Diaxial Interactions in Methylcyclohexane

more severe as the substituent(s) on the ring become(s) larger. In *t*-butylcyclohexane the equatorial conformer is 10,000 times more abundant at room temperature than the axial! It is worth remembering that there are also *attractive* interactions (dispersion forces) between diaxially opposed substituents, but unless these groups are highly polarizable, the repulsive interaction outweighs the attractive one.

An analysis of the conformations of cyclohexane rings carrying more than one substituent proceeds by an extension of the reasoning used above. It will first be desirable to master one of the nomenclature conventions. If a cyclohexane ring is viewed from above, then at each carbon atom one of the hydrogen atoms is "up" while the other one is "down." The "up" hydrogens are alternately axial and equatorial; the same applies to the "down" hydrogens. Any two hydrogen atoms that are on the same side of the ring, either "up" or "down," are said to be *cis* to one another, regardless of whether they are axial or equatorial. Similarly, two hydrogens, one of which is "up" and the other "down," are in a *trans* relationship. These designations apply to any substituents as well. When the cyclohexane ring is represented as a planar hexagon, it is usual to designate "up" substituents by drawing their bonds to the ring with heavy lines; "down" substituents are connected to the ring with broken lines (Figure 6-13).

Consider the isomers of 1,2-dimethylcyclohexane, *cis*- and *trans*-1,2-dimethylcyclohexane. Each *isomer* consists of a mixture of *conformers*. The *cis* isomer must have one methyl group axial and the other equatorial (*ae*; Figure 6-14). Furthermore, ring inversion leads to an indistinguishable equatorial-axial conformer (*ea*). *trans*-1,2-Dimethylcyclohexane, on the other

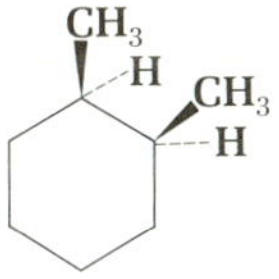

cis-1,2-dimethylcyclohexane

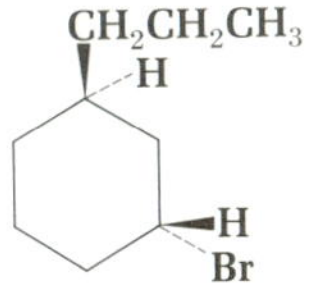

trans-3-bromopropylcyclohexane

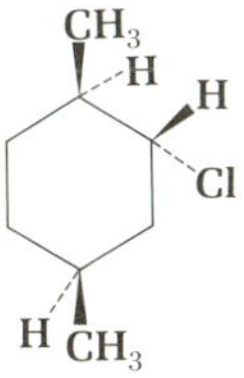

trans, trans-2-chloro-1,4-dimethylcyclohexane

FIGURE 6-13
Several *cis* and *trans* Isomers of Substituted Cyclohexanes

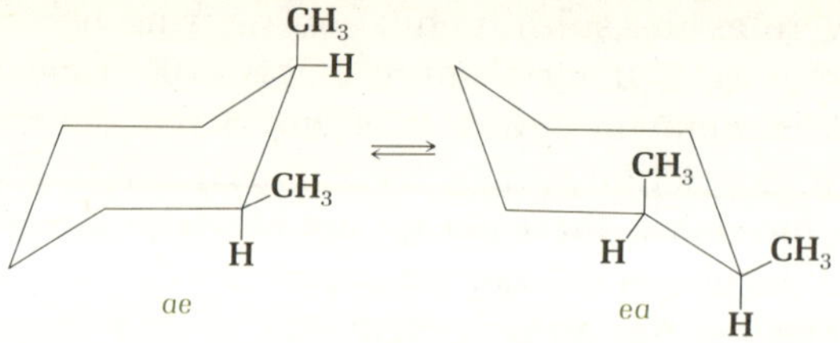

FIGURE 6-14
Conformational Equilibrium in *cis*-1,2-Dimethylcyclohexane

hand, may exist as either a diequatorial (*ee*) or diaxial (*aa*) conformer (Figure 6-15). The diequatorial is by far the less energetic of the two, and this is the conformer that predominates in *trans*-1,2-dimethylcyclohexane. If we estimated the relative stabilities of the *cis* and *trans* isomers, we would have to conclude that the *trans* isomer is more stable because an *ee* arrangement of substituents is preferable to an *ae* arrangement.

In 1,3-dimethylcyclohexane the conformational situation is the reverse of that found with the 1,2-isomer. Now it is the *cis* compound that can exist as an *ee* conformer, and it is therefore of lower energy than its *ae trans* isomer. In general, the *cis* isomer of 1,2- and 1,4-disubstituted cyclohexanes must be *ae*, which is also true of the *trans* isomer of a 1,3-disubstituted cyclohexane. The conformer that has the largest group in the equatorial position will predominate.

It would seem that *trans*-1,3-di-*t*-butylcyclohexane would have to have one axial *tert*-butyl group — despite the extreme reluctance of this group to occupy an axial position. In this case, it turns out to be most energetically favorable for the molecule to assume a twist boat conformation so that both alkyl groups can be equatorial. This molecule is one of the relatively few known to exist mainly as the boat conformer.

(6.4)

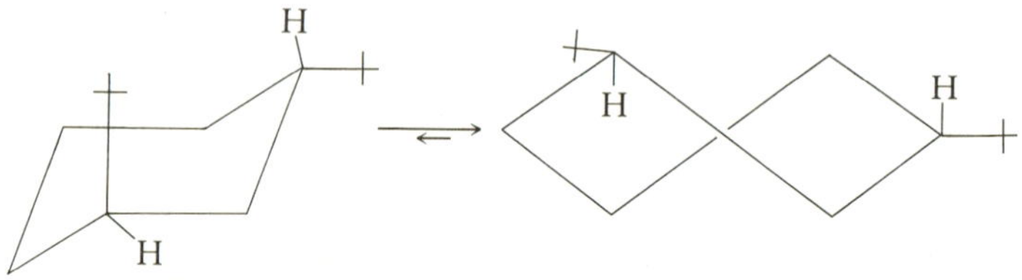

$\Delta H = 0.4$ kcal/mole

—+ = *t*-butyl

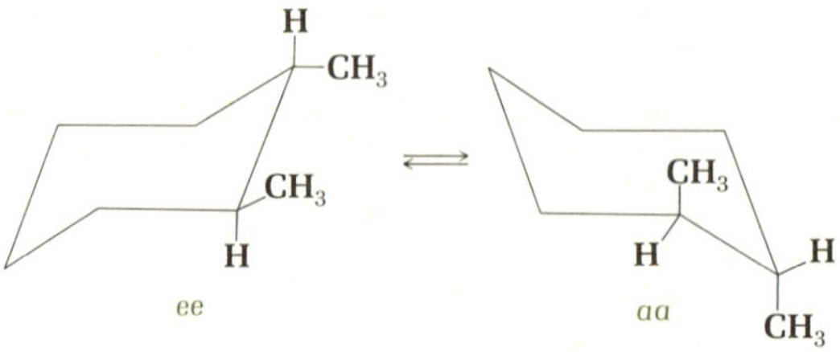

FIGURE 6-15
Conformational Equilibrium in *trans*-1,2-Dimethylcyclohexane

Of all the cycloalkanes, cyclopentane could most easily be planar with a minimum of angle strain since a regular pentagon has angles of 108°. In any planar cycloalkane however, all the hydrogens would be eclipsed by those on neighboring carbon atoms (Figure 6-16). We noted that in ethane it required about 3 kcal/mole to eclipse 3 pairs of hydrogen atoms; planar cyclopentane would require 10 pairs of eclipsed hydrogen atoms. Cyclopentane is quite flexible, and one of its energetically most favorable conformations seems to be the "envelope," in which one of the carbon atoms projects slightly above the plane defined by the other four. This enables the bond angles to remain close to 109.5° and, at the same time, relieves some of the eclipsing interactions. The molecule does not remain fixed in any one conformation, however; the various carbon atoms are constantly alternating between being "in" and "out" of the plane.

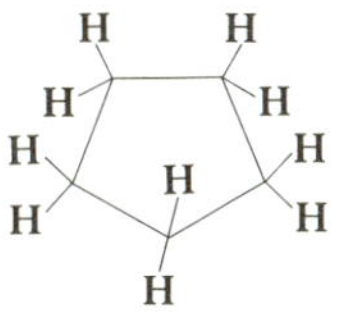

FIGURE 6-16

Eclipsing Interactions in Planar Cyclopentane

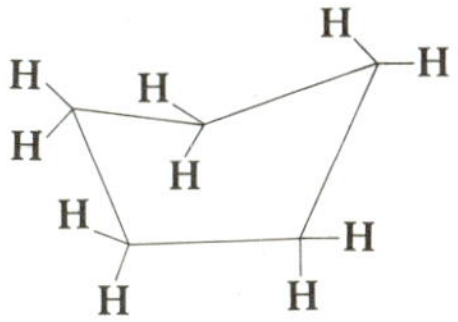

FIGURE 6-17

Envelope Conformation of Cyclopentane

Cyclobutane exists mostly in a "folded" conformation (Figure 6-18) that probably undergoes rapid flipping at room temperature. The bond angle strain in this conformation is substantial, but it is apparently more than compensated for by the relief of eclipsing interactions between neighboring hydrogen atoms.

Cyclopropane has of necessity a planar carbon skeleton and all the hydrogen atoms are eclipsed. The high degree of angle strain in this compound and in cyclobutane might be expected to result in unusual chemical and physical properties. Chemical reactivity will be discussed later in this chapter but we may consider some of the physical data here.

The protons in cyclopropane give an unsplit nmr peak at $\delta 0.22$, while those in cyclobutane resonate at $\delta 1.96$. Both values differ significantly from the absorption position of the CH_2 groups in other cycloalkanes, which seems to lie in the range of $\delta 1.40$ to $\delta 1.55$. These characteristic nmr absorptions are very useful for identifying cyclopropane and cyclobutane rings. The C—H stretching frequencies are also abnormally high for cyclopropane and somewhat low for cyclobutane.

(6.5)  $\longrightarrow \cdot CH_2CH_2CH_2 \cdot$

Finally, it has been calculated that $\Delta H = 55$ kcal/mole for the reaction shown in equation 6.5. This should be compared with the 82 kcal/mole needed to bring about the reaction shown in equation 6.6.

(6.6) $CH_3CH_2CH_2CH_3 \longrightarrow 2CH_3CH_2\cdot$

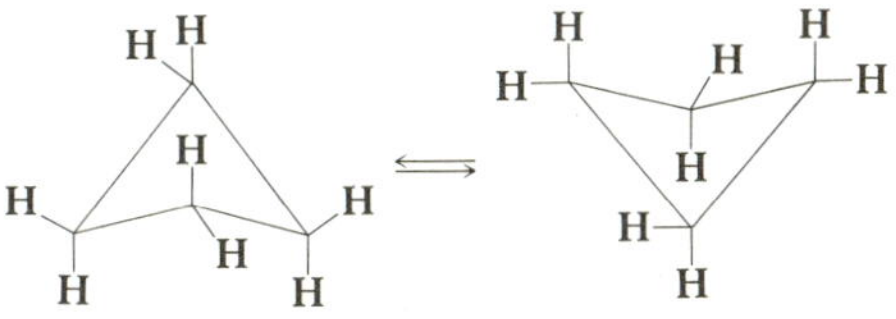

FIGURE 6-18

Ring Distortion Equilibrium in Cyclobutane

BONDING IN CYCLOPROPANE

The physical and thermochemical data just cited are strong indications that the bonding in cyclopropane is "abnormal" by comparison with other cycloalkanes. What then is the nature of the bonding in this molecule? While it is not possible to answer the question with finality, several hypotheses have been advanced that are in good accord with the observations.

One of these suggests that the carbon atoms in cyclopropane are in an unusual hybridization state and that the four atomic orbitals centered on each carbon atom are *not* equivalent to one another, as they are with typical sp^3-hybridized carbon atoms. In particular, it is proposed that the orbitals used in forming the C—C bonds have a large amount of *p*-character, while the orbitals used for C—H bonding are rich in *s*-character. The prediction would be that the C—C bonds of cyclopropane are weaker than those found in most alkanes; the prediction and the experimental data agree. The C—H bonds should, conversely, be unusually strong, and we shall see that this prediction is also borne out.

RINGS LARGER THAN CYCLOHEXANE

All monocyclic alkanes larger than cyclohexane are considerably more flexible than cyclohexane, and they all assume puckered conformations involving relatively little angular distortion or bond eclipsing. The medium rings (C_8 to C_{11}) are prone to adopt conformations in which hydrogen atoms across the ring from one another are brought too close together; repulsions between the **nonbonded** hydrogen atoms can add substantially to the energy of the molecule. Some angular distortion usually occurs in these rings in order to minimize the nonbonded repulsions, and this is reflected in the fact that the heat of combustion for medium rings is about 1 kcal/mole per CH_2 group more negative than that for cyclohexane.

Cyclooctane does not have one preferred conformation of lowest energy. Two possible conformations are indicated in Figure 6-19, in which the hydrogen atoms responsible for nonbonded interactions have been shaded.

In the large rings ($> C_{11}$) the heat of combustion per CH_2 group is very close to the cyclohexane value. These rings are sufficiently large that conformations free of nonbonded repulsions may be attained.

ORIGIN AND REACTIONS OF CYCLOALKANES

Some petroleum contains a very large percentage of cycloalkanes — this is true of the crude oil obtained from the mid-continental United States. Of

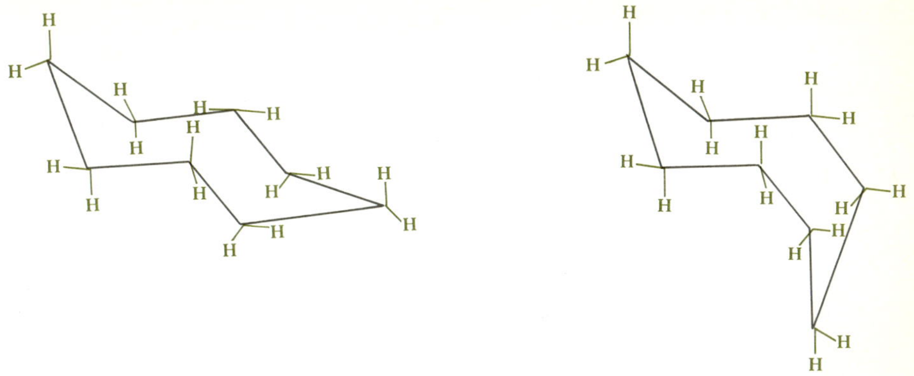

FIGURE 6-19
Two Conformations of Cyclooctane

the monocycloalkanes the principal constituents are cyclopentane, cyclohexane, and a variety of their alkylated derivatives; a small quantity of cycloheptane is also found. Industrially, the principal use for cycloalkanes lies in their dehydrogenation to aromatic compounds. In the course of this reaction, alkyl cyclopentanes are isomerized to cyclohexanes before dehydrogenation.

(6.7) 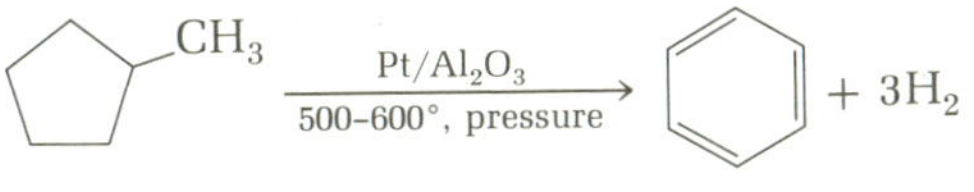

benzene

(6.8) 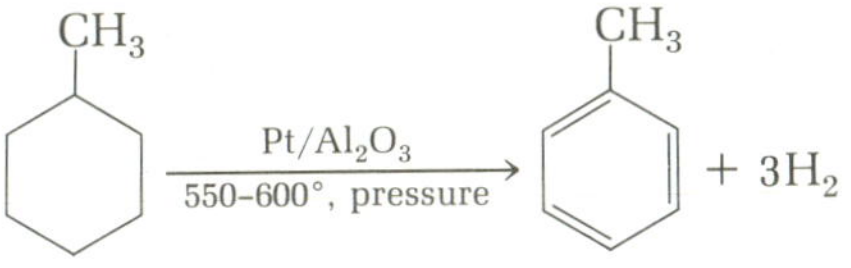

toluene

Under the usual laboratory conditions the cycloalkanes, with the exception of the small rings (C_3 and C_4), display the same lack of reactivity as their open-chain analogues. They may be halogenated by means of a free radical reaction; only one monohalogenated product is possible from each cycloalkane. The reactivity of cycloalkyl hydrogen atoms toward free radical attack is dependent on the size of the ring. If the cycloalkyl radical produced tends to have a planar (sp^2) configuration at the trivalent carbon, then the bonds around this carbon will seek to orient themselves 120° apart. In small rings the result will be a further increase in strain. This factor, plus the abnormally high C—H bond strength, explains the relatively large E_a (4.1 kcal/mole) for hydrogen abstraction from cyclopropane by chlorine atoms. (For ordinary 2° hydrogens this figure is about 0.6 kcal/mole.) Cyclobutane is considerably more reactive than cyclopropane, and cyclopentane is more reactive yet. Interestingly, cyclohexane is somewhat less reactive than cyclopentane.

Equations 6.9 to 6.11 indicate the conditions necessary for reductively cleaving cyclopropane, cyclobutane, and cyclopentane. The difference in energy released in these reactions is another measure of the qualitative difference between cyclopropane and cyclobutane, on the one hand, and the remaining cycloalkanes on the other.

(6.9)  $+ H_2 \xrightarrow[50\text{–}100^\circ]{Pt} CH_3CH_2CH_3 \qquad \Delta H = -37.6$ kcal/mole

(6.10)  $+ H_2 \xrightarrow[150\text{–}200^\circ]{Pt} CH_3CH_2CH_2CH_3 \qquad \Delta H = -36.5$ kcal/mole

(6.11) 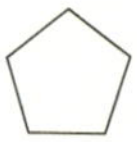 $+ H_2 \xrightarrow[250\text{–}300^\circ]{Pt} CH_3(CH_2)_3CH_3 \qquad \Delta H = -16.5$ kcal/mole

Hydrogenolysis (cleavage by hydrogen) is feasible in the laboratory only for the small rings.

Two other reactions are known that result in destruction of the cyclopropane ring — hydrolysis (equation 6.12) and brominolysis (equation 6.13). The other cycloalkanes are inert under these conditions; *in general, carbon-carbon bonds in alkanes and cycloalkanes are not broken in chemical reactions at ordinary laboratory temperatures.*

Ferric bromide functions as a Lewis acid catalyst in the brominolysis reaction (equation 6.13) and serves to polarize the bromine molecule and to increase its reactivity.

(6.12)  $+ H_2O \xrightarrow{H_2SO_4} CH_3CH_2CH_2OH$

(6.13)  $+ Br_2 \xrightarrow{FeBr_3} C_3H_6Br_2$

The reaction itself is anything but straightforward, as evidenced by the fact that $BrCH_2CH_2CH_2Br$, $CH_3\underset{\displaystyle Br}{\underset{|}{C}}HCH_2Br$, and $CH_3CH_2CHBr_2$ are all isolated as reaction products. A reaction scheme has been postulated in which **1** and **2** are intermediates. The dotted lines denote partial bonds (*less* than one electron pair) between the atoms concerned.

(6.14) 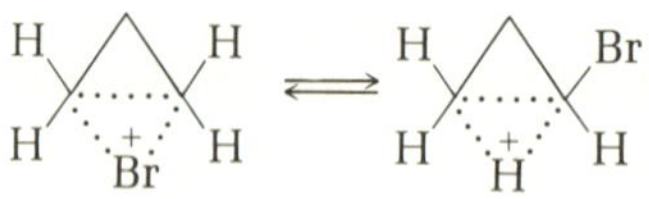

1 **2**

These are **delocalized bonds,** in which an electron pair may be strongly attracted by more than two nuclei — these types of orbitals will be encountered many times in future chapters.

The formation of delocalized bonds in the course of certain reactions of cyclopropane is rendered more plausible if we believe that the electrons in the carbon-carbon bonds are held relatively loosely. This supposition is, of course, in line with the expected properties of cyclopropane derived from other data; it certainly fits in with the bent bond picture. While intermediates such as **1** and **2** do violence to the traditional picture of electron pair bonding in organic molecules, the traditional picture cannot satisfactorily account for the product distribution in the brominolysis reaction.

The reaction of cyclopropane with bromine and with hydrogen parallels the reactions of compounds containing a double bond with these reagents (Chapters 2 and 9). Double bonds may, in fact, be viewed as "two-membered rings," which would be expected to have very high strain energies. We shall see that an alternative to the σ-π formulation of double bonds employs the bent bond approach.

POLYCYCLIC ALKANES

If two rings share two or more carbon atoms, they are said to be "fused." A large number of naturally occurring molecules, in particular many that are important physiologically, fall into this class. The bicyclo(4.4.0)decanes, known by their common name of **decalins,** are representative members.

The most highly substituted carbon atoms common to both rings are referred to as the **bridgehead** carbons. For the purposes of naming these compounds systematically, the bridgeheads are assumed to be tied together

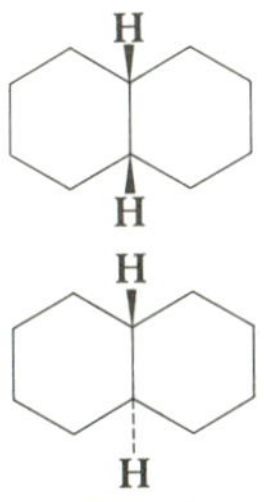

FIGURE 6-20
Planar Representation of *cis*- and *trans*- Decalin

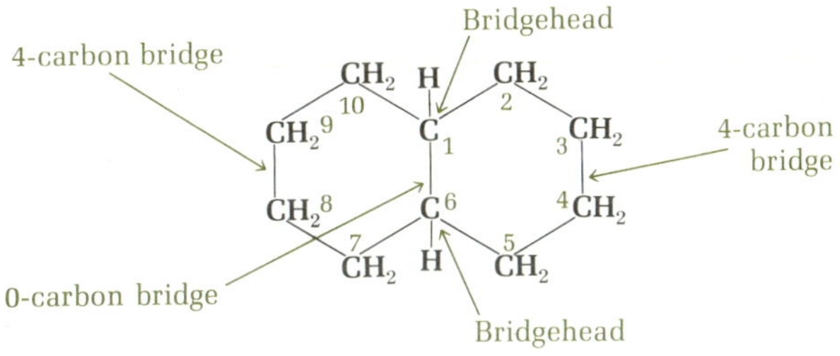

bicyclo(4.4.0)decane

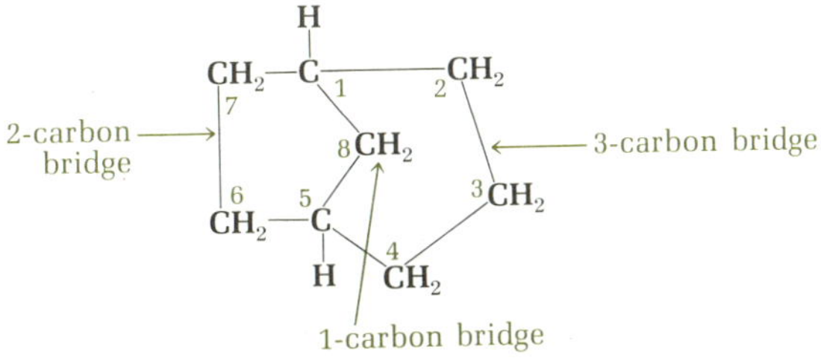

bicyclo(3.2.1)octane

FIGURE 6-21
Numbering Some Bicyclic Alkanes

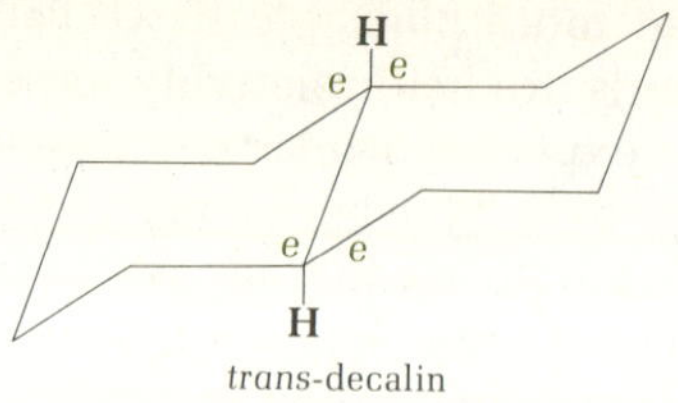

FIGURE 6-22
Chair Form of *trans*-Decalin

by bridges, which may consist of methylene groups or just bonds. The figures in brackets specify the number of carbon atoms in each separate bridge. When naming derivatives, numbering starts at one bridgehead and proceeds along the longest bridge to the other bridgehead; it then continues along the second longest bridge and so forth.

The name decalin is used in the chemical literature in preference to bicyclo(4.4.0)decane, almost without exception. A certain number of polycyclic alkanes have generally accepted nonsystematic names, although (for the majority) use of the IUPAC system just described is mandatory. With respect to the decalins, the prefixes *cis* and *trans* refer to the relative positions of the hydrogen atoms attached to the bridgehead carbon atoms. Each isomer adopts a conformation in which both of its rings are in a chair form.

cis-Decalin is conformationally flexible and the entire molecule may undergo inversion, in the course of which all axial protons become equatorial and vice versa (equation 6.15).

(6.15)

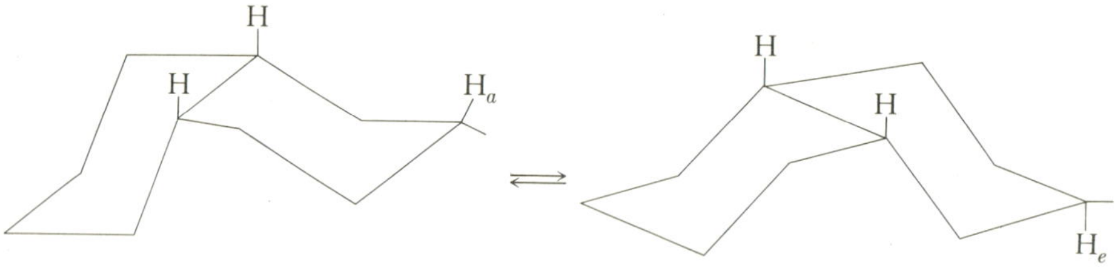

trans-Decalin, on the other hand, is conformationally rigid. It is therefore extensively used for studies of the effect of conformation on reactivity since an axial substituent, for instance, must remain in that position and cannot be transposed to an equatorial position by way of a ring inversion.

The decalin ring system is incorporated in virtually all members of the

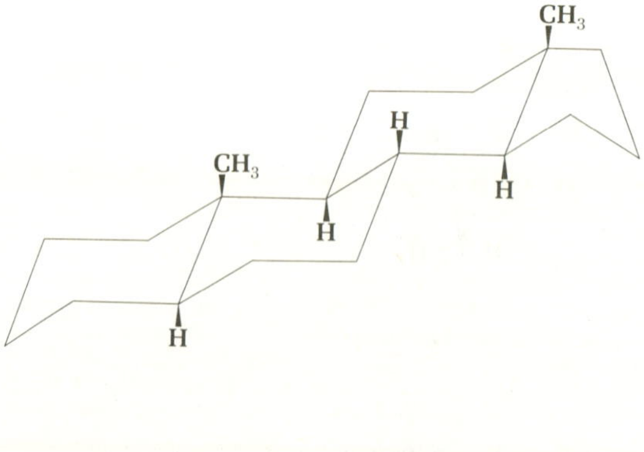

FIGURE 6-23
All *trans* Steroid Skeleton

class of biologically potent compounds known as steroids (Figure 6-23). Their skeleton consists of three six-membered and one five-membered rings all fused together, usually in a *trans* fashion. A large number of our bodily hormones that regulate growth rate, sexual activity, and innumerable other functions are steroids. These substances are particularly remarkable for the relatively small amounts that will cause a biological response (*billionths* of the weight of the organism receiving the steroid).

SUMMARY

1 If it is assumed that cycloalkanes are planar, then it would be anticipated that rings both larger and smaller than cyclopentane would exhibit angle strain. This expectation is not supported, for the larger rings at least, by the results of heat of combustion studies. These rings are all puckered; the puckering not only avoids angle strain, but it also tends to relieve eclipsing interactions between hydrogen atoms on adjacent carbons.

2 For cyclohexane there are three important puckered conformations: the chair, the true boat, and the twist boat. The chair has by far the lowest energy of the three, and cyclohexane exists almost exclusively in this conformation at room temperature despite the higher entropy of the boat forms.

There are two types of hydrogen atom in chair cyclohexane: axial and equatorial. At room temperature a given cyclohexane molecule will be equilibrating rapidly between the two possible chair conformations. In the course of this equilibration a particular hydrogen atom shifts very quickly from an axial to an equatorial conformation and back again. This process is sufficiently rapid that at room temperature the nmr spectrometer is not able to distinguish between the two types of hydrogen atom. As the temperature of the sample is lowered, the rate of interchange decreases, and it is eventually possible to obtain peaks corresponding to the axial and equatorial protons.

3 A substituent on a cyclohexane ring is also rapidly alternating between axial and equatorial orientations. In general, the conformation in which the substituent is equatorial is the preferred one. The presence of two or more substituents on a single ring results in *cis-trans* isomerism. Each of the isomers may exist in two chair conformations; the relative energy of the two isomers will usually equal the energy difference between the most stable conformers of each. The most stable conformer will be the one in which the greatest number of large substituents are oriented equatorially.

4 Cyclopentane and cyclobutane assume nonplanar conformations in order to minimize eclipsing interactions. Cyclopropane suffers from both eclipsing

interactions and angle strain. Cyclopropane and cyclobutane both possess considerable quantities of strain energy, as indicated by their relative ease of thermolysis.

In addition, cyclopropane may be cleaved by bromine in the presence of ferric bromide. The complex product mixture from this reaction suggests the intervention of some unusual intermediates containing delocalized σ-electrons. Such intermediates are plausible if one recalls that the electrons in the carbon-carbon bonds are relatively loosely held.

5 Several rings may be fused together to give polycyclic alkanes. The ring junctures may be *cis* or *trans* in these fused ring systems. One of the most important polycyclic alkanes is decalin; the *cis* isomer undergoes inversion just as cyclohexane does, but the *trans* isomer is conformationally rigid.

PROBLEMS

Note: Problem 1 appears in text.

2 The heats of combustion of three gaseous cycloalkanes are 11.9, 11.7, and 11.3 kcal/g. One of these compounds is cyclobutane, one is cyclopentane, and the third is cyclopropane. Which is which? Calculate the heat of combustion **per mole** for each of these compounds and then the heat of combustion per mole per methylene group. If the "normal" heat of combustion per methylene group is 157.4 kcal/mole, calculate the total strain energy in cyclopropane, cyclobutane, and cyclopentane.

3 Draw and label all monochlorination products of (a) ethylcyclopentane, (b) methylcyclohexane, and (c) bicyclo(3.2.0)heptane.

4 When methylcyclopentane is monobrominated, two products are obtained that result from bromination at carbon atom 2. They are *cis*- and *trans*-2-bromomethylcyclopentane. The *trans* isomer is found in greater abundance than the *cis*. Suggest an explanation for this result.

5 2-Isopropylmethylcyclohexane can exist in *cis* and *trans* isomeric forms.

a. Draw the conformers corresponding to each isomer.

b. Which is the more stable conformer of each pair?

c. Which is the more stable isomer?

d. Carry out the same analysis on the 1,3-compound.

6 The free energy difference for an axial methyl group versus an equatorial one is 1.7 kcal/mole. The corresponding free energy differences for ethyl and isopropyl are not very different — 1.9 and 2.0 kcal/mole, respectively. Axial *t*-butyl, however, is about 5.6 kcal/mole higher in free energy than its equatorial conformer. Explain these facts.

7 a. Calculate the minimum value of K for reaction (1).

$$(aa)\text{-1,2-Dimethylcyclohexane} \rightleftharpoons (ee)\text{-1,2-dimethylcyclohexane} \quad (1)$$

b. Estimate the free energy difference between *cis*- and *trans*-1,2-dimethylcyclohexane, assuming that the *trans* isomer consists solely of *ee*.

c. In an equilibrium between the *cis* and *trans* isomers mentioned, which one would be favored by entropy alone?

8 The proton attached to the halogen-bearing carbon atom of chlorocyclohexane gives rise to an easily distinguishable peak in the nmr spectrum of the compound. At 25° there is only one sharp peak in this region. If a solution of the compound is cooled to −115°, however, this single peak splits into two peaks, separated by about 0.7δ. Account for these observations.

9 Construct models of *cis*- and *trans*-decalin having both rings in the chair form. Which of the two isomers is the more stable and why?

10 Using the bond energies supplied in Chapter 5, estimate ΔH for equation (2).

$CH_3(CH_2)_3CH_3 \longrightarrow$ 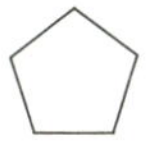 $+ H_2$ (2)

Compare your value with the heat of hydrogenation given in equation 6.11. Account for the discrepancy.

11 When cyclopropane is dissolved in a mixture of H_2O and H_2SO_4, it is converted to *n*-propyl alcohol, $CH_3CH_2CH_2OH$ (equation 6.12). If cyclopropane is reacted with D_2O-D_2SO_4 (chemically equivalent to H_2O-H_2SO_4), the product, after separation and purification, is found to be monodeuterated *n*-propyl alcohol, C_3H_6DOH.

a. Which species in the solution probably reacts first with the cyclopropane?

b. Suggest a structure for the intermediates in the reaction taking place in the deuterated system (see the bromination reaction, equation 6.13, and structures **1** and **2**).

c. Reasoning by analogy, how would you expect to find the deuterium atom distributed in the product?

d. How would you experimentally determine the position(s) of the deuterium atom in the product?

12 Bicyclo(2.2.1)heptane contains a six-membered ring in a true boat conformation, its bow and stern held in place by a methylene bridge that connects them (construct a model). If you knew the heat of combustion of this compound, would it be possible to calculate the energy of true boat cyclohexane with reference to chair cyclohexane? What additional information would you need, and what assumptions might you make in the absence of that information? There is a destabilizing interaction in boat cyclohexane that is absent in the bicyclic compound — what is it?

CHAPTER 7

ALCOHOLS AND ALKYL HALIDES—NUCLEOPHILIC DISPLACEMENT

One of the fundamental operations in organic synthesis is the conversion of one functional group into another without affecting the remainder of the molecule. Nucleophilic displacement is a widely used method of bringing about the interchange of functional groups and will be discussed at length in this chapter. The value of alcohols and alkyl halides as synthetic intermediates lies in the fact that they readily undergo a variety of nucleophilic displacement reactions.

MANUFACTURE AND USES OF ALCOHOLS AND ALKYL HALIDES

As noted in Chapter 5, alkyl halides, in particular the chlorides, are available from halogenation of alkanes. Aside from the synthetic utility of these halides, the chlorides often serve as solvents. Most are liquids at room temperature, immiscible with water, and slightly more polar than their parent alkanes. They are also less volatile than the corresponding alkanes; for a given alkyl group, the boiling point increases going from the fluoride to the iodide, reflecting increased van der Waals attractions.

Ethanol (grain alcohol) was in commercial production thousands of years before anything was known of the chemistry of its production or of its consumption. Until quite recently, the principal source of ethanol was a fermentation process usually employing corn or molasses. By using a different bacterium to carry out the fermentation, a mixture of acetone and 1-butanol can be obtained as the major product. At present, most alcohols of commercial use are derived from petroleum by the addition of various reagents to the double bonds of alkenes (Chapter 9). The first member of the series, methanol, is manufactured by allowing hydrogen and carbon monoxide to react over a catalyst at high temperatures and pressures.

$$2H_2 + CO \xrightarrow[\substack{250\text{–}300\text{ atm} \\ 300\text{–}400^\circ}]{\text{ZnO catalyst}} CH_3OH \qquad (7.1)$$

SOLVENT PROPERTIES AND HYDROGEN BONDING IN ALCOHOLS

Alcohols are extensively employed as solvents, and in this capacity they show marked differences from the alkyl halides. They are considerably more polar and can solvate anions and cations much more effectively. For this reason alcohols are able to dissolve many inorganic salts as well as many organic compounds, which less polar solvents are unable to do.

TABLE 7-1
Boiling Points of Some Alkyl Halides

Name	Alkyl Group	Boiling Point (°C) F	Cl	Br	I
Methyl	$CH_3—$	−78	−24	5	43
Ethyl	$CH_3CH_2—$	−38	12	38	72
n-Propyl	$CH_3CH_2CH_2—$	−3	47	71	102
i-Propyl	$(CH_3)_2CH—$		37	60	90
n-Butyl	$CH_3CH_2CH_2CH_2—$	32	78	101	130
s-Butyl	$CH_3CH_2\underset{\vert}{C}HCH_3$		68	91	120
i-Butyl	$(CH_3)_2CHCH_2—$		69		120
t-Butyl	$(CH_3)_3C—$		51	73	100
n-Amyl	$CH_3(CH_2)_3CH_2—$	63	108	130	155
Neopentyl	$(CH_3)_3CCH_2—$		84	91	129
n-Octyl	$CH_3(CH_2)_6CH_2—$	143	182	202	256

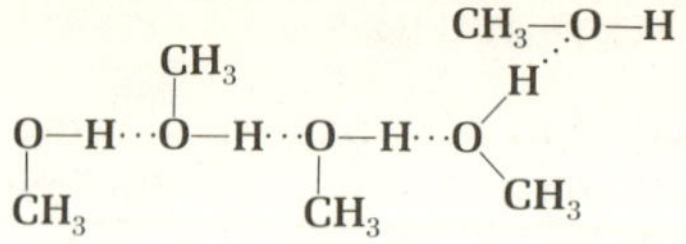

FIGURE 7-1
Hydrogen-Bonded Methanol

Like water and hydrogen fluoride, alcohol molecules are bound to one another in the liquid state by hydrogen bonds. These interactions have a profound effect on the boiling point of the alcohol, as seen by comparing the boiling points of fluoroethane and ethanol, whose molecular weights are nearly identical.

CH_3CH_2F	CH_3CH_2OH
fluoroethane (bp −32°)	ethanol (bp 80°)

Equally dramatic is the comparison between ethanol and its isomer, dimethyl ether (CH_3OCH_3), which has a boiling point of −25°; the ether molecules are incapable of hydrogen bonding with one another. It is worth reiterating with the aid of an example that elements of the second row are by far the most effective in hydrogen bonding. Sulfur is immediately beneath oxygen in the periodic table. The sulfur analogues of dimethyl ether and ethanol differ very little in their boiling points, showing how unimportant hydrogen bonding is for ethanethiol. For these reasons, alcohols are often solvents of choice when a reaction must be conducted at relatively high temperatures.

CH_3SCH_3	CH_3CH_2SH
dimethyl sulfide (bp 38°)	ethanethiol (bp 37°)

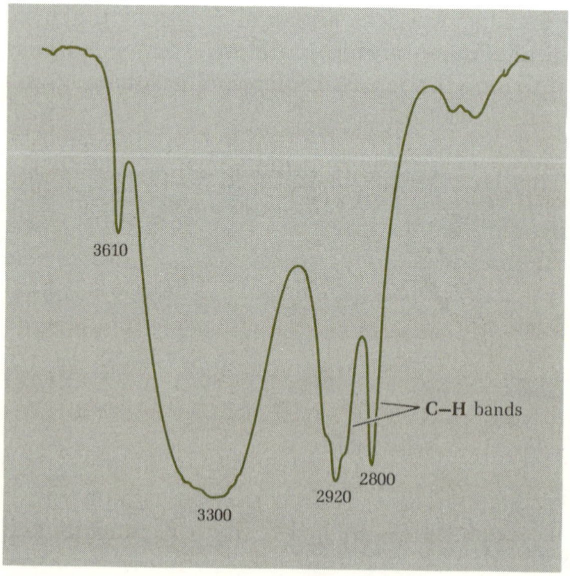

FIGURE 7-2
Partial Infrared Spectrum of Methanol

The tendency of alcohol molecules to associate by hydrogen bonding is evident from their infrared spectra. A partial spectrum of methanol (Figure 7-2) shows two bands in the —OH region of the spectrum. The sharp band centered at 3610 cm^{-1} arises from the unassociated alcohol molecules, while the broad band centered at 3300 cm^{-1} is due to hydrogen-bonded dimers and higher clusters. The relative intensity of these two bands is a function of concentration; the one due to "free" alcohol molecules gets relatively more intense as the solution is diluted and vice versa.

Association among alcohol molecules also affects their nmr spectra. Figure 7-3(a) shows the nmr spectrum of a highly purified sample of ethanol. The

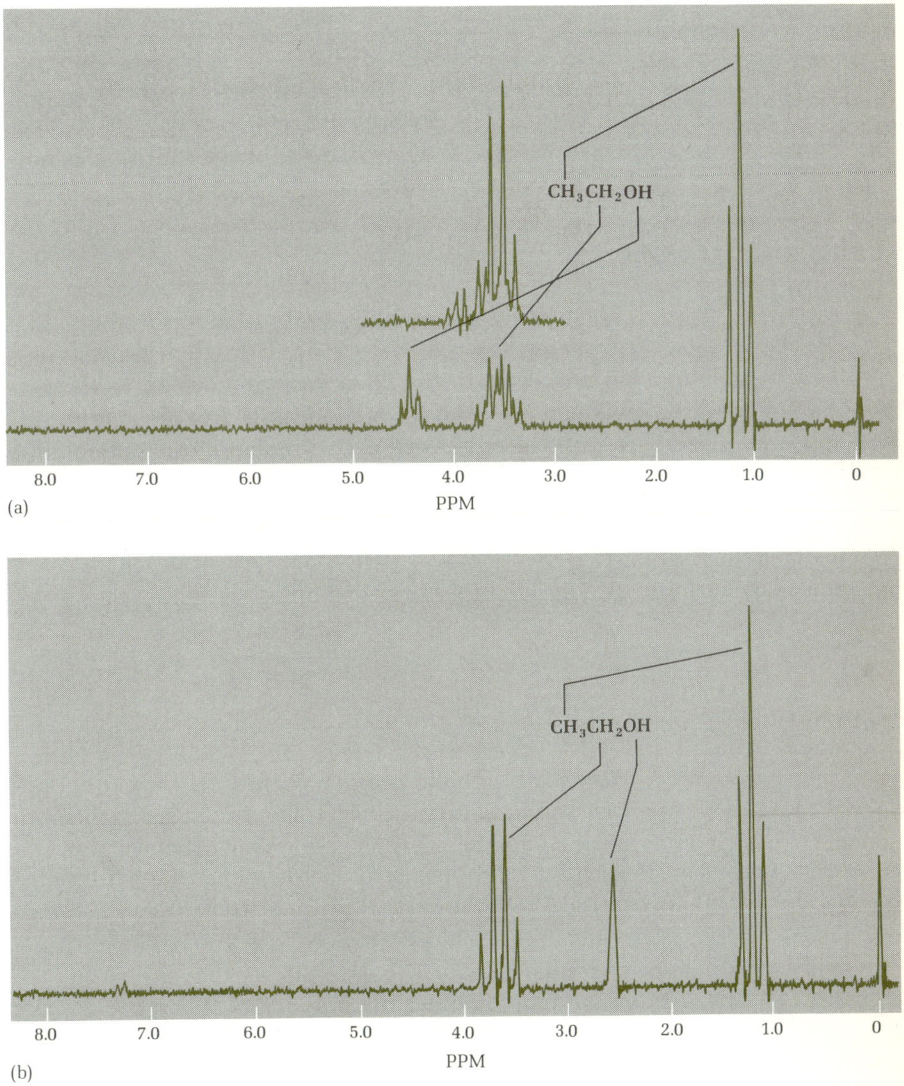

FIGURE 7-3
NMR Spectra of Two Samples of Ethanol

splitting pattern is just what would be expected. In particular, the —OH proton signal (δ4.48) is split into a triplet by the methylene protons, while the peak due to the methylene protons ($\sim\delta$3.6) appears as a multiplet because these protons are coupled to both the methyl group and the hydroxyl group. Figure 7-3(b) displays the spectrum of a slightly less pure sample. The hydroxyl proton resonance is now a singlet (δ2.6), and the methylene protons appear to be coupled only to the methyl protons.

The decoupling of the hydroxyl and methylene protons evident in Figure 7-3(b) is due to **chemical exchange** — the hydroxyl groups on different molecules swap protons among themselves. Acids and bases catalyze this process, which is very rapid compared to the nmr scan time, so that the hydroxyl protons give rise to only a single peak which reflects their average environment. Note that the hydroxyl resonances occur at different frequencies in Figures 7-3(a) and 7-3(b). The first spectrum was obtained in CCl_4 and the second in $CDCl_3$. The position of absorption for —OH protons is a function of solvents, concentration, and temperature.

ACIDITY OF ALCOHOLS

The hydrogen-bonding ability of alcohols tends to make them fairly soluble in water; those having four or less carbon atoms are miscible with water in all proportions. Their resemblance to water extends to the fact that the oxygen-bound hydrogen atom is weakly acidic, sufficiently so that alcohols will react with very active metals. Since the alcohols are weak acids, their *conjugate bases* are strong and find many uses where powerful organic bases are needed.

(7.2) $2CH_3OH + Mg \longrightarrow (CH_3O)_2Mg + H_2$

(7.3) $2CH_3CH_2CH_2OH + 2Na \longrightarrow 2CH_3CH_2CH_2O^-Na^+ + H_2$

NOMENCLATURE

As with other classes of organic compounds, there are IUPAC and semisystematic nomenclatures. The semisystematic method is identical with that used for haloalkanes — the hydrocarbon name is converted to its adjectival form by substituting -yl for -ane and this is followed by the word "alcohol." In the IUPAC system the final -e of hydrocarbon name is replaced by -ol, which denotes the presence of the functional group —OH. If there are several nonequivalent carbon atoms in the molecule, then a number is written immediately before the name to indicate the position of the hydroxyl group. The name of the alcohol is based on the longest carbon chain that contains the hydroxyl group, numbered so that the carbon atom bearing this

TABLE 7-2
Names and Physical Properties of Some Alcohols

IUPAC Name	"Alcohol" Name	Structure	Melting Point (°C)	Boiling Point (°C)
Methanol	Methyl	CH_3OH	−98	65
Ethanol	Ethyl	CH_3CH_2OH	−117	78
1-Propanol	*n*-Propyl	$CH_3CH_2CH_2OH$	−127	97
2-Propanol	*i*-Propyl	$(CH_3)_2CHOH$	−90	82
1-Butanol	*n*-Butyl	$CH_3(CH_2)_2CH_2OH$	−90	118
2-Butanol	*s*-Butyl	$CH_3CH_2CHCH_3$	−100	100
		OH		
2-Methyl-1-propanol	*i*-Butyl	$(CH_3)_2CHCH_2OH$	−108	108
2-Methyl-2-propanol	*t*-Butyl	$(CH_3)_3COH$	26	83
1-Pentanol	*n*-Amyl	$CH_3(CH_2)_3CH_2OH$	−79	138
2-Pentanol		$CH_3(CH_2)_2CHCH_3$		119
		OH		
3-Pentanol		$(CH_3CH_2)_2CHOH$		116
1,2-Ethanediol	Ethylene glycol	$HOCH_2CH_2OH$	−13	198

group has the smallest possible number. Many examples will be found in Table 7-2.

Another system of nomenclature, relatively archaic and slowly disappearing but still in use, is the **carbinol** system. All alcohols are named as derivatives of methanol, which carries the root name carbinol. The system is used most often to name symmetrical alcohols.

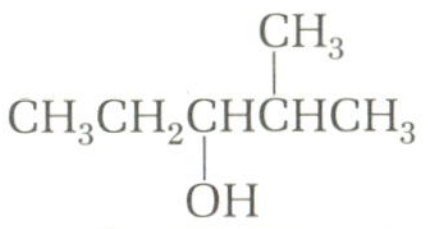
2-methyl-3-pentanol
ethyl isopropyl carbinol

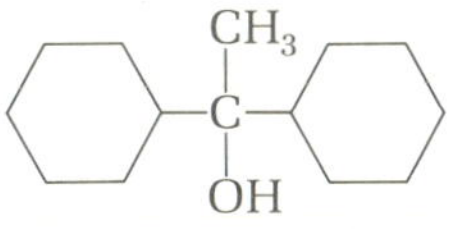
1,1-dicyclohexylethanol
dicyclohexyl methyl carbinol

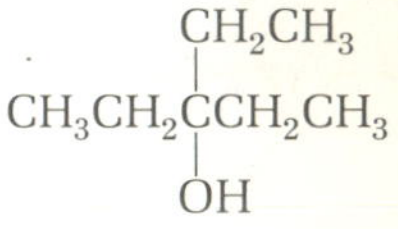
3-ethyl-3-pentanol
triethyl carbinol

THE S_N2 REACTION

An ion or molecule is defined as **nucleophilic** (nucleus-seeking) if it possesses at least one pair of electrons that it will share with another atom to form a covalent bond. Nucleophiles are therefore Lewis bases. The ion or molecule that accepts the proffered electrons is said to be **electrophilic** (electron-seeking). Just as the male-female polarity makes the world go round, so the mating of nucleophiles and electrophiles is responsible for a good deal of activity on the molecular level.

A reaction characteristic of nucleophiles is the **displacement** or **substitution** reaction. In essence, the nucleophile becomes bonded to an atom in the electrophile (or **substrate**) that had previously had another functional group attached to it; in the course of the reaction, the original functional group is displaced. The displaced group is known as the **leaving group.**

Electrophile		Nucleophile			Leaving Group	
CH_3CH_2Br	+	$:CN^- \longrightarrow$	CH_3CH_2CN	+	Br^-	(7.4)
CH_3I	+	$:NH_3 \longrightarrow$	$CH_3\overset{+}{N}H_3$	+	I^-	(7.5)
$CH_3\overset{+}{S}(CH_3)_2$	+	$:OH^- \longrightarrow$	CH_3OH	+	CH_3SCH_3	(7.6)
Common nucleophiles: I^-, Br^-, Cl^-, F^-, OH^-, OR^-, SH^-, CN^-, N_3^-, H^-, $:NH_3$, $H_2O:$, $R\ddot{O}H$						
Common leaving groups: —I, —Br, —Cl, $—\overset{+}{O}H_2$, —OTs*						
*The tosylate group, symbolized by the abbreviation OTs, will be described later in this chapter.						

Two fundamental mechanisms have been uncovered for nucleophilic substitution reactions. Although there are "borderline" reactions that are sometimes difficult to put firmly into one category or the other, we shall find it very convenient to classify most nucleophilic displacements according to this scheme. In one mechanism (equations 7.7 and 7.8) the electrophile dissociates initially by losing the leaving group along with the bonding electron pair; the cation left behind contains a carbon atom possessing only six valence electrons, known as a **carbonium ion.** The carbonium ion reacts subsequently with the nucleophile.

(7.7) $$(CH_3)_3CCl \xrightarrow{\text{slow}} CH_3—\overset{\displaystyle CH_3}{\underset{\displaystyle CH_3}{C^+}} + Cl^-$$

t-butyl cation

(7.8) $$CH_3—\overset{\displaystyle CH_3}{\underset{\displaystyle CH_3}{C^+}} + :OH^- \xrightarrow{\text{fast}} (CH_3)_3COH$$

The second mechanism (equation 7.9) involves only one continuous step — the bond between the nucleophile and the electrophile gets stronger and stronger as, simultaneously, the bond between the electrophile and the leaving group becomes weaker and weaker. It is this latter case that we shall study first.

(7.9) $$:CN^- + CH_3\ddot{\underset{\cdot\cdot}{Br}}: \longrightarrow \left[NC^{\delta-}\cdots\overset{H\ \ H}{\underset{H}{C}}\cdots Br^{\delta-} \right] \longrightarrow NCCH_3 + :\ddot{\underset{\cdot\cdot}{Br}}:^-$$

transition state

It is appropriate to recall the stereochemical conventions: Heavy, wedge-shaped lines lead to substituents above the plane of the paper, while dashed lines connect to substituents below the plane (p. 123); dotted lines depict partial bonds (pp. 77, 128).

The transition state of the reaction shown in equation 7.9 is derived from one molecule of methyl bromide and one cyanide ion. The reaction rate ought therefore to depend on the concentration of each of these, and it has been found experimentally that in processes of this type equation 7.10 applies. For this reason these reactions have been labeled S_N2 — substitution, nucleophilic, bimolecular.

(7.10) $\text{Rate} = k(\text{RBr})(\text{CN}^-)$

The term bimolecular indicates that the transition state was formed from two independent molecules or ions.

As we shall shortly demonstrate, the process depicted in equation 7.9, in which the nucleophile approaches the carbon atom from the side opposite the leaving group, has been shown to be an indispensable element of the S_N2 reaction. Kinetic studies support the idea that the approach of the nucleophile is synchronized with thc dcparture of the leaving group. The data in equation 7.11 are most easily rationalized by assuming that as one goes from methyl to *t*-butyl, there is increasing steric hindrance to the approach of the nucleophile, which must be occurring in the rate-determining step (Figure 7-4).

(7.11) $\text{RBr} + \text{Cl}^- \xrightarrow{\text{acetone}} \text{RCl} + \text{Br}^-$

If R = CH_3, the relative rate is 1.0.
If R = CH_3CH_2, the relative rate is 1.6×10^{-2}.
If R = $(CH_3)_2CH$, the relative rate is 2.1×10^{-4}.
If R = $(CH_3)_3C$, there is no measurable S_N2 reaction.

Since carbon cannot accommodate more than four electron pairs in its valence shell, the arrival of the nucleophile and the departure of the leaving group must be concerted. For these reasons, it is generally true that in S_N2 reactions primary alkyl derivatives are more reactive than the corresponding secondary ones, which are in turn more reactive than tertiary alkyl compounds.

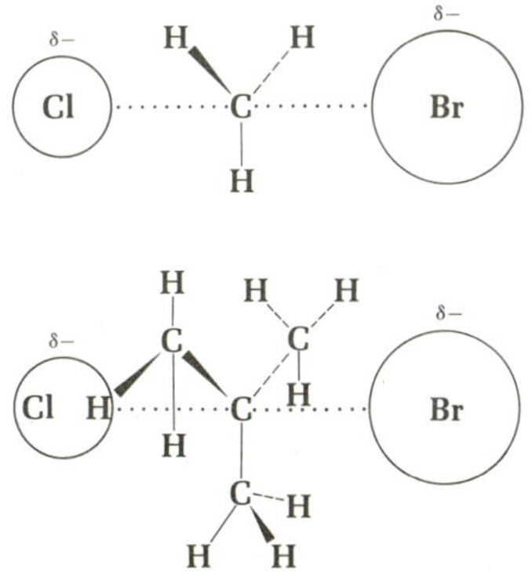

FIGURE 7-4

S_N2 Transition States in the Reactions of Methyl and *t*-Butyl Bromides with Chloride Ion

Some reactions are listed below that illustrate the synthetic utility of the S_N2 displacement.

(7.12) $$CH_3(CH_2)_{10}CH_2Br + KCN \xrightarrow[\text{reflux}]{\text{aq. ethanol}} CH_3(CH_2)_{10}CH_2CN + KBr$$
dodecyl bromide → dodecyl cyanide (90%)

(7.13) $$CH_3(CH_2)_5CH(OTs)CH_3 + CH_3CH_2OH \xrightarrow{\text{dry ethanol}} CH_3(CH_2)_5CH(OCH_2CH_3)CH_3 + HOTs$$
2-octyl tosylate → ethyl 2-octyl ether

(7.14) $$BrCH_2CH_2Br + 2H_2S + 2KOH \xrightarrow[\text{reflux}]{\text{ethanol}} HSCH_2CH_2SH + 2KBr + 2H_2O$$
1,2-ethanedithiol (60%)

The use of mixed solvents — water and an organic compound — is designed to permit both organic and inorganic reactants to dissolve. Note that in equation 7.13 ethanol is both solvent and nucleophile — an example of a *solvolysis* reaction.

MOLECULAR DISSYMMETRY

Our illustrations of S_N2 reactions have been drawn on the assumption that the nucleophile approaches the substrate on the side opposite the leaving group. While this assumption is reasonable, since it allows for minimization of interference between the two groups, none of the evidence so far presented *requires* such an interpretation. Since the **stereochemical** course of the reaction — how the atoms are arranged in space before and after reaction — is significant in both theoretical and synthetic investigations, we shall explore that question a little further. Information about the stereochemical course of a reaction is obtained through the use of **dissymmetric molecules** (those lacking all symmetry).

Methylene chloride, isopropyl bromide, and *cis*-1,3-dimethylcyclopentane all contain *a plane of symmetry* — a plane through the molecule such that a section on one side of the plane is the exact reflection of the section on the other side (Figure 7-5). If a molecule is capable of adopting a conformation in which such a plane is present, then there is one and only one spatial *configuration* that corresponds to the structural formula of that molecule. While isopropyl bromide and *cis*-1,3-dimethylcyclopentane may assume a number of different conformations, there is only one way in which their constituent atoms may be connected to one another in space. The most common though not the only indication that mirror images are superimposable is the presence of a plane of symmetry.

Examination of molecules that lack an internal symmetry plane, such as bromochlorofluoromethane, *s*-butyl bromide, or *trans*-1,3-dimethylcyclopentane, reveals that there are *two* independent spatial configurations that

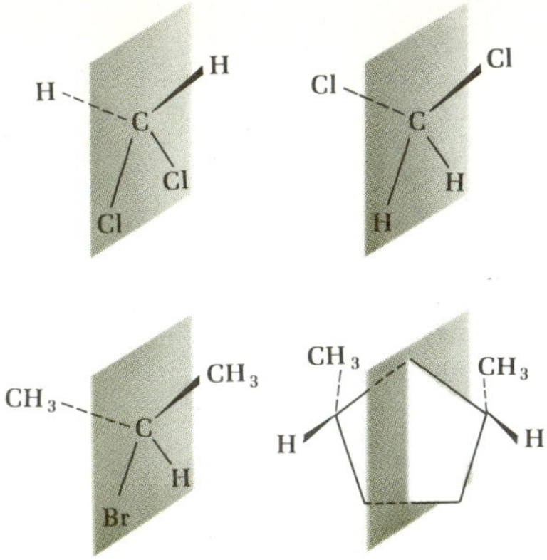

FIGURE 7-5
Some Symmetric Molecules, with the Symmetry Planes Shown

correspond to each structural formula (Figure 7-6). The relationship of these two configurations to one another is the same as the relationship of one's right and left hands, one to the other; i.e., one configuration is the **mirror image** of the other and these mirror images are *nonsuperimposable.* Thus, s-butyl bromide and *trans*-1,3-dimethylcyclopentane exist in left-handed and right-handed forms. The left- and right-handed forms of an asymmetric molecule are **enantiomers** (Greek — opposite of one another); enantiomers are also known as **optical isomers.** Isopropyl bromide by contrast is completely superimposable on its own mirror image and exists therefore in only one form (Figure 7-7).

The origin of the dissymmetry in the molecules just discussed is the presence of one or more **asymmetric atoms** — atoms that are connected to four different groups (these atoms are in color in the structural formulas).

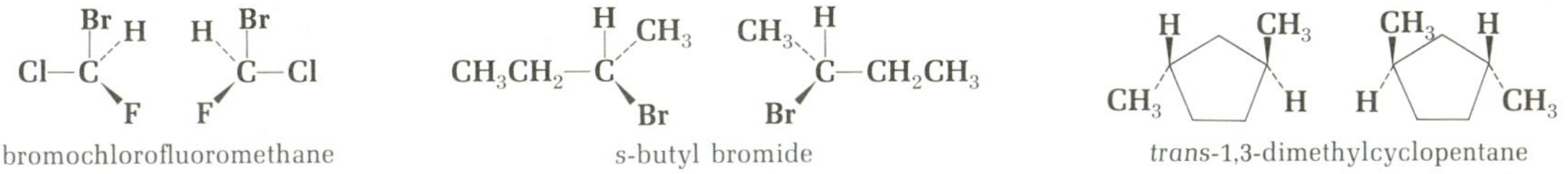

FIGURE 7-6
Some Dissymetric Molecules, Showing Both Spatial Configurations

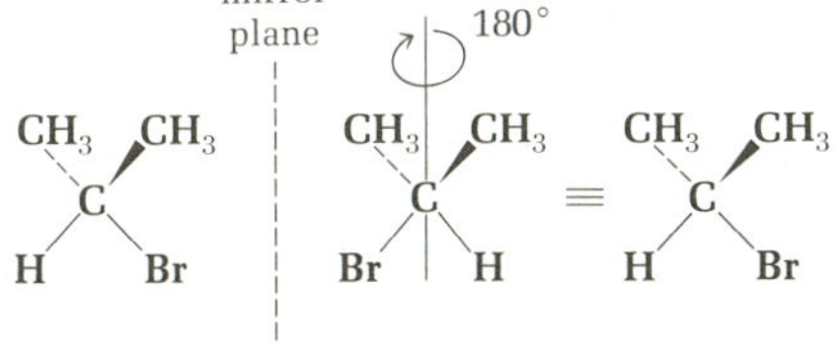

FIGURE 7-7
The Complete Identity of the Mirror Images of Isopropyl Bromide

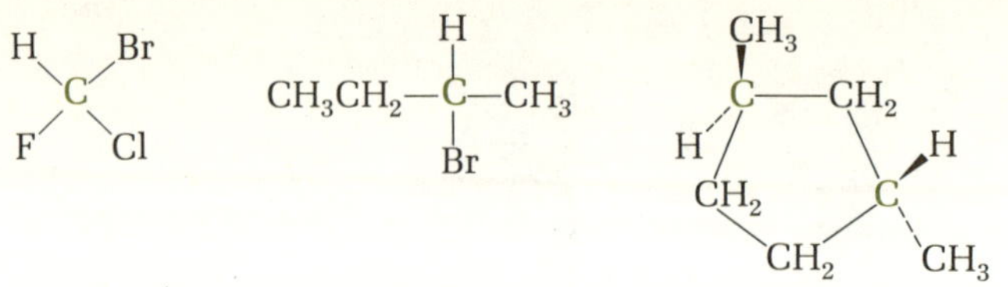

An asymmetric carbon atom is the most common although by no means the sole source of dissymmetry in organic compounds. A sampling of other compounds containing asymmetric atoms is given below.

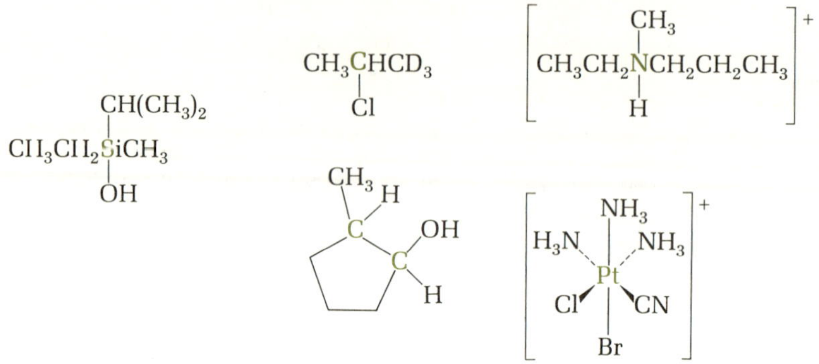

The requirements for atomic asymmetry are as follows:

1 The atom in question and those immediately attached to it must not lie in a plane.

2 Of all the groups attached to the asymmetric atom, at least three must be unique (differ structurally or spatially from the rest). If the asymmetric atom is a quadrivalent carbon atom, then all four attached substituents must be different.

The change in configuration of dissymmetric molecules as they undergo reaction is a prime source of information about the detailed mechanism of that reaction. Our understanding of nucleophilic substitution reactions has particularly benefited from such studies. In order to obtain this information, it is necessary of course to have an experimental probe to measure changes in molecular dissymmetry. Such a probe is provided by the interaction of molecules with polarized light.

MOLECULAR DISSYMMETRY AND OPTICAL ACTIVITY

A ray of light consists of a number of individual waves all traveling along the same line of propagation. The waves are planar, and the wave planes are oriented at all possible angles to one another (Figure 7-8). Several mate-

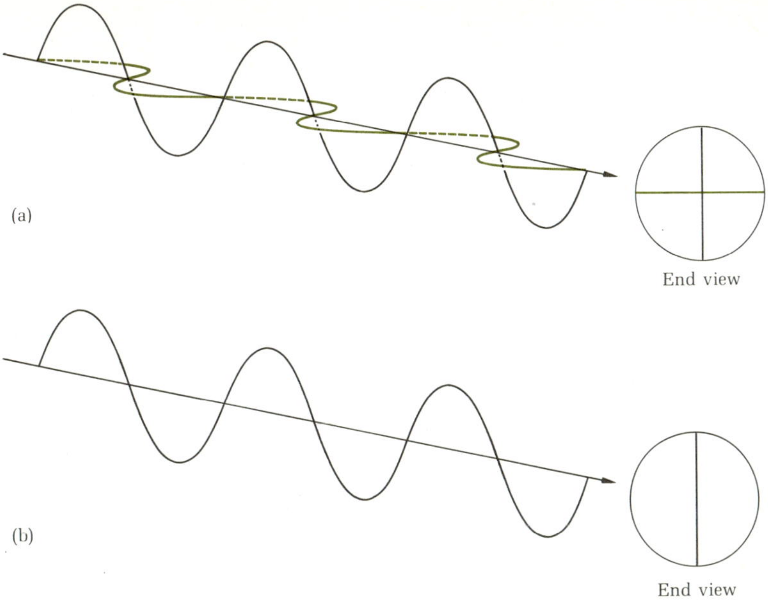

FIGURE 7-8

Beams of (a) Unpolarized Light and (b) Plane-polarized Light

rials are known that will subject light waves to **plane polarization** — a process whereby all those waves not vibrating in one particular plane are filtered out. It was discovered early in the nineteenth century that solutions of certain naturally occurring compounds could rotate the plane of polarization of light that was already polarized. The light was still polarized but its plane of polarization was now tilted at an angle to what it was before passing through the solution (Figure 7-9). Different compounds were known whose solutions would rotate the plane either to the left **(levorotatory)** or to the right **(dextrorotatory).** All compounds possessing this property were referred to as **optically active.**

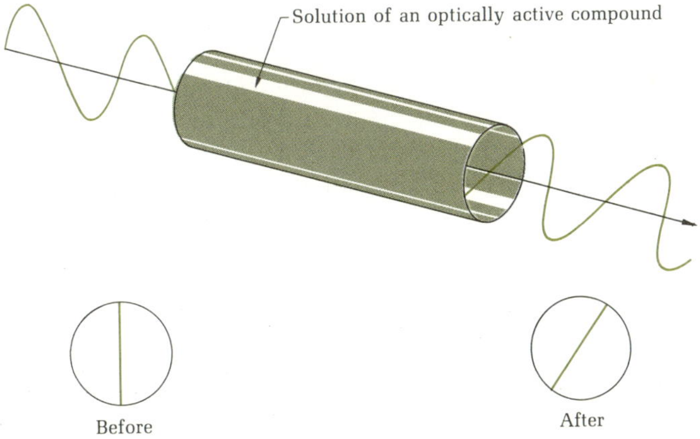

FIGURE 7-9

Plane-Polarized Light Before and After Passing Through a Solution of an Optically Active Compound

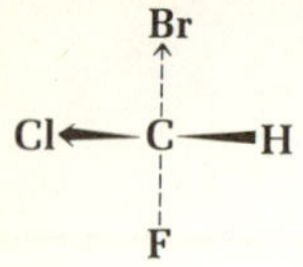

FIGURE 7-10
Projection of a Molecule of Bromochlorofluoromethane

Light rays have alternating electric fields associated with them; the plane of vibration is in fact the plane in which the electric field is oscillating back and forth. In a beam of polarized light the electric field is undergoing oscillations in one plane only. Due to the presence of this electric field, light rays may interact with the electrons in molecules. The strength of this interaction will depend, in general, on the polarizabilities of the electrons involved. (One should not confuse the two uses of polarize. When applied to light, it means restriction of the plane of vibration. The term **polarizable,** when applied to atoms or other particles, means susceptibility to the influence of an electric field.) Suppose that a plane-polarized light ray is passing through a solution of $CHFClBr$, a dissymmetric molecule. Figure 7-10 depicts a single molecule as it might appear to an approaching light ray. As the ray begins to pass through the molecule, it first encounters the hydrogen and chlorine atoms; since the electrons of the chlorine are more polarizable than those of hydrogen, the chlorine atom interacts most strongly with the light ray. At the "back" of the molecule it is the bromine atom that affects the light more strongly than the fluorine does. **As the molecule is traversed from front to back, the point of maximum polarizability undergoes an angular rotation.**

In a solution of $CHFClBr$ there will be, at any given instant, a number of molecules oriented in every possible configuration with respect to an incoming beam of light. Each encounter of a light ray with a molecule results in some displacement of the plane of polarization; since the molecules are arranged in all possible configurations, however, successive rotations of the plane may tend to cancel one another. The net effect on the plane of polarization of the entire light beam is the sum of all the molecular interactions. They may cancel out to zero, or there may be a net rotation.

If the solution contains *equal* quantities of the enantiomers of bromochlorofluoromethane, then the sum of the interactions of a polarized light beam with the solution will be zero; i.e., no rotation of the plane of polarization will be observed. Such a solution is said to be **racemic** and is optically inactive. If the solution contains an excess of one enantiomer over the other, however, then a net rotation of the plane of polarization will be observed. The solution is then *optically active.* Passage of polarized light through a liquid containing only *symmetric* molecules cannot result in any rotation of the plane of polarization. With symmetric molecules the distribution of molecular orientations in space is exactly symmetric (for a large sample), and all individual molecular interactions with the polarized light beam will sum to zero. A number of such orientations are shown for bromochloromethane.

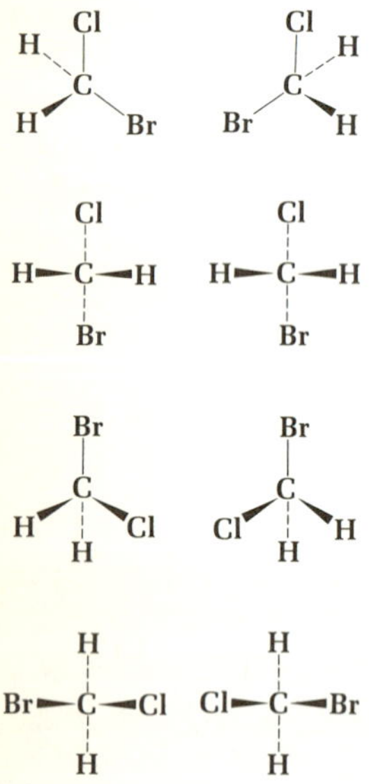

FIGURE 7-11
A Number of Projections of Bromochloromethane

There is an important distinction that has been made that must be kept clear. A compound such as bromochloromethane, which contains no asymmetric atom, is inherently optically inactive. A racemic mixture of a compound like bromochlorofluoromethane, on the other hand, happens fortuitously to contain equal numbers of dextrorotatory and levorotatory enantiomers. In the future, when a dissymmetric compound is under discussion, it will be assumed that it is the racemic mixture unless otherwise specified.

QUANTITATIVE MEASUREMENT OF OPTICAL ACTIVITY

A pure optical isomer may be characterized by its **specific rotation,** which is the rotation per unit weight per unit path length. This number is dependent on the temperature of the solution and the wavelength of light used, but once these have been fixed, it is invariant. The observed specific rotation, symbolized by $[\alpha]_\lambda^T$, may be used as a measure of the **optical purity** of the material; that is, the extent to which it is free of contamination by its enantiomer, providing that the specific rotation of a pure sample is known.

$$[\alpha]_\lambda^T = \frac{\alpha}{l \cdot c} \tag{7.15}$$

In equation 7.15, T stands for temperature; λ, for the wavelength of light; l, the path length of the light through the solution; c, the concentration in grams per milliliter; and α, the actual rotation measured under the specified conditions. Rotatory power may also be calculated on a molar rather than a weight basis. Equation 7.16 defines the **molecular rotation;** mw stands for molecular weight.

$$[M] = \frac{[\alpha] \cdot \text{mw}}{100} \tag{7.16}$$

A pair of enantiomers have all their physical properties (melting point, density, refractive index, etc.) except one in common. The absolute value of their specific rotations will be identical, but the signs will be opposed; one will be dextrorotatory (+) while the other will be levorotatory (−).

ABSOLUTE CONFIGURATION – RETENTION AND INVERSION

In order to derive information about the stereochemical course of a reaction, it is important to know what the relationship is between the spatial arrangement of groups around an asymmetric atom and the sign of rotation of a particular enantiomer. Is there, in fact, any straightforward correlation between them? The examples shown below would seem to indicate that there is not. All three compounds contain the same spatial arrangement of corresponding groups around the asymmetric carbon atom (they all have the same **absolute configuration**), but substitutions in the molecule that are remote from the asymmetric atom may change the sign of rotation.

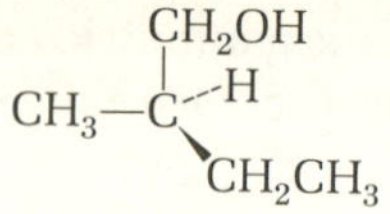

(+)-2-methyl-1-butanol

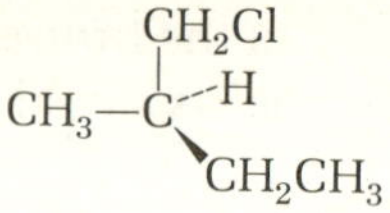

(−)-1-chloro-2-methylbutane

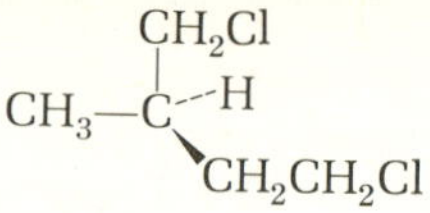

(+)-1,4-dichloro-2-methylbutane

If a strongly polarizable group is at the asymmetric atom, then the sign of rotation and the absolute configuration may well correlate.

To determine the absolute configuration of molecules, it is ultimately necessary to determine the structure of some dissymmetric compounds by means of X-ray crystallography. These investigations yield the absolute configuration of the molecule directly and unequivocally. The absolute configuration of other compounds may then be established by less direct chemical methods. If a compound of known absolute configuration is converted to a product by means of reactions that take a known **stereochemical path,** then the absolute configuration of the product is established.

How is the absolute configuration to be *specified,* even if it is known? There is a clear, unambiguous method which requires that an order of priority be assigned to the groups bonded to the asymmetric atom. Priority is assigned on the basis of the **atomic number** of the atom directly connected to the asymmetric center; the higher the atomic number, the higher the priority. If *a* denotes the substituent of highest priority, and *d* the lowest priority substituent, then bromochlorofluoromethane has the priorities shown. In 2-butanol two of the atoms directly attached to the asymmetric centers are identical; in such cases, priority is assigned by referring to the next closest atoms.

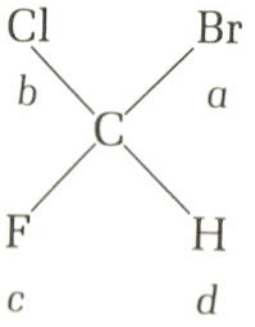

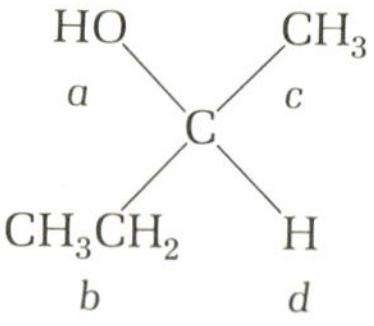

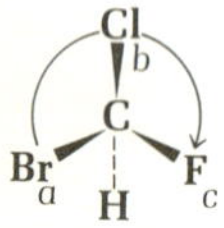

(*R*)-bromochloro-fluoromethane

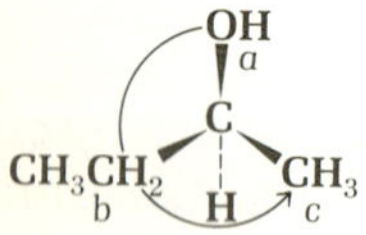

(*S*)-2-butanol

FIGURE 7-12

Absolute Configurations of One Enantiomer of Bromo-chlorofluoromethane and 2-Butanol

In this case, ethyl takes precedence over methyl since carbon has a higher priority than hydrogen. Similarly, isopropyl, $(CH_3)_2CH$–, takes precedence over ethyl, CH_3CH_2–.

Once priority has been assigned, then the projection of the molecule is manipulated so that the group of *lowest priority* is projecting behind the plane of the paper, and the other three groups are sitting above the plane of the paper (models can be a great help here). **If the order of descending priority, *a* to *b* to *c*, travels clockwise around the substituents, then the configuration at that asymmetric atom is** *R* (Latin *rectus* — right). **If the priority descends in a counterclockwise sense, the absolute configuration at that atom is** *S* (Latin *sinister* — left) (Figure 7-12). The enantiomers of the molecules shown are (*S*)-bromochlorofluoromethane and (*R*)-2-butanol. The complete process of assigning the absolute configuration to a molecule is depicted in Figure 7-13.

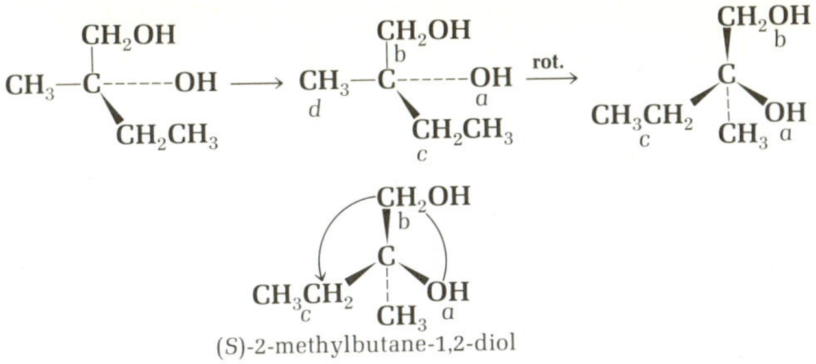

FIGURE 7-13

Assigning the Absolute Configuration to One Enantiomer of 2-Methylbutane-1, 2-diol

Now that we are able to assign absolute configurations to asymmetric atoms, we may turn back to the question of chemical interest, which is: If a nucleophilic substitution is performed at an asymmetric carbon, what is the relationship between the absolute configurations of the reactants and products? Three plausible results are shown in equations 7.17 to 7.19. The racemization shown in equation 7.19 is total only if equal amounts of the *R* and *S* enantiomers are formed; if not, then partial racemization has occurred and there has been a net retention or inversion of configuration.

(7.17) 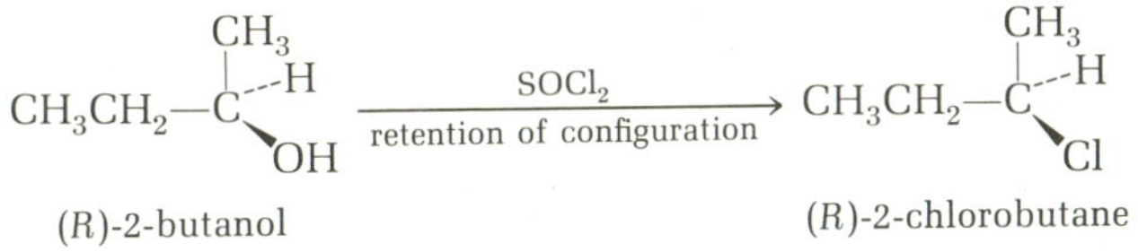

(7.18) 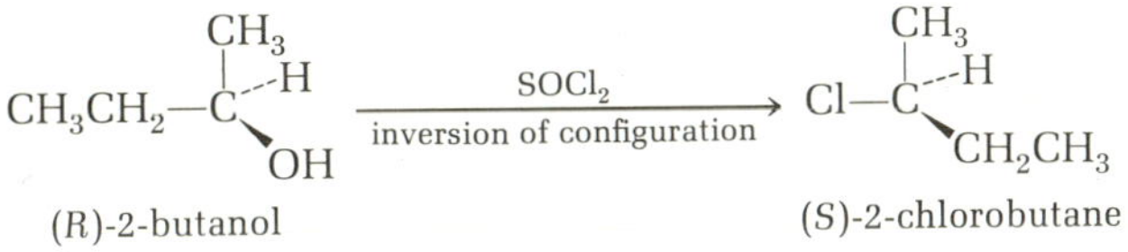

(7.19) 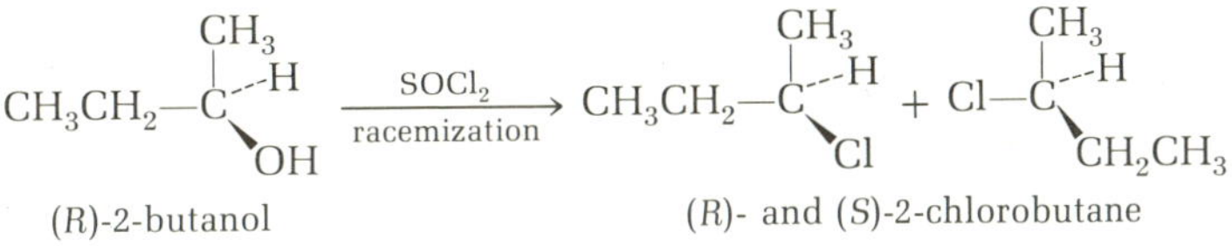

Each of the transformations shown in equations 7.17 to 7.19 may be brought about by proper choice of solvent and reaction conditions. Reactions 7.17 and 7.18 are **stereospecific;** one enantiomer of the reactant gives rise to only one enantiomer of the product.

STEREOCHEMISTRY OF THE S_N2 REACTION

The conversion of (+)-2-octanol to (+)-2-octyl tosylate does not affect the configuration of the molecules, as the carbon-oxygen bond remains undisturbed. When this tosylate undergoes reaction in aqueous acetone, the product is pure (−)-2-octanol. Thus, displacement of the tosylate anion by the water must have involved an *inversion* of configuration.

(7.20) 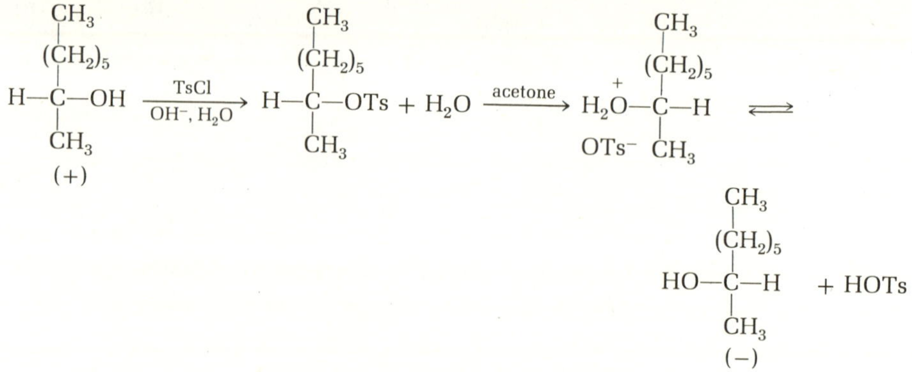

Numerous other examples are known in which nucleophilic displacements take place at asymmetric secondary carbon atoms.

(7.21) 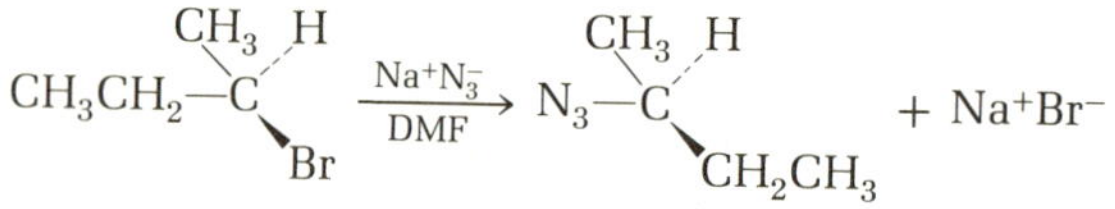

(7.22) 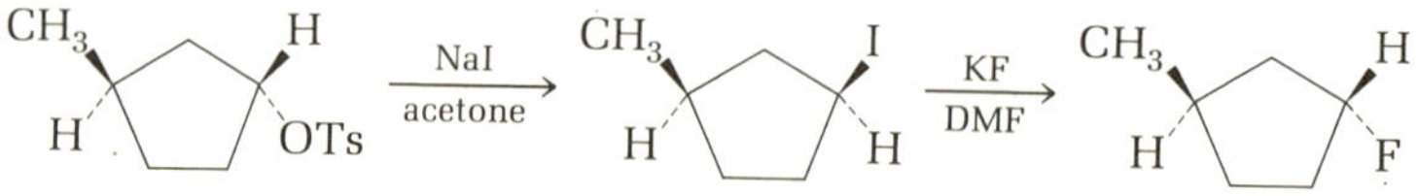

In all cases in which the kinetic data establish that the reaction is of the S_N2 type, the predominant mode of displacement is that involving inversion of configuration. This process has been likened to an umbrella turning itself inside out in a high wind. Note that at the transition state (Figure 7-14) the carbon atom undergoing inversion is essentially sp^2-hybridized. The non-reacting substituents are bonded to this carbon atom by way of normal σ-orbitals derived from sp^2-orbitals, while the leaving group and the nucleophile are each weakly bonded by overlap with one lobe of the *p*-orbital.

A primary carbon atom may be made asymmetric by substituting a deuterium atom for one of the hydrogen atoms, as in **1.** Studies of derivatives of this alcohol have shown that the reactions involve complete inversion. It may be taken as a general rule that **S_N2 reactions always occur with inversion of configuration.**

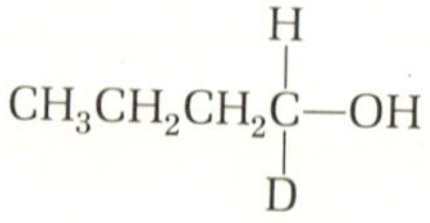

1-butanol-1-*d* (**1**)

This fact is very useful in planning stereospecific syntheses. The general nature of the process is further confirmed by the observation that many

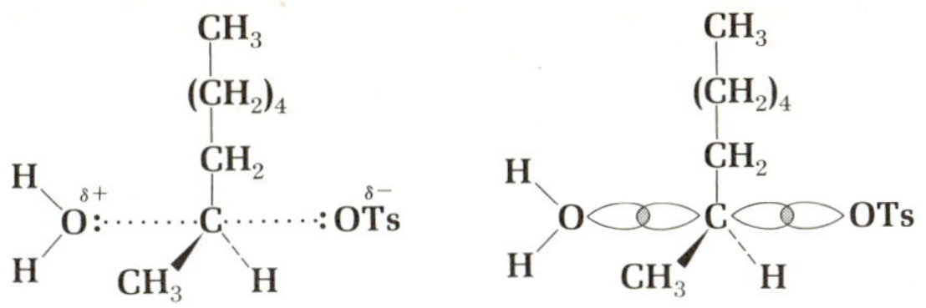

FIGURE 7-14
Transition State for Reaction of 2-Octyl Tosylate with Water

optically active silicon compounds undergo nucleophilic displacement with inversion. Silicon is directly below carbon in the periodic table.

LEAVING GROUP EFFECTS

Among the methyl halides, methyl iodide is the most reactive toward a given nucleophile and methyl fluoride is the least reactive. Since the carbon-fluorine bond is quite strong and the carbon-iodine bond is quite weak, the result is not surprising — one would expect the most weakly held leaving group to be the most readily displaced. That, however, is not the whole story.

In the series CH_3F, CH_3OH, CH_3NH_2, and CH_3CH_3, the strength of the CH_3—Y bond generally decreases from left to right; one might expect that OH^- could be more easily displaced from CH_3OH than F^- from CH_3F. In fact, although F^- can be displaced from alkyl fluorides (albeit with great difficulty), it has not proved possible to displace OH^- from alcohols or NH_2^- from **amines** (Chapter 18). The same trend is found among the elements in the next row: Cl^- may be displaced from CH_3Cl, but not SH^- from CH_3SH.

A central factor affecting the efficacy of Y as a leaving group is the stability of the anion Y^-. One measure of stability is the basicity of Y^- — the less the tendency to acquire a proton exhibited by Y^-, the greater is its stability. Thus, —Cl, —Br, and —I are effective leaving groups while —OH, —OR, —CN, and —SH, which give rise to relatively basic anions, are poor leaving groups.

NUCLEOPHILIC SUBSTITUTION OF ALCOHOLS

Alkyl halides may be easily converted to alcohols by allowing them to react with hydroxide ion or water. Treatment of alcohols with halide ions does not, however, result in the formation of alkyl halides. Since alcohols of all kinds are available in profusion, it would be desirable to have methods for replacing the hydroxyl group by other nucleophiles. Hydroxyl groups can be displaced from alcohols if a strong acid is present; the reaction is widely used for synthetic purposes.

$$\text{(7.23)}\quad CH_3CH_2CH_2CH_2OH + NaBr \xrightarrow[\text{reflux}]{\text{aq. } H_2SO_4} \underset{(90\%)}{CH_3CH_2CH_2CH_2Br} + NaHSO_4 + H_2O$$

$$\text{(7.24)}\quad CH_3CH_2\underset{\underset{OH}{|}}{C}HCH_3 + HCl \xrightarrow[\text{reflux}]{ZnCl_2} \underset{(65\%)}{CH_3CH_2\underset{\underset{Cl}{|}}{C}HCH_3} + H_2O$$

$$\text{(7.25)}\quad HO(CH_2)_6OH + 2KI + 2H_3PO_4 \xrightarrow{\text{heat}} \underset{(85\%)}{I(CH_2)_6I} + 2KH_2PO_4 + 2H_2O$$

The evidence indicates that the first step involves the reversible protonation of the alcohol, which converts it to an **oxonium ion.**

$$\text{(7.26)}\quad R\ddot{\underset{..}{O}}H + H_3O^+ \rightleftharpoons H_2O + R\underset{..}{\overset{+}{O}}\begin{smallmatrix}H\\ \\H\end{smallmatrix}$$

The oxonium ion is much more susceptible to nucleophilic attack than was the alcohol. The effective electronegativity of the oxygen atom has been increased, and the poor leaving group OH^- has been converted into the good leaving group H_2O.

$$\text{(7.27)}\quad X\!: + ROH_2^+ \longrightarrow RX + H_2O$$

This method fails if the nucleophiles in question are appreciably basic, such as SH^- or CN^-, because they are converted by acid to the much poorer nucleophiles H_2S and HCN.

A hydroxyl group may also be transformed into a much better leaving group if the hydrogen atom is replaced by a more electronegative element such as phosphorus. Halogen derivatives of trivalent and pentavalent phosphorus are extensively used to convert alcohols to alkyl halides; among them are the phosphorus trihalides (PCl_3, PBr_3, PI_3), two phosphorus pentahalides (PCl_5, PBr_5), and the dichloride and dibromide of triphenylphosphine [$(C_6H_5)_3PCl_2$, $(C_6H_5)_3PBr_2$]. (The phenyl group, C_6H_5, is extensively discussed in Chapter 14; it is highly stable and unreactive under the reaction conditions.)

$$\text{(7.28)}\quad 3CH_3CH_2\underset{\underset{OH}{|}}{C}HCH_3 + PBr_3 \xrightarrow{0\text{–}25^\circ} \underset{(80\%)}{3CH_3CH_2\underset{\underset{Br}{|}}{C}HCH_3} + H_3PO_3$$

$$\text{(7.29)}\quad P + \frac{3}{2}I_2 \rightleftharpoons PI_3 + \underset{\text{3-methyl-1-butanol}}{3(CH_3)_2CHCH_2CH_2OH} \xrightarrow{\text{reflux}} \underset{(90\%)}{3(CH_3)_2CHCH_2CH_2I} + H_3PO_3$$

$$\text{(7.30)}\quad (C_6H_5)_3PCl_2 + (CH_3)_3CCH_2OH \longrightarrow \underset{(90\%)}{(CH_3)_3CCH_2Cl} + (C_6H_5)_3P{=}O + HCl$$

The reaction conditions are quite mild, and optically active alcohols yield halides in which inversion of configuration has taken place to a large extent.

Phosphorus has a powerful affinity for oxygen, and the various phosphorus oxides are all less basic than hydroxide ion and consequently better leaving groups. A detailed mechanism for the action of the phosphorus trihalides is given in equation 7.31; note that trivalent phosphorus is converted to pentavalent phosphorus in the course of the reaction.

(7.31)

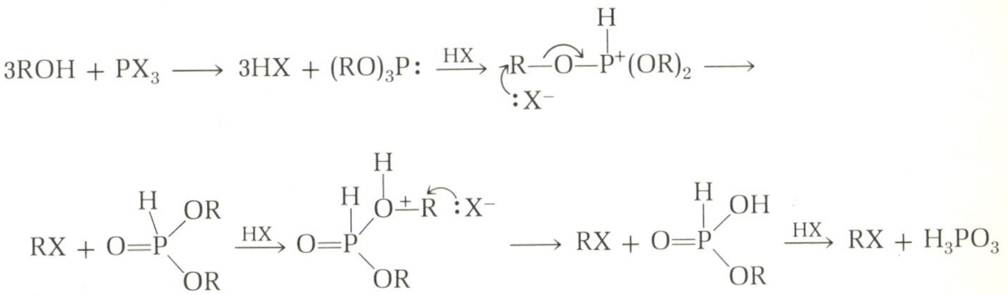

Certain sulfur halides also play a limited role in halide synthesis from alcohols. Thionyl chloride is a useful reagent for converting alcohols to chlorides.

(7.32)

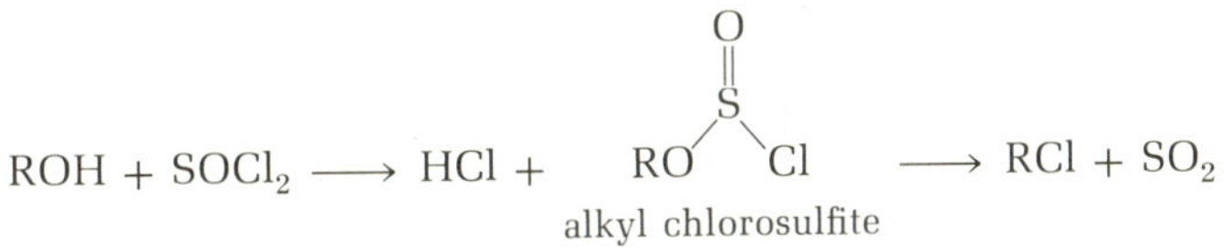

Conversion of an alcohol to a tosylate is often a very convenient way of ultimately displacing the hydroxyl group. The term "tosylate" is a useful contraction of toluenesulfonate, the correct name. Tosylates are made by the reaction of toluenesulfonyl chloride (**2**) with an alcohol in the presence of a base. Do not be disturbed by the apparently formidable structure of the toluenesulfonyl group — it is essential only that its *function* be understood. It is a very weakly basic anion and is thus a good leaving group.

(7.33)

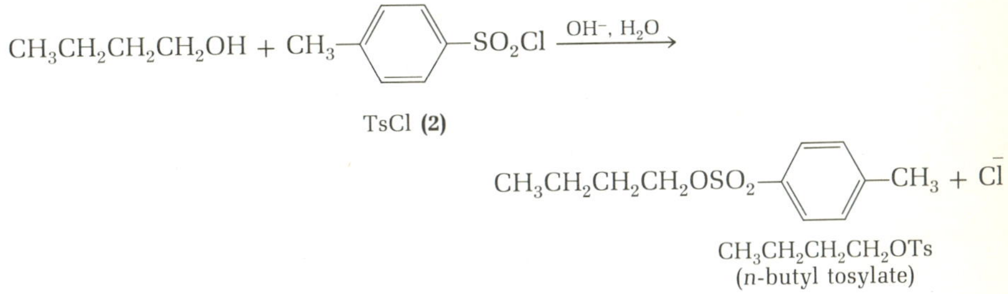

The sulfur is firmly bound to the alcohol oxygen, and all depart together as the leaving group. Formation of the tosylate involves no change at the asymmetric carbon, which is important in stereospecific syntheses; the tosylate group is displaced with inversion, however. Tosylates of tertiary alcohols usually cannot be formed in acceptable yield.

Alkyl fluorides, cyanides, and thiols (RSH) are usually not formed directly from alcohols. Rather, the alcohols are converted to halides or tosylates and the latter then allowed to react with fluoride, cyanide, or hydrosulfide ion.

RELATIVE NUCLEOPHILICITY

The relative nucleophilicity of two particular nucleophiles is not a fixed quantity. It depends strongly on the substrate and on the solvent. There are, nonetheless, a few generalizations that always hold true. Anions are always more nucleophilic than the related neutral molecules — OH^- better than H_2O, CH_3O^- better than CH_3OH, and SH^- better than H_2S. When comparing nucleophiles whose nucleophilic atoms are in the same row of the periodic table, the most reactive will have the least electronegative atom — NH_3 better than H_2O, SH^- better than Cl^-, etc. It has also appeared to be true that nucleophiles with large polarizable atoms — I^-, Br^-, SH^-, R_3P: — are generally very reactive. Their polarizability would allow them to "feel" the approach of the electrophilic substrate quite readily; on the other hand, at least part of the effect must be tied up with the solvation phenomena described below.

SOLVENT EFFECTS

The choice of solvent can be very important in nucleophilic displacement reactions. In **protic solvents** (those capable of hydrogen bonding) such as water or ethanol, the nucleophile will generally be solvated by means of hydrogen bonds. To the extent that the nucleophile is tied up by the solvent, it is less effective in displacing a given leaving group; its **nucleophilicity** is thus decreased. Hydrogen bonding is strongest when the nucleophilic atom is in the second row of the periodic table. In protic solvents, therefore, OH^- reacts more slowly than SH^-; F^-, more slowly than I^-; etc.

(7.34) $$CH_3(CH_2)_4CH_2Cl + KF \xrightarrow[200°]{\text{ethylene glycol}} \underset{(54\%)}{CH_3(CH_2)_4CH_2F} + KCl$$

All anionic nucleophiles react more rapidly in **dipolar aprotic solvents** — liquids that do not contain acidic hydrogens but that have sufficiently large dipole moments so that they are capable of dissolving salts. Some compounds that have been widely used are dimethyl sulfoxide (DMSO, **3**), dimethylformamide (DMF, **4**), acetonitrile (**5**), acetone (**6**), and hexamethylphosphoramide (HMPA, **7**). They are all characterized by highly polar double or triple bonds, and while they solvate metal cations quite well, they leave the anions relatively free.

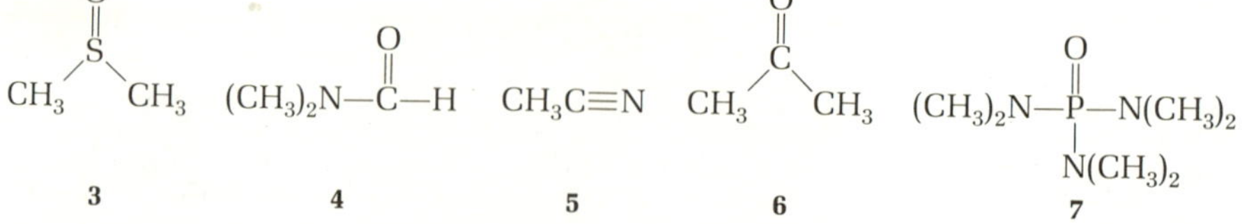

3 **4** **5** **6** **7**

It is the small anions with high charge densities (typically those derived from second-row atoms) whose nucleophilicity is most affected by a change from protic to dipolar aprotic solvents; F^-, Cl^-, and CN^- are about 10^6 times more reactive toward methyl iodide in DMF than in methanol. The nucleophilicity order $I^- > Br^- > Cl^-$ holds in water and alcohols, but it is reversed to $Cl^- > Br^- > I^-$ in DMF, DMSO, and acetone. A saturated solution of sodium iodide in acetone is often very useful in synthesizing organic iodides from the corresponding bromides or chlorides. The sodium bromide or chloride produced is insoluble in acetone, and the unfavorable equilibrium is driven in the desired direction.

(7.35) $$C(CH_2Br)_4 + 4NaI \xrightarrow[100^\circ]{\text{acetone}} \underset{(89\text{–}98\%)}{C(CH_2I)_4} + 4NaBr$$

Solvents also affect leaving groups. Since ions such as Cl^- are strongly solvated in hydrogen-bonding media, the leaving group ability of —Cl is enhanced in hydroxylic solvents. The difference in equilibrium constants between equations 7.36 and 7.37 lies mostly in the fact that bromide ion is more strongly solvated than iodide ion in water, while the difference in solvation energy is much smaller (and possibly reversed) in acetone.

(7.36) $$I^- + CH_3Br \rightleftharpoons CH_3I + Br^- \qquad K^{35} = 15.4 \text{ (water)}$$

(7.37) $$I^- + CH_3Br \rightleftharpoons CH_3I + Br^- \qquad K^{35^\circ} = 0.62 \text{ (acetone)}$$

In reaction 7.36 the solvation effect outweighs the bond energy effect.

THE S_N1 REACTION

The point has been made that replacing hydrogen atoms by alkyl groups at an atom undergoing S_N2 reaction slows down the substitution because of increasing steric hindrance to nucleophilic attack. Tertiary carbon atoms are therefore very resistant to nucleophiles. Yet *t*-butyl chloride reacts quite rapidly with water to give *t*-butyl alcohol, while the hydrolysis of *n*-butyl chloride is many orders of magnitude slower. In order to accommodate this observation along with the previously developed data, it is necessary to assume that *t*-butyl chloride is reacting by way of a different mechanism, the S_N1 mechanism (substitution, nucleophilic, unimolecular).

(7.38) $$(CH_3)_3CCl + H_2O \xrightarrow[15^\circ]{\text{very fast}} (CH_3)_3COH + HCl$$

(7.39) $$CH_3CH_2CH_2CH_2Cl + H_2O \xrightarrow[100^\circ]{\text{very slow}} CH_3CH_2CH_2CH_2OH + HCl$$

In the S_N1 mechanism the slow step is the heterolytic breaking of the bond joining the carbon atom and the leaving group; the carbonium ion formed then reacts rapidly with any nucleophiles present. The rate depends only on the concentration of the substrate.

(7.40) $\text{Rate} = k[(CH_3)_3CCl]$

Furthermore, the rate of the reaction should be unaffected by the relative nucleophilicity of the nucleophile.

(7.41) $(CH_3)_3CCl \xrightleftharpoons{\text{slow}} (CH_3)_3C^+ + Cl^-$

(7.42) $(CH_3)_3C^+ + H_2\ddot{O} \xrightleftharpoons{\text{fast}} (CH_3)_3C{-}\overset{+}{O}H_2 \xrightleftharpoons{\text{fast}} (CH_3)_3COH + H^+$

All these predictions have been verified experimentally. For instance, if 1 mole of *t*-butyl alcohol is reacted with ½ mole each of bromide ion and chloride ion in the presence of acid, about 80% of the product will be *t*-butyl bromide. But the rate of the reaction will be independent of the ratio of bromide to chloride.

(7.43) 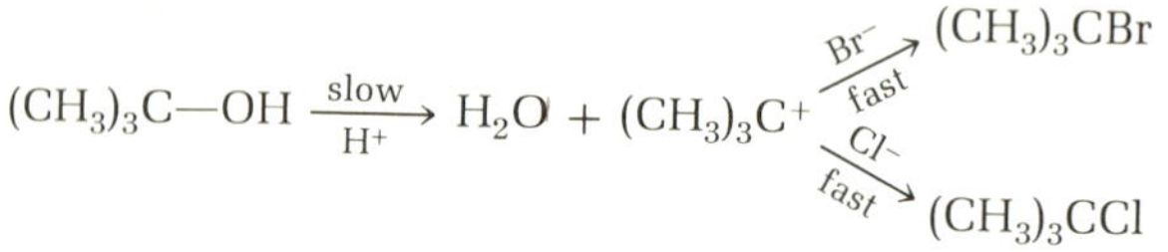

The S_N1 and S_N2 reaction mechanisms are each favored by different thermodynamic forces. The dissociation of a molecule into two particles, as in the S_N1 mechanism, is entropically favorable; on the other hand, the energetically unfavorable bond cleavage is not compensated for by simultaneous bond formation, as happens in an S_N2 reaction. Just which reaction path a given system will follow is determined by the balance of the various factors.

Increased stabilization of the carbonium ion will favor operation of the S_N1 mechanism; consequently, it usually applies to reactions occurring in polar media such as water and aqueous organic solvents, but not those taking place in dry acetone or acetonitrile. The hydrogen-bonding ability of the solvent is also important since the hydrogen bonding can assist the departure of the leaving group. The most important determinant of mechanism, however, seems to be the degree of substitution of the potential carbonium ion. S_N1 reactivity in alkyl systems follows the order $3° > 2° > 1°$.

The electrons in an alkyl group are more polarizable than those around a single hydrogen atom. An electron-deficient carbon atom in a carbonium ion will attract the electrons in the alkyl groups attached to it and thereby partially disperse its positive charge. This in turn lowers the energy of the carbonium ion and, in so doing, the energy of activation for its formation.

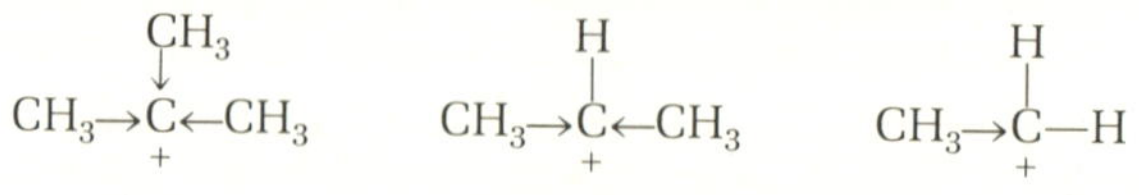

3° carbonium ion   2° carbonium ion   1° carbonium ion

The order of S_N1 reactivity of alkyl halides, tosylates, alcohols, etc., follows the order of stability of the carbonium ion intermediate. Reactions of primary derivatives in solution rarely occur by way of carbonium ions, except for certain uniquely constructed molecules. A dramatic illustration is provided by the difference in conditions necessary for converting *n*-butyl alcohol and *t*-butyl alcohol to their respective chlorides (equations 7.44 and 7.45).

(7.44) $$CH_3CH_2CH_2CH_2OH + \text{conc. } HCl \xrightarrow[\text{reflux } 3\frac{1}{2} \text{ hr}]{ZnCl_2} \underset{(65\%)}{CH_3CH_2CH_2CH_2Cl}$$

(7.45) $$(CH_3)_3COH + \text{conc. } HCl \xrightarrow[\text{temp, 20 min}]{\text{shake at room}} \underset{(85\%)}{(CH_3)_3CCl}$$

The donation of electrons from an atom or group to an electron-deficient site is known as the **inductive effect.** Several other physical models have been proposed that give rise to the same predictions as the inductive model. A few of these will be discussed in future chapters.

CARBONIUM ION REARRANGEMENTS

A general property possessed by carbonium ions is their tendency to undergo rearrangement, particularly to a more stable carbonium ion. Rearrangement takes place through the shift of an atom or group, with its pair of bonding electrons, over to the electron-deficient atom, creating thereby a new carbonium ion.

(7.46)

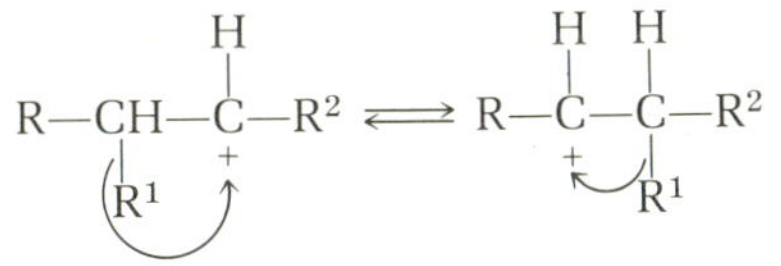

These shifts are often reversible.

In general, 1° carbonium ions will rearrange to 2° or 3° carbonium ions, if possible; 2° carbonium ions to 3°, etc. Carbonium ions of equal stability may interconvert, but under the conditions usually employed (aqueous solution, ~100°) the rearrangement of more stable to less stable carbonium ions — e.g., 2° to 1° — is too slow to be observed. If two different groups may migrate to give equally stable carbonium ions, then an alkyl group will usually migrate in preference to hydrogen. The occurrence of rearrangement in a reaction is often indicative of the intermediacy of carbonium ions.

When primary alcohols are heated with HCl, HBr, or HI at 100 to 120°, only primary halides are obtained. Primary carbonium ions are formed with difficulty, and the nucleophilic halide ions apparently react with the oxonium ions before they dissociate to carbonium ions. But the reaction of 2-

or 3-pentanol even at 20° leads to rearrangement, reflecting the greater ease of 2° carbonium ion formation.

(7.47) 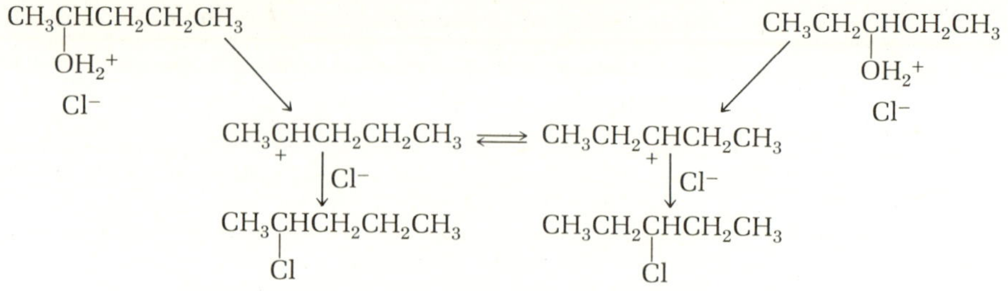

(7.48) 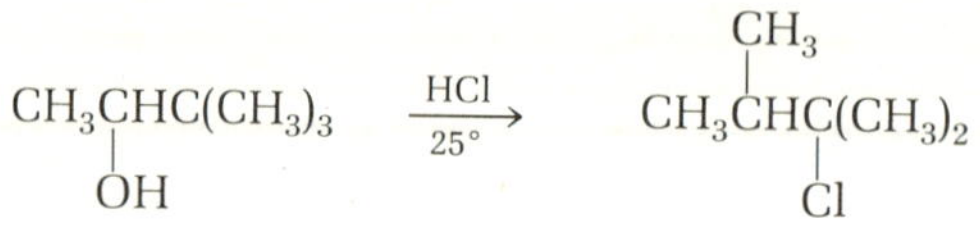

3,3-dimethyl-2-butanol 2,3-dimethyl-2-chlorobutane

Primary and tertiary alcohols, halides, etc., often provide examples of reactions that clearly proceed by S_N2 or S_N1 mechanisms, respectively. The reactions of secondary derivatives lie in a border area, and they take place by either or both of these routes, depending on the solvent, nucleophile, leaving group, etc. The reaction of (+)-2-octyl tosylate with pure water gives (−)-2-octanol by complete inversion. Displacement of Br from 2-octyl bromide by ethoxide ion ($CH_3CH_2O^-$) in dry ethanol also proceeds with complete inversion. However, 2-octyl bromide in aqueous ethanol forms, among other products, 2-octanol that is partly racemized.

The racemization can be taken as evidence of carbonium ion formation. Since the ions are planar (the electron-deficient atom is sp^2-hybridized), the nucleophile may attack equally well from either side of the plane, producing

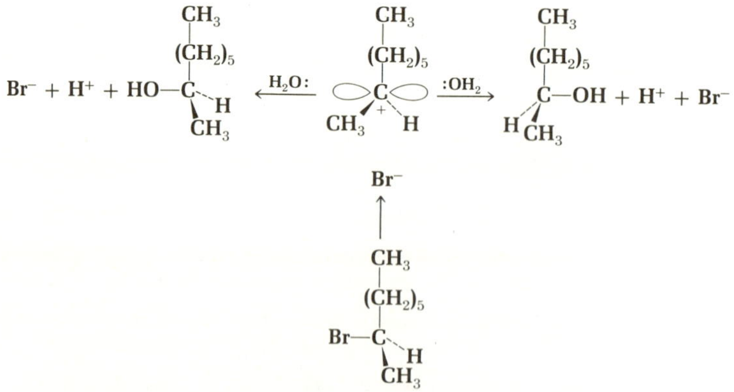

FIGURE 7-15
Racemic 2-Octanol from Optically Active 2-Octyl Bromide

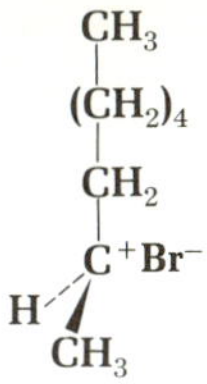

FIGURE 7-16
Octyl Bromide Ion Pair

the two enantiomers in equal amounts (Figure 7-15). Obtaining racemic products from optically active reactants is often used as a criterion of an S_N1 mechanism. Even in pure S_N1 reactions, however, one does not always obtain completely racemic product. The anion and cation may not be completely independent of one another, but they remain in close proximity as an **ion pair** (Figure 7-16), resulting in a shielding of one side of the carbonium ion.

It must be emphasized at this point that the great majority of carbonium ions are relatively high energy, reactive species. Under normal reaction conditions they cannot be obtained in sufficient concentration in solution to be observed spectroscopically. Carbonium ions are reasonably stable, however, in strongly ionizing, non-nucleophilic media; *t*-butyl cation may be preserved in concentrated sulfuric acid. The nmr spectra of a number of carbonium ions have been obtained in solution in very strong acids.

CYCLOALKYL COMPOUNDS – NONCLASSICAL IONS

The principles of nucleophilic substitution that apply to alkyl derivatives are found also to operate for cycloalkyl compounds. Substitution under S_N2 conditions takes place stereospecifically (equation 7.49), while S_N1-type reactions usually produce mixtures. Ring size is also important, and for S_N2 reactions it has been found that cyclopropyl < cyclobutyl < cyclopentyl in rate of reaction.

(7.49)
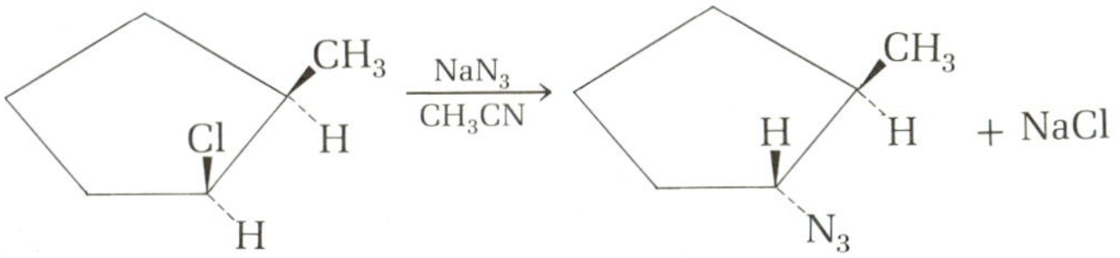

The order of reactivities is undoubtedly associated with the relative difficulty in achieving, at the transition state, bond angles of 120° around the atom undergoing substitution. The same order of reactivities is observed in S_N1 reactions since here again the three atoms attached to the reacting carbon atom must all attain coplanarity and 120° bond angles.

Some fascinating and controversial results have been obtained in studies of nucleophilic substitution. 1-Bromonorbornane (**8**) reacts with water only under extreme conditions.

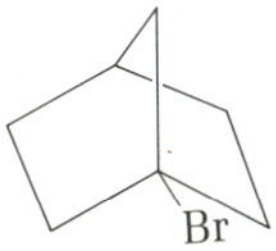

8

The distortion of the ring system necessary to create a planar carbonium ion makes the S_N1 transition state very high in energy; on the other hand, it is impossible for the nucleophile to approach the rear of the carbon atom, which precludes S_N2 substitution.

The 2-halonorbornanes behave very differently from the 1-isomers. *exo*-2-Chloronorbornane (**9**) is more reactive than one would expect of a simple secondary chloride.

(7.50) 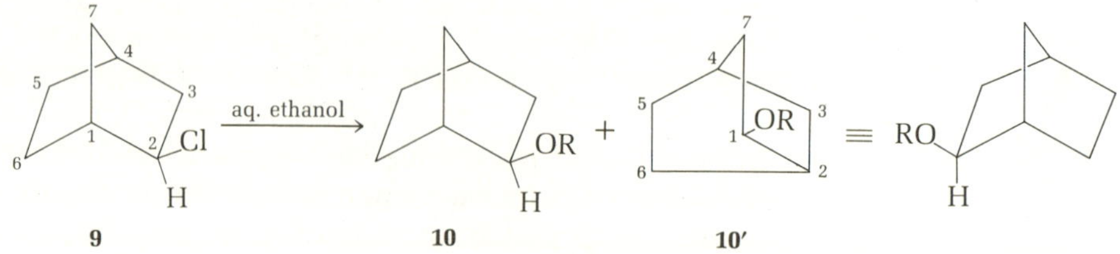

Furthermore, it undergoes an interesting rearrangement of a typical carbonium ion sort, in which the carbon atoms labeled 2 and 6 in the reactant become bonded, and the nucleophile attaches itself to carbon atom 1. This can be shown either by having a radioactive carbon atom at one of the crucial positions or by starting with optically active **9.** Reaction of a single enantiomer of **9** results in the formation of both enantiomers of the product **10** in equal amounts (equation 7.50, R = H or C_2H_5).

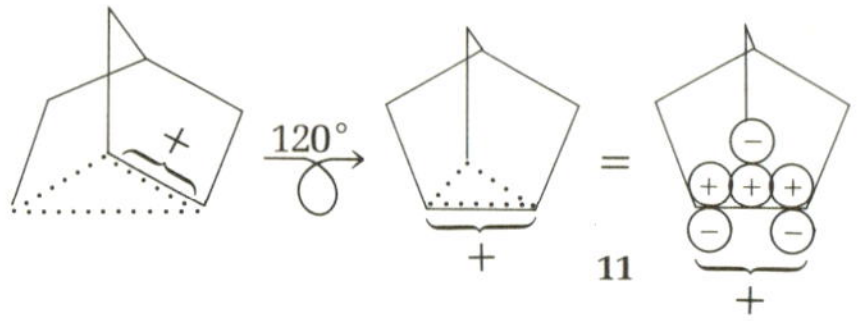

FIGURE 7-17
Dotted Line and Orbital Representations of a Non-classical Norbornyl Cation

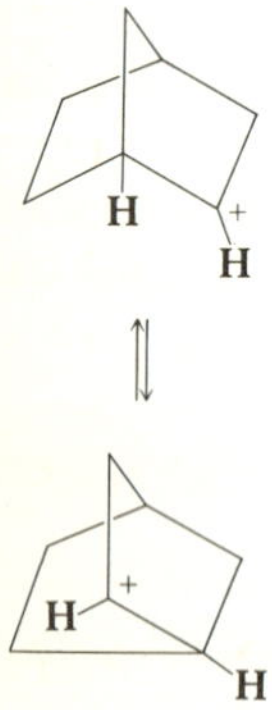

FIGURE 7-18
Equilibrating Classical Norbornyl Cations

In order to explain the rearrangement and the abnormal reactivity, it has been proposed that, as the chloride ion departs, part of the electron density from the 1-6 bond flows toward carbon atom 2 to partially relieve the charge. Eventually an intermediate is formed (**11**) in which carbon 6 is half-bonded to carbons 1 and 2. This ion has been termed a "nonclassical" carbonium ion. Figure 7-17 shows an orbital picture, obtained by rotating the dotted line representation 120° clockwise. An atomic orbital from each of the three carbons overlaps to form a molecular orbital containing a pair of electrons. It is the notion of three atoms sharing one pair of σ-electrons that accounts for the name "nonclassical;" compare the proposed intermediates in the bromination of cyclopropane (Chapter 6).

This hypothesis has not been universally accepted, and it has been maintained that all the data can be accounted for by postulating a pair of "classical" carbonium ions rapidly equilibrating by means of 1-2 bond shift (Figure 7-18). While on balance there seems to be strong evidence in favor of the nonclassical ion, the question is by no means settled. The continuing controversy has in fact been very fruitful, for in the course of attempting to settle this question chemists have performed many ingenious experiments and have learned a great deal about the chemistry of these and related systems.

NUCLEOPHILIC DISPLACEMENTS AT OTHER ELEMENTS

Although attention has naturally been focused on nucleophilic displacements at carbon, a vast number of other elements serve as targets for nucleophilic attack. Silicon has already been mentioned. As samples we list the formation of hydrazine from ammonia and chloramine plus the oxidation of iodide ion by hydrogen peroxide in the presence of acid (which probably involves nucleophilic displacements at oxygen and at iodine).

(7.51) $$\ddot{N}H_3 + \ddot{N}H_2{-}Cl \longrightarrow \overset{+}{N}H_3\ddot{N}H_2\ Cl^- \xrightarrow{NH_3} \underset{\text{hydrazine}}{\ddot{N}H_2{-}\ddot{N}H_2} + NH_4Cl$$

(7.52) $$HO{-}OH \xrightarrow{H^+I^-} HO{-}OH_2^+ \longrightarrow H_2O + HOI$$
$$:I^-$$

(7.53) $$HOI \xrightarrow{H^+I^-} I{-}OH_2^+ \longrightarrow I_2 + H_2O$$
$$:I^-$$

SUMMARY

1 Alkyl halides are nonpolar compounds having much in common physically with the alkanes. The corresponding alcohols, on the other hand, are quite polar and have high boiling points for their molecular weights. The anions of the alcohols are useful as bases and nucleophiles. Alcohols give rise to characteristic ir and nmr spectra. The ir spectra provide evidence of extensive hydrogen bonding among alcohol molecules, and nmr spectra suggest rapid exchange of hydroxyl protons among these molecules.

2 Except for the fluorides, alkyl halides are widely employed as synthetic intermediates since the halogen atom may be replaced by a variety of nucleophiles. Iodides are most reactive, chlorides are least. The nucleophiles are usually anions although several common ones are neutral molecules, such as water. They all possess unshared electron pairs.

3 The OH group in alcohols is too firmly bound to be replaced directly; it is necessary to attach other groups to the oxygen to convert the OH to a

better leaving group. This hindrance is commonly overcome by heating the alcohol with acid in the presence of the nucleophile. Phosphorus and thionyl halides convert alcohols to phosphorus and sulfur derivatives that are readily displaced by halide ions. Another useful method involves converting the alcohol to a tosylate, which may then be readily replaced.

4 In an S_N2-type displacement the nucleophile replaces the leaving group in one continuous step; when reaction takes place at an asymmetric carbon, the result is always stereochemical inversion. This mechanism is operative in the reactions of 1° and 2° alkyl derivatives but only rarely with 3° derivatives. Nucleophiles tend to be more reactive in dipolar aprotic solvents than in protic ones, particularly those most susceptible to hydrogen bonding.

5 Tertiary alkyl derivatives usually undergo substitution by way of an S_N1 mechanism that involves slow ionization of the organic compound to a carbonium ion. The ion subsequently reacts rapidly with the available nucleophiles. If an optically active compound undergoes S_N1 displacement, the product will always be at least partially racemic. Carbonium ions may rearrange to isomeric carbonium ions of equal or greater stability by way of alkyl or hydrogen shifts.

6 Compounds possessing an asymmetric carbon atom may exist in right- and left-hand forms, called enantiomers. Enantiomers are identical in physical properties except for the direction in which they rotate the plane of polarization of polarized light. A mixture containing equal quantities of two enantiomers is said to be racemic and does not affect polarized light. A mixture in which one or the other enantiomer predominates will be optically active (rotate the plane of polarized light).

In order to gain information on the stereochemistry of a reaction, it is necessary to know the absolute configurations of products and reactants. Absolute configurations may be assigned unambiguously by means of a system that ranks the substituents attached to the asymmetric atom on the basis of the atomic weight of the attached atom.

PROBLEMS

1 Explain each of the following observations.

a. For a given concentration the relative size of the hydrogen-bonding band is much larger in the ir spectrum of *n*-butyl alcohol than it is in the spectrum of *t*-butyl alcohol.

b. The ir spectrum of a dilute solution of *trans*-cyclopentane-1,2-diol shows only one peak in the —OH stretching region. A spectrum of *cis*-cyclopentane-1,2-diol at the same concentration exhibits two peaks in this region. The relative intensities of these two peaks do not change with concentration.

2 Draw the nmr spectrum of a mixture of ethanol and methanol containing a drop of acid. Label the peaks.

3 a. Anhydrous potassium hydroxide dissolves easily in dry ethanol. What species will be present in the solution?

b. Write an expression for the equilibrium constant that relates to the equilibrium among these species.

c. The ionization constant for ethanol is about 10^{-16}. Assuming that the ionization constant of water is 10^{-14} under these conditions, calculate the value of the equilibrium constant from part b.

4 Draw the structures of all the alcohols containing five carbon atoms and name them all by the IUPAC system. Classify them as primary, secondary, or tertiary alcohols.

5 How would you perform the following conversions? Specify all reagents.

a. Ethyl bromide $\longrightarrow$ ethyl iodide

b. Isopropyl chloride $\longrightarrow$ isopropyl azide

c. Ethyl chloride $\longrightarrow$ diethyl ether ($CH_3CH_2OCH_2CH_3$)

d. Cyclobutyl tosylate $\longrightarrow$ cyclobutylamine 

e. Isobutyl iodide $\longrightarrow$ isobutyl thiol

f. Cyclopentane $\longrightarrow$ cyclopentyl cyanide

6 A sample of 2-octanol had $\alpha_D = -9.9°$. When a portion of this alcohol was allowed to react with sodium and the salt then treated with ethyl iodide, ethyl 2-octyl ether was isolated that had $\alpha_D = -17.5°$. A second portion of 2-octanol was treated with tosyl chloride in base that led to the isolation of 2-octyl tosylate, $\alpha_D = -6.8°$. When this tosylate is solvolyzed in dry ethanol, ethyl 2-octyl ether is produced that has $\alpha_D = +17.5°$. What conclusions can *rigorously* be drawn (without unfounded assumptions) from these experiments?

7 If (+)-2-iodooctane is allowed to stand in a solution of sodium iodide in acetone, it is eventually converted to racemic 2-iodooctane. Explain this result. Why is the product not (−)-2-iodooctane?

8 Methyl iodide reacts with radioactive iodide ions ($^{131}I^-$) to yield radioactive methyl iodide (equation 1). The rate of formation of radioactive iodide is proportional to the concentration of methyl iodide and to the concentration of ^{131}I.

$$CH_3I + {}^{131}I^- \rightleftharpoons CH_3{}^{131}I + I^- \quad (1)$$

It would seem plausible that this is a one-step, concerted displacement. A two-step mechanism, consistent with the kinetic data, may also be envisaged for this reaction.

$$CH_3I \xrightarrow{\text{fast}} CH_3^+ + I^- \qquad (2)$$

$$CH_3^+ + {}^{131}I^- \xrightarrow{\text{slow}} CH_3{}^{131}I \qquad (3)$$

This mechanism may be immediately rejected. Why? (Remember the principle of microscopic reversibility, Chapter 4.)

9 If 1-fluoro-1-iodoethane is treated with sodium azide ($Na^+N_3^-$) in DMF, 1 mole of a sodium halide is obtained. What is the structure of the organic product? If the *R* enantiomer of the starting material had been used, what would have been the absolute configuration of the product?

10 An alcohol cannot be converted to an alkyl cyanide by heating it with a solution of sodium cyanide and acid.

a. Explain this observation.

b. Suggest two routes, each proceeding by way of stereospecific reactions, for converting 2-pentanol to 2-pentyl cyanide.

c. Suppose that you had used (*R*)-2-pentanol in part b. What would be the absolute configuration of the product obtained by each route?

11 Describe methods for effecting the following conversions. Specify all reagents and conditions.

a. 1-Butanol ⟶ 1-bromobutane (two ways)

b. Methanol ⟶ methanethiol

c. Cyclohexanol ⟶ cyclohexyl chloride

d. *trans*-3-Methylcyclopentanol ⟶ *trans*-3-methylcyclopentyl azide

e. (*S*)-3-Heptanol ⟶ (*S*)-3-chloroheptane

12 Thiosulfate ion (I) and sulfite ion (II) belong to the category of *ambident nucleophiles*, i.e., species that have more than one center of nucleophilic activity.

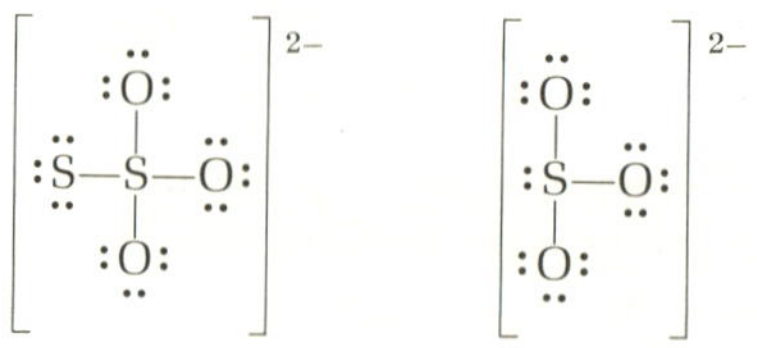

I II

For the cases cited, alkylation could take place at either oxygen or sulfur.

a. When ethyl bromide is allowed to react with sodium thiosulfate in aqueous solution, sodium ethylthiosulfate (***A***) is obtained (equation 4). Hydrolysis of ***A*** in hot aqueous acid gives sodium bisulfate and an organic product (***B***).

$$CH_3CH_2Br + Na_2S_2O_3 \longrightarrow \textbf{\textit{A}} \xrightarrow[\text{heat}]{H_3O^+} \textbf{\textit{B}} + NaHSO_4 \qquad (4)$$

B is identical to the compound obtained from the reaction of ethyl bromide with sodium hydrosulfide. What are the structures of ***A*** and ***B***? Propose a mechanism for hydrolysis of ***A***.

b. Ethyl bromide and sodium sulfite combine in aqueous solution to form sodium ethanesulfonate (***C***).

$$CH_3CH_2Br + Na_2SO_3 \longrightarrow \textbf{\textit{C}} + NaBr \qquad (5)$$

C may also be obtained by oxidizing ethanethiol with nitric acid and then neutralizing the mixture with sodium hydroxide. What is the structure of ***C***?

c. Rationalize the results presented in parts a and b. Thiosulfate ion and sulfite ion are both quite reactive nucleophiles — on the basis of your findings, how would you expect sulfate ion to compare with them?

13 Two fundamental mechanisms for nucleophilic substitution at a saturated carbon have been discovered, one in which leaving group departure precedes nucleophilic addition and a second in which the two events are concerted. When tetravalent silicon compounds are subjected to nucleophilic substitution, a third mechanism is conceivable. What is it? What properties of silicon permit the operation of this mechanism? What stereochemical results would follow from the operation of this mechanism?

$$Y\!:^- + R_3SiX \longrightarrow R_3SiY + X\!:^- \qquad (6)$$

14 Write out the structure of the product(s) from each of the following reactions, showing its (their) absolute configuration.

a. (*S*)-3-Chloroheptane $\xrightarrow[\text{DMSO}]{\text{KCN}}$

b. (*R*)-3-Methyl-3-hexanol $\xrightarrow[\text{heat}]{\text{HCl, } H_2O}$

c. (*R*)-3-Methyl-3-hexanol $\xrightarrow[\text{base}]{\text{TsCl}}$ ***A*** $\xrightarrow[\text{reflux}]{\text{aq. ethanol}}$

d. (*S*)-3,3-Dimethyl-2-butanol $\xrightarrow{PI_3}$

e. (*S*)-3,3-Dimethyl-2-butanol $\xrightarrow[H_2SO_4\text{, heat}]{\text{NaBr, } H_2O}$

f. 1-Iodo-*cis*-1,4-dimethylcyclohexane $\xrightarrow{\text{aq. reflux}}$

g. (*R,S*)-2-Butyl tosylate $\xrightarrow[\text{DMSO}]{\text{NaBr}}$

15 (+)-2,2-Dimethyl-3-hexanol is treated with concentrated hydrochloric acid at 25°. A small quantity (6%) of (−)-3-chloro-2,2-dimethylhexane is formed along with large quantities of 2-chloro-2,3-dimethylhexane and 3-chloro-2,3-dimethylhexane, the latter two being approximately equal in quantity and both racemic. Write out a mechanism that accounts for the formation of all the products.

16 In the presence of acid catalysts, straight-chain primary alcohols tend to give unrearranged substitution products. If there is a branch at the carbon atom *next to* the primary carbon atom (the β carbon), as in isobutyl alcohol, then rearrangement becomes much more probable. Neopentyl alcohol (2,2-dimethylpropanol) is notorious for the ease with which it rearranges upon attempted conversion to neopentyl halides (equation 7.30 is a noteworthy exception). If the hydrocarbon branch is further away, then the tendency toward rearrangement fails. How would you explain these findings?

17 In ethanol containing a little water, isopropyl chloride solvolyzes more rapidly than di-*t*-butylcarbinyl chloride (3-chloro-2,2,4,4-tetramethylpentane). The major product from the solvolysis of isopropyl chloride is ethyl isopropyl ether, but di-*t*-butylcarbinyl chloride gives rise to rearranged products.

When the solvent is 2,2,2-trifluoroethanol, it is found that isopropyl tosylate solvolyzes considerably more *slowly* than di-*t*-butylcarbinyl tosylate does. Indicate the mechanisms of reaction in the isopropyl and di-*t*-butylcarbinyl systems, and explain the solvent effect.

18 When alkanes are bubbled through hot solutions of very strong acids, such as HF-SbF_5 (called "magic acids"), reaction takes place. The introduction of methane into a "magic acid" results in the evolution of hydrogen gas and the formation of multicarbon species, particularly *t*-butyl cation (identified by nmr spectroscopy). It is believed that a vital intermediate in these reactions is CH_5^+, a species shown by mass spectrometry to be stable in the gas phase.

a. Draw a structure for CH_5^+ showing its geometry and its orbital structure.

b. It is thought that CH_5^+ loses H_2 and is converted to CH_3^+. Suggest a mechanism, based on the chemistry just discussed, whereby higher carbonium ions might be formed from CH_3^+.

SUPPLEMENTARY READINGS

J. O. Edwards, "Bimolecular Nucleophilic Displacement Reactions," *J. Chem. Ed.*, **45,** 386 (1968).

M. S. Puar, "Nucleophilic Reactivities of the Halide Ions," *J. Chem. Ed.*, **47,** 473 (1970).

D. F. Mowery, Jr., "Criteria for Optical Activity," *J. Chem. Ed.*, **46,** 269 (1969). Some of this article will be beyond the beginner, but the first part gets directly to the heart of the matter.

W. H. Saunders, Jr., *Ionic Aliphatic Reactions*. Englewood Cliffs, N.J.: Prentice-Hall, Inc., 1965, Chapter 3. (P)

L. B. Clapp, *The Chemistry of the OH Group*. Englewood Cliffs, N.J.: Prentice-Hall, Inc., 1967, Chapters 1 and 2. (P)

CHAPTER 8

ALKENES I—SYNTHESIS AND GEOMETRICAL ISOMERISM

NOMENCLATURE

This chapter will be concerned with the chemistry of compounds possessing a carbon-carbon double bond. The generic IUPAC name for these compounds is alk**enes,** and the presence of the suffix "ene" in the name of any compound indicates that it contains one or more such double bonds. When positional isomerism is possible, the location of the double bond is indicated by a numerical prefix; the number refers to the carbon atom carrying the smaller numerical value. The alkene chain is numbered so that the double bond is denoted by the smallest possible prefix; the root of the name is based upon the longest chain *containing the double bond.* The system is easily grasped with the aid of a few examples. If a molecule contains both a double bond and a hydroxyl group, it is named as an unsaturated alcohol (the name ends in "ol"), and the hydroxyl-bearing carbon atom is accorded the lowest number.

$CH_2{=}CH_2$
ethene

$\overset{1}{C}H_2{=}\overset{2}{C}\overset{3}{H}\overset{4}{C}H_2CH_3$
1-butene

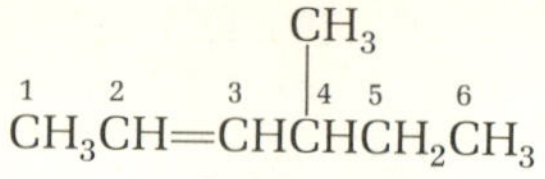
4-methyl-2-hexene

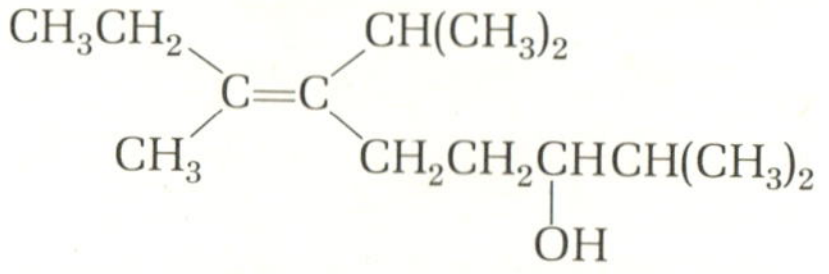
2,7-dimethyl-6-isopropyl-6-nonen-3-ol

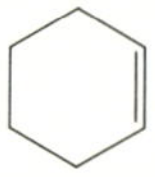
cyclohexene

As one might expect, there is a widely used semisystematic nomenclature for alkenes which is most often applied to the lower members of the series. Thus we have ethylene, propylene, the butylenes, etc. More highly branched alkenes are named as derivatives of ethylene, where the "derivative" is obtained by replacement of the hydrogen atoms on the doubly bonded carbon atoms by other groups. Some examples are listed below, with their IUPAC names in italics.

The pre-IUPAC generic name for alkenes was **olefins** because some common chemicals synthesized from ethylene were rather oily. At present the terms olefin and alkene are used interchangeably for the most part. Substances containing double bonds are also classified as unsaturated because they are capable of taking up hydrogen.

$CH_3CH{=}CH_2$
propylene
propene

$(CH_3)_2C{=}CH_2$
α,α-dimethylethylene
2-methylpropene

$(CH_3)_2C{=}CHCH_3$
trimethylethylene
2-methyl-2-butene

$CH_3CH{=}CHCH_2CH_3$
α-methyl-β-ethylethylene
2-pentene

$(CH_3)_2CHCH{=}C(CH_3)CH(CH_3)_2$
α-methyl-α,β-diisopropylethylene
2,3,5-trimethyl-3-hexene

BONDING AND GEOMETRICAL ISOMERISM

The π-bond of alkenes derives from the side-by-side overlap of two atomic *p*-orbitals in a bonding manner (Chapter 2). As Figure 8-1 illustrates, this overlap will be maximal when the axes of the *p*-orbitals are exactly parallel. As one orbital is rotated with respect to the other, their overlap decreases and the energy of the system rises; the overlap is minimal and the system most unstable when the angle between the orbital axes is 90°. For this reason, there is a barrier to rotation about double bonds of approximately 60 kcal/mole in simple alkenes. In contrast therefore to single bonds, double

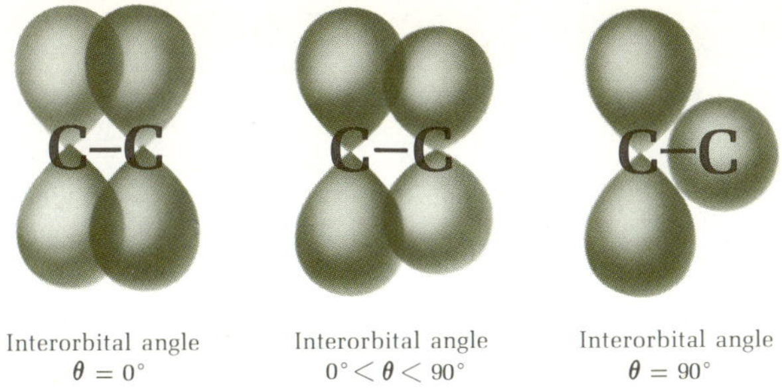

FIGURE 8-1
Rotational Angle and Orbital Overlap in Ethylene

bonds are essentially rigid at room temperature, and the geometry of the groups attached to them is fixed.

If the two groups attached to each carbon atom of the double bond are different, then two possible compounds might result, differing only in the arrangement of substituents about that bond (Figure 8-2). These compounds are **geometrical isomers** and they are differentiated by the terms *cis* and *trans*. In the case where there is a hydrogen atom on each of the doubly bonded carbon atoms, the *cis* isomer is the one with the hydrogen atoms on the same side of the double bond. More generally, for the purpose of naming, one picks out the longest continuous chain containing the maximum number of double bonds. Then the designation of double bond geometry simply depends on whether the groups defining the chain are on the same (*cis*) or opposite (*trans*) sides. Alkenes in which the double bond is at the end of the chain (terminal alkenes) cannot exhibit *cis-trans* isomerism.

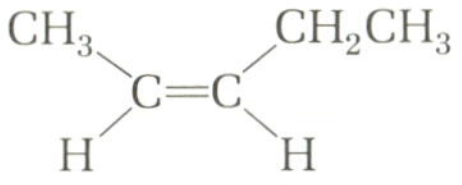

cis-2-pentene

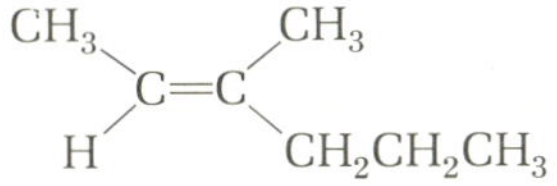

3-methyl-*trans*-2-hexene

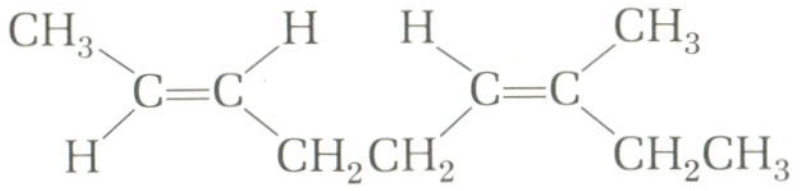

7-methyl-*trans*-2,*cis*-6-nonadiene

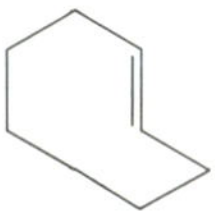

trans-cycloheptene

Cis and *trans* isomers have different physical properties and can often be separated from one another by conventional techniques — distillation, gas chromatography, etc. In this respect they are quite different from enantiomers. It might be useful at this point to comment on the relationship between *conformers* and *geometrical isomers*. Conformers are different spatial arrangements of a given molecule that may interconvert solely by way of rotations around bonds. It can immediately be seen that *cis*- and *trans*-2-pentene, for example, may be transformed one into the other by rotations about the double bond. Due to the high energy barrier, this rotation

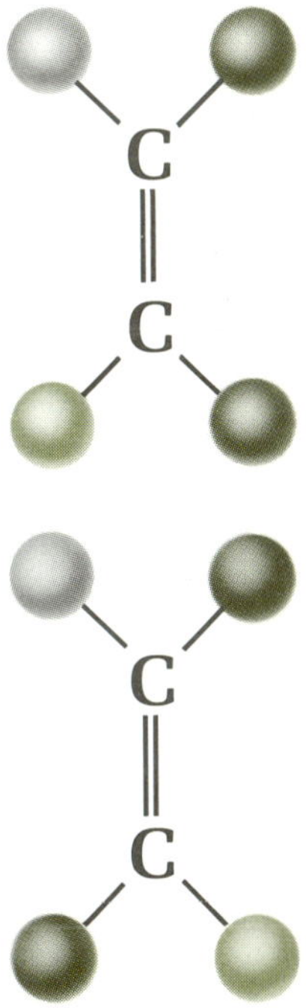

FIGURE 8-2
Geometrical Isomerism in Alkenes

TABLE 8-1
Physical Properties of Some Alkenes

	Melting Point (°C)	Boiling Point (°C)
Ethylene	−169	−104
Propylene	−185	−48
1-Butene	−185	−6.3
cis-2-Butene	−139	3.7
trans-2-Butene	−106	0.9
1-Octene	−102	121
1-Dodecene	−35	213

is negligible at room temperature and the two isomers may be isolated and kept in the pure state for very long periods of time. This is certainly not true of, say, the two different staggered conformations of 1,2-dichloroethane (*gauche* and *anti*). If one worked at sufficiently low temperatures, however, they would be conformationally stable and separable. Conversely, if room temperature were several hundred degrees hotter than it is, most *cis-trans* alkene isomers would not be separable since rotation about the double bond would be fairly rapid. The applications of the terms conformer and isomer to a given system is then seen to be a temperature-dependent procedure.

Melting points may diverge widely for a pair of *cis-trans* isomers because their different shapes will strongly influence how well they pack into a crystal lattice (Table 8-1). Their boiling points are much closer to one another because the isomers have identical molecular compositions.

By the same token, the boiling points of almost all alkenes are very close to those of the corresponding alkanes. For this reason quantitative separation of alkenes and alkanes of the same or similar molecular weight is not possible by distillation. Gas chromatography has proved invaluable for effecting these separations, however. There are several column packings known that interact much more strongly with alkenes than with alkanes and thereby slow the progress of the alkene through the column.

Among the most effective column packings are solutions of silver nitrate in certain organic solvents, which are deposited on the inert support. Silver ion complexes with the π-bond of olefins (Chapter 9) while remaining unreactive to alkanes. The stability of the complex is very sensitive to the geometry of the double bond, and it is possible to separate positional and even *cis-trans* isomers.

SPECTROSCOPIC PROPERTIES OF ALKENES

Alkene isomers have similar spectroscopic properties but there are certain differences that are extremely useful in assigning a *cis* or *trans* configuration to the molecule. The infrared stretching frequencies of the hydrogens bound to the sp^2-hybridized carbon atoms lie between 3000 and 3100 cm^{-1}, distinguishing them from absorptions due to the sp^3-bound ones that occur below 3000 cm^{-1}. The carbon-carbon stretching frequency for the double bond lies in the region 1600 to 1680 cm^{-1}. The peak is normally very weak due to the lack of a substantial polarity difference between the two carbon atoms. The carbon-carbon stretching band is totally absent in the infrared spectrum of ethylene since this motion causes no change at all in the dipole moment of the molecule. At lower frequencies the bending motions of the C—H bonds are found. These divide into two types: bending of the C—H bonds within the plane of the double bond and bending out of that plane. In-plane vibrations occur within the region 1280 to 1450 cm^{-1}; consistently strong absorption is usually found around 1400 cm^{-1} and is of value for confirming the presence of sp^2-bound hydrogens.

The out-of-plane vibrations are the most useful because the band position can be correlated with the stereochemistry around the double bond. α,β-*trans*-Disubstituted ethylenes give rise to infrared bands in the 965 to 990 cm^{-1} region, while the *cis* isomers show bands near 690 cm^{-1} (usually 675 to 730 cm^{-1}, but higher frequencies are known). α,α-Disubstituted ethylenes have a band near 890 cm^{-1} in their spectra, characteristic of the grouping $=CH_2$; this band appears at about 910 cm^{-1} in monosubstituted ethylenes. Most of these assignments are fairly reproducible and provide a reliable method of assigning stereochemistry, which is rather difficult to do by chemical means.

Not unexpectedly, the protons on the double bond have different nmr chemical shifts from those on saturated carbon atoms. Carbon atoms that are sp^2-hybridized are slightly more electronegative than sp^3-hybridized ones, and the hydrogen atoms attached to them are accordingly more deshielded. Proton absorptions arising from double bonds are found between 4.5 and 7.0δ, depending on the substituents; they are normally readily distinguishable from those due to alkane hydrogens (Figure 8-4).

FIGURE 8-3
Infrared Spectrum of *trans*-3-Hexene

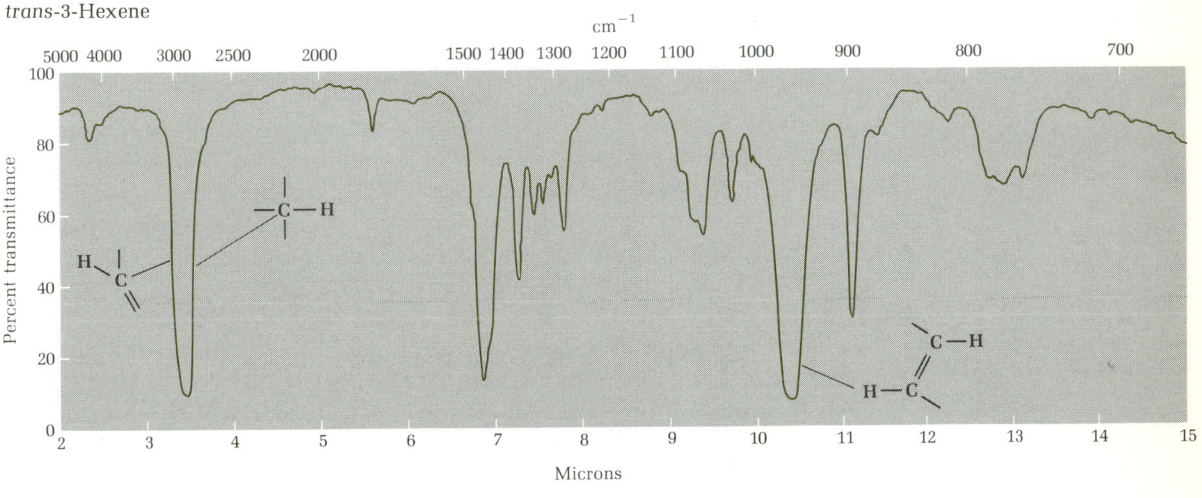

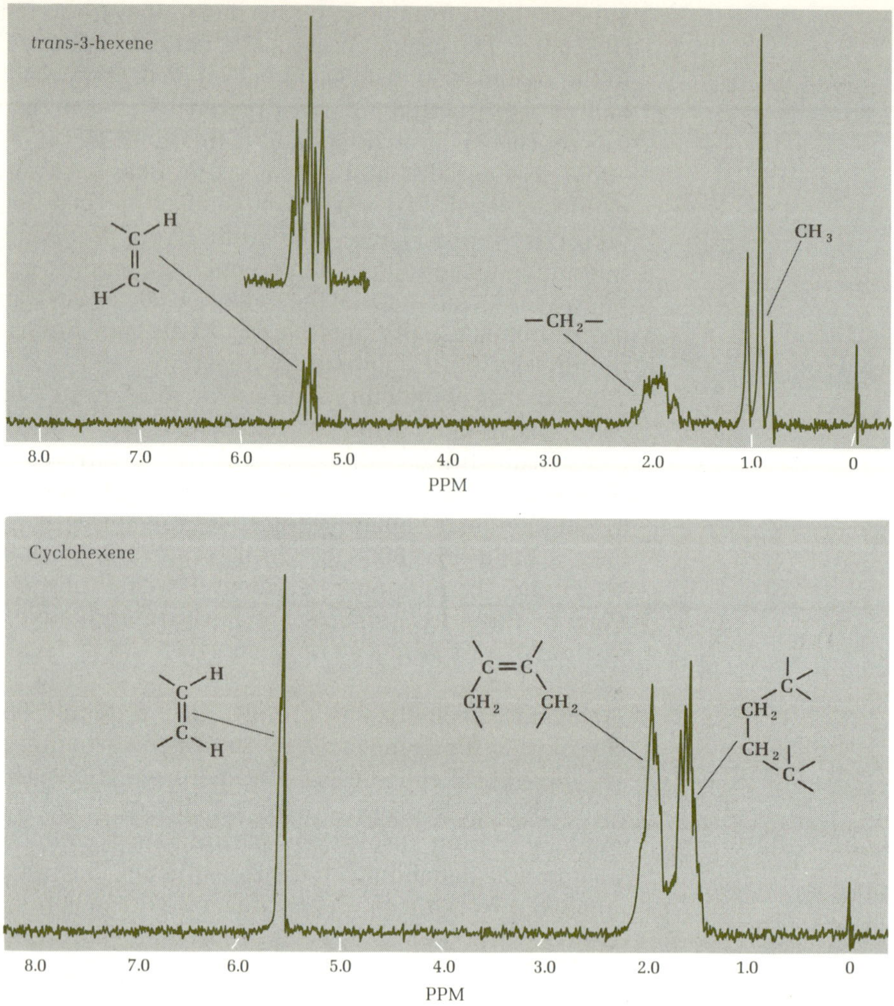

FIGURE 8-4
NMR Spectra of Cyclohexene and *trans*-3-Hexene

Particularly noteworthy is the fact that the degree of proton spin-spin coupling is markedly dependent on the geometry about the double bond (Figure 8-4). α,β-Disubstituted ethylenes with *trans* hydrogen atoms show coupling constants of 11 to 18 cps for the interaction of the protons attached to the double bond. Since the *cis* isomers have coupling constants in the range of 6 to 14 cps, it is usually possible, particularly by bringing in infrared data, to classify *cis-trans* isomers with a good deal of confidence.

The hydrogen atoms at the end of a terminal alkene chain are not equivalent unless the two α-positions are identically substituted; they may therefore show spin-spin coupling. The couplings are usually no more than a few cycles per second, despite the fact that both hydrogen atoms are attached to the same carbon atom.

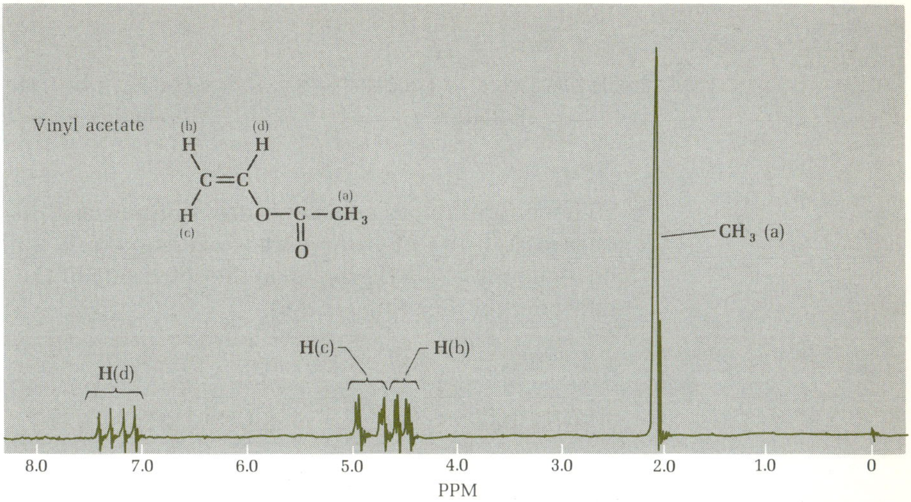

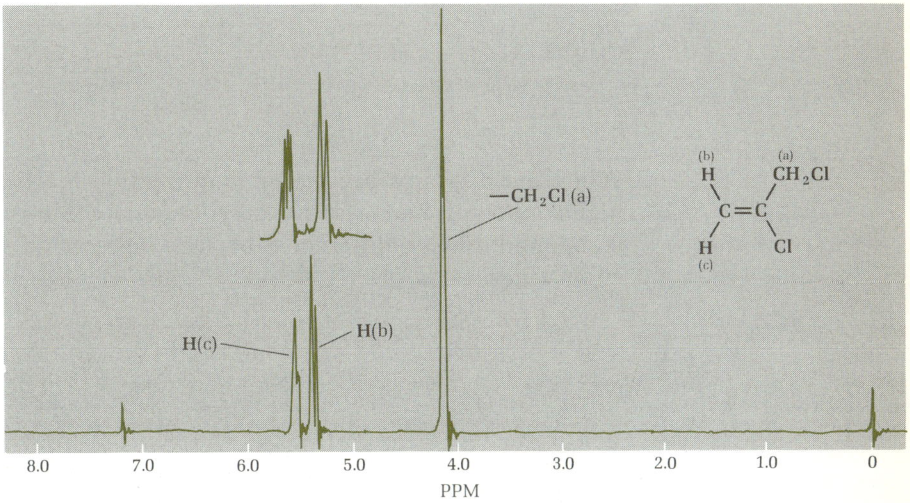

FIGURE 8-5
NMR Spectra of Vinyl Acetate and 2,3-Dichloropropene

ISOMER STABILITY AND *cis-trans* INTERCONVERSION

Alkene isomers that differ in double bond position or geometry differ also in energy. A study of the equilibrium among the various straight-chain heptenes has given some precise data that are probably applicable to most straight-chain alkenes. The actual technique used to establish equilibrium will be discussed later (Chapter 11) but it would be appropriate to note that the alkene mixtures were analyzed by gas chromatography, the column packing consisting of silver nitrate dissolved in ethylene glycol.

(8.1) 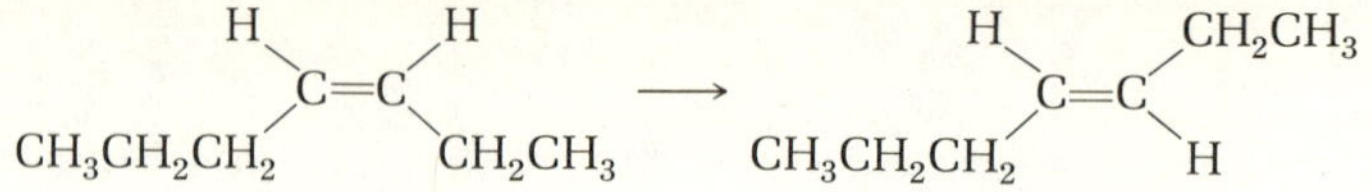

cis-3-heptene → *trans*-3-heptene

$\Delta H = -0.9$ kcal/mole

The important conclusions are summarized in the equations. The increased stability of *trans* versus *cis* isomers is understandable because of the tendency of alkyl groups on the same side of the double bond to interfere with one another (Figure 8-6).

(8.2) 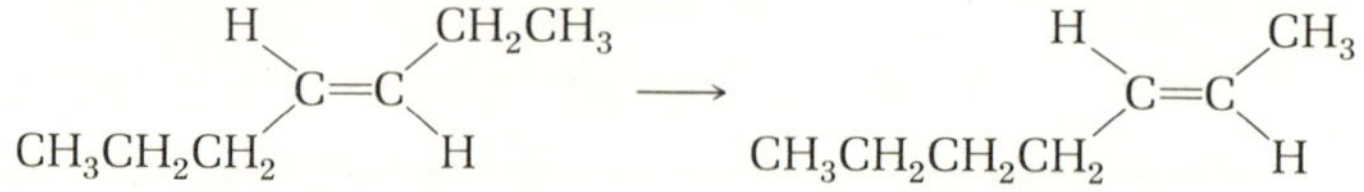

trans-3-heptene → *trans*-2-heptene

$\Delta H = 0$ kcal/mole

(8.3) 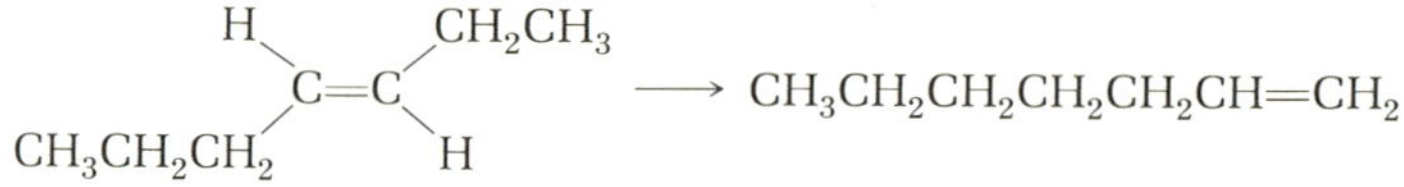

trans-3-heptene → 1-heptene

$\Delta H = -2.7$ kcal/mole

When these groups are larger than methyl, the *trans* isomer also has the higher entropy because the alkyl groups cannot rotate freely in the *cis* compound. Energy differences become more acute as branching of the alkyl chains increases.

(8.4) 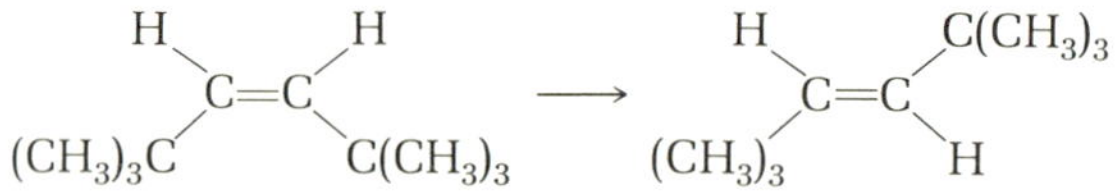

cis-di-*t*-butylethylene → *trans*-di-*t*-butylethylene

$\Delta H = -10.5$ kcal/mole

Internal olefins have a decided energetic advantage over terminal ones in the absence of substantial steric factors. Since the π-electrons are so polarizable, the dispersion forces (Chapter 4) within the alkene molecule will be much stronger than those found in saturated molecules. The greater the number of atoms clustered about the polarizable electrons, the larger are these attractive forces, and the more stable the molecule. At least part of the increased stability of internal olefins is due to dispersion forces.

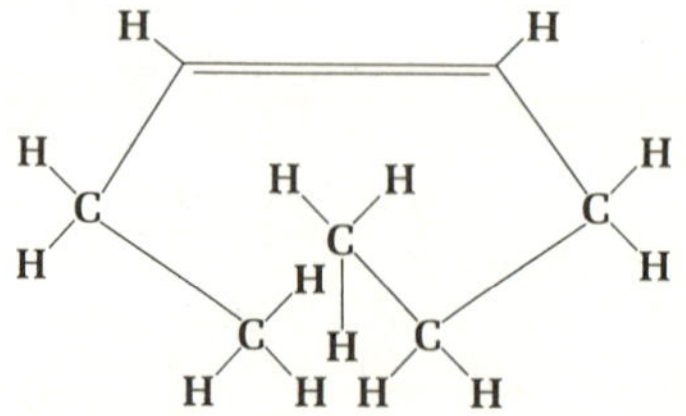

FIGURE 8-6
cis-3-Heptene, Showing Interference Between the Alkyl Groups

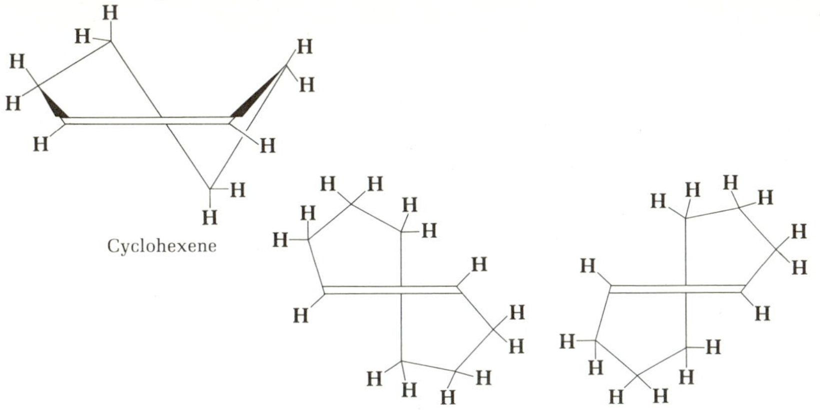

FIGURE 8-7
Preferred Conformation of Cyclohexene and of the Enantiomeric *trans*-Cyclooctenes

Cycloalkenes having fewer than eight carbon atoms in the ring are stable only in the *cis* form at room temperature. The preferred conformation for cyclohexene is the so-called "half chair," in which eclipsing interactions are minimized (Figure 8-7). *trans*-Cycloheptene has been generated as a transient, reactive species but *trans*-cyclooctene is the smallest *trans*-cycloalkene capable of being isolated. The preferred conformation of the ring is such that the entire molecule is dissymmetric, **even though it contains no asymmetric carbon atom.** Two enantiomers of *trans*-cyclooctene are therefore possible, and they have in fact each been obtained in pure form (Figure 8-7). By simply twisting the remainder of the ring around the double bond, one enantiomer may be converted to the other, i.e., racemized. In cyclooctene this motion is hindered by a substantial energy barrier because of the consequent close approach of several atoms. Larger *trans*-cycloalkenes, in which the interconversion is much easier, cannot be obtained in their pure enantiometric forms. (Ten or fifteen minutes spent with a set of molecular models will clarify this entire topic.)

The τ-BOND MODEL

At this point it might be worthwhile to put the σ-π model, which has been relied upon so heavily, into some perspective. Division of the carbon-carbon bonds into two types, σ and π, is a mathematical convenience useful in deriving molecular orbitals for alkenes and other molecules. Since many of the properties predicted by theory are found in the actual compounds, we are justified in considering the σ-π model a good one. It is, however, by no means a unique one. If the σ- and π-orbitals themselves are combined ($\sigma + \pi$, $\sigma - \pi$) in the same manner as atomic orbitals, two new hybrids, the τ-orbitals, are produced. They strongly resemble the bent bonds invoked to explain

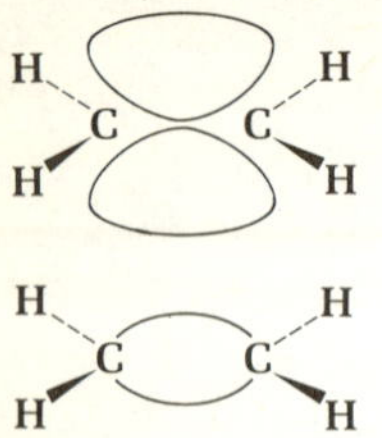

FIGURE 8-8
Orbital and Line Representations of τ-Bonds in Ethylene

the bonding in cyclopropane (Chapter 6), and they point to the double bond as a two-membered ring, which is in line with its physicochemical properties (Figure 8-8). The presence of a rotational barrier in double bonds then follows naturally.

The same electron density distribution is predicted by both the τ and σ-π models. We shall continue to rely on the σ-π formulation because it provides a very convenient framework for discussing many other more complex systems. The student should be nonetheless aware that it has no claim to absolute veracity and that an insurrection by a new group of theoreticians could relegate it to theoretical Siberia.

cis-trans INTERCONVERSION

The rotational barrier in alkenes is by no means insurmountable. In particular, less stable isomers may be largely converted to the more stable ones at appropriate temperatures. Isomer interconversion may also be accomplished photochemically; the evidence suggests that the same mechanism may sometimes operate in both cases.

(8.5)
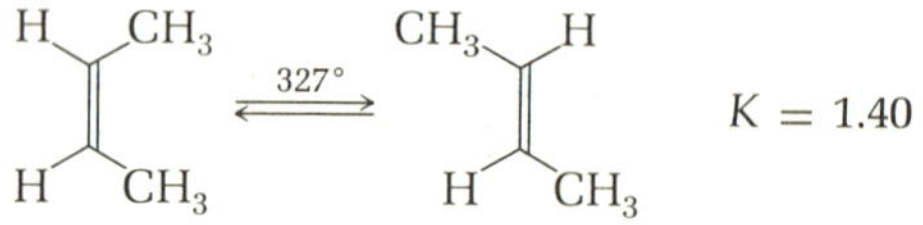

(8.6)
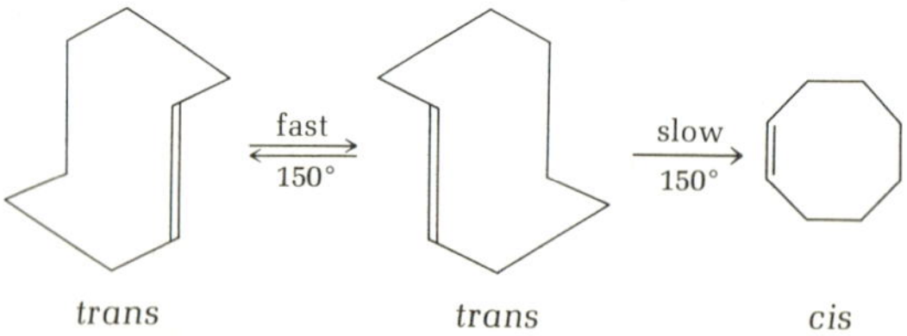

It will be recalled that two atomic *p*-orbitals can combine to produce not only a bonding π-orbital but an antibonding π^*-orbital as well (Chapter 2). Both π-electrons normally occupy the lower energy level. If the molecule acquires sufficient energy from collisions, or if it absorbs a quantum of light the energy of which equals the energy difference between the two orbitals, one electron may be promoted to the antibonding level (Figure 8-9).

(8.7)
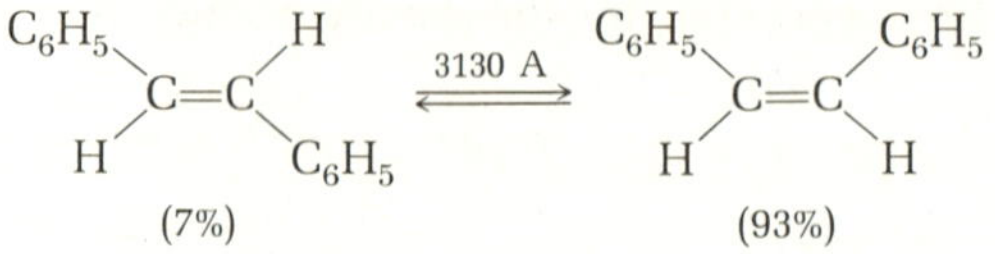

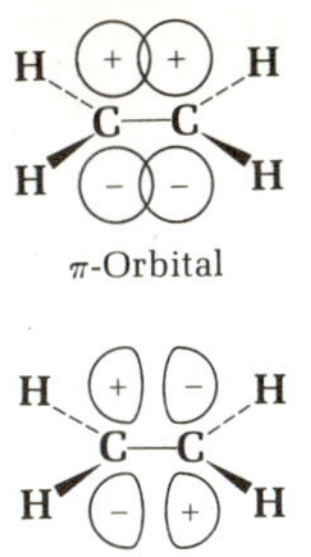

FIGURE 8-9
Atomic Orbital Representation of the π- and π*-Orbitals of Ethylene

In such an excited state the binding energy of the π-bond is essentially lost, and the molecule apparently adopts a configuration with the *p*-orbitals at right angles to one another, to minimize electron repulsions. Collapse of the excited state may then lead to either isomer. Since *cis*- and *trans*-alkenes absorb light of slightly different wavelengths (but generally in the region of 1600 to 1900 A for simple alkenes), it is possible to selectively build up the concentration of the less stable isomer.

The rationalization offered above accounts quite well for the observed phenomena, but it does contain several simplifications. There is not one but many excited states available, which is revealed by calculations using higher atomic orbitals. Even when one electron is in the lowest antibonding orbital, which is the situation under discussion, there are two energy states available to the molecule. As soon as the electrons are in different orbitals, the requirement that their spins remain paired disappears. If they continue to do so, the excited state is known as a **singlet state;** if the spins are unpaired, we have the **triplet state.** The triplet state is lower in energy than the singlet, but it is usually much more difficult to obtain by direct excitation because flipping an electron spin in a molecule is normally a forbidden process. In the presence of heavy atoms such as iodine, transitions to triplet states become much more "allowed." One can also use **sensitizers,** molecules that are capable of being excited to triplet states and that can transfer their excitation energy and triplet character to another molecule. Mercury vapor is a common triplet sensitizer. Most thermal isomerizations seem to proceed by way of the singlet states, but either singlet or triplet states may be involved in photochemical processes.

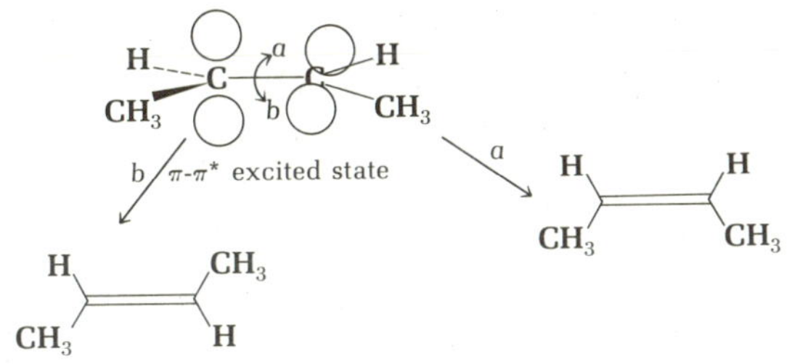

FIGURE 8-10
Excited State of 2-Butene Collapsing to *cis*- or *trans*-2-Butene

COMMERCIAL PRODUCTION OF ALKENES

Because of the ease with which they can be converted to a variety of other compounds containing many different functional groups, olefins are extremely important synthetic intermediates and prime products from petroleum conversion. Ethylene itself is produced by direct thermal cracking of ethane in tubular furnaces.

(8.8) $CH_3CH_3 \xrightarrow{800^\circ} CH_2{=}CH_2 + H_2$

The reaction is strongly endothermic but entropically favorable, so it is carried out at elevated temperatures. It has been shown to be a radical chain process, initiated by methyl radicals (Problem 14, Chapter 5). The methane and hydrogen by-products provide fuel for the plant.

(8.9) $CH_3CH_3 \longrightarrow 2CH_3\cdot$

(8.10) $CH_3\cdot + CH_3CH_3 \longrightarrow CH_4 + \cdot CH_2CH_3$

(8.11) $CH_3CH_2\cdot \longrightarrow H\cdot + CH_2{=}CH_2$ } propagation steps (8.10)–(8.12)

(8.12) $H\cdot + CH_3CH_3 \longrightarrow H_2 + \cdot CH_2CH_3$

This method of alkene production is not suited to higher alkanes due to the intervention of competing reactions that involve C—C bond breaking which requires less energy than C—H bond cleavage.

(8.13) $CH_3CH_2CH_3 \longrightarrow CH_3CH{=}CH_2 + H_2$

(8.14) $CH_3CH_2CH_3 \longrightarrow CH_2{=}CH_2 + CH_4$

It becomes necessary, therefore, to employ catalysts that promote hydrogen abstraction without encouraging C—C breakage.

(8.15) $CH_3CH_2CH_2CH_3 \xrightarrow[500^\circ,\ \text{low pressure}]{Cr_2O_3\text{-}Al_2O_3} \text{butenes} + H_2$

There are a number of important industrial processes that involve "addition" of various reagents to the double bond of alkenes. Most commercial alcohols are obtained this way. Molecules other than water may be added as well. Most of the addition reactions are conducted at relatively low temperatures, as they are exothermic and entropically unfavorable.

(8.16) $CH_2{=}CH_2 + H_2O \xrightarrow{H_3PO_4} CH_3CH_2OH$

(8.17) $CH_3CH{=}CH_2 + H_2SO_4 \longrightarrow CH_3\underset{\substack{|\\ OSO_3H}}{C}HCH_3 \xrightarrow{H_2O} CH_3\underset{\substack{|\\ OH}}{C}HCH_3$

(8.18) $CH_2{=}CH_2 + \tfrac{1}{2}O_2 \xrightarrow[250^\circ]{\text{Ag catalyst}}$ 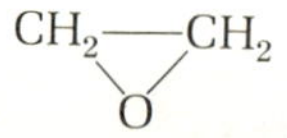

ethylene oxide

(8.19) $C_6H_6 + CH_2{=}CH_2 \xrightarrow[100^\circ]{AlCl_3} C_6H_5CH_2CH_3$

ethylbenzene

(8.20) $C_6H_5CH_2CH_3 \xrightarrow{\text{cat.}} C_6H_5CH{=}CH_2 + H_2$

styrene

Ethylene, propylene, their oxides, and styrene are all important starting materials for polymer production. These polymers may wind up as plastics, as artificial rubber, etc. Other ethylene derivatives are converted to synthetic fibers. Manufacture of polymers and their precursors represents one of the most lucrative branches of the American chemical industry.

FORMATION OF ALKENES BY ELIMINATION

Synthesis of alkenes by direct removal of H_2 is not a generally useful laboratory procedure. But **elimination** of other molecules of the type X—Y, from reactants in which the groups X and Y occupy *adjacent* carbon atoms, provides the most general route to alkenes. These elimination reactions are frequently the reverse of **additions** to the double bond (Chapter 9). Of the molecules eliminated in the course of double bond creation, the most common are water, hydrogen halides, *p*-toluenesulfonic acid, and ammonium salts.

(8.21)

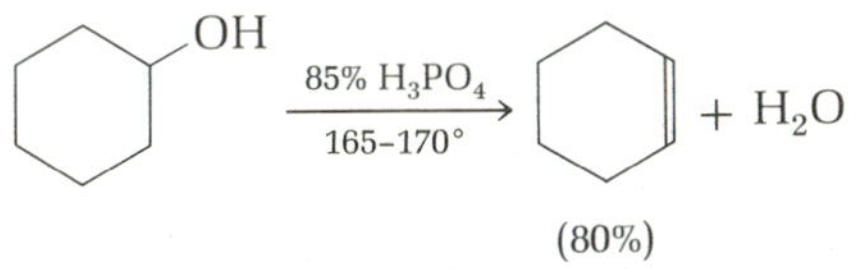

(8.22)

$$BrCH_2\underset{\displaystyle Br}{\underset{|}{C}}HCH_2Br + NaOH \xrightarrow{H_2O} CH_2{=}\underset{\displaystyle Br}{\underset{|}{C}}{-}CH_2Br + NaBr + H_2O$$

(74–84%)

(8.23)

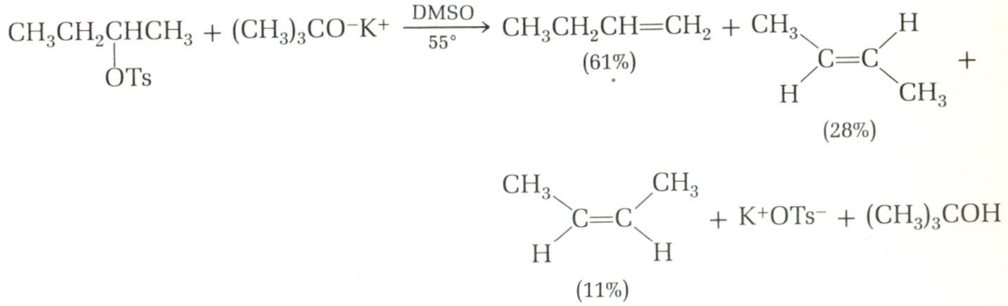

(8.24)

$$CH_3CH_2CH_2CH_2\overset{+}{N}(CH_3)_3\ OH^- \xrightarrow[130°]{\text{conc. aq. soln.}} CH_3CH_2CH{=}CH_2 + (CH_3)_3N{:} + H_2O$$

(85%)

Removal of water requires acid catalysis, while elimination of the acidic components proceeds most readily in the presence of base.

There are two fundamental mechanisms operative in elimination reactions; they are related to, and compete with, the S_N1 and S_N2 nucleophilic displacement reactions. In an *E*1 (elimination, unimolecular) reaction the slow step involves the ionization of the organic compound. The carbonium ion so formed may capture a nucleophile (S_N1) or lose a proton to produce an alkene (*E*1).

(8.25)
$$(CH_3)_3CBr \xrightarrow[\text{slow}]{C_2H_5OH} Br^- + (CH_3)_3C^+ \xrightarrow[S_N1]{C_2H_5OH} (CH_3)_3COCH_2CH_3 + H^+$$
$$(CH_3)_3C^+ \xrightarrow[E1]{C_2H_5OH} (CH_3)_2C{=}CH_2 + CH_3CH_2OH_2^+$$

The *E*2 (elimination, bimolecular) mechanism requires that both the base and the molecule undergoing reaction be present during the slow step.

(8.26)
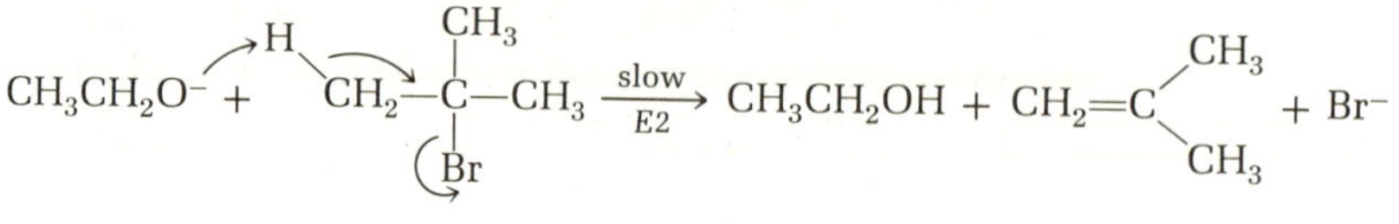

*E*1 REACTIONS

The *E*1 and S_N1 reactions share a common slow step, the formation of a carbonium ion. In the subsequent fast step the carbonium ion either captures a nucleophile or loses a proton, giving S_N1 or *E*1 products, respectively. The extent to which the carbonium ions are diverted into one or the other path depends on both internal and external factors.

If the carbonium ion is generated in a nucleophilic medium such as water or ethanol, in the absence of substantial concentrations of strong base, then nucleophilic substitution will prevail. Even under these conditions, however, strongly hindered carbonium ions eliminate a proton in preference to nucleophile capture.

(8.27)
$$(CH_3)_3CBr \xrightarrow[25^\circ]{CH_3CH_2OH} \underset{(81\%)}{(CH_3)_3COCH_2CH_3} + \underset{(19\%)}{(CH_3)_2C{=}CH_2} + HBr$$

The carbonium ion from di-*t*-butylcarbinyl chloride (equation 8.28) presents just such a case; note the typical carbonium ion rearrangement (see Chapter 7, Problem 17).

(8.28)
$$(CH_3)_3C\overset{\displaystyle Cl}{\overset{|}{C}}HC(CH_3)_3 \xrightarrow{CH_3CH_2OH,\ H_2O} (CH_3)_3\overset{+}{C}HC(CH_3)_3$$
$$\longrightarrow (CH_3)_2\overset{+}{C}\underset{\underset{\displaystyle CH_3}{|}}{C}HC(CH_3)_3 \longrightarrow CH_2{=}\overset{\overset{\displaystyle CH_3}{|}}{\underset{\underset{\displaystyle CH_3}{|}}{C}}CHC(CH_3)_3 + H^+$$

2,3,4,4-tetramethyl-1-pentene (100%)

One of the best methods for promoting E1 elimination involves heating an alcohol in excess concentrated sulfuric or phosphoric acids. The only nucleophiles present in these solutions are the HSO_4^- and $H_2PO_4^-$ ions, the nucleophilic powers of which are quite weak.

(8.29) 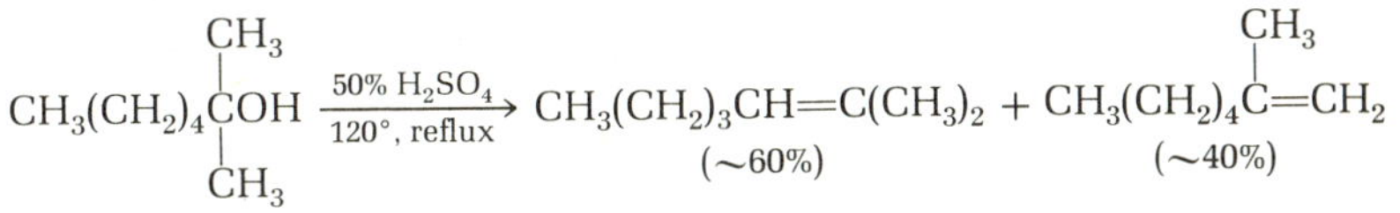

The olefin often remains in contact with the catalyst for some time after its formation.

(8.30) 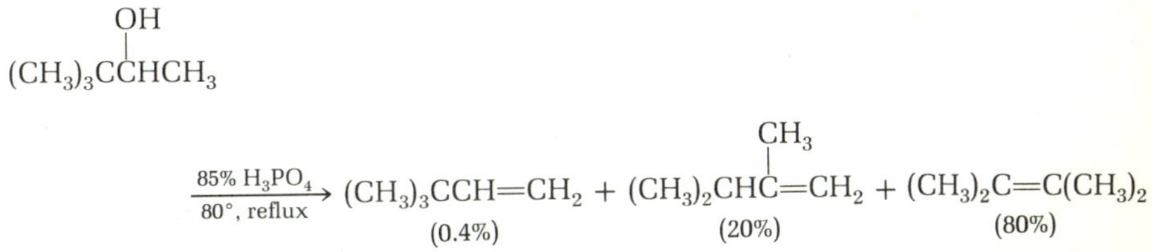

Since the olefin may be reprotonated to form a carbonium ion, which may then lose the same or a different proton, acid-catalyzed dehydration often yields equilibrium mixtures of olefins that may include substantial quantities of rearranged products. The method is not a good one for preparing terminal alkenes since these are usually the least stable isomers.

Dehydration of alcohols may also be effected with Lewis acid catalysts. These catalysts bring about much less rearrangement than the proton acids do but rather high temperatures are required unless the alcohols are tertiary.

(8.31) 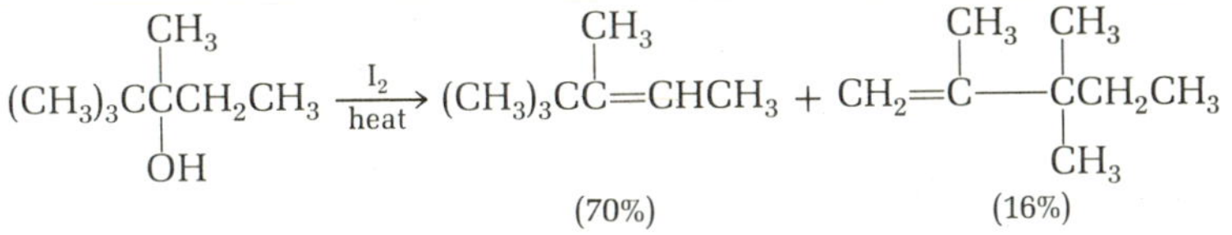

(8.32) $$CH_3(CH_2)_6CH_2OH \xrightarrow[350°]{Al_2O_3} \underset{(>90\%)}{CH_3(CH_2)_5CH{=}CH_2}$$

E2 REACTIONS

Allowing alkyl halides and tosylates to react with strong bases almost always results in substantial amounts of elimination products being formed; this statement applies even to tertiary derivatives in ionizing solvents.

(8.33) $$(CH_3)_3CBr + \underset{(excess)}{CH_3CH_2O^-Na^+} \xrightarrow[25°]{CH_3CH_2OH} \underset{(91\%)}{(CH_3)_2C{=}CH_2} + \underset{(9\%)}{(CH_3)_3COCH_2CH_3}$$

Secondary and tertiary halides and tosylates yield mostly elimination products, while primary derivatives often give rise mostly to substitution (S_N2) products since most strong bases are also effective nucleophiles.

Under the usual reaction conditions the alkene products are not interconvertible, and the proportion of each in the product mixture is a direct measure of its rate of formation. 2-Bromobutane, for instance, can lose either a primary or a secondary proton to yield 1-butene or 2-butene (*cis* or *trans*), respectively. Bases such as hydroxide and ethoxide ion tend to promote formation of the more stable products (*Zaitsev elimination*).

(8.34)
$$CH_3CH_2\underset{\displaystyle Br}{\underset{|}{C}}HCH_3 + KOH \xrightarrow[80^\circ]{CH_3CH_2OH}$$
$$CH_3CH{=}CHCH_3 + CH_3CH_2CH{=}CH_2 + KBr + H_2O$$
cis and *trans* (80%) (20%)

More powerful (and sterically hindered) bases such as *t*-butoxide ion generate higher percentages of the less substituted alkene (*Hofmann elimination*). (In order to appreciate equations 8.34 and 8.35, it must be kept in mind that a completely nondiscriminatory base would produce 60% of 2-butene and 40% of 1-butene.

(8.35)
$$CH_3CH_2\underset{\displaystyle Br}{\underset{|}{C}}HCH_3 \xrightarrow[(CH_3)_3COH,\ 70^\circ]{(CH_3)_3CO^-K^+} CH_3CH{=}CHCH_3 + CH_3CH_2CH{=}CH_2$$
cis and *trans* (47%) (53%)

The olefin distribution from any given starting material and the relative importance of elimination versus substitution are variables that are amenable to manipulation. With a given set of reactants an increase in temperature favors elimination of the expense of substitution; similarly, dipolar aprotic solvents such as DMSO increase the rate of elimination even more than they increase the rate of substitution. Even chloride ion can be an effective base in aprotic solvents.

The choice of leaving group will also exert a substantial effect on the product distribution, as shown by the examples.

(8.36)
$$CH_3(CH_2)_3\underset{\displaystyle F}{\underset{|}{C}}HCH_3 \xrightarrow[CH_3OH,\ 130^\circ]{CH_3O^-Na^+}$$
$$CH_3(CH_2)_3CH{=}CH_2 + CH_3(CH_2)_2CH{=}CHCH_3 + CH_3(CH_2)_3\underset{\displaystyle OCH_3}{\underset{|}{C}}HCH_3$$
(46%) *trans* (13%), *cis* (5%) (36%)

(8.37)
$$CH_3(CH_2)_3\underset{\displaystyle I}{\underset{|}{C}}HCH_3 \xrightarrow[CH_3OH,\ 99^\circ]{CH_3O^-Na^+}$$
$$CH_3(CH_2)_3CH{=}CH_2 + CH_3(CH_2)_2CH{=}CHCH_3 + CH_3(CH_2)_3\underset{\displaystyle OCH_3}{\underset{|}{C}}HCH_3$$
(16%) *trans* (53%), *cis* (15%) (16%)

A useful though certainly not infallible general rule is that elimination will be maximized if an alkyl bromide or iodide is exposed to a strong hindered base in an aprotic solvent.

DIASTEREOISOMERISM

The S_N2 reaction pathway is associated with a specific stereochemical consequence — inversion of configuration. In order to find out if there is a corresponding preferred stereochemical pathway in the E2 elimination, it is necessary to work with compounds containing two asymmetric atoms. The nomenclature of such compounds is outlined below.

A fairly clear case is that in which both asymmetric atoms are bonded to identical groups; two stereochemical arrangements are then possible. The absolute configuration at each of the two centers might be identically the same or identically opposite. If the second arrangement is found, then the compound must have an optical rotation of exactly 0° because it must possess a plane of symmetry. Such a compound is internally racemic and is known as a **meso** compound (Greek *mesos* — in the middle).

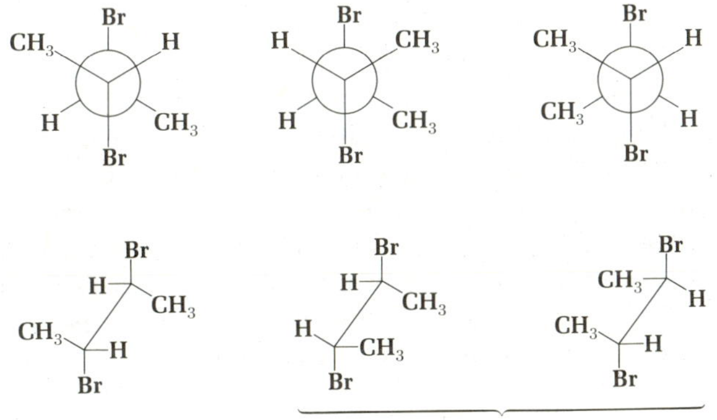

meso-2,3-dibromobutane

dl-2,3-dibromobutane

FIGURE 8-11 Line and Newman Projections of *meso*- and *dl*-2,3-Dibromobutane

A convenient example is provided by the isomers of 2,3-dibromobutane (Figure 8-11). The isomers are shown in staggered, energetically favored conformations; if they are rotated into eclipsed conformations, then it will be seen that the *meso* form may adopt a conformation in which all the identical groups eclipse one another. Such a conformation would then possess a plane of symmetry in accord with its lack of optical activity. Some *meso* compounds appear below.

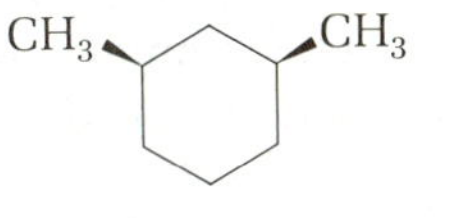

cis-1,3-dimethylcyclohexane

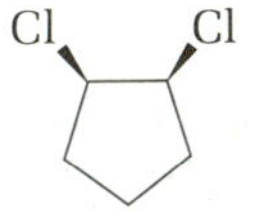

cis-1,2-dichlorocyclopentane

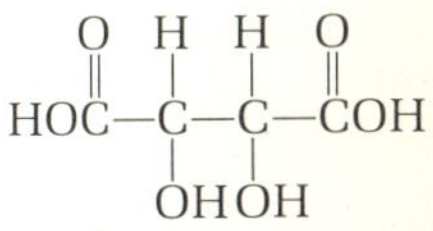

meso-tartaric acid

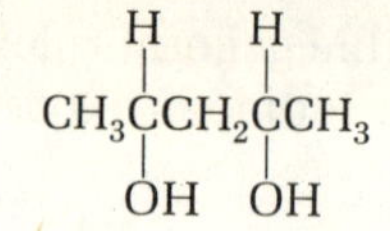

meso-2,4-pentanediol

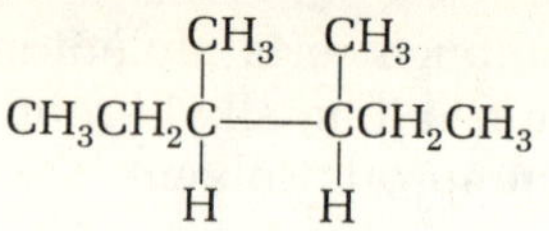

meso-3,4-dimethylhexane

The noncyclic compounds illustrated above have been depicted by means of **Fisher projections.** In this convention all horizontal bonds to an asymmetric atom are assumed to project up from the plane of the paper and all vertical bonds, down from that plane.

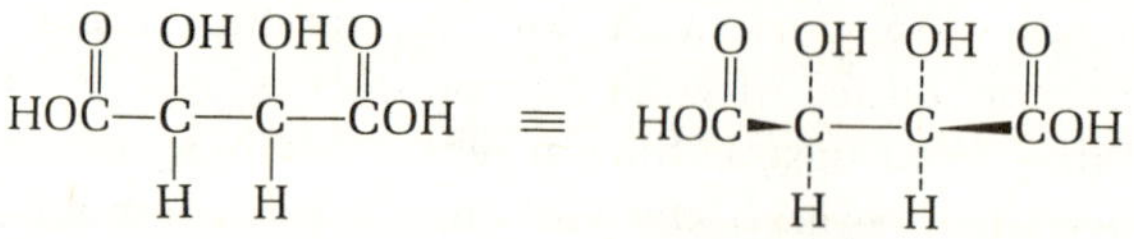

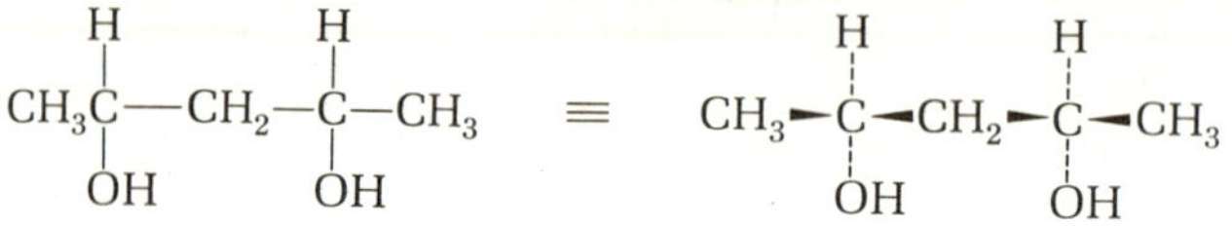

A bond that directly connects two asymmetric atoms lies in the plane of the paper.

The other isomer of 2,3-dibromobutane is one in which the absolute configuration at each asymmetric atom is the same (*S*,*S* or *R*,*R*). No matter which eclipsed conformation of this molecule is chosen, the corresponding groups on the two carbon atoms will not line up with one another (Figure 8-11). These molecules are inherently dissymmetric and exist in dextro- and levorotatory enantiomeric forms; they are therefore known as *dl*-isomers. The racemic mixture of optically active 2,3-dibromobutanes is labeled *dl*-2,3,-dibromobutane. The *dl*-isomers of the *meso* compounds shown above are included below for comparison.

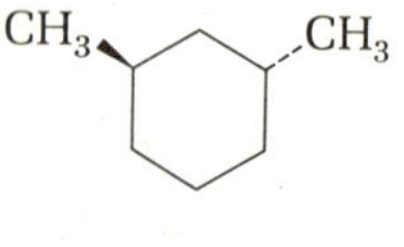

trans-1,3-dimethylcyclohexane

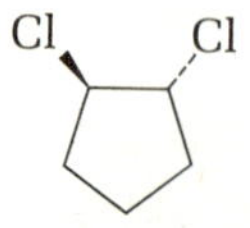

trans-1,2-dichlorocyclopentane

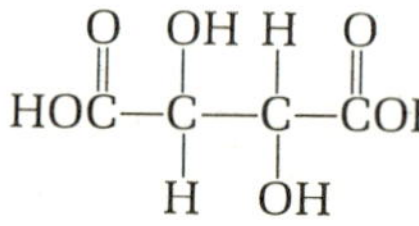

dl-tartaric acid

If only two of the groups on each asymmetric atom are identical, then it is not possible to have a *meso* form. That isomer in which the two identical groups may be opposite one another in an eclipsed conformation is the **erythro** isomer; the remaining one is the **threo** isomer (Figure 8-12). Both *erythro* and *threo* isomers exist as enantiomeric pairs, so that there are four separate compounds that correspond to the designation 3-methyl-2-pentyl tosylate, for instance. In general, if a particular structure contains *n* asymmetric atoms, then 2*n* distinct isomers of that structure are possible. If a

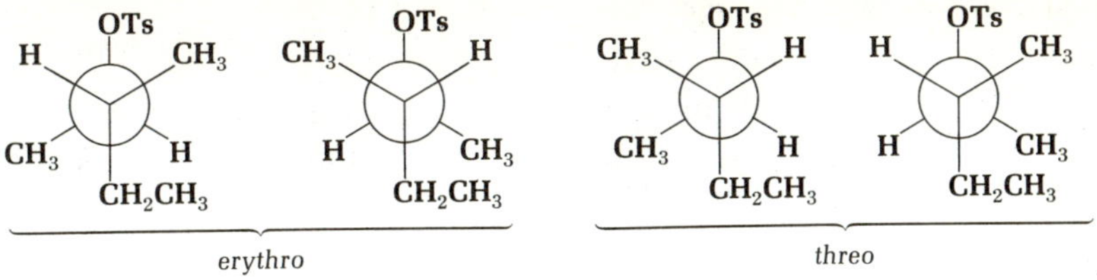

FIGURE 8-12
Enantiomers of *threo-* and *erythro-*3-Methyl-2-Pentyl Tosylate

meso form occurs, then there can be only $2n - 1$ distinct isomers since there is only one *meso* isomer.

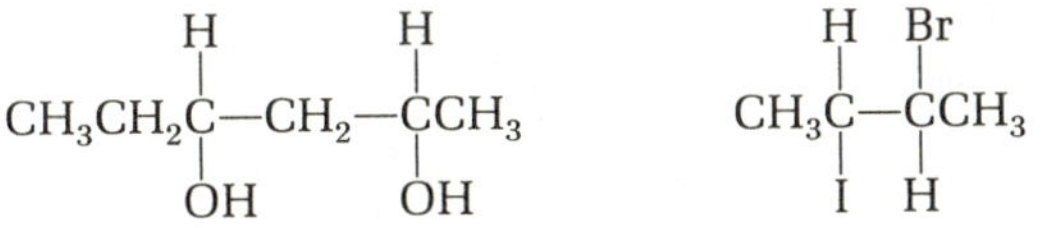

erythro-2,4-hexanediol *threo*-3-bromo-2-iodobutane

Meso isomers and *dl*-isomers are *not* mirror images (enantiomers) of one another; they are **diastereomers** (Latin *dis* — apart). *Erythro* and *threo* forms of a given compound are also diastereomers of one another, and yet they are *not* enantiomers. This situation arises because the compound possesses more than one asymmetric atom; the absolute configuration is identical at one or more of these atoms in the diastereomers, but it is also opposed at one or more of the asymmetric atoms (Figure 8-13). Diastereomers differ in all physical properties, and mixtures of them may be separated by conventional techniques (fractional distillation, fractional crystallization, glc, etc.).

To further illustrate the concept of diastereoisomerism we can make literal use of an expression that has come to denote a clumsy individual — a person with "two left feet." With most ordinary people the two halves of the body are mirror images of one another, at least insofar as the external features are concerned. Such people would fall under the category of "enantiomeric man." In the case of someone with two left feet, most parts of left and right sides of the body would retain their mirror image relationship while another section would have an *identity* relationship. A person with two left feet would thus constitute a "diastereomeric man." If an individual were missing one toe, however, he or she would qualify for neither of these categories.

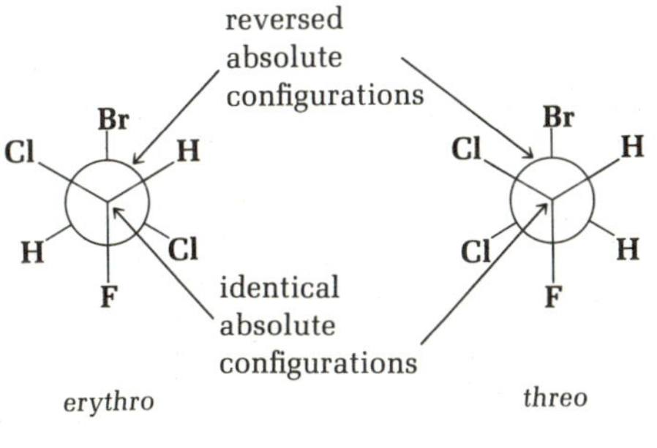

FIGURE 8-13
Comparison of Absolute Configurations Between One Enantiomer Each of *erythro-* and *threo-*2-Bromo-1,2-dichloro-1-fluoroethane

E2 REACTIONS OF DIASTEREOMERS

If a pair of diastereomers such as the 3-methyl-2-pentyl tosylates are subjected to E2 elimination conditions, then each isomer will give rise predominantly to a different alkene. The major product from the *threo* isomer is *cis*-3-methyl-2-pentene, while the *erythro* tosylate is converted primarily to *trans*-3-methyl-2-pentene (Figures 8-14 and 8-15). If the molecules undergo reaction from a staggered conformation, then the results cited lead to the following conclusions:

1 The C—H and C—X bonds (X is the leaving group) lie in the same plane in the transition state.

2 The proton being abstracted and the leaving group are on opposite sides of the molecule, i.e., *anti* to one another.

These generalizations are in accord with reasonable expectations, because of the following:

1 If C—H and C—X are coplanar, the *p*-orbitals that are forming on the adjacent carbon atoms will be oriented for maximum overlap (Figure 8-14).

2 Since the base and the leaving group are on opposite sides of the molecule, steric interference between them is minimized.

FIGURE 8-14
Anti Elimination in *threo*-3-Methyl-2-pentyl Tosylate

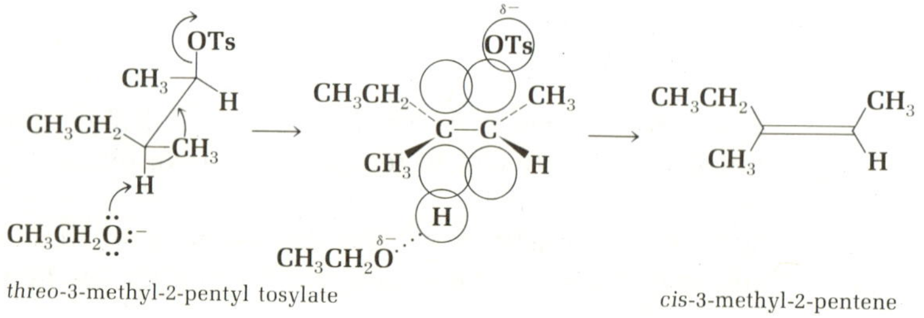

FIGURE 8-15
Anti-Elimination in *erythro*-3-Methyl-2-pentyl Tosylate

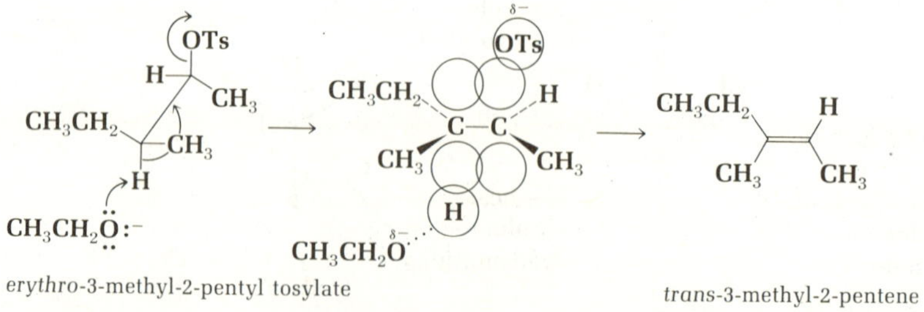

Virtually all open-chain halides and tosylates that undergo bimolecular elimination (E2) experience *anti* elimination of the proton and the leaving group. A particularly illuminating case is the E2 elimination of *erythro*-2-bromobutane-3-*d*. Three different bases were used — potassium ethoxide in ethanol, potassium *t*-butoxide in dimethyl sulfoxide, and tetrabutylammonium fluoride [$(n\text{-}C_4H_9)_4N^+F^-$] in dimethylformamide. The stereochemical results were the same in all cases. In addition to 1-butene, only *trans*-2-butene and *cis*-2-butene-2-*d* were detected.

(8.38) 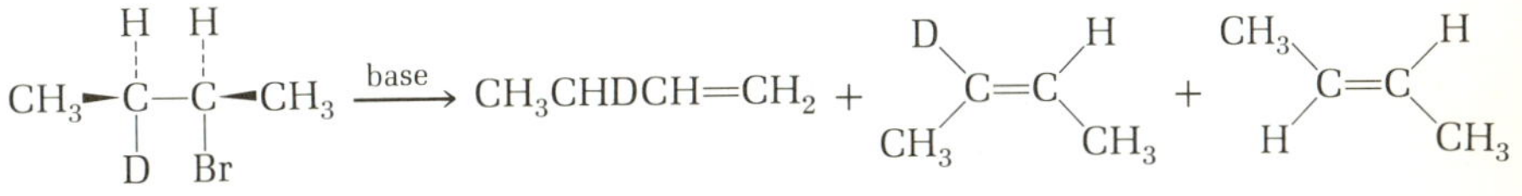

Elimination must have occurred from conformations **1** and **2**. The absence of *cis*-2-butene or *trans*-2-butene-2-*d* shows that elimination did not occur by way of conformations such as **3** (*syn* elimination).

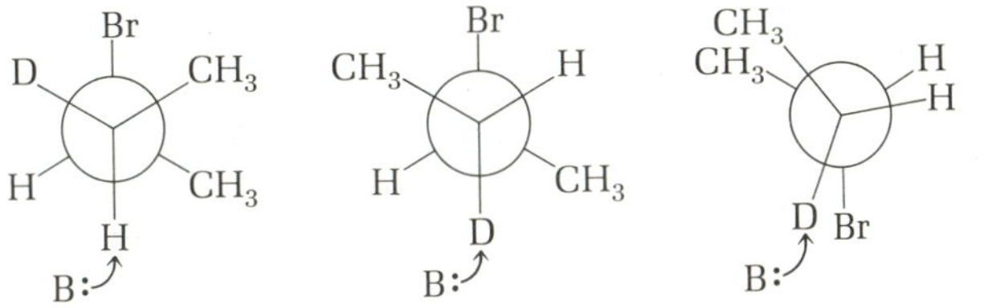

1 **2** **3**

*E*2 REACTIONS IN CYCLOALKYL SYSTEMS

Studies of E2 reactions with cycloalkyl derivatives reveal a more complex picture. Cyclohexyl halides and tosylates generally prefer to undergo *anti* elimination. This mechanism requires the proton and the leaving group to be in a *trans*-diaxial relationship to one another (Figure 8-16). Only if this conformation is prohibitively high in energy will there be a weakening of the preference for *anti* elimination.

In open-chain and cyclohexyl compounds it is energetically possible, in most cases, for the molecule to assume the conformation necessary for *anti*

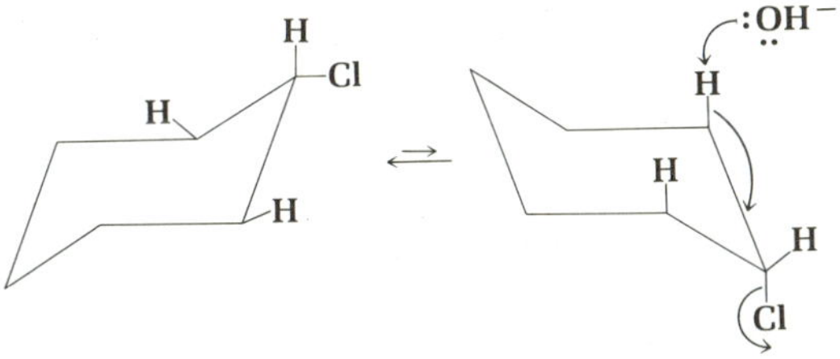

FIGURE 8-16
Conformational Equilibrium and Elimination in Cyclohexyl Chloride

elimination. This generalization is not necessarily valid for cyclic molecules that are not six-membered. Of the two stereochemical requirements previously noted, that the C—H and the C—X bonds be both parallel and *anti*, the more important one seems to be that they be parallel. This restriction ensures maximum overlap of the *p*-orbitals of the developing π-bond. If *anti*-parallel C—H and C—X orientation is difficult or impossible to achieve, then the base may remove a proton on the same side of the molecule as the leaving group, leading to *syn* elimination.

Models of cyclobutane and norbornane show that it is difficult to get a coplanar arrangement for bonds on opposite sides of the ring but relatively easy when the bonds are on the same side. Experiments have confirmed the predominance of *syn* elimination with derivatives of these cycloalkanes.

(8.39)

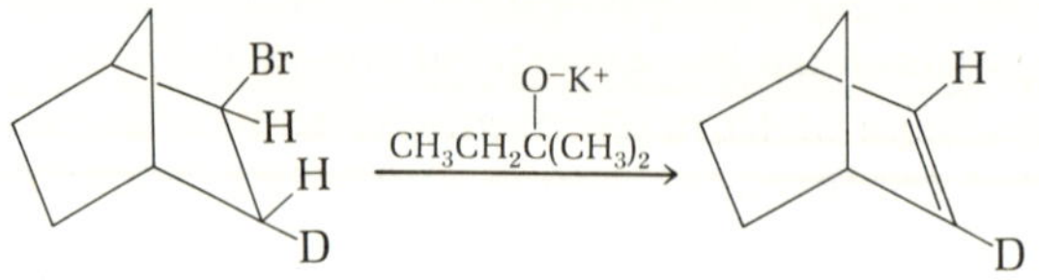

Cyclopentane and the C_7 to C_{12} cycloalkanes exhibit mixed behavior; the transition states for *syn* and *anti* elimination are apparently not too different in energy, and both may occur simultaneously in the same compound. The balance between them is dependent on the base, the solvent, etc.

HOFMANN ELIMINATIONS

One of the classical methods of introducing a carbon-carbon double bond into a molecule involves the decomposition of an organic derivative of ammonium hydroxide, known as a **quaternary ammonium hydroxide.** This technique is known as the Hofmann elimination, after its discoverer. Equations 8.40 and 8.41 give examples of this reaction.

(8.40)

$$\underset{\displaystyle (CH_3)_3\overset{+}{N}OH^-}{CH_3CH_2\underset{|}{C}HCH_3} \xrightarrow{\text{heat}} \underset{(95\%)}{CH_3CH_2CH{=}CH_2} + \underset{(5\%)}{CH_3CH{=}CHCH_3} + (CH_3)_3N\colon + H_2O$$

(8.41)

$$CH_3CH_2CH_2\overset{\displaystyle CH_3}{\underset{\displaystyle CH_3\ OH^-}{\overset{|+}{\underset{|}{N}}}}CH_2CH_3 \xrightarrow{\text{heat}}$$

$$\underbrace{CH_3CH{=}CH_2 + (CH_3)_2NCH_2CH_3}_{(2\%)} + \underbrace{CH_2{=}CH_2 + CH_3CH_2CH_2N(CH_3)_2}_{(98\%)}$$

The distribution of products found in Hofmann eliminations is rather different from that obtained from reaction of the corresponding halides or tosylates. When halides or tosylates are used, the product mixture is usually richer in the more stable products — internal olefins predominate over terminal ones and *trans* isomers, over *cis* (equations 8.1 and 8.3). Hofmann elimination with alkylammonium hydroxides tends, on the other hand, to produce high yields of the less stable alkenes. In fact, the base shows a distinct preference for abstracting a proton from a CH_3— group as compared to a —CH_2— group and a slight preference for —CH_2— over $-\overset{|}{\underset{|}{C}}-H$. This selectivity is far in excess of what would be expected if the number of each type of hydrogen atom were the sole determinant.

The reasons behind the fundamentally different behavior of alkyl ammonium salts on the one hand and most halides and tosylates on the other have not been fully elucidated by any means. Zaitsev-type eliminations seem to be governed largely by alkene stability. Hofmann eliminations seem to be controlled at least in part by the relative acidities of the protons available for abstraction (1° > 2° > 3°). Furthermore, the base strength, the nature of the solvent, and the steric effects in both base and substrate all play a part. Reactions in which an abnormal amount of terminal alkenes are produced usually also show a high *cis:trans* ratio for the internal alkenes, and a large portion of the products are all connected but the causes behind the connection remain to be defined.

In any case, the differences between Hofmann and Zaitsev elimination can be exploited for synthetic purposes (equations 8.42 and 8.43).

(8.42)

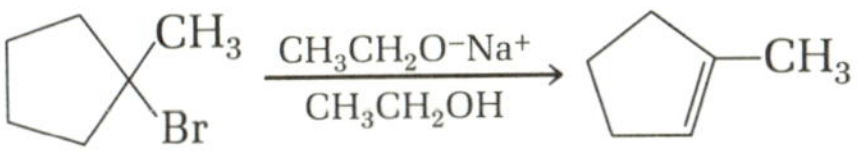

(8.43)

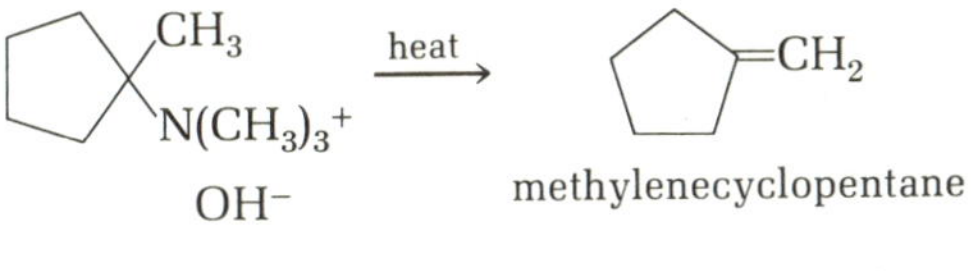

Examples of the preparation of quarternary ammonium hydroxides are given in equations 8.44 to 8.46. The formation of the C—N bonds occurs in every case by way of an S_N2 displacement. Methyl groups are used on the nitrogen because they cannot undergo the elimination reaction, which will be confined to the fourth group.

(8.44)

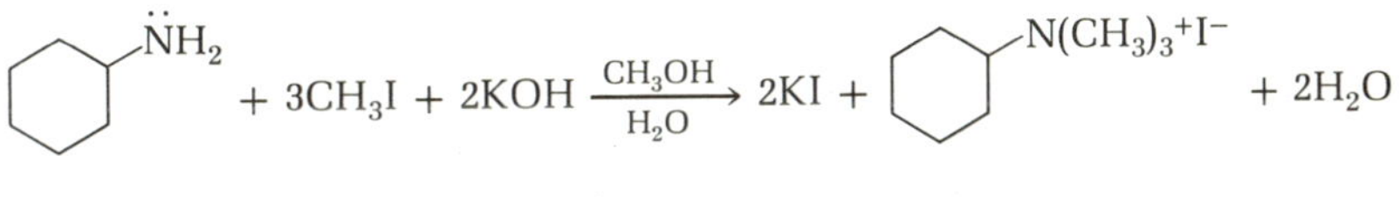

(8.45)

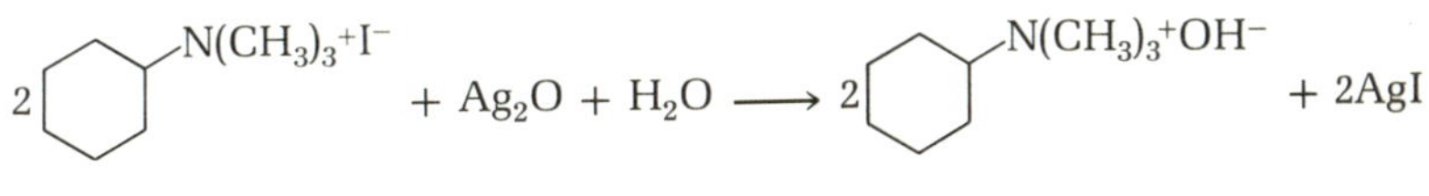

(8.46) $$CH_3CH_2CH_2I + (CH_3)_3N\!: \xrightarrow{CH_3OH} CH_3CH_2CH_2\overset{+}{N}(CH_3)_3\ I^- \xrightarrow[H_2O]{Ag_2O} CH_3CH_2CH_2\overset{+}{N}(CH_3)_3\ OH^-$$

DEBROMINATION OF VICINAL DIBROMIDES

The addition of bromine to alkenes yields alkyl dibromides in which the bromine atoms are attached to adjacent carbon atoms (*vicinal* dibromides, Chapter 9). Debromination of 1,2-dibromides by iodide ion is an example of an *E*2 elimination in which no proton removal is involved.

(8.47)

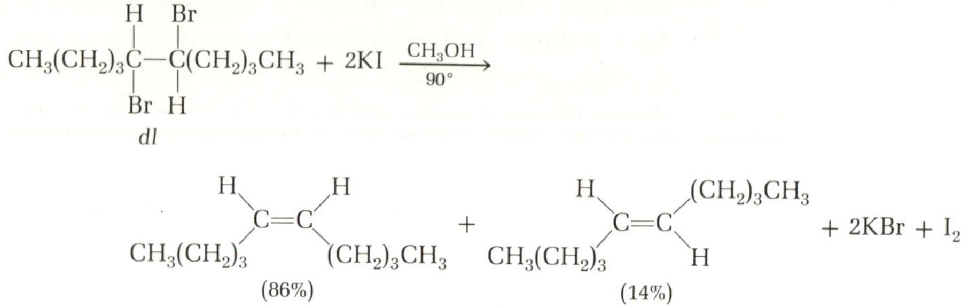

The high proportion of *cis*-alkene from *dl*-dibromide indicates that iodide ion prefers to attack that conformation of the molecule in which the bromines are *anti* to one another.

(8.48)

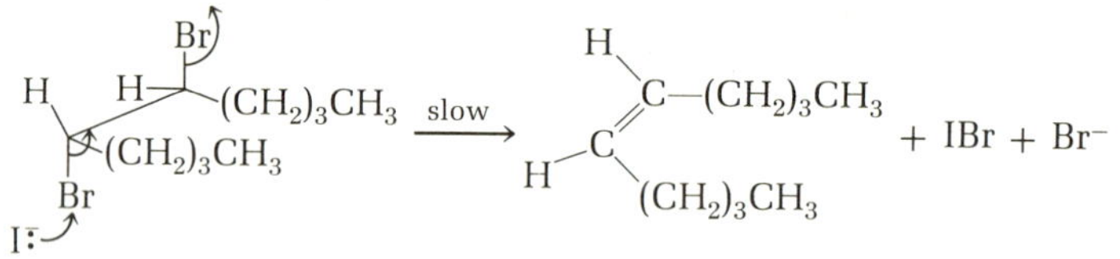

(8.49) $$IBr + I^- \xrightarrow{\text{fast}} I_2 + Br^-$$

Zinc metal can bring about the same debromination reaction, with the same predominantly *anti* stereochemistry.

(8.50)

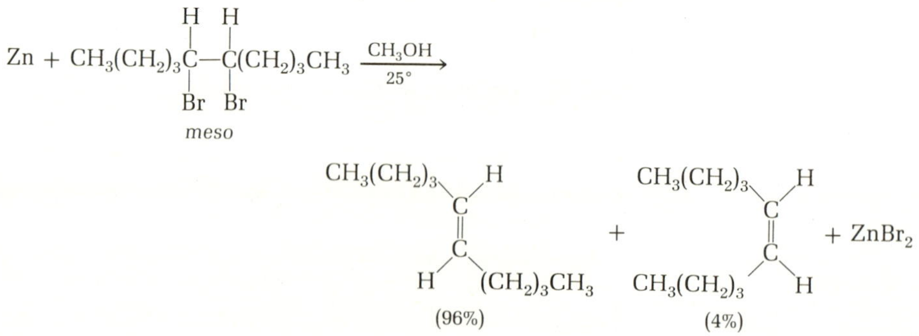

These reactions are of no preparative value because the dibromides are obtained from alkenes by addition of bromine. Since the bromine addition proceeds by an *anti* mechanism, however, one may add bromine and then remove it without affecting the stereochemistry of the double bond. This can be an effective method for **protecting** the double bond when a chemical transformation must be carried out elsewhere in the molecule with a reagent that would normally react with the double bond.

SUMMARY

1 The double bonds in alkenes are conveniently viewed as composites of σ- and π-bonds. Overlap of the *p*-orbitals comprising the π-bond is most effective when their axes are colinear; this requirement results in a substantial barrier to rotation about double bonds. It is therefore possible to isolate pure *cis* and *trans* isomers of internal alkenes at room temperature. These isomers will usually be differentiable by means of ir and nmr spectroscopy.

Cis and *trans* isomers may interconvert at sufficiently high temperatures or upon photochemical excitation. It is also possible, by chemical means, to equilibrate positional isomers. Equilibrium studies show that in general internal alkenes are less energetic than terminal ones, and *trans*-alkenes more stable than *cis*-alkenes. These isomers each have distinct physical properties.

Double bonds may also be viewed as consisting of two bent σ-bonds (τ-bonds). This picture could adequately account for the geometrical isomerism and reactivity of alkenes. Although the σ-π model will be relied on most of the time as a matter of convenience, the τ-model should be seen as complementary to rather than in competition with the σ-π model.

2 Carbonium ions may capture a nucleophile (S_N1 reaction) or lose a proton and be converted to alkenes (*E*1 reaction). A common method for effecting *E*1 reactions involves heating alcohols with concentrated, strong acids. Rearrangements are quite common, and equilibrium mixtures of alkenes are customarily obtained.

3 Treatment of halides and tosylates with strong bases brings about bimolecular elimination (*E*2 reaction) accompanied by varying amounts of nucleophilic substitution. The most stable olefin usually predominates in the product mixture; the products are not interconvertible under the reaction conditions.

Studies of the stereochemistry of elimination have been made using *erythro* and *threo* diastereomers. In open-chain and cyclohexyl halides and tosylates, eliminations almost always proceed mainly with *anti* stereochemistry. In cycloalkyl rings other than C_6, the energetic preference for *anti* elimination is by no means assured and substantial amounts of *syn* elimination may take place. The most important factor appears to be the necessity for coplanarity of the bonds being broken.

4 Eliminations involving quartenary ammonium salts follow a different path. The least stable olefin is commonly the most abundant product, the base apparently removing a proton from the least substituted carbon atom. *Syn* elimination also seems to be much more prevalent when ammonium ions are substrates.

5 Bromination of alkenes often serves to protect the double bond from numerous reagents. The double bond may be stereospecifically regenerated by allowing the dibromide to react with zinc or iodide ion.

PROBLEMS

1 Write out the structures of all the isomers, including optical and geometrical isomers, of the following compounds, and name them:

a. 1-Hexene
b. 2-Hexene
c. 2-Methyl-2-pentene
d. Chloropropene
e. 4-Methyl-2-hexene
f. 4-Isopropylcyclononene
g. 6-Methylcyclononene
h. 1,3,8-Trichloro-6-propyl-3-undecene

2 1-Methylcyclohexene is completely unreactive toward neutral methanol. If the alkene and methanol are dissolved in a photochemical sensitizer and the solution then irradiated, reaction does take place in the manner shown in equation 1.

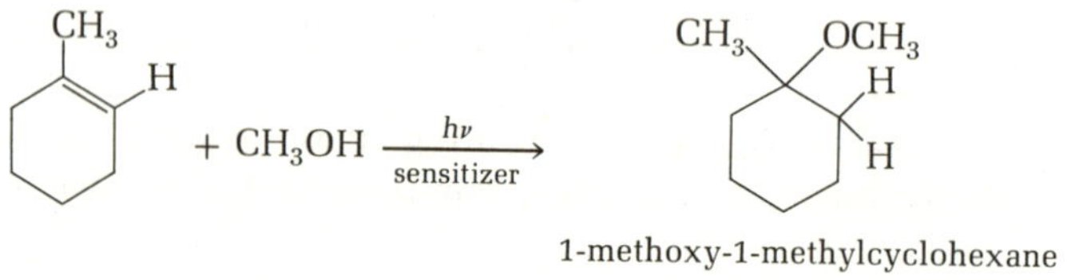

(1)

The same observations apply to 1-methylcycloheptene. Smaller cycloalkenes do not undergo this reaction, while larger ones simply experience *cis-trans* isomerization under the same reaction conditions.

What is the intermediate produced in reaction 1? How do you account for its reactivity?

3 *Bredt's rule* states that a bridgehead carbon atom in a bicyclic ring in which all the bridges are composed of atoms cannot be part of a double bond. Thus, bicyclo(2,2,1)hept-1-ene (I) and bicyclo(2,2,2)oct-1-ene (II) are unknown.

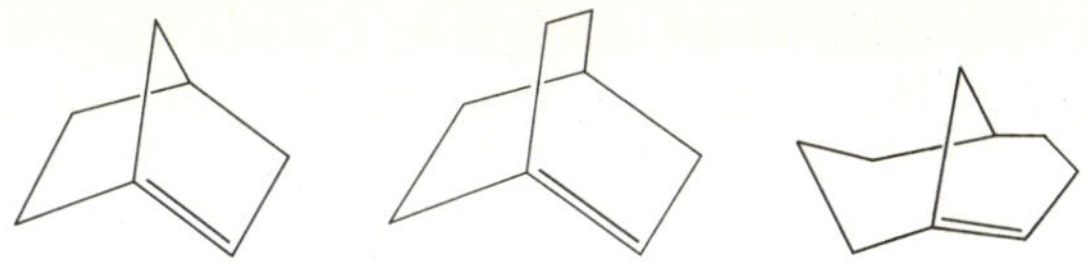

I II III

a. What is the origin of Bredt's rule?

b. The reactive bicyclo(3,3,1)non-1-ene (III) has been prepared. Why has this been possible? Propose a modification of Bredt's rule.

c. What products would be expected from the reaction of 1-chloro- and 2-chlorobicyclo(2,2,2)octane with potassium hydroxide in hot ethanol?

4 2-Chloropentane is allowed to react with sodium isopropoxide [$(CH_3)_2CHO^-Na^+$] in isopropyl alcohol.

a. Write out all the products that could be obtained from this reaction; name all the alkenes.

b. Assuming that the individual products could be separated by glc, how would they each be identified, if authentic specimens were not available for comparison?

5 The rearranged carbonium ion in equation 8.28 could have given rise to two olefinic products, although only one was observed.

a. What is the structure of the second product?

b. The reaction appears to be an exception to the rule that under E1 conditions the most stable olefin will be the predominant product. Account for this apparent discrepancy.

6 Write the structures of the probable products from the following reactions.

a. Cyclopentanol $\xrightarrow[\text{heat}]{H_2SO_4}$

b. 6-Methyloctan-3-ol $\xrightarrow[\text{heat}]{H_3PO_4}$

c. 2,2-Dimethylcyclohexanol $\xrightarrow[140°]{ZnCl_2}$

d. Cyclopentylmethyl tosylate $\xrightarrow[80°]{H_2O}$

e. Isohexylbromide $\xrightarrow[\text{DMSO}]{(CH_3)_3CO^-K^+}$

f. *threo*-2-Chloro-3-methylpentane $\xrightarrow[CH_3CH_2OH]{CH_3CH_2O^-Na^+}$

g. 2-Chloro-3,3-dimethylbutane $\xrightarrow[C_2H_5OH]{KOH}$

h. (1-Cyclohexylethyl)trimethylammonium hydroxide $\xrightarrow{\text{heat}}$

7 Suggest means of carrying out the following conversions.

a. Methylcyclohexane $\longrightarrow$ 1-methylcyclohexene

b. Isopentyl alcohol $\longrightarrow$ 3-methyl-1-butene

c. 2-Chloropentane $\longrightarrow$ 1-pentene

d. *threo*-4-Methyl-3-hexanol $\longrightarrow$ *cis*-3-methyl-3-hexene

8 How could *cis*-2-bromomethylcyclopentane be converted predominantly to

a. 1-Methylcyclopentene?

b. 3-Methylcyclopentene?

9 The reaction of *trans*-2-methylcyclohexyl tosylate with potassium *t*-butoxide in *t*-butyl alcohol at 50° gives rise to an olefin mixture that is 99% 3-methylcyclohexene and only 1% 1-methylcyclohexene. Account for the great predominance of the less stable isomer over the more stable one.

10 Propyl bromide reacts with dimethyl sulfide to produce a crystalline compound that is quite soluble in polar solvents but only slightly soluble in nonpolar solvents. When this compound is dissolved in aqueous sodium hydroxide and the solution heated, a gas is evolved. The gas is found to have a molecular weight of 42. Write out equations for the chemistry.

11 Trimethyl-2-*cis*-cyclobutylammonium-d_1 hydroxide (IV) was pyrolyzed to a mixture of cyclobutene (31%) and cyclobutene-d_1.

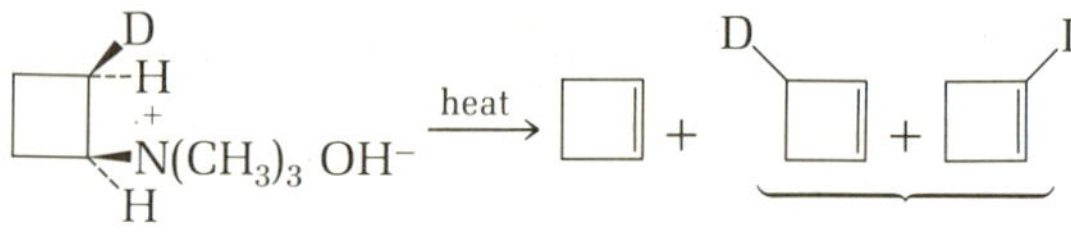

IV (31%) 69% (2)

a. How were the proportions of deuterated and nondeuterated cyclobutene in the mixture determined?

b. Assume that deuterium and hydrogen are abstracted with equal ease. What percentage of each product would you expect if the reaction proceeded solely by way of a *syn* mechanism? Using experimental figures, calculate what percentage of the molecules decompose by *syn* and *anti* mechanisms.

c. In actual fact it is approximately twice as easy to abstract a proton as a deuteron in this reaction. Using this fact, calculate the percentage of the molecules that decompose by *syn* and *anti* mechanisms.

12 Pyrolysis of trimethyl-*cis*-4-cyclooctenylammonium hydroxide (V) gives rise to two isomeric products, ***A*** and ***B***. While ***A*** is quite stable, ***B*** is reactive and unstable.

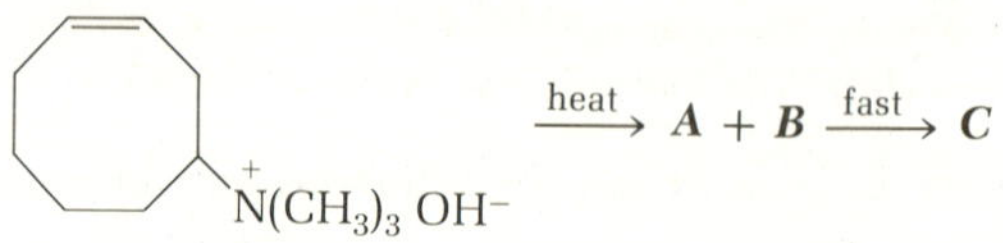

V (3)

B may be readily isomerized to ***C***, which is different from ***A***. ***A***, ***B***, and ***C*** are all converted to cyclooctane when treated with hydrogen and a catalyst. Draw structures for ***A***, ***B***, and ***C***.

13 If it is assumed that equation 8.30 represents an equilibrium distribution of products, then the equilibrium constant for conversion of 2,3-dimethyl-1-butene to 2,3-dimethyl-2-butene is about 4.

a. What is the free energy change for this process?

b. Is this free energy change almost exclusively an enthalpy difference? Explain your answer.

14 Consider qualitatively the overall entropy and enthalpy changes that accompany S_N2 and E2 reactions. Is it to be expected that the rate of an E2 reaction will rise more rapidly with temperature than the rate of the competing S_N2 reactions? Such an analysis is most easily made if it is assumed that the entropy and enthalpy changes undergone by the solvent are roughly equivalent for both reactions. Is such an assumption justified?

SUPPLEMENTARY READINGS

E. A. Walters, "Models for the Double Bond," *J. Chem. Ed.*, **43**, 134 (1966).

R. J. Baumgarten, "Substitution Products in the Hofmann Elimination," *J. Chem. Ed.*, **45**, 122 (1968).

W. H. Saunders, Jr., *Ionic Aliphatic Reactions*. Englewood Cliffs, N.J.: Prentice-Hall, Inc., 1965, Chapters 5 and 6.

CHAPTER 9

ALKENES II—ADDITIONS TO THE UNPOLARIZED DOUBLE BOND

INTRODUCTION

As noted in Chapter 2, the π-bond of alkenes is over 20 kcal/mole weaker than a carbon-carbon σ-bond. As a consequence, reactions of the type shown in equation 9.1 (**addition reactions**) are almost always thermodynamically favorable since there is a net conversion of a π-bond to a σ-bond.

(9.1) 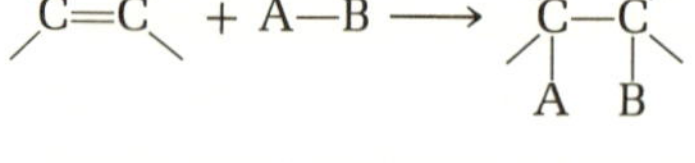

In fact, the chemistry of alkenes encompasses an enormous number of reactions of this type.

The net concern is the mechanism of addition. A concerted mechanism seems reasonable (equation 9.2), although few reactions of this type have been discovered.

(9.2) 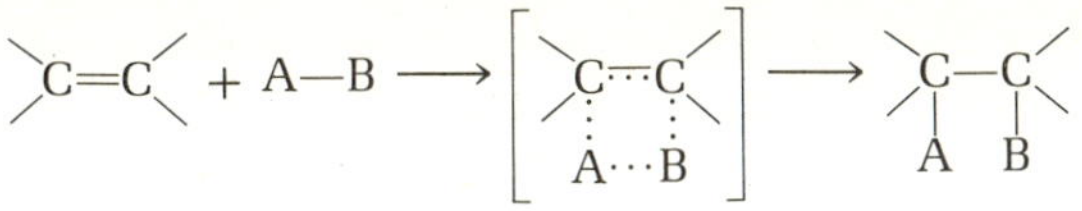

The reagent A—B usually dissociates prior to or concurrently with attack on the double bond, and the two parts of the reagent are added successively. Dissociation may be homolytic (equation 9.3) or heterolytic (equation 9.4).

(9.3) 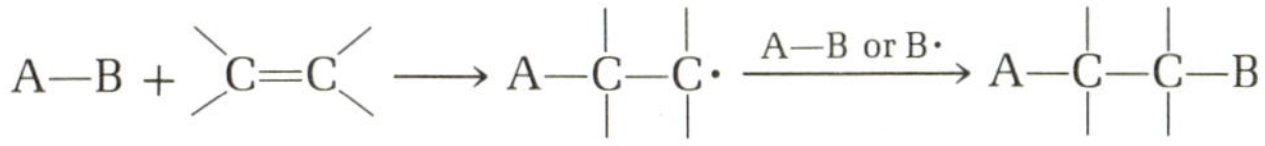

In the latter case there are two possibilities — initial reaction with either the electrophilic (reaction 9.4a) or nucleophilic portion of the reagent (reaction 9.4b).

(9.4) 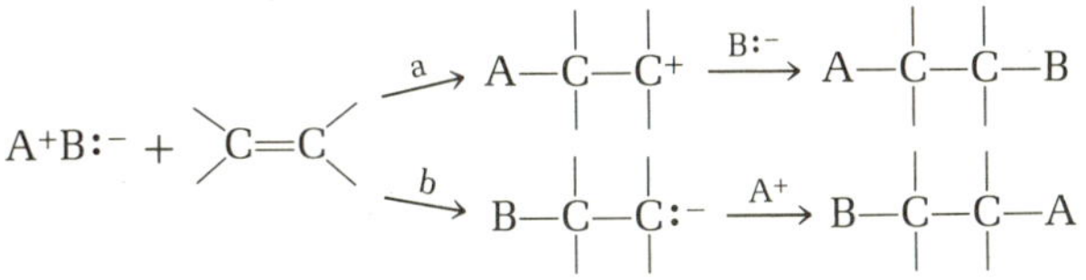

Reaction 9.4b is quite uncommon unless the carbon atom on which the charge accumulates bears fairly electronegative substituents. On the other hand, examples of reaction 9.4a are abundant and will be the subject of a sizable fraction of this chapter. The following generalization will be a useful guide. **The π-electrons of alkenes are sterically accessible and easily polarized and are therefore subject to initial attack by a wide variety of electrophiles. Concurrent or subsequent attachment of a nucleophile usually completes the reaction.**

ALKENE-METAL COMPLEXES

The complexation of alkenes with silver ion (π-complex) is a typical example of electrophilic attack on the double bond. Silver ion brings an empty orbital that overlaps with the carbon p-orbitals, forming new three-center molecular orbitals. The lowest in energy of these three new molecular orbitals is bonding and accommodates the pair of electrons (silver contributed none). From one viewpoint, the complex resembles a system containing two carbon atoms joined by one bent bond and one normal σ-bond (Figure 9-1).

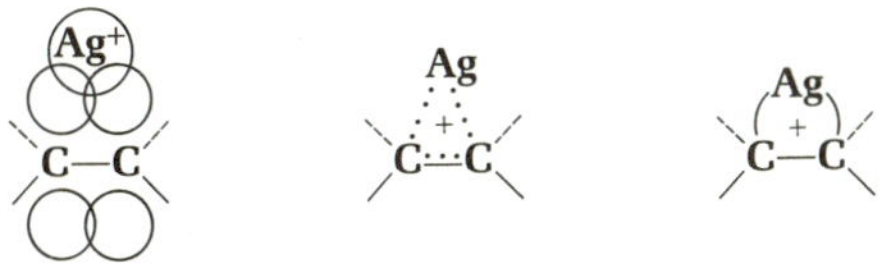

FIGURE 9-1
Several Representations of a Silver-Olefin Complex

The equilibrium constants for complex formation between silver ion and simple olefins are usually small, and the complexes are not normally isolable (they have been mentioned in connection with the glc of alkenes, Chapter 8). Many transition metals with empty orbitals form alkene complexes, and some of these are quite stable. Platinum, in particular, is capable of reacting with alkenes to form stable compounds that can be isolated; they have a fascinating chemistry of their own. Compound **1** is a complex formed from ethylene, platinum dichloride and potassium chloride.

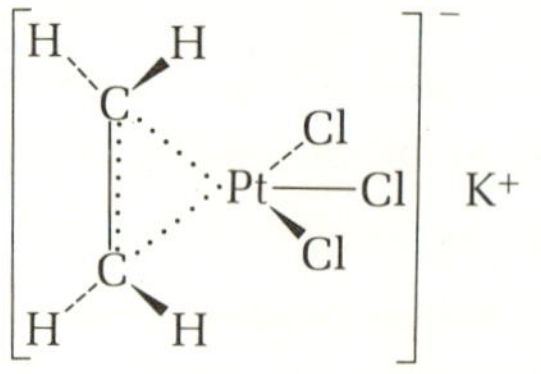

1

ACID-CATALYZED ADDITIONS

One need not search for exotic electrophiles, however. The simplest one, hydrogen ion, is quite reactive toward double bonds. In this case the initially formed π-complex is not stable as such but rearranges to a carbonium ion with concomitant formation of a normal C—H σ-bond.

(9.5)

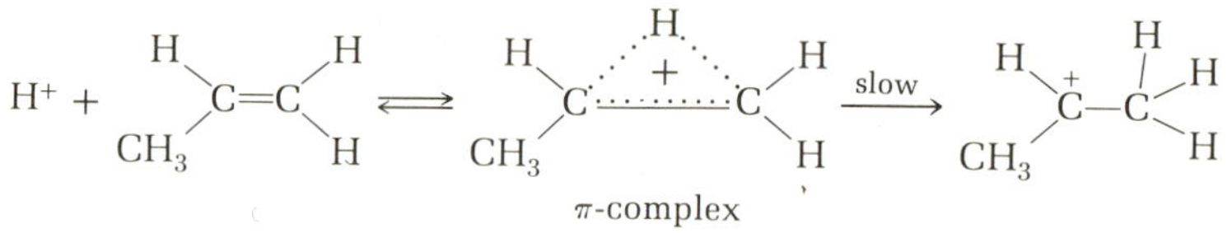

The carbonium ion is then available for reaction with a wide spectrum of nucleophiles.

(9.6)

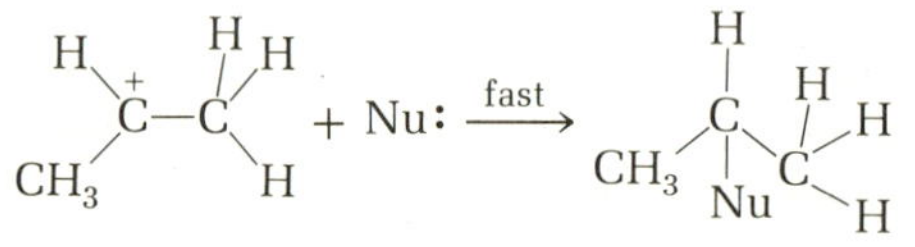

Among the large number that have been used in this reaction are F^-, Cl^-, Br^-, I^-, H_2O, ROH, RSH, etc. If the nucleophile is too basic, the reaction will not be successful because the acid catalyst will preferentially protonate the nucleophile, destroying its reactivity. Thus, $:NH_3$, CN^-, OH^-, etc., cannot be added to double bonds in this way; furthermore, it becomes necessary to add HF by using anhydrous liquid hydrogen fluoride, a rather nasty procedure.

When water is being added to a double bond (*hydration*), the acid chosen for a catalyst must contain a fairly nonnucleophilic anion (H_2SO_4, HNO_3,

$HClO_4$), so that the carbonium ion will react preferentially with water rather than the anion.

(9.7) 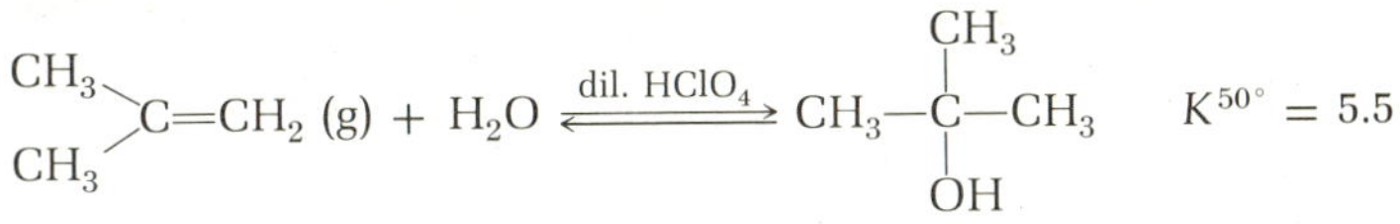

The practical limitation on this method is the limited solubility of most alkenes in dilute aqueous acid. The problem is often circumvented by dissolving the alkene in strong, cold sulfuric acid, in which it is converted to an alkyl hydrogen sulfate.

(9.8) 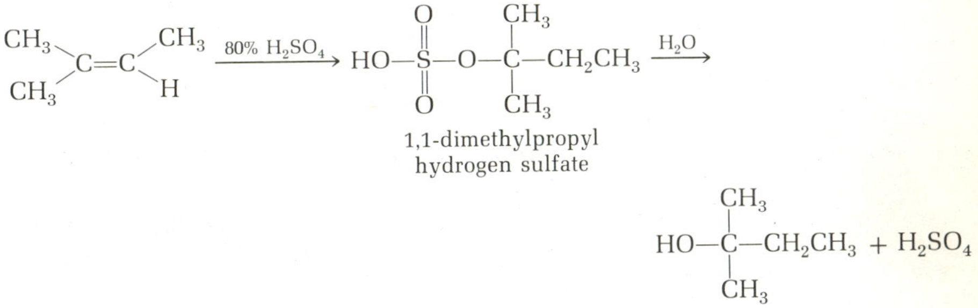

Hydrolysis of the sulfate then yields an alcohol and sulfuric acid. Both the one-step and two-step hydrations proceed by way of carbonium ions, and energetically favorable rearrangements will occur.

(9.9) 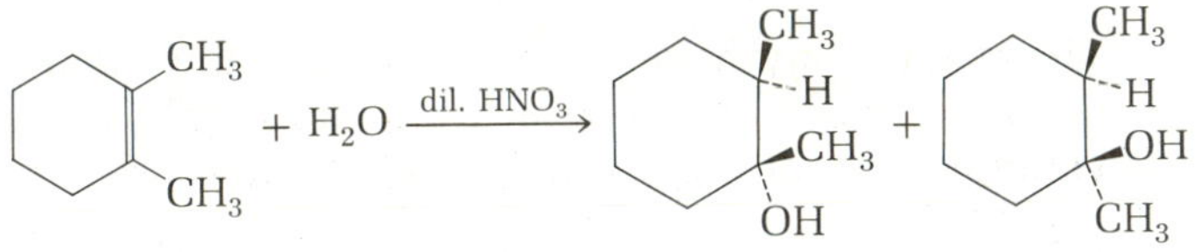

Alcohols may also be added to alkenes in the presence of acid, in complete analogy to hydration. The addition of water and alcohols is readily reversed, particularly when acid concentration is high, the nucleophile concentration low, and the temperature elevated.

Addition of the hydrogen halides to alkenes is usually carried out neat or in nonnucleophilic solvents, in order to avoid mixtures of products.

(9.10) $$(CH_3)_2C{=}CH_2 + HCl \xrightarrow{-80^\circ} \underset{(88\text{–}94\%)}{(CH_3)_3CCl}$$

Pentane, methylene chloride, and ether are often used, the dry gas (HCl or HBr) being passed into the solvent or into the neat alkene. In these nonpolar solvents it appears that truly "free" carbonium ions are not favored; the nucleophile begins to attack the π-complex before it has fully rearranged. Nucleophilic addition seems to take place on the side of the molecule opposite that point where protonation occurred.

(9.11) 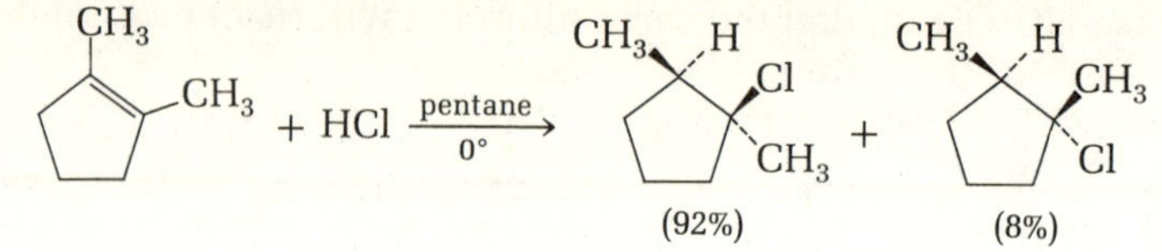

This arrangement probably minimizes interference between incoming groups (**2**).

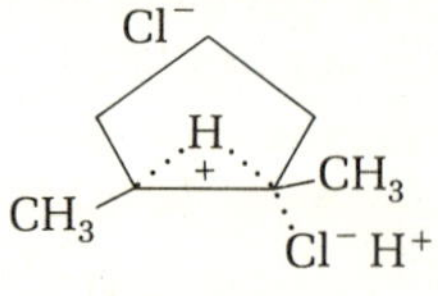

2

This method is quite suitable for adding a hydrogen halide to an olefin but not for making a stereospecific product. Mixtures result in most cases, and their composition is dependent on factors such as time, temperature, and solvent. Furthermore, reactions are known in which HCl and HBr add primarily *syn* to double bonds that carry groups particularly effective at stabilizing carbonium ions.

MARKOVNIKOV'S RULE

The reader may have noted that when unsymmetrical olefins have been reactants then essentially only one product has been obtained. This observation was made some time ago by the Russian chemist Markovnikov, who concluded that in the addition of reagents H—G to unsymmetrical double bonds the H becomes attached to the carbon already possessing the greatest number of hydrogens. Stated in modern terms, the rule says that attack of an electrophile on a double bond proceeds in such a manner as to give the most stable carbonium ion (equations 9.12 to 9.14). The last example (equation 9.14) shows that there is not much difference in rate of formation between two isomeric secondary carbonium ions.

(9.12) 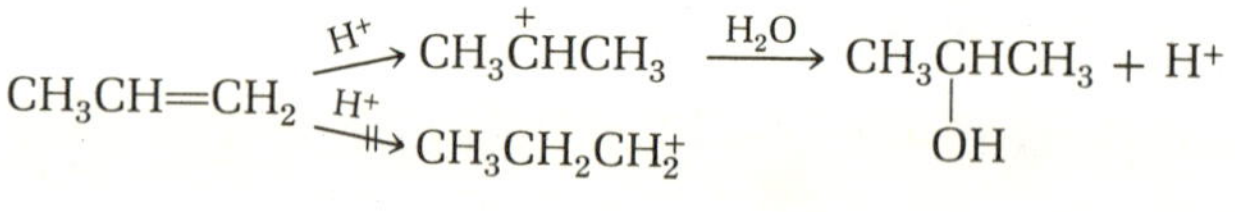

(9.13) 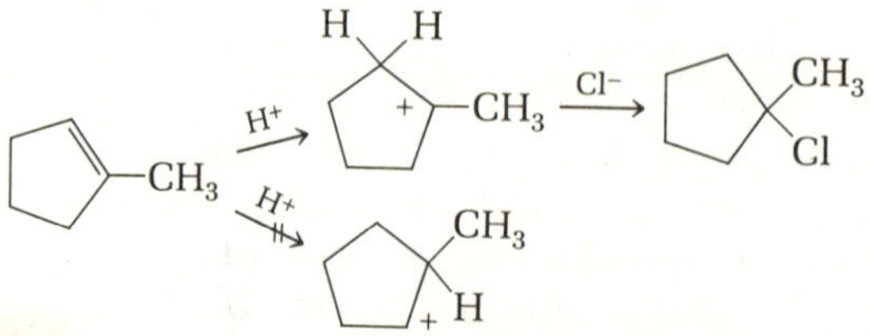

(9.14) 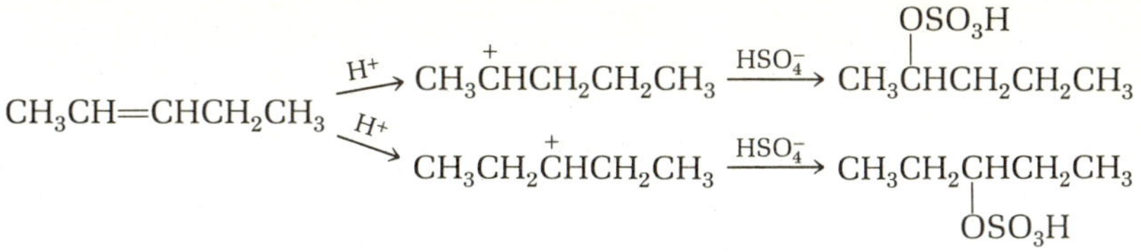

In nonpolar solvents, proton addition is generally irreversible, and the product distribution depends on the rate of formation of each product rather than on its stability.

One consequence of the Markovnikov rule is that primary alcohols or halides cannot be made by ionic additions to alkenes.

(9.15) 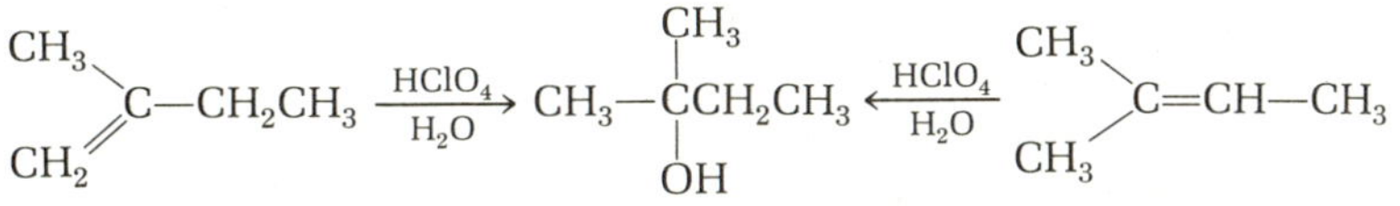

There are ways of getting around the Markovnikov rule, however, that testify to the synthetic chemist's determination to defy thermodynamics whenever possible (see hydroboration, p. 214).

Since carbonium ion stability has such a dominating effect on the orientation in addition reactions, it would be expected to affect the rate as well. A study of the reaction of hydrogen chloride with a series of alkenes uncovered the reactivity order shown below. The result is not surprising since it would be expected that the electron-supplying methyl groups would facilitate attack by the electrophilic proton.

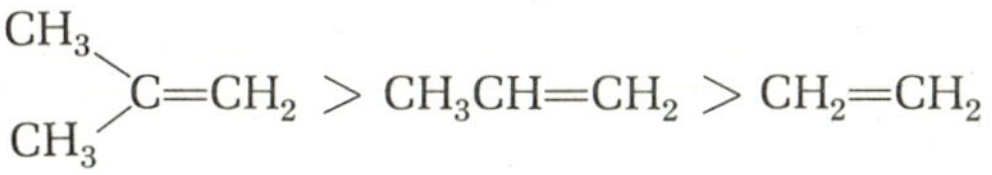

Interestingly, 2-methylpropene sometimes reacts even faster than 2-methyl-2-butene. This would seem to indicate that the polarization of the double bond in the ground state may also affect the rate of reaction.

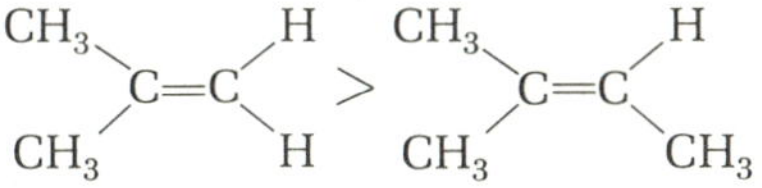

2-Methylpropene should be relatively highly polarized because the electron-supplying methyl groups are concentrated at one end of the double bond. Indeed, isobutylene is one of the most reactive alkenes in addition reactions.

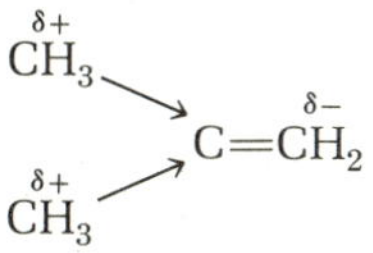

IONIC HALOGENATION

Among the most important addition reactions of olefins are those involving the halogens and their derivatives. In particular, the addition of chlorine or bromine to carbon-carbon double bonds is exothermic, goes to completion, and is readily controlled. (Oxygen is used during chlorinations to suppress free radical reactions; see the section on free radical additions to double bonds, p 216.) Direct fluorination is extremely exothermic and leads to extensive degradation of the molecule.

(9.16) 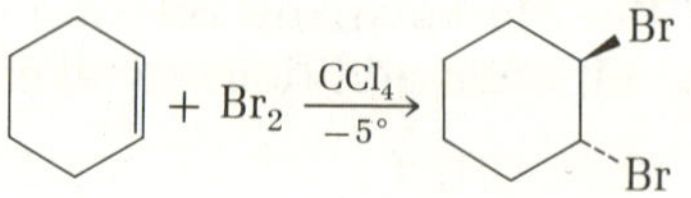

(95%)

(9.17) $$CH_3CH_2CH{=}CH_2 + Cl_2 \xrightarrow[-9^\circ]{O_2} CH_3CH_2\underset{\underset{Cl}{|}}{C}HCH_2Cl$$

(97%)

Iodine, on the other hand, is rarely used in addition reactions because they do not normally go to completion (see below).

As indicated by the reaction shown in equation 9.16, and by numerous other cases, addition of molecular halogen to a simple alkene under the conditions shown gives essentially only one dihalide. Each one of a pair of *cis-trans* alkene isomers forms only one of the two possible diastereomeric dihalides.

The stereospecificity of the reaction suggests strongly that neither free radicals (equation 9.18) nor free carbonium ions (equation 9.19) are intermediates in the halogenation. If they were, then rotation could take place about the σ-bond that was formerly part of the double bond.

(9.18) 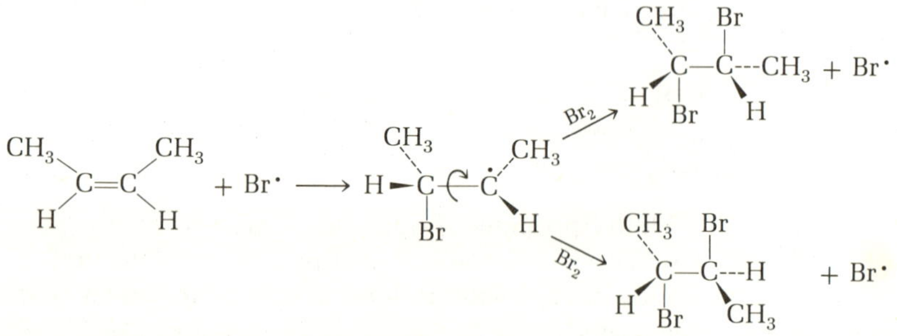

(9.19) 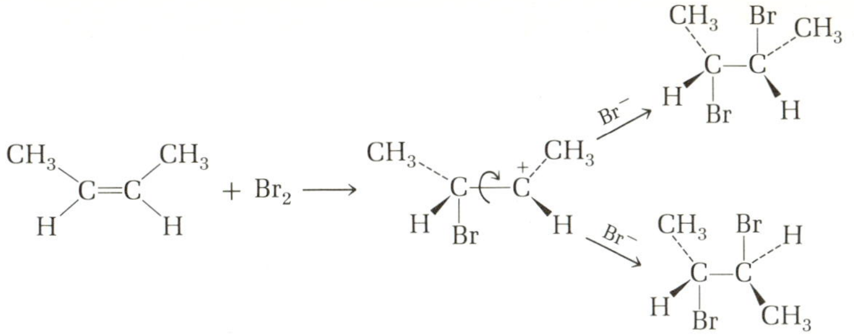

Were that to occur, then a mixture of products would result. It might be postulated alternatively that both bromine atoms become attached simultaneously to the reacting carbon atoms (*syn* addition). Such a mechanism would require that *cis*-2-butene give rise to *meso*-2,3-dibromobutane; in fact, the *dl*-dibromide is isolated in very high yield.

(9.20) 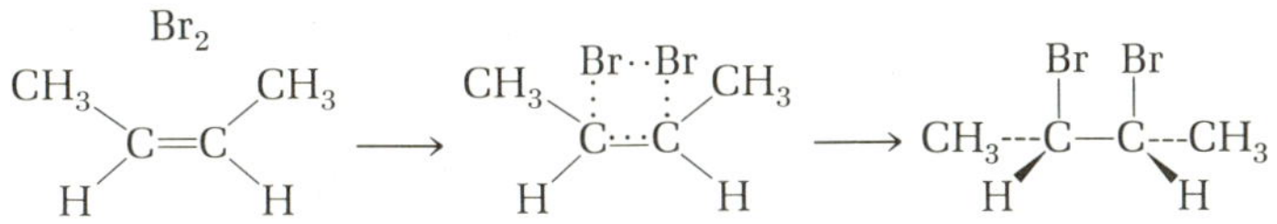

cis *meso*

The results were rationalized by means of a mechanism involving a **bromonium ion (3)** that would lead to overall *anti* addition of bromine to the double bond. (It was assumed, quite reasonably, that bromide ion would attack the bromonium ion in an S_N2 fashion, with inversion.) Several other observations are also compatible with this mechanism.

(9.21) 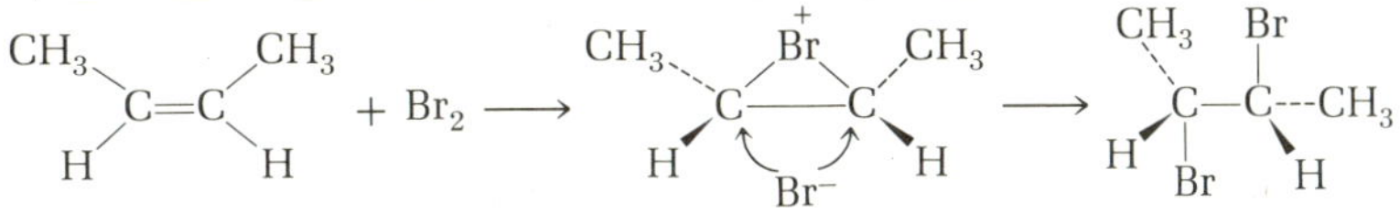

cis **3** *dl*

In CCl_4 or $CHCl_3$, the usual solvents for these brominations, bromide ion is the sole nucleophile available. If other nucleophiles are present, including the solvent itself, then they should also attack the bromonium ion. Such is indeed the case (equations 9.22 and 9.23). Finally, electron-donating groups such as methyl, when substituted for a hydrogen atom, increase the reactivity of an alkene toward bromination; electron-attracting groups such as chlorine have the opposite effect.

(9.22) $$Br_2 + CH_2{=}CH_2 \xrightarrow{CH_3OH} BrCH_2CH_2OCH_3 + BrCH_2CH_2Br + HBr$$

(9.23) $$Br_2 + CH_2{=}CH_2 \xrightarrow[NaCl]{H_2O} BrCH_2CH_2Cl + BrCH_2CH_2OH + BrCH_2CH_2Br + HBr$$

The bromonium ion has obvious similarities to the π-complex postulated for acid-catalyzed additions to alkenes. An important difference lies in the fact that bromine has nonbonding electrons and is able to form *normal single bonds* to both carbon atoms. This leads to a more stable system and also to less carbonium ion character. Characteristic carbonium ion rearrangements are not observed, even when favored (equation 9.24).

(9.24)

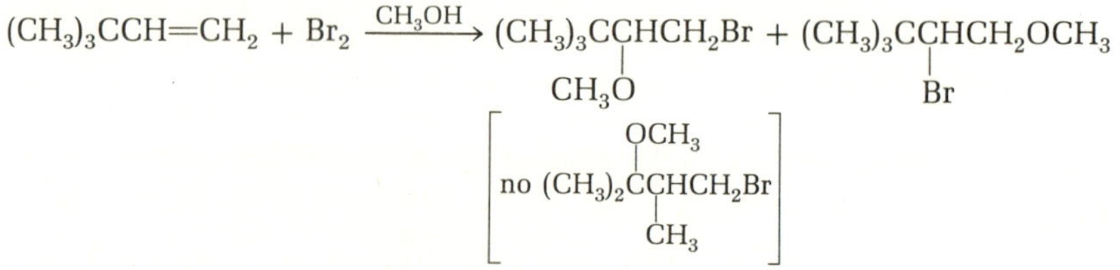

Similar mechanisms apply to chlorinations that are also stereospecific in the absence of radical chains. Chlorinations are much less common than brominations in preparative work because chlorine is rather inconvenient to handle and to measure.

(9.25)

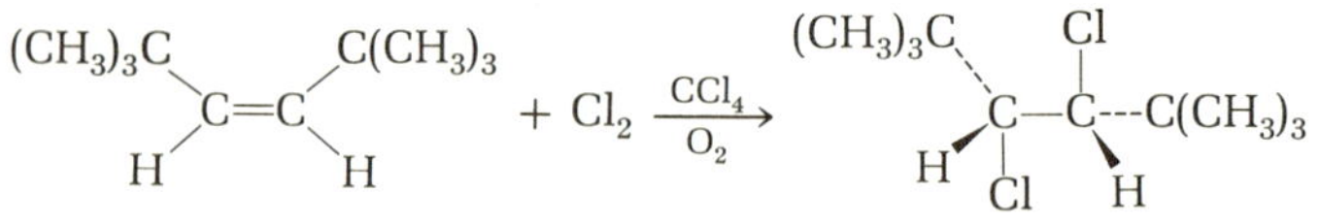

(sole product)

(9.26)

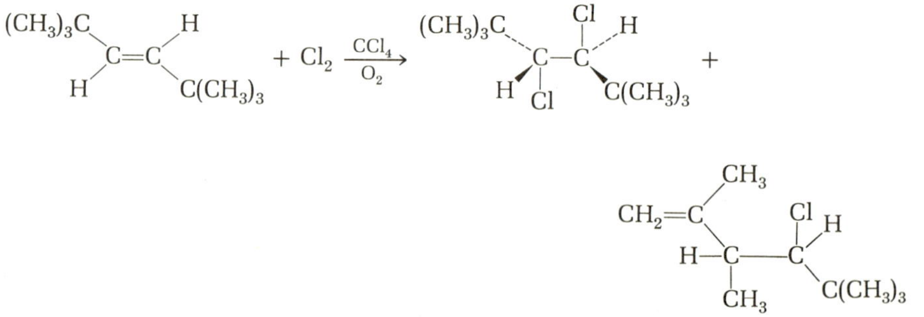

(from carbonium ion-type rearrangement)

In the chlorination examples shown, the stereospecificity of addition is remarkable, particularly with regard to the *cis* isomer (equation 9.25), since the bulky *t*-butyl groups interfere strongly with one another.

The halonium ion hypothesis was put on a firmer basis when it was found possible to generate these ions in very nonnucleophilic media at low temperatures. Direct observations of chloronium, bromonium, and iodonium ions formed by this route have been made by means of nmr spectroscopy.

(9.27) 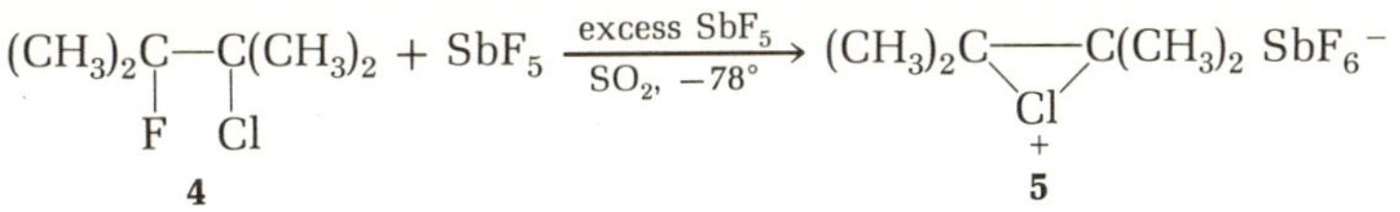

When a dihalide such as **4** is converted to a chloronium ion (**5**, equation 9.27), the methyl groups become more deshielded by about 1 ppm. In a classical rearranging carbonium ion (equation 9.28) the methyl groups are deshielded to the extent of about 2 ppm compared to covalent compounds.

$$(9.28)\quad (CH_3)_3C\overset{+}{C}(CH_3)_2 \underset{SO_2,\ -78^\circ}{\overset{SbF_5}{\rightleftharpoons}} (CH_3)_2\overset{+}{C}C(CH_3)_3$$

Addition of the ion to methanol and base resulted in the production of the expected products of nucleophilic attack.

(9.29) 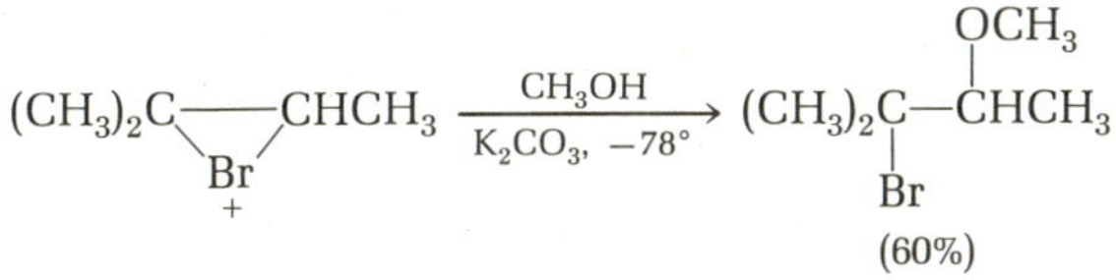

Evidence such as that noted above argues strongly for the accessibility and stability of halonium ions. Similar experiments have shown that fluoronium ions *cannot* be generated, an observation consistent with the high electronegativity of fluorine.

PROBLEM

1 When the bromofluoride I is dissolved in SbF_5-SO_2 at -60°, it gives rise to two halonium ions that appear to be *cis-trans* isomers.

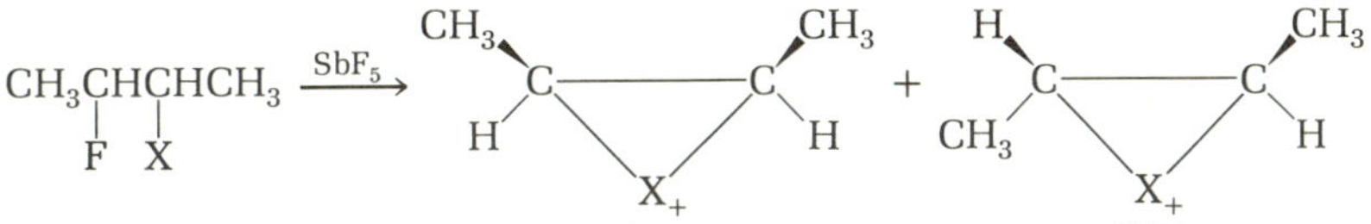

I: X = Br
II: X = I

(1)

Even if a diastereomerically pure starting material is used (*threo*-I, for instance), a mixture of products is obtained. The iodofluoride II behaves somewhat differently. *Threo*-II gives rise to only one ion under the reaction conditions, while *erythro*-II yields only one of the isomeric ions.

a. Which ion arises from each of the diastereomers of II?

b. Account for the difference of reaction stereospecificity for I and II.

CARBONIUM VERSUS HALONIUM IONS

Although halonium ions are common intermediates in ionic halogenations, their formation is not inevitable. The formation of a halonium ion may be viewed as an *intramolecular* nucleophilic attack on a carbonium ion that naturally competes with proton loss and with intermolecular reactions.

Although most simple alkenes give high yields of 1,2-dihalides (*vicinal* dihalides) by way of halonium ions, rearranged dihalides and haloalkenes are sometimes obtained (equation 9.26). These products most likely arise from normal, open carbonium ions, and they occur more frequently in chlorinations than in brominations. Since chlorine is more electronegative than bromine, it is less able to support a positive charge, so chlorine is less effective as a bridging atom. The intervention of nonbridged, open carbonium ions, as evidenced by the formation of nonstereospecific and/or rearranged products, is enhanced if a particularly stable carbonium ion can be formed. This seems to be the case in the norbornyl system (p. 160), which rearranges to a considerable degree upon bromination.

(9.30)

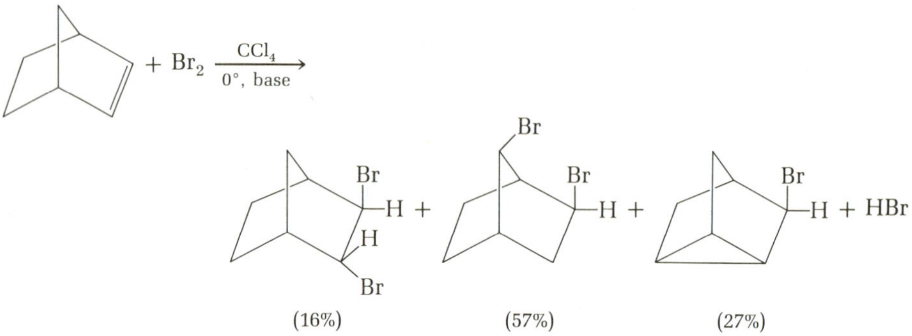

INTERHALOGENS AND HYPOHALOUS ACIDS

There are a number of interhalogen and pseudohalogen compounds that add to double bonds in exactly the same manner as the molecular halogens. A representative collection of these compounds is listed below; they are depicted with the electrophilic atom on the left.

BrCl, BrN_3, IF, ICl, IBr, IN_3, INCO

Related to these are the hypohalous acids, ClOH, and BrOH, that add to double bonds to give halohydrins (compounds containing a halogen atom and a hydroxyl group). The hypohalous acids are generated as needed from hypohalite salts and acid or from molecular halogens and water; their addition reactions are acid-catalyzed, the true reactant apparently being XOH_2^+.

(9.31) 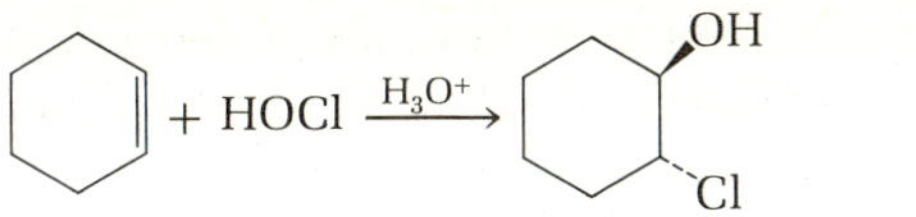

(70%)
trans-2-chlorocyclohexanol
(cyclohexene chlorohydrin)

ANALYTICAL PROCEDURES

A solution of bromine in carbon tetrachloride or chloroform is a deep red-brown color. If several drops of such a solution are added to a solution of a compound suspected of containing unsaturation, rapid discharge of the color is taken as presumptive evidence of its presence. While double bonds are not the only functional group that reacts with bromine, few others react with such rapidity and without evolution of hydrogen bromide. It must be stressed that the rate of reaction of bromine with alkenes is many, many orders of magnitude faster than with alkanes under comparable conditions.

Alkenes may be titrated with bromine as a means of determining unsaturation quantitatively. Iodine monochloride has been used for the same purpose.

THERMODYNAMICS OF HALOGEN ADDITION

To understand why iodination is not a generally useful procedure, it is helpful to calculate representative heats of halogenation, using the average bond strengths listed in Chapter 2.

(9.32) 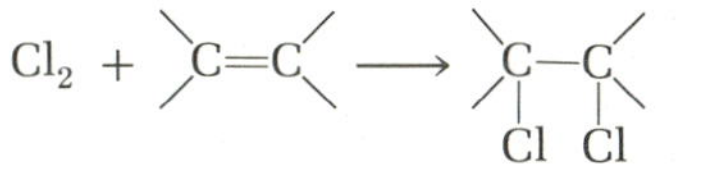 $\Delta H = -41$ kcal/mole

+58 +146 −83 − (2 × 81)

(9.33) 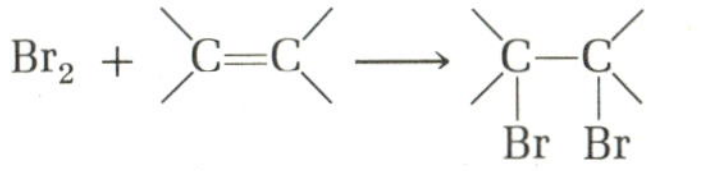 $\Delta H = -27$ kcal/mole

+46 +146 −83 − (2 × 68)

(9.34) 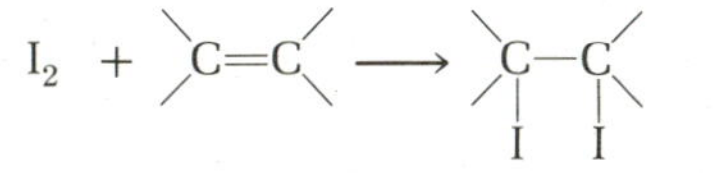 $\Delta H = -7$ kcal/mole

+36 +146 −83 − (2 × 53)

The value for iodination is sufficiently small that the unfavorable entropy term becomes important; the equilibrium constant for addition is also very sensitive to double bond substitution.

In order to appreciate the importance of thermodynamic effects in answering the practical question of whether or not a reaction will "go," it is instructive to compare carbon-carbon with carbon-oxygen double bonds. The equilibrium constant for addition of water to isobutylene is about 5 (equation 9.7). Using values of 91 and 97 kcal/moles for the average bond energies of C—O and C—H bonds, respectively, we may calculate that this reaction should be exothermic by roughly 6 kcal/mole.

(9.35)

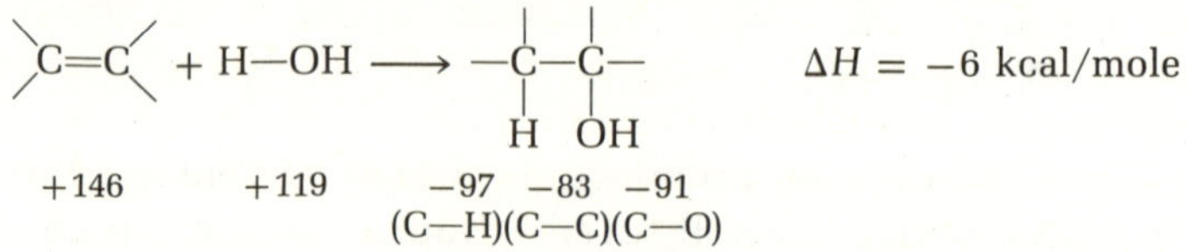

An analogous calculation may be carried out for hydration of a carbon-oxygen double bond, by taking 110 kcal/mole as the average bond energy of O—H.

(9.36)

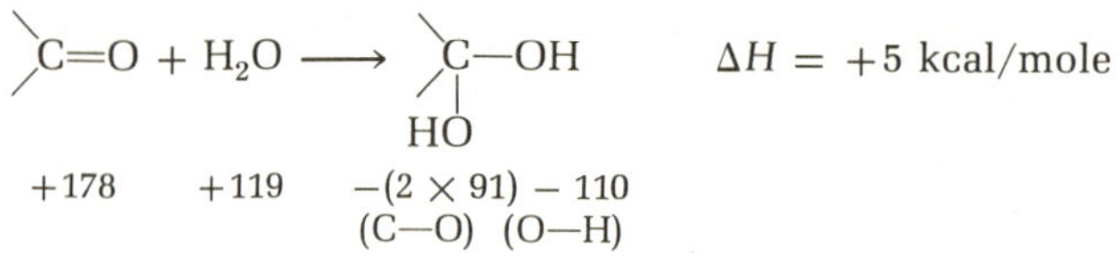

It is not surprising that the equilibrium constant for addition of water to acetone is only 0.002.

(9.37) $(CH_3)_2C{=}O + H_2O \rightleftharpoons (CH_3)_2C(OH)_2 \qquad K^{25^\circ} \sim 0.002$

While solvation effects will play some part, the 30 kcal difference in >C=C< and >C=O bond strengths is undoubtedly decisive.

OXIDATIVE ADDITION TO DOUBLE BONDS

Although halogen addition to an alkene may be viewed as an oxidation, we shall reserve that term for reactions in which oxygen atoms become directly attached to the doubly bonded carbon atoms. These reactions proceed in two stages; in the first, the π-bond is attacked and two C—O bonds replace the C=C π-bond. The presence of these oxygen atoms weakens the remaining σ-bond and it may be cleaved by further oxidation. All these reactions are driven by the tendency of an atom in the oxidant (usually, though not always, a metal) to attain a lower valence state.

Treatment of an alkene with a dilute, basic solution of potassium permanganate in water or aqueous organic solvents, at room temperature or below, results in the addition of one hydroxyl group at each end of the double bond. The addition is stereospecifically *syn*.

(9.38)

$$3(CH_3)_2CHCH{=}CH_2 + 2KMnO_4 + 4H_2O \longrightarrow 3(CH_3)_2CH\underset{\displaystyle OH}{\underset{|}{C}}HCH_2OH + 2MnO_2 + 2KOH$$

3-methylbutane-1,2-diol (50%)

The reaction involves a cyclic manganese derivative, which accounts for the stereochemistry.

(9.39)

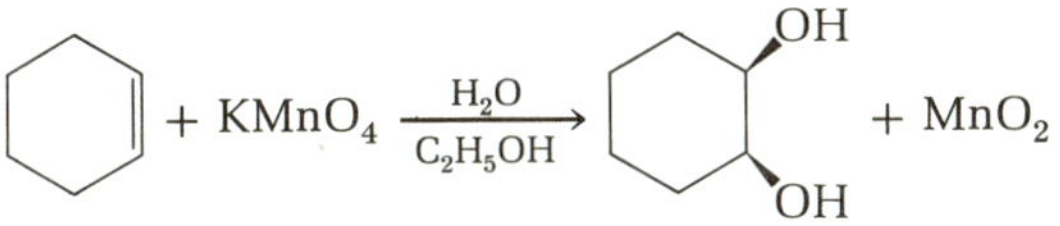

(33%)

Hydroxide ion catalyzes the decomposition of this intermediate and prevents further oxidation.

(9.40)

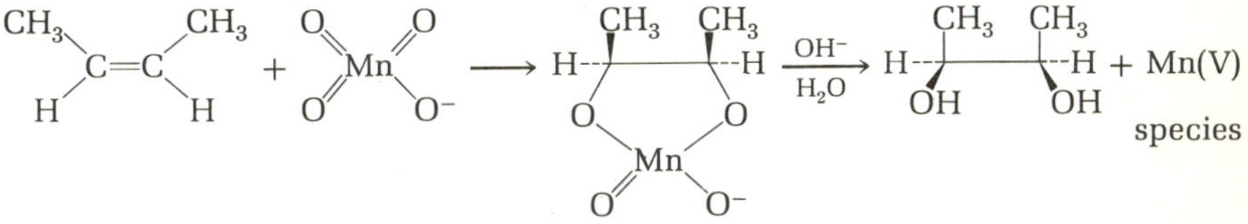

The discharge of the intense purple color of a dilute permanganate solution, with simultaneous deposition of brown manganese dioxide, has been taken as a positive test for unsaturation. Unfortunately several other functional groups, such as alcohols, will also react.

Osmium tetroxide, OsO_4, accomplishes the same type of *syn* hydroxylation that potassium permanganate does, and with greater efficiency. A cyclic intermediate is again involved. The use of osmium tetroxide has been limited because of its high cost and severe toxicity.

Diols are often known by the class name **glycols** (Greek *glykys* — sweet; many sweet tasting compounds are polyols). Individual compounds in this class take their names from the olefins from which they are derived: ethylene glycol, $HOCH_2CH_2OH$; propylene glycol, $HOCH_2\underset{\displaystyle OH}{\underset{|}{C}}HCH_3$; *etc.* A method of forming glycols by overall addition of 2 —OH groups with *anti* stereochemistry is discussed in Chapter 10.

OXIDATIVE CLEAVAGE OF DOUBLE BONDS

The carbon-carbon single bond between the two hydroxyl-bearing carbon atoms (in a glycol) is very susceptible to oxidative cleavage, in contrast to most other carbon-carbon single bonds. Reaction of the glycol with the oxidizing agent leads to a cyclic derivative that cleaves in the course of an internal rearrangement of electrons (equation 9.41).

(9.41) 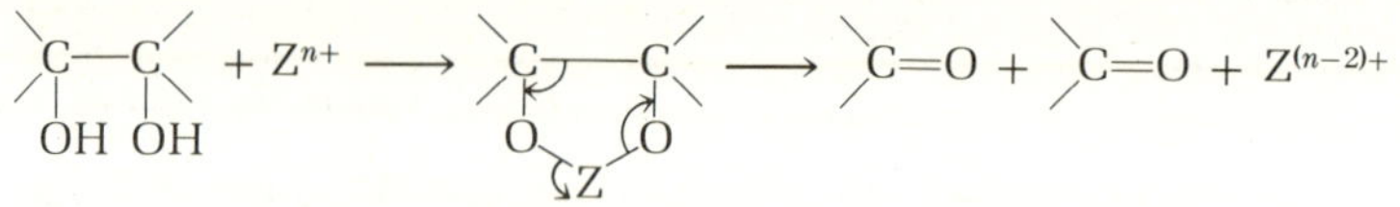

The driving force derives from the reduction in the valence state of an atom in Z by two units and by the relatively high C—O π-bond strength.

The effect of the hydroxyl groups on the intermediate carbon-carbon single bond is illustrated by the ease with which this bond is cleaved by a number of reagents. Among the most useful is periodic acid, which can be employed for quantitative determinations.

(9.42) 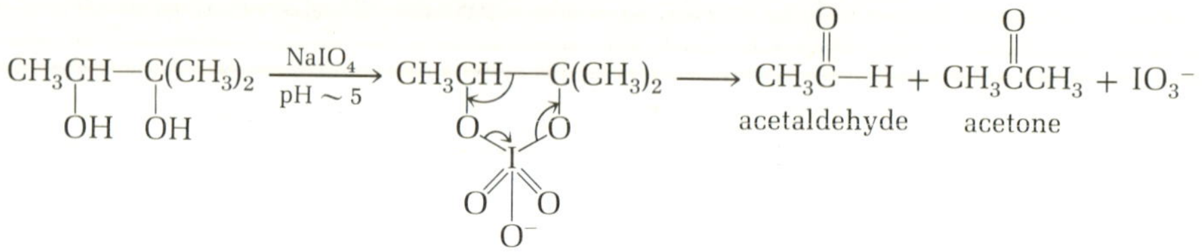

(9.43) 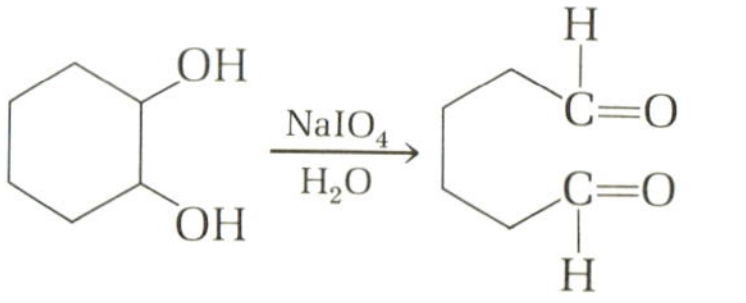

adipaldehyde
(hexanedialdehyde)

If a cleavage product contains only one alkyl group, it is an **aldehyde,** while those with two alkyl groups attached to the >C=O are called **ketones** (Chapter 11). In this way, by successive hydroxylation (addition of —OH groups) and oxidation of a double bond, and identification of the fragments, it is possible to reconstruct the structure of the original alkene. The hydroxylation and cleavage may be accomplished in one step.

The intervention of a cyclic intermediate in the oxidation reaction with periodate may be inferred from the fact that certain *trans*-glycols in which the hydroxyl groups are far apart will not undergo cleavage.

(9.44) 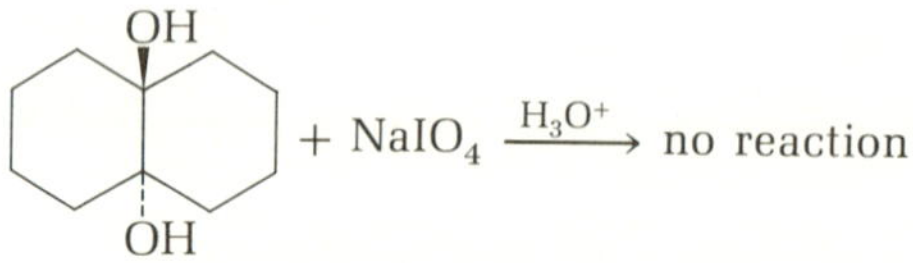

Stronger oxidizing agents will cleave both σ- and π-bonds in one step; if one of the sp^2-hybridized carbon atoms carries a hydrogen atom, then one of the products will be a **carboxylic acid.** Typical oxidizing agents are acidic or basic permanganate solution or aqueous chromic acid.

(9.45) 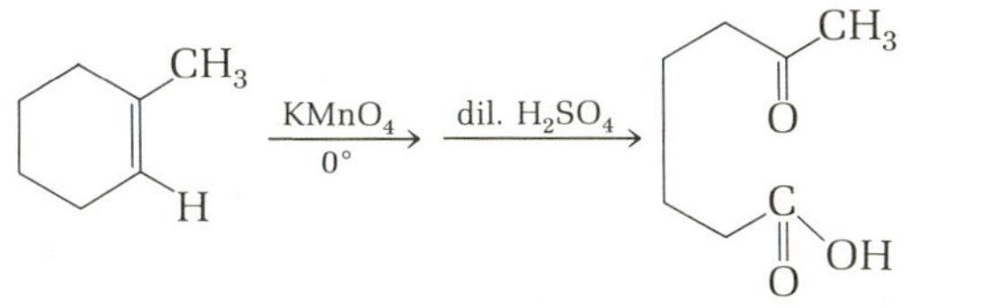

6-oxoheptanoic acid

(9.46) $$(CH_3)_2C{=}C(CH_3)_2 \xrightarrow[H_2SO_4,H_2O]{CrO_3} (CH_3)_2C{=}O + (CH_3)_3C\overset{\displaystyle O}{\overset{\|}{C}}CH_3 + (CH_3)_3C\overset{\displaystyle O}{\overset{\|}{C}}OH$$

These oxidations are often complicated by side reactions and rearrangements (equation 9.46). For specific double bond oxidation, especially for the purpose of locating the double bond in the molecule, a mixture of potassium permanganate and sodium periodate is very useful. Many of the oxidizing agents that attack double bonds do not restrict their attentions to that functional group but will also attack alcohols, aldehydes, thiols, etc.

(9.47) $$CH_3CH_2CH{=}C(CH_3)_2 \xrightarrow[25°,\ H_2O]{KMnO_4\text{-}NaIO_4} \underset{\text{propionic acid}}{CH_3CH_2\overset{\displaystyle O}{\overset{\|}{C}}OH} + (CH_3)_2C{=}O$$

(9.48) $$CH_3CH_2CH_2CH_2CH{=}CH_2 \xrightarrow[NaIO_4]{KMnO_4} \underset{\text{valeric acid}}{CH_3CH_2CH_2CH_2\overset{\displaystyle O}{\overset{\|}{C}}OH} + \underset{\text{formaldehyde}}{CH_2{=}O}$$

This will often necessitate the use of specialized reagents or, alternatively, the protection of some functional groups.

Ozone, O_3, reacts with few functional groups other than double bonds. It has been used principally as an aid in structure determination, although it can also be used for preparative purposes. It is obtained in the laboratory by passing oxygen gas into an "ozonizer," where a continuous electrical discharge produces a mixture of 3 to 8 weight percent ozone in oxygen. The gaseous mixture is then bubbled through a solution of the alkene in an inert solvent (chloroform, methanol, and ethanol are quite common), usually at low temperatures.

Present evidence indicates that ozonization proceeds through at least two intermediates. The first results from attack of ozone on the π-bond, giving a very unstable molozonide that undergoes a rearrangement involving σ-bond cleavage, leading to the far more stable ozonide.

(9.49) 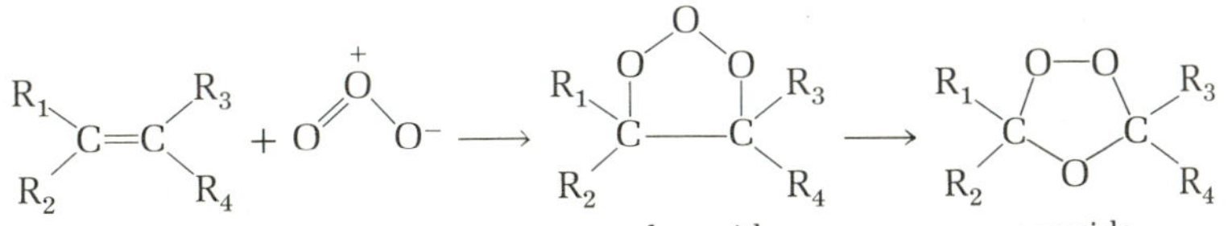

molozonide ozonide

In most cases the ozonide is not isolated but subjected directly to oxidative or reductive work-up, depending on the products desired.

(9.50)

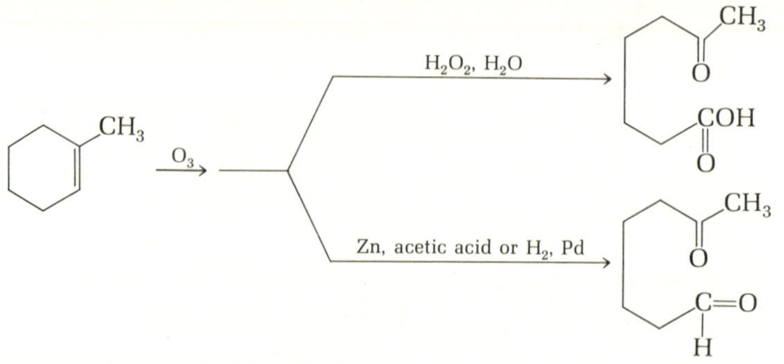

Note that none of these degradative reactions gives any information about the *stereochemistry* of the original alkenes — one must still rely on spectroscopic data. Ozonization is an oft-used method for locating the position of the double bond in an alkene.

REDUCTION OF DOUBLE BONDS

The "saturation" of double bonds by addition of hydrogen is a process that has seen repeated use in both laboratory and plant. Although the addition is strongly exothermic, mixing together hydrogen and an alkene at any reasonable temperature produces no change. The barrier to reaction lies in the 104 kcal/mole needed to dissociate molecular hydrogen. For this reason reductions involving molecular hydrogen require a catalyst.

Numerous metals are known that **adsorb** hydrogen on their surfaces; some cause the hydrogen to dissociate, so that hydrogen atoms are actually the adsorbed species. These metals adsorb organic molecules as well, particularly those with polarizable electrons, such as alkenes. The metal surface thus acts as a molecular matchmaker, bringing together the reacting species. Most hydrogenation catalysts are completely insoluble in organic solvents so that the reaction is heterogeneous, taking place on the catalyst surface.

Platinum, palladium, and nickel are the metals most often used in the hydrogenation reaction. The catalyst surface is not uniformly active; there are, rather, regions called **active sites** that seem to be responsible for most of the catalytic activity. The means by which the catalyst is prepared may

FIGURE 9-2
Adsorption and Dissociation of Hydrogen on Platinum

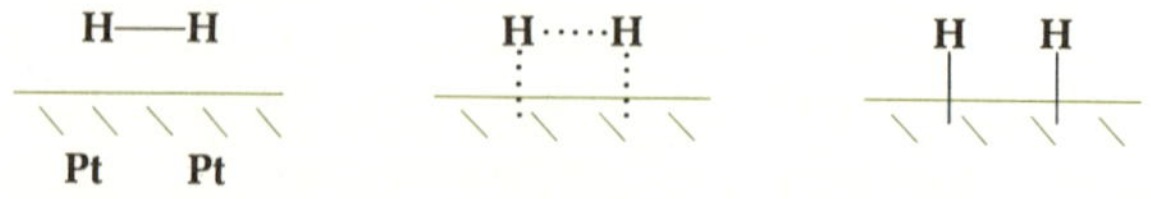

profoundly affect the number and nature of these sites. The metals may be used by themselves (platinum and nickel) or dispersed on the surface of a "support." Supporting materials have no catalytic properties of their own, but they do modify those of the dispersed metal. Common supports include activated charcoal, various metal carbonates and sulfates, and alumina (Al_2O_3). It is vital that any catalyst, supported or not, have the maximum amount of active surface area.

(9.51) $PtO_2 + 2H_2 \longrightarrow Pt + 2H_2O$

(9.52) $PdCl_2 + NaBH_4 \xrightarrow[H_2O]{\text{charcoal}} NaCl + H_3BO_3 + Pd/C$ (palladium on carbon)

(9.53) Ni-Al (Raney alloy) + NaOH $\longrightarrow$ $NaAl(OH)_4 + H_2$ + Ni (Raney nickel)

Reductions usually go to completion and give very high yields.

(9.54) $(CH_3)_2C{=}CHCH_3 + H_2\ (3\ atm) \xrightarrow[\text{ethanol, 25°}]{PtO_2} (CH_3)_2CHCH_2CH_3$ (100%)

(9.55) 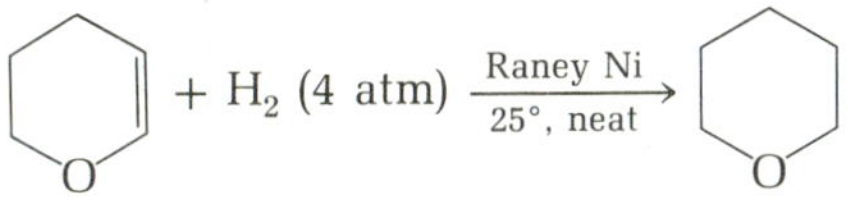

(100%)

(9.56) 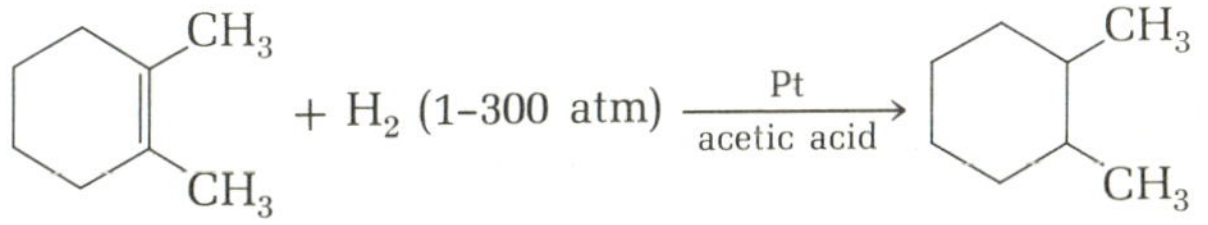

(*cis*, 82–95%)
(*trans*, 5–18%)

It is customary to use a solvent, ethanol being the most common. Hydrogen uptake may be measured with reasonable accuracy so that hydrogenation is useful for quantitative analysis. The heats of hydrogenation may also be measured, and these data provide a means of calculating the energy differences between alkenes that give the *same* alkane upon reduction. The energy differences among the butenes, as given by the hydrogenation study, are very close to those found from the equilibrium among the straight-chain heptenes (Chapter 8).

(9.57) $CH_3CH_2CH{=}CH_2 + H_2 \xrightarrow[82°]{\text{gas phase}} CH_3CH_2CH_2CH_3 \qquad \Delta H = -30.3$ kcal/mole

(9.58) 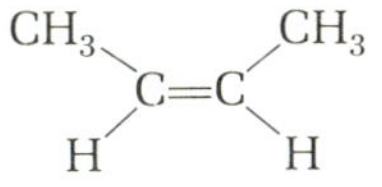

$+ H_2 \xrightarrow[82°]{\text{gas phase}} CH_3CH_2CH_2CH_3 \qquad \Delta H = -28.57$ kcal/mole

(9.59) 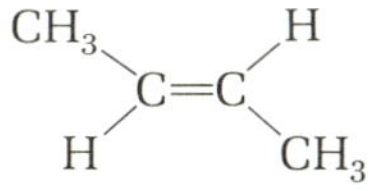

$+ H_2 \xrightarrow[82°]{\text{gas phase}} CH_3CH_2CH_2CH_3 \qquad \Delta H = -27.62$ kcal/mole

(9.60) 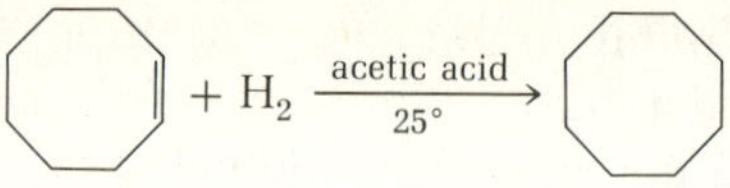 $\Delta H = -22.98$ kcal/mole

cis

(9.61) 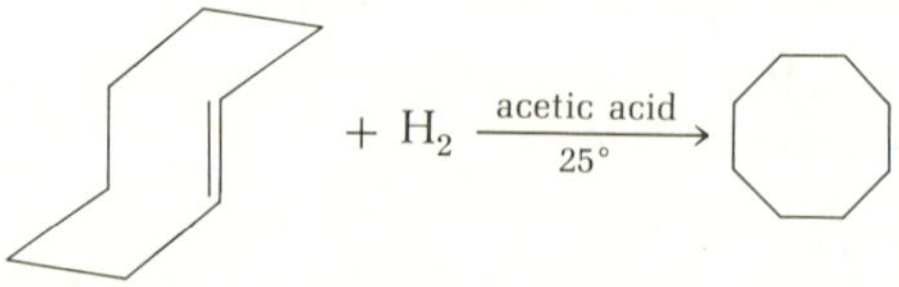 $\Delta H = -32.24$ kcal/mole

trans

The heat of hydrogenation of isobutylene cannot be directly compared to those of the straight-chain butenes since the product in this case is isobutane rather than *n*-butane. Relative energies of all the butene isomers may be arrived at, however, by combustion measurements since the same numbers of moles of water and carbon dioxide are produced from each of the isomers.

(9.62) $$C_4H_8 + 6O_2 \longrightarrow 4CO_2 + 4H_2O + \text{heat}$$

HYDROBORATION

Double bond reduction may also be accomplished indirectly, the hydrogen being added consecutively rather than simultaneously. Diborane, B_2H_6, generated from sodium borohydride and boron trifluoride in ether solvents, is very useful for this purpose.

(9.63) $$3NaBH_4 + 4BF_3 \xrightarrow[\text{solvents}]{\text{ethereal}} 2B_2H_6 + 3NaBF_4$$

This unusual molecule contains several features of great interest. Since boron has only three electrons in its valence shell, neutral boron compounds are trivalent. The hydride of boron, borane (BH_3), contains only six electrons in its valence shell and is therefore a Lewis acid. Borane is not an isolable compound; however, its electron deficiency is partly overcome by the formation of diborane that contains three center bonds composed of two boron atoms and one proton that share one pair of electrons (Figure 9-3).

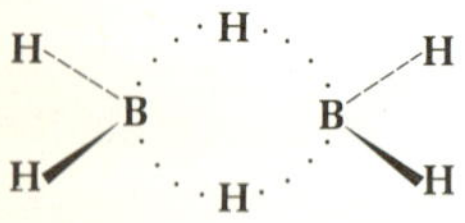

FIGURE 9-3
Molecular Structure of Diborane

The resemblance between diborane and a bent bond ethylene is striking. In fact, by removing one proton from each carbon nucleus of ethylene and placing it squarely in the middle of the bent bond, the olefin would be transformed into the isomeric diborane.

In Lewis base solvents, such as ethers, diborane dissociates into solvated borane. Most likely it is the monomeric borane that reacts with the olefin.

(9.64) $$B_2H_6 + 2R\text{—}\ddot{O}\text{—}R \rightleftharpoons 2H_3B\!:\!OR_2$$

The electrophilic boron atom initiates the reaction, and a shift of a hydride ion acting as an internal nucleophile completes the addition. The process is repeated until all the hydrides have been added.

(9.65)

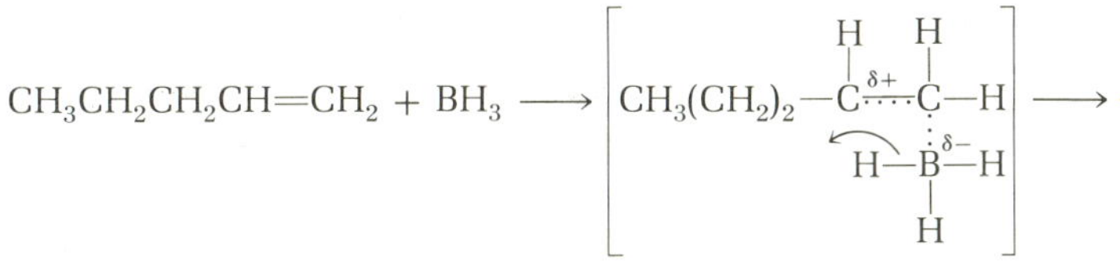

transition state

$$CH_3(CH_2)_3CH_2BH_2 \xrightarrow{C_5H_{10}} \xrightarrow{C_5H_{10}} [CH_3(CH_2)_4]_3B$$

tris(1-pentyl)borane

CLEAVAGE OF ORGANOBORANES

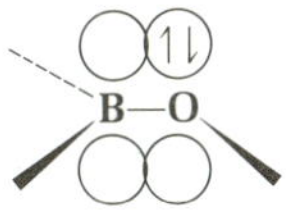

FIGURE 9-4
Overlap of Empty and Filled *p*-Orbitals on Neighboring Boron and Oxygen Atoms

Boron forms very strong bonds with atoms such as oxygen because the filled *p*-orbitals on oxygen may overlap with the empty *p*-orbital of trivalent boron (Figure 9-4). One may take advantage of this affinity to replace a boron atom bound to carbon by other elements.

When organoboranes are treated with organic acids such as acetic acid, they are cleaved to alkanes and boron triacetate, resulting in a net reduction of the starting alkene.

(9.66) $$(C_5H_{11})_3B + 3CH_3\overset{O}{\overset{\|}{C}}OH \longrightarrow 3CH_3(CH_2)_3CH_3 + (CH_3\overset{O}{\overset{\|}{C}}O)_3B$$

(9.67)

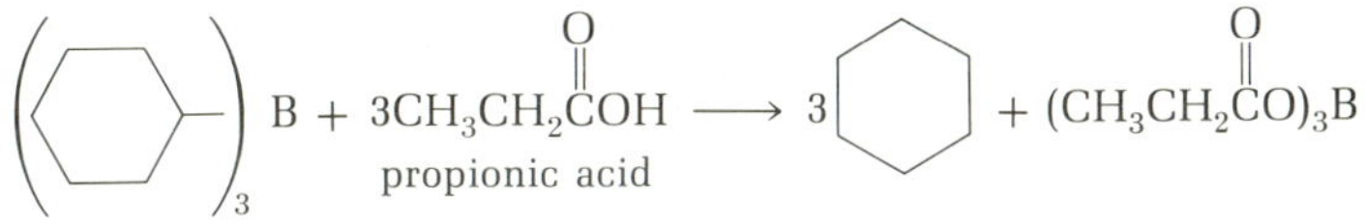

The reaction rate is enhanced by the ability of the acid to form a complex with the borane. Use of B_2D_6 and/or CH_3CO_2D allows the introduction of deuterium into a specific position in the molecule.

(9.68)

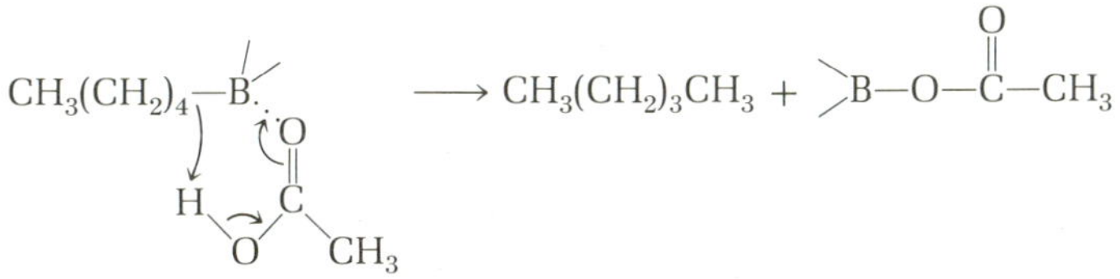

Organoboranes may be oxidatively cleaved to alcohols by basic hydrogen peroxide. This reaction is also "driven" by the oxygen affinity of boron.

(9.69) $$[CH_3(CH_2)_4]_3B + 3H_2O_2 \xrightarrow{OH^-} 3CH_3(CH_2)_4OH + \underset{\text{boric acid}}{B(OH)_3}$$

(9.70) 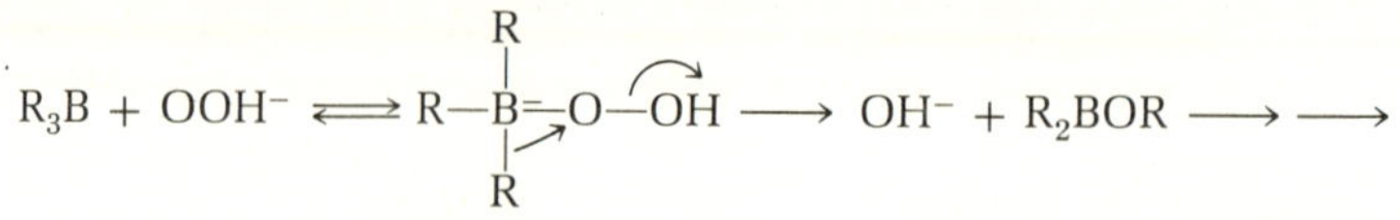

$$(RO)_3B \xrightarrow[OH^-]{H_2O} 3ROH + B(OH)_3$$

Since the borane undergoes stereospecific *syn* addition to the alkene, and replacement of the boron atom by oxygen proceeds with retention of configuration, the stereochemistry of the alcohol can be predicted with confidence.

(9.71) 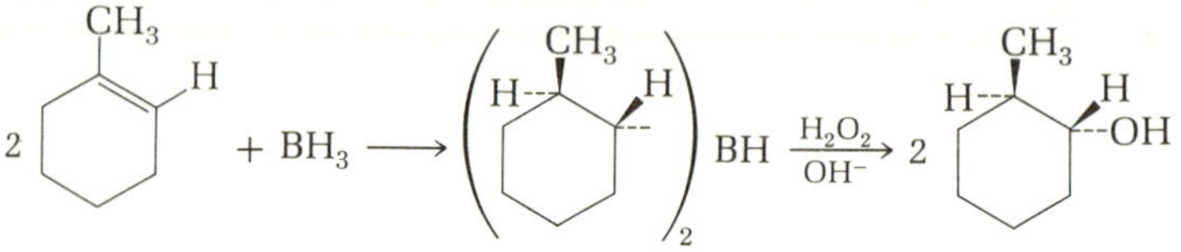

The hydroboration-oxidation method is referred to as "anti-Markovnikov" hydration because the product has the opposite orientation to that obtained by treatment of the same alkene with aqueous acid.

(9.72) 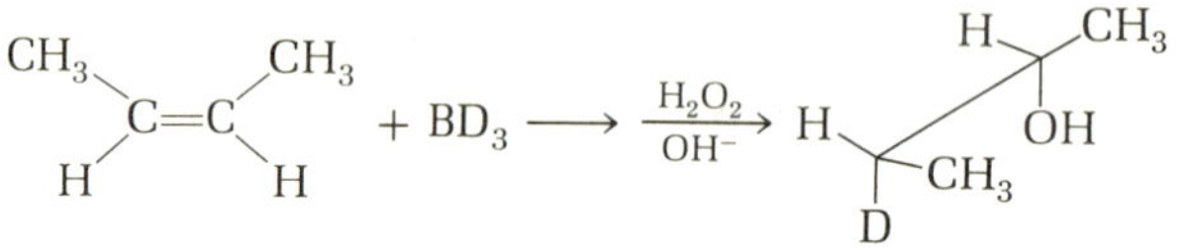

erythro

Mechanistically speaking, the Markovnikov rule is obeyed in hydroboration reactions; it must be remembered that the boron atom rather than the hydrogen atom is the electrophile.

FREE RADICAL ADDITIONS AND POLYMERIZATION

It is by no means always easy to decide whether a particular reaction proceeds by way of a radical or an ionic mechanism. Since radicals are electron-deficient species, they exhibit a marked tendency to attack carbon-carbon π-bonds, and numerous bond additions occur by way of free radical mechanisms.

Early investigators were puzzled by the reaction of hydrogen bromide with alkenes such as propene because mixtures of 1- and 2-bromopropane of variable composition were often produced. It was later shown that when pure alkenes and solvents were used, only 2-bromopropane — the product of normal ionic addition — could be found. 1-Bromopropane was apparently formed by a free radical process, initiated by impurities in the reactants, usually peroxides. The free radical chain can also be initiated by the photodissociation of hydrogen bromide, using ultraviolet light.

(9.73) $R{-}O{-}O{-}R \xrightarrow[\text{or heat}]{h\nu} 2RO\cdot$

(9.74) $RO\cdot + HBr \longrightarrow ROH + Br\cdot$

(9.75) $Br\cdot + CH_2{=}CH{-}CH_3 \longrightarrow BrCH_2\dot{C}HCH_3$

(9.76) $BrCH_2\dot{C}HCH_3 + HBr \longrightarrow BrCH_2CH_2CH_3 + Br\cdot$

Many other molecules may be added to double bonds by free radical processes. (The 2- and 3-bromohexane in equation 9.77 arise because under radical conditions some of the 1-hexene isomerizes to 2-hexene.) The principal requirement is that these molecules contain one bond that can undergo homolytic breakage with relative ease.

(9.77)
$$CH_3(CH_2)_3CH{=}CH_2 + HBr \xrightarrow[4^\circ]{h\nu} \underset{(83\%)}{CH_3(CH_2)_5Br} + \underset{(10\%)}{CH_3(CH_2)_3\overset{\overset{\large Br}{|}}{C}HCH_3} + \underset{(7\%)}{CH_3CH_2CH_2\overset{\overset{\large Br}{|}}{C}HCH_2CH_3}$$

(9.78) $$(CH_3)_2C{=}CH_2 + CH_3CH_2SH \xrightarrow[100^\circ]{\text{peroxides}} \underset{(94\%)}{(CH_3)_2CHCH_2SCH_2CH_3}$$

(9.79) $$CH_3(CH_2)_5CH{=}CH_2 + CCl_4 \xrightarrow[\text{peroxide}]{\text{acetyl}} \underset{(85\%)}{CH_3(CH_2)_5\overset{\overset{\large Cl}{|}}{C}HCH_2CCl_3}$$

Ultraviolet light is often an effective initiator for these reactions. Molecular halogen addition in the presence of ultraviolet light and in the absence of oxygen (which intercepts free radicals) occurs by way of a free radical chain mechanism. Exclusive *anti* addition then no longer prevails.

The hydrogen bromide addition products are labeled "anti-Markovnikov" products since they are the opposite of those obtained by the normal ionic process. Bromine atom attack proceeds to give the more highly substituted carbon radical, however, just as proton attack in the ionic case leads to the more substituted carbonium ion. A more useful and comprehensive statement of Markovnikov's rule might be: **The attack of an electrophilic ion or radical on a double bond will take place in such a manner that the most stable intermediate carbon radical or ion will be produced.**

PROBLEM

2 Although HBr may be added in an "anti-Markovnikov" (free radical) fashion to double bonds, neither HCl nor HI will undergo addition in this fashion. Rationalize these results, paying particular attention to the fact that the addition is a two-step process.

Under some conditions free radical additions to double bonds can display a high degree of stereospecificity. There is some evidence that bromine, for example, can act as a bridging atom in radicals (such as **6**), analogous to its role in bromonium ions.

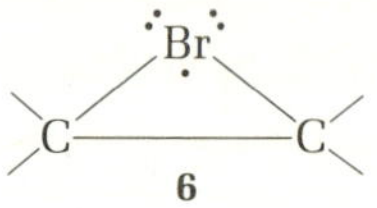

6

These bridged radicals (in which there are nine electrons in the valence shell of bromine) are much less stable than bromonium ions and open very readily to unbridged radicals.

When a free radical such as a bromine atom attacks an alkene, a carbon radical is produced. This carbon radical may abstract an atom from any reactive species present (HBr, Br_2, CCl_4, etc.) or it may attack another molecule of alkene. The second path will be favored if the concentration of alkene is high and the concentrations of other reactive substrates are low.

(9.80) $$Y\cdot + CH_2{=}CH_2 \longrightarrow Y{-}CH_2CH_2\cdot + CH_2{=}CH_2 \xrightarrow{\text{etc.}} \underset{\text{polyethylene}}{Y{+}CH_2CH_2{+}_n Z}$$

As this process continues, a **polymer** is formed (Greek *polymeres* — of many parts). The individual molecular units that comprise the polymer are known as **monomers.** The mode of polymerization illustrated by equation 9.80 is **addition polymerization,** in which the joining together of the monomer units entails no loss of atoms. Individual polymer molecules have a wide variety of molecular weights, but the average molecular weights of commercial polymers ordinarily lie in the range of 10^4 to 10^6.

Polyethylene is manufactured by several different processes. The most common involves polymerization under fairly high pressures (1500 atm) with oxygen as the free radical initiator. Not all addition polymerization is of a free radical nature, however. The initiator may be a powerful anionic nucleophile or a cationic acid.

(9.81)

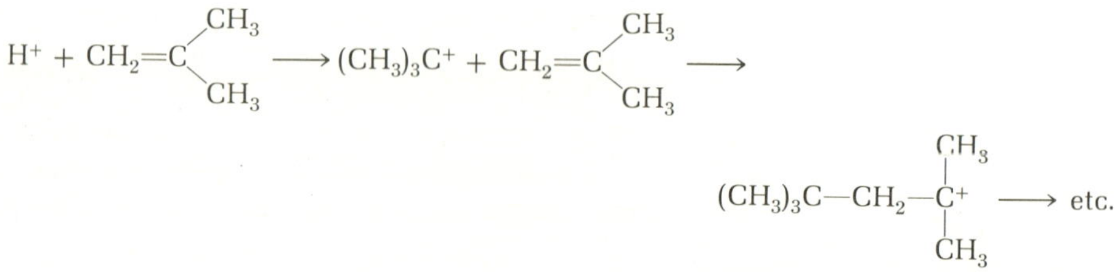

Note that most all polymerizations are of the "head-to-tail" type because the more stable intermediate radical or ion tends to be produced at every step.

STEREOCHEMISTRY OF POLYMERIZATION

Ethylene may be polymerized at rather modest temperatures and pressures by use of a so-called Ziegler catalyst; a typical catalyst is made by mixing triethylaluminum, $(CH_3CH_2)_3Al$, and titanium tetrachloride, $TiCl_4$, in a hydrocarbon such as heptane. A complex series of reactions ensues, in which much insoluble material is produced. While the mechanism of action of these catalysts is still a matter under active investigation, it seems that it encompasses aspects of both cationic and anionic polymerization.

If a terminal alkene such as propylene is polymerized, every other carbon atom in the polymer (marked by asterisks) is asymmetric.

$$-CH_2\overset{CH_3}{\underset{*}{C}}HCH_2\overset{CH_3}{\underset{*}{C}}HCH_2\overset{CH_3}{\underset{*}{C}}HCH_2\overset{CH_3}{\underset{*}{C}}H-$$

Most processes result in polymers in which there is random distribution of configurations at the asymmetric carbons (**atactic** polymers) (Figure 9-5). If Ziegler-type catalysts are used, however, polymers are formed in which all the asymmetric centers have the same configuration (**isotactic** polymers) or in which there is a regular alteration of configuration (**syndyotactic** polymers) (Greek *iso* — same; *syndyo* — two together; *taktikos* — of order). Isotactic and syndyotactic polypropylene have markedly different physical properties compared to the atactic variety. The stereoregular polymers are generally denser, higher melting, less soluble, and mechanically stiffer than their randomly polymerized counterparts. Syndyotactic and isotactic polymers are often highly crystalline.

FIGURE 9-5
Diagrams of Atactic, Isotactic, and Syndyotactic Polypropylene

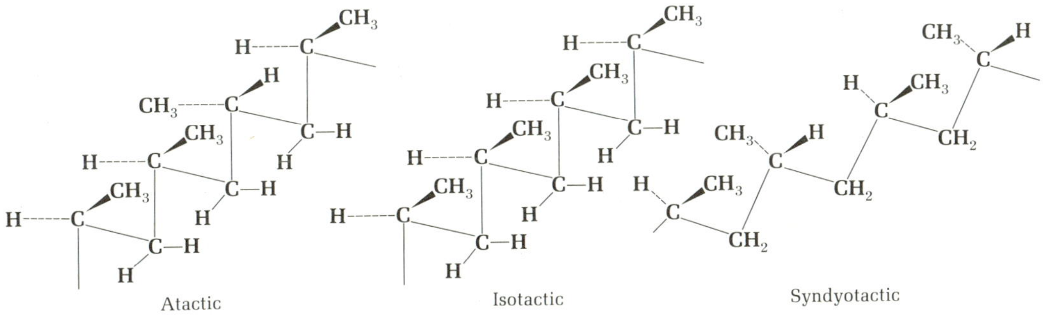

There can be no doubt that some of the most challenging and beautiful chemistry available for our scrutiny is to be found in living organisms. The systems are enormously complex, and rare indeed is the scientist who has even a general appreciation of all the forces at work. Yet the individual chemical events are relatively simple and quite analogous to those occurring in much simpler systems. It is this juxtaposition of simplicity and complexity, of unity and diversity that endows the study of biological chemistry with aesthetic as well as scientific appeal.

Biochemistry made an important advance when it was realized that the results obtained by physical organic chemists from studying reactions *in vitro* (literally "in glass") could aid in the understanding of processes *in vivo* (in the living organism). If attention was turned specifically to the functional group undergoing transformation, then the detailed reaction mechanism was often found to resemble very closely those deduced for laboratory transformations of the same functional group. The most striking difference between two similar reactions *in vitro* and *in vivo* is that the latter usually proceeds very much faster and under much milder conditions (moderate temperature, pH near neutral) and with a high degree of stereospecificity. These differences can be attributed to the intervention of **enzymes.**

Enzymes are biological catalysts. They all belong to a class of compounds known as **proteins** (biological polymers; see Chapters 17 and 25); they often have high molecular weights and usually contain a number of acidic and basic groups. The reactant(s) [always called substrate(s) in biochemical parlance] form(s) complexes with the enzyme that subsequently decompose, giving the product and regenerating the enzyme. Covalent bonds between substrate and enzyme are not necessary because the enzyme invariably contains several sites that permit hydrogen bonding and ionic, dipolar, and dispersion force attractions between itself and the substrate.

Enzymes often assume particular shapes due to intramolecular interactions, and the substrate must be able to fit itself to the enzyme. The various reactants are all attached to the enzyme and are held in the conformation most favorable for reaction. The enzyme thereby sharply lowers the free energy of activation for the reaction, with consequent acceleration of the rate. For instance, an example of an enzyme-catalyzed hydration of a double bond will shortly be discussed. The enzyme binds not only the alkene but the water molecule and the proton that initiates the reaction as well. The proton and the water molecule are probably held on opposite sides of the double bond, close to the carbon atoms to which they will become bound.

Some enzymes catalyze the reaction of only one substrate, while others are more versatile. Most of them are very sensitive to the absolute configuration of the substrate and will react with only one of an enantiomeric pair; furthermore, they form their products stereospecifically.

It is possible, though often quite difficult, to isolate an enzyme from tissue

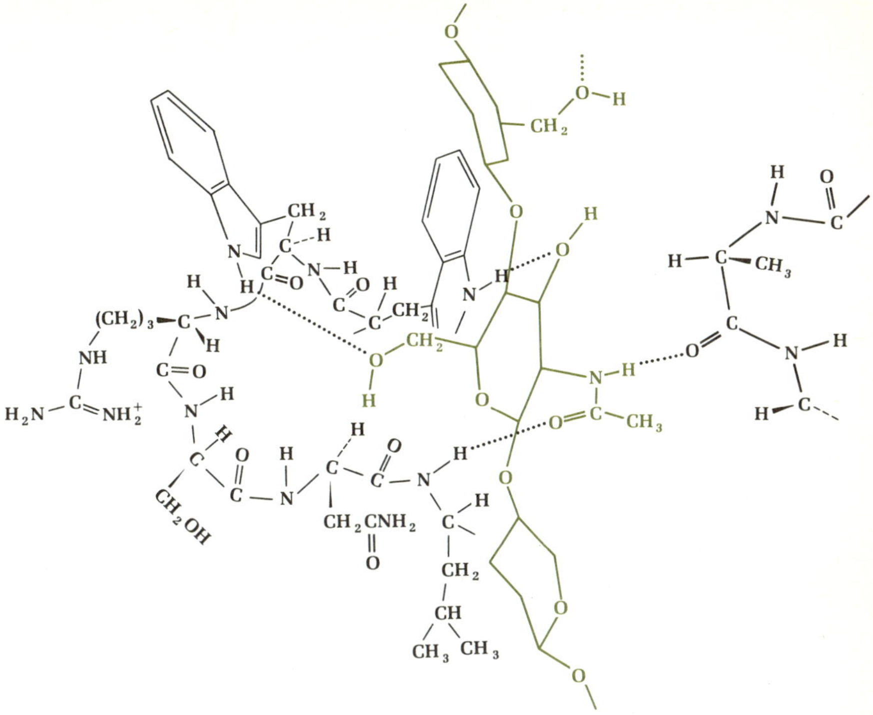

FIGURE 9-6
A Portion of the Enzyme Lysozyme and its Substrate (in color), Showing the Substrate Bound to the Enzyme by a Number of Hydrogen Bonds

with at least some of its catalytic power intact. This allows one to study the enzymatic reaction *in vitro*, provided that a suitable medium is present. The enzyme preparation may also be used for synthetic purposes since a number of transformations that are commonplace for enzymes would be rather taxing for most organic chemists. Microorganisms are attaining some importance as commercial synthesizers, particularly in the steroid field.

ENZYMATIC CHEMISTRY OF SOME ALKENES

The hydration of alkenes and the dehydration of alcohols are reactions that are of primary importance in the **metabolism** of plant and animal cells (Greek *metabole* — change). Metabolism encompasses the chemical process by which molecules are broken down and energy released (*catabolism*), as well as the chemical synthesis of compounds necessary for cell growth (*anabolism*). The compounds chosen for examples are *fatty acids*, substances vital to cellular life. They consist of hydrocarbon chains containing a carboxyl group (—CO_2H) at their terminus. A more detailed discussion of the biochemistry of the fatty acids is to be found in Chapters 17 and 22.

A bacterial enzyme has been isolated that catalyzes the addition of water to the double bond of oleic acid (equation 9.82). The addition is *regiospecific;* i.e., the hydroxyl group becomes attached to only one end of the double bond, a result impossible to achieve with this molecule in the absence of an enzyme. Furthermore, the addition is *stereospecific* since the hydroxyl group attacks from only one side of the double bond.

(9.82)
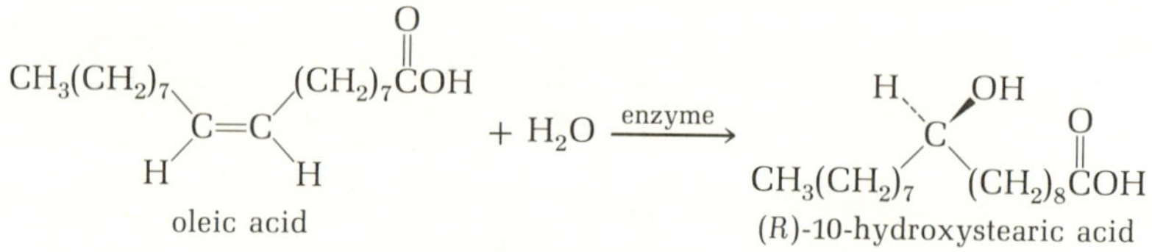

The oleic acid itself is obtained in part by the metabolic dehydrogenation of stearic acid (equation 9.83).

(9.83)
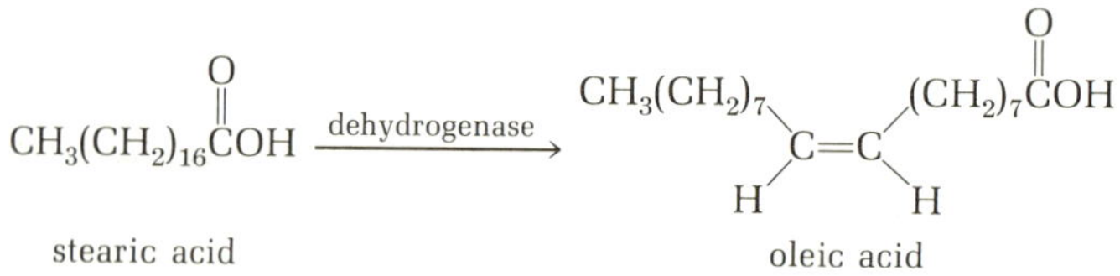

The two hydrogen atoms removed from stearic acid eventually wind up as water. (The name of the enzyme in equation 9.83 is obtained by addition of the suffix -ase to a word description of its function.)

The remarkable stereospecificity of enzymatic reactions is dramatically illustrated by the course of reaction 9.83. There are four possible combinations of pairs of hydrogen atoms that can be abstracted in the course of the transformation of stearic to oleic acid. Yet only one out of those four pairs of hydrogen atoms is in fact removed, as demonstrated by the use of specifically labeled substrate (equations 9.84 and 9.85).

(9.84)
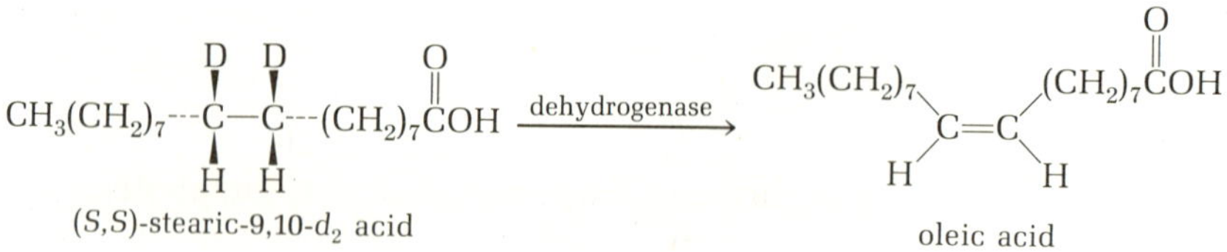

(9.85)
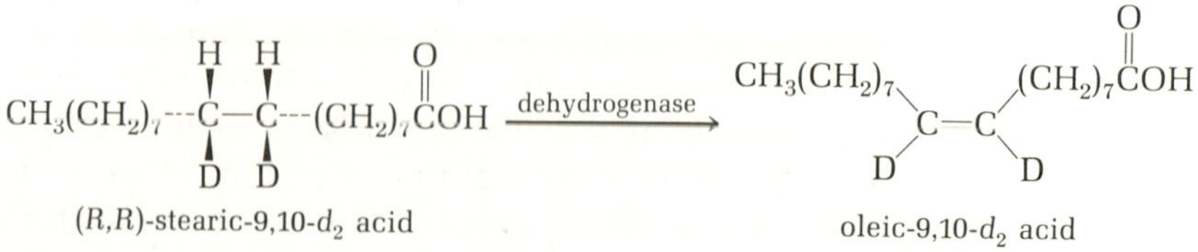

SUMMARY

1 The readily polarizable π-electrons of the carbon-carbon double bond are subject to attack by a wide spectrum of electrophiles. These include cations (protons, metal cations, carbonium ions), neutral molecules (halogens, osmium tetroxide, ozone, borane, chromic acid), and anions (permanganate ion). The intermediate may or may not be an open carbonium ion, depending on whether or not the electrophile can supply a pair of electrons.

2 Proton attack on alkenes results in carbonium ion formation. The carbonium ion may capture a nucleophile or lose a proton to yield an alkene, which may or may not be identical to the starting alkene. The usual nucleophile is the anion of the acid, but nucleophilic solvents may also react. For preparative purposes hydrogen halides are ordinarily allowed to react with alkenes in solvents like pentane and methylene chloride. The stereochemistry of addition is quite dependent on the ionizing power of the solvent.

3 Molecular halogens and pseudohalogens transfer one atom to the alkene with formation of a cyclic halonium ion. This halonium ion is then attacked by a nucleophile on the side away from the positive halogen so that the overall addition is stereospecifically *anti*. The nucleophile may be a halide or pseudohalide ion, or the solvent where appropriate.

4 The oxidation of double bonds may involve π-bond cleavage or destruction of both the carbon-carbon σ- and π-bonds. Cleavage of the π-bond alone is accomplished with dilute permanganate or osmium tetroxide. The glycol formed is the product of *syn* addition. Further oxidation of the glycol may be accomplished with sodium periodate. Complete cleavage of the double bond results when more concentrated permanganate or acidic dichromate is used.

Ozone and permanganate-periodate mixtures are used when complete scission of the double bond is required under mild conditions, for the purposes of locating the double bond positions.

5 Radicals attack carbon-carbon π-bonds with ease. If the new radical abstracts an atom from another molecule, then a net addition to the double bond has taken place. If, on the other hand, the new radical reacts with a second molecule of alkene, then polymerization will result. Polymerization is suppressed if there is available a high concentration of molecules containing readily abstracted atoms.

Addition of a radical to a double bond will proceed in such a manner as to give the most stable new radical. As a result, the products from free radical addition will usually differ in orientation from those obtained by way of an ionic mechanism.

6 Many enzyme-catalyzed reactions in biological systems strongly resemble the reactions of the same molecules under laboratory conditions. The enzymatic reactions are characterized, however, by a very high rate and complete stereospecificity, in contrast to their *in vitro* counterparts. Hydration of double bonds takes place by a path in which the O—H group is added to only one face of the double bond, resulting in an optically pure alcohol.

PROBLEMS

Note: Problems 1 and 2 appear in text.

3 Indicate all reagents and conditions necessary for carrying out the following syntheses.

a. Ethylene ⟶ ethanol

b. 1-Pentene ⟶ 2-pentanol

c. 1-Bromobutane ⟶ 2-butanol

d. 2-Bromopropene ⟶ 2,2-dibromopropane

e. 2-Bromomethylcyclohexane ⟶ 1-bromomethylcyclohexane

f. Chlorocyclopentane ⟶ *trans*-1,2-dichlorocyclopentane

g. *trans*-3-Hexene ⟶ *erythro*-4-chloro-3-iodohexane

h. 2-Methylbutane ⟶ 2-methylbutane-2,3-diol

4 The phenyl group (C_6H_5) is very effective at stabilizing carbonium ions, much more so than an alkyl group. A series of additions was made to the double bonds of *cis*- and *trans*-$C_6H_5CH{=}CHCH_3$, using the compounds INCO, Br_2, Cl_2, F_2, and HBr. In all cases products of both *syn* and *anti* addition were found. The greatest tendency for *anti* addition was found with INCO and the smallest with HBr. The others showed intermediate behavior, in the order listed. Explain these data.

5 Salts of mercury(II) react readily with alkenes in nucleophilic solvents such as water or alcohols. The reaction shows a strong preference for *anti* addition.

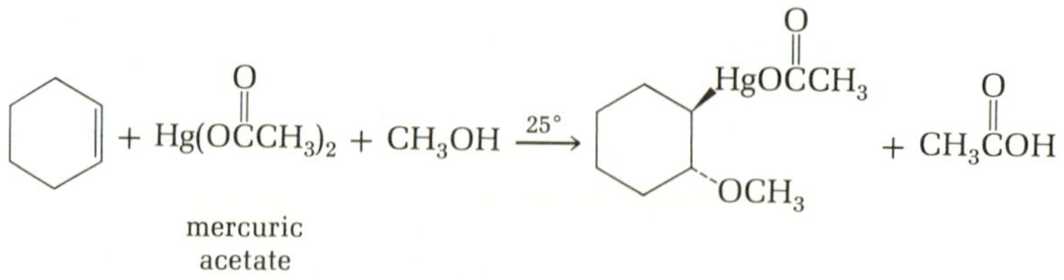

(2)

Increased substitution of the double bond tends to accelerate the rate; yet carbonium ion-type rearrangement products are not observed. Markovnikov's rule is obeyed.

$$(CH_3)_3CCH{=}CH_2 + Hg(O\overset{\overset{O}{\|}}{C}CH_3)_2 + H_2O \xrightarrow{THF} (CH_3)_3C\overset{\overset{OH}{|}}{C}HCH_2HgO\overset{\overset{O}{\|}}{C}CH_3 \qquad (3)$$

Suggest a structure for the intermediate in this reaction that accommodates the data. (Note that mercury belongs to the same class of elements that platinum and silver do — the transition metals. They are almost all reactive toward alkenes.)

The usefulness of reactions 2 and 3 lies in the fact that the mercury is readily replaced by hydrogen, resulting in a net overall addition of water or an alcohol to the double bond under mild conditions without danger of rearrangement.

$$(CH_3)_3C\underset{\underset{OH}{|}}{C}HCH_2HgO\overset{\overset{O}{\|}}{C}CH_3 \xrightarrow[OH^-]{NaBH_4} (CH_3)_3C\underset{\underset{OH}{|}}{C}HCH_3 + Hg + CH_3\overset{\overset{O}{\|}}{C}O^- \qquad (4)$$

6 Glycols are subject to oxidative cleavage by lead tetraacetate. Propose a mechanism for this reaction.

$$\underset{}{CH_3\underset{\underset{OH}{|}}{C}HCH_2OH} + \underset{\text{lead tetraacetate}}{Pb(O\overset{\overset{O}{\|}}{C}CH_3)_4} \longrightarrow \qquad (5)$$

$$CH_3\overset{\overset{O}{\|}}{C}H + CH_2{=}O + Pb(O\overset{\overset{O}{\|}}{C}CH_3)_2 + 2CH_3\overset{\overset{O}{\|}}{C}OH$$

7 What series of steps are probably involved in the cleavage of double bonds by $KMnO_4$-$NaIO_4$?

8 Write out the structures of the organic products from the following reactions:

a. 1-Pentene $\xrightarrow{OsO_4}$

b. *cis*-2-Butene $\xrightarrow[25°]{\text{dil. }KMnO_4}$

c. *threo*-Hexane-2,3-diol $\xrightarrow[pH\ 5]{NaIO_4}$

d. 2-Hexene $\xrightarrow[100°]{KMnO_4}$

e. Cyclobutene $\xrightarrow[NaIO_4]{KMnO_4}$

f. 1-Bromo-4-chloro-*cis*-2-butene $\xrightarrow{O_3}\ \xrightarrow{H_2,\ Pt}$

g. 1-Bromo-4-chloro-*trans*-2-butene $\xrightarrow{O_3}\ \xrightarrow[H_2O]{H_2O_2}$

h. 1-Methyl-2-ethylcycloheptene $\xrightarrow{O_3}\ \xrightarrow[\text{acetic acid}]{Zn}$

9 a. Listed below are some heats of hydrogenation for a series of *cis*- and *trans*-cycloalkenes. What conclusions can be drawn about the relative stabilities of the isomers?

	cis	*trans*
Cyclooctene	−22.98 kcal/mole	−32.24 kcal/mole
Cyclononene	−23.62 kcal/mole	−26.94 kcal/mole
Cyclodecene	−20.67 kcal/mole	−24.01 kcal/mole

b. How would you determine the relative stabilities of 1-pentene and 2-methyl-2-butene?

10 a. The heat of hydrogenation of 2,3-dimethyl-2-butene equals −26.63 kcal/mole. At the same temperature the heat of hydrogenation of 3,3-dimethyl-1-butene is −30.34 kcal/mole. Furthermore, it has been found from combustion studies that 2,3-dimethylbutane is more stable than 2,2-dimethylbutane by 2.079 kcal/mole. Calculate the energy difference between 2,3-dimethyl-2-butene and 3,3-dimethyl-1-butene.

b. The heat of hydrogenation of 2,3-dimethyl-1-butene equals −28.00 kcal/mole. At equilibrium at 80° the concentration of this olefin is one-fourth that of 2,3-dimethyl-2-butene (equation 8.30). Calculate the change in *entropy* that takes place when 2,3-dimethyl-1-butene is isomerized to 2,3-dimethyl-2-butene at 80° [see also Chapter 8, problem 13, and *J. Chem. Ed.*, **46,** 849 (1969)].

11 The heats of hydrogenation are commonly used to measure the relative stabilities of isomeric alkenes. In some ways it would be experimentally easier to measure the heats of bromination of these alkenes (bromination is always exothermic). Comment on the possibility of using this method.

12 Devise syntheses of the following compounds using any alkene as a starting material and any other necessary reagent, including isotopically labeled ones.

a. 1-Hexanol

b. 3-Methyl-1-butanol-2-*d*

c. Butyl methyl ether [$CH_3(CH_2)_3OCH_3$]

d. Cyclohexane-*cis*-1,2-d_2

e. 1-Bromocyclopentane-*trans*-2-*d*

13 "Anti-Markovnikov" addition of hydrogen bromide to olefins occurs because bromine atoms, rather than hydrogen atoms, are the chain-carrying radicals. Would it be possible to have a free radical, Markovnikov addition of hydrogen bromide to terminal alkenes?

14 Write the first few steps in the polymerization of propylene, initiated by a sodium atom.

15 Rubber is a natural linear polymer, containing no rings, of empirical formula $(C_5H_8)_x$. When rubber is ozonized and then treated with zinc, compound I is the only low molecular weight substance isolated.

$$\text{Rubber} \xrightarrow{O_3} \xrightarrow{Zn} CH_3\overset{\overset{O}{\|}}{C}CH_2CH_2\overset{\overset{O}{\|}}{C}H$$

I

Propose a formula for rubber, showing a section containing three of the repeating units. Is yours the only structure that will fit the data?

16 Alkene *A* has the molecular formula C_7H_{14}; its infrared spectrum contains a band at 970 cm^{-1}. When *A* is treated successively with ozone and zinc dust and water, it is transformed into a mixture of *B* and *C*. The nmr spectrum of *B*, C_3H_6O, consists of a triplet, a multiplet, and a triplet, the areas of which are in the ratio 3:2:1. The nmr spectrum of *C*, C_4H_8O, contains a doublet, a multiplet, and a doublet; area ratios are 6:1:1. Write structures for *A*, *B*, and *C*.

17 4-Methyl-2-pentanol is heated in sulfuric acid and the volatile products distilled. Analysis of the product mixture by glc with the aid of a silver nitrate-ethylene glycol column reveals no fewer than seven isomeric components. Five of the compounds were minor products, while the other two, *A* and *B*, were present in substantial quantities and could be isolated by preparative glc. In the nmr spectrum of *A* the ratio of aliphatic to olefinic protons was 5:1; in the spectrum of *B* the same ratio was found to be 11:1.

The crude product mixture was treated with platinum and hydrogen. After gas absorption had ceased, the resulting liquid was analyzed again by glc. It showed only two components, *C* and *D*, in a ratio of 50:1.

Write out the structures of *A*, *B*, *C*, and *D* as well as the structures of the other unknowns, and show how they were formed.

SUPPLEMENTARY READINGS

J. G. Traynham, "The Bromonium Ion," *J. Chem. Ed.*, **40**, 392 (1963).

W. R. Dolbier, Jr., "Electrophilic Additions to Olefins," *J. Chem. Ed.*, **46**, 342 (1969).

N. Isenberg and M. Grdinic, "A Modern Look at Markovnikov's Rule and the Peroxide Effect," *J. Chem. Ed.*, **46**, 601 (1969).

W. H. Saunders, Jr., *Ionic Aliphatic Reactions*. Englewood Cliffs, N.J.: Prentice-Hall, Inc., 1965, Chapter 2.

D. C. Phillips, "The Three-Dimensional Structure of an Enzyme Molecule," *Scientific American*, **215**, 78 (November, 1966).

CHAPTER 10

ETHERS—LEWIS BASES AS SOLVENTS (ORGANOMETALLIC CHEMISTRY)

Ethers may be thought of as hydrocarbon derivatives, obtained by replacing a —CH_2— group by —O—. This is a useful way of viewing these compounds because they have many properties in common with the corresponding hydrocarbons. They are slightly more polar, differ in molecular weight by only two units, and usually boil no more than 20° higher. The important difference between the two classes arises from the availability of nonbonding electrons in ether molecules.

For naming purposes, the word "ether" is used as a class name in the same way as "alcohol." When used by itself, it invariably refers to diethyl ether (use of the prefix "di" in naming symmetrical ethers is optional).

$CH_3OCH_2CH_3$
ethyl methyl ether

$CH_3CH_2OCH_2CH_3$
diethyl ether

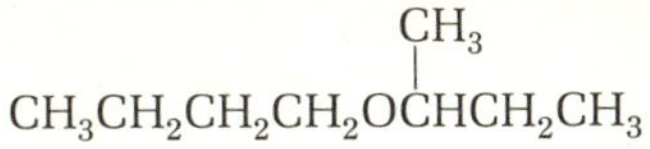
n-butyl *s*-butyl ether

When the ether is comprised of a relatively small alkyl group and a much larger one, then the small alkyl group plus the oxygen atom are treated as a substituent on the large alkyl chain. In rather complex ethers, particularly cyclic ones, the oxygen is treated as a substitute for a methylene group, its location given by a number and the prefix "oxa."

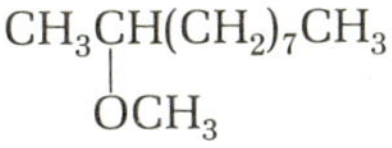
2-methoxydecane

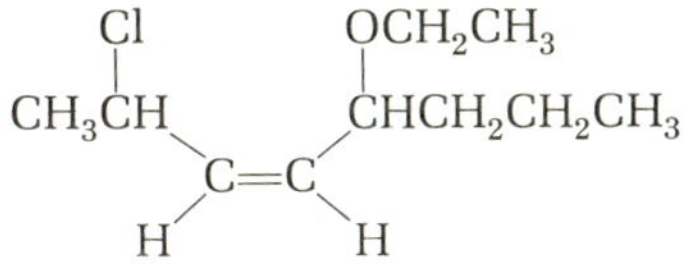
2-chloro-5-ethoxy-*cis*-3-octene

oxacycloheptane

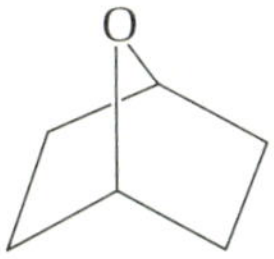
7-oxabicyclo(2.2.1)heptane

The lower molecular weight cyclic ethers have individual names, systematic and trivial, by which they are almost always known.

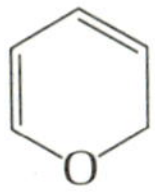
α-pyran
(2*H*-pyran)

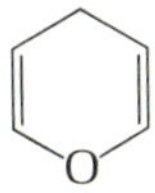
γ-pyran
(4*H*-pyran)

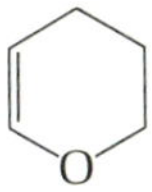
2,3-dihydropyran

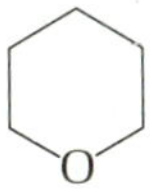
tetrahydropyran

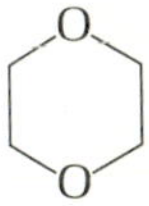
1,4-dioxane

furan

tetrahydrofuran
(THF)

oxetane

oxirane

Spectroscopically, ethers may be distinguished from alkanes by the presence of strong bands in their infrared spectra in the region 1050 to 1150 cm^{-1}, attributable to C—O stretching vibrations. The remainder of the spectrum is hydrocarbon-like (Figure 10-1). The presence of the electronegative oxygen atom causes a deshielding of the hydrogens on the adjacent carbon atoms. Peaks due to these hydrogens are found at 3.2 to 4.5δ in the nmr spectra of ethers.

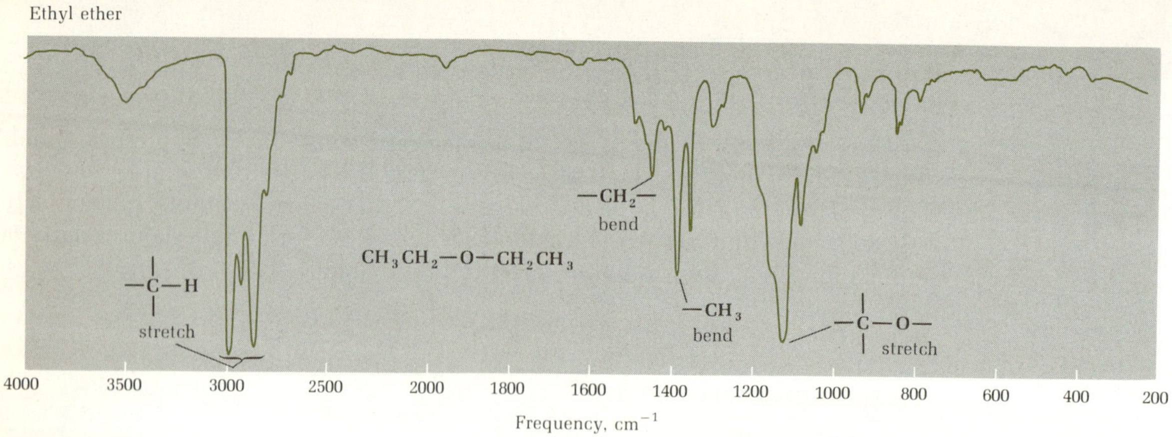

FIGURE 10-1
Infrared Spectrum of Diethyl Ether

More crudely, but quite effectively, ethers can be distinguished from hydrocarbons by their more pungent odors, very unlike the "gasoline-like" smell of the hydrocarbons. Several acyclic ethers have sweet, pleasant odors. Their solubility in concentrated sulfuric acid also sets them apart from alkanes.

BASICITY OF ETHERS

Ethers are many times more miscible with water than hydrocarbons, and some are completely miscible; hydrogen bonding between water and ether molecules accounts for this behavior. Solutions of the hydrogen halides in various ethers are readily prepared, and they provide a convenient medium for acid-catalyzed reactions in the absence of nucleophilic water. Ethers will even neutralize some mineral acids with formation of isolable oxonium salts.

(10.1) $$(CH_3CH_2CH_2CH_2)_2\ddot{O}: + HNO_3 \longrightarrow (CH_3CH_2CH_2CH_2)_2\overset{+}{\ddot{O}}H\ NO_3^-$$
dibutyloxonium nitrate

Ethers are not sufficiently strong bases to affect the pH of water, however.

Lewis acids, as well as proton acids, will coordinate with ethers. Boron trifluoride forms a complex with diethyl ether that may be distilled as a single liquid under vacuum (bp 48°/10 mm).

(10.2) $$BF_3 + (CH_3CH_2)_2\ddot{O}: \longrightarrow (CH_3CH_2)_2\ddot{O}:BF_3$$
boron trifluoride etherate

Aluminum compounds are quite soluble in most ethers and some aluminum compounds form unusually strong ether complexes. The alkali metal and alkaline earth cations often exhibit remarkable solubility in these solvents. They provide a relatively unreactive medium that will dissolve a variety of salts and a great range of organic compounds as well.

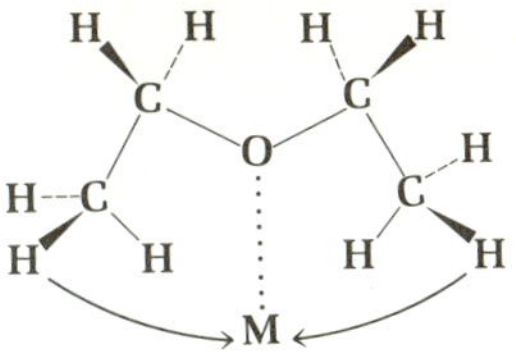

FIGURE 10-2
Steric Interactions in a Metal-Diethyl Ether Complex

Studies of basicity and complexing ability reveal that cyclic ethers are usually superior in both properties with respect to their acyclic analogues. This may be due to the fact that the normal rotations and vibrations in an acyclic ether could lead to steric interference in the complex (Figure 10-2). Salts are generally much more soluble in tetrahydrofuran than they are in diethyl ether for example, although the two ethers differ in composition by only two hydrogen atoms. Certain polyethers are particularly successful as solvents for reactions involving metals and their salts — chief among these are **glyme** and **diglyme** (acronyms for ethylene **gly**col di**m**ethyl **e**ther and **di**ethylene **gly**col di**m**ethyl **e**ther).

$CH_3OCH_2CH_2OCH_3$	$CH_3OCH_2CH_2OCH_2CH_2OCH_3$
glyme	diglyme
(1,2-dimethoxyethane)	[bis-(2-methoxyethyl)ether]

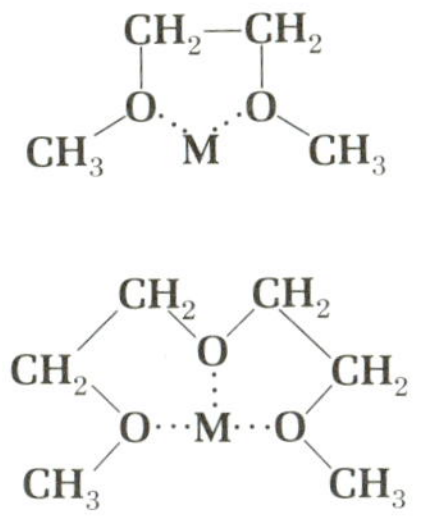

FIGURE 10-3
Solvation of a Cation by Glyme and by Diglyme

These ethers provide several pairs of solvating electrons in each molecule, requiring fewer molecules for solvation of a cation. The polyethers thus have an entropic advantage over the monoethers as complexing solvents (Figure 10-3). Complexes in which a ligand provides more than one site for attachment of the group being complexed are known as *chelates* (Greek *chele* — claw).

GRIGNARD REAGENTS

Reactions in which metal cations are formed from the zerovalent metal can be advantageously conducted in ethers. If one of the other reactants is organic, an *organometallic compound* will be a product. These substances contain a direct carbon-metal bond and are enormously useful and versatile. Their formation directly from the metal and an organic compound is a heterogenous reaction since the metal is insoluble in the solvent; a sizable proportion of the driving force for the reaction is undoubtedly the exothermic solvation of the metal cation as it is formed. Of all the organometallic compounds known, those derived from magnesium have pride of place as far as utility and ease of preparation are concerned.

They are known as **Grignard reagents,** after their discoverer, and have undoubtedly been employed by organic chemists more often than all other organometallic compounds combined. They are of sufficient importance to have earned Victor Grignard one-half of a Nobel Prize (1912).

A variety of ethers will serve for the preparation of Grignard reagents but anhydrous diethyl ether is the solvent par excellence for this reaction. These organometallic compounds react with both oxygen and moisture; diethyl ether, being quite volatile, provides a blanket of vapor that serves to protect the contents of the solution. It is also easy to remove once the reaction is over. The organic substrate is a chloride, bromide, or iodide that is increasingly reactive in that order; fluorides do not react.

$$\text{(10.3)} \quad CH_3CH_2Br + Mg \xrightarrow[35°]{\text{ether}} \underset{\text{ethylmagnesium bromide (97\%)}}{CH_3CH_2MgBr}$$

$$\text{(10.4)} \quad (CH_3)_2CHBr + Mg \xrightarrow{\text{ether}} \underset{\text{isopropylmagnesium bromide (84\%)}}{(CH_3)_2CHMgBr}$$

$$\text{(10.5)} \quad (CH_3)_3CCl + Mg \xrightarrow[\text{slow addition}]{\text{ether, 35°}} \underset{t\text{-butylmagnesium chloride (80\%)}}{(CH_3)_3CMgCl}$$

The principal by-products arise from reaction of 2 moles of organic halide and 1 mole of magnesium. For this reason, the halide dissolved in ether is slowly added to ether-covered magnesium.

It is more difficult to form vinyl Grignard reagents than it is to form alkyl Grignard reagents because a halogen atom is more strongly bound to an sp^2-hybridized carbon atom than to an sp^3-hybridized one. For those compounds it is necessary to use tetrahydrofuran as solvent since it allows higher temperatures to be attained and it is better at solvating metal cations.

$$\text{(10.6)} \quad CH_2{=}CHBr + Mg \xrightarrow[66°]{\text{THF}} \underset{\text{(85–90\%)}}{CH_2{=}CHMgBr}$$

The actual structure of the Grignard reagent in solution has been a matter for much debate and experiment. It is known from various experiments, including some with radioactive magnesium, that the equilibrium shown in equation 10.7 is established rapidly in most Grignard solutions.

$$\text{(10.7)} \quad 2RMgX \rightleftharpoons \underset{\text{dialkyl magnesium}}{R_2Mg} + MgX_2$$

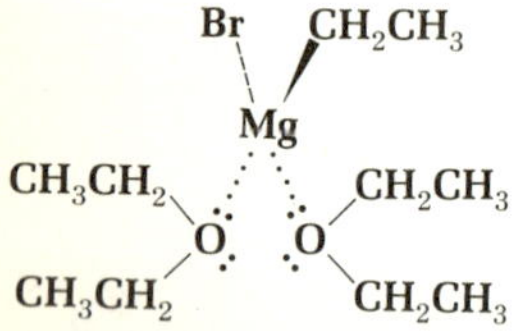

FIGURE 10-4
Ethylmagnesium Bromide Solvated by Ether Molecules

The reagent may also aggregate in dimeric (such as **1** and **2**) and higher forms; the extent of aggregation, as well as the position of equilibrium in equation 10.7, is a function of the solvent and the halogen X. But it will be sufficiently accurate for our purpose to represent the Grignard reagent as RMgX since this structure can account for most of its chemistry. It should be borne in mind that Grignard reagents are strongly bound to the solvent in solution and in the crystal (Figure 10-4).

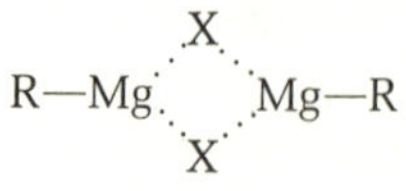

1

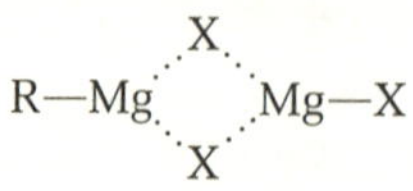

2

The mechanism by which Grignard reagents are formed is under active investigation; the current choice is a two-step radical process, as outlined in equations 10.8 and 10.9.

(10.8) $RX + Mg \longrightarrow R\cdot + \cdot MgX$

(10.9) $R\cdot + \cdot MgX \longrightarrow RMgX$

One datum that supports this mechanism is the finding that optically active halides, after conversion to Grignard reagents and subsequent reaction, give rise to racemic products. There is evidence that the reaction of a Grignard reagent and an electrophilic substrate (equation 10.10) involves *retention* of configuration at the carbon atom bound to magnesium. Any loss of optical activity must then have occurred when the halide was converted to the Grignard reagent.

(10.10) $RMgX + YZ \longrightarrow R{-}Y + ZMgX$

The magnesium-carbon bond is strongly polarized in the sense $C^{\delta-}{-}Mg^{\delta+}$, and Grignard reagents are therefore strong bases, the carbon atom becoming attached to a proton or a Lewis acid. Water and alcohols are sufficiently strong acids to take part in this reaction, and for this reason the ether solvent must be rigorously dried. This reaction has its uses, however, as shown by equation 10.13. It also forms the basis for a method of determining the amount of so-called "active hydrogen" (hydrogen bound to oxygen or nitrogen) in a substance. An excess of ethereal methylmagnesium iodide is added to the substance, and the volume of methane liberated is then measured.

(10.11) $CH_3CH_2MgBr + H_2O \longrightarrow CH_3CH_3 + HOMgBr$

(10.12)
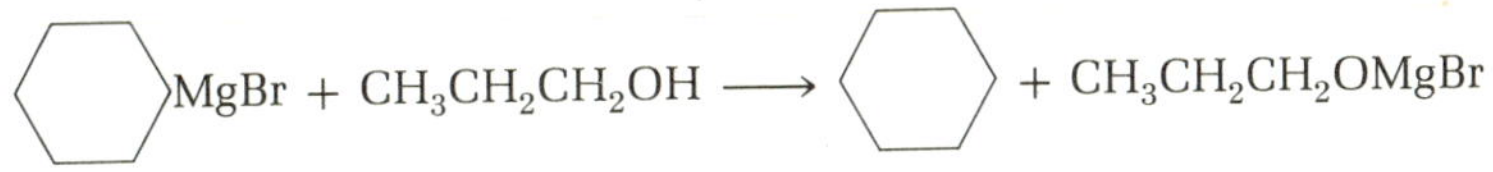

(10.13) $(CH_3)_3CMgCl + D_2O \longrightarrow \underset{\text{2-methylpropane-2-}d}{(CH_3)_3CD} + DOMgCl$

The reaction of Grignard reagents with organic compounds is covered at length in the next and subsequent chapters. It might be useful to mention here that they are widely employed for the preparation of organic derivatives of more electronegative elements.

(10.14) $2CH_3CH_2MgBr + ZnCl_2 \xrightarrow[25°]{\text{ether}} \underset{(100\%)}{(CH_3CH_2)_2Zn} + 2MgBrCl$

(10.15) $3CH_3CH_2CH_2CH_2MgBr + PCl_3 \xrightarrow{\text{ether}} \underset{\text{tributylphosphine (57\%)}}{(CH_3CH_2CH_2CH_2)_3P} + 3MgBrCl$

ORGANOMETALLIC COMPOUNDS OF OTHER ELEMENTS

The only other organometallic compounds that approach the Grignard reagents in popularity among organic chemists are those derived from lithium. Since lithium is more electropositive than magnesium, organolithium compounds cannot be made from Grignard reagents and lithium halides. They are formed by direct reaction between an organic halide and the metal.

(10.16) $CH_3I + 2Li \xrightarrow[N_2,\ 35°]{\text{ether}} CH_3Li + LiI$

methyllithium (90–95%)

(10.17) $CH_3CH_2CH_2CH_2Br + 2Li \xrightarrow[N_2,\ -10°]{\text{ether}} CH_3CH_2CH_2CH_2Li + LiBr$

n-butyllithium (80–90%)

The organolithium compounds are more reactive and less stable than their magnesium counterparts. Calculations indicate greater ionic character in the C—Li bond than in the C—Mg bond.

Vinyllithium compounds may be synthesized, as with the corresponding Grignard reagents, by reaction of the appropriate halide with metal. These reactions may be carried out in ether.

(10.18)

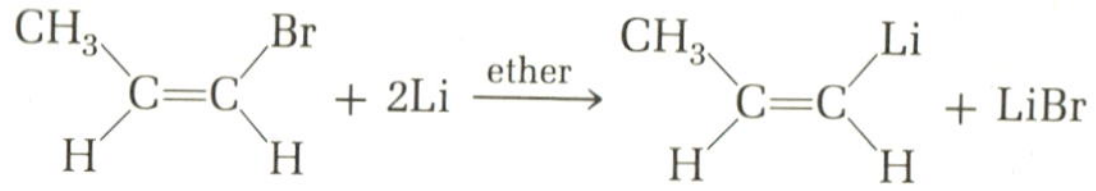

(10.19)

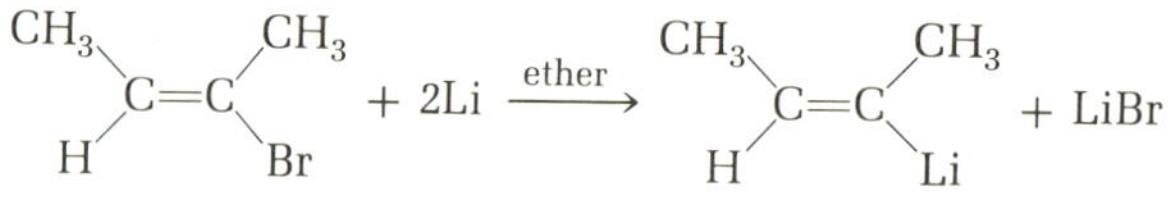

Geometrically isomeric halides give rise to isomeric vinyllithiums, which do not interconvert under the conditions of their preparation (equations 10.18 and 10.19).

One characteristic reaction of many organometallic reagents is the **coupling reaction** in which two organic residues are joined together with elimination of a metal halide. It is tempting to view this reaction as being simply a displacement of X^- by R^-.

(10.20) $RM + R'X \longrightarrow R{-}R' + MX$

The analogy is not supported, however; in many cases, for instance, the absolute configuration of R′ remains unchanged. Furthermore, there is evidence that in some cases at least a mechanism involving radicals is involved.

(10.21) $RM + R'X \longrightarrow R\cdot + R'\cdot + MX$

(10.22) $R\cdot + R'\cdot \longrightarrow R{-}R'$

Many reactions of organometallics probably involve radical intermediates but these are often not looked for because there are other, nonradical mechanisms that can account for the products.

Coupling reactions are often very desirable for synthetic purposes, particularly if two different groups are to be coupled. Organic derivatives of copper seem to be the most efficient organometallic compounds for coupling reactions, as this and subsequent chapters show. Organocopper compounds are almost always synthesized by the reaction of copper halides with an organic derivative of a more reactive metal. Addition of 1 mole of an alkyllithium to 1 mole of cuprous iodide in ether produces a precipitate of polymeric alkylcopper.

(10.23) $RLi + CuI \xrightarrow{\text{ether}} LiI + RCu\downarrow \xrightarrow{RLi} Li^+R_2Cu^-$

When a second mole of alkyllithium is added, the precipitate dissolves; the product is conveniently depicted as a lithium dialkylcopper, although no firm evidence on its structure is available.

These lithium dialkylcopper compounds are reactive toward a number of alkyl and vinyl halides; coupling takes place in reasonably good yield, with retention of geometrical configuration.

(10.24) 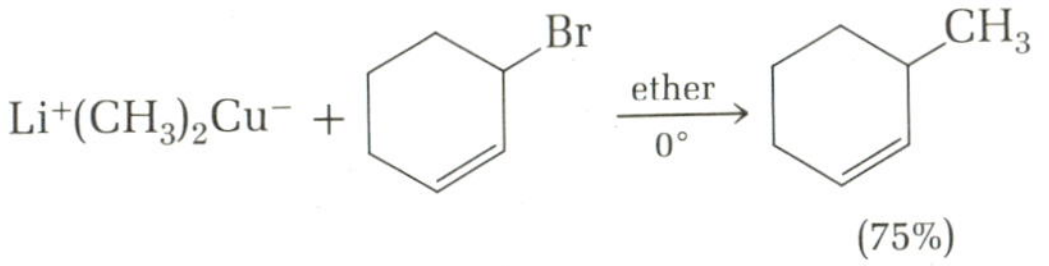

(10.25) $CH_3(CH_2)_7(H)C{=}C(H)(I) + Li^+(n\text{-}C_4H_9)_2Cu^- \xrightarrow[-95^\circ]{\text{ether}} CH_3(CH_2)_7(H)C{=}C(H)(CH_2)_3CH_3$

(71%)

OXONIUM SALTS

Protonation of water, alcohols, and ethers gives rise to the oxonium ions H_3O^+, $R\overset{+}{O}H_2$, and $R_2\overset{+}{O}H$, respectively. The last member of this series, R_3O^+, is also known. Syntheses of compounds in this class depend on the weak nucleophilic properties of ethers. Trialkyloxonium salts are very reactive, sensitive to moisture, etc.

(10.26) $CH_3CH_2F + (CH_3CH_2)_2O{:}BF_3 \longrightarrow (CH_3CH_2)_3O^+BF_4^-$

Almost any molecule with some nucleophilic capacity will displace an ether from an oxonium ion because it is energetically unfavorable to have a positive charge on the highly electronegative oxygen atom.

$$(10.27)\quad \underset{\substack{\uparrow\\ H_2\ddot{O}:}}{CH_3CH_2}-\overset{+}{O}(CH_2CH_3)_2 \longrightarrow \ddot{O}(CH_2CH_3)_2 + CH_3CH_2\overset{+}{O}H_2 \rightleftharpoons CH_3CH_2OH + H^+$$

Ethers also exhibit nucleophilic reactivity in certain *intramolecular* reactions. The rate of ethanolysis of chloromethyl methyl ether (**3**) is enormously fast compared to that of propyl chloride because in the ether the chloride ion is being displaced in large measure by the electrons on oxygen.

CH_3OCH_2Cl (**3**)	:	$CH_3CH_2CH_2Cl$	
10^{13}	:	1	ethanolysis rate

The intermediate is a cation in which the charge is shared by the carbon and oxygen atoms (these cations are described in greater detail in Chapter 12).

$$(10.28)\quad CH_3\ddot{O}-CH_2-Cl \xrightarrow{\text{slow}} CH_3\overset{+}{O}{=}CH_2 + Cl^-$$

(with curved arrows from the oxygen lone pair to the C–O bond and from the C–Cl bond to Cl)

$$(10.29)\quad CH_3\overset{+}{O}{=}CH_2 + CH_3CH_2OH \xrightarrow{\text{fast}} CH_3OCH_2OCH_2CH_3 + H^+$$

SYNTHESIS OF ETHERS

The commercially available acyclic ethers are made from the related alcohols by an acid-catalyzed process. A typical example is the commercial synthesis of diethyl ether; the rate-determining step is probably the attack of an ethanol molecule on the ethyloxonium ion.

$$(10.30)\quad CH_3CH_2OH + H^+ \rightleftharpoons CH_3CH_2\overset{+}{O}H_2$$

$$(10.31)\quad CH_3CH_2\ddot{O}H + CH_3CH_2\overset{+}{O}H_2 \rightleftharpoons CH_3CH_2\overset{\overset{H}{|}}{\underset{+}{O}}CH_2CH_3 + H_2\ddot{O}$$

$$(10.32)\quad CH_3CH_2\overset{\overset{H}{|}}{\underset{+}{O}}CH_2CH_3 \rightleftharpoons CH_3CH_2OCH_2CH_3 + H^+$$

Ether, being much more volatile than ethanol, distills from the solution immediately after formation. Control of the pot temperature is fairly critical here because an increase of 20°, from 140° to 160°, results mainly in the production of ethylene.

The same type of reaction can occur intramolecularly, leading to formation of a cyclic ether.

(10.33) $HO(CH_2)_5OH \xrightarrow[165°,\ \text{remove water}]{\text{mild acid}}$ 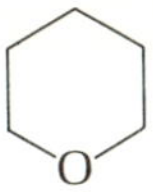

(100%)

Secondary and tertiary alcohols give rise mainly to carbonium ions in the presence of acid catalysts, and alkene formation competes strongly with ether synthesis. In essence, ether formation is either an S_N2 or S_N1 process complicated by competing E2 and E1 eliminations, respectively.

Acid-catalyzed condensations are often not an efficient method of synthesizing mixed ethers. These can be made by means of a general reaction involving a metal alkoxide and an alkyl halide or tosylate (*Williamson synthesis*). It is merely a synthetic application of the S_N2 displacement reaction.

(10.34) $RO^-M^+ + R'X \longrightarrow ROR' + M^+X^-$

The alkoxides are formed by direct reaction of the metal (usually sodium) with excess alcohol, which also serves as solvent.

The most important side reaction is E2 elimination since the alkoxides are fairly strong bases. The chemist has a good deal of control over the amount of elimination and can suppress it markedly by the correct choice of conditions. For instance, in the series X = Cl, Br, I, iodides give the most elimination and chlorides the least. Bromides have been the usual choice because they provided an acceptable compromise between reasonable rate and useful yield. Studies have shown that tosylates are more reactive than bromides; yet they give rise to less alkene.

(10.35) $(CH_3)_2CHCH_2Br + Na^+C_2H_5O^- \xrightarrow[55°]{C_2H_5OH} \underset{(40\%)}{(CH_3)_2CHCH_2OCH_2CH_3} + \underset{(60\%)}{(CH_3)_2C{=}CH_2}$

(10.36) $(CH_3)_2CHCH_2OTs + Na^+C_2H_5O^- \xrightarrow[55°]{C_2H_5OH} \underset{(56\%)}{(CH_3)_2CHCH_2OCH_2CH_3} + \underset{(44\%)}{(CH_3)_2C{=}CH_2}$

More critical is the choice of alcohol and halide (tosylate) when a mixed ether is being synthesized. This is particularly true if the carbon atoms attached to the oxygen are substituted to different extents. Elimination in alkyl halides and tosylates increases sharply in the order $1° < 2° < 3°$. Therefore, ethyl *t*-butyl ether should be synthesized as shown in equation 10.37 rather than as shown in equation 10.38.

(10.37) $(CH_3)_3CO^-K^+ + CH_3CH_2Br \longrightarrow \underset{\text{major product}}{(CH_3)_3COCH_2CH_3}$

(10.38) $(CH_3)_3CBr + CH_3CH_2O^-Na^+ \longrightarrow \underset{\text{major product}}{(CH_3)_2C{=}CH_2}$

The S_N2 reaction can also be applied intramolecularly, particularly in the case of five- and six-membered rings.

(10.39) 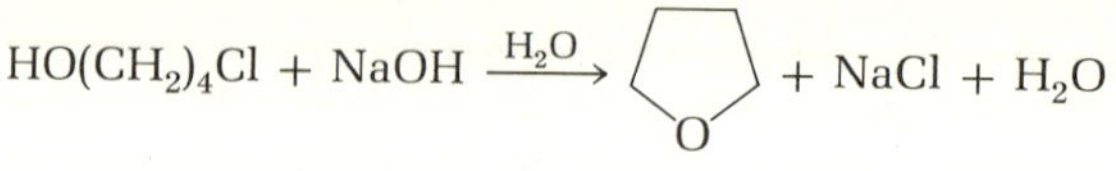

(95%)

EPOXIDES

Discussion of the three-membered cyclic ethers has deliberately been postponed since their chemical behavior is quantitatively different from that of other cyclic ethers and their syntheses somewhat specialized. These compounds are properly called epoxides or oxiranes. Since they can be derived formally (and chemically) from alkenes by insertion of an oxygen atom into the π-bond, they are often named as alkene oxides. For simple molecules, this is probably the most widely used system.

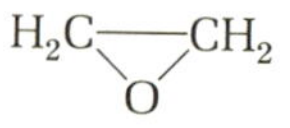

oxirane
epoxyethane
ethylene oxide

methyloxirane
1,2-epoxypropane
propylene oxide

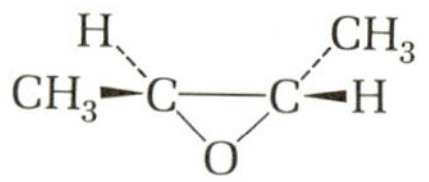

trans-2,3-dimethyloxirane
trans-2,3-epoxybutane
trans-2-butene oxide

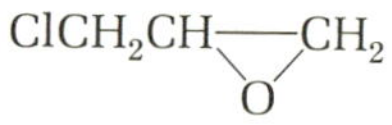

chloromethyloxirane
3-chloro-1,2-epoxypropane
epichlorohydrin

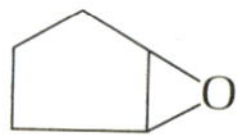

6-oxabicyclo(3.1.0)hexane
1,2-epoxycyclopentane
cyclopentene oxide

Like other cyclic ethers, epoxides can be synthesized by way of an internal nucleophilic displacement. The starting material, an α-halo alcohol, can be made by addition of a hypohalous acid to an alkene (Chapter 9); this route constitutes a two-step conversion of an alkene to its oxide.

(10.40) 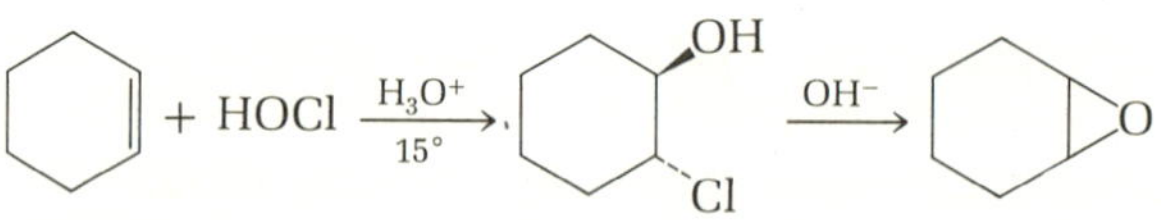

(50%)

Since the addition occurs in an *anti* manner, and the elimination is also *anti*, the stereochemistry of the alkene is retained in the epoxide.

(10.41)

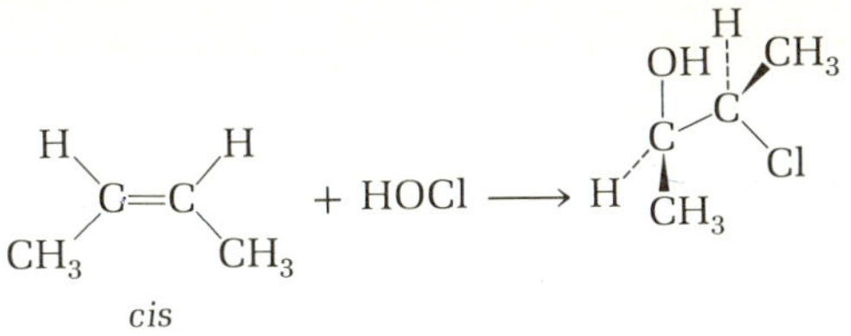

(10.42)

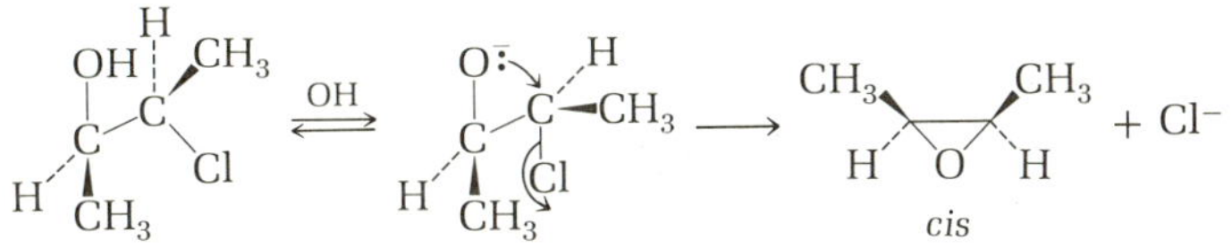

Epoxides may be obtained in one step from alkenes by insertion of an oxygen atom into the π-bond. This can be accomplished through the use of certain organic derivatives of hydrogen peroxide known as **peracids.** They contain the weak oxygen-oxygen bond and serve therefore as oxygen atom transfer reagents. The reaction of *cis*-alkenes gives rise to *cis*-epoxides and that of *trans*-alkenes to *trans*-epoxides (equations 10.43 to 10.45).

(10.43)

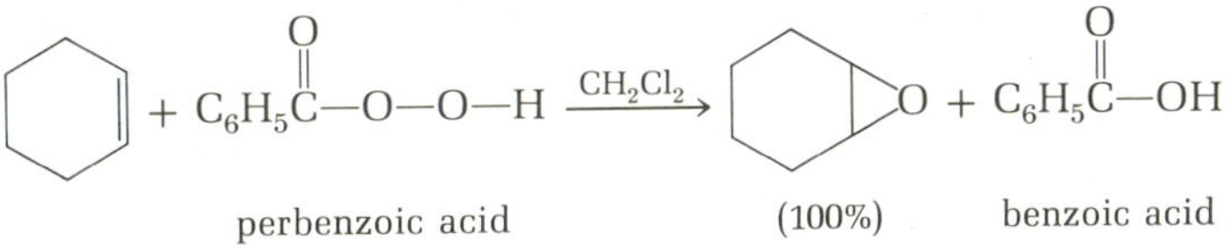

(10.44)

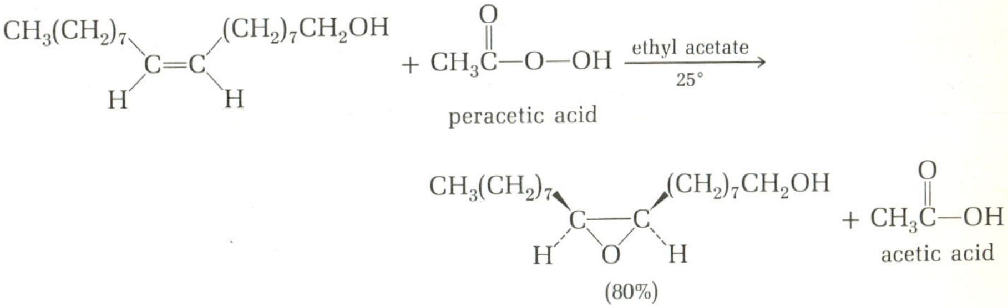

(10.45)

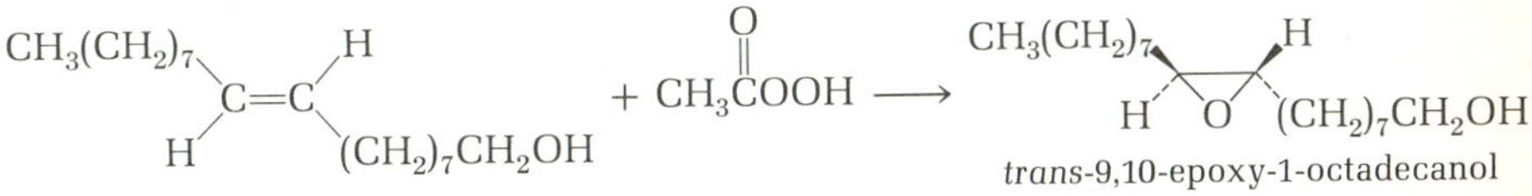

The fact that no geometrical isomerization occurs upon epoxidation with peracids, combined with other evidence, strongly indicates that both C—O σ-bonds are being formed simultaneously. In fact, the generally accepted mechanism for the reaction is one in which *all* the bond making and bond breaking takes place simultaneously.

(10.46) 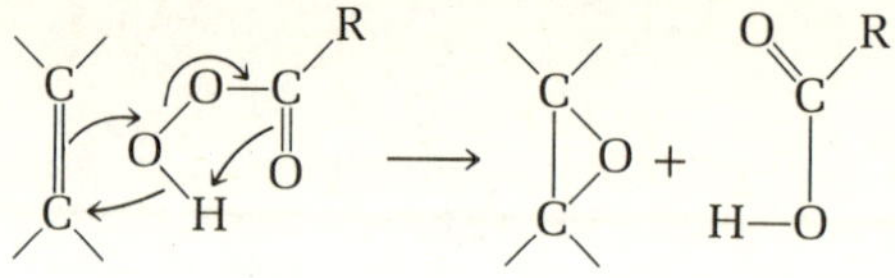

The strained three-membered ring not only confers unusual chemical reactivity upon epoxides, but it alters some physical properties as well. The C—O stretching band occurs at about 1250 cm^{-1}, and the hydrogens on the ring give rise to bands between 800 and 900 cm^{-1}, very much like the bending mode vibrations of alkenes.

NUCLEOPHILIC ETHER CLEAVAGE

Ethers are generally impervious to most nucleophiles and bases. Epoxides, however, will react with a number of ordinary nucleophiles with resultant opening of the strained oxirane ring. The characteristic properties of an S_N2 displacement are in evidence: the nucleophile preferentially attacks the least substituted carbon atom with inversion. Included among the nucleophilic attacks on epoxides would be their reduction by metal hydrides. Epoxidation followed by reduction is equivalent to addition of water to the alkene.

(10.47) 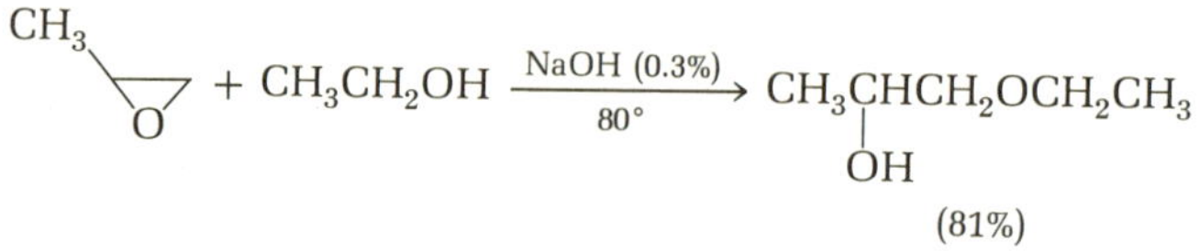

(10.48) 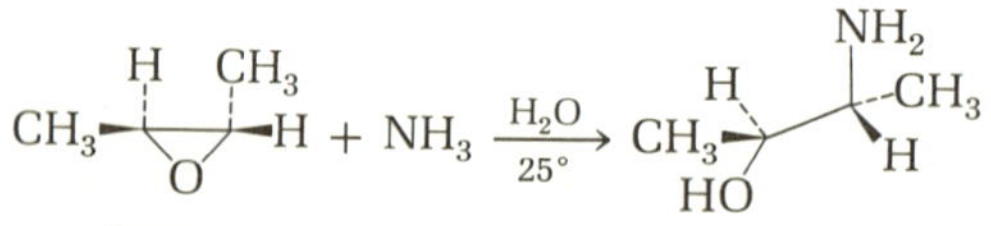

trans   *threo*-3-amino-2-butanol

(10.49) 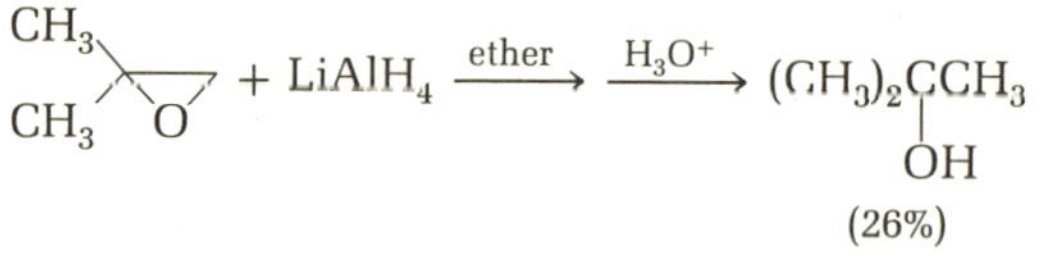

Epoxides are easily hydrated to diols in the presence of a basic catalyst. Due to the stereochemistry of this reaction, the overall result of epoxidation and hydrolysis is the *anti* addition of two —OH groups to the double bond.

(10.50) 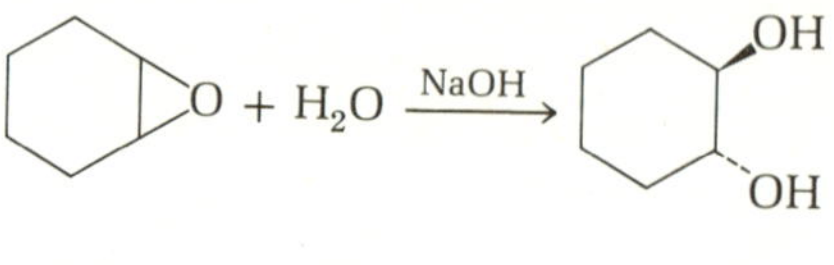

This method is complementary to the *syn* hydroxylation of double bonds with $KMnO_4$ or OsO_4. The epoxide acts in the same manner as a halonium ion does in the *anti* addition of halogens to a double bond.

Grignard reagents may also act as nucleophiles toward epoxides. Ethylene oxide is the one most commonly used, and this provides a convenient method of lengthening the carbon chain by two atoms.

(10.51) 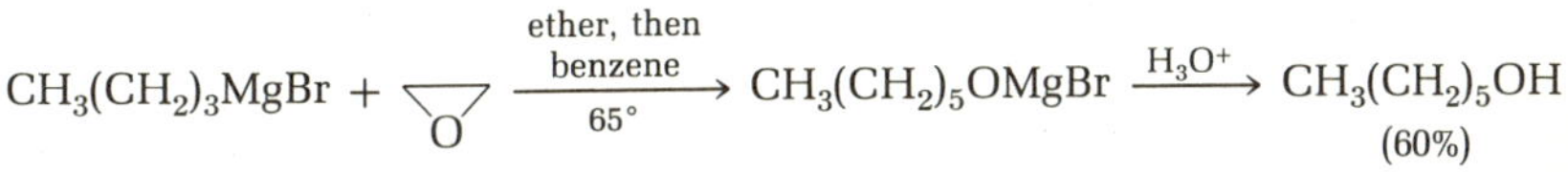

(10.52) 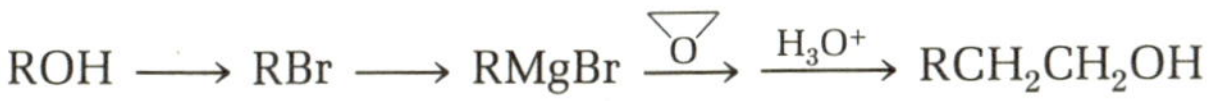

ACID-CATALYZED ETHER CLEAVAGE

Nucleophiles will react with ordinary ethers if they are first activated through complexing of an acid with the oxygen atom. Concentrated aqueous hydrogen bromide and hydrogen iodide (in particular) have long been used to cleave ethers.

$$CH_3CH_2CH_2CH_2OCH_3 \xrightleftharpoons{H^+} CH_3CH_2CH_2CH_2\overset{H}{\underset{+}{O}}CH_3 \xrightarrow[\text{reflux}]{I^-} CH_3CH_2CH_2CH_2OH + CH_3I \xrightarrow{\text{HI (excess)}} CH_3CH_2CH_2CH_2I \quad (10.53)$$

Where primary alkyl groups are involved, the least substituted is converted into the halide. This is the expected result if the reaction involves S_N2-type nucleophilic attack on an oxonium ion. On the other hand, it appears that when secondary or tertiary alkyl groups are present the reaction has much more carbonium ion character (equation 10.54).

$$(CH_3)_3COCH_2CH_3 + HBr \xrightarrow{\text{heat}} (CH_3)_3CBr + HOCH_2CH_3 \quad (10.54)$$

Lewis acids are also effective in cleaving ethers; $AlCl_3$ and BCl_3 have found much use in this respect (equation 10.55).

$$CH_3CH_2CH_2CH_2O\underset{CH_3}{\underset{|}{C}}HCH_2CH_3 + BCl_3 \rightleftharpoons CH_3CH_2CH_2CH_2\overset{BCl_3^-}{\underset{+}{\ddot{O}}}\underset{CH_3}{\underset{|}{C}}HCH_2CH_3 \longrightarrow CH_3CH_2CH_2CH_2OBCl_2 + CH_3\underset{Cl}{\underset{|}{C}}HCH_2CH_3 \quad (10.55)$$

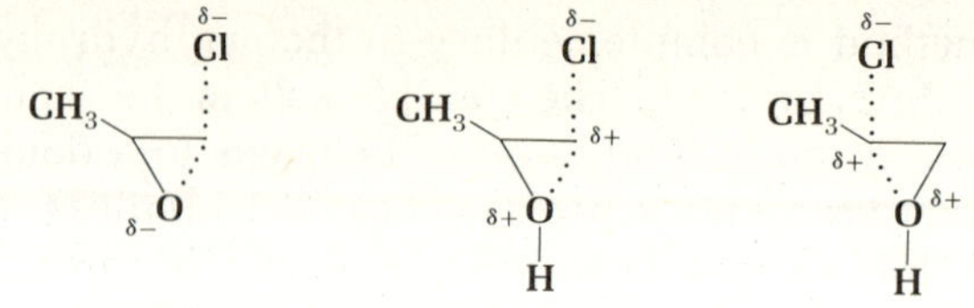

4 5 6

Neutral Solution   Acid Solution

FIGURE 10-5 Transition States for Reaction of Propylene Oxide in Neutral and Acidic Solutions

When epoxides are cleaved under acidic conditions, the pattern of substitution changes by comparison with that found under neutral or basic conditions.

Under neutral conditions, chloride ion attacks the unprotonated epoxide to give a tight transition state (**4,** Figure 10-5), and steric effects play a large part in determining the site of attack. Conversely, under acidic conditions the protonated epoxide is the substrate, and the transition state is much looser.

(10.56) $$CH_3CH(O)CH_2 + Cl^- \xrightarrow[H_2O]{pH\ 7} \underset{(86\%)}{CH_3CH(OH)CH_2Cl} + \underset{(14\%)}{CH_3CH(Cl)CH_2OH}$$

(10.57) $$CH_3CH(O)CH_2 + Cl^- \xrightarrow[H_3O^+]{pH\ 3.6} \underset{(64\%)}{CH_3CH(OH)CH_2Cl} + \underset{(36\%)}{CH_3CH(Cl)CH_2OH}$$

There is an appreciable degree of positive charge on the carbon atoms, and the transition states **5** and **6** are nearly equal in energy since the steric and electronic effects are of relatively equal importance.

The presence of Lewis acids may also exert a sizable effect on the product distribution from epoxide reduction (compare equation 10.58 with equation 10.49).

(10.58) $$(CH_3)_2C(O)CH_2 + LiAlH_4 \xrightarrow[\text{(4 moles per mole of hydride)}]{AlCl_3} \xrightarrow{H_3O^+} \underset{(3\%)}{(CH_3)_2C(OH)CH_3} + \underset{(47\%)}{(CH_3)_2CHCH_2OH}$$

ETHERS AS PROTECTING GROUPS

The insensitivity of most ethers to nucleophilic attack in the absence of acid catalysts has been put to very good use. Alcohol groups may be protected from oxidation or other transformations, while reactions are carried out on the remainder of the molecule. The hydroxyls are converted to ethers, which are stable as long as strong acid is not used. The ethers are then split and the alcohols recovered. A very useful blocking group is 2-tetrahydropyranyl because it is very easily made and removed.

(10.59)
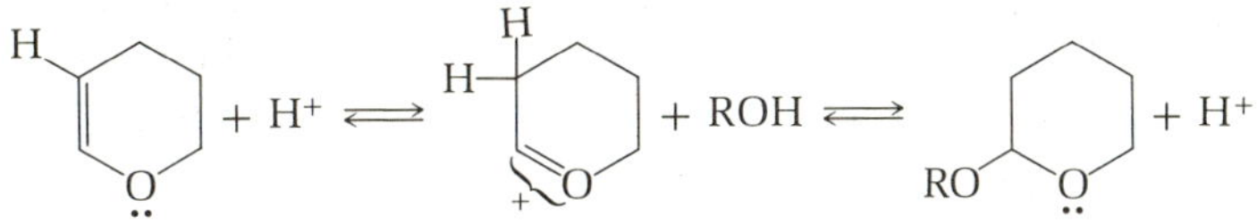

Its reactivity depends on its giving rise to a carbonium ion, stabilized by charge dispersion (see p. 288). The 2-tetrahydropyranyl group is removed by aqueous acid at room temperature.

(10.60)
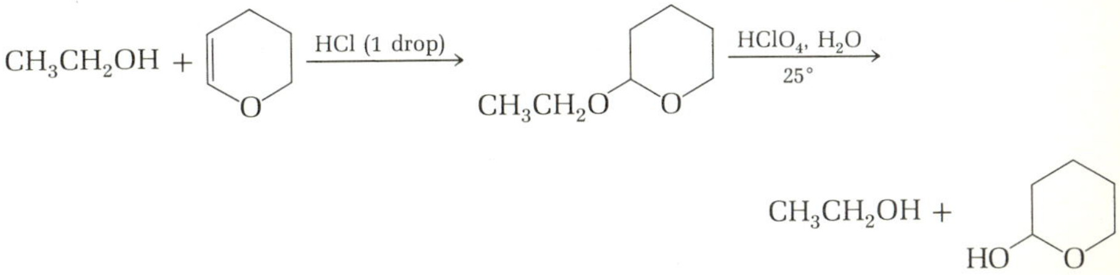

SOME EPOXIDES IN BIOLOGICAL SYSTEMS

Alkyl ethers as a class have not been found to be of great biochemical importance, but evidence has recently accumulated to support the contention that epoxides may be important biochemical substances. One example is a juvenile hormone that controls molting in certain insect larvae.

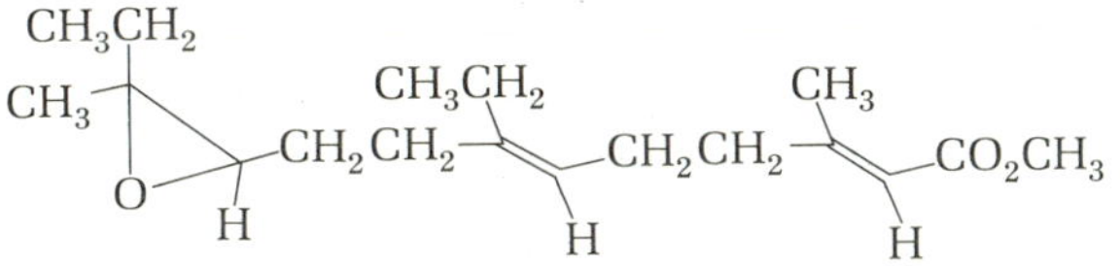
an insect juvenile hormone

It could prove to be a very potent insecticide because its presence in the larvae in excessive quantities results in the formation of an immature insect, incapable of reproducing.

The close relationship between an alkene and its derived epoxide is emphasized by the finding that the same enzyme system that catalyzes the addition of water to oleic acid (equation 9.82) also catalyzes the hydration of its epoxide (equation 10.61). Once again it is found that the enzymatic reaction is regiospecific and stereospecific.

(10.61)
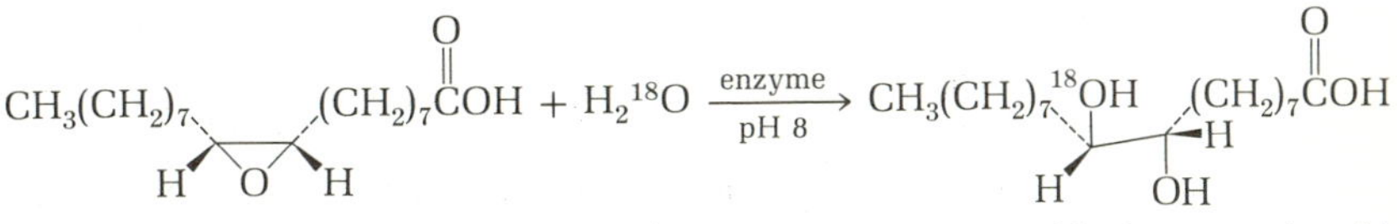

9,10-epoxystearic acid

threo-9,10-dihydroxystearic acid

One of the most important processes in which epoxide has been implicated is in the conversion of the acyclic symmetrical 30-carbon hexaene *squalene* to the tetracyclic steroid *lanosterol* (equation 10.62). (Note that in equation 10.62 squalene has been drawn in the folded conformation essential to its cyclization. Careful inspection will show that the molecule is symmetrical about the dotted line.)

The conversion of squalene to lanosterol is a critical reaction in biosynthesis since lanosterol is a precursor of many other steroids.

(10.62)

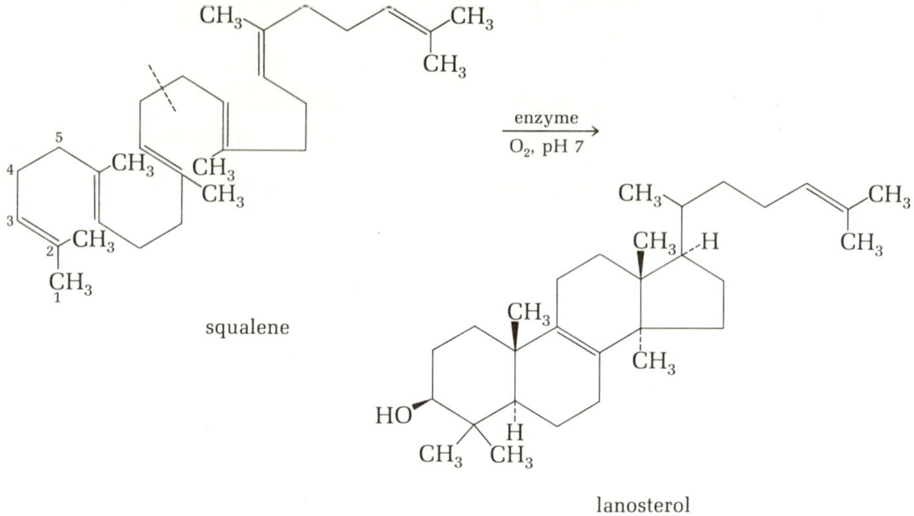

Experiments have established that the enzyme(s) involved in this transformation first convert(s) squalene to squalene 2,3-oxide, in a step requiring oxygen. The monoepoxide is then cyclized to lanosterol in one step, without isolable intermediates and without any requirement for oxygen. The epoxide oxygen ends up as the hydroxyl oxygen of lanosterol. The suggested mechanism, although formidable in appearance, is really a combination of a number of straightforward processes in concerted fashion (Figure 10-6):

1 Acid-catalyzed opening of the epoxide ring, with partial carbonium ion character at C-2.

2 Attack of this electron-deficient carbon on a π-bond, which leads to a kind of internal polymerization.

3 Concerted shifts of hydrogens and methyls to electron-deficient carbon atoms.

4 Proton loss from a carbonium ion to form a stable alkene.

Note that squalene itself has no asymmetric carbons, the epoxide has one, and lanosterol has seven. The enzyme system uses only one enantiomer of

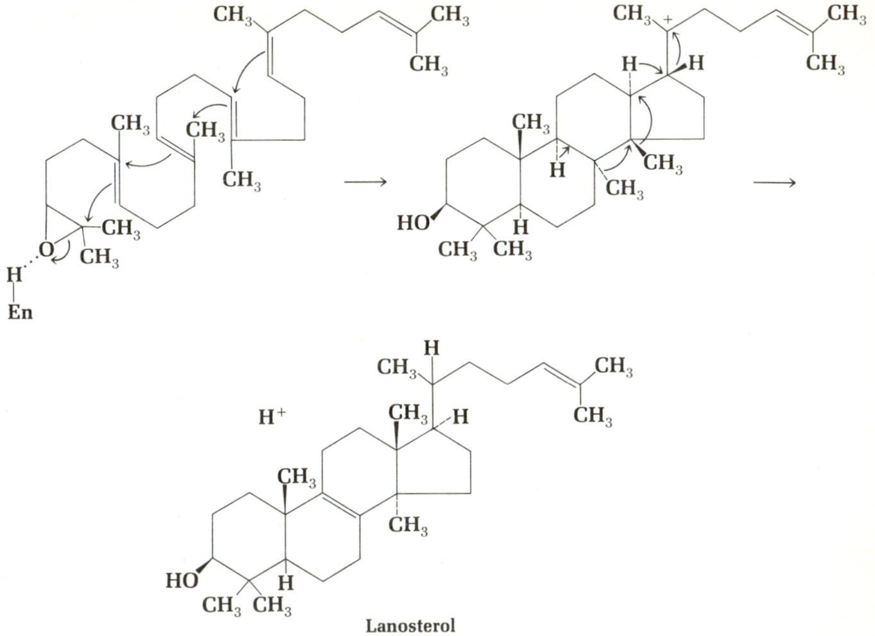

FIGURE 10-6
Suggested Mechanism for the Isomerization of Squalene Oxide to Lanosterol

the epoxide **and produces only one of the 128 possible enantiomers of lanosterol.**

If Nobel Prizes were awarded solely on the basis of achievement, without regard to the question of conscious volition, then every living cell would be a laureate.

SUMMARY

1 Ethers exhibit a strong resemblance in physical properties to the corresponding alkanes. The principal difference between them lies in the nonbonding electrons possessed by the ethers, which confer nucleophilicity and weak basicity upon the compounds in this class. Their basicity makes them effective solvents for many metal salts and organometallic compounds.

2 The reactivity of organometallic compounds is generally proportional to the electronegativity of the metal, the most reactive being the alkali metal derivatives. Organolithium compounds and Grignard reagents are made by the direct reaction between an organic halide and the metal in an ethereal solvent. These compounds are very reactive toward oxygen and protons. They may be used to prepare organic derivatives of more electronegative elements, by reaction with the halides of these elements. The organocopper

derivatives obtained by adding 2 moles of an alkyllithium compound to 1 mole of cuprous iodide will react with alkyl or vinyl halides, leading to coupling of the organic groups attached to the halogen and to the metal.

3 The most generally used method for synthesizing ethers is one involving the reaction of a sodium alkoxide with an alkyl halide. When a mixed ether is to be made, the reactants are chosen so that the least highly substituted halide is used, to minimize elimination. This synthesis can be applied intramolecularly, in order to produce cyclic ethers. With the exception of epoxides, ethers are pretty well immune to attack by nucleophiles in the absence of an acid. Alcohol groups are sometimes converted to ethers to protect them from oxidations, etc. The cleavage of most ethers requires mineral acids such as hydrogen bromide and hydrogen iodide or Lewis acids like boron trichloride and aluminum chloride.

4 Due to their highly strained rings, epoxides are readily attacked by nucleophiles such as hydroxide, hydride, and chloride ions; the attack takes place preferentially at the less hindered carbon atom. When the opening of the epoxide ring is catalyzed by an acid, then there is considerable positive charge on the carbon atoms at the transition state, and the nucleophile tends to become attached to the most highly substituted carbon atom. Epoxidation of a double bond, followed by hydride reduction, is equivalent to hydration of the double bond. If the epoxide ring is opened by water, the net effect is the *anti* addition of HO—OH to the double bond.

PROBLEMS

1 Write out the reaction of tetrahydrofuran with the following compounds (at 0°, unless otherwise indicated):

a. Sulfuric acid
b. Perchloric acid
c. Boron tribromide
d. Aluminum trichloride
e. Triethylaluminum
f. Gallium trichloride
g. Lithium perchlorate
h. Hydrogen bromide
i. Hydrogen bromide $\xrightarrow{\text{reflux}}$

2 Magnesium bromide is reasonably soluble in diethyl ether but addition of dioxane to the solution causes complete precipitation of the halide as a dioxane complex. This process is used to prepare dialkylmagnesiums by driving the equilibrium in equation 10.7 to the right.

Propose a structure for the magnesium bromide-dioxane complex that accords with its stability and insolubility.

3 Describe preparations of the following compounds from an appropriate organic halide and any inorganic reagents:

a. s-Butylmagnesium iodide

b. Cyclohexylmagnesium chloride

c. Isopropyllithium

d. *t*-Butoxymagnesium bromide [$(CH_3)_3COMgBr$]

e. Dimethylaminomagnesium bromide [$(CH_3)_2NMgBr$]

f. Lithium amide ($LiNH_2$)

g. Dimethylmercury

h. Tetravinyltin [$(CH_2{=}CH)_4Sn$]

i. Butylsodium

4 When *cis*-2-methylcyclopropyl bromide and isopropyllithium are mixed together, an exchange reaction takes place and *cis*-2-methylcyclopropyllithium is generated. This compound retains its initial geometry; most alkyllithium compounds, by contrast, are configurationally very unstable. Optically active alkyllithiums tend to racemize very quickly, for instance. Why should the cyclopropyllithium be an exception? What does this suggest about the mechanism by which alkyllithiums lose their initial configuration?

5 The reaction of 1,2-dibromoethane with magnesium does not follow the usual course. When the dihalide is added to magnesium in ether, gas is evolved *before* water has been added. The gas decolorizes a solution of bromine in chloroform. Write equations for the reactions that are occurring.

6 Suggest reagents and conditions for carrying out the following conversions:

a. Cyclohexane $\longrightarrow$ cyclohexane-*d*

b. 1-Propanol $\longrightarrow$ di-n-propylzinc

c. 3-Methyl-1-butene $\longrightarrow$ 2,3-dimethylpentane

d. 3-Methyl-1-butene $\longrightarrow$ (3-methylbutyl)cyclopentane

7 Propose a mechanism for the formation of $(CH_3CH_2)_3O^+BF_4^-$ (equation 10.26).

8 a. The acid-catalyzed addition of water to the double bond of alkyl alkenyl ethers (I, X = O) is very much more rapid than the addition of water to the double bonds of the corresponding alkenes (I, X = CH_2). Explain.

$$\underset{\text{I}}{R{-}X{-}CH{=}CHR'}$$

b. Grignard reagents do not ordinarily undergo extensive coupling with the usual alkyl halides. Yet they will couple with α-chloroethers ($RO\underset{|}{\overset{}{C}}HR'$ with Cl on C). Offer a rationalization of this finding.

9 Under what conditions might pure mixed ethers be obtained by an acid-catalyzed process?

10 Indicate all necessary conditions for effecting the following conversions:

a. 1-Butanol ⟶ dibutyl ether

b. Isopropyl bromide ⟶ diisopropyl ether

c. Isobutyl alcohol ⟶ cyclohexyl isobutyl ether

d. 2-Methyl-2-butene ⟶ trimethyloxirane (two ways)

e. *trans*-2-Pentene ⟶ *erythro*-pentane-2,3-diol

f. *trans*-2-Pentene ⟶ *threo*-pentane-2,3-diol

g. Methylenecyclohexane ⟶ 1-methyl-1-cyclohexanol

h. Cyclopentene ⟶ *trans*-2-aminocyclopentanol

i. Cyclobutanol ⟶ 2-cyclobutylethyl bromide

11 When propylene oxide is treated with boron trifluoride and the reaction mixture then worked up to remove boron fluoride, propionaldehyde is isolated (equation 1).

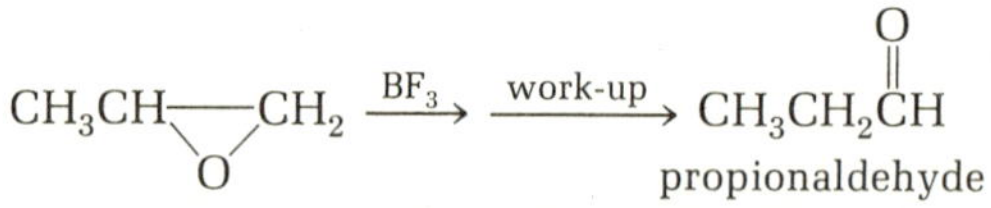

(1)

a. Propose a mechanism for this reaction.

b. What product(s) might be obtained from the boron trifluoride-catalyzed rearrangement of cyclohexene oxide?

12 The acid-catalyzed reaction between 1-butanol or isopropyl alcohol and dihydropyran yields only one ether in each case. When 2-butanol is allowed to react with dihydropyran, however, glc analysis shows the presence of two isomeric products.

a. What are the structures of the two products that are found in the 2-butanol reaction?

b. Compound II has been suggested as a replacement for dihydropyran as a reagent for protecting alcohols.

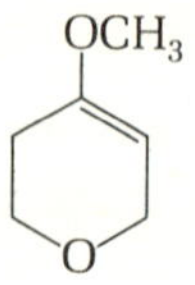

II

Show the product from reaction of compound II with 2-butanol. What advantage does compound II have over dihydropyran as a protecting agent?

CHAPTER 11

ALDEHYDES AND KETONES I— THE POLARIZED DOUBLE BOND

Many classes of organic compounds contain the **carbonyl group** — carbon doubly bonded to oxygen, >C=O. If both of the atoms attached directly to the carbonyl group are carbon atoms, then the compound is a ketone. If one of them is a hydrogen atom instead, the compound is an aldehyde. If one of the immediately attached atoms is neither hydrogen nor carbon, then the compound is neither an aldehyde nor a ketone.

NOMENCLATURE AND PHYSICAL PROPERTIES

In the IUPAC system, one identifies the longest chain containing the functional group and then names the compound as a derivative of that. The characteristic suffixes are "al" to denote an aldehyde and "one" for a ketone.

$H_2C{=}O$ — methanal, *formaldehyde*

CH_3CHO — ethanal, *acetaldehyde*

$(CH_3)_2CHCHO$ — 2-methylpropanal, *isobutyraldehyde*

$CH_3CH_2CH(CH_2CH_3)CH{=}CHCH_2CHO$ (H, H on the double-bond carbons)

cis-5-ethylhept-3-enal

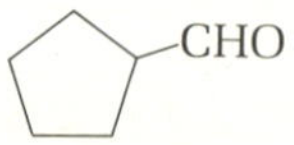

cyclopentane carboxaldehyde

The low molecular weight aldehydes are frequently known by their non-IUPAC names (Table 11-1) that are derived from those of the carboxylic acids (Chapter 16) to which they are related. The semisystematic nomenclature for ketones involves naming the alkyl groups attached to the carbonyl group, followed by the word "ketone."

Aldehydes and ketones of the same molecular weight have fairly similar physical properties. They are considerably more polar than the corresponding ethers and have correspondingly higher boiling points; they cannot act as hydrogen bond donors, however, and thus have lower boiling points than the related alcohols. Aldehydes and ketones containing up to four carbon atoms are reasonably soluble in water.

All compounds containing carbonyl groups exhibit a peak in their infrared spectra in the region 1650 to 1900 cm^{-1}, characteristic of the $>C{=}O$ stretching vibration.

$CH_3{-}C(=O){-}CH_3$ — propanone, *acetone*

$CH_3C(=O)CH_2CH_3$ — butanone, *methyl ethyl ketone*

$CH_3CH_2C(=O)CH_2CH_3$ — 3-pentanone, *diethyl ketone*

$CH_3CH_2C(=O)CH(CH_3)_2$ — 2-methyl-3-pentanone, *ethyl isopropyl ketone*

$CH_3CH_2C(=O)CH{=}CHCH_2CH_3$ — 4-hepten-3-one

TABLE 11-1
Low Molecular Weight Aldehydes and Ketones

Semisystematic Name	IUPAC Name	Boiling Point (°C)
Formaldehyde	Methanal	−21
Acetaldehyde	Ethanal	20
Propionaldehyde	Propanal	50
Butyraldehyde	Butanal	76
Isobutyraldehyde	2-Methylpropanal	61
Pivaldehyde	2,2-Dimethylpropanal	75
Valeraldehyde	Pentanal	103
Acetone	Propanone	56
Methyl ethyl ketone	Butanone	80
Methyl *n*-propyl ketone	2-Pentanone	102
Diethyl ketone	3-Pentanone	102
Methyl isopropyl ketone	3-Methyl-2-butanone	93

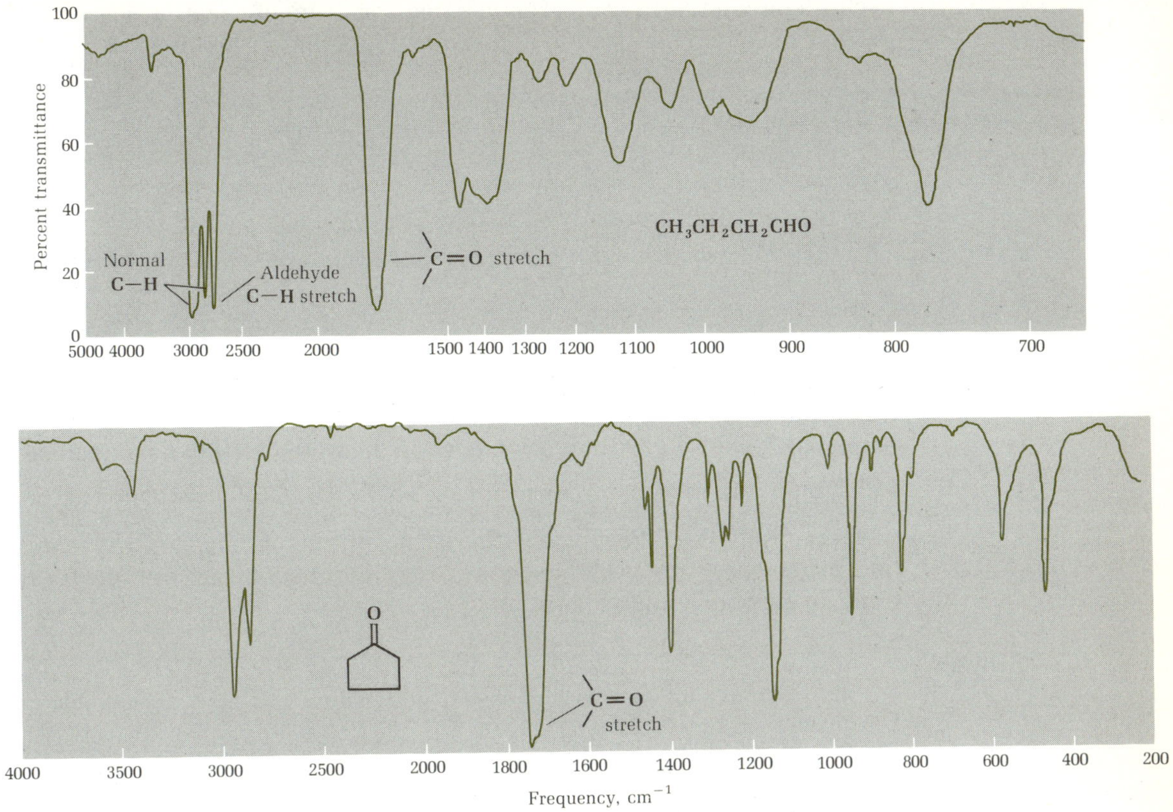

FIGURE 11-1
Infrared Spectra of Butyraldehyde and Cyclopentanone

For most common aldehydes and ketones, this band lies in the range 1690 to 1780 cm^{-1}. These peaks are very intense, and they will usually be the most intense bands in the spectra of simple molecules. Aldehydes may be differentiated from ketones by the presence in their spectra of a peak in the 2720 to 2820 cm^{-1} region due to the stretching motion of the aldehyde hydrogen atom.

As already noted in Chapter 3, the C=O stretching vibration occurs at higher frequencies than does the C=C stretching vibration, a fact attributable in part to the relatively strong carbon-oxygen π-bond. In Chapter 9 chemical evidence was introduced that also supported this contention.

Infrared spectroscopy has been very successful in the identification of the ring size of cyclic ketones. Cycloheptanones and cyclohexanones exhibit C=O stretching vibrations in the same region as open-chain ketones, 1700 to 1725 cm^{-1}. Contraction of the ring causes marked increases in the carbonyl stretching frequency: cyclopentanone, 1745 cm^{-1}; cyclobutanone, 1775 cm^{-1};

cyclopropanone, 1815 cm^{-1}. The infrared spectrum of camphor has a strong peak at 1745 cm^{-1}, indicating that the smallest ring containing the carbonyl group is a five-membered ring.

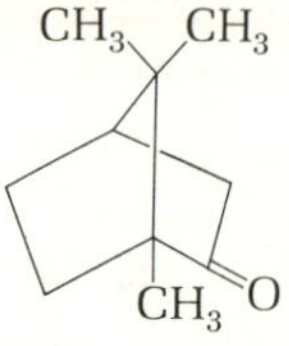

camphor

cyclopropanone

In Chapter 6 it was shown that the properties of cyclopropane could be accounted for if one assumed that the carbons were not sp^3-hybridized. The ring orbitals were taken to have increased p-character, resulting in weaker bonds; the exocyclic carbon orbitals were then predicted to have increased s-character and to form stronger bonds. The high $>$C=O frequency accords with the prediction. Presumably, there is a continuous change in hybridization as one goes from cyclohexanone to cyclopropanone, accounting for the trend in carbonyl frequencies.

PROBLEM

1 For most ketones the equilibrium shown in equation 1 is very unfavorable and lies heavily to the left.

$$\mathrm{R\overset{\overset{O}{\|}}{C}R' + HCl \rightleftharpoons R\overset{\overset{OH}{|}}{\underset{\underset{Cl}{|}}{C}}R'} \qquad \textbf{(1)}$$

An exception is the very reactive cyclopropanone, which forms a stable hydrogen chloride addition product. Suggest an explanation for this finding.

Aldehydes are usually detected with ease by means of nmr spectroscopy because the hydrogen atom attached to the carbonyl group is strongly deshielded; its resonance occurs at 9.6 to 10.0δ. The hydrogen atoms α and β to the carbonyl group are also deshielded, although not as drastically. The cause is in part the electron-attracting property of the carbonyl group, but it is also due to the induced circulation of the π-electrons of the carbonyl group, which creates a local magnetic field that can deshield protons relatively distant from the carbonyl group. Note, in particular, the nmr spectrum of cyclopentanone in which both types of methylene group are deshielded to about the same extent (Figure 11-3, p. 254).

The spectroscopic evidence certainly suggests that carbonyl groups are strongly polarized, with negative charge accumulating on oxygen; this impression is reinforced by the examination of some molecular dipole mo-

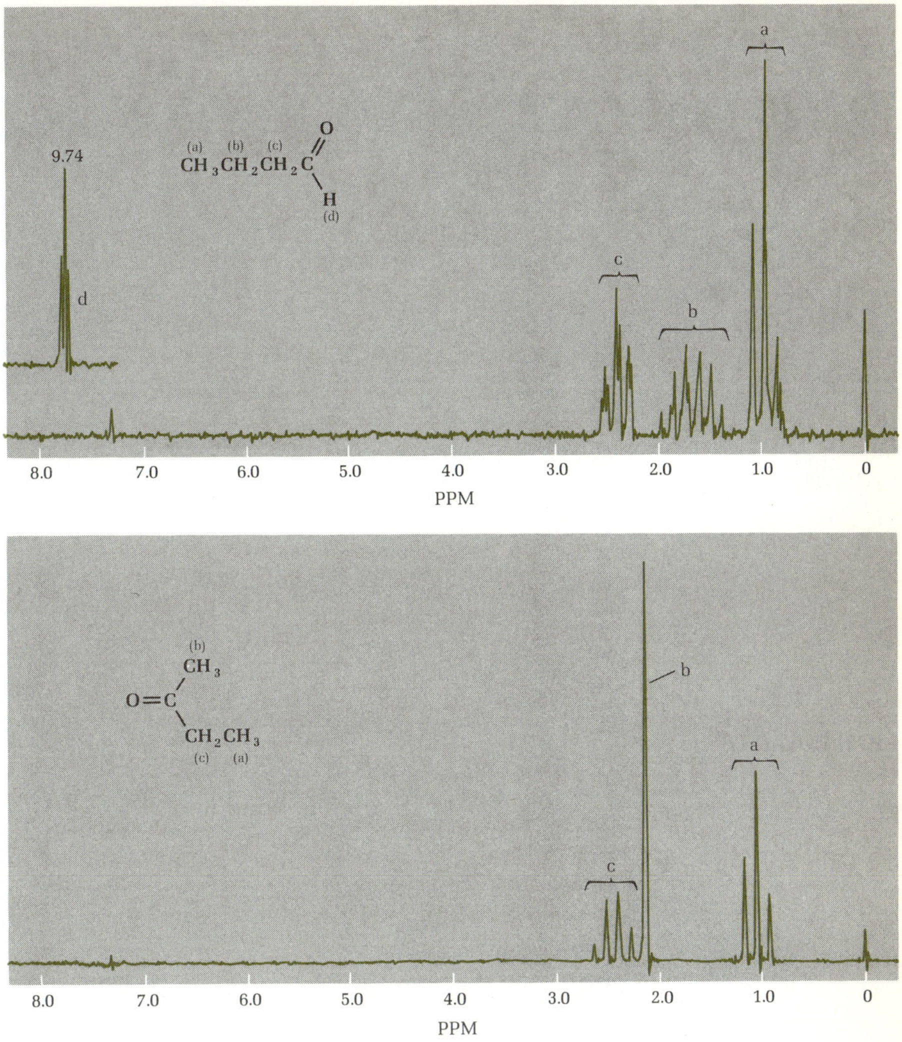

FIGURE 11-2
NMR Spectra of Butyraldehyde and Methyl Ethyl Ketone

ments. We can therefore expect that groups attached directly to a carbonyl carbon atom will be somewhat depleted in electron density.

CH_2Cl_2	CH_2F_2	$CH_2{=}O$
1.57 D	1.93 D	2.34 D

A comparison of the dipole moments of formaldehyde, acetaldehyde, and acetone reinforces our belief that methyl groups are more polarizable than hydrogen atoms.

$CH_2{=}O$	$CH_3CH{=}O$	$(CH_3)_2C{=}O$
2.34 D	2.70 D	2.88 D

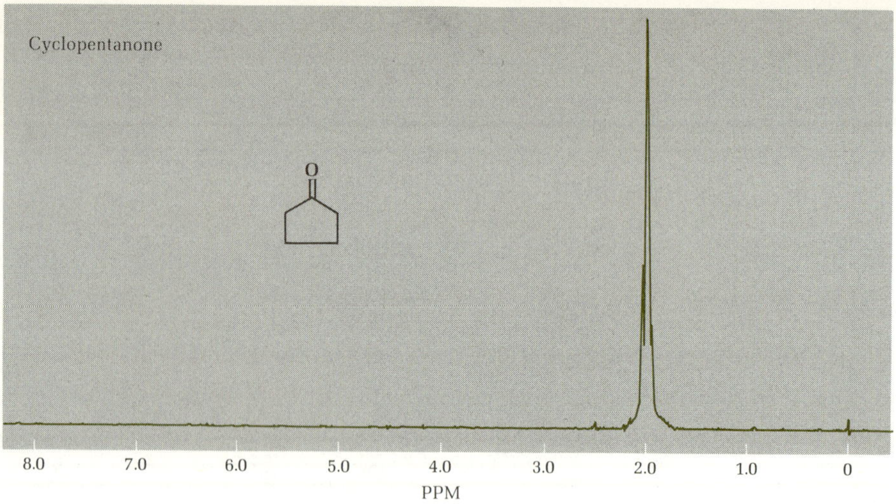

FIGURE 11-3
NMR Spectrum of Cyclopentanone

As we shall see, the chemistry of carbonyl compounds supports the view that the carbon atoms attached to the functional group (the α-carbon atoms) are relatively electron-deficient.

MOLECULAR ORBITALS AND PHOTOCHEMISTRY

The carbonyl group double bond may be formulated as a σ-π double bond, somewhat analogous to the carbon-carbon double bond. Its formation may be depicted as the combination of an sp^2-hybridized carbon atom and an unhybridized oxygen atom. Of course, simultaneously with the formation of the bonding σ- and π-orbitals, the antibonding σ^*- and π^*-orbitals are also created (Figure 11-4). In the ground state of the molecule there are two electrons each in the σ- and π-orbitals, and the high energy antibonding orbitals are empty. In addition, there are two sets of nonbonding electrons (or n-electrons) occupying an s- and a p-orbital localized on oxygen. The filled orbital of highest energy is the p-orbital. In formaldehyde it takes only 250 kcal/mole to remove an electron from the n-orbital to infinity, while it requires 332 kcal/mole to remove an electron from the π-bond.

If energy of the correct wavelength is put into the molecule, an electron may be excited from any one of the filled orbitals to one of the unfilled ones. In practice, the σ-orbital lies too low and the σ^*-orbital too high for them to be much involved. It is not difficult, however, to obtain radiation of the right wavelength to cause electrons to jump from the n- or π-orbitals to the π^*-orbital. (These events are usually labeled $n \rightarrow \pi^*$ or $\pi \rightarrow \pi^*$ transitions by organic chemists; the excited states that result are labeled n, π^* and π, π^* states.)

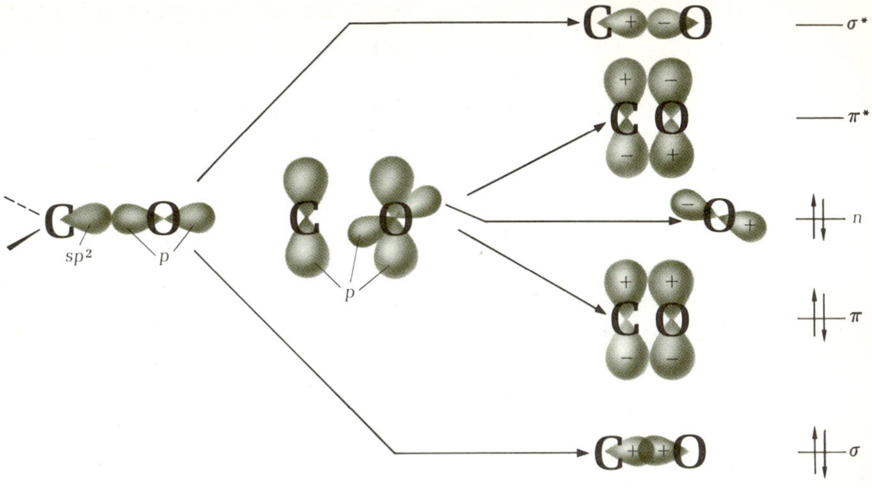

FIGURE 11-4
Combination of Carbon and Oxygen Orbitals to Produce σ- and π-Orbitals

The $\pi \rightarrow \pi^*$ transition is classified as an "allowed" transition, since there is a high probability that it will occur under favorable conditions (correct wavelength of light). The $n \rightarrow \pi^*$ transition is "forbidden," however, because the electron must "turn a corner" in order to get from the *n*-orbital to the π^*-orbital. As a consequence, the probability of an $n \rightarrow \pi^*$ transition occurring is rather low. On the other hand, the π, π^* state is higher in energy; it requires light of wavelength 1800 to 2000 A (159 to 143 kcal/mole) to effect this transition in simple aldehydes and ketones, and this is somewhat inconvenient experimentally. The $n \rightarrow \pi^*$ transitions lie in the region 2800 to 3000 A (102 to 95 kcal/mole), which is readily accessible, so that most of the photochemistry of simple aldehydes and ketones is derived from the n, π^* state.

The excited state may be a singlet or a triplet (Chapter 8). Excitation usually results in the immediate formation of a singlet state, in which all electron spins remain paired, as in the ground state. The excited singlet may spontaneously convert to a triplet, a process known as **intersystem crossing.**

(11.1) 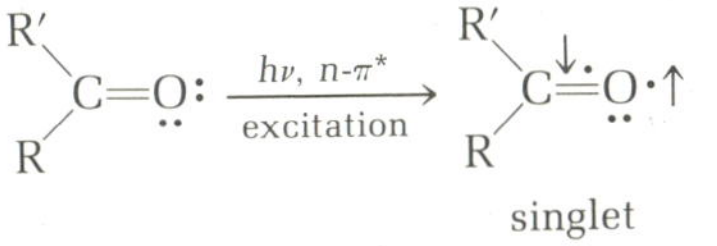

$^1(n, \pi^*)$

(11.2) 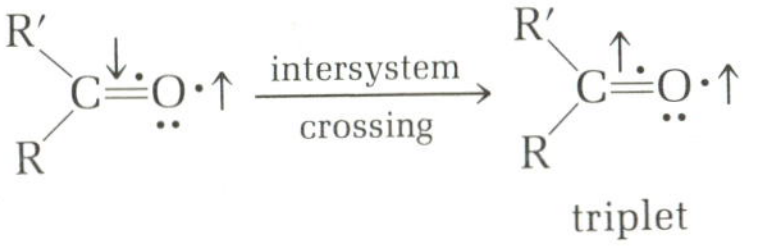

$^3(n, \pi^*)$

The organic chemist most wants to know how these transitions affect the chemical behavior of the molecule. When an $n \rightarrow \pi^*$ transition takes place, one electron that has essentially been confined to the oxygen atom is placed in an orbital in which it spends time on both the carbon and oxygen atoms. So there is a net depletion of electron density on oxygen. Furthermore, one of the oxygen orbitals is now half empty, making the oxygen atom radical-like. And lastly, the molecule contains a large amount of excitation energy.

When a carbonyl compound is irradiated in the presence of an alkene, reactions take place between the two that do not occur in the absence of light. Experiments have shown that the $^3(n, \pi^*)$ state reacts with the alkene — the reaction may be visualized as an attack of the electron-deficient oxygen atom on the alkene, attended by electron rearrangement to produce a biradical. The biradical then closes, forming an oxetane.

(11.3) 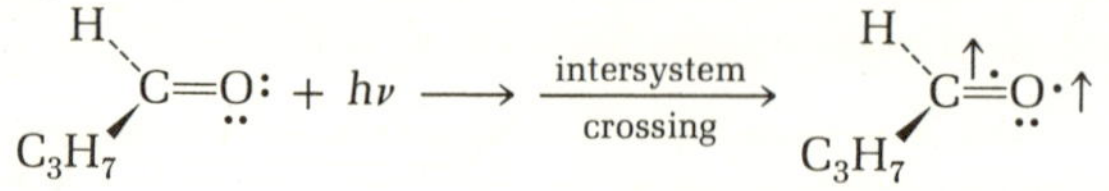

(11.4) 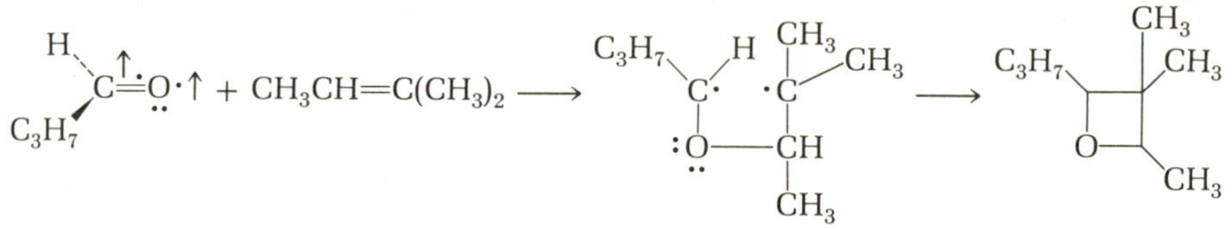

The same reaction may be induced *intramolecularly* and is a generally useful method of making exotic oxetanes. In fact, a number of molecules that would be difficult to synthesize using ordinary thermal reactions are accessible photochemically. This is especially true of cyclobutane derivatives.

(11.5) 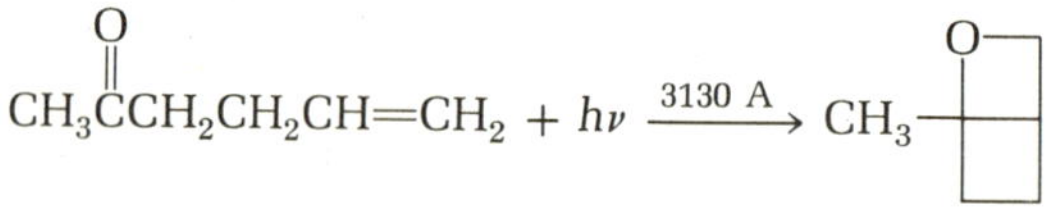

The radical character of the n, π^* state is also apparent in some intramolecular photochemical reactions of aldehydes and ketones. The electron-deficient oxygen atom may abstract a hydrogen atom from the γ-position if there is a sufficiently long alkyl chain, producing two normal molecules [in equation 11.6 the half-arrow or "fishhook" (⇀) indicates motion of a single electron].

(11.6) 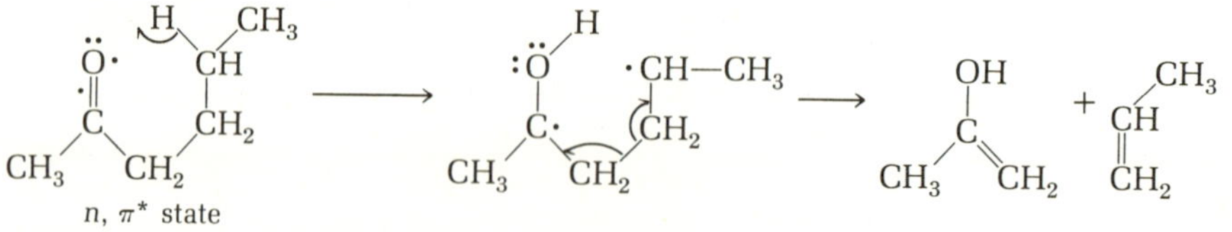

This is the Type II cleavage. The energetic advantage of six-membered cyclic transition states is indicated by the fact that only the γ-hydrogen atoms are attacked. Both reactions can involve the singlet and triplet n, π^* states, depending on conditions. Analogies to the Type II cleavage are found in the decomposition of ketones in the mass spectrometer.

GROUND STATE CARBONYL GROUP CHEMISTRY: INTRODUCTION

Carbonyl compounds are generally susceptible to a variety of reactions, and this propensity makes them key intermediates in many synthetic schemes, as well as in a variety of biochemical systems. Their reactivity can be ascribed to three factors:

1 The large electronegativity difference between carbon and oxygen results in a strong polarization of the double bond, creating an electron-deficient carbon that is susceptible to nucleophilic attack.

2 The nonbonded electrons on the electron-rich oxygen atom provide a site for electrophilic attack.

3 The planar arrangement of atoms around the sp^2-hybridized carbonyl carbon atom makes it sterically accessible to various reagents. It is worth recalling that at the transition state of a typical S_N2 reaction the carbon atom is close to sp^2-hybridized and so, of course, are the carbonium ions produced in S_N1 reactions.

As a general rule aldehydes are more reactive than ketones. A hydrogen atom is less effective than an alkyl group in relieving the electron deficiency of the carbonyl carbon atom; in addition, it offers less steric hindrance to an attacking nucleophile. This difference in reactivity is reflected in the tendency of low molecular weight aldehydes to react with themselves to form polymers, a reaction not observed in ketones. Formaldehyde exists in several polymeric forms, including two linear polymers (paraformaldehyde and polyoxymethylene) and a cyclic trimer (trioxane). Other aldehydes usually form trimers if they polymerize at all.

$$(CH_2{-}O{-}CH_2{-}O{-}CH_2{-}O)_x$$

paraformaldehyde ($x \sim 15$ to 55)
polyoxymethylene ($x \sim 300$ to 3000)

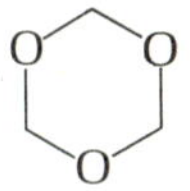

trioxane

Acetaldehyde trimerizes to paraldehyde, but propionaldehyde is essentially monomeric.

(11.7) $3CH_3CHO \underset{\text{heat, } H^+}{\overset{25°,\ H^+}{\rightleftharpoons}}$ 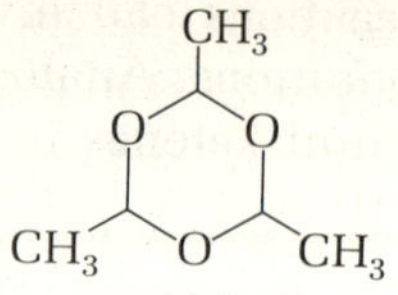

paraldehyde

Strong nucleophiles attach themselves to the carbonyl carbon atom by breaking the carbon-oxygen double bond. The oxygen atom may then combine with an electrophile such as a proton, and a stable product will result. Alternatively, the addition step may be reversed and the double bond regenerated. It depends on the interplay of several factors, the most prominent of which are the strength of the nucleophile-carbon bond, the strength of the C—O π-bond, the extent of crowding around the carbonyl carbon atom, the relative solvation energies of reactants and product, and the fact that the addition is usually entropically unfavorable.

RAPIDLY REVERSIBLE ADDITIONS

Cyanide ion will add to the carbonyl groups of most aldehydes in mildly alkaline solution to produce stable **cyanohydrins.** The equilibrium is less favorable with the more hindered ketones, although cyanohydrins of some ketones are isolable.

(11.8) $$CH_3\overset{\overset{\displaystyle O}{\|}}{C}—H + CN^- \rightleftharpoons CH_3\underset{\underset{\displaystyle H}{|}}{\overset{\overset{\displaystyle O^-}{|}}{C}}—CN \overset{H_2O}{\rightleftharpoons} CH_3\underset{\underset{\displaystyle H}{|}}{\overset{\overset{\displaystyle OH}{|}}{C}}—CN + OH^- \qquad K^{25°} = 14.3 \times 10^3$$

acetaldehyde cyanohydrin

Similarly useful derivatives may be formed by reaction of aldehydes with sodium bisulfite (equation 11.10). These "bisulfite addition compounds" are sparingly soluble crystalline solids and are readily purified by recrystallization. The aldehyde may be regenerated by treating the addition compound with a reagent that destroys the sodium bisulfite.

(11.9) $$HSO_3^- \rightleftharpoons :SO_3^{2-} + H^+$$

(11.10) $$CH_3CHO + :SO_3^{2-} \rightleftharpoons CH_3\underset{\underset{\displaystyle SO_3^-}{\backslash}}{\overset{\overset{\displaystyle O^-}{/}}{CH}} \overset{H^+}{\rightleftharpoons} CH_3\overset{\overset{\displaystyle OH}{|}}{C}HSO_3^-$$

(11.11) $$CH_3\underset{\underset{\displaystyle OH}{|}}{C}HSO_3^-Na^+ + HCl \longrightarrow CH_3CHO + NaCl + SO_2 + H_2O$$

Only unhindered ketones, such as cyclohexanone or methyl ketones, form bisulfite addition compounds to an appreciable extent at equilibrium.

Hydroxide ion attacks the carbonyl group reversibly; one cannot tell that anything has "happened" unless isotopically labeled water is used.

(11.12)

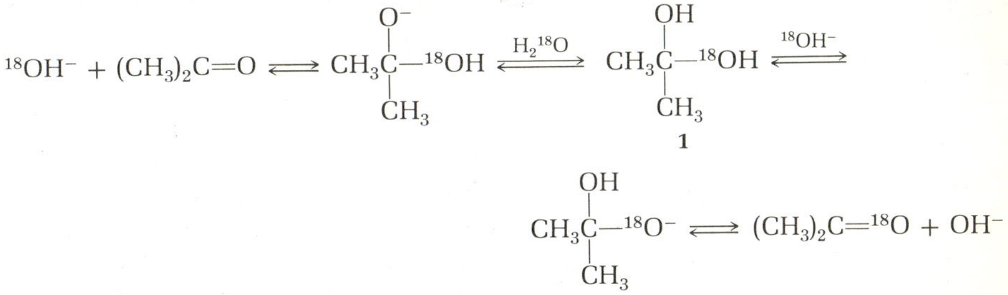

The intermediate **1** in the exchange shown above is not very stable, as indicated by the low value of the equilibrium constant for equation 11.13.

(11.13) $(CH_3)_2C{=}O + H_2O \rightleftharpoons (CH_3)_2C(OH)_2 \qquad K^{25^\circ} = 2 \times 10^{-3}$

acetone hydrate (**1**)

Here again, however, the position of equilibrium is very sensitive to steric and electronic factors (equations 11.14 and 11.15).

(11.14) $(CH_3)_2CHCHO + H_2O \rightleftharpoons (CH_3)_2CHCH(OH)_2 \qquad K^{25^\circ} = 6.2 \times 10^{-1}$

(11.15) $CCl_3CHO + H_2O \rightleftharpoons CCl_3CH(OH)_2 \qquad K^{25^\circ} = 2.8 \times 10^{4}$

CARBONYL REACTIONS WITH AMMONIA DERIVATIVES

A number of derivatives of ammonia actively attack the carbonyl group. The important characteristic of these reactions is that further transformations, in particular dehydration, are possible after the addition step. In practice the intermediate addition compound (**2**) is rarely isolated.

(11.16)

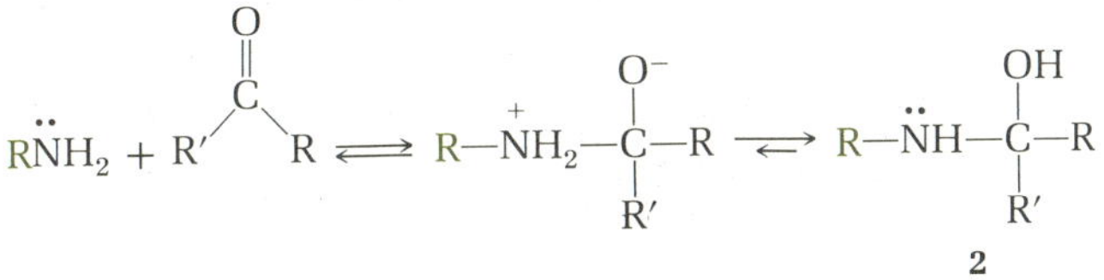

(11.17)

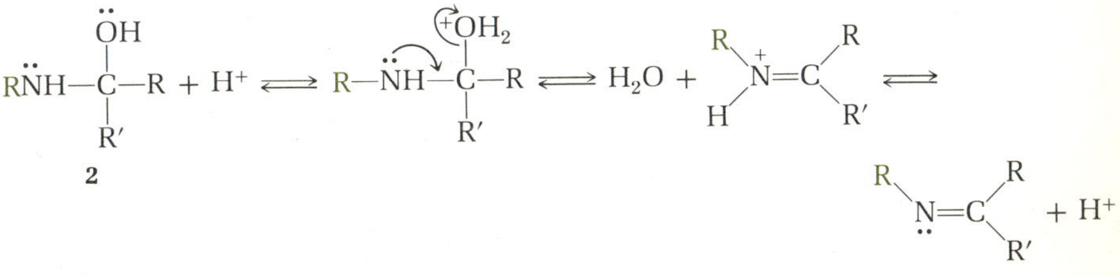

a Schiff base

Its existence may be inferred in many cases from the fact that there is a distinct time lag between the disappearance of the carbonyl group (measured spectroscopically) and the appearance of the product. The product contains the functional group $\overset{R}{\ddot{N}}{=}C\langle$ and is obviously related to an aldehyde or ketone — compounds containing this grouping are known generically as **imines.** When the imine is derivable from a carbonyl compound and a **primary amine,** RNH_2, it is almost always referred to as a **Schiff base.**

When R, R′, and R are alkyl groups (or hydrogen), formation of the Schiff base is unfavorable, and the product is very readily hydrolyzed. If any of these groups (particularly R and R′) are **aromatic** (Chapter 14), the Schiff base is stabilized considerably.

(11.18) $$CH_3NH_2 + CH_3CHO \xrightarrow[Na_2SO_4\text{ (absorbs water)}]{\text{dry ether}} CH_3N{=}CHCH_3 \quad (\sim 40\%)$$

Schiff bases are often reactive intermediates in a number of chemical and biochemical reactions. Protonation of a Schiff base (equation 11.19), which is more basic than the ketone from which it is derived, yields an ion that is even more reactive toward nucleophiles than the corresponding ketone.

(11.19) $$\mathrm{R'(R)C{=}\ddot{N}{-}R} + H^+ \rightleftharpoons \mathrm{R'(R)C{=}\overset{+}{N}(R)H}$$

The role of Schiff bases in the biologically important transaminase reaction is discussed at some length in Chapter 23 (p. 656).

There is a special class of ammonia derivatives, $G{-}\ddot{N}H_2$, in which G is connected to the $-NH_2$ group through an oxygen or nitrogen atom. These reagents condense readily with aldehydes or ketones. The products are very stable, probably because of the overlap between the *p*-orbital on the oxygen or second nitrogen atom and the carbon-nitrogen double bond.

(11.20) $$(CH_3)_2C{=}O + NH_2\ddot{O}H \xrightarrow[\text{ethanol-water}]{\text{acetic acid}} (CH_3)_2C{=}N\ddot{O}H$$

hydroxylamine — acetone **oxime**

(11.21)

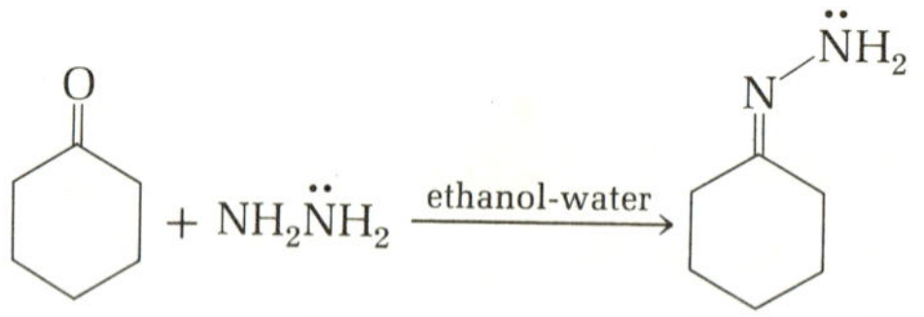

hydrazine — cyclohexanone **hydrazone**

(11.22) $$CH_3CH_2CH_2CHO + NH_2\ddot{N}H\overset{O}{\overset{\|}{C}}NH_2 \xrightarrow[\text{acetic acid}]{\text{ethanol-water}} CH_3CH_2CH_2CH{=}\ddot{N}NH\overset{O}{\overset{\|}{C}}NH_2$$

semicarbazide — butyraldehyde **semicarbazone**

These derivatives of the carbonyl compounds are almost all solids, even though the aldehydes and ketones themselves are liquids. They are readily purified by recrystallization and have characteristic, sharp melting points; for this reason, they are very useful for identifying unknown aldehydes and ketones. The carbonyl compound may also be regenerated from the derivative, so that they are occasionally used for indirect purification of an aldehyde or ketone.

The formation of a hydrazone is not really the straightforward procedure implied in equation 11.21. Reaction of 2 moles of carbonyl compound with 1 mole of hydrazine is an unavoidable side reaction; this is the major reaction of aliphatic aldehydes, even in the presence of excess hydrazine.

(11.23)

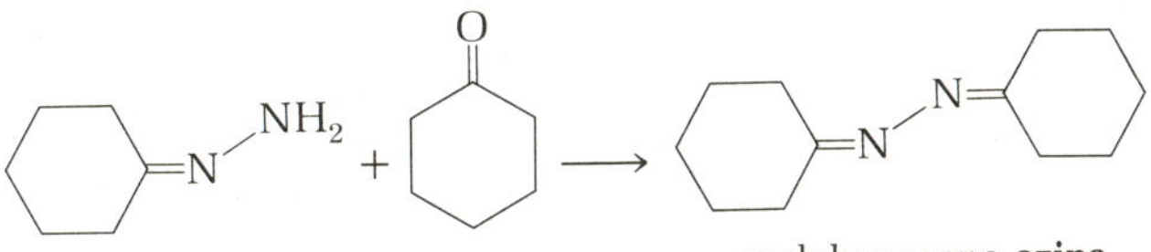

cyclohexanone **azine**

For this reason, substituted hydrazines, such as phenylhydrazine, are used to make carbonyl derivatives rather than hydrazine itself.

(11.24) $$CH_3CH_2\overset{\overset{\displaystyle O}{\|}}{C}CH_3 + NH_2NHC_6H_5 \longrightarrow CH_3CH_2\overset{\overset{\displaystyle NNHC_6H_5}{\|}}{C}CH_3$$

phenylhydrazine — butanone **phenylhydrazone**

ACID-CATALYZED CARBONYL ADDITION

If a nucleophile is relatively weak, it may attack a carbonyl group rather slowly. Such is the case with water or alcohols. If acid is added to the reaction medium, it speeds up the rate; by hydrogen-bonding with, or actually protonating, the carbonyl group, it increases the $\rangle\overset{\delta+}{C}=\overset{\delta-}{O}$ polarization and lowers the activation energy for nucleophilic attack.

(11.25)

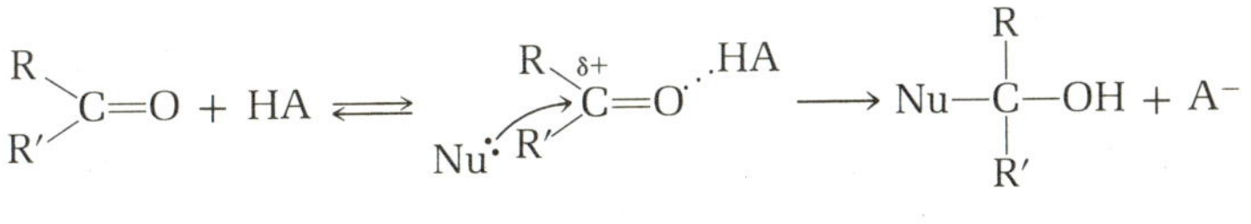

When an aldehyde or ketone is dissolved in an alcohol, 1 mole of the alcohol may add across the carbon-oxygen double bond, forming a hemiacetal or hemiketal. These are usually too unstable to be isolated (see below). If a trace of acid catalyst is present, however, then the hemiacetal or hemiketal may be dehydrated, an oxonium-carbonium ion **(3)** being produced.

(11.26)

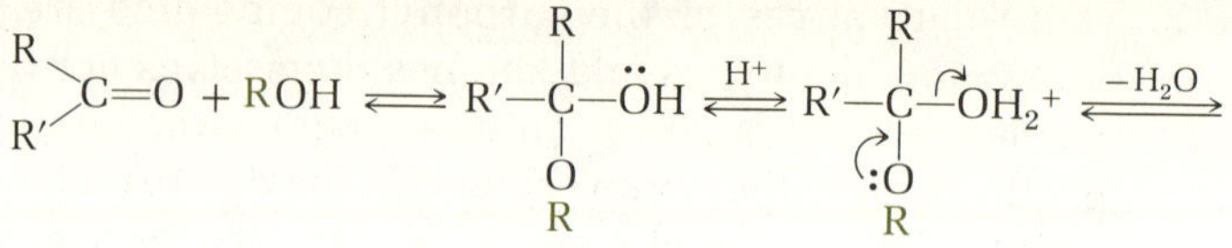

hemiacetal or
hemiketal

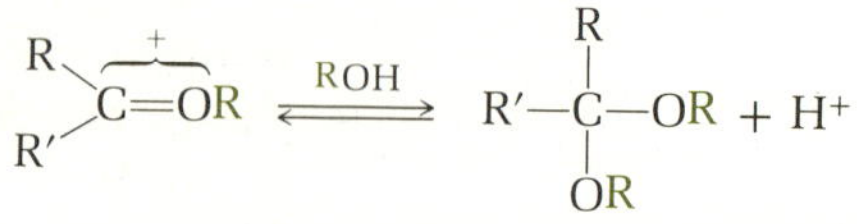

3

R′ = H: acetal
R′ = alkyl: ketal

This ion then reacts with a second mole of alcohol, yielding the acetal or ketal. The acetals and ketals are completely stable in base; they may be used to protect the carbonyl group under basic reaction conditions.

(11.27)

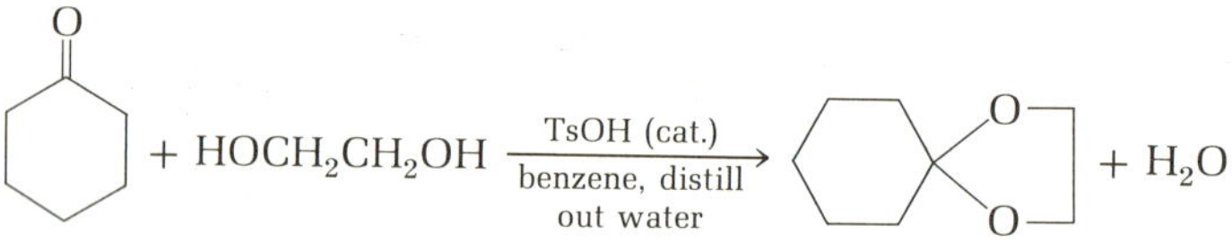

cyclohexanone
ethylene ketal (80%)

(11.28) $$CH_3CHO + \underset{\text{(excess)}}{CH_3CH_2CH_2OH} \xrightarrow[25°,\ 24\ hr]{HCl\ (cat.)} \underset{\text{acetaldehyde dipropyl acetal (100\%)}}{CH_3CH(OCH_2CH_2CH_3)_2} + H_2O$$

The formation of ketals and acetals is subject to the same factors that affect other carbonyl addition reactions, such as steric hindrance. Ketals are not formed from monoalcohols in good yield and, for this reason, cyclic ketals (derived from glycols) are almost always made. Their formation is entropically more favorable than that of acyclic ketals. Acetals and ketals may be readily converted back to the carbonyl compounds with dilute acid.

(11.29) $$CH_3CH(OCH_2CH_2CH_3)_2 + H_2O \xrightarrow{H_3O^+} CH_3CHO + 2CH_3CH_2CH_2OH$$

Hemiacetals and hemiketals are normally quite unstable and not isolable. One exception, and an important one, occurs when the hemiacetal or hemiketal is cyclic and contains a five- or six-membered ring. The entropy effect no doubt accounts for the stability difference between cyclic and acyclic hemiacetals.

(11.30)

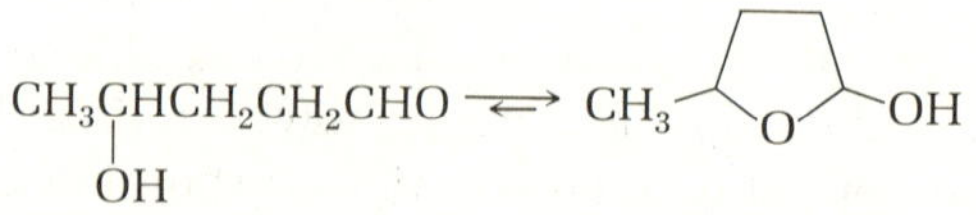

(11.31) 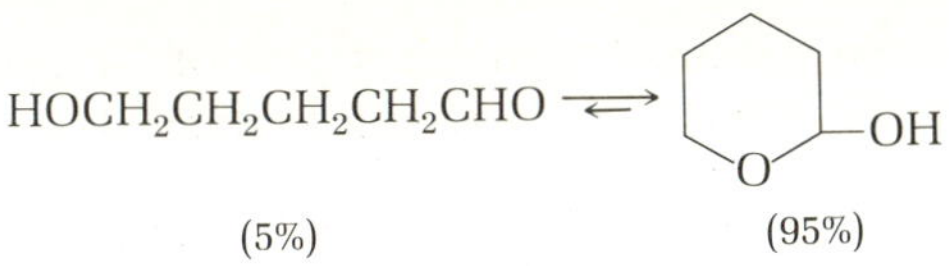

The relative proportions of the two isomers in equation 11.31 were determined spectroscopically; the aldehyde form absorbs ultraviolet radiation at 2900 A ($n \rightarrow \pi^*$ region). If one attempts to make an acetal of these compounds, only 1 mole of alcohol is consumed and a mixed acetal results. The chemistry of these stable hemiacetals supports the proposed intermediacy of unstable hemiacetals during acetal formation.

(11.32) 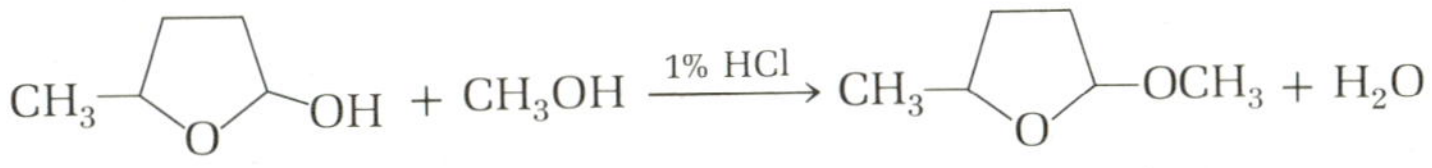

GRIGNARD REACTIONS – IRREVERSIBLE CARBONYL ADDITION

The reaction of aldehydes and ketones with Grignard reagents constitutes an extremely versatile method of synthesizing substituted alcohols.

$$(11.33)\quad (CH_3)_2CHMgBr + CH_3CHO \xrightarrow{\text{ether}} \xrightarrow{H_3O^+} (CH_3)_2CH\overset{\overset{\displaystyle OH}{|}}{C}HCH_3 \quad (54\%)$$

(11.34) 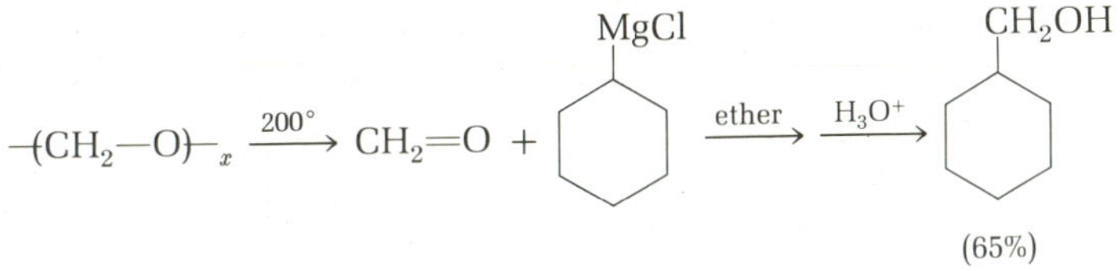

$$(11.35)\quad CH_3CH_2CH_2CH_2MgBr + (CH_3)_2C{=}O \xrightarrow{\text{ether}} \xrightarrow{H_3O^+} CH_3CH_2CH_2CH_2\overset{\overset{\displaystyle CH_3}{|}}{\underset{\underset{\displaystyle CH_3}{|}}{C}}OH \quad (92\%)$$

Depending on whether the carbonyl compound is formaldehyde (equation 11.34), another aldehyde (equation 11.33), or a ketone (equation 11.35), the product will be, respectively, a primary, secondary, or tertiary alcohol. In a great many cases there will be two or more routes to the same compound. For example, the product in equation 11.33 would also be obtained from reaction of methylmagnesium bromide with isobutyraldehyde. In general, other things being equal, one chooses the carbonyl compound with the least hindered carbonyl group.

Present evidence indicates that 2 moles of Grignard reagent and 1 mole of ketone are involved in the transition state.

(11.36)
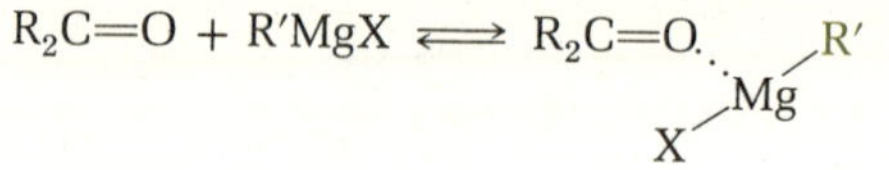

(11.37)
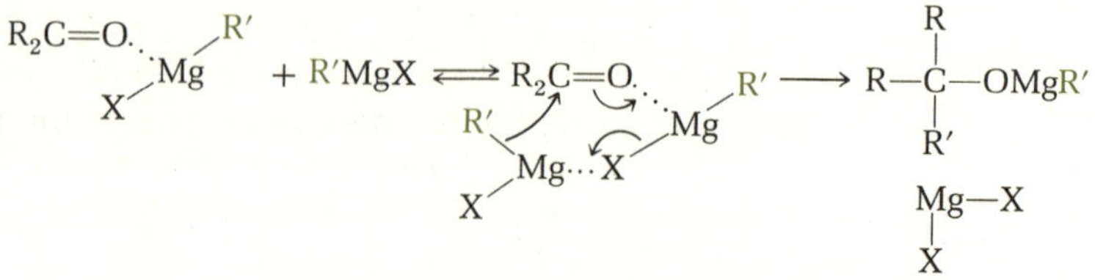

(11.38)
$$R_2C(R')\text{—}OMgR' + MgX_2 \rightleftharpoons R_2C(R')\text{—}OMgX + R'MgX$$

For instance, the rate of the reaction has been shown to be proportional to the square of the Grignard reagent concentration in several cases. The mechanism proposed is analogous to that postulated for acid-catalyzed additions to the carbonyl group. The ability of the magnesium atoms to complex with oxygen and halogen atoms would make the sterically favorable six-membered cyclic transition state energetically favorable as well (Figure 11-5).

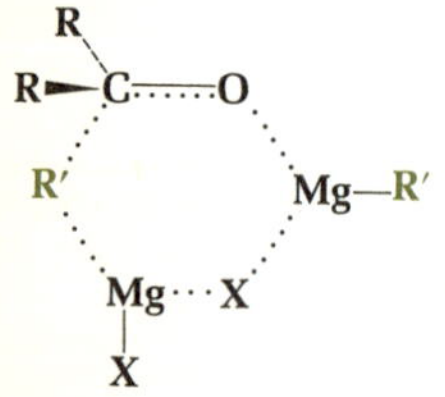

FIGURE 11-5
Cyclic Transition State Grignard Addition

It should be noted that in order to get the product alcohol it is necessary to decompose the intermediate alkoxymagnesium halide, $R_2C(R')\text{—}OMgX$, with aqueous acid. This is a separate step; water may not be present while the reaction mixture contains unreacted Grignard reagent or the reagent will be instantly destroyed.

The related **Grignard reduction** provides some support for the postulated mechanism of the normal Grignard reaction. If sterically hindered aldehydes or ketones, particularly the latter, are treated with branched alkyl Grignard reagents, appreciable amounts of reduction product are obtained.

(11.39)
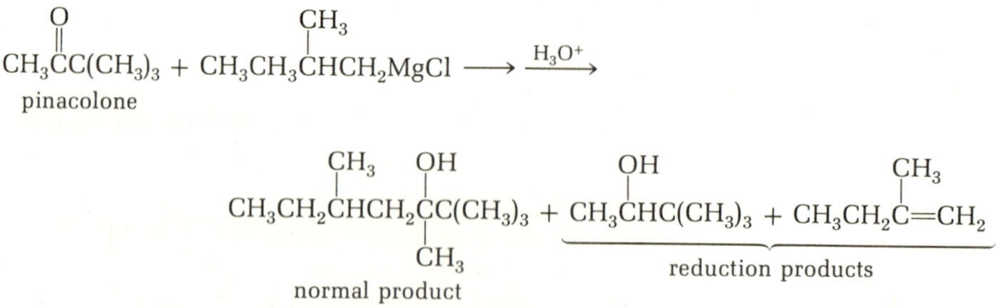

This observation is compatible with the suggestion that there is initial formation of a Grignard-ketone complex; if attack of a second mole of

Grignard reagent is slow because of steric hindrance, then transfer of a hydride ion via a six-centered transition state becomes competitive.

(11.40)

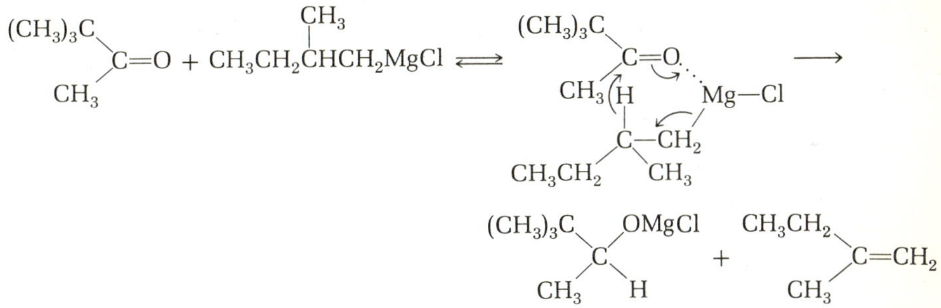

Grignard reagents will attack a wide variety of polarized multiple bonds. In the reaction with carbon dioxide the intermediate bromomagnesium salt resists further Grignard attack, in large part because the oxygen atom possesses a substantial fraction of a negative charge.

(11.41)

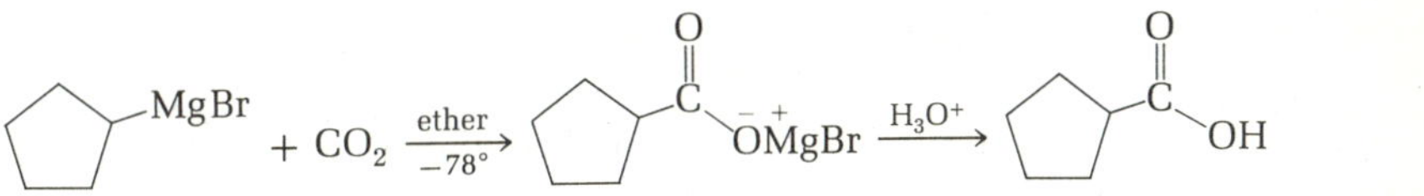

cyclopentanecarboxylic acid (85%)

The low temperature also retards addition of a second mole; the ethereal Grignard reagent is mixed with powdered solid carbon dioxide. The more reactive alkyllithium reagents will undergo double addition to carbon dioxide, forming symmetrical ketones.

(11.42)

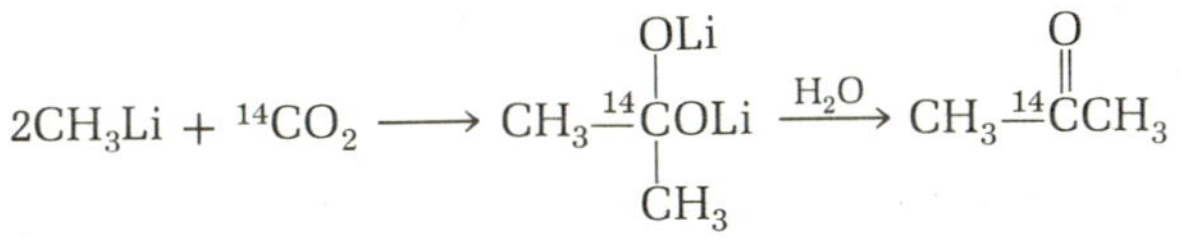

Grignard reagents will also react with $>C{=}N-$ and $-C{\equiv}N$ groups. The reaction with alkyl cyanides can generally be confined to addition of 1 mole of Grignard reagent; the hydrolysis product is a ketone. Rather low yields often result because the α-hydrogens of alkyl cyanides are relatively acidic, and an acid-base reaction competes with addition.

(11.43)

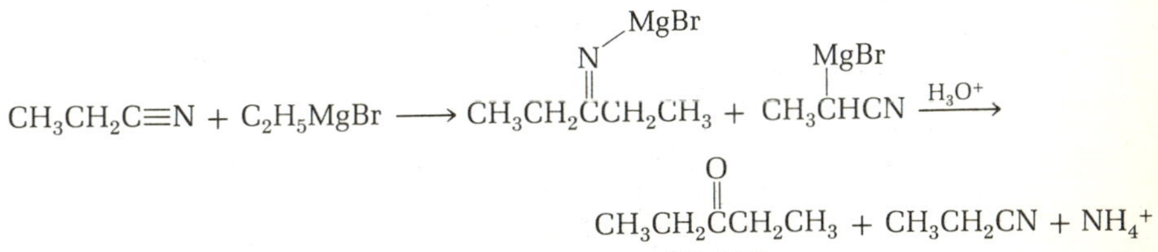

REDUCTION OF ALDEHYDES AND KETONES TO ALCOHOLS

Defining the oxidation state of an atom is often a straightforward process in inorganic chemistry, when ions are involved. It is more difficult, and more arbitrary, to assign oxidation numbers to covalently bound carbon atoms. A rule can be formulated, however, that is useful for bookkeeping purposes. In essence, it involves assigning one unit of positive charge to a carbon atom for each bond between that carbon atom and a more electronegative atom or group of atoms. Bonds between carbon and a metal or hydrogen atom confer a negative charge on that carbon atom, and carbon-carbon bonds involve a zero charge contribution.

By this criterion the following group of compounds all contain a carbon atom with the same oxidation number: **C**H_3OH, **C**H_3Cl, **C**H_3NH_2, **C**H_3CN, **C**H_3SH, etc. It can be seen that nucleophilic substitutions do not usually involve oxidation or reduction at carbon, according to our rule. Similarly, the following compounds are arranged in order of steadily increasing oxidation number for carbon: CH_4, CH_3CH_3, CH_3OH, $HC{\equiv}CH$, $CH_2{=}O$, $H\overset{\overset{O}{\|}}{C}OH$, CO_2.

Finally, addition or removal of water to or from an organic compound does not change the oxidation state *of the entire molecule* (equation 11.44). These rules cannot be pressed too hard since they are not grounded on firm physical distinctions, but they can be useful as thumbnail guides.

(11.44)

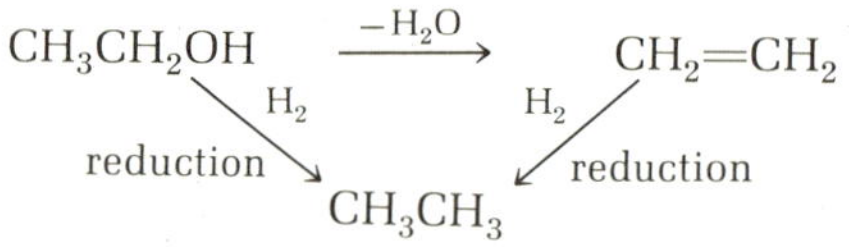

Reduction of carbonyl compounds to alcohols is most commonly done with the complex metal hydrides, lithium aluminum hydride, $LiAlH_4$ (abbreviated LAH), and sodium borohydride, $NaBH_4$. They act by transferring the strong nucleophile hydride ion to the carbonyl carbon. Their use is often complementary. Lithium aluminum hydride is very reactive, will decompose exothermically in water or alcohols (often sufficiently so to start a fire), and is soluble in diethyl and other ethers in which solvents it is commonly used. Sodium borohydride is a less vigorous reducing agent (it reduces few functional groups other than aldehyde or ketone), and it reacts only slowly with neutral or basic water and alcohols. It is often used in these solvents since it is practically insoluble in most ethers, except diglyme. A hydrolysis step is necessary to complete lithium aluminum hydride reductions, but this is not usually necessary with borohydride reductions that have been run in water or alcohols.

(11.45)

$$4CH_3\overset{\overset{O}{\|}}{C}CH_2CH_2CH_3 + Na^+BH_4^- \xrightarrow[OH^-]{CH_3OH} 4CH_3\overset{\overset{OH}{|}}{C}HCH_2CH_2CH_3 + NaOCH_3 + B(OCH_3)_3$$

(100%)

(11.46)

$$4CH_3(CH_2)_5CHO + Li^+AlH_4^- \xrightarrow{ether} \xrightarrow{H_3O^+} CH_3(CH_2)_5CH_2OH$$

(86%)

(11.47)

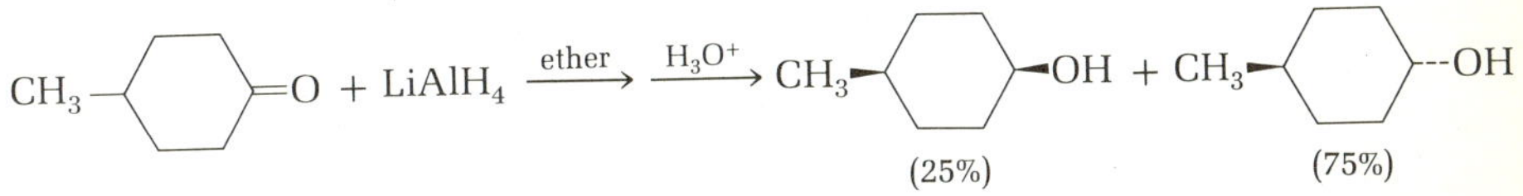

Sources of hydride ion other than metal hydrides are available. Alcohols might serve in this capacity if an equilibrium of the type shown in equation 11.48 could be achieved.

(11.48)

$$\begin{matrix}R \\ R'\end{matrix}\!\!>C{=}O + \begin{matrix}R \\ R'\end{matrix}\!\!>CHOH \rightleftharpoons \begin{matrix}R \\ R'\end{matrix}\!\!>CH{-}OH + \begin{matrix}R \\ R'\end{matrix}\!\!>C{=}O$$

Such equilibria may be established, using aluminum salts of alcohols as catalysts (**Meerwein-Pondorf-Verley reduction**). The equilibrium may be driven in either direction by the appropriate choice of conditions. For reduction, aluminum isopropoxide in isopropyl alcohol is used, and the acetone that forms is distilled from the reaction mixture.

(11.49)

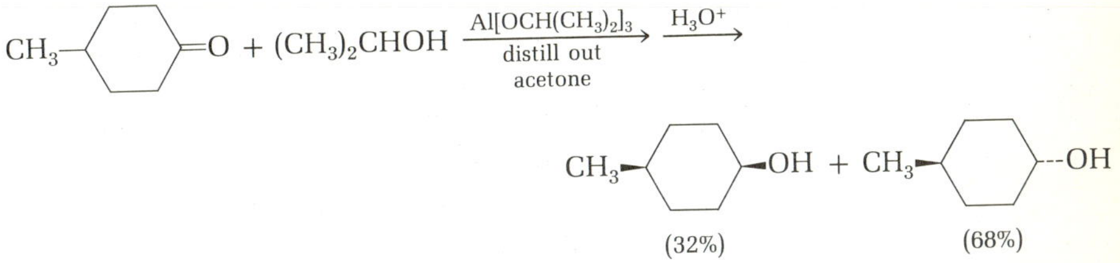

(11.50)

$$(CH_3)_2C{=}CH\overset{\overset{O}{\|}}{C}CH_3 + (CH_3)_2CHOH \xrightarrow{Al[OCH(CH_3)_2]_3} (CH_3)_2C{=}CH\underset{\underset{OH}{|}}{C}HCH_3 + (CH_3)_2C{=}O$$

(80%)

The virtue of this method of reduction is that other functional groups in the molecule remain unaffected, provided that they are not base-sensitive.

Since the reaction is reversible, one can assume that a mixture of diastereomeric products is at equilibrium. The postulated mechanism is very similar to that proposed for Grignard reduction.

(11.51)

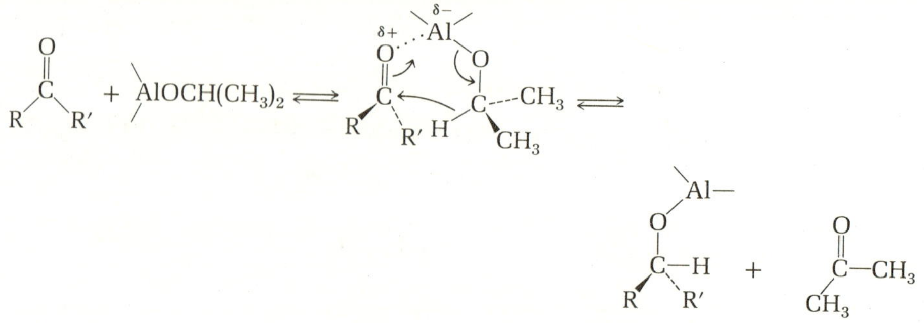

(11.52)

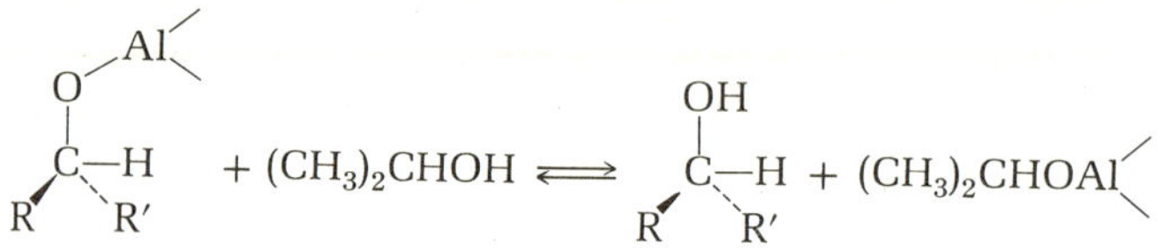

BIOLOGICAL REDUCTIONS OF CARBONYL COMPOUNDS

Hydride ion transfer to a carbonyl group is found in a number of biological reductions, many of which are indispensable for metabolism (biological synthesis and degradation). The reaction is reversible and the same system may also serve to oxidize alcohols. These reactions require, in addition to the substrate and enzyme, a coenzyme. The coenzyme acts as a source of hydride ion for reductions and as a hydride receptor during oxidations. The coenzyme thus has a reduced and an oxidized form. Two very similar cofactors are known: **nicotinamide-adenine dinucleotide (NADH, NAD⁺)** and **nicotinamide-adenine dinucleotide phosphate (NADPH, NADP⁺).** The nicotinamide ring is the site of the hydride transfers and has the structure shown in equation 11.53.

(11.53)

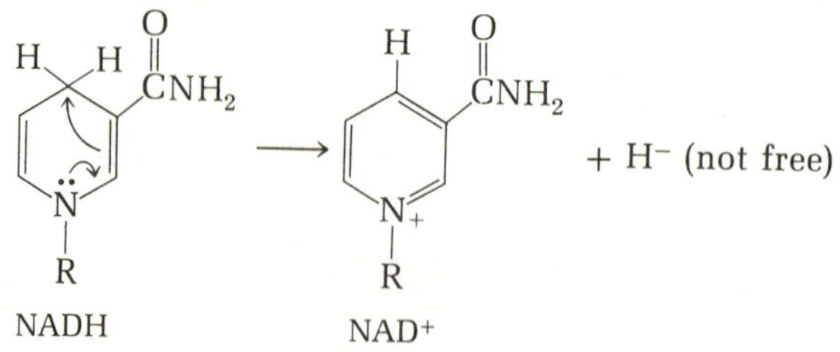

R is a complex structure that serves, among other things, to bind the coenzyme to the enzyme but does not directly affect its chemical function (see complete structure in Chapter 23).

If NAD^+ is mixed with ethanol-1,1-d_2 and an enzyme called **alcohol dehydrogenase** (isolated from yeast), acetaldehyde-1-*d*, is produced, along with NADD (equation 11.54). If this same NADD is now treated with ordinary acetaldehyde and the enzyme, ethanol-1-*d* is produced (equation 11.55).

(11.54) 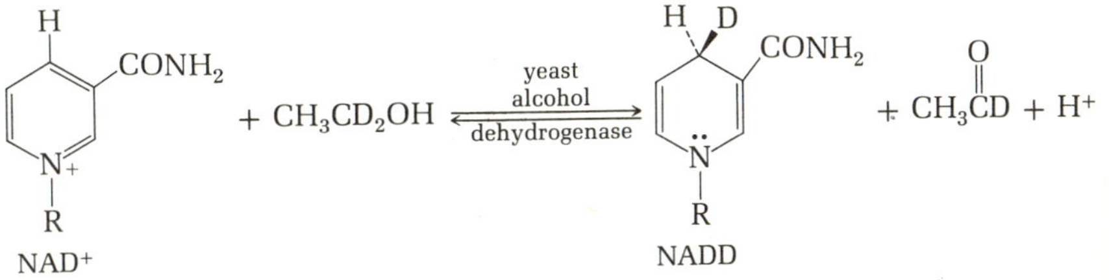

Only deuterium is transferred from the coenzyme to the aldehyde, which accepts the hydride ion on only one face of the carbonyl group. The enzyme is thus able to distinguish between the hydrogens on the two faces of the nicotinamide ring, as well as the two faces of the carbonyl group. Other enzymes are known that solely catalyze the transfer of the other hydrogen atom.

(11.55) 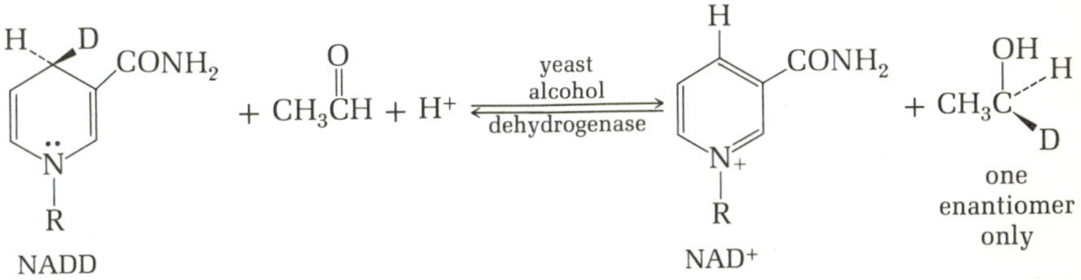

This enzymatic system has been used to synthesize optically active neopentyl-*d* alcohol. A very schematic mechanism for all these reductions is presented in Figure 11-6.

(11.56) 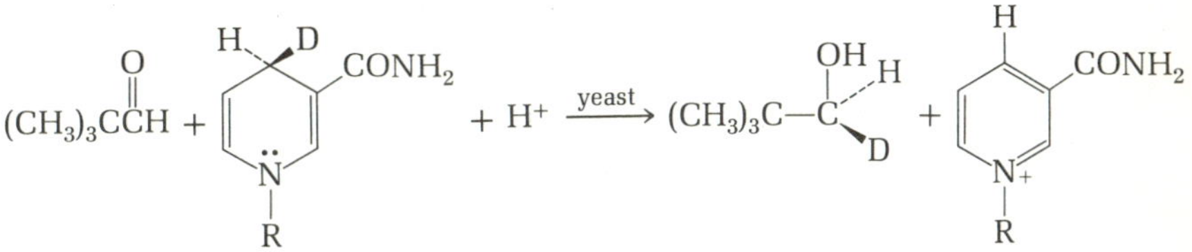

PINACOLS

If a ketone is subject to reduction by a metal in the absence of a source of protons, a product derived from 2 moles of ketone can be obtained. The general reaction is known as the **pinacol reduction,** after the trivial name of the diol derived from acetone.

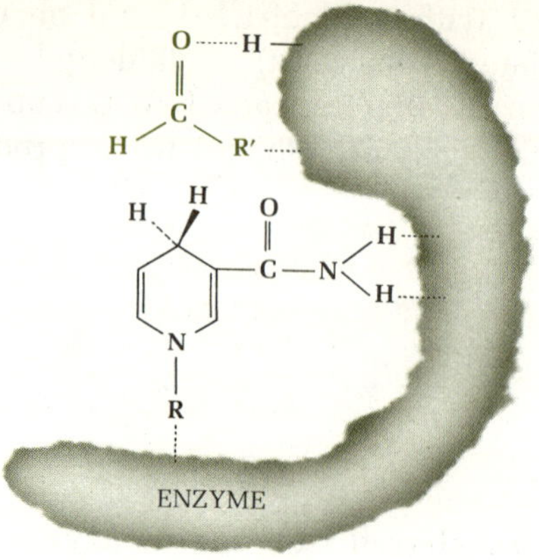

FIGURE 11-6
Schematic Mechanism for Enzymatic Aldehyde Reduction

(11.57) $$2(CH_3)_2C{=}O + Mg(Hg) \xrightarrow{\text{benzene}} \xrightarrow{H_2O} CH_3{-}\underset{\underset{\displaystyle HO}{|}}{\overset{\overset{\displaystyle CH_3}{|}}{C}}{-}{-}\underset{\underset{\displaystyle OH}{|}}{\overset{\overset{\displaystyle CH_3}{|}}{C}}{-}CH_3 \cdot 6H_2O$$

pinacol hexahydrate (42%)

The reaction is fairly general for ketones. Pinacols are subject to acid-catalyzed dehydration-rearrangement, leading to carbonyl compounds that could often be obtained only by means of multistep syntheses.

(11.58) $$(CH_3)_2\underset{\underset{\displaystyle OH}{|}}{C}{-}{-}\underset{\underset{\displaystyle OH}{|}}{C}(CH_3)_2 \xrightarrow[\text{distill}]{H_2SO_4\text{-}H_2O} CH_3\overset{\overset{\displaystyle O}{\|}}{C}C(CH_3)_3 + H_2O$$

pinacalone (100%)

The rearrangement is applicable to any 1,2-glycol, so that it provides a route for converting alkenes to aldehydes and ketones.

(11.59) $$(CH_3)_2\underset{\underset{\displaystyle OH}{|}}{C}CH_2OH \xrightarrow[H_2O,\ 80^\circ]{HClO_4} (CH_3)_2CHCHO$$

CATALYTIC REDUCTIONS

Carbon-oxygen double bonds may be reduced with hydrogen and a metal catalyst, in complete analogy with the reduction of carbon-carbon double bonds. Alcohols are products. The reaction is not nearly so rapid nor so exothermic since the $>C{=}O$ π-bond is considerably stronger than the $>C{=}C<$ π-bond.

(16.60) 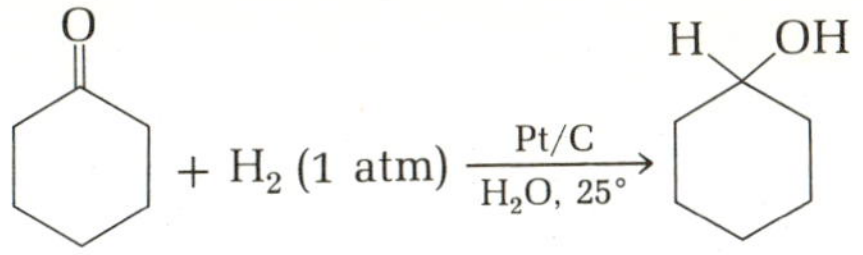

In compounds containing both types of double bond, it is possible to reduce them both or the carbon-carbon double bond alone. In order to reduce only the carbon-oxygen double bond, one must employ metal hydrides or the M-P-V method.

(11.61) 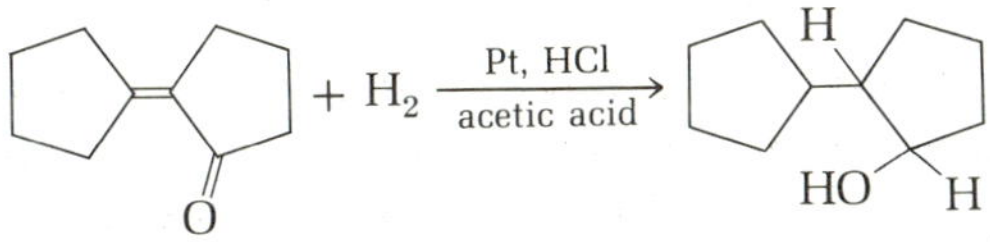

cis and *trans*

(11.62) 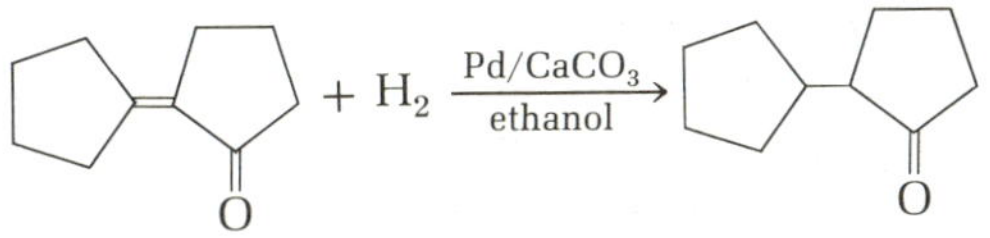

CARBONYL-TO-METHYLENE REDUCTIONS

Aldehydes or ketones may undergo reduction to the corresponding hydrocarbons in one step. Methods are available that involve acidic, basic, or neutral conditions so that it is possible to avoid serious side reactions with almost any molecule. The **Clemmensen reduction** is heterogenous, involving an organic layer, an aqueous layer containing concentrated hydrochloric acid, and solid amalgamated zinc. Since the reaction is first order in ketone, the rate-determining step cannot be the production of hydrogen atoms, which would react very rapidly if formed. Present evidence indicates that reduction occurs at the surface of the zinc, and that alcohols are not intermediates since they seem to be unreactive under Clemmensen conditions.

(11.63) $$CH_3CO(CH_2)_5CH_3 + Zn(Hg) + 2HCl \xrightarrow{\text{heat}} \underset{(51\%)}{CH_3(CH_2)_6CH_3} + ZnCl_2 + H_2O + Hg$$

The **Wolff-Kishner reduction** involves the decomposition of a hydrazone to nitrogen and a hydrocarbon; in most cases this decomposition is base-catalyzed. A simplified procedure involves mixing the ketone or aldehyde, hydrazine, and potassium hydroxide in a high boiling solvent such as ethylene glycol. The mixture is heated at 120 to 130° to form the hydrazone; excess hydrazine and water are distilled; and the reactants are heated to 190 to 200° to effect decomposition.

(11.64) 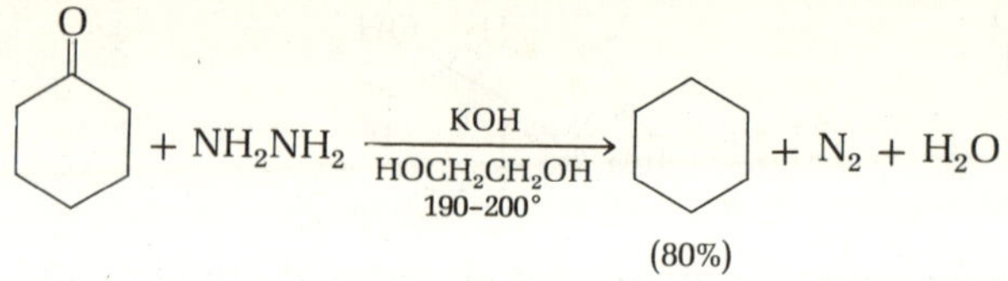

KETONES AND ALDEHYDES FROM OXIDATIONS

The oxidation of primary and secondary alcohols to aldehydes and ketones is a very important process in synthetic, commercial, and biological chemistry. Most reactions in solution seem to follow one of two general mechanisms:

Hydride Transfer Mechanism

(11.65) 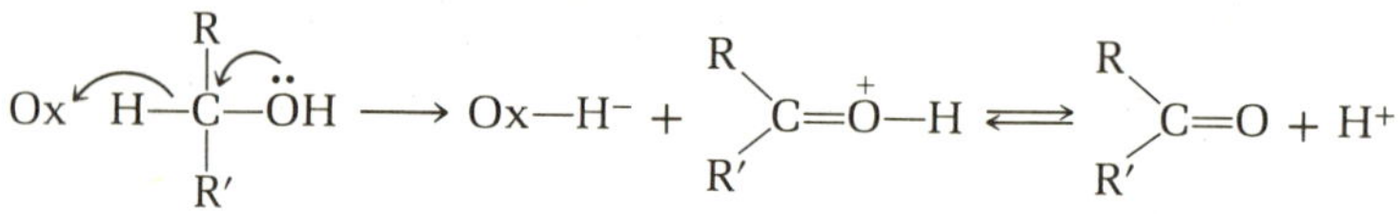

Elimination Mechanism

(11.66) 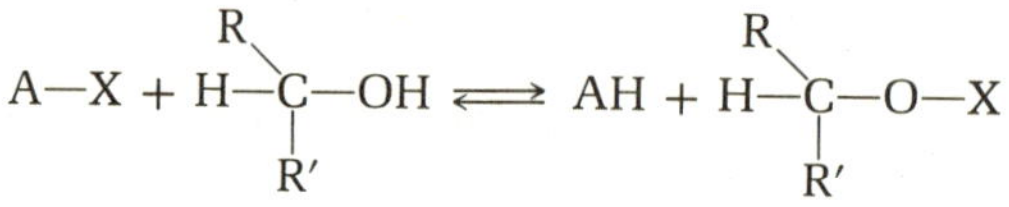

(11.67) 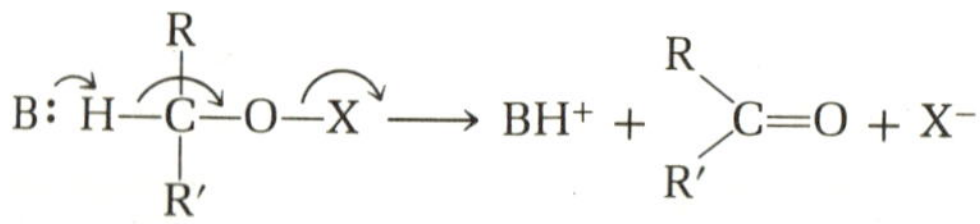

In the mechanism depicted in equation 11.65 the hydrogen atom attached to carbon is removed *as a hydride ion;* its bonding electrons go with it. In the other mechanism (equations 11.66 and 11.67) that same hydrogen atom is removed *as a proton;* its binding electrons flow toward the oxygen atom and form the π-bond. In addition, concerted loss of hydrogen can occur on the surface of metal catalysts at elevated temperatures (the reverse of hydrogenation).

(11.68) 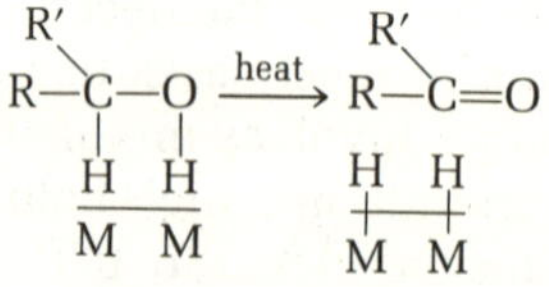

Some oxidations that proceed according to the first mechanism (equation 11.65) are those brought about by NAD^+ and $NADP^+$ in biological systems.

The Meerwein-Ponndorf-Verley reduction may also be run in the reverse direction (while simultaneously changing its name to the **Oppenauer oxidation**). Here again, hydride transfer from the alkoxide to the ketone seems to be rate-determining.

(11.69)

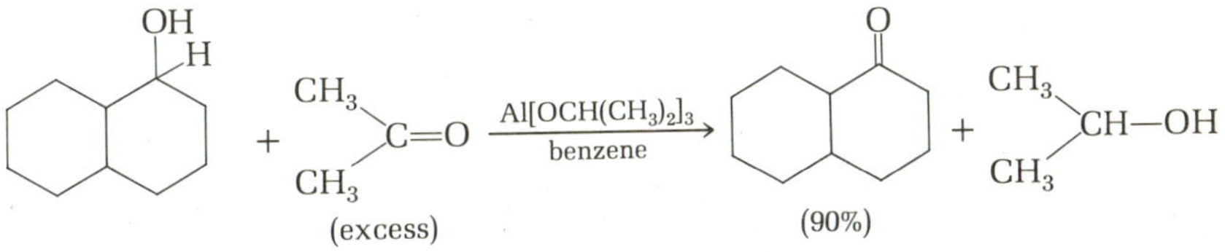

Carbon-carbon double bonds are unaffected by the oxidation conditions.

(11.70)

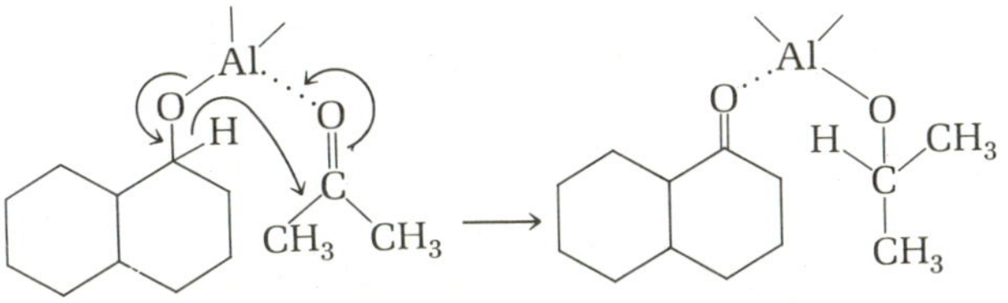

The elimination mechanism has obvious analogies to the reaction used to form carbon-carbon double bonds. In this case X, the group being reduced, is such a good leaving group that weak bases such as water will suffice to remove the proton. The most widely used oxidizing agent that reacts according to this mechanism is chromic acid, H_2CrO_4. It is usually formed *in situ* by mixing either sodium dichromate ($Na_2Cr_2O_7$) or chromium trioxide (CrO_3) with aqueous sulfuric acid.

(11.71) $6H^+ + 3CH_3\underset{\displaystyle OH}{\underset{|}{C}}H(CH_2)_5CH_3 + 2H_2CrO_4 \xrightarrow{\text{reflux}} 3CH_3\underset{\displaystyle O}{\underset{\|}{C}}(CH_2)_5CH_3 + 2Cr^{3+} + 8H_2O$

(96%)

(11.72)

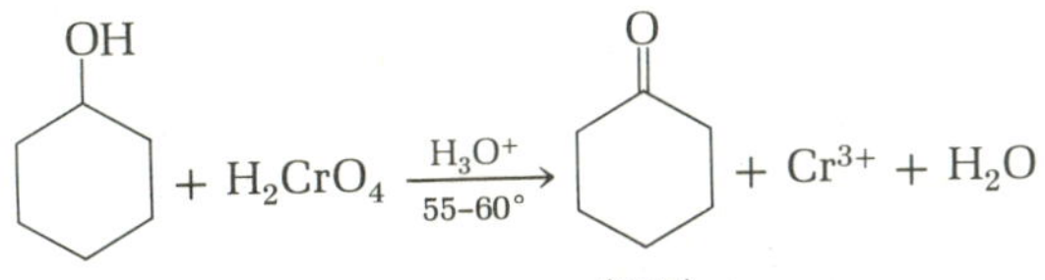

(85%)

The mechanism involves the reversible reaction of the alcohol with chromic acid to form a **chromate ester,** which subsequently decomposes in the rate-determining step.

(11.73)

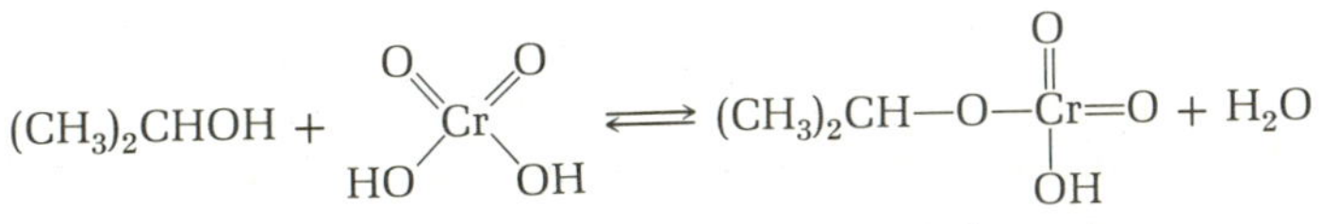

isopropyl chromate

(11.74) 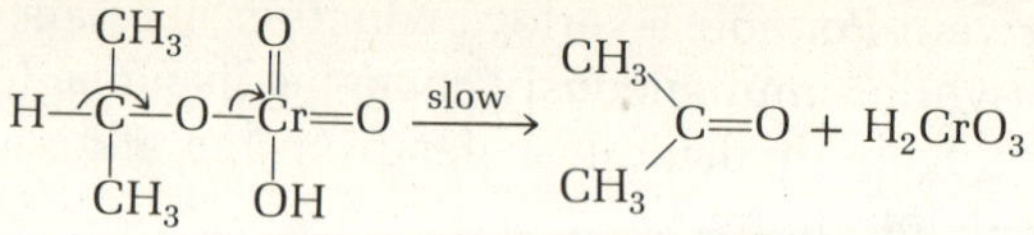

(11.75) $H_2CrO_3 + H_2CrO_4 \xrightarrow{\text{fast}} 2Cr(V)$

(11.76) $2Cr(V) + 2(CH_3)_2CHOH \xrightarrow{\text{fast}} 2Cr^{3+} + 2(CH_3)_2C{=}O$

The Cr(IV) species that results then participates in additional rapid redox steps (equations 11.75 and 11.76).

The presence of a carbon-carbon double bond in the molecule poses problems since it too is susceptible to oxidation. One solution is to use a mixed acetone-water solvent; an organic layer separates after a time and protects the product from further attack. Tertiary alcohols are normally inert to oxidation but in strongly acidic media they may dehydrate and the alkene so formed is then oxidized.

Aldehydes are somewhat tricky molecules to synthesize. They can undergo condensation polymerizations that are acid- and base-catalyzed, and they are also easily oxidized or reduced. If they are being synthesized by an oxidation or reduction reaction, therefore, there is the danger that one will "overshoot" and produce instead the corresponding acid or alcohol. For instance, chromic acid can oxidize aldehydes by the following mechanism (equation 11.77).

(11.77) 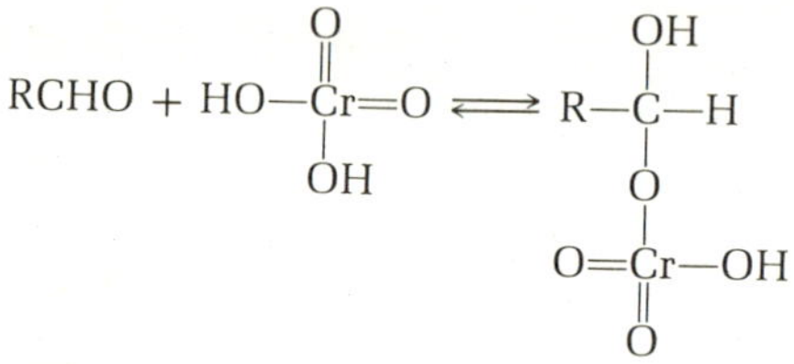

(11.78) 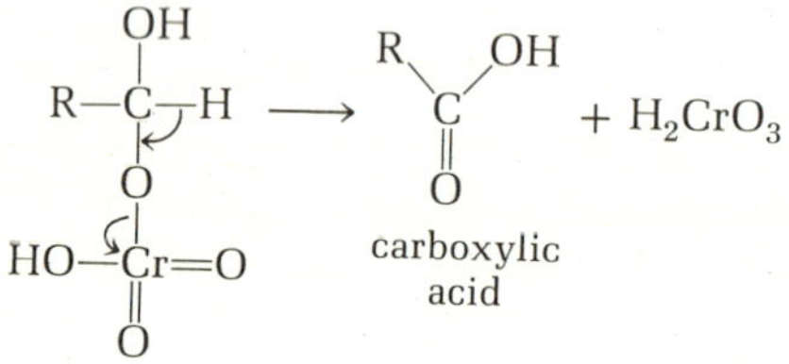

No such problem exists with ketones.

Since the presence of chromic acid is undesirable in the oxidation of alcohols to aldehydes, means have been sought to carry out the oxidation in nonaqueous media. Chemists have taken advantage of the fact that chromium trioxide forms a complex with the organic base *pyridine*, 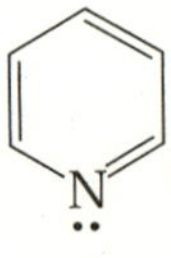, (abbreviated py), which is soluble in many organic solvents. The complex not only does not oxidize aldehydes further, but it is also unreactive toward sensitive groups such as carbon-carbon double bonds.

(11.79) $$CrO_3 + 2py \longrightarrow CrO_3 \cdot 2py$$

(11.80) $$\underset{\text{(excess)}}{CrO_3 \cdot 2py} + CH_2{=}\overset{\displaystyle CH_3}{\overset{|}{C}}(CH_2)_2CH{=}\overset{\displaystyle CH_3}{\overset{|}{C}}(CH_2)_4OH \xrightarrow[25^\circ]{CH_2Cl_2} \underset{(92\%)}{CH_2{=}\overset{\displaystyle CH_3}{\overset{|}{C}}(CH_2)_2CH{=}\overset{\displaystyle CH_3}{\overset{|}{C}}(CH_2)_3CHO}$$

Other oxidizing systems can be used in which the nature of the oxidant does not allow the possibility of further oxidation if the product is an aldehyde. Dimethyl sulfoxide is a very mild oxidizing agent for 1° and 2° alcohols.

(11.81) $$H^+ + RR'CHOH + (CH_3)_2\overset{+}{S}{-}O^- \xrightarrow[\text{agent}]{\text{dehydrating}} RR'CHO{-}\overset{+}{S}(CH_3)_2 + H_2O$$

(11.82) $$RR'\underset{\displaystyle H}{\underset{|}{C}}{-}O{-}\overset{+}{S}(CH_3)_2 \longrightarrow RR'C{=}O + (CH_3)_2S\colon + H^+$$

(11.83) $$CH_3(CH_2)_6CH_2OH + (CH_3)_2\overset{+}{S}{-}O^- \xrightarrow[\text{agent, }H^+]{\text{dehydrating}} \underset{(100\%)}{CH_3(CH_2)_6CHO} + (CH_3)_2S\colon$$

It is worth remembering that aldehydes or ketones may be obtained by the oxidation of alkenes or glycols under the appropriate conditions.

ALDEHYDE TESTS

The marked ease of aldehyde oxidation is the basis of several tests for distinguishing them from ketones. One is the silver mirror test (metallic silver is deposited as a film on a clean glass surface); the reagent is known as *Tollen's reagent* (equation 11.84).

(11.84) $$RCHO + 2Ag(NH_3)_2^+ + 3OH^- \longrightarrow R{-}\overset{\displaystyle O}{\overset{\|}{C}}{-}O^- + 2Ag\downarrow + 4NH_3 + 2H_2O$$

Another is *Fehling's test* (equation 11.85), in which a brick-red solid precipitates.

(11.85) $$2Cu^{2+} + 5OH^- + RCHO \longrightarrow RCO_2^- + Cu_2O\downarrow + 3H_2O$$

Of all the ketones only α-hydroxyketones will give positive tests with these reagents.

SYNTHESES OF SYMMETRICAL KETONES

Several special methods are available for obtaining symmetrical ketones. They are relatively simple, usually give good yields, and provide a short route to many ketones whose syntheses by conventional techniques might be rather laborious.

It has been known for some time that divalent heavy metal salts — Ba, Pb, Ca — of aliphatic carboxylic acids would give varying yields of dialkyl ketones upon thermal decomposition.

(11.86) $$(RCO_2)_2M \xrightarrow{\text{heat}} R\overset{\overset{\displaystyle O}{\|}}{C}R + CO_2 + MO \qquad (M = \text{divalent metal})$$

Similar results can be obtained by heating the acid with the metal oxide — and in some cases by slow distillation of the acid alone. The mechanism of this reaction is still obscure, but it has been shown that the ferrous salts of the acids decompose to ketones in good yields.

(11.87) $$2CH_3CH_2CH_2\overset{\overset{\displaystyle O}{\|}}{C}OH + Fe \longrightarrow (CH_3CH_2CH_2CO_2)_2Fe + H_2$$

(11.88) $$(CH_3CH_2CH_2CO_2)_2Fe \xrightarrow{\text{heat}} \underset{(69\text{–}75\%)}{CH_3CH_2CH_2\overset{\overset{\displaystyle O}{\|}}{C}CH_2CH_2CH_3} + FeO + CO_2$$

The highly versatile organoboranes can be used in ketone synthesis as well. In this reaction two of the alkyl groups from the borane are tacked on to a molecule of carbon monoxide, which apparently acts initially as a nucleophile in attaching itself to boron. The presence of water in the reaction mixture when the CO is added prevents the migration of the third alkyl group. The entire reaction can be conducted in one flask since none of the intermediates are isolated. If no water is present when the CO is bubbled in, then all three alkyl groups are transferred from boron to carbon, producing tertiary alcohols (equation 11.90).

(11.89)

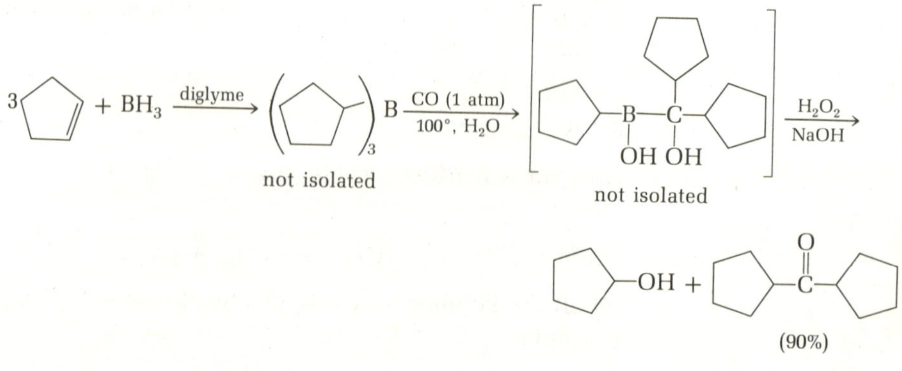

Hindered alcohols, such as the one shown, would be obtained in much poorer yields by the Grignard method.

(11.90)
$$3CH_3CH_2CH{=}CH_2 + BH_3 \xrightarrow{\text{diglyme}} (CH_3CH_2CH_2CH_2)_3B \xrightarrow[\text{1 atm}]{CO} \xrightarrow[\text{NaOH}]{H_2O_2} CH_3CH_2CH_2CH_2\underset{\underset{CH_2CH_2CH_2CH_3}{|}}{\overset{\overset{OH}{|}}{C}}CH_2CH_2CH_2CH_3 \quad (90\%)$$

The first step in all these reactions would seem to be equation 11.91.

(11.91)
$$R_3B + :\overset{-}{C}{\equiv}\overset{+}{O}: \longrightarrow R_2\overset{-}{B}\underset{\underset{R}{|}}{}{-}C{\equiv}\overset{+}{O}: \longrightarrow R_2B{-}\overset{\overset{\ddot{O}:}{\|}}{C}{-}R$$

These versatile syntheses have yet to be fully exploited.

ALDEHYDES AND KETONES AS INTERMEDIATES FOR SYNTHESIS

Aldehydes and ketones are valuable intermediates for the synthesis of complex organic molecules. The carbonyl carbon atom acts as a point of attachment for a variety of nucleophilic species; furthermore, after addition to the carbonyl group has taken place a hydroxyl group still remains, providing a "handle" for further transformations. An example of a multistep sequence is given in equation 11.92.

(11.92)

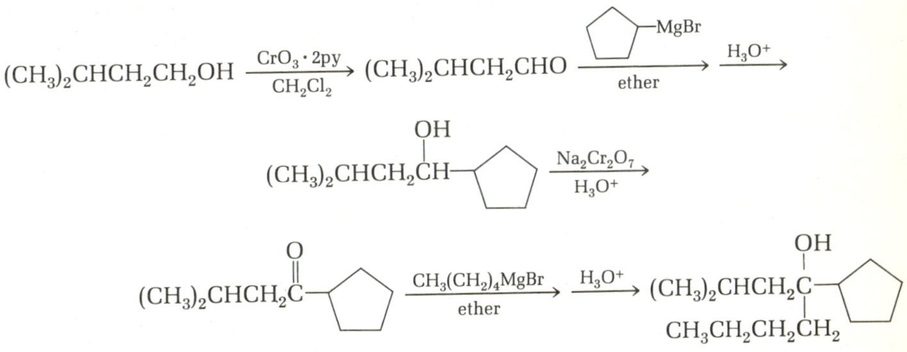

As will be made evident in Chapter 20, the α-hydrogen atoms of aldehydes and ketones are unusually reactive, a fact that further enhances the synthetic utility of these compounds.

SUMMARY

1 The chemistry of aldehydes and ketones derives principally from the properties of the carbonyl group. The carbon-oxygen double bond is quite strong and polarized so that there is an appreciable fractional positive charge on the carbonyl carbon atom. Intense bands in the region 1675 to 1775 cm^{-1} appear in the infrared spectra of aldehydes and ketones due to the $>C{=}O$ stretch. The two may be differentiated because of the characteristic aldehyde C—H peak in ir and nmr spectra.

2 Carbonyl groups absorb ultraviolet light (~3000 A), which causes a nonbonded electron localized on oxygen to be promoted to the antibonding π-orbital ($n \rightarrow \pi^*$). The electrons may remain paired in the excited state $[^1(n, \pi^*)]$ or their spins may become unpaired $[^3(n, \pi^*)]$. In the excited state the oxygen atom has considerable radical character. It may attack the double bond of an olefin, leading to an oxetane, or it may break C—C or C—H bonds intramolecularly.

3 Strong nucleophiles such as cyanide and hydroxide ion add to the carbonyl group without catalysis. The resulting oxygen anion may then acquire a proton to form a stable product; the equilibrium constant for the overall process is strongly dependent on steric and electronic factors. The attack of weaker nucleophiles is usually catalyzed by a proton or Lewis acid, which complexes with the oxygen atom and increases the polarization of the carbonyl group.

4 When derivatives of ammonia add to the carbonyl group, the intermediate hydroxy compound is dehydrated, yielding products containing a carbon-nitrogen double bond. The oximes, semicarbazones, and phenylhydrazones are stable solids, useful for characterizing carbonyl compounds.

5 Carbonyl compounds may be reduced to alcohols containing the same number of carbon atoms. Lithium aluminum hydride, sodium borohydride, another alcohol, or platinum and hydrogen may be used for these reductions. The first three react by hydride transfer, which is also the mechanism for biological reductions. These latter are reversible and are remarkable for the stereospecific hydride transfers from both the coenzyme and alcohol.

Alcohols with an increased number of carbon atoms may be synthesized by allowing carbonyl compounds to react with Grignard reagents. Primary, secondary, or tertiary alcohols may be obtained from formaldehyde, other aldehydes, or ketones, respectively.

6 Oxidation of primary or secondary alcohols leads to aldehydes or ketones. Aldehydes are very susceptible to further oxidation to acids. Several mild oxidizing systems are known that avoid this problem. They all follow an

elimination mechanism, in which H—C—O—H is first converted to H—C—O—X and HX is then eliminated.

PROBLEMS

Note: Problem 1 appears in text.

2 Note the equilibria depicted in equations 2 and 3.

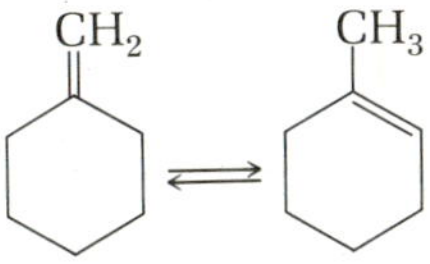

$K = 2.4 \times 10^{2}$ **(2)**

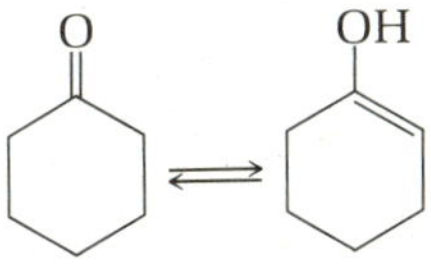

$K = 4.1 \times 10^{-6}$ **(3)**

Rationalize the size of the equilibrium constant in each case, and explain the substantial difference between the two.

3 When solutions of ketones are irradiated with ultraviolet light, they are often reduced to alcohols and symmetrical glycols (pinacols), provided that they are in a solvent that is a good source of hydrogen atoms (equation 4).

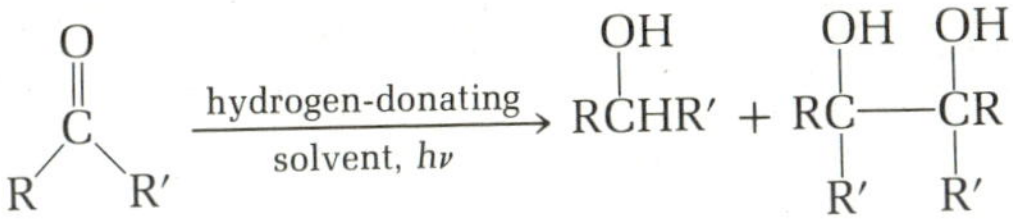

(4)

This process does not take place in the absence of light. Suggest a mechanism.

4 Write out equations for the following reactions, including the products.

a. 4-Octanone $\xrightarrow{h\nu}$

b. Isobutyraldehyde $\xrightarrow[H_2O]{KCN}$

c. 2-Hexanone $\xrightarrow{NaHSO_3}$

d. 3-Methylcyclopentanone $\xrightarrow{\text{hydroxylamine}}$

e. 5-Methyl-*trans*-2-hexenal $\xrightarrow{\text{semicarbazide}}$

f. Pentanal $\xrightarrow{\text{hydrazine}}$

g. Cyclooctanecarboxaldehyde $\xrightarrow{\text{phenylhydrazine}}$

h. 5-Ethyl-2-methyl-3-heptanone $\xrightarrow[\text{glycol, H}^+]{\text{ethylene}}$

i. *trans,cis*-2,3-Dimethylcyclobutanecarboxaldehyde $\xrightarrow[\text{H}^+]{\text{CH}_3\text{OH}}$

j. Cyclohexanone $\xrightarrow[\text{H}^+]{\text{1,3-propanediol}}$

5 2-Butanone-^{18}O may be prepared by treating the ordinary ketone with $H_2{}^{18}O$ containing a small quantity of acid. Suggest a mechanism for this reaction.

6 a. A sample of octanal is contaminated with a few percent of 2-octanol. How can the aldehyde be obtained in pure form?

b. It is desired to recover the 4-heptanone from a mixture of 2-heptanone and 4-heptanone. Suggest a procedure.

c. How can one separate a mixture of hexanal, 3-hexanol, and ethyl propyl ether into its components?

d. How could the presence of a small quantity of 2-pentanone in a sample of 2-pentanol be verified?

7 When solutions of acetone in aqueous hydrogen peroxide are treated with acids, the cyclic peroxides I and II are formed. Propose a mechanism for the formation of these products.

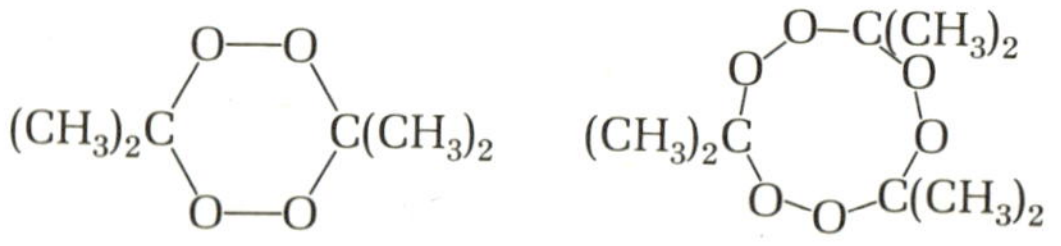

I II

8 Suggest syntheses for the following alcohols using aldehydes, ketones, and alkyl halides as starting materials. In cases where more than one possibility presents itself, propose two possible synthetic routes; and if you believe one of them to be preferable, then indicate the fact.

a. 2-Hexanol

b. 3-Methyl-3-pentanol

c. 2-Ethyl-1-butanol

d. 1,3-Dimethylcyclohexanol

e. 2,2-Dimethyl-3-pentanol

f. 4-Ethyl-7-methyl-4-nonanol

9 Indicate how 1-bromopentane might be converted to the following products, noting all reagents and essential conditions. Supply names (more than one if possible) for all the products that are either aldehydes or ketones.

a. $CH_3(CH_2)_3CHO$

b. $CH_3(CH_2)_4CO(CH_2)_3CH_3$

c. $CH_3(CH_2)_4\overset{O}{\overset{\|}{C}}OH$

d. $CH_3(CH_2)_4CO(CH_2)_4CH_3$

e. $CH_3(CH_2)_2CHO$

10 The Grignard reduction of pinacolone (equation 11.39) has been carried out with (+)-2-methyl-1-butylmagnesium bromide. In this case the reduction product, 3,3-dimethyl-2-butanol, was optically active when isolated, showing that reduction had occurred preferentially on one side of the carbonyl group. Draw the possible transition states for this reaction, paying attention to the stereochemical details. What interaction(s) might account for stereoselective reduction?

11 Write out the structures of the organic products of the following reactions.

a. Acetone $\xrightarrow[H_2O,\ OH^-]{NaBH_4}$

b. *trans*-3-Hexen-2-one $\xrightarrow[CH_3OH]{NaBH_4}$

c. 4-Methylpentanal $\xrightarrow[ether]{LiAlH_4}$ $\xrightarrow{H_3O^+}$

d. 1,4-Cyclohexanedione $\xrightarrow[ether]{LiAlH_4}$ $\xrightarrow{H_3O^+}$

e. Cyclopentanone $\xrightarrow[THF]{LiAlD_4}$ $\xrightarrow{H_3O^+}$

f. Cyclopentanone $\xrightarrow[THF]{LiAlH_4}$ $\xrightarrow{D_3O^+}$

g. 2-Cyclohexyl-3-pentanone $\xrightarrow[ether]{LiAlH_4}$ $\xrightarrow{H_3O^+}$

h. 2-Cyclohexenone $\xrightarrow[H^+,\ C_2H_5OH]{Pt,\ H_2}$

i. 2-Cyclohexenone $\xrightarrow[(CH_3)_2CHOH]{Al[OCH(CH_3)_2]_3}$

12 The rates of reduction of some cyclic ketones by sodium borohydride in isopropanol stand in the order cyclobutanone > cyclohexanone > cyclopentanone.

a. Offer an explanation for this reactivity order.

b. What would be the relative stabilities of the cyanohydrins of these ketones?

13 If acetaldehyde-1-*d* is reduced with NADH in the presence of yeast alcohol dehydrogenase, then optically active ethanol-1-*d* is isolated from the reaction mixture. What is the relationship of this ethanol-1-*d* to that obtained by way of reaction 11.55?

14 Contrary to the situation in many enzymatic systems, a bacterial enzyme system has been found that will catalyze the reduction of *both* enantiomers of *trans*-1-decalone (III) by NADH. The reduction of each enantiomer is stereospecific,

however — one gives rise only to equatorial alcohol, while the other gives rise only to axial alcohol. What conclusions might one draw about the positions of the substrate and of NADH on the enzyme?

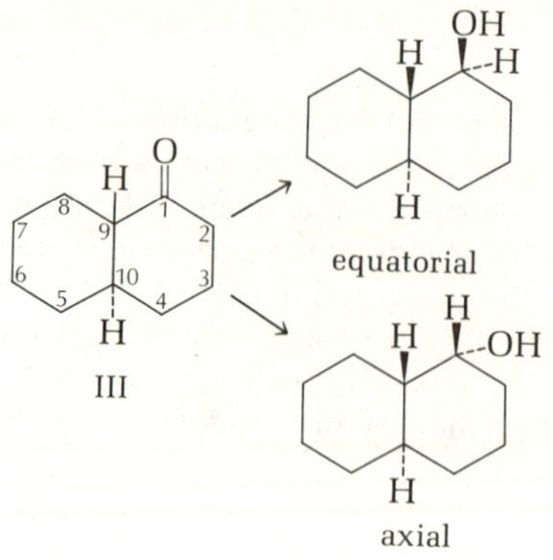

(5)

15 a. Suggest a mechanism for the pinacol rearrangement (equation 11.58).

b. Equation 11.59 depicts the pinacol rearrangement of 2-methylpropane-1,2-diol. The major product, as indicated, is isobutyraldehyde. How would you explain this? What other product might be expected — why would it be present in only minor amounts?

16 a. What products would be expected from the reduction of cyclopentanone with amalgamated magnesium, followed by aqueous work-up?

b. If the compound from part a were treated with hot acid, what result would be anticipated?

17 Suppose that you did not have the means of carrying out either the Clemmensen or Wolff-Kishner reduction; yet you had a number of aldehydes and ketones that had to be converted to hydrocarbons. What other routes could be employed? Do they entail any disadvantages?

18 Describe syntheses of the following compounds, using as starting materials any organic compound containing five or fewer carbon atoms and any needed inorganic reagents.

a. 2-Pentanone
b. 3-Pentanone
c. Propanal
d. 2-Hexanone
e. Heptanal
f. 7-Methyl-3-octanone
g. 5-Nonanone
h. 2-Methylcyclopentanone
i. Dicyclohexylmethanol
j. 2,6-Dimethyl-4-isobutyl-4-heptanol

19 An additional complication encountered in the chromic acid oxidation of primary alcohols is the formation of *esters*.

$$2RCH_2OH \xrightarrow{H_2CrO_4} \underset{\text{a carboxylic acid ester}}{R\overset{O}{\overset{\|}{C}}OCH_2R} \qquad (6)$$

Suggest a mechanism for equation 6 that accords with the known characteristics of chromic acid oxidations.

20 When 1-iodooctane is heated in chloroform with trimethylamine oxide $[(CH_3)_3\overset{+}{N}—O^-]$, a 43% yield of octanal is obtained, along with some trimethylammonium iodide $[(CH_3)_3\overset{+}{N}H\ I^-]$. Suggest a mechanism for this mild oxidation that is useful for aldehyde synthesis.

21 When 3,3-dimethyl-2-pentanol is oxidized with chromic acid in acetic acid-water, the major product is 3,3-dimethyl-2-pentanone. Also isolated, however, are various quantities of acetaldehyde and 2-methyl-2-butanol. Suggest a mechanism for the formation of these "abnormal" oxidation products. Why are they found in this particular reaction?

22 A solution of bromine in water is effective at oxidizing secondary alcohols to ketones; primary alcohols are oxidized to carboxylic acids by way of aldehydes. Ethers containing primary and secondary alkyl groups are also subject to oxidative cleavage by bromine water. Suggest one or more mechanisms for these reactions.

23 Indicate how the following conversions would be effected:

a. Cyclopentanone ⟶ cyclopentyl tosylate

b. 3-Methyl-1-pentanol ⟶ 3-methylpentanal diethyl acetal

c. Pentanal ⟶ 3-cyanoheptane

d. 2-Pentanol ⟶ 2,4-dimethyl-4-heptanol

e. 3-Hexanone ⟶ 2-ethylpentanal

f. Isobutyric acid $[(CH_3)_2CH\overset{O}{\overset{\|}{C}}OH]$ ⟶ 2,4-dimethyl-3-pentanol

g. Cyclohexyl bromide ⟶ cyclohexylacetaldehyde

h. 3-Methyl-3-hexene ⟶ 2-methyl-2-ethylbutanal

i. Cyclohexanone ⟶ 6-oxoheptanal $[CH_3CO(CH_2)_4CHO]$

(Remember that carbonyl compounds may be obtained by oxidation of appropriate alkenes and glycols.)

24 a. A compound **A,** C_8H_{16}, is oxidized with hot acidic potassium permanganate. Gas chromatographic analysis of the product reveals the presence of two compounds, **B** and **C,** which are separable by distillation. They each give a precipitate when treated with semicarbazide solution. Analytical results for **B** are compatible with the formula $C_5H_{10}O$. The nmr spectrum of **B** contains the following peaks (in

order of increasing field strength): a triplet, a singlet, a multiplet, and a triplet; area ratios are 2:3:2:3. **C** is a volatile liquid, the nmr spectrum of which consists of a single peak.

Write structures for **A, B,** and **C.**

b. Compound **D,** $C_5H_{12}O_2$, exhibits strong absorption in the 3400 to 3600 cm^{-1} region in its ir spectrum. When **D** is oxidized with chromic acid for short periods of time, one of the isolated products is **E,** $C_5H_8O_2$. The ir spectrum of **E** contains an intense band at 1720 cm^{-1}; its nmr spectrum consists of a singlet, a triplet, a multiplet, and a second triplet far downfield; area ratios are 3:2:2:1. **E** gives positive tests with phenylhydrazine (precipitate) and with Tollen's reagent. **E** may be further oxidized to **F,** $C_5H_8O_3$, which also gives a precipitate with phenylhydrazine. If **D** is treated with sulfuric acid, it is transformed into **G,** $C_5H_{10}O$, which gives no reaction with phenylhydrazine and does not decolor solutions of bromine or of potassium permanganate.

Write structures for **D, E, F,** and **G.**

c. Compound **H,** $C_7H_{14}O$, gives a crystalline precipitate when treated with a solution of hydroxylamine. **H** is allowed to react with sodium borohydride in methanol and gives rise to two compounds separable by means of glc. Both these substances (**I** and **J**) have nearly identical mass spectra in which the parent peak is found at *m/e* 116. Both **I** and **J** form crystalline tosylates when treated with tosyl chloride and base. The tosylates react with potassium *t*-butoxide to yield similar mixtures of isomers (molecular weight, 98). Ozonization of these mixtures yields two major products — acetaldehyde and **K,** which both give positive tests with Fehling's solution. Write structures for **H, I, J,** and **K.**

SUPPLEMENTARY READINGS

C. D. Gutsche, *The Chemistry of Carbonyl Compounds.* Englewood Cliffs, N.J.: Prentice-Hall, Inc., 1967, Chapters 1 (Sections 1.1–1.6, 1.12, 1.13), 2 (Sections 2.1, 2.2, 2.5), 3 (Sections 3.1, 3.2), and 4 (Sections 4.1, 4.3). (P)

C. D. Hurd, "Hemiacetals, Aldals and Hemialdals," *J. Chem. Ed.,* **43,** 527 (1966).

CHAPTER 12

ALKADIENES—CONJUGATION AND RESONANCE

ALLYL RADICAL AND DELOCALIZATION

When propene is chlorinated in the gas phase, the principal product is 3-chloropropene (**allyl chloride**), the substitution product, rather than the 1,2-dichloropropane that would be obtained by addition.

(12.1) $CH_2{=}CHCH_3 + Cl_2 \xrightarrow{400^\circ} CH_2{=}CHCH_2Cl + HCl$

That substitution can compete with addition is made plausible by the knowledge that the primary C—H bonds in propene are 13 kcal/mole weaker than the corresponding bonds in propane.

(12.2) $CH_2{=}CHCH_2{-}H \longrightarrow \underset{\text{allyl radical}}{CH_2{=}CHCH_2\cdot} + \cdot H \qquad \Delta H = 85 \text{ kcal/mole}$

(12.3) $CH_3CH_2CH_2{-}H \longrightarrow CH_3CH_2CH_2\cdot + \cdot H \qquad \Delta H = 98 \text{ kcal/mole}$

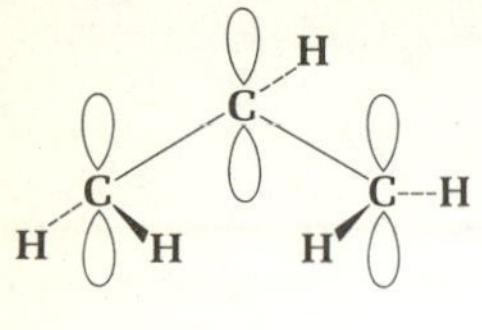

FIGURE 12-1
Atomic Orbitals of Allyl Radical

The relatively low energy requirement for the dissociation of the aliphatic C—H bond in propene is directly attributable to the stability of the allyl radical. As the hydrogen atom is lost, C-3 assumes sp^2-hybridization. The p-orbital on C-3 is then able to overlap with the neighboring p-orbital of the double bond (Figure 12-1). The interaction among the three p-orbitals results in the formation of three new molecular π-orbitals (Figure 12-2). There are bonding and antibonding orbitals, just as there are in alkenes, in which neighboring atomic orbitals are aligned with or against one another. The third molecular orbital is referred to as nonbonding and lies between the other two in energy; it is composed essentially of only two atomic orbitals. An electron in the nonbonding π-orbital has approximately the same energy as it would in a carbon $2p$-orbital. In the ground state of the allyl radical, two of the π-electrons will reside in the bonding orbital while the third, unpaired, electron will be in the nonbonding orbital.

In order to understand the nature of the allyl radical, we are forced to abandon a fiction that has generally served us very well so far. We have depicted molecules as collections of nuclei held together by electron pairs localized between pairs of nuclei. While this picture is probably incorrect, it nonetheless allows us to predict molecular behavior with surprising accuracy, except for certain notable exceptions (Chapter 6). The localized bond model will simply not do for the allyl radical, however. We must concede that the electron pair in the lowest π-orbital is delocalized over all three carbon atoms, which undoubtedly contributes to the stability of the radical. Similarly, the nonbonding electron divides its time between C-1 and C-3; the decreased spin density at each atom should lower the reactivity of the radical.

Another approach to understanding the structure of the allyl radical is as follows. One can write two different structures for the radical, **1** and **2**,

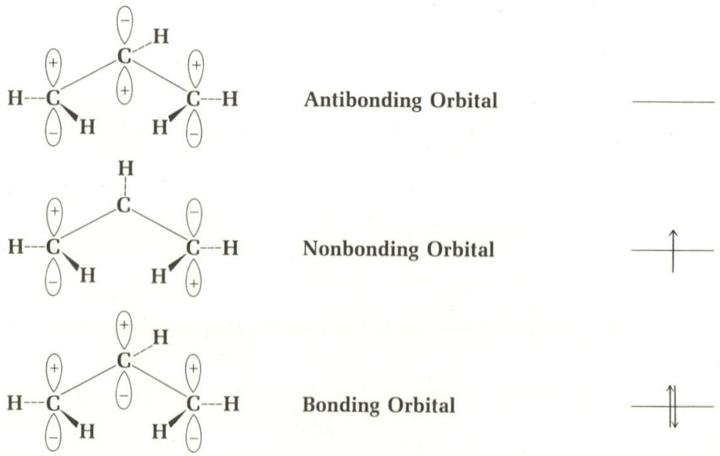

FIGURE 12-2
Molecular π-Orbitals of Allyl Radical

which are chemically identical, although they could be made distinguishable by use of isotopes.

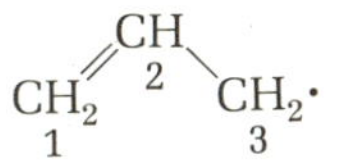

1

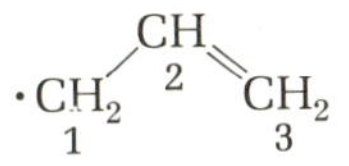

2

We might suppose that there is a rapid equilibrium between these two radicals, which differ mainly in their electron distributions (and bond lengths). **All evidence that we have for systems of this type indicates that this is not the case. Only one species is detected,** even at low temperatures.

The real allyl radical is then described as a **resonance hybrid** of the two contributing (or **canonical**) structures **1** and **2.** This implies that the real species has properties that are a blend of the individual properties of the hypothetical canonical structures, just as the optical properties of orange can be described as a blend of those of red and yellow. We would expect the C-1 to C-2 and C-2 to C-3 bonds in the allyl radical to be equal and to have a bond length intermediate between those of single and double bonds. The principal difference is that red and yellow are real, while canonical structures are purely imaginary. But just as an orange object does not oscillate back and forth between yellow and red, the real molecule, radical, or ion has fixed properties.

In order to qualify as valid canonical forms, the hypothetical structures must have the same spatial arrangement of nuclei, have the same number of unpaired electrons, and be reasonably close in total energy content. Thus **3** would not be an allowable canonical structure because of the number of unpaired electrons, while **4** can be disqualified on the grounds of its high energy.

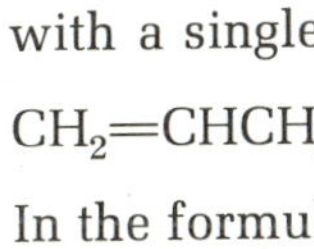

3

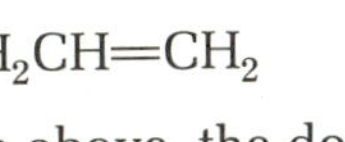

4

It should be emphasized that one reason for employing resonance pictures is our inadequacy at depicting the essential characteristics of the real species with a single drawing.

$$CH_2{=}CHCH_2\cdot \longleftrightarrow \cdot CH_2CH{=}CH_2$$

In the formulation shown above, the double-headed arrow signifies that the two structures are canonical forms and are *not* in equilibrium. A good quantitative generalization may be extracted from the resonance approach: If a species can be equally well represented by two or more classical structures, then it cannot be adequately represented by a single classical structure.

Attempts have been made to depict species such as the allyl radical with the aid of a single picture. An example is **5.**

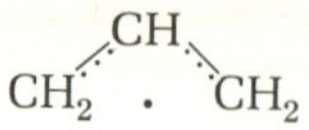

5

One postulate of resonance theory is that the greater the number of resonance forms of equal energy that can be drawn, the more stable is the real species. The difference in energy between a resonance-stabilized molecule and a comparable nonstabilized one (which may be hypothetical) is known as the **resonance energy.** We shall see a little further on how this quantity is calculated. It should be noted that as far as a species such as the allyl radical is concerned, both molecular orbital theory and resonance theory give the same qualitative predictions as far as electron delocalization, relative stability, etc., are concerned. The two methods do not always lead to identical predictions, however; molecular orbital theory is far more rigorous and is to be preferred. The resonance approach has the virtue of convenience.

When a given *p*- or π-orbital is geometrically able to overlap with a neighboring π-orbital, then the two are said to be **conjugated.** Conjugation results in more highly delocalized molecular orbitals and a net lowering of the energy of the π-electrons. Accordingly, it is sometimes stated that the unpaired electron in allyl radical is stabilized by conjugation. All those phrases are actually different ways of saying the same thing.

Just as delocalization of an unpaired electron stabilizes a radical, so delocalization of charge stabilizes an ion. Some examples can be cited from previous chapters. The isomerization of hydrazones, an essential process in the Wolff-Kishner reduction (Chapter 11), proceeds by way of a delocalized anion.

(12.4)

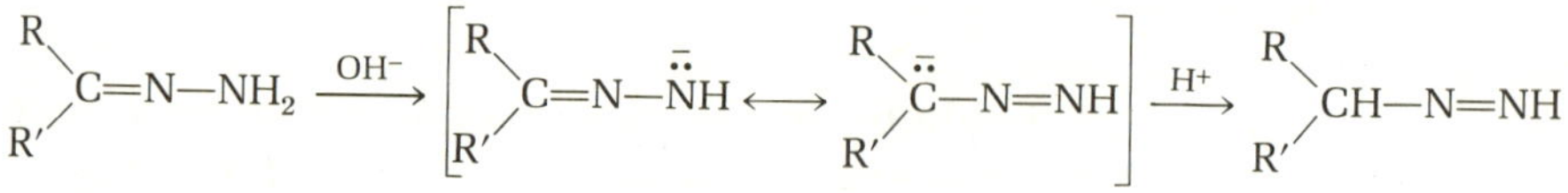

Carbonium ions are stabilized by adjacent oxygen atoms (Chapter 10) because the empty carbon 2*p*-orbital can overlap with a filled nonbonding oxygen orbital.

PROBLEM

1 The hydrolysis of Schiff bases (I, equation 1) usually proceeds quite readily.

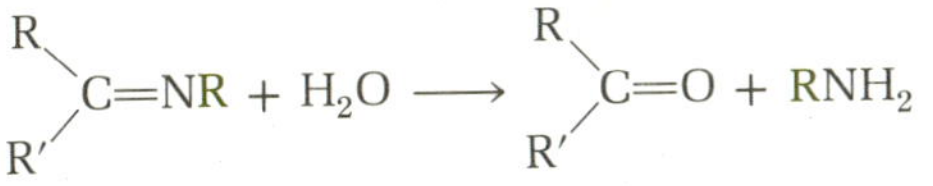

I **(1)**

By contrast, the hydrolysis of the corresponding oximes (II) and hydrazones (III) is generally a slow process.

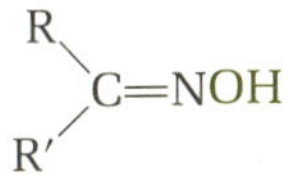

II

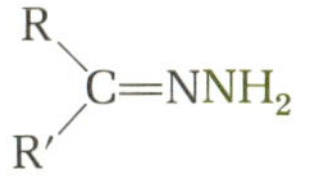

III

How would you account for this difference?

ESR SPECTRUM OF ALLYL RADICAL

One method of studying free radicals, which has given important evidence about the nature of allyl radical, is **electron spin resonance (esr)** spectroscopy. Electrons have a spin of ½, just as protons do. If an unpaired electron is placed in a magnetic field, it can assume two orientations: Its magnetic moment is either parallel or antiparallel to the applied field. Radiation of the correct frequency, for a given magnetic field strength, can cause a flip of the electron spin.

The esr spectrum of a hydrogen atom consists not of one peak but two. The reason is that the nucleus of the hydrogen atom is also oriented in the applied magnetic field and it produces a small additional field that either adds to or substracts from the applied field felt by the electron. Since the hydrogen nucleus has two possible orientations, the esr spectrum of a hydrogen atom consists of two lines (Figure 12-3). In general, if an electron spends some time on a group of atoms bearing n equivalent hydrogen atoms, its esr spectrum will consist of $n + 1$ lines. If there are n hydrogen atoms of one kind and m of another, the spectrum consists of $(n + 1)(m + 1)$ lines. The effect of protons on the electron transition is completely analogous to the effect of neighboring protons on a nuclear transition, which gives rise to spin-spin splitting in nmr spectra.

The esr spectrum of methyl radical ($CH_3\cdot$) consists therefore of four lines. Since only *unpaired* electrons can absorb energy, methyl cation (CH_3^+) and methyl carbanion ($CH_3{:}^-$) do not give rise to esr spectra. At liquid nitrogen temperature the esr spectrum of allyl radical consists essentially of five lines, which means that the electron is interacting with four equivalent protons.

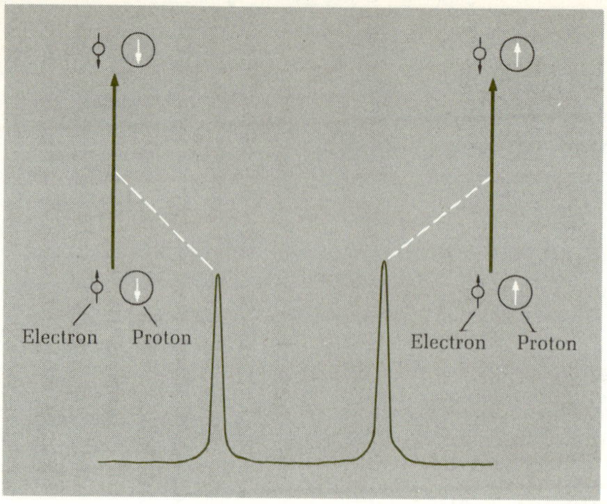

FIGURE 12-3
Splitting of an Electron Spin Resonance Signal by Interaction of the Electron with Proton Spins

While not conclusive, this tends to rule out the idea of an equilibrium between radicals **1** and **2** (p. 287) since the two would interconvert only slowly at this temperature. The anticipated spectrum would then be that of the bond-localized species, which should consist of three lines. The esr spectrum also confirms our expectation that the unpaired electron spends no time on C-2.

Electron spin resonance spectroscopy has been introduced at this time particularly because of the light it could shed on the structure of allyl radical. It should be clear, however, that theoretically any free radical is amenable to study by esr.

ALLYLIC HALOGENATION

Because of the low dissociation energy of allylic C-H bonds, these hydrogen atoms are readily attacked by a variety of free radicals. They may be replaced by halogen atoms; the usual reagents used for this purpose are *t*-butyl hypochlorite, **6** (Chapter 5), and *N*-bromosuccinimide (NBS) (**7**).

$(CH_3)_3COCl$

6

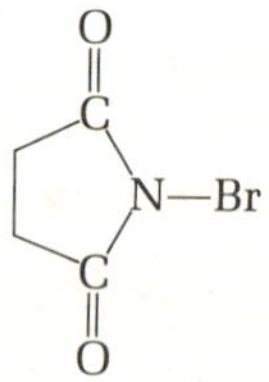

7

t-Butyl hypochlorite dissociates into chlorine atoms and *t*-butoxy radicals; the latter initiate the chain by hydrogen abstraction.

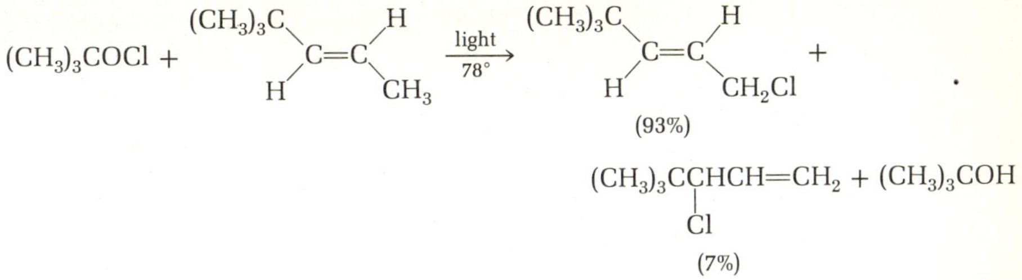

(12.5)

PROBLEM

2 Write out a mechanism for equation 12.5, showing all steps including the formation of both products. Draw the stable conformations of the allylic radical intermediate.

NBS functions slightly differently. It reacts with HBr to form succinimide and molecular bromine. Since NBS is only slightly soluble in carbon tetrachloride, the usual reaction medium, the bromine concentration is quite low.

The bromine atoms may add to the double bond or abstract in allylic hydrogen atom. Hydrogen abstraction is irreversible, while bromine atom addition is not (equation 12.6).

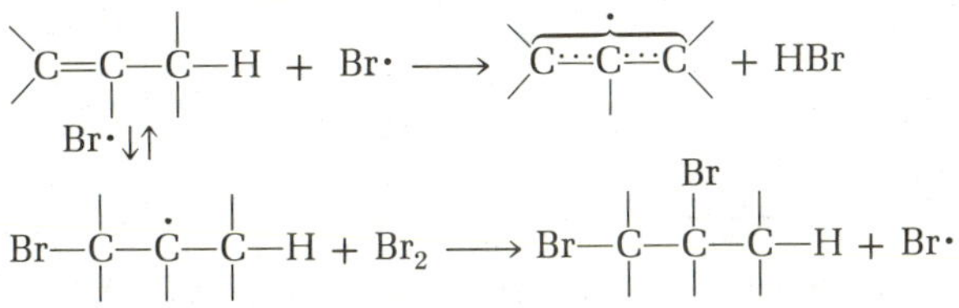

(12.6)

Since the radical arising by addition must react with a molecule of bromine in order to be irreversibly converted to product, low bromine concentrations favor allylic substitution (equation 12.7).

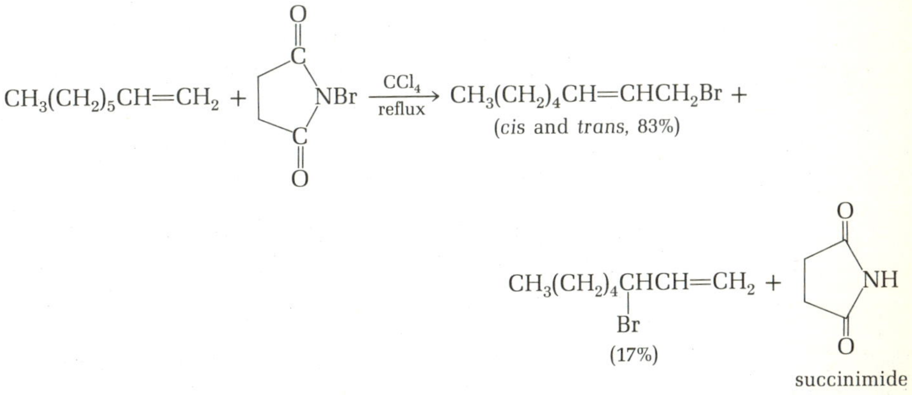

(12.7)

A mixture of isomeric products is to be expected if a substituted allyl radical is an intermediate.

The reaction of iodine with alkenes at elevated temperatures does not result in any appreciable amount of halogenated product because the equilibrium constant is very small. Reactions do occur, nevertheless, leading to isomerization of the alkene. This method was used to establish equilibrium among all the possible straight-chain hexene isomers (Chapter 8).

(12.8) 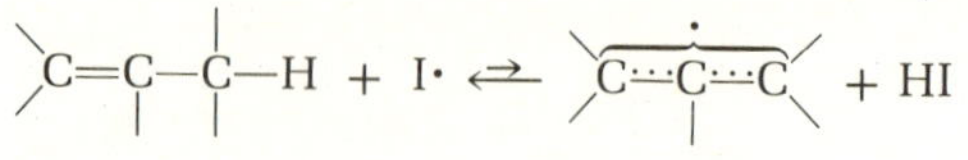

(12.9) 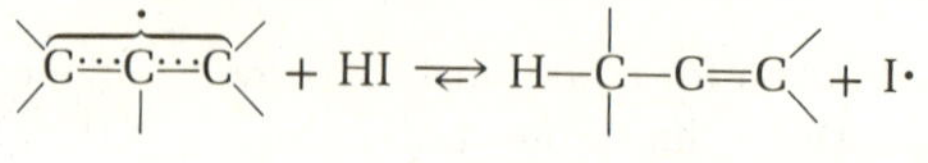

ALLYL CATION – ALLYLIC REARRANGEMENT

Allyl cation is an unusually stable carbonium ion; the positive charge is delocalized over the two terminal carbon atoms. It is possible to prepare stable solutions of allylic cations in very strong acids, such as $H_2SO_4-SO_3$ or SbF_5, and measure their spectroscopic properties.

$$CH_2{=}CH{-}\overset{+}{C}H_2 \longleftrightarrow \overset{+}{C}H_2{-}CH{=}CH_2 = \overbrace{CH_2 \cdots CH \cdots CH_2}^{+}$$

PROBLEM

3 a. The ir spectra of symmetrically substituted allylic cations contain a very intense peak near 1530 cm^{-1} and *no* other peaks between 1450 and 2850 cm^{-1}. What conclusions might one draw about the structure of allyl cations from this data?

b. The nmr spectra of the presumed allyl and 2-methylallyl (IV) cations were obtained at $-60°$ in SbF_5-SO_2 solution.

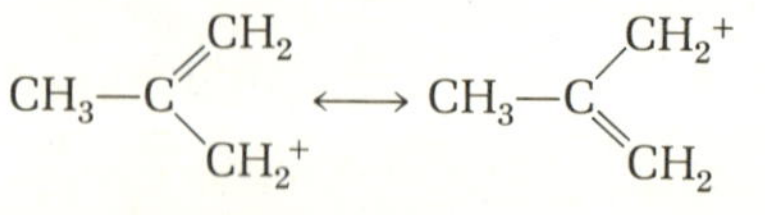

IV

The allyl cation spectrum consists of a broad singlet at δ9.64 (1 H) and a broad doublet at δ8.97 (4 H). Two singlets were observed in the spectrum of IV: δ3.85 (3 H) and δ8.95 (4H). How do these observations accord with the model of allylic systems so far presented?

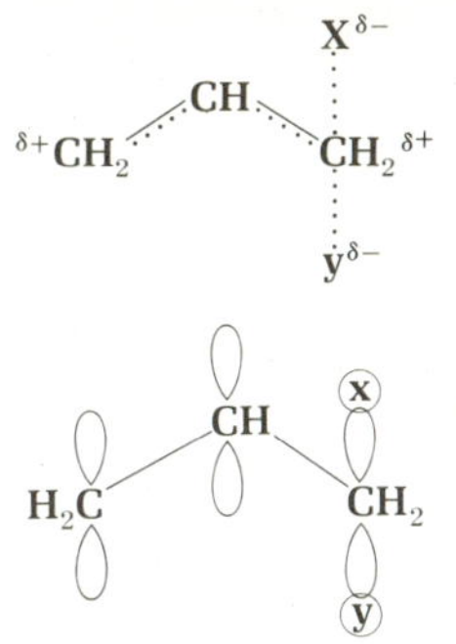

FIGURE 12-4
Transition State for an S_N2 Displacement in an Allylic System — Two Representations

Allylic compounds show somewhat enhanced reactivity in nucleophilic substitution reactions. For instance, they undergo S_N2 reactions at a rate notably greater than that of the corresponding saturated compounds. The stability of the allyl system eases bond breaking to the leaving group at the transition state and simultaneously means that less bond making is required by the nucleophile.

$$CH_3CH_2CH_2Cl + CH_3CH_2O^- \xrightarrow{CH_3CH_2OH} CH_3CH_2CH_2OCH_2CH_3 \qquad k^{rel} = 1 \tag{12.10}$$

$$CH_2{=}CHCH_2Cl + CH_3CH_2O^- \xrightarrow{CH_3CH_2OH} CH_2{=}CHCH_2OCH_2CH_3 \qquad k^{rel} = 4 \tag{12.11}$$

Since the carbon atom undergoing substitution is sp^2-hybridized in the transition state, some conjugative stabilization may be present (Figure 12-4).

Rearranged products may result from the reactions of allylic compounds with nucleophiles. If the substitution takes place under the customary conditions for bimolecular reactions, the paths that lead to unrearranged and rearranged products are labeled S_N2 and S_N2', respectively.

$$\underset{\displaystyle Cl}{CH_3CHCH{=}CH_2} + CH_3CH_2O^- \xrightarrow{CH_3CH_2OH} \underset{\substack{\displaystyle OCH_2CH_3 \\ (95\%) \\ S_N2 \text{ product}}}{CH_3CHCH{=}CH_2} + \underset{\substack{(5\%) \\ S_N2' \text{ product}}}{CH_3CH{=}CHCH_2OCH_2CH_3} \tag{12.12}$$

FIGURE 12-5
Electronic Displacements in an S_N2' Reaction

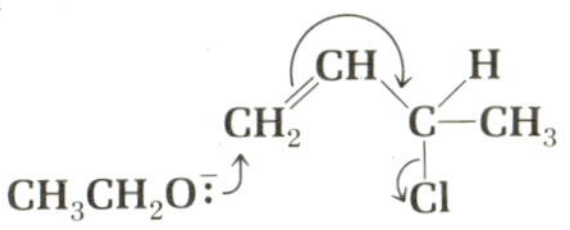

FIGURE 12-6
Energy Diagram for Reaction of 1,1-Dimethylallyl Cation with Water

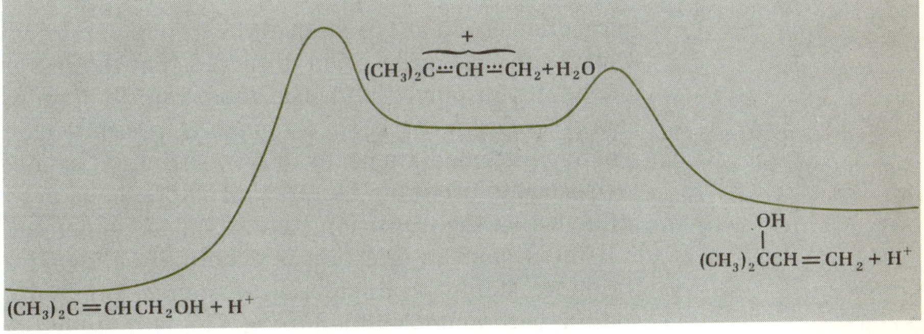

(12.13)

$$\underset{\substack{|\\ Cl}}{(CH_3)_3CCHCH{=}CH_2} + CH_3CH_2O^- \xrightarrow{CH_3CH_2OH} \underset{\substack{(100\%)\\ S_N2'\ \text{product}}}{(CH_3)_3CCH{=}CHCH_2OCH_2CH_3}$$

It has been proposed, and until recently widely accepted, that the S_N2' reaction involves attack of the nucleophile on one end of the double bond, the π-electrons of that bond simultaneously displacing the leaving group (Figure 12-5). The steric effect on the relative proportions of S_N2 and S_N2' products (equations 12.12 and 12.13) is in accord with this proposal.

Some chemists now doubt the existence of this pathway and propose instead that the allylic compound undergoes an internal rearrangement, followed by a normal S_N2 reaction of the rearranged derivative (equation 12.14).

(12.14)

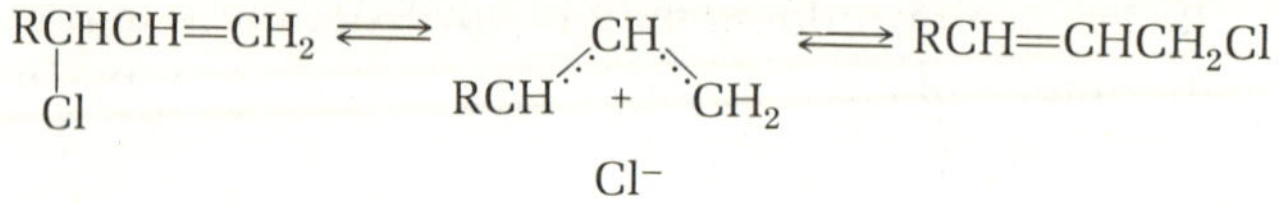

Such rearrangements are known to be relatively facile and the two competing theories are therefore quite difficult to disentangle experimentally.

Under conditions such that the rate-determining step is ionization of the allyl halide, a mixture of products is invariably found. These are sometimes referred to as S_N1 (unrearranged) and S_N1' (rearranged) products. Equation 12.15 provides a characteristic example.

(12.15)

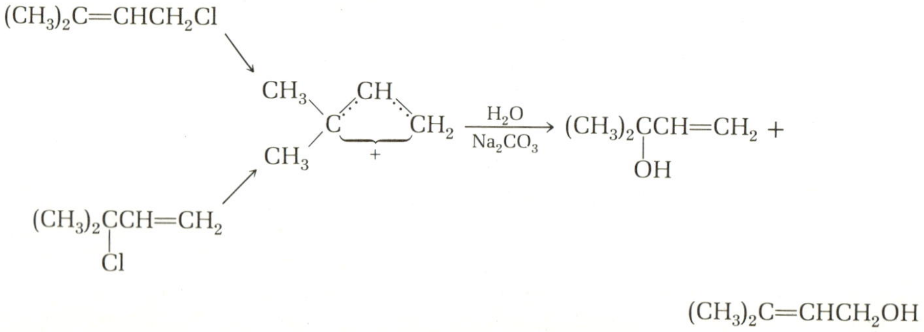

In reactions such as equation 12.15, *in which a highly reactive intermediate intervenes,* it is quite often observed that the less stable product (terminal alkene, for instance) is formed more rapidly than the more stable one. The energy diagram for such a system is presented in Figure 12-6. Since the reactive intermediate needs only to surmount small energy barriers in order to proceed to products, factors that stabilize the product will *not* exert much influence on the transition state (refer to Hammond's postulate, Chapter 5).

If the system in question is eventually allowed to come to equilibrium, the more stable product will, of course, predominate. In equation 12.15, sodium carbonate is added to remove HCl, which would catalyze just such an equilibration.

PROBLEM

4 By what mechanism does an acid catalyze the equilibration of the products of equation 12.15?

As a final illustration of the reactivity of allyl halides, we may mention their facile reaction with Grignard reagents, a form of coupling reaction. Grignard reagents do not react with ordinary alkyl halides except under forcing conditions.

(12.16) $$CH_2{=}CHCH_2Cl + (CH_3)_2CHCH_2CH_2MgCl \xrightarrow{\text{ether}} \underset{(60\%)}{(CH_3)_2CH(CH_2)_3CH{=}CH_2}$$

The condensation reaction can often be useful in the preparation of alkenes. Because of the ease with which coupling occurs with allyl halides, special care must be taken in the preparation of allyl Grignard reagents.

(12.17) $$Mg + \underset{(3\text{ moles})}{CH_2{=}CHCH_2Cl} \xrightarrow[0°,\ 10\text{ hr}]{\text{ether}} \underset{(85\%)}{CH_2{=}CHCH_2MgCl}$$

It is useful to compare the reactivities of allyl halides with those of *vinyl* halides, $CH_2{=}CHX$, in which the halogen atom is directly bound to one of the sp^2-hybridized carbon atoms. While the allyl halides are more reactive toward nucleophiles than the corresponding alkyl halides, the vinyl halides are much less so.

$$\underset{\text{reactivity}}{XCH_2CH{=}CH_2 > XCH_2CH_2CH_3 \gg XCH{=}CHCH_3}$$

The carbon-halogen bond is unusually strong in vinyl halides, partly because of the compact carbon orbital and partly because of the conjugation between the π-bond and the nonbonding *p*-orbitals on the halogen. In addition, there seems to be no low energy pathway for nucleophilic attack.

$$CH_2{=}CH\ddot{C}l \longleftrightarrow \overset{-}{\ddot{C}}H_2{-}CH{=}\overset{+}{C}l$$

In vinyl halides, conjugation stabilizes the reactant, lessening its tendency to undergo nucleophilic displacements. With allyl halides, conjugation lowers the energy of the transition state, thereby promoting reaction.

CUMULATED, CONJUGATED, AND NONCONJUGATED DIENES

There are three possible open-chain alkenes having the empirical formula C_5H_8: 1,2-pentadiene, 1,3-pentadiene, and 1,4-pentadiene. They each have distinct chemical and physical properties.

$CH_2{=}C{=}CHCH_2CH_3$ — 1,2-pentadiene

$CH_3CH{=}CHCH{=}CH_2$ — 1,3-pentadiene

$CH_2{=}CHCH_2CH{=}CH_2$ — 1,4-pentadiene

In 1,4-pentadiene the two double bonds do not interact and a normal $>C{=}C<$ peak is observed at about 1650 cm^{-1} in the infrared spectrum. The two double bonds in 1,3-pentadiene are conjugated and the resulting interaction is reflected in the infrared spectrum. Two peaks now appear in the double bond region, the more intense of the two occurring at 1600 cm^{-1}. The intensity of this peak is considerably enhanced compared to that of an isolated double bond.

The double bonds in 1,2-pentadiene are said to be **cumulated** — that is, one carbon atom is a part of two different double bonds. These **allenes** are readily distinguishable from their isomers on the basis of their spectra (1,2-propadiene is called allene and compounds having the same functional group are named as its derivatives). A band at 1900 cm^{-1}, well outside the normal double bond region, is characteristic of compounds with cumulated double bonds. The π-systems are not conjugated because they are at right angles to one another (Figure 12-7). In addition, the planes of the two $>C{=}$ groups at each end of the molecule are also at right angles to one another. As a result, if X and Y are nonidentical on both carbon atoms, the entire molecule will be dissymmetric *even though it contains no asymmetric atom.* This is illustrated with the enantiomers of 1,3-dimethylallene (**8** and **9**).

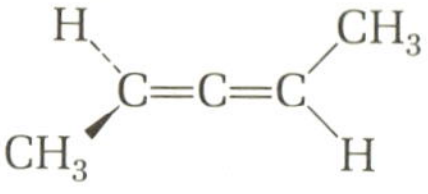

8

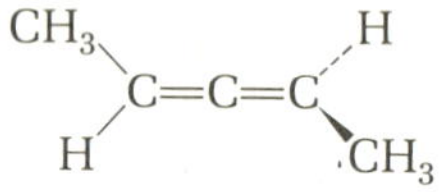

9

A few allenes have been resolved and some optically active ones have been obtained from natural sources. Their chemistry is discussed in Chapter 13.

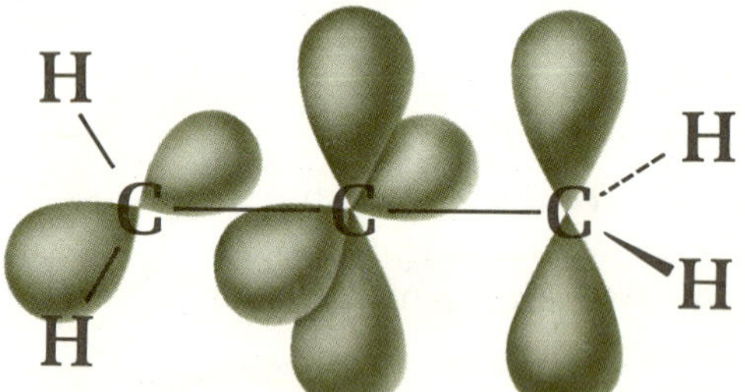

FIGURE 12-7
π-Bonds in Allene

ULTRAVIOLET SPECTROSCOPY

1,3-Butadiene is the simplest conjugated diene. In order for conjugation to be maximized, the molecule must be planar. There are then two possible conformations, the *s-cis* and the *s-trans* (which predominates at room temperature).

s-*cis*-butadiene s-*trans*-butadiene

These are not true *cis* and *trans* isomers since they cannot be isolated at room temperature; the barrier to rotation around the single bond is only about 5 kcal/mole. Butadiene has four molecular π-orbitals, two bonding and two antibonding (Figure 12-8). Normally the four π-electrons are in the two bonding orbitals; by providing approximately ~130 kcal/mole of energy, it is possible, however, to place the molecule in a π, π^* state in which an electron has been promoted from the highest bonding orbital to the lowest antibonding orbital.

Light of wavelength 2170 A contains the required energy and irradiation of a sample of butadiene with it results in strong absorption. It is experimentally convenient to measure quantitatively the absorption of light of

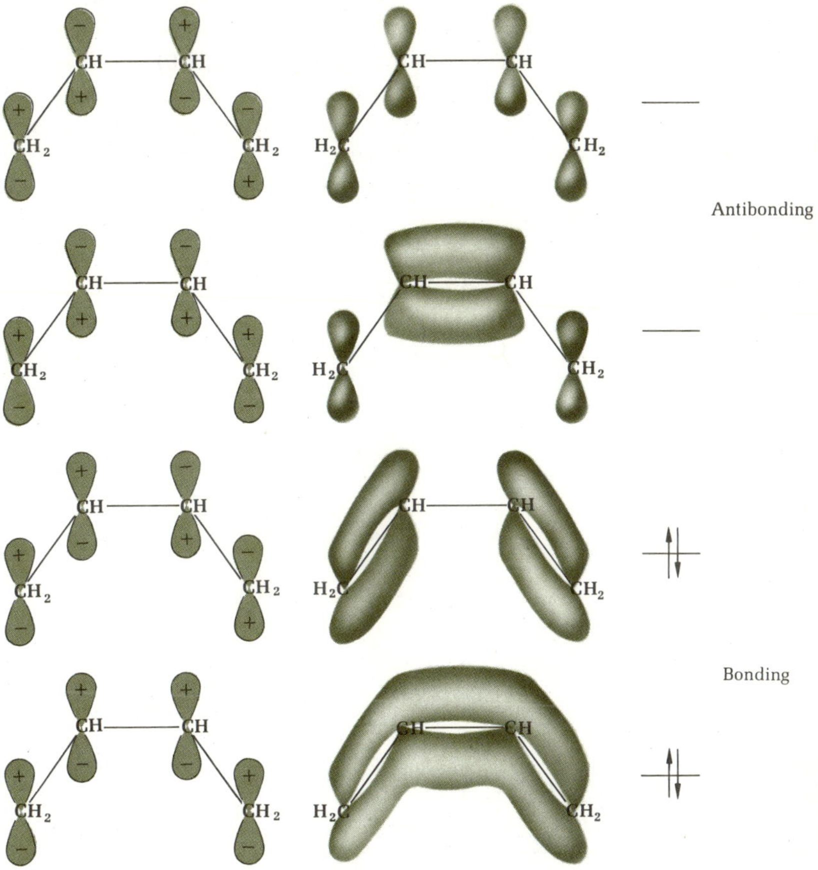

FIGURE 12-8
Diagram of the π-Orbitals of 1,3-Butadiene

wavelength >2000 A for any substance, particularly in solution. Simple alkenes do not fit into this category because of the large energy gap between their bonding and antibonding orbitals, but conjugated dienes and higher polyenes do.

These measurements are made with an **ultraviolet** (uv) **spectrometer** — the instrument usually operates in the visible region as well. Its operation is basically the same as that of an infrared spectrometer. A beam of light, the wavelength of which is continuously varied, is split and each half sent through a cell. One cell contains a solution of the sample and the other contains pure solvent. The difference in absorption between the two cells is automatically and continuously recorded on a chart, yielding the uv spectrum.

From previous discussions one might be led to expect that a compound would exhibit sharp absorption peaks at certain wavelengths in the ultraviolet, separated by broad regions of essentially no absorption. In fact, ultraviolet spectra as normally measured consist of very broad absorptions containing fairly well-defined maxima. The main cause for this is the fact that the molecules could be in several different vibrational levels at the moment of excitation and they may also be in one of several vibrational levels after excitation (see Figure 12-9). Each of these transitions occurs at a slightly different wavelength; the individual vibrational transitions are further blurred by rotation and by solution effects, and a broad band results (the spectrum of benzene shows some of the individual vibrational absorptions, Figure 12-10). (One mμ (millimicron) = 10^{-7} cm = 10 A.)

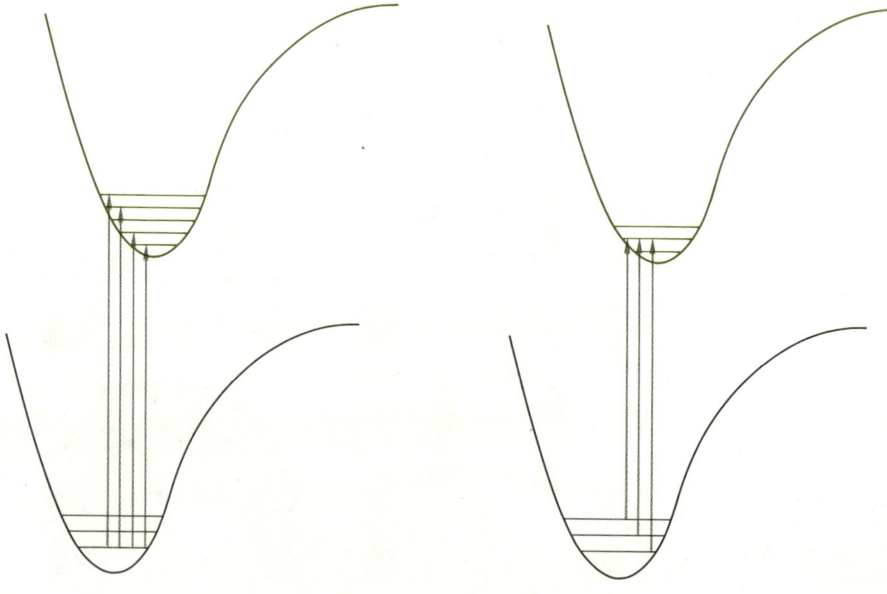

FIGURE 12-9
Electronic Excitation of a Diatomic Molecule, from Various Ground State Vibrational Levels to Various Excited State Levels

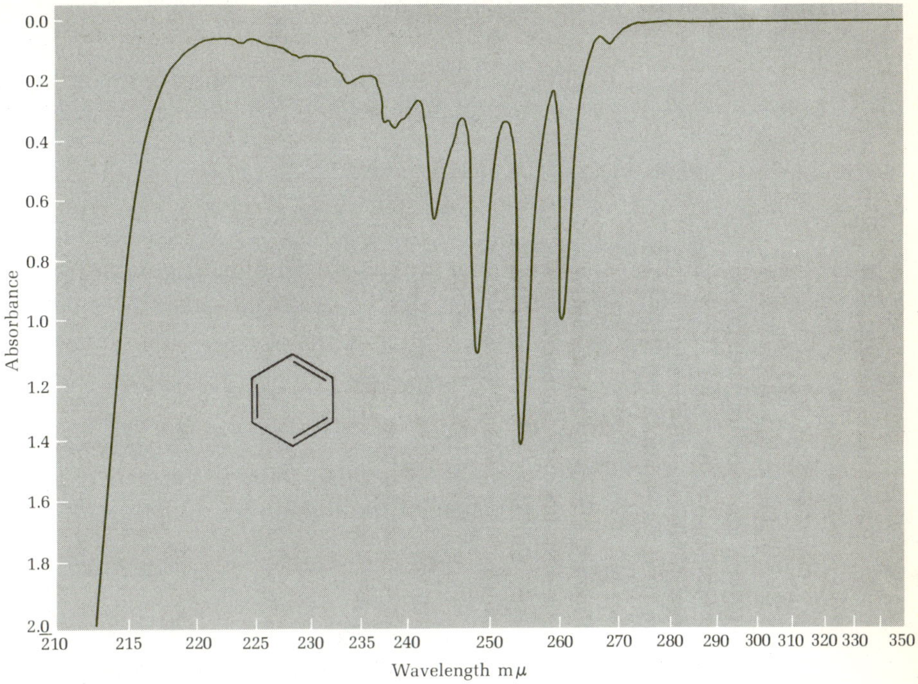

FIGURE 12-10
Ultraviolet Spectrum of Benzene Showing Vibrational Bonds

It is often useful to know just how intensely any compound absorbs radiation of a given wavelength. This information is conveyed by use of the symbol ϵ, defined by equation 12.18 [the absorbance was formerly known as the optical density (D)].

(12.18) $$\log \frac{I_0}{I} = \text{absorbance } (A) = \epsilon \cdot c \cdot l$$

I_0 is the intensity of the light before absorption by the sample; I, the intensity afterward; c, the sample concentration (M); and l, the thickness of the sample cell in centimeters. The molar extinction coefficient, ϵ, is then a function only of the compound, the wavelength of light, the solvent, and the temperature. When giving spectroscopic data for a compound that absorbs in the ultraviolet or visible region, it is customary to specify the wavelength of the maxima and the extinction coefficients at those maxima. Thus, 1,3-butadiene is characterized by one peak in the accessible region of the ultraviolet spectrum, λ_{max}^{hexane} 2170 A (ϵ 21,000). Peaks arising from allowed transitions usually have sizable extinction coefficients, $\epsilon = 10^3 - 10^5$, while "forbidden transitions" give peaks of low intensity, $\epsilon = 0 - 10^2$.

PHOTOCHEMISTRY OF CONJUGATED DIENES

If butadiene is irradiated with short wavelength ultraviolet light, it gives rise to two principal photoproducts, cyclobutene and bicyclo(1,1,0)butane.

(12.19) 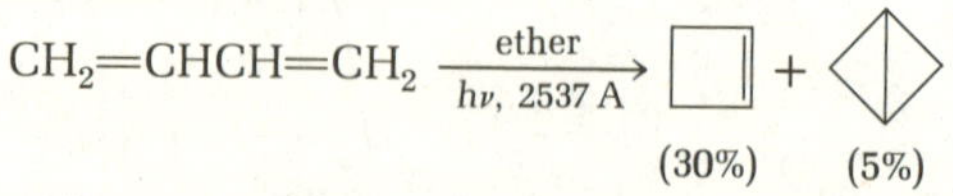

It is likely that the second product originates from the diradical intermediate **10.**

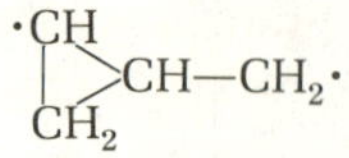

10

Cyclobutene, on the other hand, is formed in a concerted process from the *s-cis* conformer of butadiene.

(12.20) 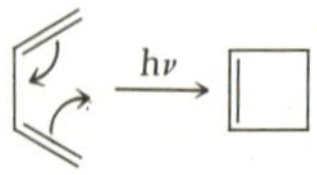

Irradiation of *trans,trans*-2,4-hexadiene under the same conditions leads to *cis*-3,4-dimethylcyclobutene as the only cyclobutene product. In general, irradiation of a single isomer of a substituted butadiene leads to only *one* of the possible cyclobutene isomers.

(12.21) 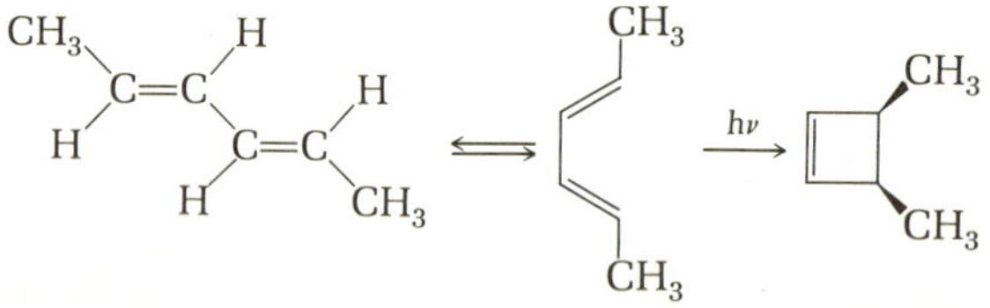

THE WOODWARD-HOFFMANN RULES

This remarkable stereospecificity was explained by Woodward and Hoffmann, who formulated a set of general rules to cover all **electrocyclic reactions** (reactions in which a ring is created by σ-bond formation between the ends of one or more π-electron systems). Their work was based on the hypothesis that as the reactants in any reaction are converted to products, the reactant molecular orbitals are continuously converted to product molecular orbitals. This is reasonable since both the reactant and product molecular orbitals are derived from the same set of atomic orbitals (since the same *atoms* are present in the reactant and product).

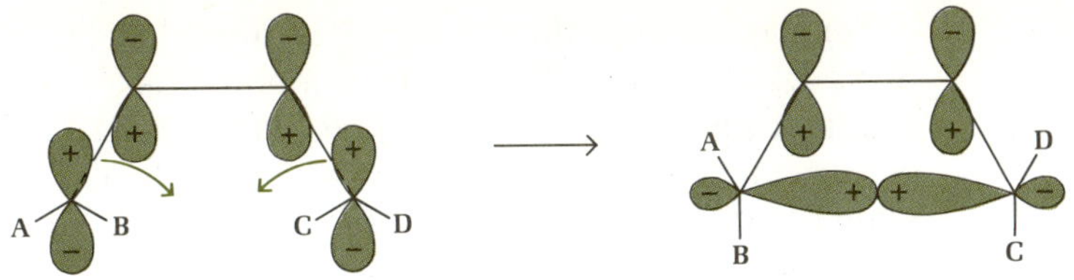

FIGURE 12-11
Cyclobutene Formation and Transformation of the Third π-Orbital of Butadiene

The critical point is that there are restrictions on just which reactant molecular orbitals may be transformed into a particular product molecular orbital — it is not a random process. Certain reactant orbitals convert to particular product orbitals — they are said to **correlate.** Because of these restrictions the stereochemistry of the reaction may be highly controlled. If the orbitals involved contain some symmetry elements, then it is much easier to decide how the reactant and product orbitals correlate. **In these cases, the symmetry of the occupied molecular orbital of highest energy often seems to "control" the stereochemistry of the reaction.**

Butadiene and its derivatives provide good examples for illustrating the operation of these principles. When butadiene is photochemically excited, one electron is promoted from the second to the third π-orbital (Figure 12-11). In order for the ring to be closed, the CH_2 groups must each undergo a 90° rotation. In the course of this rotation the *p*-orbitals at carbon atoms 1 and 4 overlap to form the new σ-bond. Woodward and Hoffmann made the reasonable assumption that the rotations would occur in such a manner that orbital lobes of the *same sign* would overlap. Due to the symmetry of the third π-molecular orbital of butadiene, the CH_2 groups must rotate in opposite directions — this is known as **disrotatory motion.**

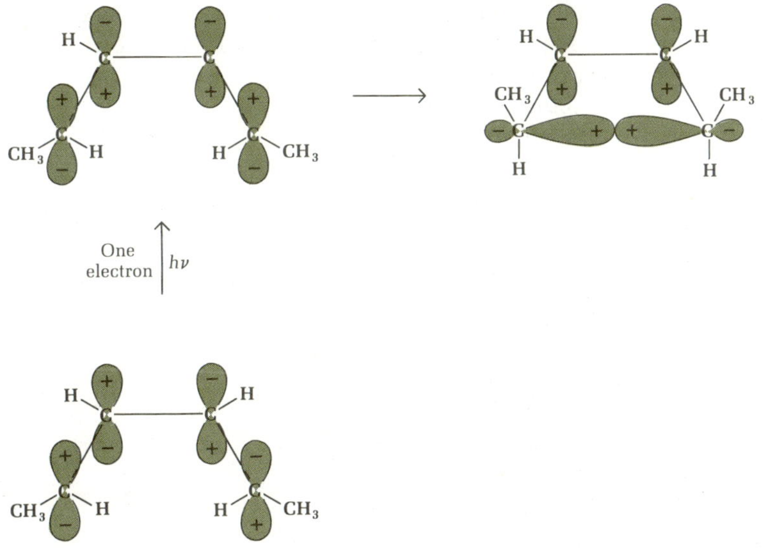

FIGURE 12-12
Photochemical Cyclization of *trans, trans*-2,4-Hexadiene: Orbital Diagram

Confirmation of this proposal is provided by the cyclization of *trans,trans*-2,4-hexadiene. When this compound is arrayed in the *s-cis* conformation and photochemically excited (equation 12.21), then a disrotatory ring closure should produce *cis*-3,4-dimethylcyclobutene (Figure 12-12)—which is exactly what is found.

THERMAL REACTIONS OF CONJUGATED DIENES

Cyclobutenes may be converted back to the related butadienes thermally. Heating *cis*-1,2,3,4-tetramethylcyclobutene to 200° leads to its clean conversion to *cis,trans*-3,4-dimethyl1-2,4-hexadiene. The ring opening is exclusively **conrotatory** (rotation of both termini in the same direction).

(12.22)

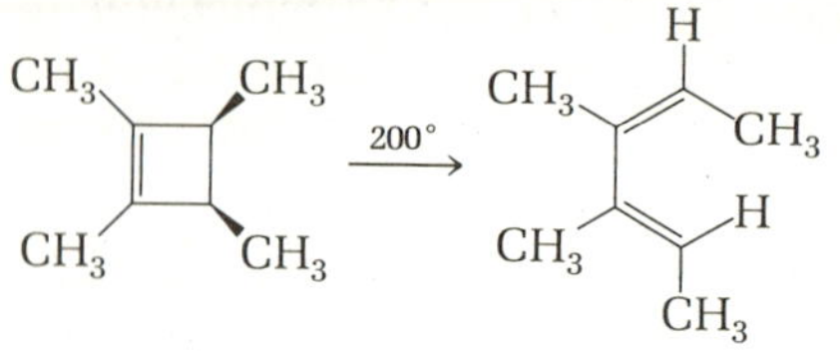

The thermal reaction involves no electronic excitation so that the highest occupied π-orbital in the product is the second. Its symmetry demands that conversion of the diene to a cyclobutene, or the reverse, involves conrotatory motion at the ends of the π-system (Figure 12-13).

These analyses are of general applicability, and cover the cyclization of 1,3,5-hexatrienes to 1,3-cyclohexadienes, etc. A number of such electrocyclic reactions may be reversed; the same orbital symmetry arguments apply no matter in which direction the reaction is proceeding. From analysis of these and many other examples two very simple rules have been derived which allow one to predict the stereochemistry of ring closure for any **intramolecular** electrocyclization.

The rules are:

1 **If the number of π-electrons equals 4*n*** ($n = 0,1,2, ...$)**, then thermal electrocyclization proceeds with conrotatory motion, while photochemical electrocyclization proceeds with disrotatory motion.**

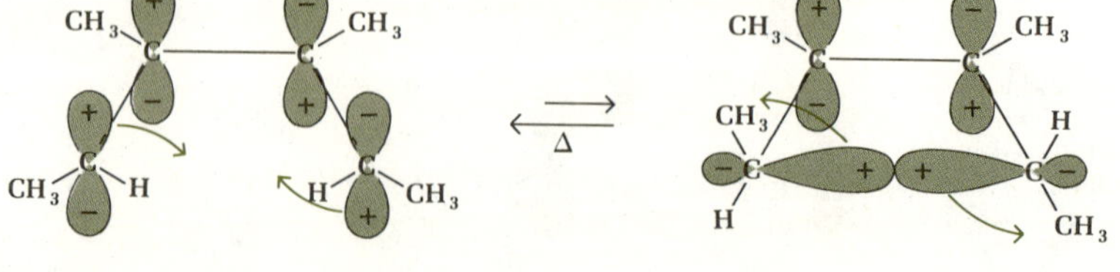

FIGURE 12-13
Thermal Interconversion of Tetramethylcyclobutene and Dimethyl-2,4-hexadiene

2 **If the number of π-electrons equals $4n + 2$ ($n = 0,1,2, \ldots$), then thermal electrocyclization requires disrotatory motion while photochemical electrocyclization requires conrotatory motion.**

INTERMOLECULAR ELECTROCYCLIC REACTIONS

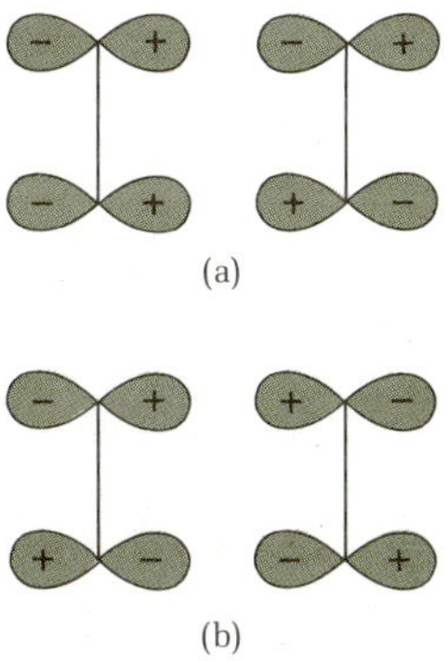

FIGURE 12-14
(a) Unfavorable Concerted Overlap of Ethylene π- and π^*-Orbitals;
(b) Favorable Concerted Overlap of Two π^*-Orbitals

Woodward-Hoffmann analysis also permits predictions to be made about *intermolecular* reactions between two or more unsaturated reactants. In particular, one may decide whether it is feasible for the reactants to bond together concertedly, in a single step, or whether the reaction is required to follow a multistep course. In a concerted reaction, the various bond-breaking processes that are taking place are largely compensated for by the bond-forming steps that are occurring simultaneously. One would therefore expect the concerted process to be favored by reason of its low activation energy, *provided* that there are not other constraints such as those arising from orbital symmetry considerations.

The reactions of two ethylenes to yield a cyclobutane offers a convenient example (equation 12.23). This reaction is very difficult to induce by thermal means but examples are abundant in which one of the reactants has been photochemically excited. Application of the Woodward-Hoffmann rules leads to the conclusion that two ethylenes may clip together concertedly under photochemical activation but not under thermal activation.

(12.23)
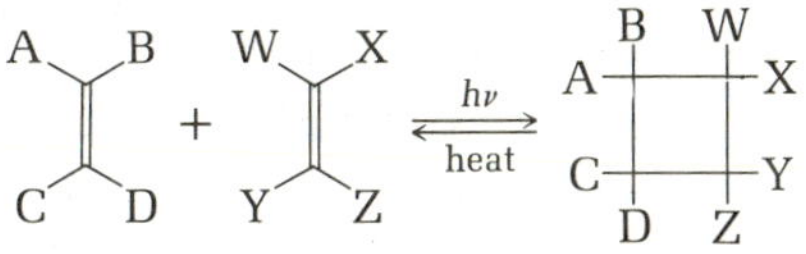

The system may be naively analyzed as follows. As the two reactants approach one another, the filled molecular orbitals of one begins to feed electron density into appropriate empty molecular orbitals of the other. (The interactions of two filled orbitals would lead to strong repulsions, as per Pauli's principle.) For ethylene, the highest filled orbital is the π-orbital, while the lowest empty orbital available from the co-reactant is the π^*-orbital. Figure 12-14 makes clear the fact that these two orbitals are not of the correct symmetry to allow complete overlap. If, on the other hand, one of the ethylene molecules had been photochemically excited, its highest occupied orbital would be a π^*-orbital and the chemical situation would be one involving the favored overlap of two ethylene π^*-orbitals.

PROBLEM

5 The reaction of a butadiene and an ethylene to yield a cyclohexene is an important and useful one (**Diels-Alder reaction;** Diels and Alder subsequently shared a Nobel

Prize). A great many examples are known, many of which result in high yields of product and mild reaction conditions.

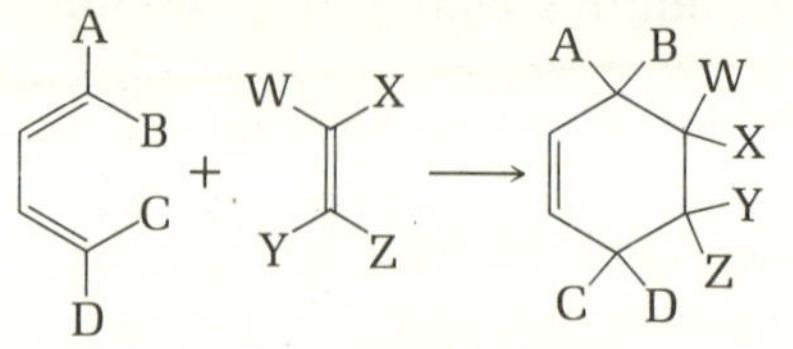

(2)

There is a sizable body of evidence that argues for the concertedness of the reaction.

a. Is this Diels-Alder reaction predicted to be an allowed concerted process under thermal conditions?

b. Could it be a concerted photochemical process?

c. Two butadienes might react in Diels-Alder fashion to give vinylcyclohexene or they might alternatively combine to yield 1,5-cyclooctadiene [equation (3)].

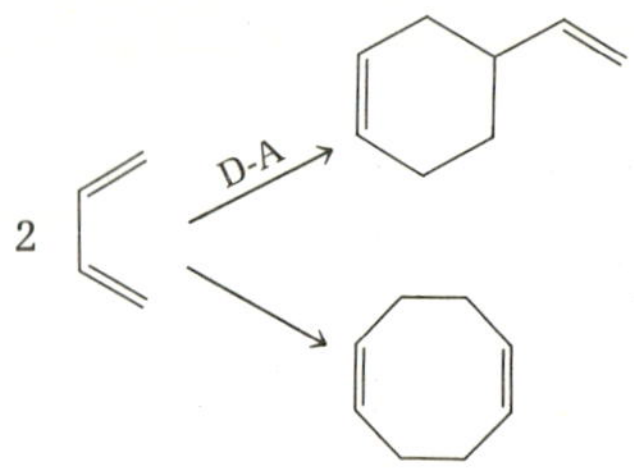

(3)

What about the thermal and photochemical allowedness of these competing paths?

d. Based on the examples discussed so far, could you extrapolate and propose a general rule to govern the possibility of a concerted reaction between two conjugated π-systems under thermal and photochemical conditions?

THE DIELS-ALDER REACTION

This *thermal* reaction provides a prime route to all manner of substituted six-membered cyclic compounds. Normally, the rate of reaction is enhanced if the alkene (**dienophile**) is substituted with electron-withdrawing groups.

(12.24)
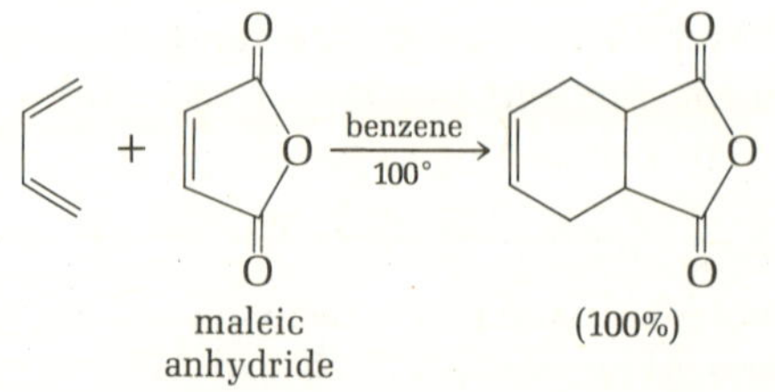

(12.25) 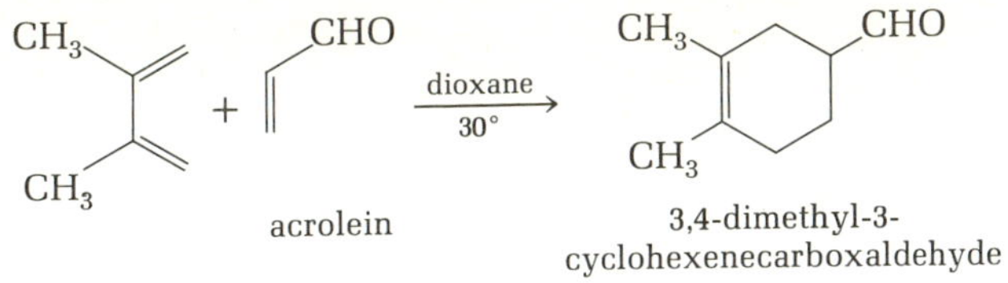

2,3-Dimethylbutadiene is about five times as reactive as butadiene toward maleic anhydride, showing that electron-supplying groups activate the diene. On the other hand, ethylene is a very poor dienophile and requires elevated temperatures to undergo reaction, while tetracyanoethylene, $(NC)_2C{=}C(CN)_2$, reacts instantaneously with many dienes at 0^0. This confirms the activating influence of electron-withdrawing groups on the dienophile.

Steric relationships that are present in the dienophile are retained in the product (equations 12.26 and 12.27).

(12.26) 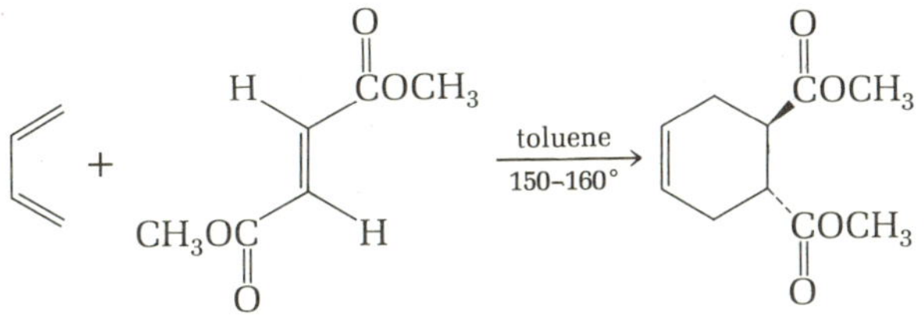

(12.27) 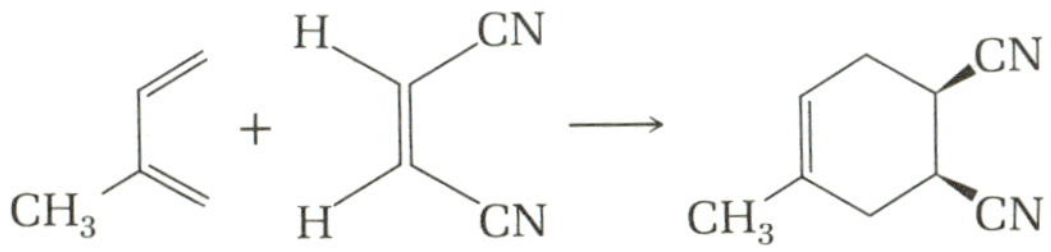

Furthermore, the diene appears to react only in the *s-cis* form. A diene such as 3-methylenecyclohexene, **11**, does not undergo Diels-Alder reactions, while cyclopentadiene, which is locked in a *cis*-conformation, is a very reactive diene.

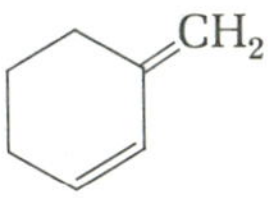

11

Cyclopentadiene is not stable as a monomer at room temperature but exists as a Diels-Alder dimer.

(12.28) 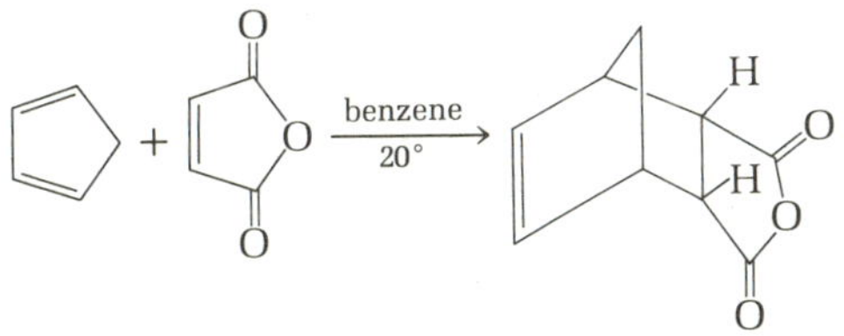

(100%)

(12.29)

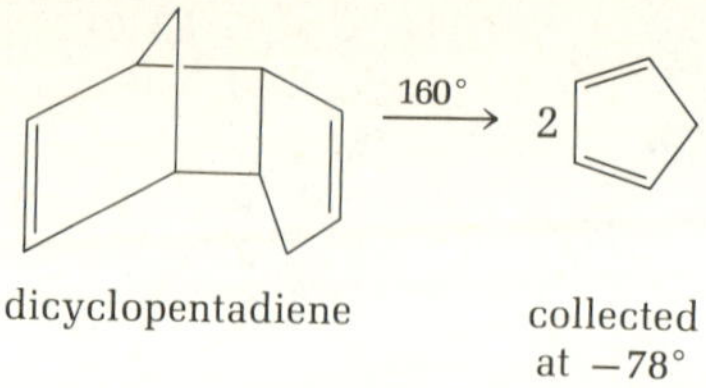

It is obtained by heating the dimer, which then undergoes a **retro Diels-Alder reaction.** The reaction of cyclopentadiene with a number of dienophiles offers a very convenient and versatile route to norbornane derivatives (Chapter 7, p. 159).

Many **heterocycles** (cyclic compounds containing ring atoms other than carbon) are available by way of the Diels-Alder reaction.

(12.30)

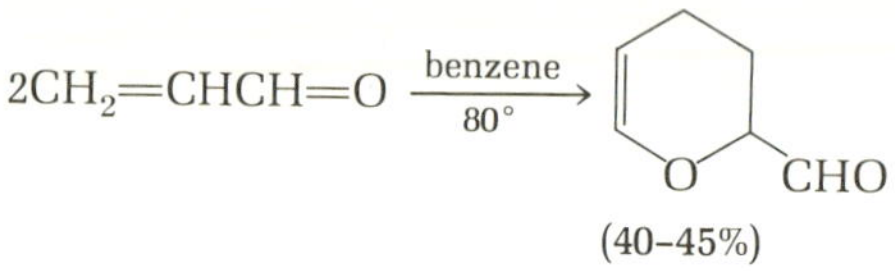

(12.31)

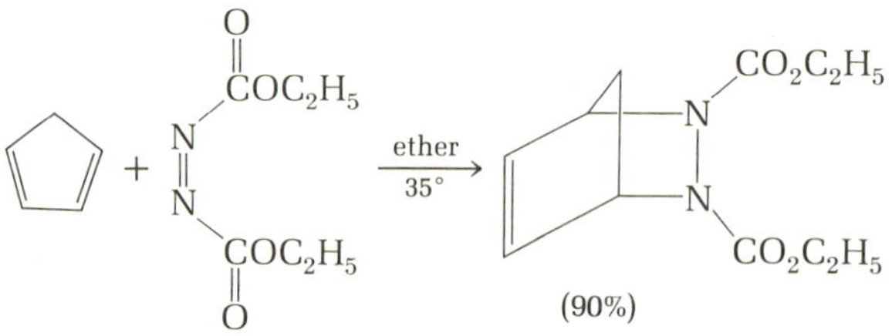

A discussion of the mechanism of this versatile, and apparently straightforward, reaction is deferred until the last chapter.

The ethylene dimerization and Diels-Alder reaction are specific examples of more general reactions in which two π-systems interact to yield a cyclic product. General rules may be derived which allow one to predict whether these **intermolecular** cyclizations may proceed in a concerted manner or whether they will be constrained to follow a multistep path.

1 **If the total number of π-electrons equals $4n$** ($n = 0,1,2, ...$), **then a concerted cyclization is allowed photochemically but not thermally.**

2 **If the total number of π-electrons equals $4n + 2$** ($n = 0,1,2, ...$), **then a concerted cyclization is allowed thermally but not photochemically.**

IONIC ADDITIONS TO CONJUGATED DIENES

The same reagents that add to simple alkenes also add to conjugated dienes. These additions are initiated by electrophilic attack in both cases. The electrophile always attaches itself to one of the *terminal* atoms of the π-system since the intermediate carbonium ion so produced is an allylic one (equation 12.32).

(12.32) 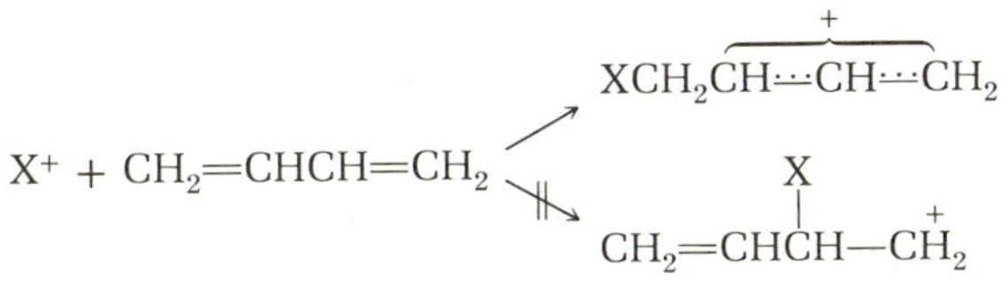

Reaction of this delocalized carbonium ion with a nucleophile can lead to two different products (neglecting *cis-trans* isomerism), deriving from overall 1,2- or 1,4-addition.

$$\text{(12.33)}\quad XCH_2\overbrace{CH\overset{...}{-}CH\overset{...}{-}CH_2}^{+} + Y^- \longrightarrow XCH_2\underset{\displaystyle Y}{CH}CH{=}CH_2 + XCH_2CH{=}CHCH_2Y$$

Slow addition of 1 mole of bromine to 1 mole of 1,3-butadiene under conditions where the products do not isomerize leads to roughly equal amounts of 3,4-dibromo-1-butene and *trans*-1,4-dibromo-2-butene, and no *cis* isomer.

(12.34) 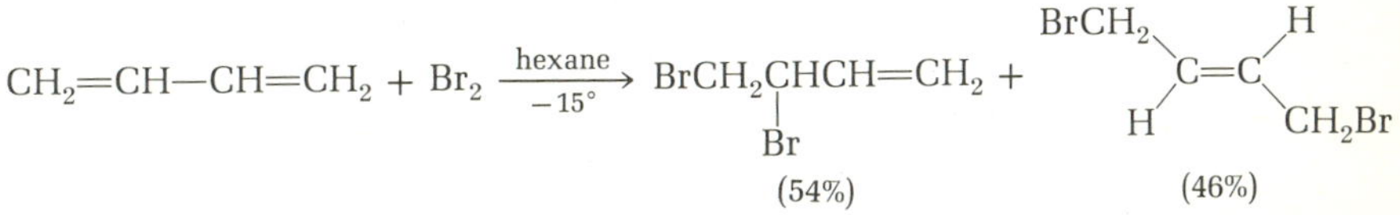

Since additions are much faster with conjugated dienes than with simple alkenes, it is possible to get solely the dibromides. The relative proportion of the products certainly does not reflect their relative stabilities, as shown when they are allowed to equilibrate (equation 12.35).

(12.35) 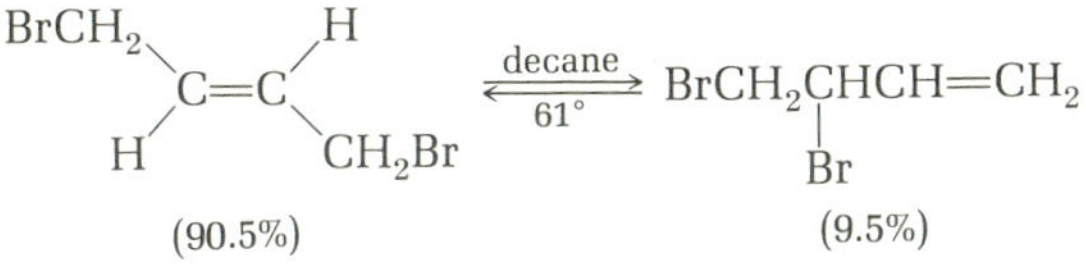

The preference for the thermodynamically less stable isomer is attributed to unequal charge distribution in the 4-bromobutenyl cation **12**, the secondary carbon atom presumably carrying more than 50% of the positive charge and the bromide ion preferentially attacking that carbon atom.

$$BrCH_2\overbrace{CH\overset{...}{-}CH\overset{...}{-}CH_2}^{+}$$

12

Note that, due to the stability of the delocalized carbonium ion, a typical cyclic bromonium ion is apparently not formed. Chlorination of butadiene brings about similar results.

HYDROGENATION OF CONJUGATED DIENES

Conjugated dienes may also be hydrogenated by the usual catalytic methods. It is very difficult to stop at the monohydrogenation stage with platinum but relatively easy with palladium. Complete reduction of 1,3-butadiene to butane in the gas phase results in the release of 57.1 kcal/mole.

(12.36) $$CH_2{=}CHCH{=}CH_2 + 2H_2 \xrightarrow[80°]{\text{gas phase}} CH_3CH_2CH_2CH_3 \qquad \Delta H = -57.1 \text{ kcal/mole}$$

This value is somewhat less than twice the heat of hydrogenation of either ethylene or propylene. The difference is ascribed to stabilization of the conjugated molecule by resonance and is often labeled the empirical resonance energy.

(12.37) $$CH_2{=}CH_2 + H_2 \xrightarrow[80°]{\text{gas phase}} CH_3CH_3 \qquad \Delta H = -32.8 \text{ kcal/mole}$$

(12.38) $$CH_3CH{=}CH_2 + H_2 \xrightarrow[80°]{\text{gas phase}} CH_3CH_2CH_3 \qquad \Delta H = -30.1 \text{ kcal/mole}$$

Depending on whether one chooses ethylene or propylene as the appropriate standard, the resonance energy of butadiene calculates out as $(2 \times 32.8) - 57.1 = 8.5$ kcal/mole or $(2 \times 30.1) - 57.1 = 3.1$ kcal/mole. The latter is probably a more realistic estimate. This example points up the fact that the meaningfulness of these empirical resonance energies is very much a function of the standard chosen for comparison. Most such figures should be treated with caution.

Even the existence of a substantial resonance effect in butadiene is open to question. Upon complete reduction of the diene, the central carbon-carbon σ-bond changes from sp^2-sp^2 to sp^3-sp^3, an endothermic process. It is quite possible that this change accounts for the greater part of the apparent resonance energy.

The hydrogenation of allene contrasts sharply with that of butadiene, the heat release being abnormally large. This presumably reflects the repulsion

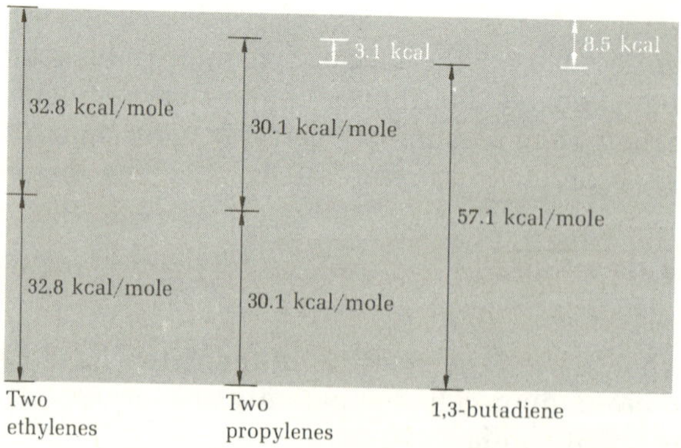

FIGURE 12-15
Heat of Hydrogenation of 1,3-Butadiene, Two Ethylenes and Two Propylenes

between nonconjugated interleaved π-bonds. Allenes are distinctly unstable with respect to the corresponding conjugated dienes.

(12.39) $$CH_2{=}C{=}CH_2 + 2H_2 \xrightarrow[80^\circ]{\text{gas phase}} CH_3CH_2CH_3 \qquad \Delta H = -71.3 \text{ kcal/mole}$$

SYNTHESES OF DIENES

Dienes can be synthesized in the laboratory by the same methods used for the synthesis of simple alkenes, such as dehydration of diols. Dehydrohalogenation of dihaloalkanes is also feasible. If the halogen atoms are on adjacent carbon atoms, however, elimination of 2 moles of hydrogen halide leads preferentially to formation of alkynes (Chapter 13) rather than dienes (equation 12.40). When structural features block the more favorable path, diene synthesis is feasible.

(12.40)

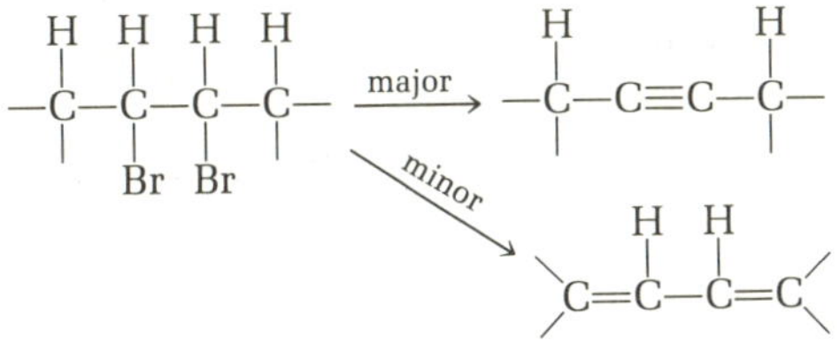

(12.41)

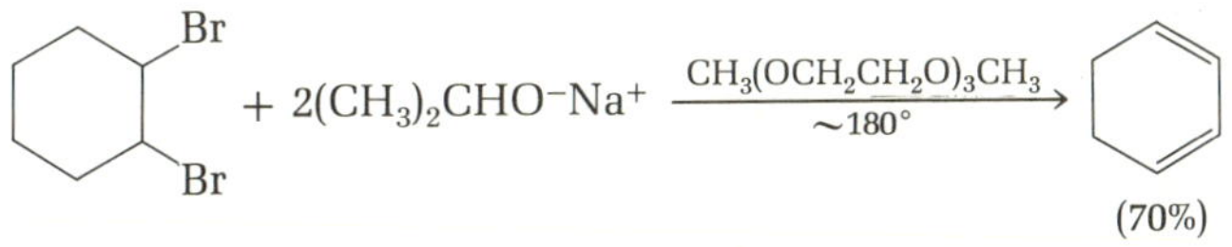

(12.42)

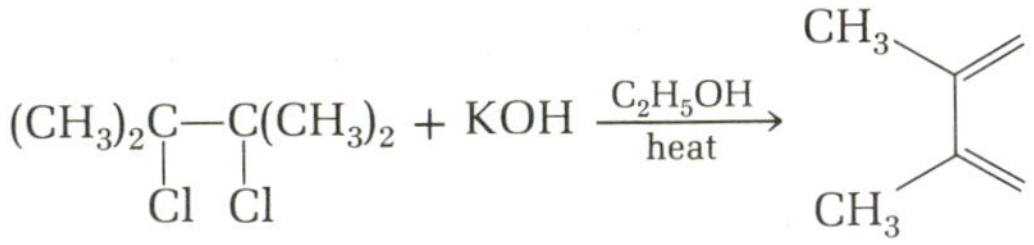

NATURAL AND ARTIFICIAL RUBBER

The chief use of the considerable quantity of butadiene produced each year is for the synthesis of artificial rubber and rubber-like polymers, known collectively as **elastomers.** Latex, the form in which *natural* rubber is first obtained from rubber trees, is basically an aqueous colloidal solution of a polymer of **isoprene** (2-methylbutadiene). Isoprene is an extremely important natural building block; for instance, squalene, the precursor of the steroids, can be envisaged as resulting from the condensation of a number of isoprene units. In natural rubber the isoprene units are polymerized in a head-to-tail, all *cis* fashion (Figure 12-16).

FIGURE 12-16
Section of a Natural Rubber Molecule

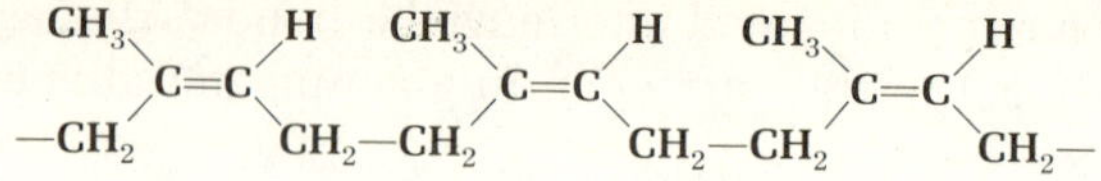

Prodded by commercial and military necessity, several nations undertook studies of the feasibility of producing synthetic rubber. It seemed logical to start by polymerizing butadiene or one of its derivatives, in imitation of Nature. The most satisfactory recipe, widely used during World War II, turned out to be a free radical-initiated copolymer of butadiene and styrene. Polymerization was completely random; about 60% of the butadiene was polymerized *trans*-1,4; 20%, *cis*-1,4; and another 20% underwent 1,2-addition.

FIGURE 12-17
Section of a Butadiene-Styrene Polymer Molecule

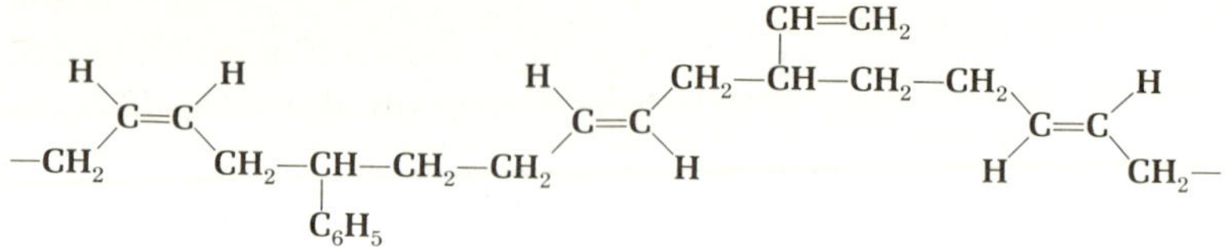

It has been found that many of the catalysts that were capable of bringing about stereoregular polymerization of simple alkenes (Chapter 9) could do the same for dienes. Both *cis*-1,4-polybutadiene and *cis*-1,4-polyisoprene are manufactured in this country. They have many of the desirable features of natural rubber not found in the randomly polymerized artificial rubbers.

All forms of rubber, natural and artificial, must be modified considerably before they are suitable for use. One of the most critical alterations is **vulcanization** (*Vulcan* — Roman god of fire), in which sulfur is blended into the raw rubber and the mixture then heated. The raw rubber is very plastic, does not wear well, and has mechanical properties that are very temperature-dependent. Vulcanization renders it tougher, most elastic, and "rubbery." What is happening chemically is that some of the allylic hydrogen atoms are being replaced by sulfur atoms and different polymer molecules then become cross-linked to one another by means of chains of sulfur atoms (Figure 12-18).

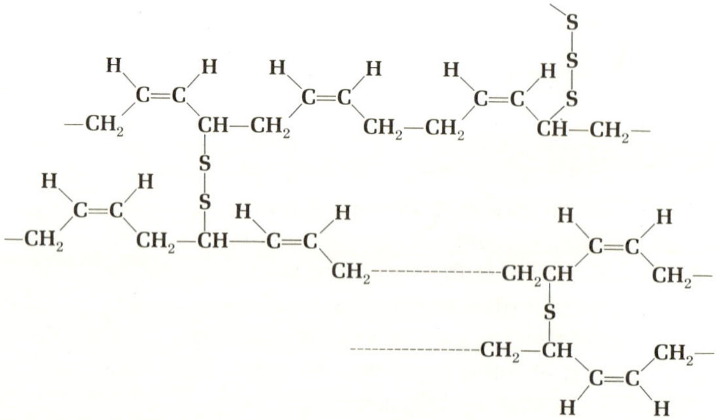

FIGURE 12-18
Cross-Linked Polybutadiene Chains

The linear hexaene squalene plays a vital role in biological chemistry since it may be cyclicized to lanosterol, a predecessor of numerous other steroids (Chapter 10, p. 244). Squalene, in turn, is constructed from two 15-carbon triene units, called *farnesyl* groups, joined head-to-head (a speculative mechanism for this process is presented in Chapter 20). As made explicit by structure **13,** farnesol itself is the end product of the linking together of three isoprene moities.

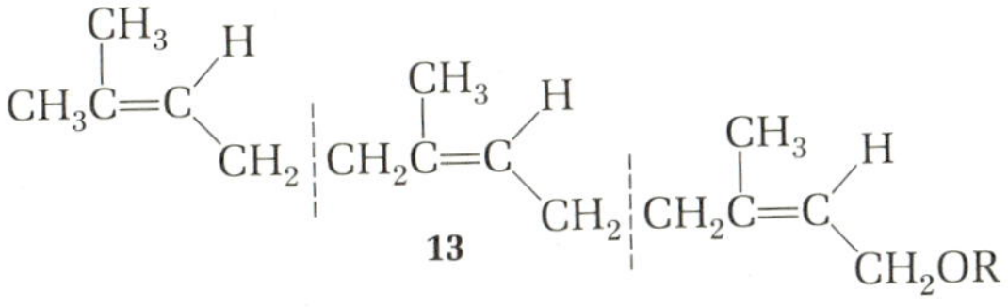

13

R = H, farnesol

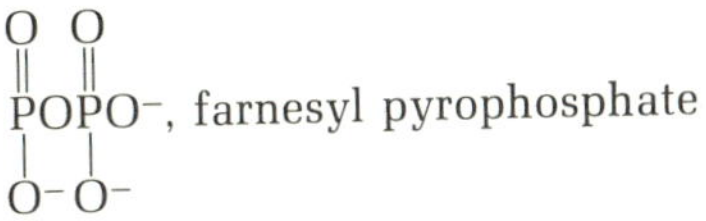

R = POPO⁻, farnesyl pyrophosphate

The isoprene unit is present in tissue as isopentenyl pyrophosphate (**14**). Its precursor is mevalonic acid (**15**), which is transformed into **14** after suffering enzymatic dehydration and decarboxylation.

(12.43)

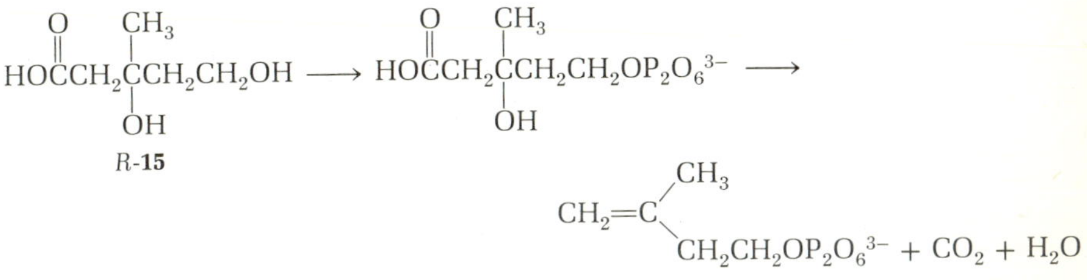

R-**15**

14

Current hypotheses presume that the biosynthesis begins with the rearrangement of one molecule of **14** to its allylic isomer. Nucleophilic displacement of the allylically placed pyrophosphate anion then follows, with the double bond of another molecule of **14** acting as the nucleophile.

(12.44)

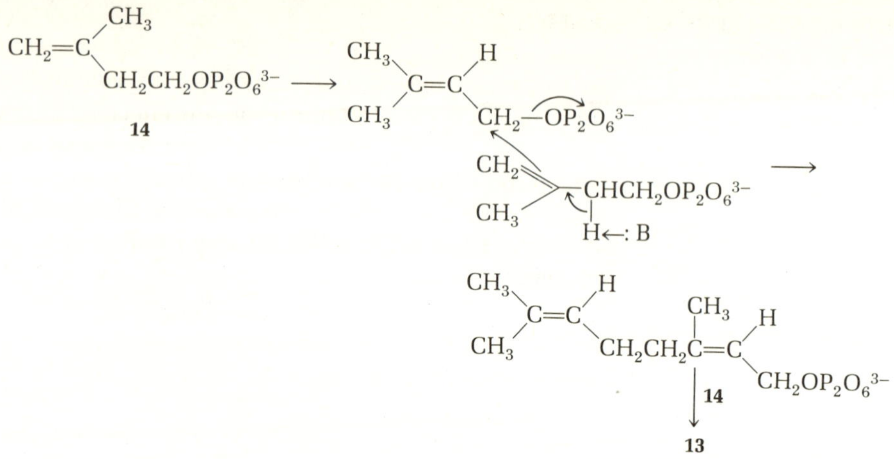

A basic group in the enzyme abstracts a proton, thus shifting the position of the double bond in the second isopentenyl pyrophosphate unit. A repetition of the substitution step leads finally to **13.**

SUMMARY

1 Allylic radicals or ions are notably more stable than their saturated counterparts. As a result of conjugation in these species, the charge or unpaired spin is delocalized over the terminal atoms of the allylic system. This delocalization is predicted by molecular orbital theory, as well as by resonance theory. The ions or radicals are in static electronic states — they do not oscillate back and forth between different discrete structures.

2 The consequences of delocalization may be observed directly. The electron spin resonance spectrum of allyl radical indicates that the four terminal hydrogen atoms are all equivalent. The infrared spectrum of allyl cation suggests that the carbon-carbon bonds are somewhere between normal single and double bonds. Furthermore, nucleophilic substitution of allylic halides, by both S_N1 and S_N2 mechanisms, proceeds with varying degrees of rearrangement. The rates of both S_N1 and S_N2 reactions, particularly the former, are enhanced in allylic systems compared to their saturated analogues.

3 Dienes may be unconjugated, conjugated, or cumulated. Cumulated dienes (allenes) are relatively unstable. They may exist in enantiomeric forms even though they contain no asymmetric atoms. Unconjugated polyalkenes behave substantially like monoalkenes.

4 Conjugated dienes are of the greatest chemical interest. The two double bonds interact, resulting in some additional delocalization and concomittant stabilization of the π-electrons. This is reflected in the ultraviolet spectra and heats of hydrogenation of butadiene and its derivatives.

5 The symmetry of the molecular orbitals of these conjugated dienes determines which electrocyclic reactions will occur by way of energetically favorable transition states and which will not. A general conclusion is that if a particular reaction path is favored thermally, then it will be disfavored photochemically, and vice versa. One thermally favorable reaction of great synthetic utility is the Diels-Alder reaction, in which an ethylene derivative and a butadiene derivative unite to produce a substituted cyclohexene.

6 Ionic or free radical additions to conjugated dienes can give both 1,2- and 1,4-addition products, by way of an allylic intermediate. It is often found that 1,2-addition is faster, even though the 1,4-addition product is the more stable.

7 Isopropene is a fundamental building block of nature and its oligomers are very widely distributed in both plants and animals. Natural rubber is a stereoregular *cis*-polymer of isoprene. The open-chain precursor of steroids in plants and animals, squalene, is a symmetric molecule formed by the head-to-head linkage of two farnesyl units. Farnesyl pyrophosphate, in turn, is constructed by the condensation of three isopentenyl pyrophosphate molecules. The condensation appears to involve the nucleophilic displacement of allylic pyrophosphate ions and concurrent allylic double bond rearrangements in the nucleophilic molecule.

PROBLEMS

Note. Problems 1 through 5 appear in text.

6 Pyrolysis of butane involves its initial dissociation to either two ethyl radicals or a propyl radical plus a methyl radical. What would you expect the predominant dissociation process(es) to be when 1-butene is pyrolyzed?

7 Do you consider cyclopropyl radical and allyl radical to be distinct species or merely canonical forms of the same structure? Defend your answer; reference to the rules for resonance structures may help.

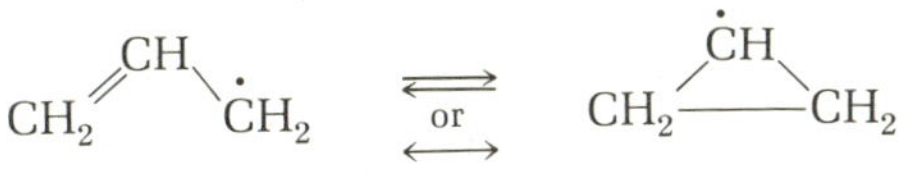

8 Addition of hydrogen chloride to vinyl chloride yields 1,1-dichloroethane as the predominant product. Account for this result with the aid of resonance structures.

9 Provide an explanation for each of the following observations, using appropriate canonical structures to illustrate your argument.

a. When 1,4-pentadiene is exposed to strong bases, it isomerizes to 1,3-pentadiene.

b. Boron forms unusually strong single bonds with nitrogen and oxygen.

c. Despite the electronegativity of fluorine, compounds containing the CF_3— group are characterized mass spectrometrically by the appearance of intense peaks corresponding to CF_3^+.

10 Write out reaction sequences that would allow one to bring about the conversions shown below. Include all important reagents and conditions.

a. Cyclohexanol ⟶ 3-chlorocyclohexene

b. Acetone ⟶ 3-chloro-2-methylpropene

c. Methyl isopropyl ketone ⟶ 1-bromo-2,3-dimethyl-2-butene

d. Pinacolone (methyl *t*-butyl ketone) ⟶ 1-chloro-2,3-dimethyl-2-butene

e. Cyclopentylmethyl bromide ⟶ 1-bromomethylcyclopentene

f. 1-Iodo-2-methylpentane ⟶ 2-methyl-2-pentene

11 *N*-Bromosuccinimide reacts only very slowly with 3,3-dimethyl-1-butene. How do you account for this observation?

12 What products would be anticipated from the reaction of 1-methylcyclopentene with *t*-butyl hypochlorite?

13 *cis*-3-Hexene was subjected to allylic bromination with *N*-bromosuccinimide in carbon tetrachloride. The reaction was interrputed before completion and the unbrominated hexene recovered. It was found that a portion of the starting material had isomerized to *trans*-3-hexene. Explain this finding.

14 a. 2-Methyl-2-butene-1-ol and 3-methyl-2-butene-1-ol are each converted to chlorides under ionizing conditions (concentrated hydrochloric acid containing zinc chloride). Which one should react more rapidly? Explain.

b. 3-Chloro-1-butene and 3-chloro-2-methylpropene are each reacted with sodium iodide in acetone (S_N2 conditions). Which compound would be expected to be converted most rapidly to iodide? Explain.

15 *cis*-2-Buten-1-ol (*cis*-crotyl alcohol) was warmed with dilute sulfuric acid. Gas chromatographic analysis showed the appearance of two isomeric alcohols, **A** and **B.** The nmr spectrum of **A** contained two peaks in the aliphatic proton region, a doublet (3 H) and a quartet (1 H) that showed signs of further splitting. There was a close resemblance between the nmr spectrum of **B** and that of *cis*-crotyl alcohol,

although there were clearly differences in several of the coupling constants. Write out structures for **A** and **B** and explain how they were formed.

16 A chemist sought to make a sample of *trans*-2-heptene free of isomers, which are often difficult to remove. The method he used involved a reaction between propylmagnesium bromide and pure 1-bromo-*trans*-2-butene. Is it likely that he was successful? Justify your answer.

17 How might *trans,trans*-3,5-octadiene be converted stereospecifically to *cis,trans*-3,5-octadiene?

18 The third and fourth π-orbitals of 1,3,5-hexatriene have the symmetry shown. *trans,cis,trans*-2,4,6-Octatriene (V) may be converted to 5,6-dimethylcyclohexadiene, either thermally or photochemically.

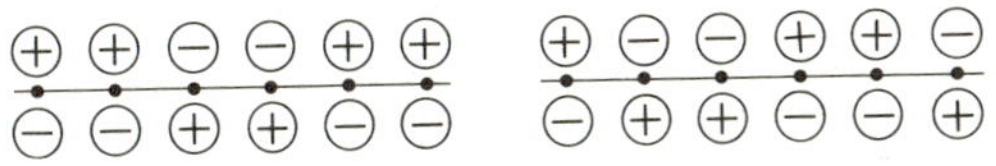

third π-orbital   fourth π-orbital

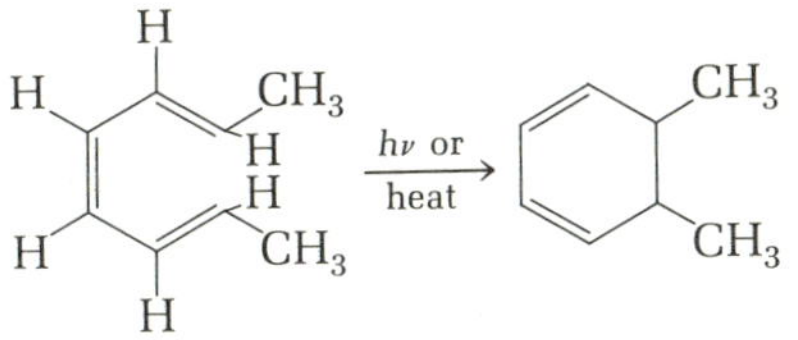

V

(4)

What will be the stereochemistry of the product in each case?

19 The "allowedness" or "forbiddenness" of competing paths in electrocyclic reactions (such as conrotatory or disrotatory ring closures) must arise from a difference in the free energies of activation of the two routes. It is not uncommon to find that at least 99.5% of the isolated product is what would be expected on the basis of Woodward-Hoffmann predictions of the allowed reaction route. What is the *minimum* free energy of activation difference required between the allowed and forbidden reactions to produce this degree of stereoselectivity at 25°? It has been estimated that the difference in *enthalpy* of activation between allowed and forbidden paths amounts to 10 to 15 kcal/mole.

20 The following cyclohexane derivatives may be synthesized by means of Diels-Alder reactions, followed by reduction of the double bond in the cyclohexene ring. Write out all the possible Diels-Alder reactions that may lead to each product and comment on their relative merits.

a. Cyclohexanecarboxaldehyde.

b. 3,4-Dimethylcyclohexanecarboxaldehyde.

c. *trans*-2-Cyanomethylcyclohexane.

21 How would one free a sample of 3-methylenecyclohexene from an impurity of isomeric 1-methylcyclohexa-1,3-diene?

22 1,2-Dimethylenecyclobutane (VI) reacts with one molecule of maleic anhydride at room temperature to yield a product, **A.** When **A** is heated to 150°, it isomerizes to **B,** which is capable of reacting with a second molecule of maleic anhydride to give **C.** Write out structures for **A, B,** and **C.**

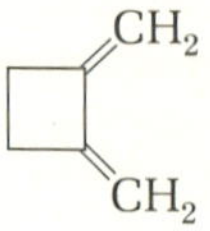

VI

23 The enthalpy difference between *s-cis-* and *s-trans*-butadiene amounts to 2.4 kcal/mole.

a. Calculate the equilibrium percentage of each conformer at 25°, assuming no significant entropy difference between them.

b. *trans*-1,3-Pentadiene reacts 3 times more rapidly than butadiene with maleic anhydride at 30°. Cyclopentadiene, however, reacts 1400 times more rapidly than butadiene under the same circumstances. To what factors may the high reactivity of cyclopentadiene by ascribed?

c. While *trans*-1,3-pentadiene is slightly more reactive than butadiene, *cis*-1,3-pentadiene is considerably *less* so. Explain this difference.

24 Simple alkenes may react with those compounds that function as strong dienophiles in the Diels-Alder reaction; equation 5 provides an example of this so-called "ene" reaction.

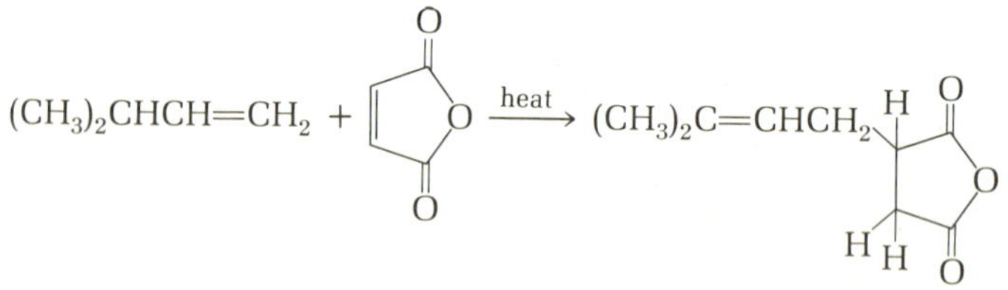

(5)

a. Propose two or three plausible mechanisms for this reaction.

b. It has been found that an optically active alkene can give rise to an optically active product (equation 6). Which of the mechanisms, if any, now seem most plausible in the light of this finding?

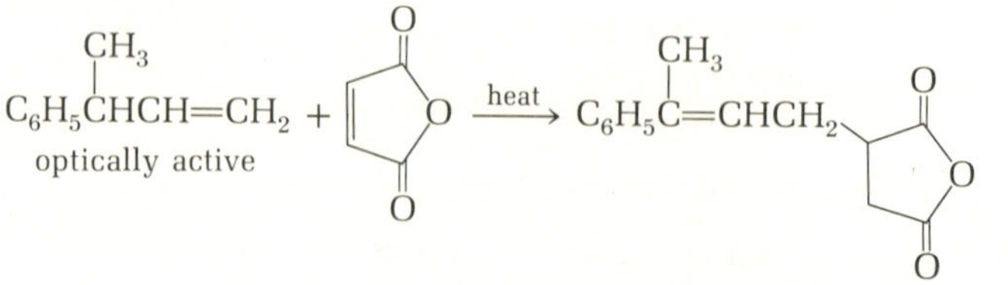

(6)

25 a. The fact that monobromination of butadiene leads to much more 1,2-addition than 1,4-addition has been explained on the basis of asymmetric charge distribution in the 4-bromobutenyl cation **12** (p. 307). It was presumed that more of the positive charge resides on C-2 than on C-4. Is this reasonable?

b. Monobromination of isoprene (2-methylbutadiene) under nonequilibrium conditions may lead to a number of products. What are they? Which will be the major and which will be the minor products?

26 Bromination of cyclopentadiene results in the formation of substantial quantities of *cis*-3,5-dibromocyclopentene. How does this result accord with the results from the bromination of butadiene? Is the same mechanism operative in the bromination of both compounds?

27 Equations 7 and 8 describe the conversion of butyraldehyde and crotonaldehyde (2-butenal) to their respective methylacetals.

$$CH_3CH_2CH_2CHO + 2\ CH_3OH \longrightarrow CH_3CH_2CH_2CH(OCH_3)_2 + H_2O \qquad \Delta H = -8.8 \text{ kcal/mole} \tag{7}$$

$$CH_3CH{=}CHCHO + 2\ CH_3OH \longrightarrow CH_3CH{=}CHCH(OCH_3)_2 + H_2O \qquad \Delta H = -3.7 \text{ kcal/mole} \tag{8}$$

What is the cause of the energy difference between these two related processes?

SUPPLEMENTARY READINGS

P. W. Allen, D. Barnard, and B. Saville, "A New Look at Natural Rubber," *Chem. Brit.*, **6,** 382 (1970).

M. Orchin and H. H. Jaffé. *The Importance of Antibonding Orbitals*. Boston, Mass.: Houghton Mifflin Co., 1967, Chapter 5.

J. S. Swenton, "Photochemistry of Organic Compounds. I. Selected Aspects of Olefin Photochemistry," *J. Chem. Ed.*, **46,** 7 (1969).

J. J. Vollmer and K. L. Servis, "Woodward-Hoffmann Rules: Electrocyclic Reactions," *J. Chem. Ed.*, **45,** 214 (1968).

CHAPTER 13

ALKYNES—BIFUNCTIONAL SYNTHETIC INTERMEDIATES

All the diene hydrocarbons discussed in Chapter 12 adhere to the general empirical formula C_nH_{2n-2}. There is one other class of compounds to which this formula applies, the **alkynes,** also known as acetylenes after the first member of the class. They each contain a carbon-carbon triple bond, composed of one σ-bond and two π-bonds.

In the IUPAC naming system, the suffix -yne is appended to a root denoting the number of carbon atoms, while a number denotes the position of the triple bond. The older system treated acyclic alkynes as derivatives of acetylene. Naming alkynes by either method is quite straightforward, particularly since there is no possibility of *cis-trans* isomerism.

$HC{\equiv}C(CH_2)_5CH_3$
1-octyne
n-hexyl acetylene

$CH_3C{\equiv}CCH_2CH_3$
2-pentyne
methyl ethyl acetylene

cyclooctyne

$CH_2{=}CHC{\equiv}CCH_3$
pent-1-en-3-yne
methyl vinyl acetylene

$HC{\equiv}CCH(OH)C{\equiv}CH$
1,4-pentadiyn-3-ol

ORBITAL DESCRIPTION AND SPECTROSCOPIC PROPERTIES

In acetylene (ethyne) itself the carbon atoms are in the *sp*-hybridization state. One *sp*-orbital on each carbon atom is involved in the carbon-carbon σ-bond and the other one in the carbon-hydrogen bond. That leaves two orthogonal *p*-orbitals on each carbon atom that overlap with their opposite numbers to form two orthogonal π-orbitals. The resulting π-electron density around the carbon-carbon bond is cylindrically symmetrical. Due to the *sp*-hybridization of the carbon atoms, the entire molecule is linear.

The $—C{\equiv}C—$ stretching vibration occurs at a much higher frequency, 2100 to 2250 cm^{-1}, than do $>C{=}C<$ vibrations, as expected. In a dialkyl acetylene, such as 2-pentyne, the stretching of the triple bond causes almost no change in dipole moment and the corresponding ir band is therefore quite weak. In a conjugated alkyne, such as 4,4-dimethyl-2-butynal (Figure 13-2), this band is considerably more intense. Terminal alkynes are recognizable because of the characteristic sharp C—H band at 3300 cm^{-1}. Almost no other functional group gives a band at precisely this position. The alkyne C—H band is relatively intense.

The nmr spectra of alkynes are distinctive as well, due to the cylindrical symmetry of the π-orbitals. If the axis of the triple bond is parallel to an applied magnetic field, then the π-electrons may circulate around the carbon-carbon bond as if it were a solenoid (Figure 13-3). The induced magnetic field caused by this electronic motion opposes the applied field

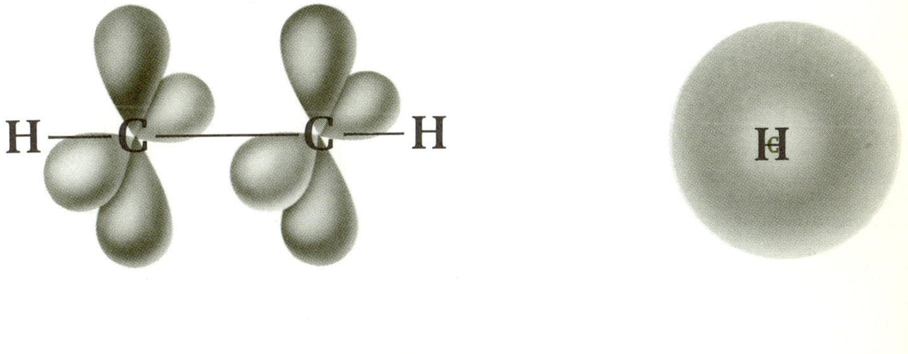

FIGURE 13-1
p-Orbitals in Acetylene. π-Electron Density in Acetylene — End View

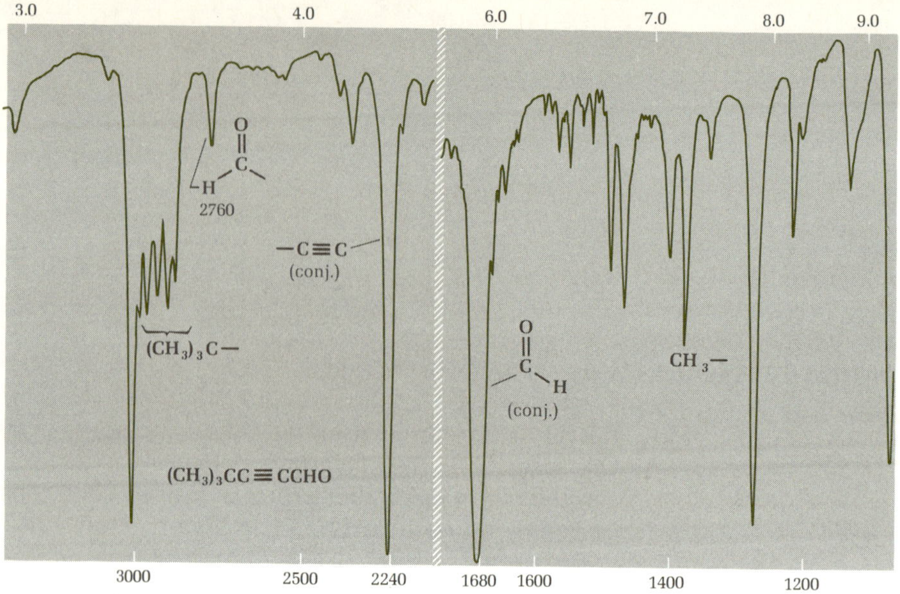

FIGURE 13-2
IR Spectrum of 4,4-Dimethyl-2-butynal

in the center of the "solenoid." Of course other molecular orientations are possible and some of these give rise to the opposite magnetic effect, but the most powerful electron circulation occurs when the triple bond is parallel to the applied field. As a result, hydrogen atoms attached to the triply bonded carbon atoms are heavily shielded; this effect even overcomes the deshielding due to the electronegative *sp*-hybridized carbon atoms. The characteristic alkyne proton resonance occurs at $\delta 2.0$ to 3.0, well upfield of the alkene proton peaks.

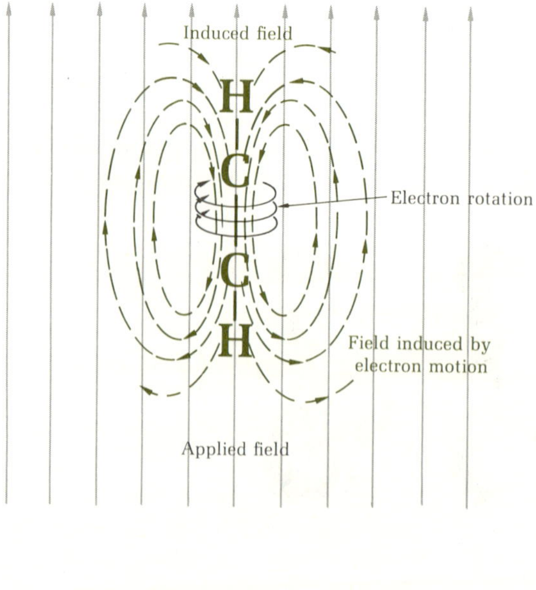

FIGURE 13-3
Motion of π-Electrons in an Acetylene Molecule Parallel to an Applied Field

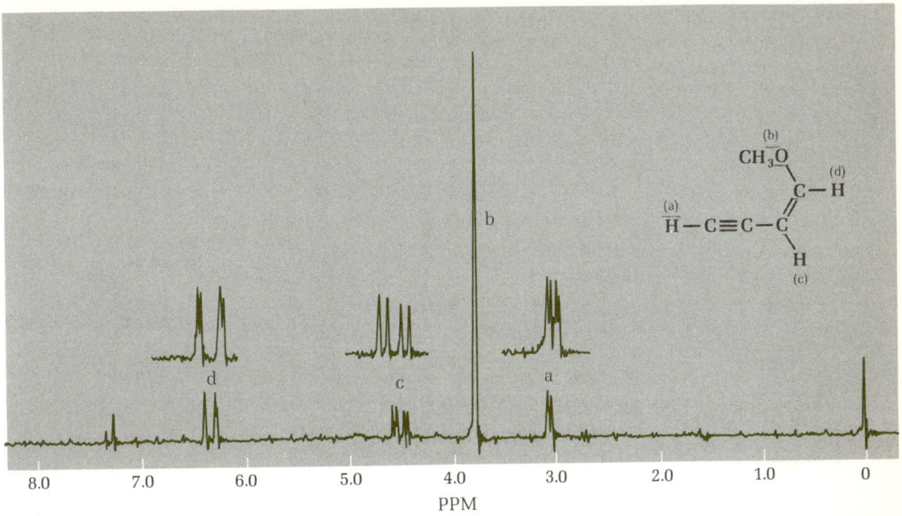

FIGURE 13-4
NMR Spectrum of 1-Methoxy-1-buten-3-yne

The nmr spectrum of *cis*-1-methoxy-1-buten-3-yne (Figure 13-4) is an instructive one. Note first of all that the signal from the alkynyl proton (*a*) is further upfield than the signals from the vinylic protons (*c* and *d*). The latter two are readily differentiated because proton *c* is relatively strongly coupled to both proton *d* and proton *a* (through the triple bond), while proton *d* is only weakly coupled to proton *a*.

REPLACEMENT OF ALKYNYL HYDROGEN

In Chapter 11, in a discussion of carbonyl frequencies, it is noted that there is a correlation between carbon-X bond strengths and the amount of *s*-character in the carbon bond orbital. *s*-Electrons are, on the average, closer to the nucleus than *p*-electrons in the same shell; as a result, they are more tightly bound by that nucleus. Of all the carbon-hydrogen bonds, ≡C—H is close to being the strongest (~125 kcal/mole). It is also the most strongly polarized in the sense

$$\equiv\overset{\delta-}{C}—\overset{\delta+}{H}$$

This strong bond polarization permits acetylene (pK_a~26) and its monosubstituted derivatives to act as weak acids, intermediate in strength between water and ammonia (pK_a~40).

(13.1) $$RC\equiv CH + NaOH \overset{H_2O}{\rightleftarrows} RC\equiv C{:}^{-}Na^{+} + H_2O$$

(13.2) $$RC\equiv CH + NaNH_2 \overset{NH_3}{\rightleftarrows} RC\equiv C{:}^{-}Na^{+} + NH_3$$

Acetylene itself may be generated by hydrolysis of its calcium salt (calcium carbide), which is manufactured commercially.

(13.3) $$CaO + 3C \xrightarrow{2000^\circ} CO + CaC_2 \xrightarrow{2H_2O} C_2H_2 + Ca(OH)_2$$

Alkynes are sufficiently acidic to react very readily and irreversibly with alkyl Grignard reagents, this being the only practical method of generating alkynyl Grignard reagents.

(13.4) $$HC{\equiv}CH + C_2H_5MgBr \xrightarrow[\text{dilute}]{\text{THF}} C_2H_6 + \underset{\text{ethynylmagnesium bromide}}{HC{\equiv}CMgBr} \xrightarrow{C_2H_5MgBr} \underset{\text{ethynyldimagnesium dibromide}}{BrMgC{\equiv}CMgBr}$$

The metal salts of the alkynes, easily generated in liquid ammonia or in ethereal solvents, are powerful nucleophiles and find considerable synthetic use. Both monosubstitution and disubstitution can be effected, depending on the starting materials.

(13.5) $$HC{\equiv}CH + Na \xrightarrow{\text{liq } NH_3} \frac{1}{2} H_2 + \underset{\text{sodium acetylide}}{HC{\equiv}C^-Na^+} \xrightarrow{CH_3(CH_2)_3Br} \underset{\text{1-hexyne (75\%)}}{HC{\equiv}C(CH_2)_3CH_3} + NaBr$$

(13.6) $$HC{\equiv}CH + 2NaNH_2 + 2CH_3CH_2Br \xrightarrow{\text{liq } NH_3} \underset{\text{3-hexyne (75\%)}}{CH_3CH_2C{\equiv}CCH_2CH_3}$$

(13.7) $$2ROCH_2C{\equiv}CH + Br(CH_2)_6Br \xrightarrow[\text{ether}]{\text{Na, liq } NH_3,} \xrightarrow{H_3O^+} \underset{\text{2,10-dodecadiyn-1,12-diol}}{HOCH_2C{\equiv}C(CH_2)_6C{\equiv}CCH_2OH}$$

R =
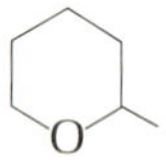

These alkyne derivatives not only displace halogens, but they can add to carbonyl groups as well. Note that only primary halides may be used in the synthesis of substituted acetylenes.

(13.8)
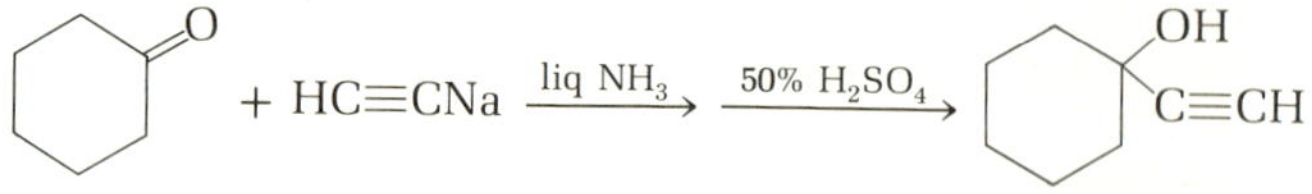

(65–75%)

(13.9) $$(CH_3)_3CC{\equiv}CH + CH_3CH_2MgBr \xrightarrow{\text{ether}}$$

$$(CH_3)_3CC{\equiv}CMgBr \xrightarrow{(CH_3)_2C{=}O} \xrightarrow{H_3O^+} (CH_3)_3CC{\equiv}C{-}\overset{\displaystyle CH_3}{\underset{\displaystyle CH_3}{C}}{-}OH$$

2,5,5-trimethyl-hex-3-yn-2-ol (81%)

The alkyne salts are strong bases as well as nucleophiles and they bring about considerable elimination with secondary and tertiary halides. Acetylenes containing secondary and tertiary alkyl groups must be synthesized by means of an elimination reaction (see below).

PROBLEM

1 2-Methyl-1-buten-3-yne is an inexpensive commercial chemical. Suggest a simple synthesis for this compound based on inexpensive starting materials.

Heavy metal alkyne derivatives are also known; they have a chemistry substantially different from that of the lighter metal salts. They are covalent compounds, ordinarily very weak bases and nucleophiles, and usually insoluble in both protic and aprotic solvents. Formation of cuprous and silver **acetylides** constitutes a useful test for the presence of a terminal alkyne.

(13.10) $$RC{\equiv}CH + Cu(NH_3)_2^+ \xrightarrow{H_2O} RC{\equiv}CCu\downarrow + NH_4^+ + NH_3$$

(13.11) $$RC{\equiv}CH + 2AgNO_3 \xrightarrow[C_2H_5OH]{95\%} RC{\equiv}CAg \cdot AgNO_3\downarrow + HNO_3$$

(13.12) $$RC{\equiv}CR' + Cu(NH_3)_2^+ \text{ or } AgNO_3 \longrightarrow \text{no reaction}$$

The alkynes may be liberated from their copper and silver salts; this provides a means of purifying them and freeing them from their nonterminal isomers.

(13.13) $$RC{\equiv}CCu + 2CN^- \xrightarrow{H_2O} RC{\equiv}CH + Cu(CN)_2^- + OH^-$$

COUPLING REACTIONS

The most important use of the copper acetylides is in **coupling reactions.** The heavy metal acetylides are usually not isolated but rather generated in the course of the reaction, during which they undergo further transformation. The coupling reaction is run by merely shaking a cuprous chloride-pyridine or cuprous chloride-ammonium chloride solution of the terminal alkyne in an oxygen atmosphere; the yields are quite good.

(13.14) $$2(CH_3)_2C(OH)C{\equiv}CH \xrightarrow[CH_3OH,\ O_2]{CuCl\text{-pyridine}} (CH_3)_2C(OH)C{\equiv}CC{\equiv}CC(OH)(CH_3)_2$$

(70–80%)

(13.15) $CH_3C{\equiv}CCu \xrightarrow[O_2,\ 25^\circ]{NH_3,\ CH_3OH} CH_3C{\equiv}CC{\equiv}CCH_3$ (74%)

Cyclic compounds may also be obtained by the coupling of terminal diynes. It is possible to get very large rings, which would be difficult to obtain by other methods, by the method of cyclic polymeric coupling.

(13.16) 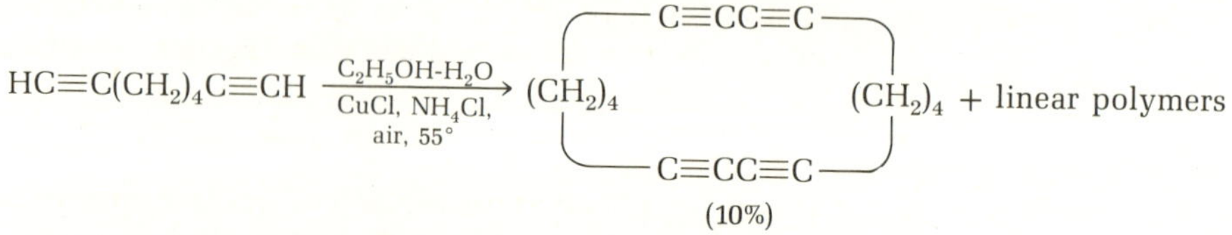

Experiments have indicated that the principal function of the oxygen is to oxidize the cuprous acetylide to a cupric derivative. What happens thereafter is still unknown; it is possible that the cupric acetylides decompose to cuprous ions and alkynyl radicals, which then dimerize.

(13.17) $RC{\equiv}CCu + O_2 \longrightarrow RC{\equiv}CCu^+$

(13.18) $RC{\equiv}CCu^+ \longrightarrow Cu^+ + RC{\equiv}C\cdot \longrightarrow RC{\equiv}CC{\equiv}CR$

In any event, it seemed natural to introduce cupric ion into the reaction mixture from the start. Heating the alkyne with a mixture of cupric acetate and pyridine leads to coupling. This method is very effective for synthesizing cyclic polymers because it minimizes formation of linear polymers.

(13.19) 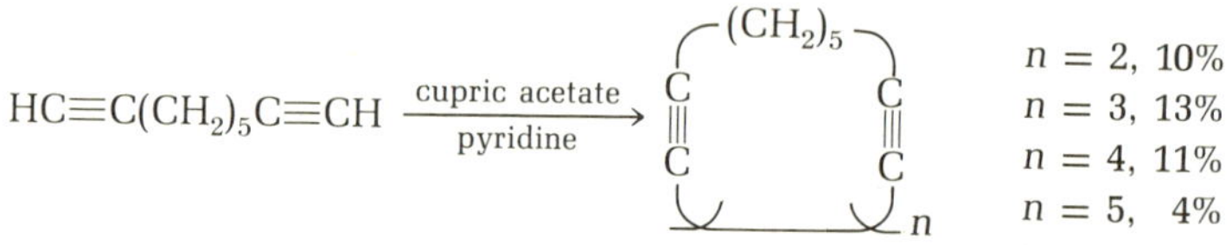

ELIMINATION REACTIONS – SYNTHESIS AND ISOMERISM

Alkynes may be synthesized by a double elimination reaction on a dihaloalkane. The most common materials are 1,2-dihalides, obtained by halogen addition to alkenes; 1,1-dihalides and even 1,3-dihalides may also be used. The 2 moles of hydrogen halide are eliminated consecutively, so that a haloalkene is an intermediate in the reaction; these are often formed first in a separate step. Elimination of the first mole of hydrogen halide is quite easy and may be effected by hydroxide or alkoxide ions at relatively low temperatures.

(13.20) 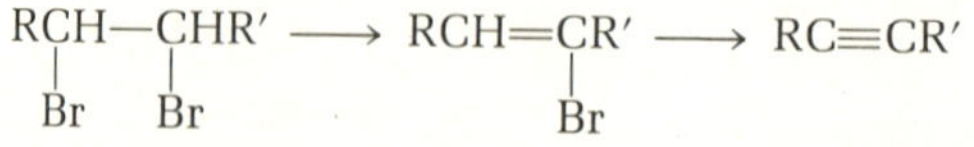

The second elimination is quite difficult; either a strong base such as sodium amide or an elevated temperature is required.

(13.21)
$$CH_3\underset{\large Cl}{\underset{|}{C}}{=}CHCH_2Cl \xrightarrow[165\text{–}170^\circ]{KOH,\ HOCH_2CH_2OH} \underset{(43\text{–}48\%)}{CH_2{=}CHC{\equiv}CH}$$

(13.22)
$$CH_3(CH_2)_7CH{=}CH_2 + Br_2 \xrightarrow{CCl_4} \underset{(95\text{–}100\%)}{CH_3(CH_2)_7\underset{\large Br}{\underset{|}{C}}H{-}CH_2Br} \xrightarrow[\text{or NaH, DMSO}]{NaNH_2,\ NH_3} \underset{(95\%)}{CH_3(CH_2)_7C{\equiv}CH}$$

(13.23)
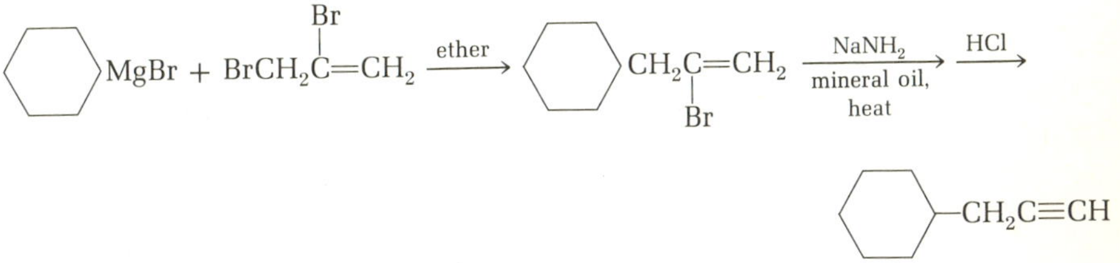

The alkynes obtained by these processes are sometimes isomers of the expected product. This is due to the acidity of the hydrogens *adjacent* to the triple bond. Abstraction of one of these hydrogens leads to a resonance-stabilized anion, not unlike an allyl anion; this anion may be protonated at the other end to yield an allene.

(13.24)
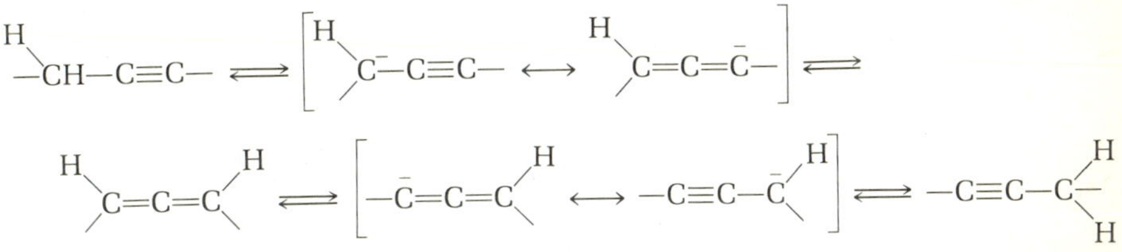

A second deprotonation-reprotonation step completes the triple bond migration.

PROBLEM

2 When allene is allowed to react with lithium amide in liquid ammonia, and the reaction mixture worked up by addition of D_2O, propyne-1-*d* is obtained. Show in detail how this product was formed.

In open-chain systems, internal alkynes are more stable than allenes or terminal alkynes.

(13.25) 1-Alkyne $\longrightarrow$ 2-alkyne $\Delta H = -4.8$ kcal/mole

Isomerizations carried out in the presence of sodium amide at high temperatures often result in the formation of high yields of the terminal alkynes, however. This result is apparently caused by the selective precipitation of the sodium salt of the terminal alkyne.

CYCLIC ACETYLENES

Cycloalkynes are capable of existence, just as cycloalkenes are; in general, cycloalkynes show the same chemical properties as their acyclic analogues. Due to the hybridization of the triply bonded carbon atoms, however, the atoms C—C≡C—C must lie in a straight line; this imposes restrictions on the ring size. Cyclooctyne is the smallest member of this series that is sufficiently unstrained to allow its isolation at 25°. Smaller homologues may be generated, but they undergo instantaneous reactions with other compounds or with themselves, unless the temperature is quite low.

The mixture of products obtained when a cyclic haloalkene is dehydrohalogenated is very different from that obtained from reactions of acyclic compounds (equation 13.26).

(13.26)
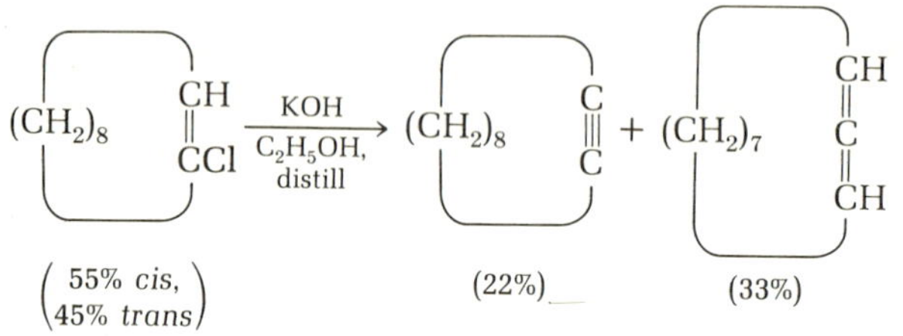

The cyclic allene is now favored by the fact that it is less strained since only the atoms CH=C=CH must be colinear. Thermodynamic measurements indicate that this effect diminishes in importance with increasing ring size, as expected.

(13.27)
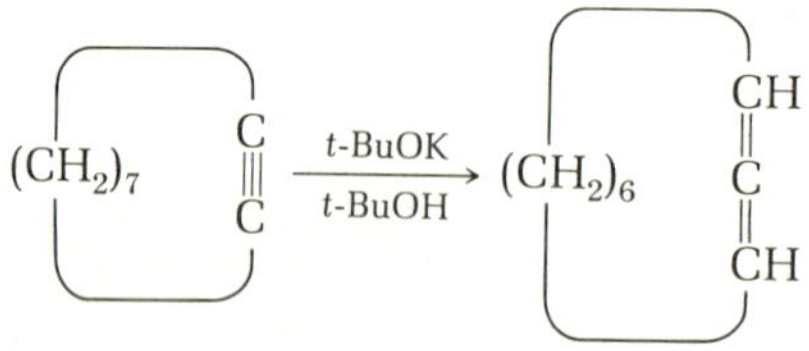

$\Delta H = -2.5$ kcal/mole

(13.28)
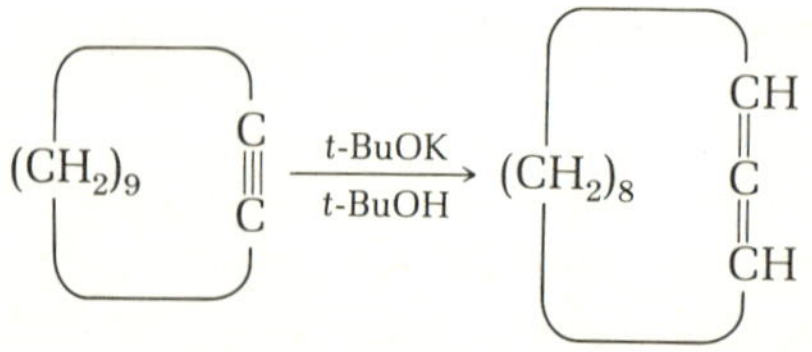

$\Delta H = +1.3$ kcal/mole

ADDITIONS TO THE TRIPLE BOND

The triple bond in acetylenes is susceptible to attack by the same electrophilic reagents that add to the double bond of olefins. The alkynes react much more slowly, however, than the corresponding alkenes. It is not obvious why this should be so. A possible explanation depends on the fact that the carbon-carbon triple bond is appreciably shorter than the carbon-carbon double bond; the π-electrons may be closer, on the average, to the carbon nuclei and less polarizable. The chemical consequences are illustrated by equation 13.29.

$$\text{HC}\equiv\text{CCH}_2\text{CH}=\text{CH}_2 \xrightarrow[-20^\circ]{\text{Br}_2,\ \text{CCl}_4} \text{HC}\equiv\text{CCH}_2\underset{\underset{\text{Br}}{|}}{\text{C}}\text{HCH}_2\text{Br} \quad (90\%) \tag{13.29}$$

Electrophilic additions to alkynes may be *envisaged* as proceeding by way of vinylic cations; one would expect to see, and in fact one does observe, the Markovnikov rule in operation.

$$-\text{C}\equiv\text{C}- \xrightarrow{\text{X}^+} -\overset{+}{\text{C}}=\text{C}\begin{smallmatrix}\diagup\text{X}\\ \diagdown\end{smallmatrix} \xrightarrow{\text{Y}^-} -\text{CY}=\text{CX}- \tag{13.30}$$

The mechanism of addition of halogens to the triple bond seems to vary with the substrate and with the conditions. Under different circumstances, addition may require vinylic cations, bromonium ions, a concerted *anti* attack (see below), or combinations of these. Simple alkynes (e.g., 3-hexyne) give almost exclusively the products of *anti* addition, either by way of bromonium ions or as a result of a concerted process.

The addition of 1 mole of a halogen, a halogen acid, an alcohol, etc., leads to a vinyl derivative that may react further with the reagent. In some cases, it is comparatively easy to isolate the monoaddition product; in other cases, it is impractical to do so. Hydrogen halides and halogens may be added essentially 1 mole at a time, since the product of the first addition is less reactive than the starting material.

$$\text{CH}_3(\text{CH}_2)_3\text{C}\equiv\text{CCH}_2\text{OH} \xrightarrow[0^\circ]{\text{Br}_2,\ \text{CCl}_4} \text{CH}_3(\text{CH}_2)_3\text{CBr}=\underset{\underset{\text{Br}}{|}}{\text{C}}\text{CH}_2\text{OH} \quad (80\%) \tag{13.31}$$

$$\text{CH}_3(\text{CH}_2)_3\text{C}\equiv\text{CH} + \text{HBr} \xrightarrow[\text{inhibitors, } 15^\circ]{\text{radical}} \text{CH}_3(\text{CH}_2)_3\text{CBr}=\text{CH}_2 \xrightarrow{\text{HBr}} \text{CH}_3(\text{CH}_2)_3\overset{\overset{\text{Br}}{|}}{\underset{\underset{\text{Br}}{|}}{\text{C}}}\text{CH}_3 \tag{13.32}$$

The stereochemistry of addition seems rather dependent on the compound, the solvent, and in particular the presence of additional quantities of the nucleophilic part of XY. The addition can be sterospecifically *anti* if the last condition is met; this suggests that the addition is a concerted process in which the electrophile and nucleophile bond simultaneously to the alkyne.

(13.33) 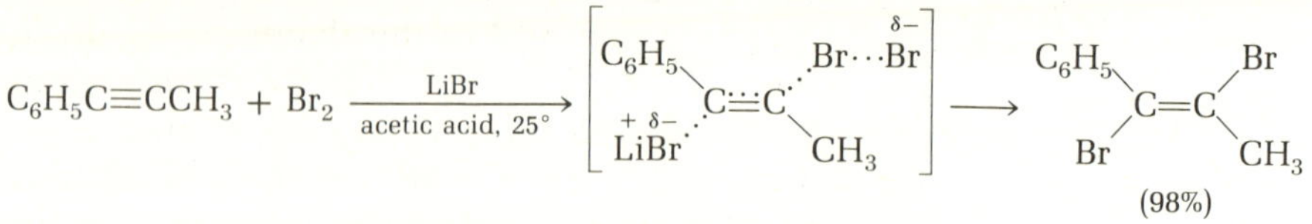

(13.34) 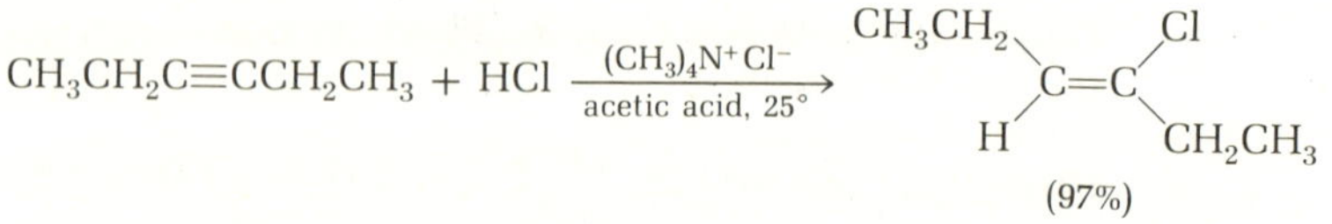

True vinylic cations seem to be ruled out in these cases; they probably intervene in some acid-catalyzed additions to alkynes containing substituents capable of conjugating with the carbonium ion.

Acid-catalyzed additions of alcohols to acetylenes are difficult to arrest at the monoaddition stage because the intermediate vinyl ether is very susceptible to further electrophilic attack.

(13.35) 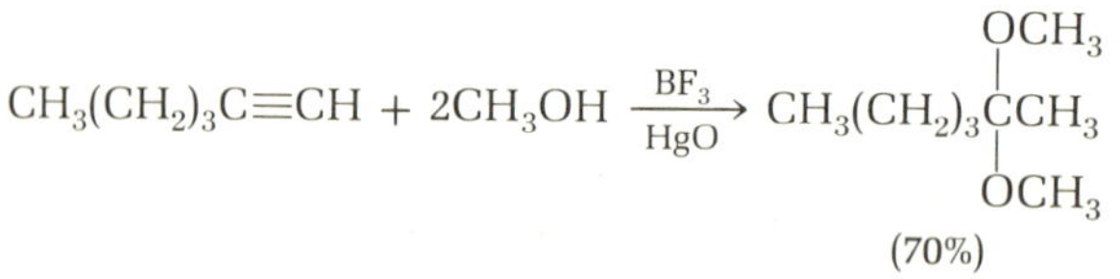

Ketals are invariably the major product. Water may also be added, Markovnikov fashion, to the triple bond of alkynes. The intermediate vinyl alcohol isomerizes completely to the isomeric ketone. The vinyl alcohol-ketone equilibrium is explored in Chapter 20; suffice it to say at this point that the principal energy factor influencing the equilibrium is the difference in carbon-oxygen and carbon-carbon double bond strengths.

(13.36) 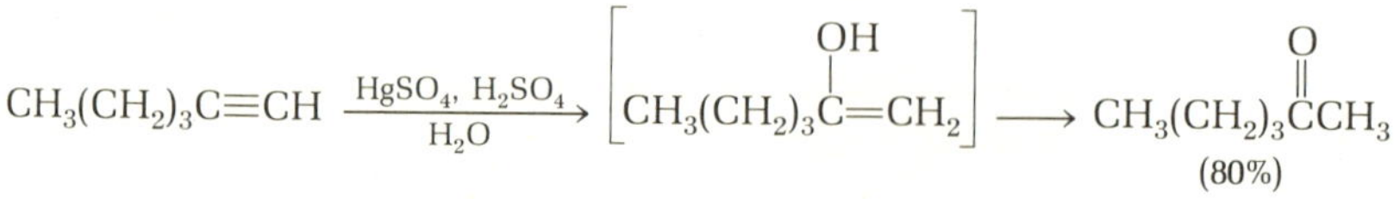

The effectiveness of mercuric salts, such as the sulfate and acetate, as catalysts for the hydration is probably due to complex formation between the cation and the alkyne. This would serve to bring the alkyne into solution and also render it more susceptible to nucleophiles. An organomercury compound is a likely intermediate in these reactions since these compounds are known to undergo acid-catalyzed hydrolysis. The reaction shown in equation 13.37 is used commercially for the production of acetaldehyde.

Like carbon-carbon double bonds, triple bonds are also attacked by radicals. Hydrogen bromide addition is subject to a peroxide effect, leading to anti-Markovnikov addition.

(13.37) 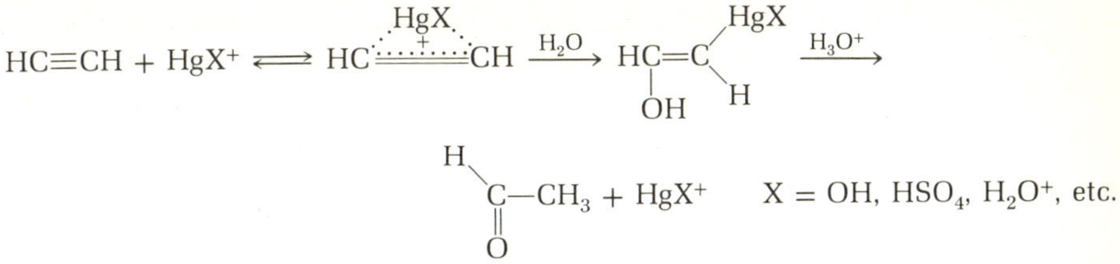

These radical additions are much faster than the corresponding ionic additions, confirming the unreactivity of the triple bond toward electrophilic attack.

(13.38) $$CH_3(CH_2)_3C{\equiv}CH + HBr \xrightarrow{\text{peroxides}} \underset{(74\%)}{CH_3(CH_2)_3CH{=}CHBr}$$

REDUCTION

Alkynes may be reduced to either alkenes or alkanes, or a mixture of the two. For synthetic purposes it is important that the degree of reduction be capable of control, as well as the stereochemistry of reduction. Although the addition of the first mole of hydrogen is more exothermic than the addition of the second, most catalysts are sufficiently active that reduction of the intermediate alkene is nearly as fast as that of starting alkyne.

(13.39) $$CH_3(CH_2)_3C{\equiv}CH + 2H_2 \xrightarrow{25^\circ} CH_3(CH_2)_4CH_3 \qquad \Delta H_{gas} = -69.5 \text{ kcal/mole}$$

(13.40) $$CH_3(CH_2)_3CH{=}CH_2 + H_2 \xrightarrow{25^\circ} CH_3(CH_2)_4CH_3 \qquad \Delta H_{gas} = -30.0 \text{ kcal/mole}$$

Thus, after addition of 1 mole of hydrogen the product consists of a mixture of alkyne, alkene, and alkane.

Raney nickel is sufficiently discriminatory that it is possible to obtain mostly the alkene; the hydrogenation is stereospecifically *cis*. The catalyst most preferred for partial reduction of alkynes is palladium-on-calcium carbonate that has been "poisoned" with lead acetate and quinoline (an organic base) (*Lindlar catalyst*). The purpose of the poisons is to reduce the activity of the catalyst and thereby make it more selective.

(13.41) 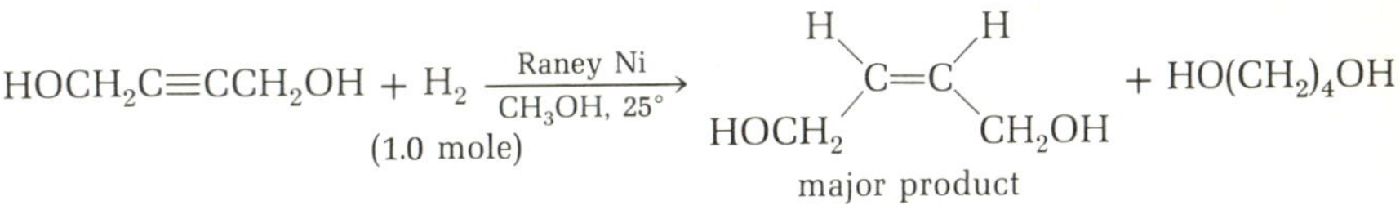

(13.42) 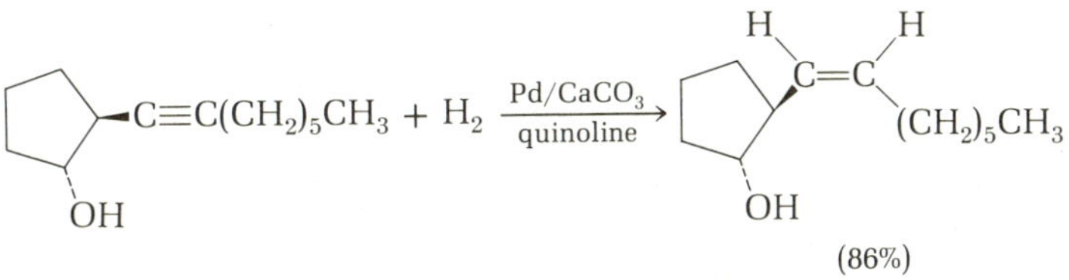

These partial reductions are very useful, for they provide a convenient route to a number of thermodynamically disfavored *cis*-alkenes.

Internal alkynes may be transformed to *trans*-alkenes as desired if reduction is accomplished by chemical means. Sodium metal in liquid ammonia is one reducing agent for this purpose; alkenes are unaffected by it. The same result may be brought about by lithium aluminum hydride in an ethereal solvent. Both lithium aluminum hydride and sodium possess the undesirable feature that they attack many other functional groups, which must be protected if possible or entirely absent.

(13.43)

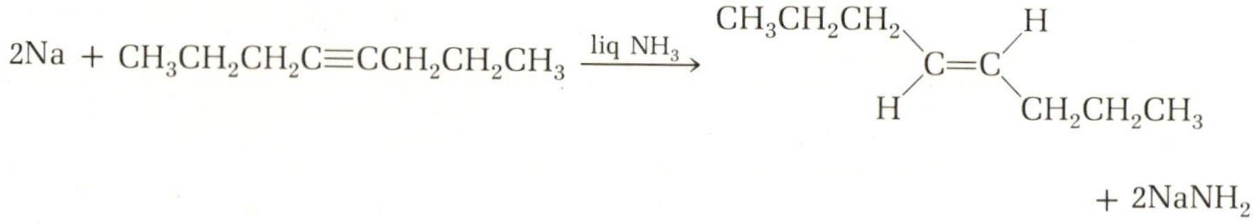

(13.44)

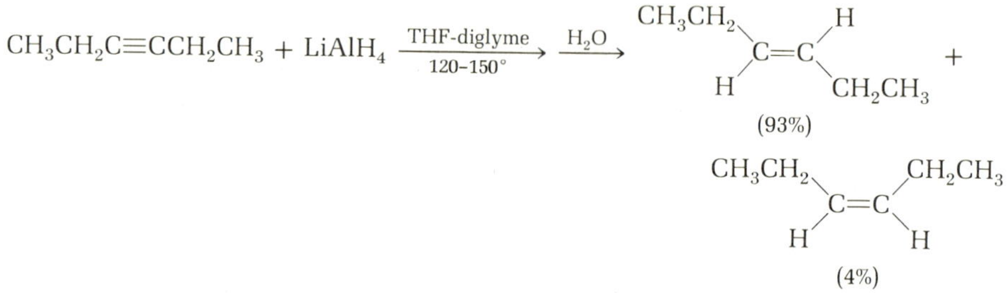

DIELS-ALDER REACTIONS

The similarities between alkenes and alkynes carries over to the Diels-Alder reaction, in which both types of compound may participate as dienophiles. As in the case with alkenes, the presence of electron-withdrawing substituents attached to the π-bond enhances the dienophilic activity of an alkyne. Acetylenic dienophiles are used frequently in the synthesis of aromatic compounds (Chapter 14).

(13.45)

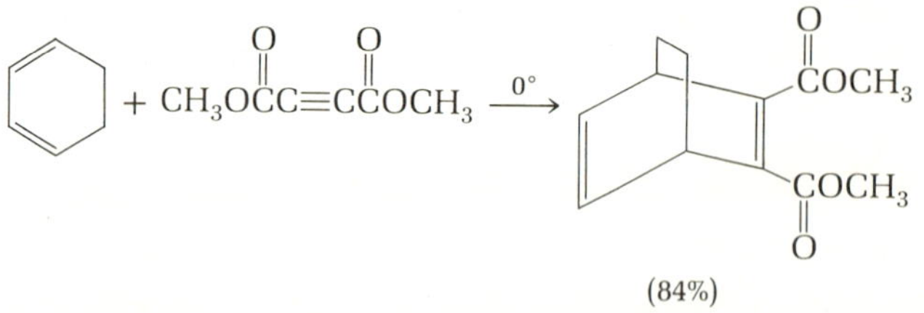

(13.46)

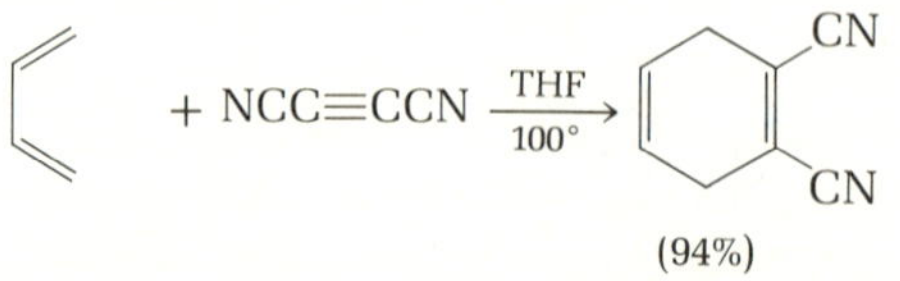

SCISSION OF THE TRIPLE BOND

Oxidative fission of the carbon-carbon triple bond may be brought about by the same reagents that attack the double bond. The most commonly used are ozone and potassium permanganate; the products are acids.

(13.47) $$CH_3(CH_2)_7C{\equiv}C(CH_2)_7CO_2H \xrightarrow[CCl_4]{O_3} CH_3(CH_2)_7CO_2H + HO_2C(CH_2)_7CO_2H$$

Since the acids are usually easy to identify, it is possible to calculate the structure of the original alkyne.

(13.48) $$CH_3CH_2C{\equiv}C(CH_2)_7CH_3 \xrightarrow[\text{pH 12, 25°}]{KMnO_4} CH_3CH_2CO_2H + HO_2C(CH_2)_7CH_3$$

There is much evidence that diketones are intermediates in both types of oxidation; by adjusting the pH, it is possible to stop the oxidation at the diketone stage.

(13.49) $$CH_3(CH_2)_7C{\equiv}C(CH_2)_7CO_2H \xrightarrow[\text{pH 7.0–7.5}]{\text{dil. }KMnO_4} CH_3(CH_2)_7\overset{O}{\overset{\|}{C}}\overset{O}{\overset{\|}{C}}(CH_2)_7CO_2H$$

This procedure is occasionally of synthetic value.

SYNTHETIC UTILITY OF ACETYLENES

Acetylene and the terminal alkynes are very useful as synthetic intermediates because they are *bifunctional*. One or two molecules may be added onto them by replacement of the acidic hydrogen(s) and the resulting compound subjected to further transformations by reagents that attack the triple bond. The coupling reaction, for instance, is obviously a good tool for building up all sorts of carbon chains and rings. Some other examples are given below.

(13.50)

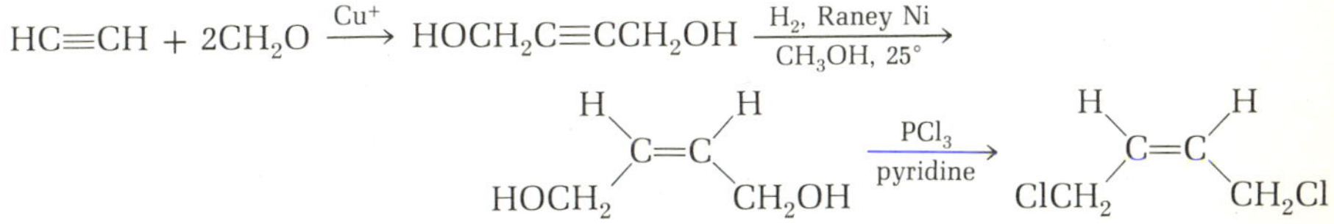

Note that the final product in equation 13.50, *cis*-1,4,-dichloro-2-butene, *cannot* be obtained by addition of chlorine to butadiene; only the 1,2- and *trans*-1,4-isomers are obtained by that method.

(13.51)

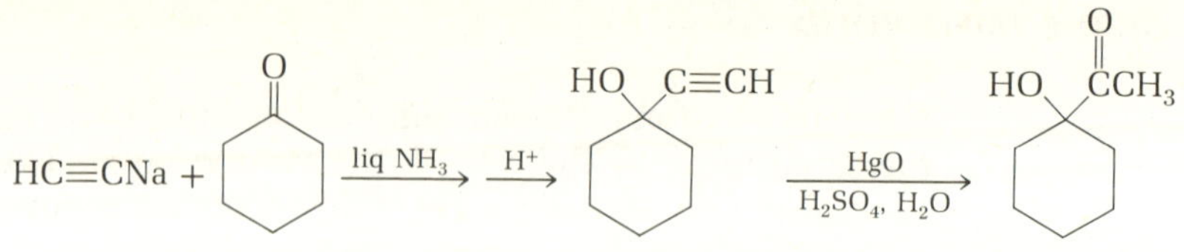

1

Compound **1** cannot be made by reaction of cyclohexanone with $\mathrm{CH_3\overset{\overset{\displaystyle O}{\|}}{C}MgBr}$ because the proposed Grignard reagent cannot be made. It is impossible to make a Grignard derivative of a halide that also has a carbonyl group in it since it would react with another molecule immediately upon formation. Thus, the $HC{\equiv}C$— group may be used as CH_3CO— group "in disguise," in cases where it is not feasible to have the ketone present during reaction. Reactions such as equation 13.51 are of great importance in the steroid field, where they are used in the synthesis of a wide variety of compounds including several with contraceptive activity.

Equation 13.52 shows how double bonds may be placed at given positions in a long chain, with no danger of isomerization. Most syntheses based on elimination reactions would give a mixture of positional isomers, complicated further by *cis-trans* isomerism.

(13.52)

$$CH_2{=}CHCH_2CH_2CH{=}CH_2 \xrightarrow[CCl_4]{2Br_2} \xrightarrow[NH_3]{4NaNH_2} HC{\equiv}CCH_2CH_2C{\equiv}CH \xrightarrow[NaNH_2]{C_2H_5I}$$

$$CH_3CH_2C{\equiv}CCH_2CH_2C{\equiv}CH \xrightarrow[THF]{C_2H_5MgBr} \xrightarrow{CH_2{=}O}$$

$$CH_3CH_2C{\equiv}CCH_2CH_2C{\equiv}CCH_2OH \xrightarrow[Pd/CaCO_3]{H_2}$$

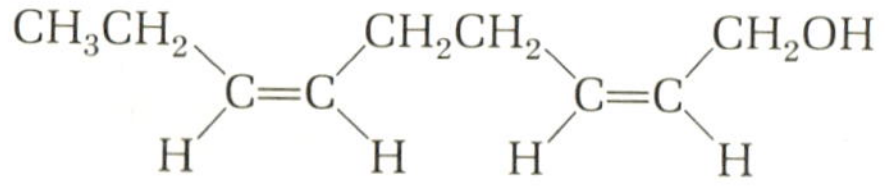

SUMMARY

1 The triple bond in alkynes consists of one σ- and two orthogonal π-bonds. The π-electron distribution is cylindrically symmetrical about the carbon-carbon σ-bond; the shielding of protons adjacent to the triple bond is understandable in terms of the movement of electrons in the π "solenoid." Hydrogen atoms attached directly to *sp*-hybridized carbon atoms stand out in both ir and nmr spectra, which makes it possible to distinguish terminal from internal alkynes.

2 Due to the relatively high electronegativity of the *sp*-orbitals on carbon, terminal alkynes are comparatively quite acidic hydrocarbons. They may

be converted to metallic derivatives by bases; the most important are the sodium, halomagnesium, and copper derivatives. The sodium salts enter into S_N2 reactions with primary alkyl halides to yield alkynes of increased chain length. They also add to carbonyl groups, which is the principal use made of alkynyl Grignard reagents.

The cuprous acetylides may be formed in aqueous solution in which they are insoluble. This constitutes a convenient chemical test for terminal alkynes. If oxygen is bubbled through a suspension of a cuprous acetylide, a coupling reaction takes place producing a symmetrical linear or cyclic diyne. These may in turn be converted to many other compounds.

3 Alkynes may be synthesized by double elimination from 1,2-dihalides. Elimination of the first mole of hydrogen halide requires only relatively mild conditions, while more stringent conditions are necessary to bring about loss of the second mole. The final product may undergo base-catalyzed isomerization by way of delocalized anions. Allenes are intermediates in these isomerizations. The allenes are usually thermodynamically disfavored with respect to the corresponding alkynes, except in small cyclic compounds where ring strain effects are very important. Internal alkynes also predominate over terminal alkynes in these isomerizations, except in the presence of very strong bases that convert terminal alkynes to their salts.

4 The triple bond adds most of those electrophilic reagents that also attack the double bond, albeit at a much lower rate. Markovnikov addition prevails. It is usually possible to stop the reaction after only 1 mole of the electrophile has been added. The vinyl alcohol that is the initial product of addition of water to alkynes isomerizes completely to the isomeric ketones. Mercuric ions catalyze the latter reaction.

Reduction of the triple bond may be halted after the alkene stage is reached. The use of special catalysts and hydrogen results in the formation of *cis*-alkenes almost exclusively; chemical reduction yields the *trans* isomers.

5 Terminal alkynes are useful synthetic intermediates because they contain two sites that are avilable for further transformation.

PROBLEMS

Note: Problems 1 and 2 appear in text.

3 Indicate whether you expect the following equilibria to have constants much less than, much greater than, or substantially equal to unity.

a. $CH_3C{\equiv}CH + CH_3Li \rightleftharpoons CH_3C{\equiv}CLi + CH_4$

b. $CH_2{=}CH_2 + CH_3CH_2Na \rightleftharpoons CH_2{=}CHNa + CH_3CH_3$

c. $CH_3Na + CH_3CH_3 \rightleftharpoons CH_4 + CH_3CH_2Na$

d. $CH_3CH_2CH_3 + CH_2{=}CHCH_2Li \rightleftharpoons CH_3CH_2CH_2Li + CH_2{=}CHCH_3$

e. $CH_4 + NaNH_2 \rightleftharpoons CH_3Na + NH_3$

f. $CH_3C{\equiv}CBr + CH_3CH_2MgBr \rightleftharpoons CH_3C{\equiv}CMgBr + CH_3CH_2Br$

4 Show how acetylene might be converted to each of the following compounds. Indicate all reagents and conditions.

a. 1-Hexyne
b. 3-Decyne
c. 2,7-Dimethyl-4-octyne
d. 1,6-Heptadiyne
e. 4,6-Decadiyne
f. 1,3-Cyclododecadiyne
g. Dodeca-3,9-diyne-2,11-diol
h. Trichloroethylene
i. 4-Bromo-*trans*-4-octene
j. 3-Hexanone
k. 2-Methyl-7-ethyl-*cis*-4-nonene
l. *cis*-1,6-Octadiene
m. *trans*-9-Methyl-5-tridecene

5 Compound **A** exhibited a strong band in its ir spectrum at 3320 cm^{-1}, a weak band at 2200 cm^{-1}, and no bands between 1500 and 2200 cm^{-1}. What would you expect to happen when **A** is exposed to the following reagents.

a. Br_2 in CCl_4
b. $Ag(NH_3)_2^+$ in C_2H_5OH
c. Aqueous $KMnO_4$
d. NH_2OH in C_2H_5OH-H_2O
e. Concentrated H_2SO_4

Carry out the same analysis for compound **B**, the ir spectrum of which contains a weak band at 2180 cm^{-1}, a strong band at 1720 cm^{-1}, and no absorption bands above 3000 cm^{-1}.

Compound **C** gives rise to a spectrum that contains a strong band at 1740 cm^{-1}, no bands between 1740 and 2800 cm^{-1}, and none between 1500 and 1740 cm^{-1}. Predict its reactions with the reagents listed above.

6 When 3-pentanone reacts with phosphorus pentachloride (PCl_5), it is converted in part to 3,3-dichloropentane. This compound is treated with a large excess of sodium amide and the reaction mixture then subjected to acid work-up and distillation. A liquid is obtained that gives the correct analysis for C_5H_8. Examination of this liquid by glc shows, however, that it contains a number of components.

a. Draw the structures of all compounds that could reasonably be expected to be present in the mixture.

b. Suppose that all the components could be separated by glc. What chemical or physical data could you acquire that would aid in assigning the correct structure to each component?

7 The elimination of 1 mole of hydrogen halide from an open-chain 1,2-dihalide (equation 13.20) seems to proceed almost solely to give the haloalkene rather than the isomeric allylic halide. Expalin this observation.

8 When 1,2-dibromodecane is allowed to react with sodium hydroxide in boiling ethanol, three products are generated — *cis*- and *trans*-1-bromo-1-decene and 2-bromo-1-decene. They are separated and each converted to 1-decyne by sodium amide in liquid ammonia; it is found that *trans*-1-bromo-1-decene reacts distinctly more slowly than the other two isomers. What inferences might one draw about the stereochemistry of elimination reactions with haloalkenes?

9 3-Bromo-1-hexene was allowed to react with sodium acetylide in liquid ammonia. The hydrocarbon product isolated showed no characteristic acteylene bands in its ir spectrum and it was found by mass spectrometry to have a molecular weight of 82. Suggest a structure for this product and account for the failure of the anticipated reaction to take place.

10 a. 1-Methoxy-2-pentyne was treated with 2 moles of sodium amide in liquid ammonia. After acidic work-up, two products were isolated: methanol and 3-penten-1-yne. Show how these compounds were produced.

b. If only 1 mole of sodium amide is used, then one-half of the starting material is left unreacted. Explain why 2 moles of base are required.

11 When 1-ethynylcyclohexanol is treated with acid, it rearranges in part to cyclohexylideneacetaldehyde.

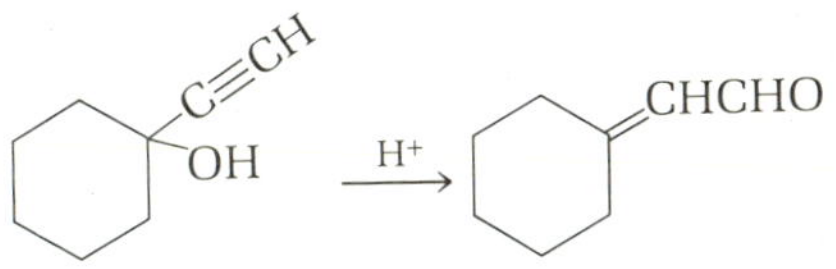

(1)

Write out a mechanism for this process.

12 3-Hexyne is treated with hydrogen and poisoned $Pd/CaCO_3$ catalyst; the product of the reduction is brominated in chloroform and then allowed to react with potassium *t*-butoxide in *t*-butyl alcohol at room temperature. Write out the structures of the product and the intermediates, taking care to indicate without ambiguity their stereochemistry.

13 Suggest a method whereby *cis*-cyclododecene may be converted to *trans*-cyclododecene.

14 Under base catalysis, one molecule of an alcohol may be added to acetylene to form a vinyl ether. The reaction goes no further, however, because alcohols do not add to alkenes under basic conditions.

$$ROH + HC{\equiv}CH \xrightarrow{OH^-} ROCH{=}CH_2 \qquad (2)$$

Write out a mechanism for the reaction, and explain why alkynes but not alkenes are susceptible.

15 Propose structures for **A** to **E**.

$$HC{\equiv}CH \xrightarrow{C_2H_5MgBr} \mathbf{A} \xrightarrow{(CH_3)_2C{=}O} \xrightarrow{H_3O^+} \mathbf{B} \xrightarrow{PBr_3} \mathbf{C}\ (C_5H_7Br) + \mathbf{D}\ (C_5H_6)$$

$$\mathbf{C} \xrightarrow[\text{ether}]{LiAlH_4} \mathbf{E}\ (C_5H_8;\ \text{shows no acetylenic bands in its ir spectrum}).$$

16 a. The standard heats of formation of ethane, ethylene, and acetylene are −20.2, 12.5, and 54.2 kcal/mole, respectively. The standard heats of formation of carbon dioxide and water are −94.0 and −57.8 kcal/mole. Which of the three hydrocarbons is the most efficient fuel per unit weight?

b. Containers of pure acetylene have been known to explode after being subject to mild shock. What are some likely reactions that may take place during these detonations?

SUPPLEMENTARY READINGS

T. F. Rutledge, *Acetylenic Compounds*. New York: Reinhold Book Corp., 1968.

T. F. Rutledge, *Acetylenes and Allenes*. New York: Reinhold Book Corp., 1969.

CHAPTER 14

BENZENE AND THE CONCEPT OF AROMATICITY

"Let us learn to dream, gentlemen, then perhaps we shall find the truth ... but let us beware of publishing our explanations before they have been put to the proof by the waking understanding."

Kekulé

HISTORY OF THE BENZENE PROBLEM

In 1823 Michael Faraday undertook to analyze the volatile fraction of lighting fuel that was produced by the thermal cracking of whale oil. After repeated fractional distillations and fractional crystallizations of this material he obtained, among other products, a pure hydrocarbon that melted at 5.5° and that he named "bicarburet of hydrogen." It should be noted that this was a feat of some magnitude — a recent investigation with the aid of gas chromatography has shown that the mixture with which Faraday worked contained literally hundreds of components. One factor that undoubtedly proved very fortunate for the isolation of "bicarburet of hydrogen" was the melting point of the material; it was remarkably high for a hydrocarbon of low molecular weight.

Faraday reported the empirical formula of "bicarburet of hydrogen" as C_2H, based upon the then current belief that the atomic weight of carbon was six. After the problem of atomic and molecular weights was clarified in the course of the next three or four decades, the molecular formula of Faraday's compound was established as C_6H_6. It had earlier been renamed *benzene*, by Mitscherlich, who showed that it could be obtained by heating benzoic acid with calcium hydroxide. As chemical data about benzene and its derivatives began to accumulate, the stage was set for the intervention of Friederich August Kekulé.

Kekulé, one of the seminal thinkers in organic chemistry, started out as a student of architecture. After his interest turned to chemistry, he began to immerse himself in the study of a number of the most important chemical questions of the day. He played a major part in establishing the quadrivalency of carbon and he was the first to advance the hypothesis that carbon atoms could combine with one another to form chains. In Kekulé's time most chemists viewed chemical formulas as symbols that depicted only atomic groupings that underwent transformation in the course of a reaction. These symbols were not credited with having any *structural* significance; i.e., they did not and could not represent the physical arrangement of the atoms in space.

Kekulé at first adhered to the traditional view; he later came to believe that chemists could deduce, and symbolically represent, the structural features of molecules. The architect in him was perhaps expressing itself. Kekulé was not completely alone in his views but his papers were probably more influential than anyone else's in advancing these hypotheses. Considering his interests, it was natural that he should concern himself with the "benzene problem."

In Kekulé's time the concept of unsaturation was being formulated and evidence was accumulating concerning the chemical behavior of unsaturated compounds. The molecular formula of benzene indicated that it should be highly unsaturated but its relative lack of reactivity belied this expectation. In the absence of light or catalysts, benzene reacts very slowly with bromine or chlorine and it is impervious to the usual oxidizing agents. When halogenation is effected, it usually leads to substitution of a halogen atom for one of the hydrogen atoms rather than addition of a halogen molecule that is so characteristic of unsaturated compounds. These substitution reactions also led to an isomer pattern that was quite confusing.

Kekulé's solution was to postulate (in 1865) that the carbon atoms of benzene formed a ring, lying at the corners of a regular hexagon. He first proposed a regular plane hexagon as a symbol for benzene **(1)** but later added a system of conjugated double bonds to this formula **(2)** in order to make explicit the quadrivalency of carbon. (It was made clear that each carbon atom bore one hydrogen atom).

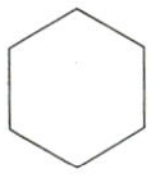

1

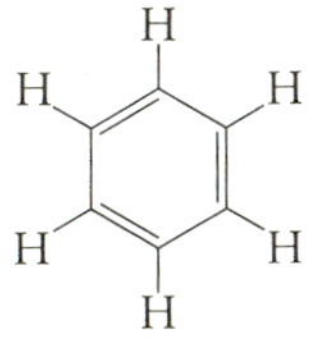

2

In an often repeated story, Kekulé related how the hexagon formula came to him one day while he was dozing; chains of carbon atoms disported themselves like snakes before his eyes, until one snake seized its own tail with its mouth, and Kekulé had the vital clue. (It is interesting to note that Kekulé's inspiration for the theory of carbon chains also came to him in a reverie — while he was atop a London bus.) He was one of chemistry's most fruitful dreamers.

The novelty of Kekulé's solution lay in part in his proposal of a cyclic structure in the first place. Chemists worked almost entirely with linear structures and theories of cyclic or unsaturated molecules were largely absent. For instance, the Russian chemist, Buterlov, who originated the term *molecular structure,* depicted (in 1867) the dehydration of 1-butanol, in part, as shown in equation 14.1. The primes represent one-half of an unsaturation; there is not yet a suggestion that these half unsaturations may be joined to give rise to double bonds or rings.

(14.1) 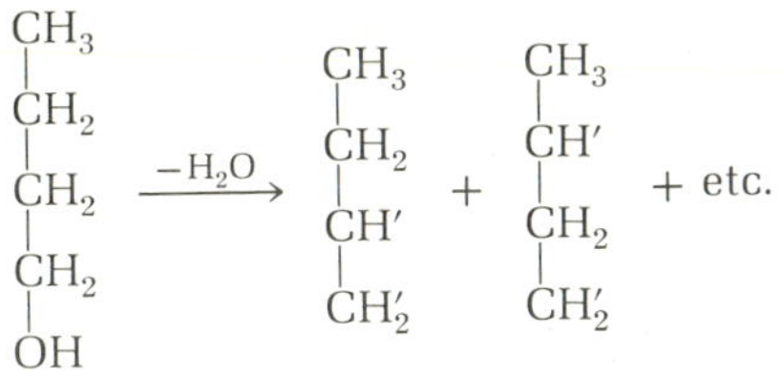

In order to fully reconcile theory with experiment insofar as the isomer problem was concerned, Kekulé made one more modification of his original hypothesis. Using **2** as the formula for benzene, one would predict four possible symmetrically disubstituted benzenes: two 1,2-disubstituted compounds **(ortho** isomers, **3** and **4)**, one 1,3-disubstituted compound **(meta** isomer, **5)**, and one 1,4-disubstituted compound **(para** isomer, **6)**.

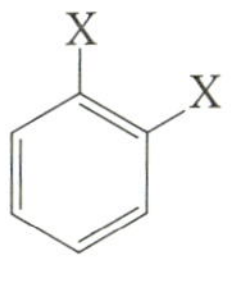

3

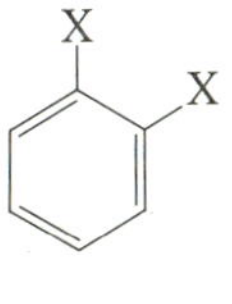

4

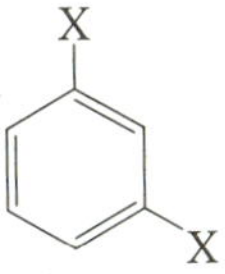

5

6

In fact, no more than three isomers of a given disubstituted benzene have ever been found, regardless of the nature of X. Kekulé disposed of this problem by postulating that the benzene molecule constantly oscillated between the two different but indistinguishable conjugated bond arrangements (equation 14.2).

(14.2) 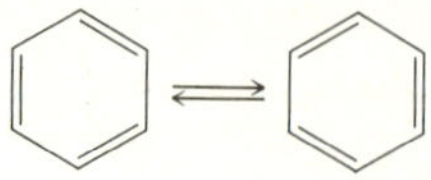

This oscillation was a consequence of the vibrational motion of the carbon atoms; its net effect was to make all the carbon-carbon bonds identical. In such a regular plane hexagon there could be only one *ortho*, one *meta*, and one *para* isomer of any disubstituted benzene.

(14.3) 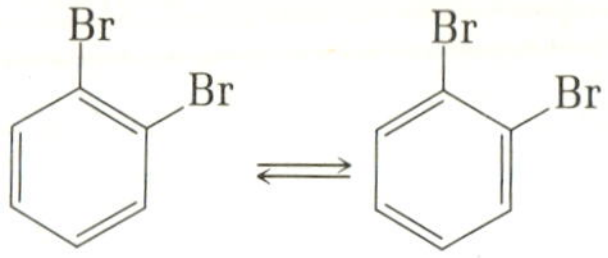

3a **4a**

Kekulé's proposals, effective as they were in rationalizing most of the available data, were not universally accepted. For one thing, it was by no means apparent from the formula why benzene did not behave like an alkene. Various alternatives were proposed as chemists strove to utilize the fourth valence of each carbon atom in some way other than by incorporating it into a double bond. Claus proposed several formulas, chief among these a ring with diagonal bonds **(7)**; there were a number of so-called "centric" formulas, in which the fourth valence was directed toward the center of the ring **(8)**; Dewar proposed a model with a long *para* bond **(9)**; finally, a number of proposals came from Ladenburg, the one now associated with his name being the prism formula **(10)**. The Dewar and Ladenburg formulas are still of interest to modern chemists (see below).

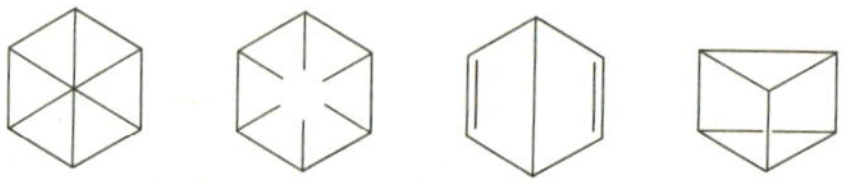

7 **8** **9** **10**

PROBLEM

1 Kekulé's choice of the cyclic model for benzene derived in part from the necessity of finding a structure of high symmetry in order that the small number of disubstituted isomers be accounted for. Draw some linear structures that correspond to C_6H_6. Do any of them fit the experimental data as far as number of monosubstituted and disubstituted isomers is concerned?

AROMATICITY – CHEMICAL CRITERIA

The generic name "aromatic compounds" was bestowed by Kekulé on benzene, a number of its derivatives, and several related compounds because of their aroma. After a time, however, the terms "aromatic" and "aromaticity" acquired purely chemical connotations. One of the most striking facts about benzene had been its chemical inertness — its resistance to uncatalyzed halogenation, its resistance to oxidation, and, perhaps most importantly, its tendency to undergo substitution rather than addition, thereby maintaining the integrity of the carbon skeleton. Aromaticity thus denoted the unusual chemical stability of benzene, and other compounds were judged to be aromatic to the extent that they emulated benzene in chemical behavior. This criterion was often difficult to apply in practice since a given compound might appear to be "aromatic" toward one set of reagents but distinctly "nonaromatic" toward another set.

Near the end of the nineteenth century the belief became current that the aromaticity of benzene could be ascribed to its system of cyclic, conjugated double bonds. Other compounds possessing this property were likewise expected to exhibit aromatic behavior. The synthesis of cyclooctatetraene **(11)** in the first decade of this century and the demonstration of its olefin-like, nonaromatic behavior undermined this theory.

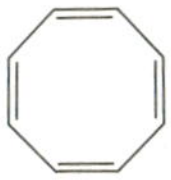

11

Robinson proposed in the 1920's that aromaticity was peculiarly the result of a compound's containing a *sextet* of cyclic conjugated electrons; compounds displaying aromatic properties were known in which the ring size was other than six but the electron sextet was always present. Robinson was substantially correct and, as we shall now see, his criterion was really a special case of a more general definition of aromaticity.

MOLECULAR ORBITAL THEORY OF BENZENE

One might attempt to explain the unusual stability of the benzene ring by invoking resonance arguments (Chapter 12). If equation 14.1 is rewritten so that the equilibrium arrows are replaced by the double-headed arrow, it would follow (from the resonance approach) that the stability of benzene was a consequence of the contribution from two canonical forms of identical energy.

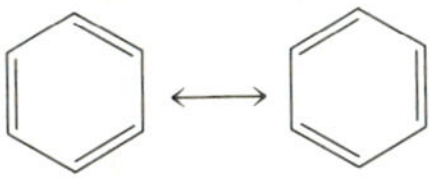

The larger the number of classical canonical structures that may be written for a molecule, the lower is its energy because the existence of these structures *implies* extensive electron delocalization in the actual molecule. This treatment would lead to the prediction that cyclobutadiene and cyclooctatetraene would also be aromatic since exactly analogous canonical forms can be drawn in these cases.

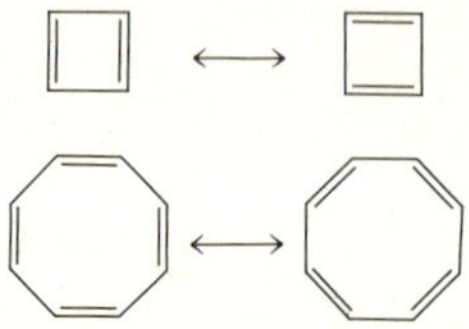

As previously noted, cyclooctatetraene behaves as a classical alkene; cyclobutadiene is so reactive that it cannot be isolated in uncomplexed form since the free compound undergoes an extremely rapid Diels-Alder reaction with itself. The resonance explanation thus seems to be of little help; the molecular orbital description seems to hold out more hope of rationalizing the observed phenomena.

The six π-molecular orbitals of benzene, three bonding and three antibonding, are formed by combining six $2p$-orbitals, one from each carbon atom; these molecular orbitals are depicted in Figure 14-1. The π-electron distribution is completely symmetrical about the ring. Half the electron density is in an "electron cloud" above the ring and the other half in an "electron cloud" below the ring (Figure 14-2). The energy released when six carbon $2p$-orbitals combine to form the π-system of benzene may be estimated from molecular orbital calculations and this number is sizably greater than the energy released when six $2p$-orbitals combine to form three ordinary π-bonds, like those in ethylene.

Benzene is a regular hexagon with a fixed carbon-carbon bond length of 1.40 A, as determined by X-ray analysis. This value is appropriately intermediate between that calculated for a sp^2-sp^2 single bond (1.49 A) and that found for the double bond of ethylene (1.34 A). But making use of the X-ray data and of the molecular orbital calculations, we might construct a picture of benzene such as **12,** in which the dots represent the π-electrons. There is effectively one π-electron, *on the average,* between each pair of carbon atoms. Compare this structure with that of cyclohexatriene, the imaginary isomer of benzene with three fixed double bonds (1.34 A) and three single bonds (1.49 A) **(13).**

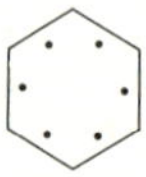

12

13

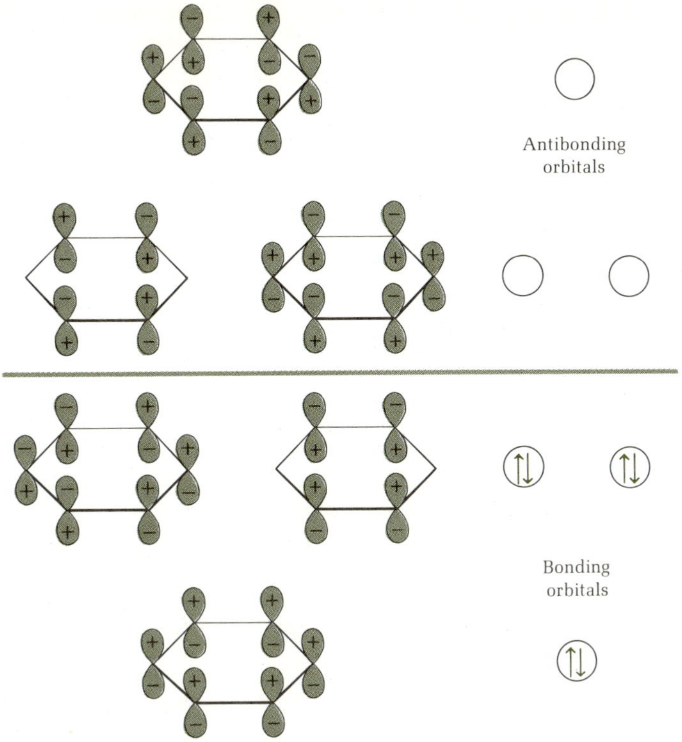

FIGURE 14-1
Molecular π-Orbitals of Benzene

It can be readily seen that in benzene, as compared to cyclohexatriene, the interelectronic repulsion must be much smaller due to the more symmetrical electron distribution, with a consequent increase in stability. Furthermore, since the π-electron density in **12** is nowhere as great as in the double bonds of **13**, **12** should react much more slowly with electrophiles than **13**. To answer the question of why the same reasoning does not apply to cyclobutadiene and cyclooctatetraene, we must also consider the wave properties of electrons (see aromaticity and antiaromaticity, p. 351).

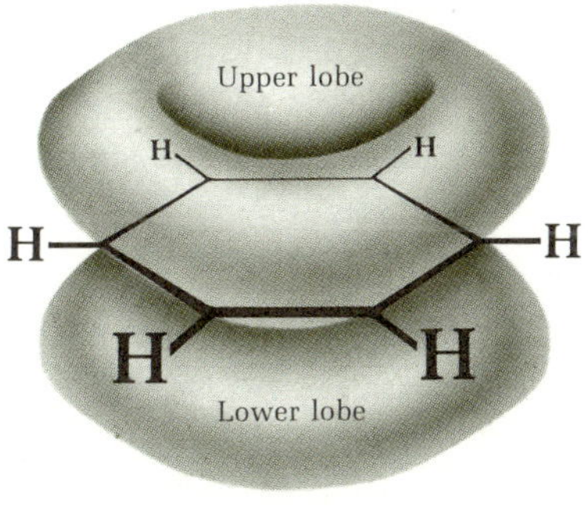

FIGURE 14-2
π-Cloud Picture of Benzene

The resonance energy, or delocalization energy, of benzene (its "extra" stabilization) is simply the difference in the heat of formation between benzene and cyclohexatriene, its hypothetical isomer.

(14.4)

benzene cyclohexatriene

Resonance energy (benzene) = ΔH_f (benzene) − ΔH_f (cyclohexatriene)

Since cyclohexatriene does not exist, however, its heat of formation must be estimated and every method will be open to criticism because of the assumptions made in its use.

An empirical measure of the resonance energy of benzene may be made in the following way: The heat of hydrogenation of cyclohexene is −28.6 kcal/mole; cyclohexatriene might therefore be expected to have a heat of hydrogenation of $3 \times (-28.6) = -85.8$ kcal/mole.

(14.5)
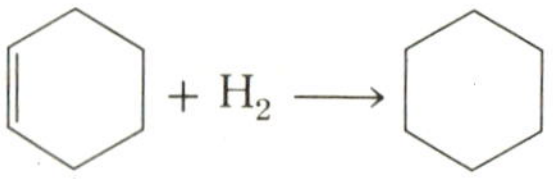
$\Delta H = -28.6$ kcal/mole

When benzene is reduced to cyclohexane, the heat released amounts to only 49.8 kcal/mole. (In particular, the first step of the reduction is actually endothermic, which is quite unusual for a hydrogenation (equation 14.7).)

(14.6)
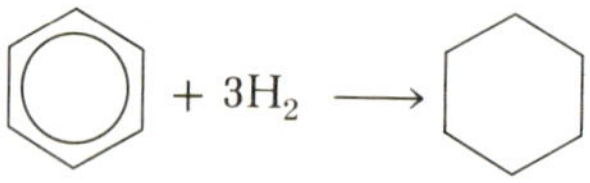
$\Delta H = -49.8$ kcal/mole

(14.7)
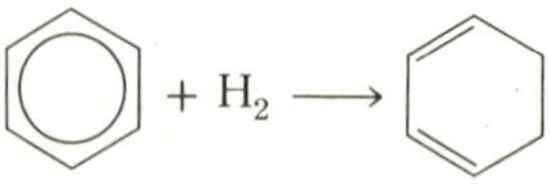
$\Delta H = +5.6$ kcal/mole

The conclusion might be drawn that the resonance energy of benzene is $(85.8 - 49.8) = 36.0$ kcal/mole. This conclusion ignores, however, the changes that take place in *single bonds* as hydrogenation proceeds (for instance, sp^2-sp^2 going to sp^2-sp^3), and that would also be taking place when the classical cyclohexatriene undergoes hydrogenation. Estimates of the resonance energy of benzene that attempt to take account of single bond changes range from 18 to 63 kcal/mole.

PROBLEM

2 Based on the heat of hydrogenation data, calculate the following quantities:
a. The enthalpy change when 1,3-cyclohexadiene is reduced to cyclohexene.

b. The resonance energy of cyclohexadiene.

c. The enthalpy change when 2 moles of 1,3-cyclohexadiene react to give 1 mole of benzene and 1 mole of cyclohexene.

SPECTROSCOPY AND PHOTOCHEMISTRY OF BENZENE

Due to the extremely high symmetry of the molecule, the infrared spectrum of benzene is comparatively simple. Peaks due to the C—H stretching motions are found above 3000 cm^{-1} as expected for hydrogen bound to sp^2-hybridized carbon. Peaks of weak to medium intensity assigned to stretching vibrations of the aromatic ring lie in the 1450 to 1600 cm^{-1} region. The most intense band in the spectrum is found near 700 cm^{-1} and is due to the out-of-plane bending motions of the hydrogen atoms. Bands in this region are very characteristic of the aromatic ring and their number and frequency are dependent on the number and location of the substituents attached to the ring.

The benzene protons being all equivalent, the nmr spectrum of benzene consists of a single peak at $\delta 7.27$. It is significant that this figure is 0.3 to 2.8δ *below* that usually found for hydrogen atoms connected to doubly bonded carbon atoms. This additional deshielding throws light on the electronic structure of benzene. The carbon ring can be viewed as a conducting wire or solenoid. When this ring is placed in a magnetic field, the highly mobile, delocalized π-electrons circulate around the ring, much as electrons do in a wire, giving rise to a **ring current.** This ring current in turn results in a local magnetic field being induced around the ring (Figure 14-3).

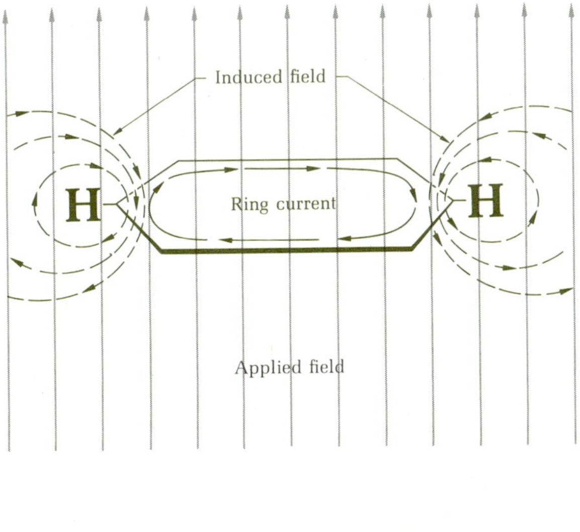

FIGURE 14-3

Ring Current and Induced Magnetic Field in Benzene

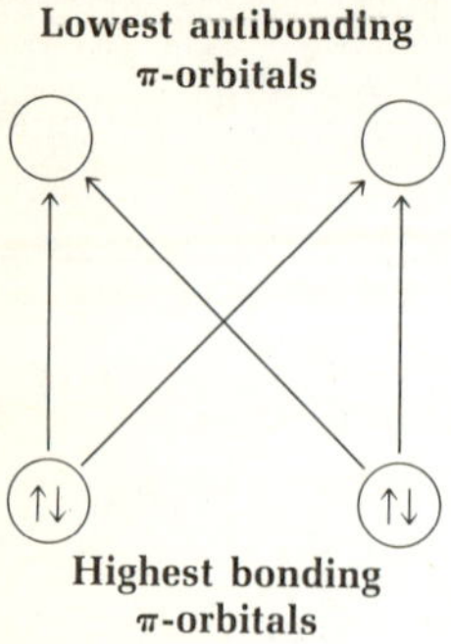

FIGURE 14-4

Low Energy Electronic Transitions in Benzene

The induced magnetic field is opposed to the applied field inside the ring but it reinforces the applied field outside the ring. The benzene protons, which are on the outer periphery of the ring, are therefore deshielded by the induced magnetic field. If there were any protons inside the ring, they would be abnormally *shielded* by the induced magnetic field. The ability of a compound to sustain a ring current is one useful criterion of aromaticity and it is easily determined because of the effect of the ring current on the resonance absorption of the protons. The application of this criterion is illustrated on p. 349.

The π-orbitals are so arranged that the two highest filled orbitals have the same energy (they are **degenerate**). Similarly, the two unfilled orbitals of lowest energy are also degenerate (Figure 14-4). The electronic transition requiring the least energy would involve the promotion of an electron from the highest occupied orbital to the lowest empty one. (The matter is not so simple as it is made to appear — there are four equivalent ways of bringing about this transition, as the figure shows, and they actually mix together. This discussion should be taken as very approximate.)

Benzene can absorb ultraviolet light in the region 2280 to 2680 A, simultaneously with the occurrence of the low energy electronic transition. This transition, which has its maximum at 2540 A, is forbidden and has a low ϵ (Figure 14-5). The low transition probability is a consequence of the sym-

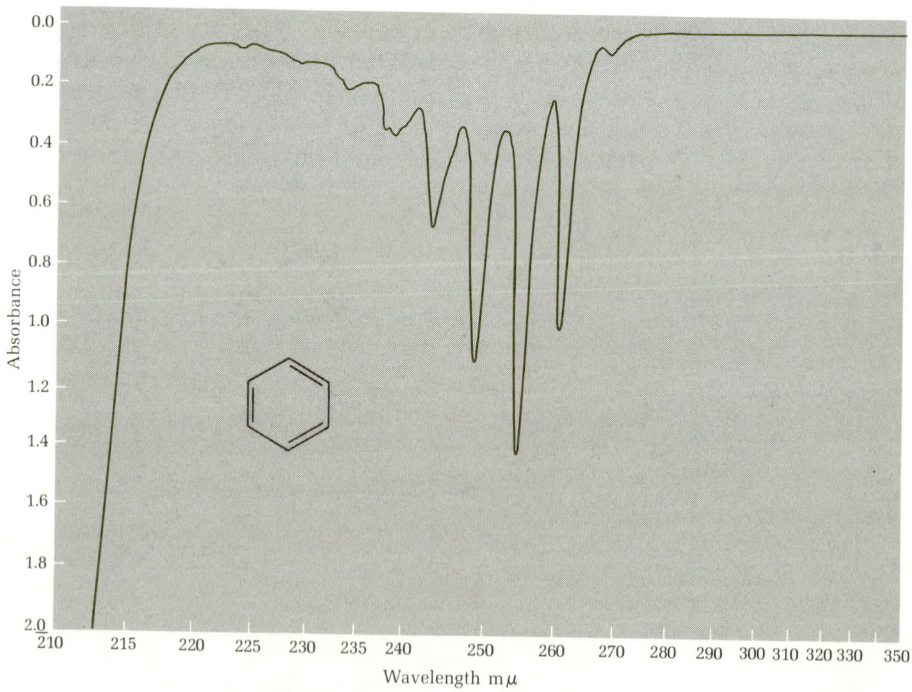

FIGURE 14-5

Near Ultraviolet Spectrum of Benzene

metry of the π-orbitals of benzene and confirms our picture of the benzene molecule. The low intensity uv band is characteristic of the benzene ring; its position and strength may be somewhat altered by substituents attached to the ring. There are very intense absorptions in benzene at shorter wavelengths but these are usually beyond the convenient range of many uv spectrophotometers.

Upon irradiation, benzene may be transformed into a number of isomers, some of which would be difficult to obtain by other routes. The product distribution is dependent on the wavelength of light, the temperature, and the physical phase.

(14.8)
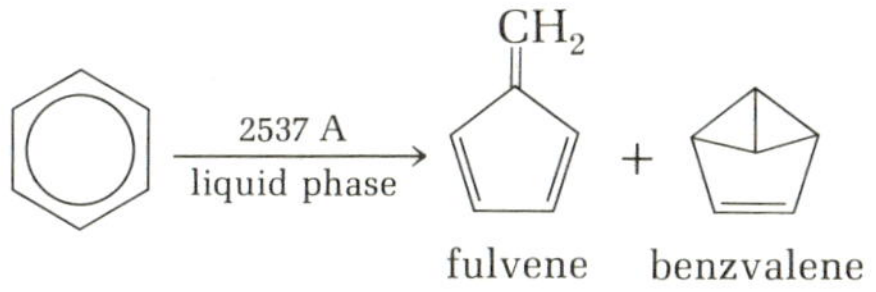

(14.9)
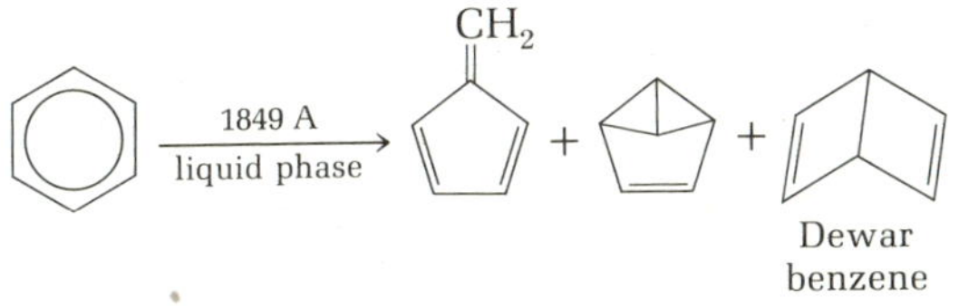

The isomers may be further interconverted by the radiation. These reactions show that excited states of benzene may have appreciable bonding between nonadjacent atoms.

(14.10)
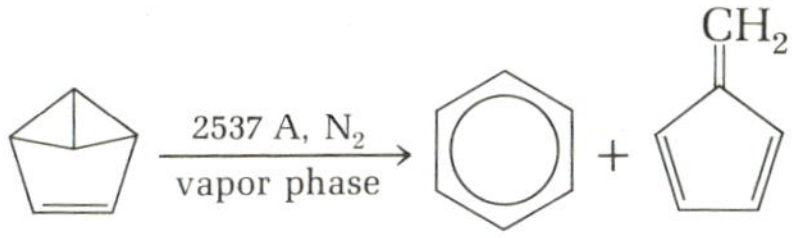

PROBLEM
3 When benzene-1,3,5-d_3 is irradiated in the vapor phase with 2537 A light, it is partially converted to benzene-1,2,4-d_3. The evidence indicates that free radicals are not involved in this isomerization. Suggest a mechanism for the reaction.

Isolation of Dewar benzene from the photolysis of benzene is very intriguing. Dewar benzene and several of its substituted derivatives may be obtained by nonphotolytic routes. In one respect the Dewar benzenes differ in structure from the original proposal.

(14.11)

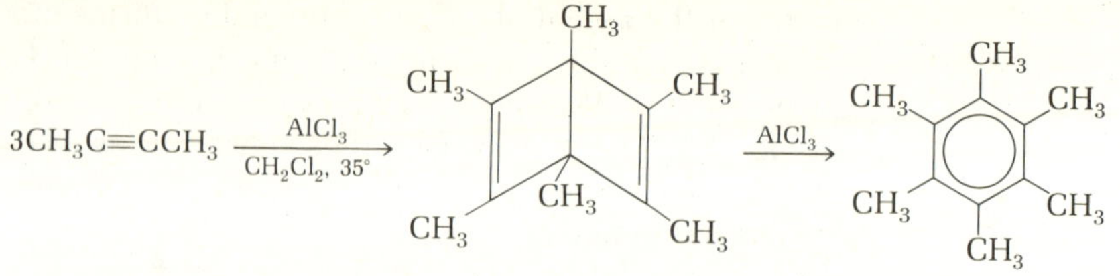

The molecule is not planar, with a long single bond, but rather V-shaped and contains bonds of the expected lengths (Figure 14-6). Due to the "tilting" of the double bonds, there is some overlap between their π-orbitals, even though they are not conjugated. When Dewar benzene derivatives are irradiated, they are partially converted to prismanes, or Ladenburg benzenes.

(14.12)

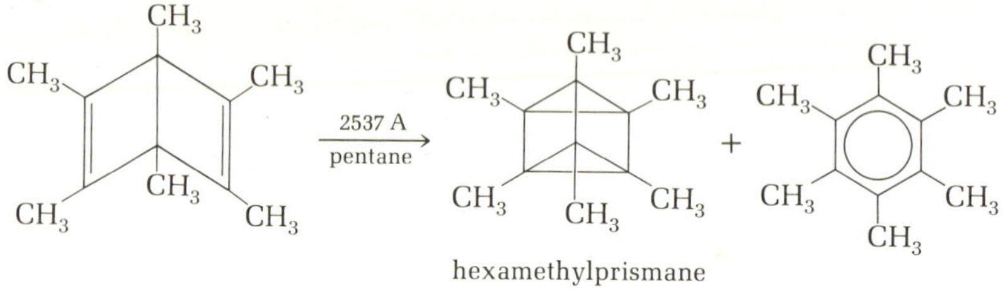

The synthesis of various Dewar and Ladenburg benzenes, and the demonstration that they are less stable isomers of their "normal" benzene counterparts, constitutes in many ways the final word on the classical benzene controversy.

OTHER AROMATIC SYSTEMS – HÜCKEL'S RULE

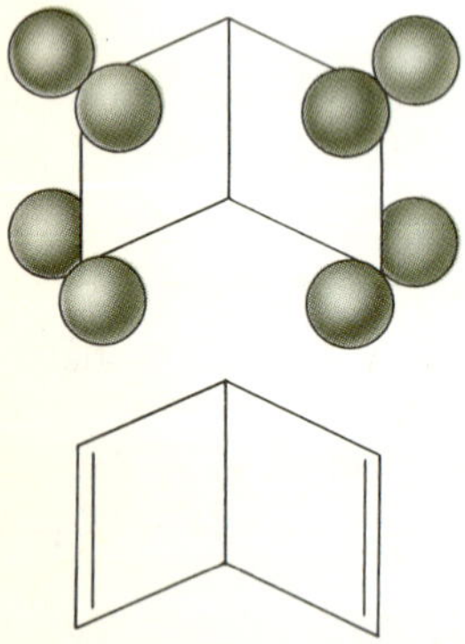

FIGURE 14-6
Geometry of Dewar Benzene

Shortly after the development of the molecular orbital theory, the theoretician Hückel proposed a general rule for deciding whether or not appropriate molecules would be aromatic — that is, if they would be unusually stable and unreactive. In any monocyclic planar hydrocarbon composed only of sp^2-hybridized carbon atoms, in which each atom contributed one p-orbital to the π-orbital system, the molecule would be aromatic if the number of *π-electrons* equaled $(4n + 2)$, where n is a positive integer or zero.

It is clear that benzene meets the criterion ($n = 1$). It is also clear that cyclooctatetraene and cyclobutadiene do not — they are $4n$ π-electron systems. Cyclooctatetraene is not even planar — it is tub-shaped, which avoids the bond angle distortion that would be present in a planar molecule. Cyclobutadiene has been isolated only in the form of its complexes with various metals and metal ions.

The fact that a cyclic molecule contains $(4n + 2)$ π-electrons does not in itself guarantee its aromatic behavior — other factors may intervene. Cyclodecapentaene **(14)**, for instance, a compound that satisfies Hückel's rule for $n = 2$, has never been isolated. The form with the least amount of bond angle distortion would suffer from the interference of two internal hydrogen atoms **(14)** that would prevent molecular planarity. A related material has been synthesized in which the offending hydrogen atoms have been removed and replaced by a methylene bridge **(15)**. This molecule is substantially planar and satisfies most of the operational criteria for aromaticity.

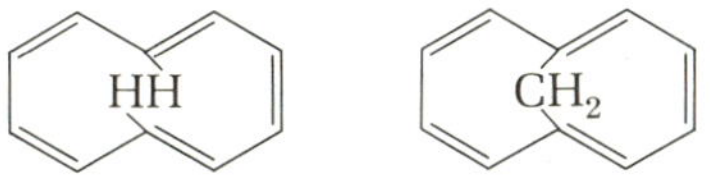

14 **15**

When placed in a magnetic field, it apparently develops a ring current, for the peripheral protons are definitely deshielded. The protons on the methylene group, on the other hand, lie over the central cavity of the molecule and are shielded — they resonate 0.5δ *upfield* from tetramethylsilane.

The nmr criterion of aromaticity correlates very nicely with Hückel's rule. It is found that $4n$ π-electron systems behave in reverse fashion from $4n + 2$ systems. In cyclic $4n$ molecules the outer peripheral protons are abnormally *shielded*, while those within the inner cavity are abnormally deshielded.

The Hückel rule says nothing about the number of carbon atoms present in the ring — it requires only that there be $(4n + 2)$ π-electrons. This leaves open the possibility of creating aromatic systems in which the number of π-electrons would differ from the number of carbon atoms — in other words, the aromatic rings would be anions or cations. The first aromatic ion discovered was the **cyclopentadienide** anion **(16)**. Cyclopentadiene exhibits remarkable acidity for a hydrocarbon.

(14.13) 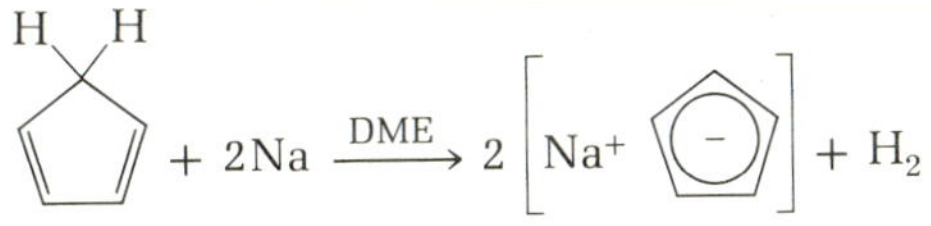

16 ($n = 1$)

Removal of a proton from cyclopentadiene leaves behind a negatively charged six π-electron system; the charge is equally distributed over the five carbon atoms.

Hückel aromatic anions that satisfy the equation for values of n other than unity are also known (equation 14.14).

(14.14) 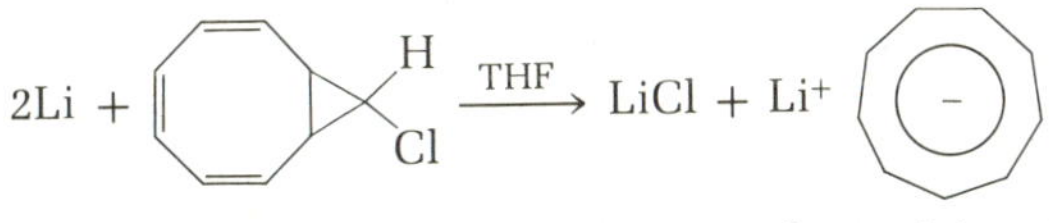

cyclononatetraenide ion
($n = 2$)

Aromatic *cations* have also been synthesized that satisfy Hückel's rule for various values of n. The first to be made, and the most widely studied, is the seven-carbon, six π-electron cation known as the **tropylium ion.** Its fluoborate (equation 14.15) may be isolated and stored for indefinite periods. The fact that salts with three-membered aromatic cations are capable of isolation (equation 14.16) is a tribute to the stabilization which comes with aromaticity because the ring strain in this cation must be considerable.

(14.15)

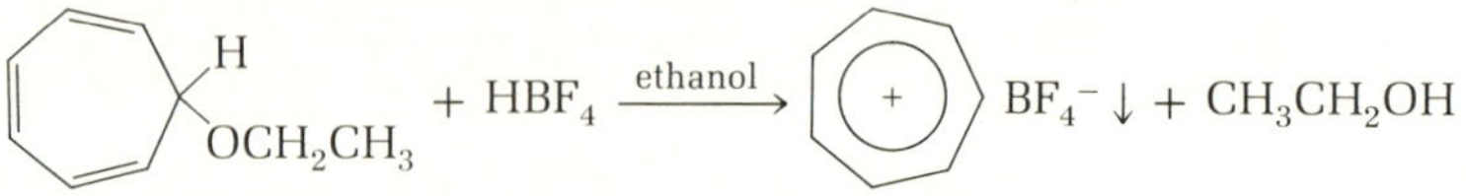

tropylium fluoborate
($n = 1$)

(14.16)

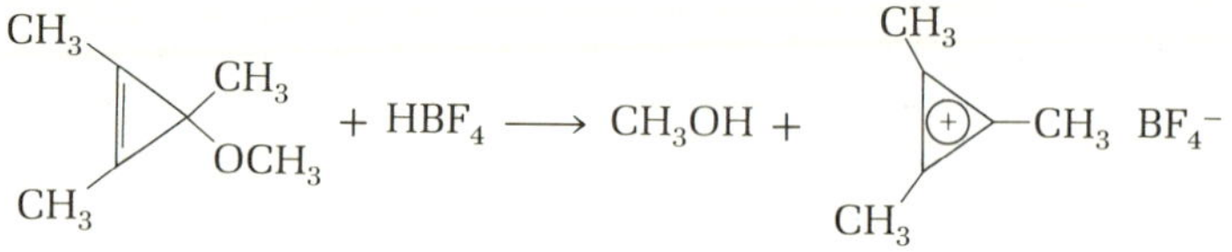

trimethylcyclopropenium
fluoborate ($n = 0$)

It is even possible to generate aromatic dianions and dications. Recognition should be made of the fact that the aromatic cations are much more reactive than uncharged aromatic systems — the cations are susceptible to attack by strong nucleophiles, while the anions are sensitive to acids, oxygen, etc.

(14.17)

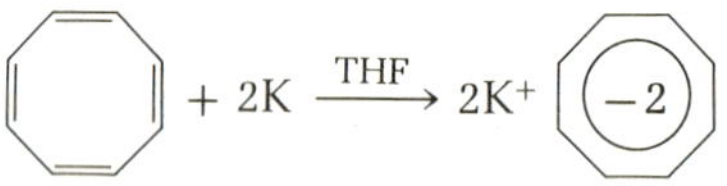

($n = 2$)

(14.18)

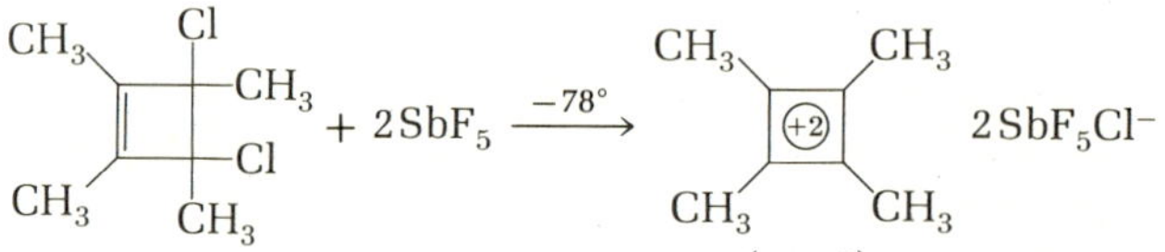

($n = 0$)

Nonetheless, the aromatic ions are usually perfectly stable in the absence of substances with which they can react. Furthermore, they exhibit many physical properties that we associate with aromaticity — identical C—C bond lengths, equal distribution of the charge over all the carbon atoms, and the ability to sustain a ring current when placed in a magnetic field.

We close this section with another view of aromaticity that sheds some light on the "stabilizing force." Imagine a linear chain of sp^2-hybridized carbon atoms carrying a certain number of π-electrons that may or may not equal the number of carbon atoms. Suppose now that we snip one hydrogen atom from each end of the chain and join up the ends to form a ring. It is a relatively easy matter to calculate the change in energy of the π-electrons as this process is carried out. If the energy decreases — that is, the π-electrons are more stable in the ring than in the linear system — then the ring is aromatic. If, on the other hand, the energy increases upon formation of the ring, then it is **antiaromatic.** If essentially no change takes place, then the ring is **nonaromatic.**

A good example is the transformation of hexatriene to benzene — there is a very substantial decrease in π-electron energy.

(14.19)
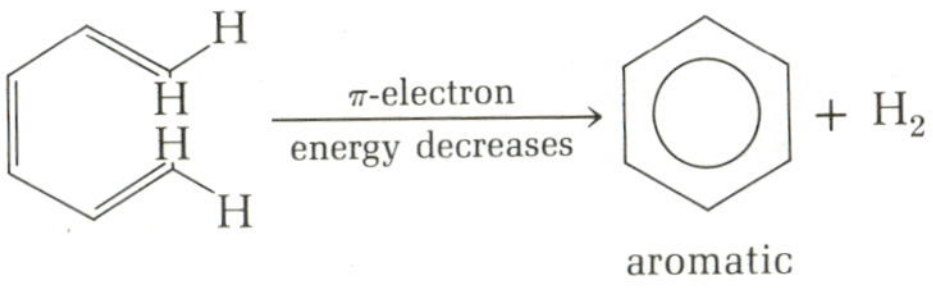

Chemical experience tells us that the opposite must be true of the butadiene-cyclobutadiene conversion, based on the relative reactivity and isolability of the two compounds.

(14.20)
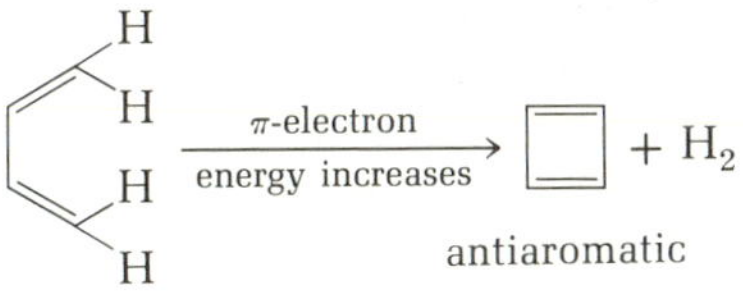

In similar fashion, cyclizing allyl cation to cyclopropenyl cation lowers the energy of the π-electrons, while the reverse result obtains when allyl anion is converted to cyclopropenyl anion.

(14.21)
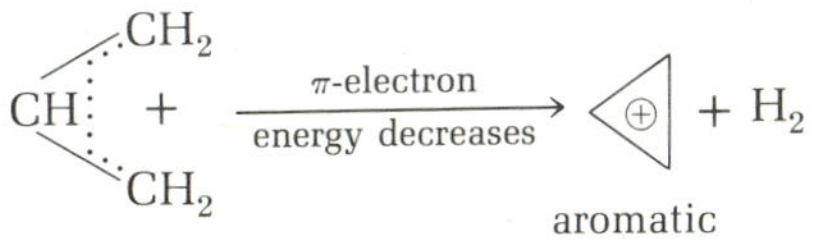

(14.22)
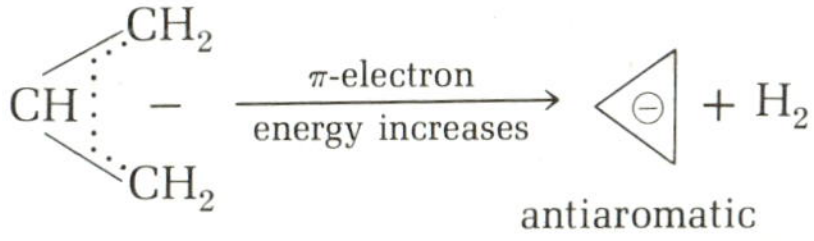

When the number of π-electrons exceeds about 20, the energy of the linear and cyclic forms is about the same for both $4n$ and $4n + 2$ molecules. These rings are nonaromatic.

PROBLEM

4 Substituted 1*H*-azirines (I) are very unstable compounds that tend to rearrange readily to 2*H*-azirines (II). What might be the driving force for this facile rearrangement?

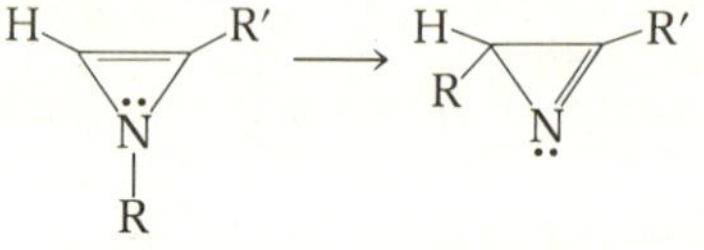

I II (1)

In order to understand, in a qualitative way, the origin of aromaticity and antiaromaticity, we must fall back on the molecular orbital diagrams that we utilized in deriving the Woodward-Hoffman rules. When the ends of an acyclic polyene are connected, two consequential forces come into play that exert opposing effects on the π-electron energy. On the one hand, increased electron delocalization may occur that would result in a lower π-electron energy. On the other hand, by bringing the ends together, additional electronic repulsions are made possible that serve to raise the energy.

The predominant effect is different for electrons in different orbitals, depending on their symmetry, and the *net* effect for the entire molecule is the sum of the energy decreases or increases for the electrons in each occupied molecular orbital. Take, as an example, the lowest three molecular orbitals of 1,3,5-hexatriene (the only ones occupied in the ground state) and follow their energy progression as the triene is transformed into benzene (Figure 14-7). The lowest molecular orbital (Ψ_1) is symmetric, and when C-1 and C-6 are joined, there will be a bonding interaction, electron delocalization will increase, and the energy of the electrons in that orbital will fall. The same effect is observed with Ψ_3 but the opposite result is experienced by Ψ_2; an antibonding interaction is generated and the electron energy rises. (The *magnitude* of the energy change will not be the same for each orbital.) There is an overall energy lowering accompanying the cyclization of 1,3,5-hexatriene, and benzene is therefore stable and "aromatic."

A few calculations will show that $(4n + 2)\,\pi$-electron systems always experience a net lowering of energy upon cyclization, while for $4n$ π-electron systems the number of bonding interactions exactly equals the number of antibonding interactions and the total π-electron energy actually rises. This

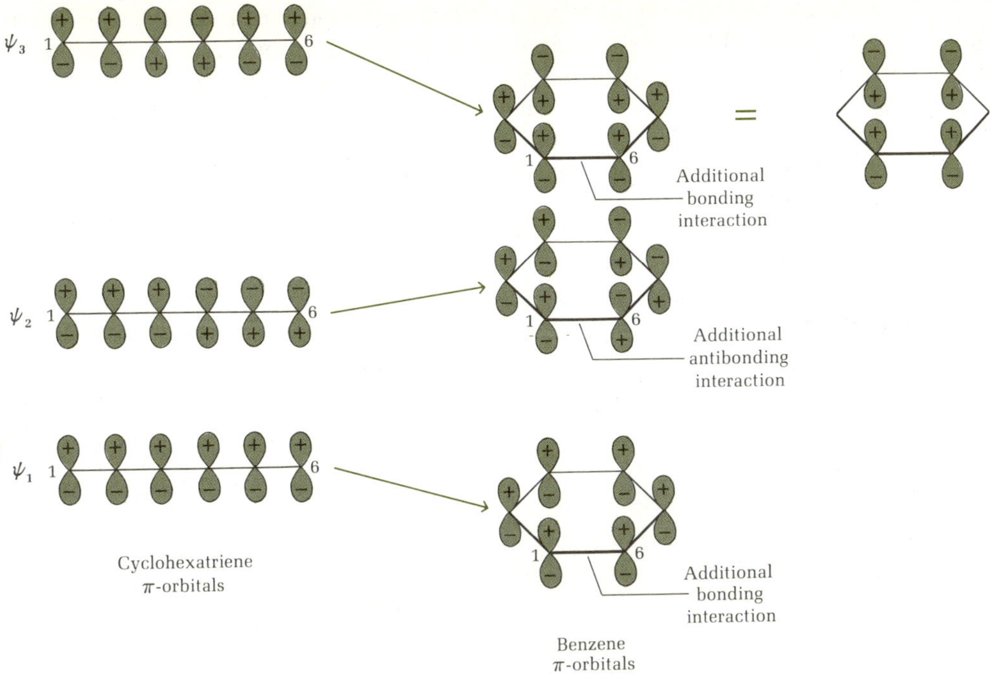

FIGURE 14-7
Orbital Energy Changes Upon Conversion of Cyclohexatriene to Benzene

analysis then amounts to a restatement of Hückel's rule and hopefully makes it more physically comprehensible by introducing the idea of a standard against which the cyclic compound is to be compared — its open-chain analogue.

THE BENZYL SYSTEM

Interesting effects occur if we replace one of the hydrogen atoms on the benzene ring by an sp^2-hybridized carbon atom so that the *p*-orbital on this atom is conjugated with the π-orbitals of the ring. This is the **benzyl** system. Depending on whether that *p*-orbital originally held zero, one, or two electrons, the product will be a benzyl cation, benzyl radical, or benzyl anion, respectively.

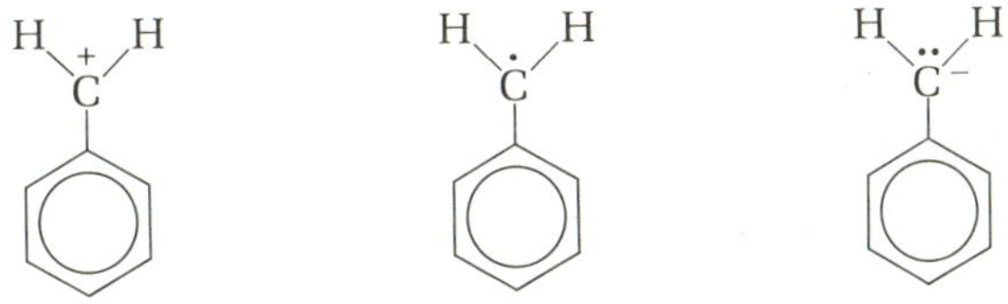

benzyl cation   benzyl radical   benzyl anion

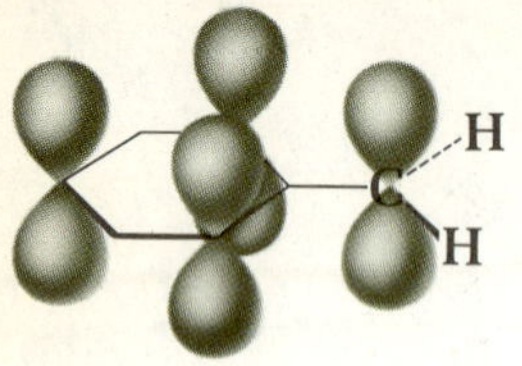

FIGURE 14-8
Benzyl Nonbonding Molecular Orbital

In the benzyl system there are a total of seven carbon 2*p*-orbitals and there must consequently be seven molecular π-orbitals. In benzene itself there are three bonding and three antibonding π-orbitals; the same is true of the benzyl system, the seventh orbital being a nonbonding orbital. The nonbonding orbital is composed of the atomic orbitals shown in Figure 14-8. If there is an electron in that orbital, as in benzyl radical, then the electron spends most of its time on the exocyclic (out-of-the-ring) carbon atom and on the ring carbon atoms *ortho* and *para* to it. Similarly, in the benzyl cation the positive charge is distributed over the exocyclic atom and the *ortho* and *para* positions. These points may be illustrated by means of resonance structures.

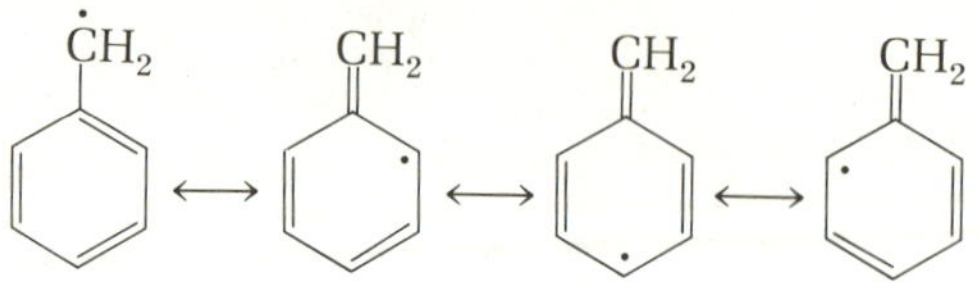

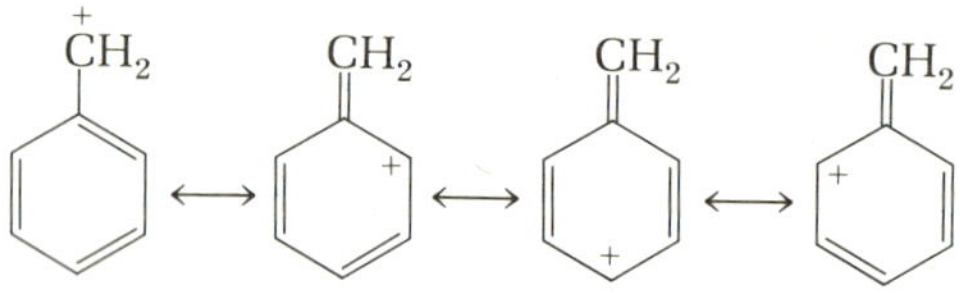

One would expect the benzyl radical and ions to be stabilized because of the additional electron delocalization. This stability is reflected in the value of the dissociation energy for the methyl C—H bonds of toluene; this quantity is 13 kcal/mole smaller than the C—H bond dissociation energy in ethane.

(14.23) 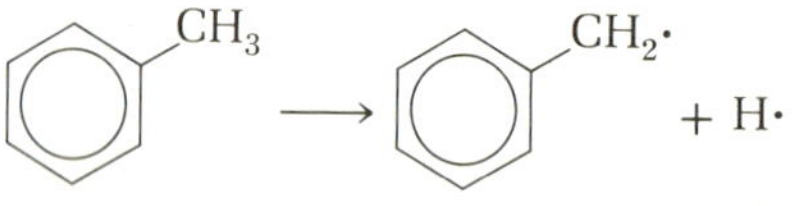 $\Delta H = 85$ kcal/mole

The presence of a **phenyl group** also enhances the ease of carbonium ion formation at the carbon atom to which it is attached. (When the benzene ring itself is treated as a substituent on another atom, the adjectival form of its name is "phenyl," from the word *phène*, the name given to benzene by the French chemist Laurent.) A comparison of equations 14.24 and 14.25, and also 14.26 and 14.27, indicates that phenyl is much more effective than methyl in promoting carbonium ion formation.

(14.24) $CH_3CH_2Br + H_2O \xrightarrow{\text{aq. acetone}} CH_3CH_2OH + HBr$ definitely S_N2

(14.25) 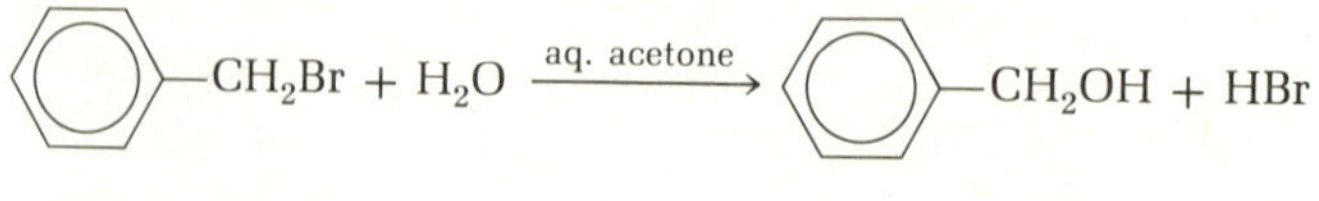 S_N1-S_N2 borderline

(14.26) $(CH_3)_2CHBr + H_2O \xrightarrow{\text{aq. acetone}} (CH_3)_2CHOH + HBr$ S_N1-S_N2 borderline

(14.27) 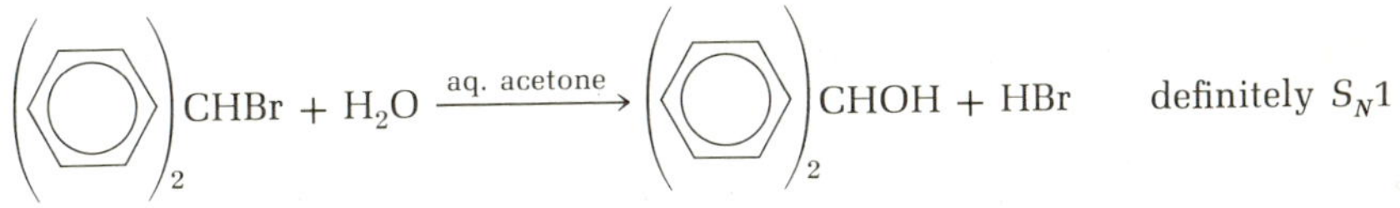

diphenylmethyl bromide diphenylmethanol

Triphenylmethyl chloride, 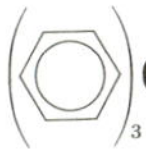 CCl (usually known as **trityl chloride**), hydrolyzes at an appreciable rate even upon exposure to moist air.

TRITYL CATION AND RADICAL

Trityl cation is sufficiently stable to be isolable as its fluoroborate or perchlorate salt, $(C_6H_5)_3C^+BF_4^-$ or $(C_6H_5)_3C^+ClO_4^-$. The stability of trityl cation may be ascribed to three causes: partial delocalization of the positive charge over the three rings, steric crowding that favors bond angles of 120° around the central atom, and the absence of certain reaction paths normally open to carbonium ions, such as loss of a hydrogen ion to form an alkene.

Just as trityl cation is stabilized by charge dispersion so it would be expected that trityl radical, $(C_6H_5)_3C\cdot$, would be stabilized by delocalization of the unpaired spin. This is in fact so and, because of its relative stability, trityl radical has played an important part in chemical history. It was the first organic free radical whose existence was supported by convincing experimental evidence.

The American chemist Gomberg reported in 1900 on his synthesis of hexaphenylethane, in which he coupled two trityl groups with the aid of a metal. When he carried out the reaction under normal conditions, he obtained a product that did not give the correct analysis for hexaphenylethane; the fraction of the material that was not carbon or hydrogen was equivalent in weight to 1 mole of oxygen per mole of hexaphenylethane.

(14.28) $2(C_6H_5)_3CCl + Zn \xrightarrow[N_2]{\text{benzene}} (C_6H_5)_3C\text{—}C(C_6H_5)_3 + ZnCl_2$

When Gomberg repeated his reaction and work-up in the absence of air, he produced a white crystalline solid, *from a yellow reaction solution,* that gave the correct analysis for hexaphenylethane. It was found that the color of the reaction solution could be discharged by admitting a small volume of air to the reaction flask; upon standing, the solution then regained its color. This cycle could be repeated many times until finally the color no longer reappeared; if the solution was then worked up, the oxygen-containing compound was formed.

Gomberg interpreted the results in a very convincing manner. He assumed that hexaphenylethane is in equilibrium with a small concentration of trityl radical. The radical is yellow in color and stable in the absence of oxygen.

(14.29) $(C_6H_5)_3C{-}C(C_6H_5)_3 \rightleftharpoons 2(C_6H_5)_3C\cdot$

When oxygen is admitted to the system, the radicals are converted irreversibly to triphenylmethyl peroxide.

(14.30) $2(C_6H_5)_3C\cdot + O_2 \longrightarrow (C_6H_5)_3C{-}O{-}O{-}C(C_6H_5)_3$

More dissociation of hexaphenylethane then takes place in order to maintain equilibrium.

Modern investigations have tended to fully substantiate Gomberg's hypothesis. Electron spin resonace spectroscopy (Chapter 12) has confirmed the delocalization of the unpaired electron and the fact that it resides principally at the *ortho* and *para* positions. Recent nmr studies have cast doubt on what was the least controversial part of Gomberg's original hypothesis — the structure of hexaphenylethane itself! The structure **17** has been proposed for this compound; such a suggestion is not unreasonable because of the sizable steric repulsion expected between the phenyl group on carbon atoms 1 and 2 in the "normal" structure.

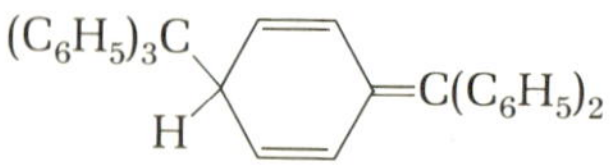

17

The existence of compound **17** would also bolster our ideas on electron delocalization.

SYNTHESIS AND PHYSICAL PROPERTIES OF BENZENE

Until fairly recently all the benzene used commercially and in the laboratory was derived from coal tar (next section). The tremendous commercial demand for benzene and other aromatics has resulted in their being synthesized by the dehydrogenation of assorted cycloalkanes (Chapter 6, pp. 126–127). It has also been known, since Kekulé's time, that acetylene would trimerize in the gas phase in a hot tube; this fact was used as evidence in favor of Kekulé's original structure for benzene.

(14.31) 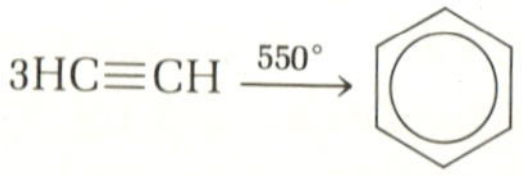

Although renewed interest in this reaction was evinced by the Germans during World War II and a number of catalysts have been found that promote it, this synthesis is not now of commercial importance.

Benzene melts at 5.5° and boils at 80°, which gives it a very short liquid range of only 75°. The meaning of these facts is illuminated by an examination of the physical constants of toluene in which a ring hydrogen atom has been replaced by a methyl group. The boiling point rises to 111° as expected, but the melting point drops to −95°! We must therefore view the 5.5° melting point of benzene as being abnormally high, a consequence of its hexagonal symmetry, which allows it to pack very efficiently into a crystal lattice. As mentioned at the beginning of the chapter, this abnormal melting point was an important factor in the early discovery of benzene.

AROMATIC COMPOUNDS FROM COAL TAR – STRUCTURE AND PROPERTIES

The conversion of iron ore to metallic iron consumes large quantities of coke, the material that remains after the destructive distillation of coal. The distillate consists of **coal gas** (light hydrocarbons, ammonia, hydrogen sulfide, etc.) and **coal tar.** Further refinement of the coal tar by extraction and by distillation yields a number of aromatic compounds of commercial importance: benzene, toluene, the three xylenes (dimethylbenzenes), phenol (hydroxybenzene) and its methylated derivatives, and pyridine (in which a CH group of benzene is replaced by N) and its methylated derivatives.

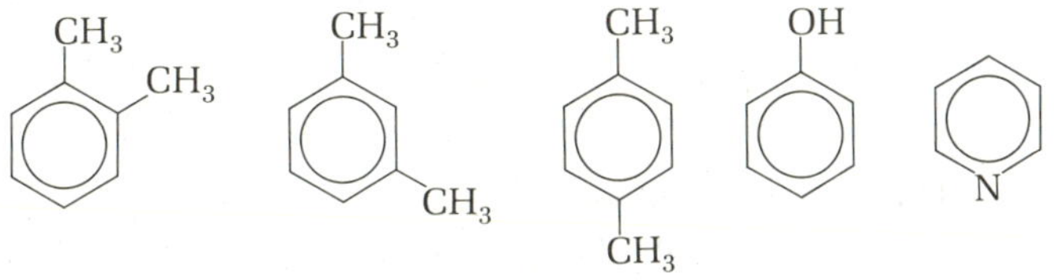

o-xylene *m*-xylene *p*-xylene phenol pyridine

It would be difficult to overestimate the importance of these compounds to the synthetic organic chemical industry. Also isolated are aromatic compounds containing *condensed rings,* in which two or more aromatic rings share one or more edges. Of those compounds shown, by far the most important are naphthalene and anthracene.

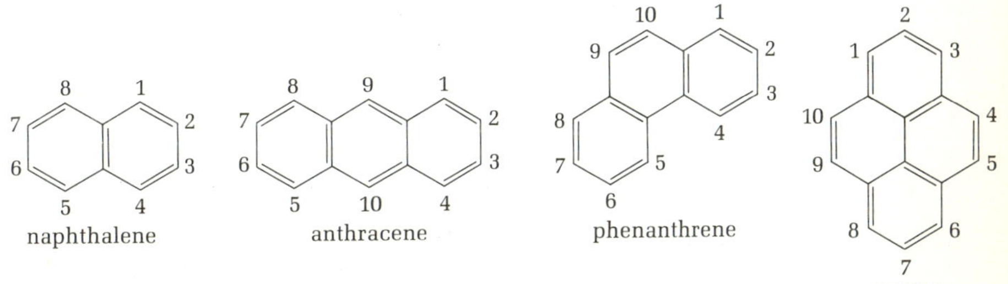

naphthalene anthracene phenanthrene pyrene

Although the chemical behavior of many condensed aromatic compounds resembles that of benzene, particularly insofar as electrophilic substitution reactions are concerned (next chapter), many anomalies have been discovered. In fact, it soon becomes clear that although all the individual rings in a given compound may appear to be aromatic, some are more aromatic than others. For instance, one ring of naphthalene may be reduced under conditions that leave the second ring unaffected.

(14.32)

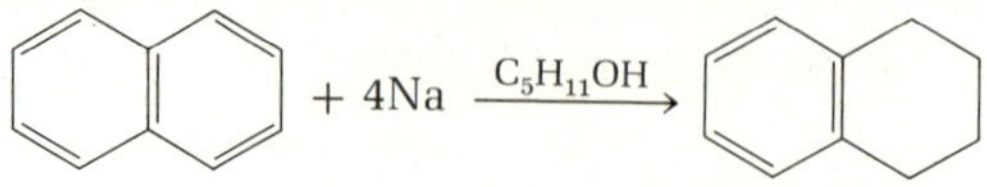

1,2,3,4-tetrahydronaphthalene
(tetralin)

Similarly, benzene does not undergo the Diels-Alder reaction, while certain polymethylbenzenes combine with only the most powerful dienophiles. These facts accord with aromatic, nonolefin nature of benzene. Naphthalene, by contrast, combines with a moderate dienophile like maleic anhydride under forcing conditions, and substituted naphthalenes react with this dienophile under nonextreme conditions.

(14.33)

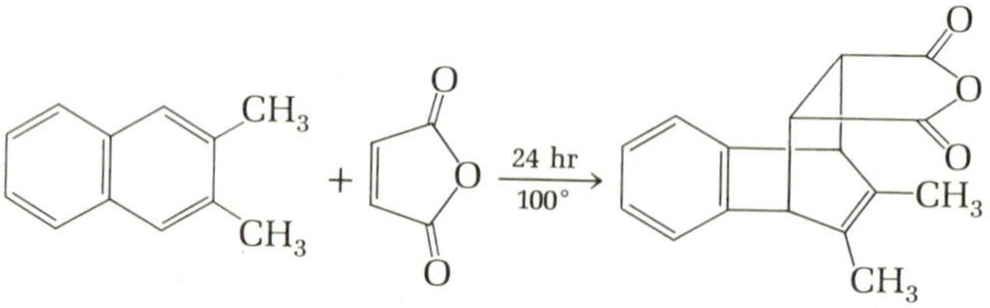

The tricyclic molecules provide even more graphic examples. Anthracene undergoes Diels-Alder additions very readily in the central ring, at the 9 and 10 positions. Phenanthrene suffers many addition reactions characteristic of alkenes at the 9,10-double bond.

(14.34)

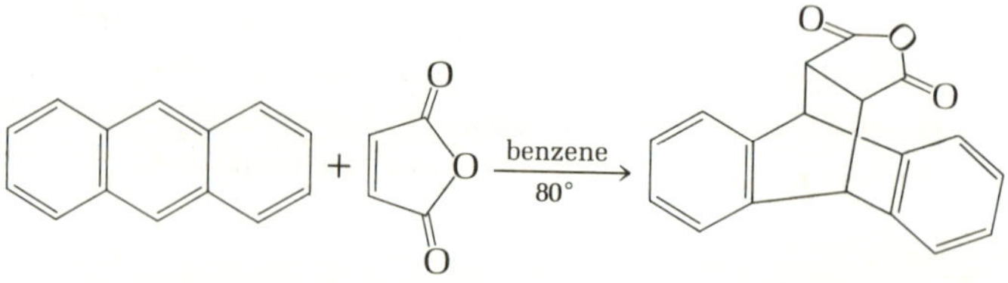

(100%)

(14.35)

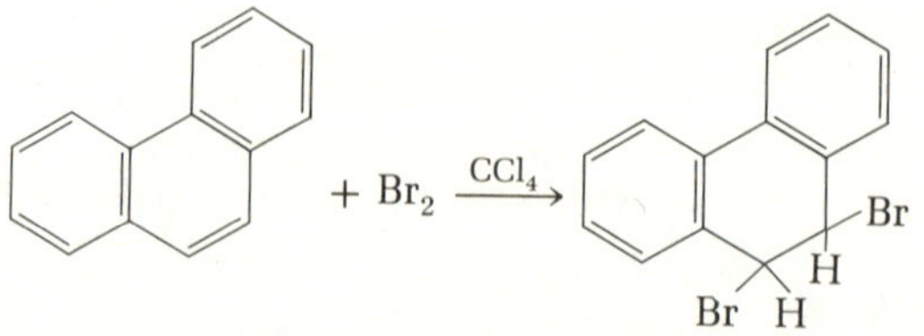

Note that in the two examples cited the destruction of the conjugation in one ring results in the formation of two completely benzene-like rings; *i.e.*, they each have their own sextet of π-electrons that need no longer be shared with a neighboring ring. One might regard this result in a very qualitative way as providing part of the driving force for these reactions.

SUMMARY

1 Benzene is a hydrocarbon with molecular formula C_6H_6. The formula suggests extensive unsaturation but benzene resists halogenation and oxidation under conditions that give rapid results with alkenes. Only one monosubstitution product of benzene with a given functional group is known and only three disubstitution products with any particular pair of substituents. These data suggest that benzene is highly symmetrical. The hexagon and the triangular prism were two geometrical forms that had been suggested that satisfied the isomerism criterion. Kekulé proposed a hexagon with alternating single and double bonds that were ceaselessly interchanging positions. Modern investigations point strongly to a structure with fixed equilibrium bond lengths and the *absence* of any tendency to oscillate between equivalent cyclohexatriene structures.

2 X-Ray analysis of benzene indicates that the carbon atoms lie at the vertices of a regular plane hexagon, each one bearing a hydrogen atom. The carbon-carbon distance is intermediate between that found in single bonds and in double bonds. Molecular orbital calculations correctly predict the geometry of benzene based on a model in which there is continuous overlap of neighboring carbon 2*p*-orbitals. The π-orbitals so formed allow extensive electron delocalization, which accounts in part for the relative stability of benzene. This delocalization makes itself felt when the molecule is placed in a magnetic field. A ring current is induced that leads in turn to the appearance of a local magnetic field around the benzene ring. The local field deshields the hydrogen nuclei, causing a low field proton resonance for benzene.

3 "Aromaticity" (enhanced stability and diminished reactivity) is not a phenomenon unique to benzene. Theory predicts that monocyclic, planar conjugated hydrocarbons having $(4n + 2)$ π-electrons will be aromatic; this prediction has been verified for the lower members of the series. These compounds are more stable than their open-chain analogues. The opposite is true in the $4n$ π-electron series; the cyclic members are antiaromatic. The criteria used for verifying aromaticity and antiaromaticity are often spectroscopic ones.

4 Ions as well as neutral molecules may be aromatic. The best known aromatic ions are the CH-homologues of benzene, cyclopentadienyl anion and cycloheptatrienyl cation (tropylium ion). There are several tropylium salts that are easily isolable and quite stable. Aromatic ions having *n* less than and greater than unity are also known; some doubly charged aromatic ions have even been generated. As expected, the aromatic ions are more reactive than the corresponding neutral molecules with the same number of π-electrons.

5 Coal tar, after appropriate work-up, yields a number of substituted benzenes as well as a variety of condensed aromatic systems. These latter are generally more reactive than benzene and as the number of rings increases, parts of the molecule become rather alkene-like in behavior.

PROBLEMS

Note: Problems 1 through 4 appear in text.

5 As would be expected from the known structure of benzene, all disubstituted benzene derivatives C_6H_4XY have been found to be *meso* compounds. Which of the various cyclic structures previously proposed for benzene would be placed seriously in doubt by this finding?

6 It has been shown that benzene will undergo ozonolysis, albeit quite slowly compared with ordinary alkenes. After hydrolytic work-up, the product obtained was glyoxal, $H\overset{\overset{O}{\|}}{C}\overset{\overset{O}{\|}}{C}H$. What products would be expected from ozonolysis of *o*-, *m*-, and *p*-xylene (1,2-, 1,3-, and 1,4-dimethylbenzene)? Could this method be used to distinguish among *ortho, meta,* and *para* isomers?

7 a. From the bromination of benzene three separate dibromobenzenes ($C_6H_4Br_2$) have been isolated. In the absence of spectroscopic methods there was no unambiguous method for assigning the relative positions of the bromine atoms in the isomers. Upon chlorination of each of the pure isomers, however, they are converted to dibromochlorobenzenes ($C_6H_3Br_2Cl$). By analysis of the chlorination products it is possible to assign structures to the dibromobenzene isomers (*Korner's method*). Outline how this could be done. What *practical* problems might interfere with the application of this technique?

b. With the aid of nmr spectroscopy the three dibromobenzenes could be readily differentiated. Describe the appearance of the spectra of each of the three isomers.

8 Tris(methylene)cyclopropane (III) is an isomer of benzene that can yield only one monosubstituted product.

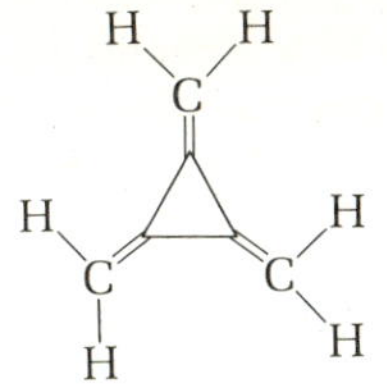

III

a. How many symmetrically disubstituted derivatives of III may be formed? How many unsymmetrical ones?

b. Does structure III qualify as a competitor of the hexagonal structure of benzene *on symmetry grounds*?

9 a. Write out the Diels-Alder dimerization of cyclobutadiene ($2C_4H_4 \longrightarrow C_8H_8$), showing the structures of the reactants and the product. Which molecule is being treated as the diene and which as the dienophile (Chapter 12)? Draw the two stereoisomeric forms of the product.

b. Irradiation of one of the product isomers could yield another isomer that contains *no* unsaturation (cubane). Draw the structure of this product.

10 *o*-Difluorobenzene could be photolyzed to difluoro(Dewar benzene), which upon further photolysis would yield difluoroprismane. Heating the prismane would convert it back, in part, to difluorobenzene. Draw the structures of the intermediate Dewar benzenes and prismanes, and show which difluorobenzene isomers could be derived from each of the latter.

11 The compound *trans*-15,16-dimethyldihydropyrene (IV) was synthesized several years ago by a multistep route.

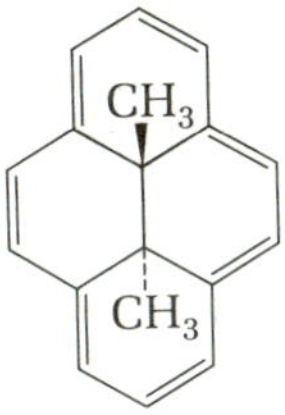

IV

a. How many peaks (neglecting splitting) would one expect to find in the nmr spectrum of IV? What would be their relative areas?

b. Assign approximate chemical shifts to the peaks. How would you classify this compound?

12 Pentalene (V) has so far defied attempts at synthesis.

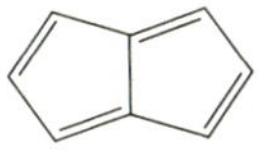

V

When dihydropentalene (VI) is treated with 2 moles of strong base, however, it is readily converted into pentalene dianion (equation 2). How do you explain the observations?

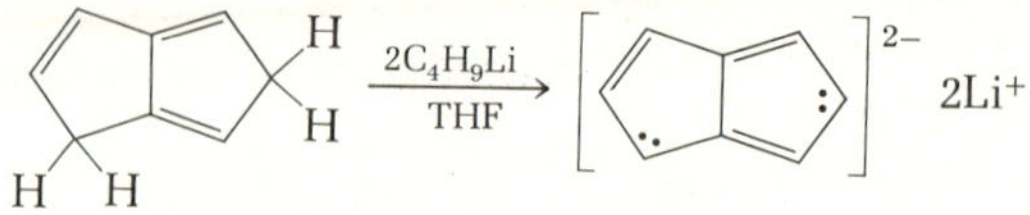

VI

(2)

13 a. Cyclopropanone may be generated and stored in solution at $-78°$ but it decomposes quite rapidly when warmed to $0°$. Cyclopropenone, on the other hand, is a much more stable material despite the fact that it should be suffering more from angle strain than its saturated analogue. Cyclopropenones are also quite susceptible to protonation on oxygen and form salts with strong acids. Similar behavior toward acids is exhibited by tropone (cycloheptatrienone). The stretching frequency of the carbonyl group is unusually low in tropone. With the aid of canonical structures, explain these observations.

b. By contrast, cyclopentadienone is a very reactive molecule. It has never been isolated because it undergoes a very facile Diels-Alder reaction with itself; only heavily substituted derivatives have been obtained. Rationalize this finding.

14 a. Addition of bromine to *cis*-1-phenylpropene produces a mixture of *erythro*- and *threo*-dibromides (write out their structures). How do you account for this nonstereospecific addition reaction?

b. Addition of hydrogen chloride to 3-phenylpropene and 1-phenylpropene leads to a different product in each case (write out their structures). Which addition would you expect to be the more rapid?

15 The polymer derived from styrene (phenylethylene) is an extremely important commercial material. Styrene is usually polymerized with the aid of a peroxide initiator. Write out the first three steps in the polymerization of styrene.

16 Trityl alcohol (triphenylmethanol) is a white, crystalline solid. When a small quantity of the compound is dissolved in concentrated sulfuric acid, an intensely red solution is formed. As water is added to this solution, the color begins to fade and a white solid precipitates. Eventually the solution becomes completely colorless; the white solid is collected and upon examination proves to be trityl alcohol. Rationalize these observations.

17 The dissociation constant (equation 14.29) of hexaphenylethane in benzene is $2 \times 10^{-4}M$ at 25°. If the *para* positions in four of the benzene rings (two on each side) are substituted by methyl groups, the dissociation constant increases to $1 \times 10^{-3}M$. If the four methyl groups are placed at *ortho*, rather than *para*, positions in the same rings, the dissociation constant becomes 2. Explain all these data.

18 a. 1,3-Cyclohexadiene + $CH_3O\overset{O}{\overset{\|}{C}}C{\equiv}C\overset{O}{\overset{\|}{C}}OCH_3 \longrightarrow$ **A** $(C_{12}H_{14}O_4) \xrightarrow{\text{heat}} C_2H_4 +$ **B**

b. Naphthalene $\xrightarrow[H_2SO_4,\ heat]{CrO_3}$ **C** ($C_8H_6O_4$) $\xrightarrow[H^+]{CH_3OH}$ **B.** Draw structures for **A, B** and **C.**

c. The type of reaction sequence outlined in part a is of general value for synthesizing substituted benzenes. Why does it work so well, and why should it be less successful with cyclopentadiene instead of 1,3-cyclohexadiene?

SUPPLEMENTARY READINGS

"Kekulë Symposium," *Advances in Chemistry Series,* No. 61, American Chemical Society, Washington, D.C., 1966.

R. Kaiser, "The Discovery of Benzene," *Angewandte Chemie International Edition in English,* **7**, 345 (1968).

G. M. Badger, *Aromatic Character and Aromaticity.* London, England: Cambridge University Press, 1969.

D. Ginsburg, ed., *Non-benzenoid Aromatic Compounds.* New York: Interscience Pub., 1959.

CHAPTER 15

ELECTROPHILIC SUBSTITUTION—DISRUPTION AND REFORMATION OF THE AROMATIC SEXTET

AROMATIC π-COMPLEXES

Chapter 14 presents a picture of benzene in which the π-electrons are shown to be relatively low in energy, delocalized, and highly mobile. It seems perhaps to suggest that benzene is qualitatively different from other unsaturated molecules. In this chapter we modify that conception somewhat; under some conditions the chemical behavior of benzene and its derivatives is quite olefin-like, while in other circumstances the resemblance is more remote. Overall, the chemical differences between benzene and the alkenes would seem to be mostly quantitative — certain reactions characteristic of the alkenes are infrequently encountered in aromatic chemistry only because they occur slowly compared to the rates of competing processes.

One property of benzene and many of its derivatives that is reminiscent of the alkenes is their ability to form complexes. In the majority of cases the aromatic ring is the electron donor, complexing with substances more electrophilic than itself. If the aromatic ring is substituted with a number of electron-withdrawing groups, however, it may then function as the electrophile.

As is true of the alkenes, the electrophiles are bound to the aromatic ring principally by way of the π-electrons and the resultant complexes may be classified as π-complexes. They may in turn be subdivided into two groups:

1 Weak complexes that are readily dissociable.

2 Strong complexes in which the electrophile is tightly bound to the ring, which can best be thought of as compounds in the usual sense.

Typical examples of complexes of the first kind are those that form between benzene derivatives and the molecular halogens. As noted in Chapter 14, solutions of chlorine, bromine, or iodine in benzene are quite stable in the absence of catalysts, coreactants, or ultraviolet light.

(15.1) $I_2 + C_6H_6 \rightleftharpoons C_6H_6 \cdot I_2 \qquad K^{25°} = 0.15$

There is spectroscopic evidence that molecular complexes are formed reversibly in these solutions, however. Moreover, at low temperatures it is possible to obtain these complexes in crystalline form but they are only stable within the crystal lattice.

(15.2) $C_6H_6 + Br_2 \xrightarrow{-40°} \underset{(mp\ -\ 14°)}{C_6H_6 \cdot Br_2 \downarrow}$

In the crystal the geometry of these complexes is such that the axis of the halogen molecule is perpendicular to the plane of the ring and lies directly above its center. The halogen atom nearest the ring is thus symmetrically coordinated to all the π-electrons. The crystal consists of molecular chains formed from alternating benzene rings and halogen molecules (Figure 15-1).

The solubility of the hydrogen halides in aromatic hydrocarbons substantially exceeds their solubility in related saturated hydrocarbons. Vapor pressure data, as well as infrared measurements, support the picture of the hydrogen halide (HF, HCl) weakly complexed with the aromatic ring.

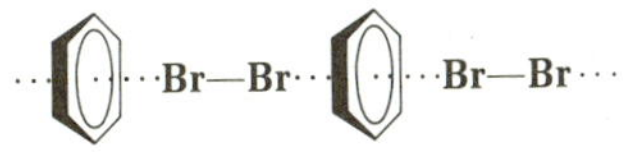

FIGURE 15-1

Benzene-Bromine Complex in the Solid State

(15.3) 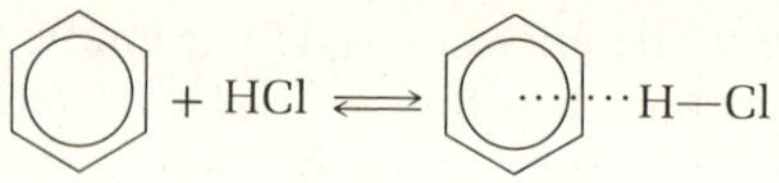

FIGURE 15-2
Benzene-Silver Ion Complex

They are held together by a hydrogen bond involving the π-electrons of the ring as a whole.

Certain metals ions, such as Cu(I) and Ag(I), form weak aromatic π-complexes in which the ion is situated directly over a particular C—C bond. They resemble typical silver-olefin complexes (Figure 15-2). X-Ray analysis of these complexes shows that the ring no longer retains hexagonal symmetry; inequalities have appeared in the C—C bond lengths.

(15.4) 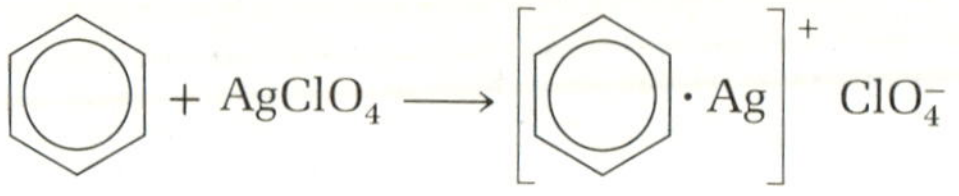

This is significant, for it shows that an electrophile can effectively distort the symmetrical π-electron distribution characteristic of many aromatic systems.

Aromatic rings form relatively stable complexes with a wide variety of transition metals in which the metal atom is symmetrically disposed with respect to the π-electrons. Dibenzenechromium is an example of a complex containing a zerovalent metal. It has the structure shown in Figure 15-3 and belongs to the group of "sandwich complexes."

(15.5) $$C_6H_6 \text{ (excess)} + CrCl_3 + Al \xrightarrow{\text{heat}} (C_6H_6)_2Cr + AlCl_3$$

The aromatic cyclopentadienide anion is much more effective than benzene in sandwich complex formation; the first of these ever prepared was dicyclopentadienyliron **(ferrocene),** by the reaction of the anion with a ferrous salt in a nonaqueous solvent.

(15.6) $$2Na^+C_5H_5^- + FeCl_2 \xrightarrow{\text{THF}} \underset{\text{ferrocene}}{(C_5H_5)_2Fe} + 2NaCl$$

The cyclopentadienide rings retain their aromatic character in the complex, undergoing many substitution reactions characteristic of benzene.

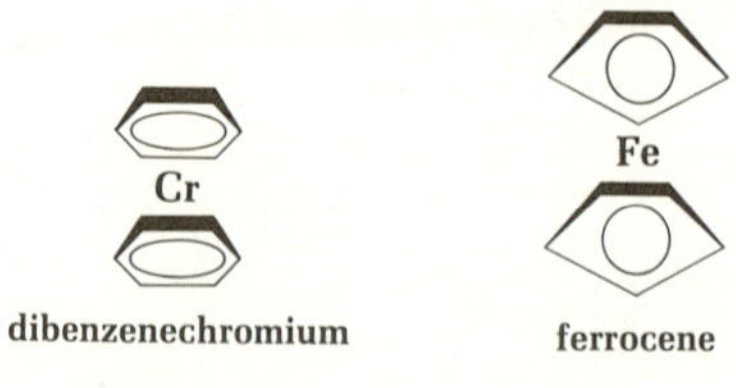

FIGURE 15-3
Structure of Dibenzenechromium and Ferrocene

These rings are, in fact, less reactive in the complex than they are when free, due to the reduced negative charge present in the complexed ring. Cyclopentadienyl π-complexes of numerous other transition metals are known, as well as complexes involving a variety of other aromatic systems.

σ-COMPLEXES AND ELECTROPHILIC SUBSTITUTION

Solutions of HCl or HF in aromatic hydrocarbons at low temperatures give every indication of being just π-complexes; there is no evidence of reaction between the two components. Upon addition of a Lewis acid such as $AlCl_3$ or BF_3, which promotes dissociation of the hydrogen halide, the character of the solution changes markedly. It becomes colored and acquires the ability to conduct an electric current. Clearly ions are being formed; all the evidence so far obtained supports the view that the acid proton has become attached to one of the ring carbon atoms by means of a σ-bond, giving rise to a **σ-complex.**

(15.7)

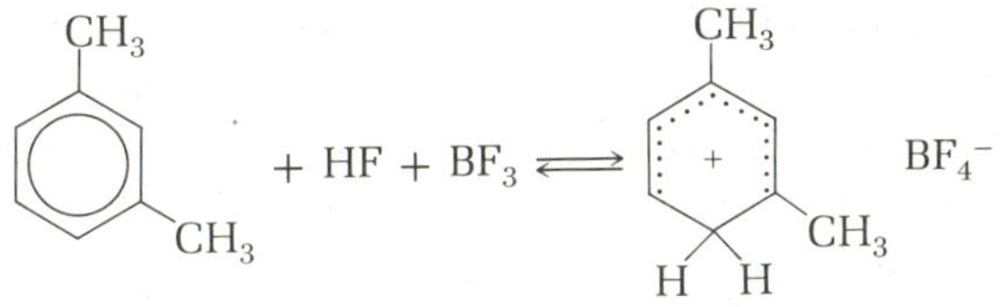

σ-complex (mp −55°)

In order to attach a hydrogen ion to the ring, two of the π-electrons must be used to create the new σ-bond. The carbon atom to which the new bond is formed will then have sp^3-hybridization; a total of four π-electrons remain to be distributed over the remaining five sp^2-hybridized carbon atoms. The formation of the dimethyl-substituted **benzenium ion (1)** is an endothermic process since it entails the disruption of the cyclic conjugation of the aromatic ring.

(15.8)

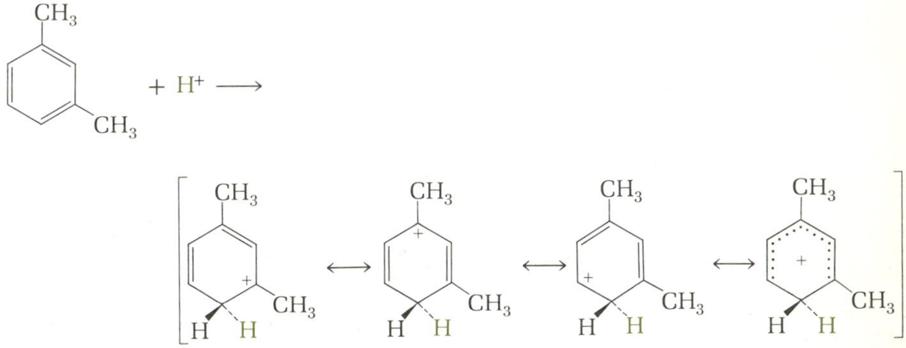

1

The ion is energetically accessible as an intermediate in part, however, because of the delocalization of the remaining π-electrons.

When a σ-complex of the type shown in equation 15.7 is allowed to warm up, it decomposes with evolution of HF and BF_3. If DF rather than HF is employed in the formation of the σ-complex, then the decomposition of the latter will lead to partial incorporation of deuterium into the aromatic compound.

(15.9)

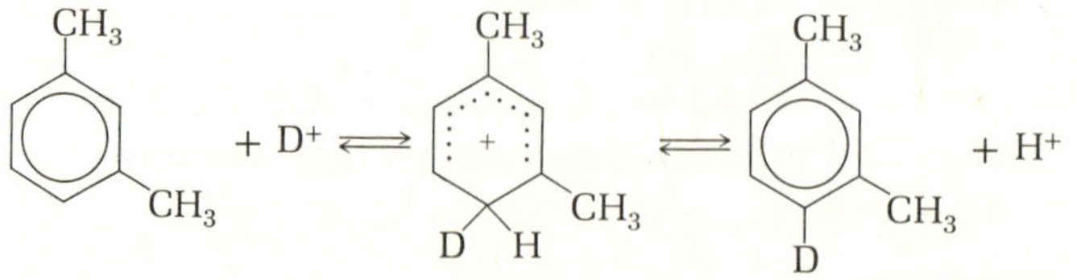

Decomposition of the benzenium ion intermediate is, of course, favored because the resonance stabilization of the ring is regained. Deuterium exchange in benzene and its derivatives can be accomplished just by shaking a mixture of the hydrocarbon and a strong mineral acid, such as D_2SO_4-D_2O, at room temperature.

The dissociation energies of the C—H and C—D bonds are nearly identical so that the energy profile for the hydrogen-deuterium reaction in benzene resembles that show in Figure 15-4. Note that the transition states are close to the intermediate in structure; there must be a substantial fraction of charge in the ring at the transition state.

It would be worthwhile to look a bit harder at the benzenium ion in molecular orbital terms. There are five carbon 2p-orbitals available to form five π-orbitals. Two of these are bonding (and contain the four π-electrons); two are antibonding; the fifth π-orbital is nonbonding. In a neutral benzenium radical there would be one electron in this nonbonding orbital. Since the nonbonding orbital is empty in the ion, those carbon atoms whose atomic

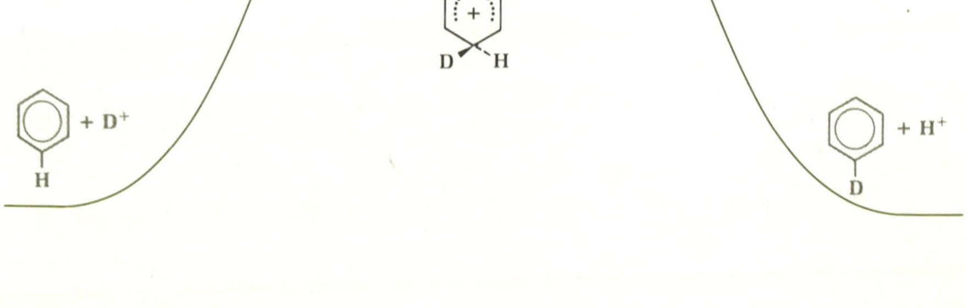

FIGURE 15-4

Energy Profile for the Isotope Exchange Reaction of Benzene

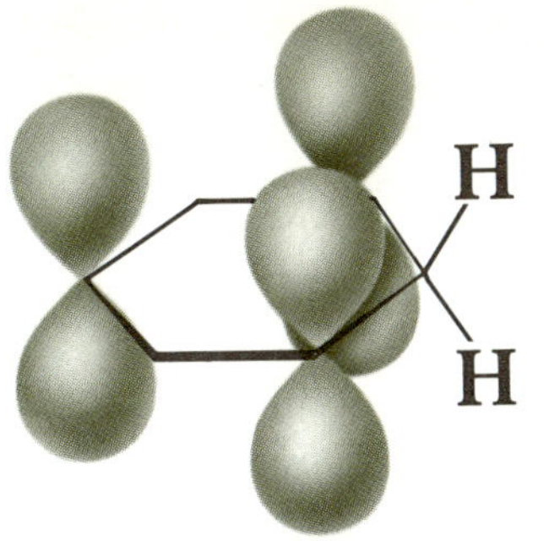

FIGURE 15-5
Nonbonding Orbital in Benzenium Ion

orbitals contribute to it (Figure 15-5) will be the ones that are most electron-deficient. The distribution of these orbitals confirms our expectation from the resonance structures (equation 15.8) that the charge will be borne by the carbon atoms *ortho* and *para* to the tetrahedral atom.

Although proton exchange reactions in aromatic compounds are not intrinsically very important, the mechanistic pattern that has been established for these reactions provides a prototype for *all* electrophilic aromatic substitution reactions, many of which are subsequently discussed. That pattern is as follows:

1 An electrophile becomes attached to one of the carbon atoms of the ring, converting the aromatic compound into a benzenium ion (this is usually the rate-determining step of the reaction).

2 The benzenium ion rearomatizes by ejecting the proton from the carbon atom carrying the electrophile.

It is noteworthy that step 2 is apparently much faster than reaction of the benzenium ion with a nucleophile. Here is a real difference in behavior between aromatic compounds and alkenes.

ELECTROPHILIC SUBSTITUTION OF BENZENE

Electrophilic substitution reactions are ubiquitous in aromatic chemistry — they hold the same position of importance that nucleophilic substitutions do in aliphatic chemistry. A few of the more common ones are set out in equations 15.10 to 15.15.

(15.10)

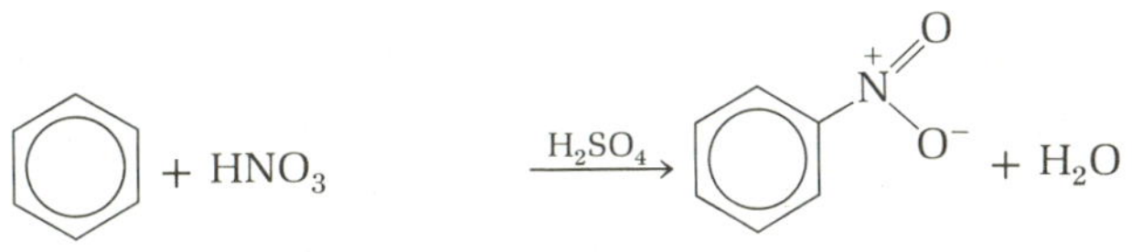

nitrobenzene

(15.11) 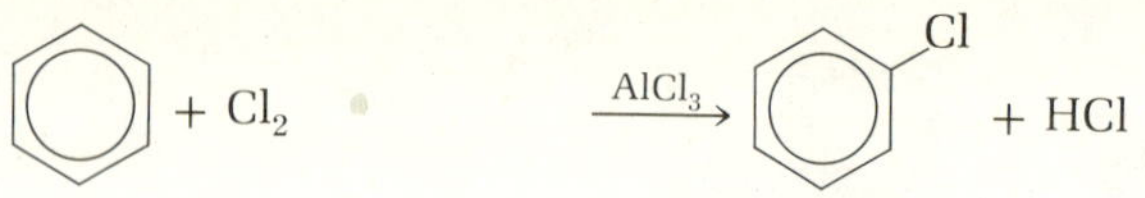

chlorobenzene

(15.12) 

bromobenzene

(15.13) 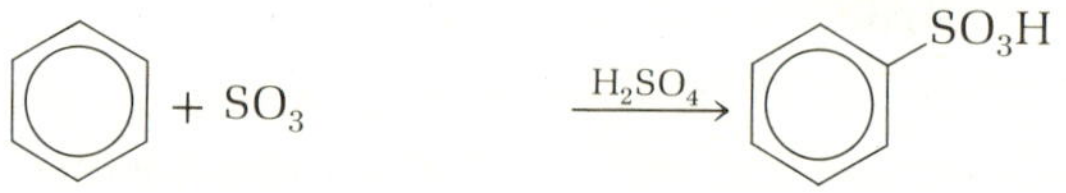

benzenesulfonic acid

(15.14) 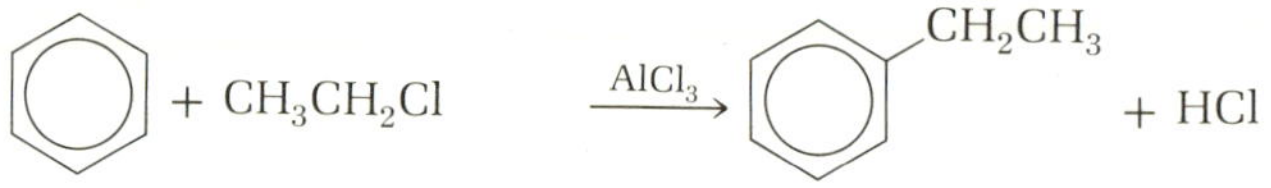

ethylbenzene

(15.15) 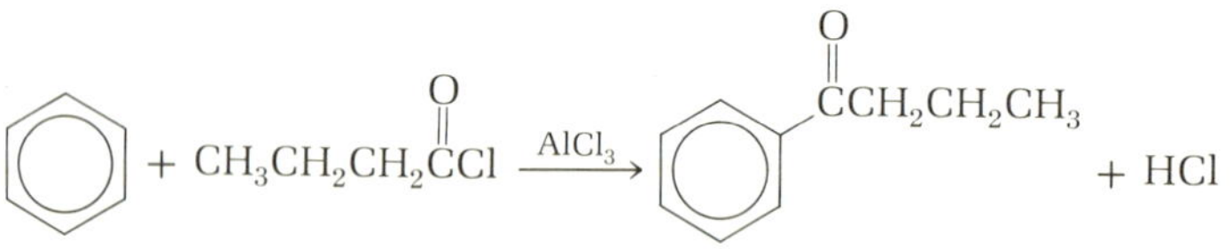

butyrophenone

In all cases cited the reactions require an acidic catalyst, either of the proton or of the Lewis type. The function of these catalysts is to generate electrophilic species sufficiently reactive to attack the benzene ring at a convenient rate. In the last two examples cited, the electrophiles are carbonium ions; these **Friedel-Crafts reactions** are discussed in detail in Chapter 16. It should be borne in mind, however, that they are not mechanistically distinct from the other reactions.

NITRATION

A good deal of effort has been expended in identifying the electrophilic species in each reaction. Chapter 25 details the investigations that have been made into the mechanism of nitration. It is sufficient to state at this time that nitrations with "mixed acid" (a mixture of concentrated nitric acid and sulfuric acid) take place by way of **nitronium ion,** $O{=}\overset{+}{N}{=}O$. This ion is isoelectronic with carbon dioxide, from which it may be obtained (on paper) by substituting a nitrogen atom for a carbon atom.

(15.16) $HNO_3 + 2H_2SO_4 \rightleftharpoons NO_2^+ \, HSO_4^- + H_3O^+ \, HSO_4^-$

Salts of the nitronium cation containing nonnucleophilic anions may be prepared and these salts may be used as nitrating agents.

(15.17) $BF_3 + HF + HNO_3 \text{ (red fuming)} \xrightarrow{CH_2Cl_2} NO_2^+ \, BF_4^- + H_2O$

This method of nitration is much less convenient than use of the mixed acid but it provides an alternative in cases when the aromatic ring carries an acid-sensitive substituent.

(15.18)

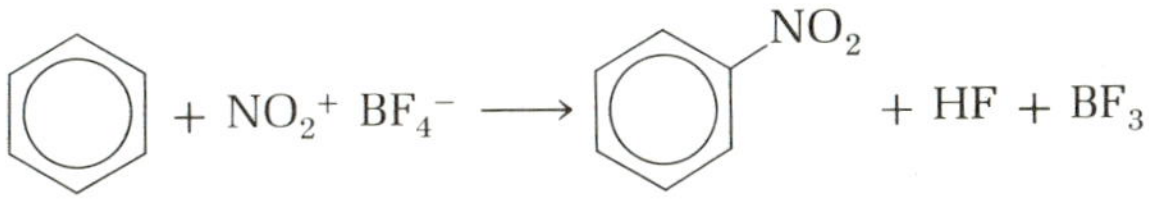

The mechanism of nitration meets our expectations based on the discussion of proton exchange (in equation 15.20, B: represents a weak base, such as HSO_4^-).

(15.19)

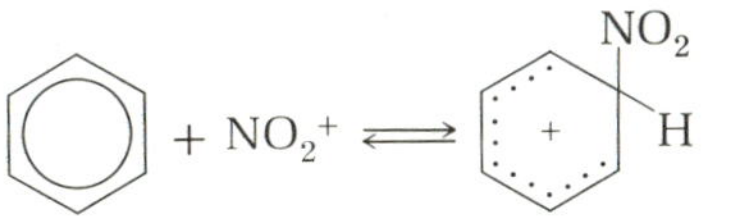

(15.20)

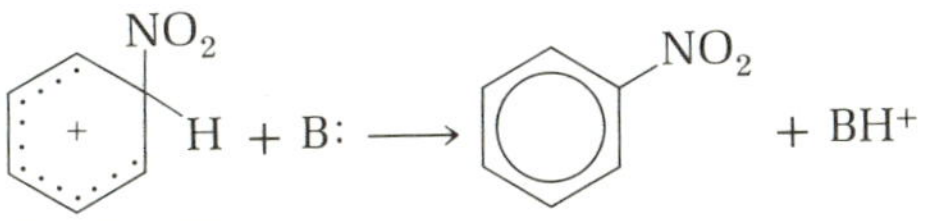

The **nitro group** is attached to the benzene ring by way of the positively charged nitrogen atom, the negative charge being delocalized over both oxygen atoms.

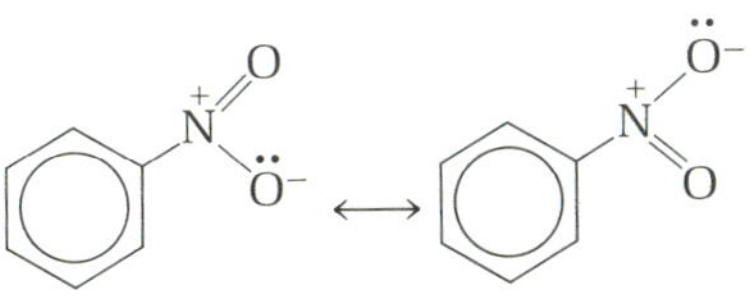

The nitro group as a whole is electrically neutral. It is important not to confuse nitrobenzene with its isomers, such as phenyl nitrite **(2)**, which have very different chemical properties.

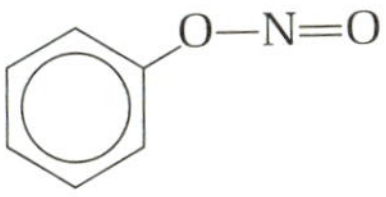

2

HALOGENATION

Bromination and chlorination of aromatic compounds require the use of a Lewis acid as catalyst — $AlCl_3$, $AlBr_3$, $FeCl_3$, and $FeBr_3$ are the most common. The most likely mechanism is one in which the molecular halogen is polarized by complex formation with the metal halide, and the complex then transfers $:\ddot{Cl}^+$ or $:\ddot{Br}^+$ to the aromatic ring. There is no firm evidence that free chloronium (Cl^+) or bromonium (Br^+) ions are formed.

(15.21)

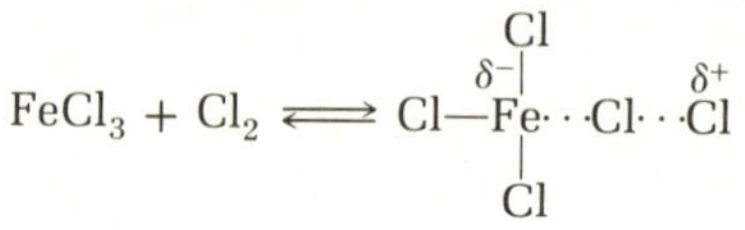

(15.22)

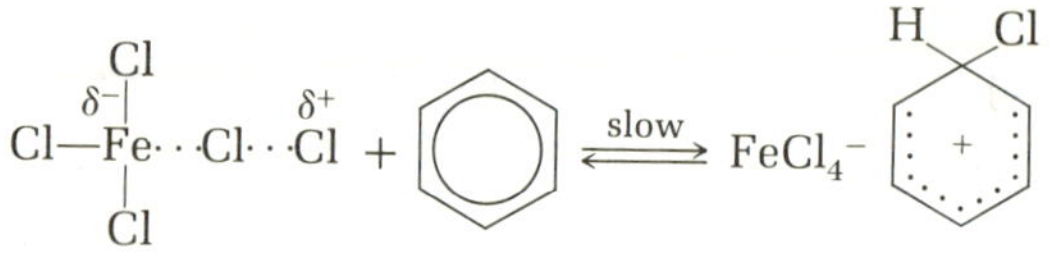

(15.23)

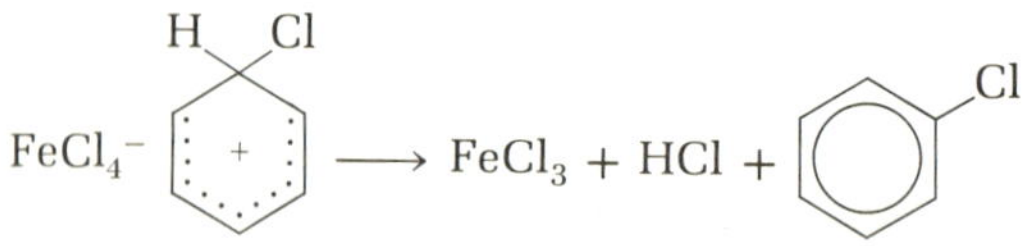

Aromatic rings may also be halogenated by a mixture of a hypohalous acid and a mineral acid; the electrophile in these cases is probably H_2OBr^+ and H_2OCl^+.

As usual fluorine is so reactive that fluorbenzene cannot be obtained by the methods that served for the production of chlorobenzene and bromobenzene. When fluorine and benzene react, a variety of products are obtained, most of which are formed by way of free radical processes. On the other hand, the equilibrium is unfavorable for the direct formation of iodobenzenes, and many iodobenzene derivatives are formed by indirect methods (Chapter 21).

(15.24)

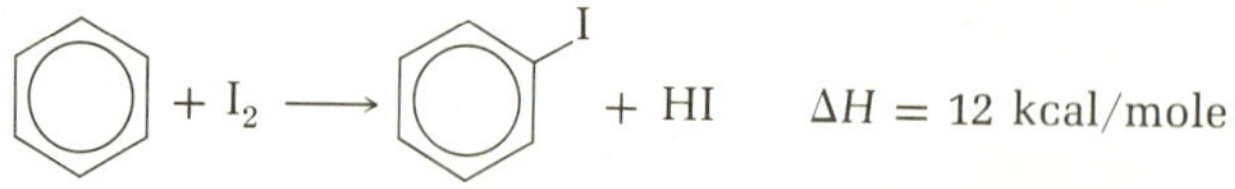

Direct iodination is feasible if the hydrogen iodide formed is removed chemically; its oxidation with nitric acid is a common alternative to the method shown in equation 15.25.

(15.25)

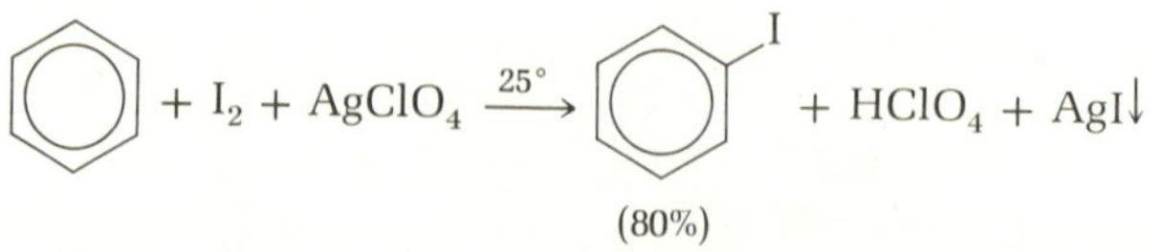

SULFONATION

The **sulfonation** of aromatic compounds can be brought about by ordinary concentrated sulfuric acid or by fuming sulfuric acid (H_2SO_4-SO_3), the latter being very much more reactive. In either case the electrophilic species is probably sulfur trioxide, SO_3. The intermediate formed in these substitutions is then neutral overall, in contrast to the cationic intermediates present in the other electrophilic aromatic substitution reactions.

(15.26)
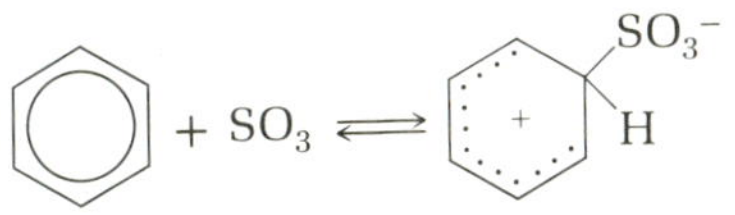

(15.27)
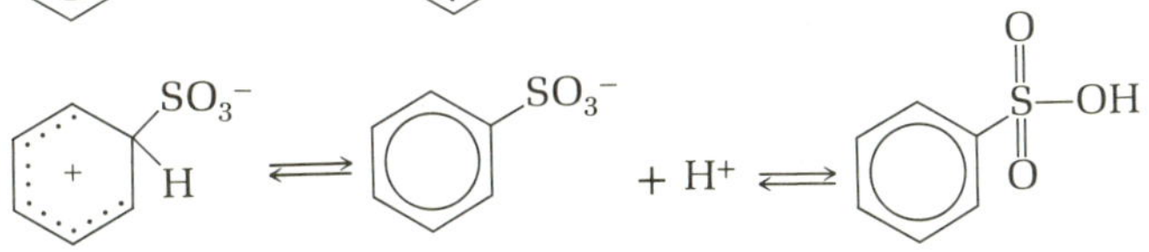

Benzenesulfonic acid is a reasonably strong acid, its K_a being slightly *larger* than the first K_a for sulfuric acid.

(15.28)
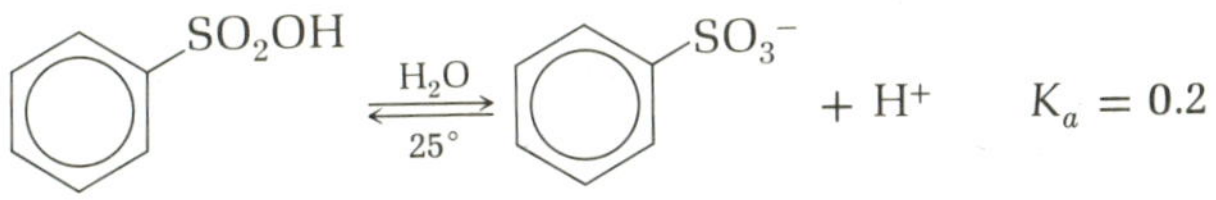

This allows the neutral intermediate to be regenerated by ring protonation of the benzenesulfonate anion (reverse of equation 15.27); the result is a relatively easy reversal of the sulfonation reaction.

REVERSIBLE ELECTROPHILIC SUBSTITUTIONS

Benzenesulfonic acid itself may be desulfonated by boiling it with aqueous sulfuric acid. The sulfonic acid group may be replaced by other electrophiles in the course of an aromatic substitution, particularly when the reaction is run in water.

(15.29)
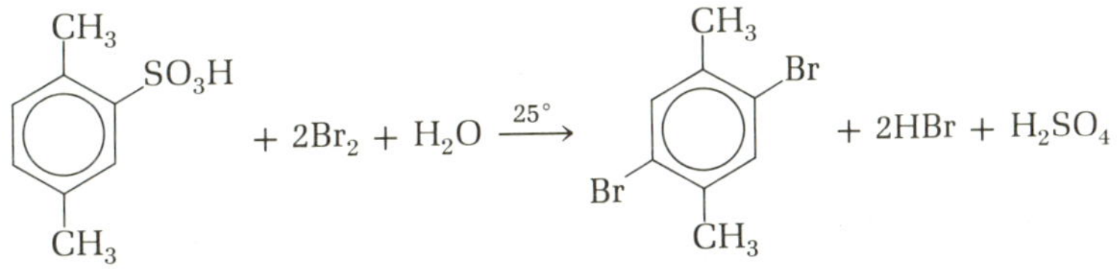

One of the most graphic instances of reversible sulfonation occurs with naphthalene. When this hydrocarbon is sulfonated at 40°, the preponderant product is naphthalene-1-sulfonic acid (α-naphthalenesulfonic acid); the

α-acid is formed much more rapidly than the β-acid. Heating the reaction mixture to 160°, however, or carrying out the sulfonation at that temperature from the start, gives a product mixture rich in β-naphthalenesulfonic acid.

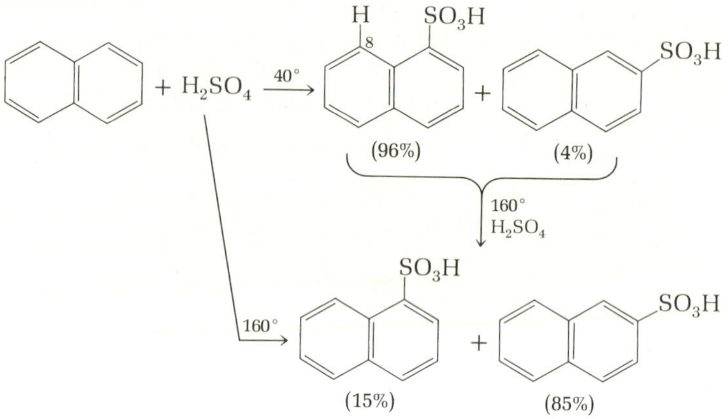

(15.30)

The β-acid is the more stable isomer because in the α-compound there is interference between the sulfonic acid group and hydrogen atom at the 8 position (*peri*-hydrogen). Thus, at 40° the product mixture is kinetically determined but at 160° equilibrium is rapidly attained by way of a desulfonation-resulfonation process and the most stable isomer predominates (Chapter 12, p. 294).

As noted above, iodination of aromatic compounds is quite readily reversible; bromination is reversible if "activating groups" (see below) are present on the ring (equation 15.31).

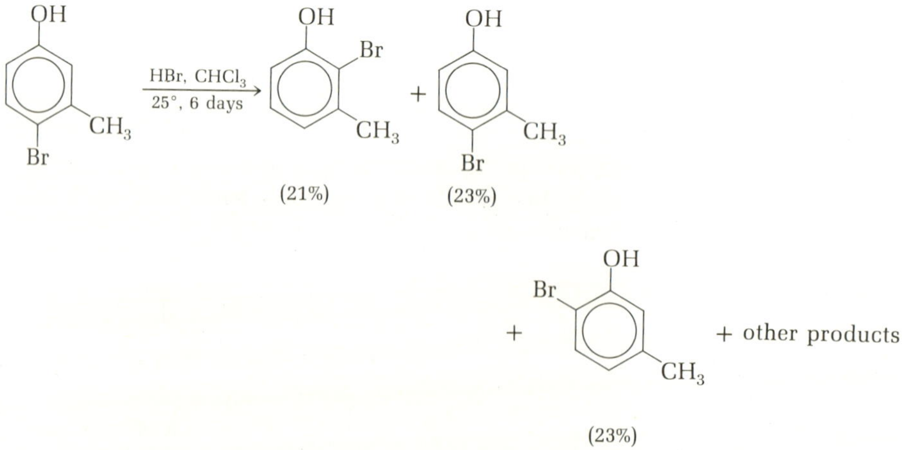

(15.31)

There is no evidence that chlorination or nitration are measurably reversible. The reversibility of electrophilic substitution seems to be related to the reactivity of the electrophile itself. Nitronium ion is probably the most powerful electrophile that is commonly employed in aromatic chemistry; molecular bromine and iodine, on the other hand, are relatively weak electrophiles.

EFFECT OF SUBSTITUENTS ON RATE

It had been observed some time ago that the presence of certain substituents on a benzene ring tended to hinder the introduction of a second group by way of an electrophilic substitution reaction. These substituents have been labeled **deactivating substituents** since they reduce the rate of any given substitution reaction when they are present in the aromatic ring, compared to the rate of the same reaction involving the unsubstituted ring (equations 15.32 and 15.33).

(15.32)

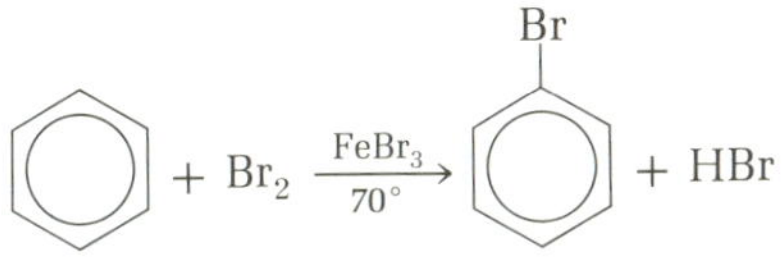

(15.33)

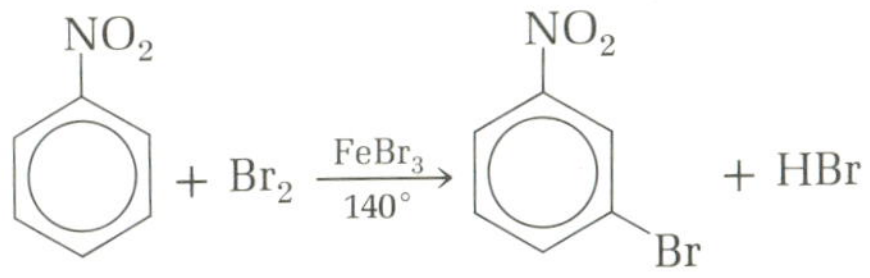

A partial list of deactivating substituents is given in Table 15-1 (the term **aryl** means an aromatic ring acting as a substituent).

TABLE 15-1

Deactivating Substituents in Electrophilic Aromatic Substitution Reactions

	No R	R=H	R=alkyl	R=aryl
—F	Fluoro			
—Cl	Chloro			
—Br	Bromo			
—I	Iodo			
$-NO_2$	Nitro			
$-\overset{O}{\overset{\Vert}{C}}-R$		Formyl	Acyl	Aroyl
$-\overset{O}{\overset{\Vert}{C}}-OR$		Carboxylic acid	Alkoxy-carbonyl	Aryloxy-carbonyl
$-SO_3H$	Sulfonic acid			
—CN	Cyano			
$-NR_3^+$		Ammonium	Mono-, di-, or trialkyl-ammonium	Mono-, di-, or triaryl-ammonium
$-CF_3$	Trifluoromethyl			
$-CCl_3$	Trichloromethyl			

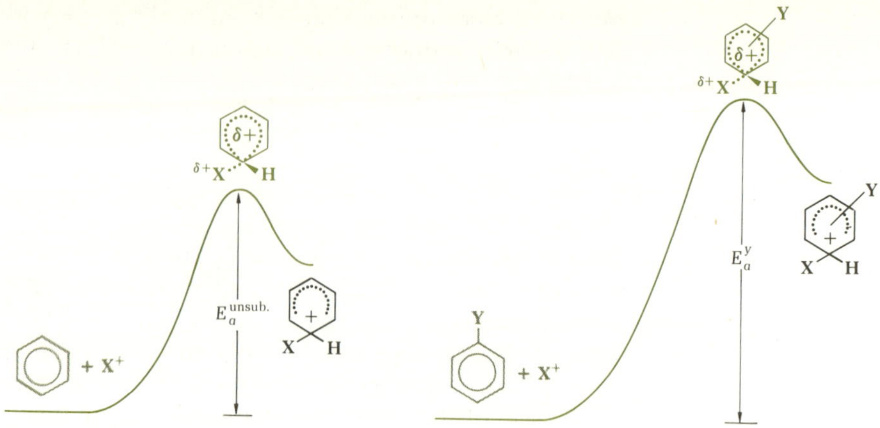

FIGURE 15-6

Activation Diagrams for Electrophilic Attack on Benzene and on a Deactivated Benzene

In general a substituent that decreases the reactivity of the ring is one that raises the free energy of activation. The substituent could exert its influence in two ways: by stabilizing the ground state of the molecule and/or by destabilizing the transition state. In our analysis we concentrate mostly on transition state effects since this procedure seems to give a coherent and self-consistent rationale of the observed data.

All the substituents listed in Table 15-1 contain electronegative atoms and would be expected to be *electron-withdrawing* when compared to a hydrogen atom. Such a substituent would be likely to destabilize an electron-deficient transition state leading to a positively charged benzenium ion and thereby raise the energy of activation (Figure 15-6).

The substituents may be divided into two categories. The first category embraces those groups whose effect on the reaction is exerted primarily by way of an inductive effect — the halogens, trifluoromethyl (CF_3) and trichloromethyl (CCl_3), and the ammonium (NR_3^+) substituents. In a molecule with highly polarizable electrons, such as π-electrons, the inductive effect might be considerable. An alternative, or complementary, explanation focuses on the bond dipole between the substituent and the ring and is discussed later in this chapter.

The other class of deactivating substituents have one feature in common: They contain the partial grouping —X=Y attached directly to the benzene ring, in which Y is an atom more electronegative than carbon. This permits reasonable canonical structures in which a negative charge is placed on Y and a positive charge is distributed around the ring.

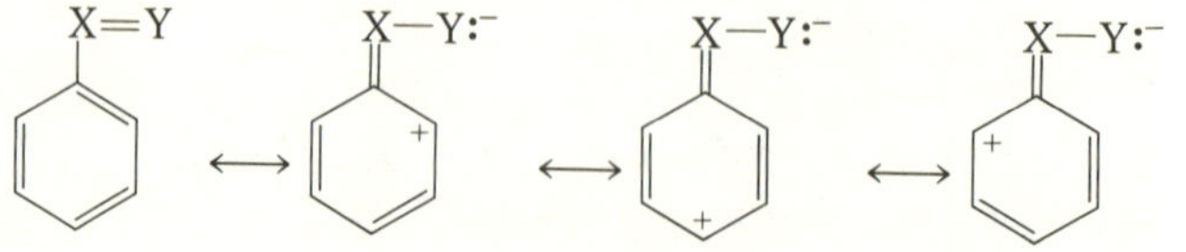

Such resonance interactions stabilize the *ground state* of the molecule but they will be much less important in stabilizing the benzenium ion since they would require removal of electron density from an already electron-deficient ring. Resonance energy is thus *lost* as one goes from the ground state to the transition state and this loss materially increases the free energy of activation.

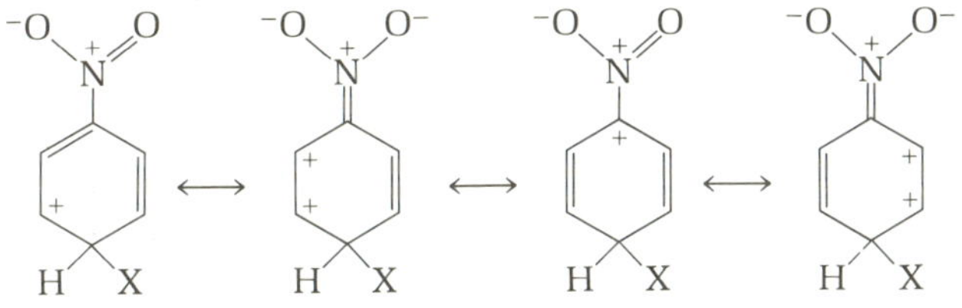

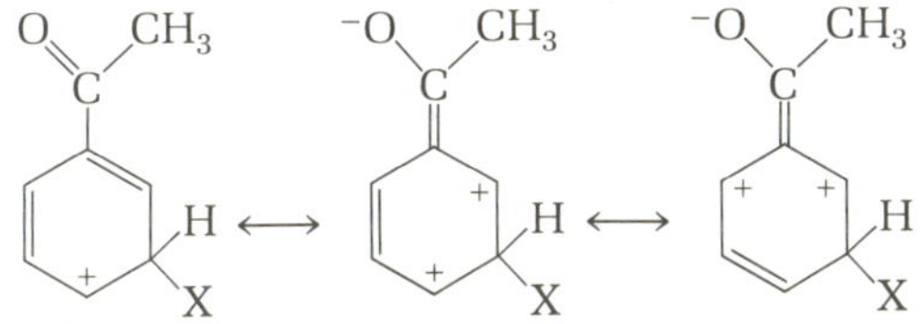

There is a body of substituents that have the opposite effect. Comparison of equations 15.34 and 15.32 leads one to the conclusion that the hydroxyl group has activated the ring toward electrophilic attack.

(15.34) 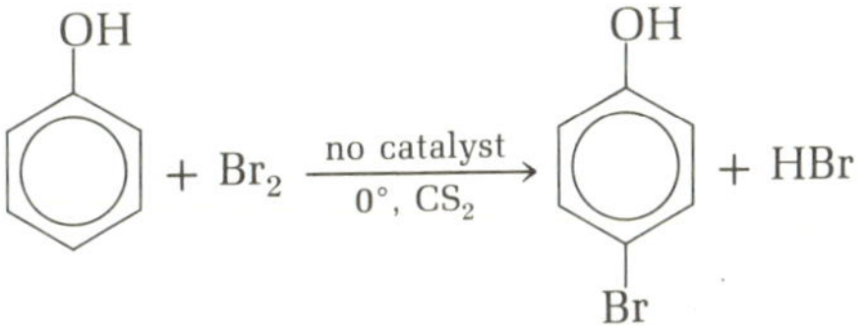

It is in fact one of a number of **activating substituents** (Table 15-2), the presence of which on an aromatic ring increases the rate at which that ring undergoes electrophilic attack.

TABLE 15-2
Activating Substituents in Electrophilic Aromatic Substitution Reactions

	No R	R=H	R=alkyl	R=aryl
—R			Alkyl	Aryl
$—\ddot{N}R_2$		Amino	Mono- or dialkylamino	Mono- or diarylamino
$—\ddot{N}H—\overset{O}{\overset{\Vert}{C}}CH_3$	Acetamido			
$—\ddot{O}R$		Hydroxy	Alkoxy	Aryloxy
$—\ddot{S}R$		Thiol	Alkylthio	Arylthio
$—\ddot{O}\overset{O}{\overset{\Vert}{C}}—R$		Formyloxy	Acyloxy	Aroyloxy

The alkyl substituents are mildly activating and undoubtedly stabilize the electron-deficient transition state by inductively supplying electron density to it. The other substituents listed in Table 15-2 are all more powerfully activating than the alkyl groups and they all share one feature in common: In each one, the atom immediately attached to the ring possesses an unshared electron pair. Conjugation of this electron pair with the benzenium ring greatly stabilizes the ion and decreases the activation energy required for its formation.

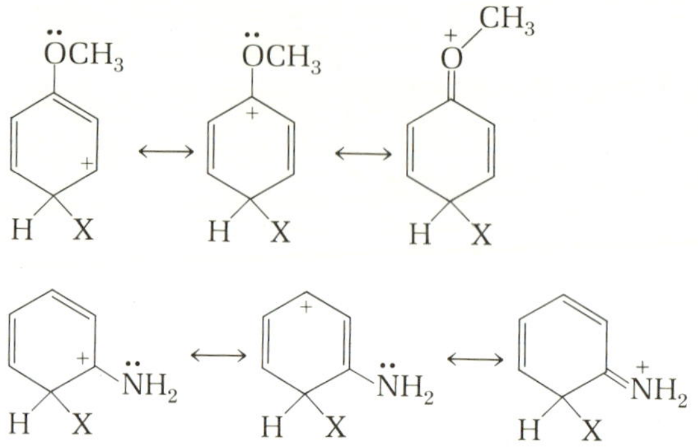

There is undoubtedly some conjugation between the substituent and the ring in the ground state but this interaction becomes much more important in the transition state.

In general, substituents that can conjugate with the ring, both electron-supplying and electron-withdrawing, exert a more pronounced effect on reactivity and orientation (see below) than do substituents that operate purely by induction.

The reactivity of a particular aromatic nucleus determines the reagents used to introduce a given substituent. The nitration of benzene, for instance, is usually carried out with a mixture of concentrated nitric and sulfuric acids but the nitration of *m*-dinitrobenzene involves fuming nitric and sulfuric acids. On the other hand, some phenols and anilines may be nitrated by dilute nitric acid alone. Similarly, the halogenation of unreactive aromatic rings requires $X_2 \cdot FeX_3$ or $H^+ + HOX$ but more reactive compounds may be conveniently halogenated by the molecular halogen without a catalyst.

ORIENTATION IN ELECTROPHILIC ATTACK ON MONOSUBSTITUTED BENZENES

It had been known for some time that certain substituents, when already present in an aromatic ring, exert a powerful directive effect on an incoming electrophile. In many cases a correlation was noted between directive influence and activating or deactivating ability. The bromination of nitrobenzene (equation 15.33), for example, results in the almost exclusive formation of *m*-bromonitrobenzene. By contrast, the predominant product obtained from bromination of phenol (equation 15.34) is *p*-bromophenol, along with lesser amounts of *o*-bromophenol. In order to understand these directive effects, it is most instructive if we examine the transition states for the substitution reactions.

The properties of the transition states cannot, of course, be uncovered by direct observation. There is, however, evidence that indicates that for most electrophilic substitution reactions the transition state resembles the benzenium ion intermediate. **Any substitution that lowers the energy of the benzenium ion also lowers the energy of the transition state leading to it, and vice versa.**

Suppose, for example, that an electrophile A^+ attacks a benzene ring *para* to the substituent B. The positive charge in the intermediate is partially centered on the carbon atom carrying the substituent (those atoms bearing most of the charge are asterisked).

(15.35)

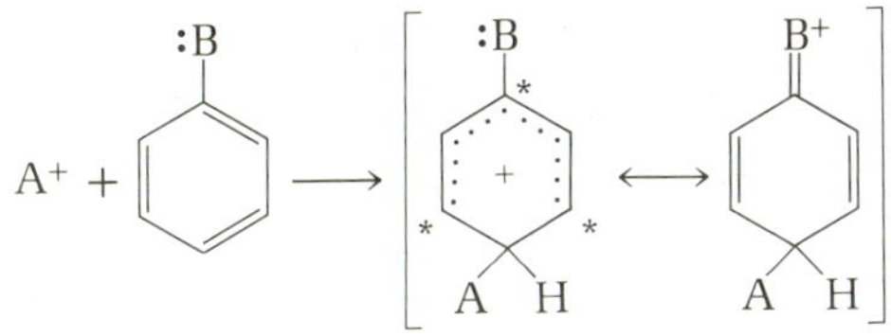

(15.36)

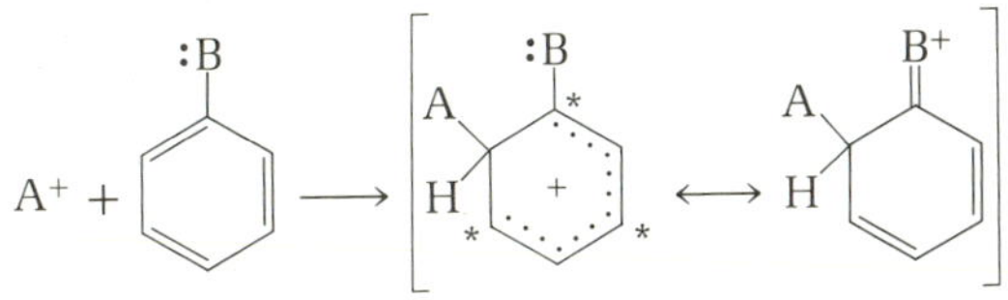

If the substituent carries a nonbonded electron pair, it may share it with the carbon atom and further delocalize the charge, thus lowering the energy of the intermediate (equation 15.35). The same result obtains when the electrophile attacks the carbon atom *ortho* to the substituent (equation 15.36). If a *meta* attack occurs, however, the substituent cannot stabilize the positive ion by conjugation (equation 15.37).

(15.37)

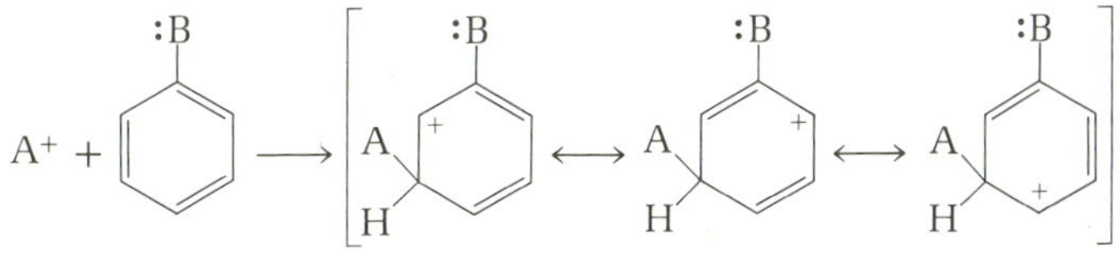

Substituents with nonbonded electrons adjacent to the ring therefore favor *ortho* and *para* substitution; they are known as **ortho,para-directors.** They stabilize the benzenium ion intermediates (and their predecessor transition states) in the manner shown. Alkyl groups may stabilize the benzenium ions by means of an inductive interaction; although this is not as effective as conjugative stabilization, it is sufficiently strong to make alkyl groups *ortho, para*-directors.

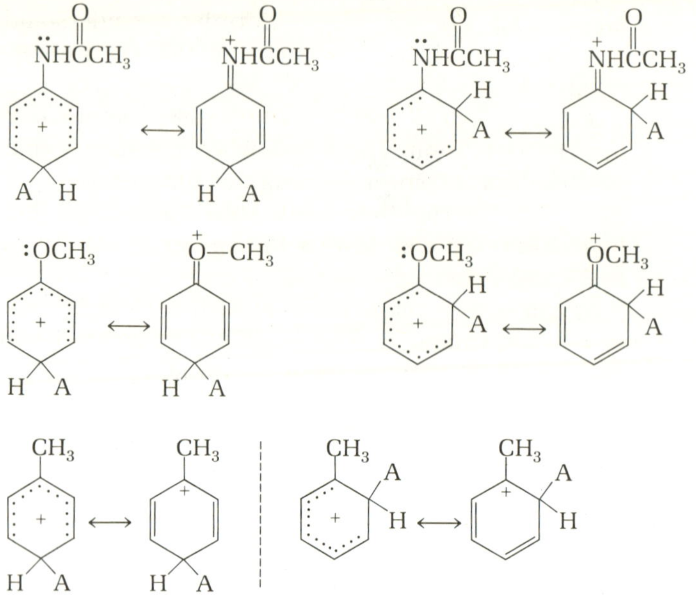

All the activating groups listed in Table 15-2 — R—, $R_2\ddot{N}$—, etc., — are strong *ortho,para*-directors. They are able to donate electron density to the ring and are further able to do so most effectively when they are *ortho* or *para* to the attacking electrophile. Some examples are given in equations 15.38 to 15.40.

(15.38)

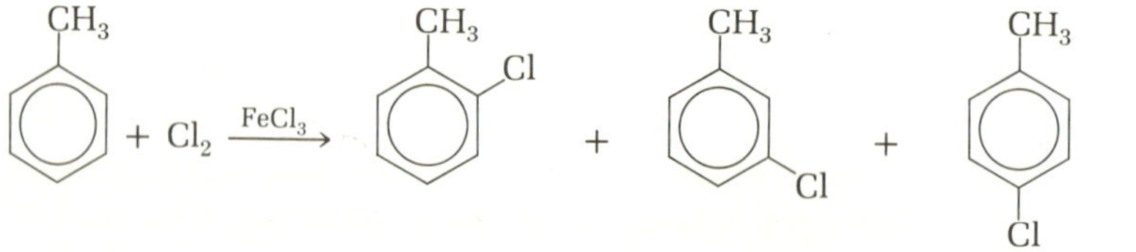

toluene | *o*-chlorotoluene (75%) | *m*-chlorotoluene (2%) | *p*-chlorotoluene (23%)

(15.39)

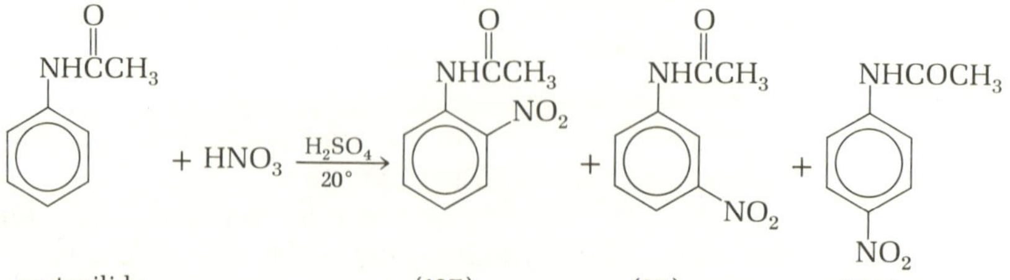

acetanilide | (19%) | (2%) | (79%)

(15.40) 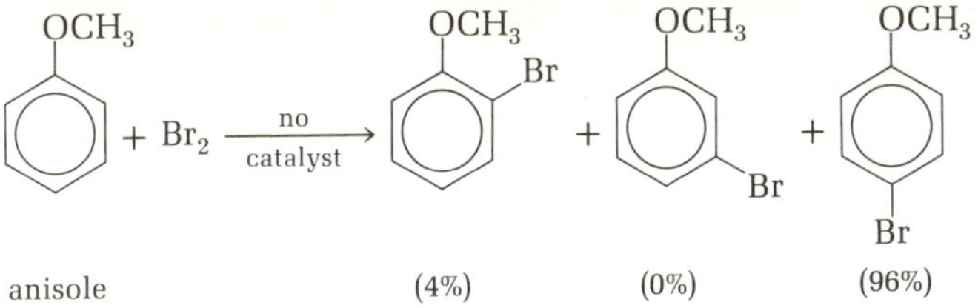

The halogens, X—, are also *ortho,para*-directors even though they tend to deactivate aromatic rings toward electrophilic attack. The general interpretation offered for this apparent anomaly is that although the halogens deactivate aromatic rings by inductive electron withdrawal, they may partially compensate by conjugative electron donation when they are located *ortho* and *para* to the point of electrophilic attack. Conjugation must play an important part in determining the chemical character of the molecule. For instance, fluorobenzene is generally the most reactive of the halobenzenes toward electrophilic attack and fluorine is the strongest *ortho, para*-directing halogen atom, despite the high electronegativity of fluorine.

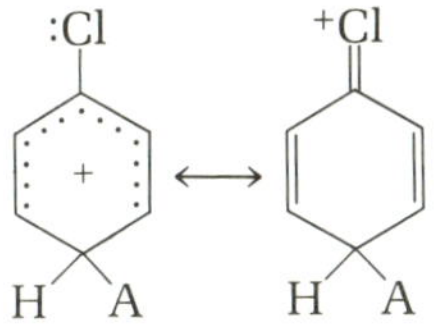

Additions to vinyl halides parallel the additions to aromatic halides. Hydrogen chloride reacts much more slowly with vinyl chloride than it does with ethylene. The intermediate carbonium ion, however, is the one in which the charge resides on the carbon atom bearing the chlorine. Here again chlorine is deactivating the molecule toward electrophilic attack while stabilizing a particular transition state for that attack.

(15.41) $$CH_2{=}CHCl + HCl \xrightarrow{\text{slow}} CH_3CHCl_2$$

All the deactivating substituents listed in Table 15-1, except for the halogens, are **meta-directors.** The transition state leading to **3,** in which the electron-withdrawing substituent is attached to the carbon atom bearing little positive charge, is lower in energy than the transition state leading to **4,** where the substituent is drawing charge away from an already deficient carbon atom.

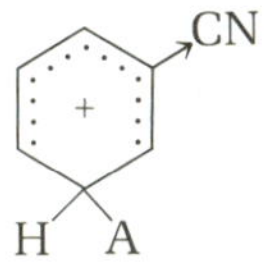

3

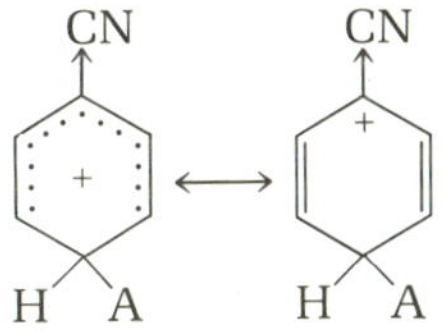

4

Some illustrative examples follow.

(15.42) 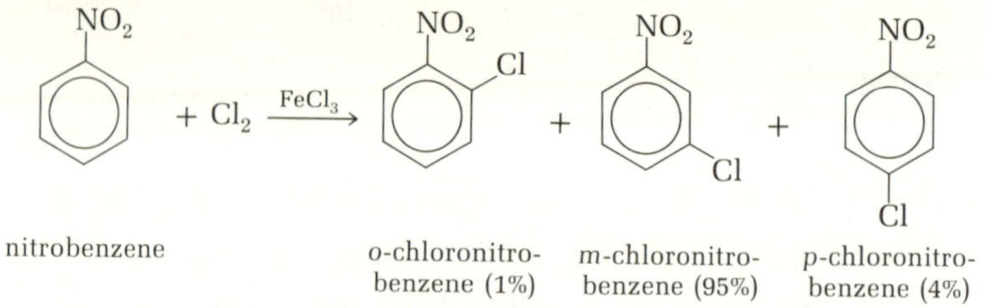

(15.43) 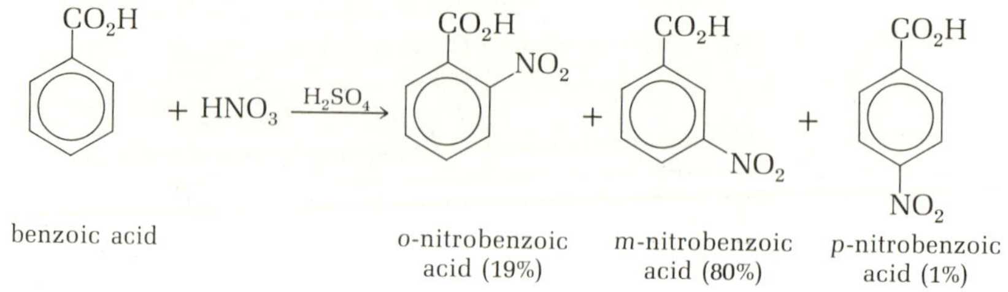

(15.44) 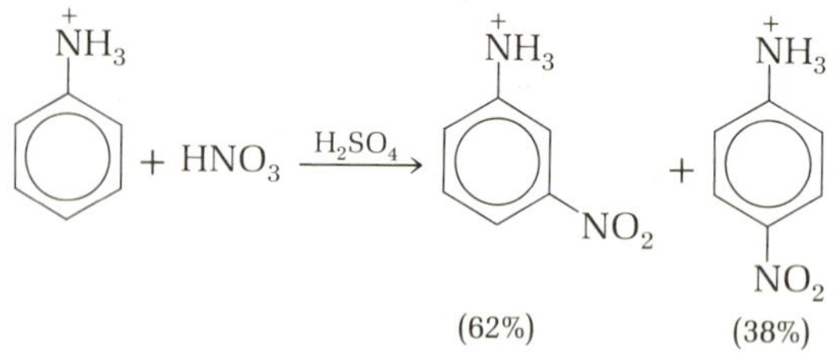

FIELD EFFECTS

The product distribution from the nitration of **anilinium ion** (equation 15.44) is curious since 38% of the *para*-substitution product is found and *none* of the *ortho* isomer. The explanation offered is that a field effect is operative — the positive electric field emanating from the substituent repels the positively charged electrophile. Since virtually all substituents will contain bond dipoles, and the charge centers of these dipoles will give rise to electric fields, the field effect will play some part in determining the orientation in any electrophilic aromatic substitution.

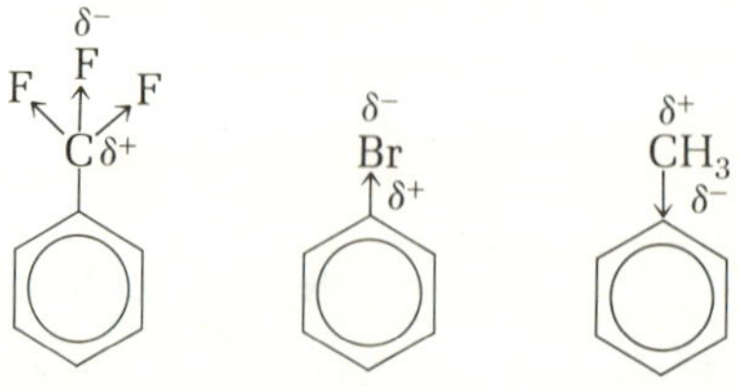

5 6 7

It should be noted that in the great majority of cases the field and inductive effects would lead to the same results: The substituents in compounds **5** and **6** would be expected to discourage *ortho*-substitution, while the opposite result would be expected in the case of compound **7**.

SUBSTITUTION OF PHENOLS AND ANILINES

The two most powerfully activating groups are the amino ($-\ddot{N}H_2$) and hydroxyl groups (along with their alkyl derivatives). Their presence in an aromatic ring that is to be substituted may lead to complications. The $-\ddot{N}H_2$ group is so powerfully activating that oxidation reactions, particularly in the presence of nitric acid, intervene to a serious extent; degradation, coupling of rings, etc., may divert a large proportion of the starting material. The amino group is also basic and upon protonation it changes character completely; the product distribution then becomes sensitive to the acidity of the medium.

(15.45) 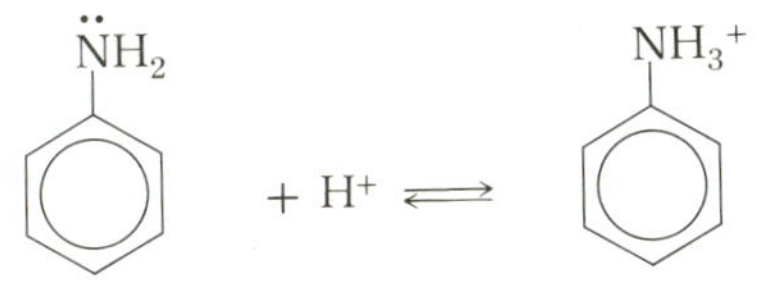

activated *o,p*-substitution — deactivated *m*-substitution

One method of controlling the substitution of anilines involves the *acetylation* of the amino group. Due to the electron-withdrawing properties of the *acetyl group* ($CH_3\overset{O}{\overset{\|}{C}}-$), the basicity and activating power of the amino group are considerably reduced.

(15.46) 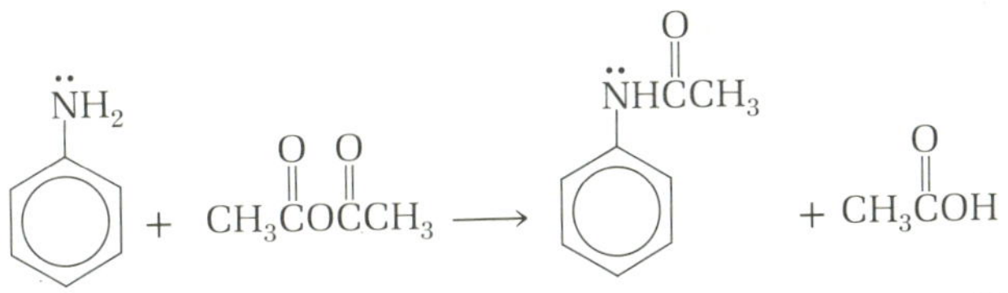

aniline — acetic anhydride — *N*-acetylaniline (acetanilide) — acetic acid

The *acetamido* group ($CH_3\overset{O}{\overset{\|}{C}}NH-$) is still a powerful *ortho,para*-director but it is not nearly so strongly activating as the amino group; yet it is one of the most powerful substituents in this regard. The acetyl group may be removed from the nitrogen atom with relative ease upon completion of the desired substitution (Chapter 17).

The hydroxy group in phenol makes that molecule and its derivatives susceptible to oxidation and other side reactions. Furthermore, phenol is, unlike aliphatic alcohols, appreciably acidic and partially ionized in polar solvents. The —O⁻ group of the phenolate ion is the most powerfully activating, *o,p*-directing substituent known.

(15.47)

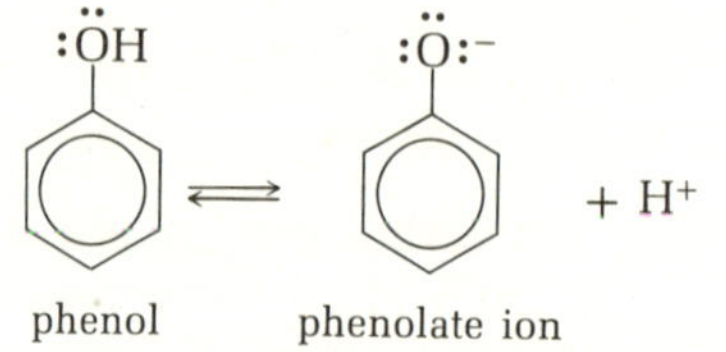

It is sometimes difficult to prevent multiple substitution of the anion, even when a deficiency of the electrophilic reagent is used. For these reasons phenols are often subjected to electrophilic substitution as their ethers; *anisole* (methoxybenzene) is the most common ether or phenol. The alkyl group may then be removed if desired. Further data on the electrophilic substitution of anilines and phenols is presented in Chapters 18 and 22, respectively.

SYNTHETIC APPLICATIONS

The products of the various electrophilic substitution reactions are very useful materials in themselves and are, in addition, important intermediates in the synthesis of other aromatic compounds. Bromobenzenes may, for instance, be converted to Grignard reagents that are then able to undergo all the expected reactions.

(15.48)

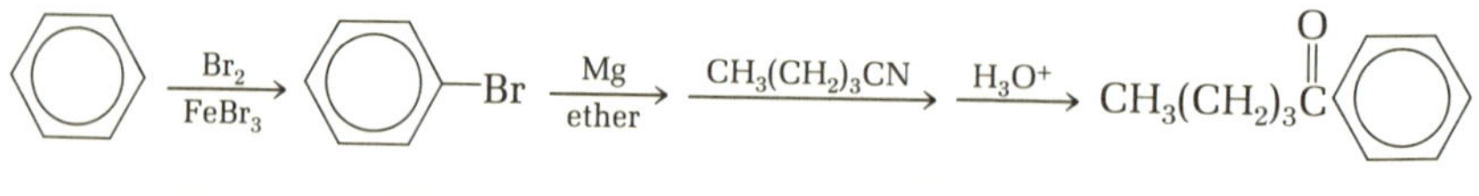

(15.49)

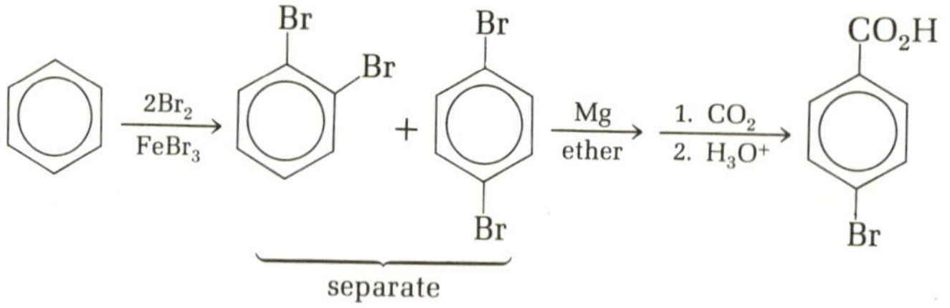

Chlorobenzene is resistant to Grignard formation in ether (THF must be used); iodobenzenes are very reactive, on the other hand, but give higher yields of side products. One may take advantage of the differing reactivity of the halogens.

(15.50) 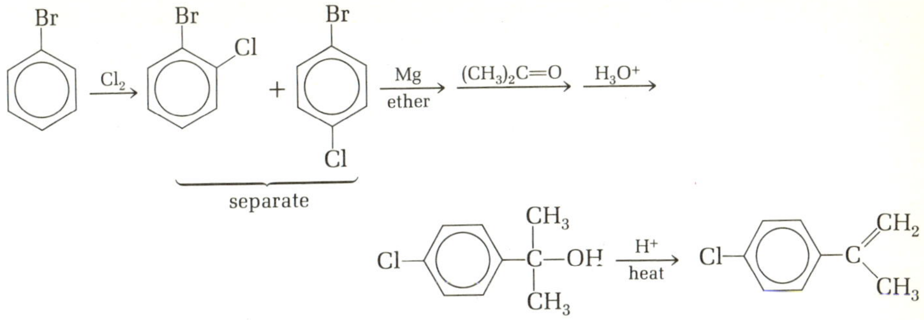

The nitro group may be reduced to an amino group, which is susceptible to a variety of subsequent transformations (Chapters 18 and 19).

(15.51) 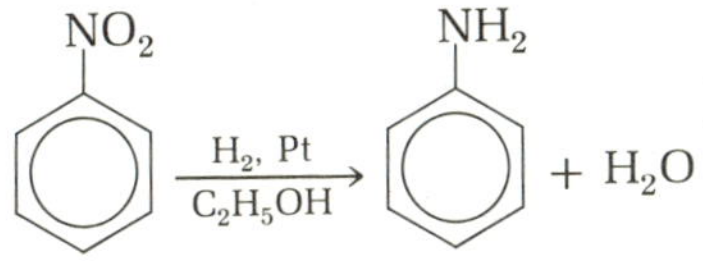

The sulfonic acid group has some unique and important properties. Arylsulfonic acids are rather strong acids, considerably stronger than the related carboxylic acids (Chapter 16), and they often find use amid circumstances where one requires a strong acid that is soluble in solvents of low polarity. For synthetic purposes, advantage is taken of the reversibility of the sulfonation reaction. When an aromatic ring is sulfonated, the most reactive position is occupied (blocked) by the sulfonic acid group. Subsequent electrophilic substitution (halogenation, nitration, etc.) of the sulfonic acid results in placement of the new electrophile at a less reactive position, after which the sulfonic acid group is removed (compare equations 15.52 and 15.53).

(15.52) 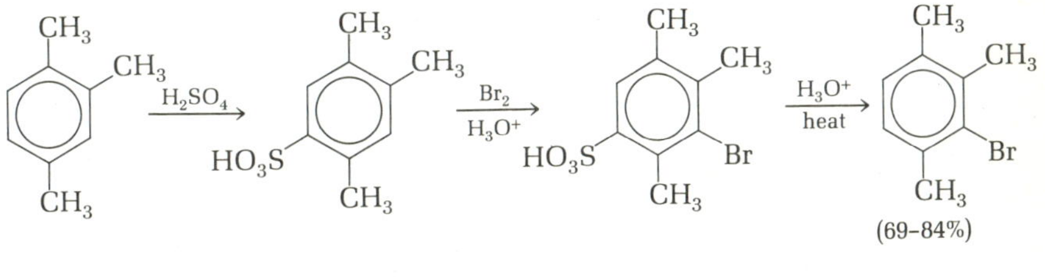

(15.53) 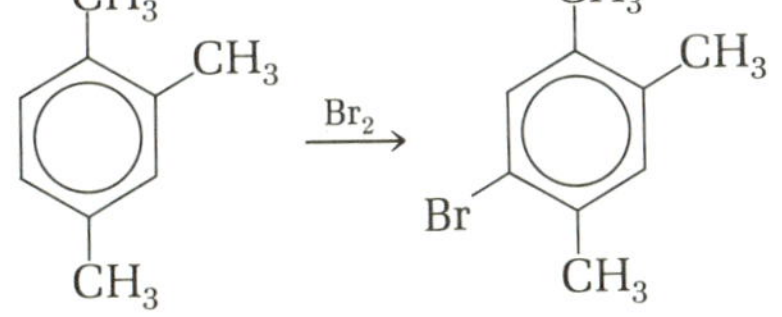

Table 15-3 lists the boiling points of some sets of disubstituted benzene isomers. When neither substituent is strongly polar (alkyl, alkoxy, or halogen), the three isomers have very similar boiling points. Although mixtures of these isomers may often be satisfactorily analyzed by means of gas-liquid chromatography, their fractionation by means of distillation on a preparative scale is impractical. This restriction applies especially to isomers present in small amounts. The isomers may sometimes have sufficiently different solubility properties to make them amenable to separation by fractional crystallization but this can be a tedious procedure.

TABLE 15-3
Boiling Points of Some Disubstituted Benzene Isomers

Substituents	Boiling Points (°C)		
	ortho	*meta*	*para*
Ethyl, ethyl	184	182	182
Chloro, methyl	159	162	162
Bromo, iodo	257	252	252
Methyl, nitro	220	233	238
Bromo, cyano	252	225	236
Acetamido, methyl	296	303	307

If the compound contains at least one polar substituent (those with polar double bonds and/or hydrogen bound to oxygen or nitrogen), then a greater spread of boiling points is usually found among the isomers and their separation by fractional distillation becomes a practical procedure. Since nitro and amino groups may be ultimately converted to halogen and halogen-like substituents (Chapter 21), this route is often to be preferred for the preparation of pure polysubstituted benzene isomers.

The nomenclature rules for polysubstituted benzenes are quite simple. The substituents are numbered so as to minimize the sum of the position numbers. Placing the substituent names in alphabetical order is the most convenient method; the substituent at position 1 may be placed at the end of the list and its number omitted. Methylbenzene, aminobenzene, and hydroxybenzene are known, respectively, as toluene, aniline, and phenol; for the purpose of numbering substituted compounds, the methyl, amino, and hydroxy groups are considered to be attached to position 1. If two of these groups should appear on the same ring, the order of preference in classification is phenol > aniline > toluene.

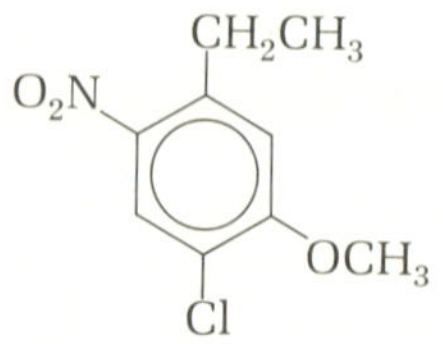

4-ethyl-2-methoxy-5-nitrochlorobenzene

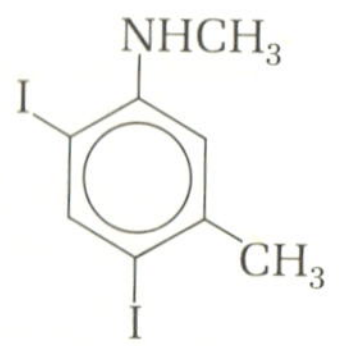

2,4-diiodo-N,5-dimethylaniline

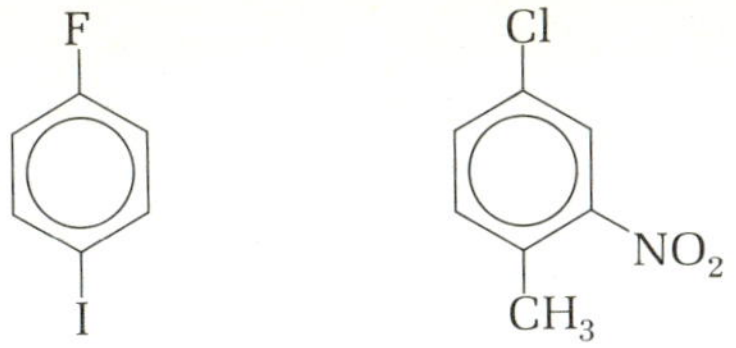

p-fluoroiodobenzene
4-fluoroiodobenzene

4-chloro-2-nitrotoluene

3-methylphenol
(*m*-cresol)

PARTIAL RATE FACTORS

In order for chemists to compare the activating and orienting effects of a given substituent in various electrophilic aromatic substitution reactions, **partial rate factors** have been defined and calculated for a number of groups. The partial rate factor is the rate at which a particular position in a substituted benzene undergoes a given substitution reaction relative to the rate at which a single benzene position would undergo the same reaction. The calculation of partial rate factors requires that one know the relative rate of overall reaction of the substituted benzene compared to benzene itself and that the product distribution also be known.

A few examples quickly illustrate the method of calculation. Monochlorination of toluene with a mixture of hypochlorous acid and a mineral acid ($HOCl + H^+$) proceeds at a rate 60 times as great as that for the monochlorination of benzene under the same conditions. The products are shown in equation 15.54.

(15.54)

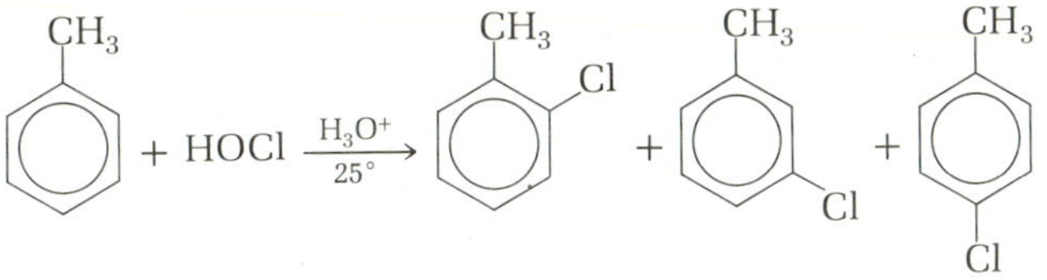

(74.6%) (2.2%) (23.2%)

Three partial rate factors may now be calculated: those for *ortho* methyl ($o_f^{CH_3}$), *meta* methyl ($m_f^{CH_3}$), and *para* methyl ($p_f^{CH_3}$).

In order to calculate the partial rate factor for chlorination of toluene in the *ortho* position, the overall rate factor (60.0) must be multiplied by $6/2$ because there are six equivalent positions in benzene but only two *ortho* positions in toluene. Since only 74.6% of the product was *o*-chlorotoluene, however, the rate figure derived above must be multiplied by 0.746.

(15.55) $o_f^{CH_3}(HOCl + H^+) = 6/2 \times 60.0 \times 0.746 = 134$

The resulting partial rate factor, 134, (equation 15.55) is a shorthand way of saying that toluene chlorinates at a single *ortho* position 134 times as rapidly as benzene chlorinates at a single position, when acidic aqueous hypochlorous acid is the reagent. The partial rate factors for *meta*- and *para*-substitution may be similarly calculated; it must be kept in mind that there is only one *para* position.

(15.56) $m_f^{CH_3}(HOCl + H^+) = \frac{6}{2} \times 60.0 \times 0.022 = 4.0$

(15.57) $p_f^{CH_3}(HOCl + H^+) = 6 \times 60.0 \times 0.232 = 82$

The toluene results tell us that the methyl group activates all positions of the ring since they are all chlorinated faster than a single benzene carbon atom. The *ortho* and *para* positions are far more activated than the *meta* position is, however; the partial rate factors provide a quantitative measure of what we know qualitatively. Accurate measurement of the rate of substitution of strongly deactivated aromatic rings is difficult but it has been possible to estimate partial rate factors for the nitration of nitrobenzene:

$o_f^{NO_2} = 1.8 \times 10^{-6}$ $m_f^{NO_2} = 2.8 \times 10^{-5}$ $p_f^{NO_2} = 2 \times 10^{-7}$

The nitro group deactivates all positions severely but the *meta* position is a little less strongly affected and is therefore the favored position of attack.

Table 15-4 contains the partial rate factors for the nitration of the four halobenzenes with nitronium fluoborate ($NO_2^+BF_4^-$) at 25°. The data in this table provide support for our mechanistic interpretation of the behavior of the halogens as substituents.

TABLE 15-4
Partial Rate Factors for the Nitration of the Halobenzenes

	o_f^x	m_f^x	p_f^x
Fluorobenzene	0.115	0.000	2.47
Chlorobenzene	0.093	0.0029	0.643
Bromobenzene	0.093	0.0040	0.526
Iodobenzene	0.034	0.000	1.07

In chlorobenzene and bromobenzene all the ring positions are deactivated but the *ortho* and *para* positions are less severely affected. Reaction is favored at the *para* position where the inductive (or field) effect of the halogen atom would be least severe, while its conjugation effect would be fully operative. Fluorobenzene, in which the conjugative effect should be strongest, is the most reactive halobenzene.

ORIENTATION OF ATTACK ON DISUBSTITUTED BENZENES

There are some simple qualitative rules that allow one to predict, with a fair degree of certainty, the major product(s) from electrophilic attack on disubstituted benzenes. The easiest cases to treat are those in which the two substituents are so placed that their orienting effects reinforce one another. This is the case when one *o,p*-director and one *m*-director are either *ortho* or *para* to one another **(8)**. If two *o,p*-directors **(9)** or two *m*-directors **(10)** are *meta* to one another, then their orienting powers also reinforce one another.

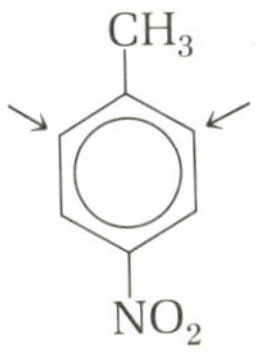

8

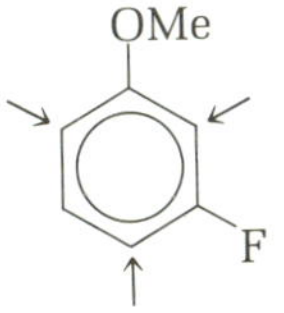

9

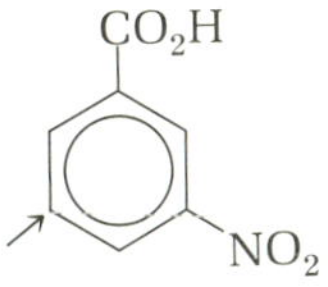

10

The interesting situations arise when the orienting effects of the two substituents are in opposition, as when two *o,p*-directors are *para* to one another. A qualitative conclusion can be reached even in some of these cases if one remembers that the attacking species is electrophilic; should one of the substituents be substantially more activating than the other, it will determine the substitution pattern.

(15.58)

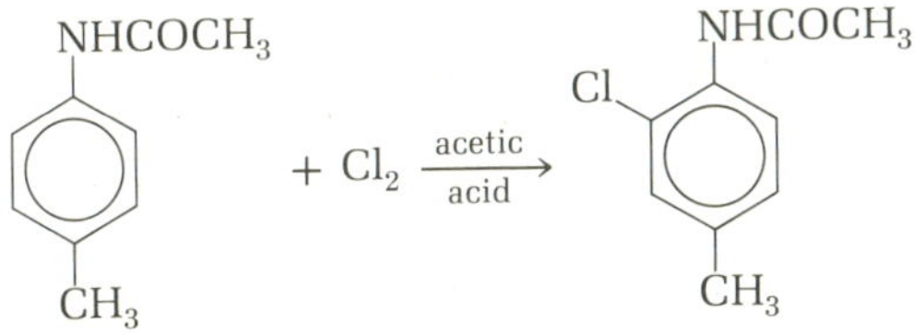

(15.59)

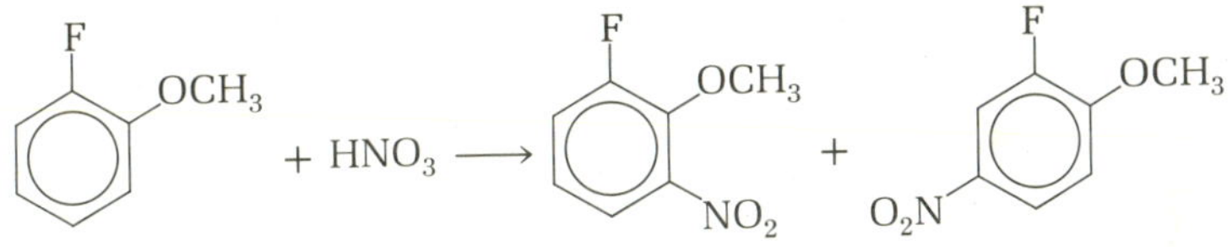

(66%) (31%)

The order of decreasing activating ability among activating substituents is $R_2N— > HO— > RO— > CH_3CONH— \gg$ alkyl. All these substituents except alkyl are powerful activators. The order of increasing deactivating ability among the deactivating substituents is:

halogen $< -\underset{\underset{\mathrm{O}}{\|}}{\mathrm{C}}-\mathrm{Z} < -\mathrm{CN} < -\mathrm{NO_2}$

(15.60)

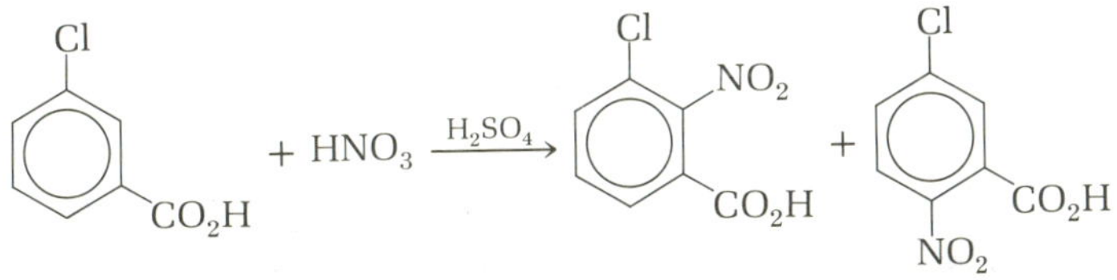

(8%) (92%)

Qualitative predictions of substitution in disubstituted benzenes fail when the two substituents do not differ greatly in activating ability. This is the case with alkyl groups and halogen atoms. Structure **11** shows the relative amount of each isomer obtained upon nitration of *o*-chlorotoluene. No set of qualitative rules would allow one to predict such a product distribution.

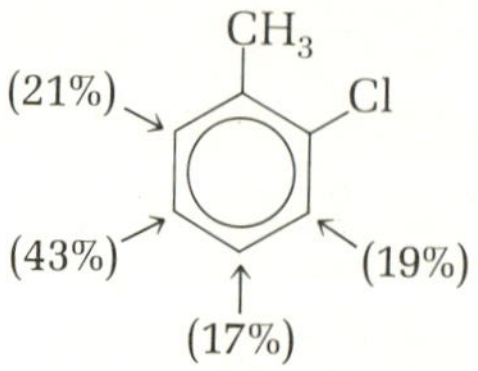

11

FREE RADICAL ATTACKS ON AROMATIC SYSTEMS

It was stated previously that benzene reacts only very slowly with the halogens in the dark and in the absence of Lewis acid catalysts. If the halogen-hydrocarbon mixture is irradiated with ultraviolet light, however, a vigorous free radical reaction sets in. Hexachlorocyclohexane ("benzene hexachloride") may be isolated as a mixture of stereochemical isomers.

(15.61) 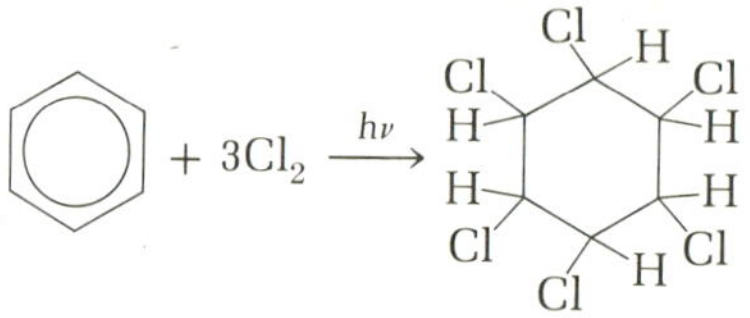

This mixture has some value as an insecticide. By use of an iodine catalyst, it has been possible to isolate the intermediate tetrachlorocyclohexenes as an isomeric mixture; free radical chlorination of this mixture gives rise to the various "benzene hexachloride" isomers. When the photochemical chlorination is carried out at low temperatures, dichlorocyclohexadiene may be detected.

(15.62) 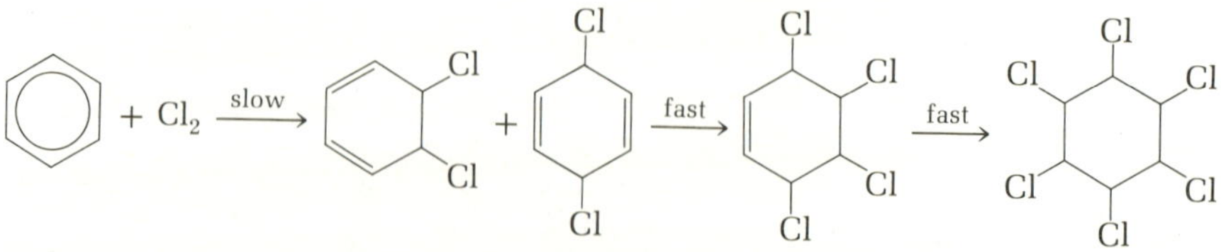

The first step is undoubtedly the attack of a radical on the aromatic π-system to form a pentadienyl radical. The unpaired electron occupies the nonbonding orbital (Figure 15-5) and therefore spends its time on the carbon atoms *ortho* and *para* to the point of attack.

(15.63)

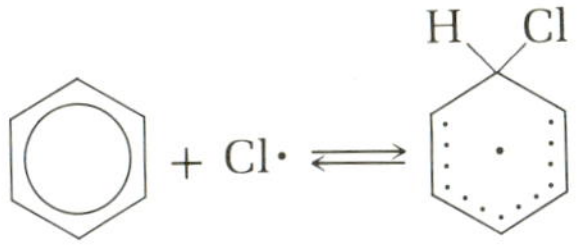

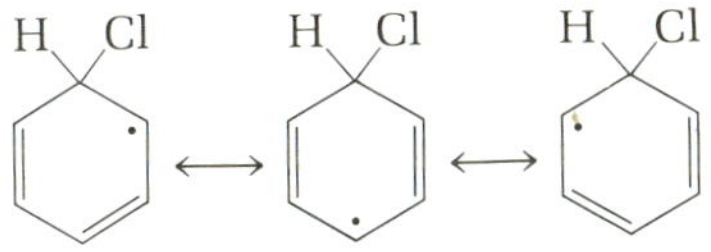

Reaction of the radical intermediate with molecular chlorine would then produce dichlorocyclohexadienes and a chlorine atom to continue the chain.

(15.64)

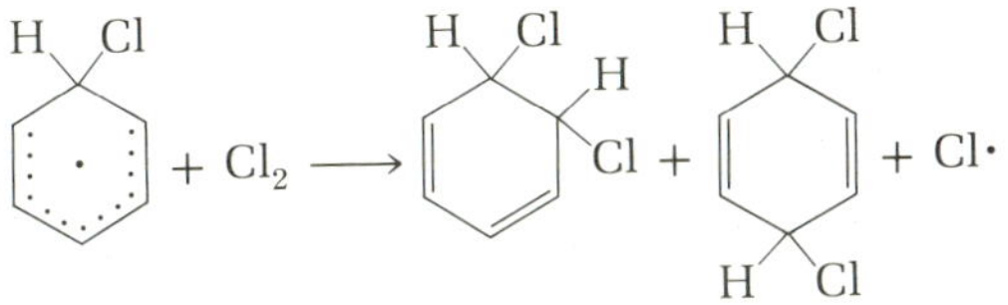

Some chlorobenzene is always formed in the course of these free radical chlorinations. While it may arise by the customary ionic path, at least part of it must result from hydrogen abstraction from the radical intermediate. This mechanism is a very general one in aromatic free radical chemistry.

(15.65)

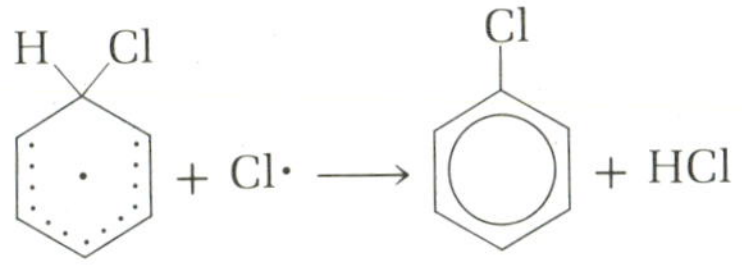

There are many free radical reactions in which a reagent AB reacts with benzene (and other aromatics) to give C_6H_5B and HA. Direct hydrogen abstraction from benzene by A· is apparently not favored, as the C—H bond of benzene has a dissociation energy of 112 kcal/mole. Rather, B· adds to the ring first and hydrogen abstraction follows; the C—H bond undergoing cleavage is much weaker in the intermediate than in benzene itself.

(15.66)

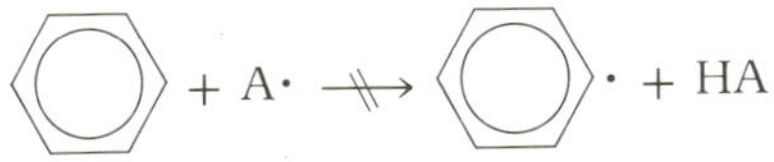

(15.67)

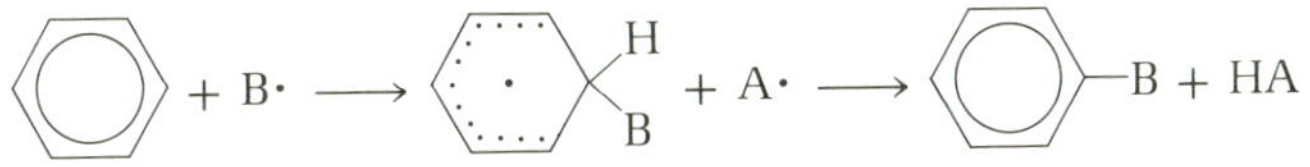

SUMMARY

1 Benzene and other aromatic compounds may act as donors in the formation of π-complexes with a number of electron acceptors. When the acceptors are molecular halogens or transition metals, the complexes are usually symmetrical, the acceptor lying over the center of the aromatic ring. Halogen and hydrogen halide complexes are quite weak but many transition metal complexes are fairly strong and capable of being isolated. The cyclopentadienyl anion is a very effective π-donor, forming a number of sandwich complexes with various transition metals.

Another class of π-electron acceptors, notably Ag(I) and Cu(I), form asymmetrical complexes with benzene and its derivatives, destroying the original hexagonal symmetry of the benzene ring in the process.

2 A further distortion of the π-complex, so that the acceptor is bound to a single carbon atom by means of an electron pair bond, results in the creation of a σ-complex. This is the course that the reaction takes when a Lewis acid is added to a solution of HF or HCl in benzene (or its derivatives). The σ-complex is a benzenium ion in which the five remaining sp^2-hybridized carbon atoms share only four electrons. The electron deficiency is concentrated in a nonbonding orbital that has an appreciable amplitude only at the three carbon atoms *ortho* and *para* to the point of attack. Loss of one of the two groups attached to the sp^3-hybridized carbon atom of the benzenium ion regenerates an aromatic ring.

3 Electrophilic aromatic substitution proceeds by way of a σ-complex formed by attack of an electrophile on the aromatic ring. Formation of the σ-complex is usually, although not invariably, the rate-determining step. In nitration the electrophile is the nitronium ion, usually generated *in situ* by the dehydroxylation of nitric acid. Halogenation may be accomplished by the molecular halogen, by a complex of the molecular halogen and a metal halide, or by protonated hypohalous acid. Sulfonation seems to involve sulfur trioxide as the electrophile, with concentrated or fuming sulfuric acid as its source.

Of the common electrophilic substitution reactions, iodination, bromination, sulfonation, and proton exchange are reversible. The equilibrium constant for iodination is unfavorable and the hydrogen iodide must be destroyed if decent yields of iodinated products are to be obtained.

4 Deactivating substituents are groups that cause a substituted benzene to undergo electrophilic substitution more slowly than benzene itself does. These substituents withdraw electrons from the electron-deficient transition state leading to the benzenium ion, thereby substantially raising the energy of activation. Some of the deactivating groups are also conjugated with the benzene ring and a substantial amount of the conjugative stabilization is lost upon electrophilic attack.

There are also substituents that increase the reactivity of the benzene ring. These activating groups donate electron density to the benzenium ring, either inductively or by conjugation (which is the most effective means of benzenium ion stabilization).

5 An activating substituent directs incoming electrophiles to those carbon atoms *ortho* and *para* to itself. These substituents stabilize the transition states leading to *ortho*- and *para*-substitution. Stabilization may occur by an inductive or conjugative mechanism; conjugative interactions are much stronger than inductive ones. Deactivating substituents tend to raise the energy of the transition states leading to *ortho*- and *para*-substitution more than they affect the transition states leading to *meta*-substitution and hence encourage *meta* orientation of the incoming electrophile. The halogens, although deactivating, stabilize the transition states for *o,p*-substitution.

A useful general rule, *to which the halogens are the principal exception,* is that **activating substituents activate the** *ortho,para*-**positions preferentially, while deactivating substituents deactivate the** *ortho,para*-**positions preferentially.**

6 Substituted benzenes are important synthetic intermediates. Bromobenzenes are easily converted to Grignard reagents that are able to undergo the full range of characteristic reactions. Nitro groups are readily reduced to amino groups that are, in turn, convertible to a number of other functionalities. Sulfonic acid substituents are very useful because of the relative ease with which they may be removed; this makes them valuable as blocking groups.

7 The positional isomers of disubstituted benzenes may have very similar boiling points if neither substituent is appreciably polar. Separation of the isomers by fractional distillation becomes a difficult matter. A more advantageous spread of boiling points among the isomers is encountered if at least one of the substituents is reasonably polar. In such cases the individual isomers may be obtained in pure form. These compounds may often be chemically transformed into pure isomers of other compounds that would be difficult to separate into isomers by physical means.

8 The effect of a substituent on the rate of a particular reaction at a particular position may be quantitatively assessed by calculation of its partial rate factor. The prf is defined as the rate of the reaction in question at the given position relative to the rate of the same reaction at a single position of benzene. In order to calculate partial rate factors, one needs to know the overall rate of reaction of the aromatic compound and the distribution of the isomeric products. The partial rate factors give quantitative proof of the qualitative assertions made above. One may also compare the extent to which different reagents discriminate among the *ortho, meta,* and *para* positions and relate the differences to electronic and steric effects.

9 The pattern of electrophilic attack in a disubstituted benzene is controlled by one of the substituents if it is distinctly more activating (or less deactivating) than the other one. When the two substituents exert comparable influences, then all possible isomers are usually formed.

10 Free radicals may add to the benzene ring to give a delocalized cyclohexadienyl radical. This is the first step in the chlorination of benzene promoted by ultraviolet light; the usual products are the various isomers of hexachlorocyclohexane. Aromatic rings may also undergo substitution by way of a free radical mechanism. The substituting radical just adds to the benzene ring and another radical then abstracts the hydrogen atom from the carbon atom undergoing substitution.

PROBLEMS

1 Ferrocene has been offered as an example of a transition metal π-complex. Suppose that you had obtained a sample of this compound and had been asked to determine whether in fact it was a π-complex or, rather, if it had a structure in which the cyclopentadiene rings were bound to the iron atom by σ-bonds (I). What sort of data would you seek?

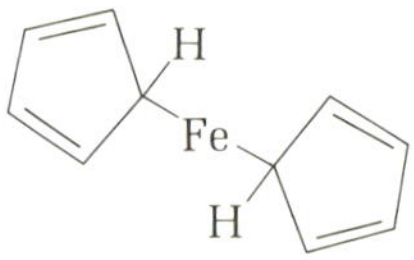

I

2 The nmr spectrum of anthracene (II) consists of three peaks in the aromatic region, between $\delta 7.2$ and $\delta 8.5$. When anthracene is treated with HCl gas and $AlCl_3$ in an inert solvent, a new spectrum appears.

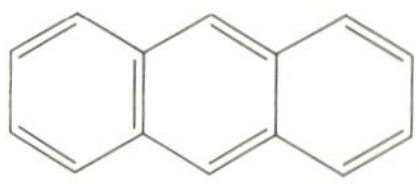

II

It consists of a singlet at $\delta 4.05$, a multiplet centered at $\delta 7.6$, and a small multiplet at $\delta 8.83$; these peak areas are in the ratio 2:8:1.

a. Show the three types of protons in anthracene.

b. Draw a structure for the new species that is consistent with the data that you have.

3 In polar solvents favoring ionization, such as trifluoroacetic acid, 2-phenylethyl tosylate (IIIa) is much more reactive than ethyl tosylate (IIIb).

$$RCH_2CH_2OTs + CF_3CO_2H \longrightarrow RCH_2CH_2O\overset{\overset{\large O}{\|}}{C}CF_3 + HOTs \qquad (1)$$

IIIa: $R = C_6H_5$
b: $R = H$

Furthermore, when deuterium atoms are substituted at C-1 in the substituted reactant, they are found to be completely scrambled between C-1 and C-2 of the product.

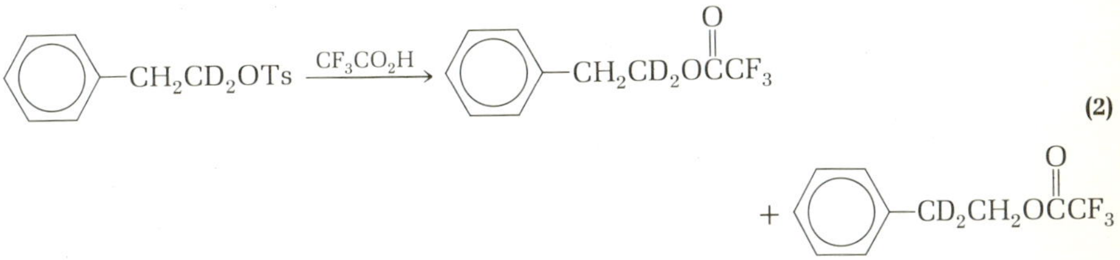

(2)

Write out a mechanism that is consistent with the rate data and the isotope scrambling studies. What sort of intermediate might intervene in this reaction?

4 Ethyl fluoride and boron trifluoride form a colorless complex at $-80°$ that exhibits only weak conductivity. Addition of toluene to this solution results in the appearance of a yellow color and a marked increase in conductivity. A crystalline solid may be obtained from the mixture at low temperatures. When allowed to warm up, the solid decomposes with evolution of gas. Write out equations that account for all these phenomena.

5 In the silver perchlorate-assisted iodination of benzene (equation 15.25) the added salt has another function in addition to preventing the reversal of the reaction. What is this second function?

6 If an aromatic compound that is not strongly deactivated is allowed to react with mercuric acetate in acidic solution, it can be *mercurated*.

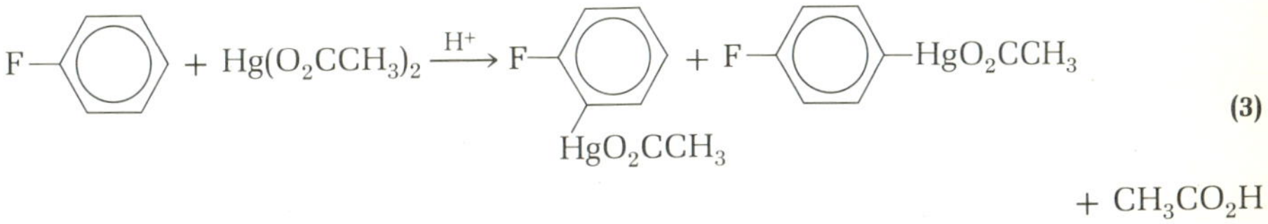

(3)

This reaction can be of preparative use since the products are readily *demercurated* by a variety of electrophiles under mild conditions.

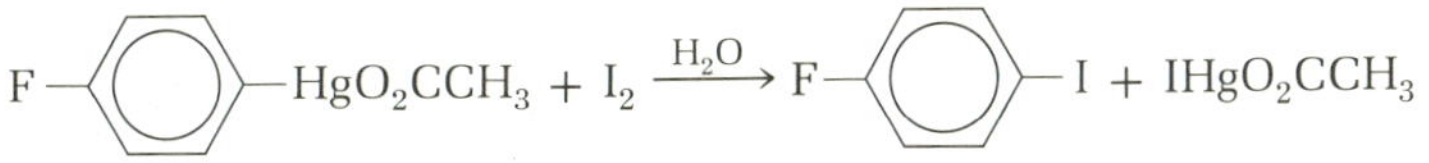

(4)

Write out plausible mechanisms for the mercuration and demercuration reactions.

7 a. Aromatic compounds have been brominated with a variety of brominating agents, among them BrCl, Br_2, and ($HOBr + H^+$). Rank them in order of increasing effectiveness and explain your reasoning.

b. While iodine is a very weak electrophile, iodine chloride is considerably more reactive and has been used to iodinate aromatics. To what is this difference in behavior due?

8 Due to the strong orienting effects of most substituents, the order in which groups are introduced into an aromatic molecule is very critical in the synthesis of polysubstituted benzenes. Devise efficient syntheses of the following compounds that do not involve isolation and purification of the minor product of a reaction. Benzene, toluene, or benzoic acid may be used as starting materials.

a. *m*-Chloronitrobenzene

b. *p*-Chloronitrobenzene

c. *p*-Iodobenzenesulfonic acid

d. *o*-Bromochlorobenzene

e. *m*-Nitrobenzoic acid

f. 2-Bromo-4-nitrotoluene

g. 4-Chloro-3-nitrobenzenesulfonic acid

h. 3-Bromo-5-nitrobenzoic acid

i. 2,4-Dinitrochlorobenzene

j. *p*-Nitroaniline

k. *m*-Diaminobenzene

l. 2-Chloro-6-nitrotoluene

9 Would you expect the following compounds and ions to undergo predominant *m*- or *o,p*-substitution? Provide a brief explanation in each case:

a. 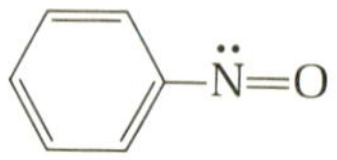

(nitrosobenzene)

b. 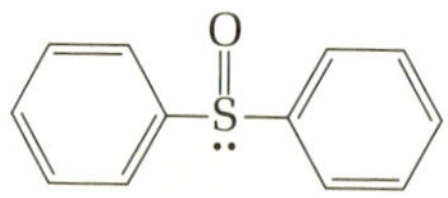

(diphenyl sulfoxide)

c. 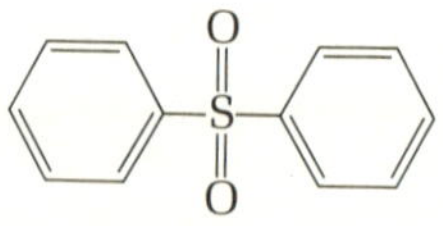

(diphenyl sulfone)

d. 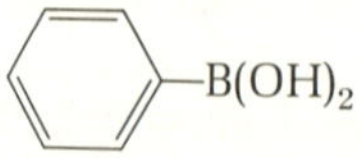

(phenylboronic acid)

e. 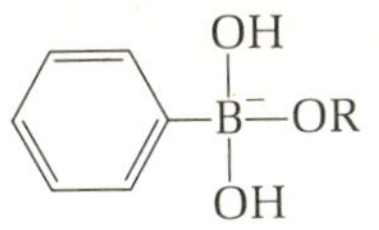

(phenylboronate anion)

10 Treatment of phenol with an excess of aqueous bromine leads to the precipitation of a crystalline solid, the analysis of which corresponds to $C_6H_2Br_4O$. The ir spectrum of the compound has a peak in the carbonyl region and the nmr spectrum consists of a single line. Draw a structure for the compound, and propose a mechanism for its formation.

11 The separation of the three xylenes is a somewhat complex procedure, due to their similar boiling points (*o*-, 144°; *m*-, 139°; *p*-, 138°). By careful distillation one can separate the *m*- and *p*-isomers from the *o*-isomer. The former two isomers can be separated in pure form by taking advantage of the fact that one of them can be sulfonated at a much lower temperature than the other; the resulting sulfonic acid is then isolated and desulfonated. Which isomer is the more reactive? Explain.

12 Propose structures for the unknowns in the equations listed below. In all cases the unknowns represent the major product(s) of the indicated reaction.

a. $C_7H_8O \xrightarrow{Br_2} \mathbf{A} + \mathbf{B}$ (both C_7H_7BrO)

$\mathbf{A} \xrightarrow{HNO_3} \mathbf{C}\ (C_7H_6BrNO_3)$

$\mathbf{B} \xrightarrow{HNO_3} \mathbf{D} + \mathbf{E}$

b. $C_7H_6O \xrightarrow{H_2SO_4} \mathbf{F}\ (C_7H_6O_4S) \xrightarrow[H_2SO_4]{HNO_3} \mathbf{G}\ (C_7H_5NO_6S)$

c. Nitrobenzene $\xrightarrow[FeCl_3]{Cl_2} \mathbf{H} \xrightarrow{H_2SO_4\text{-}SO_3} \mathbf{I} \xrightarrow{Pt,\ H_2} \mathbf{J}\ (C_6H_6ClNO_3S)$

$\mathbf{H} \xrightarrow{Pt,\ H_2} \mathbf{K}\ (C_6H_6ClN) \xrightarrow{H_2SO_4} \mathbf{L} + \mathbf{M}$

d. Acetanilide $\xrightarrow{Cl_2} \mathbf{N} + \mathbf{O}$

$\mathbf{N} \xrightarrow{H_2SO_4} \xrightarrow{H_3O^+} \mathbf{P} + \mathbf{Q}$ (both $C_6H_6ClNO_3S$)

$\mathbf{O} \xrightarrow{H_2SO_4} \xrightarrow{H_3O^+} \mathbf{R}$

13 The reaction of benzyl Grignard reagents with alkyl tosylates or sulfates (equation 5) leads to the synthesis of homologues of toluene. When the reaction was carried out as shown, there was obtained, together with the desired propylbenzene, an isomer of the product.

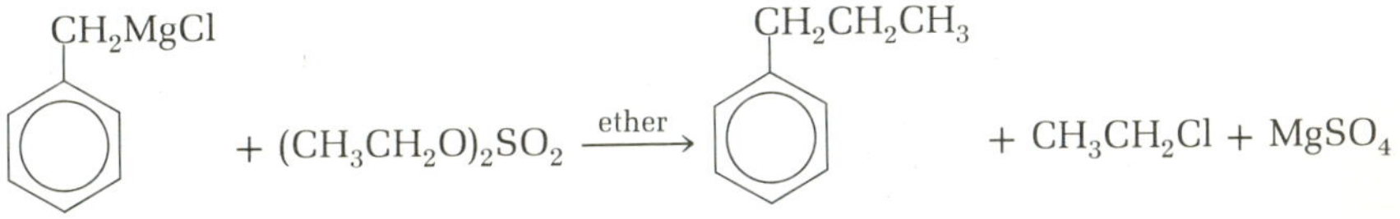

(5)

The nmr spectrum of this isomer contained four peaks: δ1.2 (triplet, 3 H); δ2.3 (singlet, 3 H); δ2.6 (quartet, 2 H); and δ7.1 (symmetrical multiplet, 4 H). What is the structure of this compound? How would you rationalize its formation?

14 a. The hydrolysis of cumyl chloride (IV) is accelerated by the presence of a *p*-methoxyl substituent, but a *m*-methoxyl group *retards* the hydrolysis. Explain these substituent effects.

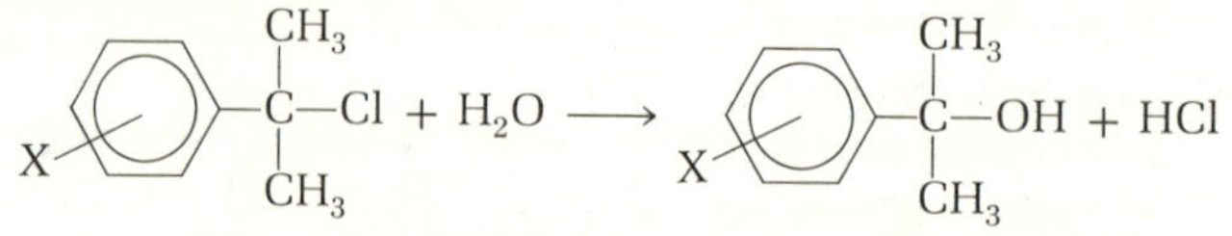

IV

b. 3,5-Dimethoxybenzyl phosphate (V) is quite stable in aqueous solution but upon ultraviolet irradiation it rapidly hydrolyzes to 3,5-dimethoxybenzyl alcohol, apparently by way of a carbonium ion.

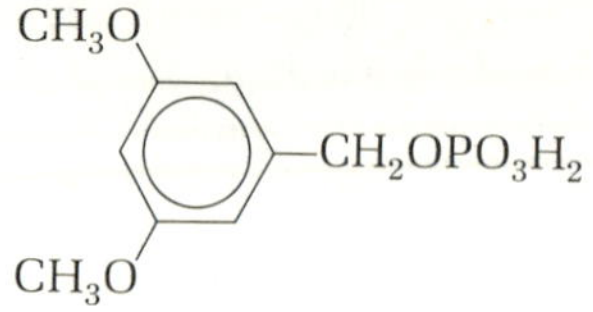

V

Propose resonance structures for the photoexcited carbonium ion that explain the stabilizing effects of the *m*-methoxyl groups. (For suggestions, consult the section on benzene photochemistry, Chapter 14.)

15 Aniline and its nitrogen-substituted derivatives undergo substitution almost exclusively at the *ortho* and *para* positions. Benzoquinuclidine (VI), on the other hand, is substituted exclusively at C-7. To what do you ascribe the differences in reactivity pattern between aniline and VI?

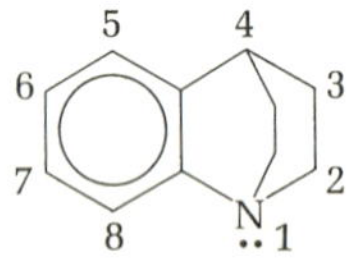

VI

16 Propose syntheses of the following compounds using benzene, toluene, and any needed aliphatic and inorganic reagents as starting materials.

a. 1-Phenylethanol

b. 1-(*p*-Methylphenyl)cyclopentene

c. *p*-Chlorophenyl *n*-butyl ketone

d. *m*-Chlorophenyl *n*-butyl ketone

e. 1-Phenyl-1-butyne

f. Toluene-2-*d*

17 Nitration of toluene gives a mixture of 58% *o*-isomer, 5% *m*-isomer, and 37% *p*-isomer. Under the same conditions the isomer proportions from benzyl chloride are 32%,

16%, and 52%; from benzal chloride (dichloromethylbenzene), 23%, 34%, and 43%; and from benzotrichloride (trichloromethylbenzene), 7%, 64%, and 29%.

a. How would you rationalize the trends in *o*- and *m*-substitution?

b. At what point, as hydrogen atoms are replaced by chlorine atoms, does the methyl group change from being an *o,p*-director? (What would be the distribution of products if the substituent exerted no directive power?)

c. Would the change in directive character of the methyl group come more or less quickly if the hydrogen atoms were being replaced by fluorine rather than chlorine? Explain your answer.

18 One mole of chlorine is allowed to react with a large excess of an equimolar mixture of benzene and toluene; $FeCl_3$ is the catalyst. Using the partial rate factors for chlorination of toluene given in equations 15.55 to 15.57, calculate the yields of chlorobenzene and *o*-, *m*-, and *p*-chlorotoluene after all the chlorine has been consumed.

19 a. Nitration of biphenyl (VII) yields principally 2-nitro- and 4-nitrobiphenyl. How would you classify phenyl as a substituent? Illustrate its mode of action with the aid of canonical structures.

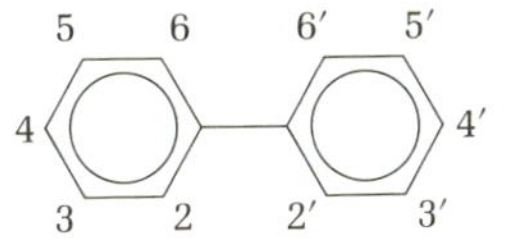

VII

b. If an activating substituent is attached to one ring of biphenyl, then another substituent is directed only into that ring. On the other hand, the presence of a deactivating substituent in one ring ensures that an incoming electrophile will attack the unsubstituted ring. What does this reactivity preference suggest about the extent of conjugation between the two rings? What structural feature of the molecule might be responsible for the observed behavior?

20 a. Under mildly acidic conditions naphthalene is nitrated about 350 times faster than benzene. The product consists of 90% 1-nitronaphthalene and 10% 2-nitronaphthalene. Calculate partial rate factors for the 1 and 2 positions of naphthalene.

b. Naphthalene undergoes electrophilic substitution much more rapidly at C-1 than at C-2. Can you explain this preference with the aid of the appropriate resonance structures?

21 a. Treatment of mesitylene (1,3,5-trimethylbenzene) with acidic hydrogen peroxide results in the formation of mesitol (2,4,6-trimethylphenol). Write out a mechanism for this reaction.

b. When hexamethylbenzene is treated with the same reagent, 2,3,4,5,6,6-hexamethyl-2,4-cyclohexadienone (VIII) is isolated from the reaction mixture.

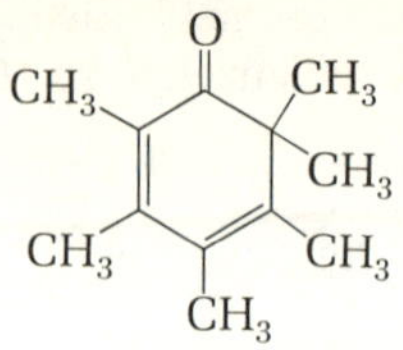

VIII

What is the probable mechanism for this reaction?

22 a. How many isomers are possible for the dichlorocyclohexadiene, tetrachlorocyclohexene, and hexachlorocyclohexane that arise during the free radical chlorination of benzene (equation 15.62)?

b. Which hexachloro isomers arise from which tetrachloro derivatives?

SUPPLEMENTARY READINGS

L. M. Stock, *Aromatic Substitution Reactions.* Englewood Cliffs, N.J.: Prentice-Hall, 1968. (P)

R. O. C. Norman and R. Taylor, *Electrophilic Substitution in Benzenoid Compounds.* New York: Elsevier Publishing Co., 1965.

CHAPTER 16

FRIEDEL-CRAFTS SYNTHESES OF ALKYLBENZENES — AROMATIC AND ALIPHATIC CARBOXYLIC ACIDS

NOMENCLATURE

This chapter is concerned in large part with the chemistry of **carboxylic acids** and to a lesser extent with one class of carboxylic acid derivatives, the **acid chlorides.** Carboxylic acids are distinguished by the presence of the functional group $-\overset{\overset{\displaystyle O}{\|}}{C}-OH$. In the IUPAC system, open-chain compounds are named by dropping the final "-e" from the name of the hydrocarbon corresponding to the longest chain and replacing it with the suffix "*-oic acid*." The carbon atom of the carboxyl group is counted as part of the longest chain and is understood to be numbered one.

ethanoic acid

4-methylpentanoic acid

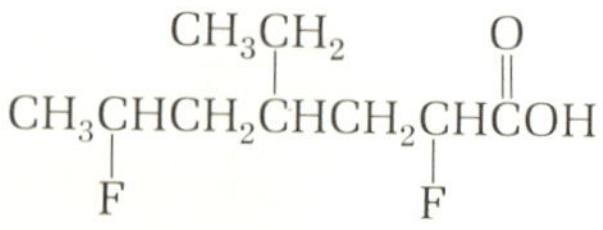
2,6-difluoro-4-ethylheptanoic acid

The carboxylic acids of low molecular weight have been known for some time and they have acquired nonsystematic names that derive from their original source (Table 16-1). Thus, formic acid was obtained by pyrolysis of red ants (Latin *formica* — ant); acetic acid is the characteristic ingredient of vinegar (Latin *acetum* — vinegar, sour wine); butyric acid is released when butter turns rancid (Latin *butyrum* — butter); and caproic, caprylic, and capric acids owe their names to their particularly penetrating odors (Latin *caper* — goat). The whole series of acids is reknowned for its pungency. Many straight-chain monocarboxylic acids are present in, and were first isolated from, animal fats and these compounds are often known as **fatty acids.** Those fatty acids containing five or fewer carbon atoms are invariably referred to by their trivial names.

TABLE 16-1
IUPAC and Trivial Names of Some Fatty Acids

	IUPAC Name	Trivial Name
HCO_2H	Methanoic	Formic
CH_3CO_2H	Ethanoic	Acetic
$CH_3CH_2CO_2H$	Propanoic	Propionic
$CH_3(CH_2)_2CO_2H$	Butanoic	Butyric
$(CH_3)_2CHCO_2H$	2-Methylpropanoic	Isobutyric
$CH_3(CH_2)_3CO_2H$	Pentanoic	Valeric
$(CH_3)_3CCO_2H$	2,2-Dimethylpropanoic	Pivalic
$CH_3(CH_2)_4CO_2H$	Hexanoic	Caproic
$CH_3(CH_2)_6CO_2H$	Octanoic	Caprylic
$CH_3(CH_2)_8CO_2H$	Decanoic	Capric
$CH_3(CH_2)_{16}CO_2H$	Octadecanoic	Stearic

There is also a well-known series of straight-chain dicarboxylic acids; these compounds are named by adding the suffix *-dioic acid* to the alkane name denoting the longest carbon chain *containing both carboxyl groups.* For denoting the position of substituents in the mono- and dioic acids, two systems are available. When systematic nomenclature is used, substituent positions are identified by numbers, the carboxyl group always occupying position one. If the trivial name of an acid is being used, then the position immediately *adjacent* to the carboxyl group is the α-position, the next one is the β-position, and so on. The dioic acids are almost universally referred to by their trivial names; many chemists are aided in remembering these names by the following mnemonic: Oh my, such good apple pie, sweet as sugar. Not as colorful as the resistor code mnemonic perhaps, but quite useful (Table 16-2).

TABLE 16-2
Nomenclature of Aliphatic Dicarboxylic Acids

	IUPAC Name	Trivial Name
HO_2CCO_2H	Ethanedioic	Oxalic
$HO_2CCH_2CO_2H$	Propanedioic	Malonic
$HO_2CCH_2CH_2CO_2H$	Butanedioic	Succinic
$HO_2C(CH_2)_3CO_2H$	Pentanedioic	Glutaric
$HO_2C(CH_2)_4CO_2H$	Hexanedioic	Adipic
$HO_2CCH(CH_2)_3CO_2H$ CH_3CH_2	2-Ethylhexanedioic	α-Ethyladipic
$HO_2C(CH_2)_5CO_2H$	Heptanedioic	Pimelic
$HO_2C(CH_2)_6CO_2H$	Octanedioic	Suberic
$HO_2C(CH_2)_7CO_2H$	Nonanedioic	Azelaic
$HO_2C(CH_2)_8CO_2H$	Decanedioic	Sebacic

When more than two carboxyl groups are present in the molecule, then it is often convenient to treat them as substituents on the principal chain. In that event the carboxyl carbon atoms are not counted as part of the chain. The same situation applies when carboxyl groups are attached directly to an alicyclic or aromatic ring.

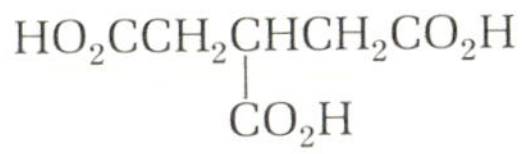

propane-1,2,3-tricarboxylic acid

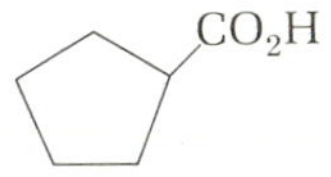

cyclopentanecarboxylic acid

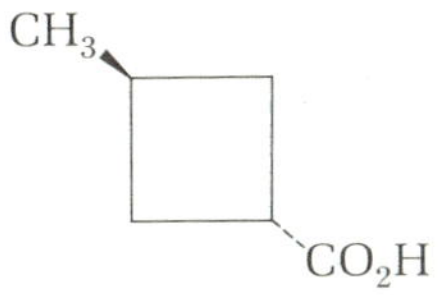

trans-3-methylcyclobutanecarboxylic acid

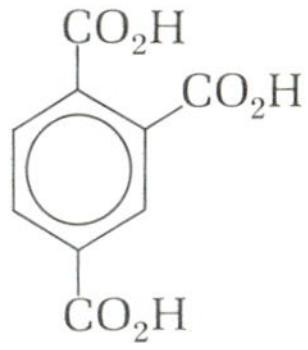

benzene-1,2,4-tricarboxylic acid

Benzene rings carrying one or two carboxyl groups are known by their trivial names, as shown below.

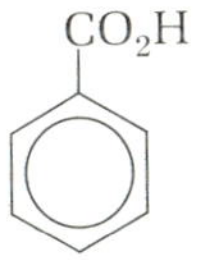

benzoic acid

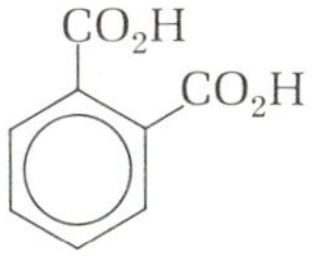

phthalic acid

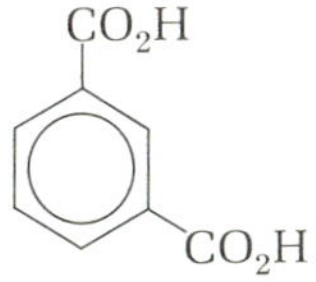

isophthalic acid

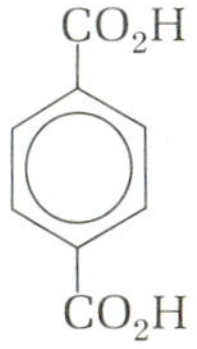

terephthalic acid

Compounds in which the hydroxyl group of the acid is replaced by a chlorine atom are labeled *acid chlorides*. The individual species are named by taking the name of the acids from which they are derived and replacing *-ic* acid by *-yl chloride*. If the acid name contains the phrase *carboxylic acid*, the latter is replaced by *carbonyl chloride*.

$CH_3C(=O)Cl$

acetyl chloride

$CH_3(CH_2)_4C(=O)Cl$

hexanoyl chloride

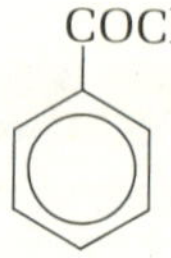

benzoyl chloride

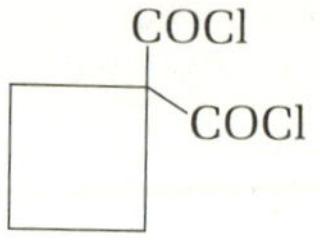

cyclobutane-1,1-dicarbonyl chloride

$ClOC{-}CH_2{-}COCl$

malonyl chloride

$ClOC{-}CH_2{-}CH_2{-}CO_2H$

β-carboxypropionyl chloride
(succinic acid monoacid chloride)

The only acid halides of general importance are the chlorides. Acid fluorides are occasionally useful; they may be prepared by allowing the corresponding acid chloride to react with hydrogen fluoride.

Acid chlorides are invariably prepared from the corresponding carboxylic acids. The most common reagents are inorganic chlorides such as phosphorus trichloride, phosphorus pentachloride, and thionyl chloride.

(16.1) $$CH_3CH_2CH_2CO_2H + SOCl_2 \xrightarrow{80^\circ} \underset{(85\%)}{CH_3CH_2CH_2\overset{O}{\overset{\|}{C}}Cl} + SO_2 + HCl$$

(16.2) 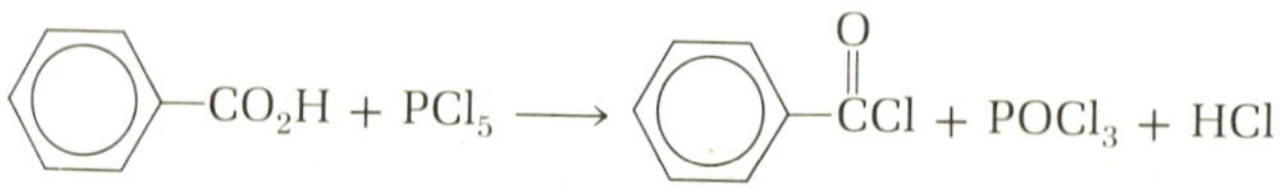

(90%)

FRIEDEL-CRAFTS ACYLATIONS

The abstraction of a halide ion from an acid halide by a Lewis acid gives rise to an **oxocarbonium ion** salt, which may be isolated under favorable conditions. The relative stability of these carbonium ions derives from the conjugation of the empty *p*-orbital on carbon with the filled oxygen *p*-orbital.

(16.3) $$CH_3\overset{O}{\overset{\|}{C}}F + SbF_5 \longrightarrow \left[CH_3\overset{+}{C}{=}\ddot{O}\!: \longleftrightarrow CH_3C{\equiv}\overset{+}{O}\!:\right] SbF_6^-$$

methyloxocarbonium hexafluoroantimonate

PROBLEM

1 When triphenylacetyl chloride (I) is treated with aluminum chloride in an inert solvent, a gas is evolved. Aqueous work-up of the reaction mixture leads to the isolation of triphenylmethanol.

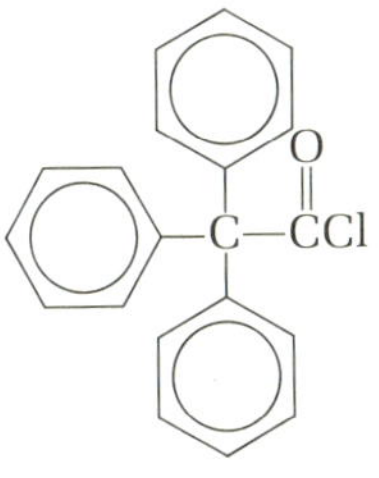

I

Show the steps leading to formation of this product. What accounts for the difference in behavior of this compound and acetyl chloride (or fluoride)?

If the carbonium ion is generated in the presence of a sufficiently reactive aromatic ring, an electrophilic substitution takes place, the carbonium ion acting as the electrophile.

(16.4)

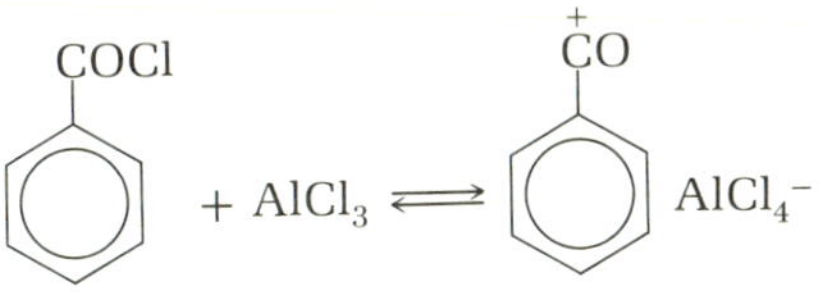

(16.5)

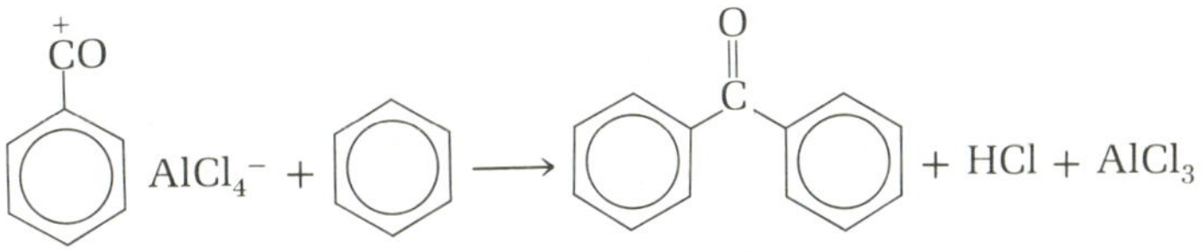

benzophenone

Although aluminum chloride is regenerated in the second step, an amount equal to that of the acid chloride must be used because it forms a complex with the ketonic product that must be decomposed with aqueous acid (see equation 16.6).

(16.6)

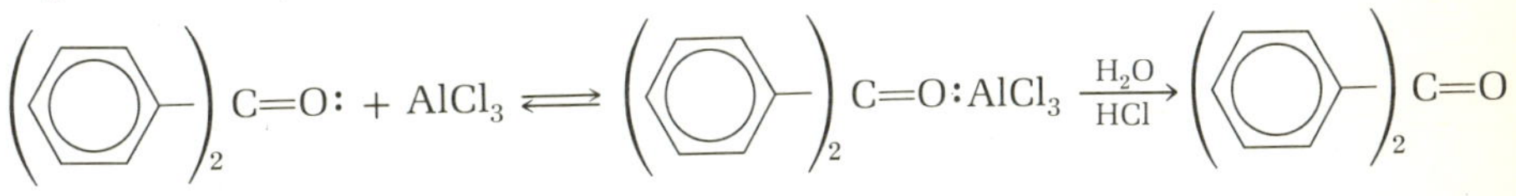

The carbonium ion salt is not the only reactive species in the mixture. Various Lewis acids form complexes with acid halides by attaching themselves to the carbonyl oxygen atom. Complex formation increases the reactivity of the carbonyl atom, although the complex is not as reactive as the oxocarbonium ion.

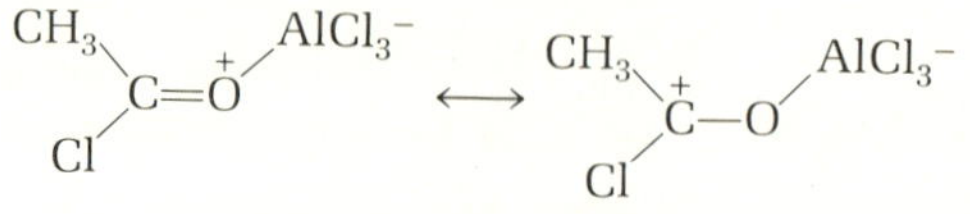

The **Friedel-Crafts acylation** (*acyl* is the generic name for R—C(=O)—) may be carried out using aliphatic or aromatic acid chlorides, under intra- or intermolecular conditions.

(16.7)

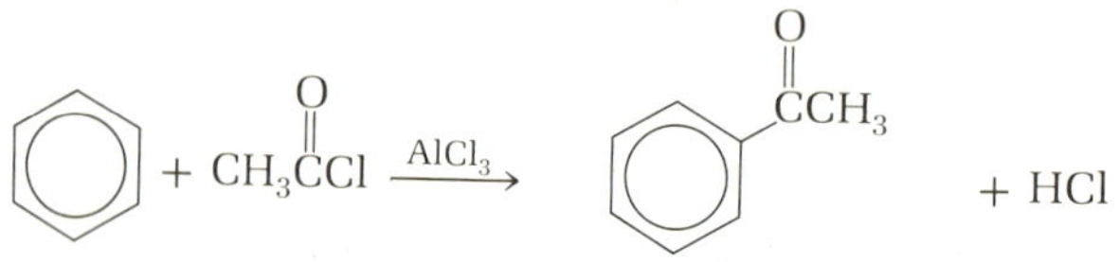

acetophenone (97%)
(methyl phenyl ketone)

(16.8)

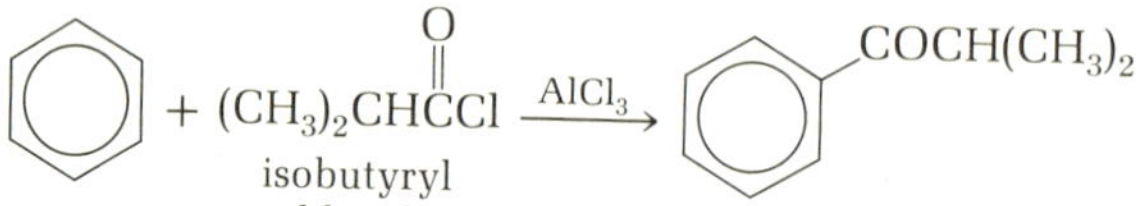

isobutyryl chloride

isobutyrophenone (76%)

(16.9)

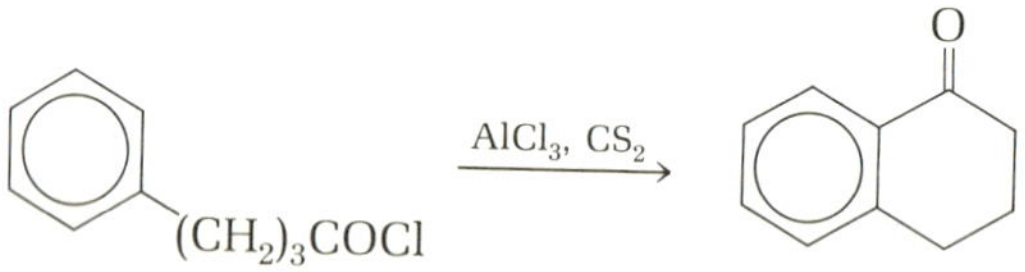

γ-phenylbutyryl chloride

α-tetralone (75–90%)

Intermolecular acylations are normally run using an excess of the substrate as the solvent (if it is inexpensive and a liquid). If a solvent is required, two of the most common are carbon disulfide (CS_2) and nitrobenzene.

Acid chlorides are the most common source of **acylonium cations** (oxocarbonium ions) and acyl complexes for Friedel-Crafts acylations but they are not the sole source of these species. Carboxylic acid **anhydrides** are also used in this capacity, the other product being a carboxylic acid.

(16.10)

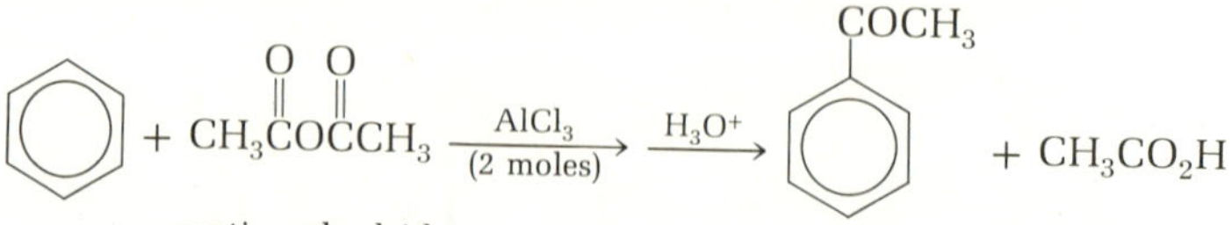

acetic anhydride

(85%)

(16.11)

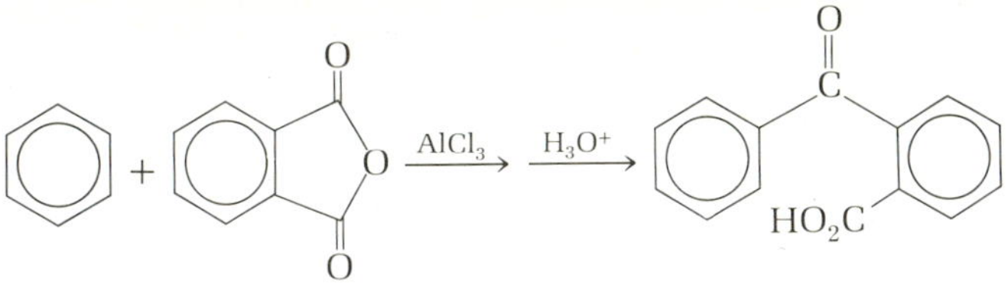

phthalic anhydride

o-benzoylbenzoic acid (88%)

Acyl cations may also be generated from carboxylic acids by abstracting the —OH group with the aid of a strong acid.

(16.12)

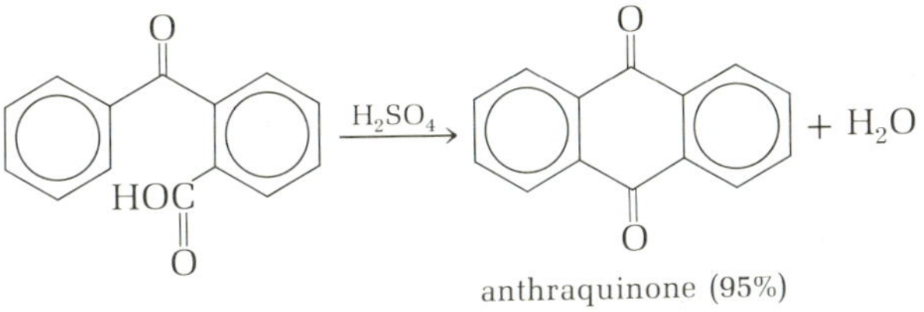

anthraquinone (95%)

(16.13)

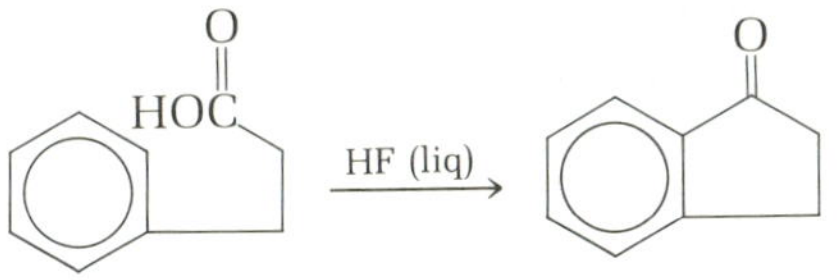

β-phenylpropionic acid

1-indanone (75%)

PROBLEM

2 a. Aldehydes may be synthesized by a procedure that is related to the Friedel-Crafts acylation. (Formyl chloride, HCOCl, is not stable enough to be isolated, while the stable formyl fluoride, HCOF, is not widely used.) Carbon monoxide and hydrogen chloride are bubbled into a mixture of an aromatic compound (not strongly deactivated) and a Lewis acid.

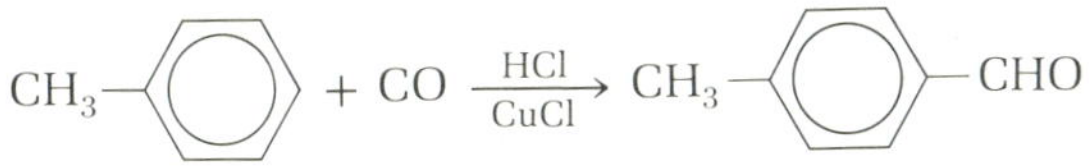

(1)

p-tolualdehyde (85%)

Write out a plausible mechanism for this reaction.

b. Hydrogen cyanide (*Gattermann reaction*) and alkyl cyanides (*Hoesch reaction*) react with activated aromatic nuclei, in the presence of hydrogen chloride and a Lewis acid, to give aldehydes and ketones, respectively. Write out a mechanism for these processes.

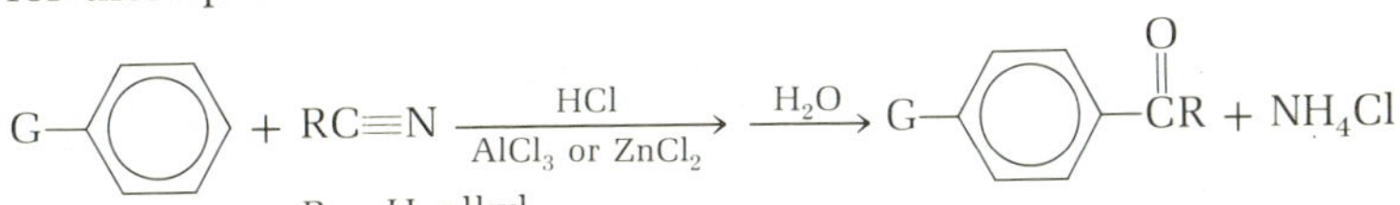

(2)

R = H, alkyl

Since the carbonyl group is electron-withdrawing, the incoming acyl group deactivates the ring, allowing the introduction of only one substituent at a time. The acylating agents are highly selective in their attack. Aromatic rings containing nitro, cyano, carbonyl, or carboxyl groups cannot normally be acylated unless they also contain strongly activating groups. Furthermore, monosubstituted benzenes that are capable of being acylated are attacked almost exclusively in the *para* position. The partial rate factors for acetylation of toluene with acetyl chloride and aluminum chloride are

$o_f^{CH_3}$, 4.5 $m_f^{CH_3}$, 4.8 $p_f^{CH_3}$, 749

Aluminum chloride is by far the most widely used catalyst, and one of the most active, for Friedel-Crafts acylations. Other metal halides may serve, however, such as the halides of boron, iron, and zinc, as well as stannic chloride ($SnCl_4$). Stannic chloride and zinc chloride are rather weak catalysts and are normally used only with very reactive aromatic substrates.

Anilines and phenols form complexes with aluminum chloride in which an unshared electron pair on nitrogen or oxygen is coordinated with the aluminum atom. In the case of aniline, these complexes are usually resistant to Friedel-Crafts reaction conditions. Anilines are usually acylated on nitrogen first and then subjected to Friedel-Crafts acylation in the ring.

(16.14)

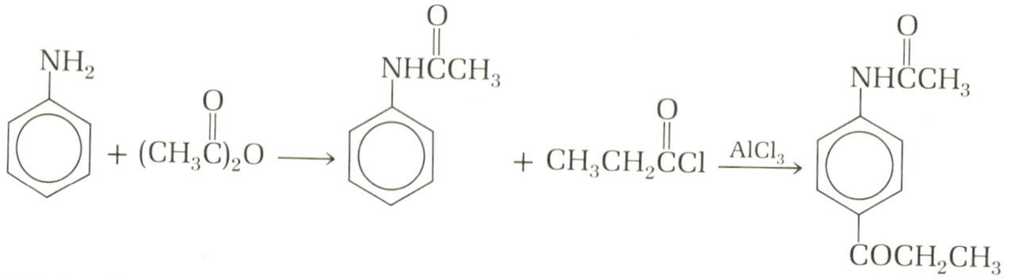

With phenols, as with anilines, acylating agents attack the atom with the unshared pair (oxygen) faster than they attack the ring. The products from acylation of phenols may be rearranged by the action of aluminum chloride, the acyl group moving from the oxygen atom to the *ortho* or *para* position of the ring (*Fries rearrangement*).

(16.15)

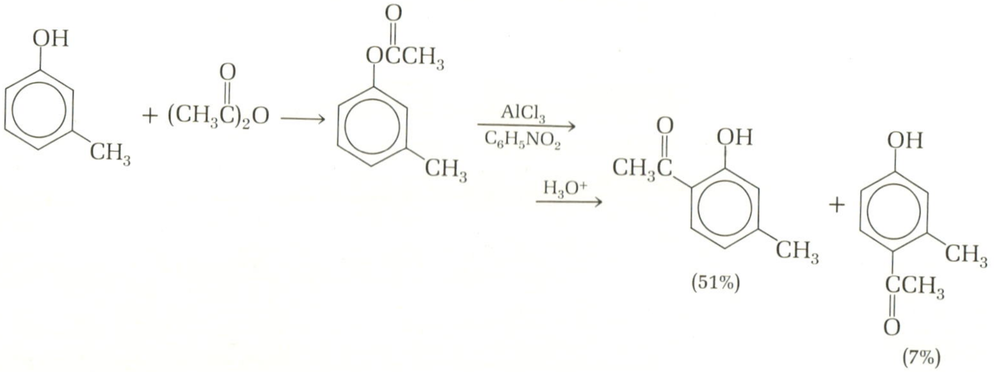

(16.16)

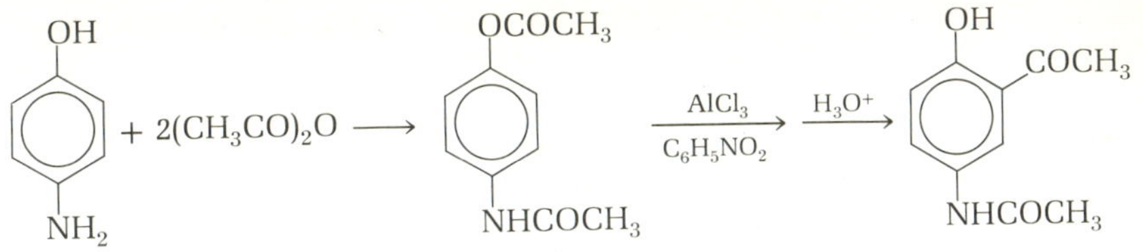

The *ortho* and *para* rearrangements apparently follow different mechanistic paths and the ratio of *ortho* to *para* product is dependent on the amount of catalyst, its concentration, the nature of the acyl group, etc.

Various alkenes may also be acylated under Friedel-Crafts conditions. These reactions must be more closely controlled to suppress side reactions and they do not generally possess the same synthetic utility as the aromatic acylations.

FRIEDEL-CRAFTS ACYLATIONS – SYNTHETIC APPLICATIONS

The molecules obtained by means of the Friedel-Crafts acylation reaction are important intermediates and end products in synthetic chemistry. As we see below, the acyl groups may be reduced to alkyl groups, which is a highly useful reaction. The carbonyl group also provides a focus for further chemical transformations. Note that the 1-phenylpropanol-1-*d* (equation 16.18) can be obtained by way of propionaldehyde-1-*d* but that the latter compound is relatively difficult to come by.

(16.17)

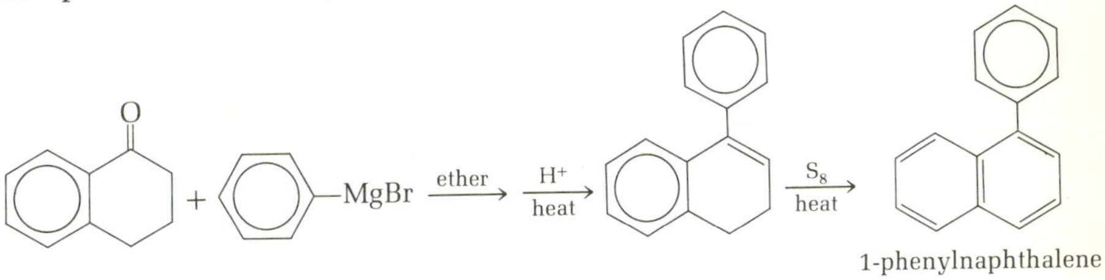

(16.18)

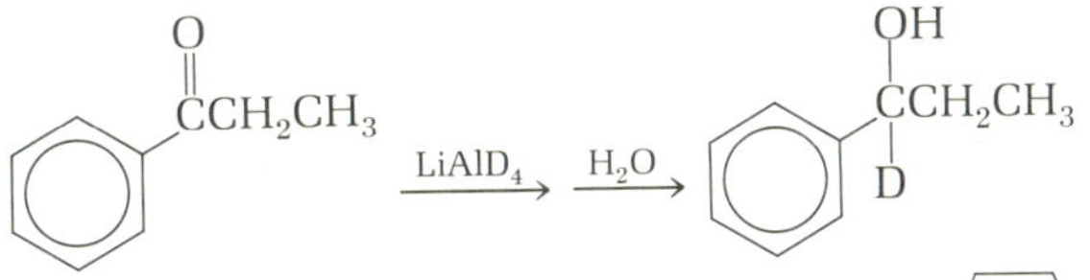

(16.19)

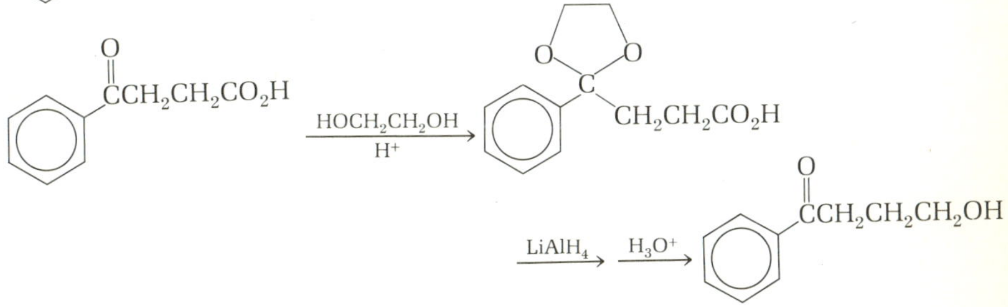

The synthesis of cyclic species, as exemplified by equations 16.9, 16.12, and 16.13, is one of the most important applications of the Friedel-Crafts acylation. In all the cases cited the acylation takes place *ortho* to the point of attachment of the side chain. As noted above, however, acylation takes place preferentially *para* to an alkyl substituent and this preference is indeed observed in intramolecular reactions provided that a sufficiently long alkyl chain is available. The *cyclophanes* are of substantial interest to chemists because they permit one to study the effect of chain length and strain on aromatic spectroscopic properties, the effect of the aromatic ring current on nmr spectra of the methylene groups, etc.

(16.20)

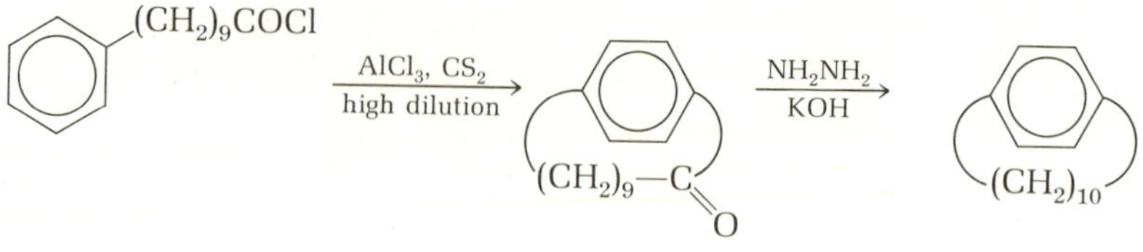

(25%) (10)paracyclophane

FRIEDEL-CRAFTS ALKYLATION

Acyl halides are not the only sources of carbonium ions and electrophilic complexes. Friedel and Crafts also discovered that *alkyl* halides would react with aromatic rings in the presence of Lewis acids such as aluminum chloride.

(16.21)

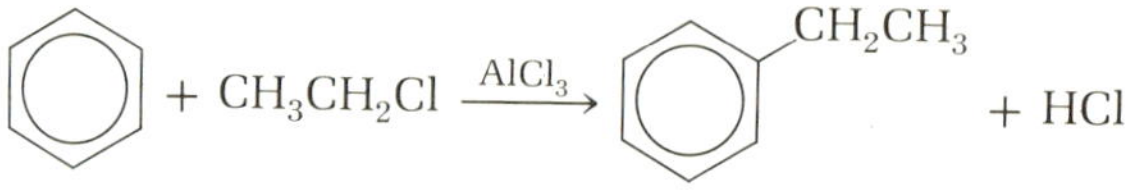

Although toluene and other methylbenzenes are available from natural sources, many higher alkylbenzenes are only obtainable from Friedel-Crafts syntheses.

The Friedel-Crafts alkylation reaction has a number of limitations that restrict its general applicability. The mechanistic evidence obtained so far suggests that secondary and tertiary halides dissociate into carbonium ions with the aid of the acid catalyst.

(16.22) $(CH_3)_2CHCl + AlCl_3 \rightleftharpoons (CH_3)_2CH^+AlCl_4^-$

(16.23)

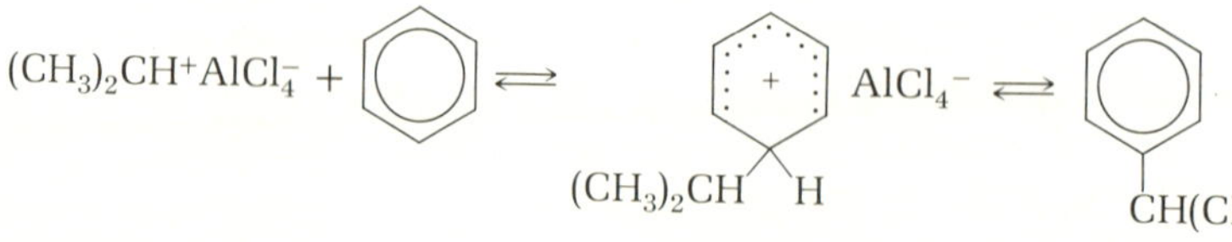

isopropylbenzene

+ HCl + $AlCl_3$

Primary halides seem rather to form electrophilic complexes that lose halide ion simultaneously with a nucleophilic attack by the aromatic ring.

(16.24) $$CH_3CH_2Cl + AlCl_3 \rightleftharpoons \overset{\delta+}{CH_3CH_2}—\overset{\delta-}{Cl\cdots AlCl_3}$$

(16.25) 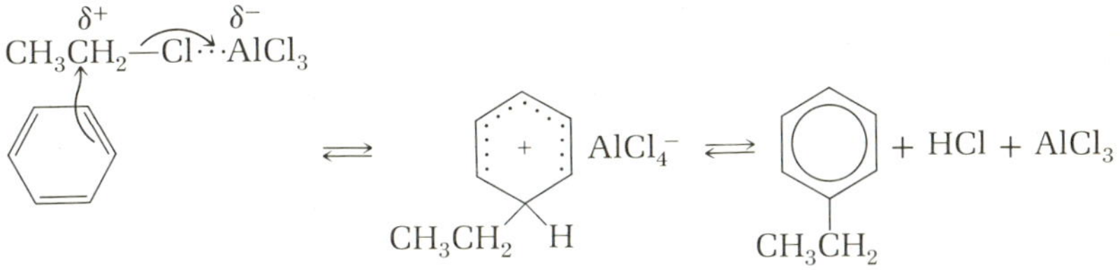

The aluminum chloride is always regenerated so that only catalytic amounts are required (0.1 to 0.25 mole per mole of alkyl halide).

Since the alkylations proceed by way of carbonium ions and carbonium ion-like complexes, a good deal of rearrangement may take place in the alkyl group, with mixtures of isomeric products resulting.

(16.26) 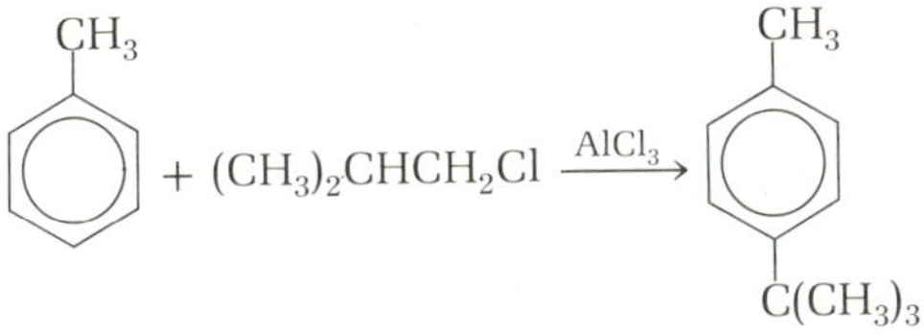

(16.27) 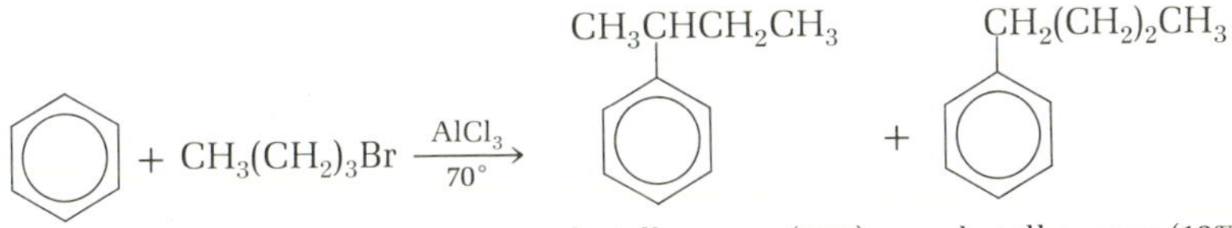

s-butylbenzene (39%) *n*-butylbenzene (13%)

(16.28) 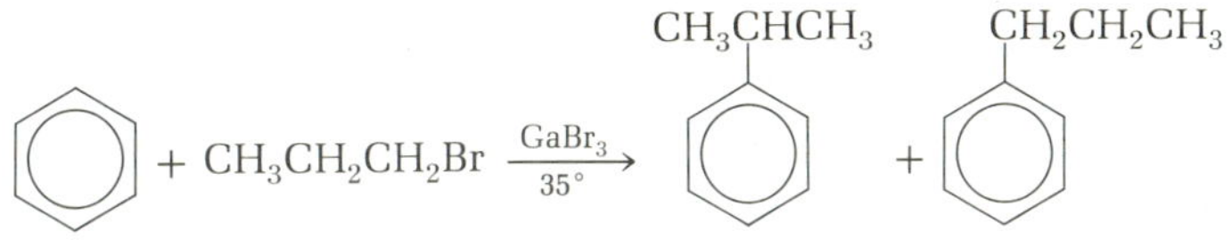

(72%) (18%)

The isomerization problem is especially serious with primary halides; the extent of isomerization will be a function of the catalyst, the temperature, and the nucleophilicity of the aromatic ring. The more nucleophilic the aromatic ring, the better are its chances of reacting with the alkyl halide-catalyst complex before rearrangement occurs.

Friedel-Crafts alkylations are further complicated by the mildly activating effect of the alkyl group once it is attached to a ring. This activation tends to encourage further substitution in previously substituted rings. Equation 16.29 gives some idea of the product mixture obtained in the course of these alkylations.

(16.29)

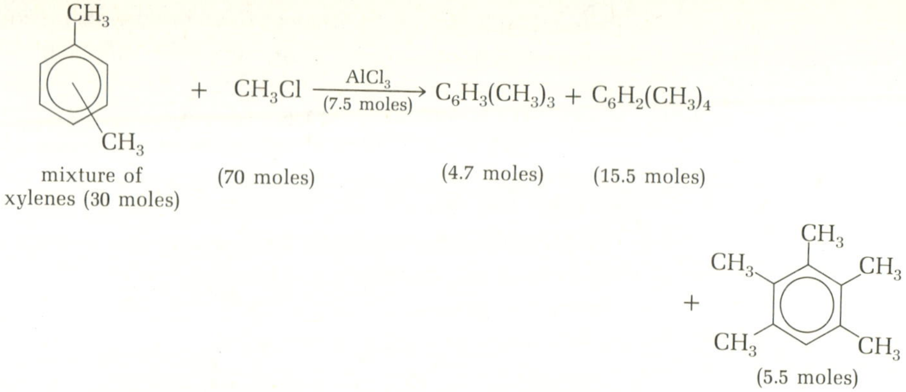

(16.30)

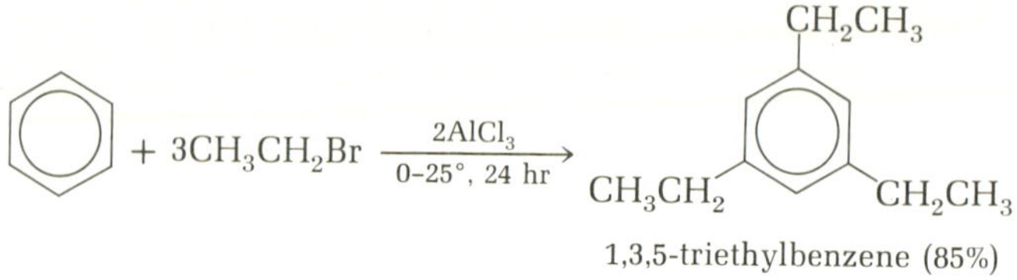

The object of this synthesis was to prepare durene (1,2,4,5-tetramethylbenzene); it may be obtained from the tetramethylbenzene fraction in yields of 4.0 to 4.6 moles.

Alkylation is usually reversible under the reaction conditions, particularly if substantial amounts of catalyst are present and some of the hydrogen halide released remains in the reaction mixture, which is almost always the case. As a consequence, alkyl groups that are present on the ring at the start of the reaction may be removed or transferred to other rings, particularly if they are secondary or tertiary. Equilibration of positional isomers also sets in; the compounds initially formed have the alkyl groups predominantly in *ortho* and *para* relationships (kinetic control) but *meta* isomers accumulate as the reaction period lengthens (thermodynamic control).

PROBLEM

3 Electrophilic substitution reactions on *t*-butylbenzenes often involve partial replacement of one or more *t*-butyl groups by an incoming electrophile, as in equation 3.

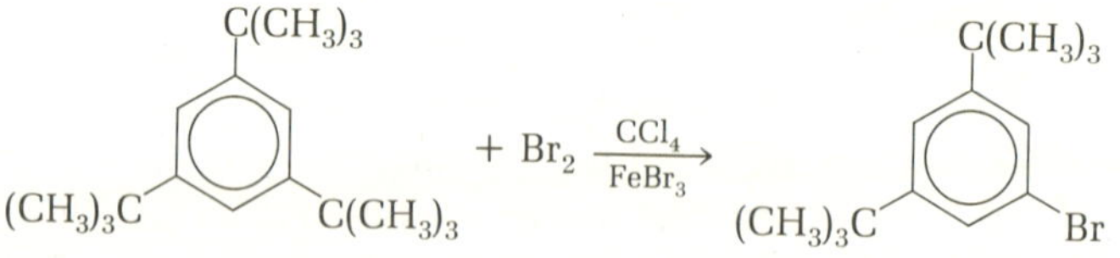

(3)

Write a reasonable mechanism for this process. What properties of *t*-butyl make it particularly prone to replacement when compared with other alkyl groups?

Halides in which the halogen atom is attached to an aromatic or vinylic carbon atom cannot normally take part in Friedel-Crafts reactions. Chlorobenzene is sometimes employed as a solvent in both acylations and alkylations. Ethylene chloride, $ClCH_2CH_2Cl$, finds use as a solvent for Friedel-Crafts acylations, which seems to indicate that acylations take place much more readily than alkylations, particularly when the latter involve primary halides.

Alkyl halides are not the only source of carbonium ions and electrophilic complexes for alkylation reactions; olefins and alcohols may also serve. Either Lewis or proton acids are catalysts for these alkylations.

(16.31) 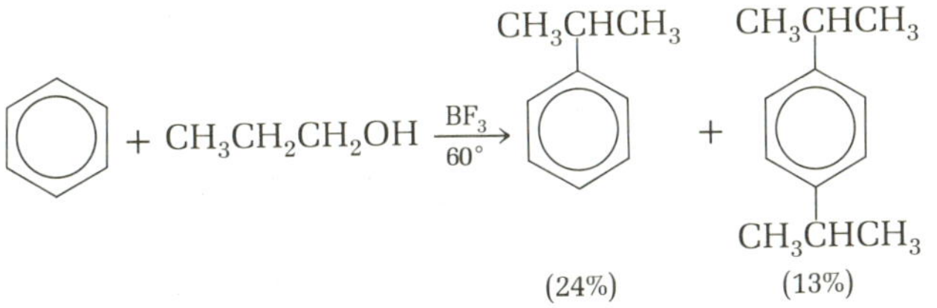

(16.32) 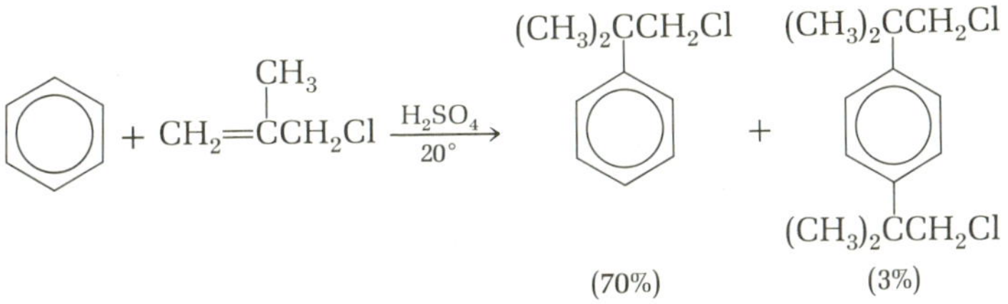

SYNTHETIC LIMITATIONS OF THE FRIEDEL-CRAFTS ALKYLATION

Because of the rearrangements, isomerizations, and polyalkylations that usually accompany Friedel-Crafts alkylations, plus the difficulty posed by the presence in the product mixture of difficulty separable isomers, this reaction has limited synthetic value. This stricture applies particularly to the synthesis of primary alkylbenzenes. These compounds are best obtained by a two-step route, in which the aromatic ring is first acylated (the acylation reaction being substantially free of the complications noted above); the ketone group is then reduced to a methylene group, usually by the Clemmensen method.

(16.33) 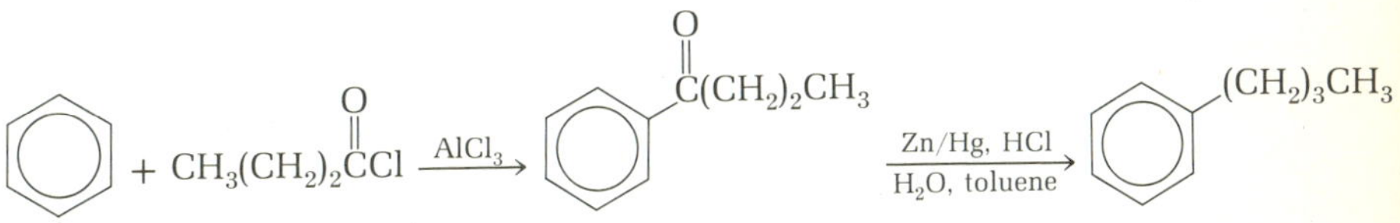

(16.34)

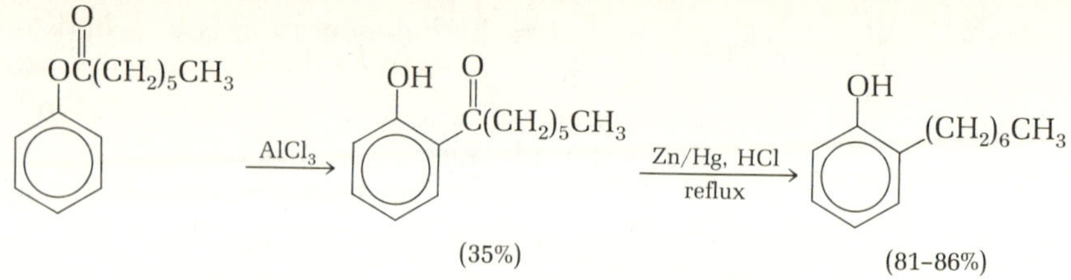

(16.35)

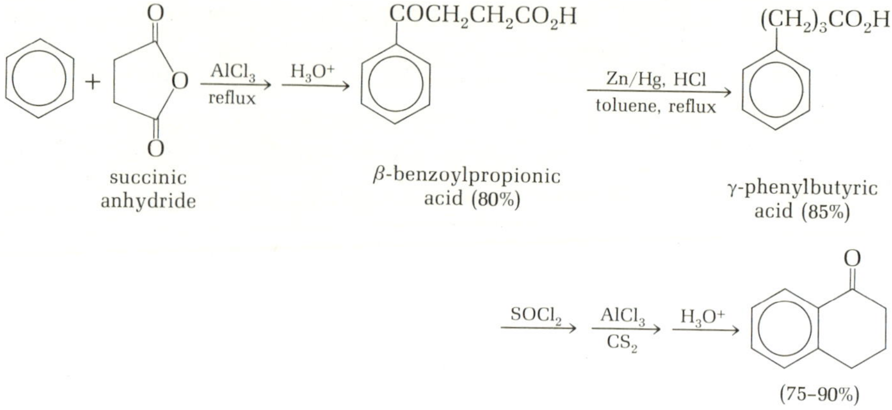

PROBLEM

4 The superiority of the Friedel-Crafts acylation over the corresponding alkylation rests primarily upon the fact that acyl cations do not rearrange while alkyl cations are often quite prone to do so. How would you explain this difference in behavior?

REACTION OF ORGANOCADMIUM COMPOUNDS WITH ACYL HALIDES

The reaction between an acid chloride and an aromatic ring almost always requires an electrophilic catalyst because the aromatic ring is too weak a nucleophile to displace the chloride ion without some assistance. The same pattern appears in certain other reactions of acid chlorides that may seem at first to be of a different character.

Cadmium chloride reacts with aryl or *primary* alkyl Grignard reagents to produce organocadmium compounds that are stable and may be isolated. The pure organocadmium compounds are rather unreactive toward ketones, acid chlorides, etc.

(16.36) $$2C_6H_5MgBr + CdCl_2 \xrightarrow[\text{reflux}]{\text{ether}} (C_6H_5)_2Cd + MgBr_2 + MgCl_2$$

If the original reaction mixture, *which still contains the magnesium halides,* is used instead, then the organocadmium compound reacts with acid chlorides to produce ketones in fair to excellent yields.

(16.37) $$(CH_3CH_2CH_2CH_2)_2Cd + ClCH_2\overset{O}{\overset{\|}{C}}Cl \xrightarrow[MgX_2,\ 20^\circ]{\text{benzene}} \xrightarrow{H_3O^+} CH_3CH_2CH_2CH_2\overset{O}{\overset{\|}{C}}CH_2Cl \quad (51\%)$$

(16.38)
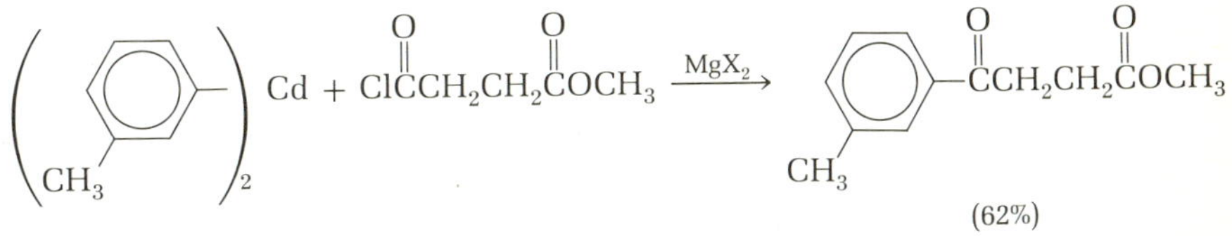

The magnesium halide complexes with the acid chloride and increases its reactivity toward the sluggishly reactive organocadmium compound.

(16.39) $$R\overset{O}{\overset{\|}{C}}\!-\!Cl + MgX_2 \rightleftharpoons R\!-\!C\begin{smallmatrix}\nearrow O\!:\!MgX_2 \\ \searrow Cl\end{smallmatrix} \quad \text{and/or} \quad RC\begin{smallmatrix}\nearrow O \\ \searrow Cl\!:\!MgX_2\end{smallmatrix}$$

The latter is still sufficiently unreactive so that it does not attack the product ketones.

Organocadmium compounds do not react with a number of functional groups ($-NO_2$, $-COR$, $-CN$) that are reactive toward Grignard reagents. Since Grignard reagents are used to prepare organocadmium compounds, however, some care must be exercised in planning syntheses of compounds containing these groups. For instance, in equation 16.40, the two aromatic functionalities could *not* be interchanged in the synthetic sequence.

(16.40)
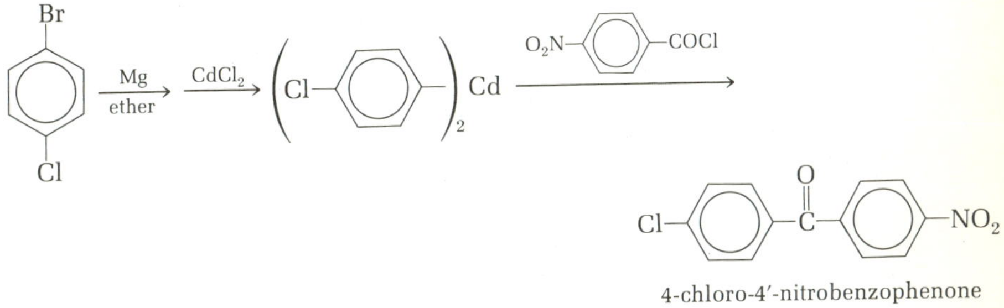

4-chloro-4′-nitrobenzophenone

HALOGENATION OF SIDE CHAINS OF ALKYLBENZENES

Reaction of alkylbenzenes with chlorine in the absence of electrophilic catalysts (i.e., under free radical conditions) results in both chlorine atom addition to the ring and hydrogen atom abstraction from the side chain. Chlorination of the side chain becomes the predominant reaction when the halogen is bubbled through the hydrocarbon under ultraviolet illumination, often at elevated temperatures.

(16.41) 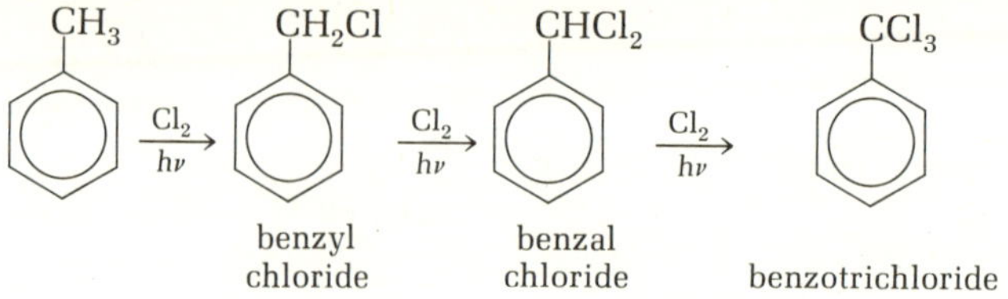

The extent of chlorination can usually be controlled, particularly if halogen atom sources other than the molecular halogens are used, and these reactions are often useful.

(16.42) 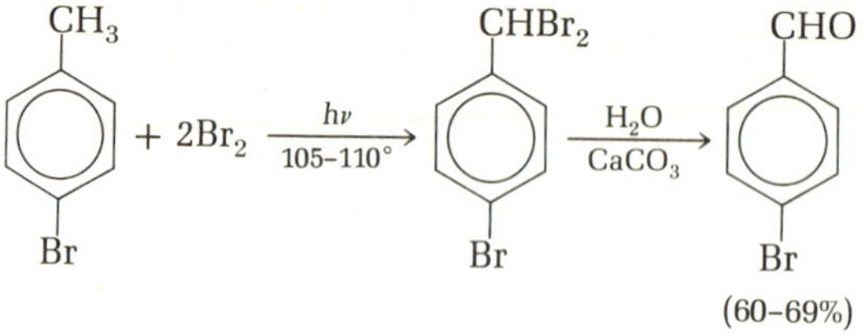

(16.43) 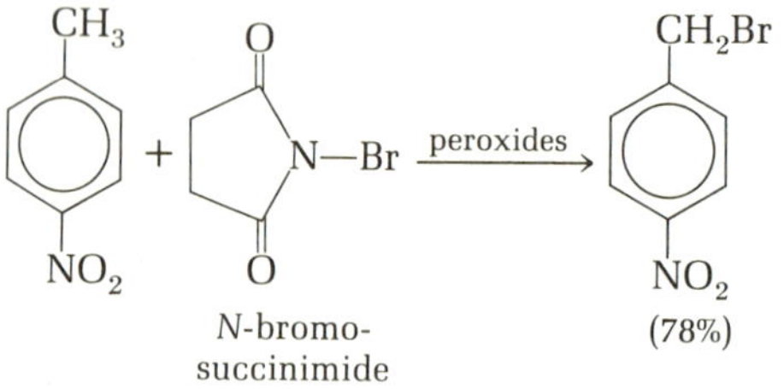

Halogenation of alkylbenzenes possessing longer chains might be expected to lead to reaction primarily at the position α to the ring since the phenyl group is known to weaken adjacent C—H bonds by stabilization of the related radical (Chapter 14). When chlorine is the reactant, this expectation is fulfilled at low temperatures; at higher temperatures the chlorine atom is much less discriminating and seems in fact to prefer positions removed from the ring.

(16.44) 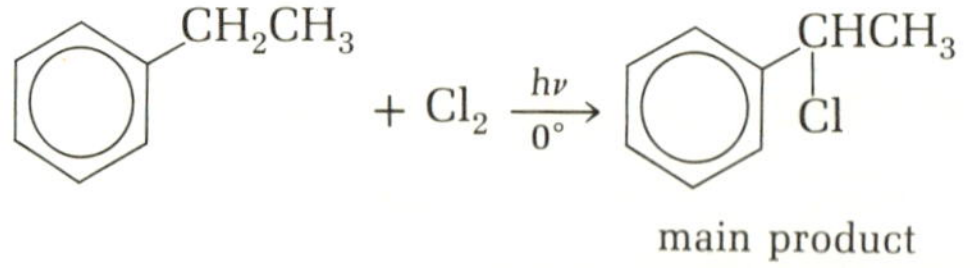

(16.45) 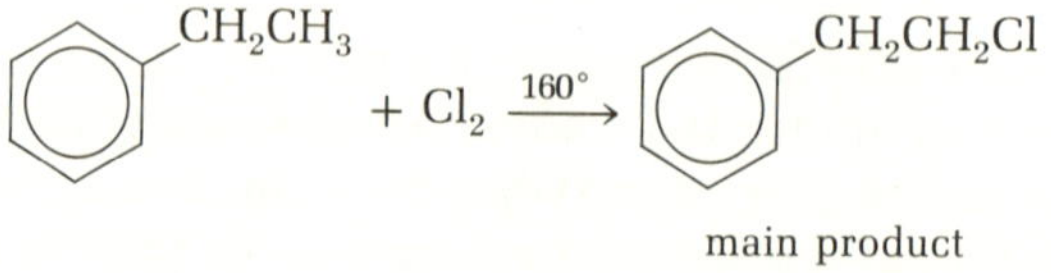

Bromine is quite selective however, and overwhelmingly favors attack at the α-position.

(16.46) 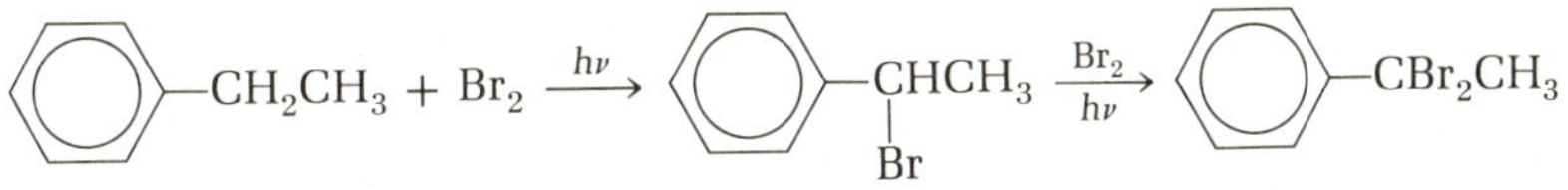

(16.47) 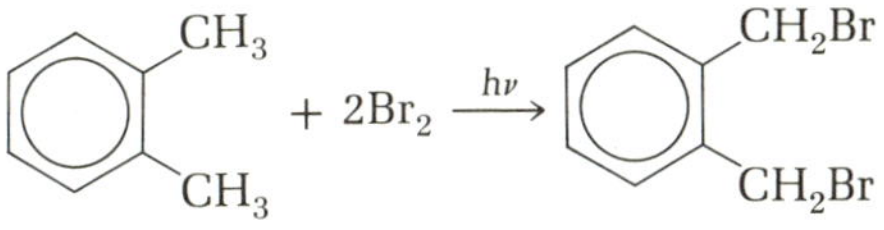

OXIDATION OF SIDE CHAINS OF ALKYLBENZENES

Most alkanes and cycloalkanes are attacked rather sluggishly by the common oxidizing agents — chromic acid and its derivatives and potassium permanganate. Furthermore, the oxidation is usually random, in that all secondary hydrogen atoms are attacked at about the same rate. If a benzene ring is attached to one carbon atom of the chain, however, then the hydrogen atoms at that position become the focal point for oxidative attack.

Strong aqueous oxidizing solutions oxidize alkyl side chains to carboxyl groups; if the side chain is longer than methyl, it will be degraded, with the carbon atoms beyond the α-position being sliced off.

(16.48) 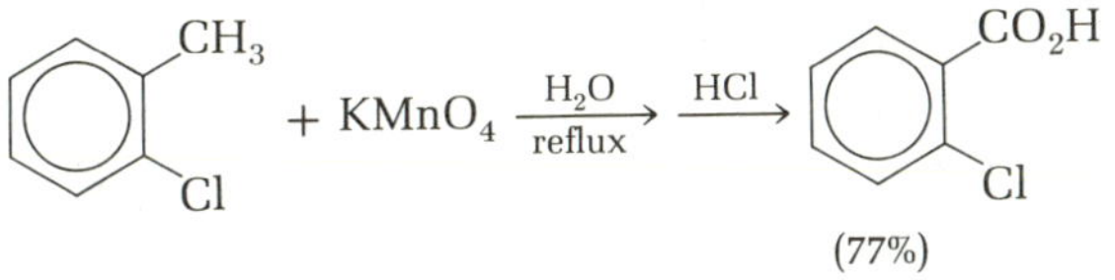

(16.49) 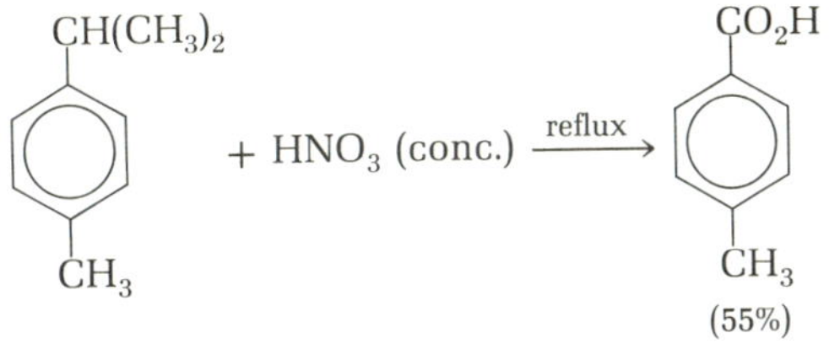

Equation 16.49 shows the degradation of such a chain and also illustrates the selective character of nitric acid — the introduction of an electron-withdrawing carboxyl group inhibits attack at the second alkyl group.

PROBLEM

5 Benzene and ethyl chloride react in the presence of aluminum chloride to yield ethylbenzene. It is of some interest to know whether or not the ethyl group underwent a carbonium ion-type rearrangement in the process. Devise an experiment to

answer this question; be specific about the materials to be used and the experiments to be performed.

The initial step in these oxidations appears to involve the abstraction of a hydrogen atom by the oxidizing agent, yielding an organic radical. It is thus not surprising that oxidation and halogenation prefer the same point of attack on arylalkanes. Once a hydroxyl group has been attached to the molecule, then the oxidation follows familiar paths, with an alkene (Chapter 9) or an alcohol (Chapter 11) as the intermediate. Alkyl benzenes containing no hydrogen atoms α to the ring (e.g., *t*-butylbenzene) are not oxidized in this manner.

(16.50)

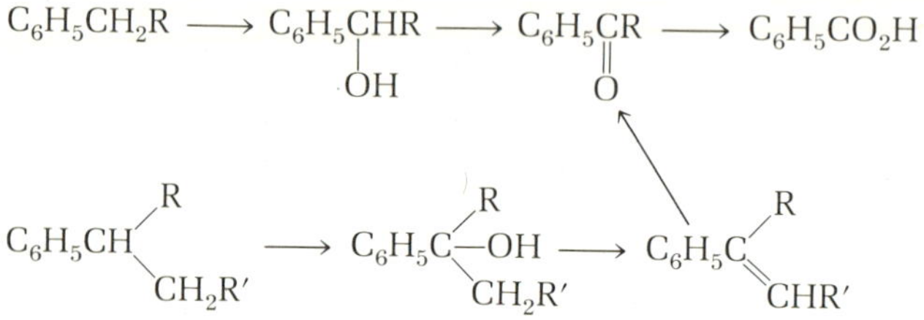

Note that oxidation converts an *o*, *p*-directing alkyl group to a *m*-directing carboxyl group. This transformation allows the synthesis of compounds such as *o*-chlorobenzoic acid (equation 16.48), which would be a very minor product of chlorination of benzoic acid.

Oxidation of a methyl group may be halted before the carboxylic acid stage is reached, provided that the reaction takes place in the absence of strong aqueous acid. The technique provides a useful route to aromatic aldehydes. Derivatives of chromium(VI) oxide, CrO_3, are the most widely used oxidants for these reactions.

(16.51)

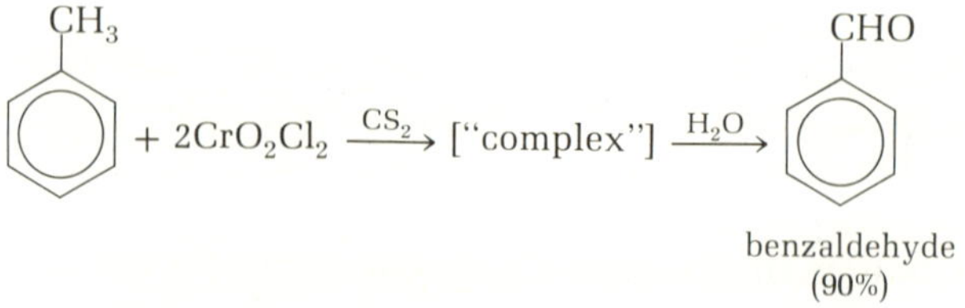

(16.52)

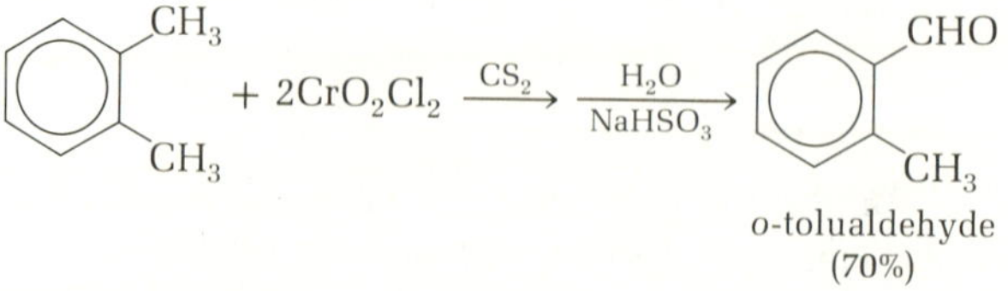

For example, chromyl chloride, $O{=}CrCl_2{=}O$, is mixed with the appropriate toluene in carbon disulfide or carbon tetrachloride; a mysterious "complex" precipitates and when the complex is decomposed with water (often in the presence of a mild reducing agent), the aldehyde is liberated. The "complex" contains 2 moles of chromyl chloride per mole of aromatic and a reasonable structure that has been proposed for it is **1.**

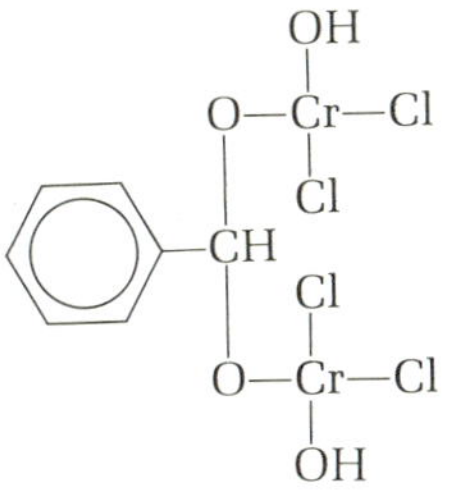

1

A similar technique utilizes chromic oxide, acetic anhydride, and concentrated sulfuric acid. The diacetate obtained (equation 16.53) is then hydrolyzed to the aldehyde.

(16.53)

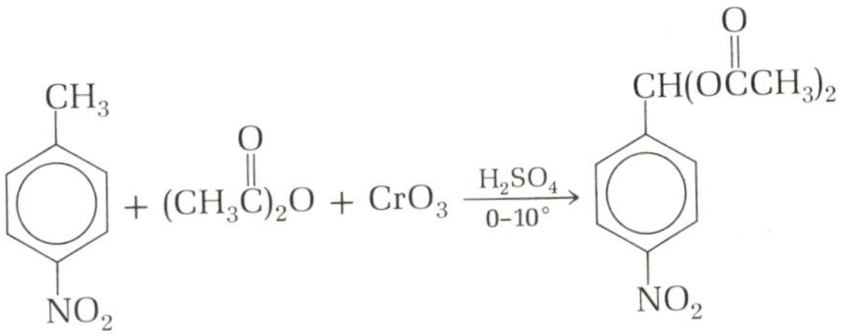

(16.54)

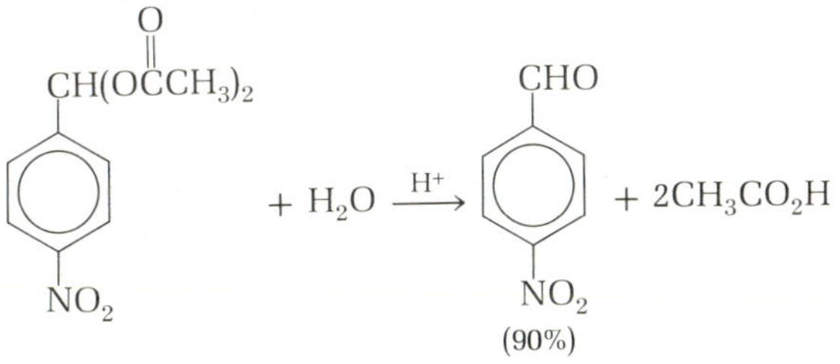

In all the cases cited, precipitation of the dioxygenated intermediate may be vital to the prevention of further oxidation.

The activating effect of a benzene ring on an adjacent methyl group, insofar as oxidation is concerned, is also discernible in biological systems. Because of its great stability, benzene is not subject to metabolic degradation by the human body. Since benzene is very slightly water-soluble, it is excreted only slowly and tends to accumulate in the liver; due to these factors, and the permeability of skin and lungs to benzene, this hydrocarbon is quite poisonous. Toluene, on the other hand, is transformed into benzoic acid by the human metabolism, albeit slowly. Benzoic acid is many times more water-soluble than benzene and may be excreted without difficulty. Toluene is therefore to be preferred to benzene in those instances when an aromatic solvent is required and there is a reasonable probability of human exposure to its vapors.

While oxidation of alkylbonzenes is a valuable method for synthesizing benzoic acids (equations 16.48 to 16.50), the oxidation of simple alkanes is not a generally useful technique for obtaining aliphatic carboxylic acids. In order to synthesize an aliphatic carboxylic acid by oxidation, it is necessary that one carbon atom already be oxygenated (alcohol or aldehyde) or that a double bond be present (alkene or ketone; see Chapters 9 and 11). Alcohols, aldehydes, and alkenes are readily oxidized, while ketones are more difficult. Conditions can always be found under which ketones are not attacked while the other three functional groups are.

Oxidation of alcohols and aldehydes leads to carboxylic acids with the same number of carbon atoms; oxidation of alkenes and ketones gives rise to acids with a smaller number of carbon atoms, unless the starting material is cyclic. The most widely used oxidants are permanganate, chromic acid and related compounds, and nitric acid.

(16.55) $$ClCH_2CH_2CH_2OH + HNO_3 \xrightarrow{25°} \underset{(55\%)}{ClCH_2CH_2CO_2H} + H_2O + NO + NO_2$$

(16.56) $$CH_3(CH_2)_5CHO + KMnO_4 \xrightarrow[20°]{H_3O^+} \underset{(77\%)}{CH_3(CH_2)_5CO_2H} + MnO_2$$

It should also be recalled that reaction of Grignard reagents with carbon dioxide provides a convenient synthesis of carboxylic acids (Chapter 11).

(16.57) 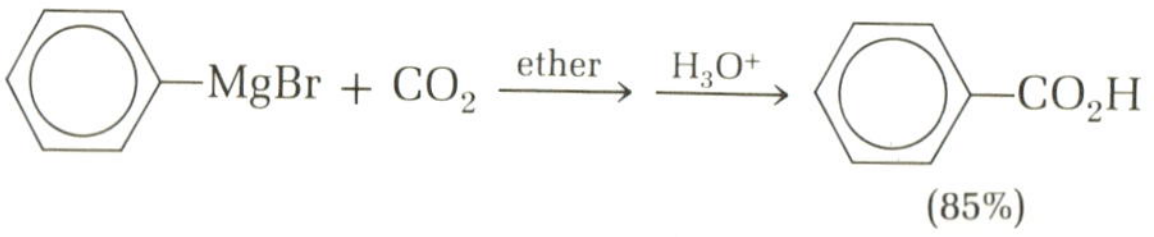

The straight-chain monocarboxylic acids up through nonanoic acid are liquids at room temperature; all substituted benzoic acids are solids, however. The pure liquid acids exist almost entirely in the form of molecular dimers, in which two molecules are held together by a pair of hydrogen bonds. Carboxylic acids have very small dielectric constants, considering the fact that they contain the highly polar $>C{=}O$ and —OH groups: acetic,

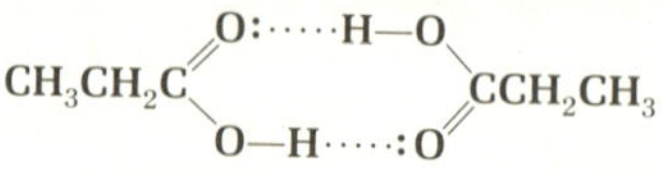

FIGURE 16-1
Propionic Acid Dimer

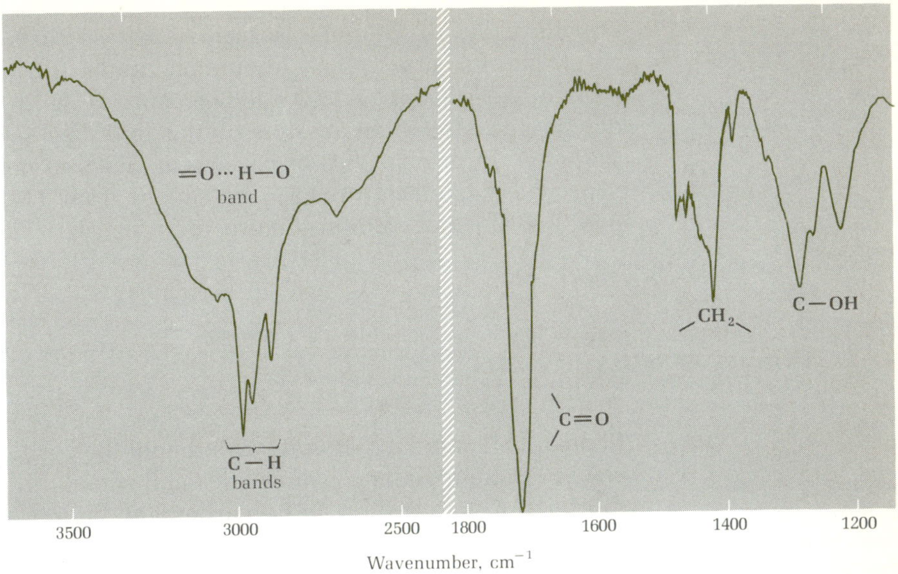

FIGURE 16-2

Partial Infrared Spectrum of Valeric Acid in Carbon Tetrachloride

6.2; propionic, 3.3; isobutyric, 2.7. (The glaring exception is formic acid, $\epsilon = 58.5$.) The polar nature of the individual carboxylic acid molecule and the low polarity of the bulk liquid can be reconciled if the liquid consists principally of dimers in which most of the bond dipoles cancel one another.

These dimers persist even in the vapor state, resulting in a low entropy of vaporization for the acids, and they are also found in solutions of car-

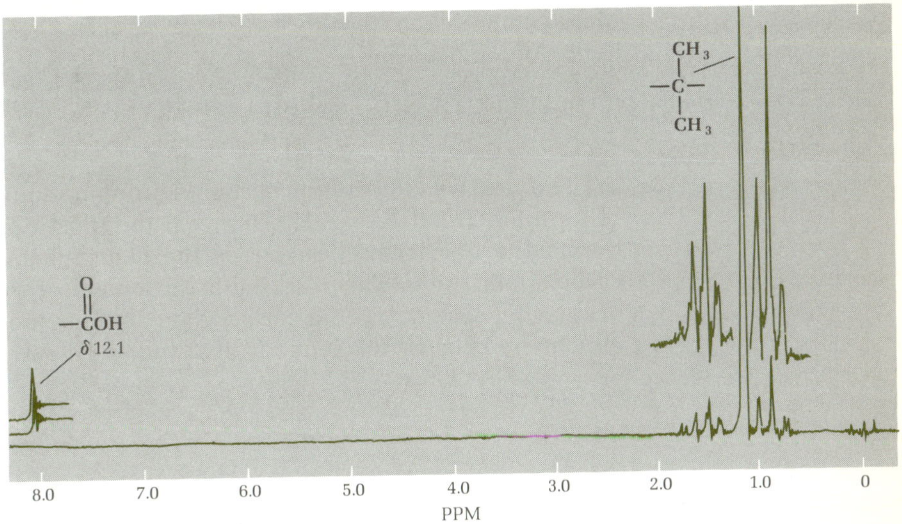

FIGURE 16-3

NMR Spectrum of 2,2-Dimethylbutanoic Acid

boxylic acids in nonpolar solvents such as carbon tetrachloride. The ir spectrum of a carboxylic acid obtained under the usual conditions contains a broad peak between 2500 and 3500 cm^{-1} (Figure 16-2). This absorption is due to the stretching motion of the hydrogen atom in the =O···H—O grouping. In nmr spectra of carboxylic acids in nonpolar solvents there is usually a singlet 11.0 to 13.0δ downfield from TMS, corresponding to the highly deshielded proton (Figure 16-3).

IONIZATION OF CARBOXYLIC ACIDS

In polar solvents such as water or ethanol the carboxylic acid dimers dissociate into the monomeric molecules and some of the latter in turn undergo dissociation to a proton and a **carboxylate anion.**

(16.58) $CH_3CO_2H + H_2O \rightleftharpoons CH_3CO_2^- + H_3O^+ \qquad K_a^{25} = 1.8 \times 10^{-5}$

acetate ion

Although most carboxylic acids are weak acids with pK_a's in the range 3.5 to 5.0, they are many orders of magnitude more acidic than the corresponding alcohols containing the same number of carbon atoms. (The pK_a equals $-\log K_a$, and the smaller the pK_a, therefore, the more dissociated is the acid at equilibrium. Very strong acids have negative pK_a's.) An aqueous solution of a carboxylic acid turns litmus red; it can be neutralized with weak bases such as ammonia and its derivatives; it liberates carbon dioxide from sodium bicarbonate; it reacts with a number of active metals with the release of hydrogen; and so on.

(16.59) $(CH_3)_2CHCO_2H + NH_3 \rightleftharpoons (CH_3)_2CHCO_2^-\ NH_4^+$

ammonium isobutyrate

The inductive effect of the carbonyl group undoubtedly promotes the loss of a proton from the —OH group but the most decisive factor bearing on the acidity of carboxylic acids is the delocalization of the charge in the carboxylate anion. The π-orbital arrangement of this anion resembles that of the allyl system but the carboxylate anion is much more stable than the allyl anion because its charge is distributed over two electronegative atoms.

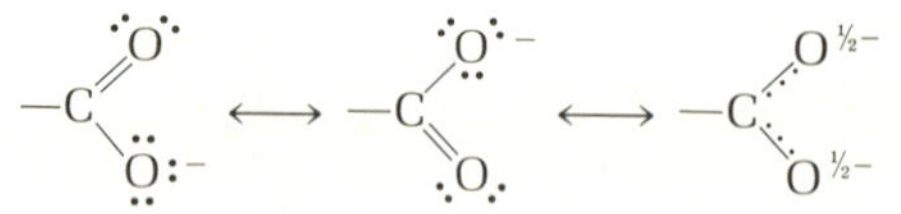

X-Ray studies of salts of carboxylic acids have shown that the two C—O bond distances are identical and intermediate in value between those found for carbon-oxygen single and double bonds. The infrared spectra of carboxylates also lack the characteristic C=O and C—O bands of the acids (Figure 16-2) and contain instead a new band at an intermediate frequency (1550 to 1610 cm^{-1}).

There is some conjugative interaction between the carboxyl group and the hydroxyl oxygen atom even in the unionized acid and this interaction tends to stabilize the acid.

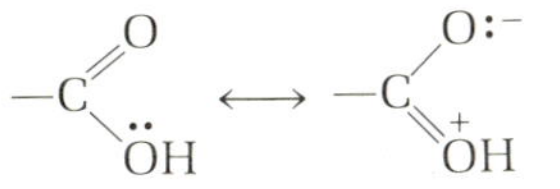

Resonance interactions are nonetheless much less important in the acids, where two canonical forms of unequal energy are involved, then they are in salts where the two canonical forms are identical in energy. As a consequence there is a net gain in resonance energy upon ionization of a carboxylic acid.

Substituents can exert a marked effect on the pK_a of a carboxylic acid; the most sizable alterations occur when the substituent is in close proximity to the carboxyl group. Inductive electron withdrawal apparently facilitates proton loss, as evidenced by the trend in pK_a's of the chloroacetic acids.

CH_3CO_2H	$ClCH_2CO_2H$	Cl_2CHCO_2H	Cl_3CCO_2H
4.75 (pK_a)	2.85	1.48	0.70

A hydroxyl group also increases the acidity of an alkanoic acid by an inductive (or field) effect; donation of electron density from the hydroxyl to the carboxyl group by conjugative interaction is not possible.

$HOCH_2CO_2H$

3.83

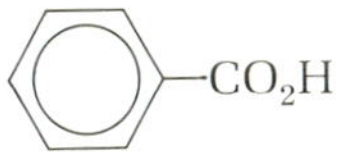

4.19

The phenyl group may simultaneously withdraw from the carboxyl group (inductive effect) and donate electrons to it (resonance effect). The inductive effect seems to predominate.

$CH_3CH_2CO_2H$	$(CH_3)_2CHCO_2H$	$(CH_3)_3CCO_2H$
4.87	4.84	5.03

$CH_3CH_2CH_2CO_2H$	$(CH_3)_2CHCH_2CO_2H$
4.81	4.77

Replacing the α-hydrogen atoms of acetic acid successively by methyl groups lowers the acidity of the compound. This result has commonly been ascribed to the electron-donating inductive effect of methyl, which should make proton loss more difficult. Since the organic portion of the acid acquires a negative charge upon ionization, it has been assumed that electron-withdrawing substituents would stabilize the ion and thereby promote ionization, while electron-donating substituents would have the opposite effect. If this interpretation were correct, then the *enthalpy* of ionization for trichloroacetic acid should be lower than that for acetic acid, which should in turn be lower than that for trimethylacetic acid. Such a relationship is *not* found; enthalpies of ionization vary only slightly, and unpredictably, among various aliphatic carboxylic acids.

There are sizeable differences in the *entropies* of ionization of these acids, however, and these differences are due in large part to solvation effects. Ionization usually results in a decrease in entropy, because the polar solvent clusters about the ions. The difference in solvation between the unionized molecule and the ionic species is then very important, and the substituent can affect this difference in multiple ways. Thus, although the free energies of ionization (equilibrium constants) of aliphatic carboxylic acids vary in an apparently straightforward way, it is not always easy to ascribe these variations to one or two simple changes in molecular properties.

THE HAMMETT EQUATION

The effect of a substituent on the chemical behavior of a compound can be a very useful parameter, both for altering the course of a given reaction and as an aid in understanding it. These substituent effects would clearly be most valuable if they could be expressed in a quantitative manner. One instance of such an approach is the use of partial rate factors in electrophilic aromatic substitution (Chapter 15).

A very general equation that quantitates substituent behavior is the **Hammett equation** (16.60 or 16.61).

(16.60) $$\log\left(\frac{k}{k_0}\right) = \sigma\rho$$

(16.61) $$\log\left(\frac{K}{K_0}\right) = \sigma\rho$$

In these equations σ is a constant characteristic of the substituent, ρ is characteristic of the reaction and measures its sensitivity to changes in substituent, k and K are rate or equilibrium constants involving the substituted compound, and k_0 and K_0 are rate or equilibrium constants for the reaction of the unsubstituted compound.

It was known that substituted benzoic acids had pK_a's that differed from that of benzoic acid itself. Not unexpectedly, electron-withdrawing substituents increased the acidity of benzoic acid, while a decrease in acidity was measured when an electron-donating substituent was present. By defining $\rho = 1.000$ for the ionization of benzoic acids and taking the logarithm of appropriate acidity constant ratios, a set of σ's could be defined.

It was found, for instance, that the dissociation constant of benzoic acid in water at 25° was equal to 6.27×10^{-5}. The corresponding values for *p*-anisic (*p*-methoxybenzoic) and *m*-fluorobenzoic acids were 3.38×10^{-5} and 1.365×10^{-4}, respectively. σ values for *p*-methoxy and *m*-fluoro were then calculated as shown in equations 16.62 and 16.63 (the denominator is the value of ρ, defined as 1.000 for the ionization of benzoic acids).

(16.62) $$\sigma^{p\text{-OCH}_3} = \frac{\log\left(\dfrac{3.38 \times 10^{-5}}{6.27 \times 10^{-5}}\right)}{1.000} = -0.27$$

(16.63) $$\sigma^{m\text{-F}} = \frac{\log\left(\dfrac{13.65 \times 10^{-5}}{6.27 \times 10^{-5}}\right)}{1.000} = 0.34$$

A representative list of σ values calculated in this way is contained in Table 16-3. Not surprisingly, the σ values for a given substituent are heavily dependent on its position relative to the reacting center.

TABLE 16-3
A Representative List of Substituent σ Values

Substituent	σ	Substituent	σ
m-$N(CH_3)_3^+$	0.88	*m*-OH	0.12
p-NO_2	0.78	*m*-OCH_3	0.12
m-NO_2	0.71	*p*-F	0.06
p-CN	0.66	H (defined)	0.000
m-CN	0.56	*m*-CH_3	−0.07
p-CF_3	0.54	*m*-$C(CH_3)_3$	−0.10
p-$COCH_3$	0.50	*m*-NH_2	−0.16
m-Br	0.39	*p*-CH_3	−0.17
m-Cl	0.37	*p*-OCH_3	−0.27
m-F	0.34	*p*-OH	−0.37
p-Br	0.23	*p*-NH_2	−0.66
p-Cl	0.23	*p*-$N(CH_3)_2$	−0.83

The more positive the σ value, the greater the acidity of the substituted benzoic acid and vice versa. Positive substituents presumably decrease the electron density at the reaction site, compared to hydrogen, while substituents with negative σ's presumably increase it. Those uncharged substituents capable of direct conjugation with the ring carbon atom connected to the substituent (*p*-NO_2, *p*-CN) have the greatest effect in increasing the acidity. They are less effective from the *meta* position.

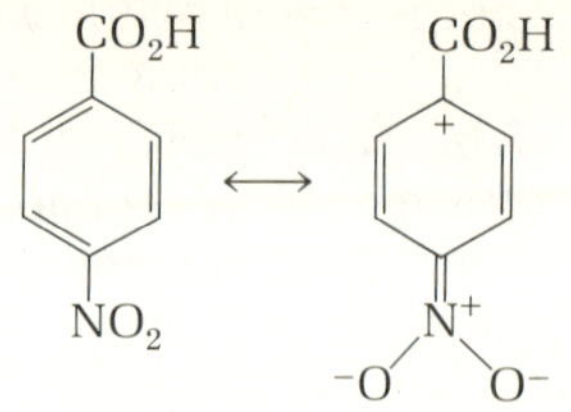

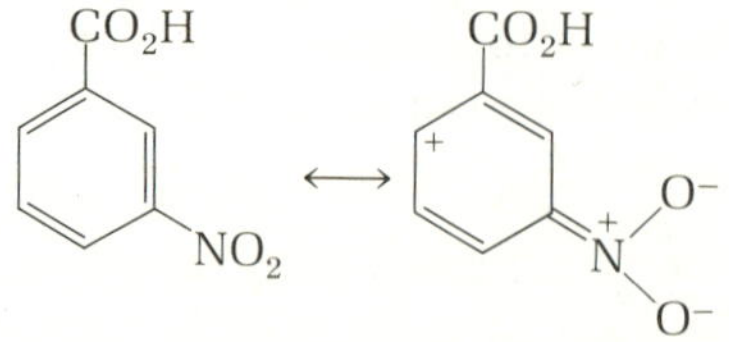

Substituents that influence the reaction "purely" by way of an inductive or field interaction are usually more effective in the *meta* position, where they are closer to the reaction site.

Hydroxyl and methoxyl have negative σ values when in the *para* position (conjugative effect) but positive σ values when *meta* to the carboxyl group [primarily inductive (field) effect]. The halogens exhibit positive σ values in both positions but they are reduced in magnitude at the *para* position. The largest decrease on going from *meta* to *para* is exhibited by fluorine, which can most effectively conjugate with the π-system of the ring.

When the dissociation constants of a number of substituted phenylacetic acids (**2**) are plotted according to equation 16.60, using the σ constants given in Table 16-3, a good linear correlation results, indicating that the values of the σ constants are valid for more than one reaction alone. We have then a **linear free energy relationship** — the effect of a set of substituents on the free energy of one reaction is directly proportional to the change in free energy of a second reaction as a function of substituent.

2

From the Hammett plot for the ionization of the phenylacetic acids, it is found that $\rho = 0.47$. Since ρ is positive, electron-withdrawing substituents enlarge the equilibrium constant as expected; but since ρ is smaller for phenylacetic than for benzoic acid, the equilibrium constants for the phenylacetic acids are less sensitive to a change in substituent. This conclusion is reasonable since there is a methylene group interposed between the ring and the carboxyl group in these acids.

Rate data may also be correlated by means of the Hammett equation. The reaction of various *m*- and *p*-substituted *N*,*N*-dimethylanilines with methyl iodide (equation 16.64) was studied quantitatively.

(16.64) 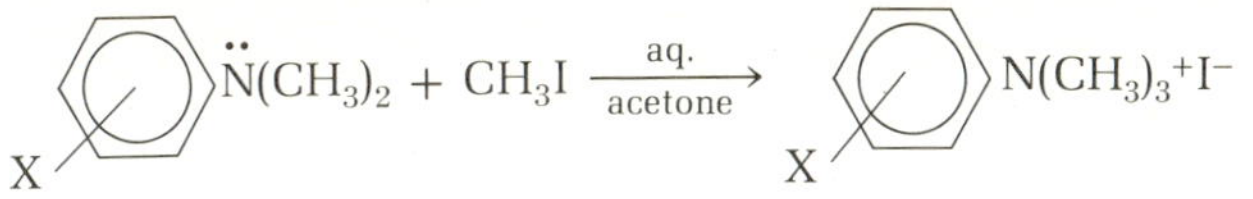

It was found, for example, that *p*-methyl-*N,N*-dimethylaniline ($k = 7.06 \times 10^{-2}$ l/mole sec) reacted more rapidly than the unsubstituted compound ($k = 2.69 \times 10^{-2}$ l/mole sec). From this data a value of ρ could be computed for the entire reaction (equation 16.65; -0.17 is the σ for *p*-methyl).

(16.65)
$$\rho = \frac{\log\left(\frac{7.06 \times 10^{-2}}{2.69 \times 10^{-2}}\right)}{-0.17} = -2.46$$

The data for a half dozen compounds were treated in this way and the best value of ρ was found to be -2.74. Negative ρ values suggest that electron-donating substituents accelerate the reaction, which is consistent with our view of the reaction as one in which nucleophilic nitrogen donates electrons to an electrophile.

Some success has been achieved in correlating the reactivities of purely aliphatic reactants by means of the Hammett equation. The best results are obtained when σ_{meta} values are used, which tends to reinforce the hypothesis that substituents in the *meta* position act principally by way of inductive or direct field mechanisms; resonance effects are absent in aliphatic compounds.

While the rate data from a surprisingly large number of reactions may be fitted to equations 16.60 and 16.61 by using the σ values in Table 16-3, there are some significant exceptions. One category of exceptions is that type of reaction in which a powerful electron deficiency is created at the transition state; equation 16.66 provides a well-studied example.

(16.66) 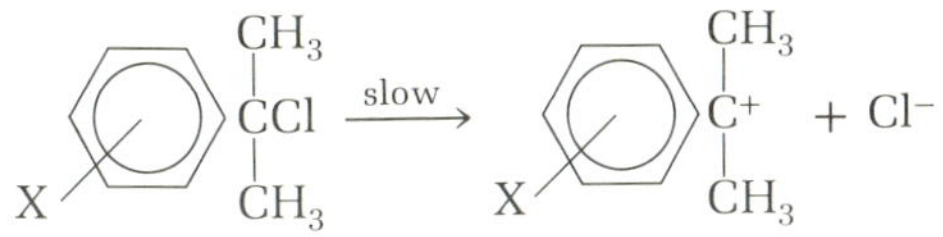

For reactions of this type a new set of substituent constants, σ^+, has been devised. A comparison of Tables 16-3 and 16-4 shows that the biggest differences between σ and σ^+ values occur for those substituents that may conjugate directly with the empty orbital such as *p*-NH_2.

TABLE 16-4
Some Representative Values of σ^+

Substituent	σ^+	Substituent	σ^+
p-NO_2	0.79	*p*-F	−0.07
m-NO_2	0.67	*m*-CH_3	−0.07
p-CN	0.66	*m*-NH_2	−0.16
m-CN	0.56	*p*-CH_3	−0.3
p-Cl	0.11	*p*-NH_2	−1.3

The Hammett equation may be used to predict the rate of reaction for a certain compound provided that ρ is known for the general reaction. More importantly, the sign and magnitude of ρ often provide valuable clues to the nature of the slow step in the reaction and to the structure of the transition state. Of significance is the question of whether the data are better correlated by σ or σ^+; this information is particularly valuable when attempts are made to draw conclusions about transition state structure.

PROBLEM

6 Sulfonic acid chlorides may participate in Friedel-Crafts **sulfonylations** (equation 4).

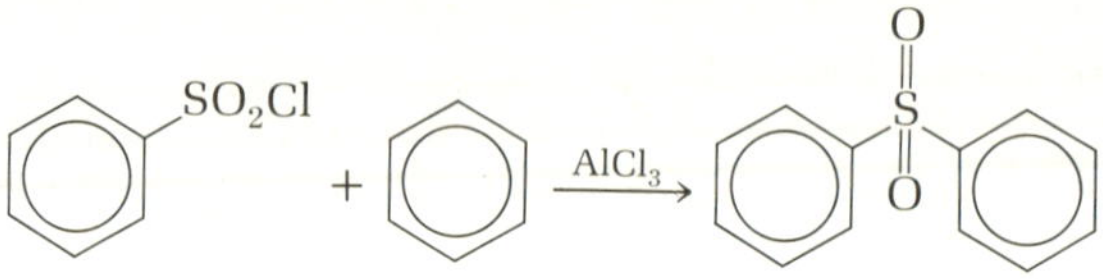

(4)

When the reaction rates of substituted benzenesulfonyl chlorides (equation 5) are put on a Hammett plot, it is found that $\rho = -1.80$.

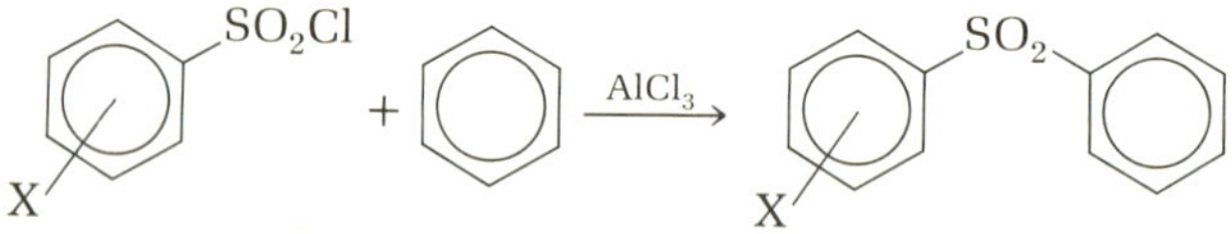

(5)

Suggest a mechanism for the reaction consistent with this finding, including an indication of the rate-determining step.

Substituent constants for *ortho* substituents have been conspicuous by their absence in the treatment outlined above. Steric factors would contribute a great deal to such substituent constants and since these would vary widely from one reaction to the next, it would be most difficult to formulate σ's of wide applicability. As an example of such "*ortho* effects" one may cite the unusually large dissociation constant of *o*-hydroxybenzoic acid (equation 16.67).

(16.67)

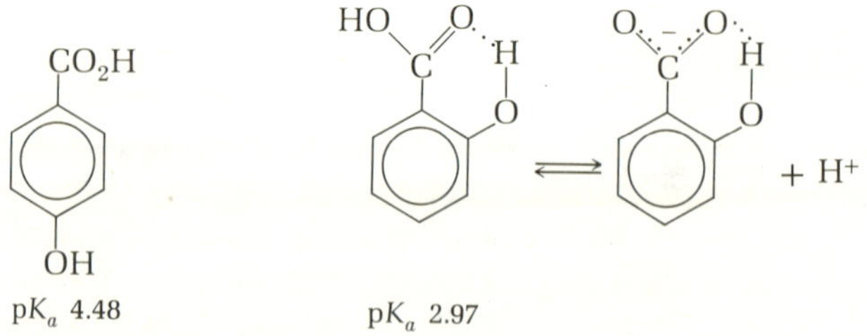

The enhanced acidity of this acid may be laid to stabilization of its anion by internal hydrogen bonding.

SULFONIC ACIDS

Although both aromatic and aliphatic sulfonic acids are known, it is the aromatic acids that are most often encountered in the laboratory, due to their ready availability by the sulfonation of various aromatic compounds. The only aliphatic acid common in the laboratory is methanesulfonic acid, CH_3SO_3H. Sulfonic acids have much larger dissociation constants than their carboxylic acid counterparts do because the negative charge in the anion may be distributed over three oxygen atoms.

(16.68) 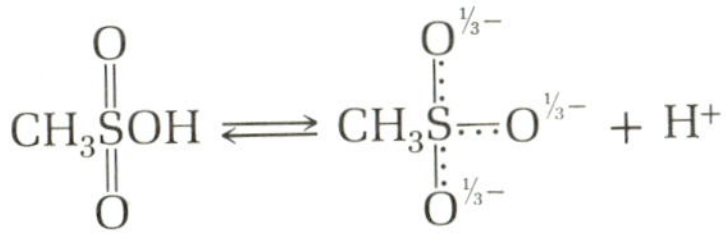

(16.69) 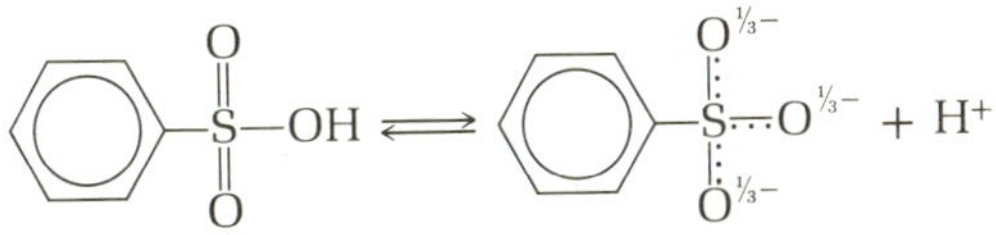 $K_a = 0.21$

Because of their high dissociation constants, sulfonic acids are much more water-soluble than the corresponding carboxylic acids. At the same time they retain their ability to dissolve in much less polar solvents. *p*-Toluenesulfonic acid is often used when a strong acid catalyst is needed for a reaction being run in a nonpolar solvent such as benzene.

SUMMARY

1 The reaction of an acid chloride with a Lewis acid such as aluminum chloride leads in many cases to the formation of acylonium ions. If these ions are generated in the presence of an aromatic molecule, electrophilic substitution occurs. Acid anhydrides may also serve as acylonium ion sources; carboxylic acids may be used in the presence of a strong mineral acid.

2 Acylonium ions only attack monosubstituted benzenes that contain *o,p*-directing substituents. The reaction shows a very heavy preference for the *para* position. Because of the deactivating effect of the acyl group, the reaction may easily be controlled to give monosubstitution only. It is usually necessary to protect anilines and phenols by acylating their nitrogen or oxygen atoms before acylating the ring. Phenols acylated on oxygen rearrange in the presence of aluminum chloride to *o*- and *p*-acylated phenols.

3 Alkyl halides react with aromatic compounds under the influence of Lewis acids. Carbonium ions are intermediates when secondary or tertiary halides

are reactants but primary halides alkylate by way of electrophilic complexes. The alkyl groups are very prone to undergo rearrangement to stable carbonium ions. Since an alkyl group exerts a mildly activating effect on an aromatic ring, it is very difficult to prevent polyalkylation, even when an excess of the aromatic reactant is present.

4 The Friedel-Crafts alkylation reaction invariably leads to mixtures of isomers alkylated to various extents. A preferable procedure, where applicable, involves acylating the aromatic ring and then reducing the carbonyl group to a methylene group. Acylation with a cyclic anhydride, followed by reduction of the ketone group and then intramolecular acylation, provides a standard route for building a six-membered ring onto an aromatic ring.

5 Many organometallic reagents displace chloride ion from acid chlorides and form ketones. Organocadmium compounds react very sluggishly with acid chloride but in the presence of catalysts such as magnesium halides the reaction takes place at a convenient rate. Only primary alkyl or aryl groups may be attached to the cadmium atom. The organocadmium reagent is unreactive toward many groups that would react with a Grignard reagent ($-NO_2$, $-CN$, $-CO_2R$, etc.).

6 Under conditions of high radical selectivity (low temperatures, ultraviolet light) bromine and chlorine attack the side chains of alkylbenzenes, principally at the α position. The extent of halogenation may be controlled and the product halides hydrolyzed to alcohols or carbonyl compounds. Oxidizing agents also attack alkylbenzenes at the α positions, eventually cleaving off the remainder of the side chain and leaving behind benzoic acid or a derivative. Special oxidizing agents, such as chromyl chloride or chromium trioxide in acetic anhydride, may be used to oxidize methyl side chains to aldehydes.

7 Carboxylic acids are weak acids with pK_a's usually in the range 5 to 2; the presence of a number of electron-withdrawing substituents may, however, lower the pK_a to values characteristic of mineral acids. The fatty acids tend to exist as hydrogen-bonded dimers in the pure liquid or in nonpolar solvents; in polar solvents the dimers dissociate to monomers, which may in turn undergo ionization. The principal driving force for ionization seems to be the formation of the charge-delocalized carboxylate anion.

8 If the dissociation constants of a number of *meta*- and *para*-substituted benzoic acids are plugged into the equation $\log(K/K_0) = \sigma\rho$, and ρ is set equal to 1.000, then a set of σ constants may be derived that are characteristic of the substituents. The rate and equilibrium constants for the reactions of a number of substituted benzene derivatives may be fitted to the Hammett equation using that same set of σ's. Substituents with a positive σ value

decrease the electron density at a reaction site, compared to hydrogen, in either the ground state or the transition state.

A given functional group will often have markedly different σ values depending on whether it is located *meta* or *para* to the reaction site. *para*-Substituents that can conjugate directly with an electron-deficient reaction center often deviate badly on a Hammett plot when the standard σ values are used. For reactions of this type a different set of substituents, σ^+, has been devised.

The magnitude and sign of ρ for a given rate process may give much information on the nature of the activation step.

PROBLEMS

Note: Problems 1 through 6 appear in text.

7 Indicate all steps in the syntheses of the following compounds, using benzene, toluene, or benzaldehyde as starting materials.

a. Propiophenone (two methods)

b. α-Methylbutyrophenone,

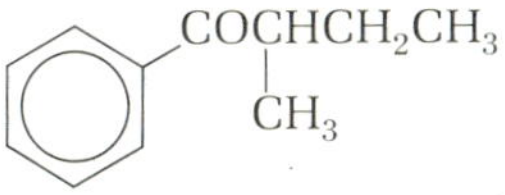

(two methods)

c. *p*-Methylvalerophenone

d. Cyclohexyl phenyl ketone

e. Benzophenone (two ways)

f. 4-Nitrobenzophenone

g. 4,4′-Dibromobenzophenone

h. γ-Benzoylbutyric acid

i. 5-Phenylpentanoic acid

j. 1-Methyl-1,2,3,4-tetrahydronaphthalene

8 Propose syntheses of the following compounds based on benzoic acid as a starting material.

a. Butyrophenone

b. β-Methylbutyrophenone

c. *m*-Nitroacetophenone

d. 4-Methylbenzophenone

e. 3,4′-Dichlorobenzophenone

f. 1-Naphthyl phenyl ketone

g. 2-Naphthyl phenyl ketone

9 Ethylene is bubbled through a solution of acetyl chloride and aluminum chloride in nitrobenzene. After work-up, a compound of molecular formula C_4H_7ClO is obtained. Propose a structure for this compound, and write a mechanism that explains its formation.

10 When benzene is treated with a mixture of formaldehyde and concentrated hydrochloric acid, benzyl chloride is produced in good yield (the *chloromethylation* reaction). This synthesis can also be applied to other aromatics, as long as their reactivity equals or exceeds that of benzene. Write a detailed mechanism for the reaction, showing all important intermediates.

11 a. Toluene is benzoylated (with benzoyl chloride and aluminum chloride) 106 times faster than benzene is in nitromethane at 25°. The products consist of 3.2% *o*-methylbenzophenone, 0.9% *m*-methylbenzophenone, and 95.9% *p*-methylbenzophenone. Calculate partial rate factors for benzoylation.

b. The relative rates of propionylation (propionyl chloride and aluminum chloride) of *o*-, *m*-, and *p*-xylene are 54:15:1, respectively. Products from the *ortho* and *meta* isomers are 3,4-dimethylpropiophenone and 2,4-dimethylpropiophenone. How would you rationalize this reactivity order? What does it suggest about the transition state for acylation?

12 Propose syntheses for the following compounds using benzene, toluene, or one of the xylenes as the starting material.

a. Ethylbenzene
b. *p*-Isopropyltoluene
c. *p*-*t*-Butylethylbenzene
d. *o*-Cyclohexyltoluene
e. 1,2,4-Trimethylbenzene
f. 3,4-Dimethylethylbenzene
g. *n*-Hexylbenzene
h. *p*-Chloropropylbenzene
i. *p*-Nitroisopropylbenzene

13 When toluene is methylated with methyl chloride and aluminum chloride, a small quantity of benzene is found among the products. Suggest a mechanism for its formation.

14 Polyalkylation of benzene, as in equation 16.29, leads to *meta*-substituted products after equilibration in the presence of a large quantity of aluminum chloride, provided that the hydrogen chloride formed in the reaction is retained in the mixture. Account for this observation.

15 Ethylbenzene was treated with one equivalent of bromine in the presence of ferric bromide, which yields two isomeric compounds, **A** and **B**. Vigorous oxidation with hot potassium permanganate converts **A** to **C** and **B** to **D**. Neither **C** nor **D** are appreciably soluble in water but both are soluble in aqueous sodium carbonate solution. When treated with a mixture of nitric and sulfuric acids, **C** is transformed into **E** ($C_7H_4BrNO_4$). When **D** is subjected to the same conditions, it gives rise to two compounds, **F** and **G**, both of which are isomers of **E**.

a. Write out structures for all compounds.

b. How could **C** and **D** be distinguished by nonchemical means and in the absence of a table of physical constants?

16 If 4-chlorobutanoic acid were as readily available as succinic anhydride, would it not constitute a more desirable starting material for the synthesis of 1-tetralone? Explain your answer.

17 Mesitylene (1,3,5-trimethylbenzene), upon treatment with dichloromethyl methyl ether (CH_3OCHCl_2) and titanium tetrachloride in methylene chloride, yields **A** ($C_{11}H_{15}ClO$). Upon hydrolysis **A** is converted to mesitaldehyde ($C_{10}H_{12}O$). Write out mechanisms for both steps of this process, which is a general one for converting highly activated aromatic rings to the corresponding aldehydes.

18 a. Compound II was treated with methylmagnesium iodide and the product dissolved in aqueous acid (for a synthesis of II, see Chapter 15, Problem 21).

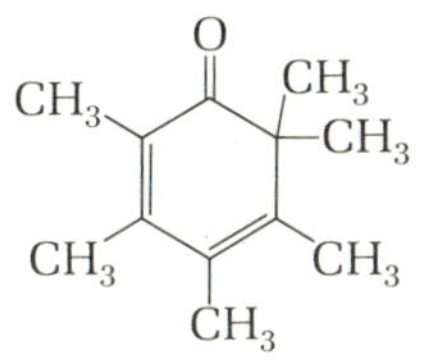

II

A stable organic cation was formed that generated an nmr spectrum containing singlets at $\delta 2.20$, $\delta 3.01$, $\delta 3.31$, and $\delta 3.51$, relative areas 2:2:2:1. The same cation is produced when hexamethylbenzene is allowed to react with methyl chloride and aluminum chloride. What is its structure? Assign the nmr peaks for this ion.

b. When the solution of the cation is neutralized with sodium bicarbonate, a hydrocarbon is obtained that gives the following nmr spectrum: singlets at $\delta 1.09$, $\delta 1.78$, and $\delta 4.77$; area ratios, 3:6:1. Propose a structure for this hydrocarbon, and assign the nmr peaks.

19 Show how the following conversions can be accomplished efficiently, listing all essential conditions and reagents.

a. Toluene ⟶ benzyl bromide

b. Toluene ⟶ α-bromo-α-chlorotoluene

c. Toluene ⟶ *p*-iodobenzyl alcohol

d. *p*-Xylene ⟶ 1,4-bis(hydroxymethyl)benzene

e. Toluene ⟶ benzaldehyde-*o*-sulfonic acid

f. Benzene ⟶ *m*-bromobenzaldehyde

g. Toluene ⟶ acetophenone-*p*-carboxylic acid

h. *t*-Butyl alcohol ⟶ trimethylacetic acid

i. Cyclohexene ⟶ cyclohexanecarboxylic acid

j. Cyclohexyl chloride ⟶ adipic acid

k. Benzene ⟶ *p*-nitrobenzoic acid

20 Which compound of each of the following pairs is the more acidic? Explain.

a. HNO_2 or HNO_3

b. $HClO_3$ or $HClO_4$

c. 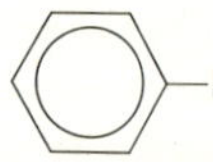$-SO_3H$ or 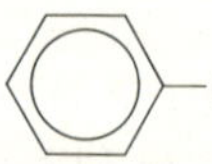$-SO_2H$ (benzenesulfinic acid)

d. FCH_2CO_2H or $ClCH_2CO_2H$

e. FCH_2CO_2H or $FCH_2CH_2CO_2H$

f. $CH_3OCH_2CO_2H$ or CH_3CO_2H

g. CH_3CO_2H or $CH_3\overset{\overset{\displaystyle O}{\|}}{C}OOH$ (peracetic acid)

h. CH_3O-$-CO_2H$ or $-CO_2H$

21 a. The first dissociation constants for oxalic, malonic, and succinic acids are 5.9×10^{-2}, 1.5×10^{-3}, and 6.9×10^{-5}, respectively. How do you account for this trend?

b. The ratio of first and second dissociation constants is 920 for oxalic acid and 740 for malonic acid. Why do the two dissociation constants differ so markedly? As a qualitative estimate, what ratio would you expect between the first and second dissociation constants of adipic acid?

22 When a series of monosubstituted benzenes are chlorinated, the reaction rates for *meta* and *para* substitution may be correlated by a Hammett plot. Will the rate at each position correlate best with σ or with σ^+? Explain.

23 a. The benzylation of a series of substituted benzenes with benzyl chloride and a Lewis acid catalyst has been studied quantitatively. When the logarithms of the rate ratios were plotted against σ^+, a straight line was obtained.

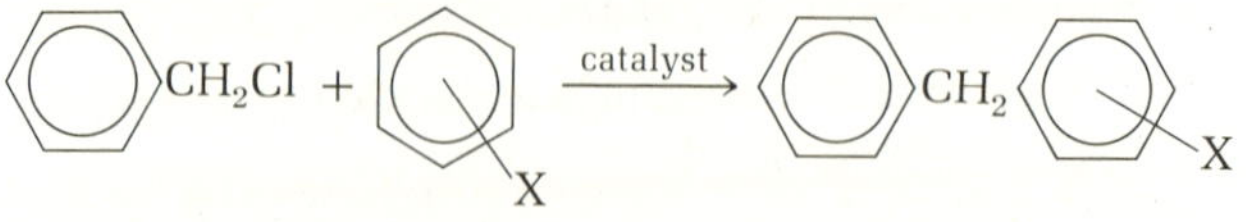

(6)

Was the slope of this line positive or negative? Explain your answer.

b. Reactions between substituted benzyl chlorides and benzene were also studied.

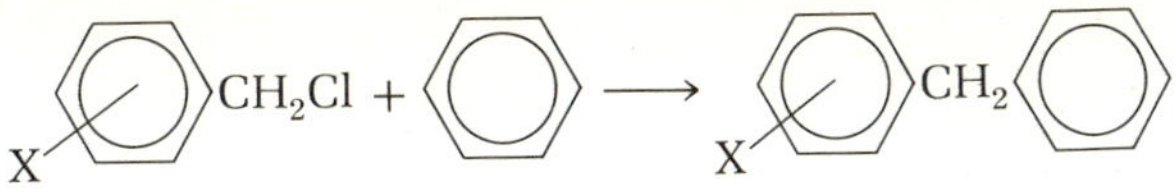 (7)

Would you expect the ρ for this reaction to be positive or negative? (What is the rate-determining step in the reaction? Draw the complete mechanism.)

SUPPLEMENTARY READINGS

R. O. C. Norman and R. Taylor, *Electrophilic Substitution in Benzenoid Compounds.* New York: Elsevier Pub. Co., 1965.

K. L. Marsi and S. H. Wylen, "Friedel-Crafts Alkylation," *J. Chem. Ed.*, **40,** 214 (1963).

J. Shorter, "Linear Free Energy Relationships," *Chemistry in Britain,* **5,** 269 (1969).

G. V. Calder and T. J. Barton, "Actual Effects Controlling the Acidity of Carboxylic Acids," *J. Chem. Ed.*, **48,** 338 (1971).

CHAPTER 17

DERIVATIVES OF CARBOXYLIC ACIDS — NUCLEOPHILIC DISPLACEMENT AT THE CARBOXYL CARBON ATOM

CLASSES OF CARBOXYLIC ACID DERIVATIVES

There are several classes of compounds that may be derived formally (and often in practice) from carboxylic acids by replacing the hydroxyl group with some other functional group. Acid chlorides represent one example of such carboxylic acid derivatives. A number of important derivatives are listed in Table 17-1. These compounds are very widely used in the laboratory

TABLE 17-1
Some Important Carboxylic Acid Derivatives

General Formula	Name	Specific Example
$-\overset{\text{O}}{\overset{\Vert}{\text{C}}}-\text{X}$	Acid (or acyl) halides	$CH_3(CH_2)_3COF$ pentanoyl fluoride
$-\overset{\text{O}}{\overset{\Vert}{\text{C}}}\text{OR}$ R = alkyl, aryl	Esters	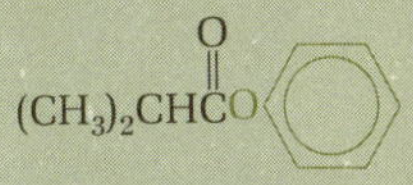 phenyl isobutyrate
$-\overset{\text{O}}{\overset{\Vert}{\text{C}}}\overset{\text{R}}{\overset{\vert}{\text{N}}}-\text{R}'$ R,R′ = H, alkyl, aryl	Amides	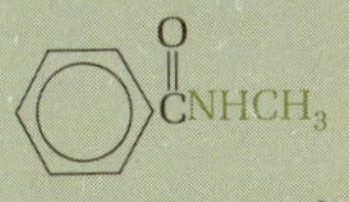 *N*-methylbenzamide
$-\overset{\text{O}}{\overset{\Vert}{\text{C}}}\text{O}\overset{\text{O}}{\overset{\Vert}{\text{C}}}\text{R}$ R = H, alkyl, aryl	Acid anhydrides	$CH_3CH_2\overset{\text{O}}{\overset{\Vert}{\text{C}}}\text{O}\overset{\text{O}}{\overset{\Vert}{\text{C}}}CH_2CH_3$ propionic anhydride

and in commercial operations and some of them are of supreme importance in biochemical systems.

RELATIVE REACTIVITY OF CARBOXYLIC ACID DERIVATIVES

The common carboxylic acid derivatives form a convenient hierarchy in terms of reactivity and stability. Elaboration of this hierarchy can be a very useful exercise, for it suggests a theoretical framework for understanding the chemistry of these compounds. Moreover, it provides a scheme for classifying the large number of individual reactions.

The reactivity of the carboxylic acid derivatives decreases in the following order:

$$\underset{\text{acid chlorides}}{\text{R}\overset{\text{O}}{\overset{\Vert}{\text{C}}}\text{Cl}} > \underset{\text{anhydrides}}{\text{R}\overset{\text{O}}{\overset{\Vert}{\text{C}}}\text{O}\overset{\text{O}}{\overset{\Vert}{\text{C}}}\text{R}} > \underset{\text{esters}}{\text{R}\overset{\text{O}}{\overset{\Vert}{\text{C}}}\text{OR}'} > \underset{\text{amides}}{\text{R}\overset{\text{O}}{\overset{\Vert}{\text{C}}}\text{NR}'_2}$$

An important consequence of this order is that **the less reactive derivatives may be synthesized from the more reactive ones but not vice versa.** Acid chlorides are starting materials for the synthesis of anhydrides, esters, and amides and both esters and amides are conveniently made from anhydrides, but they never serve as starting materials for the synthesis of anhydrides or acid chlorides. As noted in Chapter 16, acid chlorides are synthesized with the aid of inorganic chlorides such as phosphorus pentachloride and thionyl chloride. These compounds may be viewed as acid chlorides of phosphoric and sulfurous acids, respectively, thus suggesting the conclusion that in order to make carboxylic acid chlorides one must treat the acid with an even more reactive acid chloride.

One feature that is common to all these derivatives is that their reactions are principally those that involve nucleophilic substitution at the carbonyl carbon atom (equation 17.1).

(17.1) $$\mathrm{R\overset{\overset{\displaystyle O}{\|}}{C}Y + Nu\!:^- \longrightarrow R\overset{\overset{\displaystyle O}{\|}}{C}Nu + Y\!:^-}$$

As we go down the reactivity order, we can see immediately that the leaving group gets progressively poorer. This fact alone would account for the decreasing reactivity along the series.

(17.2) $$\mathrm{RCOCl + Nu\!:^- \longrightarrow RCONu + Cl^-}$$

(17.3) $$\mathrm{R\overset{\overset{\displaystyle O}{\|}}{C}O\overset{\overset{\displaystyle O}{\|}}{C}R + Nu\!:^- \longrightarrow RCONu + RCO_2^-}$$

(17.4) $$\mathrm{RCOOR' + Nu\!:^- \longrightarrow RCONu + R'O^-}$$

(17.5) $$\mathrm{RCONR'_2 + Nu\!:^- \longrightarrow RCONu + R'_2N^-}$$

There is another factor that is complementary to the leaving group effect and that works in the same direction. As one goes down the series from acid chlorides to amides, the delocalization energy of the functional group increases, which contributes to the decreasing reactivity of the compounds. Since the atom directly attached to the carbonyl carbon in all these derivatives has an unshared electron pair, this pair may overlap with the carbonyl π-bond, thereby providing a certain amount of delocalization energy. Since the extent of overlap is different for different groups, however, the degree of stabilization that arises from this effect varies substantially.

Overlap between chlorine p-orbitals and the carbonyl π-orbital is relatively inefficient ($2p$-$3p$ bonding) and the resonance energy of the chlorocarbonyl group would be expected to be small.

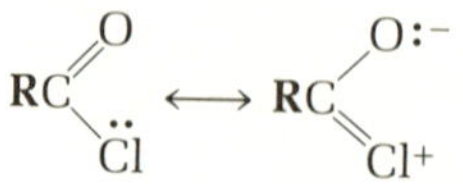

The interaction of oxygen p-orbitals with the π-orbital, however, should be much stronger. In anhydrides the extent of this interaction should be smaller *per carbonyl group* than it is in esters, leading to the conclusion that esters should be more stable than anhydrides, a conclusion in agreement with experience.

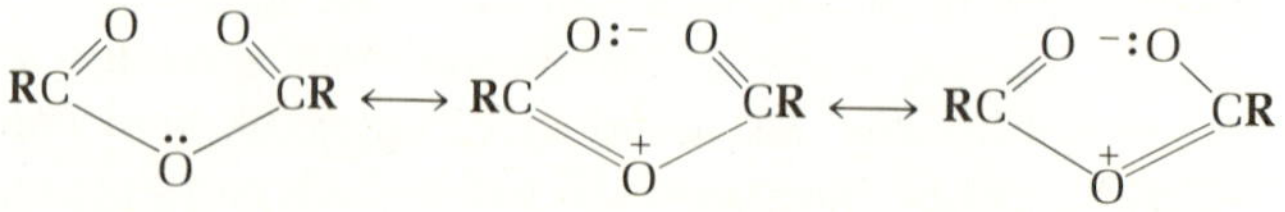

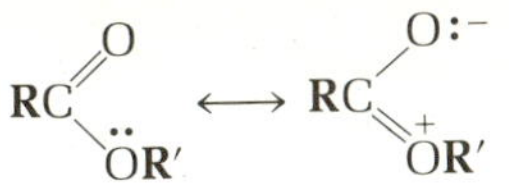

Finally, resonance interactions should be strongest in amides since nitrogen is less electronegative than oxygen and can more readily relinquish some electronic charge.

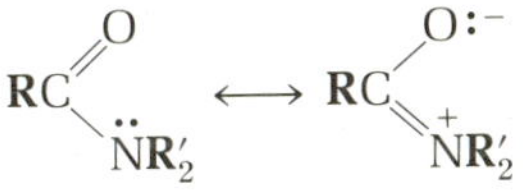

The conclusions about the reactivity-stability order derived from a qualitative consideration of resonance effects is in complete agreement with the order obtained from an analysis of leaving group effects.

PROBLEM

1 The nmr spectrum of N,N-dimethylacetamide (I), taken at room temperature, contains *three* singlets of equal area.

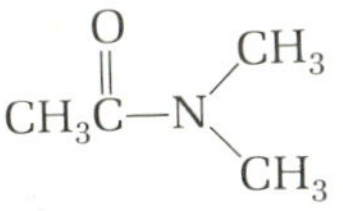

I

This finding indicates that the two N-methyl groups are not equivalent on the nmr time scale under these conditions. Suggest a reason for this phenomenon. What spectral changes might occur if the sample were heated?

Carboxylic acids and esters should be about equal in stability and reactivity; their resonance energies are very similar, and OH^- and OR^- are roughly equivalent leaving groups. When attempting to assess the reactivity of carboxylic *acids* in relation to that of their derivatives, however, complications intervene.

Many of the nucleophiles to be discussed in this chapter are also bases and the fastest reaction between them and an acid is neutralization. The neutralization product, the carboxylate anion, is extremely unreactive toward nucleophiles, due both to its negative charge and to its relatively high resonance energy.

$$\mathrm{R\overset{\overset{\displaystyle O}{\|}}{C}OH + Nu\colon \longrightarrow RCO_2^- + Nu{-}H^+} \quad (17.6)$$

As a result, carboxylic acids do not undergo nucleophilic substitution under basic conditions. As we see below, however, they can take part in this reaction in the presence of acid.

CARBONYL FREQUENCIES OF ACID DERIVATIVES

All the carboxylic acid derivatives discussed above give rise to ir spectra that contain intense peaks attributable to >C=O stretching. The position of the carbonyl band in the spectra of these compounds varies in a manner that can be correlated with their chemical properties. Thus, the highest frequency carbonyl peaks are found in the spectra of acid chlorides (1780

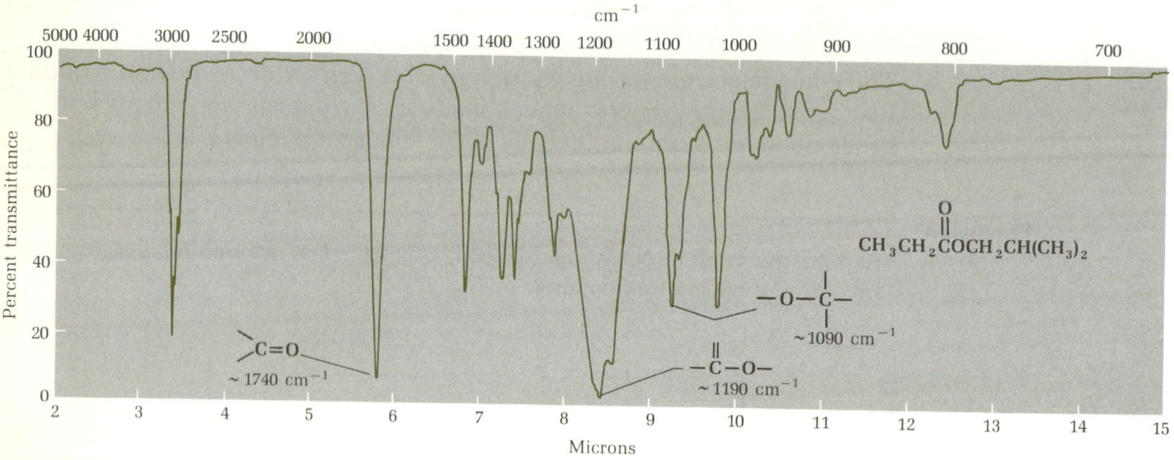

FIGURE 17-1

Infrared Spectrum of Isobutyl Propionate

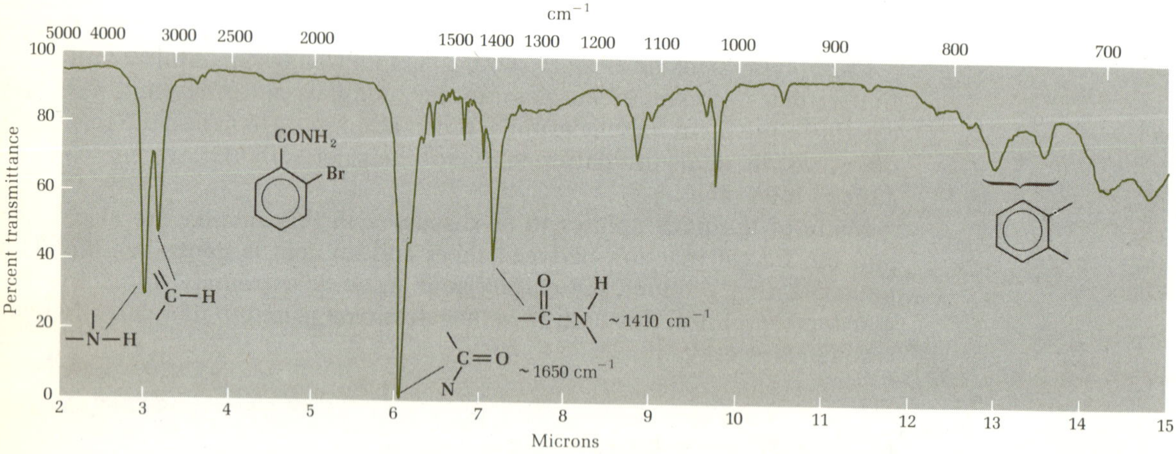

FIGURE 17-2

Infrared Spectrum of o-Bromobenzamide

to 1815 cm^{-1}) and anhydrides (two bands, 1800 to 1850 and 1740 to 1790 cm^{-1}). Ester carbonyl frequencies are somewhat lower (1735 to 1750 cm^{-1}) and those of amides are markedly lower (1630 to 1700 cm^{-1}).

One explanation that has been offered for the observed trend is that the carbonyl frequencies reflect the extent of "pure" double bond character in the carbonyl groups. The greater the double bond character in this group, the higher its stretching frequency should be. It follows then that a strong resonance interaction between the carbonyl group and an attached atom should lower the stretching frequency. The empirical results are in accord with the analysis of resonance interactions presented above.

PROBLEM

2 Saturated carbonyl derivatives invariably exhibit higher stretching frequencies than do the corresponding α,β-unsaturated or aromatic compounds. For instance, the carbonyl bands for saturated esters are normally found in the range 1735 to 1750 cm^{-1}, while the range for conjugated unsaturated and aromatic esters is 1715 to 1730 cm^{-1}. Offer an explanation for this difference.

NOMENCLATURE

The names of esters consist of two words, the first of which denotes the alcohol from which the ester is derived and the second of which denotes the carboxylic acid. Substituents in the alcohol portion are noted in the alcohol name and are separate from the substituents in the acid portion.

ethyl acetate

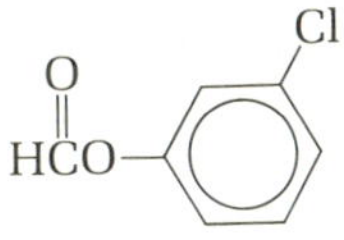

m-chlorophenyl formate

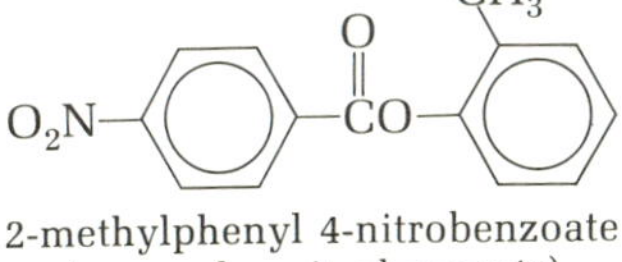

2-methylphenyl 4-nitrobenzoate
(*o*-cresyl *p*-nitrobenzoate)

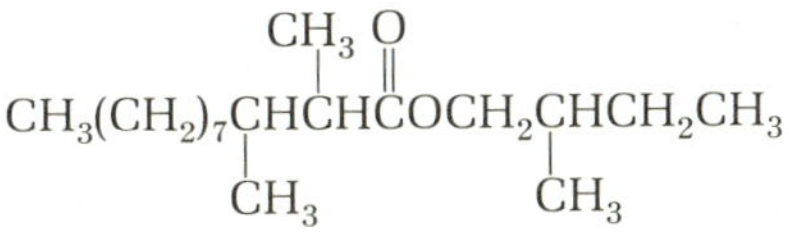

2-methylbutyl 2,3-dimethylundecanoate

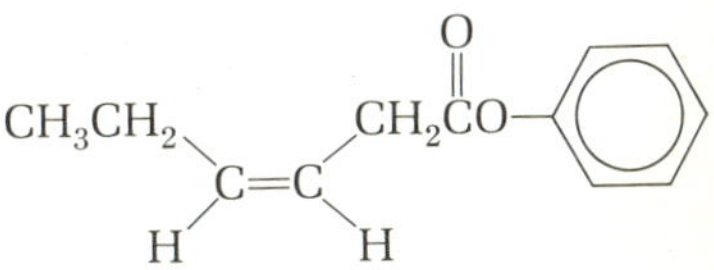

phenyl *cis*-3-hexenoate

Amide names are derived from the corresponding carboxylic acid names. The *-ic* or *-oic acid* ending is replaced by *-amide*. If the suffix *-carboxylic acid* appears in the name, it is replaced by *-carboxamide*.

$$(CH_3)_2CH\overset{\overset{\displaystyle O}{\|}}{C}NH_2$$

isobutyramide

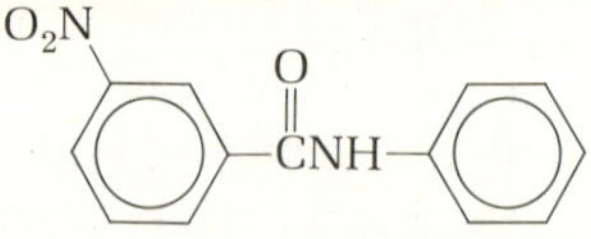

N-phenyl-*m*-nitrobenzamide
(*m*-nitrobenzanilide)

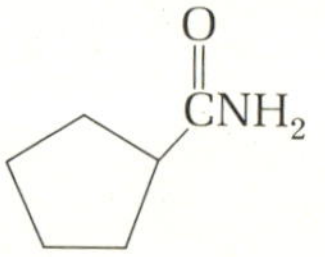

cyclopentanecarboxamide

$$CH_3CH_2\overset{\overset{\displaystyle CH_3}{|}}{C}H(CH_2)_4\overset{\overset{\displaystyle O}{\|}}{C}NHCH_2CH_3$$

N-ethyl-6-methyloctanamide

PREPARATION OF ACID CHLORIDES AND ANHYDRIDES

Acid chlorides are synthesized by allowing carboxylic acids to react with the chlorides of inorganic acids; thionyl chloride and phosphorus pentachloride are the most common inorganic reactants (Chapter 16). Anhydrides may also serve as starting materials.

(17.7) 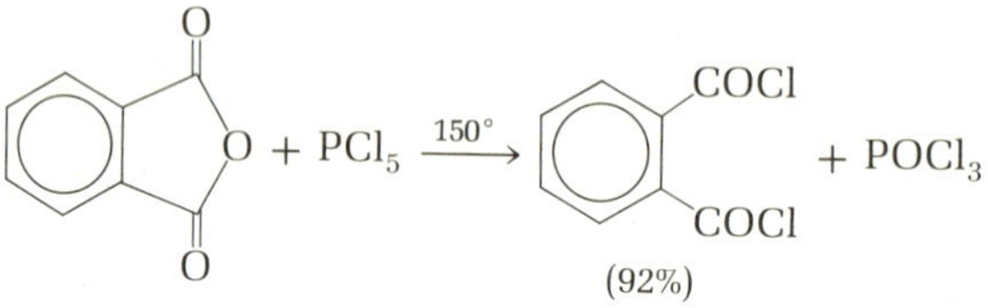

(92%)

Anhydrides are formally derived from 2 moles of acid by intermolecular elimination of 1 mole of water. Acyclic anhydrides are actually prepared by the reaction of 1 mole of acid chloride with 1 mole of carboxylate salt. Another useful method involves the equilibrium between a given acid and acetic anhydride on the one hand and the acid anhydride and acetic acid on the other. The reaction is driven to completion by distilling off the volatile acetic acid.

(17.8) $$CH_3CH_2\overset{\overset{\displaystyle O}{\|}}{C}Cl + CH_3CH_2CO_2^-Na^+ \longrightarrow (CH_3CH_2\overset{\overset{\displaystyle O}{\|}}{C})_2O + Na^+Cl^-$$

propionic anhydride

(17.9) 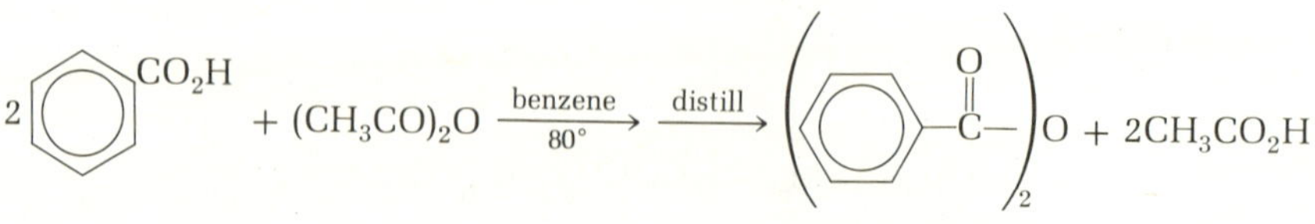

benzoic anhydride
(81%)

Mixed anhydrides are not easily obtained because on distillation they tend to disproportionate to the symmetrical anhydrides. The only mixed anhydride of any importance is formic acetic anhydride, which is used as a formylating agent because formic anhydride is unknown.

Dibasic acids can give rise to cyclic anhydrides by *intramolecular* loss of water; succinic acid is the smallest acid that will form a cyclic anhydride. Succinic and glutaric anhydrides are the easiest to prepare because they possess five- or six-membered rings, respectively. Glutaric anhydride is formed merely by heating the free acid (equation 17.10).

(17.10)

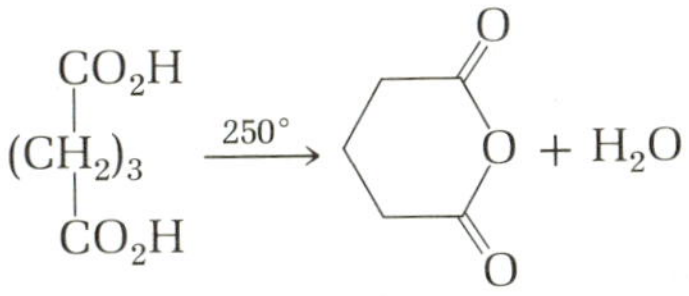

Cyclic anhydrides up to and including sebacic anhydride may be prepared by heating the diacid with acetic anhydride; the anhydrides of larger diacids tend to be formed as linear polymers.

AMIDES AND ESTERS FROM ANHYDRIDES AND ACID CHLORIDES

When making esters by way of acid chlorides, it is customary to add a weak base to neutralize the hydrogen chloride liberated. As noted before, this method and the related anhydride technique are very useful for making esters of teritiary alcohols.

(17.11)

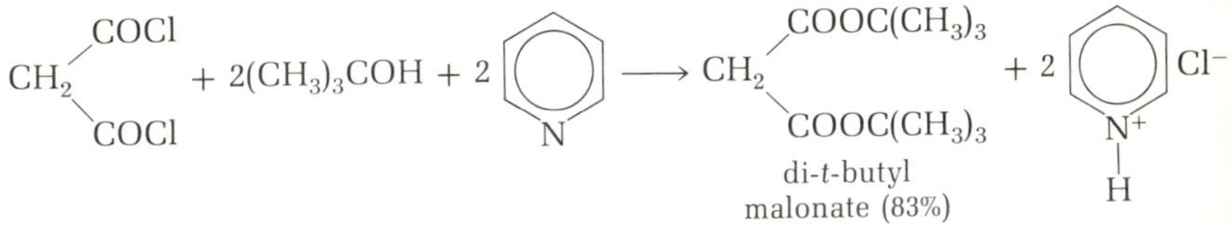

di-*t*-butyl malonate (83%)

(17.12)

$$(CH_3CH_2\overset{O}{\overset{\|}{C}})_2O + CH_3OH \xrightarrow{\text{heat}} CH_3CH_2\overset{O}{\overset{\|}{C}}OCH_3 + CH_3CH_2CO_2H$$

Note that when a cyclic anhydride is a reactant (equation 17.13), then the product is the half ester of a dicarboxylic acid.

(17.13)

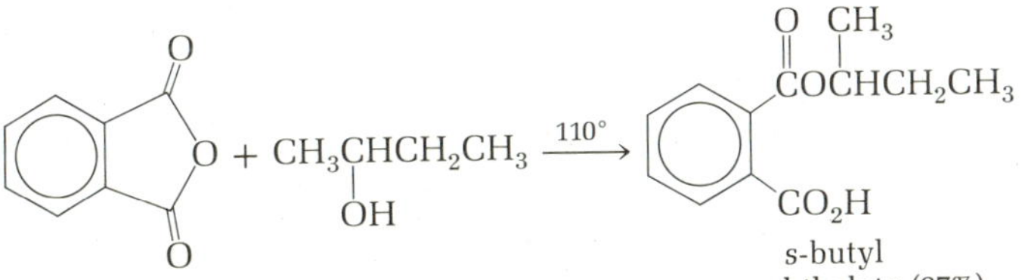

s-butyl monophthalate (97%)

When applicable, this method of synthesizing half esters is preferable to reacting the diacid with 1 mole of alcohol. The latter procedure generally yields a mixture of products.

Ammonia and its derivatives (*amines*) react very readily with acid chlorides and anhydrides to yield amides. Since an acid (hydrochloric or carboxylic) is released, 1 mole of amine will be consumed as a base for each mole of amide formed. If the amine cannot be spared, then 1 mole of an external base, such as sodium hydroxide or sodium acetate, must be added.

Ammonia is often introduced by use of the concentrated aqueous solution since the ammonia is much more nucleophilic than water. The amines are reacted directly with the acylating agent (if it is a liquid) or in solvents such as benzene and ether.

(17.14) $$CH_3\overset{O}{\overset{\|}{C}}Cl + 2NH_3 \xrightarrow[0^0]{H_2O} CH_3\overset{O}{\overset{\|}{C}}NH_2 + NH_4^+\ Cl^-$$

acetamide

(17.15) 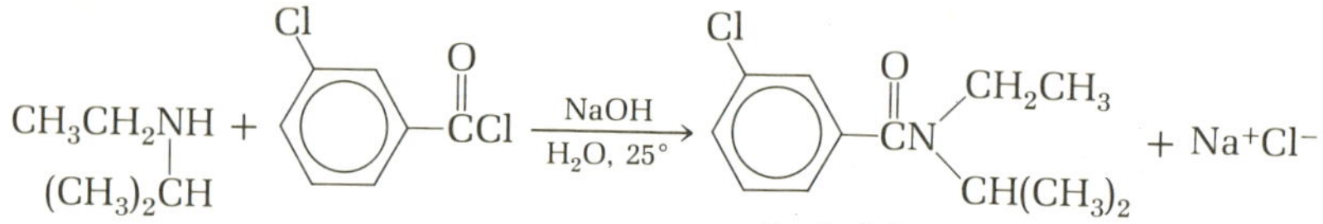

N-ethyl-*N*-isopropyl-*m*-chlorobenzamide

(17.16) 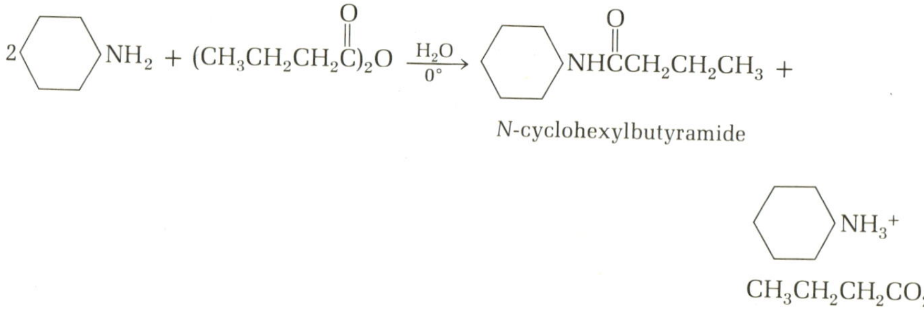

N-cyclohexylbutyramide

The products of equations 17.14 to 17.16 are representative of the three different classes of amides. Acetamide, which has two hydrogen atoms attached to the nitrogen atom, is a *primary amide;* *N*-cyclohexylbutyramide is a *secondary amide* because it has only one nitrogen-bound proton; and *N*-ethyl-*N*-isopropyl-*m*-chlorobenzamide is a *tertiary amide*. It should be noted that an amine must possess at least one proton bound to nitrogen in order for it to be convertible to an amide. Amines such as trimethylamine, $(CH_3)_3N$, do not react with acylating agents to give stable amide products.

Acid chlorides and anhydrides are readily hydrolyzed to carboxylic acids, water being sufficiently nucleophilic to react rapidly with both. The mechanisms of all these reactions involves nucleophilic addition to the carbonyl group, followed by expulsion of the leaving group. In the following two sections, the fundamental mechanisms characteristic of reactions of carboxylic acid derivatives are discussed in some detail.

REVERSIBLE ESTER FORMATION

The formation of esters and amides from acid chlorides and anhydrides is essentially irreversible. Esters may also be obtained by the direct interaction of alcohols and carboxylic acids. Since esters and acids are approximately equal in stability, it would be expected that there would be no great driving force for the *esterification* of carboxylic acids with alcohols. This reaction was one of the first in organic chemistry to be studied quantitatively.

Equilibrium between the products and reactants is attained only slowly in esterification reactions unless the acid and alcohol are both fairly high in molecular weight, thereby allowing the reaction to be conducted at elevated temperatures. The size of the equilibrium constant is quite dependent on steric factors in the alcohol; in most cases, the reaction seems to stop well short of completion.

(17.17) $$CH_3CO_2H + CH_3OH \overset{155°}{\rightleftharpoons} CH_3CO_2CH_3 + H_2O \qquad K = 5.2$$

(17.18) $$CH_3CO_2H + (CH_3)_3COH \overset{155°}{\rightleftharpoons} CH_3CO_2C(CH_3)_3 + H_2O \qquad K = 4.9 \times 10^{-3}$$

Adding a few drops of a strong mineral acid, such as hydrochloric, perchloric, or sulfuric acid, has a tremendous accelerating effect on the rate at which equilibrium is reached in an esterification, although the equilibrium constant is unaffected. Acid also catalyzes the reverse of the esterification reaction, hydrolysis of the ester. It is known from labeling studies involving use of ^{18}O that the oxygen atom in the alcohol is retained in the ester (equation 17.19).

(17.19) $$RCO_2H + H^{18}OR \rightleftharpoons R\overset{O}{\overset{\|}{C}}{-}^{18}OR + H_2O$$

This finding (which applies to primary alcohols) rules out a mechanism involving a nucleophilic attack by the carboxylic acid on the protonated alcohol (equation 17.20).

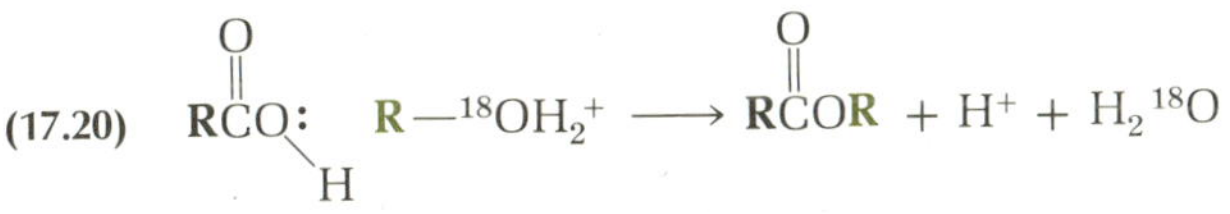

All the evidence so far uncovered is compatible with a mechanism in which there is first addition of the alcohol to the protonated carbonyl group, followed by expulsion of a molecule of water (equation 17.21).

(17.21)

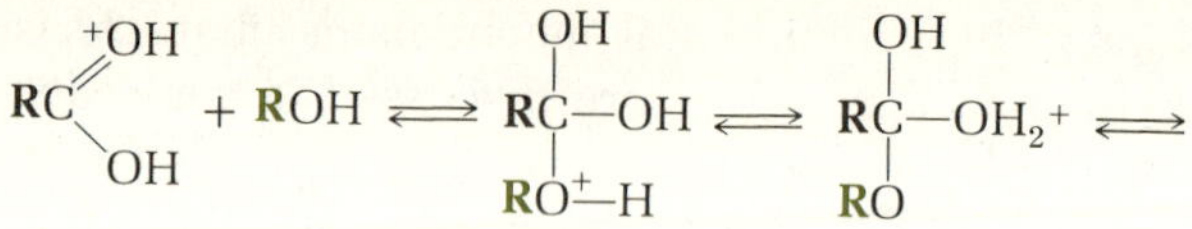

1 2

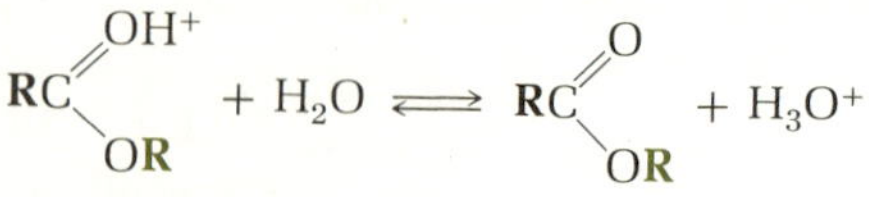

3

The first step, acid-catalyzed addition of an alcohol molecule to the carbonyl group, is completely analogous to the corresponding reaction of an aldehyde or ketone (Chapter 11). The principal difference in this instance between aldehydes and ketones on the one hand and carboxylic acids on the other is that with the acids it is possible to expel water (a good leaving group) from the intermediate (**2**).

Expulsion of water is of course favored by entropy; from the energetic point of view, the breaking of the C—O σ-bond is compensated for by the reestablishment of the strong C—O π-bond. This is particularly advantageous under the present circumstances. Intermediates **1** and **2** have the charge localized on a single oxygen atom, while intermediate **3** is capable of charge delocalization; the charge delocalization is only possible after the carbon-oxygen double bond has reappeared.

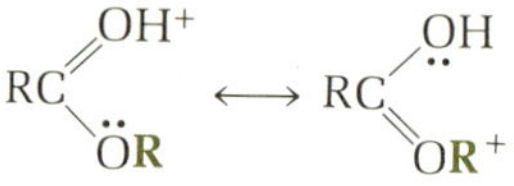

3

Since esterification is an equilibrium process, and a slowly established one at that, numerous devices are used to maximize the yield. When making methyl or ethyl esters, one normally uses large excesses of the alcohol since it is both volatile and inexpensive.

(17.22) $$\underset{\text{(5 moles)}}{CH_3CH_2OH} + \underset{\text{(1 mole)}}{CH_3CO_2H} \rightleftharpoons \underset{\text{(97\%)}}{CH_3CO_2CH_2CH_3} + H_2O$$

After the mixture has been allowed to reflux, the unreacted acid is removed by extraction with base and the products are separated by fractional distillation. With less volatile alcohols it is customary to remove the water continuously by azeotropic distillation; benzene or toluene may be used for the purpose or even just the azeotrope of water with the ester and/or alcohol.

Sterically hindered alcohols react only slowly under normal esterification conditions; if they happen to be tertiary alcohols, they are also prone to suffer acid-catalyzed dehydration. These alcohols are best converted to esters by allowing them to react with acid chlorides or anhydrides.

The conversion of acid to ester is conveniently monitored by means of infrared spectroscopy. The broad band due to the carboxyl —OH group disappears; the carbonyl peak moves to slightly higher frequencies and a very intense peak due to the C—O single bond stretch appears in the 1100 to 1200 cm^{-1} region.

Amides cannot be obtained directly by the reaction between carboxylic acid and amines because neutralization (yielding a salt) is the faster reaction. If the salt is isolated and pyrolyzed, however, then amides may be isolated in satisfactory yields. The amides may be hydrolyzed back to acids in the presence of aqueous acid, the amine being converted to a salt.

(17.23)

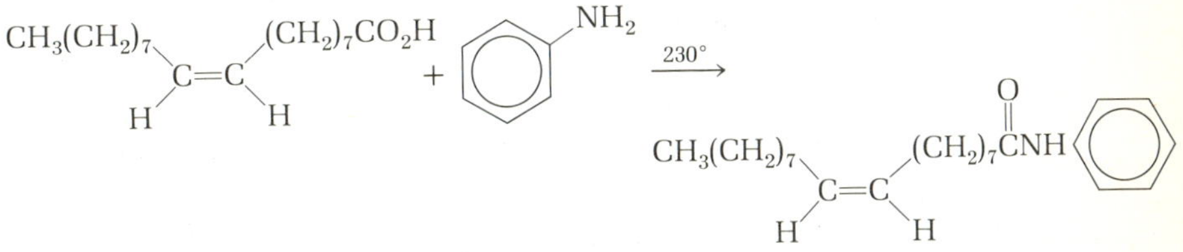

(17.24)

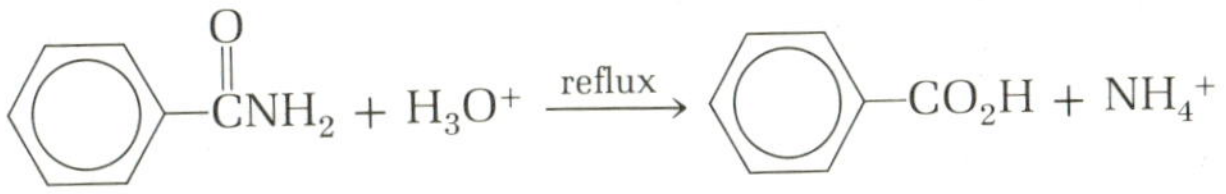

REACTIONS OF ESTERS AND AMIDES UNDER BASIC CONDITIONS

The base-catalyzed hydrolysis, or *saponification,* of carboxylate esters is a very important reaction and its mechanism is prototypical for the mechanism of reaction of many acid derivatives with strong nucleophiles. (Soaps, which are basically the salts of long-chain fatty acids, are obtained by basic hydrolysis of fats that contain large amounts of these acids as esters.) The mechanism set out below is supported by the results of labeling experiments.

(17.25)
$$\mathrm{R\overset{\overset{O}{\|}}{C}OR} \underset{}{\overset{OH^-}{\rightleftharpoons}} \mathrm{R\overset{\overset{O^-}{|}}{\underset{\underset{OH}{|}}{C}}OR} \rightleftharpoons \mathrm{R\overset{\overset{O}{\|}}{C}OH} + \mathrm{RO^-} \rightleftharpoons \mathrm{RCO_2^-} + \mathrm{ROH}$$

4

The tetrahedral intermediate **4** can expel either a hydroxide ion or an alkoxide ion, with roughly equal probability. If the second path is followed, then the products (carboxylic acid and alkoxide ion) neutralize one another, this final step serving to drive the hydrolysis to completion. The hydrolysis is said to be base-catalyzed, although the reaction involves consumption of the strong base hydroxide ion and substitution of the weak base carboxylate anion.

(17.26)
$$\mathrm{CH_3CH_2\overset{\overset{O}{\|}}{C}OCH_2CH(CH_3)_2} + \mathrm{OH^-} \xrightarrow[\text{reflux}]{H_2O} \mathrm{CH_3CH_2CO_2^-} + \mathrm{(CH_3)_2CHCH_2OH}$$

Almost all carboxylate esters are hydrolyzed by way of the mechanism depicted in equation 17.25; sterically hindered esters, particularly those hindered in the acyl portion (**5**), are saponified only very slowly.

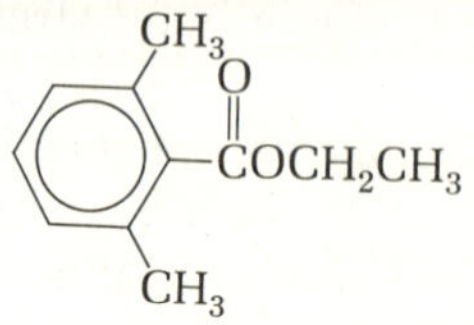

5

The facts enumerated above prompt the conclusion that nucleophiles such as hydroxide ion attack the carbonyl carbon atom, in preference to displacing the entire carboxylate anion by attacking the α-carbon atom of the alcohol portion of the molecule.

It is notable that nucleophiles that are quite reactive toward sp^3-hybridized carbon atoms — I^-, Br^-, SH^-, etc. — are almost totally unreactive toward carbonyl carbon atoms. The nucleophiles that show the greatest preference for the latter type — OH^-, CN^-, NH_3, etc. — are those that are the most basic.

Nucleophilic attack on carboxylic acids is not possible under basic conditions because the bases react much more rapidly with the proton than they do with the carbonyl carbon atom. Carboxylate anions are, not unexpectedly, almost completely immune to nucleophilic attack.

(17.27) 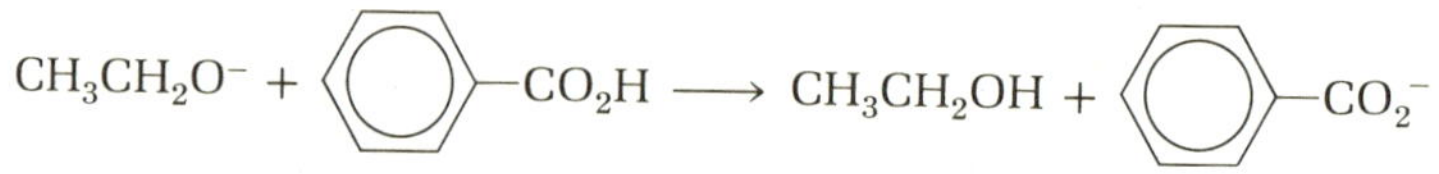

(17.28) $NH_3 + CH_3CH_2CO_2H \longrightarrow CH_3CH_2CO_2^-\ NH_4^+$

Ammonia is a weak nucleophile compared to hydroxide ion and its attack on an ester may be enhanced by simultaneous rather than subsequent proton loss (equation 17.29). The proton could be lost to a molecule of water, ammonia, or another appropriate base. This process is known as *general base catalysis*.

(17.29) 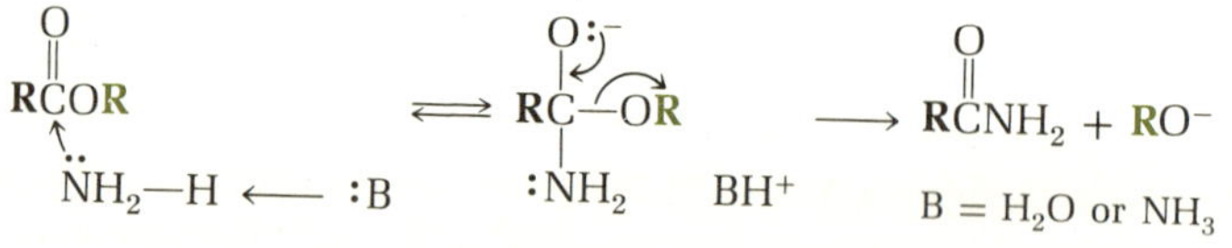

Such catalysis may be important in many other reactions of carboxylic acid derivatives; the determining factors are the leaving group, the pH of the medium, and the nucleophile.

The amide ion, NH_2^-, and its substituted derivatives are very poor leaving groups. For that reason the expulsion of this group during the base-catalyzed hydrolysis of amides may be synchronized with proton transfer from a water molecule. The reaction is therefore *general acid-catalyzed.*

(17.30)

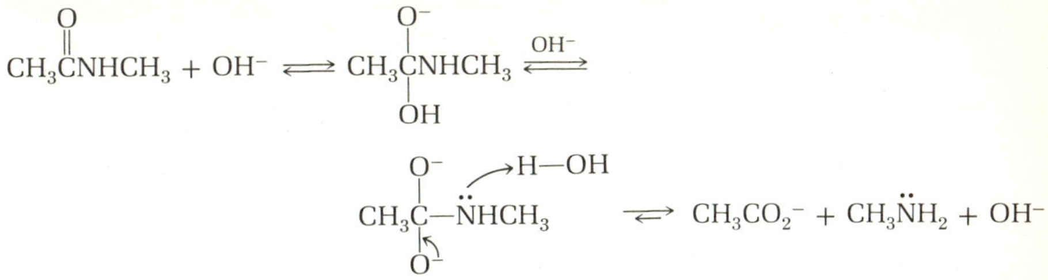

Note that due to the poor leaving ability of the methylamido group, it has been proposed that the reaction proceeds by way of a relatively high energy dianion. For this and other reasons, amides are usually the most difficult carboxylic acid derivatives to decompose.

The relative unreactivity of amides toward hydrolysis and other cleavage reactions may be vividly illustrated by comparing the hydrolysis of an amide and an ester under comparable conditions.

(17.31)

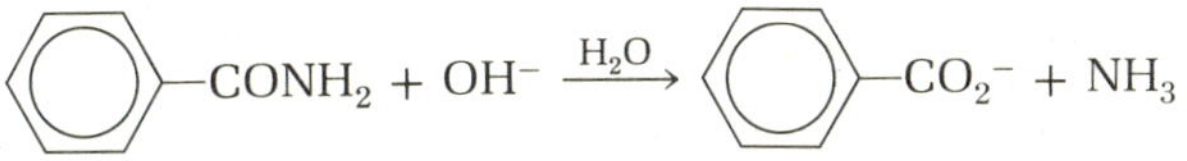

$k^{40^\circ} = 4.0 \times 10^{-5}$ l/mole sec

(17.32)

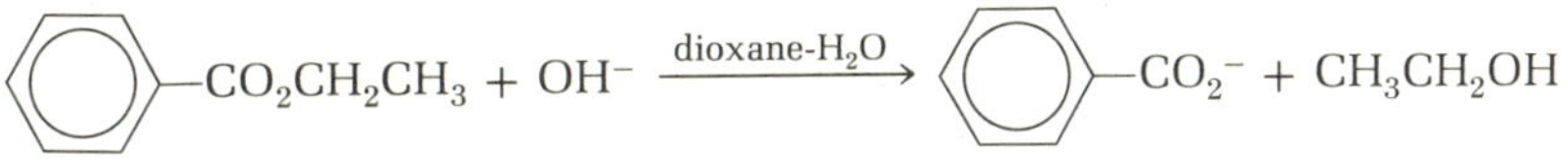

$k^{40^\circ} = 2.58 \times 10^{-2}$ l/mole sec

The special character of amides manifests itself in other ways. In nucleophilic attacks on acid chlorides, anhydrides, and most esters the addition of the nucleophile to the carbonyl group is the rate-determining step; with many amides, however, expulsion of the leaving group becomes the rate-limiting process. The amide nitrogen atom is much less basic than the amine from which it is derived.

(17.33)

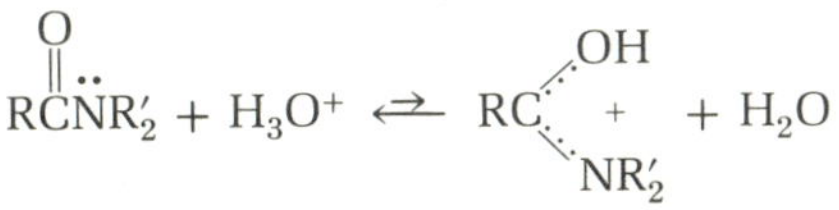

Finally, amides are identifiable by at least two strong infrared bands, both of which arise from modes in which O, C, and N are all involved. The higher frequency band at 1630 to 1700 cm^{-1} is most like a carbonyl stretching mode but lies at lower frequency than most carbonyl bands do. The second amide frequency, at 1200 to 1400 cm^{-1}, involves a mostly N—C motion and would seem from its position to imply considerable double bond character between these atoms.

The degree of substitution of the nitrogen atom may be deduced from the infrared spectrum of the amide. The presence of two, one, or no bands in the N—H stretching region denotes a 1°, 2°, or 3° amide, respectively.

Amides and esters may form *intramolecularly* if the molecule contains an amino or hydroxyl group, respectively, as well as a carboxyl group. These cyclic derivatives are known as **lactams** (amides) and **lactones** (esters).

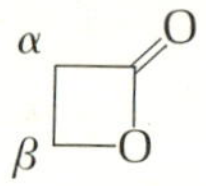

β-propiolactone

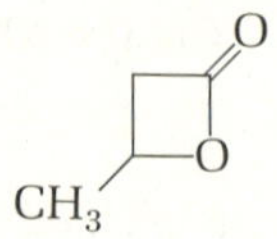

β-butyrolactone

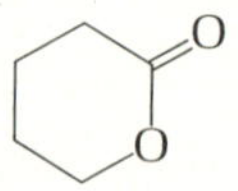

δ-valerolactone

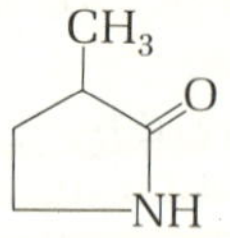

α-methyl-
γ-butyrolactam

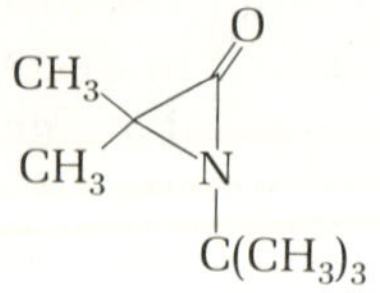

N-*t*-butyl-α-methyl-
α-propiolactam

In one system of nomenclature the word lactam or lactone is preceded by a root that denotes the acid from which it derives and that, in turn, is preceded by a Greek letter denoting the position of attachment of the nitrogen or oxygen atom.

In line with much other data on cyclic compounds, five- and six-membered lactones and lactams are far more readily formed than any others. It is, in fact, often difficult to prevent lactone or lactam formation in those cases in which γ- or δ-derivatives result.

(17.34)

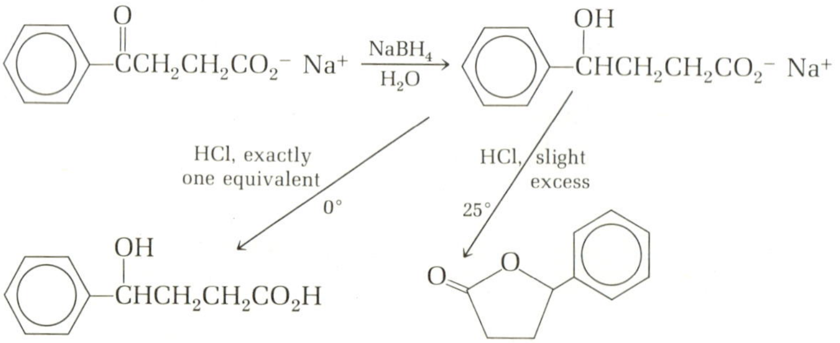

β-Lactones and β-lactams are quite reactive in ring opening reactions, as a result of ring strain. The chemotherapeutic action of the various penicillins (**6**) is thought to center largely in the β-lactam ring.

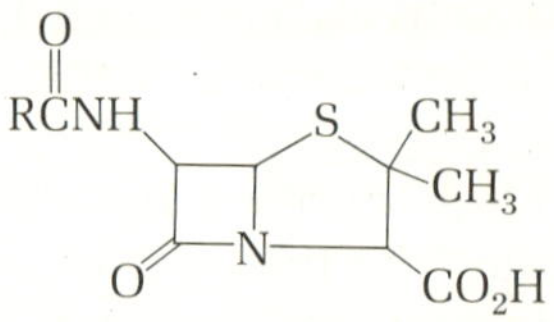

6

A few α-lactams have been isolated, a necessary prerequisite being the presence of a bulky group on the nitrogen atom as well as substitution of the α-carbon atom. α-Lactams and β-lactones exhibit an unusual property that may be understood in terms of their high reactivity. When they are subjected to ring opening by nucleophilic attack, the nucleophile may attach itself to either the carbonyl carbon atom or the aliphatic carbon atom.

(17.35)

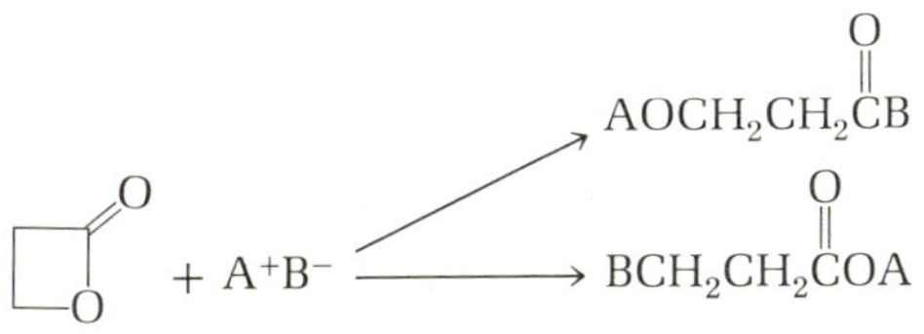

TAUTOMERISM

By moving a proton from the nitrogen atom of an amide to the carbonyl oxygen and adjusting the electron distribution, one obtains an isomer known as an **imide.** The two forms are **tautomers** of one another (Greek — *to auto* the same + *mer* unit).

(17.36)

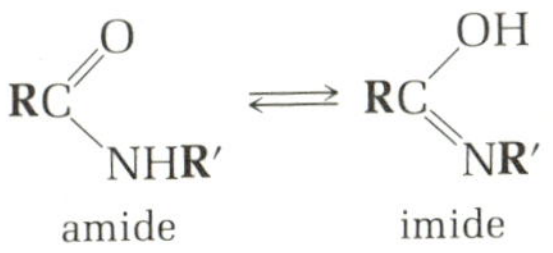

The term tautomer was originally used interchangeably with the word isomer but it is now most often used to denote isomers that may interchange by shifting a proton (electron rearrangements normally accompany these shifts). Tautomeric equilibria are rapidly established when atoms at either terminus are nitrogen or oxygen; such is the case with the equilibrium in equation 17.36.

The amide-imide equilibrium is a very unequal one, the amide form being overwhelmingly predominant under most conditions. The explanation for this inequality may lie in the fact that canonical form **7** must contribute much more to the amide than **8** contributes to the imide. Another common type of tautomerism involves no displacement of atoms, but only a change in electron distribution, with corresponding alterations of bond lengths and angles. This phenomenon is known as *valence tautomerism;* butadiene and cyclobutene, for instance, would be valence tautomers.

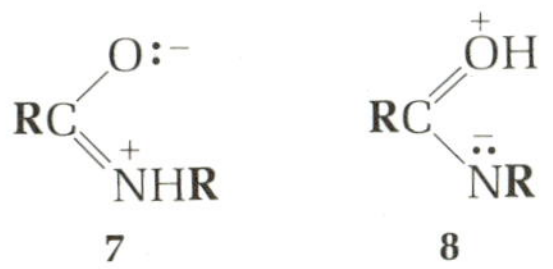

An example of a tautomeric equilibrium with an equilibrium constant equal to one is that between benzoic acid-*carbonyl*-^{18}O and benzoic acid-*hydroxyl*-^{18}O. The rate of interconversion of these two forms is so rapid that they cannot be separately isolated.

(17.37) 

Hydrolysis of benzoyl chloride in $H_2{}^{18}O$, followed by immediate isolation of the product, yields only an equimolar mixture of the two tautomers.

NITRILES

Organic cyanides all belong to the generic class, **nitriles.** They are usually named accordingly, although names such as ethyl cyanide and cyanobenzene may still be found in the literature. In the nitrile nomenclature the word is added as a suffix either to the name of the principal carbon chain (including the carbon atom of the cyano group) or to the name of the carboxylic acid containing the same number and arrangement of carbon atoms. The second system is usually used when the subject is a nitrile related to a common carboxylic acid.

CH_3CN	$CH_3(CH_2)_4CN$	
ethanenitrile acetonitrile	hexanenitrile capronitrile	benzonitrile

Nitriles and carboxylic acids are in fact interconvertible; the interconversion involves the addition or removal of water and ammonia, the common carbon atom undergoing no change in oxidation state. The hydrolysis of nitriles to amides is catalyzed by either acids or bases and these reactions point up the similarity between the strongly polarized cyano group and the carbonyl group. Equation 17.38 gives the probable mechanism for the base-catalyzed hydrolysis of a nitrile to an amide; note that the last step is an imide-amide tautomerization.

(17.38)
$$RC{\equiv}N: + OH^- \rightleftharpoons \underset{\displaystyle OH}{RC}{=}N:^- \xrightleftharpoons{H_2O} \underset{\displaystyle OH}{RC}{=}NH \rightleftharpoons \underset{\displaystyle O}{\overset{}{RC}}NH_2$$

There is a corresponding mechanism for acid-catalyzed hydrolysis in which the first step involves the reversible protonation of the nitrogen atom. Under both acidic and basic conditions the amide may be further hydrolyzed, of course, to the carboxylic acid. It is usually easier to stop the hydrolysis at the amide stage under basic conditions.

A method of selective hydrolysis of nitriles to amides employs the base-catalyzed reaction between hydrogen peroxide and the cyano compound. The virtue of this method is that very mild conditions can be used because the attacking nucleophile is the hydroperoxide anion, HOO^-, which is 10^4 to 10^5 times more nucleophilic than hydroxide ion.

(17.39) 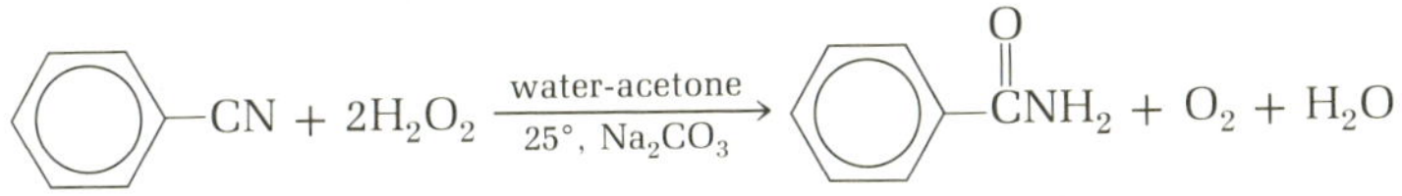

(17.40) 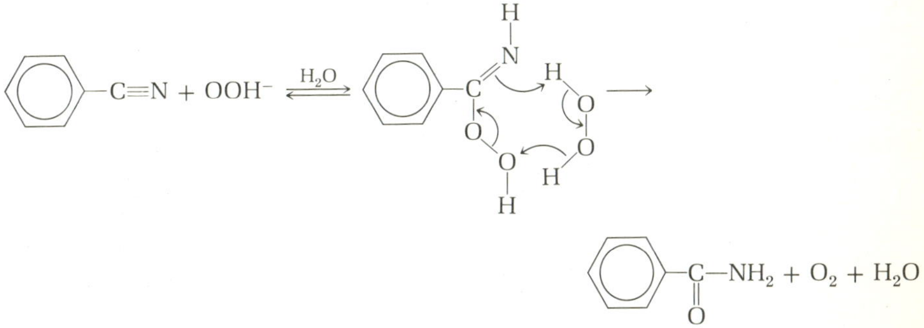

This enhanced nucleophilicity is thought to be due to the repulsions among the pairs of nonbonded electrons on different oxygen atoms and is termed the *α-effect.* The nucleophilicity of hypochlorite ion, hydroxylamine, and other related species is also unusually high.

Amides may be used to synthesize nitriles if a dehydrating agent is also present. The most popular dehydrating agents are phosphorus pentoxide and its derivatives, although acetic anhydride, thionyl chloride, and others have been used. Although the conditions may appear drastic, the yields are usually quite good.

(17.41) $$(CH_3)_2CHCONH_2 \xrightarrow[200\text{–}220^\circ]{P_2O_5} \underset{(69\text{–}86\%)}{(CH_3)_2CHCN}$$

A more common method of aliphatic nitrile synthesis has already been encountered (Chapter 7); the cyano group is introduced by means of an S_N2 displacement reaction.

(17.42) $$CH_3(CH_2)_4Cl + NaCN \xrightarrow[\text{reflux}]{\text{aq. ethanol}} CH_3(CH_2)_4CN + NaCl$$

This approach is excellent when a primary halide or tosylate is the substrate; the yields fall considerably when secondary halides are used and they are uselessly small when a tertiary derivative is employed. In such cases other approaches are required (cf. the synthesis of isobutyronitrile, equation 17.41).

Aromatic nitriles may be synthesized by several methods; the nucleophilic substitution approach delineated above is of course not applicable. One generally useful method revolves around the reaction of cuprous cyanide with aromatic bromides and iodides. This technique is discussed at greater length in Chapter 22.

(17.43)

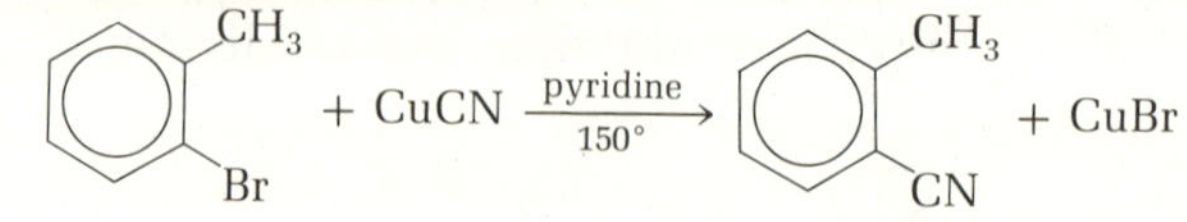

GRIGNARD REACTIONS AND REDUCTIONS OF ACID DERIVATIVES

Grignard reagents add to the carbonyl groups of carboxylic acid derivatives just as they do to the carbonyl groups of aldehydes and ketones. The important difference is that the intermediate from the carboxylate derivative contains a good leaving group and decomposes to an aldehyde or ketone. This latter compound is subjected to further attack by a second mole of Grignard reagent, finally yielding an alcohol.

(17.44)
$$\mathrm{R\overset{O}{\overset{\|}{C}}{-}Y + R'MgX \longrightarrow R\underset{R'}{\underset{|}{\overset{OMgX}{\overset{|}{C}}}}{-}Y \longrightarrow R\overset{O}{\overset{\|}{C}}R' + YMgX} \qquad \mathrm{Y = Cl, OR, O\overset{O}{\overset{\|}{C}}R}$$

(17.45)
$$\mathrm{R\overset{O}{\overset{\|}{C}}R' + R'MgX \longrightarrow \xrightarrow{H_3O^+} R\underset{R'}{\underset{|}{\overset{OH}{\overset{|}{C}}}}R'}$$

Since the two Grignard reactions (equations 17.44 and 17.45) proceed at roughly the same rate, it is impractical to try and stop the reaction at the ketone stage.

PROBLEM

3 There is some evidence that ketones actually react more rapidly than related esters with Grignard reagents. Suggest a reason for this reactivity order.

The most frequent situation involves the use of an ester to synthesize a tertiary alcohol in which at least two of the attached groups are identical. (Symmetrical secondary alcohols result when formate esters are used.)

(17.46)

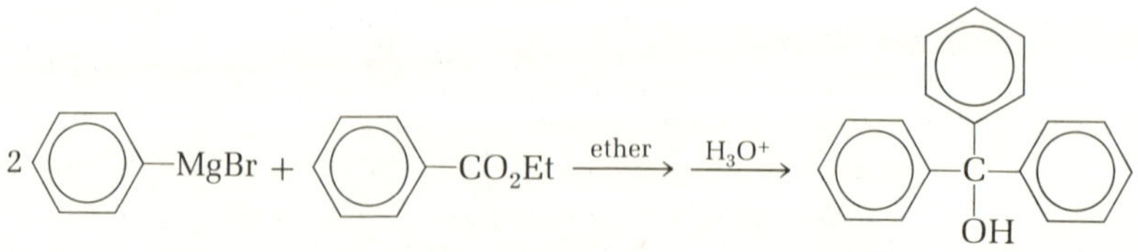

triphenylmethanol (90%)

(17.47) $$CH_3CO_2CH_2CH_3 + 2CH_3(CH_2)_3MgBr \xrightarrow{ether} \xrightarrow{H_3O^+} CH_3(CH_2)_3\overset{\overset{\displaystyle OH}{|}}{\underset{\underset{\displaystyle CH_3}{|}}{C}}(CH_2)_3CH_3 \quad (79\%)$$

A very rapid proton transfer is the outcome when a Grignard reagent is mixed with a carboxylic acid. The halomagnesium carboxylate that is a product reacts only very slowly with another molecule of Grignard reagent.

(17.48) $$CH_3MgBr + CH_3CH_2CO_2H \longrightarrow CH_4 + CH_3CH_2\overset{\overset{\displaystyle O}{\|}}{C}OMgBr$$

Hydride ion from lithium aluminum hydride acts toward carboxylic acid derivatives in the same fashion as Grignard reagents. With esters, acid chlorides, and anhydrides the intermediate formation of aldehydes occurs but these are immediately further reduced to alcohols.

(17.49) $$\mathbf{R}\overset{\overset{\displaystyle O}{\|}}{\mathbf{C}}\mathbf{X} + H{-}\overset{|}{\underset{|}{Al}}{-} \longrightarrow \mathbf{R}\overset{\overset{\displaystyle O\overset{|}{Al}{-}}{|}}{\underset{\underset{\displaystyle H}{|}}{\mathbf{C}}}\mathbf{X} \longrightarrow \mathbf{R}\overset{\overset{\displaystyle O}{\|}}{\mathbf{C}}H + \mathbf{X}\overset{|}{\underset{|}{Al}}{-}$$

(17.50) $$\mathbf{R}\overset{\overset{\displaystyle O}{\|}}{C}H + H{-}\overset{|}{\underset{|}{Al}}{-} \longrightarrow \mathbf{R}CH_2O\overset{|}{\underset{|}{Al}}{-}$$

Although esters can be reduced by sodium in alcohol, and also catalytically, lithium aluminum hydride is by far the most widely used reductant.

(17.51) $$CH_3CH{=}CHCH_2\overset{\overset{\displaystyle O}{\|}}{C}OCH_3 + LiAlH_4 \xrightarrow{ether} \xrightarrow{H_3O^+} \underset{(75\%)}{CH_3CH{=}CHCH_2CH_2OH} + CH_3OH$$

Carboxylic acids are also reduced to primary alcohols, although somewhat more slowly than the corresponding esters. The method is also slightly wasteful since the acid and aluminohydride ion react very rapidly to liberate molecular hydrogen. That the reduction may be carried out in the first place is due to the strength of the aluminum-oxygen bond, which has partial double bond character, enabling the aluminum atom to pull the oxygen atom away from carbon.

(17.52)

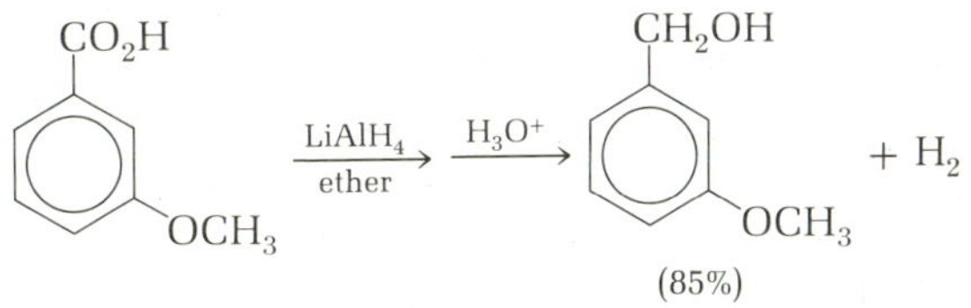

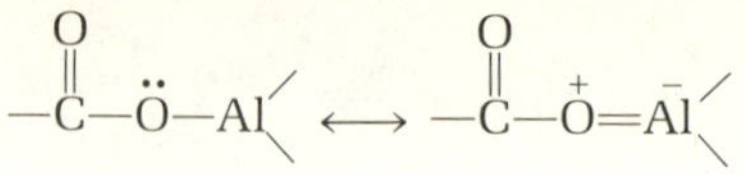

Boron has a similarly large oxygen affinity and diborane is an effective reductant for carboxylic acids.

(17.53)
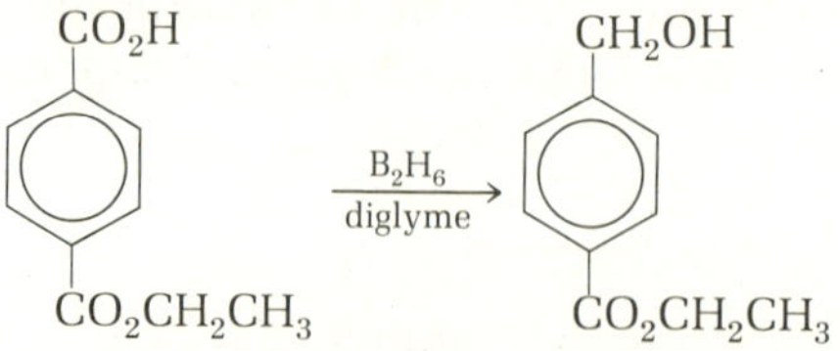

Unlike lithium aluminum hydride, diborane does not reduce esters as well so it can be used for selective reduction of mixed ester acids.

Reduction of amides with excess lithium aluminum hydride, under normal conditions, also involves the removal of an oxygen atom, the products being amines. In this case the expulsion of the oxygen atom is aided by the nonbonding electrons on nitrogen.

(17.54)
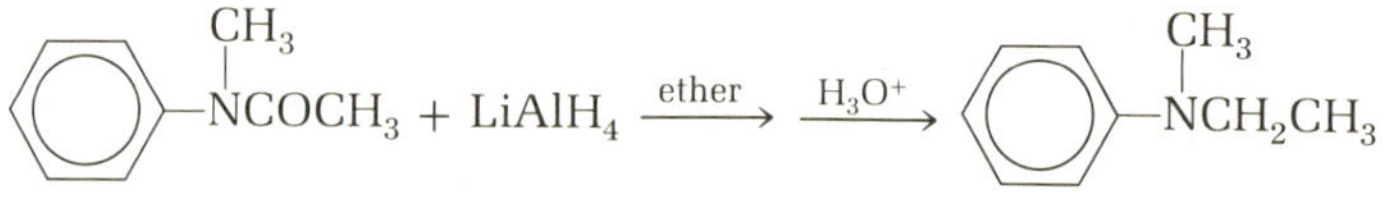

(91%)

(17.55)
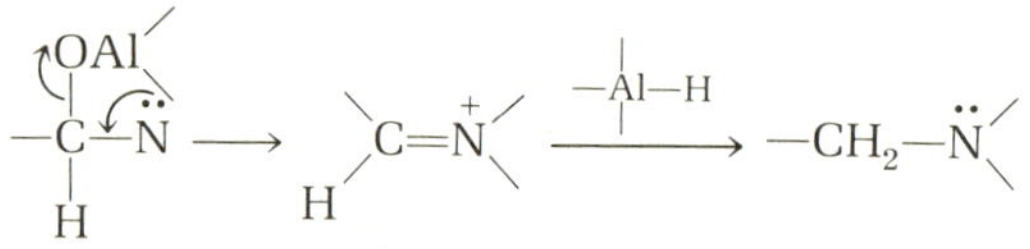

Under special conditions the reduction of acid chlorides may be arrested at the aldehyde stage. One method employs a special derivative of lithium aluminum hydride at low temperatures.

(17.56)
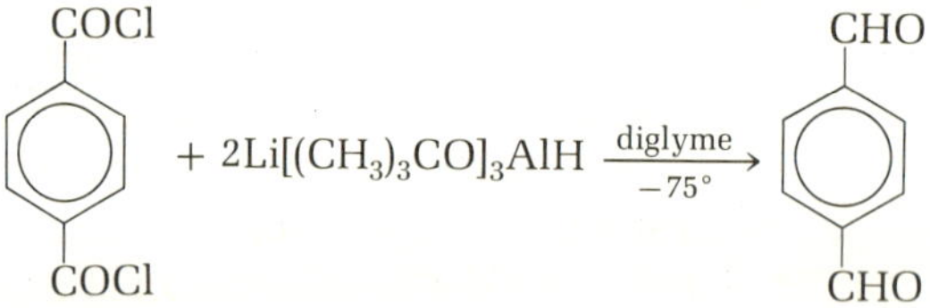

lithium tri(*t*-butoxy)-aluminohydride

(85%)

Another approach to conversion of acid chlorides to aldehydes employs molecular hydrogen and a special catalyst, which is often partially poisoned (*Rosenmund reduction*). The poisoned catalyst is capable of catalyzing the reduction of the acid chloride but not of the aldehyde.

(17.57) $$CH_3O_2CCH_2CH_2COCl + H_2\ (1\ atm) \xrightarrow[\text{xylene, }110^\circ]{Pd/BaSO_4} CH_3O_2CCH_2CH_2CHO\ (65\%)$$

Partial reduction of nitriles and amides to aldehydes is also possible through use of specially modified lithium aluminohydrides.

(17.58) $$CH_3(CH_2)_2CN + LiAlH(OCH_2CH_3)_3 \xrightarrow{\text{ether}} \xrightarrow{H_3O^+} CH_3(CH_2)_2CHO\ (68\%)$$

ESTER FORMATION VIA S_N2 ROUTES

Since carboxylate ions are nucleophiles, esters may also be formed by way of S_N2 reactions in which one oxygen atom of the carboxylate group displaces a leaving group from an sp^3-hybridized carbon atom. Carboxylate ions may be classified as weak nucleophiles and their participation in S_N2 displacements is only practical if a good leaving group is available. In the example shown, complexation with silver ion makes the iodine an unusually good leaving group.

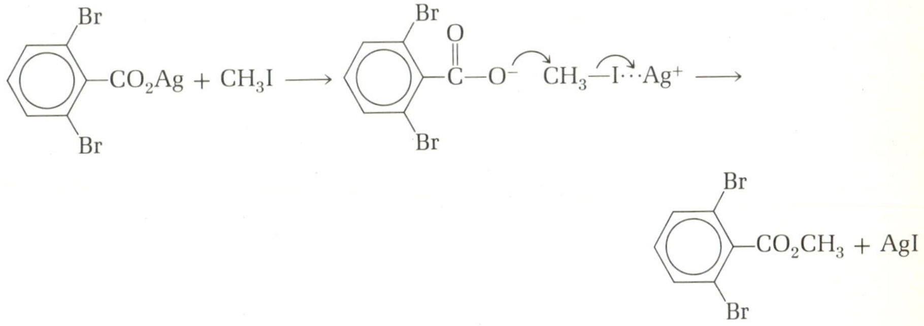

(17.59)

A useful method of converting acids to methyl esters involves the reaction

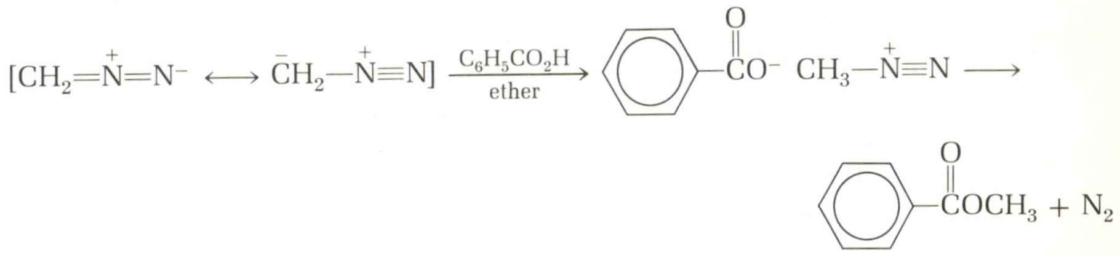

(17.60)

of these acids with *diazomethane*, $CH_2{=}\overset{+}{N}{=}N{:}^-$. Its efficacy depends on the fact that after protonation it contains the **diazonium group** $—\overset{+}{N}{\equiv}N$. Due to the great thermodynamic stability of molecular nitrogen the diazonium group is a very good leaving group indeed (Chapter 19).

CARBOXYLATES AND SULFONATES AS LEAVING GROUPS

As already noted, hydrolysis and alcoholysis of most carboxylate esters goes by way of a mechanism in which the nucleophile attacks the carbonyl carbon atom. There are some exceptions, however; esters of tertiary alcohols, and benzylic and allylic esters, may rather dissociate into carboxylate anions and carbonium ions. Such a path would be favored by the relatively weak C—O bond in these compounds as well as steric hindrance to nucleophilic attack in the case of tertiary esters.

(17.61)

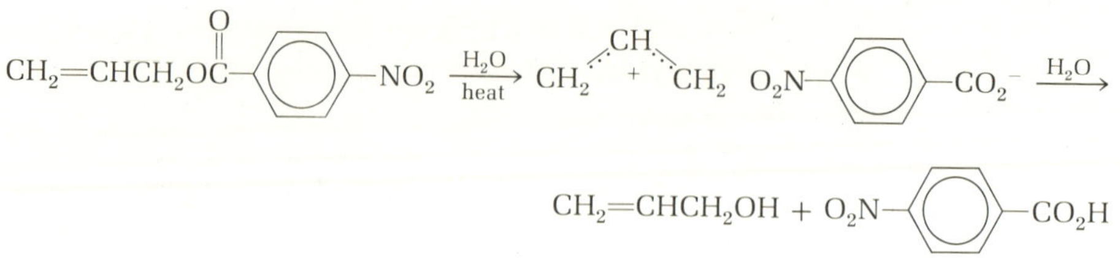

The rate of these S_N1 dissociations of carboxylate esters is greatly increased by the presence of electron-withdrawing groups in the acid portion. Thus, esters of 4-nitrobenzoic, 3,5-dinitrobenzoic, and trifluoroacetic acids dissociate fairly readily.

A particularly instructive case of mechanism change has been uncovered in studies of the uncatalyzed hydrolysis of methyl, ethyl, isopropyl, and *t*-butyl esters of trifluoroacetic acid.

(17.62) $$CF_3\overset{\overset{O}{\|}}{C}OR + H_2O \xrightarrow{\text{slow}} CF_3\overset{\overset{O^-}{|}}{\underset{\underset{OH_2^+}{|}}{C}}OR \qquad R = CH_3,\ CH_3CH_2,\ (CH_3)_2CH$$

The reactions of the first three esters require modest activation energies ($\Delta H^{\ddagger} = 10.5$ kcal/mole) and strongly negative entropies of activation ($\Delta S^{\ddagger} = -35$ eu); these data are compatible with a mechanism in which water attacks the carbonyl group during the slow step (equation 17.62). Very different parameters are found for the reaction of the *t*-butyl ester: $\Delta H^{\ddagger} = 15$ kcal/mole and $\Delta S^{\ddagger} = +14$ eu. A reasonable interpretation is that in this case the slow step consists of the unimolecular dissociation of the ester.

(17.63) $$CF_3\overset{\overset{O}{\|}}{C}OC(CH_3)_3 \xrightarrow{\text{slow}} CF_3CO_2^- + (CH_3)_3C^+$$

"Abnormal" behavior in carboxylate esters (breaking of the alkyl-oxygen bond) is quite normal for sulfonate esters. These compounds almost never undergo nucleophilic attack at the sulfur atom, even when powerful nucleophiles are being used.

(17.64) 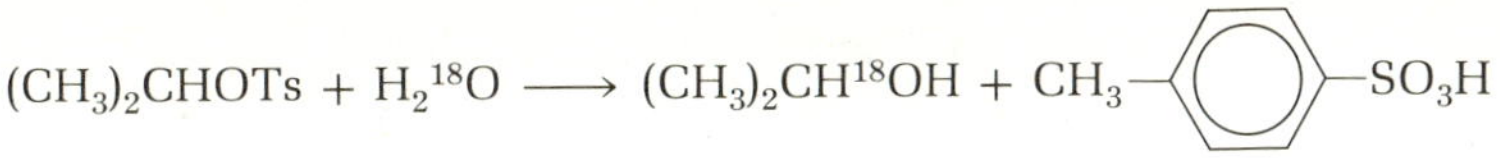

(17.65) $CH_3(CH_2)_3OTs + LiAlH_4 \xrightarrow{\text{ether}} CH_3(CH_2)_2CH_3 + Li^+OTs$

Two reasons may be adduced for this behavior. In the first place, the benzenesulfonyl group, 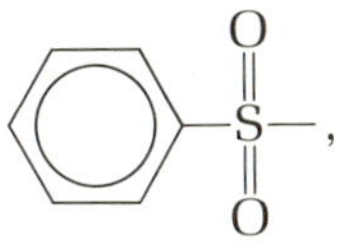, is powerfully electron-withdrawing. Secondly, the sulfur atom is surrounded by four ligands, in contrast to the carbonyl carbon atom, which bears only three. The sulfur atom is undoubtedly much less sterically accessible than the analogous carbon atom.

POLYAMIDES AND POLYESTERS

Polymers in which an ester or amide linkage is incoporated into the backbone of the polymer are of great commercial importance, as well as being very significant in living systems. Man-made fibers are very often polyesters, which are synthesized from dicarboxylic acids and diols.

(17.66) 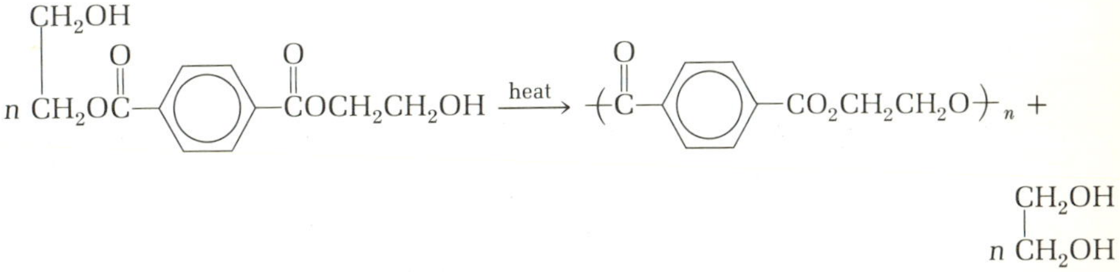

Commercial polyamides are usually synthesized by one of two methods. The first is the self-condensation of an amino acid or its lactam (equation 17.67), while the second involves the condensation of a diamine with a dicarboxylic acid.

(17.67) 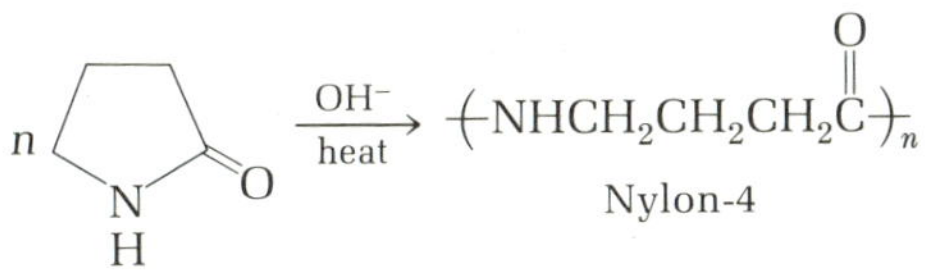

(17.68) $n\, HO_2C(CH_2)_4CO_2H + nH_2N(CH_2)_6NH_2 \xrightarrow{200\text{–}300^\circ} \left(\overset{O}{\overset{\|}{C}}(CH_2)_4\overset{O}{\overset{\|}{C}}NH(CH_2)_6NH \right)_n + 2nH_2O$

Nylon 6-6

The regular synthetic polyamides are known as nylons and the number following the name denotes the number of carbon atoms in the repeating unit(s) in the backbone. They are hydrolytically very stable, more so than the polyesters, and often quite high melting, due probably in part to hydrogen bonding between —N—H and O=C .

There are almost no analogues to the polymeric esters to be found in nature. Very much in abundance, however, are the triesters of **glycerol** (propane-1,2,3-triol). These esters (**9**) constitute the fats and oils found in both plants and animals.

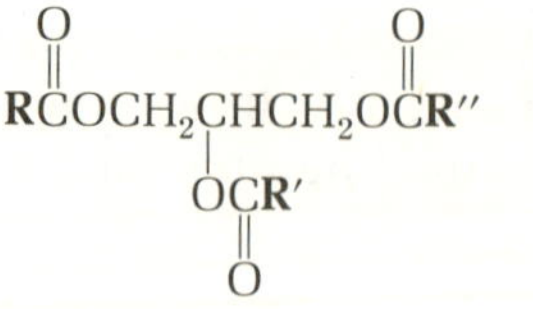

9

The esterifying acids are usually C_{12}, C_{16}, and C_{18} straight-chain saturated fatty acids, as well as several C_{18} unsaturated acids. In some cases one of the carboxylic acids is replaced by a molecule of phosphoric acid, which is monoesterified.

While polymeric esters are of little importance biochemically, the very opposite is true of polyamides. Peptides and proteins, among the most vital classes of compounds found in living organisms, are polymers of α-amino acids. In particular, cellular structural elements and enzymes are proteins. The structure, chemistry, and function of these molecules is the subject of Chapter 25.

It is interesting to speculate on the reasons why chemical evolution has favored the selection of polyamides rather than polyesters. The greater hydrolytic stability of amides compared to esters must be important since many proteins are continually in contact with aqueous solutions of substances that are active as catalysts for hydrolysis. Furthermore, many polypeptides and proteins possess so-called secondary and tertiary structures that refer to the conformation of the entire chain. It may be wound in a helix, folded back and forth on itself, etc. Intramolecular hydrogen bonds are responsible for maintaining this structure which is intimately connected with the functioning of these molecules, particularly the enzymes. Such structuring would not be available in polyesters that lack hydrogen-bond donors.

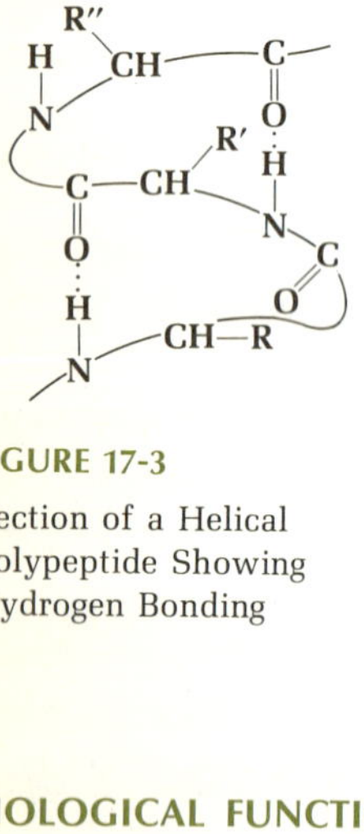

FIGURE 17-3

Section of a Helical Polypeptide Showing Hydrogen Bonding

BIOLOGICAL FUNCTIONS OF THIOL ESTERS

The sulfur analogues of alcohols, ROH, are the thiols, RSH. These compounds may be esterified, in complete analogy to alcohols, giving rise to thiol esters. Although not of great moment in the laboratory, thiol esters occupy

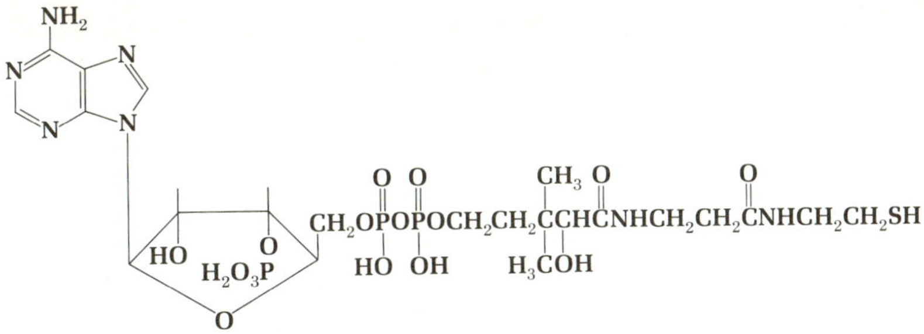

FIGURE 17-4
Structure of Coenzyme A

a position of central importance in biochemical systems. The synthesis and degradation of carbon chains in living matter often proceed by the stringing together or uncoupling of acyl groups. These acyl groups are transferred from one chemical cycle to another in the form of their thiol esters (Chapter 21). The thiol esters are also subject to reaction with a variety of other nucleophiles.

The thiol that most commonly performs this function is coenzyme A, the structure of which is shown in Figure 17-4. Ignoring the complex structure of the coenzyme, we may focus our attention on the thiol group that is the "business end" of the molecule. This fact is highlighted by the practice of abbreviating the structure of the molecule as CoASH; its acetate ester, known as acetyl coenzyme A, is represented as $\text{CoAS}\underset{\text{O}}{\underset{\|}{\text{C}}}\text{CH}_3$.

Again adopting a teleological tone, we may ask why Nature "chose" to use thiol esters rather than oxygen esters for these functions. The answer undoubtedly lies in the greater reactivity of thiol esters, particularly toward neutral nucleophiles such as amines. Thiol esters are less stabilized by resonance than oxygen esters are, due to the weakness of $2p$-$3p$ π-bonds, and are correspondingly more rapidly attacked by nucleophiles.

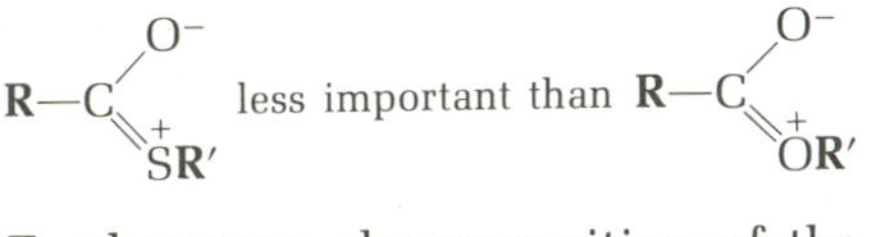

Furthermore, decomposition of the tetrahedral intermediate derived from a thiol ester (**10**) gives product much more often than decomposition of the intermediate from an oxygen ester (**11**).

10 **11**

When **10** decomposes, it can eject either the attacking nucleophile or $R'S^-$, while the choices for **11** are either the nucleophile or $R'O^-$. Since the C—S bond is considerably weaker than the C—O bond, decomposition of **10** with product formation will occur far more frequently.

SUMMARY

1 The common carboxylic acid derivatives fall into a distant reactivity sequence. The acid chlorides are the most reactive, followed by anhydrides, esters, and amides. This sequence parallels the decreasing leaving group ability of the carbonyl substituent. It also parallels the increasing resonance energy of the functional group. The less reactive derivatives may be synthesized from the more reactive ones but not vice versa.

2 Acid chlorides are synthesized by allowing carboxylic acids to react with inorganic acid chlorides, notably thionyl chloride and phosphorus pentachloride. Generally speaking, the symmetrical anhydrides are the more stable and the more widely used. They may be made by allowing the desired acid to react with the anhydride of a more volatile acid, such as acetic anhydride. Carboxylate anions and acid chlorides also combine to form anhydrides.

3 Amines and alcohols react rapidly and irreversibly with acid chlorides and anhydrides. A mole of acid is liberated in the process and it is customary to neutralize this acid with an added base. Excess amine can also serve to neutralize the acid.

4 Alcohols add reversibly to the carbonyl group of carboxylic acids, forming intermediates in which the carbonyl carbon atom is tetrahedrally substituted. In the presence of catalytic amounts of a strong acid the intermediate may lose water and become converted to an ester. In order to drive the reaction to completion one either uses a large excess of one reactant (usually the alcohol) or one continuously removes the water as formed (commonly by means of azeotropic distillation).

5 The hydrolysis of amides and esters is catalyzed by hydroxide ion, which is sufficiently nucleophilic to attack these compounds at a convenient rate. Addition of the hydroxide ion to the carbonyl double bond is followed by expulsion of the leaving group. The departure of the amide leaving group requires its prior, or simultaneous, protonation. Ammonia and amines may also react with esters, albeit slowly, to yield amides.

Nucleophilic attacks on acid chlorides, anhydrides, and esters involve a rate-determining nucleophilic attack; with amides as substrates, however, loss of the departing group is often rate-determining. These reactions are driven to completion by formation of the carboxylate ion.

6 Intramolecular formation of esters and amides may occur, leading to the cyclic lactones and lactams. These reactions are particularly favored if the resulting ring is five- or six-membered. α-Lactams and β-lactones and β-lactams are all rather reactive species due to their considerable angle strain. Under favorable circumstances nucleophiles will attack these compounds at the aliphatic rather than the carbonyl carbon atom.

7 Amides may be dehydrated to nitriles, usually with the aid of phosphorus pentoxide; nitriles, in turn, are hydrolyzed to amides with the aid of acid or base catalysts. Imides are intermediates in these hydrolyses but they tautomerize almost immediately to amides.

8 Grignard reagents add to the carbonyl double bond of acid chlorides, anhydrides, and esters, the products of addition and leaving group loss being ketones. These compounds in turn take part in further reactions with the Grignard reagent; the final products after hydrolytic work-up are tertiary alcohols containing two identical groups. Hydride reduction of the same classes of compounds follows an identical course, as the intermediate aldehydes are further reduced to primary alcohols. It is possible to stop the reduction of an acid chloride at the aldehyde stage, through use of special hydrides [lithium tri(*t*-butoxy)aluminohydride] or by employing molecular hydrogen and a specially prepared catalyst.

Amides may be reduced to amines with lithium aluminum hydride, while carboxylic acids are converted to primary alcohols by both lithium aluminum hydride and diborane. These reactions depend in part on the Lewis acid character of boron and aluminum, enabling them to abstract oxygen from the compound being reduced.

9 Carboxylate ions may act as nucleophiles in displacing good leaving groups attached to primary carbon atoms, forming esters thereby. Carboxylate ions may also function as leaving groups, especially in those cases in which stable carbonium ions are left behind after their departure. The likelihood of the entire carboxylate group's departing is increased by the presence of electron-withdrawing substituents in this group. Sulfonate esters, in contrast to carboxylate esters, almost never undergo sulfur-oxygen cleavage but rather suffer oxygen-carbon bond breaking. This difference is probably due to the very favorable charge delocalization in the sulfonate anion and to the steric crowding around sulfur.

10 Polyesters and polyamides are substances of great commercial importance; they are made by allowing diols and diamines, respectively, to react with dicarboxylic acids. The polyamides tend to be higher melting and of greater hydrolytic stability. Triesters of glycerol are widely distributed in Nature, the esterifying acids being saturated and unsaturated C_{12}, C_{16}, and C_{18} carboxylic acids. Polymers of α-amino acids, the peptides and proteins, are among the most vital classes of compound for sustaining life.

11 Biological transformations of acyl groups are often performed on their thiol esters. These compounds are particularly susceptible to nucleophilic attack because of their small resonance stabilization compared with that of oxygen esters and because of the ease of carbon-sulfur bond cleavage in the tetrahedral intermediate.

PROBLEMS

Note: Problems 1 through 3 appear in text.

4 a. When amides and acyl chlorides are allowed to react in the presence of pyridine, diacylamides (*imides*) are formed.

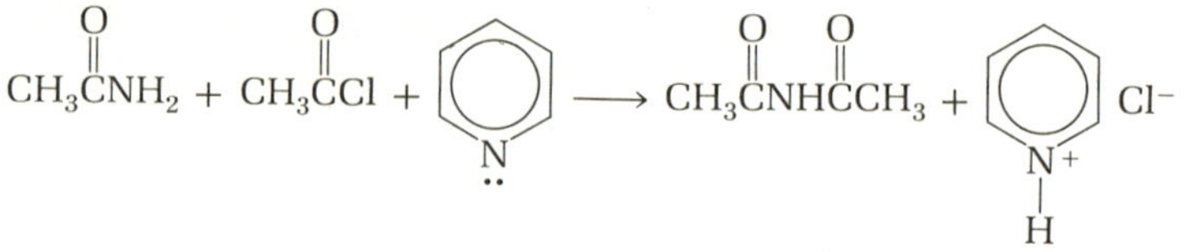

(1)

Where would you place these compounds in the reactivity-stability scale shown on p. 437? Upon further acylation, triacylamides may be obtained. Would you expect these latter derivatives to be more or less reactive than diacylamides?

b. Acyl cyanides, $R\overset{O}{\overset{\|}{C}}C{\equiv}N$, are quite reactive acylating agents toward nucleophiles. How would you rationalize this finding?

azanonatetraene

c. Azanonatetraene (see at left) is relatively quite stable for a molecule of its type; it remains unchanged for long periods of time at 35°. Its *N*-carbethoxy derivative, on the other hand, undergoes a rapid intramolecular electrocyclization at 25° (equation 2).

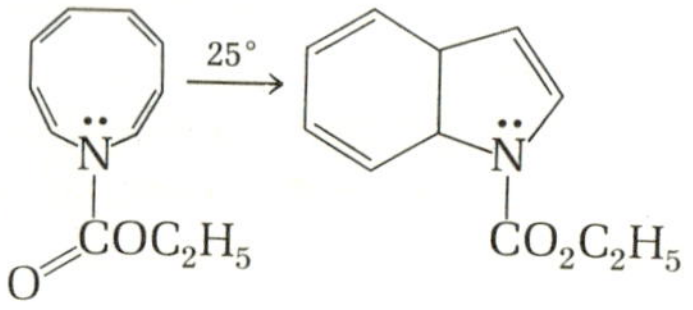

(2)

To what is the stability of azanonatetraene due (see Chapter 14), and how is it affected by the carbethoxy group?

5 a. Succinic acid $\xrightarrow{\text{heat}}$ **A** ($C_4H_4O_3$) $\xrightarrow{CH_3NH_2}$ **B** ($C_5H_9NO_3$) $\xrightarrow{\text{heat}}$ **C** ($C_5H_7NO_2$; *N*-methylsuccinimide)

b. Phthalic acid $\xrightarrow[\text{anhydride}]{\text{acetic}}$ **D** ($C_8H_4O_3$) $\xrightarrow{CH_3OH}$ **E** ($C_9H_8O_4$) $\xrightarrow{SOCl_2}$ **F** ($C_9H_7ClO_3$) $\xrightarrow{NH_3}$ **G** ($C_9H_9NO_3$)

c. Phosgene ($COCl_2$) $\xrightarrow{CH_3OH}$ **H** ($C_2H_3ClO_2$) $\xrightarrow{(CH_3CH_2)_2NH}$ **I** ($C_6H_{13}NO_2$)

Propose structures for compounds **A** to **I**.

6 When ester reactions have to be conducted in a polar solvent, alcohols are often the media chosen. Methyl esters are dissolved in methanol; ethyl esters, in ethanol; etc., but methyl esters are usually not dissolved in ethanol, whether under acidic or basic conditions. Explain.

7 Freshly fermented wine or cider is often rather raw, due in part to the presence of carboxylic acids in the mixture. If the liquid is "aged" (allowed to stand in a sealed container), it gradually becomes much more mellow and acquires its characteristic fragrance as well. What chemical process might be occurring during the aging process?

8 Write structures for all the anticipated products, if any, from the reaction of ethyl benzoate with the following reagents:

a. H_3O^+, reflux

b. OH^-, heat

c. $CH_3CH_2CH_2OH$, H^+

d. NH_3, H_2O

e. $(CH_3)_2NH$

f. NH_2NH_2

g. CH_3O^-, CH_3OH

h. $CH_3(CH_2)_3MgBr$, then H_3O^+

i. $LiAlH_4$, then H_3O^+

9 a. Compound **A** ($C_5H_{10}O_2$) was heated in aqueous base and the solution extracted with ether. 1-Propanol was isolated from the ether extract. After acidification the aqueous layer yields compound **B.** Suggest a structure for **B.**

b. When **C** ($C_6H_{13}NO$) is heated in aqueous base, a gas is evolved. The solution is then acidified and extracted with ether, from which **D** may be isolated. **D** is heated with acidic methanol to give **E,** which can be reduced with lithium aluminum hydride. The reduction products are methanol and 4-methyl-1-pentanol. Write structures for **C, D,** and **E.**

c. **F** ($C_7H_4Cl_2O$) is treated with isobutyl alcohol and **G** ($C_{11}H_{13}ClO_2$) is isolated. When **G** is heated with *n*-butylamine, **H** is produced, which may be transformed by aqueous acid to a solid, **I** ($C_7H_5ClO_2$). **I** may be heated with a mixture of nitric and sulfuric acids, giving rise to but a single product, **J** ($C_7H_4ClNO_4$). Propose structures for **F** to **J.**

10 The rate of the base-catalyzed hydrolysis of methyl benzoate is accelerated by electron-withdrawing substituents in the phenyl ring and decelerated by electron-donating substituents. In other words, the reaction exhibits a positive ρ value.

Which step of the mechanism shown in equation 17.25 is the rate-determining one? Explain how your conclusion is supported by the data just cited.

11 Write the structures of the products to be expected if the following compounds are allowed to react in the manner indicated:

a. Sodium *p*-methylbenzoate and *p*-methylbenzoyl chloride.

b. Benzoic acid dissolved in acidic $H_2{}^{18}O$.

c. Isobutyric acid with aqueous ammonia.

d. *m*-Nitrophenylacetic acid heated with ethylamine.

e. Butyric anhydride and *N*,4-dimethylaniline.

f. Heptanamide refluxed in acidic methanol.

g. Heating 5-hydroxypentanoic acid.

h. Phenyl acetate with excess methyllithium, followed by aqueous acid.

12 Most lactams, like acyclic amides, are relatively resistant to hydrolysis. Compound II, however, undergoes hydrolysis at a surprisingly rapid rate. Offer an explanation for this finding.

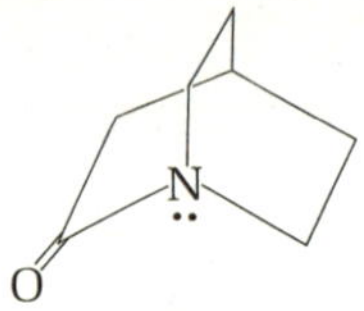

II

13 *Ketenes* are intramolecular anhydrides of carboxylic acids. The first member of the series, ketene itself ($CH_2{=}C{=}O$), may be viewed as the anhydride of acetic acid. Ketenes are conveniently prepared by the dehydrohalogenation of acid chlorides with tertiary amines.

$$(CH_3)_2CH\overset{O}{\overset{\|}{C}}Cl + (CH_3CH_2)_3N \xrightarrow{\text{ether}} \underset{\text{dimethylketene}}{(CH_3)_2C{=}C{=}O} + (CH_3CH_2)_3NH^+Cl^- \quad (3)$$

Ketenes are powerful acylating agents. Draw the structures of products you expect from the reaction of dimethylketene with the following compounds.

a. Water

b. Deuterium oxide

c. 1-Propanol

d. 1-Methylcyclopentanol

e. Aniline

f. Methylethylamine

g. 2-Aminoethanol

h. Isobutyric acid

i. Ammonia-d_3

j. Hydrogen fluoride

k. Methanethiol (CH_3SH)

14 Write out a mechanism for the acid-catalyzed hydrolysis of benzamide.

15 a. Suggest mechanisms for the operation of the equilibrium shown in equation 17.36 under both acidic and basic conditions.

b. Amides may be converted to nitriles by treating them with tosyl chloride in pyridine. Show all the steps in this process. (Pyridine acts as a base.)

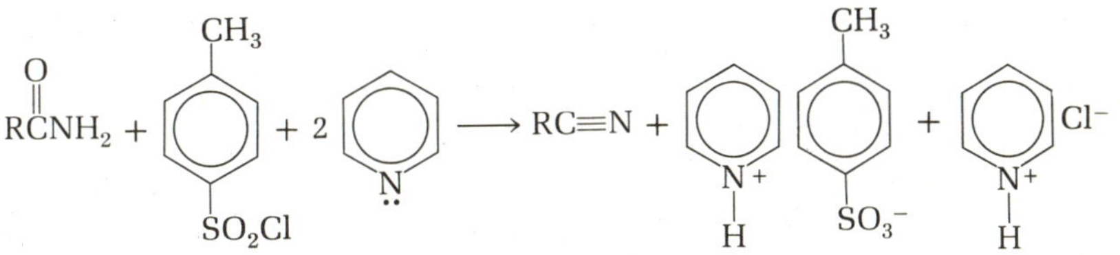

(4)

16 Write out structures for compounds **A** to **K**, showing their absolute configurations. Label the configurations for **B, E, G, I,** and **K.**

a. (S)-(+)-2-Butanol $\xrightarrow{PBr_3}$ **A** $\xrightarrow[DMF]{KCN}$ **B** $\xrightarrow[H_2O,\ \Delta]{H_2SO_4}$ (+)-**C** ($C_5H_{10}O_2$)

(S)-(+)-2-Butanol $\xrightarrow[pyridine]{TsCl}$ **D** $\xrightarrow[DMSO]{NaCN}$ **E** $\xrightarrow[H_2O,\ \Delta]{H_2SO_4}$ (−)-**C**

b. (+)-**C** $\xrightarrow[H^+]{CH_3OH}$ **F** $\xrightarrow[2.\ H_3O^+]{1.\ LiAlH_4}$ (−)-**G** ($C_5H_{12}O$) $\xrightarrow{PBr_3}$ **H** $\xrightarrow{Li^+(CH_3)_2Cu^-}$ **I** (C_6H_{14})

c. **H** $\xrightarrow[DMF]{KCN}$ **J** $\xrightarrow[H_2O,\ \Delta]{H_2SO_4}$ (+)-**K**

17 Indicate all reagents and conditions necessary for effecting the following conversions. As sources of deuterium, D_2O and $LiAlD_4$ may be used.

a. Isobutyl alcohol ⟶ ethyl isobutyrate

b. Bromobenzene ⟶ benzamide

c. *n*-Propyl bromide ⟶ 4-ethyl-4-heptanol

d. Anisole ⟶ 4,4′-dimethoxytriphenylmethanol

e. 2-Methylbutanoic acid ⟶ 1-bromo-2-methylbutane

f. Butyric acid ⟶ valeric acid-α,α-d_2

g. Aniline ⟶ ethylphenylamine

h. Ethyl acetate ⟶ ethanol-*O*-*d*(CH_3CH_2OD)

i. Cyclohexanol ⟶ cyclohexanecarboxaldehyde

j. Mesitylene ⟶ 2,4,6-trimethylphenylacetaldehyde

18 Write the structures of the organic products from the following reactions.

a. Methyl benzoate $\xrightarrow{LiAlH_4}\xrightarrow{H_3O^+}$

b. Ethyl cyclobutanecarboxylate $\xrightarrow{LiAlH_4}\xrightarrow{D_3O^+}$

c. Methyl isobutyrate $\xrightarrow{LiAlD_4}\xrightarrow{H_3O^+}$

d. Diphenyl malonate $\xrightarrow{LiAlD_4}\xrightarrow{D_3O^+}$

e. Hexanoyl chloride $\xrightarrow{Li[(CH_3)_3CO]_3AlD}$

19 a. What would be the product from reaction of excess propylmagnesium bromide with diethyl carbonate ($CH_3CH_2O\underset{\underset{O}{\|}}{C}OCH_2CH_3$), followed by acid work-up?

b. Ethyl *N*, *N*-dimethylcarbamate was allowed to react with two moles of Grignard reagent at −10° (equation 5). The solution was then treated with aqueous acid, and it yielded the products shown. Propose a mechanism for this process. How is this reaction related to that shown in part a?

$$(CH_3)_2N\overset{\overset{O}{\|}}{C}OC_2H_5 + 2RMgBr \xrightarrow[-10^\circ]{ether} \xrightarrow{H_3O^+} R\overset{\overset{O}{\|}}{C}R + C_2H_5OH + (CH_3)_2NH_2^+ \quad (5)$$

20 a. When α-bromopropionate ion is dissolved in dilute base, it is converted to α-hydroxypropionate ion. Through the use of optically active α-bromopropionate ion it was shown that the bromine is replaced by hydroxyl with *retention* of configuration. Suggest a mechanism that accounts for the stereochemical result. The reaction kinetics are first order in propionate ion and *zero order* in hydroxide ion.

b. The hydrolysis of monophenyl maleate ion (III) occurs 10,000 times more rapidly than the hydrolysis of monophenyl glutarate ion (IV) under comparable conditions. Monophenyl succinate ion (V) also hydrolyzes faster than IV but only by a factor of about 200. Propose a mechanism for the accelerated hydrolyses, and comment on the reactivity order.

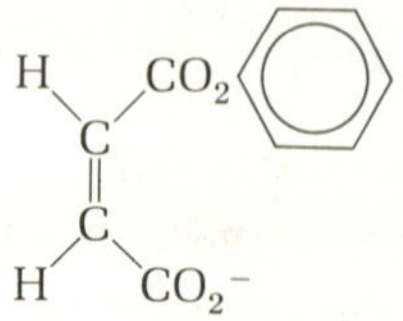
III

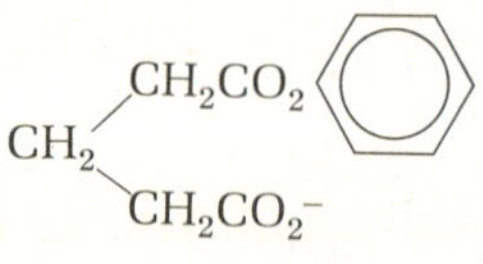
IV

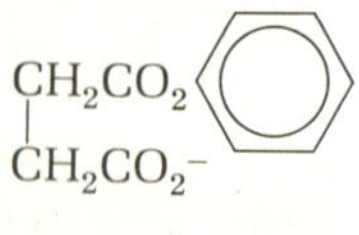
V

This mechanism may be significant in enzyme-catalyzed ester hydrolyses.

21 Draw the structures of all unknowns.

a. 2,6-Dibromo-4-methoxytoluene $\xrightarrow{\text{CuCN}}$ **A** ($C_{10}H_8N_2O$) $\xrightarrow{\text{OH}^-}$ $\xrightarrow{\text{H}^+}$ **B** $\xrightarrow[\text{H}^+]{\text{CH}_3\text{OH}}$

C $\xrightarrow[\text{2. H}_3\text{O}^+]{\text{1. LiAlH}_4}$ **D** ($C_{10}H_{14}O_3$) $\xrightarrow{\text{PBr}_3}$ **E** $\xrightarrow[\text{DMSO}]{\text{NaCN}}$ **F** $\xrightarrow[\text{2. H}^+]{\text{1. OH}^-}$ **G** ($C_{12}H_{14}O_5$)

b. **C** $\xrightarrow[\text{CCl}_4]{N\text{-bromosuccinimide}}$ **H** ($C_{12}H_{13}BrO_5$) $\xrightarrow[\text{DMSO}]{\text{NaCN}}$ **I** $\xrightarrow[\text{OH}^-]{\text{H}_2\text{O}_2}$ **J** $\xrightarrow{\text{BH}_3}$

K ($C_{13}H_{17}NO_5$) $\xrightarrow{\text{heat}}$ **L** ($C_{12}H_{13}NO_4$)

22 Mesitoic acid (2,4,6-trimethylbenzoic acid) is esterified only very slowly under the usual conditions (excess alcohol and a trace of mineral acid). If mesitoic acid is dissolved in concentrated sulfuric acid and this solution poured into an excess of cold methanol, methyl mesitoate may be recovered in high yield. Why is mesitoic acid so unreactive under the usual conditions? What is the mechanism of the sulfuric acid-catalyzed esterification (see Chapter 16)?

23 a. One mole of ammonia is allowed to react with 1 mole of *O*-methyl-*S*-ethyl thiocarbonate (VI). What is the structure of the product? Explain.

$$CH_3O\overset{\overset{\displaystyle O}{\|}}{C}SCH_2CH_3$$

VI

b. It requires high temperatures to convert a mixture of acetic acid and aniline to *N*-phenylacetamide. Thiolacetic acid (VII), on the other hand, acylates aniline at room temperature.

$$CH_3\overset{\overset{\displaystyle O}{\|}}{C}SH$$

VII

What factors account for this reactivity difference?

SUPPLEMENTARY READINGS

C. D. Gutsche, *The Chemistry of Carbonyl Compounds*. Englewood Cliffs, N. J.: Prentice-Hall, Inc., 1967.

S. Patai, Ed., *The Chemistry of Carboxylic Acids and Esters*. New York: Interscience Publishers, 1969, Chapters 3, 9 and 11.

S. Patai, Ed., *The Chemistry of Amides*. New York: Interscience Publishers, 1970, Chapters 2, 3 and 13.

CHAPTER 18

AMINES AND SULFIDES — BASICITY AND NUCLEOPHILICITY

NOMENCLATURE

As noted several times in the preceding chapters, **amines** (organic derivatives of ammonia) are both bases and nucleophiles. Our aim in this chapter is to inquire more deeply into the origin and the exploitation of these properties.

We shall be dealing with two valence states of nitrogen, N(III) and N(V), as exemplified by amines and ammonium salts, respectively. In naming these substances one enumerates the organic groups attached to the nitrogen atom, and as many hydrogen atoms as are necessary to fulfill valence requirements are also assumed to be attached to that nitrogen atom. The nomenclature is best illustrated by way of some examples. Amines that contain only one organic group bonded to the nitrogen are referred to as *primary amines,* those with two organic groups are *secondary amines,* etc. The terms primary, secondary, and tertiary, as applied to amines, have no bearing on the structure of the *organic groups.*

$CH_3CH_2NH_2$ — ethylamine

$\underset{\displaystyle CH_2CH_2CH_2CH_3}{CH_3NCH_2CH(CH_3)_2}$ — methylbutylisobutylamine

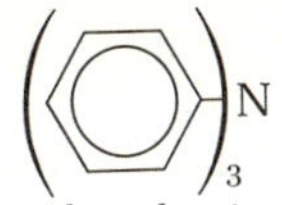

triphenylamine

Amines that carry one phenyl substituent on nitrogen are usually named as derivatives of aniline.

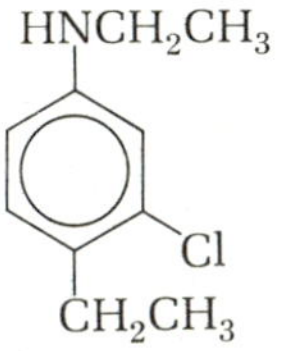

3-chloro-*N*,
4-diethylaniline

$CH_3CH_2\overset{\displaystyle CH_3}{N}H_2^+\ BF_4^-$

methylethylammonium
fluoborate

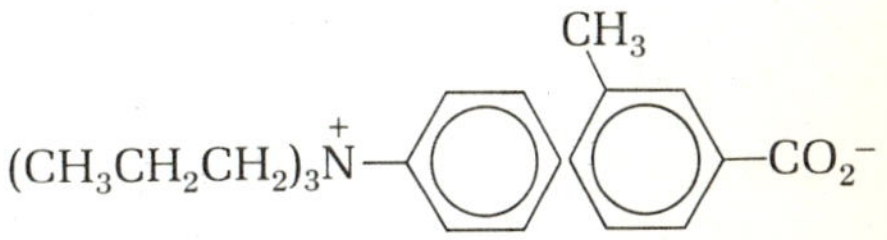

N,*N*,*N*-tripropylanilinium
3-methylbenzoate

If the nitrogen atom is attached to a nonterminal carbon atom of a chain, then the naming depends on whether the amine is primary or not. If it is primary, then the amine is named in the same manner as alcohols. Otherwise, it is named as described above, with substituents on the longest alkyl chains denoted by name and number.

$CH_3\overset{\displaystyle CH_3}{C}HCH_2\underset{\displaystyle NH_2}{C}HCH_2CH_3$

5-methyl-3-hexanamine

$CH_3CH_2CH_2\underset{\displaystyle NHCH_3}{C}HCH(CH_3)_2$

(1-isopropylbutyl)methylamine

Many cyclic amines are universally known by their nonsystematic names.

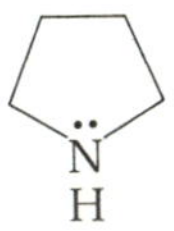

pyrrolidine

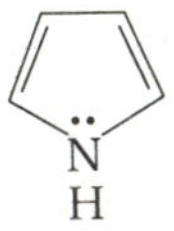

pyrrole

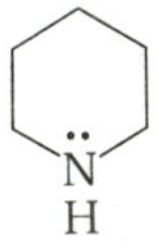

piperidine

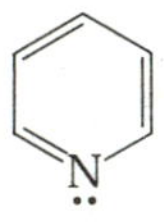

pyridine

In cases where a cyclic amine does not have a nonsystematic name, the prefix "aza" indicates that —NH— has been substituted for $—CH_2—$ in a hydrocarbon ring.

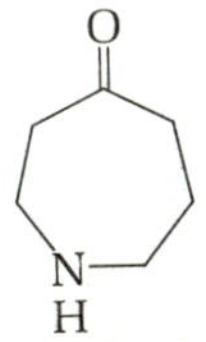

4-azacycloheptanone

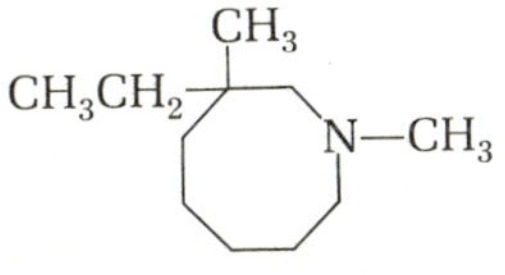

1,3-dimethyl-3-ethylazacyclooctane

$CH_3(CH_2)_3NH(CH_2)_4CONH_2$

6-azadecanamide

The aza nomenclature may also be used with complex acyclic amines.

Like ammonia, from which they are formally derived, amines are weak bases. In order to compare the basicity of different amines, it is useful to define an equilibrium constant, K_b, according to equation 18.1 (B stands for the amine).

$$(18.1) \quad K_b = \frac{(BH^+)(OH^-)}{(B)} \qquad B + H_2O \rightleftharpoons BH^+ + OH^-$$

We can then use the pK_b in a manner exactly analogous to use of the pK_a. The smaller the value of the pK_b, the more basic the amine.

Ammonia itself has a pK_b of 4.75 at 25°. Replacing the hydrogen atoms of ammonia by electron-releasing alkyl groups increases the basicity of the amine; the alkyl groups undoubtedly tend to stabilize the ammonium ion.

$CH_3CH_2NH_2$	$(CH_3CH_2)_2NH$	$(CH_3CH_2)_3N$
pK_b 3.35	pK_b 3.06	pK_b 3.35

Solvation effects are very significant, however, as shown by the fact that addition of the third alkyl group leads to a decrease in basicity.

The amine basicity may be altered by substituents in the organic groups, in a predictable manner. The markedly lower basicity of aniline, compared to typical aliphatic amines, is not due to only the greater electronegativity of sp^2-hybridized carbon atoms compared with sp^3-hybridized ones.

$CF_3CH_2NH_2$

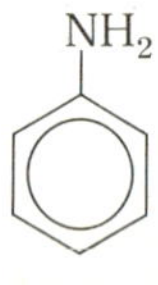

pK_b 5.70 $\qquad$ pK_b 9.40

Conjugation of the nonbonded nitrogen electrons with the π-system of the ring lowers the energy of these electrons and makes them less available to an attacking proton.

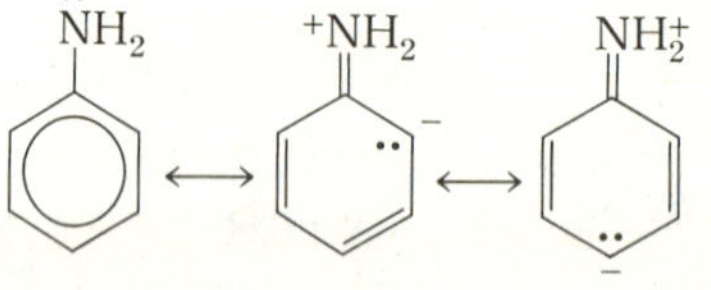

The delocalization of the nitrogen lone pair also means that the basicity of anilines will be very sensitive to substitution in the ring. Indeed, the basicity of substituted anilines may be correlated by means of a Hammett plot, the pK_b's ranging from 14.3 (*o*-nitroaniline) to 8.3 (*p*-aminophenol).

PROBLEM

1 a. The difference in acidity between acetic and benzoic acids is less than a factor of 4, while aniline is less basic than methylamine by a factor of 10^6. Why does the substitution of an aromatic substituent have such a different impact in the two systems?

b. *N,N,N'*-Trimethylnaphthalene-1,8-diamine (I) is a "normal" aromatic amine base, with a pK_b of 7.57. Addition of a fourth methyl group attached to nitrogen (II) has a dramatic effect on the pK_b — it drops to 1.66.

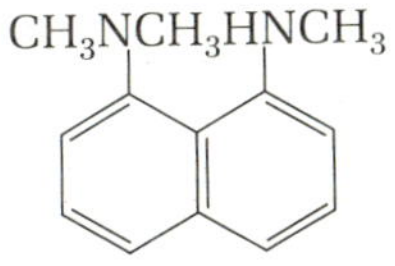

I

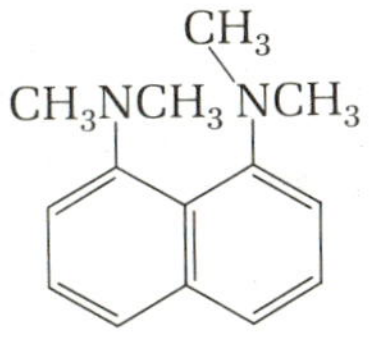

II

What is the mechanism by which the addition of the last methyl group so markedly alters the pK_b?

NUCLEOPHILIC PROPERTIES OF AMINES

As previous examples have shown (see particularly Chapters 7, 8, and 17), amines are effective as nucleophiles, attacking sp^2-hybridized and sp^3-hybridized carbon atoms. When the nitrogen atom acquires a fourth ligand, it becomes positively charged and it also loses its nucleophilic power.

(18.2) $(CH_3)_3N\colon + CH_3(CH_2)_3Br \longrightarrow (CH_3)_3N(CH_2)_3CH_3{}^+Br^-$

Whether or not further reaction is possible depends on whether or not the nitrogen atom has protons bonded to it.

Hydrogen atoms attached to positively charged nitrogen atoms are very rapidly detached from the nitrogen atom by weak bases such as amines or water. Ammonium ion, for instance, exchanges its protons very rapidly with those of heavy water (Equation 18.3) by way of a series of very rapid deprotonations and reprotonations (Equation 18.4).

(18.3) $NH_4{}^+ + 4D_2O \xrightarrow{\text{fast}} ND_4{}^+ + 4HDO$

(18.4) $NH_4{}^+ + D_2O \rightleftharpoons NH_3 + HD_2O^+ \rightleftharpoons NH_3D^+ + HDO$

Amines and ammonium ions may also exchange protons rapidly among themselves.

(18.5) $CH_3NH_3{}^+ + CH_3CH_2NH_2 \rightleftharpoons CH_3NH_2 + CH_3CH_2NH_3{}^+$

The most straightforward method of synthesizing amines would appear to be the one in which ammonia is alkylated with an alkyl halide or alkyl tosylate. Such reactions invariably produce mixtures of primary, secondary, and tertiary amines, as well as some quaternary ammonium salt.

$$CH_3CH_2Br + NH_3 \xrightarrow{CH_3CH_2OH} CH_3CH_2NH_2 + (CH_3CH_2)_2NH + (CH_3CH_2)_3N + (CH_3CH_2)_4N^+Br^- \quad (18.6)$$

This problem arises because the initially formed alkylammonium ion is deprotonated to form an alkylamine, which may be further alkylated. In contrast to the rapid transfer of protons, alkyl groups are removed only very slowly from ammonium ions.

$$RX + NH_3 \longrightarrow RNH_3^+\ X^- \quad (18.7)$$

$$RNH_3^+\ X^- + NH_3 \rightleftharpoons RNH_2 + NH_4^+\ X^- \quad (18.8)$$

$$RNH_2 + RX \longrightarrow R_2NH_2^+\ X^-, \text{ etc.} \quad (18.9)$$

The nucleophilicity of amines is usually increased when alkyl groups are substituted for hydrogen atoms because of the electron-donating character of the alkyl groups. This fact tends to increase the likelihood of polysubstitution at nitrogen. If the primary amine is the desired product, its yield may be increased by increasing the concentration of ammonia, although it is usually not possible to avoid some secondary and tertiary amine formation.

Reactions 18.2, 18.7, and 18.9 are all typical S_N2 processes; consequently, they work quite well with primary or secondary halides but not with tertiary halides. Furthermore, one cannot synthesize aromatic amines by the displacement reaction unless the aromatic halide is activated by the presence of electron-withdrawing substituents of the ring.

SYNTHESIS OF PRIMARY AND SECONDARY AMINES

Direct alkylation of ammonia or of an amine leads to mixtures of products because the nucleophilic $—NR_2$ group is susceptible to further alkylation (cf. the problem of multiple alkylation during Friedel-Crafts alkylations, Chapter 16). This problem may be met if the nitrogen atom is introduced in a considerably less nucleophilic state and the resulting functional group is then transformed into an amino group in a *subsequent step,* in the absence of alkylating agents. In practice this often means that the nitrogen atom is effectively present in a higher oxidation state when initially attached to the molecule and the second step is therefore a reduction.

A typical example of this approach involves the displacement of halide ion from an alkyl halide by azide ion, followed by reduction of the alkyl azide to a primary amine. Tosylates may also be used as substrates. The displacement is an S_N2 process and proceeds with inversion.

$$\text{(18.10)}\quad CH_3\overset{\overset{Br}{|}}{\underset{\underset{H}{|}}{C}}CO_2H + NaN_3 \xrightarrow{H_2O} CH_3\overset{\overset{H}{|}}{\underset{\underset{N_3}{|}}{C}}CO_2H \xrightarrow[Pt]{H_2} CH_3\overset{\overset{H}{|}}{\underset{\underset{NH_2}{|}}{C}}CO_2H$$

(18.11) 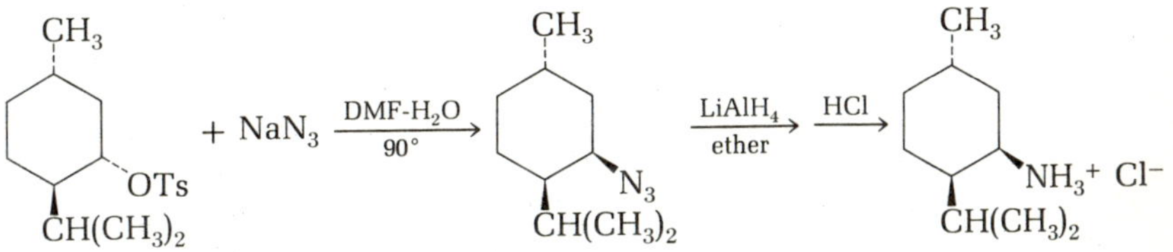

Reduction may be accomplished with lithium aluminum hydride, diborane, or hydrogen and a catalyst.

Related mechanistically to this azide path is the cyanide path, in which a nitrile is reduced to an amine; only primary alkyl primary amines are available by this route.

(18.12) 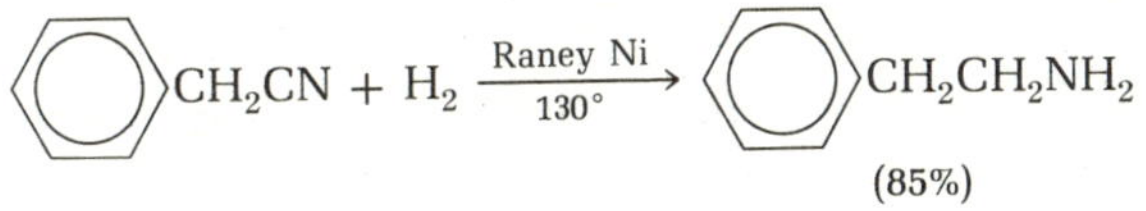

$$\text{(18.13)}\quad CCl_3CN + LiAlH_4 \xrightarrow{ether} \xrightarrow{H_2O} \underset{(61\%)}{CCl_3CH_2NH_2}$$

Nitriles may be synthesized by attack of cyanide ion on primary and secondary alkyl halides and tosylates (Chapter 7) or by treating aryl halides with cuprous cyanide (Chapter 17).

Another primary amine synthesis that revolves around a nucleophilic displacement is the **Gabriel synthesis.** Phthalimide (**1**) is too weak a nucleophile to react with alkyl halides at reasonable rates under ordinary conditions; the nucleophilicity of the nitrogen atom is severely reduced by the electron-withdrawing carbonyl groups.

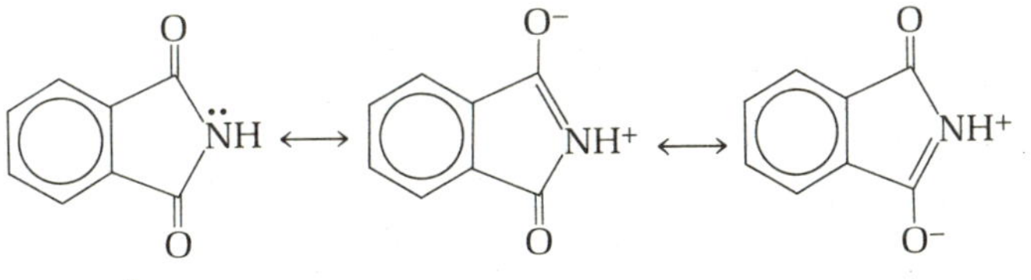

1

In fact, phthalimide is a weak acid ($pK_a = 8.3$) and is easily converted to its alkali metal salts. The phthalimide anion (**2**) is reasonably nucleophilic and may be alkylated by primary (and a few secondary) alkyl halides to give *N*-alkylphthalimides (equations 18.15 and 18.16).

(18.14)

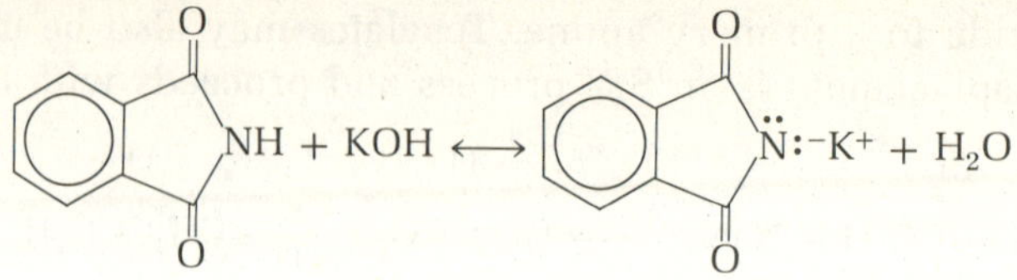

1 2

(18.15)

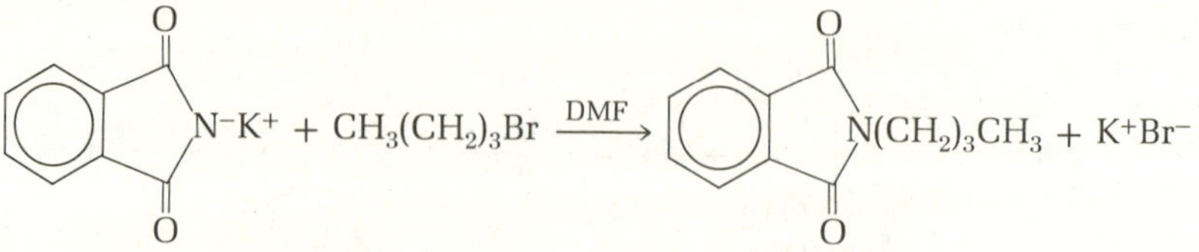

2

(18.16)

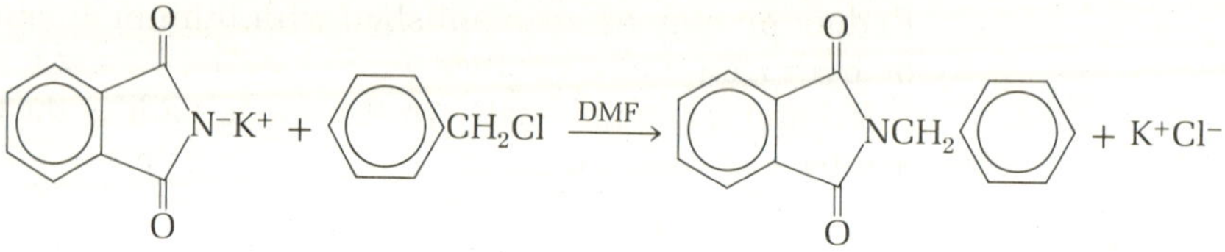

2

The *N*-alkylphthalimides are nothing more than double amides and it is necessary only to break the CO—N amide bond to recover the desired amine. Both acidic and basic hydrolyses have been used, although the conditions are often severe because the cyclic imide is quite stable. The mildest method available, which is widely used, involves the reaction between the imide and hydrazine; the effectiveness of hydrazine is due to the fact that it contains two nucleophilic $—NH_2$ groups.

(18.17)

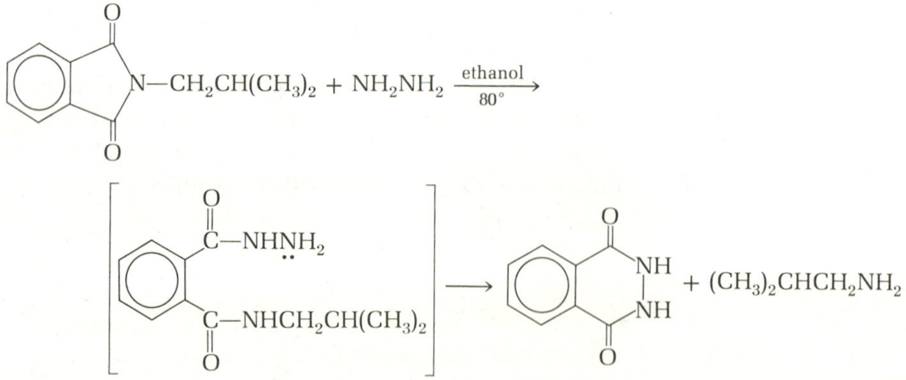

Finally, there is a method available for synthesizing primary amines that is very promising because it permits the attachment of a wide variety of alkyl groups to the nitrogen atom. It depends on the reaction of an alkylborane with hydroxylamine-*O*-sulfonic acid and retains all the characteristics of borane reactions, particularly the stereospecificity.

(18.18) 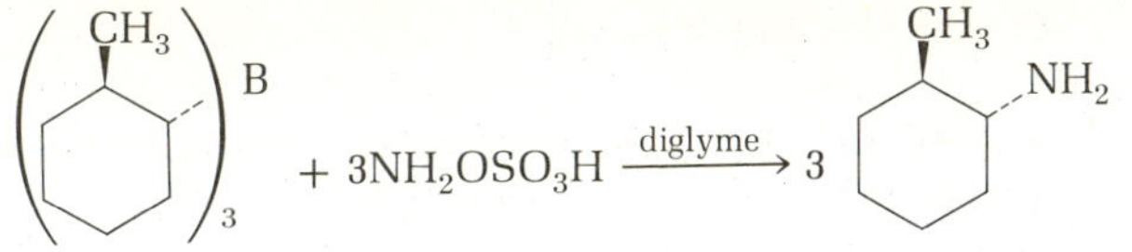

(50%)

Note that the reaction (equation 18.18) involves a nucleophilic displacement on *nitrogen*, the leaving group being bisulfate anion.

All the synthetic methods so far described lead to the formation of primary amines. There is not quite such a profusion of pathways for synthesizing secondary amines. The principal technique is one in which an imine, the condensation product of a carbonyl compound and a primary amine, is subjected to catalytic reduction.

(18.19)
$$\mathrm{RR'C{=}NR + H_2 \xrightarrow{\text{catalyst}} RR'CH{-}NHR}$$

One of the alkyl groups in the final product must be primary or secondary.

Imines are often quite unstable, particularly when the groups attached to the doubly bonded carbon and nitrogen atoms are aliphatic. The usual procedure involves mixing together the appropriate amine and carbonyl compound in the presence of the hydrogenation catalyst and then exposing the reaction mixture to hydrogen gas.

(18.20)
$$\mathrm{(CH_3)_2C{=}O + H_2NCH_2CH_2OH \rightleftharpoons H_2O + (CH_3)_2C{=}NCH_2CH_2OH \xrightarrow[C_2H_5OH]{H_2,\ Pt} (CH_3)_2CHNHCH_2CH_2OH}$$

(18.21) 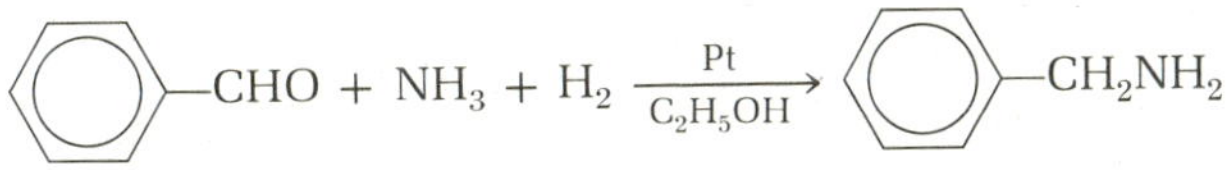

The reader is reminded that an excellent general method for amine synthesis, particularly secondary and tertiary amines, revolves around the reduction of amides with lithium aluminum hydride or diborane (Chapter 17).

(18.22) 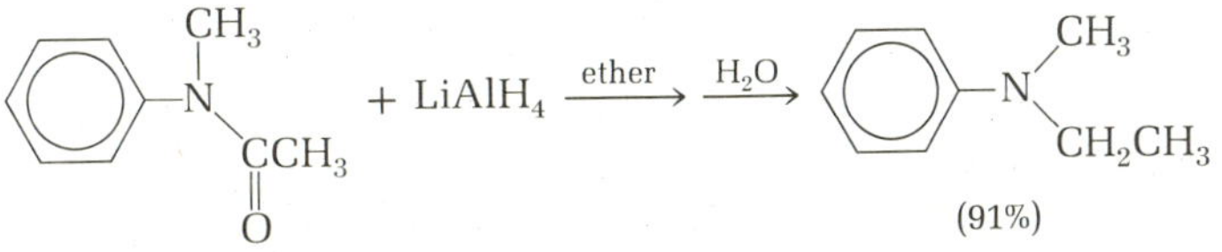

(91%)

(18.23)
$$\mathrm{(CH_3)_3C\overset{O}{\overset{\|}{C}}NHCH_3 + B_2H_6 \xrightarrow[65^\circ]{THF} \xrightarrow{H_2O} (CH_3)_3CCH_2NHCH_3}$$

(83%)

All amine syntheses that depend on an S_N2 substitution at carbon by a nitrogen-containing nucleophile are inapplicable if the carbon atom is a tertiary one. On the other hand, tertiary carbonium ions are very readily generated and one may expect that they would readily attach themselves to any available nonbonding electrons, such as those on a nitrogen atom. This approach has one hitch, however. Carbonium ions are usually present in significant concentration only in highly acidic media; in such media, most nitrogen atoms will be protonated and therefore completely nonnucleophilic. The nitrogen atoms of nitriles, however, seem to remain at least partially unprotonated in concentrated sulfuric acid.

The *Ritter reaction* involves the interaction of a carbonium ion source and hydrogen cyanide in sulfuric acid (equation 18.24). The immediate product formed upon hydrolysis of the reaction mixture is a formamide, but the amide may be further hydrolyzed to formic acid and an amine.

(18.24)

$$\text{ROH} \xrightarrow{\text{H}^+} \text{R}^+ + \text{HC}{\equiv}\text{N:} \longrightarrow \text{HC}{\equiv}\overset{+}{\text{N}}\text{—R} \xrightarrow{\text{HSO}_4^-} \text{H}\overset{\overset{\displaystyle\text{OSO}_3\text{H}}{|}}{\text{C}}{=}\text{NR} \xrightarrow{\text{H}_2\text{O}}$$

$$\text{H}\overset{\overset{\displaystyle\text{O}}{\|}}{\text{C}}\text{NHR} \xrightarrow{\text{OH}^-} \text{HCO}_2^- + \text{RNH}_2$$

Thus, any alcohol that gives rise to a stable carbonium ion, such as a tertiary or benzylic alcohol, can be converted to the corresponding amine. Alkenes may also serve as sources of carbonium ions.

(18.25)

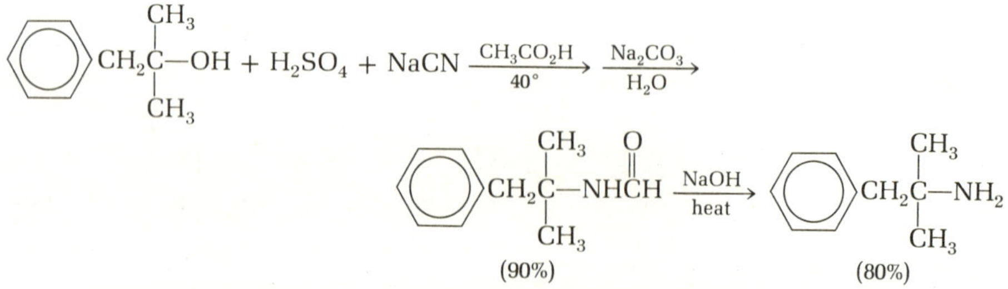

PROBLEM

2 The success of the Ritter reaction depends on the fact that $HC{\equiv}N$: is not completely converted to $HC{\equiv}NH^+$ by sulfuric acid. This implies that hydrogen cyanide and nitriles are considerably less basic than ammonia and amines, a supposition amply verified by other studies. What is the source of the decreased basicity of nitriles? (What is the hybridization state of nitrogen in hydrogen cyanide?)

A nitrogen atom usually cannot be attached to an aromatic ring by means of a nucleophilic substitution reaction, except in those special cases in which the ring is activated toward nucleophilic attack by the presence of electron-withdrawing substituents (Chapter 22). It is quite easy, however, to form a ring-nitrogen bond by way of an electrophilic aromatic substitution, when nitronium ion is the electrophile (nitration, Chapter 15). Since the nitro group is readily reducible to an amino group, the sequence aromatic ⟶ nitroaromatic ⟶ aminoaromatic is the prime route for aromatic amine production.

A very mild method of reduction is the catalytic one, which usually gives high yields and does not disturb most other functional groups.

(18.26) 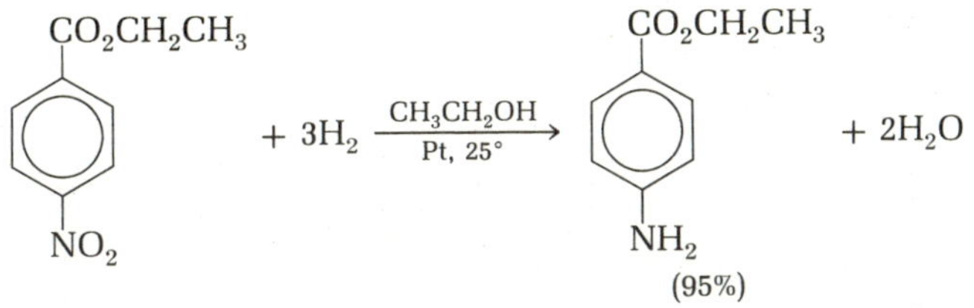

Metals such as zinc, iron, and tin may also be used to reduce the nitro group, usually in the presence of acid.

(18.27) 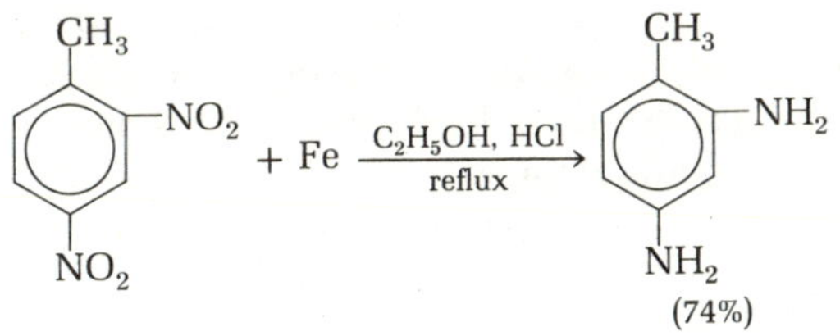

Although the conditions of reaction 18.27 are more severe than those of reaction 18.26, the latter has the advantage of requiring a less sophisticated apparatus and of being less expensive. Indeed, the commercial production of aniline from nitrobenzene utilizes scrap iron as the reductant. It should be noted that the reduction of nitro groups by metals may be stopped at intermediate valence states of nitrogen, depending on the pH of the medium.

Since the nitro group is a powerful *m*-director, while the amino group is a strong *o,p*-director, different positional isomers of substituted anilines may be obtained, depending on whether reduction precedes or follows the introduction of the substituent.

(18.28) 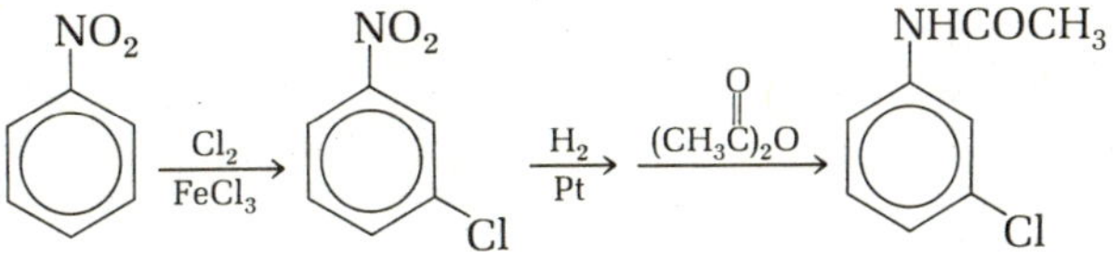

(18.29) 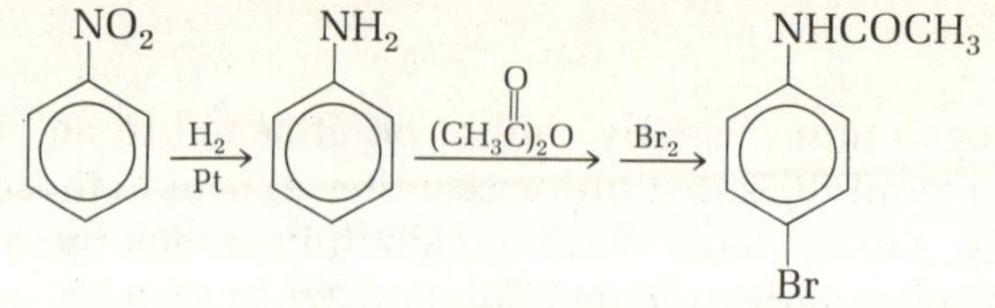

AMINE BASICITY AND NUCLEOPHILICITY

Since the behavior of amines as bases and as nucleophiles involves the donation of electrons from nitrogen to an electrophile, we would expect some parallel between amine basicity and nucleophilicity. Such a correlation is indeed often observed (equations 18.30 and 18.31).

(18.30) $$(CH_3CH_2)_3N\!: + CH_3CH_2I \xrightarrow{\text{acetone}} (CH_3CH_2)_4N^+I^- \qquad k = 6.77 \times 10^{-5}\ \text{l/mole sec}$$
pK_b 3.24

(18.31) 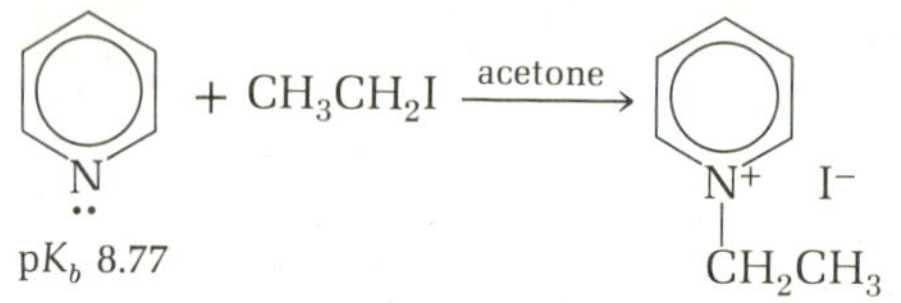
$k = 9.26 \times 10^{-6}$ l/mole sec

But that relationship is far from a straightforward one because the reactions are very different in their sensitivity to steric effects.

The proton is very small and its attachment to the nitrogen atom is fairly insensitive to the spatial environment. Alkyl groups are much larger, however, and the greater their bulk, the more sensitive will be the rate of the alkylation reaction to the structure of the amine. To take two extreme examples, both 1,2,2,6,6-pentamethylpiperidine (**3**) and 2,6-lutidine (2,6-dimethylpyridine, **4**) are stronger bases than their unmethylated precursors, but they are much inferior as nucleophiles.

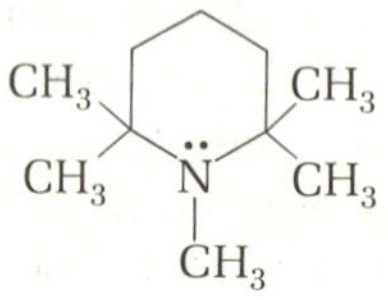
3

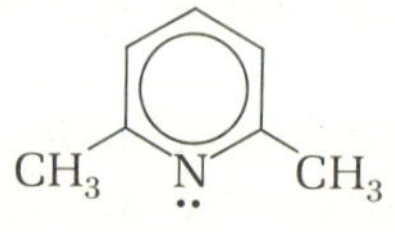
4

Advantage has been taken of this steric effect and hindered amines have been used as bases during the alkylation of *less* basic amines.

(18.32) 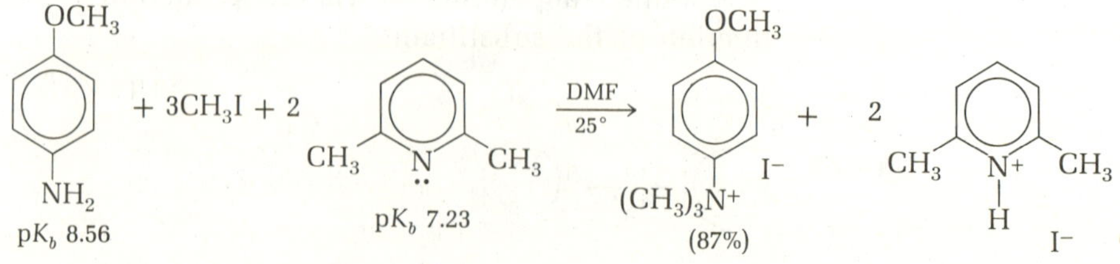

Synthesis of primary amines containing one more carbon atom than the starting material is possible with the cyanide as an intermediate. It is also possible to chop one carbon atom out of the chain if one starts with an amide. This technique involves treating the amide with an alkaline sodium hypobromite (or hypochlorite) solution and is known as the **Hofmann amide degradation** (not to be confused with the Hofmann elimination in ammonium salts).

(18.33) $$CH_3(CH_2)_4CONH_2 + OBr^- + OH^- \longrightarrow \underset{(88\%)}{CH_3(CH_2)_4NH_2} + Br^- + HCO_3^-$$

(18.34) 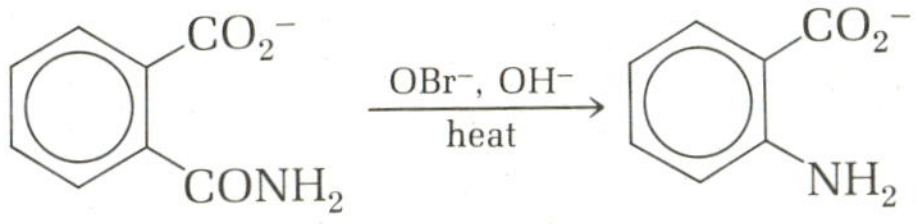

(85%)

The reaction is synthetically useful and has also been used for structure elucidation, in common with many degradative reactions.

The reaction course encompasses bromination of the amide on nitrogen and then expulsion of the bromide ion accompanied by a migration of the group attached to the carbonyl carbon atom.

(18.35) $$RC(=O)NH_2 + OBr^- \rightleftharpoons RC(=O)\ddot{N}H^- + BrOH \rightleftharpoons RC(=O)NHBr + OH^-$$

(18.36) 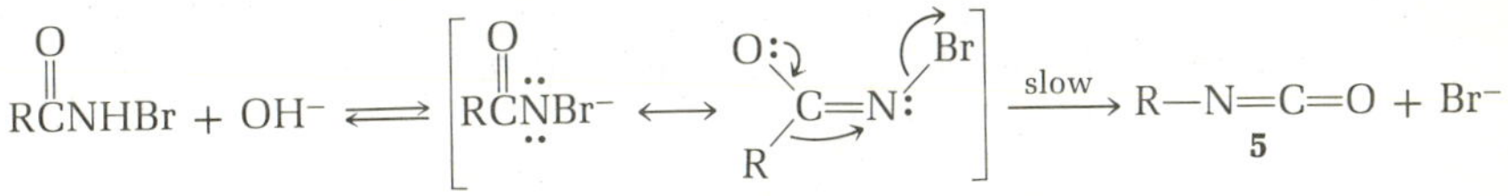

(18.37) $$\underset{\mathbf{5}}{R{-}N{=}C{=}O} \xrightarrow{H_2O} \underset{\mathbf{6}}{[RNHCO_2H]} \longrightarrow RNH_2 + CO_2$$

An organic isocyanate (**5**) is an intermediate; in aqueous base this is very quickly converted to the unstable carbamic acid (**6**), which decomposes in turn and liberates the amine.

A functional and mechanistic close cousin of the Hofmann degradation is the **Schmidt reaction.** One starts with a carboxylic acid that is converted by hydrazoic acid to an acyl azide (**7**).

(18.38) $$CH_3(CH_2)_{16}CO_2H + HN_3 \xrightarrow[\text{benzene}]{H_2SO_4,\ H_2O} \underset{(46\%)}{CH_3(CH_2)_{16}NH_2} + N_2 + CO_2$$

(18.39) $$RCO_2H + HN_3 \longrightarrow H_2O + \left[RC(=O){-}\overset{-}{N}{-}\overset{+}{N}{\equiv}N \longrightarrow \underset{\mathbf{7}}{RC(=O){-}N{=}\overset{+}{N}{=}\overset{-}{N}}\right]$$

Upon protonation the acyl azide decomposes by losing molecular nitrogen, accompanied by migration of R from carbon to nitrogen.

(18.40) 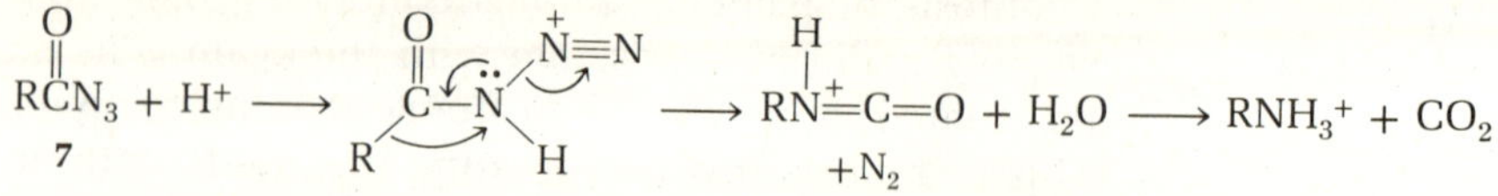

In both equations 18.36 and 18.40 it is the nitrogen lone pair electrons that facilitate the migration of R from nitrogen to carbon. The rearrangements described are seen to be very general for molecules with the proper structural features — in particular, a nitrogen atom carrying a good leaving group and bonded to a trigonal carbon atom (equation 18.41).

(18.41) 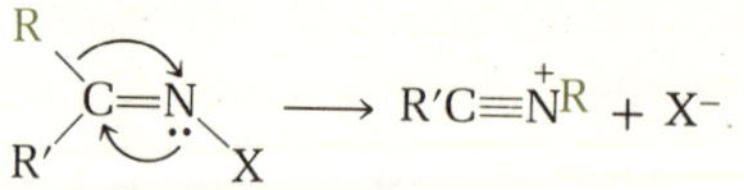

All these rearrangements seem to be concerted: if the carbon atom undergoing migration is asymmetric, its configuration will remain unaltered in the course of the rearrangement.

The mechanistic pattern illustrated here is, with many variations, quite widespread in organic chemistry. A case in point, which *appears* to be very different from the Schmidt and Hofmann rearrangements, is the oxidation of alkyl boranes to alcohols with basic hydrogen peroxide (Chapter 9). The critical step is depicted in equation 18.42; the similarity to equation 18.41 should be apparent.

(18.42) 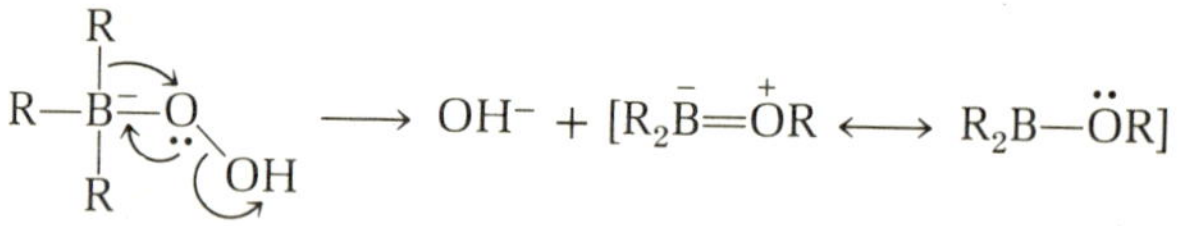

Loss of electronegative leaving groups from the electronegative N and O atoms seems to be quite facile compared with loss of the same groups from carbon.

PROBLEM

3 *The Beckmann rearrangement* involves the acid-catalyzed rearrangement of oximes to amides; an example is given in equation 1.

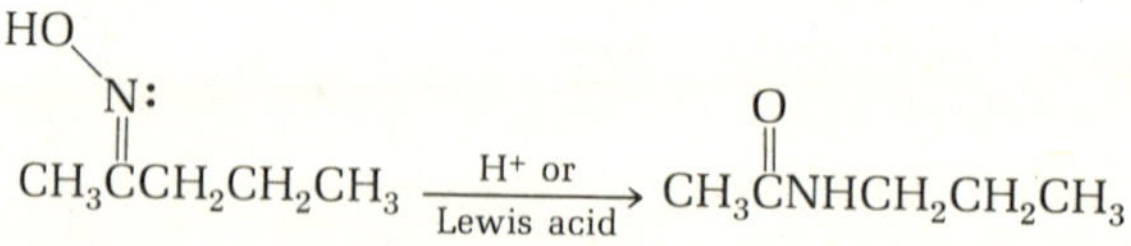 (1)

ϵ-Caprolactam, made commercially by the Beckmann rearrangement of cyclohexanone oxime, is polymerized to Nylon-6 (equation 2).

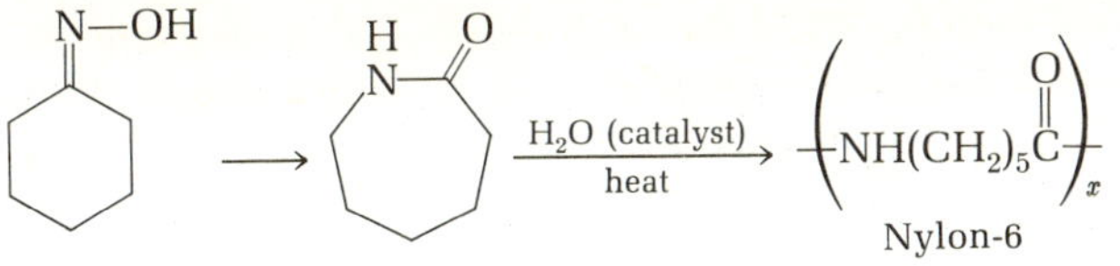

(2)

Propose a mechanism for the rearrangement.

NITROGEN INVERSION AND OPTICAL ACTIVITY

Ammonia is a pyramidal molecule that undergoes very rapid inversion. A tertiary amine carrying three different alkyl groups, such as methylethylpropylamine, has an asymmetric nitrogen atom and can exist in two enantiomeric forms. Inversion of the nitrogen lone pair converts one enantiomer into the other; since the energy barrier for inversion is quite low, it is normally not possible to isolate enantiomerically pure amines.

(18.43)

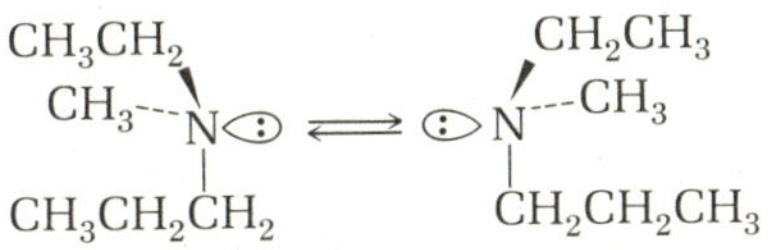

Under special circumstances one may slow or even completely inhibit nitrogen inversion. Firstly, if the nitrogen atom is incorporated into a three-membered ring (*aziridine*) and is bonded to an electronegative atom, it undergoes inversion quite slowly even at moderate temperatures (equation 18.44).

(18.44)

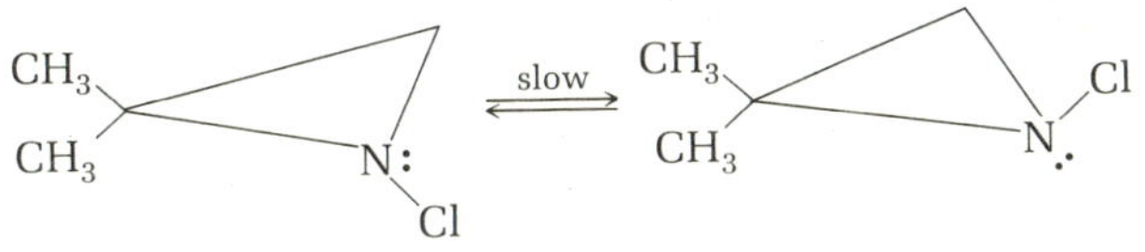

Such cases have been studied by nmr spectroscopy. Secondly, if the nitrogen atom is placed at the bridgehead position of a small polycyclic molecule, then it will no longer be capable of inverting. If that position should be asymmetric as well, then the enantiomers of the amine are in principle isolable.

When methylethylpropylamine is treated with hydrochloric acid and the salt then isolated, it consists of a racemic mixture of the two enantiomeric salts (equation 18.45). The tetracoordinate cations *cannot* undergo conversion.

(18.45)

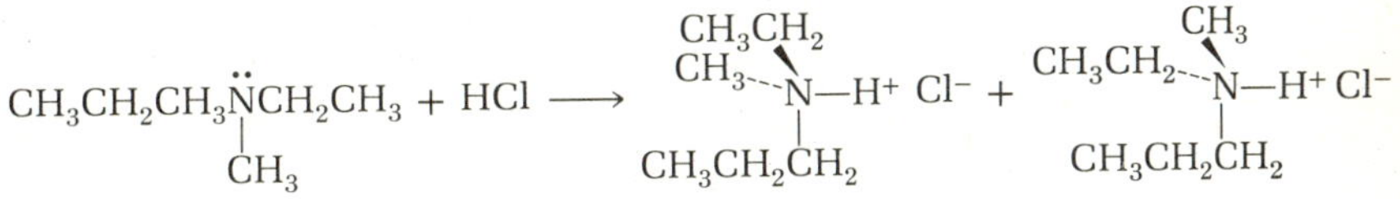

Attempts to separate these enantiomers would be very difficult, however, because if the cation dissolved in a medium in which it was able to dissociate, even for a brief period, inversion would take place and the ion would be racemized.

It is clear from the foregoing discussion that the best chance for isolating, in enantiomerically pure form, a compound containing an asymmetric nitrogen atom lies with substances containing four nondissociable groups attached to the nitrogen atom. In fact, both the *R* and *S* forms of methylallylphenylbenzylammonium iodide (**8**) were isolated at the turn of the century.

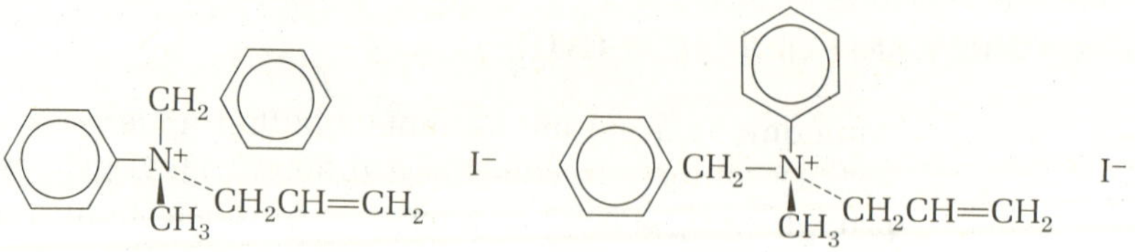

8

RESOLUTION OF ENANTIOMERS

The process of separating a racemic mixture of optical isomers into its enantiomeric components is known as **resolution.** Since enantiomers have identical physical properties, except insofar as their interaction with circularly polarized light is concerned, they cannot be separated by any purely physical process such as distillation or crystallization. The strategy is rather to have the mixture interact with another optically pure substance, such that a mixture of compounds (or complexes) is formed.

(18.46) $(+)\text{-A}, (-)\text{-A} + 2(+)\text{-B} \longrightarrow (+)\text{-A}\cdot(+)\text{-B} + (-)\text{-A}\cdot(+)\text{-B}$

These compounds (or complexes) are diastereomers since they are neither identical to nor mirror images of one another.

Diastereomers have different physical properties and are therefore theoretically capable of being separated by conventional methods. Thus, if A · B is either a molecular compound or a salt, then the diastereomers may be separated by means of fractional distillation or crystallization. However, A · B may only be a weak molecular complex. If a mixture of enantiomers is subjected to chromatography on a stationary phase (B) that is optically active, the enantiomers being chromatographed complex with the stationary phase but to different extents. The two enantiomers therefore travel at different rates along the stationary phase and may thereby be separated.

In order to resolve a mixture of enantiomers, one must have in hand an optically active resolving agent; thus, one must ultimately come to Nature. Among the first racemic mixtures resolved by man were those of acids because Nature had provided a number of optically active bases. These compounds were virtually all **alkaloids** (nitrogen-containing bases isolated from plants and usually physiologically active); brucine and quinine are alkaloids that have often been used as resolving agents.

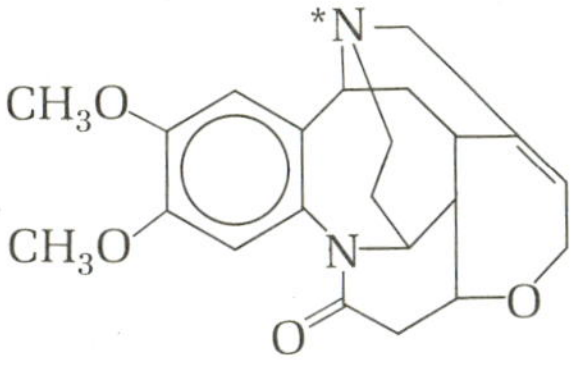

brucine

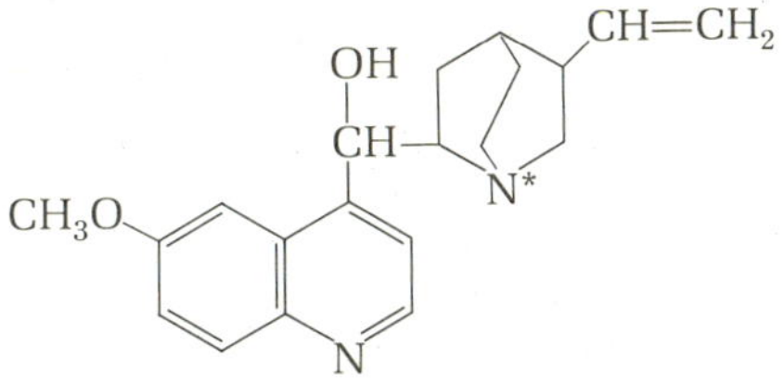

quinine

In both these compounds the amine nitrogen atom (marked by asterisks) is part of a polycyclic ring and is therefore incapable of inverting. As a result, each base gives only *one* ammonium salt upon protonation. Nature supplies these alkaloids in enantiomerically pure form.

A general scheme for resolving a racemic acid (such as 2-methylbutanoic acid) involves (1) salt formation with the optically pure base; (2) separation of the diastereomeric salts by fractional crystallization; (3) treatment of each salt separately with hydroxide ion, to free the organic base; and (4) addition of inorganic acid to the aqueous solution to precipitate the resolved organic acid.

(18.47)

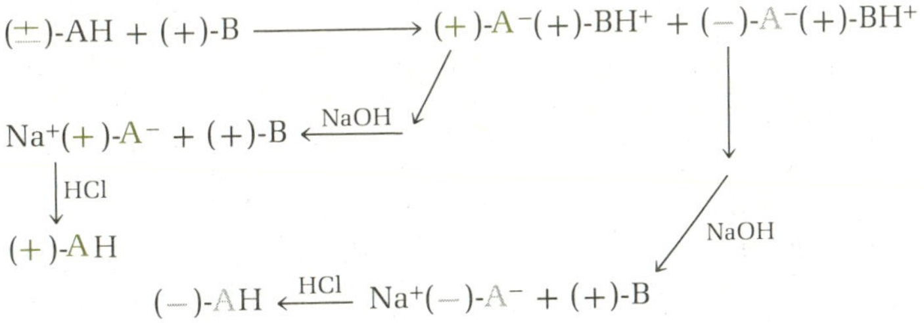

Once enantiomerically pure acids are available, they may be used in turn for the resolution of enantiomeric amines, etc. Nature also provides a few optically pure acids in plentiful quantities, the most widely used of which is (+)-tartaric acid (**9**).

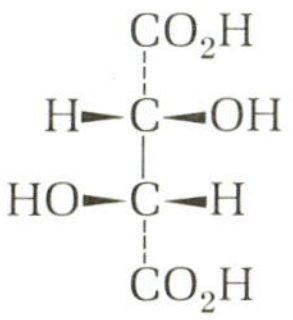

9

Resolution of a racemic alcohol, such as 2-butanol or *cis*-3-methylcyclohexanol, may be accomplished by esterifying the racemate with an optically pure acid.

(18.48) $(+)\text{-RCO}_2\text{H} + (\pm)\text{-R'OH} \longrightarrow (+)\text{-RCO}_2\text{R'-}(-) + (+)\text{-RCO}_2\text{R'-}(+) + \text{H}_2\text{O}$

A slightly more subtle method involves the reaction of the racemic alcohol with a symmetric diacid to form a monoester, the free acid group of which is subsequently neutralized with an optically active amine (equation 18.49). The resulting ester salt is a mixture of diastereomers, which may hopefully be separable by the standard techniques.

(18.49)

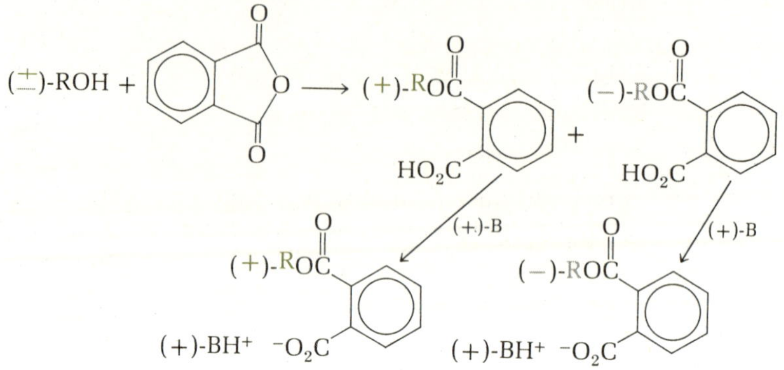

The method of equation 18.49 has an advantage over that of equation 18.48, in that salts are virtually always solids while esters are often liquids. It is usually easier to separate diastereoisomers by means of fractional crystallization than by means of fractional distillation.

INTRODUCTION TO ORGANOSULFUR COMPOUNDS

Sulfur is directly below oxygen in the periodic table and there are several classes of organic compounds that are the sulfur analogues of common oxygen-containing organic compounds, such as alcohols, ethers, and esters. In several respects the analogous oxygen and sulfur compounds exhibit very similar modes of behavior but their chemistry also discloses numerous differences that are theoretically significant and pragmatically useful.

Among the properties of sulfur atoms in organic molecules that are central to an understanding of their chemistry, three should be borne in mind particularly:

1 Sulfur is much less electronegative than oxygen is and somewhat less electronegative than nitrogen.

2 A sulfur atom has vacant 3*d*-orbitals.

3 Protons bound to sulfur are less tightly held than those bound to nitrogen or oxygen (of the same charge). A sulfide ion RS^- is more stable than an oxide ion RO^- since in the former the charge is dispersed throughout a larger volume.

Hydrogen sulfide is considerably more acidic than water: $pK_a^1 = 7.04$; $pK_a^2 = 11.96$. Addition of excess hydrogen sulfide to an aqueous solution of sodium hydroxide results in consumption of virtually all the hydroxide ion and concomittant formation of hydrogen sulfide ion.

(18.50) $$OH^- + H_2S \rightleftharpoons H_2O + SH^-$$

The hydrogen sulfide ion is a very good nucleophile compared with hydroxide ion, due to both its more polarizable electrons and its indifference to hydrogen bonding (sulfur is a poor hydrogen-bond acceptor; see Chapter 7, p. 136).

The hydrogen sulfide ion is allowed to react with primary or secondary alkyl halides to give the sulfur analogues of alcohols, the **thiols** (also known as mercaptans).

(18.51) $$CH_3(CH_2)_3Br + KOH + H_2S \text{ (excess)} \xrightarrow[\text{heat}]{C_2H_5OH} CH_3(CH_2)_3SH + KBr + H_2O$$

1-butanethiol (54%)
(butyl mercaptan)

(18.52) $$CH_3(CH_2)_5\underset{\displaystyle Br}{\underset{|}{C}}HCH_3 + K^+SH^- \xrightarrow{C_2H_5OH} CH_3(CH_2)_5\underset{\displaystyle SH}{\underset{|}{C}}HCH_3$$

2-octanethiol (66%)
(2-octyl mercaptan)

In the presence of base, thiols are converted to their sodium salts, which may be further alkylated to yield **sulfides,** the sulfur analogues of ethers.

(18.53) $$CH_3CH_2SH + CH_3(CH_2)_{15}Cl + CH_3CH_2O^-Na^+ \xrightarrow[H_2O]{C_2H_5OH} CH_3CH_2S(CH_2)_{15}CH_3$$

ethyl hexadecyl sulfide

(18.54) $$2CH_3S^-Na^+ + BrCH_2CH_2Br \xrightarrow[\text{heat}]{C_2H_5OH} CH_3SCH_2CH_2SCH_3$$

1,2-bis(methylthio)ethane (85%)

There is a very reliable two-step synthesis of thiols. An alkyl halide or tosylate first reacts with thiourea to form an isothiouronium salt; decomposition of the salt with aqueous base gives rise to the thiol.

(18.55) $$CH_3CH_2Br + H_2N\overset{\displaystyle S}{\overset{\|}{C}}NH_2 \xrightarrow{C_2H_5OH}$$

thiourea

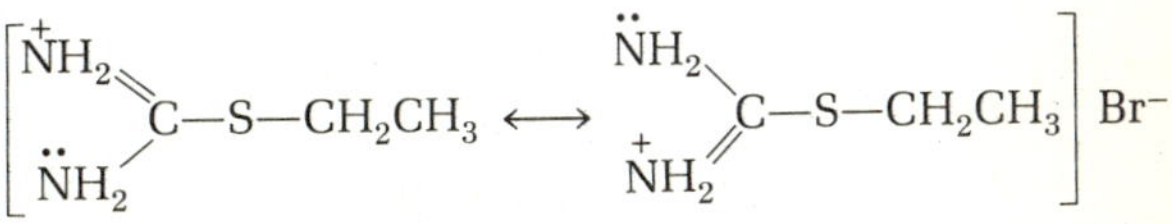

S-ethylisothiouronium bromide (95%)

$$\text{(18.56)}\quad \left[(NH_2)_2\overset{+}{C}SCH_2CH_3\right]Br^- \xrightarrow[H_2O]{NaOH} \underset{\text{urea}}{H_2NCNH_2} + \underset{(90\%)}{CH_3CH_2SH}$$

(In urea, C=O: $H_2N\overset{O}{\overset{\|}{C}}NH_2$.)

Crystalline isothiouronium salts are very useful as derivatives of alkyl halides (usually primary and secondary), as they are easily purified and have sharp melting points.

Sulfides themselves may act as nucleophiles, reacting with alkyl halides and tosylates to give rise to **sulfonium salts,** analogous to ammonium salts.

$$\text{(18.57)}\quad (CH_3)_2S + CH_3I \xrightarrow[25^\circ]{C_2H_5OH} (CH_3)_3S^+I^-$$

trimethylsulfonium iodide (99%)

$$\text{(18.58)}\quad (CH_3)_2S + BrCH_2CO_2CH_2CH_3 \xrightarrow{\text{acetone}} (CH_3)_2\overset{+}{S}CH_2CO_2CH_2CH_3\quad Br^-$$

carbethoxymethyldimethylsulfonium bromide (90%)

Sulfonium salts are electronically analogous to amines since the sulfur atoms contain three bonded electron pairs and one nonbonded electron pair in their valence shells. Furthermore, the four electron pairs are approximately tetrahedrally disposed. Unlike amines, however, sulfonium salts do not undergo inversion and enantiomers such as **10** are optically stable.

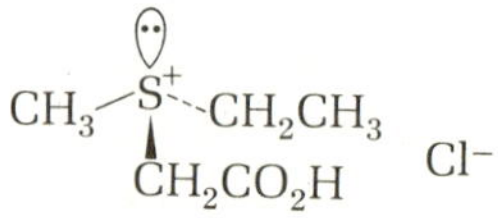

10

Sulfides are much more nucleophilic than oxides (ethers), which is to be expected since oxygen is considerably more electronegative than sulfur. While sulfonium salts are quite stable and easily synthesized by use of mild alkylating agents such as alkyl iodides, trialkyloxonium salts are extremely reactive and can only be synthesized by special methods. (Sulfide anions, RS^-, are also about 10^3 times more nucleophilic than are oxide anions, RO^-.) Sulfides are comparable in nucleophilicity to amines and sulfonium salts are in many ways reminiscent of ammonium salts; for example, they undergo Hofmann elimination.

$$\text{(18.59)}\quad \underset{(CH_3)_2S^+\ Br^-}{CH_3CH_2\underset{|}{C}HCH_3} \xrightarrow[C_2H_5OH]{C_2H_5ONa} \underset{(26\%)}{CH_3CH{=}CHCH_3} + \underset{(74\%)}{CH_3CH_2CH{=}CH_2} + (CH_3)_2S{:}$$

BASICITY OF SULFIDES

While sulfides and thiols are fairly good nucleophiles (they rapidly bond to carbon), they are not very good bases (low affinity for protons). What is being compared here is a *kinetic* property (nucleophilicity) to a *thermodynamic* property (basicity).

As some of the following examples show, factors that promote one of these properties do not necessarily promote the other.

Equation 18.60 illustrates graphically the difference in basicity between oxygen and sulfur.

(18.60) 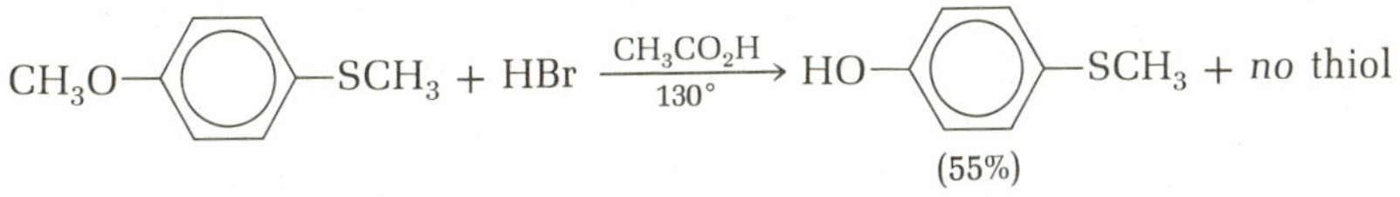

The first step in the ether cleavage involves equilibrium protonation of the heteroatom (equation 18.61); the extent of protonation is apparently substantial when X = O but negligible when X = S.

(18.61) 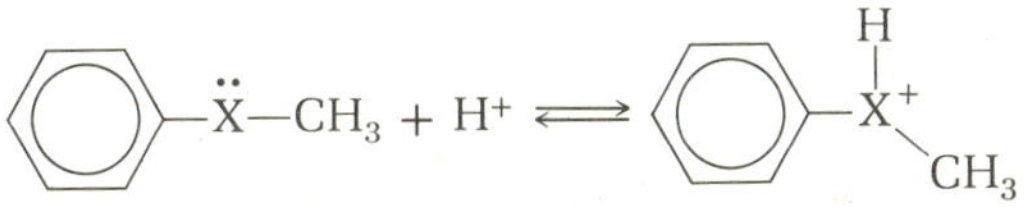

Sulfur also exhibits relatively little tendency to form complexes with those Lewis acids that are strongly attached to oxygen; these include salts of the lighter metals such as magnesium and aluminum. On the other hand, sulfides are very susceptible to strong complex formation with the salts of heavy metals such as lead and mercury.

(18.62) 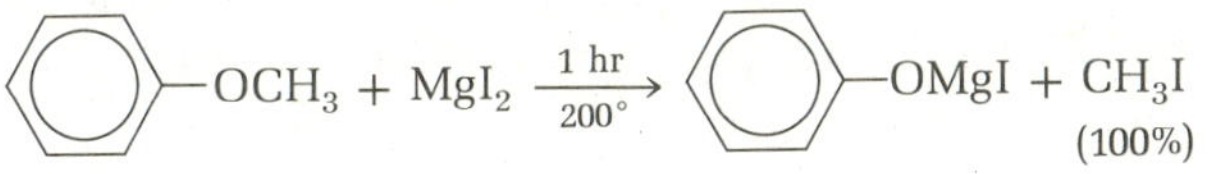

(18.63) 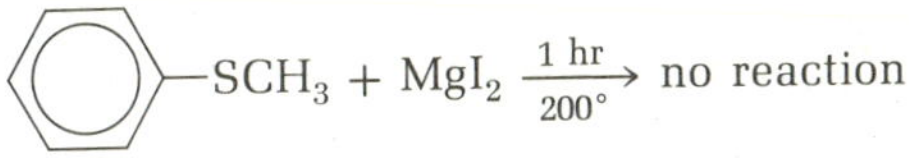

Sulfur is a nucleophilic atom because its valence electrons are highly polarizable and not tied up by hydrogen bonds in hydroxylic solvents. These attributes do not mean, however, that the bonds formed by these electrons will necessarily be *strong* ones. Thus, although amines and sulfides are comparable in nucleophilicity, the C—N bond is distinctly stronger than the C—S bond, as shown by the fact that the equilibrium depicted in equation 18.64 lies almost completely to the right.

(18.64) $$(CH_3)_3N + (CH_3)_3S^+NO_3^- \underset{60°}{\overset{C_2H_5OH}{\rightleftharpoons}} (CH_3)_4N^+NO_3^- + (CH_3)_2S$$

SULFUR AS π-ELECTRON DONOR

Divalent sulfur readily forms a third σ-bond to produce a sulfonium salt; the nonbonded electron pair is much less effective at forming π-bonds, however. This fact is a specific example of the generality put forward near the beginning of this book to the effect that strong π-bonds are generally only to be found between members of the second row (lithium through fluorine) of the periodic table.

If we accept this idea, then the relative unreactivity of thioanisole toward Friedel-Crafts acylation, when contrasted with the high reactivity of anisole itself, is immediately comprehensible. The activation of the anisole occurs by way of π-electron donation from the oxygen atom to the ring.

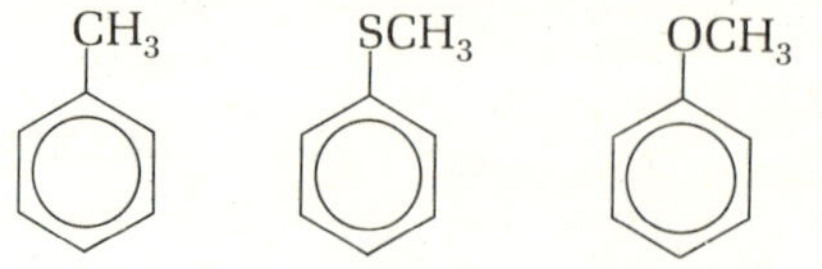

1 56 2.3×10^3

Friedel-Crafts acylation by acetyl chloride

SULFUR AS π-ELECTRON ACCEPTOR

Much evidence has accumulated that tends to indicate that sulfur atoms have a stabilizing influence on adjacent carbanionic centers. A representative example is provided by the relative hydrogen-deuterium exchange rates of a series of methyl groups bonded to various electronegative elements (the rates are per hydrogen atom in equation 18.65).

(18.65) 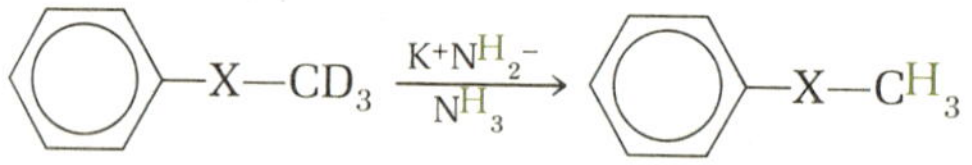

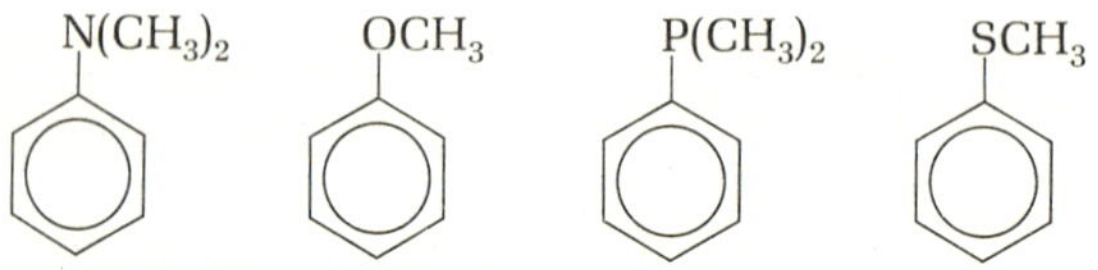

1 40 3×10^4 2×10^8

The exchange rates show only a partial correlation with electronegativity. Oxygen is more electronegative than nitrogen and sulfur is more electronegative than phosphorus; within each pair the compound having the more electronegative element shows the greater tendency to exchange hydrogens. But sulfur and phosphorus are each less electronegative than nitrogen; yet they promote exchange orders of magnitude more effectively than either nitrogen or oxygen do. The explanation most likely lies in the fact that sulfur (and phosphorus) has available low energy, vacant *d*-orbitals able to accept part of the negative charge, while oxygen (and nitrogen) does not.

(18.66) 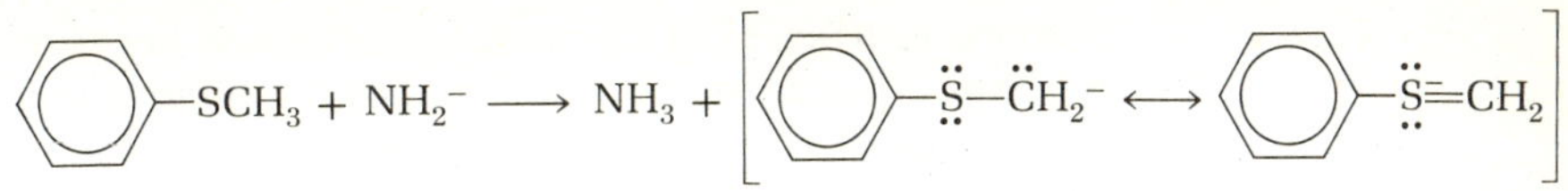

If a neutral divalent sulfur atom can stabilize an adjacent negative charge, then certainly the positively charged sulfur atom of a sulfonium ion would be markedly more efficient in this role.

(18.67) $$(CH_3)_3S^+I^- \xrightarrow[62°,\ 3\ hr]{OD^-,\ D_2O} (CD_3)_3S^+I^- \quad \text{(98\% exchange)}$$

(18.68) $$(CH_3)_4N^+I^- \xrightarrow[62°,\ 504\ hr]{OD^-,\ D_2O} \text{no exchange}$$

The conclusion is supported by the fact that exchange in sulfonium ions can be catalyzed by hydroxide ion, while sulfide exchange requires amide ions. It is important to note that the acidity of the protons undergoing exchange is not solely due to the positive charge on the heteroatom; equation (18.68) shows how sluggishly ammonium salts exchange their protons.

The exchange reaction takes place by way of an intermediate known as an **ylid** (**11**).

(18.69) $$(CH_3)_3S^+ + OD^- \longrightarrow HOD + [(CH_3)_2\overset{+}{\ddot{S}}-\overset{-}{C}H_2 \underset{\mathbf{11}}{\longleftrightarrow} (CH_3)_2\ddot{S}{=}CH_2]$$

An ylid is a neutral molecule that can be represented by a canonical structure in which a negatively charged carbon atom is bonded directly to a positively charged heteroatom. Ylids are extremely useful reactive intermediates for syntheses and are the subject of a good deal of research effort at the present time.

Dimethyl(methylene)sulfurane (**11**) has a reasonable half-life if generated in nonacidic solvents. (Ylids such as **11** are named as derivatives of the hypothetical tetravalent sulfur hydride *sulfurane*, SH_4.) It is useful as a "methylene transfer agent" and can efficiently convert ketones and aldehydes to the corresponding epoxides.

(18.70) 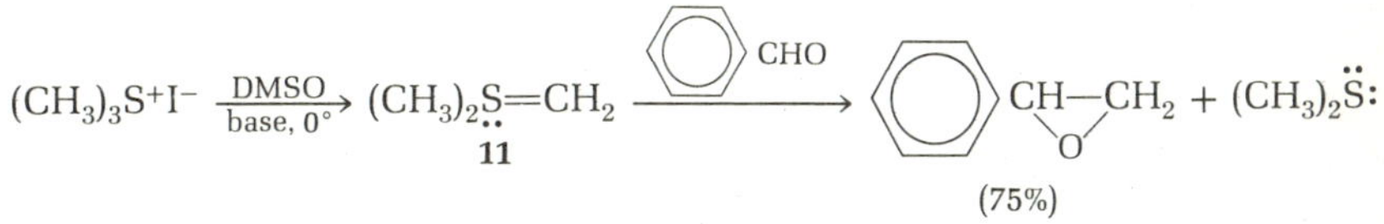

PROBLEM

4 Phosphonium salts also react with bases to give ylids that in turn react with carbonyl compounds to yield olefins (*Wittig reaction*). This process, which lends itself to extensive variation, has become one of the more important tools of the modern synthetic chemist.

Write a structure for **A**, and suggest reasons for the differing behavior of sulfonium and phosphonium ylids toward carbonyl compounds (compare equation 3 with equation 18.70).

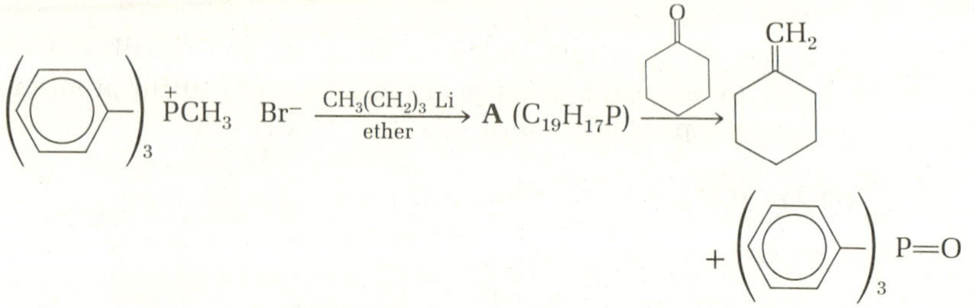

(3)

The π-bond in ylids such as **11** is certainly relatively weak; furthermore, the π-electrons are not equally shared by both atoms, the carbon atom retaining the lion's share of the π-electron density. The presence of an electron-withdrawing group on this carbon atom both increases the ease of ylid formation and also enhances the stability of the product.

(18.71)

$$(CH_3)_2S + BrCH_2CO_2CH_3 \xrightarrow{\text{acetone}} (CH_3)_2\overset{+}{S}CH_2CO_2CH_3\ Br^- \xrightarrow[H_2O\text{-}CHCl_3]{OH^-}$$

$$\left[\underset{(95\%)}{(CH_3)_2S{=}CH\overset{O}{\overset{\|}{C}}OCH_3} \longleftrightarrow (CH_3)_2\overset{+}{S}\overset{\cdot\cdot -}{C}H\overset{O}{\overset{\|}{C}}OCH_3 \longleftrightarrow (CH_3)_2\overset{+}{S}CH{=}\overset{O^-}{\overset{|}{C}}OCH_3\right]$$

Ylids such as **12** are extremely stable and unreactive and may be isolated and stored without any special precautions.

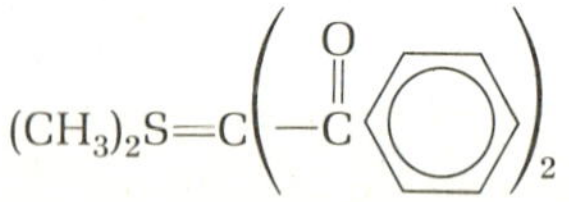

12

REARRANGEMENT OF SULFUR YLIDS

The reactions of sulfur ylids are many and various and their diversity alone precludes any general discussion of them. There is one rearrangement, however, that seems to be very general in its scope and may play an important role in a very vital biosynthesis. Equations 18.72 and 18.73 depict two variations of this rearrangement in nonspecific terms.

(18.72)

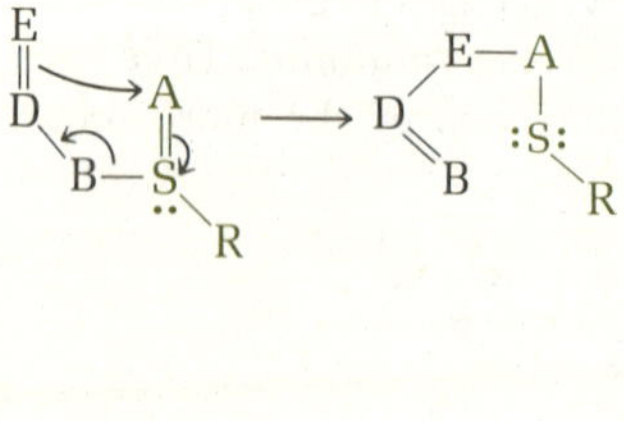

(18.73) 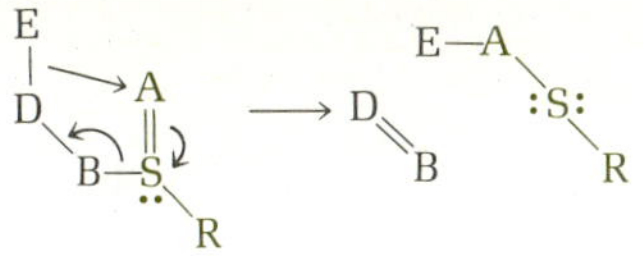

Both variations are five-center, six-electron cyclic processes (A, B, D, and E may be atoms or groups of atoms). Examples of these rearrangements are provided in equations 18.74 and 18.75; equation 18.74 is a proposed mechanism for the oxidation of alcohols to aldehydes by dimethyl sulfoxide (equations 11.81 and 11.82).

(18.74) 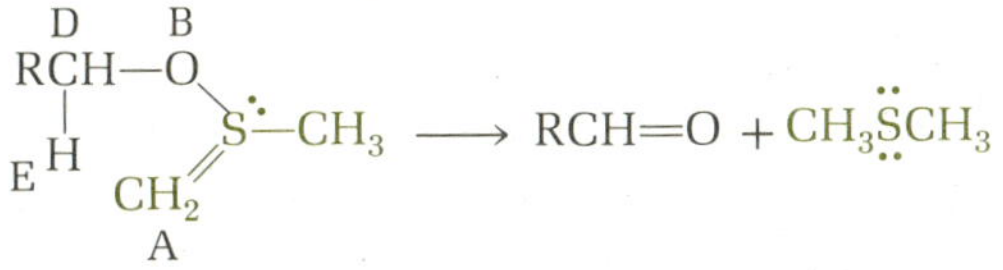

(18.75) 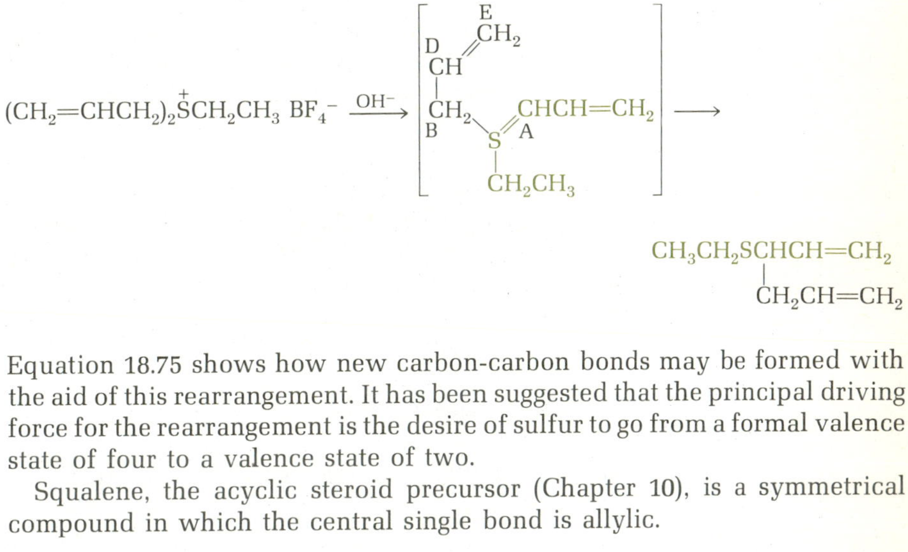

Equation 18.75 shows how new carbon-carbon bonds may be formed with the aid of this rearrangement. It has been suggested that the principal driving force for the rearrangement is the desire of sulfur to go from a formal valence state of four to a valence state of two.

Squalene, the acyclic steroid precursor (Chapter 10), is a symmetrical compound in which the central single bond is allylic.

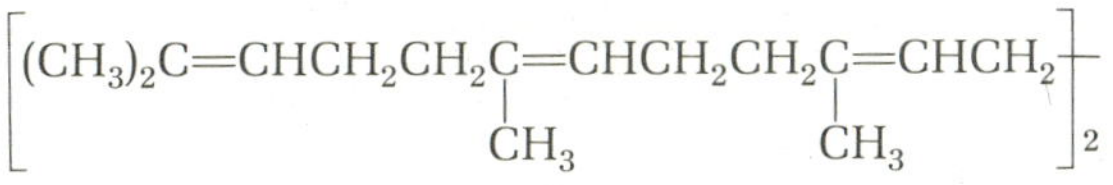

squalene

There is evidence to the effect that, in living systems, the two halves of the squalene molecule are joined by means of a sulfur ylid rearrangement (equation 18.76). The enzyme system that is responsible for synthesizing squalene is known, for instance, to require two sulfur-containing coenzymes in order for the synthesis to take place.

(18.76) 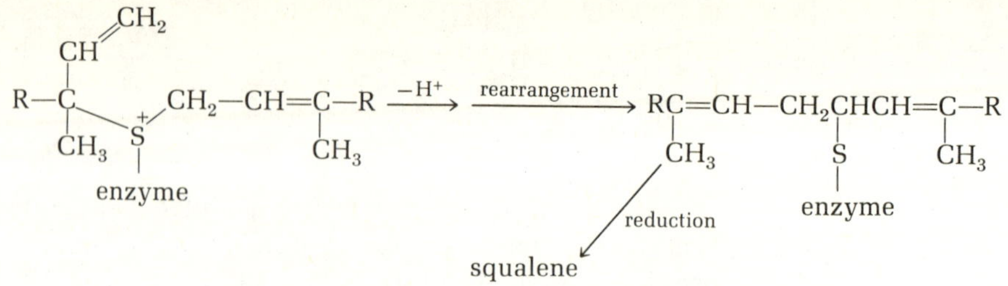

SUMMARY

1 Amines are weak bases in water, the pK_b's usually lying in the range 3 to 10. Amines in which one or more of the attached organic groups are aromatic are markedly less basic than those in which all the groups are aliphatic. Delocalization of the nitrogen lone pair seems to be the primary reason for this difference. Upon protonation the amine is converted to an ammonium ion, in which the positive charge is localized on the nitrogen atom. These ammonium ions are protonated and deprotonated very rapidly in polar media.

2 Amines are nucleophilic toward carbon atoms as well as being basic toward protons. They will displace good leaving groups, such as halide or tosylate ions, from primary and secondary alkyl derivatives. Ammonia and amines will, upon alkylation, usually give rise to a mixture of products unless either an excess of amine or of alkylating agent is used. In the latter case quaternary ammonium salts are produced.

3 Primary amines may be synthesized by way of reactions in which a good leaving group is displaced by a nitrogen-containing nucleophile such as azide ion or phthalimide anion. If the leaving group is replaced by cyanide ion and the resulting nitrile then reduced, a primary amine is obtained that contains one carbon atom more than the starting material. Degradations of acids with hydrazoic acid, or of amides with alkaline hypobromite, lead to amines containing one less carbon atom. These reactions proceed by way of a very common rearrangement, in which a substituent migrates from carbon to nitrogen.

Equilibrium formation of imines, from primary amines and carbonyl compounds, in the presence of hydrogen and a catalyst, leads to the formation of mixed secondary amines in good yield. Amines containing tertiary alkyl groups may be synthesized by allowing carbonium ions to react with hydrogen cyanide; aqueous work-up yields the tertiary alkyl formamide. Hydrolysis of the amide liberates the desired amine. Aromatic amines are usually obtained by reduction of the appropriate nitroaromatic precursor.

4 In the absence of appreciable steric effects, there is a parallel between amine basicity and nucleophilicity. That relationship breaks down, however, when there is substantial steric crowding in the vicinity of the nitrogen atom. The proton is very small, and the basicities of amines are in general but little affected by spatial variations. Alkylating agents, by contrast, are much bulkier and their approach to a nucleophilic nitrogen atom is very much governed by steric parameters. Advantage has been taken of this sensitivity to steric effects and sterically hindered amines can be used as bases in alkylation reactions.

5 If the three substituents on the nitrogen atom of an amine are all different, then that nitrogen atom is asymmetric. However, the rate of inversion of trivalent nitrogen atoms is very rapid in the usual temperature range in which organic chemists work; since each inversion converts the dissymmetric molecule into its own enantiomer, isolation of enantiomeric amines containing asymmetric nitrogen atoms is not normally possible. Ammonium salts cannot, however, undergo inversion and quaternary ammonium salts have been obtained optically pure.

In order to resolve mixtures of enantiomeric acids, the usual procedure is to add an enantiomerically pure amine. The salt that is formed exists in two diastereomeric forms that are separable, usually by means of fractional crystallization. Treatment of the resolved salt with inorganic base frees the amine and, upon acidification of the aqueous salt solution, the optically pure organic acid is recovered. The acid may in turn be used to resolve racemic amines, alcohols, etc.

6 Like trivalent nitrogen, divalent sulfur is a nucleophilic atom. Thiols, sulfides, and sulfonium salts are isolated when alkylating agents react with hydrosulfide ion, thiolate ions, and sulfides, respectively. Hydrogen sulfide and thiols are considerably more acidic than water and alcohols are and may be converted to their salts with the aid of sodium hydroxide.

The nucleophilicity of sulfides is not matched by a corresponding basicity toward protons. The sulfur valence electrons are highly polarizable and therefore very effective in kinetic competitions, but since the hydrogen-sulfur and carbon-sulfur bonds are relatively weak, the sulfur compounds do not fare well in equilibrium processes.

7 Sulfur can donate its unshared electron pairs by means of π-bonds. The overlap between the *p*-orbitals of sulfur and those of carbon is relatively inefficient and π-bonds between these atoms are quite weak. Sulfur is much weaker than oxygen or nitrogen as a π-donor. Carbon-sulfur σ-bonds are also weaker than carbon-oxygen σ-bonds but due to the polarizability of the sulfur electrons they participate readily in σ-bond formation.

8 Sulfur may also act as a π-bond acceptor by making use of its unfilled 3*d*-orbitals. The acceptor properties of sulfur are much enhanced if it bears a positive charge. Thus, the α-protons of sulfonium salts exchange very readily with the protons of the solvent under basic conditions; ylids are intermediates in these processes.

Due to the weakness of the carbon-sulfur π-bond, there is a good deal of negative charge on the ylid carbon atom. The presence of an electron-withdrawing substituent on this carbon atom greatly stabilizes the ylid. Sulfur ylids undergo a number of useful inter- and intramolecular reactions, all of which are characterized by a reduction in the formal valence of sulfur.

PROBLEMS

Note: Problems 1 through 4 appear in text.

5 Write structures for the products of the following reactions:

a. Dimethylamine and hydrochloric acid.

b. Diethylisobutylamine and hydrobromic acid.

c. *N*,3-Diethyl-*N*-hexylaniline and dilute nitric acid.

d. Pyridine and sulfuric acid (1 mole).

e. *N*-Isopropylpyrrolidine and sulfuric acid (0.5 mole).

f. *m*-Aminobenzoic acid and perchloric acid.

g. *N*-Methyl-*N*-(2-butyl)amine and acetic acid.

h. *o*-Methoxyaniline and *p*-nitrobenzoic acid.

i. Ethylamine and boron trifluoride.

6 Arrange the compounds in each group in *decreasing* order of basicity.

a. Aniline, cyclohexylamine, *p*-nitroaniline.

b. *p*-Chloroaniline, *p*-methoxyaniline, *p*-toluidine (*p*-methylaniline), *p*-cyanoaniline.

c. Aniline, *N*-methylaniline, *m*-methylaniline.

d. Aniline, benzylamine, 2-phenylethylamine.

e. Ethylamine, 2-aminoethanol, 2-fluoroethylamine.

a. The addition of two methyl groups to the nitrogen atom of aniline causes a modest increase in basicity.

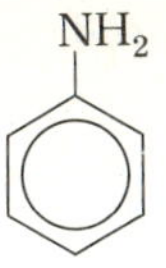 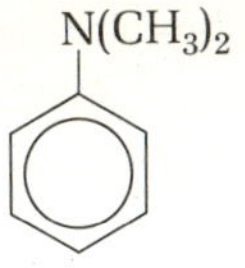

pK_b 9.40 pK_b 8.96

When the same operation is performed on 2,4,6-trinitroaniline, the increase in basicity exceeds 10^4. Explain the difference in behavior.

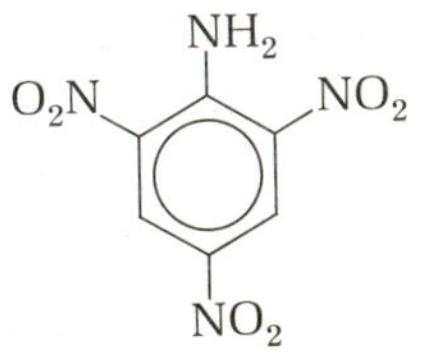 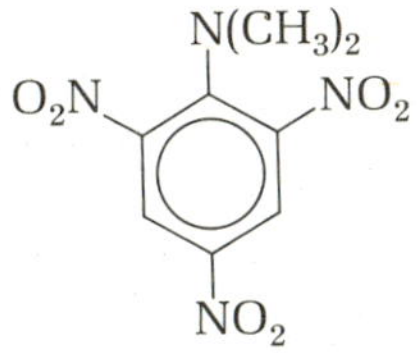

pK_b 23.3 pK_b 18.7

b. *m*-*t*-Butylaniline (I) and *m*-(trimethylsilyl)aniline (II) are essentially equal in base strength (pK_b 9.35). But *p*-*t*-butylaniline (pK_b 9.05) is a stronger base than *p*-(trimethylsilyl)aniline (pK_b 9.64), despite the fact that silicon is *less* electronegative than carbon.

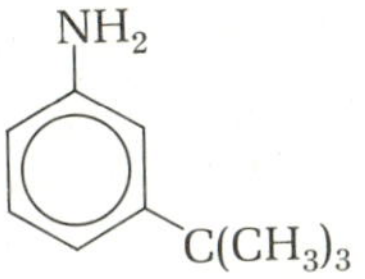

I

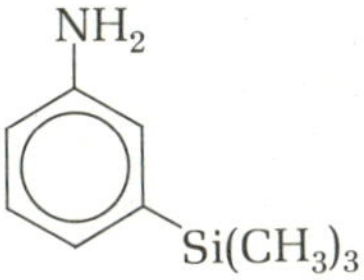

II

Explain this apparent anomaly.

8 Complete the following reactions; show all products, including isomers.

a. $CH_3CH_2NH_2$(excess) + $(CH_3)_2CHCH_2OTs \longrightarrow$

b. 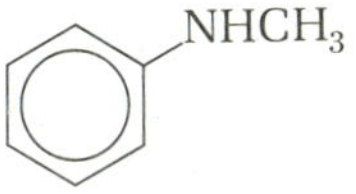 + CH_3CH_2I (excess) $\xrightarrow{\text{base}}$

c. $(CH_3)_3N$ + (*R*)-$CH_3CH(Cl)CH_2CH_3 \longrightarrow$

d. $CH_3CH_2N(CH_3)CH_2CH_2CH_3$ + (*R*)-$CH_3CH_2CH(CH_3)CH_2Br \longrightarrow$

e. $CH_3CH_2SCH_3$ + (*S*)-$CH_3CH(Br)CH_2CH_2CH_3 \longrightarrow$

9 Amines are effective nucleophiles and are usually alkylated with ease. Amides, on the other hand, are very resistant to alkylation and the reaction requires much sterner conditions. Account for this difference in reactivity.

10 A supply of *cis*-4-methylcyclohexylamine is needed for a research project. The following related compounds are available:

a. *trans*-4-Chloromethylcyclohexane

b. *cis*-4-Methylcyclohexanol

c. *cis*-4-methylcyclohexanecarboxylic acid

Show how each compound could be converted to the desired amine.

11 Describe efficient syntheses of the following amines, using appropriate alkyl halides, alcohols, aldehydes, ketones, and aromatic hydrocarbons as starting materials, plus other needed reagents.

a. Ethylamine

b. Diethylamine

c. *o*-Toluidine (*o*-methylaniline)

d. *m*-Bromoaniline

e. Ethyl *m*-aminobenzoate

f. Benzylamine

g. *p*-Iodobenzylamine

h. Diisopropylamine

i. Ethyl-*t*-butylamine

j. Methycyclopentylbenzylamine

12 Write structures for the expected organic products when each of the following compounds is first treated with sodium cyanide and sulfuric acid, followed by a work-up with aqueous base.

a. 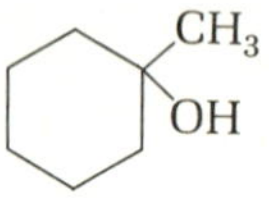

b. $(CH_3)_2\underset{|}{C}CH_2CH_3$
OH

c. $(CH_3)_3C\underset{|}{C}HCH_3$
OH

d. 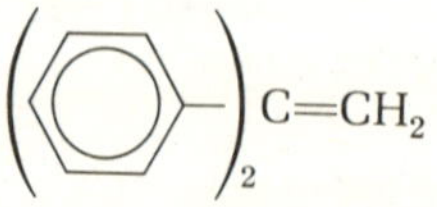

e. $CH_3CH_2\overset{CH_3}{\overset{|}{C}}{=}CHCH_3$

13 A plant that was fed sodium acetate labeled with ^{14}C in the methyl group was found to produce radioactive toluene. It was desired to know how much of the total radioactivity was present in the methyl group of toluene. Devise an experiment whereby this information could be obtained.

14 A compound $C_5H_{11}N$ **(A)** is very soluble in water, giving a basic solution. The infrared spectrum of **A** shows absorption in the N—H stretching region. When **A** is treated with benzoyl chloride in base, it is converted to **B,** $C_{12}H_{15}NO$. Reduction of **B** with lithium aluminum hydride results in the formation of **C,** $C_{12}H_{17}N$. **C** reacts with one equivalent of methyl iodide in acetone to give a crystalline precipitate **D,** $C_{13}H_{20}NI$. When **D** is treated with an equimolar quantity of the silver salt of (*R*)-2-methylbutanoic acid, in aqueous solution, there is an immediate precipitation of silver iodide. From the aqueous supernatant it is possible to separate two salts, **E** and **E′**, by fractional crystallization. Both salts have identical molecular formulas but different melting points and molecular rotations. **E** and **E′** are each separately dissolved in water and the aqueous solutions treated with hydrochloric acid. Each solution is then extracted several times with ether, and the aqueous layer evaporated to dryness. Two substances are obtained, **F** and **F′**. They both have the same molecular formula, $C_{13}H_{20}NCl$, the same melting point, density, etc. Their molecular rotations are identical in magnitude but opposite in sign. Write a plausible structure for **A,** and show the chemistry involved in its transformation.

15 Devise syntheses for the following compounds, using easily available organic starting materials and any necessary inorganic reagents.

a. $(CH_3)_2CHCH_2SH$

b. 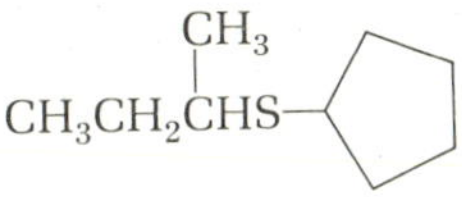

c. 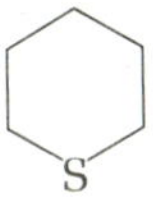

d. $HOCH_2CH_2SH$

e. 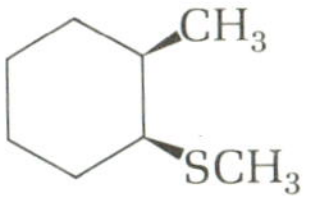

f. $CH_3CH_2\overset{+}{\underset{|\atop CH_3}{S}}CH(CH_3)_2 \quad I^-$

16 The acid-catalyzed addition of methanol to  gives 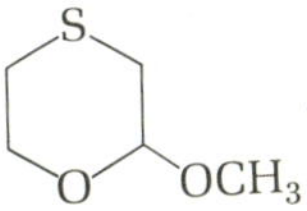 as the *only* product, in 87% yield.

What is the other logical product, and why is it not formed?

17 Reaction 4 leads almost exclusively to the rearranged product shown. Propose a mechanism that accounts for the course of this reaction.

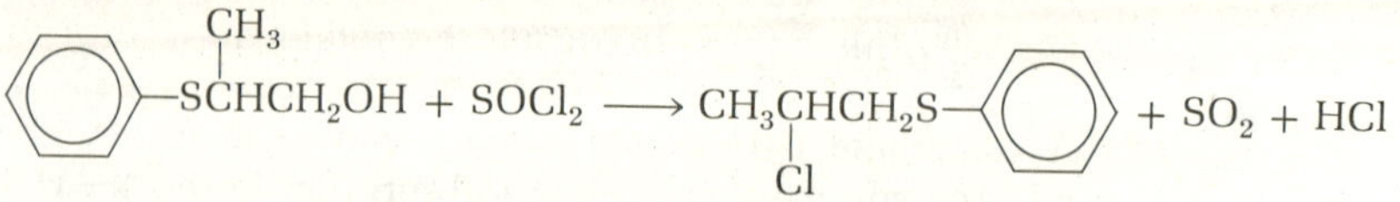

(4)

18 a. Dimethylsulfoxide can be described as a resonance hybrid of the two canonical structures V and VI. Dimethylsulfoxide is found to be appreciably more polar than acetone.

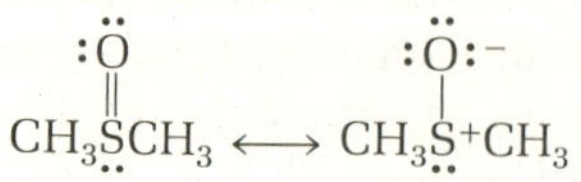

V VI

Explain.

b. Arrange dimethyl sulfoxide, dimethyl sulfide, and dimethyl sulfone (VII) in increasing order of ability to exchange protons with D_2O. Justify your order.

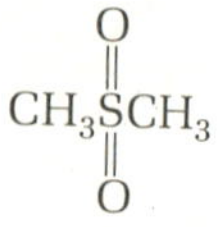

VII

c. Dimethyl sulfoxide reacts with methyl iodide to produce two isomeric compounds, C_3H_9OSI. One or the other may be preferentially produced by appropriate choice of conditions. Both are soluble in water and insoluble in benzene. Write their structures.

d. Arrange the following compounds in order of increasing ease of hydrogen-deuterium exchange *per exchangeable hydrogen atom*. Underline those hydrogen atoms that you think will exchange with solvent in a D_2O-OD^- solution.

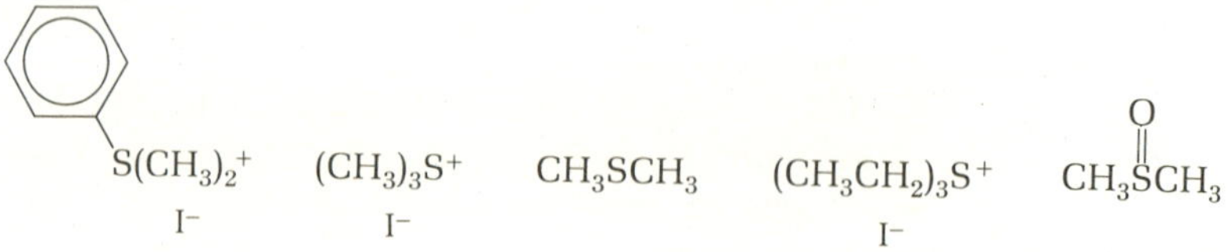

SUPPLEMENTARY READING

D. Ginsburg, *Concerning Amines*. New York: Pergamon Press, 1967.

S. Patai, ed., *The Chemistry of the Amino Group*. New York: Interscience Publishers, 1968.

C. C. Price and S. Oae, *Sulfur Bonding*. New York: The Ronald Press, 1962.

CHAPTER 19

DIAZONIUM IONS AND DIAZO COMPOUNDS—HOT CARBONIUM IONS AND DIVALENT CARBON

The nonbonded electron pair on nitrogen in amines is susceptible to attack by a wide variety of electrophiles; ammonium salts are the usual products of such reactions. An interesting and useful class of compounds is formed when amines are allowed to react with electrophiles containing the **nitroso group,** —N̈=O. The nitroso group becomes attached to the nitrogen atom and the resulting *N*-nitrosamine may, if the structure permits, undergo further transformation by way of various tautomeric shifts.

NITROSATION OF ALIPHATIC AMINES

When dimethylamine hydrochloride is treated with aqueous sodium nitrite, a yellow oil separates in high yield; this proves to be nitrosodimethylamine (equation 19.1).

(19.1) $(CH_3)_2NH_2^+ + NO_2^- \xrightarrow[75°]{H_3O^+} (CH_3)_2\ddot{N}{-}N{=}O + H_2O$ (90%)

This reaction is characteristic of secondary amines, including arylamines, and may be used as a qualitative test for identifying these compounds. Even secondary amides may be nitrosated in this way.

(19.2) 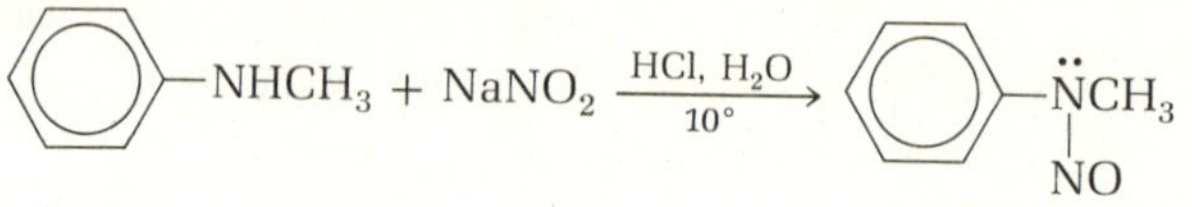

N-nitroso-N-methylaniline (90%)

(19.3) 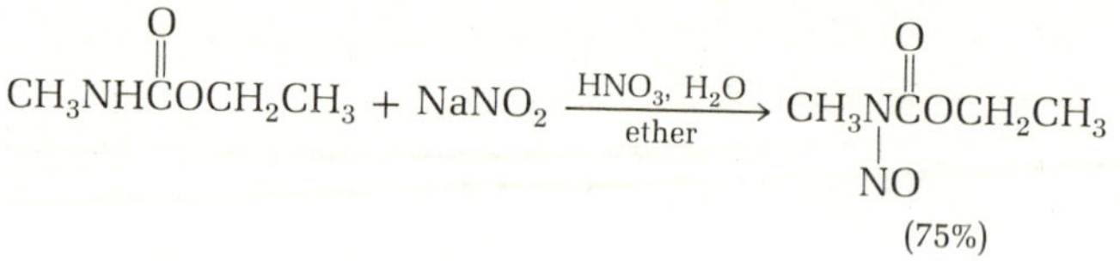

Adjustment of the pH can be important in these reactions, particularly with aliphatic amines. Acid conditions are essential for the nitrite ion must be converted to nitrous acid, which is only slightly dissociated in solution.

(19.4) $NO_2^- + H_3O^+ \rightleftarrows HON{=}O + H_2O$

If the solution becomes too acidic, however, then nitrosation does not take place because essentially all the amine has been protonated. The presence of a free electron pair on nitrogen is essential to the nitrosation reaction.

The nature of the nitrosating agent has been the subject of much controversy and investigation. In water, nitrous acid is in equilibrium with dinitrogen trioxide, N_2O_3. This compound is an anhydride of nitrous acid and it may be attacked by nucleophiles in a manner strongly analogous to the attack of nucleophiles on a carboxylic anhydride (equations 17.13 and 17.16; $RC(=O)$– is equivalent to O=N–).

(19.5) $2HONO \rightleftharpoons O{=}NON{=}O + H_2O$

The equilibrium shown in equation 19.7 lies well to the right because of the partial delocalization of the amino nitrogen lone pair.

(19.6) 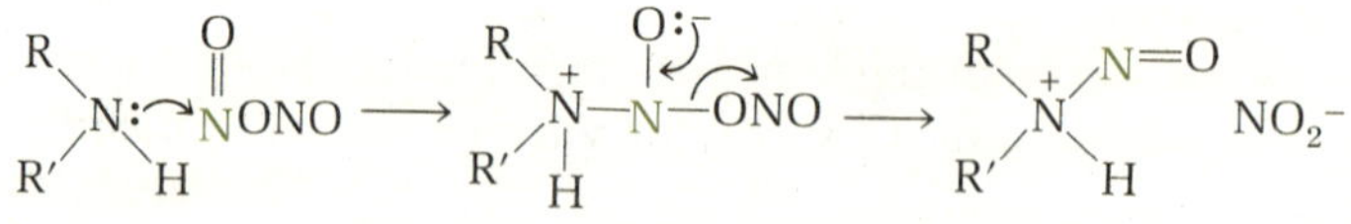

(19.7) 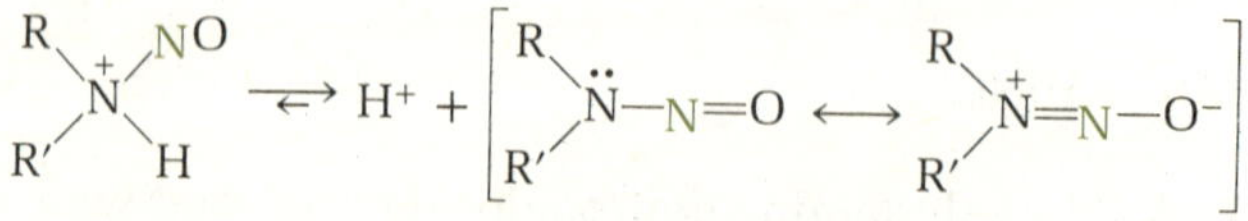

If primary amines are used instead of secondary ones, then the initial reaction steps are the same (equations 19.6 and 19.7). The nitrosamine from a primary amine can undergo a tautomeric shift, however, and eventually wind up as an **alkyldiazonium ion (1).**

(19.8) $$R\ddot{N}H{-}N{=}O \xrightarrow{H^+} R{-}\overset{+}{\underset{|\atop H}{N}}{=}N{-}OH \longrightarrow R\ddot{N}{=}N{-}OH$$

(19.9) $$R{-}\ddot{N}{=}N{-}OH \xrightarrow{H^+} H_2O + \underset{\mathbf{1}}{R{-}\overset{+}{N}{\equiv}N} \longrightarrow R^+ + N_2\uparrow$$

Alkyldiazonium ions are notorious for the ease with which they decompose to carbonium ions and molecular nitrogen (next section). The reaction of primary amines with nitrous acid (or other nitrosating agents) is known as **diazotization.**

Tertiary amines may also be nitrosated but at low temperatures the nitrosoammonium ions remain in equilibrium with amine and nitrous acid. A qualitative test for distinguishing the three classes of aliphatic amines revolves around their differing behavior in the presence of nitrous acid. A sample of the amine is dissolved in ice-cold aqueous hydrochloric acid and a solution of sodium nitrite is then added. The rapid evolution of gas (nitrogen) signifies the presence of a primary amine; the separation of an oily, yellow-green layer (nitrosamine) is diagnostic of a secondary amine; the absence of either of these manifestations, over a short period of time, may be taken to mean that the amine is tertiary.

At higher temperatures the nitrosoammonium ions decompose irreversibly, as shown in equations 19.10 to 19.12, provided only that a hydrogen atom is available on one of the carbon atoms α to the nitrogen atom.

(19.10)
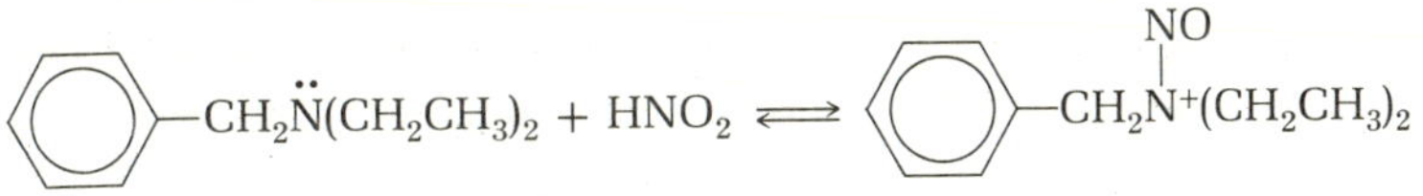

(19.11)
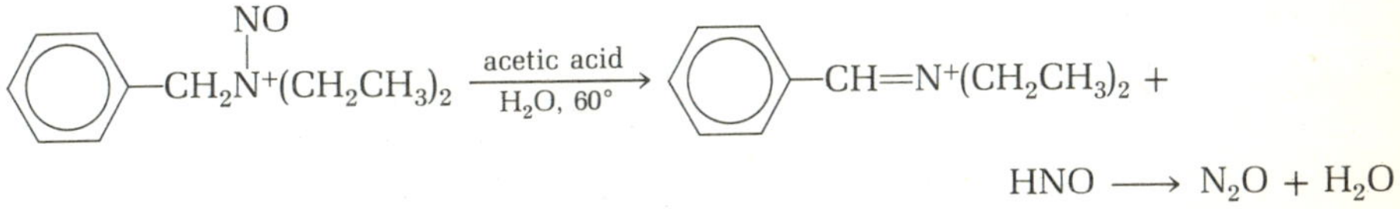

(19.12)
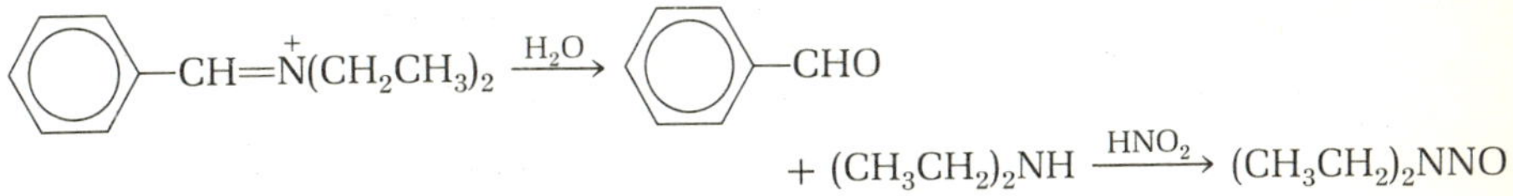

DECOMPOSITION OF ALKYLDIAZONIUM IONS

As already indicated in equation 19.9, alkyldiazonium ions decompose readily to carbonium ions. Consequently, when an alkyl primary amine is subjected to diazotization conditions, we would expect the product distribution to reflect what we already know about carbonium ion behavior from the reactions of alkyl halides and tosylates. Our expectations are generally borne out but there are important quantitative, and even qualitative, differences that are significant.

It has been known for some time that *deamination* of amines by way of the diazotization reaction led to considerable quantities of rearranged products (equations 19.13 and 19.14). The multiplicity of products spawned by these reactions made the process useless for preparative or degradative purposes.

(19.13)

$$CH_3CH_2CH_2NH_2 + NaNO_2 + HCl \xrightarrow{H_2O} \underset{(7\%)}{CH_3CH_2CH_2OH} + \underset{(28\%)}{CH_3\underset{\displaystyle OH}{\underset{|}{C}}HCH_3} + \underset{(34\%)}{CH_3CH{=}CH_2} + N_2 + NaCl$$

(19.14)

$$CH_3CH_2CH_2CH_2NH_2 \xrightarrow[HCl]{HNO_2} \underset{(25\%)}{CH_3(CH_2)_3OH} + \underset{(5\%)}{CH_3(CH_2)_3Cl} +$$

$$\underset{(13\%)}{CH_3\underset{\displaystyle OH}{\underset{|}{C}}HCH_2CH_3} + \underset{(3\%)}{CH_3\underset{\displaystyle Cl}{\underset{|}{C}}HCH_2CH_3} + \underset{(26\%)}{CH_2{=}CHCH_2CH_3} + \underset{(10\%)}{CH_3CH{=}CHCH_3} + N_2$$

When compared to the solvolysis of, say, primary alkyl tosylates, deamination of primary amines gives rise to far more rearrangement. Recent investigations have disclosed even more interesting findings. Diazotization of 1-^{14}C-1-propylamine in aqueous perchloric acid leads to labeled 1-propanol, among other products. Degradation of this material reveals that some of the ^{14}C has migrated to *both* the 2 and 3 positions of the alcohol.

(19.15)

$$CH_3CH_2{}^{14}CH_2NH_2 \xrightarrow[HClO_4]{HNO_2} \underset{(96\%)}{CH_3CH_2{}^{14}CH_2OH}$$

$$+ \underbrace{CH_3{}^{14}CH_2CH_2OH + {}^{14}CH_3CH_2CH_2OH}_{(2\%)} + N_2$$

Furthermore, careful analysis of the hydrocarbon fractions from deamination of several aliphatic amines has shown that *cyclopropanes* are formed in low yields. Thus, decomposition of propylamine in aqueous nitrous acid leads to 4% of cyclopropane; isobutylamine serves analogously as progenitor for methylcyclopropane.

A currently popular rationalization for both the label scrambling *and* the cyclopropane formation invokes *protonated cyclopropanes* (Chapter 6, p.

128) as reaction intermediates. Proton scrambling could then take place within the protonated cyclopropane, which would subsequently react with water to produce 1-propanol or lose a proton to yield cyclopropane.

(19.16) 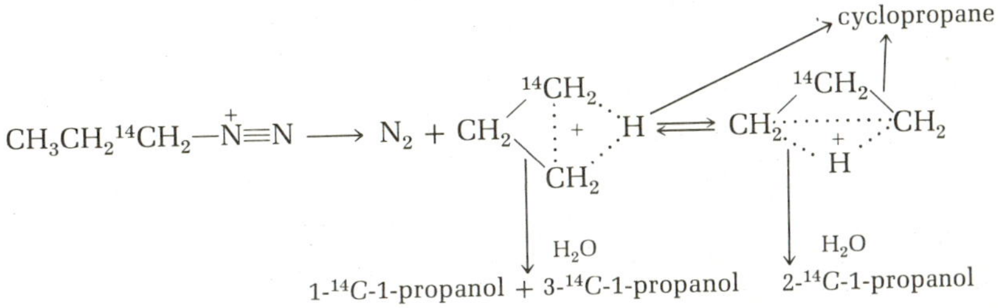

Although only a small proportion of the total products from deamination trace their origin to protonated cyclopropanes, the fact that these unusual intermediates are produced is very significant and serves to expand our vision of carbonium ion chemistry.

The difference in product distribution from primary halides and tosylates, on the one hand, and primary alkyldiazonium ions, on the other, may be explained away by noting that there is no firm evidence that the solvolysis reactions go by way of carbonium ions. Such an explanation cannot be used with reference to secondary and tertiary alkyl halides and toslyates, which can almost certainly be made to generate carbonium ions.

Why then should the carbonium ions derived by nitrogen loss from alkyldiazonium ions behave differently from those that originate by other paths? One clue is provided by the energetics of the processes involved. Most reactions in which alkyl carbonium ions are produced are endothermic and have activation energies of 25 to 30 kcal/mole. But the decomposition of alkyldiazonium ions appears to be *exothermic* because of the high thermodynamic stability of molecular nitrogen. Furthermore, it has been estimated that the activation energy for this decomposition amounts to only 3 to 5 kcal/mole.

Carbonium ions generated by the deamination of alkyl amines are often referred to as "hot" carbonium ions because of their unusually high reactivity. There are several hypotheses that attempt to account for this "hotness;" none, some, or all may be valid. One plausible explanation holds that some of the energy released in the decomposition of the diazonium ion appears as vibrational energy in the carbonium ion and that the vibrationally excited carbonium ion is relatively prone to rearrange. Another theory is that due to low activation energy and high rate for nitrogen loss the carbonium ion product is not very well solvated, unlike its counterpart in solvolytic reactions, which is heavily dependent on solvation for its formation to be at all feasible. The unsolvated ion would then have considerably more chance to rearrange before interacting with solvent.

NITROSATION OF AROMATIC AMINES

Aromatic amines may be nitrosated in much the same manner as aliphatic amines; since the aromatic amines are weaker bases than their aliphatic counterparts, more acidic solutions may be employed. Some of the results are very similar with both types of amine, and those that are different are understandable in light of what we know about aromatic rings.

Secondary alkyl aryl amines, such as N-substituted anilines, are nitrosated in the standard way when treated with aqueous solutions of nitrous acid.

(19.17)

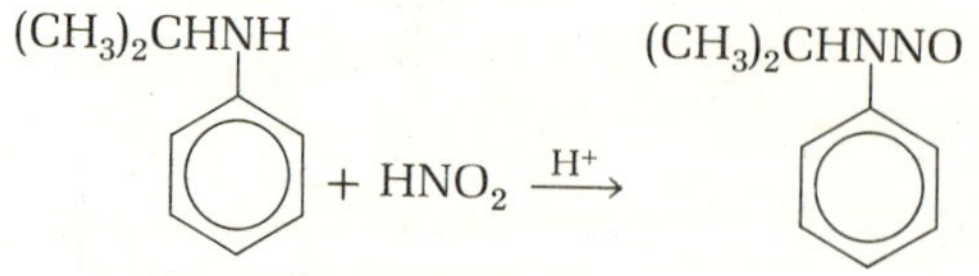

Tertiary aromatic amines, one substituent of which is an aromatic ring, may be nitrosated at the *para* position *of the ring*. The ring is strongly activated by the electron-releasing amino group and is thus readily attacked by even weak electrophiles; furthermore, attack at the nitrogen atom is readily reversible in this case.

(19.18)

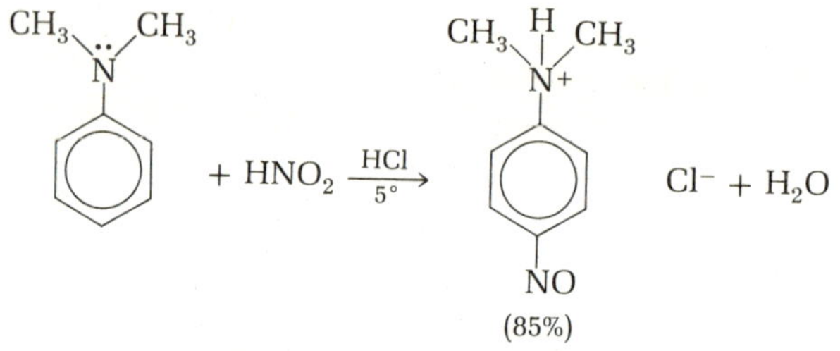

When primary aromatic amines are subjected to the action of nitrous acid, they are converted to diazonium salts in the same way that aliphatic amines are. The aromatic diazonium salts are markedly more stable than the aliphatic ones, however.

(19.19)

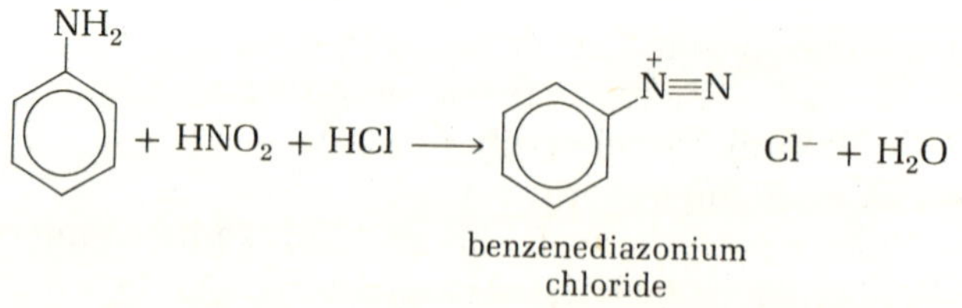

Although most aromatic diazonium salts are explosive when dry, it is possible to isolate the stable and water insoluble diazonium fluoborates.

(19.20)

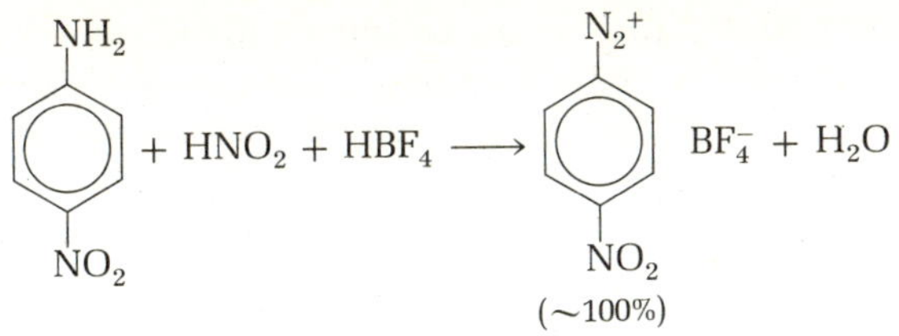

The stability of the benzenediazonium ion may be ascribed to two important effects. In the first place, conjugation between the diazonium group and the ring tends to strengthen the carbon-nitrogen bond.

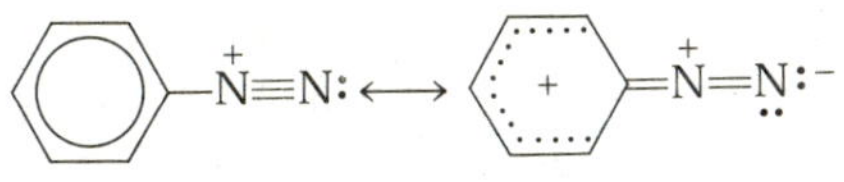

Indeed, electron-donating substituents on the ring tend to lower the nitrogen-nitrogen bond order. Secondly, loss of nitrogen from benzenediazonium ion leads to the formation of phenyl cation (equation 19.21), in which there is an unoccupied carbon sp^2-orbital.

(19.21)

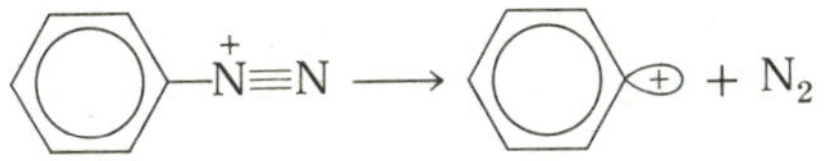

phenyl cation

Ordinary alkyl cations, such as those obtained from alkyldiazonium ions, contain unoccupied *p*-orbitals. Since *s*-electrons have higher ionization potentials than related *p*-electrons, phenyl cation is higher in energy than alkyl cations and correspondingly more difficult to generate.

PROBLEM 1

a. *m*-Methylbenzenediazonium ions lose nitrogen more rapidly than the unsubstituted ion, while the *m*-chloro and *m*-nitro ions are less reactive than the parent ion. How would you rationalize this trend?

b. When placed at the *para* position, both electron-donating *and* electron-withdrawing substituents retard the ionic decomposition of benzenediazonium ion. Explain.

REPLACEMENT OF THE DIAZONIUM GROUP

It is possible to replace the diazonium group by a number of other functional groups and, as a result, aromatic amines are important synthetic intermediates. In practice, the diazonium salt is not isolated but is rather generated *in situ* and then subjected to further reaction.

If an aqueous solution containing a diazonium ion is heated, then nitrogen is evolved and the corresponding phenol may be isolated. Since the overall process involves the conversion of the *m*-directing nitro group to the *o,p*-directing hydroxy group, this route permits substituents to be juxtaposed in a manner not available by other reaction paths.

(19.22)
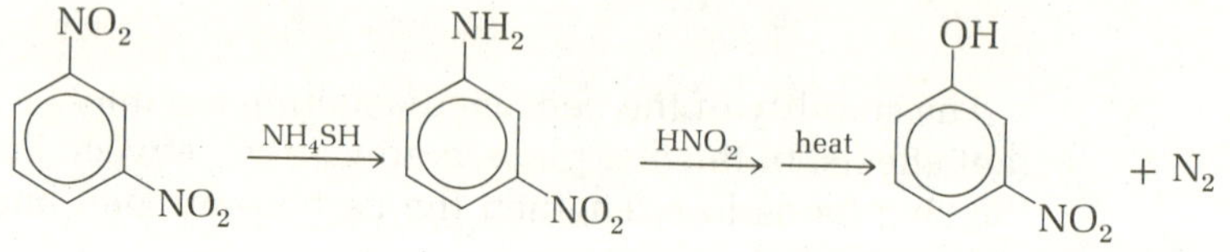

When the phenyl cation is formed in solution, it will react to some extent with each nucleophile present and not with water alone. If hydrochloric acid is used in the preparation of the diazonium ion, then a small amount of chlorobenzene will be formed. Due to its relatively high energy (see above), the phenyl cation is not very discriminatory, however. Since in aqueous solutions the nucleophile present in greatest concentration is water, phenols are usually the major products from ionic decomposition of diazonium ions. The method is therefore not very satisfactory for replacing the diazonium group by halide, cyanide, etc.

When a solution of potassium iodide is added to a cold solution of diazonium salt, nitrogen evolution begins at once and at the end of the reaction a good yield of the aromatic iodide may be isolated.

(19.23)
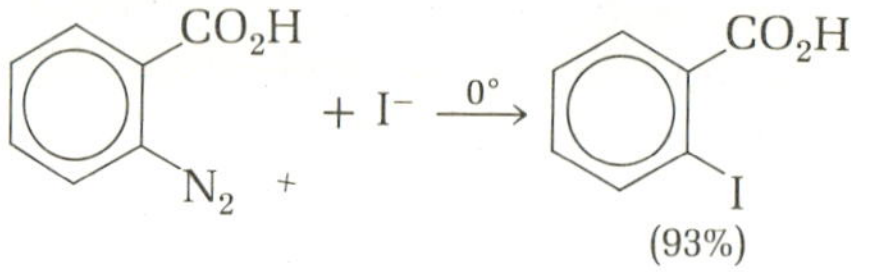

In light of the lack of discriminatory ability on the part of the phenyl cation, as noted above, how can we explain the high yields of iodo substitution product? The answer lies in the fact that the reaction apparently proceeds by way of a *free radical*, rather than an ionic, mechanism.

(19.24)
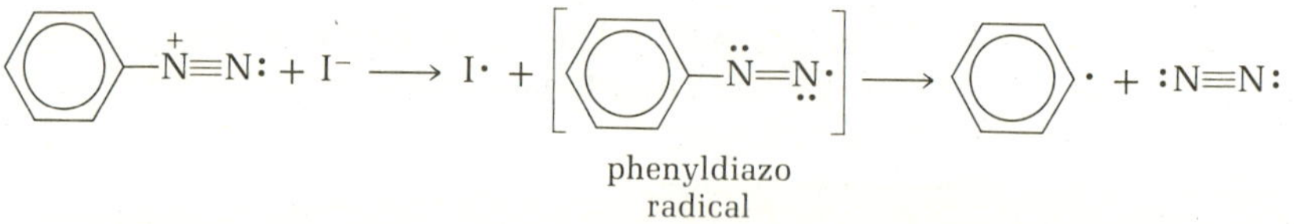

(19.25)
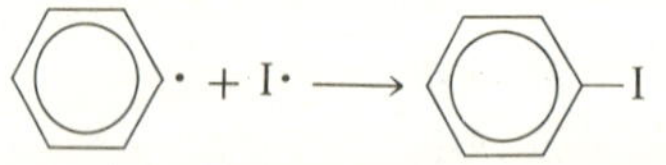

In contrast to benzenediazonium ion, the phenyldiazo radical decomposes with great ease and the phenyl radical so produced reacts preferentially with iodine. The intervention of a radical mechanism is made possible by the low oxidation potential of iodide ion. Since most potential nucleophiles are rather more difficult to oxidize, this method would appear not to be of general applicability. There is a valuable procedure, however, that involves the use of a metal cation as an electron transfer agent.

THE SANDMEYER REACTION

The replacement of a diazonium group by chloride, bromide, or cyanide can be accomplished with ease and in good yield, provided that the appropriate cuprous halide is present as a catalyst (*Sandmeyer reaction*).

(19.26)

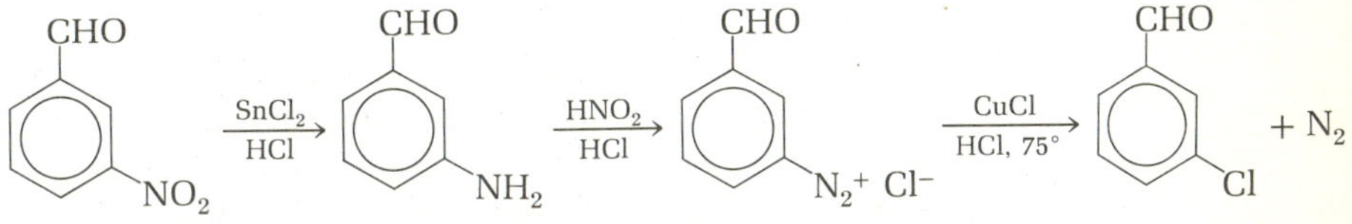

(19.27)

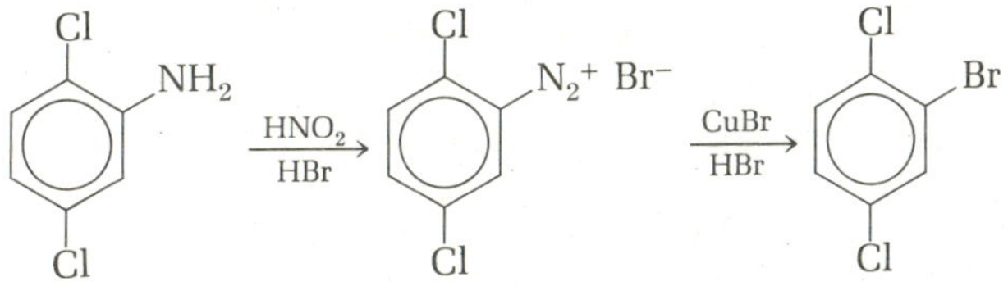

These reactions are also free radical processes but since the anions cannot be directly oxidized by the diazonium ion, the cuprous ion acts as the source of electrons (we shall represent the cuprous halides in solution as CuX, although they may in fact be more complex).

(19.28)

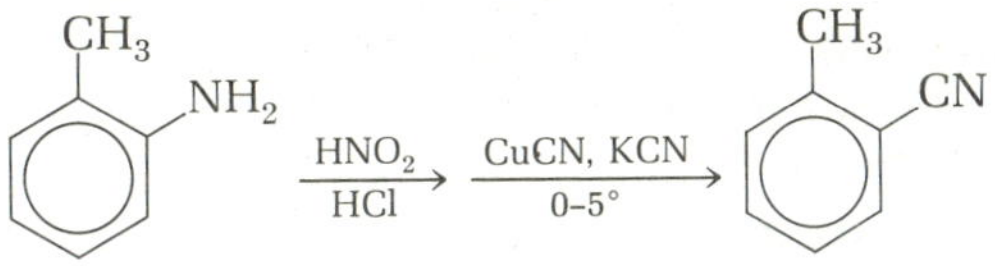

o-toluonitrile (70%)
(*o*-methylbenzonitrile)

The success of copper in this role is due in part to the ease with which it can be interconverted between the two oxidation states Cu(I) and Cu(II) (equations 19.30 and 19.31).

Once the phenyl radical is formed by Cu(I) reduction, it tends to attack the cupric halide preferentially, reducing it back to cuprous halide.

(19.29) $CuCl + Cl^- \rightleftharpoons CuCl_2^-$

(19.30)

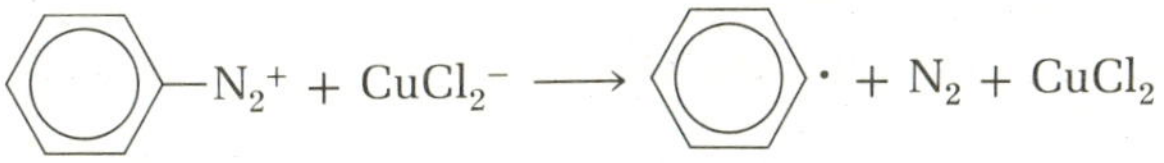

(19.31) 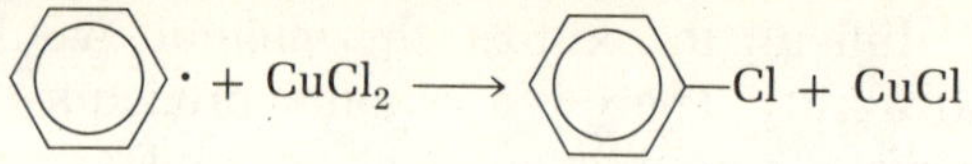

One observes small quantities of radical coupling products such as biphenyl produced in these reactions, which tends to support the proposed radical mechanism.

In synthetic aliphatic chemistry the alkyl halides play an important role because the relatively facile displacement of halide ion permits the introduction of a number of other functional groups. By contrast, aryl halides are not at all reactive except under special circumstances (Chapter 22) and aromatic derivatives with a more effective leaving group must be employed. Aromatic diazonium ions fulfill this requirement and they play somewhat the same role in synthetic aromatic chemistry that alkyl halides do in synthetic aliphatic chemistry. The halides and nitriles obtainable by the Sandmeyer reaction are themselves versatile intermediates, giving rise to Grignard reagents, carboxylic acid derivatives, etc.

DISPLACEMENT OF THE DIAZONIUM GROUP BY HYDROGEN

It is possible, by reacting a diazonium ion with *hypophosphorous acid*, H_3PO_2, to replace the diazonium group by a hydrogen atom. It thus becomes feasible to dispose of an amino group, and therefore a nitro group, which is present in the molecule. At first blush this might seem rather like a technique for turning gold into lead since one is usually attempting to put nitro or amino groups *into* molecules rather than taking them out. In a number of cases, however, it is desirable to have a nitro or amino group present temporarily in order for advantage to be taken of its directive influence. After all the other functionalities have been introduced, the now superfluous nitro or amino group may be replaced by hydrogen.

An excellent example is the synthesis of *m*-bromotoluene. Due to the directive effect of the methyl group, direct bromination of toluene yields only small quantities of this compound, which would have to be tediously separated from its more abundant isomers. Bromination of *p*-toluidine, on the other hand, puts the bromine atom in the correct position and the amino group is then easily disposed of.

(19.32) 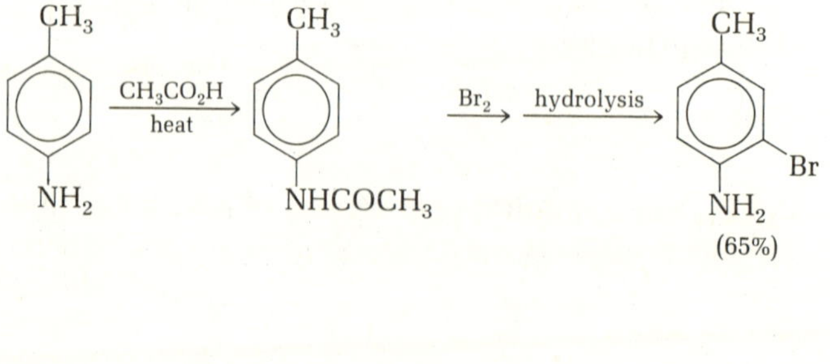

(19.33) 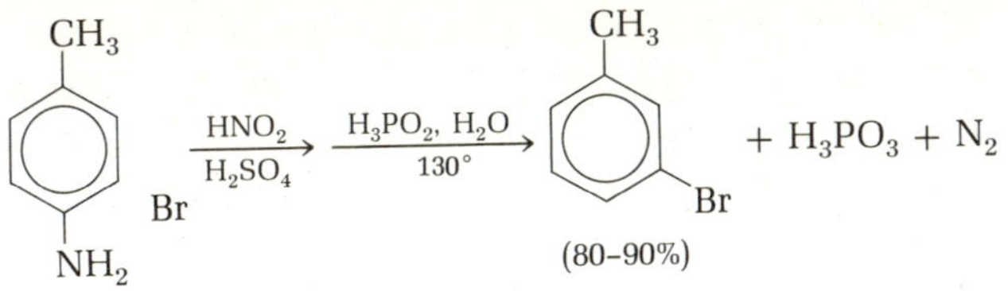

The synthesis of 2,4,6-tribromobenzoic acid is another case in point. Tribromination of the ring is an easy affair and gives the correct orientation because of the presence of the strongly activating amino group.

(19.34) 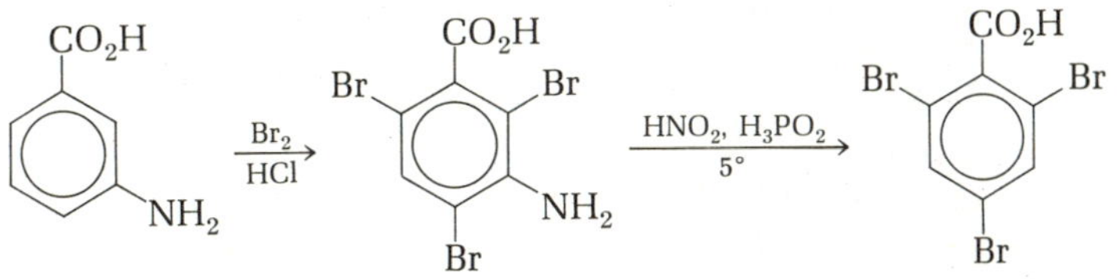

Tribromination of toluene, which could provide another route to 2,4,6-tribromobenzoic acid, is not nearly so simple a process nor does it yield so pure a product because of the weakly activating effect of methyl.

A final example concerns the use of a nitro group to block a particular ring position. Direct chlorination of 2,4-dihydroxybenzoic acid (equation 19.35) would undoubtedly lead to the 5-chloro compound as the major product.

(19.35) 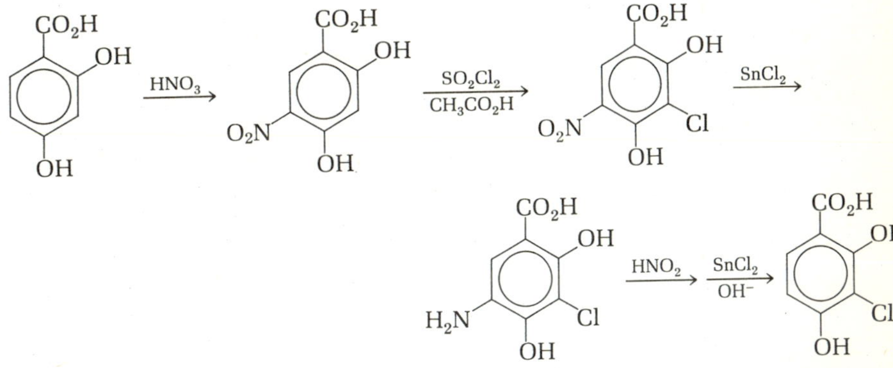

The 5 position was therefore blocked by nitrating the acid, forcing the chlorine to occupy the 3 position upon subsequent chlorination. Note that in equation 19.35 a derivative of Sn(II) [in practice, probably $Na_2Sn(OH)_4$] is used to reduce the diazonium ion, a relatively rare procedure.

AZO COUPLING

There are a number of reactions in which the diazonium ion itself acts as an electrophile toward various aromatic molecules. Both nitrogen atoms are then retained in the product. Compounds of the type R—N=N—R′ are known as **azo compounds.** Azo compounds are the most common type of synthetic dye and the discovery and exploitation of the *azo coupling reaction* (equation 19.36) fathered the rise of the enormous synthetic dye industry, particularly in Germany, toward the end of the nineteenth century.

(19.36) $ArN_2^+ + Ar'H \longrightarrow Ar{-}N{=}N{-}Ar' + H^+$

While we will not deal with the properties of dyes per se, it is worth noting a few of the characteristics that make azo compounds particularly suitable as dyes. In the first place, the azo group serves as a conjugative link between the two aromatic systems, resulting in a rather extensive conjugated system. It is *generally* true that the more extensive the conjugation is, the smaller the energy gap is between the highest filled π-orbital and the lowest empty π^*-orbital. If this gap is small enough, then even low energy light in the visible range may be sufficient to excite the molecule. The absorption of visible light by a substance will make that substance appear to possess a color *complementary* to that of the wavelength actually absorbed. Secondly, the azo link is easily formed from inexpensive precursors and it is quite stable, particularly toward hydrolysis.

Since azo coupling is an electrophilic aromatic substitution reaction, it follows that the rate of the process is greatly increased by the presence of electron-donating substituents in the ring being attacked. Thus, most of the aromatic rings used in coupling reactions contain either amino or hydroxy groups.

(19.37)

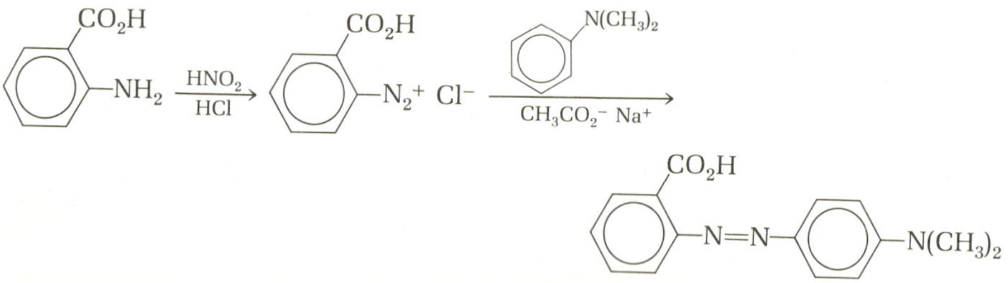

methyl red (65%)

The diazonium ion attacks preferentially *para* to the substituent but if the *para* position is blocked, it will go *ortho.*

PROBLEM

2 What would you expect the effect of substituents in the *diazonium ion* to be on the rate of diazo coupling with a given substrate?

Control of pH during the coupling reaction is of great importance. If an aromatic amine is the substrate, then one wishes to have it present as the free amine rather than as the ammonium ion, which is very much less reactive. The solution should therefore be acidic enough to keep the amine in solution but not very much more. There are, however, complications that arise if the pH is *too* high. The diazonium ion will attack the nitrogen atom of amines and if the amines are primary or secondary, then a proton may be lost and a stable diazoamino compound formed (equation 19.38).

(19.38)

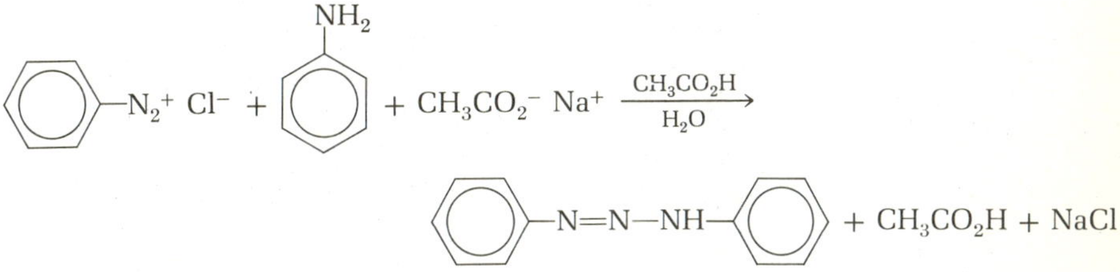

diazoaminobenzene (85%)

Formation of the diazoamino compound is reversible in the presence of mineral acid, with the normal coupling product being the more stable product.

(19.39)

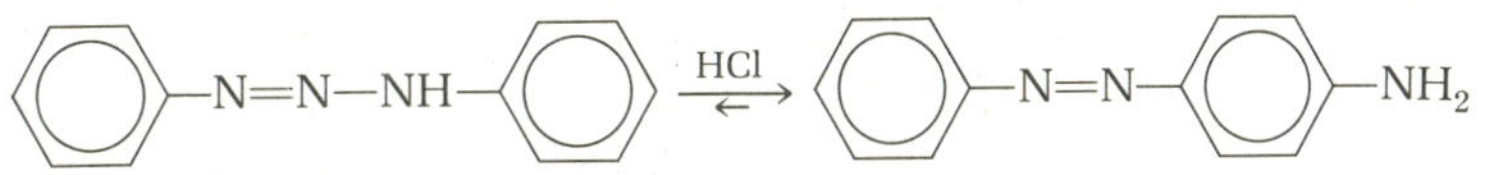

It is thus necessary to find a pH such that a substantial portion of the amine is unprotonated and, at the same time, the diazoamino compound will not accumulate.

Azo coupling reactions with phenols involve somewhat different pH problems, which nonetheless revolve around the same objectives— maximizing the rate of the coupling reaction while avoiding side reactions. Phenoxide ions are much more nucleophilic than the undissociated phenols so that phenols and naphthols are allowed to react with diazonium ions at distinctly alkaline pH's.

(19.40)

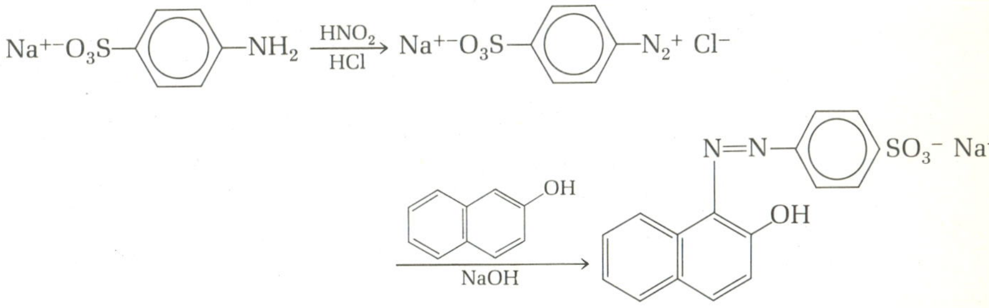

1-(*p*-sulfobenzeneazo)-2-naphthol
orange II (80%)

The danger in going to an excessively high pH lies in the fact that the diazonium ions may be converted to *diazotate ions*, which are inactive as coupling reagents.

(19.41)

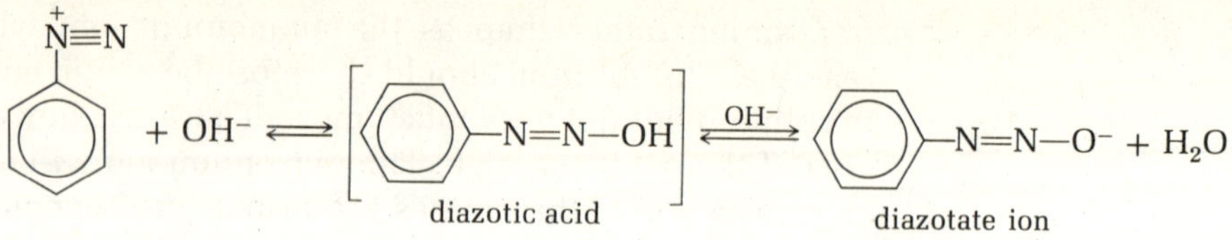

Not surprisingly the position of the diazonium ion-diazotate ion equilibrium is *very* pH dependent.

Commercially, the compounds used for coupling purposes usually contain one or more sulfonic acid groups. This group is dissociated in aqueous solution, making the reactants and the dye product water-soluble, which is obviously highly desirable. Tetrazotized diamines may be coupled with 2 moles of amine or phenol (equation 19.42); coupling may also be accomplished to *both* amines and phenols if the pH is suitably controlled.

(19.42)

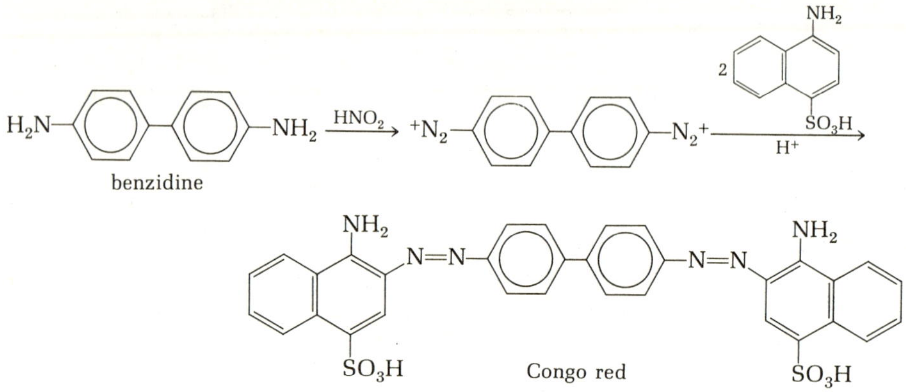

(19.43)

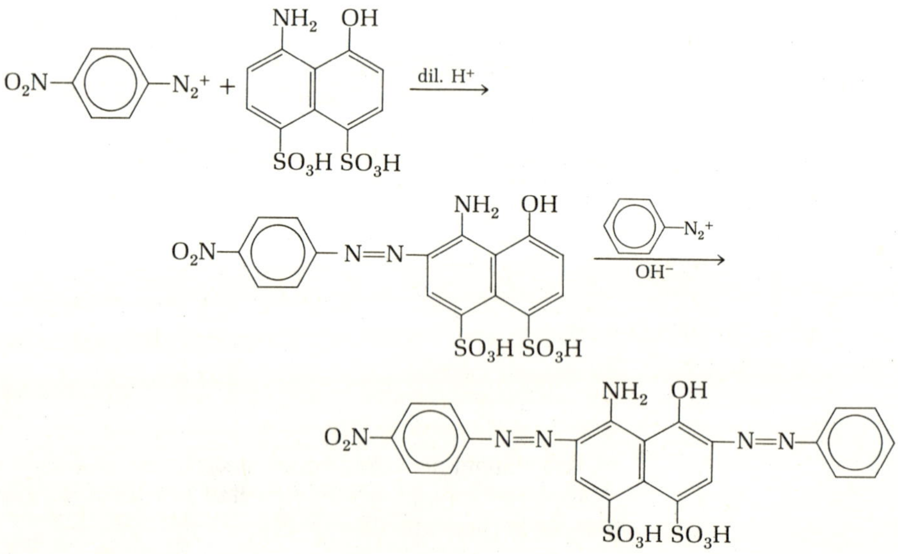

PROBLEM

3 Under the given reaction conditions (equation 19.43), why does the first azo coupling take place in the amino-substituted ring? How could this first reaction be directed into the hydroxyl-containing ring?

DIAZO COMPOUNDS

Turning our attention back to the aliphatic diazonium ions, we might note that these ions may undergo reactions other than nitrogen loss. If, for instance, the diazonium ion contains a hydrogen atom bound to the α-carbon atom, this hydrogen atom could be lost as a proton, giving rise to a neutral **diazo compound** (equation 19.44).

(19.44)

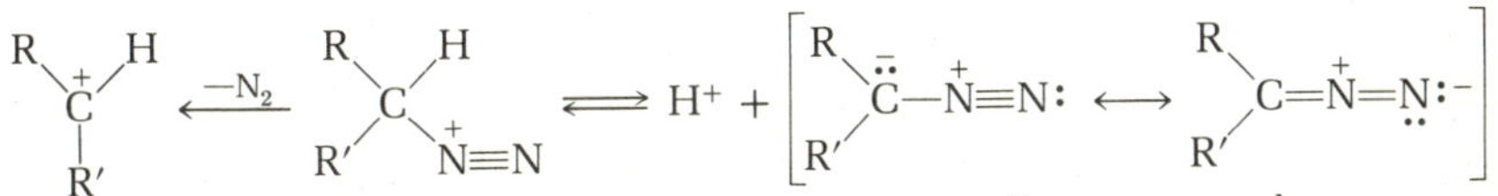

diazo compound

In the presence of proton donors (even water or alcohol), simple diazonium ions tend to decompose in the manner previously described, i.e., by losing molecular nitrogen and forming a carbonium ion. If the diazonium ion carries strongly electron-withdrawing groups, however, then it may well lose a proton preferentially and be converted to a relatively stable diazo compound.

(19.45) $$H_2NCH_2CO_2CH_2CH_3 + HNO_2 \xrightarrow{\text{ether-}H_2O} N_2CHCO_2CH_2CH_3 + H_2O$$

ethyl diazoacetate (85%)

A typical case involves the diazotization of the ethyl ester of glycine (aminoacetic acid). The carbethoxy group promotes the loss of a proton from the α-carbon atom and it also stabilizes the diazo compound. Note also that the reaction is run in a two-phase system.

(19.46)

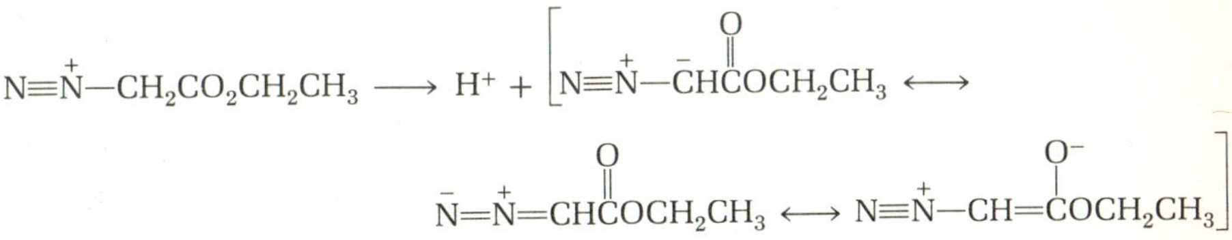

The diazo compound is not very water-soluble and immediately upon formation it is taken up into the ether layer where it is protected from the action of the acid.

This method of synthesis is applicable to diazo compounds carrying electron-withdrawing substituents but not to others. Diazo compounds not possessing such substituents must be formed under nonacidic conditions since they are very susceptible to acid-catalyzed decomposition by way of diazonium ions.

Disubstituted diazo compounds have traditionally been made, and still are made, by oxidation of the hydrazone of the corresponding ketone. The syntheses of diphenyldiazomethane (equation 19.47) and dimethyl diazomalonate (equation 19.48) are typical. Mercuric and silver oxides and lead tetraacetate are common oxidants.

(19.47)

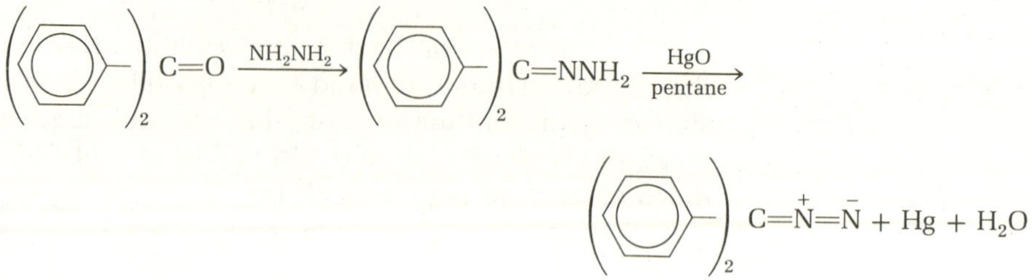

(90%)

(19.48) $H_2NN{=}C(CO_2CH_3)_2 \xrightarrow[\text{THF}]{Ag_2O} N_2C(CO_2CH_3)_2$ (89%)

An increasingly popular method of forming diazo compounds starts with *p*-toluenesulfonylhydrazine (tosylhydrazine, **2**).

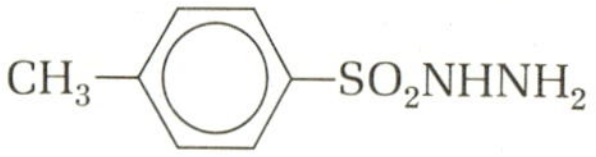

2

This compound is allowed to react with an aldehyde or ketone in the usual manner to produce a substituted hydrazone. The hydrazone is coverted to its salt by a strong base and the salt is pyrolyzed to yield the diazoalkane, which may be trapped before it decomposes further.

(19.49)

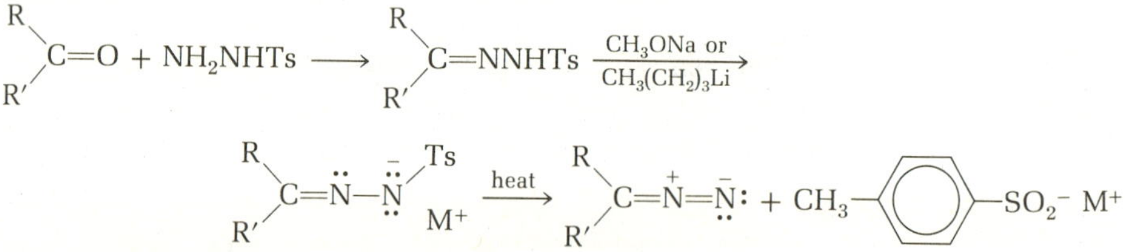

The reaction depends on the ability of a derivative of a sulfonic acid ($-SO_3H$) to be reduced to a derivative of a sulfinic acid ($-SO_2H$).

(19.50)

$$(CH_3)_2CHCH{=}NNHTs \xrightarrow{C_4H_9Li} (CH_3)_2CHCH{=}N\bar{N}{-}Ts\ Li^+ \xrightarrow[100\text{–}125^\circ]{0.3\text{ mm, dry}} (CH_3)_2CHCH{=}N_2$$

(65–75%)

The simplest diazoalkane is diazomethane, CH_2N_2. This compound was apparently prepared for the first time in the last decade of the nineteenth century. It is a yellow gas at room temperature, quite toxic, and prone to undergo explosive decomposition. The explosions may be catalyzed by nothing more sinister than a sharp surface. Diazomethane is almost always handled as an ethereal solution since it codistills with ether and is more stable in solution than in pure form.

Diazomethane is usually synthesized by decomposing an *N*-nitroso-*N*-methylamide in a basic medium. The reaction shown in equation 19.51 was used in the original preparation of the compound.

(19.51) $$CH_3N(NO)C(=O)OCH_2CH_3 + 2KOH \xrightarrow[\text{distill}]{\text{ether}} \underset{(50\%,\ \text{in ether})}{CH_2N_2} + CH_3CH_2OH + K_2CO_3$$

Some other variants are shown in the following equations; several of these methods may be adapted to the synthesis of homologues of diazomethane.

(19.52) $$CH_3N(NO)C(=O)NH_2 \xrightarrow[\text{ether}]{\text{KOH, }H_2O} \underset{(65\%)}{CH_2N_2} + KNCO + 2H_2O$$

(19.53) 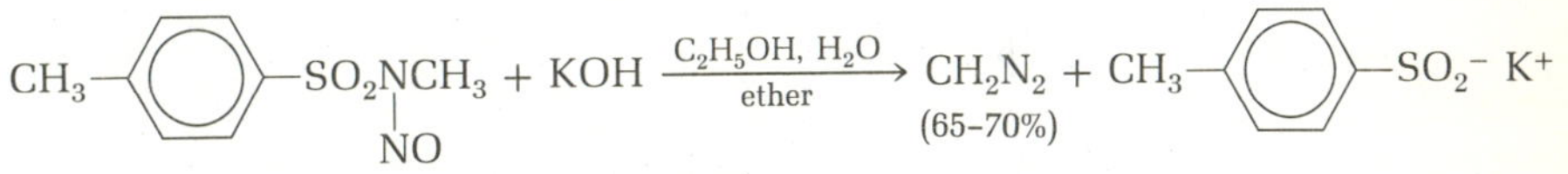

The structure of diazomethane is shown below, indicating that it has a linear C—N—N skeleton. This geometry was supported by electron diffraction studies (like X-ray diffraction, but performed on gaseous molecules) and by dipole moment measurements.

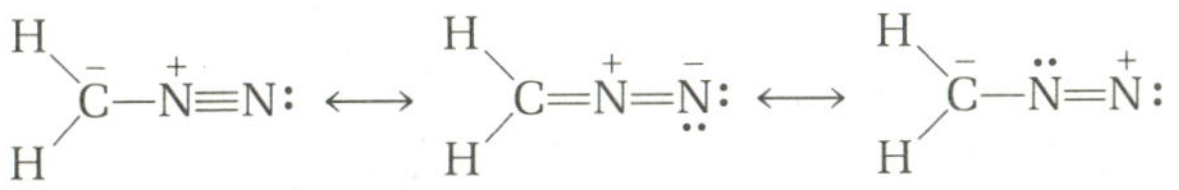

For a number of years the cyclic structure **3** was favored for diazomethane, for it avoided the necessity of writing dipolar structures.

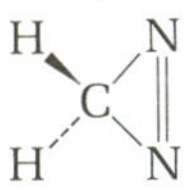

3

Compound **3**, now called **diazirine**, has been synthesized by an unambiguous route. It is definitely *not* diazomethane and while it does share some properties with diazomethane, it also exhibits important differences that can be logically related to the structural differences between the two compounds.

MODES OF REACTION OF DIAZO COMPOUNDS

Diazo compounds can react by several fundamentally different reaction paths, although an examination of the products does not always reveal the reaction mechanism. Two very important reaction modes are outlined below for diazo compounds

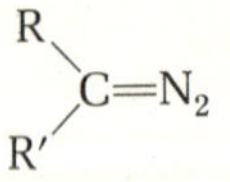

1 The compound may act as a base or as a nucleophile, with the carbon atom being the reaction site. This aspect is emphasized by the canonical structure $RR'\overset{-}{C}\text{—}\overset{+}{N}\equiv N$. The nature of R and R′ exerts a powerful effect on this reaction pathway.

2 Photochemical or thermal activation of diazo compounds may cause them to lose molecular nitrogen and form **carbenes,** highly reactive divalent carbon species.

(19.54) 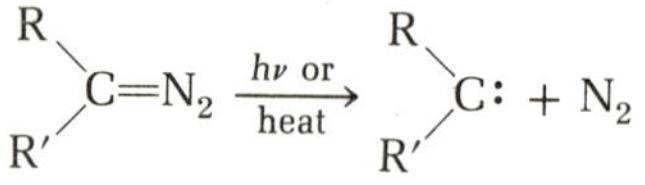

DIAZO COMPOUNDS AS NUCLEOPHILES

Diazoalkanes are not very powerful nucleophiles and they do not displace halide or tosylate ions from alkyl halides or tosylates. They react readily with carbonyl compounds, however. Equation 19.55 depicts a plausible general scheme for the reaction of diazomethane with aldehydes or ketones.

(19.55) 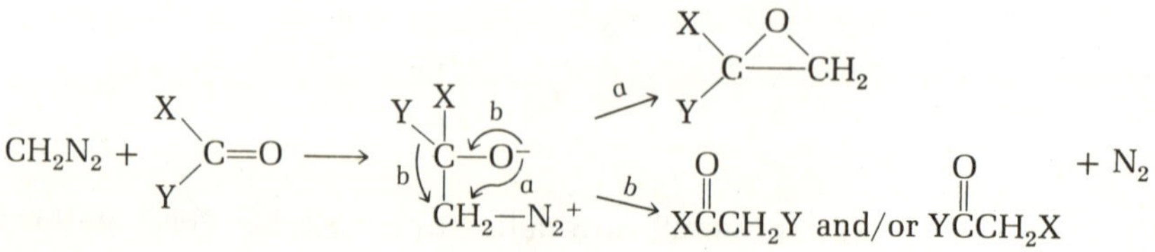

4

The postulated intermediate **4** is a diazonium ion and loses nitrogen very readily; nitrogen expulsion is concerted with either C—O bond formation or 1,2-group migration.

(19.56)

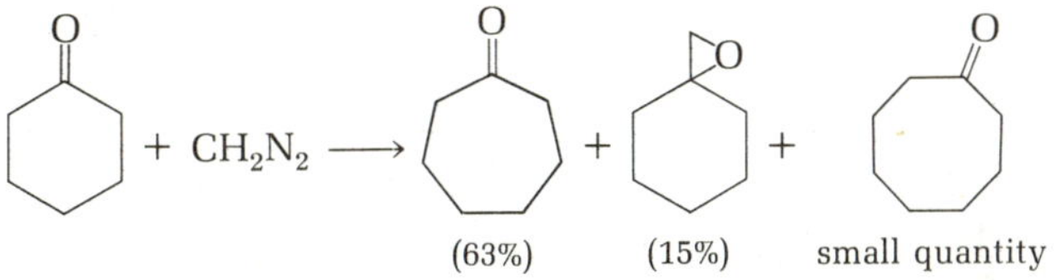

Diazomethane reacts efficiently with acid chlorides, displacing chloride ion and then losing a proton from the acyl diazonium ion. The hydrogen chloride liberated is consumed by a second mole of diazomethane, which can act as a base as well as a nucleophile.

(19.57)

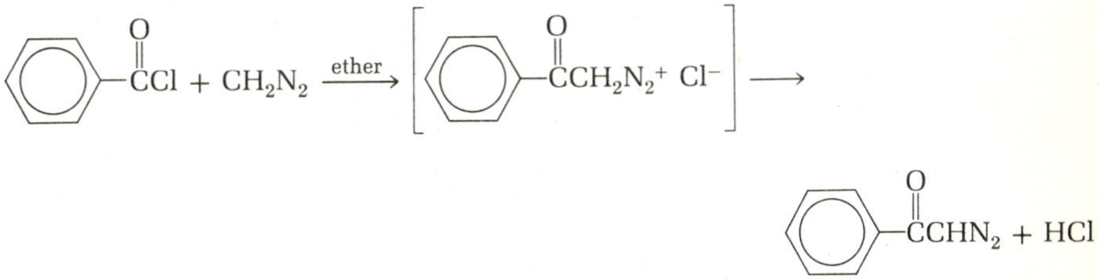

(19.58) $HCl + CH_2N_2 \longrightarrow [CH_3N_2^+\ Cl^-] \longrightarrow CH_3Cl + N_2$

The diazoketones that are the products of the acid chloride-diazomethane reaction are intermediates in the *Arndt-Eistert synthesis,* in which an acid is converted to its next higher homologue. This synthesis depends on a facile rearrangement to which diazoketones are prone (the *Wolff rearrangement;* silver is a catalyst). The mechanism of this rearrangement and related ones was extensively analyzed in Chapter 18.

(19.59)

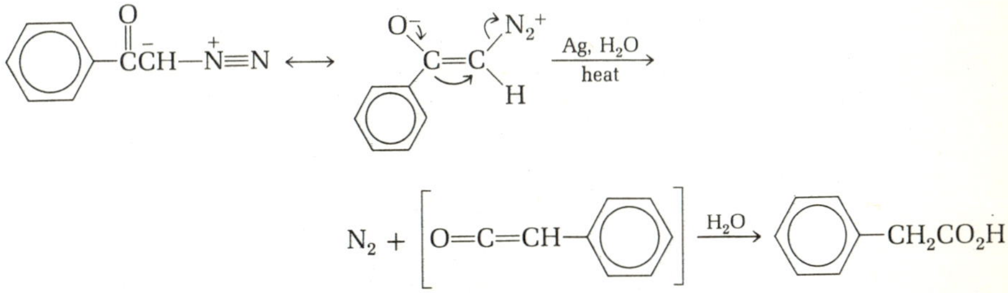

The reaction of diazomethane with carboxylic acids, to produce methyl esters, has been illustrated in Chapter 17. Here again, the basicity of diazomethane is apparent.

ELECTRONIC STRUCTURES OF CARBENES

Perhaps the most interesting role in which diazo compounds find themselves is as precursors of carbenes. All diazo compounds can be induced to eject nitrogen, either thermally or photochemically, thereby giving rise to species in which one carbon atom is formally divalent (equation 19.54). These carbenes are uncharged, the carbon atom in question carrying two ligands and having a total of six electrons in its valence shell. Not surprisingly, carbenes are highly reactive substances.

There is one well-known stable compound that may be considered to contain a divalent carbon atom; that is carbon monoxide. Experience suggests, and theoretical calculations support, the conclusion that carbon and oxygen share six electrons in carbon monoxide, one pair having been "contributed" by the oxygen atom alone.

$$:C{=}\ddot{O}: \longleftrightarrow :\overset{-}{C}{\equiv}\overset{+}{O}:$$

Such electron donation by oxygen would result in positive charge accumulating on oxygen and negative charge accumulating on carbon, in opposition to the normal charge distribution arising from electronegativity differences. Carbon monoxide has a very small dipole moment, in accord with our picture of a resonance effect in opposition to the electronegativity effect.

To return to carbenes proper, the simplest member of the series is methylene, CH_2, which is obtainable from diazomethane. It can also be obtained by photolyzing ketene, $CH_2{=}C{=}O$, in the gas phase.

(19.60) $$CH_2{=}\overset{+}{N}{=}N^- \xrightarrow[\text{heat}]{h\nu \text{ or}} :CH_2 + N_2$$

Methylene itself is an intriguing molecule, and despite its simplicity its existence poses a number of interesting theoretical questions.

Every carbon atom has four available atomic orbitals in its valence shell. In methylene two of these orbitals participate in the formation of σ-bonds to the hydrogen atoms. That leaves two nonbonded electrons and two orbitals to accommodate them; both electrons may be paired in a single atomic orbital or each of the two remaining orbitals may instead be singly occupied. If both electrons are paired, then we have a situation similar to that found in a carbonium ion in which there are three electron pairs in the valence shell of carbon. We might then expect the carbon atom to adopt sp^2-hybridization (**5**), although the H—C—H bond angle would not necessarily equal 120°. Should the nonbonded electrons elect to occupy different orbitals, then there would be only two electron pairs in the valence shell. These pairs would probably seek to be as distant from one another as possible, so that the carbon atom would be sp-hybridized. The nonbonded electrons could still have their spins paired, as in **6**, but they need not; the electron spins could be parallel, as in **7**.

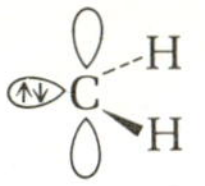

5

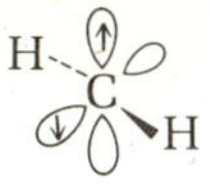

6

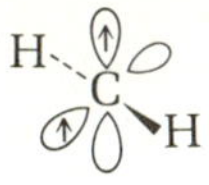

7

Carbenes **5** and **6** are *singlet* carbenes (Chapter 8) since all their electrons are spin paired. Carbene **7** is a triplet, however, since it has a total spin quantum number of one. We shall see that singlet and triplet species often exhibit characteristically different chemical behavior.

REACTIONS OF METHYLENE

Although methylene normally has a half-life of only fractions of a second, it has proved possible to study it spectroscopically in the gas phase. These studies have disclosed that the gas-phase methylene is actually a mixture of two species. They are unequal in energy and the high energy form undergoes slow conversion to the low energy species.

The high energy molecule was shown to be bent and is undoubtedly the singlet **5**. The lower energy species was found to be linear and when it was generated in an unreactive solid matrix at very low temperatures, it gave an esr signal. It must therefore be the triplet **7**. This finding would explain why **5** is converted only slowly to **7**, even though the process is exothermic. Changes in spin multiplicity are forbidden processes and are therefore quite slow, particularly in small molecules.

By adjusting the experimental conditions, it is possible to produce gas-phase methylene samples rich in either the singlet or triplet forms. (Decomposition of diazomethane in solution seems to yield mostly singlet methylene.) When triplet methylene is formed in the presence of butane, its sole reaction seems to be the abstraction of a hydrogen atom; in other words, triplet methylene behaves as if it were a diradical, which is compatible with its structure. The final reaction products are those expected from the mixture of radicals shown in equation 19.61.

(19.61) $$CH_2N_2 \xrightarrow[\text{sensitizer}]{h\nu} {}^3{:}CH_2 \xrightarrow{CH_3(CH_2)_2CH_3} CH_3CH_2CH_2CH_2\cdot + CH_3CH_2\dot{C}HCH_3 + CH_3\cdot$$

Singlet methylene, on the other hand, does not display radical traits. Its most characteristic reaction in this system is its tendency to **insert** directly into a carbon-hydrogen bond, without the formation of any intermediates.

(19.62) $$CH_2N_2 \xrightarrow[\text{heat}]{h\nu \text{ or}} {}^1{:}CH_2 \xrightarrow{CH_3CH_2CH_2CH_3} CH_3(CH_2)_3CH_3 + CH_3\underset{\displaystyle CH_3}{\underset{|}{C}}HCH_2CH_3$$

The insertion is visualized as a three-center process.

(19.63) 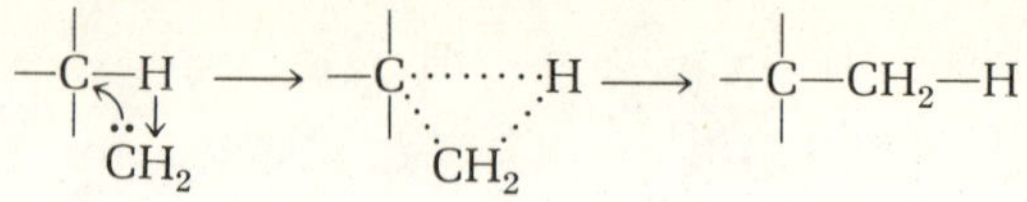

The reaction is very exothermic and methylene discriminates only very slightly among primary, secondary, and tertiary hydrogen atoms.

If the reacting molecule contains a double bond, then singlet methylene inserts preferentially into the π-bond. This reaction is extremely important in cyclopropane synthesis.

(19.64) 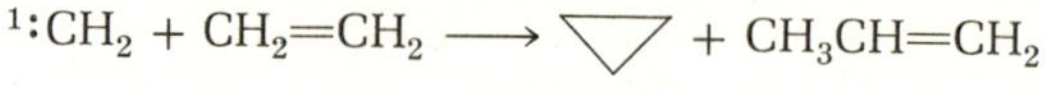

STRUCTURES OF OTHER CARBENES

Other carbenes are named as derivatives of methylene. In the preferred system, methylene is called "carbene" for the purposes of substitutive nomenclature. Carbenes are sometimes named as "methylenes" rather than as carbenes.

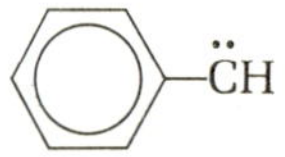

$:C(CO_2CH_3)_2$

$Br\ddot{C}OCH_3$

phenylcarbene | bis(carbomethoxy)carbene | bromomethoxycarbene

Phenylcarbene, diphenylcarbene, and most alkylcarbenes seem to have triplet ground states, although they are usually generated initially as singlets. Diphenylcarbene appears able to undergo spin inversion with unusual rapidity, reaching its ground state in a very short period of time. The ground state multiplicity of carbenes with keto and ester groups attached is still a subject for investigation.

Virtually all carbenes containing a halogen atom (fluorine, chlorine, or bromine) appear to have singlet ground states. Overlap between the filled *p*-orbitals on the halogens and the empty *p*-orbital of the carbene strongly stabilizes the singlet state. Difluorocarbene is, in fact, one of the most stable and least reactive carbenes.

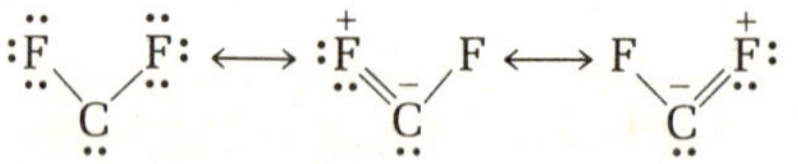

Halocarbenes cannot normally be obtained from diazo precursors but they can be obtained by decomposition of diazirines.

(19.65) 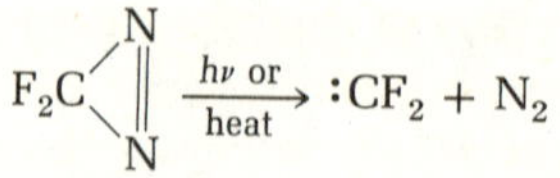

These various carbenes are useful in synthesis, particularly of compounds containing three-membered rings. For cyclopropane syntheses it is usual to employ catalysts during the decomposition of the diazo compounds.

(19.66)

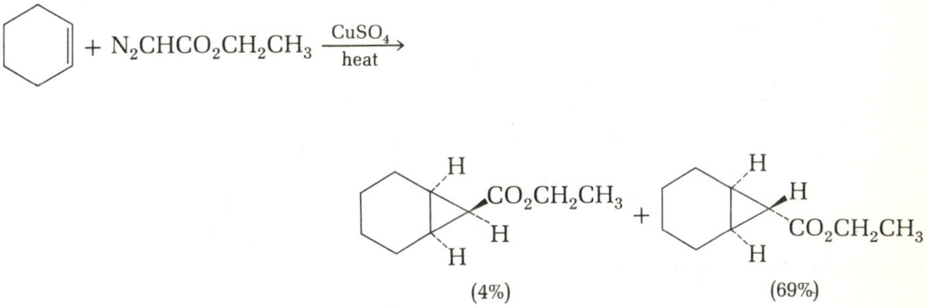

(19.67)

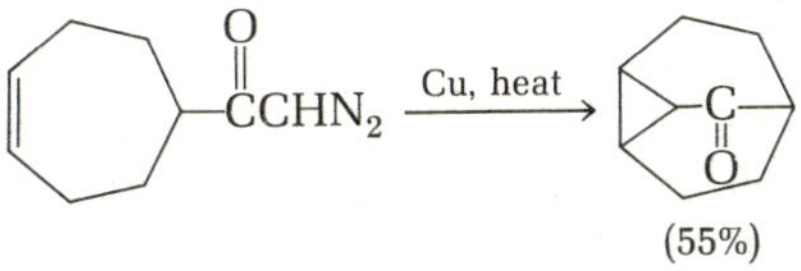

(19.68)

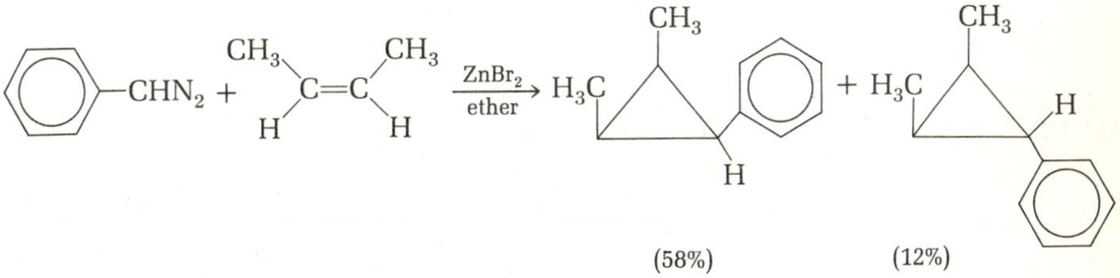

These catalysts actually complex with the carbene and alter its reactivity. The complexed carbenes tend to behave as singlets; they exhibit little tendency to insert in C—H bonds and they are more selective than their uncomplexed analogues. Copper and its salts are the most often used catalysts.

Equation 19.68 illustrates the marked stereospecificity observed when singlet carbenes add to alkenes. The methyl groups were *cis* in the starting olefin and they remained *cis* in the product cyclopropanes. Such stereospecificity is reminiscent of Diels-Alder additions to alkenes. The behavior of triplet carbenes in double bond additions is in marked contrast to that of their singlet counterparts. The new σ-bonds must be formed one at a time, with an intervening spin inversion so that the electrons in all the bonds will be spin paired. Since spin inversion is a slow process, there is time for rotation to take place about the single bonds of the diradical intermediate.

(19.69)

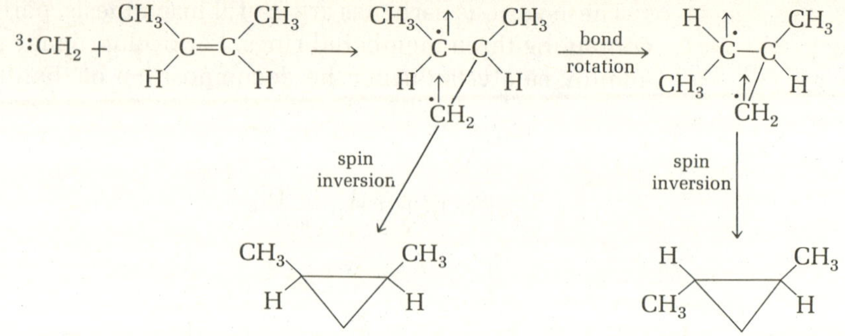

CARBENES BY α-ELIMINATION

We close this chapter by noting that carbenes are available by a variety of routes that do not require diazo compounds or diazirines. One approach is the *α-elimination* reaction (equation 19.70). The principal requirement is that AB be a reasonably stable molecule, although A and B need not be lost simultaneously. Most halocarbenes are generated in this manner.

(19.70)

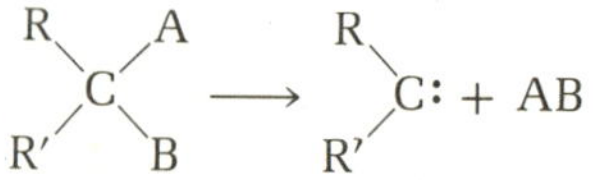

(19.71) $(CH_3)_3CO^-K^+ + CHCl_3 \rightleftharpoons (CH_3)_3COH + K^+{:}CCl_3^-$

(19.72) $K^+{:}CCl_3^- \xrightleftharpoons{\text{slow}} K^+Cl^- + {:}CCl_2$

(19.73)

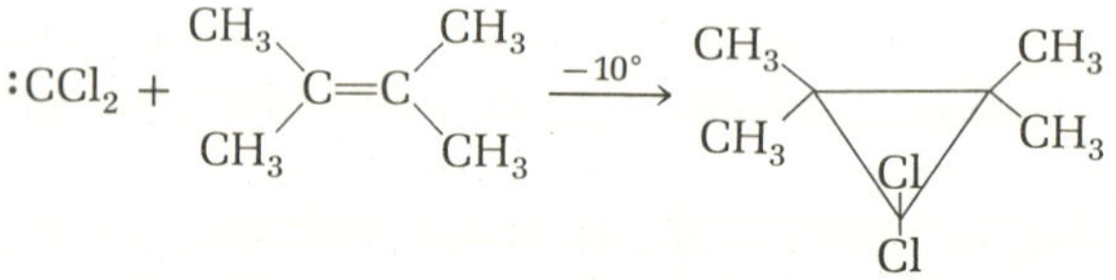

An important and nagging question is the following: Is the "free" carbene $R_2C{:}$ really present or is it still complexed with AB as it begins to attack the olefin? There is evidence that *both* situations obtain; the complexed carbenes are labeled *carbenoids*, a term that also applies to the copper- and zinc-complexed carbenes (equations 19.66 to 19.68).

(19.74)

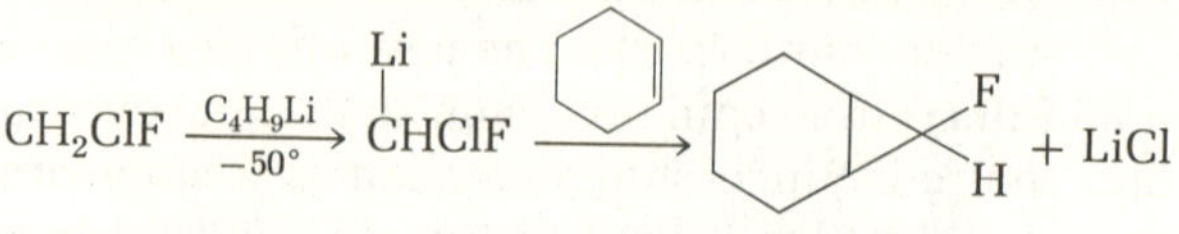

mixture of stereoisomers

SUMMARY

1 Solutions of nitrous acid are able to transfer the nitroso group, —N=O, to appropriate substrates. The nitrogen atoms of amines may be nitrosated under these conditions. Secondary amines are converted to stable nitrosamines that are only slightly soluble in dilute acid. Under the conditions usually employed for these reactions, the principal nitrosating agent is dinitrogen trioxide.

Trialkyl amines react rapidly and reversibly with N_2O_3. The nitrosoammonium ions so produced decompose only slowly at low temperatures.

2 The nitroso derivatives of alkyl primary amines undergo rapid tautomeric shifts and eventually lose hydroxide to form alkyldiazonium ions. These ions lose nitrogen with great ease, giving rise to energy-rich carbonium ions that are prone to undergo extensive rearrangement. A wide spectrum of products is observed, including cyclopropanes in some cases.

3 Tertiary amines carrying an aryl substituent can usually be nitrosated in the ring, while secondary aryl amines are nitrosated on nitrogen, just as their alkyl counterparts are. Primary aromatic amines are converted to diazonium salts that are very much more stable than their aliphatic analogues. Conjugation between the diazonium group and the ring, as well as the sp^2-hybridization of the carbon atom, is responsible for a substantial energy of activation for nitrogen loss.

4 Aromatic diazonium ions decompose in hot aqueous acid to give phenols in good yield. Phenyl cation appears to be an intermediate in these decompositions; it displays very little selectivity toward nucleophiles. Because of the high energy content and indiscriminate nature of phenyl cation, a great many reactions in which the diazonium group is replaced actually proceed by way of radical intermediates. Such is the case with the reaction of diazonium ions with iodide ions and also for the replacement of the diazonium group by chloride, bromide, or cyanide, catalyzed by the appropriate cuprous salt. The catalyst serves essentially as an electron transfer agent between anion and diazonium ion.

5 The diazonium group may also be replaced by hydrogen again by way of a free radical process. Hypophosphorus acid is generally the most effective reductant. Because of the ease of disposal of the diazonium group, nitro and amino groups are customarily used in aromatic syntheses as blocking and directing substituents, respectively. When they have served their purpose, they are converted to diazonium groups and ejected.

6 Diazonium ions may attack suitably activated aromatic rings to give azo compounds. The activating substituent is most commonly either an amino or a hydroxyl group. Amino compounds are coupled in acidic media; the

pH is controlled to avoid either complete protonation of the amine or formation of stable diazoamines. Phenolic compounds are coupled under basic conditions in order to take advantage of the very high reactivity of phenoxide ions. Extremely high pH's must be avoided in order that the diazonium ion not be converted to a diazotate anion.

7 Diazotization of a primary amine carrying an electron-withdrawing substituent on the α-carbon atom can result in proton loss being faster than nitrogen loss. Under such circumstances diazo compounds may be isolated. Oxidation of hydrazones provides another route to diazo compounds; this method makes available compounds not available by way of the diazotization reaction. Diazomethane is usually synthesized through the decomposition of *N*-methyl-*N*-nitroso compounds under basic conditions.

8 Diazo compounds react as nucleophiles, particularly toward carbonyl compounds. Aldehydes and ketones yield products formally derived by C—C and C—O insertion. Acid chlorides react with diazomethane to form diazoketones that may be rearranged to carboxylic acids having one more carbon atom than the starting acid chloride.

9 Most importantly, diazo compounds decompose thermally or photolytically to yield carbenes. The two most common ground states of carbenes are a triplet, with the two unshared electrons occupying different orbitals, and a singlet, in which the spin-paired electrons share the same orbital. The chemical properties of singlet and triplet carbenes are noticeably different, the triplet tending to behave as a diradical. Singlets, on the other hand, give the impression of being strongly electrophilic species. In addition, they display the unusual feature of being able to insert directly into carbon-hydrogen bonds. Singlet carbenes also insert into the π-bonds of olefins, all steric relationships that existed in the olefin being retained in the cyclopropane product.

10 Carbenes are accessible by way of α-elimination reactions; this is the most common method for generating halocarbenes. Halocarbenes undoubtedly have singlet ground states that are stabilized by overlap of the vacant *p*-orbital on carbon with the nonbonding electron pairs of the halogen atoms.

PROBLEMS

Note: Problems 1 through 3 appear in text.

4 Nitrosation is normally effected in hydroxylic solvents, such as water or ethanol, with nitrous acid as the reagent. It is possible, however, to diazotize amines in

nonhydroxylic solvents such as chloroform, using an alkyl nitrite, RON=O, as the diazotizing agent. One equivalent of acetic acid is usually added as well (equation 1). Write a mechanism for this reaction.

$$R'NH_2 + RONO + CH_3CO_2H \longrightarrow R'N{\equiv}N\ CH_3CO_2^- + ROH + H_2O \qquad \textbf{(1)}$$

5 The reaction of monosubstituted hydrazines, $RNHNH_2$, with nitrous acid leads to the formation of organic azides, RN_3. Write a detailed mechanism for this reaction.

6 Indicate what the likely products would be from treatment of the following amines with nitrous acid in the cold.

a. Methylethylamine
b. *N*-Methyl-2-butylamine
c. Diethylisopropylamine
d. *N*,4-Diethylaniline
e. *t*-Butylamine
f. 2-Butylamine
g. *N*-Methyl-*N'*-propylpentane-2,4-diamine
h. Cyclobutylamine

7 When *p*-aminobenzylamine (I) is subjected to the action of one equivalent of nitrous acid under strongly acidic conditions (pH ~ 1), only the aromatic amino group is diazotized.

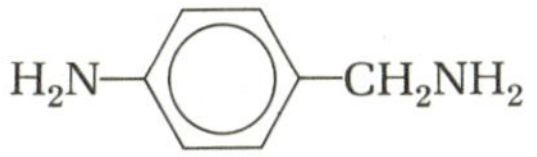

I

One can isolate *p*-hydroxybenzylamine in good yield after the reaction mixture is heated. Account for the selectivity of the diazotization process.

8 Describe efficient methods for converting aniline to the following compounds.

a. Bromobenzene
b. Phenol
c. 1-Phenylethanol
d. 4-Chlorodiphenylmethanol
e. *p*-Nitrobenzonitrile
f. *p*-Nitropropiophenone
g. *p*-Cyanobenzophenone
h. Ethyl *m*-bromobenzoate
i. Benzaldehyde-*m*-sulfonic acid
j. *m*-Nitroacetophenone
k. *m*-Dichlorobenzene
l. 5-Bromo-2-chloronitrobenzene
m. *m*-Propylbenzoyl chloride

9 At relatively high pH (still under 7), the rate of diazotization of aniline is proportional only to the square of the concentration of nitrous acid. When lower pH's are reached, the reaction becomes third order; it depends on the square of the nitrous acid concentration *and* on the concentration of amine.

a. What is the origin of the second-order dependence of the rate on nitrous acid concentration?

b. Why is the rate independent of amine concentration at higher pH's but proportional to the amine concentration at lower pH's?

c. When dilute hydrochloric or hydrobromic acids are used to prepare the nitrous acid, a new term appears in the rate equation (equation 2); the rate now becomes proportional to the concentrations of protons, halide ions, nitrous acid, and amine.

$$\text{Rate} = k(C_6H_5NH_2)(HNO_2)^2 + k'(C_6H_5NH_2)(HNO_2)(H^+)(X^-) \quad (2)$$

What are the nitrosating agents available under these conditions?

10 Describe syntheses of the following compounds that minimize the necessity of difficult isomer separations. Use any monosubstituted aromatic compound as a starting material.

a. *m*-Nitrophenol

b. *o*-Iodofluorobenzene

c. *p*-Nitrocyanobenzene

d. *m*-Chlorobenzylamine

e. 3-Bromo-4-cyanobenzoic acid

f. 3,5-Dibromochlorobenzene

g. 1,4-Benzenedicarboxylic acid (terephthalic acid)

h. 1-Acetyl-4-bromonaphthalene

i. Biphenyl-3-sulfonic acid

j. *p*-Methoxyacetophenone

11 Write a detailed mechanism for the acid-catalyzed conversion of diazoaminobenzene to *p*-aminoazobenzene (equation 19.39).

12 The decomposition of azomethane, $CH_3N{=}NCH_3$, in the gas phase, induced by either heat or light, results in the formation of ethane and nitrogen. Decomposition of azoethane under the same conditions yields ethylene, ethane, and butane. Write mechanisms for these reactions.

13 Show in detail how diazomethane converts heptanal to 2-octanone.

14 How might the following compounds be synthesized?

a. Ethyl cyclopropanecarboxylate

b. Cyclopropyl methyl ketone

c. Ethyl 2,2-diphenylcyclopropanecarboxylate

d. Cycloheptatriene

e. 1,1-Dichloro-*trans*-2,3-diphenylcyclopropane

f. 6,6-Difluorobicyclo(3,1,0)hexane

15 One class of organic compounds that appears to contain a divalent carbon atom is the isocyanides, RNC. Nonetheless, these compounds are reasonably stable and not unusually reactive. What is the electronic structure of these compounds, and how does it account for their properties?

16 Diphenylcarbene, unlike most other carbenes, reacts very readily with molecular oxygen, and benzophenone can be isolated from the reaction mixture. Why is diphenylcarbene so easily oxidized? Write a mechanism for the reaction. (*Hint:* What is the ground state of diphenylcarbene?)

17 a. When phenol is dissolved in strong aqueous base and then treated with chloroform, a good yield of salicylaldehyde (*o*-hydroxybenzaldehyde) may be isolated (after acidification). Show in detail how this product might plausibly have been formed. A small quantity of *p*-hydroxybenzaldehyde is also isolated.

b. When *p*-cresol is treated with chloroform and aqueous base, then a reaction very similar to that described above takes place. A 25% yield of 2-hydroxy-5-methylbenzaldehyde is isolated and along with it 12% of another compound of molecular formula $C_8H_8Cl_2O$. The spectra of this second compound indicate that it is not aromatic and that it contains a carbonyl group. Propose a structure for this material.

18 Ethyl azidoformate (II) may be decomposed either thermally or photochemically. When the decomposition is allowed to proceed in cyclohexene, the products III, IV, and V are isolated.

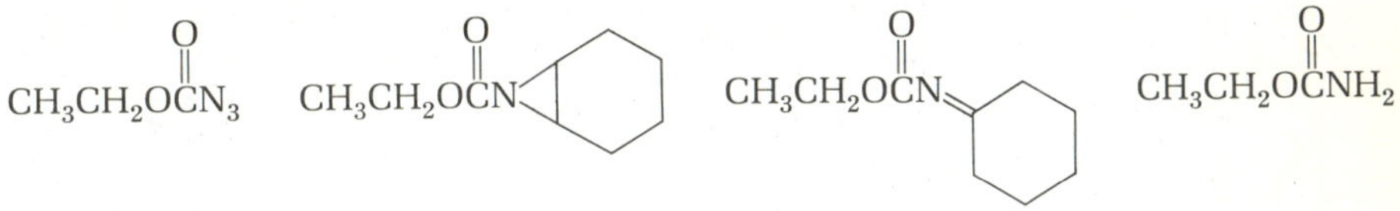

II III IV V

Write a mechanism that accounts for the formation of these products.

19 Diphenylacetylene and benzal chloride (dichlorophenylmethane) are dissolved together in benzene and the solution then allowed to react with potassium *t*-butoxide. Work-up of the benzene layer affords a white crystalline material **A**, $C_{25}H_{24}O$. The nmr spectrum of this compound shows absorption due to aromatic protons, as well as a singlet in the methyl region. The ratio of peak areas is 5:3. When **A** is dissolved in ether and dry hydrogen bromide then passed in, compound **B** precipitates, $C_{21}H_{15}Br$, mp, 270°. **B** is insoluble in benzene, ether, and chloroform but soluble in aqueous ethanol. Write structures for **A** and **B**, and account for the properties of **B**. (*Hint:* Check back to Chapter 14.)

SUPPLEMENTARY READINGS

S. Patai, ed., *The Chemistry of the Amino Group*. New York: Interscience Publishers, 1968, Chapter 8.

H. Zollinger, *Azo and Diazo Chemistry*. New York: Interscience Publishers, 1961.

J. Hine, *Divalent Carbon*. New York: The Ronald Press Co., 1964.

CHAPTER 20

ALDEHYDES AND KETONES II — THE REACTIONS OF ENOLS AND ENOLATES

PROTON EXCHANGE PROCESSES

If acetone is dissolved in neutral D_2O at 25°, no reaction is observed; the acetone may be recovered unchanged. If the solution is made slightly basic, however, then the acetone and the D_2O begin to react. Hydrogen atoms are exchanged between the solute and the solvent, a process that may be followed conveniently with the aid of nmr spectroscopy. Furthermore, it has been found that acid catalyzes the exchange of protons between acetone and water as well. Using either catalyst, one may easily prepare acetone-d_6 from ordinary acetone and D_2O.

(20.1) $$6D_2O + CH_3\overset{\overset{O}{\|}}{C}CH_3 \xrightarrow[D_3O^+]{OD^- \text{ or}} CD_3\overset{\overset{O}{\|}}{C}CD_3 + 6HDO$$

It is interesting to compare the behavior of acetone under these conditions with the behavior of some related compounds. When isopropyl alcohol is dissolved in D_2O, its hydroxyl proton exchanges almost instantaneously with the solvent; the carbon-bound protons resist exchange even at very high or very low pH. Propane is completely inert toward exchange with water.

(20.2) $$CH_3\overset{\overset{OH}{|}}{C}HCH_3 + D_2O \longrightarrow CH_3\overset{\overset{OD}{|}}{C}HCH_3 + HDO$$

There is thus presumptive evidence that the carbonyl group is performing a key role in these exchange reactions. One does not normally find that protons attached to carbon atoms are very labile but acetone seems to provide evidence that exceptions do exist. A series of experiments might then be imagined that would aim at establishing the limits of the influence of the carbonyl group.

Dissolution of acetaldehyde in slightly basic or acidic D_2O results in rapid exchange of the methyl protons but *not* the aldehyde hydrogen atom. When 2-butanone is subjected to the same treatment, nmr spectroscopy shows that only the protons α to the carbonyl group are undergoing exchange.

(20.3) $$CH_3\overset{\overset{O}{\|}}{C}H \xrightarrow[D_3O^+]{OD^- \text{ or}} CD_3\overset{\overset{O}{\|}}{C}H$$

(20.4) $$CH_3\overset{\overset{O}{\|}}{C}CH_2CH_3 \xrightarrow[D_3O^+]{OD^- \text{ or}} CD_3\overset{\overset{O}{\|}}{C}CD_2CH_3$$

The carbonyl group thus seems to labilize only those protons that are bound to carbon atoms immediately adjacent to the carbonyl function.

ENOLS AND ENOLATE IONS

As we shall discuss below, aldehydes and ketones are readily halogenated at the positions α to the carbonyl group. Under a given set of conditions the rate of halogenation of a particular ketone is equal to its rate of hydrogen exchange.

Organic chemists have in fact postulated that two intermediates play crucial roles in the chemistry of the α-positions of carbonyl compounds. They are known as **enols** (acidic or neutral conditions) and **enolate ions** (basic conditions). As the name implies, an enol contains a double bond and a hydroxyl group that is attached to one of the sp^2-hybridized carbon atoms.

The simplest enol is vinyl alcohol, $CH_2{=}CHOH$. This compound, which is a tautomer (Chapter 17) of acetaldehyde, is not isolable because the acetaldehyde form is much more stable and the tautomers are rapidly interconverted under normal conditions (see p. 451).

(20.5) $$CH_2{=}C(H)(OH) \rightleftharpoons CH_3CH{=}O$$

The greater stability of the carbonyl isomer over the enol isomer is most readily rationalized by noting the much greater strength of the C—O π-bond compared with that of the C—C π-bond.

(20.6) 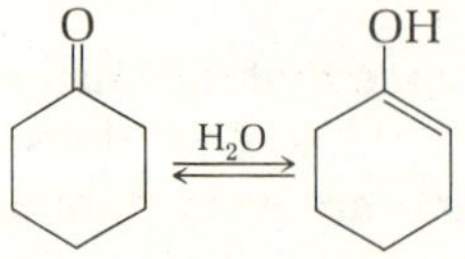 $K^{25^\circ} = 4.1 \times 10^{-6}$

In virtually all simple aldehydes and ketones the equilibrium concentration of enol is miniscule. Under strictly neutral conditions the carbonyl and enol forms interconvert only slowly, but the presence of a small quantity of either acid or base speeds up the attainment of equilibrium quite dramatically. Thus enols, which are present in only very small quantities in aldehydes and ketones, may nonetheless serve as intermediates in the reactions of these compounds, provided that a catalyst is present that will ensure that the tautomers are always in equilibrium.

In order to appreciate the role of enols in proton exchange reactions, it is useful to examine the generally accepted mechanism of acid-catalyzed enolization (equations 20.7 and 20.8).

(20.7) $$CH_3C(=\ddot{O}{:})CH_3 + H_3O^+ \rightleftharpoons CH_3C(=\dot{O}H^+)CH_3 + H_2O$$

(20.8) $$H_2\ddot{O} + H{-}CH_2{-}C(=\ddot{O}H^+)CH_3 \longrightarrow CH_2{=}C({:}\ddot{O}H)CH_3 + H_3O^+$$

acetone enol

If we then write the equations in reverse order and use D_2O as the solvent, it is readily apparent how the deuterium becomes incorporated into the ketone (equations 20.9 and 20.10). The process repeats itself until the hydrogen isotopes are nearly statistically distributed between ketone and solvent.

(20.9) $$CH_2{=}C({:}\ddot{O}H)CH_3 + D_3O^+ \longrightarrow CDH_2{-}C(=OH^+)CH_3 + D_2O$$

(20.10) $$CDH_2{-}C(=OH^+)CH_3 + D_2O \longrightarrow CDH_2C(=O)CH_3 + HD_2O^+$$

The mechanism is slightly different when basic catalysts are used; initial removal of an α-proton by the base leads to the charge-delocalized enolate anion.

(20.11) 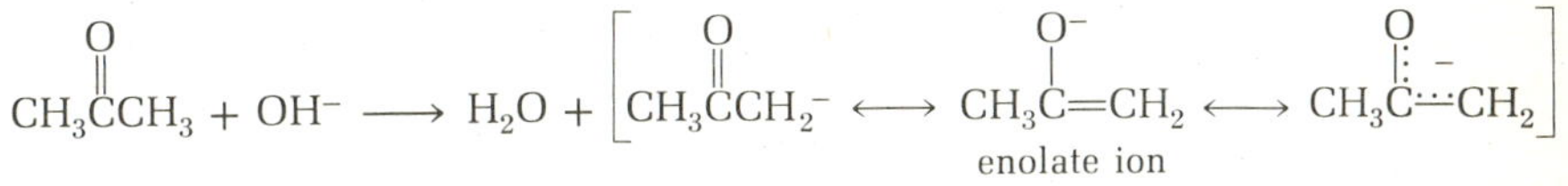

The anion may then react with water to put a proton on either the oxygen or the carbon atom; the latter path leads to isotope exchange.

(20.12) 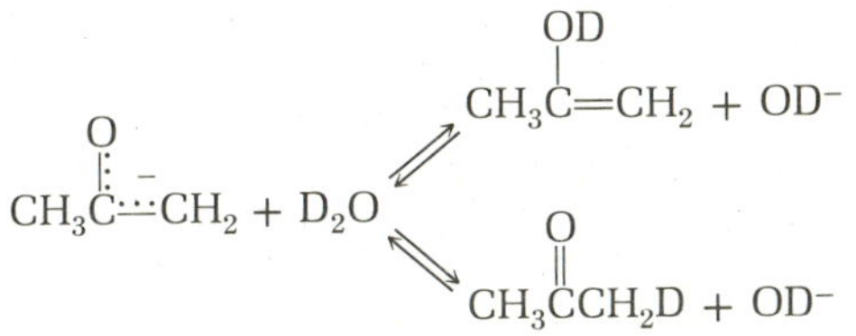

Although both enols and enolate ions are present under basic conditions, the enolate is usually the species participating in any reaction at the α-position because it is so much more reactive than the enol.

PROBLEM 1

a. When nitroethane is treated with sodium acetate in D_2O, it is converted to nitroethane-1,1-d_2. Write a mechanism for this reaction, and show with the aid of resonance structures what features of the reactant promote the reaction.

b. Addition of acetonitrile to a solution of methyllithium results in the evolution of considerable quantities of gas. After addition of D_2O to the solution, one may isolate acetonitrile-*d*, among other products. Write the reaction. What is the structure of the intermediate?

c. Butyric-α,α-d_2 acid may be obtained by heating ordinary butyric acid with DCl-D_2O in a sealed tube at 200° for 3 days. Write a mechanism for this reaction. Why are the conditions so stringent for acid-catalyzed isotope exchange in acids compared with ketones?

Aldehydes and ketones are certainly not strong acids — the pK_a of acetone is about 20. Nevertheless, these carbonyl compounds are certainly more acidic than virtually any hydrocarbon and there can be little doubt that this enhanced acidity is due to the fact that charge is delocalized in the enolate ion, in large part onto the electronegative oxygen atom (equation 20.11).

Carbon-carbon double bonds are normally weak nucleophiles. The carbon-carbon double bond of an enol is a stronger nucleophile than a plain carbon-carbon double bond because of the conjugation of the π-electrons with the *n*-electrons on oxygen.

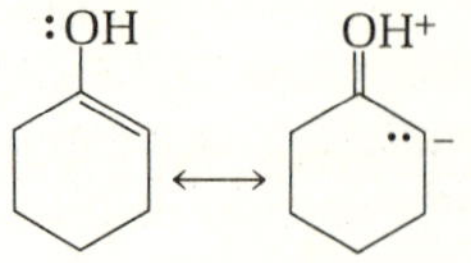

An enolate anion is, of course, a very much stronger nucleophile because the system carries a negative charge.

Enols and enolates are therefore subject to attack by a number of electrophilic species, chlorine and bromine being prominent among these. Bromination is more common because of the relative ease of handling bromine (ketones may also be iodinated).

(20.13) $$CH_3\overset{O}{\overset{\|}{C}}CH_3 + Br_2 \xrightarrow[H_2O,\ 65^\circ]{CH_3CO_2H} \underset{(50\%)}{CH_3\overset{O}{\overset{\|}{C}}CH_2Br} + \text{dibromoacetones} + HBr$$

Acetic acid, either glacial or aqueous, is a common solvent for this reaction, as it promotes enolization.

(20.14) $$CH_3\overset{O}{\overset{\|}{C}}CH_3 \xrightleftharpoons{H^+} CH_3\overset{OH}{\overset{|}{C}}{=}CH_2$$

(20.15) $$CH_3\overset{:OH}{\overset{|}{C}}{=}CH_2 + Br{-}Br \longrightarrow Br^- + CH_3\overset{OH^+}{\overset{\|}{C}}CH_2Br \rightleftharpoons CH_3\overset{O}{\overset{\|}{C}}CH_2Br + HBr$$

The reaction of bromine with enols is extremely rapid — so rapid, in fact, that the enol content of ketones can be determined by rapid titration using very small concentrations of bromine. Reaction of the enol with bromine is complete long before there is any appreciable reestablishment of the keto-enol equilibrium.

PROBLEM

2 Kinetic studies of the halogenation of ketones have shown that, except at very low concentrations of halogen, the reaction kinetics are first order in ketone concentration and *zero order* in halogen concentration. Furthermore, under identical conditions, halogenation and isotope exchange have been found to proceed at the same rate. How do these data fit the mechanism proposed in equations 20.14 and 20.15? What is the rate-determining step under normal conditions?

In the case of acid-catalyzed halogenation of unsymmetrical ketones, the halogen seems to prefer the more highly substituted position.

(20.16)

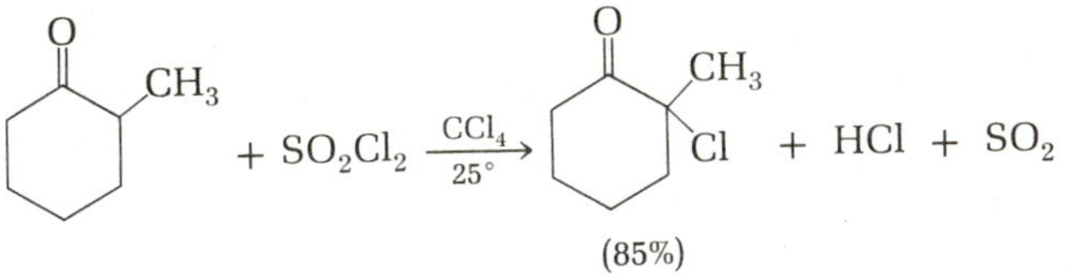

The result may be explained by assuming the greater stability of the more highly substituted of the two possible enols (equation 20.17); the same rule applies to enols as to ordinary alkenes.

(20.17)

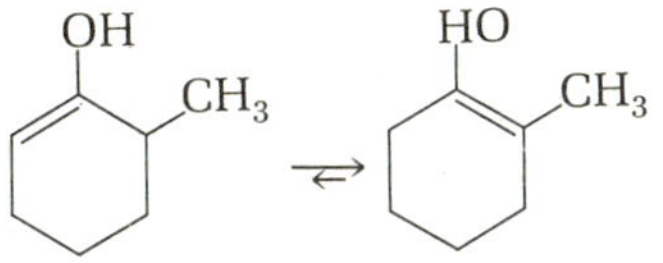

A halogen atom already in the molecule does not, however, serve to promote acid-catalyzed enolization so it is possible to selectively introduce one halogen atom per molecule.

Base-catalyzed halogenation of ketones presents a somewhat different picture. First of all, the halogens react with aqueous base to produce hypohalite ions and it is very likely that the hypohalite ions are responsible for at least part of the halogenation.

(20.18) $$Br_2 + 2OH^- \longrightarrow OBr^- + Br^- + H_2O$$

In fact, chlorinations under alkaline conditions are usually carried out with sodium hypochlorite.

(20.19) $$CH_3\overset{\overset{O}{\|}}{C}CH_3 + OCl^- \rightleftharpoons CH_3\overset{\overset{O}{\vdots}\,-}{C}{\cdots}CH_2 + ClOH \longrightarrow CH_3\overset{\overset{O}{\|}}{C}CH_2Cl + OH^-$$

Secondly, the presence of a halogen atom on a carbon atom α to a carbonyl group seems to hasten proton removal from that carbon atom. Monohalogenation of a carbonyl compound is therefore difficult under conditions of base catalysis.

(20.20) $$CH_2{\cdots}\overset{\overset{O}{-\vdots}}{C}{-}CH_2Br \rightleftharpoons CH_3\overset{\overset{O}{\vdots}\,-}{C}{\cdots}CHBr$$

The tendency of halogen atoms to accumulate on one α-carbon atom has its uses in the **haloform reaction,** in which a methyl group next to a carbonyl group is completely halogenated and finally split off as the haloform.

(20.21) $$CH_3CH_2\overset{\overset{O}{\|}}{C}CH_3 + 3Br_2 + 3OH^- \longrightarrow CH_3CH_2\overset{\overset{O}{\|}}{C}CBr_3 + 3Br^- + 3H_2O$$

$$\text{(20.22)}\quad CH_3CH_2\overset{\overset{\displaystyle O}{\|}}{C}CBr_3 + OH^- \longrightarrow CH_3CH_2CO_2^- + \underset{\text{bromoform}}{CHBr_3}$$

The haloform reaction provides a very mild method of degrading methyl ketones to carboxylic acids, by use of an oxidizing agent that does not disturb many other sensitive functional groups. More importantly, perhaps, the reaction is used as a qualitative test for the presence of the grouping $R{-}\underset{\underset{\displaystyle O}{\|}}{C}{-}CH_3$ in a ketone.

(20.23)

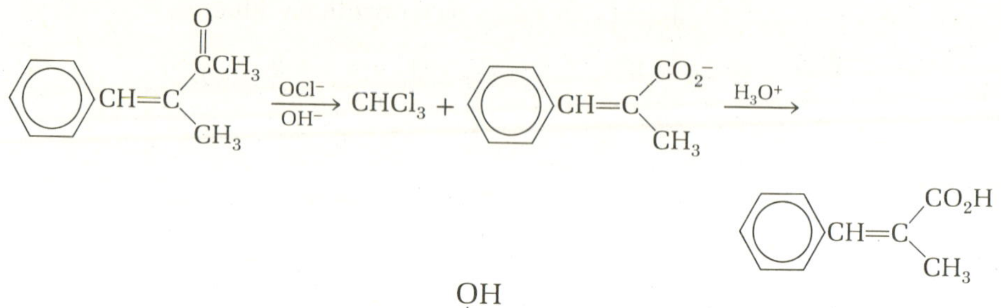

Secondary alcohols of the type $R{-}\overset{\overset{\displaystyle OH}{|}}{C}H{-}CH_3$ also give a positive test because they are first oxidized to ketones under the reaction conditions. When a functional group test is being performed, iodine is the halogen of choice because the product, iodoform (CHI_3), is a crystalline yellow solid, easily detectable in small amounts because of its pungent odor (it has found use as a mild general disinfectant in hospitals — once it has been smelled in those surroundings, the odor is never forgotten).

The success of the haloform reaction rests on the fact that the trichloromethyl ($:CCl_3^-$), tribromomethyl ($:CBr_3^-$), and triiodomethyl ($:CI_3^-$) anions are relatively stable so that they can be ejected from the trihalomethyl ketones by hydroxide ion.

$$\text{(20.24)}\quad R\overset{\overset{\displaystyle O}{\|}}{C}CX_3 + OH^- \rightleftharpoons R\underset{\underset{\displaystyle OH}{|}}{\overset{\overset{\displaystyle O^-}{|}}{C}}CX_3 \longrightarrow RCO_2H + CX_3^- \rightleftharpoons RCO_2^- + CHX_3$$

It still remains to be seen why halogenation apparently proceeds almost exclusively at a methyl group when one is present. We know that under basic conditions the attachment of one halogen atom to a particular carbon atom encourages further halogenation at that position. But does the very first introduction of halogen into the molecule go very much faster at the methyl group than at, for instance, a methylene group?

The results of some proton exchange studies would not seem to support this last hypothesis. When 2-butanone is placed in basic D_2O, the methylene position incorporates deuterium slightly *faster* than the methyl group does.

Since exchange and halogenation both seem to proceed by way of the enolate ion, one might expect initial halogenation to be slightly faster at the methylene group.

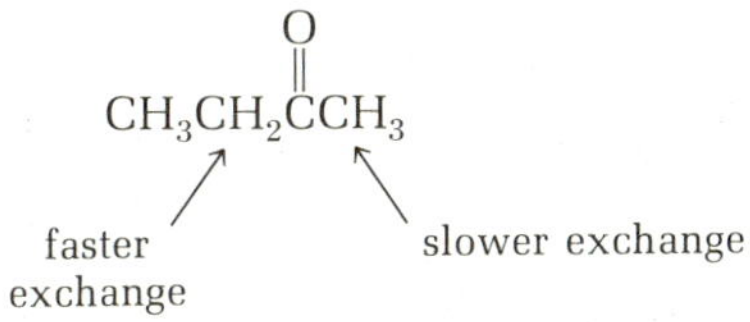

There are strong indications that halogenation, particularly with bromine and iodine, is reversible. Herein may lie the answer to the haloform "riddle." Halogenation of both the methylene and methyl groups of 2-butanone may be taking place competitively and reversibly. Complete halogenation of the methyl group, however, leads to *irreversible* cleavage to carboxylate ion and haloform; it may be the irreversibility of this last step that drives the reaction along the observed path. This suggestion still does not account for the apparent absence of products such as $CH_3CHBrCO_2^-$. One would require the additional postulate that accumulation of bromine at the methyl position actively promotes debromination of the methylene position, perhaps through operation of an inductive effect.

(20.25)

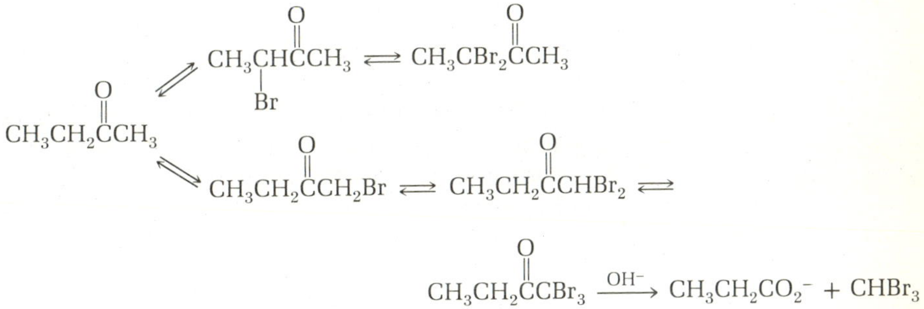

HALOGENATION OF CARBOXYLIC ACIDS

The bromination of an aliphatic carboxylic acid under the usual free radical conditions is slow and leads to a mixture of products. Addition of a small quantity of red phosphorus speeds up the reaction considerably and results in virtually exclusive bromination at the position α to the carboxyl group. Under the reaction conditions the phosphorus is converted to PBr_3, which in turn converts the acid to an acid bromide.

(20.26)

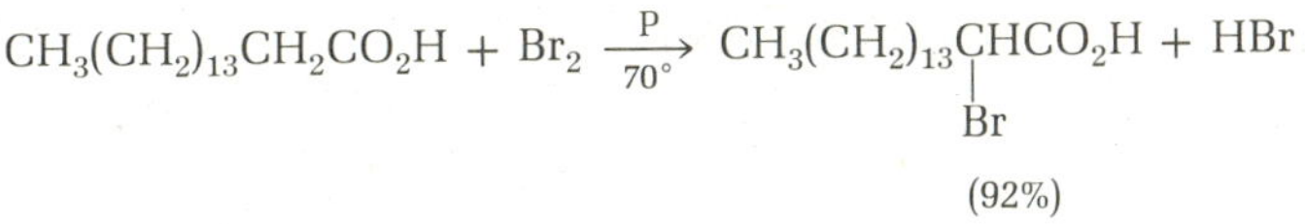

One may in fact substitute PCl_3 or PBr_3 for phosphorus, or discard them altogether by using an acid chloride or bromide directly as the reaction substrate.

(20.27) $$3RCO_2H + PBr_3 \longrightarrow 3R\overset{\overset{\displaystyle O}{\|}}{C}Br + H_3PO_3$$

α-Halogenation of carboxylic acids in this way is known as the *Hell-Volhard-Zelinsky* (HVZ) *reaction*. The exclusive reaction at the α-position is reminiscent of the acid-catalyzed halogenation of ketones, and the facts so far known about the HVZ reaction support the idea of an enol or enol-like intermediate in the halogenation. It then remains to explain why acid chlorides and bromides will take part in the reaction, while the acids themselves (and their esters) are unreactive.

We have previously noted that carboxylic acids and their esters are stabilized by conjugation between the carbonyl group and the hydroxyl oxygen atom (p. 438). When the molecule is converted to its enolic form, much of this stabilization energy is lost and therefore contributes to the ΔH of enolization.

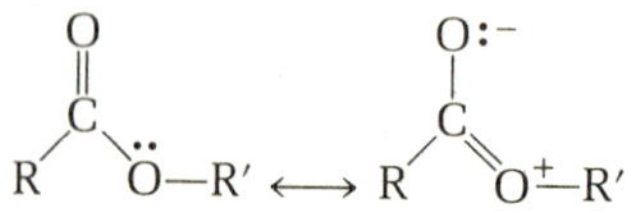

In acid chlorides and bromides there is very little conjugation between the halogen atoms and the carbonyl group to begin with and, as a result, little is lost upon enolization. We would therefore expect that reaction 20.28 would be more endothermic than reaction 20.29 [cf. Problem 1(c)].

(20.28)
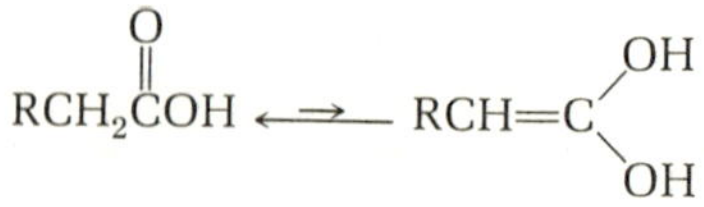

(20.29)
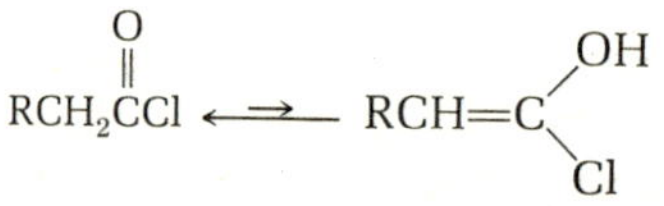

The effect of a small quantity of phosphorus halide on the course of the halogenation is dramatically highlighted by a comparison of equations 20.30 and 20.31. Equation 20.31 corresponds to a typical free radical chain process.

(20.30)
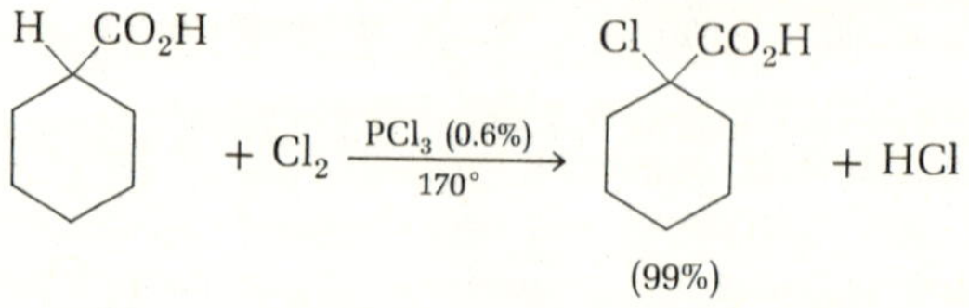

(20.31)

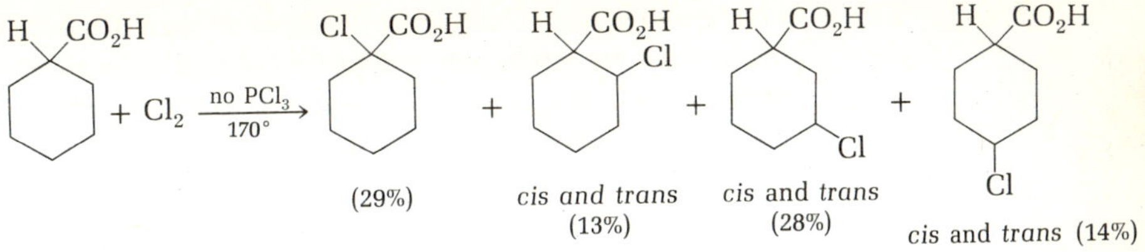

It is interesting that below about 140° the radical chain process dominates, whether PCl_3 is present or not, due to its apparently lower enthalpy of activation.

α-Halocarboxylic acids are useful as precursors to other desirable molecules, in particular, α-amino acids, the raw material for construction of peptides and proteins (Chapter 25).

(20.32)
$$CH_3(CH_2)_{13}\underset{\displaystyle Br}{\underset{|}{C}}HCO_2H + KOH \xrightarrow[\text{reflux}]{C_2H_5OH} \xrightarrow{H_3O^+} CH_3(CH_2)_{13}\underset{\displaystyle OH}{\underset{|}{C}}HCO_2H$$
(93%)

(20.33)
$$CH_3\underset{\displaystyle Br}{\underset{|}{C}}HCO_2H + NH_3\ (\text{conc.}) \xrightarrow[25^\circ]{H_2O} CH_3\underset{\displaystyle NH_3^+}{\underset{|}{C}}HCO_2^-$$
(70%)

THE ALDOL CONDENSATION

We have shown that enols and enolate ions are nucleophilic species, susceptible to attack by electrophiles such as molecular halogens. The carbonyl form of a ketone or aldehyde may also act as an electrophile toward an enol or enolate ion so that aldehydes and ketones may dimerize under the catalytic influence of acid or base.

(20.34)
$$2CH_3\overset{\displaystyle O}{\overset{\|}{C}}H \xrightarrow[5^\circ]{\text{NaOH (10\%)}} \xrightarrow[\text{then distill}]{\text{neutralize,}} CH_3\overset{\displaystyle OH}{\overset{|}{C}}HCH_2\overset{\displaystyle O}{\overset{\|}{C}}H$$
aldol (50%)

When acetaldehyde is dimerized in the presence of acid or base, 3-hydroxybutanal is obtained; this compound is usually referred to by its trivial name, aldol (*ald*ehyde + alcoh*ol*). Such dimerizations of aldehydes and ketones are known collectively as **aldol condensations.**

The equilibrium constant for equation 20.34 is favorable, although not terribly large. The overall entropy change is obviously negative and the C—O π-bond that is broken is relatively strong so that the reaction is not highly exothermic. As a consequence, the position of equilibrium in aldol condensations is sensitive to rather small changes in molecular structure. For instance, treatment of acetone with aqueous base leads to the formation of only a few percent of the aldol dimer (diacetone alcohol). In this case there is more steric congestion around the new carbon-carbon bond being formed than there was in the case of acetaldehyde dimerization.

(20.35) 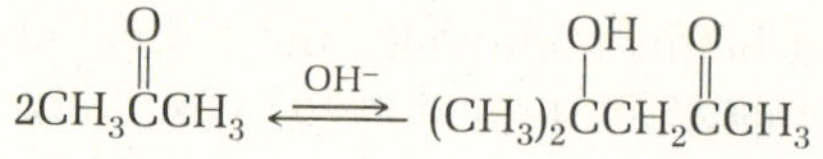

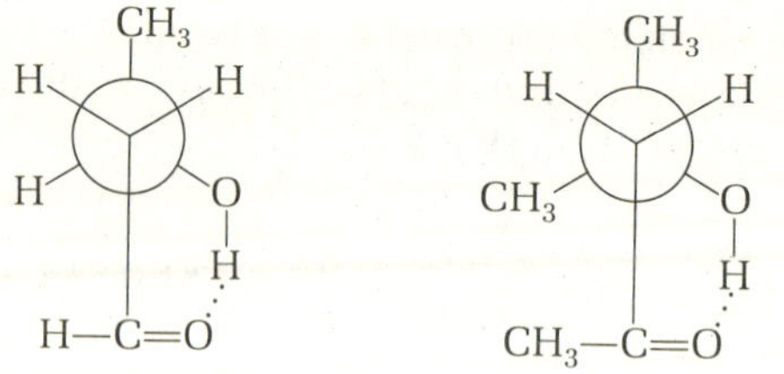

aldol — diacetone alcohol

For base-catalyzed aldol condensations the reactive intermediate is undoubtedly the enolate anion. Acetaldehyde may also be dimerized with the aid of an aqueous acid catalyst.

(20.36) 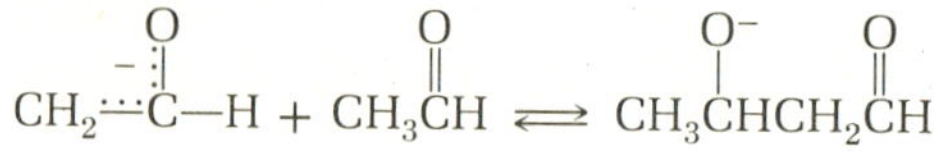

(20.37) 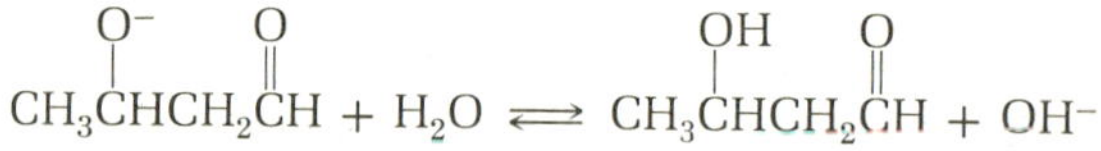

Under these circumstances the enol, rather than the enolate ion, is the nucleophile. Since it is a weak nucleophile, it appears to react not with a molecule of aldehyde but rather with a molecule of protonated aldehyde.

(20.38) 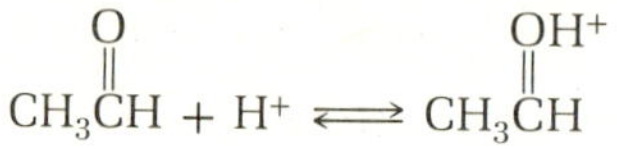

(20.39) 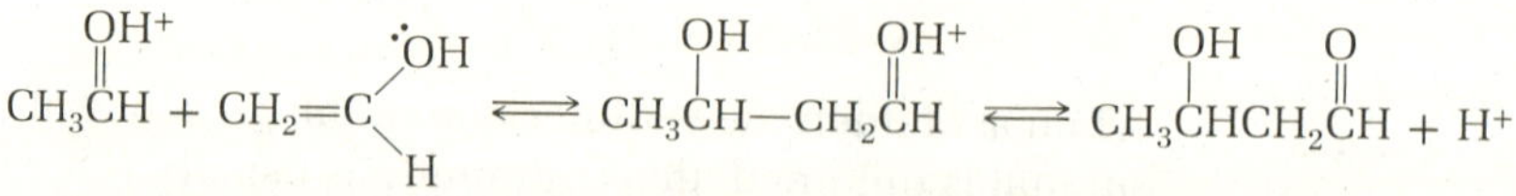

Equilibrium is usually reached much more quickly with basic catalysts than with acidic ones. For this reason one has to be fairly scrupulous about neutralizing a basic catalyst before working up an aldol condensation reaction mixture. If the dimer is heated in the presence of base, it will partly revert back to its precursor **(retroaldol cleavage)** that, being more volatile than the dimer, is continuously distilled as it is formed.

Aldols and ketols are sensitive compounds tending not only to undergo retroaldol cleavage but also to lose water, giving rise to α,β-unsaturated aldehydes and ketones. Since these conjugated unsaturated carbonyl compounds are valuable synthetic intermediates, the aldols and ketols are usually dehydrated immediately after formation. The most efficient method for dehydration involves distilling the aldol dimer in the presence of a trace of acid; even a very weak Lewis acid such as molecular iodine will serve.

(20.40)

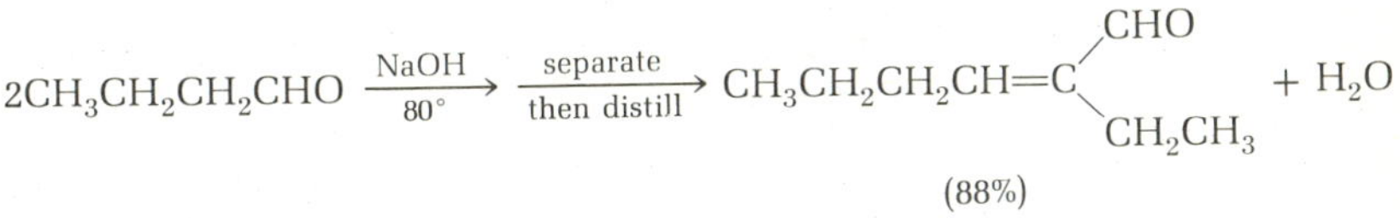

(20.41)

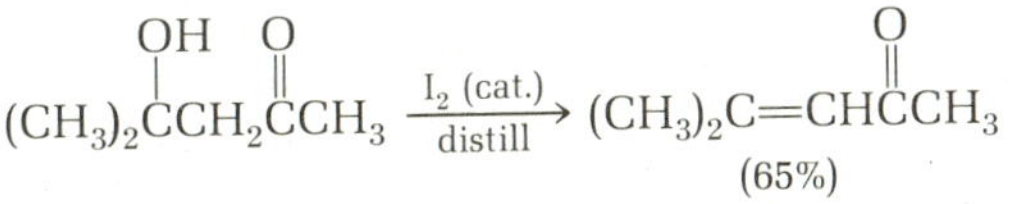

In actuality, dehydration may be brought about in the original reaction mixture if the solution is sufficiently strongly basic. Complications will often ensue, however, because under these conditions the initial product may suffer further aldol condensations, leading to polymeric products.

(20.42)

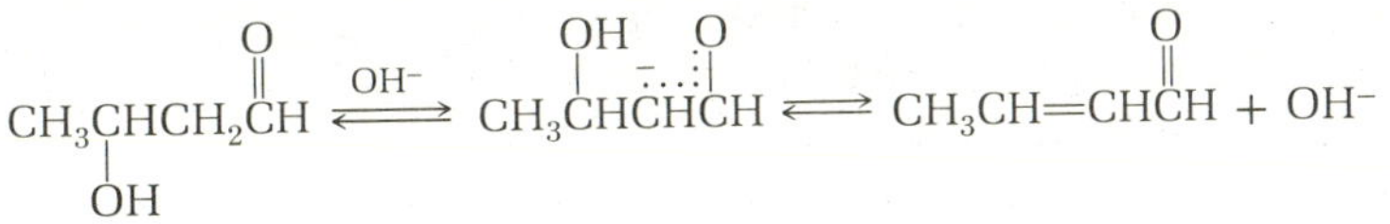

Note that the anion **1** is a vinylogous enolate ion in which there is a double bond interposed between the acidic carbon atom and the carbonyl group. The resulting anion is therefore able to delocalize charge onto the oxygen atom.

(20.43)

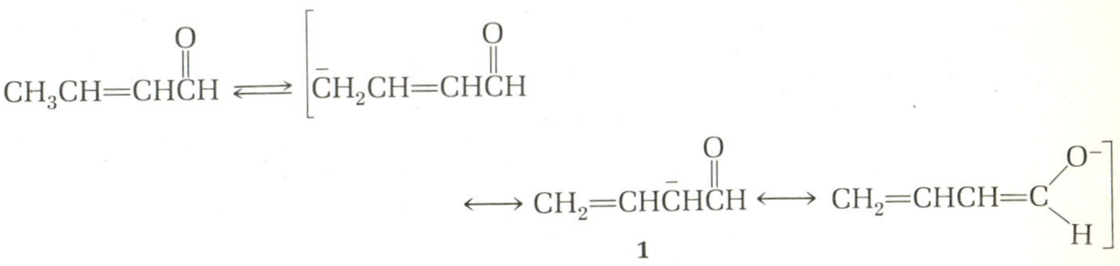

(20.44)

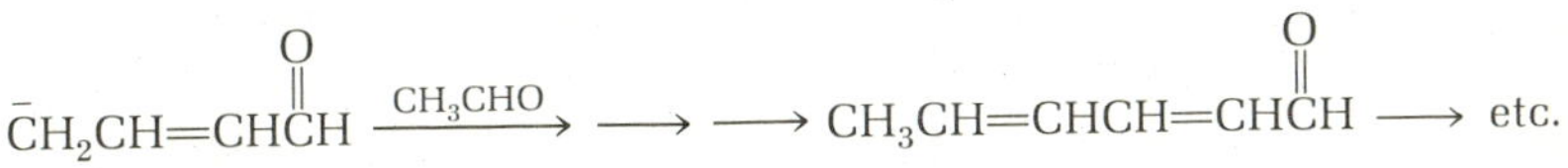

PROBLEM

3 Compound I (Chapter 15, Problem 21) was treated with potassium *t*-butoxide in dimethyl sulfoxide-d_6. Six protons of the substrate were exchanged for deuterium from the solvent.

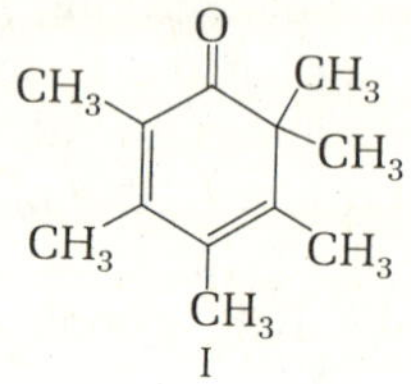

I

Which six were they? Show with the aid of canonical structures why these six were the reactive ones.

MIXED ALDOL CONDENSATIONS

If two different aliphatic aldehydes are mixed and the mixture treated with base, four possible aldol dimers are to be anticipated. The same conclusions would apply if two different aliphatic ketones were subjected to the reaction conditions. Such mixed aldol condensations are not normally of much preparative value. A slightly better situation is encountered if an aldehyde and a ketone are mixed. Ketone self-condensation is often minimal and when the aldehyde and ketone react, it is invariably the enol of the ketone that adds to the carbonyl group of the aldehyde. The only important side reaction is then self-condensation of the aldehyde, which may be suppressed by use of excess ketone.

$$CH_3CH_2CHO + CH_3\overset{\overset{\displaystyle O}{\|}}{C}CH_3 \longrightarrow \underset{(60\%)}{CH_3CH_2\overset{\overset{\displaystyle OH}{|}}{C}HCH_2\overset{\overset{\displaystyle O}{\|}}{C}CH_3} \tag{20.45}$$

Mixed aldol condensations present fewer difficulties if one of the reactants lacks α-hydrogen atoms. Formaldehyde is an exemplary molecule for this purpose since it also possesses a very reactive carbonyl group. Due to the reactivity of the formaldehyde, it is usually difficult to condense only one molecule of formaldehyde with the aldehyde or ketone in question, and it is common to use enough formaldehyde to react with all the α-hydrogen atoms in the enolizable reactant.

$$CH_3CH_2CH_2CHO + 2CH_2O \xrightarrow[H_2O]{K_2CO_3} \underset{(95\%)}{CH_3CH_2\overset{\overset{\displaystyle CH_2OH}{|}}{\underset{\underset{\displaystyle CH_2OH}{|}}{C}}{-}CHO} \tag{20.46}$$

THE CANNIZZARO REACTION

When two carbonyl compounds are mixed under basic conditions, then one or more aldol condensations will take place, *provided* that at least one of the carbonyl compounds contains an enolizable hydrogen atom. If neither carbonyl compound possesses α-hydrogen atoms, then the reaction pursues a different course.

The prototypical case is the reaction of formaldehyde with itself. Under the influence of base, formaldehyde undergoes an internal redox reaction, disproportionating to methanol and formic acid (which is present in the basic solution as formate ion).

(20.47) $$2CH_2{=}O + OH^- \longrightarrow CH_3OH + HCO_2^-$$

The mechanism apparently involves the addition of hydroxide ion to a molecule of formaldehyde and subsequent transfer of a hydride ion from this anion to another formaldehyde molecule.

(20.48) $$OH^- + CH_2{=}O \rightleftharpoons \overset{O^-}{\overset{|}{C}}H_2OH$$

(20.49)

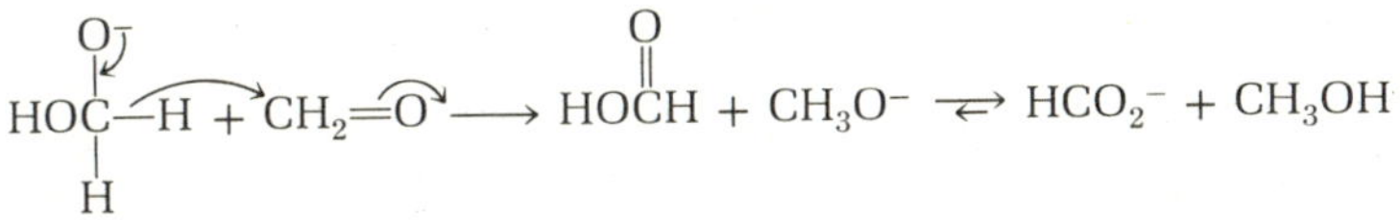

Other nonenolizable aldehydes also take part in the Cannizzaro reaction; aromatic aldehydes such as benzaldehyde are prone to undergo this disproportionation.

(20.50)

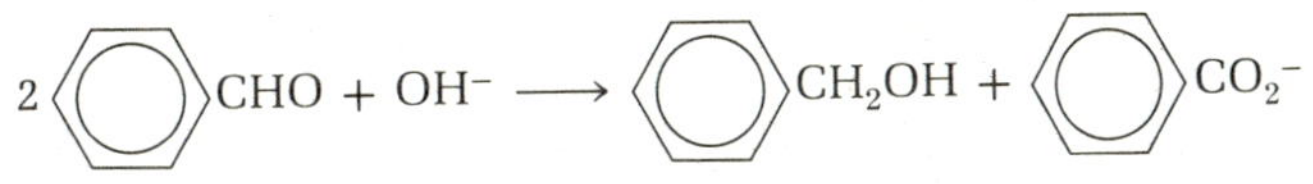

If two different nonenolizable aldehydes are mixed, then a mixture of products is possible (crossed Cannizzaro reaction). When one of the aldehydes is formaldehyde, however, then only one set of products is obtained because formaldehyde always acts as the reducing agent.

(20.51)

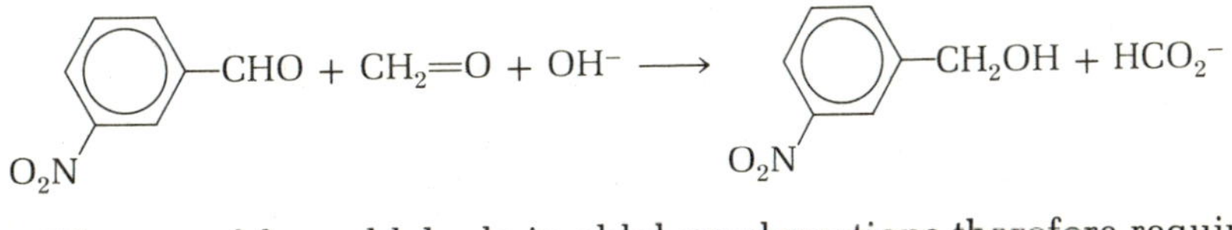

The use of formaldehyde in aldol condensations therefore requires some caution. If excess formaldehyde is present, it may, after replacing all the α-hydrogen atoms of the other aldehyde or ketone, reduce its carbonyl group as well. This result is sometimes desired, as in the synthesis of pentaerythritol.

(20.52)

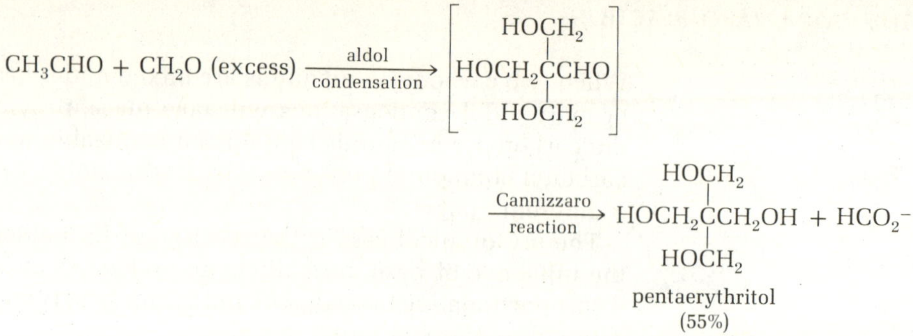

THE CLAISEN-SCHMIDT REACTION

As noted previously, when two different carbonyl compounds are combined under aldol conditions and only one of them contains enolizable hydrogen atoms, then only two products are possible. If, furthermore, the nonenolizable component is aromatic, then the reaction becomes even simpler and the mixed aldol product becomes the dominant one.

(20.53)

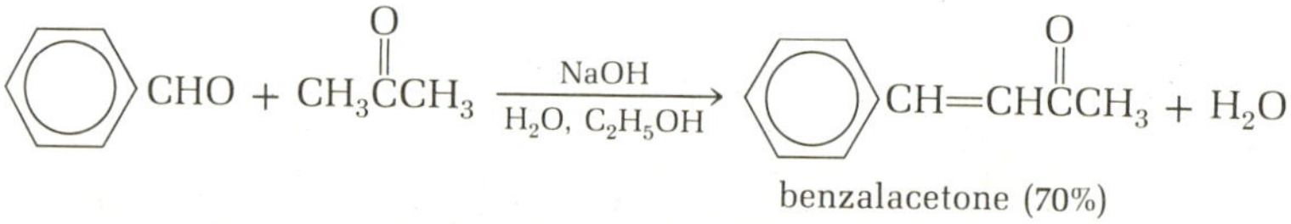

This class of condensations is known as the **Claisen-Schmidt reaction.**

Under these conditions the intermediate aldol or ketol is dehydrated in the course of the reaction, the overall entropy of reaction now becoming much more favorable than in the simple aldol dimerization (equation 20.36). Formation of the carbon-carbon double bond is aided by the fact that it is part of an extended conjugated system so that the molecule acquires considerable resonance energy upon creation of the C—C π-bond.

(20.54)

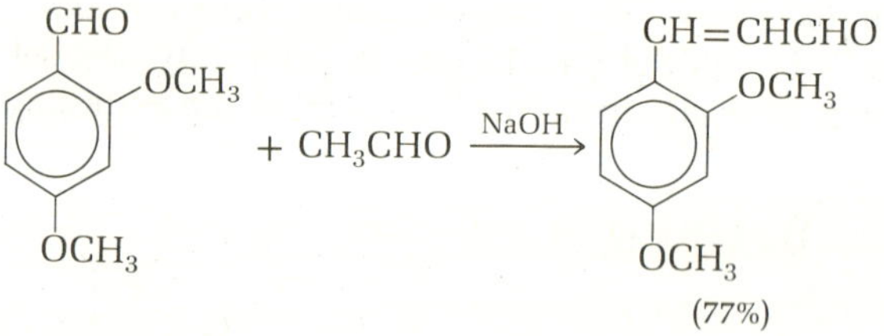

(20.55)

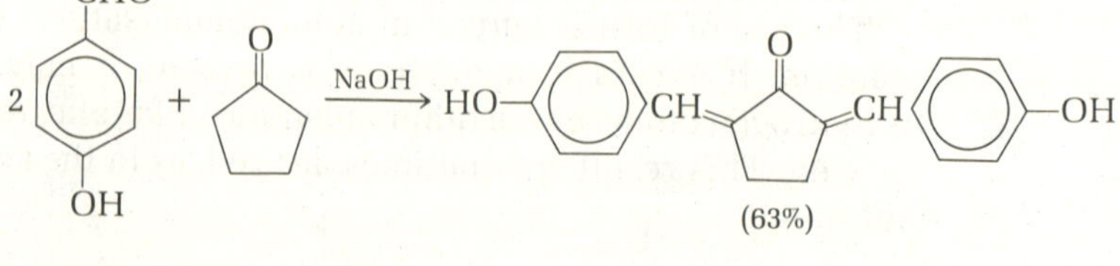

(20.56) 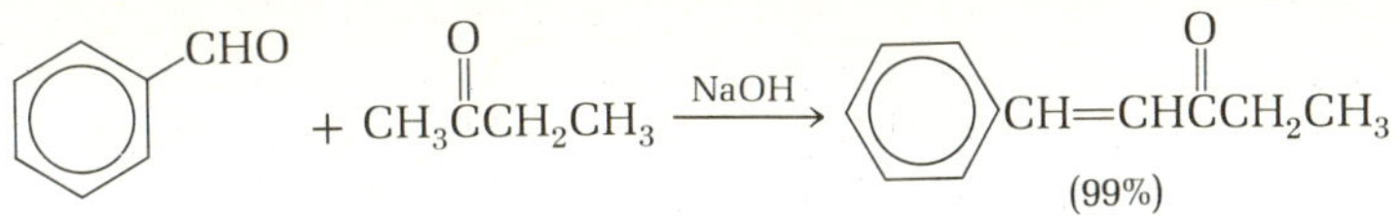

Equation 20.56 illustrates a very important point. If an unsymmetrical ketone is subjected to Claisen-Schmidt reaction with a basic catalyst, then the condensation occurs almost exclusively at the methyl group, if one is available. The selectivity is reminiscent of that which occurs in base-catalyzed halogenation. The acid-catalyzed Claisen-Schmidt condensation shows the opposite isomeric preference. Acid catalysis induces a bias toward condensation at an α-methylene group.

(20.57) 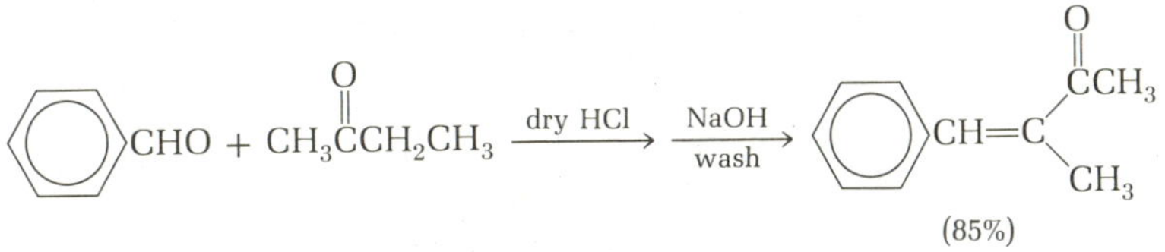

The synthetic chemist may now choose his condensation site at will by judiciously selecting the proper catalyst.

The relationship of the catalyst to the site selectivity has been brilliantly illuminated by some studies of the kinetics of the individual reaction steps. From our discussion of base-catalyzed halogenation and isotope exchange in 2-butanone, we would expect that benzaldehyde would condense competitively at the 1 and 3 positions to produce the intermediate ketols **2** and **3**, respectively.

(20.58) 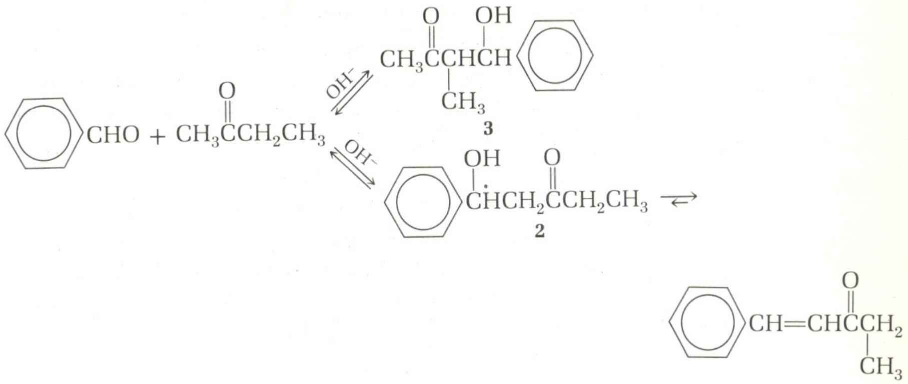

These intermediates are in rapid equilibrium with starting materials.

Because of the steric congestion in **3**, it tends very rapidly to fall apart to starting materials, and its equilibrium concentration is low. Ketol **2**, on the other hand, is much more stable and its dehydration rate is nearly equal to the rate of its reversion to starting materials. Since the dehydration is nearly irreversible, the final step determines the overall reaction path.

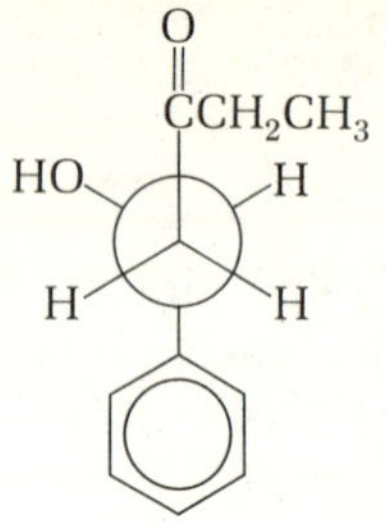

2

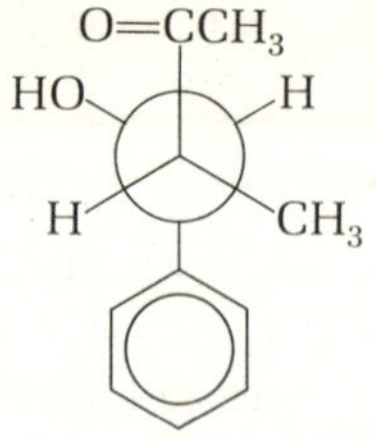

3

Under acidic conditions, a different set of kinetic relationships obtains. The reactants are the two enols, **4** and **5**, of which **5** will predominate (more highly substituted double bond).

(20.59) 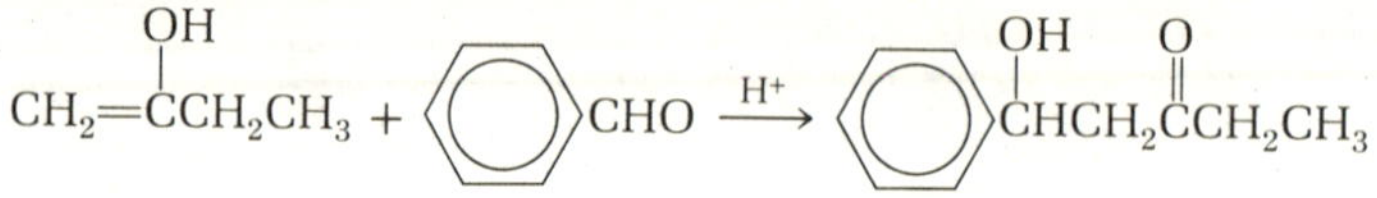

4 2

(20.60) 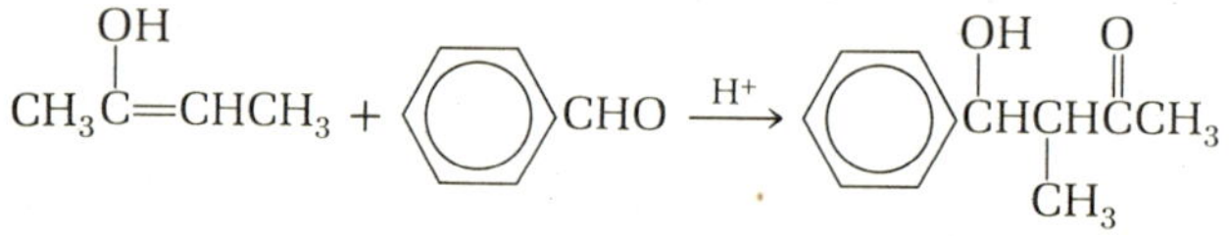

5 3

The condensation of these enols with benzaldehyde is virtually irreversible, since the dehydration of **2** and **3** under acidic conditions is much more rapid than their retroaldol cleavage.

Thus under basic conditions we have a rapid equilibrium among the two intermediates, with the product distribution being determined by the relative rapidity of the second step. Under acidic conditions, the first step is effectively irreversible and it is the relative concentration of the two enols that determines the product distribution.

Note that no matter where the condensation takes place, the product in each case will potentially be a mixture of *cis-trans* isomers. In fact, one usually isolates only that product in which the phenyl group is *trans* to the larger substituent on the other carbon atom (more stable product).

1,3-DICARBONYL COMPOUNDS

If a methylene group is flanked by *two* strongly electron-withdrawing groups (—C≡N, —COR, —COOR), then it reacts much more readily with bases than do the corresponding protons in a simple aldehyde, ketone, or nitrile. Furthermore, the equilibrium enol content of compounds containing this structural grouping is abnormally high.

The nmr and ir spectra of 2,4-pentanedione confirm the presence of some compound other than the diketone. The data are consistent with an enol structure **(6)** for this other compound, and quantitative measurements demonstrate that in nonpolar solvents the enol tautomer predominates.

(20.61) 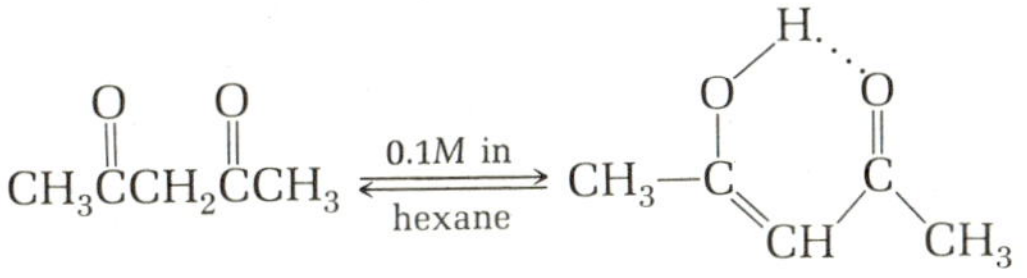 $K^{25°} = 11$

6

Another well-studied example is ethyl 3-oxobutanoate (**7**; universally known as *ethyl acetoacetate*).

(20.62) 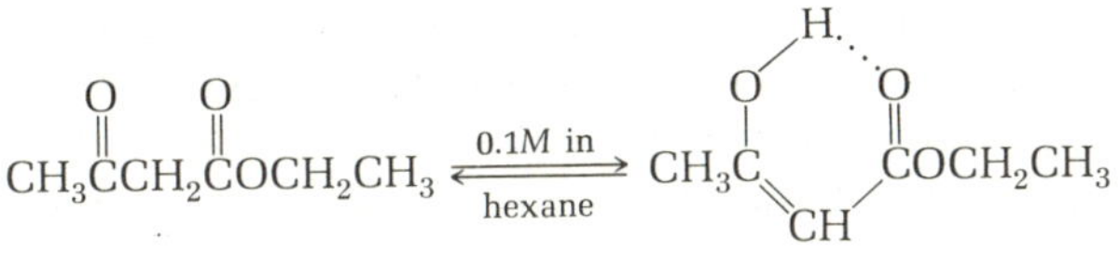 $K^{25°} = 1$

7

The driving force for enol formation in these systems must be largely due to the conversion of nonconjugated to conjugated unsaturated systems. In this respect the smaller equilibrium constant for equation 20.62 than for equation 20.61 is perfectly comprehensible. We have previously remarked that ester groups are stabilized by internal conjugation (p. 538). They therefore exhibit a diminished tendency (compared with keto groups) to conjugate with other π-electron systems and they promote enolization to a lesser extent than keto groups do. The enol tautomer of diethyl malonate [$CH_2(CO_2CH_2CH_3)_2$] is not normally detectable by ir or nmr spectroscopy.

As implied by equations 20.61 and 20.62, the enol forms of ethyl acetoacetate and of 2,4-pentanedione are stabilized by intramolecular hydrogen bonding through a six-membered ring. Hydrogen bonding must play a significant role because the equilibrium constants for enolization are very solvent-dependent. In polar solvents, which can compete as hydrogen bond acceptors and donors with the intramolecular sites, enolization is much less favored (equation 20.63).

(20.63) 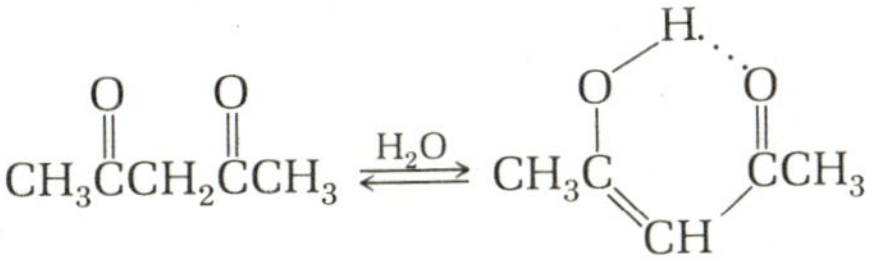 $K^{20°} = 0.17$

The stability of the enol forms of these 1,3-dicarbonyl compounds is attested to by the fact that in many cases it has proved possible to isolate this tautomer in its pure state. The keto and enol forms of ethyl acetoacetate, for example, may be separated by means of fractional distillation in a very clean quartz apparatus. Storage of the pure tautomers for substantial periods of time is possible if acids and bases are rigorously excluded; there is sufficient alkali in ordinary soft glass, for example, to promote rapid tautomeric equilibrium.

The pK_a's of ethyl acetoacetate and 2,4-pentanedione are 11 and 9, respectively, both significantly smaller than that for water. There is much more extensive charge delocalization in the enolate ions from these compounds than there is in the enolate ion from a simple aldehyde or ketone.

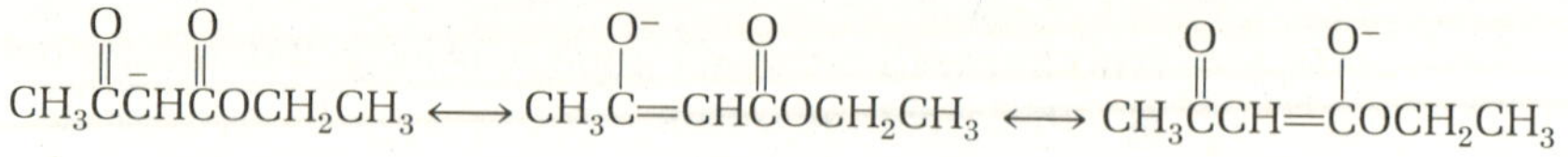

It is therefore possible to use these substances in aldol condensations under very mild conditions **(Knoevenagel reaction).**

The Knoevenagel reaction is usually run with a weak base, such as an amine, and a carboxylic acid as cocatalysts. Benzene and toluene are common solvents, and the reaction is driven to completion by removing the water by means of azeotropic distillation. One great advantage of using such mild conditions is that simple aliphatic aldehydes and ketones may be used as the carbonyl components of the aldol condensation; there is no danger of these compounds condensing with themselves.

(20.64)

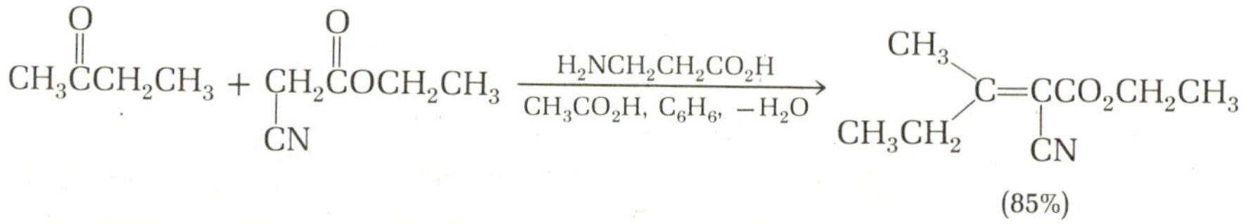

(20.65)

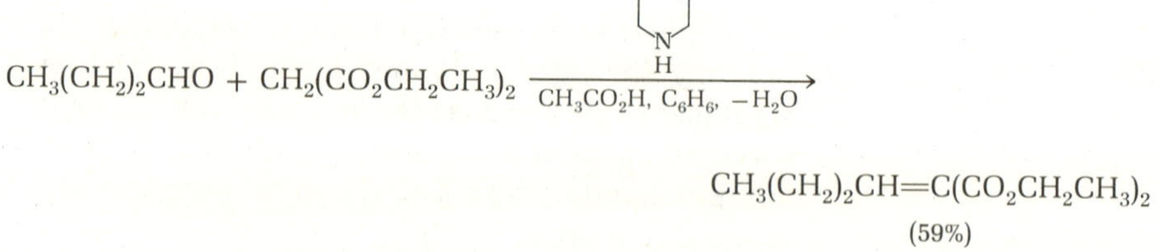

(20.66)

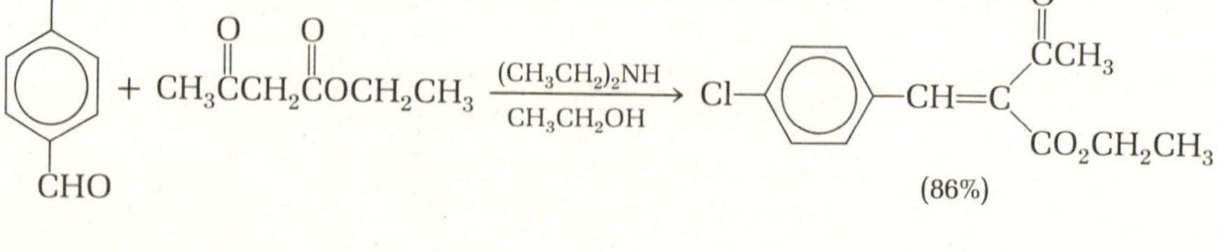

Compared to the number of methods available for transforming functional groups, there are a relatively small number of techniques for creating new carbon-carbon bonds. The aldol condensation is one of the most important of these, rivaling the Grignard reaction. One of the more actively exploited applications of the aldol condensation has been the synthesis of five- and six-membered rings, extremely important operations in the total synthesis of steroids.

(20.67)

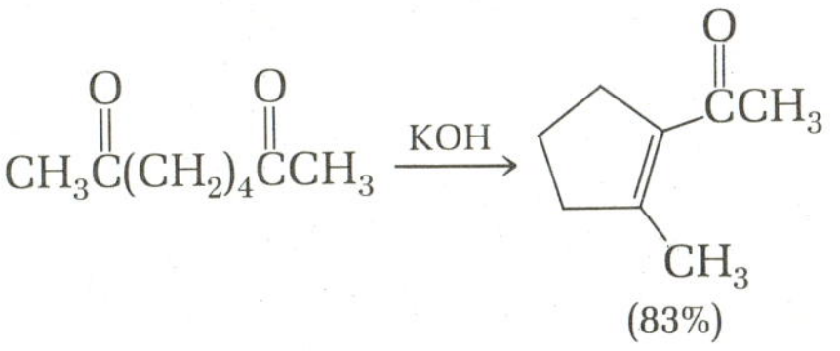

(20.68)

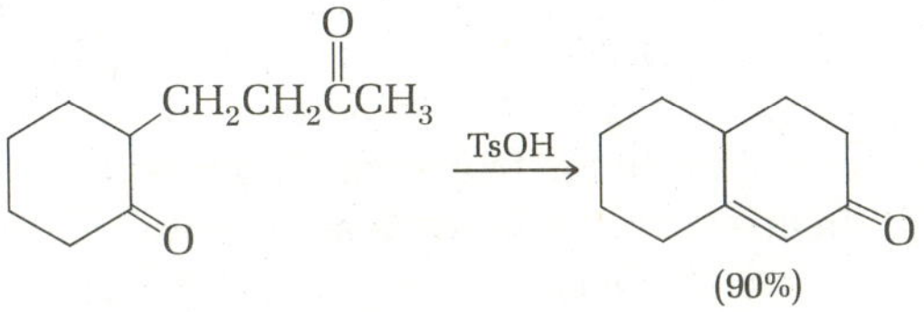

The aldol products are versatile synthetic intermediates, their double bonds being very reactive toward a variety of nucleophiles (Chapter 21). Two particular addition reactions that are noted at this time are the Diels-Alder and photoaddition reactions. The Diels-Alder addition provides a method of synthesizing substituted six-membered rings of known stereochemistry. Photochemical processes represent perhaps the most versatile and important route to cyclobutanes.

(20.69)

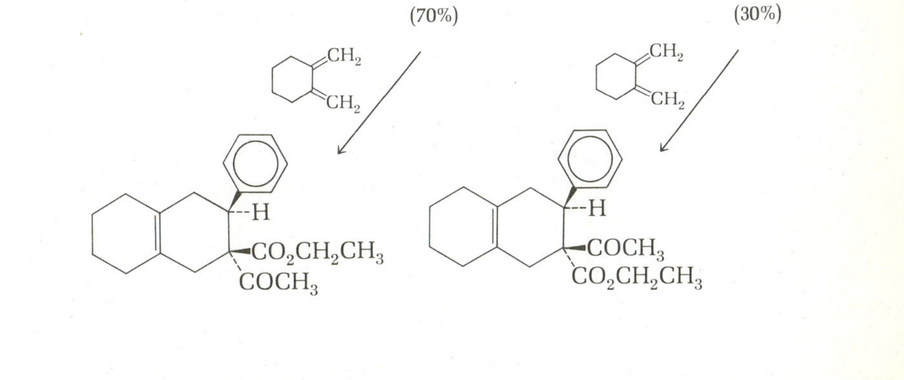

(20.70) 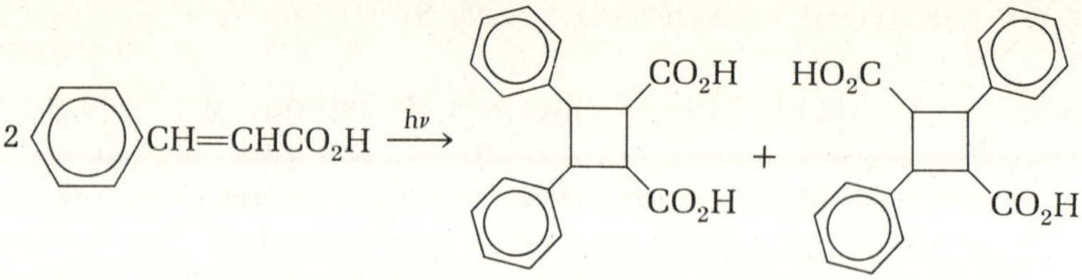

ALKYLATION OF ENOLATE IONS

Since enolate ions are such powerful nucleophiles, one would expect that they would react not only with carbonyl compounds but with almost any electrophilic species. For example, it would be reasonable to suppose that an enolate ion would displace halide ion from a primary or secondary halide, thereby forming an alkyl-substituted aldehyde, ketone, or ester. Such reactions are indeed known but their efficiency depends markedly on the nature of the carbonyl compound from which the enolate ion was derived.

When 1,3-dicarbonyl compounds are employed, few complications occur. Due to the acidity of these compounds, they react completely with a base such as sodium ethoxide, producing very stable enolate ions (equation 20.71).

(20.71) $$CH_2(CO_2CH_2CH_3)_2 + CH_3CH_2O^-Na^+ \xrightleftharpoons{CH_3CH_2OH} Na^+\ \bar{C}H(CO_2CH_2CH_3)_2 + CH_3CH_2OH$$

Addition of an alkyl halide or tosylate to a solution of the enolate results in its being alkylated, usually in good yield.

(20.72) $$Na^+CH(CO_2CH_2CH_3)_2 + CH_3(CH_2)_3Br \longrightarrow \underset{(80\text{–}90\%)}{CH_3(CH_2)_3CH(CO_2CH_2CH_3)_2} + NaBr$$

(20.73) $$CH_3\overset{O}{\overset{\|}{C}}CH_2\overset{O}{\overset{\|}{C}}CH_3 + CH_3I \xrightarrow[\text{acetone}]{K_2CO_3} \underset{(75\%)}{CH_3\overset{O}{\overset{\|}{C}}\underset{CH_3}{\underset{|}{C}H}\overset{O}{\overset{\|}{C}}CH_3}$$

Should the carbonyl compound contain only one activating group, however, then the picture changes. A base such as ethoxide ion will not convert more than a few tenths of a percent of a simple aldehyde or ketone to its respective enolate ion. The reaction between enolate and alkyl halide is therefore slow; it is very likely to be minor in extent compared with reactions between the base and the alkyl halide or between the enolate ion and unionized carbonyl compound (equation 20.74).

(20.74)

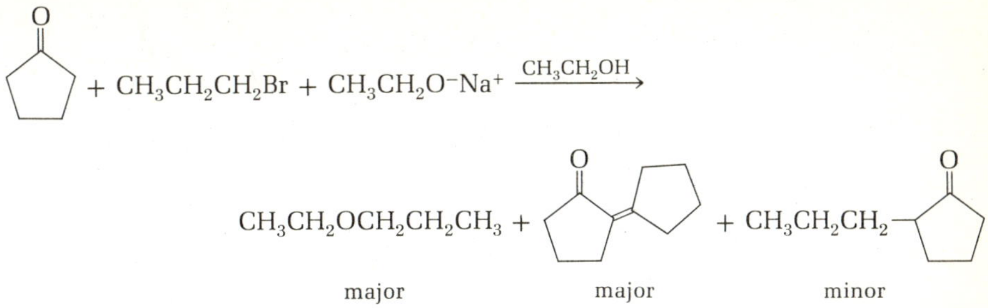

In order to provide appreciable concentrations of enolate ions, it is necessary that stronger bases be used. Potassium *t*-butoxide, $(CH_3)_3CO^-K^+$, is one of the weakest that is of value in alkylation reactions; sodium and potassium amide are very frequently employed and are sufficiently strong to completely neutralize a ketone. (Because of their tendency to suffer very rapid aldol condensations, aldehydes cannot be exposed to these strong bases.)

(20.75)

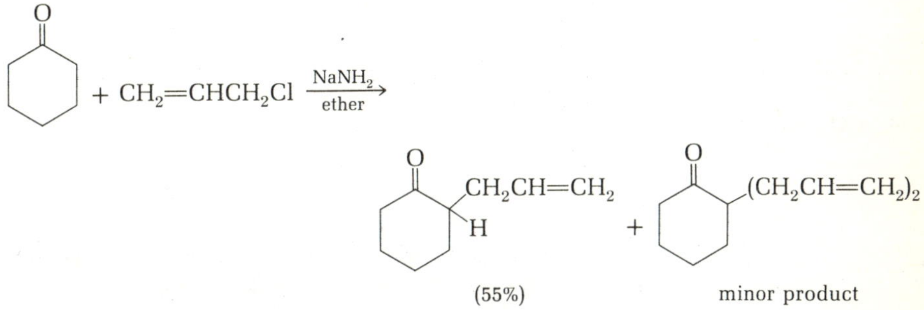

Aside from the problem of self-aldol condensation, there is also the complication of multiple alkylation. Equilibrium among the various possible enolates may be partially or completely established, leading to a nasty mixture of products (equations 20.76 to 20.78).

(20.76)

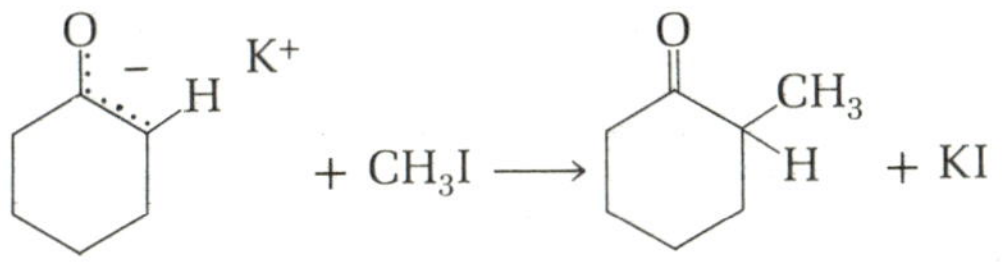

(20.77)

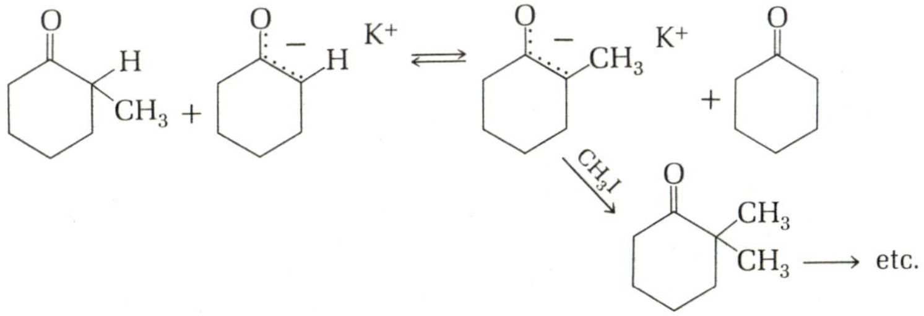

(20.78)

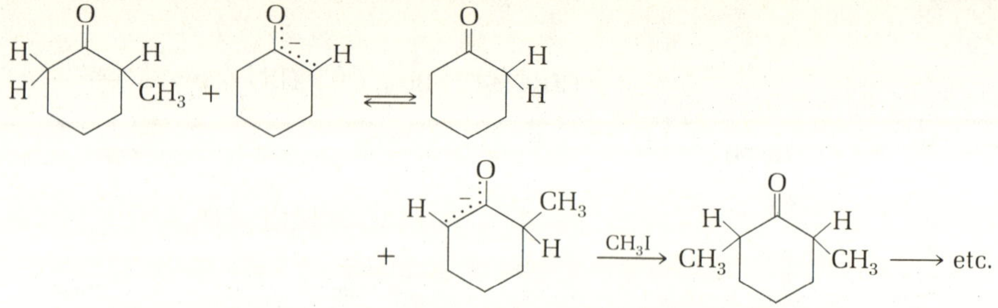

Enolate ions of monocarbonyl compounds are therefore difficult customers to handle. One solution is to use ester groups to help stabilize the enolate ions and to then subsequently dispose of the ester group (Chapter 21). Chemists have also sought some aldehyde or ketone derivative, readily prepared from the parent compound, which has the nucleophilic double bond characteristic of enols and is yet less reactive than enolate ions. These criteria have been largely satisfied by **enamines** (alk*ene* + *amine*).

ALKYLATION OF ENAMINES

These compounds are synthesized by allowing a *secondary* amine to react with an aldehyde or ketone, often with a trace of acid present. The reaction is driven to completion by removal of the water, either with the aid of a drying agent or by azeotropic distillation.

(20.79)

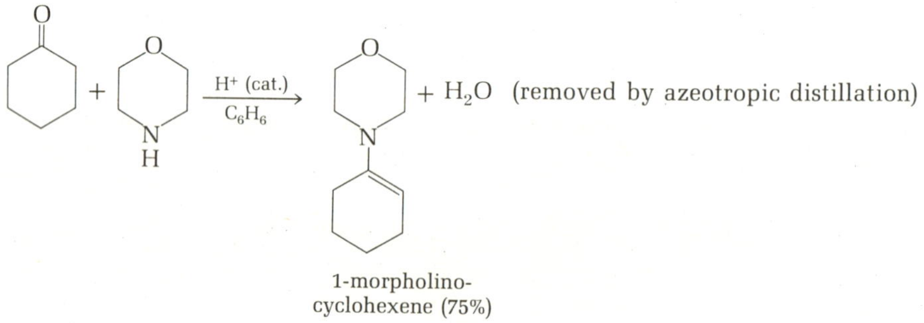

(20.80)

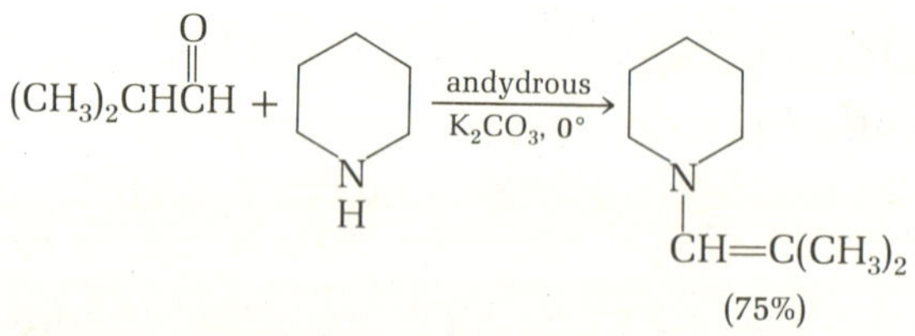

Cyclic amines such as pyrrolidine, piperidine, and morpholine are often used in the preparation of enamines.

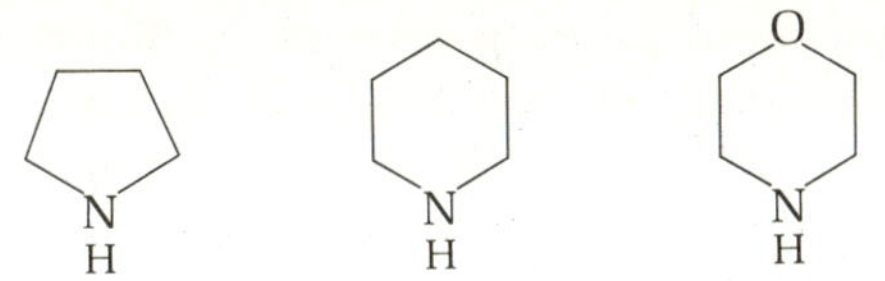

pyrrolidine piperidine morpholine

The conjugation of the nonbonded nitrogen electrons with the π-electrons confers an unusual degree of nucleophilicity on the double bond.

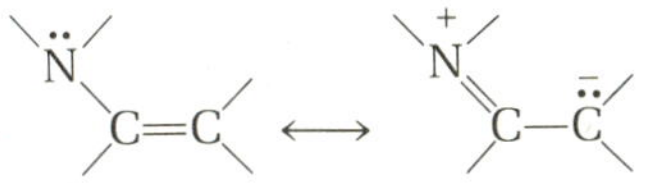

Since nitrogen is less electronegative than oxygen, enamines are more nucleophilic than enols. They are still far less reactive than enolate ions, which have a full negative charge. For this reason enamines will only react to a useful extent with reactive alkylating agents such as alkyl iodides and allyl and benzyl halides.

(20.81)

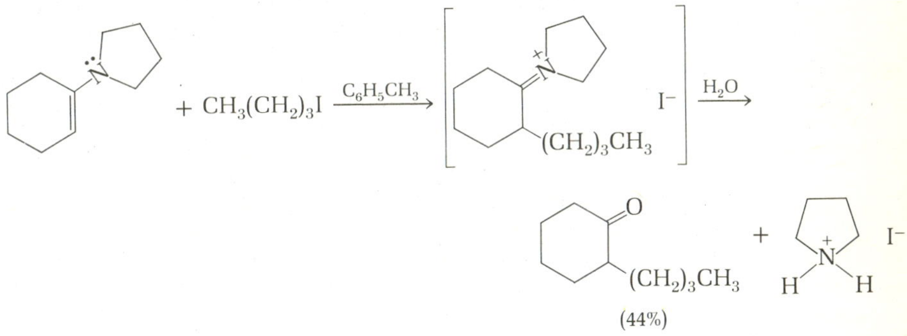

(20.82)

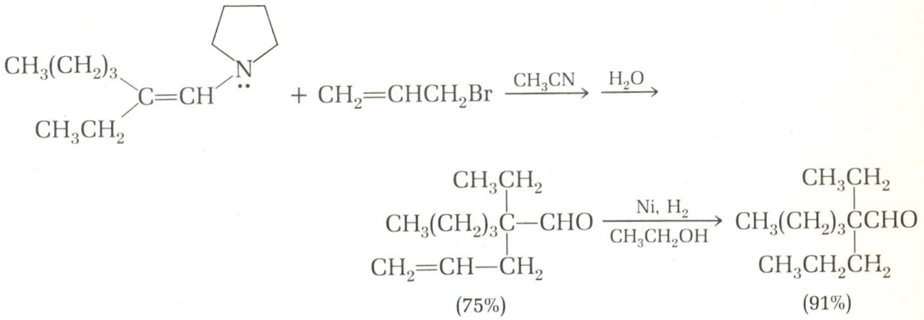

ENOLS IN BIOLOGICAL SYSTEMS

Enol chemistry is encountered very extensively in enzymic systems, and the aldol condensation in one form or another occupies an important place as a carbon-carbon bond-forming process in biological chemistry. For illustration we choose one detailed example from mammalian physiology.

The body can convert the three-carbon fragment dihydroxyacetone to glucose, which is in turn converted to a polymer known as glycogen. Glycogen is stored in the body and serves as a reserve source of fuel.

(20.83)

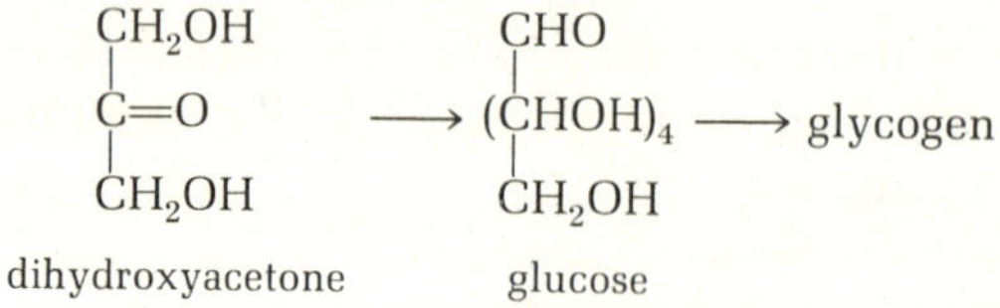

The entire sequence is reversible. Investigations of the mechanism of formation of glucose suggest that two three-carbon units are joined by way of an aldol condensation, after one of them has undergone an internal rearrangement for which its enol is an intermediate.

Many hydroxy compounds undergo physiological processes not in their "free" form but rather as esters of phosphoric acid. An enzyme has been found that catalyzes the equilibration between dihydroxyacetone phosphate and (*R*)-glyceraldehyde-3-phosphate. Nothing more than an internal redox reaction is involved; carbon atom 1 is oxidized, while carbon atom 2 is simultaneously reduced.

(20.84)

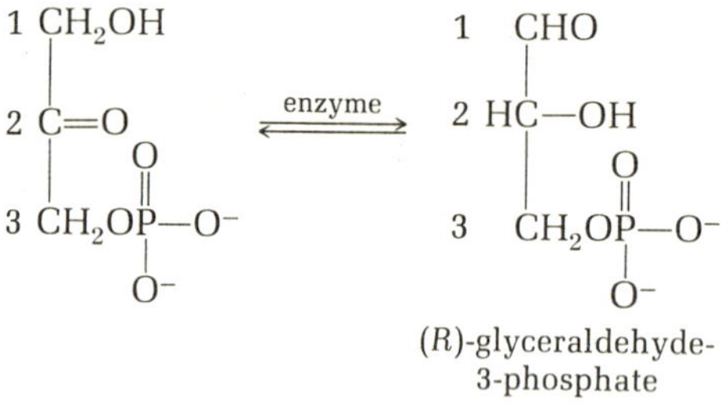

By analogy with reactions known in the laboratory, there is every reason to believe that the interconversion goes by way of the enol **8.**

(20.85)

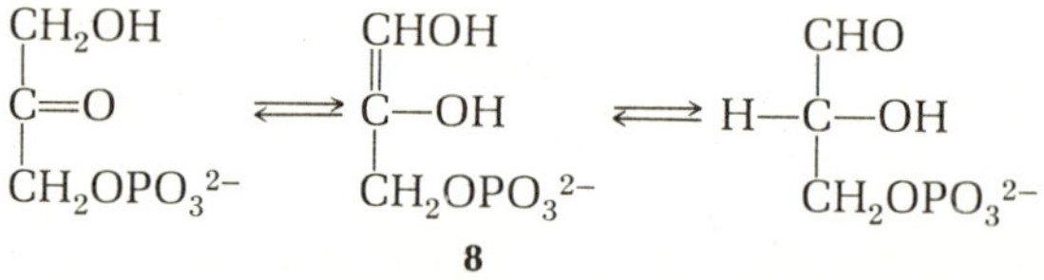

The intermediate is actually an enediol, protonation of which can take place at either carbon atom 1 or 2 and leading to ketone or aldehyde, respectively. Interchanging hydroxyl and carbonyl groups is a very facile process in α-hydroxy aldehydes and ketones, catalyzed by either acid or base. The difference between the *in vitro* and the enzymatic reaction lies in the stereospecificity of the latter.

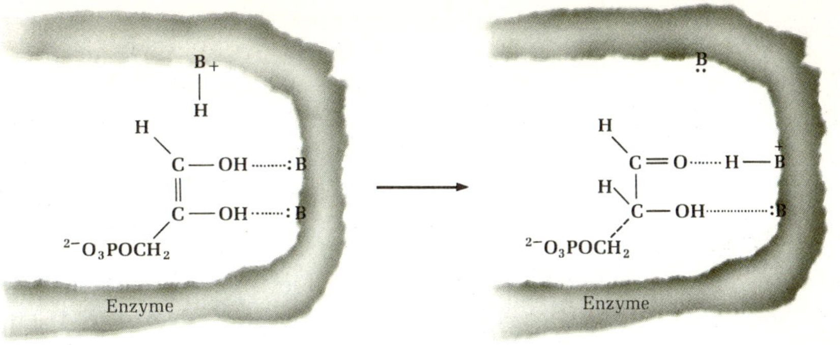

FIGURE 20-1
Schematic Diagram of the Enzymatic Protonation of Dihydroxyacetone Enol

Dihydroxyacetone phosphate is a symmetrical molecule but when it is converted to glyceraldehyde phosphate, an asymmetric atom is created at C-2. The enediol **8** is an essentially planar molecule and in the absence of enzymatic control would be protonated on either side with equal probability, giving rise to racemic glyceraldehyde phosphate. Under enzymatic catalysis, protonation occurs from one side of the molecule only and the *R*-enantiomer of the hydroxyaldehyde is the only one found.

Figure 20-1 suggests how such steric control might be achieved. If the enediol were formed as the *cis* isomer, then it could be held rigidly to the enzyme by hydrogen bonds, facilitating stereochemical specificity.

The next step in the glucose synthesis is the joining of one glyceraldehyde phosphate unit to one dihydroxyacetone phosphate unit to produce a molecule of fructose-1,6-diphosphate. This is clearly an aldol condensation between C-3 of the acetone and C-1 of glyceraldehyde. New asymmetric centers are created at C-3 and C-4 of fructose and, as before, only one absolute configuration is found at each asymmetric center.

(20.86)

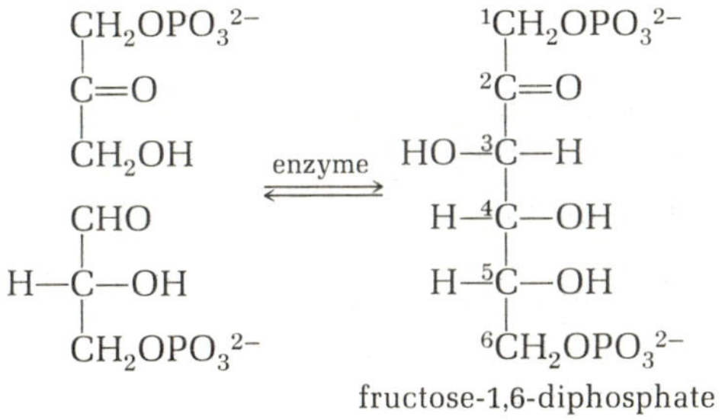

In the enzyme that catalyzes the aldol condensation there is one amino acid that specifically interacts with the dihydroxyacetone. As this amino acid contains a free amino group, which is known to condense with the carbonyl group of dihydroxyacetone, the discovery has led to the hypothesis that an enamine is an intermediate in the aldol condensation.

(20.87) 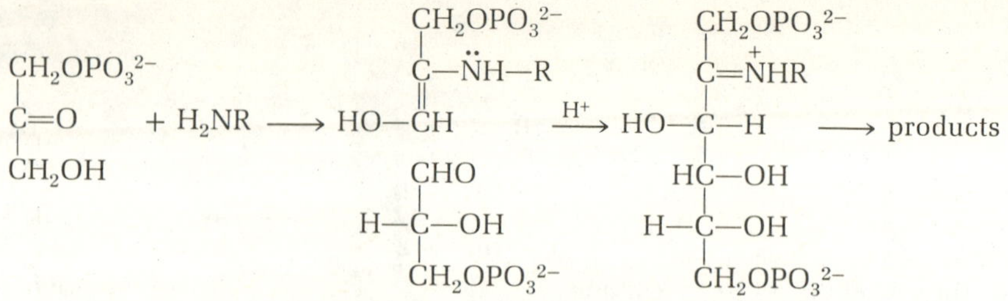

The final sequence involves hydrolysis of the phosphate at C-1 and then internal carbonyl-hydroxyl interchange at C-1 and C-2. This final step is exactly analogous to the dihydroxyacetone-glyceraldehyde isomerization and takes place with the (now) expected stereospecificity.

(20.88) 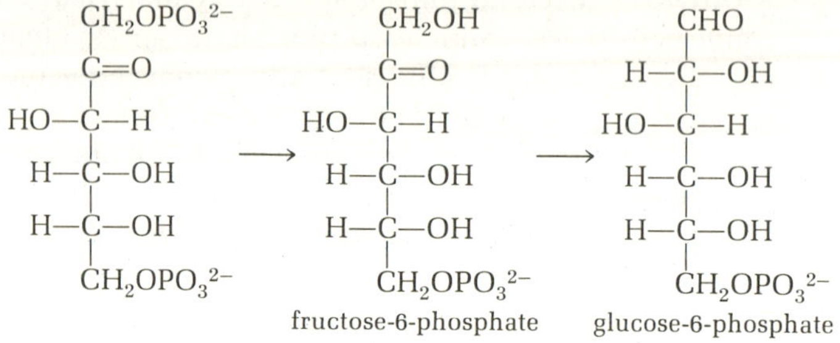

Dihydroxyacetone phosphate, a symmetrical molecule, has been converted to glucose-6-phosphate, which contains four asymmetric centers. Yet of the sixteen possible isomers, only one was produced.

SUMMARY

1 Protons attached to carbon atoms α to carbonyl groups are unusually labile. In the presence of acids and bases these protons, and only these protons, exchange rapidly with those of the solvent. Under basic conditions, enolate ions are intermediates in these exchange reactions; under acidic conditions, enols are the vital intermediates. For simple aldehydes and ketones the equilibrium concentration of enols and enolate ions is quite small but the equilibrium is rapidly established with the aid of acid or base.

2 Ketones and aldehydes are very readily halogenated in the presence of acid or base. Enols and enolate ions seem to be the reactive organic intermediates in this reaction as well. Under acidic conditions, unsymmetrical ketones seem to halogenate preferentially at the more highly substituted α-carbon atom. The greater stability of the more highly substituted enol presumably accounts for the selectivity.

Basic catalysts, on the contrary, encourage halogenation at a methyl group if one is present next to the carbonyl group. After complete halogenation of the methyl group it is cleaved off by base as a molecule of haloform, the remainder of the ketone emerging as a carboxylate ion. Under basic conditions the introduction of one halogen on a carbon atom appears to encourage further halogenation at that atom.

3 Carboxylic acids and their esters cannot be easily halogenated under non-radical conditions. If the acid is first converted to the acyl chloride or bromide, however, then it is readily halogenated at the carbon atom adjoining the carboxyl group. The acid halides are presumably more prone to enolization than the free acids themselves.

4 In basic or acidic media, aldehydes condense with themselves to yield aldols, while ketones react to give ketols. The equilibrium constants are usually favorable for aldol formation but unfavorable for ketol formation, presumably for steric reasons.

Mixed aldol condensations are usually of limited preparative value when the reactants are all aliphatic. If one reactant is an aromatic aldehyde, however, then it can be readily condensed with a variety of aliphatic aldehydes and ketones. In these cases the intermediate aldol or ketol is dehydrated in the course of the condensation, the product being a conjugated α,β-unsaturated aldehyde or ketone.

The aromatic aldehyde condenses with the methyl group of methyl alkyl ketones if a basic catalyst is employed. Acidic catalysts, on the other hand, cause condensation to take place at the methylene group.

5 Mixed aldol condensations may be carried out successfully if formaldehyde is one of the reactants, since formaldehyde possesses an extremely reactive carbonyl group but no enolizable hydrogen atoms. If formaldehyde alone or a mixture of formaldehyde and another nonenolizable aldehyde or ketone is treated with base, then a redox reaction takes place. When formaldehyde alone is used, the products are methanol and formate ion. If formaldehyde is mixed with another carbonyl compound, then the latter is always reduced and the formaldehyde always oxidized to formate.

6 The equilibrium enol content of 1,3-dicarbonyl compounds is many orders of magnitude greater than the enol content of simple ketones and aldehydes. A keto group is much more effective than an ester group in promoting enol formation. These enols are stabilized by conjugation and by intramolecular hydrogen bonding.

1,3-Dicarbonyl compounds are considerably more acidic than typical aldehydes and ketones and are completely neutralized by one equivalent of a metal alkoxide. The fairly stable enolate ions that result are useful intermediates for synthetic purposes; for instance, they are very readily alkylated by alkyl halides and tosylates.

7 Alkylation of monoketones is a much less straightforward procedure. Strong bases are required in order that there be an appreciable concentration of enolate ions. The alkylation is bedeviled by side reactions, chief among them being the aldol condensation. These problems can be partly circumvented by condensing the aldehyde or ketone to be alkylated with a secondary amine and converting it to an enamine. The enamine contains a relatively nucleophilic double bond, although it still requires fairly reactive alkylating agents. Hydrolytic work-up of the reaction mixture yields the alkylated carbonyl compound.

8 The hydroxyl and carbonyl groups of α-hydroxy aldehydes and ketones are facilely interconverted by acid and base. The interconversion is a very prominent one in enzymatic systems. The aldol condensation is also a prominent reaction in many cells. Both reactions take place with complete stereochemical control when subject to enzymic catalysis.

PROBLEMS

Note: Problems 1 through 3 appear in text.

4 Write structures for the following aldehydes and ketones, and indicate which protons you expect to exchange with D_2O under acidic or basic conditions. Write formulas for all the enols involved in these exchange reactions.

a. Cyclopentanone

b. 3-Hexanone

c. Isobutyraldehyde

d. 3,3-Dimethyl-2-butanone (pinacolone)

e. Benzaldehyde

f. Di-*t*-butyl ketone

5 The ketone II has been obtained optically pure. When treated with aqueous base, it undergoes racemization. Furthermore, the rate of racemization at a given pH is equivalent to the rate of proton exchange at the same pH.

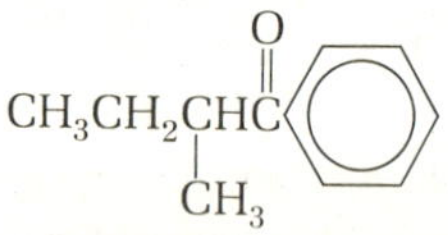

II

Rationalize these data. What do they suggest about the geometry of enolate ions?

6 Norbornanone (III) was subjected to the action of a solution of NaOD in D_2O and the ketone was then recovered from the solution and its nmr spectrum obtained. The spectrum revealed that the proton at C-1 had *not* been exchanged with solvent. Furthermore, although the protons at C-3 had been undergoing exchange, the *exo* proton was exchanging much more rapidly than the *endo* proton.

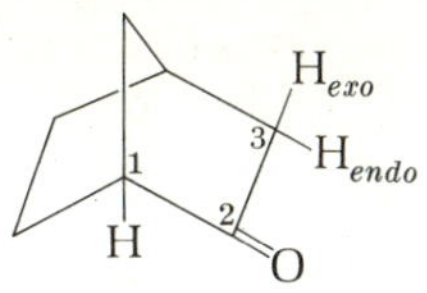

III

What conclusions can one draw about the stereochemistry of enolate ion formation and protonation? (Molecular models would be very helpful in answering this question.)

7 The reaction of hindered ketones with Grignard reagents often leads to a low yield of tertiary alcohol. When methylmagnesium bromide is added to 3,3-diethyl-2-pentanone, considerable gas evolution takes place. After virtually all the Grignard reagent has been consumed, the reaction mixture is subjected to aqueous work-up in the usual manner. The starting ketone is recovered in 94% yield. Offer an explanation for this observation.

8 Suggest efficient syntheses of the following deuterium-labeled compounds. The sources of deuterium are D_2O and $LiAlD_4$. Any other organic and inorganic reagents may be used.

a. Ethanol-2,2,2-d_3
b. Ethanol-1,2,2,2-d_4
c. Ethanol-1,1-d_2
d. 2-Propanol-1,1,1,3,3,3-d_6
e. *t*-Butyl-d_6 bromide
f. Cyclohexene-1,3,3-d_3
g. Cyclohexene-1,2,3,3-d_4
h. Acetic-d_3 acid
i. 3-Hexanol-2,2-d_2
j. 1-Pentanol-2,2-d_2

9 The following carbonyl compounds were treated with 1 mole of bromine in acetic acid. Show the possible monobromination products, and indicate which isomer(s) of a group will be favored.

a. 3-Pentanone
b. Cycloheptanone
c. Phenylacetaldehyde
d. 2-Hexanone
e. 3-Methylcyclohexanone
f. Norbornanone (III)

10 The haloform reaction depends in part on the stability of the trihalomethyl anions, CX_3^- (X = Cl, Br, I). An indication of the stability of these anions is provided by the observation that the corresponding haloforms, CHX_3, undergo rapid hydrogen exchange when dissolved in basic D_2O. Fluoroform, CHF_3, exchanges its proton much more *slowly* than the other three haloforms, despite the fact that fluorine is the most electronegative of the halogens. Write a reasonable mechanism for the exchange reaction, and offer an explanation for the low reactivity of fluoroform.

11 A compound $C_{10}H_{14}O$ absorbs infrared radiation in the 3400 to 3600 cm^{-1} region of the spectrum and also absorbs at 700 cm^{-1} and at other frequencies characteristic of an aromatic compound. When the substance is treated with alkaline sodium

hypochlorite and the reaction mixture then acidified, a crystalline acid is obtained. The acid is allowed to react successively with thionyl chloride and with lithium tri-*t*-butoxyaluminohydride. A compound is obtained that on solution in weakly basic D_2O exchanges two of its protons with the solvent. Write all the chemistry and propose one (several) structure(s) that is (are) compatible with the data presented.

12 Show all the products expected from the following reactions; write out one or two in detail showing the reactive intermediates.

a. Propionaldehyde $\xrightarrow{\text{dil. } H_3O^+}$

b. Cyclohexanone $\xrightarrow[\text{then distill}]{\text{dry HCl,}}$

c. Isobutyraldehyde $\xrightarrow{\text{dil. } OH^-}$

d. Acetaldehyde + propionaldehyde $\overset{\text{dil. } OH^-}{\rightleftarrows}$

e. Pentanal + 3-pentanone $\xrightarrow{\text{dil. } H_3O^+}$

f. 2,2-Dimethylpropanal + acetone $\xrightarrow{\text{dil. } OH^-}$

g. 2,2-Dimethylpropanal + formaldehyde $\xrightarrow{OH^-}$

h. Benzaldehyde + cyclopentanone (excess) $\xrightarrow{OH^-}$

i. *p*-Nitrobenzaldehyde + formaldehyde $\xrightarrow{OH^-}$

j. *m*-Chlorobenzaldehyde + 2-hexanone $\xrightarrow{H_2SO_4}$

k. *m*-Methoxybenzaldehyde + methylcyclobutyl ketone $\xrightarrow{OH^-}$

l. 

13 The equilibrium constant for aldol dimerization of two molecules of a ketone is usually small. However, reaction 1 gives a satisfactory yield of the dehydrated intramolecular aldol condensation product.

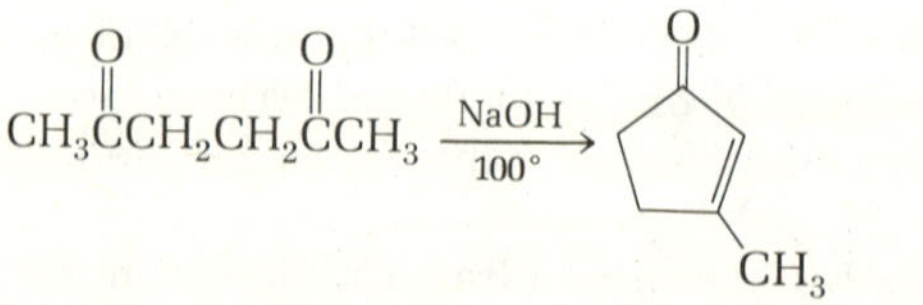
(1)

Explain why the intramolecular condensation is so much more successful.

14 When cyclopentanone is treated with isopropylmagnesium chloride and the reaction mixture then worked up with aqueous acid, the only organic product that is isolated is cyclopentylidenecyclopentanone (IV).

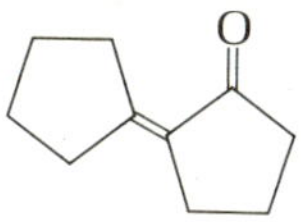

IV

Propose a mechanism for this reaction and contrast the results with those discussed in Problem 7.

15 The following compounds will all completely neutralize an equivalent quantity of sodium ethoxide. Show by means of resonance structures why their anions are relatively stable.

a. Diethyl malonate
b. Ethyl cyanoacetate
c. Malononitrile (dicyanomethane)
d. Nitromethane
e. Tribenzoylmethane
f. 2-Carbethoxycyclopentanone
g. 4-Carbethoxycyclohex-2-en-1-one (V)

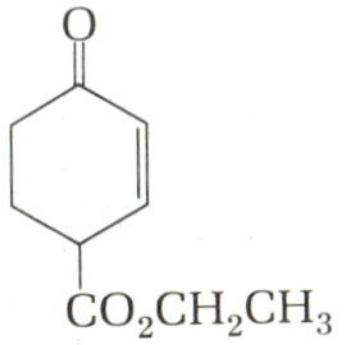

V

16 Draw the structures of the products expected from the following reactions.

a. Acetone + malononitrile $\xrightarrow[CH_3CO_2H,\ C_6H_6]{\text{piperidine}}$

b. Benzaldehyde + diethyl malonate $\xrightarrow[C_6H_6]{\text{piperidine}}$

c. 3-Methylcyclohexanone + ethyl cyanoacetate $\xrightarrow[\text{acetate, benzene}]{\text{ammonium}}$

d. Acetophenone + ethyl acetoacetate $\xrightarrow[\text{benzoic acid, toluene}]{C_4H_9NH_2}$

17 When 1,3-cyclohexanedione is treated successively with sodium methoxide and methyl iodide, two organic products are formed. One of these proves to be the desired 2-methyl-1,3-cyclohexanedione. The other product, isomeric with the first, exhibits the following nmr spectrum: a multiplet in the methylene region (6 H), a vinyl proton (1 H), and a singlet (3 H). Propose a structure for this compound, and rationalize its formation.

18 Ethyl acetoacetate is neutralized with sodium ethoxide and a solution of the resulting sodium salt is cooled to −78°. The solution is acidified at that temperature, causing two products to precipitate. One is an oil that reacts instantaneously with bromine, while the second is crystalline and unreactive toward bromine. The compounds are isomers. Explain.

19 a. Acetaldehyde $\xrightarrow{OH^-}$ **A** ($C_4H_8O_2$) $\xrightarrow[\text{heat}]{KHSO_4}$ **B** (C_4H_6O)

Anisole $\xrightarrow{Br_2}$ **C** (C_7H_7BrO) $\xrightarrow[\text{ether}]{Mg}$ $\xrightarrow[\text{2. } H_3O^+]{\text{1. } \mathbf{B}}$ **D** ($C_{11}H_{14}O_2$) $\xrightarrow[\text{heat}]{H^+}$ **E** ($C_{11}H_{12}O$)

$CH_3\overset{O}{\overset{\|}{C}}CO_2CH_2CH_3$ + malonitrile $\xrightarrow{C_2H_5OH}$ **F** ($C_8H_8N_2O_2$) $\xrightarrow{\mathbf{E}}$ **G** ($C_{19}H_{20}N_2O_3$)
ethyl pyruvate — dicyanomethane

b. Benzaldehyde + acetophenone $\xrightarrow{OH^-}$ **H** ($C_{15}H_{12}O$) $\xrightarrow[CHCl_3]{Br_2}$ **I** ($C_{15}H_{12}Br_2O$)

$\xrightarrow[C_2H_5OH]{KOH}$ **J** ($C_{15}H_{11}BrO$) $\xrightarrow[NH_3\text{ (liq)}]{NaNH_2}$ **K** ($C_{15}H_{10}O$)

Propose structures for compounds **A** to **K**.

SUPPLEMENTARY READINGS

H. O. House, *Modern Synthetic Reactions.* New York: W. A. Benjamin, Inc., 1965, Chapters 7 and 8.

D. C. Ayres, *Carbanions in Synthesis.* New York: American Elsevier Publishing Co., 1966, Chapters 4 and 5.

CHAPTER 21

ACYLATION AND MICHAEL ADDITION — FURTHER EXTENSIONS OF ENOLATE CHEMISTRY

In the preceding chapter we concentrated our attention on a reaction in which an enolate ion adds to a carbonyl group, yielding a hydroxy compound (the aldol condensation). This is the first step in a number of related reactions in which the carbonyl compound is an aldehyde or ketone. Enolate anions may also react with carboxylic acid derivatives. In this case the net reaction is a substitution rather than an addition. The enolate ion behaves toward carboxylic acid derivatives in the same manner as other nucleophiles do — after the initial addition to the double bond, the leaving group is expelled so that there is a net substitution (Chapter 17).

(21.1) 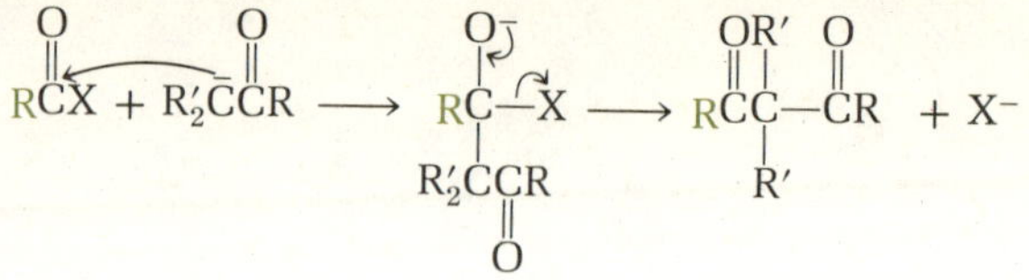

FORMATION OF β-KETOESTERS

The earliest recorded acylation of an enolate anion concerns the base-catalyzed self-condensation of ethyl acetate. This reaction was discovered by Claisen, and the entire class of reactions is known as the **Claisen condensation** (not to be confused with the Claisen-Schmidt reaction, which is an aldol condensation). The product obtained from the ethyl acetate reaction is ethyl acetoacetate (in the older literature, acetoacetic ester).

(21.2)
$$2CH_3\overset{O}{\overset{\|}{C}}OCH_2CH_3 \xrightleftharpoons{CH_3CH_2O^-Na^+} \underset{(30\%)}{CH_3\overset{O}{\overset{\|}{C}}CH_2\overset{O}{\overset{\|}{C}}OCH_2CH_3} + CH_3CH_2OH$$

The yields in this reaction are not spectacular but it is rather surprising that any product is formed at all. (Yields can be improved considerably if the ethanol is distilled continuously from the reaction mixture.) The entropy change must be very small and we can expect the reaction to be endothermic because the resonance energy of one of the ester groups is being lost. In fact, it appears that in the reaction mixture the ethyl acetoacetate is actually present as its anion, and it is the neutralization of ethoxide ion by the ester that drives the condensation forward.

(21.3) 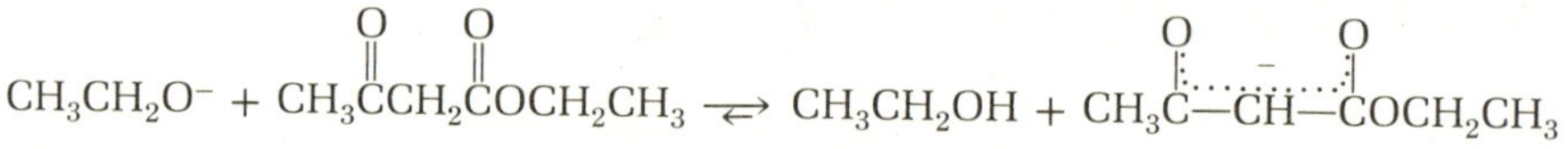

The neutral ester is obtained only after an acid work-up.

If we seek to identify the "driving force" for equation 21.2, we must conclude that it is the conversion of the strong base sodium ethoxide to the weaker base ethyl acetoacetate sodium salt. Ethoxide ion is thus not truly a catalyst. The acidity of the product ester is due to its being a β-dicarbonyl compound (p. 548) in which a methylene group is flanked by two carbonyl groups. A plausible mechanism for the first part of the condensation is given in equations 21.4 and 21.5. The entire process is readily reversible.

(21.4) 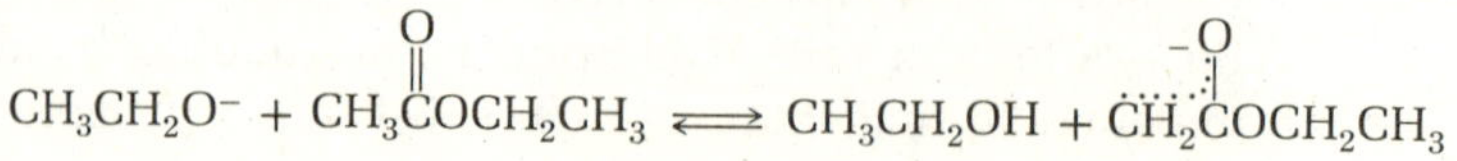

(21.5)

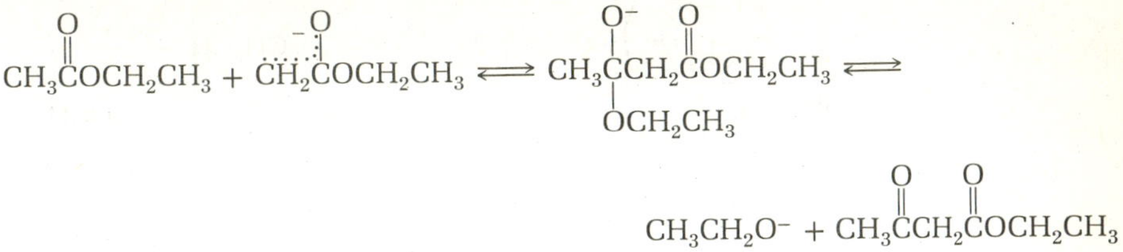

(21.6)

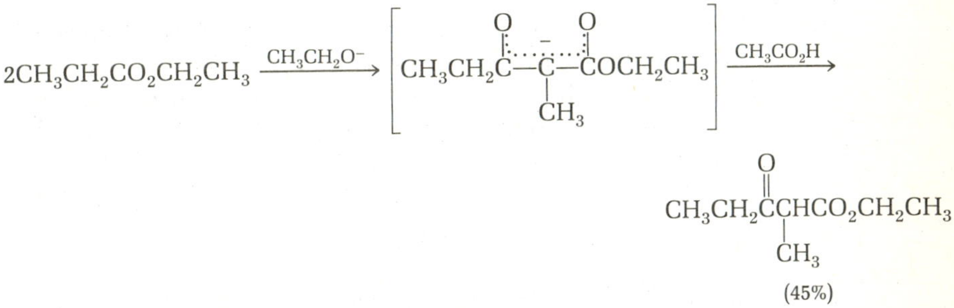

(21.7)

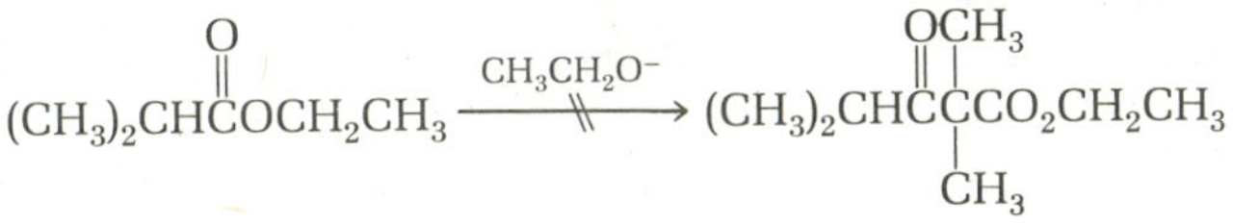

1

Ethyl propionate may also be condensed in the same manner, but ethyl isobutyrate does not undergo Claisen condensation in the presence of sodium ethoxide (equation 21.7) because the product, ethyl α,α-dimethylacetoacetate (**1**), cannot be converted to its anion as it lacks an acidic hydrogen atom. This finding emphasizes the importance of the ester-ethoxide reaction (equation 21.3) to the overall equilibrium.

PROBLEM

1 Alkyl thiolacetates, CH_3COSR, are more reactive in Claisen condensation and aldol-type condensations than are their oxygen counterparts, CH_3CO_2R. Offer an explanation for this finding.

Ethyl isobutyrate will undergo self-condensation if a very powerful base is used. One which has been employed often is the salt of triphenylmethane, sodium triphenylmethide $[(C_6H_5)_3C^-Na^+]$. Triphenylmethane is a very much weaker acid than any alcohol or even any simple ester or ketone. The condensation reaction therefore involves the conversion of triphenylmethide ion to the weaker base ethoxide ion.

(21.8) 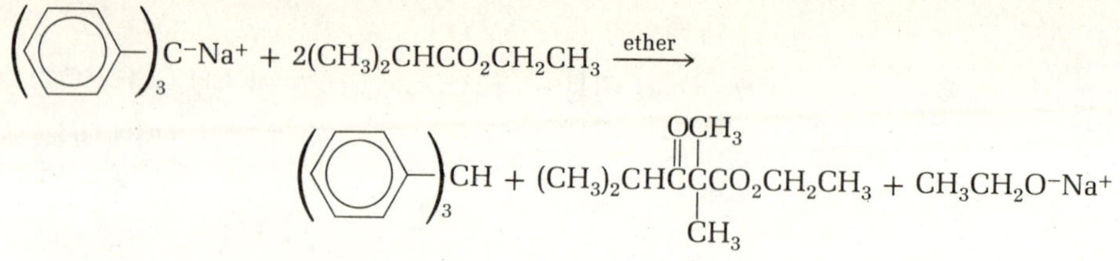

(35%)

The attempted cocondensation of two different esters usually leads to a mixture of all four possible products and is therefore of limited preparative value. If one of the two esters possesses no α-hydrogen atoms, however, then the situation is much simplified. In such cases, mixed condensations are successful, the esters of formic and benzoic acid and the diesters of carbonic and oxalic acids being the most common acylating agents.

(21.9)
$$\begin{matrix} CH_2CO_2CH_2CH_3 \\ | \\ CH_2CO_2CH_2CH_3 \end{matrix} + \underset{\text{ethyl formate}}{HC(=O)OCH_2CH_3} \xrightarrow[\text{Na metal}]{CH_3CH_2O^-Na^+} \xrightarrow{H^+} \begin{matrix} HC(=O)CHCO_2CH_2CH_3 \\ | \\ CH_2CO_2CH_2CH_3 \end{matrix}$$

(60–70%)

(21.10) 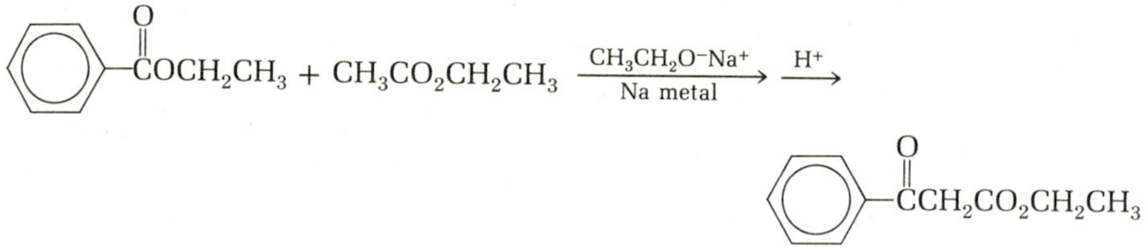

ethyl benzoylacetate
(60%)

THE DIECKMANN CONDENSATION

The intramolecular acylation of diesters to produce cyclic ketoesters is merely a special case of the Claisen condensation but it has its own name (the *Dieckmann condensation*). These acylations are usually effected with relative ease because the entropy of reaction is much more favorable than it is with acyclic reactants (equation 21.11).

(21.11) 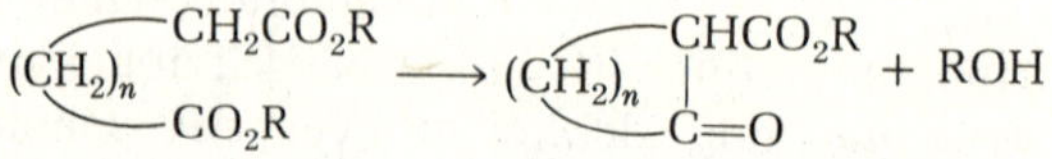

Note the use of the base sodium hydride, Na^+H^-, in equation 21.12.

(21.12)

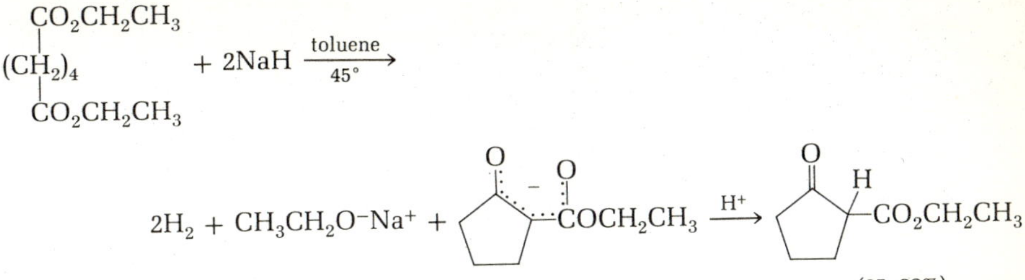

(65–80%)

Sodium hydride is an ionic, crystalline solid insoluble in all organic solvents, which reacts vigorously with acidic protons. In many reactions where it is employed, a catalytic quantity of an alcohol is added; the alcohol is immediately converted to alkoxide and this ion serves to transport protons from the carbonyl compound to the sodium hydride.

(21.13)

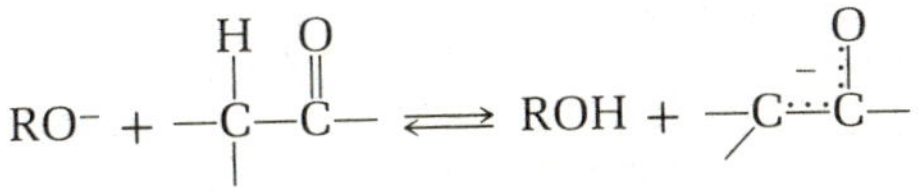

(21.14) $ROH + NaH \longrightarrow RO^-Na^+ + H_2$

The protons abstracted from the organic acids eventually go off as hydrogen gas, thus making the condensation irreversible. This irreversibility makes sodium hydride a very desirable reactant for all kinds of base-promoted acylations.

Formation of five- and six-membered rings are the processes most favored and most often encountered under Dieckmann conditions. Larger rings may be created in satisfactory yields, however, if the solution is quite dilute so that intermolecular reactions will be suppressed. When the two α-positions of a diester are not equivalent, then acylation will take place at the less substituted α-carbon atom since this route allows the formation of the stable ketoester enolate ion (equation 21.15).

(21.15)

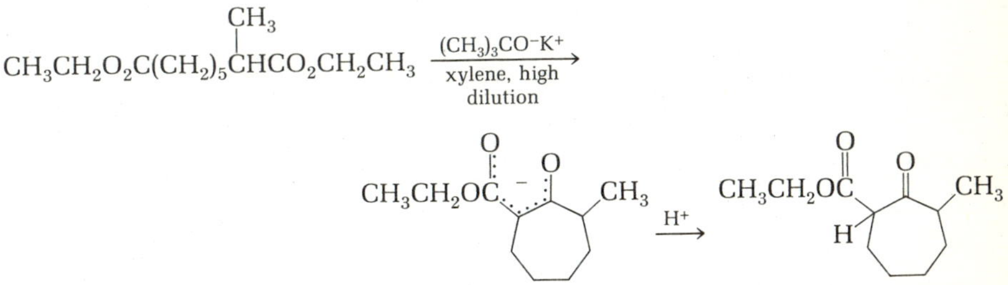

PROBLEM

2 Methyl 2-ethylcyclohexanone-2-carboxylate (I), upon being heated with aqueous base and then acidified, is converted to α-ethylpimelic acid.

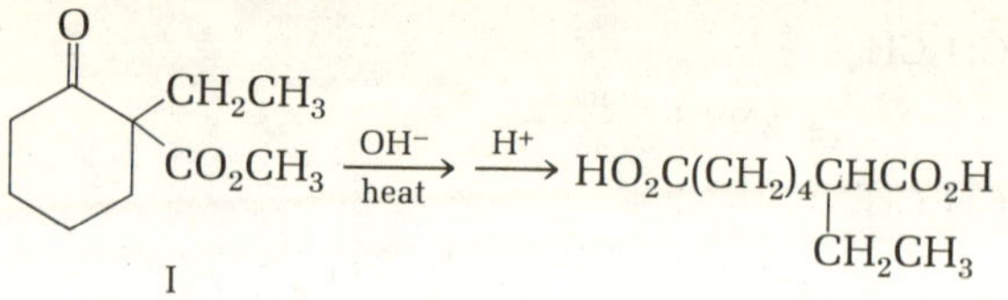

I

(1)

Propose a mechanism for reaction 1.

ACYLATION OF KETONES

The acylation of ketones may be undertaken in a manner analogous to the acylation of esters. The same basic considerations apply in both cases.

(21.16)

$$CH_3\overset{\overset{O}{\|}}{C}CH_3 + 2\,\overset{\overset{CO_2CH_2CH_3}{|}}{C}O_2CH_2CH_3 \xrightarrow[CH_3CH_2OH]{CH_3CH_2O^-Na^+} \xrightarrow{H_3O^+} C_2H_5O\overset{\overset{O}{\|}}{C}\overset{\overset{O}{\|}}{C}CH_2\overset{\overset{O}{\|}}{C}CH_2\overset{\overset{O}{\|}}{C}\overset{\overset{O}{\|}}{C}OC_2H_5 \quad (85\%)$$

(21.17)

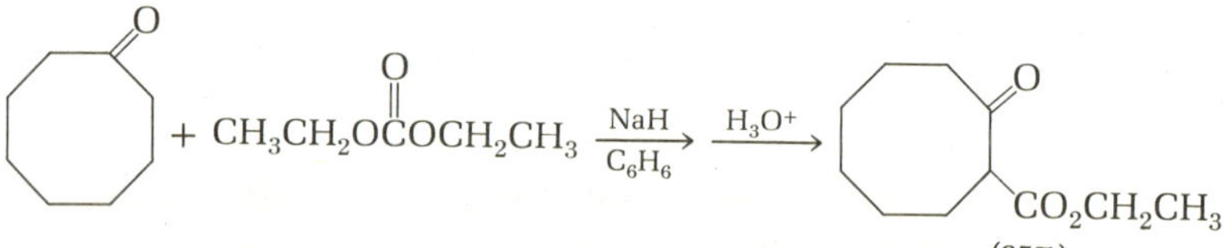

Ketones may also be acylated by way of their enamines, thus avoiding the use of strong bases that can promote side reactions such as aldol condensation. Furthermore, with enamines as substrates, one may use acid chlorides and anhydrides as acylating agents; since the chloride and carboxylate ions are weak nucleophiles, the acylation reaction is then essentially irreversible.

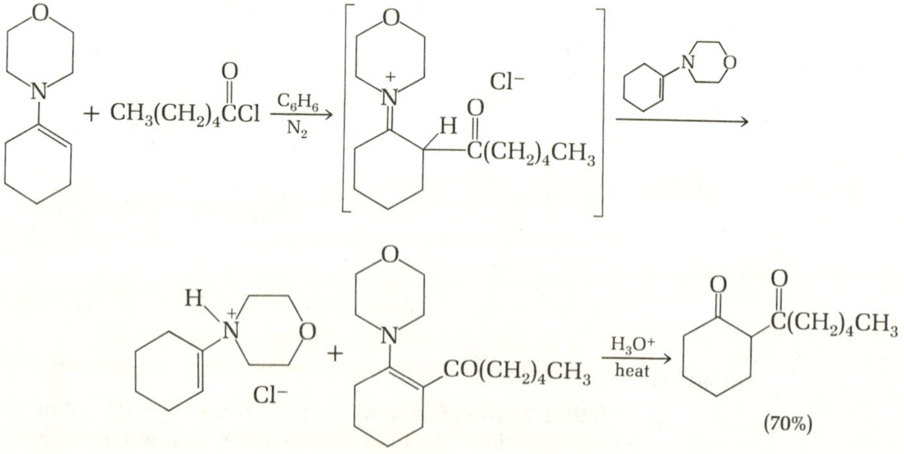

(21.18)

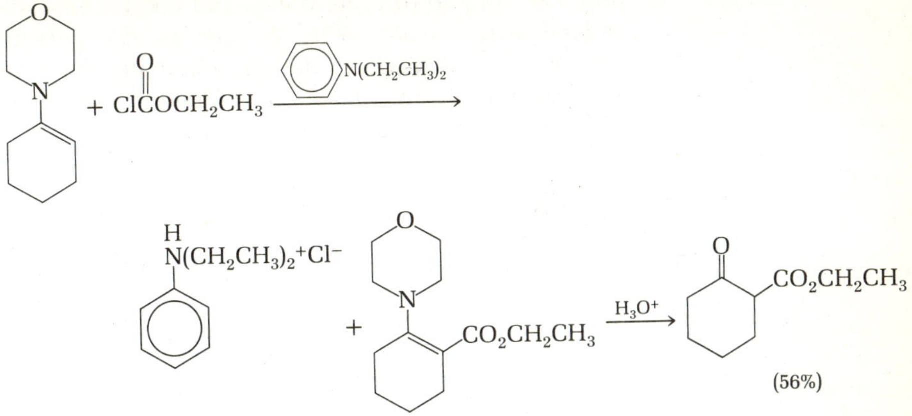

(21.19)

One equivalent of acid is released in the course of the acylation. If an external base is not added, this acid will react with the basic enamine, seriously limiting the maximum absolute yield of the reaction.

SYNTHESES WITH β-KETOESTERS AND β-DIKETONES

As already observed, 1,3-dicarbonyl compounds participate in aldol condensations under quite mild conditions, due to the lability of their acidic protons (p. 548). The stable anions derived from these compounds are also readily alkylated (equations 20.72 and 20.73). These facts by themselves establish the utility of 1,3-dicarbonyl compounds and emphasize the value of the acylation reactions used to prepare them.

(21.20) 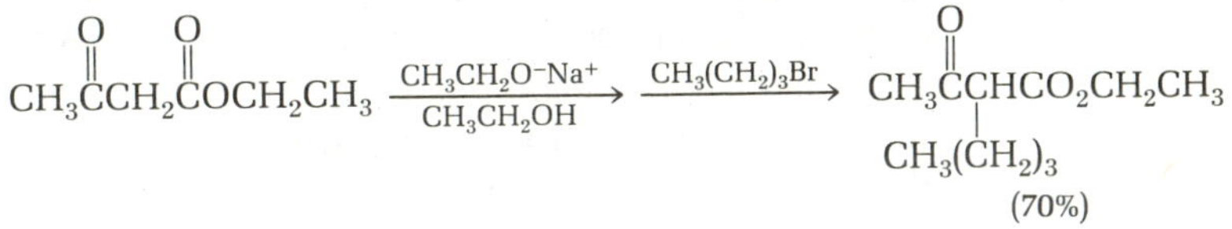

(21.21) 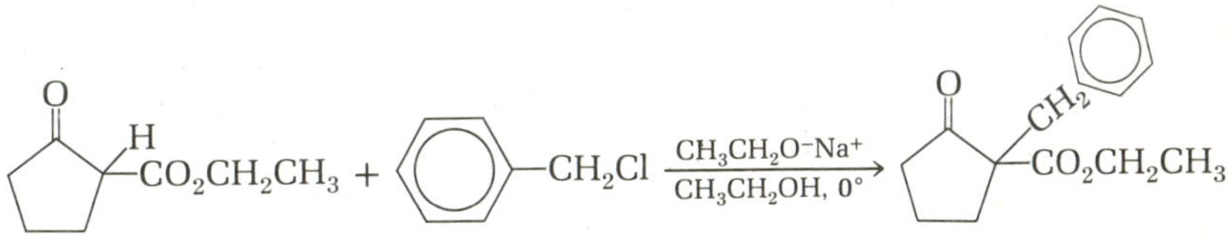

Furthermore, the acylation reaction is reversible, and one of the acyl groups may be eliminated after the alkylation or aldol condensation has taken place (see below), which considerably enhances the scope of the entire procedure.

One complication arises when the anions of 1,3-dicarbonyl compounds are subjected to alkylation. The charge in these species is distributed over both carbon and oxygen atoms, and alkylation could take place on oxygen as well as on carbon. If the oxygen atom is part of an ester group, then it will not be alkylated; diethyl malonate, for example, undergoes only *C*-alkylation. The oxygen atoms of ketones and aldehydes may be alkylated, however, as illustrated in equation 21.22.

(21.22)

$$CH_3\overset{\overset{\displaystyle O}{\|}}{C}CH_2CO_2C_2H_5 + CH_3(CH_2)_3Cl \xrightarrow[100^\circ]{K_2CO_3} \underset{\displaystyle CH_3(CH_2)_2CH_2}{CH_3\overset{\overset{\displaystyle O}{\|}}{C}\underset{|}{C}HCO_2C_2H_5} + CH_3\overset{\overset{\displaystyle O(CH_2)_3CH_3}{|}}{C}{=}CHCO_2C_2H_5$$

Solvent:			
	acetone	90%	10%
	DMF	54%	46%

This equation points up the decisive effect of the solvent in influencing *C*- versus *O*-alkylation, dipolar aprotic solvents of high polarity favoring oxygen attack. The product distribution is also sensitive to the temperature, the metal cation in the base, and the leaving group in the alkylating agent (equation 21.23).

(21.23)

$$CH_3\overset{\overset{\displaystyle O}{\|}}{C}CH_2CO_2C_2H_5 + CH_3(CH_2)_3I \xrightarrow[DMF]{K_2CO_3} \underset{\displaystyle CH_3(CH_2)_2CH_2}{CH_3\overset{\overset{\displaystyle O}{\|}}{C}\underset{|}{C}HCO_2C_2H_5} \quad (100\%)$$

DECARBOXYLATION OF β-KETOCARBOXYLIC ACIDS

Malonic acid is a perfectly stable solid at room temperature but when it is heated to about 100°, it begins to evolve gas and undergo decomposition. The product of this decomposition proves to be acetic acid. Similarly, if one attempts to obtain acetoacetic acid by hydrolyzing one of its esters under acidic conditions, one recovers only acetone.

(21.24)

$$CH_2(CO_2H)_2 \xrightarrow{heat} CH_3CO_2H + CO_2$$

(21.25)

$$CH_3\overset{\overset{\displaystyle O}{\|}}{C}CH_2CO_2CH_2CH_3 \xrightarrow[heat]{H_3O^+} CH_3CH_2OH + \left[CH_3\overset{\overset{\displaystyle O}{\|}}{C}CH_2CO_2H\right] \longrightarrow CH_3\overset{\overset{\displaystyle O}{\|}}{C}CH_3 + CO_2$$

The *decarboxylation* reaction is very characteristic of molecules in which a carboxylic acid group is β to another carbonyl group. If that carbonyl group is part of a ketone or aldehyde, then the decarboxylation takes place with particular ease, often proceeding at a fast rate even at 25°. If the carbonyl group is part of a carboxylic acid or its ester, then the decarboxylation is not quite so facile and heat must often be applied in order that carbon dioxide be lost at a convenient rate.

It is believed that decarboxylation proceeds by way of a concerted mechanism in which there is a cyclic, six-membered transition state. The immediate product would then be the *enol* of the compound actually isolated; this enol presumably tautomerizes very rapidly to the carbonyl compound.

(21.26) 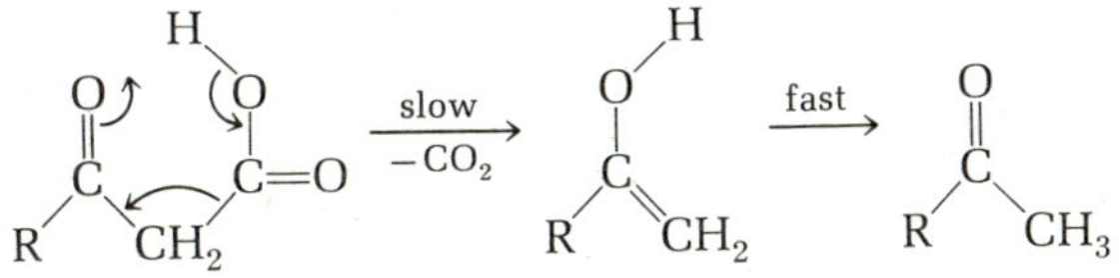

Carbonyl-containing functional groups are not the only groups that promote decarboxylation. Nitro and cyano are also effective, and cyanoacetic acid (easily prepared from chloroacetic acid and sodium cyanide) and its esters are valuable synthetic intermediates.

(21.27) $$\underset{\displaystyle CN}{\overset{}{CH_2CO_2H}} \xrightarrow{\text{heat}} CO_2 + CH_2{=}C{=}NH \rightleftharpoons CH_3CN$$

The occurrence of enols as intermediates in decarboxylations seems fairly well-established. The enols have been intercepted by halogens under conditions in which their carbonyl tautomers would have reacted much more slowly with halogen. Furthermore, the stability of the ketoacid **2** toward decarboxylation can reasonably be ascribed to the high energy of the anticipated enol **3**, with its very strained double bond.

(21.28) 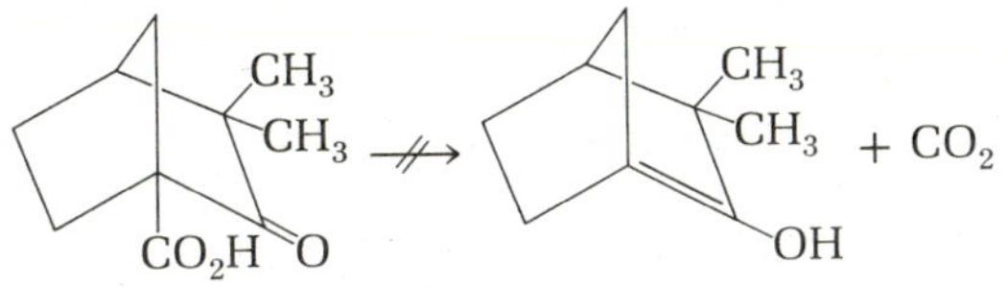

2 **3**

CARBOXYL AS AN ACTIVATING GROUP

We have been exploring two facets of carboxyl group chemistry, namely, its ability to markedly increase the acidity of neighboring protons and the ease with which it may be ejected as long as there is a suitable group in the β-position. These two properties combined make the carboxyl group a very valuable functionality since it can behave as a disposable activating group, in analogy to the role played by the amino group in aromatic chemistry (Chapter 19).

Diethyl malonate is a common starting material for the synthesis of aliphatic monocarboxylic acids (equation 21.29). It is possible to obtain α-substituted carboxylic acids by alkylating the malonic ester twice (equation 21.30).

(21.29)

$$CH_2(CO_2CH_2CH_3)_2 + (CH_3)_2CHCH_2Br \xrightarrow[CH_3CH_2OH]{CH_3CH_2O^-Na^+} (CH_3)_2CHCH_2CH(CO_2CH_2CH_3)_2 \xrightarrow{KOH} (CH_3)_2CHCH_2CH(CO_2^- K^+)_2 \xrightarrow[reflux]{H_3O^+} (CH_3)_2CHCH_2CH_2CO_2H$$

(21.30)

$$CH_3CH(CO_2CH_2CH_3)_2 + CH_3(CH_2)_7I \xrightarrow[CH_3CH_2OH]{CH_3CH_2O^-Na^+} CH_3(CH_2)_7C(CH_3)(CO_2CH_2CH_3)_2 \xrightarrow[2.\ H_3O^+]{1.\ OH^-} CH_3(CH_2)_7CH(CH_3)CO_2H$$

One could presumably have obtained the same products as those shown in equations 21.29 and 21.30 by directly alkylating ethyl acetate. This approach would have necessitated the use of a strong base, and the alkylation would have been plagued by side reactions such as Claisen condensation of the ethyl acetate. Use of a dicarbonyl compound leads to a relatively unreactive enolate anion so that side reactions are minimized. After the alkylation has been completed, the extra carboxyl group is easily disposed of.

(21.31)

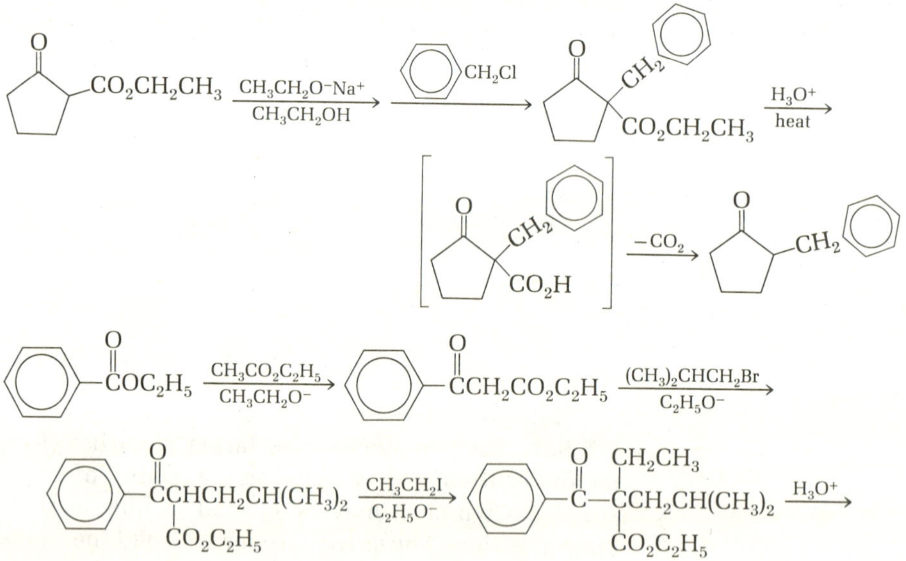

(21.32)

The diacids obtained from aldol condensation of malonic acid or its esters are difficult to decarboxylate thermally, because the enol that would be the intermediate in this reaction contains a system of cumulated double bonds and would therefore be quite high in energy.

(21.33) 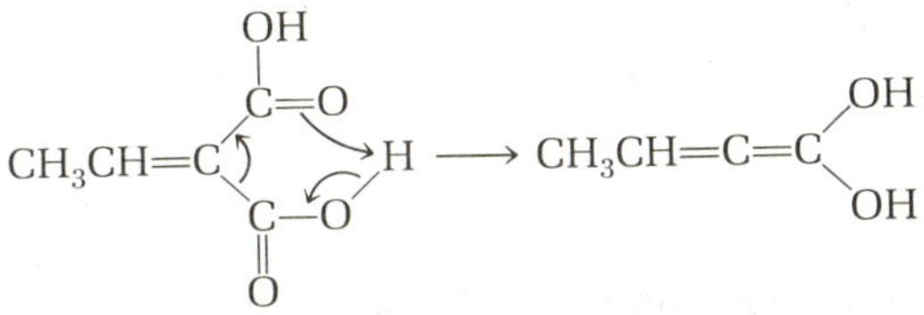

In cases of this kind there is good evidence that decarboxylation is preceded by isomerization of the double bond from the α,β-position to the β,γ-position.

It is possible, however, to condense aldehydes and ketones with malonic acid under conditions such that 1 mole of carbon dioxide is eliminated *in the course of the condensation.*

(21.34) $$CH_3CH_2CHO + CH_2(CO_2H)_2 \xrightarrow[25^\circ]{C_5H_5N} \underset{(85\%)}{CH_3CH_2CH{=}CHCO_2H} + CO_2 + H_2O$$

(21.35) 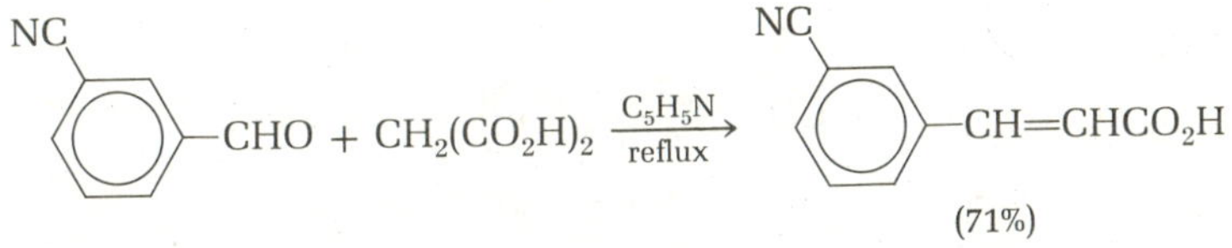

Pyridine is a very effective catalyst for this condensation (*Doebner reaction*) and a possible mechanism that accounts for its action is shown in equation 21.36.

(21.36) 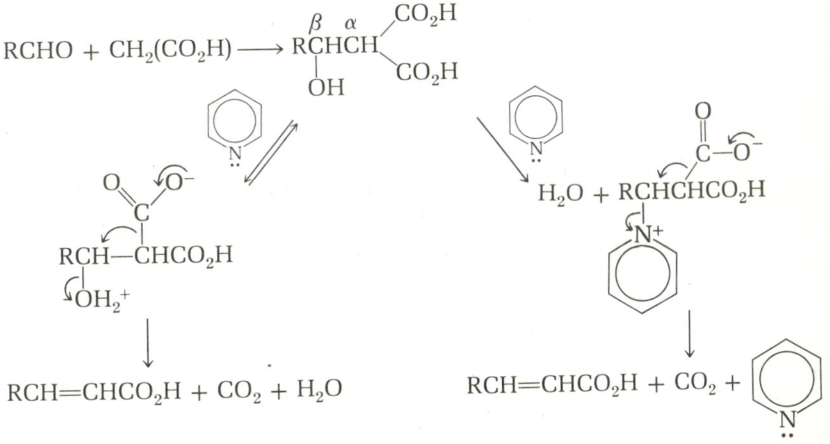

If we compare this reaction with the decarboxylation of β-ketoacids, we see that there is one common feature that they share, namely, the presence of an electron "sink" in the β position. The departure of the carbon dioxide molecule leaves a pair of electrons that must be accommodated. In β-ketoacids the ketone group constitutes the electron sink, its electrophilicity being enhanced by proton transfer during decarboxylation. For the intermediates shown in equation 21.36, the electron sinks are the positively charged oxonium and pyridinium groups. Proton transfer is not necessary and it is the anion of the acid that decarboxylates.

CARBOXYLATION OF ENOLATES

Under normal conditions the decarboxylation of β-ketoacids proceeds irreversibly, the carbon dioxide escaping to the atmosphere. The reverse reaction ought to be feasible, however, and in a few cases it has been achieved.

A commercially important process is the carboxylation of phenol and substituted phenols; salicylic acid (o-hydroxybenzoic acid), which is the starting material for the manufacture of the mountains of aspirin consumed each year, is made in this way. Sodium phenoxide may be considered to be the very stable enolate of cyclohexa-2,4-dienone.

(21.37)

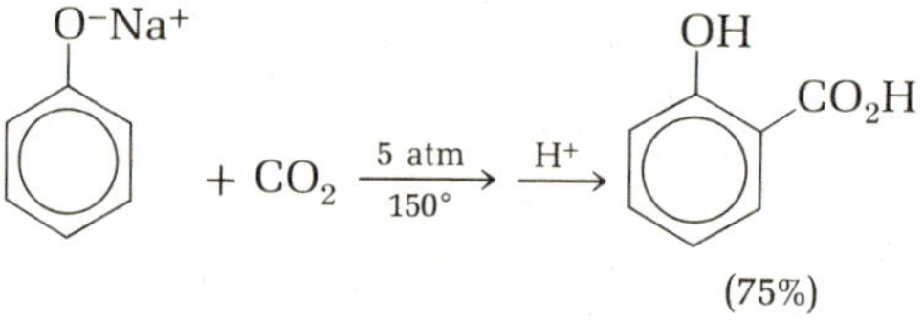

The mechanism involves the carbon acylation of phenoxide ion by carbon dioxide, followed by tautomerization to regenerate the aromatic ring.

(21.38)

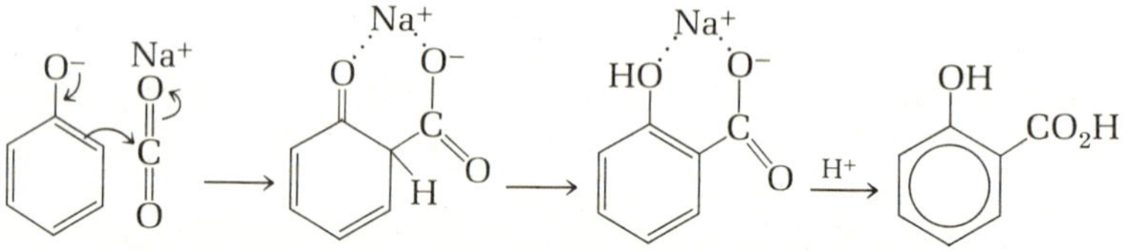

Interestingly, when the potassium salt is used, a mixture of o- and p-hydroxybenzoic acids is obtained. The high yield of *ortho* product obtained from the sodium salt is ascribed to the greater chelating power of the sodium ion.

(21.39)

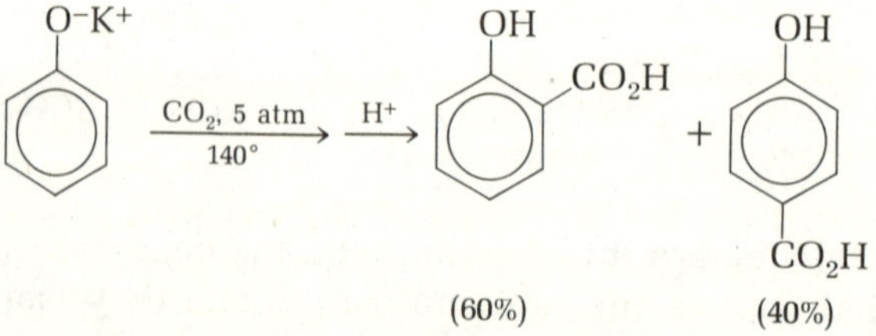

The sodium is attracted to the oxygen atoms of both the phenoxide and the carboxylate and thus stabilizes the *ortho* isomer.

Polyvalent metal cations such as those of magnesium and aluminum are even stronger chelating agents for oxygen and advantage is taken of this fact for carboxylating less stable enols. The β-ketoacids so formed are actually isolable in good yield.

(21.40) $$Mg(OCH_3)_2 + CO_2 \xrightarrow{DMF} CH_3OMgO\overset{O}{\overset{\|}{C}}OCH_3$$

methyl methoxy-magnesium carbonate

(21.41)

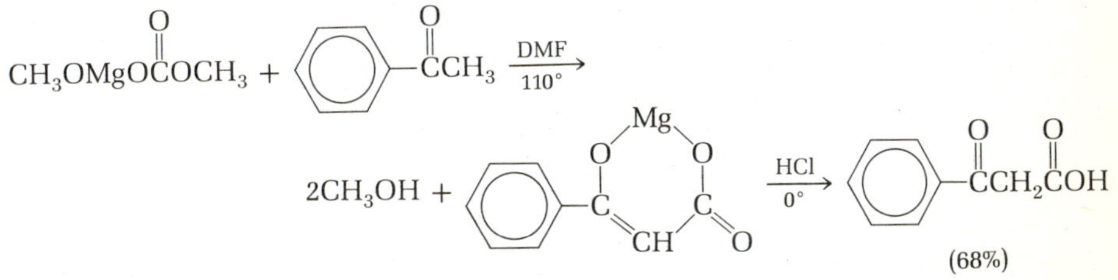

An important driving force for the reaction must be the formation of the very stable magnesium chelate.

The metal ion-promoted carboxylation of enolates is not only of synthetic value; it also throws light on a process that is extremely important in animal metabolism. The carboxylation of pyruvic acid, followed by reduction of the intermediate ketodiacid by NADPH (Chapter 11) yields malic acid.

(21.42) $$CH_3\overset{O}{\overset{\|}{C}}CO_2H + CO_2 \underset{}{\overset{enzyme}{\rightleftharpoons}} HO_2CCH_2\overset{O}{\overset{\|}{C}}CO_2H + NADPH \underset{enzyme}{\overset{H^+}{\rightleftharpoons}} HO_2CCH_2\overset{OH}{\overset{|}{C}}HCO_2H + NADP^+$$

oxalacetic acid; malic acid

The reduction helps drive the reaction to completion since the removal of the ketone group prevents reversal of the carboxylation reaction.

Capture of atmospheric carbon dioxide by plants during photosynthesis also involves carboxylation of an enolate.

ORGANOMETALLIC ADDITIONS TO POLARIZED CARBON-CARBON DOUBLE BONDS

Grignard reagents normally attack polarized double bonds ($>C{=}O$, $-C{\equiv}N$) avidly and display no tendency to react with carbon-carbon double bonds. If the carbon-carbon double bond is conjugated with a carbonyl (or similar) group, however, then the picture changes dramatically. For one thing, the electrophilic character of the carbonyl carbon atom is partly relieved by conjugation; it is therefore less reactive. Simultaneously, part of the electron deficiency of the carbonyl carbon atom is transmitted to the β-carbon atom.

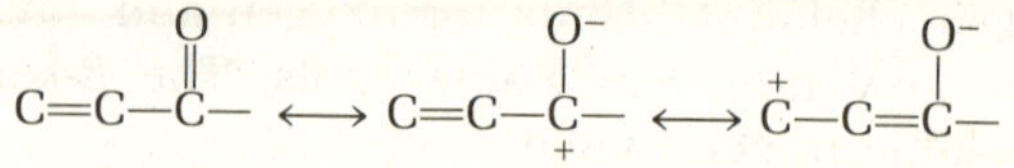

A polarized reagent A^+B^- may therefore add to the carbonyl group (1,2-addition) or its components may attach themselves to the oxygen atom and the β-carbon atom (1,4-addition, or *conjugate addition*).

(21.43)

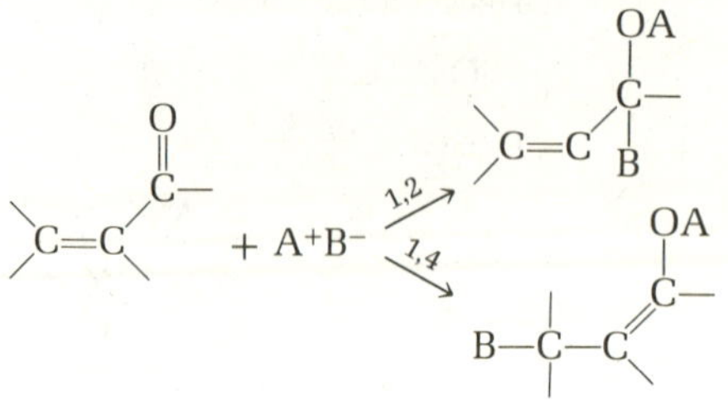

If the double bond is strongly polarized, then 1,4-addition will predominate and it will often be the exclusive mode of reaction. For instance, carbonyl attack is not detected in the course of reaction 21.44, even when the Grignard reagent is in excess.

(21.44)

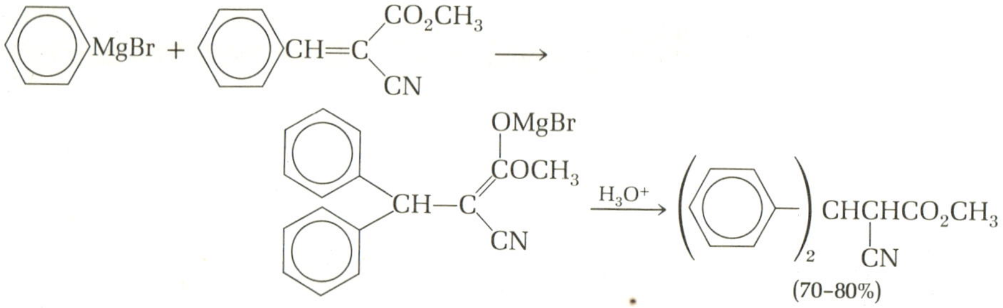

The initial product of 1,4-addition, as in equation 21.44, is an enolate, and the stability of this particular enolate makes 1,4-addition energetically very attractive compared to 1,2-addition. One need not have two polarizing groups to promote 1,4-addition, however, as shown in equation 21.45. Ethylmagnesium bromide gives even higher yields of 1,4-addition products.

(21.45)

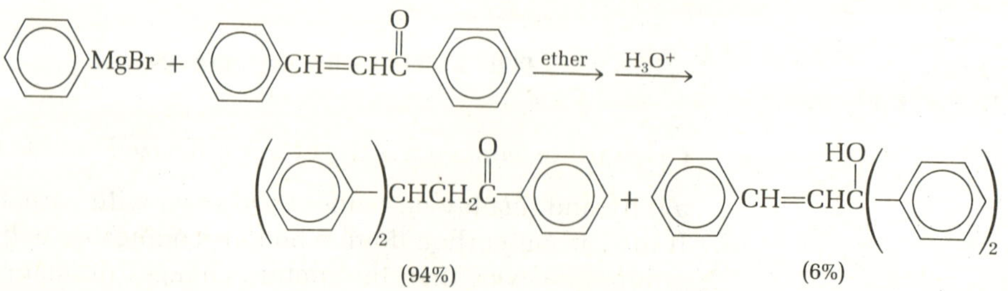

Since addition of organometallic reagents is essentially irreversible, the proportions of products obtained measure directly the relative rates of addition. The balance between 1,2- and 1,4-addition is sensitive to many factors but we shall draw attention to only two. Firstly, there is the relative steric hindrance at either end of the conjugated system. If the unsaturated compound is an aldehyde rather than a ketone or ester, the tendency toward 1,2-addition rises sharply (equation 21.46).

(21.46)

$$CH_3CH_2MgBr + CH_3CH{=}CH\overset{O}{\overset{\|}{C}}H \xrightarrow{\text{ether}} \xrightarrow{H_3O^+}$$

$$\underset{(70\%)}{CH_3CH{=}CH\overset{OH}{\overset{|}{C}}HCH_2CH_3} + \underset{(0.1\%)}{CH_3CH_2\underset{CH_3}{\underset{|}{C}}HCH_2\overset{O}{\overset{\|}{C}}H}$$

Secondly, the metal atom of the organometallic reaction exerts a decisive effect on the direction of addition. Consider the reaction pairs 21.45 plus 21.47 and 21.48 plus 21.49. In reaction 21.49 the organometallic species can be considered to be either CH_3Cu or $(CH_3)_2Cu^-$ (or their complexes).

(21.47)

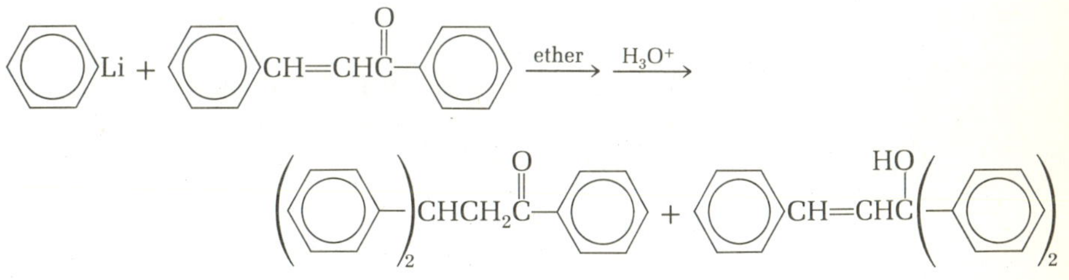

(13%) (69%)

1,2-Addition is at a maximum when lithium compounds are used; it steadily declines in importance as one proceeds to magnesium and copper.

(21.48)

$$(CH_3)_2C{=}CH\overset{O}{\overset{\|}{C}}CH_3 + CH_3MgBr \longrightarrow \xrightarrow{H_3O^+} \underset{(90\%)}{(CH_3)_2C{=}CH\overset{OH}{\overset{|}{C}}(CH_3)_2}$$

(21.49)

$$(CH_3)_2C{=}CH\overset{O}{\overset{\|}{C}}CH_3 + CH_3MgBr \xrightarrow{\text{CuCl (1\%)}} \xrightarrow{H_3O^+}$$

$$\underset{(7\%)}{(CH_3)_2C{=}CH\overset{OH}{\overset{|}{C}}(CH_3)_2} + \underset{(82\%)}{(CH_3)_3CCH\overset{O}{\overset{\|}{C}}CH_3}$$

MICHAEL ADDITIONS

Organometallic compounds are by no means the only nucleophilic reagents that will attack conjugated carbon-carbon double bonds. Most of the other nucleophiles are more stable and less reactive than the organometallics, however, and their additions are usually reversible under the reaction conditions. The product distributions obtained are therefore subject to thermodynamic control. It is not surprising then to find that 1,4-addition is usually the exclusive mode of addition since it involves destruction of a C—C π-bond rather than the stronger C—O π-bond.

(21.50)

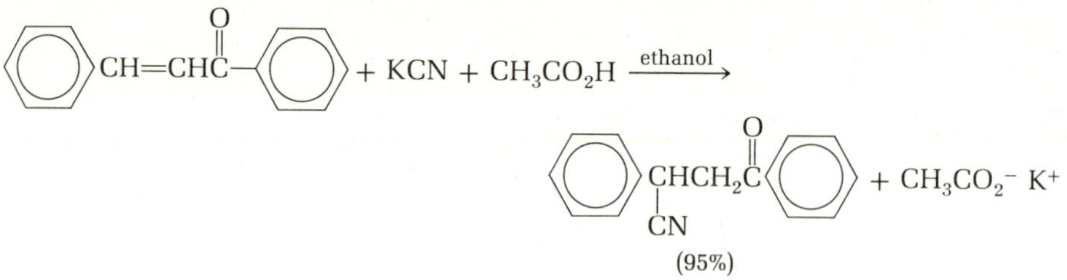

(21.51) $$H_2S + 2CH_2{=}CHCO_2CH_3 \xrightarrow{OH^-\ (cat.)} CH_3O_2CCH_2CH_2SCH_2CH_2CO_2CH_3 \quad (75\%)$$

(21.52) $$(CH_3)_2C{=}CHCCH_3\ (C{=}O) + CH_3NH_2 \xrightarrow{H_2O} (CH_3)_2C(NHCH_3)CH_2CCH_3\ (C{=}O) \xrightarrow{HNO_2} (CH_3)_2C(N(NO)CH_3)CH_2CCH_3\ (C{=}O) \quad (70\text{–}80\%)$$

(21.53) $$CH_2{=}CHCHO + CH_3CH_2OH \xrightarrow[80^\circ]{NH_4^+\ Cl^-} [CH_3CH_2OCH_2CH_2CHO] \longrightarrow CH_3CH_2OCH_2CH_2CH(OCH_2CH_3)_2 + H_2O \quad (31\text{–}39\%)$$

Derivatives of acrylic acid, $CH_2{=}CHCO_2H$, are widely used in these addition reactions.

$CH_2{=}CHCHO$ acrolein	$CH_2{=}CHCN$ acrylonitrile
$CH_2{=}CHCO_2CH_3$ methyl acrylate	$CH_2{=}C(CH_3)CO_2CH_3$ methyl α-methacrylate

These compounds are easily polymerized and one therefore uses catalytic amounts of acids or bases and mild conditions.

Enolate anions, particularly relatively stable ones, may take part in conjugate additions. This specific variation is known as the *Michael reaction* (after the American chemist A. Michael, not the saint). The catalysts are usually hydroxide or alkoxide ions, although in many cases amines will serve.

(21.54)

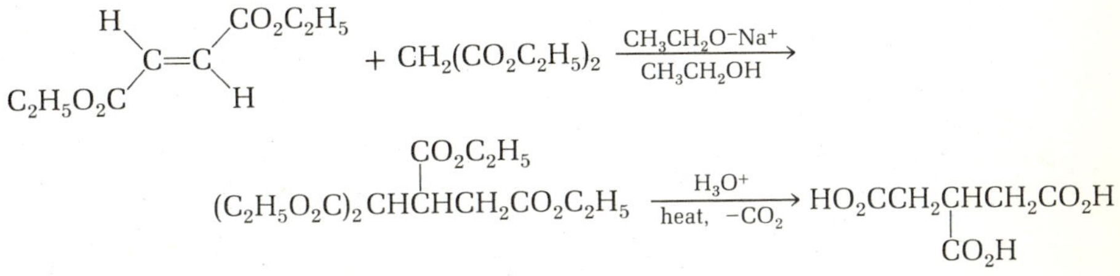

(21.55)

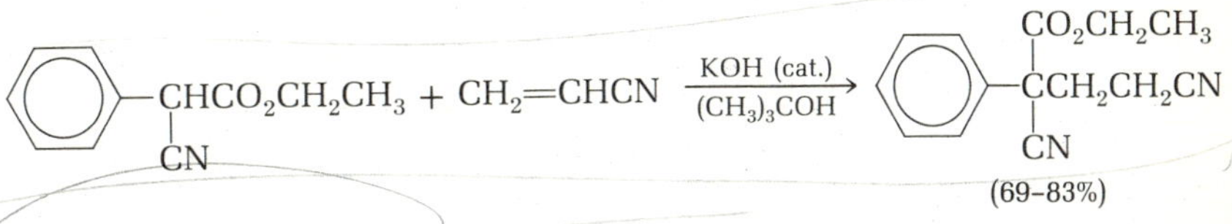

Enolates of simple aldehydes, ketones, and esters will usually not be usable because very strong bases are then required that can lead to all manner of undesirable side reactions. This problem can be circumvented by the use of the enamines of the aldehydes and ketones.

(21.56)

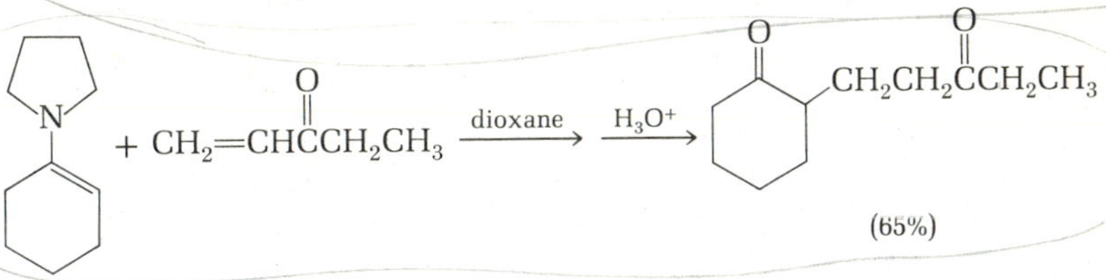

(21.57)

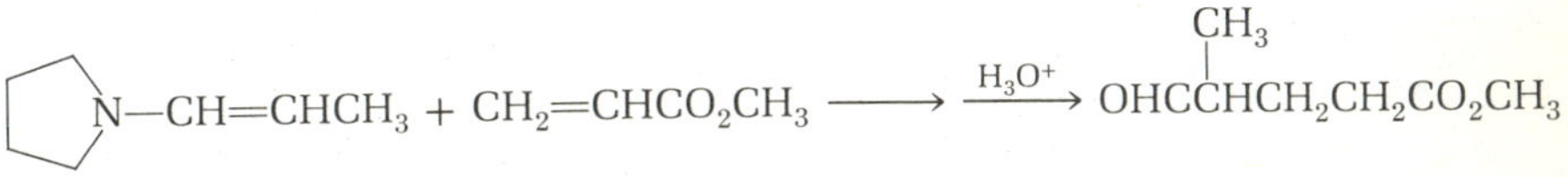

SYNTHESIS BY MICHAEL ADDITION

Michael adducts contain a number of reactive functional groups and many of them readily undergo further intramolecular transformation. Acylation and aldol condensation are the two most common reactions that occur in Michael adducts and these reactions give rise to a number of useful cyclic products.

(21.58)

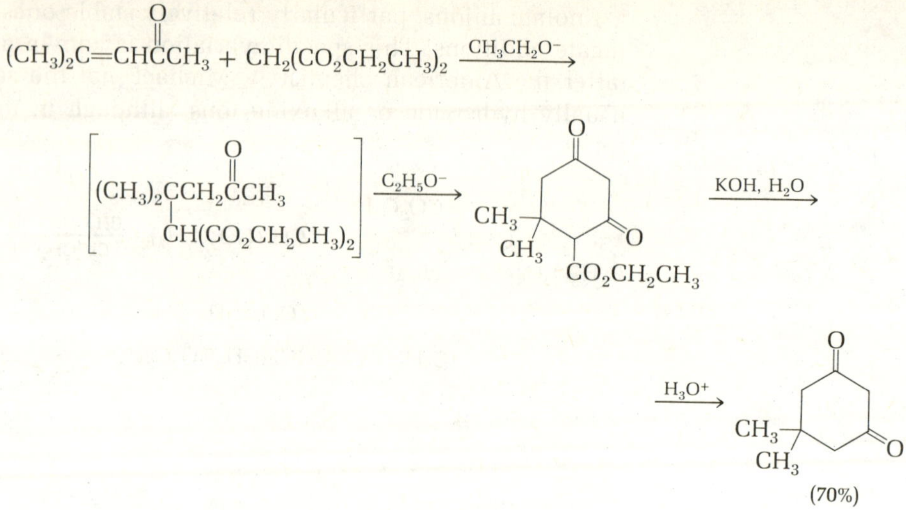

(21.59)

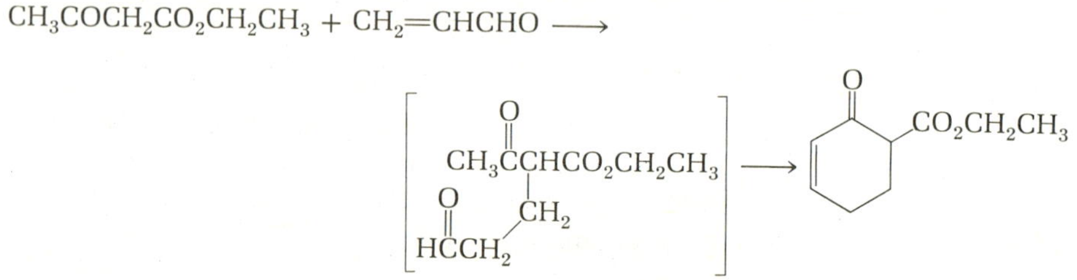

(21.60)

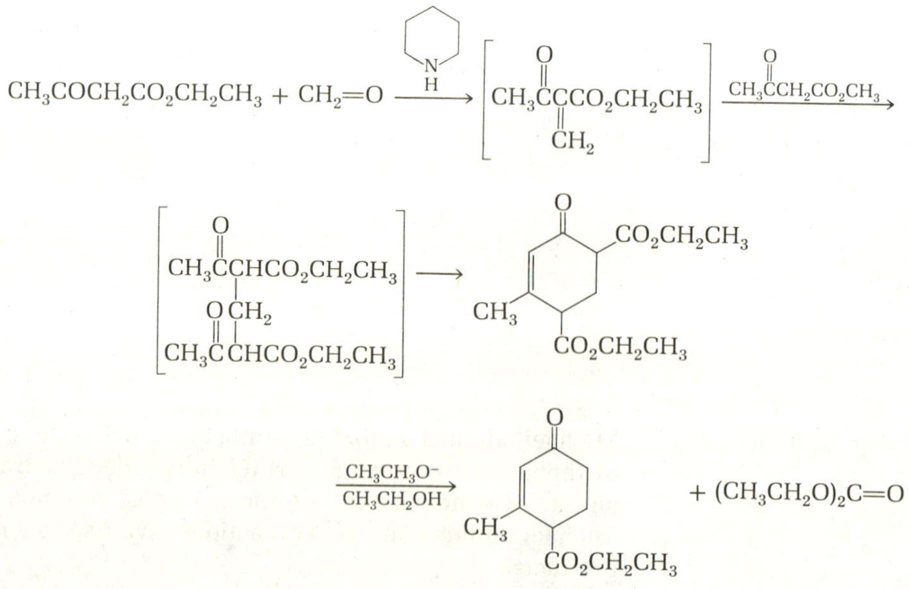

Equation 21.58 involves successively a Michael addition and an intramolecular acylation of the ketone by one of the ester groups. The remaining ester group is removed by saponification and decarboxylation. Equation 21.59 illustrates a Michael condensation followed by an intramolecular aldol condensation. Finally, equation 21.60 has something for everyone: an aldol condensation, then conjugate addition of a second mole of acetoacetic ester to the aldol product, an intramolecular aldol condensation, and to cap it off a reverse acylation reaction.

The Michael addition-aldol condensation sequence has been used innumerable times as a means of building one ring on to another, a process that is very important in steroid synthesis.

(21.61)

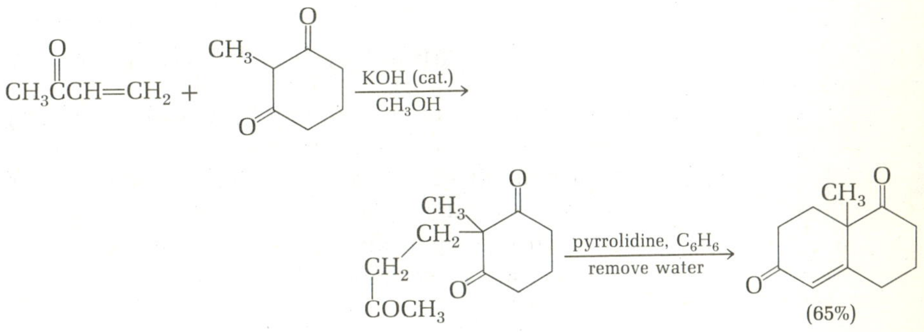

PROBLEM

3 Conjugated alkenes are popular substrates in free radical addition reactions, such as equation 2.

$$CH_3CH{=}CHCO_2CH_3 + CCl_3Br \xrightarrow{\text{peroxide}} CCl_3CH(CH_3)CH(Br)CO_2CH_3 \qquad (2)$$

Propose a mechanism for this reaction, and explain the observed orientation of addition.

THE TRICARBOXYLIC ACID CYCLE

The glucose polymer glycogen is stored in the muscle and in the liver. When an animal is exercising, the glycogen in the muscle is broken down to glucose that is, in turn, converted to carbon dioxide and water with consumption of oxygen. This reaction is very exothermic and provides the energy to power the exercising muscle.

Glucose is first isomerized to fructose, and the latter then undergoes reverse condensation to give the three-carbon fragments glyceraldehyde and dihydroxyacetone (equations 20.84, 20.86, and 20.88). These compounds are in turn oxidized to pyruvic acid, $CH_3\overset{\overset{O}{\|}}{C}CO_2H$. Pyruvic acid then enters the **tricarboxylic acid cycle** (*Krebs cycle*) in the course of which the actual oxidation to carbon dioxide takes place. The principal *chemical* events in this series of reactions are aldol condensations, decarboxylations, and conjugate additions.

The first step in the process is oxidative decarboxylation of pyruvic acid to acetic acid, which is actually present as its coenzyme A ester (p. 461).

(21.62)

$$CH_3\overset{\overset{O}{\|}}{C}CO_2H + NAD^+ + CoASH \xrightarrow{\text{enzymes}} CH_3\overset{\overset{O}{\|}}{C}SCoA + CO_2 + NADH + H^+$$

The ultimate oxidant is the hydride abstracting reagent NAD^+ (p. 268). The biochemical process is quite complex, involving several different enzymes and a number of intermediate stages. Suffice it to say that α-ketoacids are subject to facile oxidative decarboxylation nonenzymatically in solution, with hydrogen peroxide as the oxidant. In equation 21.63 the departing hydroxyl ion provides an electron sink for the decarboxylation.

(21.63)

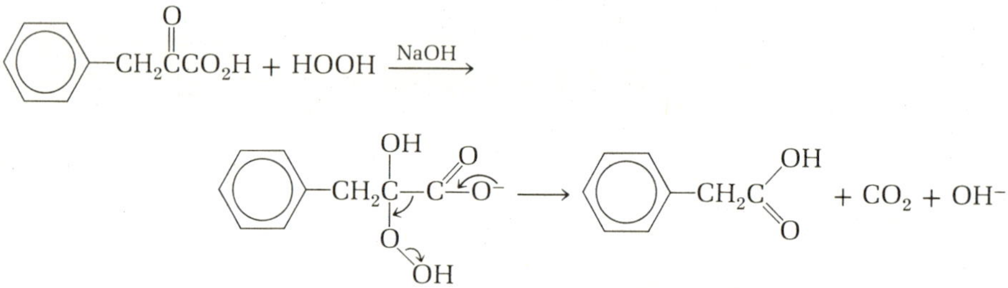

While the biochemical mechanism is quite different, there is a similarly situated, positively charged nitrogen atom that performs the same function.

Acetyl coenzyme A then takes part in an aldol condensation with oxalacetic acid (2-oxosuccinic acid) to give citric acid, which is very widespread in Nature.

(21.64)

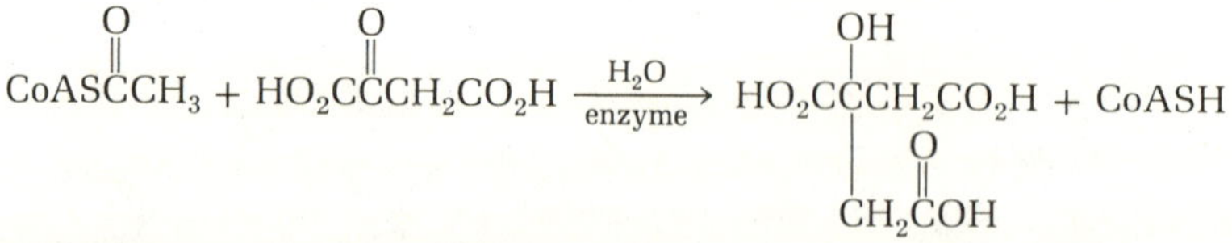

oxalacetic acid citric acid

Citric acid is dehydrated to the conjugated triacid *cis*-aconitic acid that is subsequently rehydrated to isocitric acid (equation 21.65). Thiol esters, such as acetyl coenzyme A, are expected to be more reactive in aldol and Claisen condensations than are the corresponding oxygen esters because the methyl hydrogen atoms are more labile. Since the thiolester group has a lower resonance energy than the ordinary ester group, it can more readily be converted to an enol or enolate (Chapter 17).

(21.65)

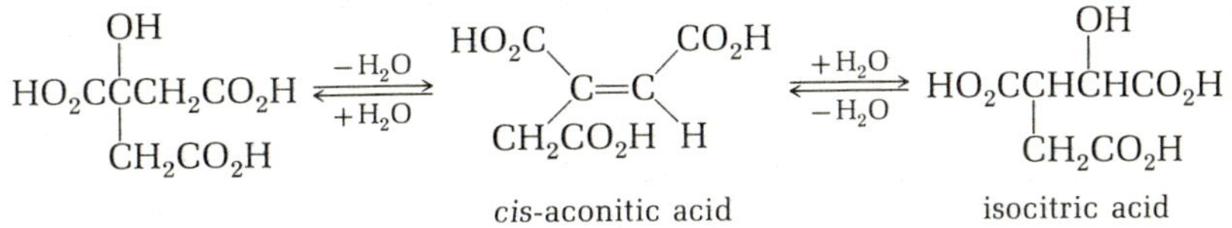

The addition of water to *cis*-aconitic acid to produce *either* citric or isocitric acids is a conjugate addition since there is an activating carboxyl group at either end of the double bond. Neither mode of addition would therefore seem to be decisively favored and, indeed, both isomers are present in appreciable amounts (along with aconitic acid) at equilibrium. Isocitric acid does contain a secondary hydroxyl group, however, and it may be carried further along the metabolic path by oxidation of this hydroxyl group.

(21.66)

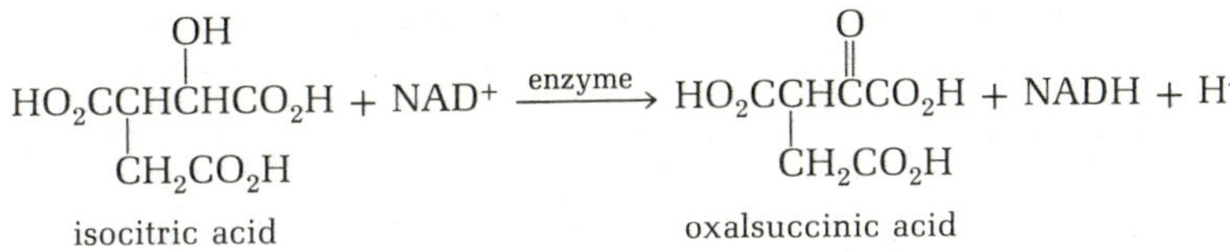

Oxalsuccinic acid contains a β-ketoacid grouping and will therefore undergo decarboxylation with ease. The decarboxylation product, α-ketoglutaric acid, is in turn subjected to oxidative decarboxylation in exactly the same manner as pyruvic acid (equation 21.62).

(21.67)

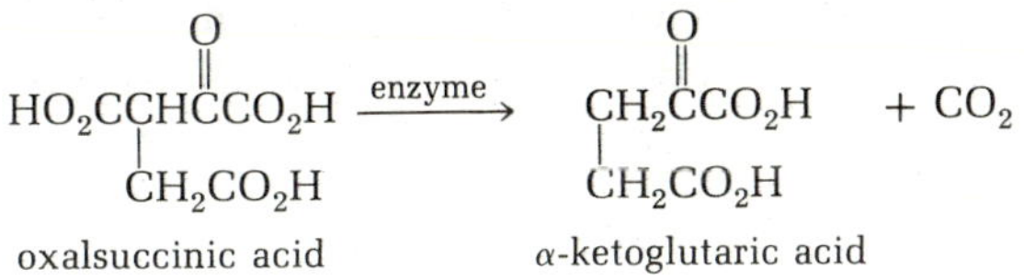

(21.68)

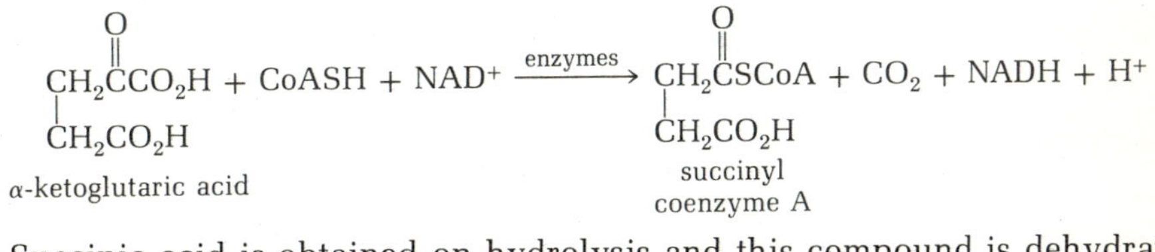

Succinic acid is obtained on hydrolysis and this compound is dehydrated to fumaric acid. Conjugate addition of water to fumaric acid gives malic acid that may be oxidized to oxalacetic acid.

(21.69)

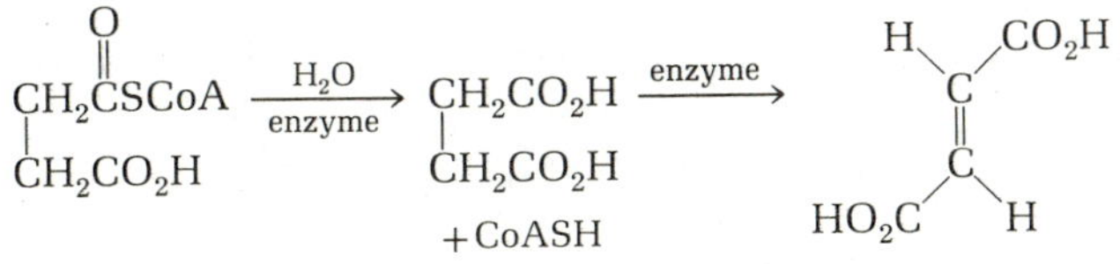

(21.70)

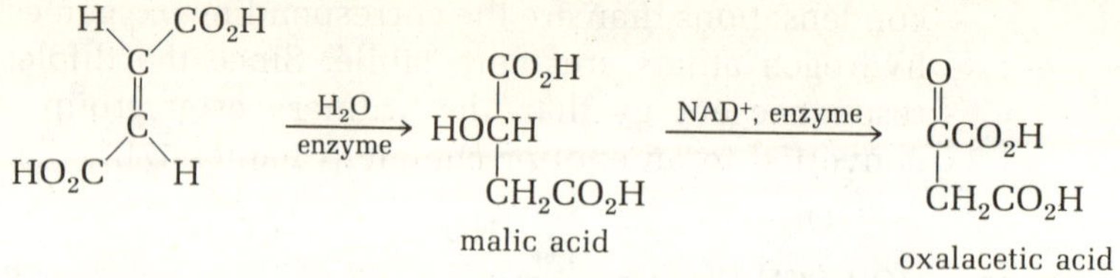

Oxalacetic acid may then condense with a molecule of acetyl coenzyme A (equation 21.64) to initiate another turn of the cycle.

In the course of this set of reactions, 3 moles of carbon dioxide were produced (equations 21.62, 21.67, and 21.68), thus completely accounting for the skeleton of pyruvic acid. Numerous NAD^+ molecules were reduced to NADH that would be reoxidized elsewhere in the cell, the hydrogen atoms eventually becoming bound to oxygen. The oxalacetic acid, which is the starting material for the cycle, was conserved. In conformity with other biochemical experience, it is found that all the intermediates that contain an asymmetric center (isocitric, oxalsuccinic, and malic acids) are present in one enantiomeric form only.

FATTY ACID SYNTHESIS

The synthesis of fatty acids, $CH_3(CH_2)_nCO_2H$ (n = even integer), is another metabolic process of great importance and it may be analyzed in terms of the chemistry discussed in this chapter. The acylation reaction is the key to this process, which takes place in animals, plants, and microorganisms.

The starting point for the synthesis is the activation of acetyl coenzyme A by carboxylation, thereby turning it into malonyl coenzyme A.

(21.71)

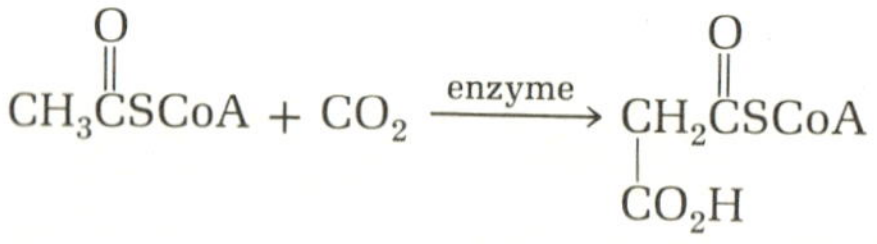

One is tempted to ask why the organism acylates first with carbon dioxide (which will subsequently be lost) rather than directly with another molecule of acetyl coenzyme A. Carboxylation may be a particularly facile form of acylation because of the possibility of stabilizing the product by chelation.

The production of acetyl coenzyme A from acetate ion is a chemically interesting process in itself. Acetate ion reacts first with the phosphorylating agent adenosine triphosphate (ATP; p. 619) to yield a mixed phosphoric-carboxylic anhydride.

(21.72)

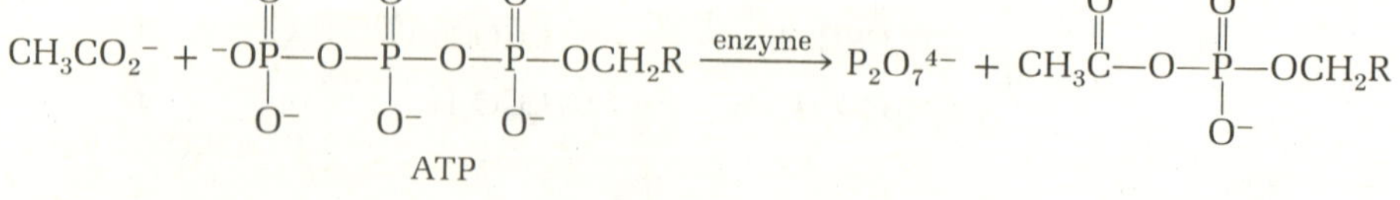

The mixed anhydride reacts with coenzyme A at the carboxyl carbon to give acetyl coenzyme A. The preference for reaction at carbon rather than at phosphorus is reminiscent of the fact that many nucleophiles attack carboxylic acid derivatives much more rapidly than they attack sulfonic acid derivatives.

(21.73)
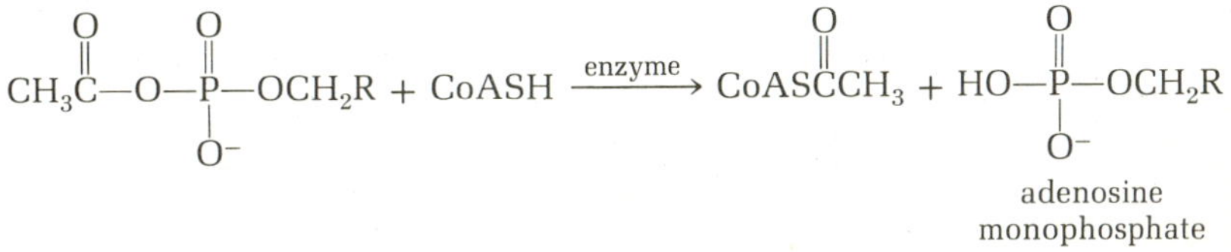

The next step in the biosynthesis is the acylation of malonyl coenzyme by acetyl coenzyme A, producing acetoacetyl coenzyme A.

(21.74)
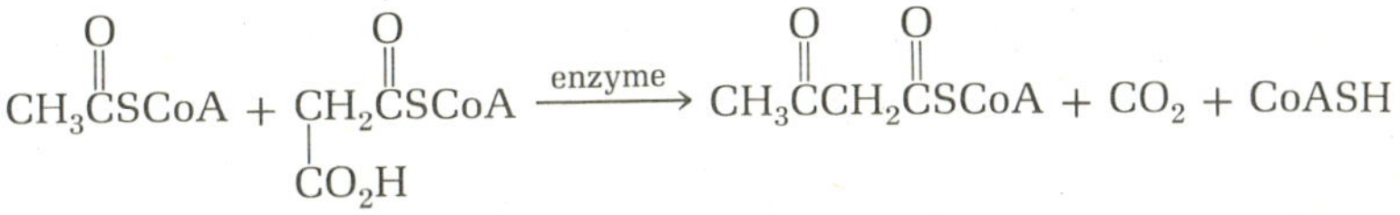

Release of carbon dioxide either precedes or occurs simultaneously with the acylation. The presence of a second carboxyl group would obviously activate the methylene group toward attack. It has, however, been hypothesized that the malonyl coenzyme A decarboxylates first to give the *enol* of acetyl coenzyme A and that this enol, stabilized by the enzyme, is the species that is being acylated.

(21.75)
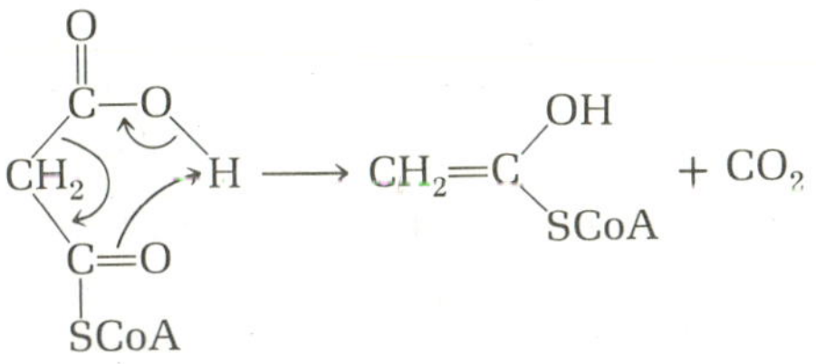

Hydride reduction of the β-ketoester, followed by dehydration of the β-hydroxyester so produced, leads to a *trans*-α,β-unsaturated ester. This product is then reduced to a saturated ester.

(21.76)
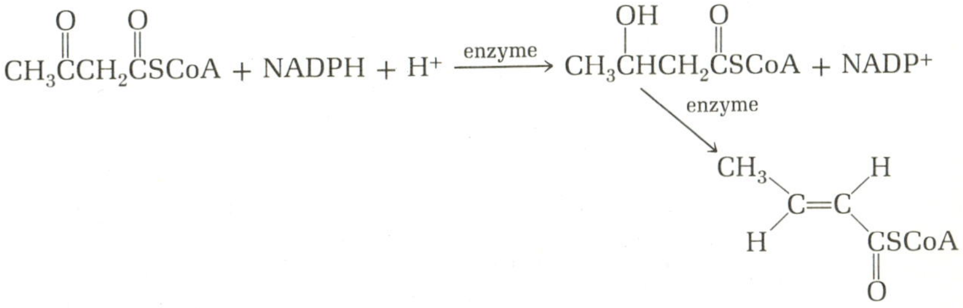

The reductant is the hydride transfer coenzyme NADPH so that one might view the reduction as a conjugate addition of H^+ and H^- to the conjugated system (equation 21.77).

(21.77)

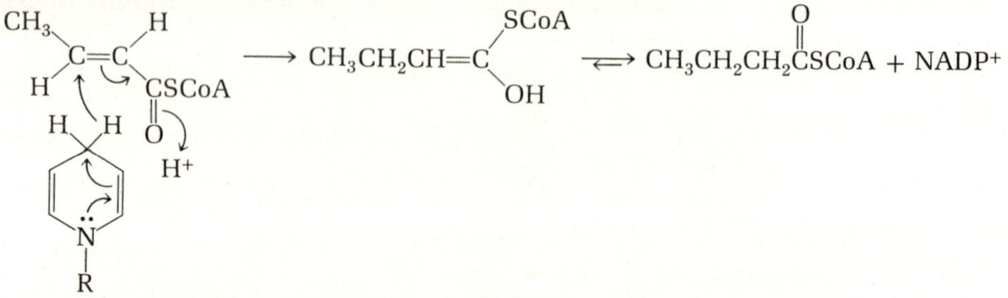

The completely saturated acyl coenzyme A ester will condense with another molecule of malonyl coenzyme A and the β-ketoester that results will be subjected to the reduction-dehydration-reduction sequence.

(21.78)

$$CH_3(CH_2)_2\overset{O}{\overset{\|}{C}}SCoA + CH_2\begin{matrix}\diagup \overset{O}{\overset{\|}{C}}SCoA \\ \diagdown CO_2H\end{matrix} \xrightarrow{\text{enzyme}} CH_3(CH_2)_2\overset{O}{\overset{\|}{C}}CH_2\overset{O}{\overset{\|}{C}}SCoA + CO_2 + CoASH$$

In this manner, saturated carboxylic acids *with an even number of carbon atoms* are synthesized in Nature. In plants and animals the most common ones are palmitic [$CH_3(CH_2)_{14}CO_2H$] and stearic [$CH_3(CH_2)_{16}CO_2H$] acids.

DEGRADATION OF FATTY ACIDS

The oxidative degradation of fatty acids bears a close resemblance to, but is not identical with, their biosynthesis. The coenzyme A ester of the saturated acid is dehydrogenated at the α- and β-positions, followed by conjugate addition of water to the double bond. Oxidation of the hydroxy ester to the β-ketoester is next and reaction of the latter with coenzyme A leads to loss of a two-carbon fragment as acetyl coenzyme A. This last product may then be chewed up by the tricarboxylic acid cycle.

(21.79)

$$R(CH_2)_nCO_2H \longrightarrow R(CH_2)_{\overline{n-2}}CH{=}CHCO_2H \longrightarrow R(CH_2)_{\overline{n-2}}\overset{OH}{\overset{|}{C}}HCH_2CO_2H \longrightarrow$$

$$R(CH_2)_{\overline{n-2}}\overset{O}{\overset{\|}{C}}CH_2CO_2H \xrightarrow{\text{CoASH}} R(CH_2)_{\overline{n-2}}CO_2H + CH_3\overset{O}{\overset{\|}{C}}SCoA$$

Unlike the biosynthesis of fatty acids, the degradation does not involve malonyl coenzyme A. The most tantalizing difference between the two sequences centers around the β-hydroxyesters that are intermediates common to both paths. The β-hydroxy esters that are created in the biosynthetic process all have the same absolute configuration at their hydroxyl-bearing carbon atom. β-Hydroxy esters that are intermediates in the degradation of fatty acids have the opposite configuration, however.

SUMMARY

1 Aliphatic esters may be condensed under basic conditions to yield β-ketoesters. The loss of the resonance energy of one of the two condensing molecules opposes the condensation, but the fact that the product can be converted to a relatively stable anion helps drive it forward. When condensing agents such as alkoxide ions are used, it is essential therefore that the product contain at least one fairly acidic hydrogen atom. Claisen condensations may also be driven toward product by the use of very strong bases, such as sodium triphenylmethide and sodium hydride.

2 The intramolecular Claisen condensation, known as the Dieckmann condensation, is thermodynamically much more favorable than its acyclic analogue when five- and six-membered rings are the products.

3 The α-carbon atoms of ketones may be acylated in the same way that esters are. In particular, esters that lack α-hydrogen atoms themselves are effective acylating agents. A mild method for acylating ketones and aldehydes involves the reaction of their enamines with acid chlorides and anhydrides. This method solves the problem of the reversibility of the acylation.

β-Ketoesters, malonic esters, and β-diketones are very useful synthetic intermediates because they can undergo facile alkylation and aldol condensation at their reactive methylene positions.

4 Ester groups are particularly valuable as activating substituents for adjacent protons. After its activating function has been utilized, the ester may be hydrolyzed to a carboxylic acid and, provided that the acid group is located β to an electron-withdrawing, conjugating substituent, decarboxylation can be induced under mild conditions. Decarboxylation usually proceeds by way of a cyclic six-membered transition state and results in an enol being produced as the primary product. Decarboxylation may also take place if the β-position carries a group that can be readily ejected and will carry a pair of electrons with it. In that case it is the carboxylate anion that is the reactant. The decarboxylative condensation of malonic and cyanoacetic acids with aldehydes, catalyzed by pyridine, fits this category.

5 Under some circumstances the decarboxylation reaction may be reversed. Phenoxide ions, for instance, may be carboxylated to give hydroxybenzoic acids. Ordinary ketones may be carboxylated with methyl methoxymagnesium carbonate because the β-ketoacid produced is stabilized as the magnesium chelate of its enolate.

6 While most organometallic reagents react only very slowly with alkenes, they will add readily to carbon-carbon double bonds if those bonds are conjugated with electron-withdrawing substituents. If two polarizing groups are located at the same end of the double bond, then addition is almost the only observable reaction. α,β-Unsaturated aldehydes, ketones, esters, and nitriles, when allowed to react with organometallic reagents, undergo a mixture of carbon-carbon and carbon-oxygen (or nitrogen) addition. The balance is sensitive to the nature of the electron-withdrawing group, the steric hindrances at the 2 and 4 positions of the conjugated system, and the metal atom in the organometallic reagent. In general, the most reactive organometallics lead to the most 1,2-addition.

7 Nucleophiles may add to polarized double bonds under equilibrium conditions, in which case one observes only C=C addition. The usual run of nitrogen, oxygen, and sulfur nucleophiles will participate in these additions. Enolate ions will participate in conjugate additions if they are generated in the presence of a suitable reactant (Michael addition); enamines are also subject to electrophilic attack by these molecules. In a number of cases the Michael addition products participate in intramolecular aldol condensations and acylations, leading to a variety of useful cyclic products.

8 Pyruvic acid is derived metabolically from the breakdown of glucose. It is converted by oxidative decarboxylation to acetyl coenzyme A. The acetyl group is then eventually converted to carbon dioxide and water by the operation of the tricarboxylic acid cycle. Aldol condensation of acetyl coenzyme A with oxalacetic acid generates citric acid, which is dehydrated. Rehydration of the double bond in the opposite sense yields isocitric acid; isocitric acid is oxidized to oxalsuccinic acid that promptly decarboxylates to α-ketoglutaric acid. Oxidative decarboxylation of the latter produces succinic acid, which is converted in turn to fumaric acid, malic acid, and oxalacetic acid. The molecule of oxalacetic acid is now available to conduct another acetate unit through the tricarboxylic acid cycle.

9 Fatty acids are built up two carbon atoms at a time, the carbon-carbon bond-forming step being acylation. An acyl coenzyme A condenses with malonyl coenzyme A, a decarboxylation occurring in the process. The β-ketoacyl coenzyme A is converted to a saturated ester by reduction, dehydration, and further reduction. Fatty acids may be degraded by a route that is nearly, but not exactly, the reverse of the biosynthetic path.

PROBLEMS

Note: Problems 1 through 3 appear in text.

4 Propose efficient syntheses of the following compounds, starting with readily available esters.

a. Ethyl α-butyrylbutyrate

b. Methyl benzoylacetate

c. Ethyl *p*-chlorobenzoylacetate

d. Isopropyl α-(*m*-nitrobenzoyl)propionate

e. Methyl α,γ-diphenyl-α-methyl-β-ketovalerate

f. Diethyl phenylmalonate

g. Ethyl α-formylphenylacetate

h. Diethyl 2-ketosuccinate

i. Diethyl cyclohexane-1,4-dione-2,5-dicarboxylate

j. Methyl 4-methylcyclopentanone-2-carboxylate

k. Diethyl cyclopentan-1,2-dione-3,5-dicarboxylate

5 When ethyl 2-methylcyclopentanone-2-carboxylate is heated with sodium ethoxide, and the reaction mixture then neutralized, ethyl 5-methylcyclopentanone-2-carboxylate is recovered. Write a plausible mechanism for this rearrangement. What is the driving force?

6 a. Propose a synthesis of diethyl oxalacetate (II).

$$C_2H_5O_2C\overset{\overset{\displaystyle O}{\|}}{C}CH_2CO_2C_2H_5$$

II

b. Ester II was hydrolyzed to the free acid and the crude product recrystallized. Depending on the solvent chosen for recrystallization, crystals melting at either 152 or 184° may be obtained. The crystalline substances are isomeric and the ir spectra of both show the absence of a peak in the ketone carbonyl region. The higher melting isomer is soluble in ether, but the lower melting one is only sparingly so. Propose structures for the isomers.

7 How would you synthesize the following compounds?

a. Ethyl 6-methylcyclohexanone-2-carboxylate

b. Ethyl 2-methylcyclohexanone-2-carboxylate

c. 2,6-Dimethylcyclohexanone

d. 2-Heptanone

e. Diethyl acetylsuccinate

f. 2,6-Dimethylheptanoic acid

g. Ethyl cyclopentylacetate

h. (2-Cyanoethyl)benzene

8 When benzaldehyde condenses with malonic acid in pyridine, the product is *trans*-cinnamic acid. The acid forms a dibromide when treated with bromine in chloroform. When the dibromoacid salt is heated in nonaqueous solvents, it loses carbon dioxide and *cis*-2-bromo-1-phenylethylene may then be isolated. Show the course of this reaction in detail, paying particular attention to stereochemical details.

9 Oxalacetic acid is decarboxylated to pyruvic acid (equation 3). The decarboxylation is accelerated by cupric ion, and the catalytic effect of the metal cation is maximal at a pH where the acid is present mainly as its dianion. The monoester of the acid is insensitive to metal catalysis.

$$HO_2CCH_2\overset{\overset{\displaystyle O}{\|}}{C}CO_2H \longrightarrow CH_3\overset{\overset{\displaystyle O}{\|}}{C}CO_2H + CO_2 \qquad (3)$$

Indicate clearly the course of the catalyzed decarboxylation, and explain the effect of the metal ion.

10 2*R*,3*R*-Mevalonic-2-*d* acid (III) is converted by an appropriate enzyme system to *trans*-isopentenyl-4-*d* pyrophosphate (IV). Compound IV (without deuterium) is a vital intermediate in the biosynthesis of squalene (Chapter 12), which is itself the precursor of the steroids (Chapter 10). Write a mechanism for this transformation (ignoring the phosphorylation step).

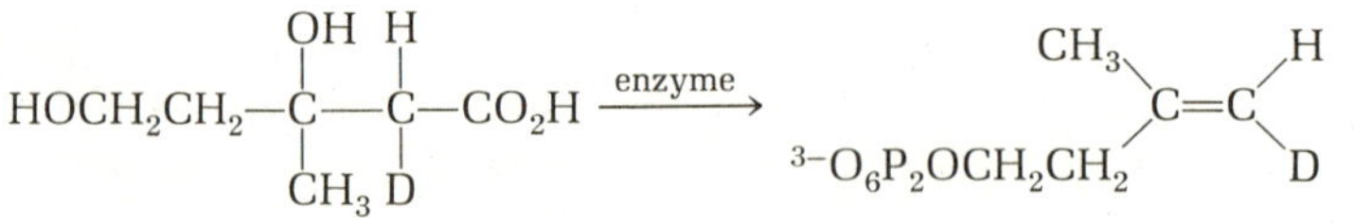

III IV (4)

11 Describe the preparations of each of the following compounds from easily accessible reagents.

a. 4-Penten-3-ol

b. Diethyl 2,2-diphenylethane-1,1-dicarboxylate

c. 4-Phenyl-2-heptanone

d. 3-Methyl-5-phenyl-3-heptanol

12 a. When cinnamaldehyde (β-phenylacrolein) is treated with potassium cyanide in the presence of acid, a product is obtained, the ir spectrum of which indicates the presence of —CN, >C=C<, —OH, and an aromatic ring. What is the structure?

b. When cinnamaldehyde is exposed to an aqueous solution of sodium bisulfite, a monoaddition product rapidly forms. If the monoaddition product is allowed to remain in contact with the bisulfite solution, a more stable product — incorporating 2 moles of sodium bisulfite — is formed. Draw structures for both products, and write the equilibria involved. (The reaction of simple carbonyl compounds with sodium bisulfite is outlined in Chapter 11.)

13 Compare the compounds comprising each pair as to their relative reactivity as electrophiles in a Michael addition.

a. $CH_2{=}CHCO_2CH_3$ versus $CH_2{=}C(CO_2CH_3)CHO$

b. $CH_2{=}CHCO_2CH_3$ versus $CH_3CH{=}CHCO_2CH_3$

c. $CH_2{=}CHCO_2CH_3$ versus $CH_2{=}C(CH_3)CO_2CH_3$

d. $CH_3CH{=}CHCO_2CH_3$ versus 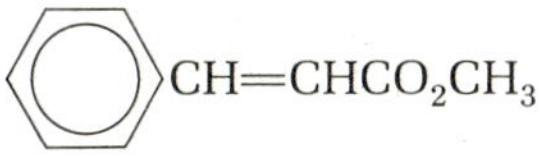

e. $CH_2{=}CHCO_2CH_3$ versus 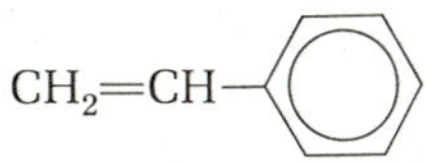

f. $CH_2{=}CHCO_2CH_3$ versus $CH_2{=}CHCOCH_3$

14 A standard method for the epoxidation of α,β-unsaturated ketones and esters involves treating them with an alkaline solution of hydrogen peroxide (equations 5 and 6). Write a plausible mechanism for the general reaction.

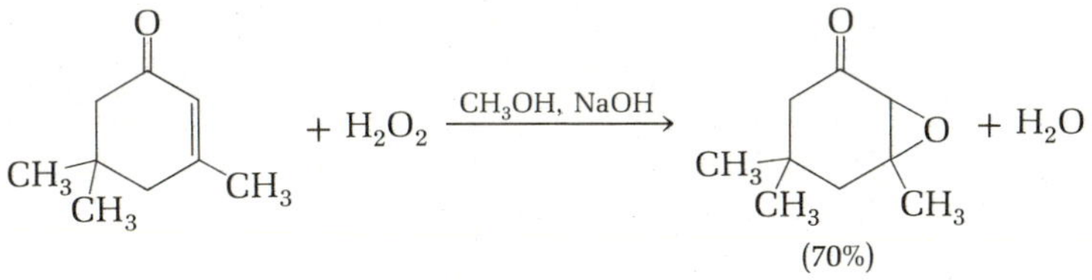

(5)

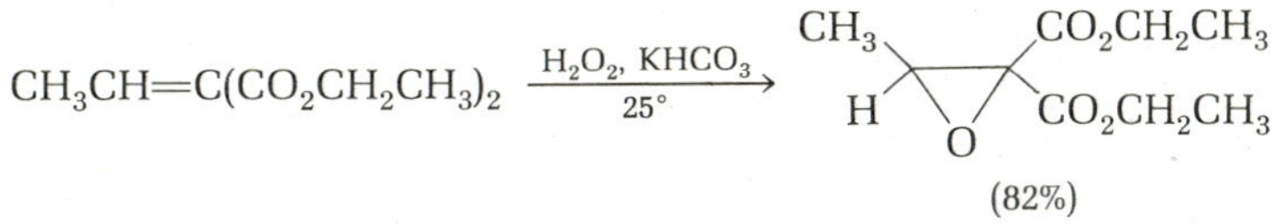

(6)

15 Reaction 7 ($R = CH_3, C_6H_5$; $Z = O, S$) proceeds at a convenient rate at 0°. By contrast, 2-chloropropene reacts at an undetectable rate with the same nucleophiles at 0°.

a. Write several plausible mechanisms for this reaction.

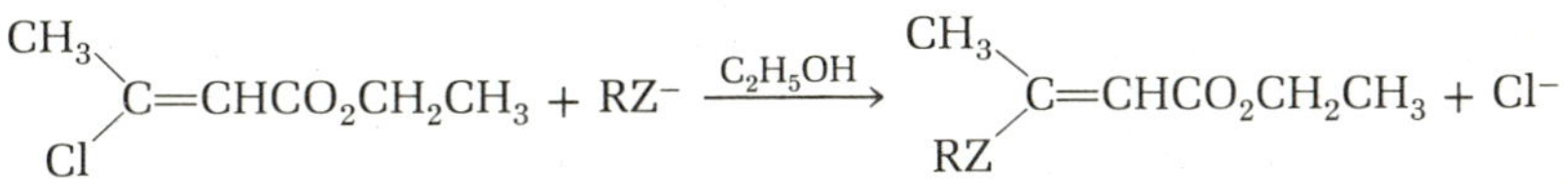

(7)

b. The ester $CH_3(Cl)C{=}C(CH_3)CO_2CH_2CH_3$ reacts similarly to its nonmethylated counterpart. Which mechanism is specifically eliminated by this finding?

c. When the sulfur-containing nucleophiles react with one isomer of the chloroester (the *trans* isomer, for instance), then both isomers of the product are obtained, although the *trans* isomer predominates. Which mechanism(s) seem(s) to be the most likely?

16 Base-catalyzed addition of diethylmalonate to 1-cyanocyclohexene yields two addition compounds that are isomers of one another. What are they? How would you attempt to determine the structures of these isomers?

17 The reaction of *t*-butyl chloride with diethyl malonate yields very little diethyl *t*-butylmalonate because elimination is the predominant reaction.

$$(CH_3)_3CCl + \bar{C}H(CO_2C_2H_5)_2 \longrightarrow \underset{\text{minor product}}{(CH_3)_3CCH(CO_2C_2H_5)_2} + \underset{\text{major product}}{(CH_3)_2C{=}CH_2} \qquad \textbf{(8)}$$

Suggest an alternative route to this compound, based on an aldol condensation product of diethyl malonate. (Keep in mind the orienting effects of the electron-withdrawing groups.)

18 The compound nootkatone (V) occurs in grapefruit and other citrus species, and its isomer isonootkatone (VI) is important in the manufacture of perfume. The construction of the carbon skeletons of these compounds illustrates the combined use of aldol condensations, acylations, and conjugate additions in organic synthesis. We shall concentrate on the carbon skeleton and ignore the stereochemical problems.

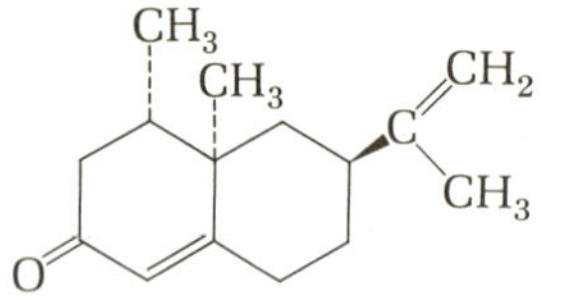

V

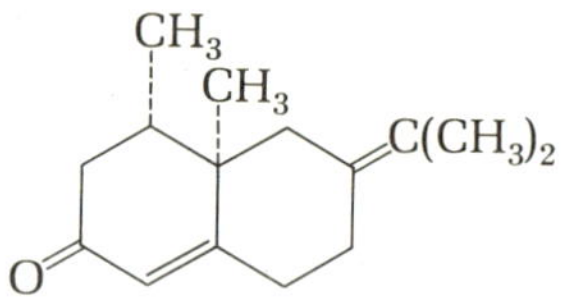

VI

Fill in either the structures of the compounds not shown or the necessary reagents and conditions over the appropriate arrows.

a. $CH_2(CO_2CH_2CH_3)_2 \longrightarrow (CH_3)_2C{=}C(CO_2CH_2CH_3)_2 \xrightarrow[\text{then } H_3O^+]{LiAlH_4,} \xrightarrow{PBr_3}$

$$\textbf{A}\ (C_6H_{10}Br_2) \longrightarrow (CH_3)_2C{=}C\begin{matrix} CH_2CH(CO_2CH_2CH_3)_2 \\ CH_2CH(CO_2CH_2CH_3)_2 \end{matrix} \xrightarrow[\text{heat}]{H_3O^+} \xrightarrow[H^+]{CH_3OH}$$

B $(C_{12}H_{20}O_4) \longrightarrow$ 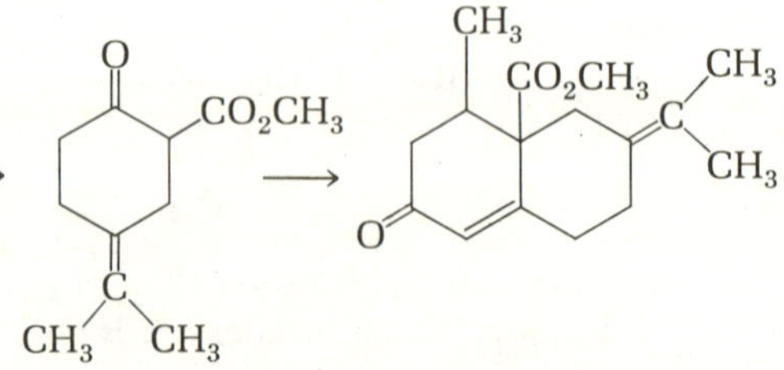

(an intermediate in the synthesis of VI)

b.

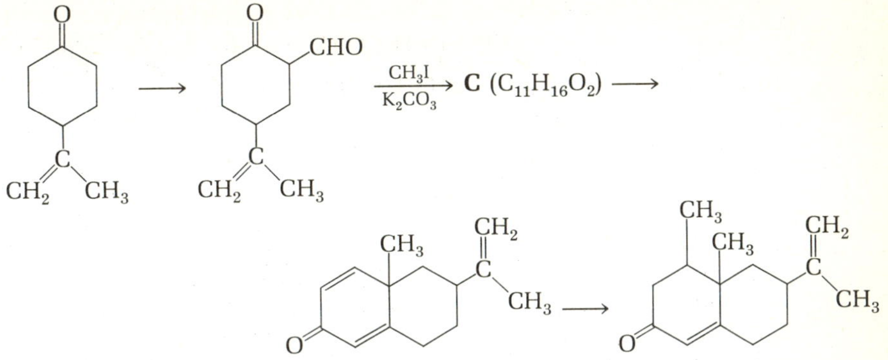

(Comment on the selectivity of the last step.)

c. $CH_2(CO_2CH_2CH_3)_2 \longrightarrow (NCCH_2CH_2)_2C(CO_2CH_2CH_3)_2 \xrightarrow[\text{heat vigorously}]{H_3O^+} \xrightarrow[H^+]{CH_3OH}$

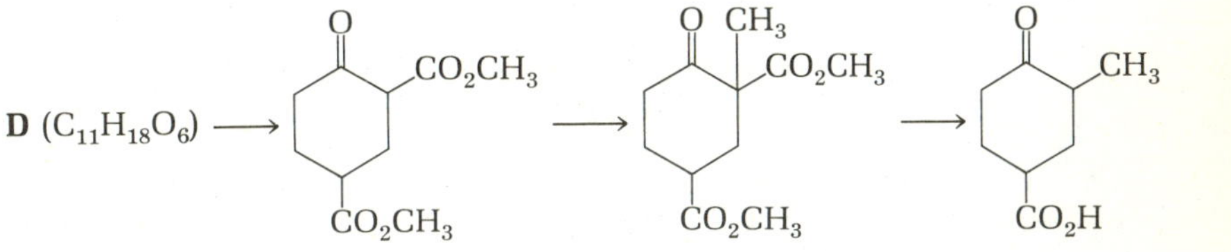

(an intermediate in the formation of V)

19 The following sequence of reactions was employed in the total synthesis of one of the antibiotics of the tetracycline series. Supply the missing structures and reagents.

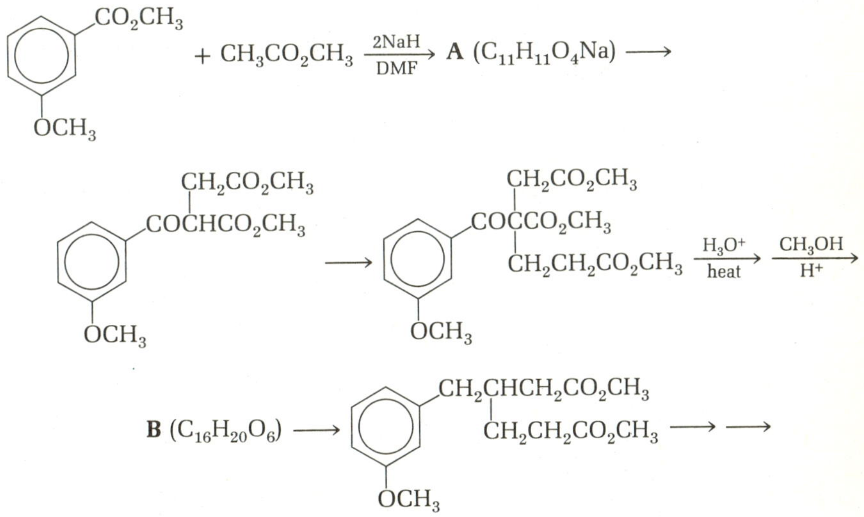

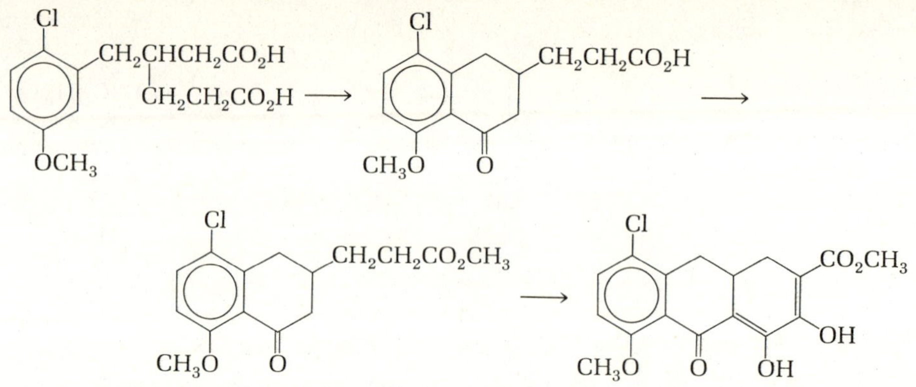

20 The last step of equation 21.60 involves the removal of a carbethoxyl group. This is accomplished by heating the ketodiester with two equivalents of sodium ethoxide in dry ethanol.

a. Why are two equivalents of base employed?

b. Why is the ester group at the 6 position removed selectively in preference to the one at the 4 position?

SUPPLEMENTARY READING

H. O. House, *Modern Synthetic Reactions*. New York: W. A. Benjamin, Inc., 1965, Chapters 7 and 9.

E. M. Kosower, *Molecular Biochemistry*. New York: McGraw-Hill Book Co., 1962.

P. Karlson, *Introduction to Modern Biochemistry*. New York: Academic Press, 1965, Chapters 11 and 12.

CHAPTER 22

PHENOLS AND NUCLEOPHILIC AROMATIC SUBSTITUTION

PHENOL ACIDITY

Aliphatic alcohols are known to be quite weak acids, their self-ionization constants being even smaller than that of water. By contrast, aqueous solutions of phenol are acid to litmus.

(22.1) $CH_3CH_2OH \rightleftharpoons CH_3CH_2O^- + H^+ \quad K_a = 10^{-17}$

Phenols are still to be considered weak acids since the K_a of phenol itself is 10^5 times smaller than the K_a of acetic acid.

(22.2) 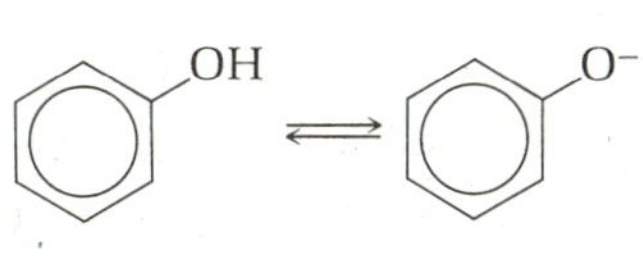

$+ H^+ \quad K_a = 1.28 \times 10^{-10}$

Phenol is even weaker than carbonic acid ($K_a = 10^{-7}$), thus allowing most phenols to be easily separated from carboxylic acids. An aqueous solution of sodium bicarbonate will dissolve carboxylic acids as their sodium salts but will not affect most phenols.

Phenol acidity may be ascribed to two principal causes. Of lesser importance is the inductive effect of the phenyl ring, arising from the greater electronegativity of sp^2-hybridized carbon atoms compared to sp^3-hybridized ones. More significant is the fact that the negative charge may be partially delocalized into the benzene ring.

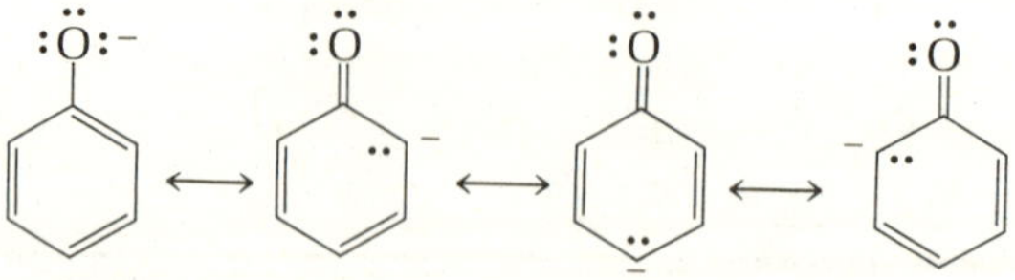

The pK_a of phenol may be modified by attaching substituents to the aromatic ring. Predictably, electron-donating substituents suppress the acidity and electron-withdrawing substituents exalt it. Direct conjugation with the oxygen atom results in the greatest substituent influence on phenolic acidity. *p*-Nitrophenol is more acidic than phenol by a factor of 500 and 2,4,6-trinitrophenol (picric acid) is as strong as several strong mineral acids.

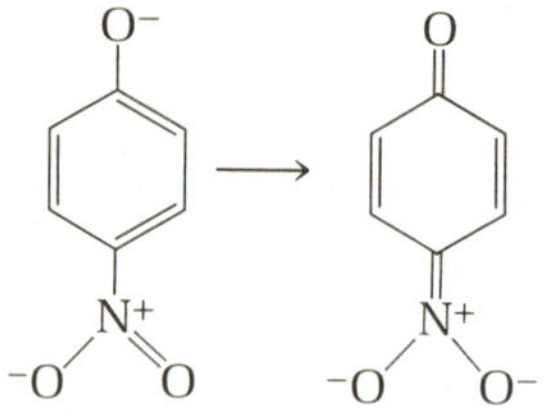

It must be kept in mind that phenol is actually an enol — it is the enol form of 2,4-cyclohexadienone. While simple ketones contain only a small percentage of enol at equilibrium, phenols are entirely enolic because of the gain in resonance energy attendant upon enolization.

(22.3)

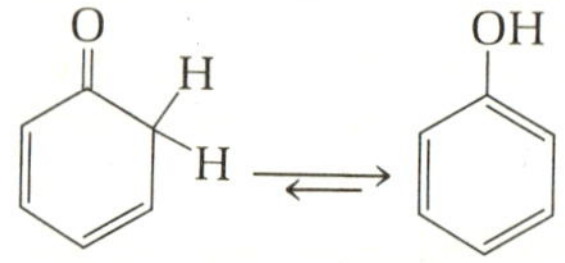

Nonetheless, there is chemical evidence to indicate that a nonnegligible concentration of the keto form is present at equilibrium with naphthols and polyhydroxybenzenes.

PROBLEM

1 a. Phenol does not react with hydroxylamine, nor do the dihydroxy- and trihydroxybenzenes. An exception is *phloroglucinol* (1,3,5-trihydroxybenzene),

which reacts with 3 moles of hydroxylamine. What is the likely structure of the product? What accounts for the difference in behavior between phloroglucinol and the other phenols?

b. *Hydroquinone* (1,4-dihydroxybenzene) labeled with ^{18}O shows little tendency to exchange its label with alkaline water. By contrast, 1,4-dihydroxynaphthalene exchanges ^{18}O fairly rapidly in alkaline solution. Suggest a mechanism for the exchange reaction, and explain the difference in reactivity of the two compounds.

PHENOL IONIZATION AND ELECTROPHILIC SUBSTITUTION

As previously noted (p. 384), phenols are prone to undergo polysubstitution by electrophilic reagents. In order to understand why, it must be borne in mind that in a polar solvent there are two phenol-derived species present — the undissociated compound and the phenoxide anion. The anion, although present to perhaps only a few tenths or hundredths of 1%, could be a million times more reactive than the parent phenol. Most of the product could therefore be formed by way of the anion despite its low concentration at any given instant.

Bromination of phenol in water results in the rapid formation of 2,4,6-tribromophenol, even in the presence of excess phenol. A plausible explanation is as follows.

(22.4)

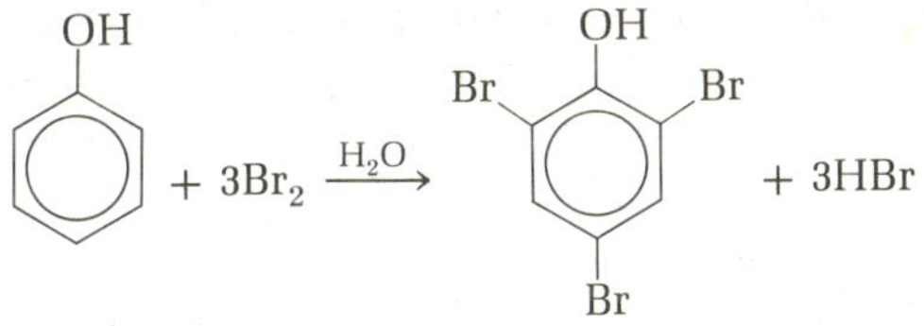

Monobromophenol, formed very rapidly, is more acidic than phenol and is dissociated to a greater extent at equilibrium. In fact, the acidity of the solution will repress the ionization of phenol strongly. Therefore, most of the organic anions present will be monobromophenoxide ions, and the bromine will attack these preferentially. The dibromophenol that is formed is even more acidic than monobromophenol and so the previous arguments apply even more strongly to this compound. Tribromophenol precipitates from the reaction medium, and indeed the formation of this precipitate constitutes a qualitative chemical test for phenol.

Monobromination of phenol is possible if an extremely nonpolar solvent such as carbon disulfide is used, in which case virtually no phenoxide ions are present. The reaction then follows the expected course — introduction of one halogen atom deactivates the ring toward further attack and monosubstitution is easily accomplished.

(22.5)

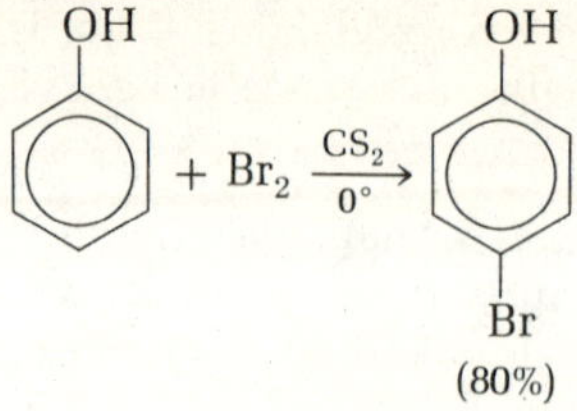

REACTIONS OF THE PHENOLIC HYDROXYL GROUP

Although the hydroxyl group of phenols is considerably more acidic than that of most alcohols, phenols will nonetheless undergo many reactions characteristic of aliphatic alcohols. Phenols are much less nucleophilic than alcohols; if a phenol and a carboxylic acid are refluxed together with a trace of mineral acid, only a small quantity of ester is formed. If water is removed continuously, however, then the phenol may be satisfactorily esterified by this method.

(22.6)

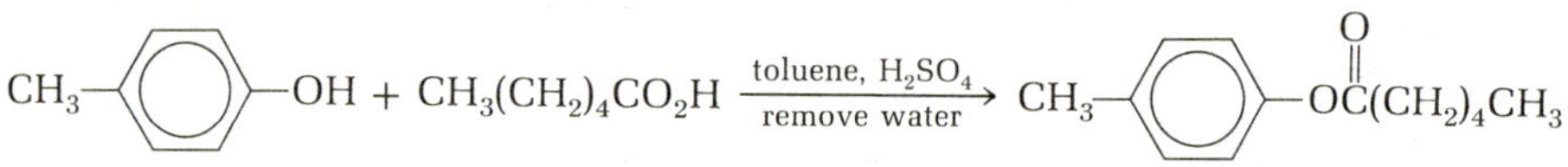

p-cresol

p-cresyl hexanoate (99%)

More commonly, phenyl esters are prepared by reaction of the appropriate phenol or its salts with the desired acid chloride or anhydride. These latter reactions are often conducted in the presence of aqueous base, which converts the phenol to the much more reactive phenoxide ion.

(22.7)

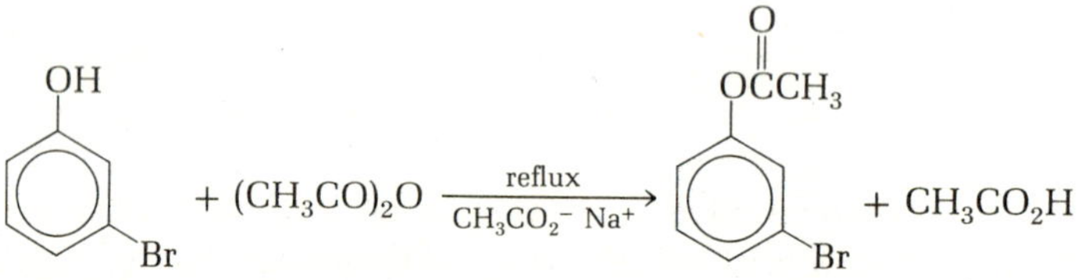

(22.8)

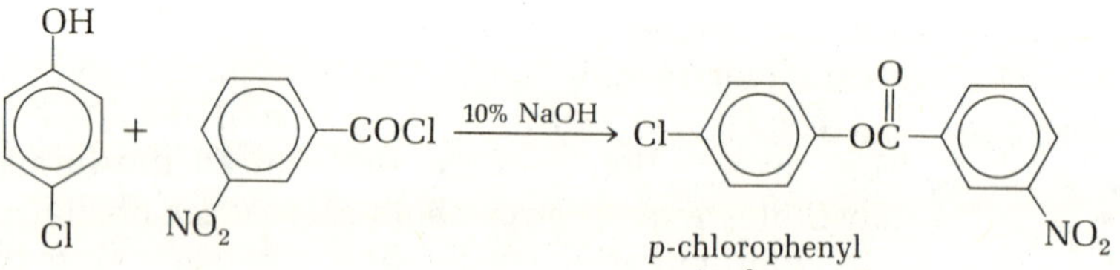

p-chlorophenyl
m-nitrobenzoate

Since phenoxide ions are much less basic than alkoxide ions, phenyl esters are much more readily hydrolyzed than alkyl esters.

Phenols may also be etherified. In the great majority of cases the reactants are a phenoxide salt and an alkyl halide (equation 22.9); the other combination (alkoxide salt and aryl halide) is much less commonly used because halogen atoms bound to aromatic rings are usually very difficult to displace.

(22.9)

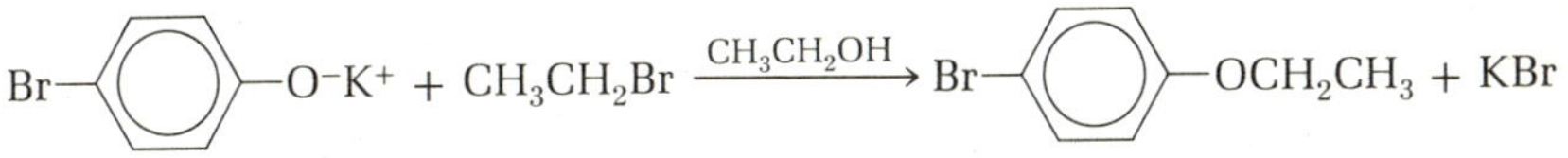

potassium
p-bromophenoxide

It is possible to induce alkylation at carbon as well as at oxygen, particularly when reactive alkylating agents are used. Dipolar aprotic solvents in which the salt is soluble promote attack at oxygen; very nonpolar solvents (heterogeneous conditions) or very good hydrogen-bonding solvents encourage reaction at carbon.

(22.10)

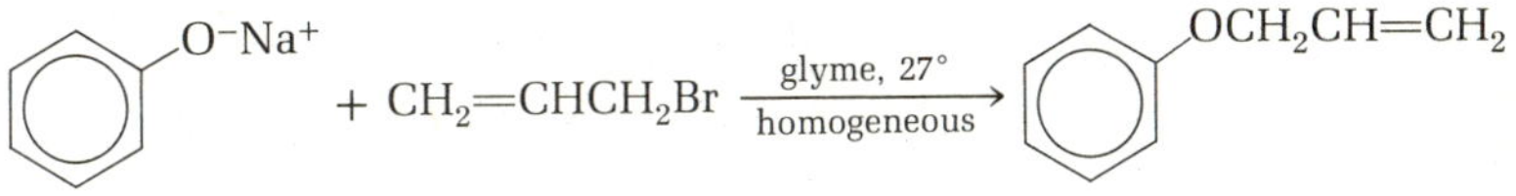

(99%)

The mechanism of formation of *o*-allylphenol (equation 22.11) is as shown in equation 22.12.

(22.11)

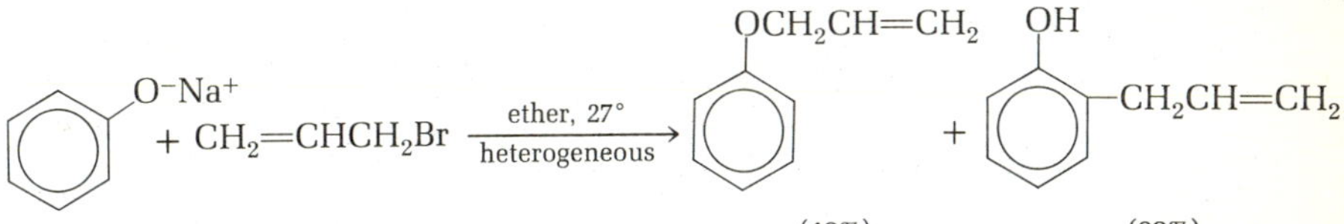

(40%) (60%)

(22.12)

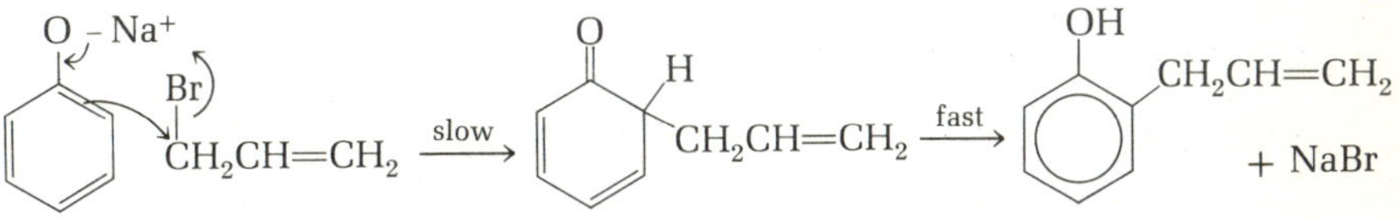

Alkyl phenyl ethers may be cleaved by the same reagents that attack dialkyl ethers (HI, BCl_3, etc.; see Chapter 10). Due to the very strong phenyl-oxygen bond, only alkyl-oxygen fission is observed.

(22.13)

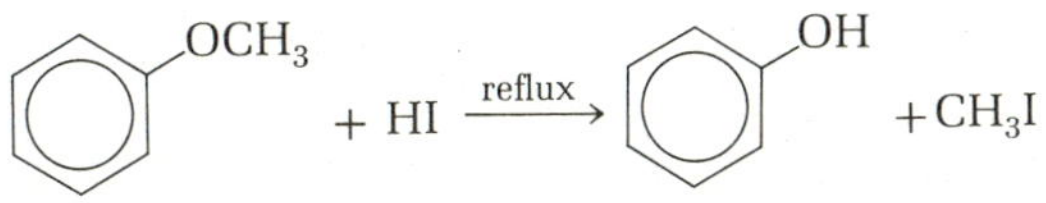

Allyl ethers of phenols, such as the one shown in equation 22.10, undergo some very interesting and useful thermal reactions. Heating allyl phenyl ether to 180 to 200° results in its conversion to *o*-allylphenol in high yield.

(22.14)

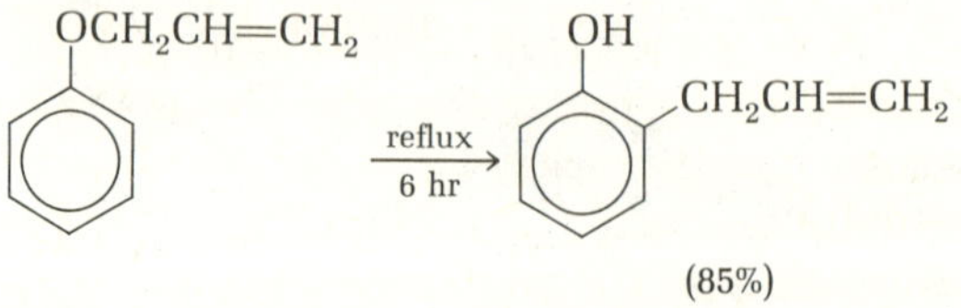

If an ether carrying an unsymmetrical allyl group is heated, the phenolic product contains a *rearranged* allyl group in the *ortho* position.

(22.15)

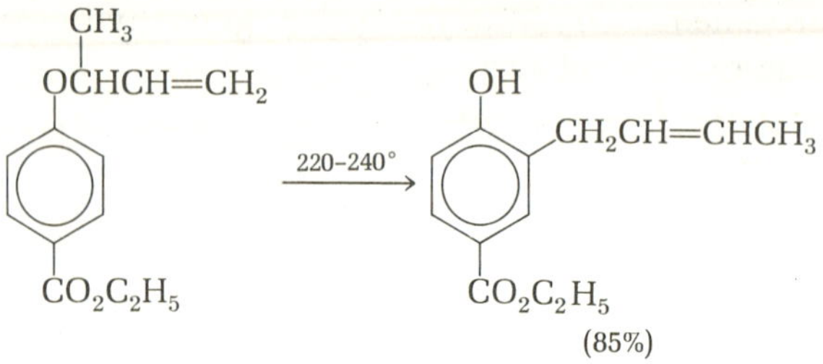

Experiments like these and a number of others have established fairly well the intramolecular nature of the rearrangement and the likelihood that it goes by way of a cyclic, six-centered transition state.

(22.16)

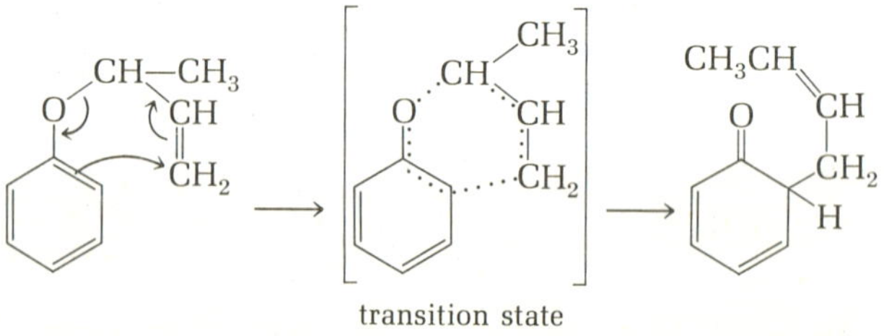

transition state

If the ether contains a group other than hydrogen in the *ortho* position, then there will be some allyl migration to the *para* position.

(22.17)

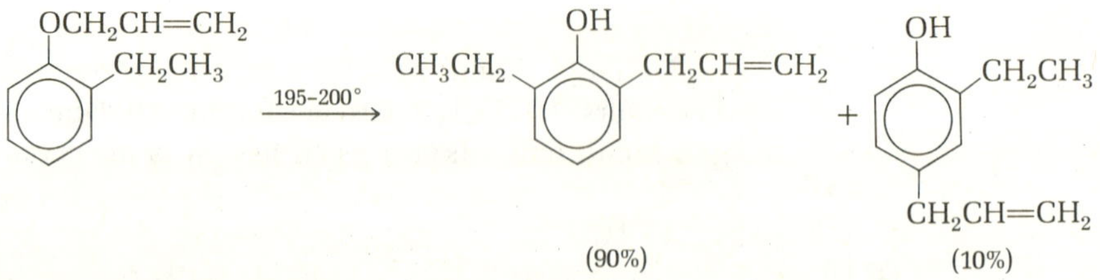

If both *ortho* positions are blocked, then the allyl group will eventually move completely to the *para* position.

(22.18)

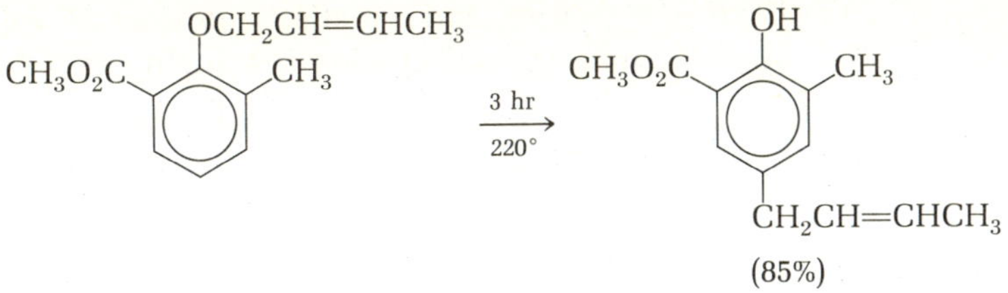

In the *para* rearrangement the allyl group has the same structure in the product that it had in the starting material. The apparent retention of structure is actually the result of two successive rearrangements, in which the 2,4-cyclohexadienone **1** is an intermediate and is subsequently transformed into a 2,5-cyclohexadienone. The conversion of **1** to products also involves a six-membered, cyclic transition state in which concerted bond making and bond breaking are taking place and serves to emphasize the fact that the *ortho* Claisen rearrangement is but one example of a more general phenomenon.

(22.19)

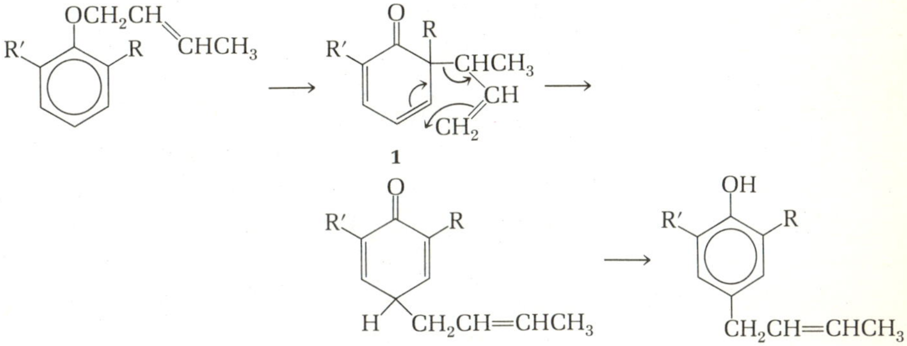

Claisen rearrangements are not confined to phenol ethers; they may also be induced in open-chain allyl vinyl ethers and the reaction is of use in preparing certain unsaturated aldehydes.

(22.20)

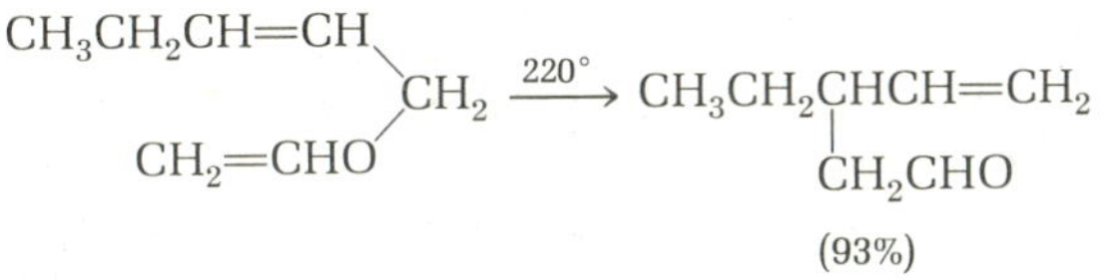

The ethereal oxygen atom is not necessary and the all-carbon analogues of the ethers also rearrange (*Cope rearrangement*).

(22.21)

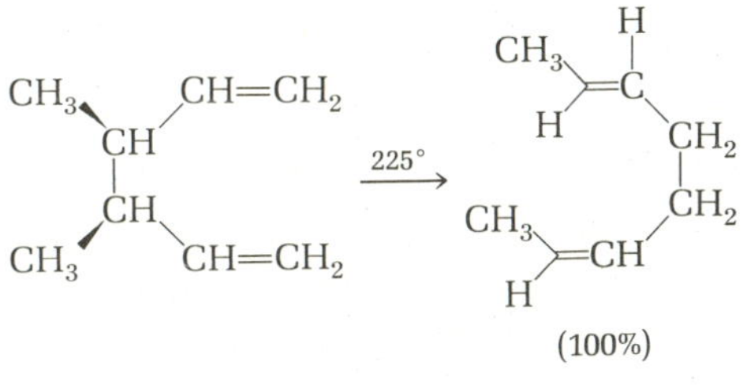

We have seen before, and we shall continue to see, evidence that the six-membered cyclic transition state is a highly favorable configuration.

NUCLEOPHILIC AROMATIC SUBSTITUTION – GENERAL CONSIDERATIONS

Halogen atoms and other active leaving groups are much more difficult to displace when bound to sp^2-hybridized carbon atoms than are those attached to sp^3-hybridized carbon atoms (p. 295). Electrons in an sp^2-orbital are more tightly bound (p. 321) so that the $\gt C{=}$—X bond strength is greater than that for —$\overset{|}{\underset{|}{C}}$—X. Furthermore, there may be conjugative interactions between X and the π-system that will be lost when X is displaced.

$$C{=}C{-}\ddot{X} \longleftrightarrow \ddot{\bar{C}}{-}C{=}\overset{+}{X}$$

As a result, very few halides of this type will undergo substitution by way of the S_N1 or S_N2 mechanisms so prevalent among aliphatic halides.

Halobenzenes devoid of strongly electron-withdrawing substituents react very sluggishly with nucleophiles, as evidenced by the extreme conditions necessary for converting chlorobenzene into phenol.

(22.22) 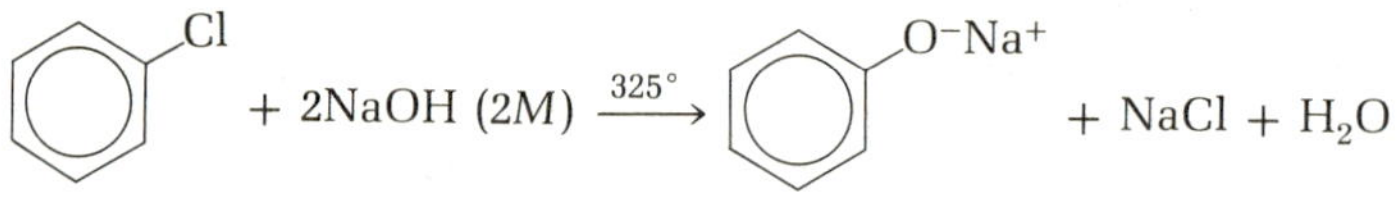

Three different processes have been discovered whereby aromatic halides and related molecules may be subjected to nucleophilic displacement; none of these corresponds to the S_N1 or S_N2 mechanisms of aliphatic chemistry. The operation of one or the other process will be a function of the reactants and the conditions.

We are omitting here a discussion of the replacement of the diazonium group in aromatic diazonium ions by a variety of nucleophiles (Chapter 19), a reaction of great importance in synthetic chemistry. The process appears, in most cases, to take place by way of a free radical mechanism and is therefore of a qualitatively different character from the reactions described below. It has been discussed previously at some length.

A. The S_NAr, or Addition-Elimination, Mechanism

The basic pattern here is one in which the nucleophile first attacks the aromatic ring, causing the formation of a delocalized carbanionic intermediate that subsequently expels the leaving group. The operation of this mechanism is greatly facilitated by the presence of *electron-withdrawing* groups on the aromatic ring.

(22.23) 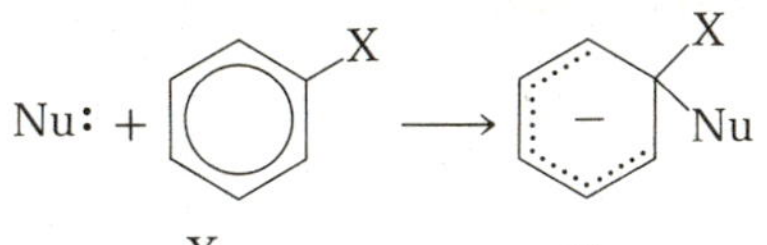

(22.24) 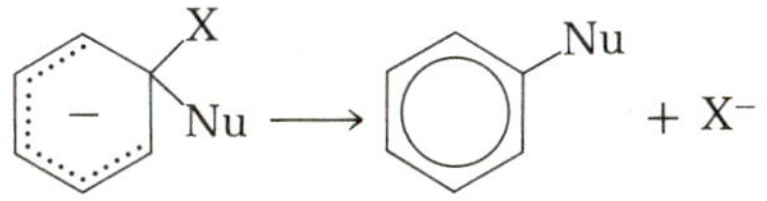

B. The Benzyne, or Elimination-Addition, Mechanism

If the nucleophile in question is also a strong base, it is possible to have an elimination reaction in which HX is lost from the aromatic ring. The resulting organic species is termed a **benzyne** and is usually depicted as 1,3-cyclohexadien-5-yne.

(22.25) 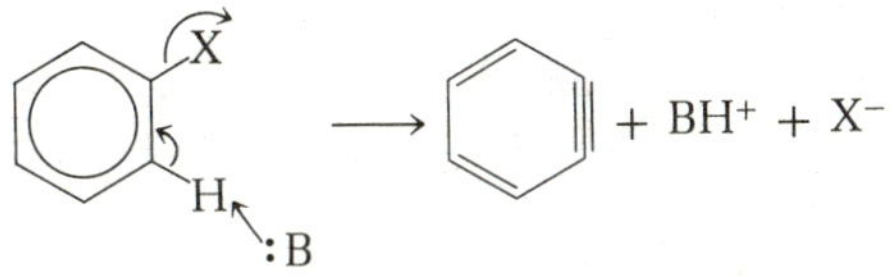

Needless to say, such a molecule is highly reactive and will undergo addition from virtually any nucleophile present. Protonation of the intermediate carbanion completes the substitution process.

(22.26) 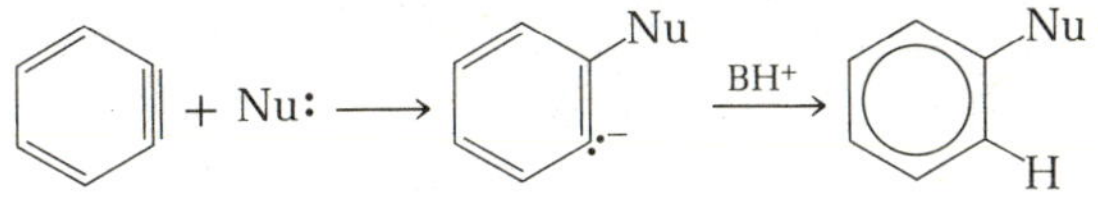

(B: and Nu: may or may not be identical.)

C. The Coupling Route

This is simply a particular variant of a process that we have encountered several times in the past. It depends on the uncanny utility of organocopper derivatives in promoting formation of carbon-carbon bonds with concomitant creation of Cu(I) halides.

(22.27) $ArX + RCu \longrightarrow ArR + CuX$

There are several ways of forming the organocopper compounds.

SUBSTITUTION VIA THE S_NAr PATH

One of the earliest studies of substitution of this type was undertaken by Meisenheimer, who found that a red salt was formed when 2,4,6-trinitroanisole was treated with potassium ethoxide. Meisenheimer further found that the same salt could be isolated from the reaction of 2,4,6-trinitroethoxybenzene with potassium methoxide.

(22.28)

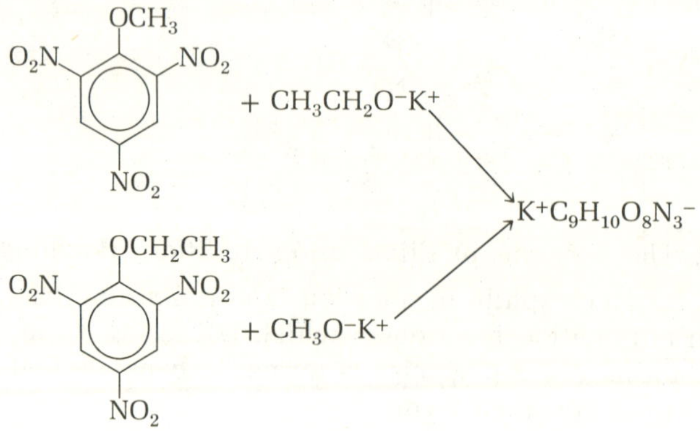

He proposed a structure for this salt that has been confirmed by subsequent investigations and that can be represented as **2** (using modern symbolic notation).

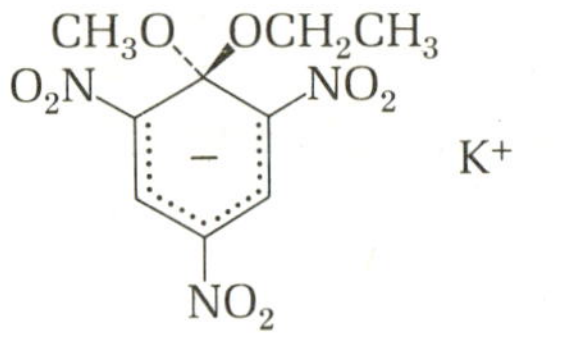

2

The stability of the anion is greatly enhanced by the three nitro groups since they are attached to the three carbon atoms of the ring over which the negative charge is primarily delocalized.

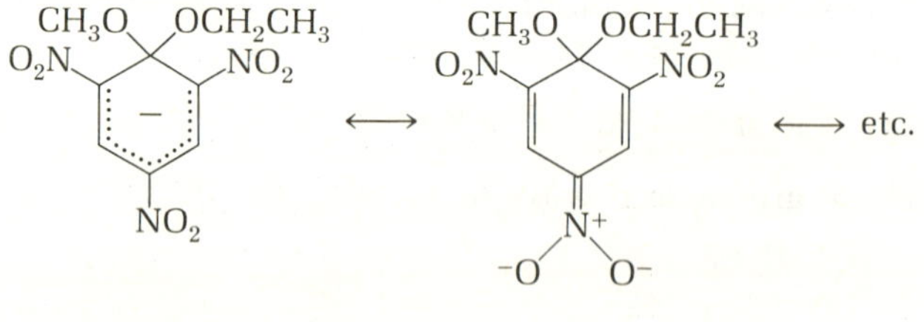

When the salt **2** is treated with acid, it decomposes into a mixture of trinitroanisole and trinitroethoxybenzene.

(22.29)

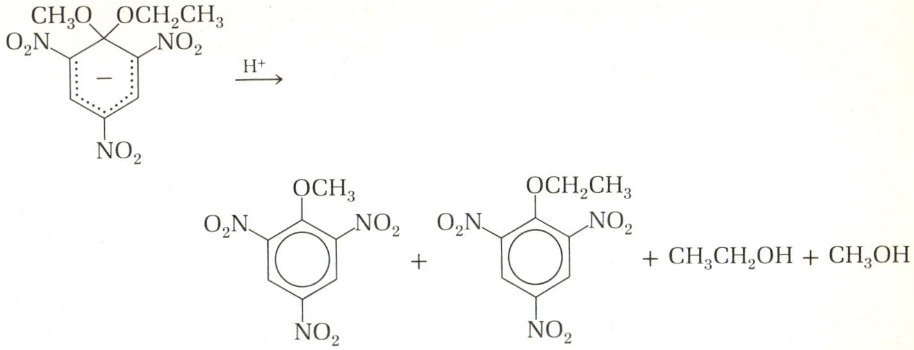

These and other experiments led to the suggestion that Meisenheimer-type complexes were intermediates in a large number of nucleophilic aromatic substitutions, particularly if the reactants satisfy one or both of the following criteria:

1 The aromatic nucleus carries electron-withdrawing groups.

2 The nucleophile is not a very strong base.

The effect of electron-withdrawing groups on the ease of replacement of an aryl halogen is brought out by a comparison of equations 22.22, 22.30, and 22.31.

(22.30)

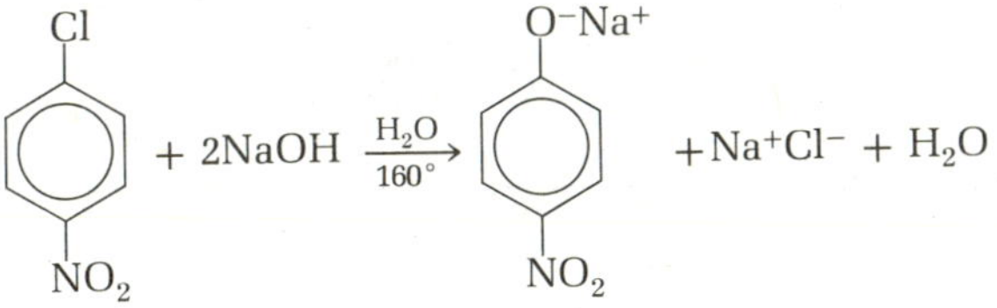

(22.31)

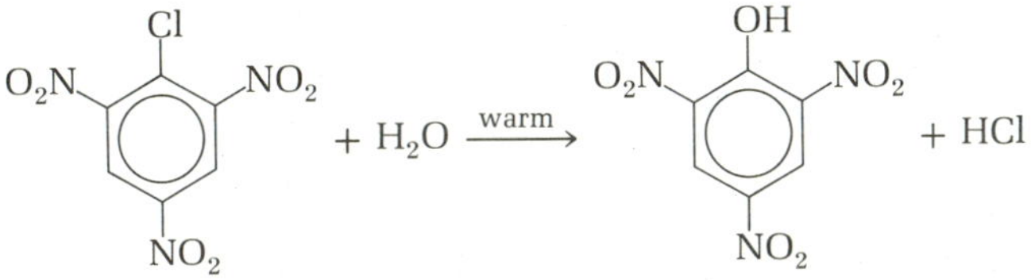

In accord with the idea that Meisenheimer-type complexes are intermediates in these reactions, the probable mechanism for reaction 22.30 is depicted in equations 22.32 and 22.33.

(22.32)

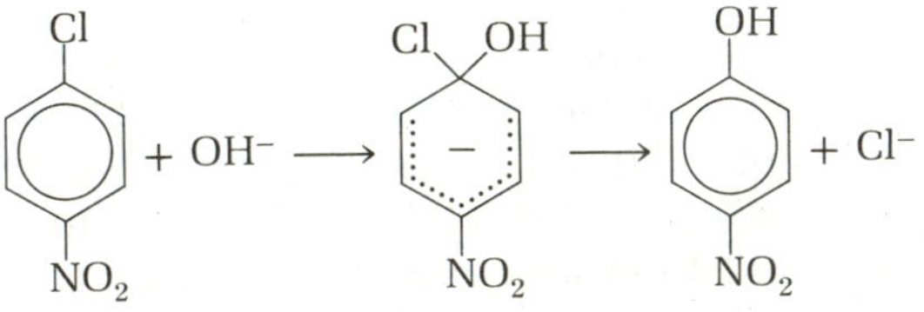

(22.33)

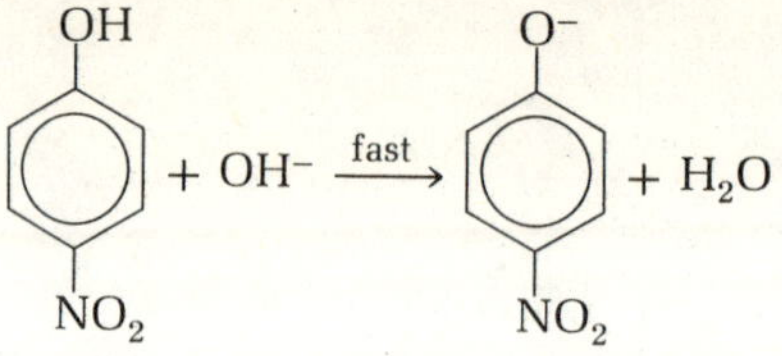

Alkoxide and hydroxide ions are not the only nucleophiles used in nucleophilic aromatic substitutions. Ammonia and the amines are very common reactants with activated aryl halides. Equations 22.34 to 22.36 not only dramatize the accelerating effects of electron-withdrawing substituents, they also show how much more pronounced these effects are from *ortho*- and *para*-substituents, in line with the mechanistic arguments already presented.

(22.34)

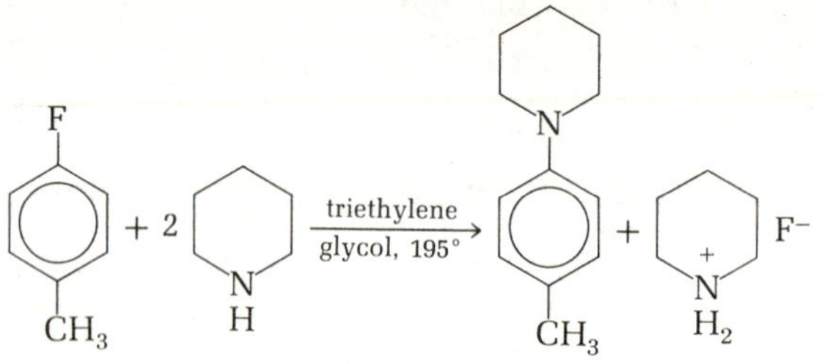

$k = 5.6 \times 10^{-8}\ M^{-1}\ \text{sec}^{-1}$

(22.35)

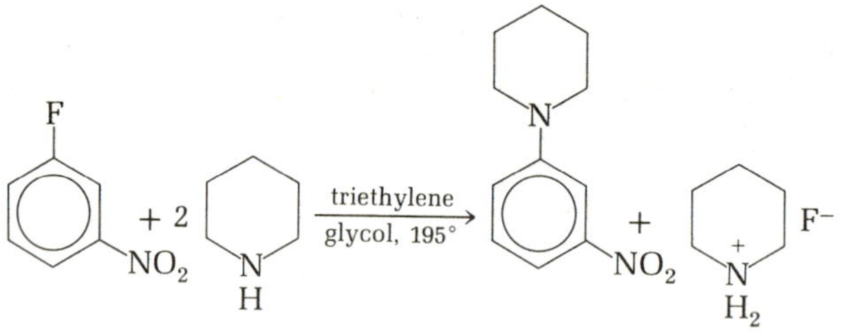

$k = 2.5 \times 10^{-4}\ M^{-1}\ \text{sec}^{-1}$

(22.36)

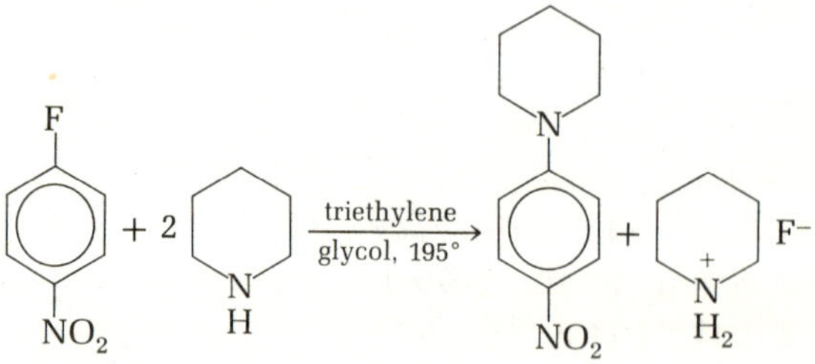

$k = 1.8 \times 10^{-1}\ M^{-1}\ \text{sec}^{-1}$

The substituents —NO_2 and —CN are most effective at promoting nucleophilic substitution by the S_NAr mechanism since conjugative withdrawal of electronic charge should have an enormous effect on the reaction rate.

The possibility of converting an aryl halide to a phenol or phenyl ether has its uses. Direct nitration of phenol is a rather poor procedure since the ring is very susceptible to oxidation. As a consequence, picric acid is synthesized by the route depicted in equation 22.37; once a deactivating group has been attached to the ring, it is much less easily oxidized.

(22.37) 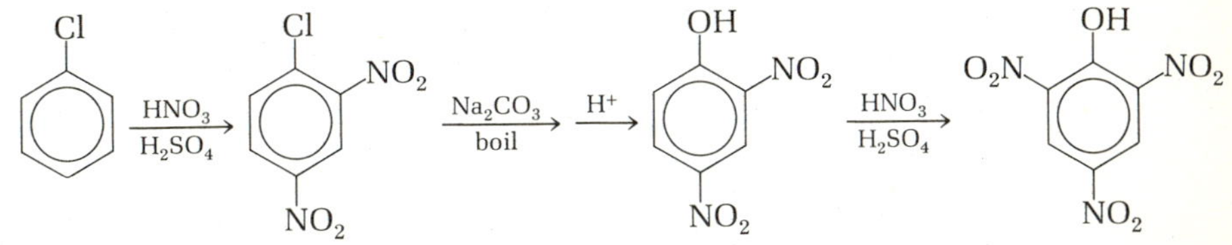

In many such phenol syntheses the actual product of the nucleophilic displacement step is a phenoxide ion, which must be converted to the phenol with the aid of a strong acid.

SUBSTITUTION BY WAY OF BENZYNES

The displacement of chloride from chlorobenzene by hydroxide ion is a slow process, requiring rather extreme conditions. It was shown in 1936 that halogen atoms of chlorobenzene, bromobenzene, and iodobenzene could be replaced by $—NH_2$ if the powerful base potassium amide ($K^+NH_2^-$) were used in liquid ammonia.

(22.38) 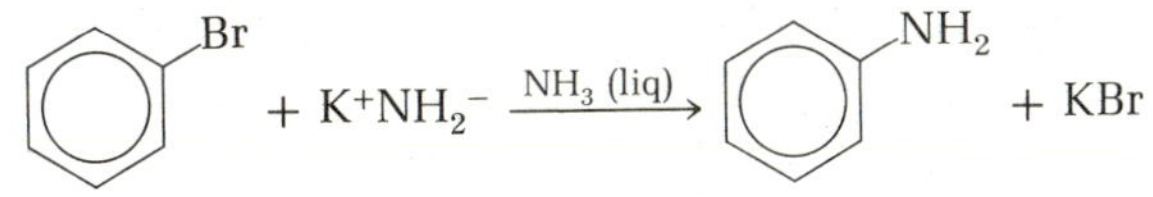

A reinvestigation of this reaction with ^{14}C-labeled chlorobenzene revealed a very startling fact: About half the amino groups in the final product were attached to carbon atoms that had *not* been bonded to chlorine in the starting material.

(22.39) 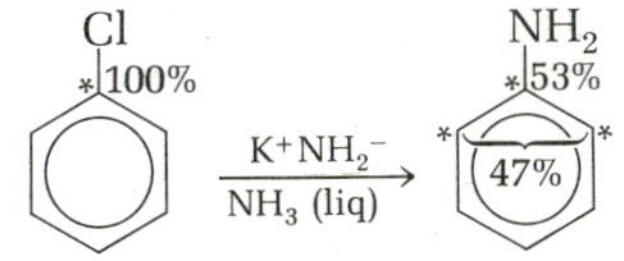

The most satisfactory explanation for these findings was that hydrogen chloride had been eliminated and that substitution had taken place by way of a benzyne intermediate. Iodobenzene reacts most rapidly and fluorobenzene does not undergo substitution, although it does exchange its *ortho*-hydrogens with those of the solvent, as shown with the use of ND_3.

(22.40)

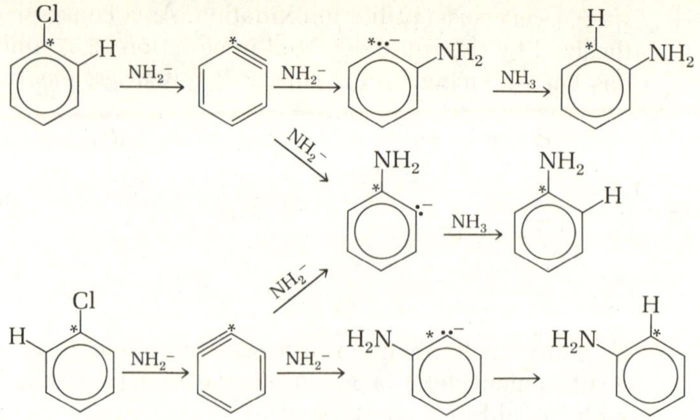

A great deal of evidence began to accumulate that was compatible with this mechanism. It was shown, for instance, that it was necessary to have a hydrogen atom *ortho* to the halogen atom in order for substitution to take place (equation 22.41).

(22.41)

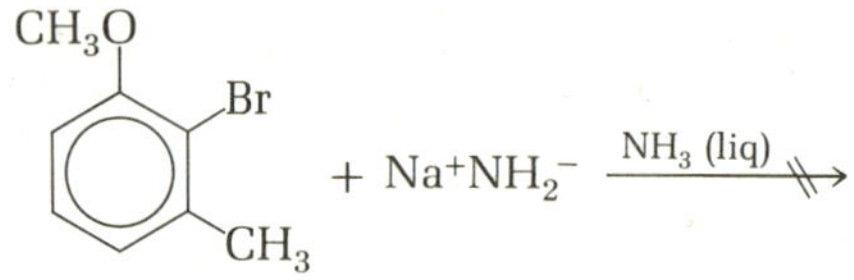

Reactions of substituted aryl halides led to rearranged products.

(22.42)

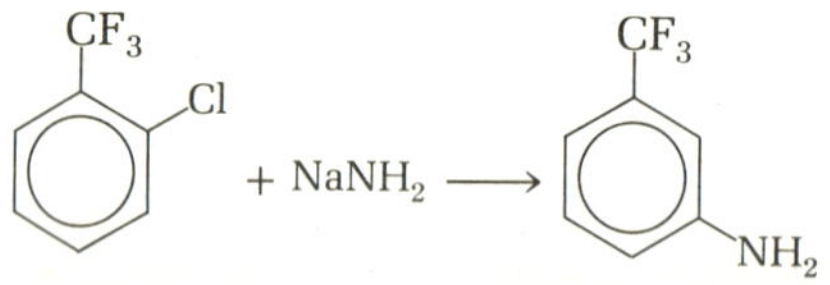

The orientation of the amino group in the product in equation 22.42 is also consistent with the mechanism shown in equation 22.40. Attack of the nucleophile on the benzyne ring leads to a phenyl anion as an intermediate; it is quite reasonable that the anion be formed so that the charge is next to the inductively electron-withdrawing trifluoromethyl group.

(22.43)

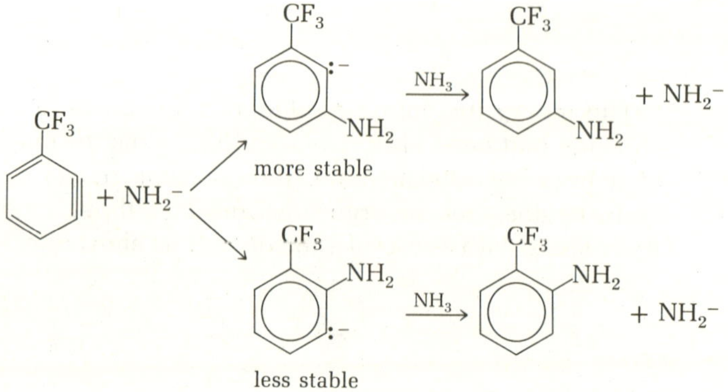

When bases weaker than amide ion are used as nucleophiles, then both the benzyne and S_NAr mechanisms may operate simultaneously; the balance between the two will be dependent on the temperature, among other factors. Lower temperatures favor the S_NAr reaction.

(22.44)

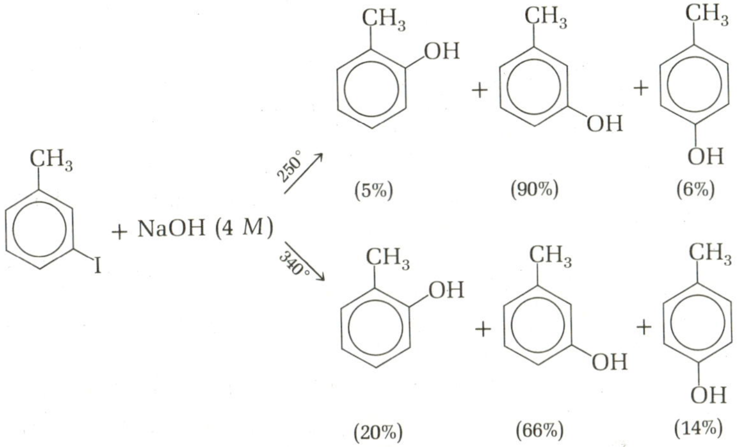

PROBLEM

2 Phenol is made commercially by the Dow process, which involves heating chlorobenzene with sodium hydroxide solution at 325° (equation 22.22). A by-product of this reaction is diphenyl ether, 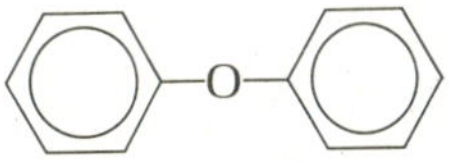. Offer an explanation for its formation.

Elimination of hydrogen halide from haloaromatics is by no means the only way of generating arynes. Removal of a pair of adjacent halogen atoms with the aid of a reactive metal (usually lithium or magnesium) is also an effective method. Equation 22.45 also shows that benzynes are active dienophiles.

(22.45)

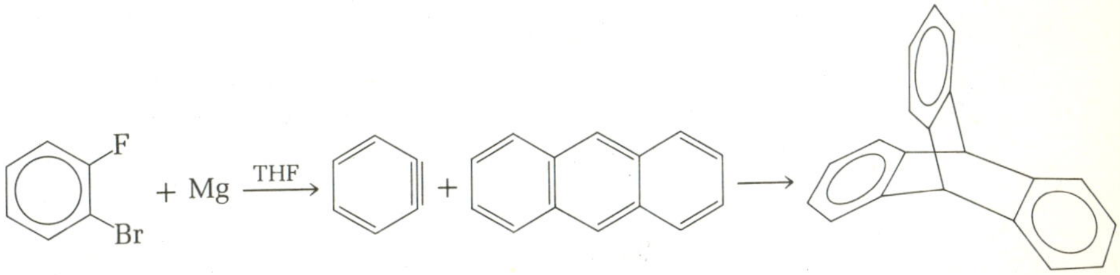

triptycene (28%)

Finally, benzynes may be created in purely thermal reactions if two stable molecules are extruded from adjacent positions on an aromatic ring.

(22.46) 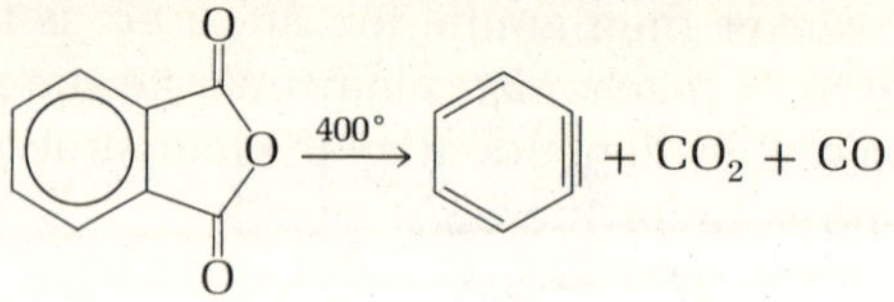

(22.47) 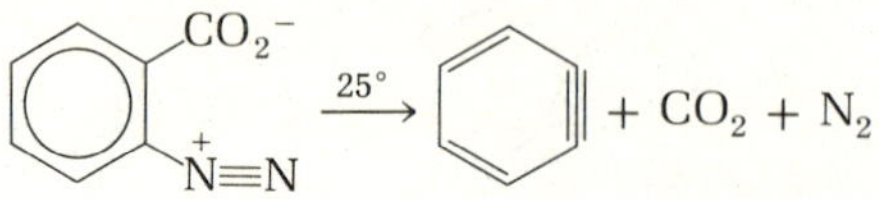

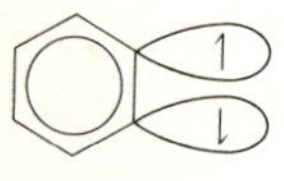

Benzyne

FIGURE 22-1

Benzyne

Benzynes are extremely reactive and none have ever been isolated, although one has been trapped in a metal complex. By use of special fast techniques, it has been possible to make spectroscopic measurements on benzyne itself. The sum of chemical and physical evidence is most readily accommodated by a molecule in which the aromatic π-system is essentially undisturbed. A weak additional π-bond arises from the partial overlap of adjacent sp^2-orbitals (Figure 22-1). The representation of benzyne that we have been using in the equations is somewhat misleading since it implies the presence of a triple bond in the molecule. In actuality, benzyne is much more likely to be a distorted hexagon.

SUBSTITUTION BY COUPLING

The formation of carbon-carbon bonds through reaction of an organocopper compound and an organic halide is very applicable to aromatic syntheses. One of the simplest cases involves the treatment of aryl halides (particularly the bromides) with cuprous cyanide, aryl nitriles being the products (Chapter 17). The reaction is promoted by solvents capable of complexing with the metal atom, such as pyridine; dimethylformamide (DMF) is particularly effective.

(22.48) 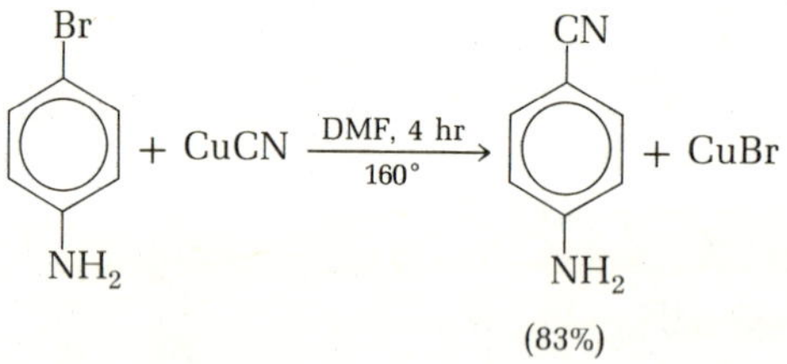

(22.49) 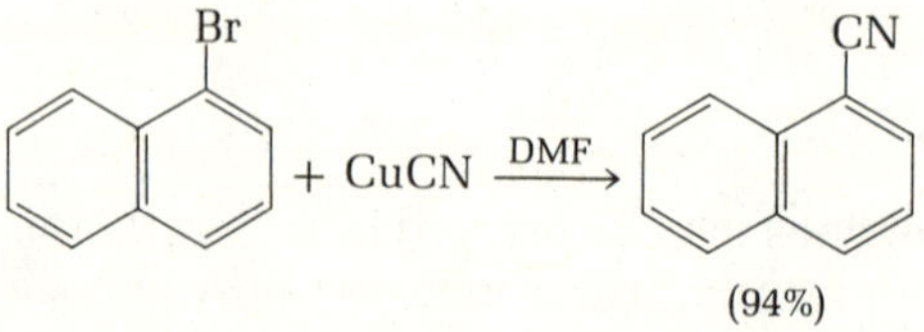

Since cyano groups may be converted to amide and carboxylic acid groups, the cuprous cyanide-aryl halide reaction represents a route to substituted benzoic, naphthoic, and other aromatic acids (and derivatives). This particular synthetic method derives its utility in part from the fact that an *o,p*-directing halogen atom can be subsequently converted to a *m*-directing cyano or carboxyl group.

(22.50) 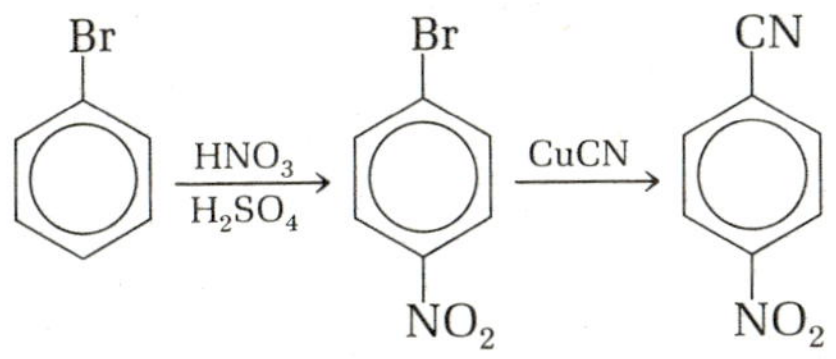

(22.51) 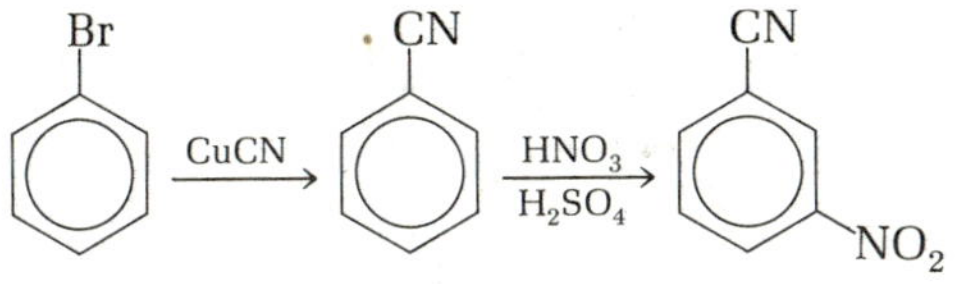

The last coupling reaction that will be covered here is the Ullman reaction in which an aryl halide or a mixture of two aryl halides is heated with copper powder added in small portions. Substituted biaryls (such as biphenyl) are the desired products. Occasional use is made of solvents such as DMF but most of the time the neat reactants are combined at fairly high temperatures.

(22.52) 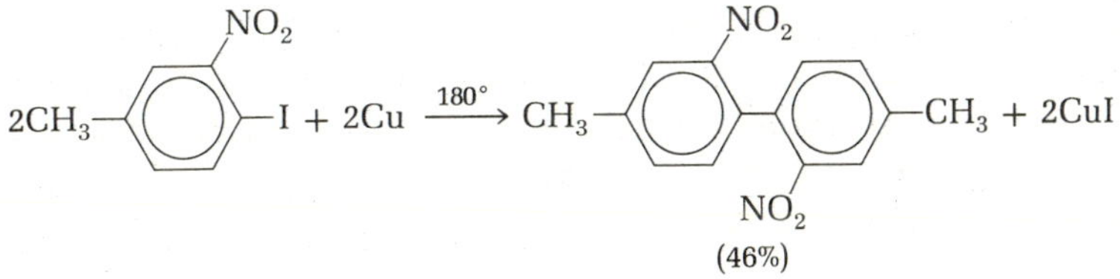

(22.53) 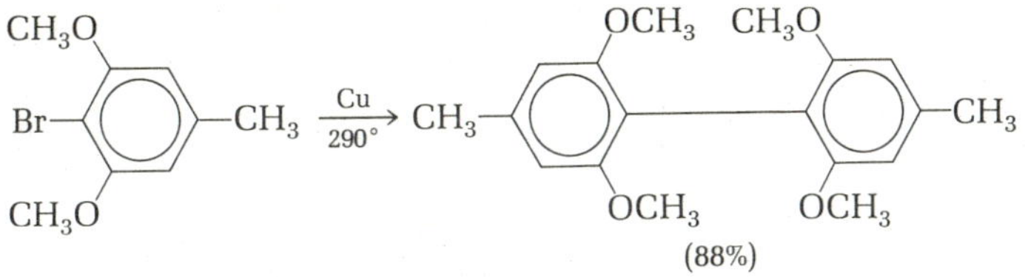

It has been shown that conjugatively electron-withdrawing substituents such as the nitro and ester groups definitely accelerate the reaction, the *ortho*-nitro group having by far the most pronounced activating effect. Among the halides the reactivity order is I > Br > Cl; fluorides are inert. Free hydroxyl, amino, and carboxyl groups suppress biphenyl formation.

A likely reaction intermediate is an arylcopper compound that then couples with a molecule of aryl halide.

(22.54) $$\mathrm{ArX + 2Cu \longrightarrow CuX + ArCu \xrightarrow{Ar'X} Ar{-}Ar'}$$

The organocopper may be complexed or part of a polymeric species, rather than free. Such a hypothesis would rationalize the fact that it is possible to prepare unsymmetrical biphenyls in reasonable yields providing that the two aryl halides differ markedly in reactivity. The more reactive halide presumably forms the organocopper compound, which then reacts with the other halide. The only side product is the symmetrical biaryl from the more reactive component; formation of this latter compound is suppressed by using a reactive aryl bromide rather than an iodide.

PHENOXY RADICALS

One-electron oxidation of a phenol is rather easily accomplished because the resulting phenoxy radical is considerably stabilized by delocalization of the unpaired electron.

(22.55)

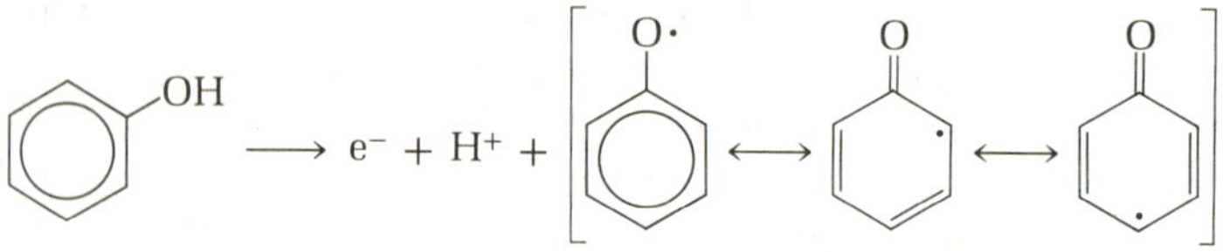

If the *ortho* and *para* positions of the ring are occupied by bulky groups, then it is possible to obtain the radical as a stable species. Such one-electron oxidations are usually carried out at neutral or basic pH, and the most common oxidizing agents are ferric chloride, potassium ferricyanide, and silver oxide.

(22.56)

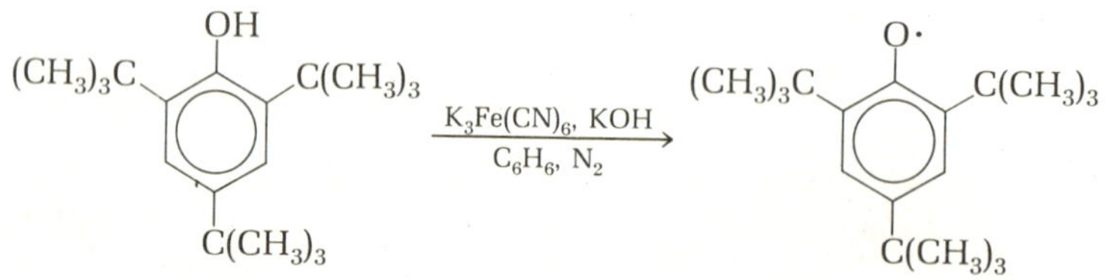

These radicals lend themselves to detailed study by means of electron spin resonance spectroscopy (Chapter 12). Such studies confirm that the unpaired spin density is found mainly *ortho* and *para* to the oxygen substituent. They also suggest that substituents at these sites stabilize the radical by helping to delocalize the lone electron.

If the *ortho* and *para* positions of the phenol are not sterically hindered, then two phenoxy radicals may couple. These couplings involve C—C and/or C—O bonding, but not O—O bonding.

(22.57) 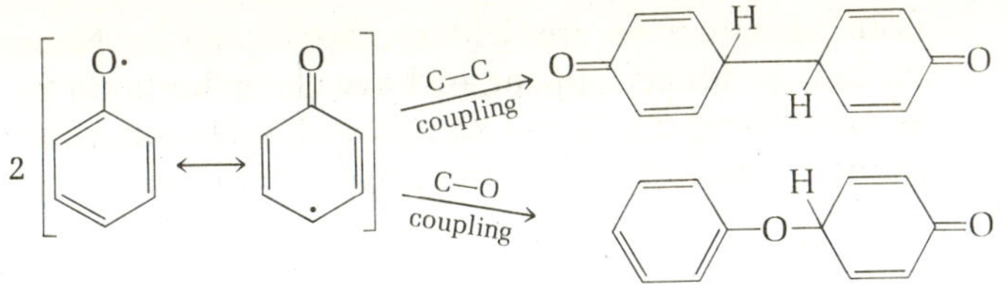

(22.58) 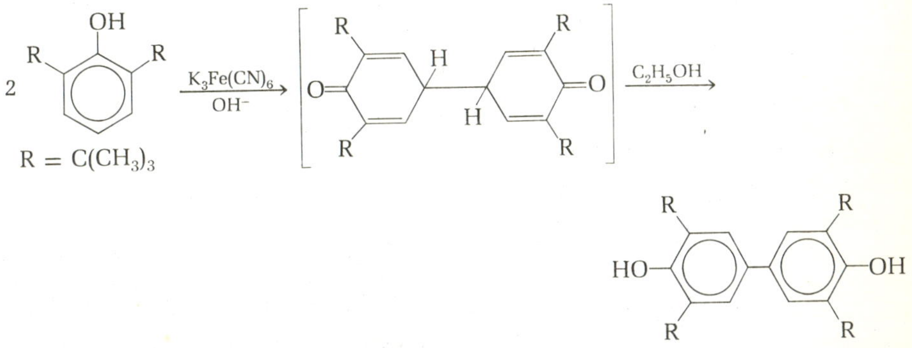

The type of coupling that takes place can be markedly influenced by the oxidant and by the reaction conditions. Intramolecular coupling between two different phenolic systems is very advantageous if the rings are properly situated with respect to one another.

(22.59) 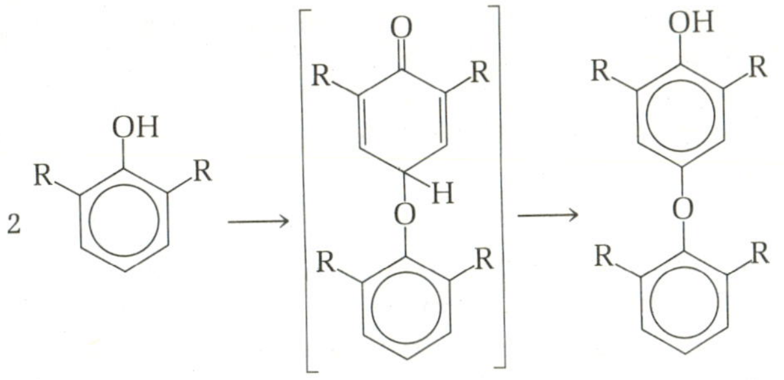

(22.60) 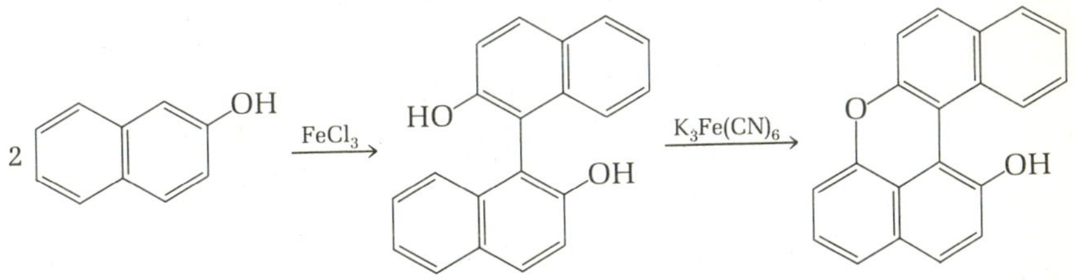

These laboratory oxidations are of interest not only for themselves but also because they provide models for biological oxidation processes. Phenolic compounds are widely distributed in Nature, and many complex natural products appear to have been built up by oxidative coupling of separate phenols. Two examples of such substances are the alkaloid morphine **(3)** and the antibiotic griseofulvin **(4)**.

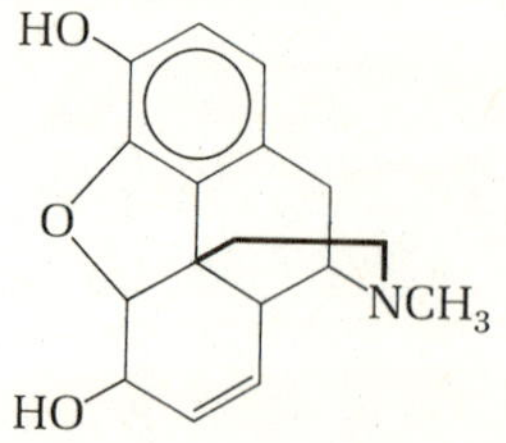

3

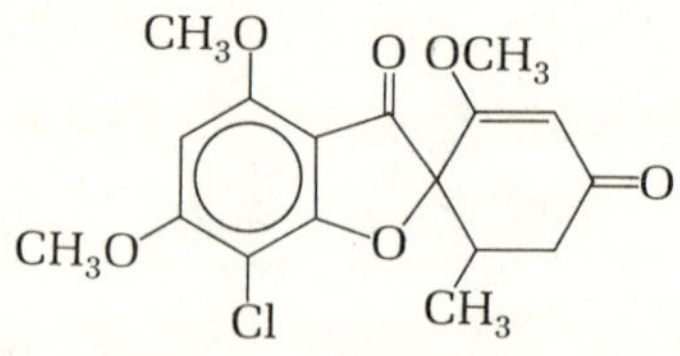

4

Because of the susceptibility of the phenolic ring to oxidation, phenols are often intermediates in biological degradation reactions in which the ring itself is eventually opened. The metabolic degradation of aromatic compounds often depends on their first being converted to phenols and then cleaved by further oxidation.

(22.61)

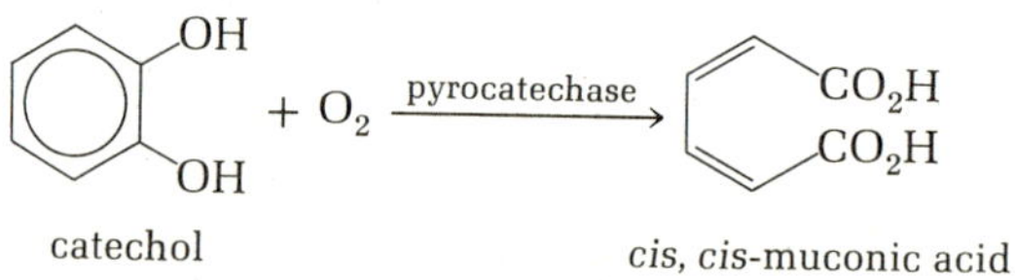

QUINONES

Oxidation of phenols under acidic conditions leads to different products than oxidation in basic solution. Coupling is suppressed and extensive degradation of the ring may occur; there is evidence that ionic intermediates intervene in these oxidations. The reaction mixtures will usually contain, in varying yield, **p-quinones** (1,4-benzoquinones).

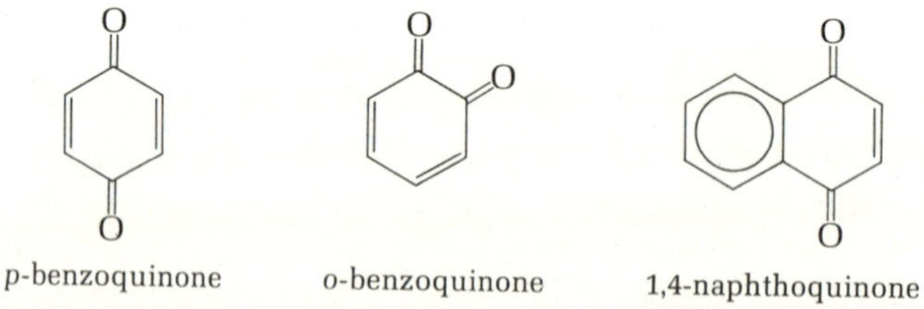

Equation 22.62 gives an example of such a reaction, which is of rather variable value as a preparative method.

(22.62) 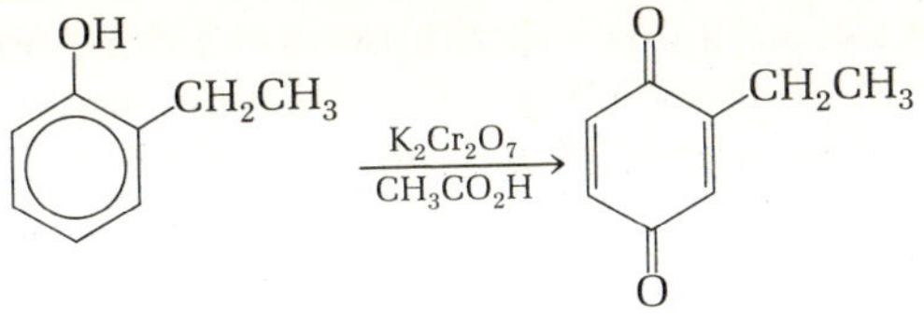

2-ethyl-*p*-benzoquinone

Anilines may also be oxidized to *p*-quinones in acid.

(22.63) 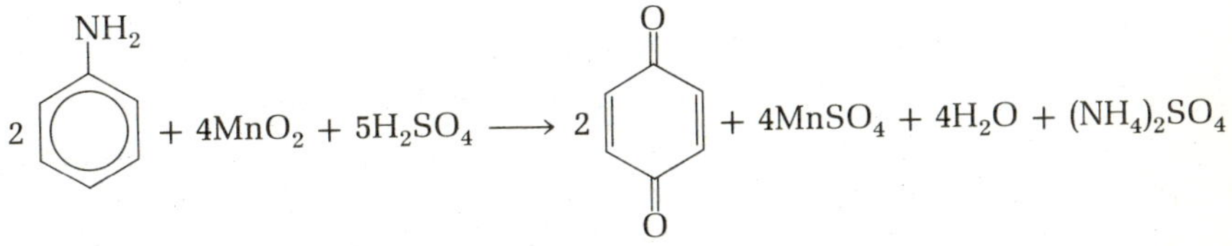

In this case polymeric products obtained from the coupling of radicals are intermediates. Under the relatively vigorous conditions employed, these polymers are further oxidized to *quinoneimines* that yield *p*-quinones upon hydrolysis (equation 22.64). The *p*-phenylenediamine released in the example shown will itself be rapidly oxidized to quinone.

(22.64) 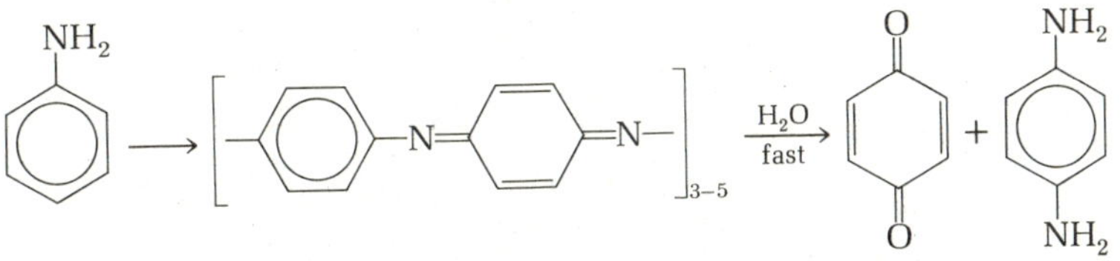

A more reliable and productive method of obtaining derivatives of *p*-benzoquinone depends on the oxidation of substituted *p*-aminophenols or *p*-dihydroxybenzenes **(hydroquinones).** With these starting materials one can use relatively mild oxidants.

(22.65) 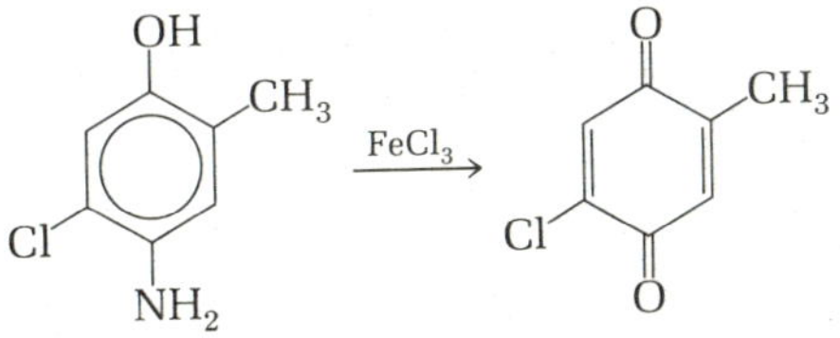

2-chloro-5-methyl-
p-benzoquinone (60%)

(22.66) 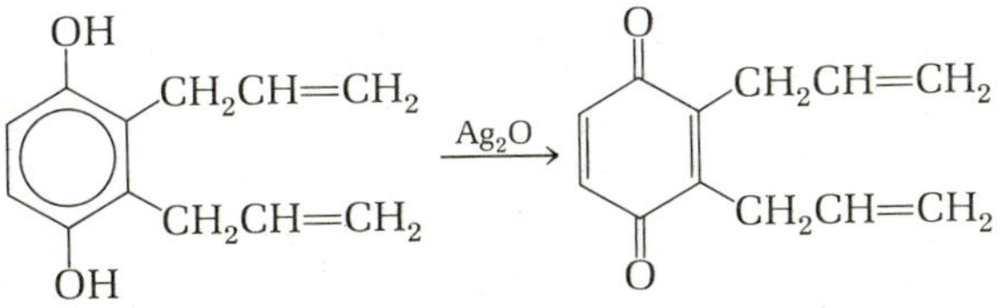

2,3-diallyl-*p*-benzoquinone (96%)

If an aniline or phenol is substituted in the *para* position by a group other than —OH, but one which is a good leaving group, then the latter may be eliminated in the course of oxidation.

(22.67)

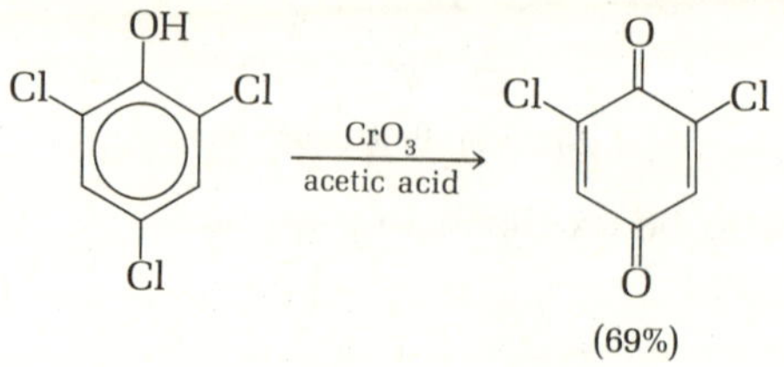

When hydroquinone itself is oxidized, a very dark green material precipitates early in the reaction; however, it is eventually consumed and the color changes to the characteristic yellow of benzoquinone. The green crystals, which may be isolated and stored, prove to consist of a molecular complex of one molecule of hydroquinone and one of benzoquinone held together by hydrogen bonds. The complex is known as **quinhydrone;** a number of substituted quinhydrones have been generated by partial oxidation of the appropriate hydroquinones.

(22.68)

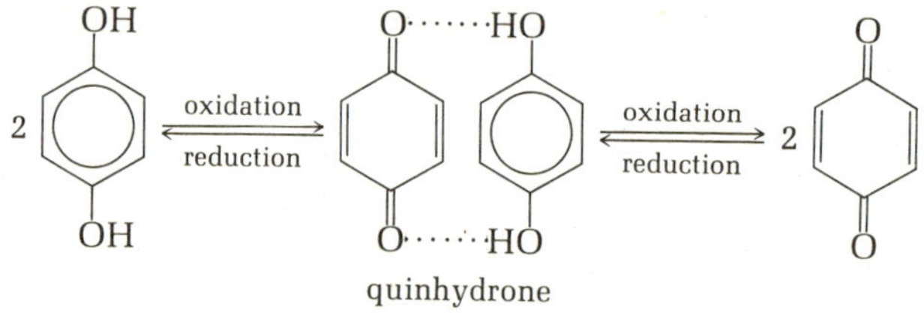

o-Benzoquinones are also known, although these compounds are invariably less stable than the corresponding *p*-quinones. The *o*-quinones are always obtained by oxidation of *o*-dihydroxybenzenes (*catechols*) or *o*-aminophenols.

(22.69)

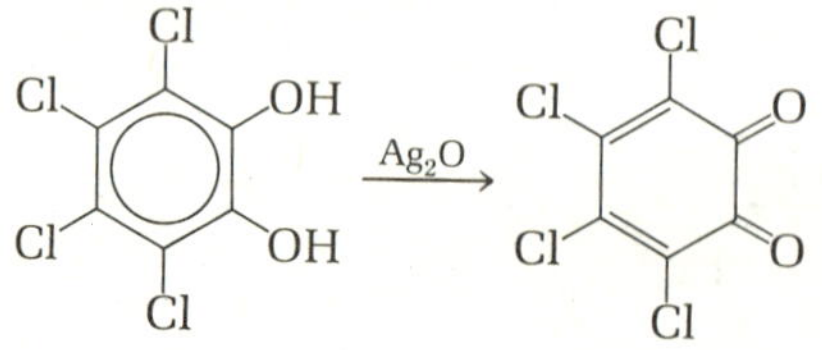

m-Quinones are unknown since it is not possible to have such an arrangement without having bonds crossing the ring.

REACTIONS IN QUINONE-HYDROQUINONE SYSTEMS

The oxidation of hydroquinone to quinone and the related reduction are completely reversible in acid, a rare phenomenon in organic chemistry. Quinhydrone is used to set up standard half-cells; since the emf of this half-cell is pH-dependent, quinhydrone electrodes can be used to measure hydrogen ion concentration. Equation 22.70 defines the *reduction potential* of *p*-benzoquinone.

(22.70) 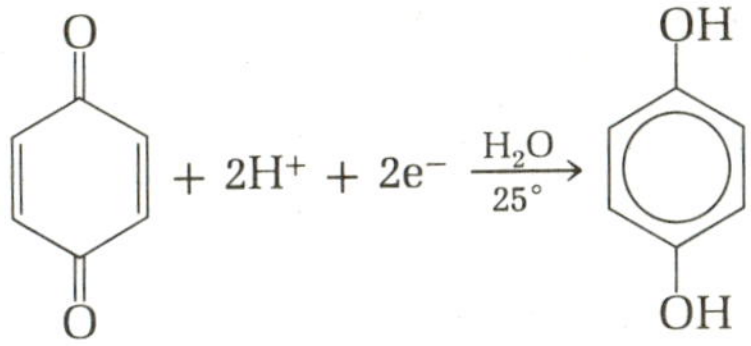 $E^\circ = 0.699$ v

If hydroquinone is oxidized with a one-electron oxidizing agent in a strongly basic medium, then the oxidation may be arrested after only one electron has been removed from the hydroquinone dianion, the product being a **semiquinone radical anion.** This unusual species derives its stability from the fact that both the negative charge and the unpaired spin are delocalized over all six ring atoms, as well as over the oxygen atoms (as confirmed by esr measurements).

(22.71) 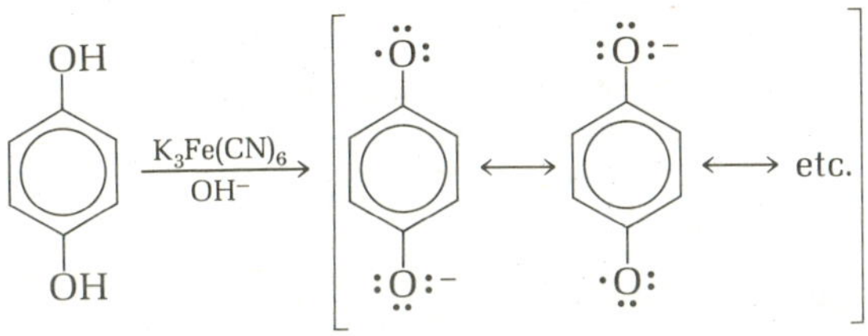

semiquinone

All benzoquinones have positive reduction potentials in water or ethanol. The presence of electron-withdrawing substituents on the ring tends to raise this potential relative to that for benzoquinone, while electron-donating substituents lower the reduction potential. Substances such as 2,3-dichloro-5,6-dicyano-1,4-benzoquinone (DDQ) are very powerful oxidizing agents that are particularly useful at dehydrogenating certain types of organic compounds.

(22.72) 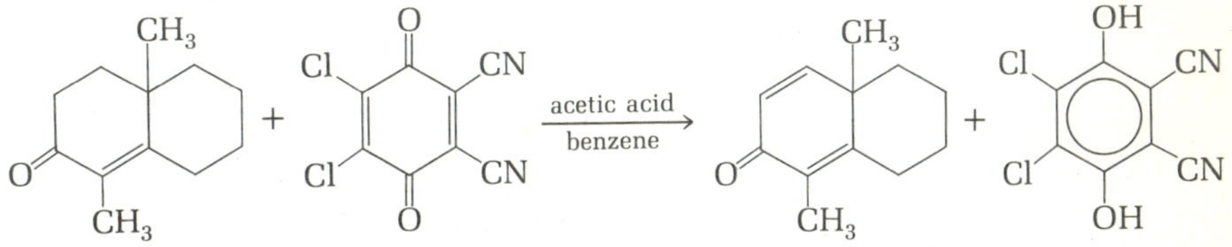

DDQ

If a quinone A and a hydroquinone BH_2 are mixed, then a redox reaction will take place until the system has come to equilibrium. The position of equilibrium will be determined by the relative reduction potentials of the quinones A and B and by the relative concentrations of the species involved.

(22.73) $A + BH_2 \rightleftharpoons AH_2 + B$

NUCLEOPHILIC ADDITIONS TO QUINONES

A number of nucleophiles such as water, alcohols, and halide ions will attack quinones under acid catalysis. These reactions are nothing more than acid-catalyzed 1,4-additions to an α,β-unsaturated ketone (Chapter 21), followed by rapid aromatization of the intermediate. For instance, addition of hydrogen chloride to benzoquinone leads to chlorohydroquinone.

(22.74) 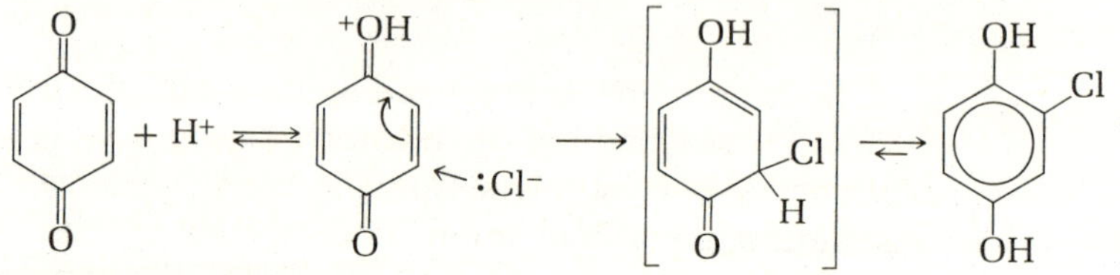

This product may in part be oxidized by the starting material to chloro-*p*-benzoquinone, which can add hydrogen chloride to give some dichlorohydroquinone.

An excellent example that illustrates the operation of the various factors set forth in the preceding paragraphs is the synthesis of 2,3-dicyanohydroquinone from benzoquinone.

(22.75) 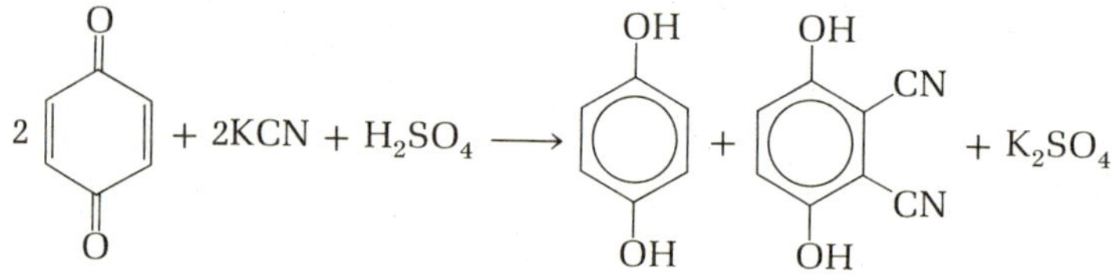

A mechanism consistent with previous observations is set out in equations 22.76 to 22.78.

(22.76) 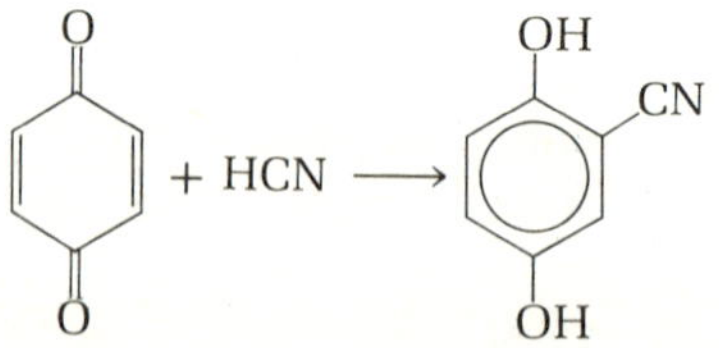

(22.77) 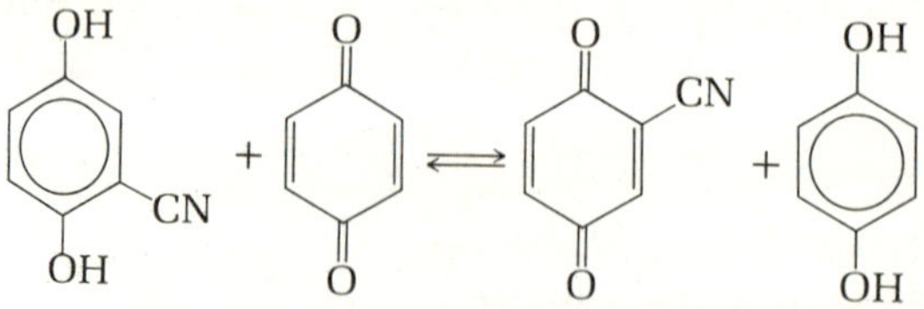

(22.78) 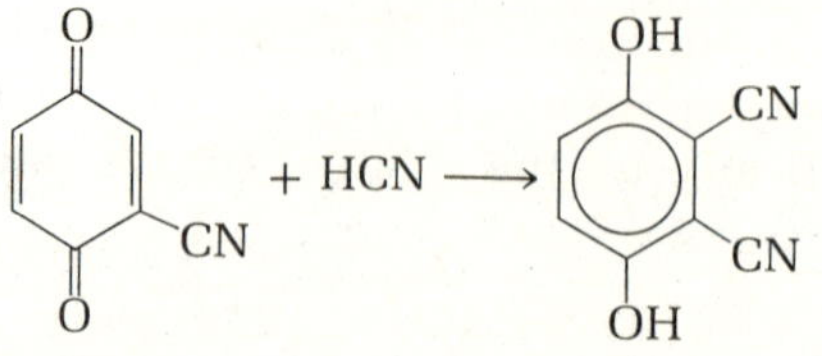

The product is used for the synthesis of DDQ (equation 22.79).

(22.79) 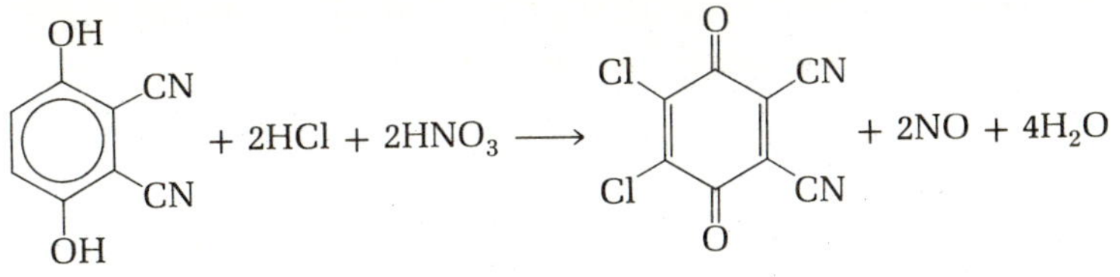

QUINONES IN BIOLOGICAL SYSTEMS

Quinones and hydroquinones are widely distributed in Nature and their functions are undoubtedly connected with their facile reduction and oxidation. As one example we take the family of coenzymes Q_n (**5**, $n = 6$ to 10), which may play an important role in the vital process known as **oxidative phosphorylation.**

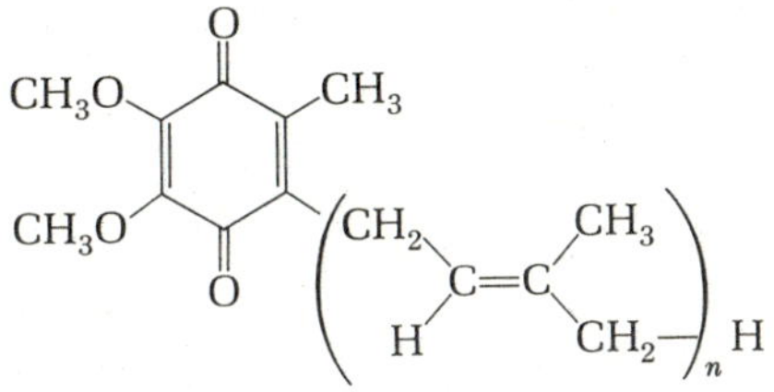

5

The construction and hydrolysis of polyphosphates is an integral part of the basic metabolic processes in cells. The phosphate must be suitably "activated" for reaction 22.80 to take place, for it is certainly not a spontaneous process.

(22.80) 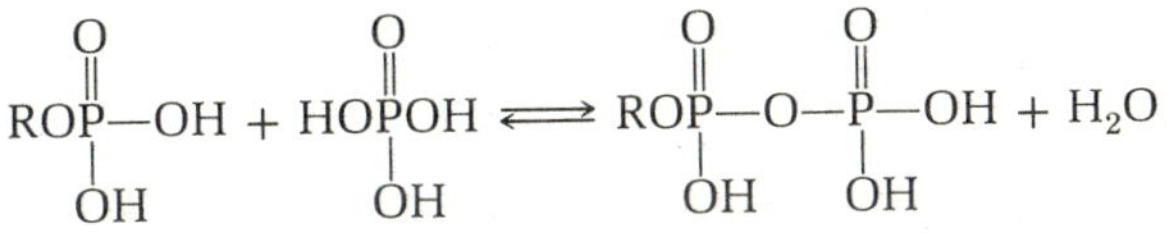

In cell mitochondria (cellular substructures specializing in metabolism), adenosine diphosphate (ADP) is converted to adenosine triphosphate (ATP, **6**), which in turn phosphorylates a number of compounds that are metabolic intermediates.

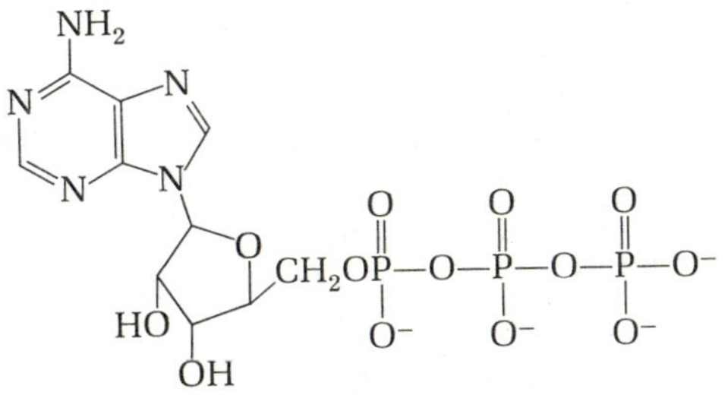

6

The synthesis of ATP allows the organism to store energy that is later released when ATP phosphorylates some suitable substrate.

The activation of phosphate by the coenzyme is thought to take place as follows. A monophosphate ester of the reduced form of a coenzyme is subjected to oxidation.

(22.81)

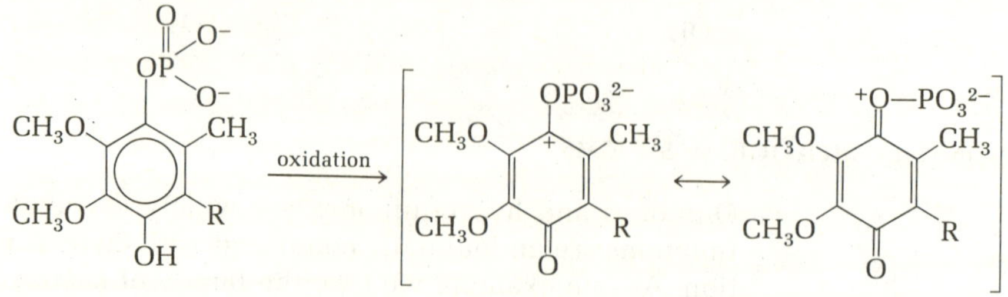

A phosphoquinonium ion is produced in which the phosphate group is now very susceptible to nucleophilic attack. The ion reacts with ADP to give ATP and the neutral quinone.

(22.82)

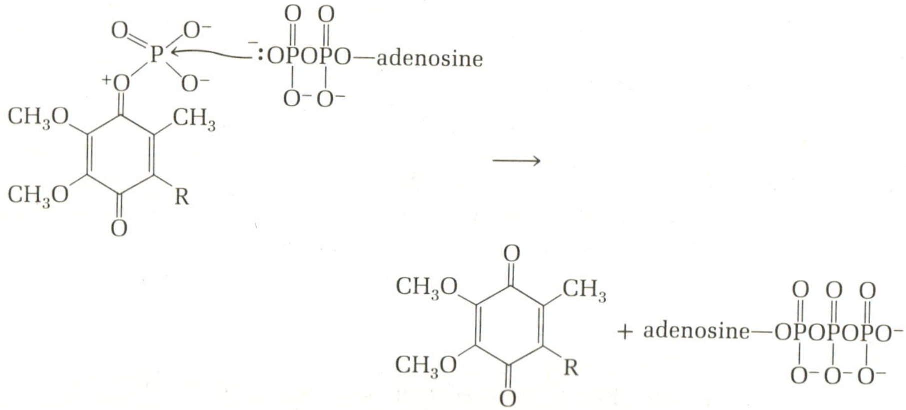

The proposed scheme is supported by the results of studies on simpler model hydroquinone monophosphates. Upon oxidation under favorable conditions they can give rise to diphosphates, triphosphates, and higher polyphosphates.

SUMMARY

1 Phenols are many orders of magnitude more acidic than aliphatic alcohols because when they dissociate, the negative charge may be delocalized into the aromatic ring. The acidity is much enhanced by the addition of electron-withdrawing substituents to the ring, particularly at the positions *ortho* and *para* to the hydroxyl group.

Electrophilic substitution of phenols often proceeds by way of the phenoxide anion, which is much more reactive than the neutral phenol. The reactivity can be suppressed by esterifying or etherifying the phenol or by dissolving it in a very nonpolar solvent.

2 Phenols are much less nucleophilic than alcohols are and they are therefore most conveniently esterified by way of their salts. Acid chlorides and anhydrides are the usual acylating agents. Phenoxide ions will also react with alkyl halides to give alkyl phenyl ethers. If a reactive alkylating agent is used, then it is possible to obtain C- as well as O-alkylation, particularly in heterogeneous media.

Allyl phenyl ethers undergo a thermal rearrangement in which the allyl group is transferred to the *ortho* position of the phenol ring. The side chain undergoes allylic rearrangement in the course of the transfer. If both *ortho* positions are blocked, then the allyl group will eventually find its way to the *para* position, after undergoing two successive rearrangements. These Claisen rearrangements take place concertedly by way of cyclic, six-centered transition states that are very common in organic chemistry.

3 Due to (a) the strength of aryl-halogen bonds, (b) the high energy of carbonium ions in which the charge is localized in sp^2-orbitals, and (c) the steric inaccessibility of the "back side" of the carbon atoms in an aromatic ring, aryl halides do not undergo nucleophilic displacement by way of either the S_N1 or S_N2 mechanisms. If the nucleophile is not a powerful base and if the temperature is not too high, then the nucleophile will first add to the ring to give a delocalized carbanion. Electron-withdrawing groups *ortho* and *para* to the point of attack speed up the process of carbanion formation considerably. The delocalized anion then expels the leaving group, thereby rearomatizing the ring and completing the substitution process.

4 When an aryl halide is exposed to a strong base, a mole of hydrogen halide may be eliminated. The resulting benzyne is essentially an intact benzene ring minus two adjacent hydrogen atoms; there is weak overlap between the two sp^2-orbitals made available by loss of the hydrogens. This weak bond is very readily attacked by nucleophiles and substitution is completed upon subsequent addition of a proton. The operation of this mechanism can lead to overall substitution with rearrangement.

5 Cuprous cyanide is a very effective reagent for synthesizing aromatic nitriles, starting from aryl bromides. The aromatic ring may carry virtually any type of substituent. The coupling of two aromatic halides by copper metal, leading to formation of a biaryl, probably proceeds by way of an organocopper compound.

6 Oxidation of phenols at neutral or basic pH's results in the formation of phenoxy radicals. These radicals are relatively long-lived and unreactive,

due to the delocalization of the unpaired electron. Substitution of the *ortho* and *para* positions of the radical by bulky groups materially increases its lifetime. Phenoxy radicals react with other radicals and radical sources; one of their most frequent methods of consumption is by dimerization. These dimers are created by C—O and C—C bond formation. The dimer may initially be in the form of a cyclohexadienone but the latter will rapidly tautomerize to a phenol, if possible.

7 More vigorous oxidation of phenols, particularly at acidic pH's, leads to degradation of the ring and also to the formation of *p*-benzoquinones. The quinones are more efficiently made by oxidation of hydroquinones or *p*-aminophenols. Oxidation of catechol derivatives leads to the much less stable *o*-benzoquinones. Quinones and hydroquinones may be reversibly interconverted by reduction or oxidation, respectively. In some cases a complex of intermediate oxidation state, called a quinhydrone, appears; it contains one molecule each of the quinone and the hydroquinone.

Oxidation of hydroquinones in basic media results in the formation of the relatively stable semiquinone radical anions, in which both the charge and the unpaired spin are delocalized over the entire ring.

Attaching electron-withdrawing substituents to the quinone ring increases its reduction potential. Some quinones that are heavily substituted by electron-withdrawing groups find use as oxidizing agents. The ready oxidation-reduction interconversion of hydroquinones and quinones seems to confer biochemical importance upon these compounds.

PROBLEMS

Note: Problems 1 and 2 appear in the text.

3 Arrange the following phenols in increasing order of acidity.

a. *m*-Cresol, *p*-cresol, *m*-bromophenol, *p*-bromophenol, phenol

b. *p*-Cresol, *p*-methoxyphenol (hydroquinone monomethyl ether), *p*-cyanophenol, *p*-trifluoromethylphenol

c. Phenol, 1-naphthol, *p*-hydroxybenzoic acid

d. *p*-Aminophenol, hydroquinone (*p*-dihydroxybenzene), catechol (*o*-dihydroxybenzene)

4 Rationalize the following facts:

a. The pK_a of 2,4-dinitrophenol is 3.96, while the pK_a of 3,5-dinitrophenol is 5.15.

b. The pK_a of 2,4,6-trichlorophenol is 6.00, while the pK_a of 2,4,6-trinitrophenol is 0.38.

5 One approach to the monobromination of phenol involves carrying out the reaction in a nonpolar solvent so that phenoxide ion cannot be formed. Can you suggest another way of dealing with this problem?

6 *p*-Aminophenol was treated with a deficiency of acetyl chloride (a) at low pH, (b) at pH 7, and (c) at high pH. What was the predominant product in each case? Explain.

7 Primary and (some) secondary phenyl ethers can be synthesized by allowing the appropriate alkyl halides to react with phenoxide ion. The reaction of phenoxide ion with tertiary halides leads to almost no ether as product. Explain.

8 Arrange the following compounds in order of increasing rate of base-catalyzed hydrolysis.

a. Ethyl acetate, phenyl acetate, *m*-cresyl acetate

b. *m*-Chlorophenyl butyrate, phenyl butyrate, 4-*t*-butylphenyl butyrate, 3,4-dichlorophenyl butyrate, *p*-nitrophenyl butyrate

c. Phenyl *p*-toluate, *o*-cresyl *p*-toluate, *p*-cresyl *p*-toluate

d. *p*-Aminophenyl succinate, *p*-trimethylammoniophenyl succinate, phenyl succinate

9 The addition of 1 mole of sodium methoxide to 1 mole of 2,4,6-trinitroanisole in DMSO-methanol leads to the formation of two new species, **A** and **B**, which may be detected by means of their proton nmr spectra. The spectrum of **A** consists of two peaks in the area ratio 3:1. Substance **B** gives rise to a four-peak spectrum; the areas of these peaks are in the ratio 3:3:1:1. After 15 min the peaks due to **B** have disappeared, while the **A** peaks have become more intense. Propose structures for **A** and **B**, and offer an explanation for this last observation.

10 Addition of sodium azide to a solution of *p*-nitrofluorobenzene in dry DMF produces a reaction — the ultraviolet spectrum of *p*-nitrofluorobenzene disappears and is replaced by a new absorption at 397 mμ. When water is added to the mixture, fluoride ion appears, the peak at 397 mμ disappears, and *p*-nitroazidobenzene may be detected. Explain all these observations, and draw the appropriate structures.

11 For a number of nucleophilic displacement reactions on nitrohalobenzenes the fluoride has been found to be the most reactive and the iodide the least reactive substrate. This order is opposite to the reactivity sequence found in S_N1 and S_N2 reactions, in which iodides are the most rapidly substituted halides. How can the experimental data be accommodated by the mechanism proposed for nucleophilic aromatic substitution (equation 22.32)?

12 The *von Richter* reaction involves the displacement of nitrite ion from nitrobenzenes by cyanide ion. This process is far from straightforward, as the cyanide becomes attached to the carbon atom *adjacent* to the position vacated by the nitro group (equation 1).

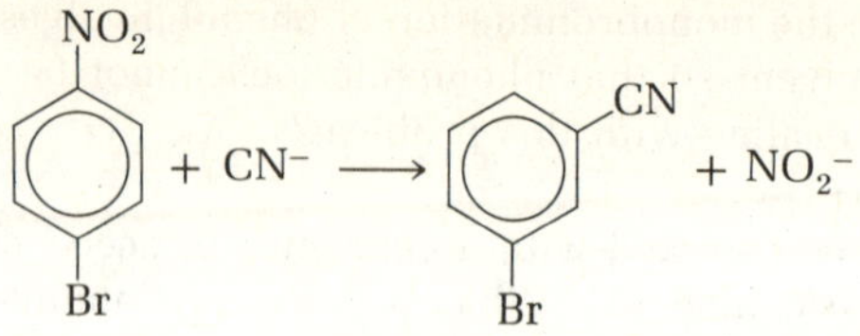

(1)

Suggest a mechanism for equation 1. (Remember that cyano is also efficient at delocalizing negative charge.)

13 Write syntheses for the following compounds, using benzene, toluene, phenol, or naphthalene as starting materials and any other necessary reagents.

a. *p*-Bromophenyl isobutyrate
b. *p*-Phenoxyacetophenone
c. *t*-Butyl 4-cyanophenyl ether
d. 4-Nitro-1-naphthol
e. 4-Cyano-2-allylphenol
f. *m*-Bromobenzonitrile
g. Methyl *p*-cyanobenzoate
h. 3,3′-Dinitrobiphenyl
i. 2,2′-Dinitrobiphenyl-4,4′-dicarboxylic acid
j. 4-Nitro-1-(*p*-methylphenyl)naphthalene

14 Equation 2 is an example of a reaction that has considerable potential for heterocyclic synthesis. What is the probable mechanism for the reaction?

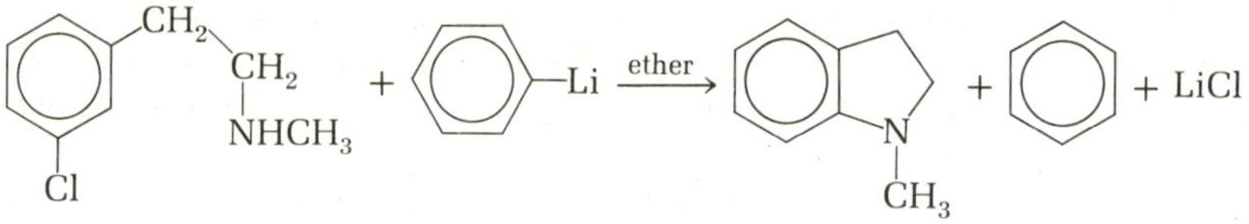

(2)

15 Indicate whether the equilibrium constants for the following reactions will be substantially greater than, substantially less than, or essentially equal to one. Justify your answers.

a.

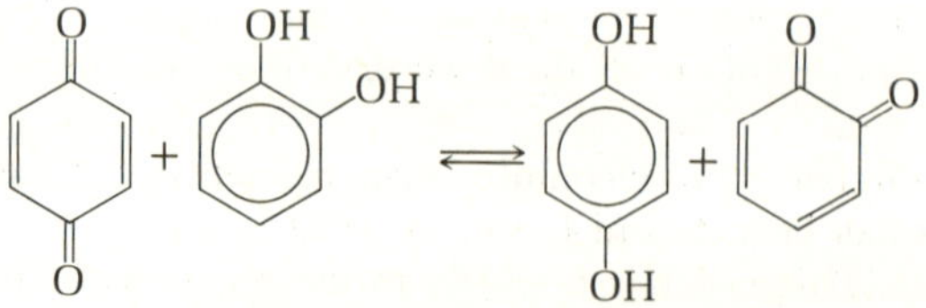

b.

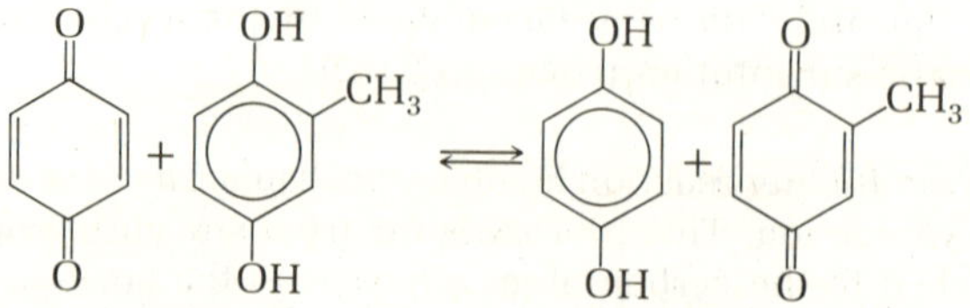

c. 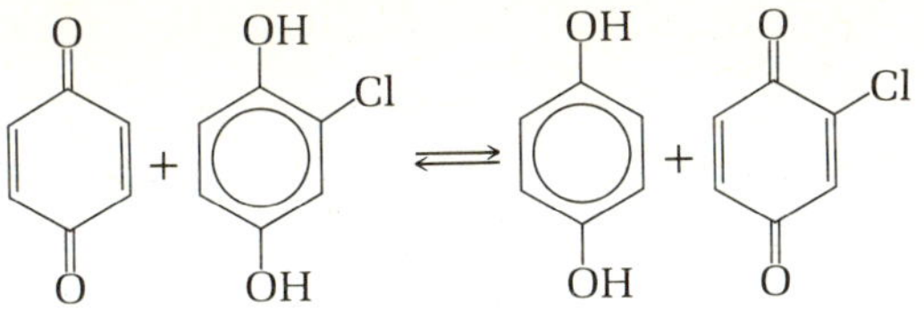

d. 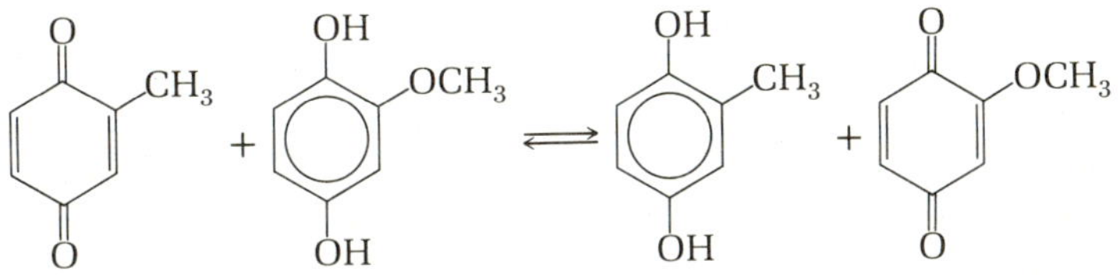

e. 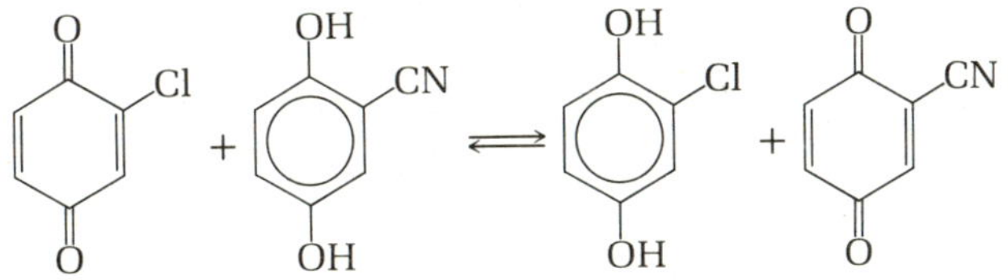

f. 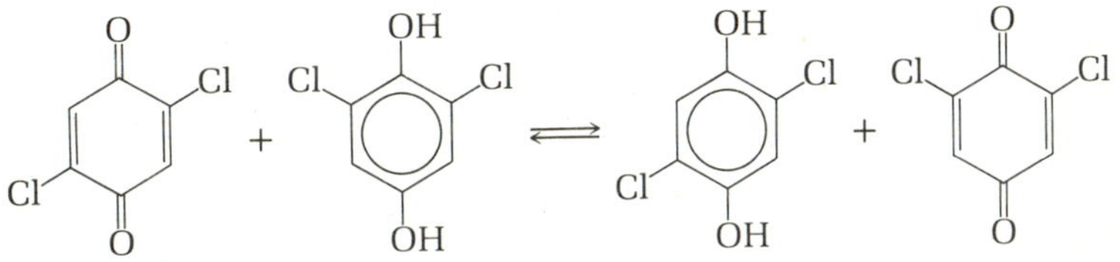

g. 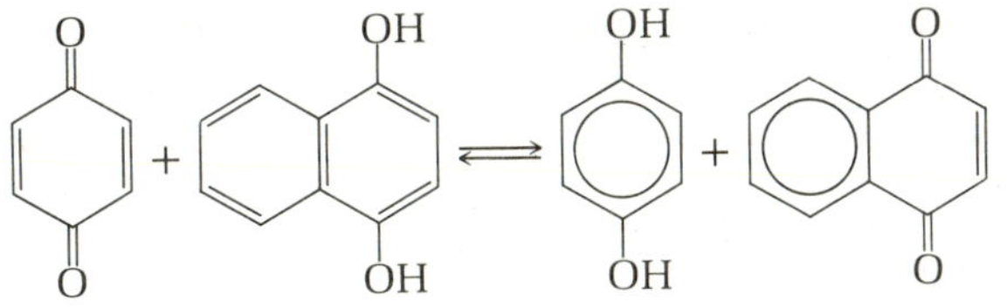

h. 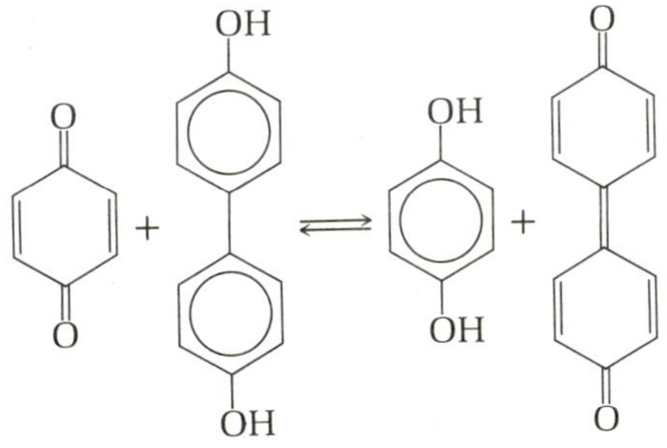

16 *trans*-,7,8-Diphenylbenzocyclobutene (I) reacts with tetracyanoethylene to give II in the stereospecific manner shown. Kinetically, the reaction is first order in I and *zero order* in tetracyanoethylene.

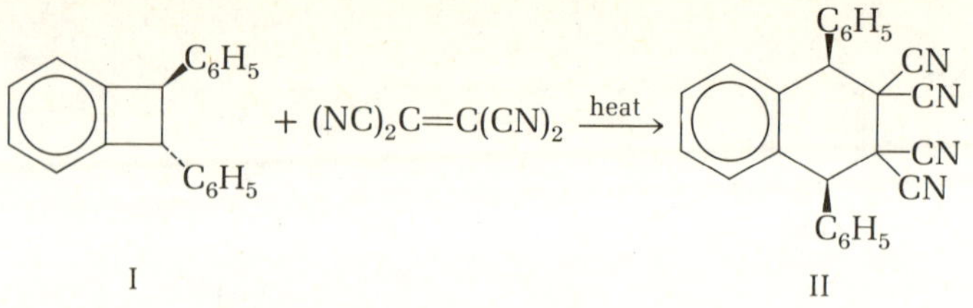

(3)

Propose a mechanism that is consistent with these facts. What is the driving force for the reaction?

SUPPLEMENTARY READING

H. J. Shine, *Aromatic Rearrangements*. New York: Elsevier Publishing Co., 1967, Chapter 2, "Claisen rearrangement."

J. F. Bunnett and R. E. Zahler, "Aromatic Nucleophilic Substitution Reactions," *Chemical Reviews*, **49**, 273 (1951).

R. W. Hoffmann, *Dehydrobenzene and Cycloalkynes*. New York: Academic Press, 1967.

L. F. Fieser and M. Fieser, *Organic Chemistry*. Boston, Mass: D. C. Heath and Co., 1944, Chapter 31, "Quinones;" *Textbook of Organic Chemistry*. Boston, Mass.: D. C. Heath and Co., 1954, Chapter 27, "Quinones."

CHAPTER 23

THE CHEMISTRY OF HETEROCYCLES

It has been traditional to treat heterocyclic chemistry on the whole as a separate branch of organic chemistry. There is some justification for this separation since there are important differences in properties between heterocyclic molecules and their carbocyclic analogues. Similarities are also very plentiful, however, and it is well to remember that the same physicochemical forces are at work in all compounds and that the mechanistic rules gleaned from the study of carbocyclic compounds will certainly serve to clarify the reactions of heterocyclic substances.

A very large number of cyclic molecules are known in which a wide variety of elements are to be found in skeletal positions. The most common heterocycles are those in which one or more atoms in the ring skeleton are nitrogen, oxygen, or sulfur. Chemical differences between these compounds and their all-carbon analogues may usually be traced to one or both of the following facts:

1 The electronegativity difference between the heteroatom and the carbon atom results in the presence of polarized bonds in the molecule.

2 The heteroatoms possess nonbonding electron pairs.

The compounds of greatest interest are those that contain five- and six-membered rings. Smaller heterocyclic rings behave as one might expect if one extrapolated from the chemistry of the related carbocycles. For instance, derivatives of aziridine, oxirane, and thiirane are very susceptible to ring-opening reactions; numerous examples are known in oxirane chemistry in particular (Chapter 10).

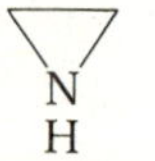
aziridine (ethyleneimine)

oxirane (ethylene oxide)

thiirane (ethylene sulfide)

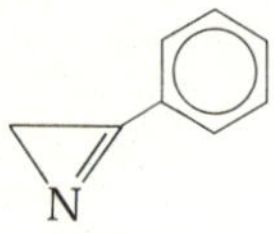
2-phenylazirine

Ring strain is substantial in all these compounds and unsaturated derivatives of aziridine alone (called azirines) have been isolated.

FIVE- AND SIX-MEMBERED SATURATED HETEROCYCLES

The chemistry of the five- and six-membered saturated heterocycles offers few surprises; these compounds resemble very strongly their acyclic analogues.

pyrrolidine

tetrahydrofuran

tetrahydrothiophene (tetramethylene sulfide)

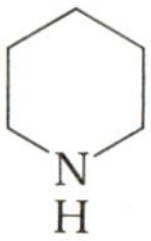
piperidine

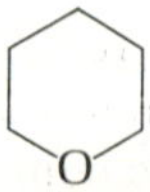
tetrahydropyran

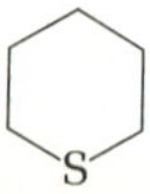
tetrahydrothiopyran (pentamethylene sulfide)

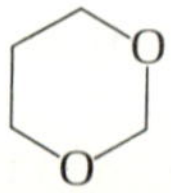
1,3-dioxane

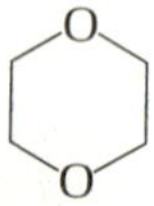
1,4-dioxane

Pyrrolidine and piperidine are typical amine bases, slightly stronger than open-chain secondary amines. Tetrahydrofuran and tetrahydropyran are representative ethers. By far the most attention recently shown these compounds has come from those interested in their conformational behavior.

The replacement of one or more atoms in a carbocyclic ring by a heteroatom will usually not bring about vast structural changes; most six-membered heterocycles still prefer the chair conformation, for instance. But the energetic balances and the conformational equilibria do not coincide between carbocycles and heterocycles and when we seek to rationalize the differences, we are forced to draw upon considerations that are usually of secondary importance with carbocycles. In particular, certain types of polar effects are much more significant in heterocyclic than in homocyclic species. The study of heterocyclic conformational analysis serves as a reminder that our conceptions of conformational behavior should not be based solely on the extremely well-studied cyclohexane derivatives.

Piperidine exists almost totally in the chair conformation at 25°. The principal question of interest with respect to this molecule is whether the proton or the nitrogen lone pair shows a marked preference for the equatorial position. The tentative answer seems to be that neither does; while there appears to be some preference for equatorial hydrogen **(2)**, both conformers **1** and **2** are present in appreciable quantity.

(23.1)

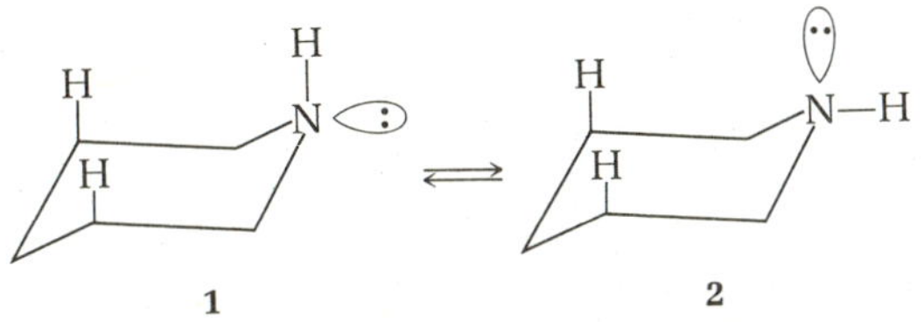

This finding, if confirmed, would seem to indicate that the 1,3-diaxial interactions between the axial protons and the nitrogen lone pair (as in **2**) are at least comparable to the proton-proton diaxial repulsions (as in **1**). Interestingly, the conjugate acid of tetrahydrothiopyran seems to favor heavily the conformation in which the proton is axial and the lone pair is equatorial. Sulfur lone pairs are probably much "larger" and more diffuse than those on nitrogen.

(23.2)

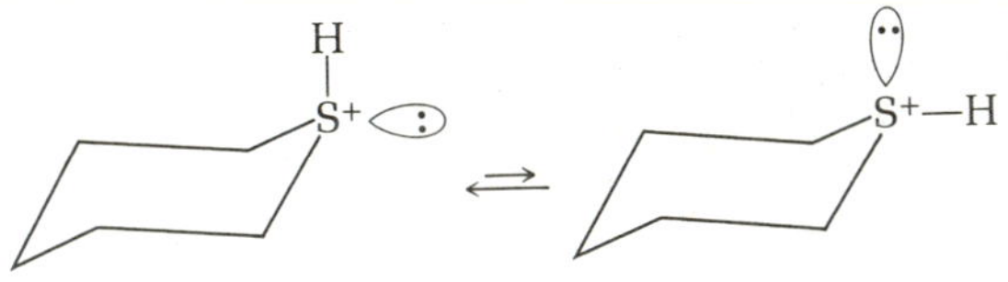

The most readily available tetrahydropyrans are those obtained by acid-catalyzed addition to dihydropyran. Surprisingly, it was found that in 2-alkoxy and 2-acyloxytetrahydropyrans the substituent prefers the *axial* position.

(23.3)

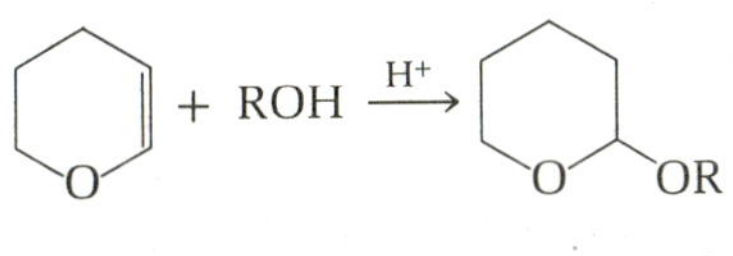

The result has been rationalized in the following way. When the alkoxy group is equatorial, its dipole is roughly perpendicular to the dipole of the C—O—C ring atoms. If the alkoxy group migrates to the axial position, then the two dipoles tend strongly to cancel one another, which is an energetically favorable occurrence (equation 23.4). 2-Halotetrahydropyrans also display a marked preference for an axial substituent.

(23.4)

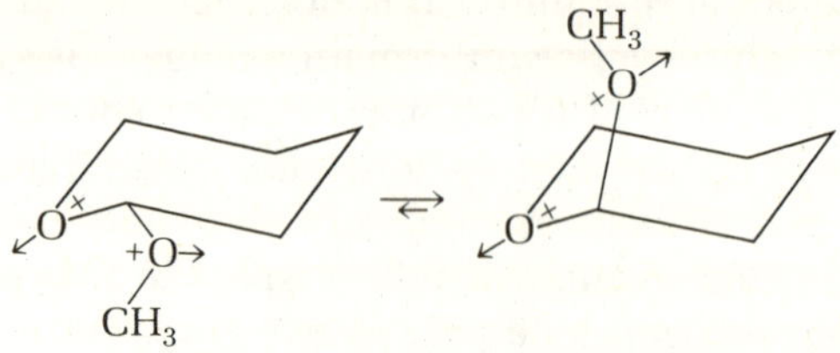

The dipolar effect, if real, must be of substantial magnitude because it overrides the 1,3-diaxial interactions that promote an equatorial arrangement for the alkoxy group. Support for the dipolar interaction hypothesis is provided by studies of the solvent effect on conformational preferences — the axial conformer is most abundant in nonpolar solvents in which dipole cancellation should be most favorable. Since many sugars exist principally as substituted 2-hydroxytetrahydropyrans, the conformational problems just discussed are of great relevance to the discussion of sugar chemistry (Chapter 24).

PROBLEM

1 Conformational equilibria have also been investigated in six-membered heterocycles that contain more than one heteroatom. The 1,3-dioxanes, in particular, have been extensively studied. For 5-methyl-1,3-dioxane the free energy change for the equilibrium shown in equation 1 amounts to about −0.8 kcal/mole.

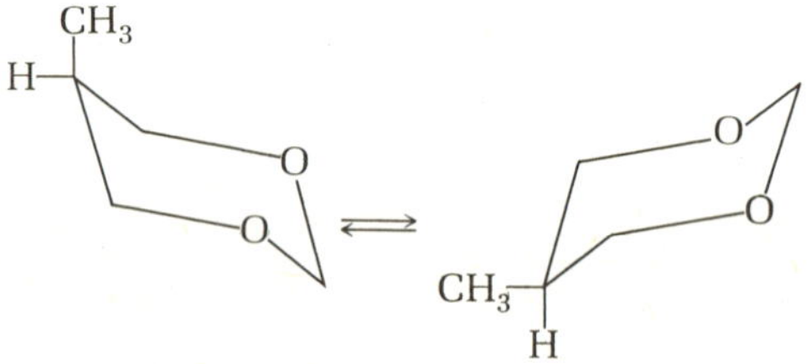

(1)

With methylcyclohexane the analogous free energy of equilibrium equals −2.7 kcal/mole. What accounts for the sizeable difference in conformational free energies between the heterocycle and the carbocycle?

FIVE-MEMBERED AROMATIC MONOHETEROCYCLES: ACIDITY AND BASICITY

The most important five-membered unsaturated heterocycles are furan, pyrrole, and thiophene.

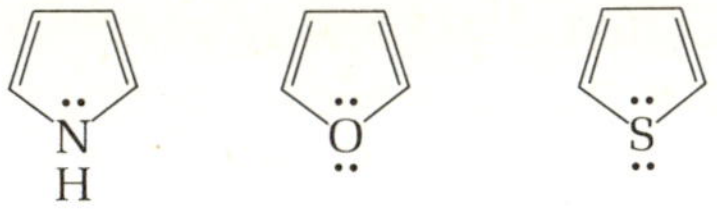

pyrrole furan thiophene

Numerous polyheterocyclic compounds containing five-membered rings are known; we shall concentrate our attention on two of them, imidazole and thiazole. In the simple rings the heteroatom is always given the number one; in the polyheterocyclic systems the atom of highest atomic number is given the number one, and the rest of the molecule numbered in such a way as to place the other heteroatoms at the lowest numbered positions.

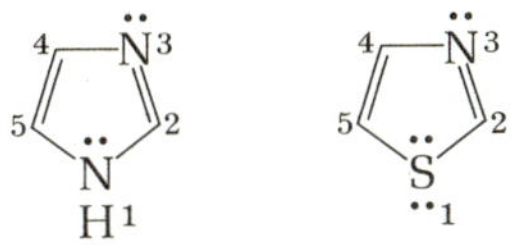

imidazole thiazole

These compounds all contain six π-electron systems in which one of the heteroatoms contributes two electrons to the sextet. Aromatic behavior is therefore to be expected of them and this expectation is borne out by a good deal of their chemical behavior. They are not so highly aromatic as benzene, however; for one thing, the π-electron density will not be symmetrically distributed. Examination of the canonical forms for these heterocycles reveals that there is a net drift of electron density away from the heteroatom and toward the carbon atoms.

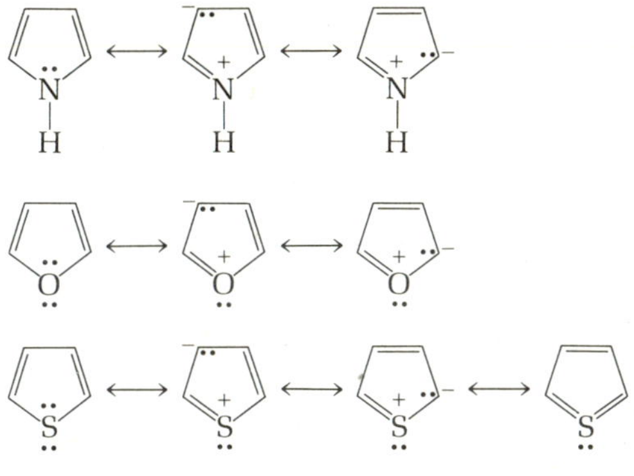

Due to the availability of the *d*-orbitals of sulfur, thiophene may be represented by a canonical form without charge separation in which sulfur contains 10 electrons in its valence shell. Calculations indicate that thiophene has the greatest resonance energy of any of the monoheterocyclic aromatics, as much as 80% of the resonance energy of benzene, while furan is the least stable and has less than 50% of the resonance energy of benzene. The relative instability of furan is understandable because electron delocalization in this molecule requires the removal of electron density from the highly electronegative oxygen atom.

Electron delocalization in the aromatic heterocycles manifests itself in a variety of ways. Unlike common sulfides (Chapter 18), thiophene will not form sulfonium salts, except when treated with very powerful alkylating agents such as trimethyloxonium fluoborate (p. 235). When sulfur in a thiophene ring is deprived of both its nonbonding electron pairs, then the ring loses its characteristic stability; thiophene 1,1-dioxide is a highly reactive molecule that cannot be isolated because it undergoes Diels-Alder dimerization.

(23.5) 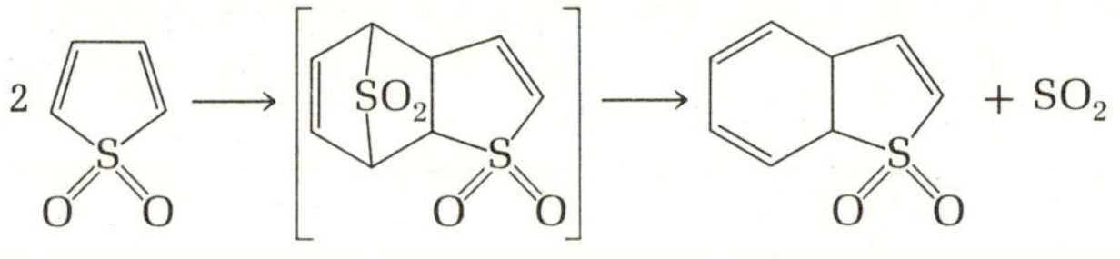

Pyrrole shows none of the characteristics of a typical secondary amine. It is actually less basic than water, in which it is only sparingly soluble. Under neutral or basic conditions pyrrole seems often to react by way of its anion, which is generated by proton removal from the nitrogen atom.

(23.6) 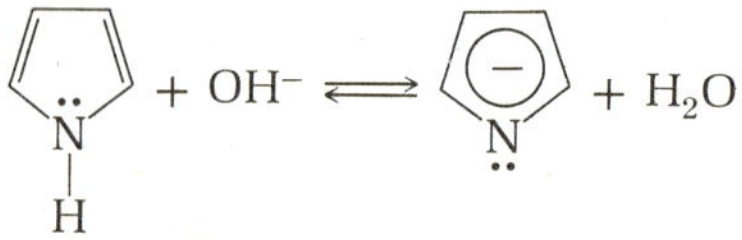

Imidazole, on the other hand, is a weak base with a pK_b of 7. The site of protonation of imidazole is *not* N-1, the electron pair of which is part of the aromatic ring; it is rather N-3, which possesses an electron pair that is orthogonal to the π-system (Figure 23-1).

(23.7) 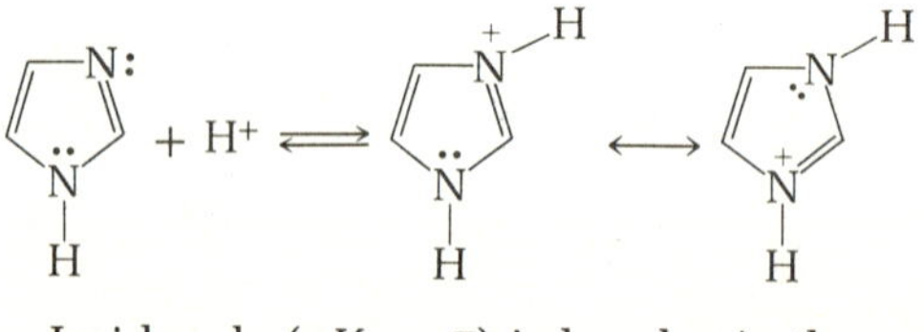

Imidazole ($pK_b = 7$) is less basic than are ordinary alkyl amines ($pK_b \sim 4$) because the electron pair being protonated is in an sp^2-orbital rather than an sp^3-orbital. Imidazole is more basic than amines such as pyridine, how-

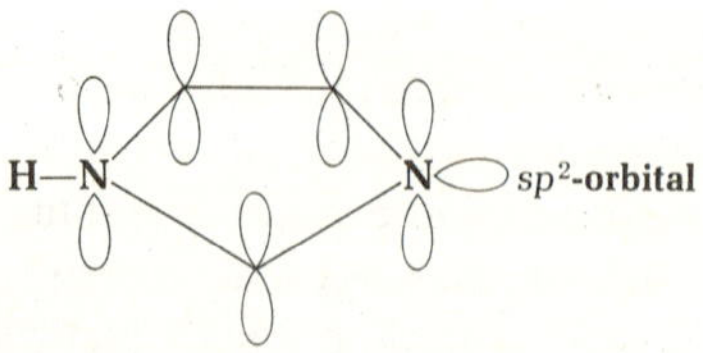

FIGURE 23-1
Atomic *p*-Orbitals of Imidazole

ever, which are also protonated on an sp^2-orbital. The reason for this latter difference is that the imidazolium cation is a symmetrical ion in which the positive charge is equally distributed over two nitrogen atoms. As we shall see, the imidazole ring is an important nucleophile and base in numerous enzymatic systems.

ELECTROPHILIC SUBSTITUTION OF FIVE-MEMBERED HETEROCYCLES

In accord with our predictions of aromatic properties for the five-membered unsaturated heterocycles, they have all been found to undergo electrophilic substitution. In the monoheterocycles there are only two distinguishable positions for monosubstitution, the 2 and 3 positions (often called the α- and β-positions). Almost without exception it is found that substitution at the 2 position is very much faster than the corresponding reaction at the 3 position. A good example is provided by acid-catalyzed deuterium exchange in thiophene (equation 23.8).

(23.8)

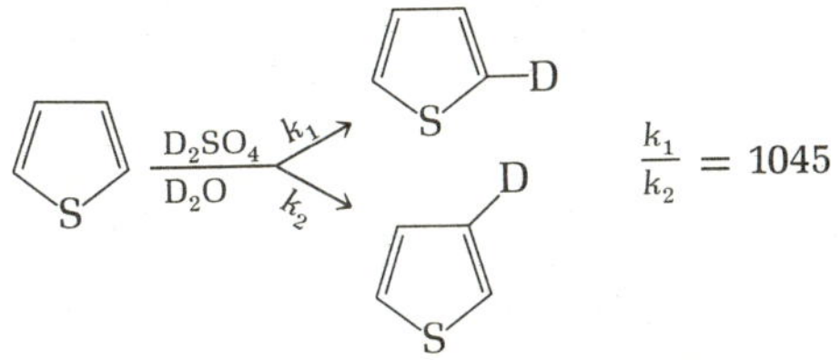

The preference for 2 substitution becomes more plausible if we examine the intermediate cations that are produced in the course of the reaction.

(23.9)

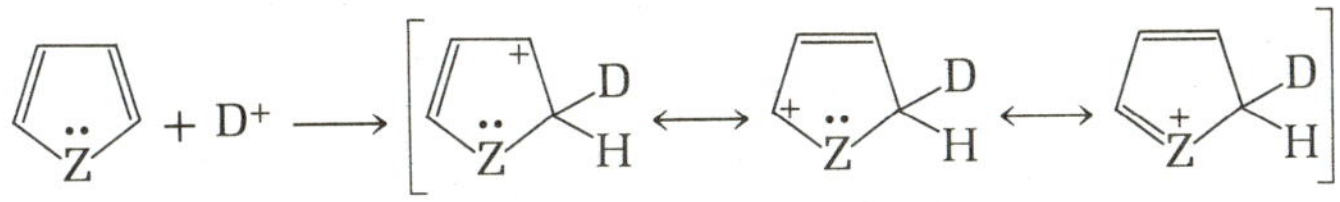

Z = N—H, O, S

3

(23.10)

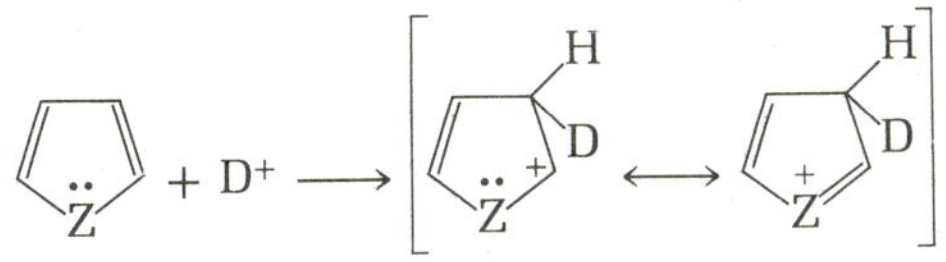

4

The intermediate arising from attack at C-2 **(3)** is capable of greater charge delocalization than is the intermediate produced by attack at C-3 **(4)**.

Due no doubt to their lower resonance energies, the aromatic heterocycles are more reactive than benzene is toward electrophilic substitution (equation 23.13) and the use of acid catalysts is often not necessary. Since the rings are susceptible to degradation, it is advantageous to use the mildest possible conditions.

(23.11) 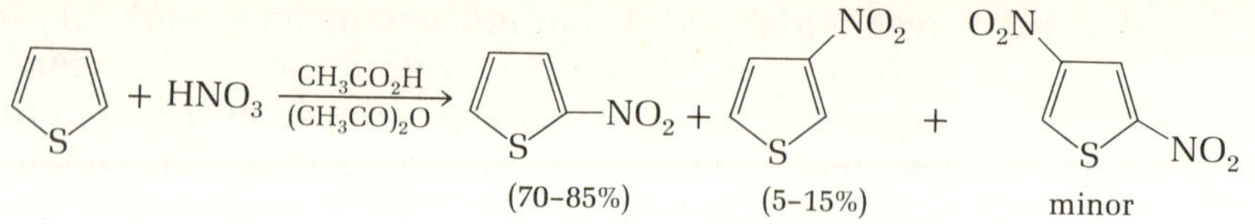

(23.12) 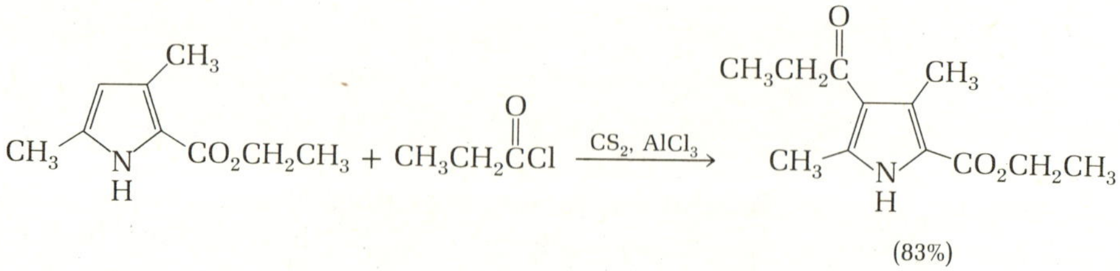

(23.13) 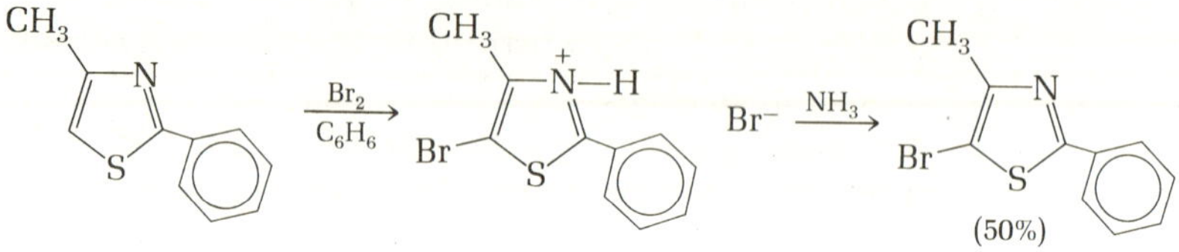

SUBSTITUENT EFFECTS

The effects of substituents on the orientation of incoming electrophiles can be understood and usually predicted with the aid of principles developed for rationalizing electrophilic substitution in benzene and its derivatives. It is important to keep in mind, however, that unlike benzene the heteroaromatic rings exert preferences of their own in addition to the influence of any attached substituents.

The heteroatoms promote reaction at the α-positions of the rings so that if one α-substituent is already present in the ring, reaction tends to take place at the free α-position. This behavior is observed with substituents that vary widely in electrical character.

(23.14) 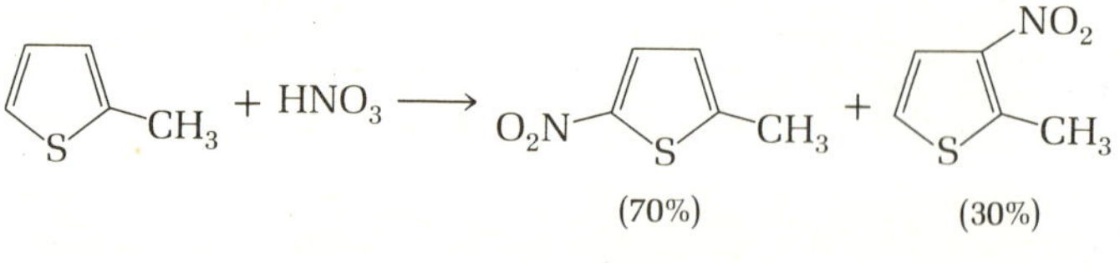

(23.15) 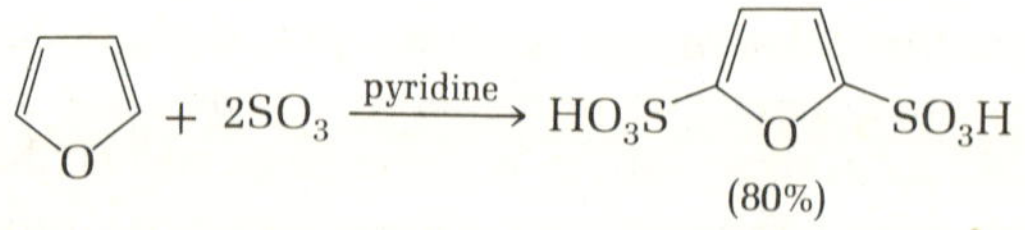

If a C-2 substituent is very powerfully electron-withdrawing, then it may sufficiently deactivate the 5 position (see canonical structures) so that substitution is encouraged at C-4 (equations 23.16 and 23.17).

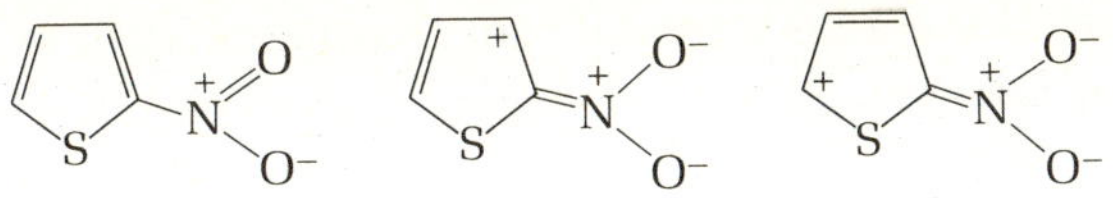

(23.16)
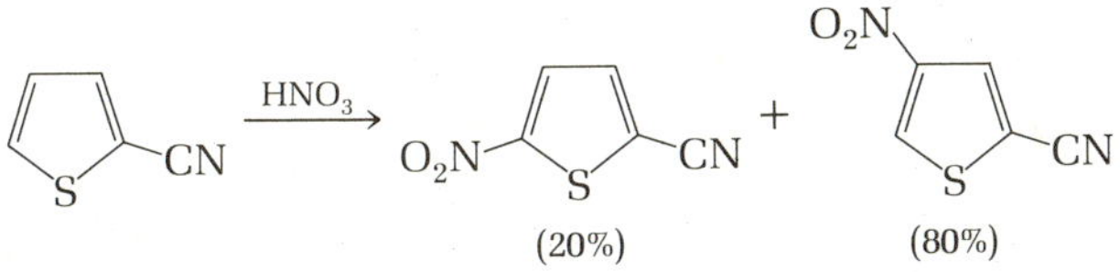

(23.17)
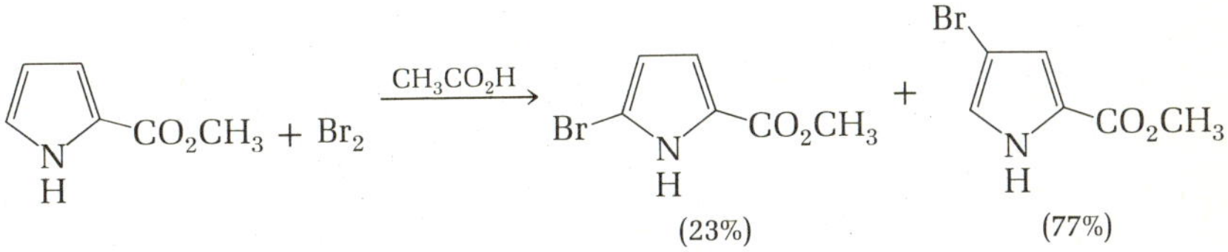

A substituent at the 3 position leaves both α-positions open for attack. If the C-3 substituent is electron-supplying (*o*,*p*-directing), then attack at C-2 will be favored; otherwise, C-5 will be the site of reaction.

(23.18)
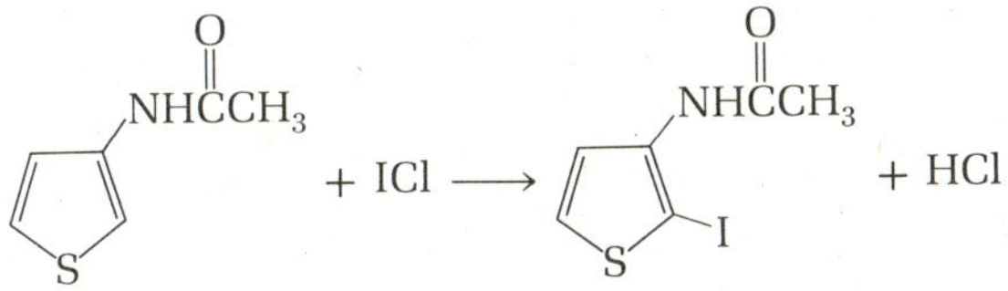

(23.19)
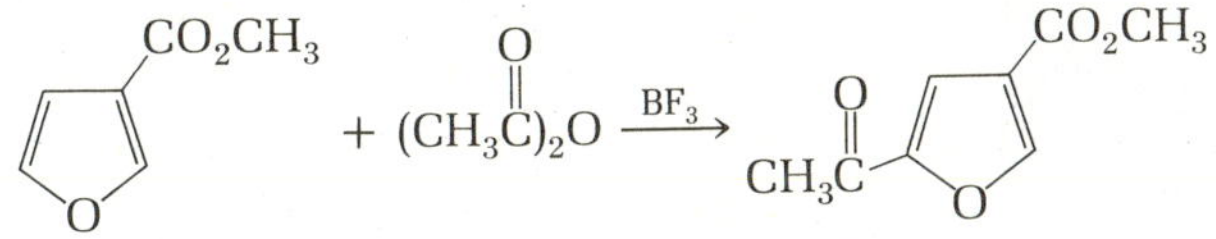

It is clear from the substitution patterns so far elucidated that it would be difficult to prepare a simple 3-substituted aromatic heterocycle. This problem has been circumvented in a clever way. Thiophene is tribromonated, and advantage is then taken of the greater reactivity of the bromines at C-2 and C-5, which are selectively removed (equation 23.20). 3-Bromothiophene may be converted to 3-thienyllithium by exchange, and the lithium compound then used to prepare a host of other derivatives.

(23.20)
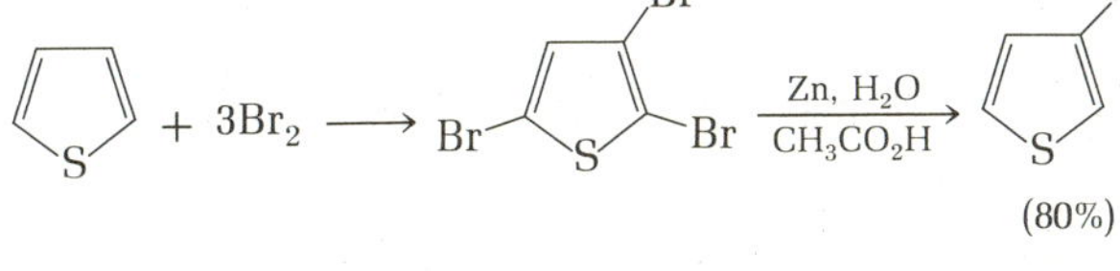

(23.21)
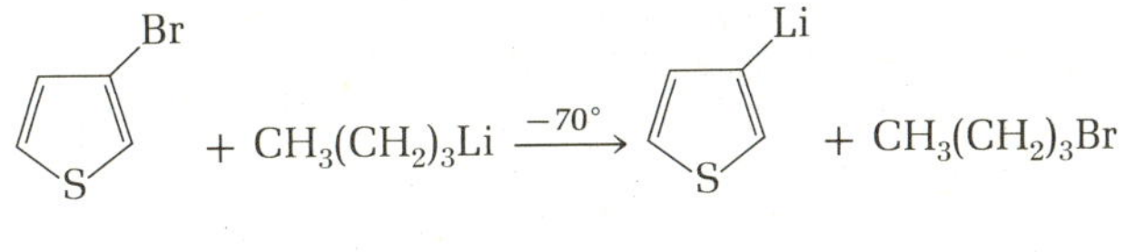

(Thienyl is the adjectival form of thiophene, just as phenyl is the adjectival form of benzene.)

ADDITION REACTIONS OF FURAN

Electrophilic attack at a five-membered heterocycle must proceed initially by way of the cyclic carbonium ion **5**. If **5** loses a proton, as does benzene-onium ion in similar circumstances, then the substitution product is obtained directly.

(23.22)

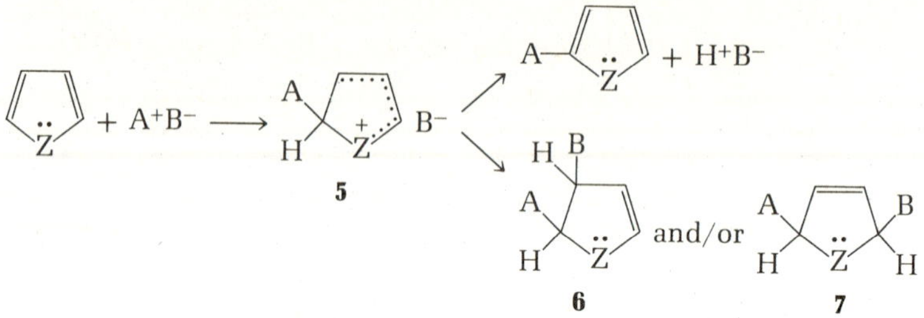

Carbonium ion **5** may also react with one or more nucleophiles present in solution to give one or both of the addition products **6** and **7**. A net substitution may occur in this case as well, if the addition product then undergoes a facile elimination (equation 23.23).

(23.23)

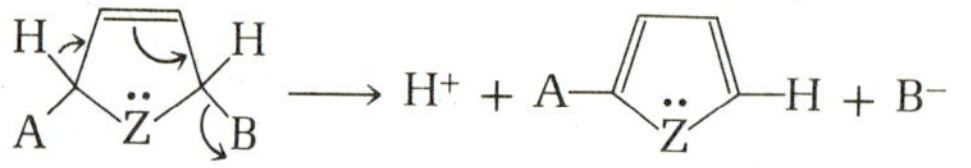

The balance between addition of a nucleophile and proton loss must depend to a large extent on the difference in resonance energy between **5** and the aromatic molecule. In the case of benzene, that difference is substantial and addition products have not been detected under ionic conditions. Among the aromatic five-membered heterocycles the resonance energies are substantially lower, particularly in the case of furan. We might expect to find addition reactions occurring with these nuclei; in fact, a large number of furan *substitutions* apparently proceed by way of *addition-elimination*, rather than by direct substitution.

The reaction of furan and several substituted furans with nitric acid in acetic anhydride leads to the precipitation of an unstable crystalline solid (**8**, equation 23.25). The intermediate may be isolated and, upon treatment with a weak base such as acetate ion or pyridine, is converted to a nitrofuran.

(23.24)

$$(CH_3\overset{O}{\overset{\|}{C}})_2O + HONO_2 \rightleftharpoons \underset{\text{acetyl nitrate}}{CH_3\overset{O}{\overset{\|}{C}}ONO_2} + CH_3CO_2H$$

(23.25) 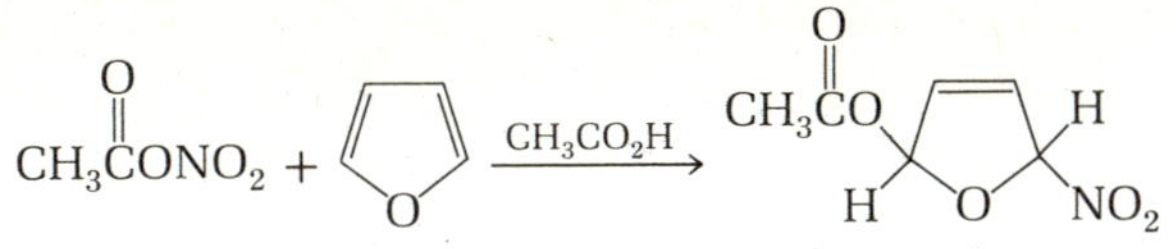

8

(23.26) 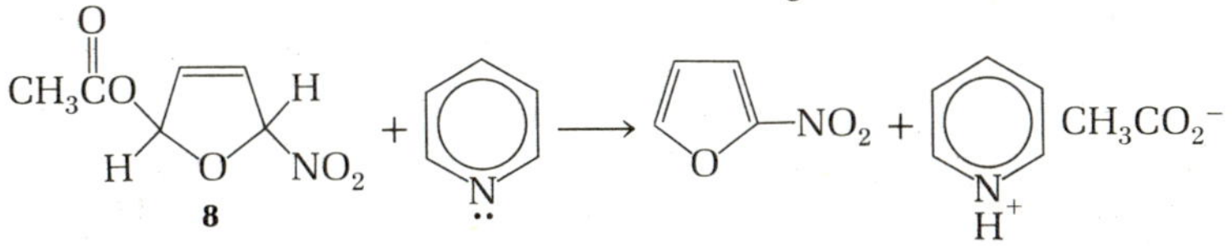

The relatively diminished aromaticity of furan is also highlighted by its capacity to act as a diene in Diels-Alder reactions. Neither thiophene nor pyrrole are active as dienes except under forcing conditions.

(23.27) 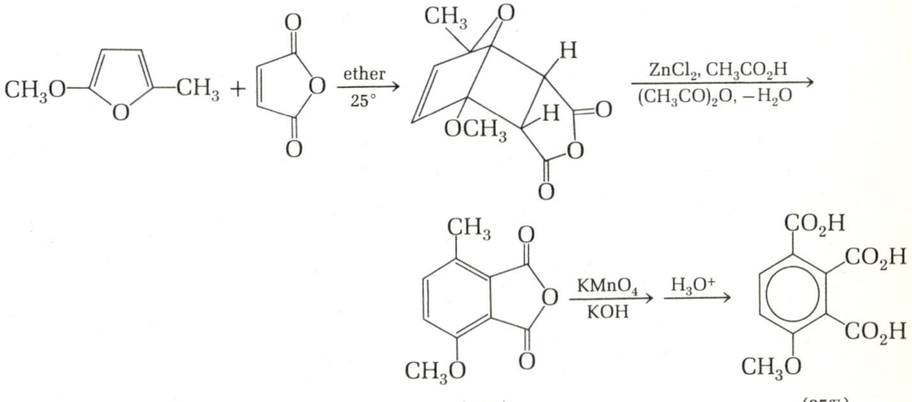

(48%) (85%)

PROBLEM

2 There is a great structural similarity between the Diels-Alder adducts of furan and those of cyclopentadiene, . The cyclopentadiene adducts are considerably more stable, however; the furan adducts suffer retro Diels-Alder cleavage to starting materials at relatively low temperatures. To what may this difference in stability be ascribed?

FUSED-RING HETEROCYCLES

Compounds consisting of heterocyclic rings fused to carbocyclic ones are abundant and well-known: indole (2,3-benzopyrrole), thionaphthene (2,3-benzothiophene), and coumarone (2,3-benzofuran). Dibenzo heterocycles have also been prepared and their chemistry has been studied.

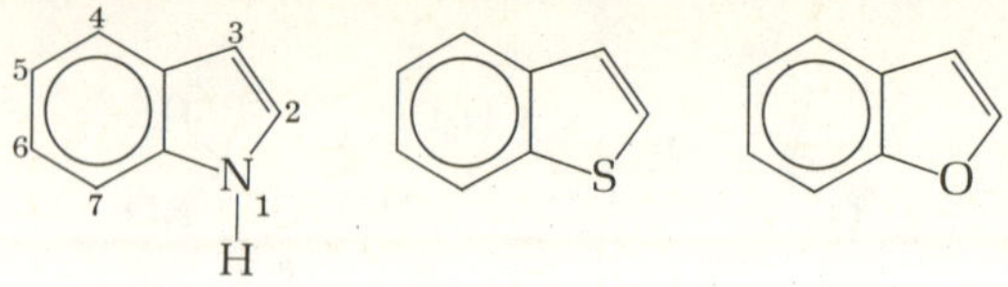

indole thionaphthene coumarone

Of all these ring systems, indole is far and away the most important because of the physiological activity of so many of its derivatives.

Electrophilic substitution of indole takes place almost exclusively at the 3 position.

(23.28)

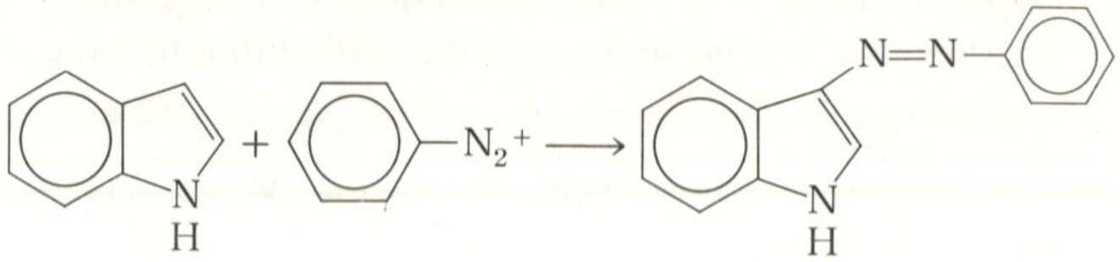

(23.29)

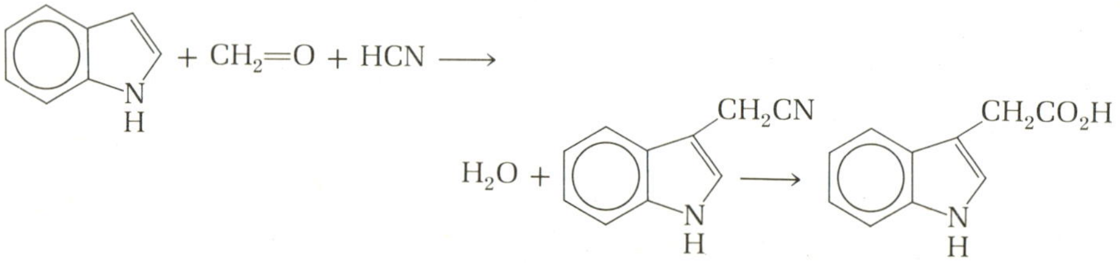

3-indoleacetic acid

The presumed greater stability of intermediate **9** over intermediate **10** is ascribed to the fact that charge delocalization may be accomplished in **9** without disturbing the aromaticity of the benzene ring, while that is not possible in **10.**

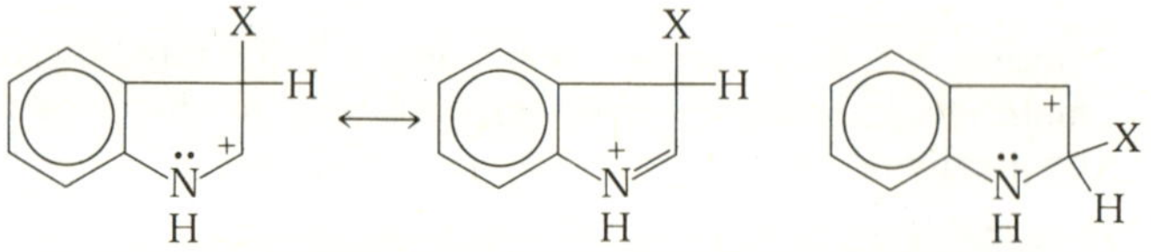

9 **10**

The ubiquity of physiologically important indole derivatives may be suggested by the following sample. 3-Indoleacetic acid (*heteroauxin*, equation 23.29) is an important growth factor in seeds. Tryptophan is one of the naturally occurring α-amino acids, and skatole is its physiological decomposition product, reknowned for its penetrating fecal odor.

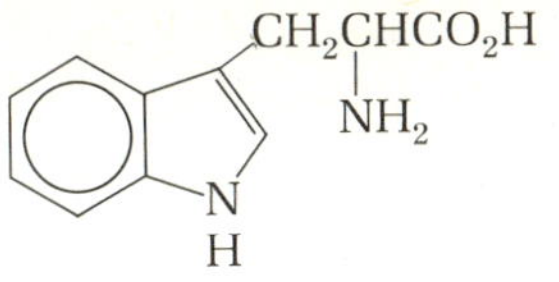

tryptophan

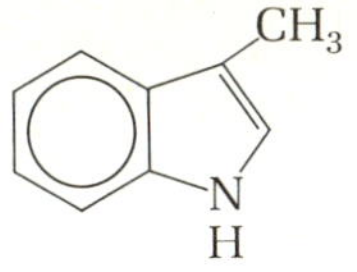

skatole

Yohimbine is one member of an important subclass of the indole alkaloids that are quite numerous. Certain members of the yohimbine family are very active in the brain and nervous system. The same is true of lysergic acid, which is a constituent of several venerable homeopathic medicines. Its diethylamide is of course notorious and needs no further publicity.

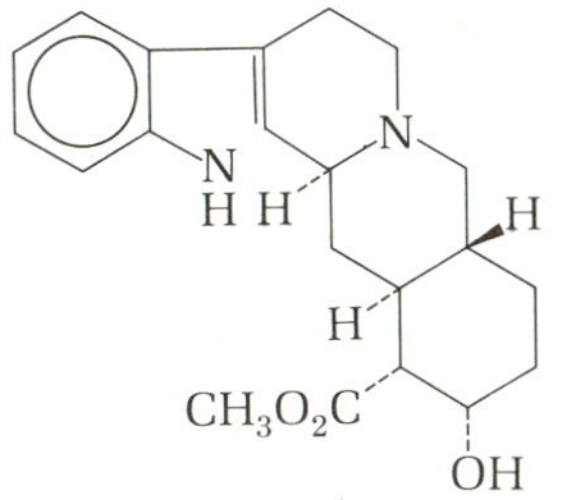

yohimbine

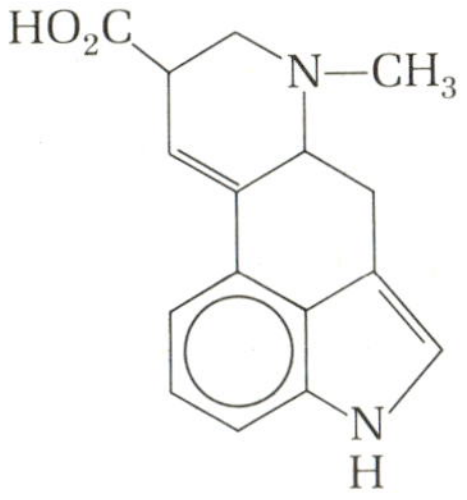

lysergic acid

The most general method of obtaining substituted indoles is the *Fischer indole synthesis*, for which the phenylhydrazones of aldehydes and ketones constitute the starting materials. Either Lewis or proton acid catalysts are usually necessary to effect the transformation (equations 23.30 and 23.31).

(23.30)

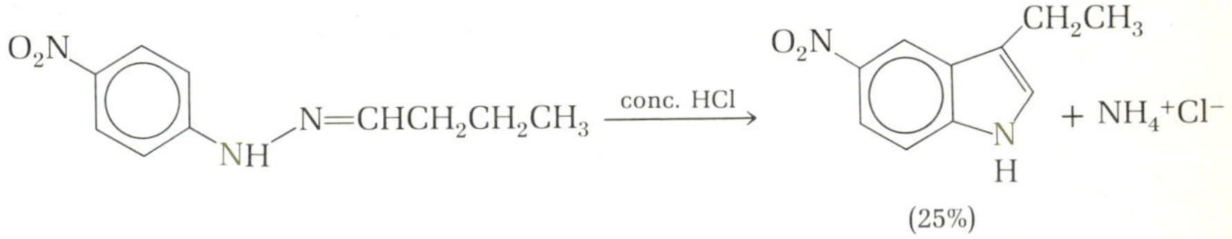

(23.31)

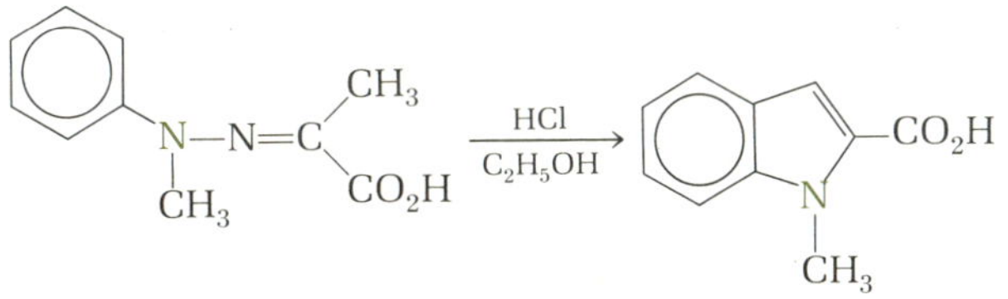

The mechanism has been fairly well established and encompasses a tautomeric equilibrium between the hydrazone and its hydrazo isomer, followed by six-center, six-electron cyclic rearrangement favored in part by the scission of the weak N—N bond (equation 23.32).

(23.32)

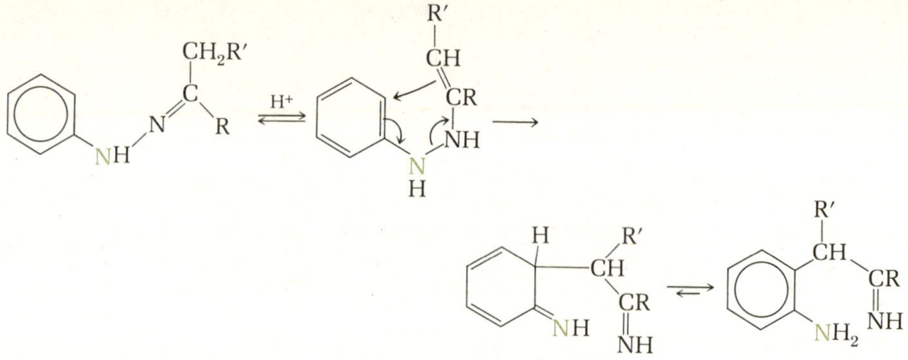

The rearrangement strongly resembles the Claisen rearrangement (Chapter 22). After the rearrangement there is an internal cyclization with elimination of ammonia.

(23.33)

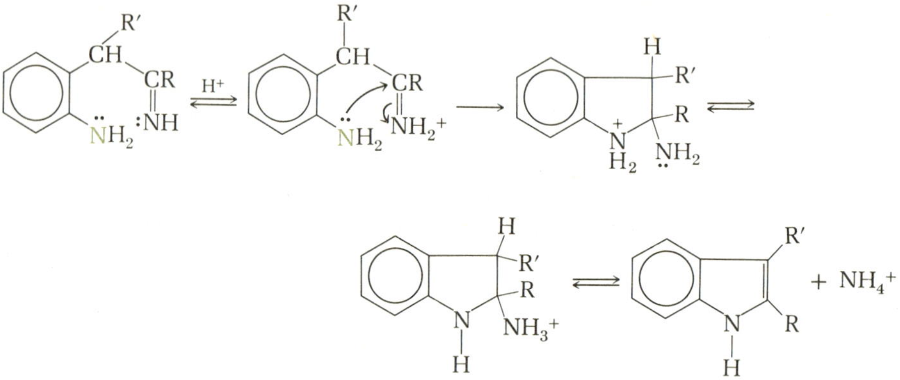

Equations 23.34 and 23.35 illustrate the synthesis of β-(3-indolyl)propionic acid, an intermediate in the total synthesis of lysergic acid. Note in particular (1) the reaction of diazonium ions with readily enolizable compounds other than phenols, and (2) the facile thermal decarboxylation of the intermediate **11.**

(23.34)

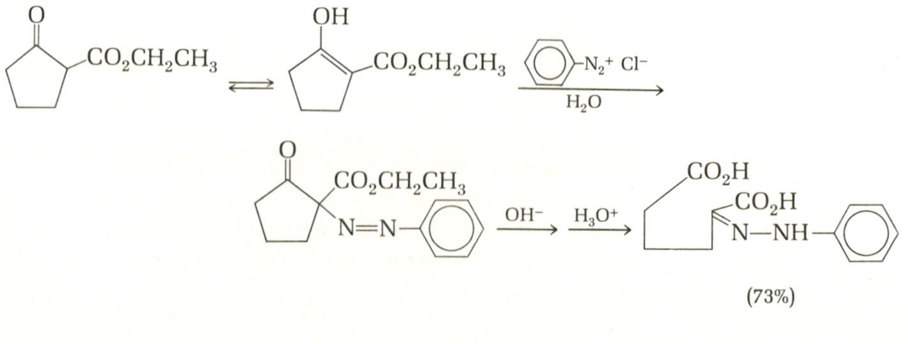

(23.35)

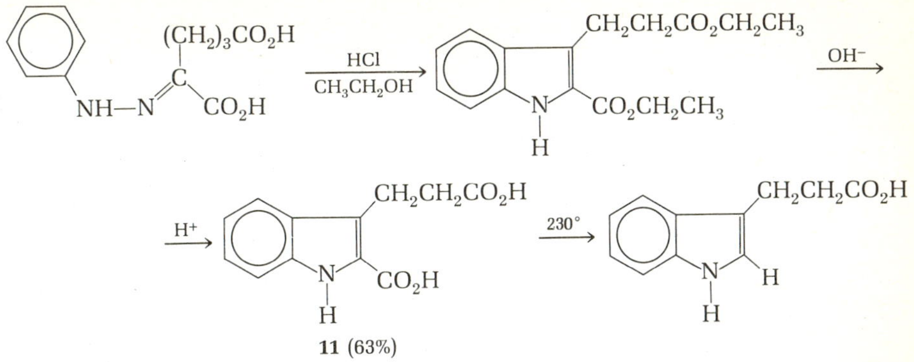

All five-membered heterocyclic 2-carboxylic acids are prone to decarboxylate easily, possibly by way of the mechanism depicted in equation 23.36.

(23.36)

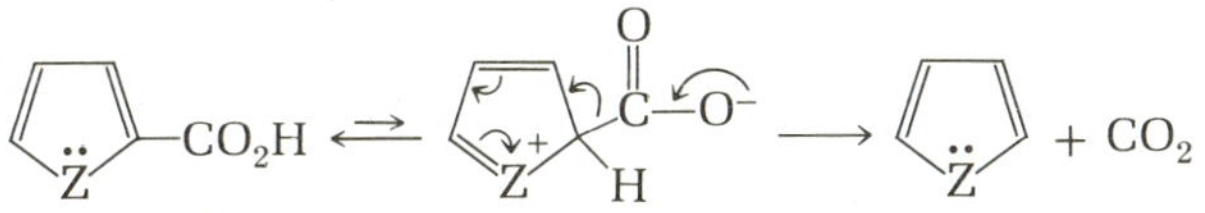

AVAILABILITY OF FIVE-MEMBERED HETEROCYCLES

Pyrrole and thiophene may be found in coal tar and in addition pyrrole may be obtained from several plant sources. Furan is produced commercially by decarbonylation of furfural (2-furanaldehyde), which is in turn obtained from the acid degradation of oat hulls and corncobs. These substances contain sizable quantities of five-carbon sugars that are dehydrated to furfural. Furan serves as the source of the considerable quantities of tetrahydrofuran produced each year.

(23.37)

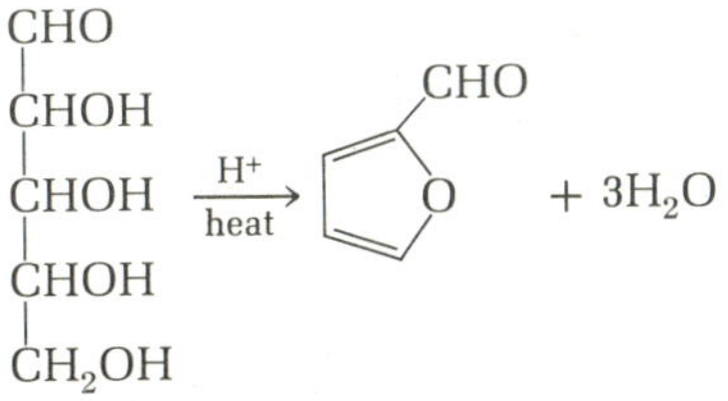

Alkyl-substituted five-membered aromatic heterocycles are usually synthesized directly in the laboratory rather than by substitution of the parent ring. 1,4-Dicarbonyl compounds are invaluable starting materials for these syntheses. They may be converted to furans, pyrroles, or thiophenes upon dehydration, reaction with a primary amine or ammonia, or reaction with a phosphorus sulfide.

(23.38) 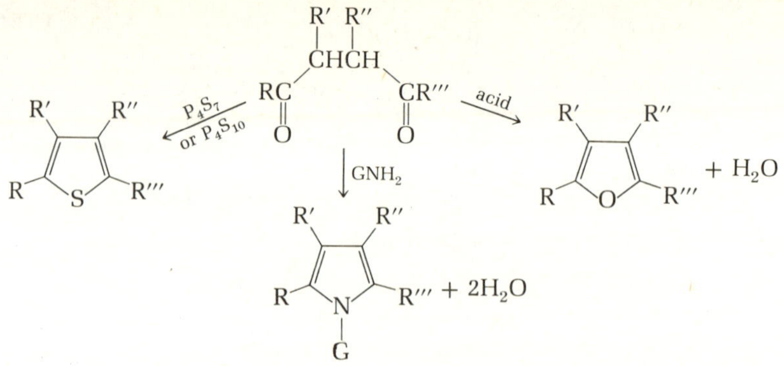

(23.39) 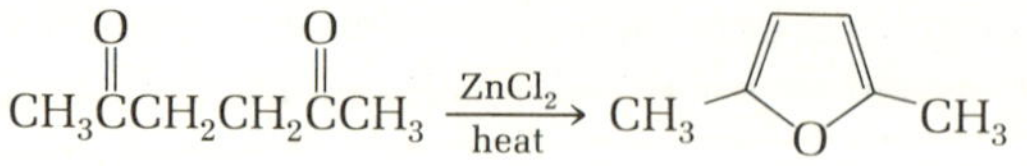

(23.40) 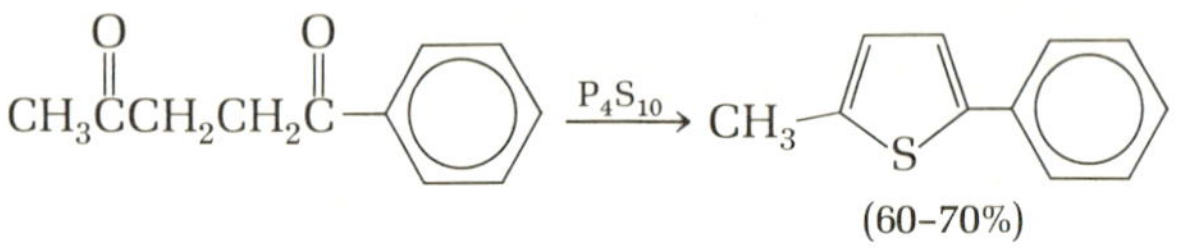

CORRINS AND PORPHYRINS

There are several very important ring systems that incorporate pyrrole and pyrroline nuclei as their basic components. Two of these are the **corrin** and **porphin** ring systems. Porphins that carry substituents on the perimeter are known as **porphyrins.** Corrins and porphyrins usually act as ligands for metal cations.

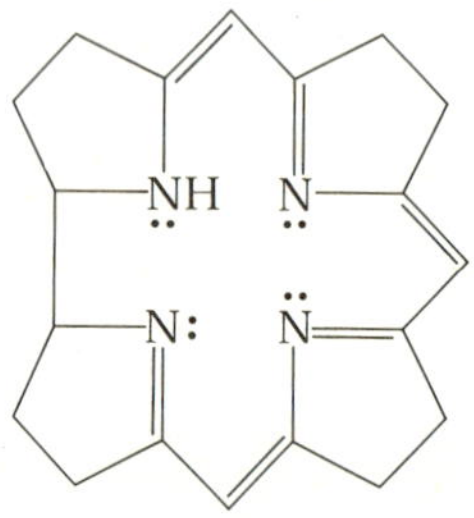

corrin

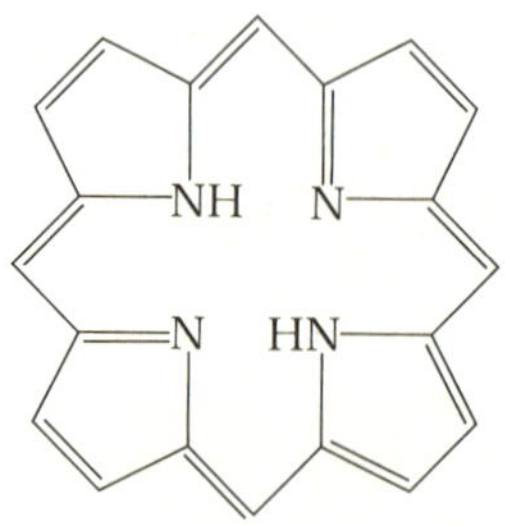

porphin

In Nature, corrins are always found complexed with cobalt ions. Vitamin B_{12} is a striking example of such a complex (Figure 23-2). The cobalt(III) ion is formally σ-bonded to only one of the nuclear nitrogen atoms, while it is complexed with the other three. Due to electron delocalization over the ring, the extent of bonding of all four nitrogen atoms to cobalt is probably

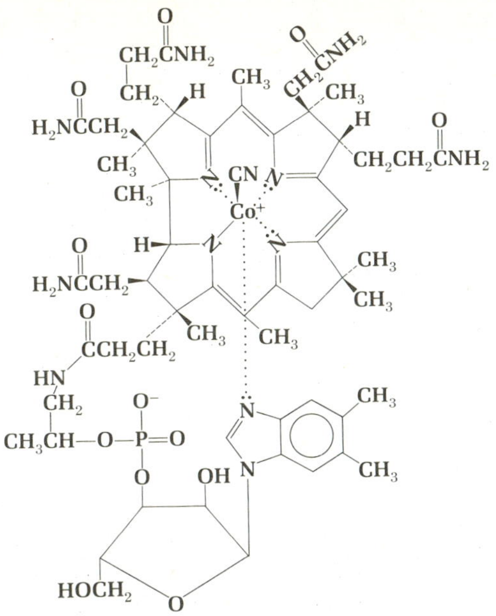

FIGURE 23-2
Vitamin B_{12}

nearly equal. Note that the cobalt atom is also coordinated to a nitrogen atom of a benzimidazole ring in the side chain.

Vitamin B_{12} is an important coenzyme for several reactions in which alteration of a carbon skeleton takes place, such as the isomerization of methylmalonyl coenzyme A to succinyl coenzyme A. This step is very important in the metabolic degradation of α-branched and odd-numbered fatty acids (Chapter 21) since methylmalonyl coenzyme A is obtained by carboxylation of propionyl coenzyme A.

$$CH_3CH(CO_2H)\overset{O}{\overset{\|}{C}}SCoA \xrightarrow[\text{vitamin } B_{12} \text{ coenzyme}]{\text{enzyme}} CH_2(CO_2H)CH_2\overset{O}{\overset{\|}{C}}SCoA \tag{23.41}$$

In its coenzyme form, vitamin B_{12} has a much more complicated group than cyanide ion as its sixth ligand; the presence of the cyanide ion is an artifact of the isolation procedure.

Heme and chlorophyll incorporate a porphyrin ring and a dihydroporphyrin (chlorin) ring, respectively, into their structures (Figure 23-3). Chlorophyll is of course the principal light-absorbing pigment in green plants that provides energy for the photosynthetic cycle. Heme is found in the blood combined with a peptide chain. Four of these heme-peptide units are held together by electrostatic bonds to form *hemoglobin,* the substance responsible for oxygen transport.

The Fe(II) in hemoglobin is six-coordinate. Four coordination sites are occupied by the nitrogen atoms of the porphyrin rings. A fifth coordinating

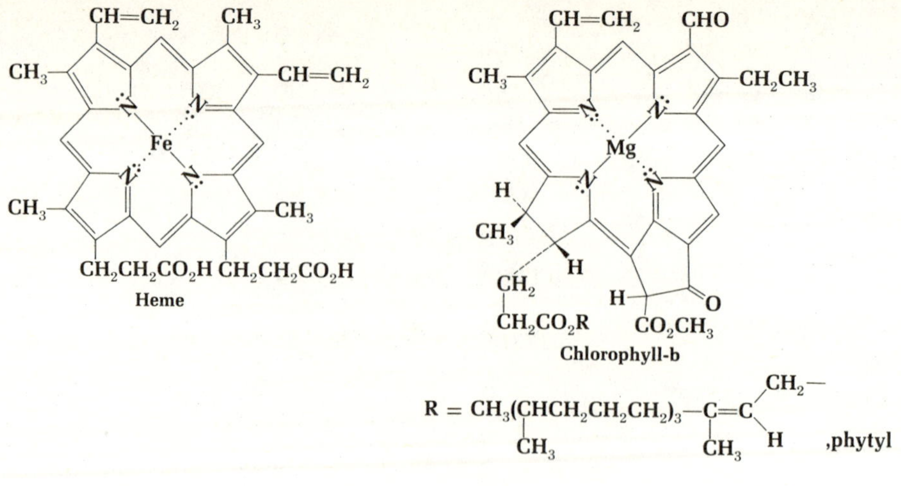

FIGURE 23-3

Heme and Chlorophyll

nitrogen atom is provided by an imidazole ring from a histidine unit in the peptide side chain. In the oxygenated form of hemoglobin (oxyhemoglobin) a molecule of oxygen is bound at the sixth coordination site (Figure 23-4). After delivering oxygen to a cell, the hemoglobin also helps to transport carbon dioxide away from the cell.

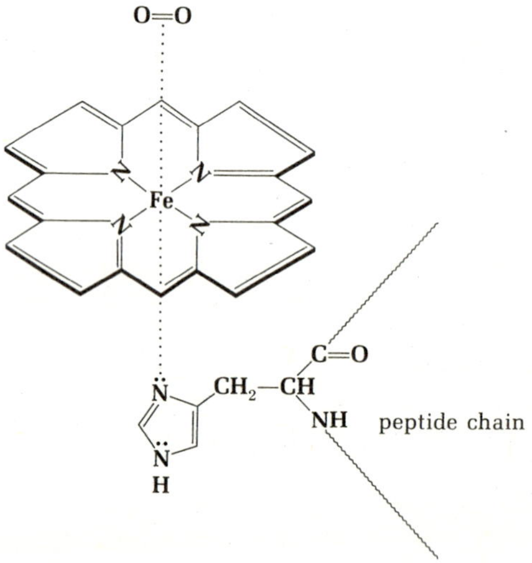

FIGURE 23-4

Partial Structure of Oxyhemoglobin

THIAZOLIUM ION CATALYSES

The thiazolium salt **12a** is an essential vitamin (vitamin B_1, thiamine) and its pyrophosphate **12b** is a coenzyme for a number of metabolic processes.

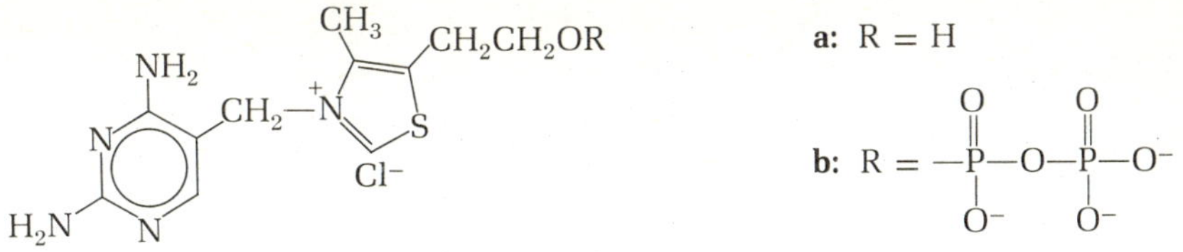

12

It serves to catalyze aldol condensation reactions and to promote the decarboxylation of pyruvic acid to acetaldehyde and its oxidative decarboxylation to acetyl coenzyme A (p. 582).

A powerful clue to the catalytic action of thiamine pyrophosphate was provided by the observation that 3,4-dimethylthiazolium bromide rapidly exchanged its C-2 proton with D_2O *in neutral solution* at room temperature. Since it is very unlikely that a deuterium ion would attack the positively charged ring, the exchange must be base-catalyzed and the ylid **13** the critical intermediate.

(23.42)

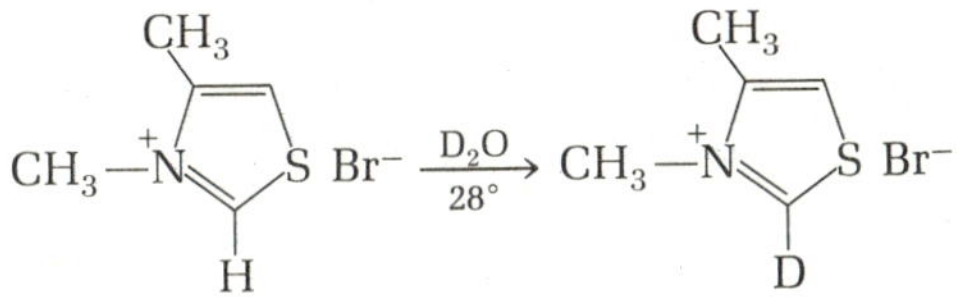

(23.43)

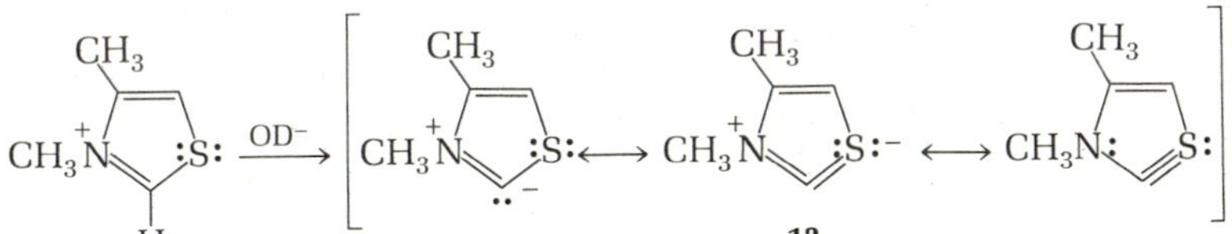

13

Sulfur plays an important part in thiazolium ion proton exchange; 1,3-dimethylimidazolium ion **(14)** exchanges much more slowly than the thiazolium ion does. Thus, the canonical form of **13** in which there are 10 electrons in the valence shell of sulfur must make a significant contribution to the description of the molecule (see formulas for thiophene, p. 631).

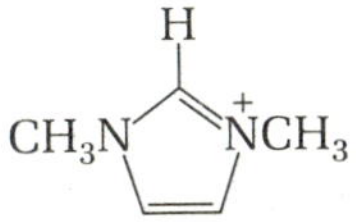

14

It has been experimentally verified that there are far fewer geometrical restrictions on carbon-sulfur multiple bonds than there are on carbon-carbon multiple bonds. This additional freedom is due both to the symmetry characteristics of the 3*d*-orbitals and also to their greater diffuseness.

It is highly likely that the ylid form of thiamine pyrophosphate is the true catalyst. The decarboxylation of pyruvic acid is illustrative; note how the positively charged nitrogen atom provides an excellent electron sink for the decarboxylation (equation 23.44).

(23.44)

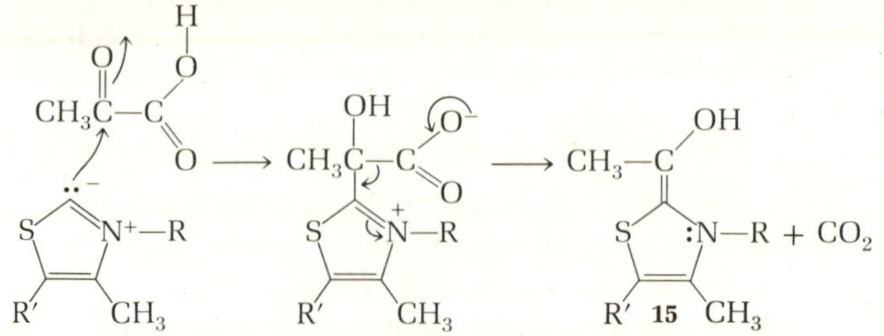

Protonation of the enamine **15**, followed by its dissociation, gives acetaldehyde and regenerates thiamine (equation 23.45).

(23.45)

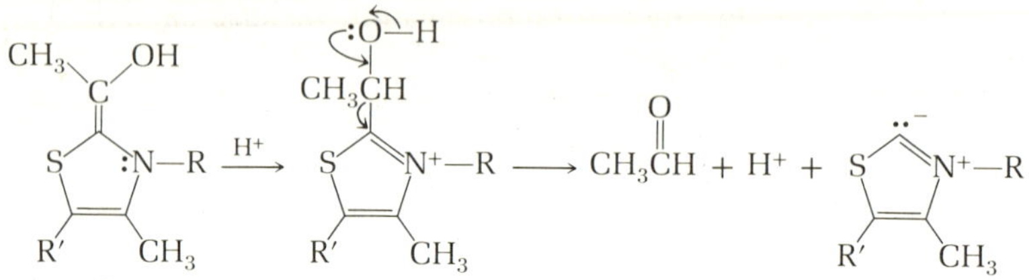

15

In the oxidative reaction, enamine **15** is oxidized by lipoic acid (equation 23.46) and the oxidation product subsequently dissociates to give thiamine and a thioacetate ester.

(23.46)

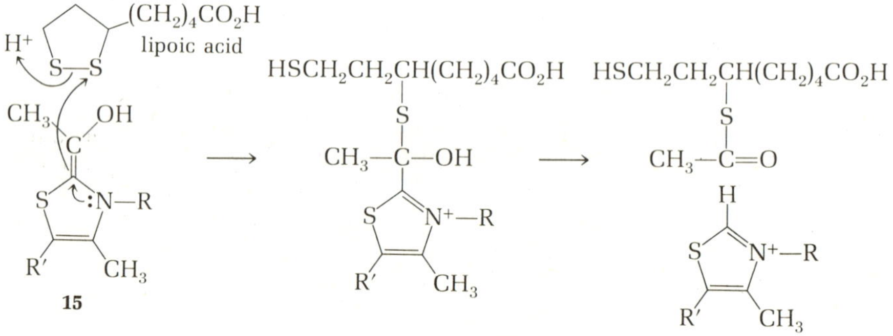

The latter then undergoes transesterification with coenzyme A, and the acetyl coenzyme A is then ready to enter the tricarboxylic acid cycle. Reduced lipoic acid is eventually oxidized back to the cyclic form.

(23.47)

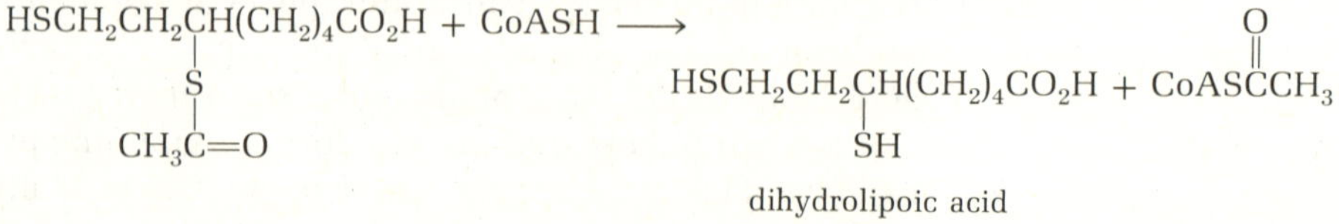

PYRIDINE AND RELATED COMPOUNDS: GENERAL CONSIDERATIONS

Of all the six-membered monoheterocyclic compounds, pyridine is by far the most important; its benzo derivatives, quinoline and isoquinoline, are also valuable substances.

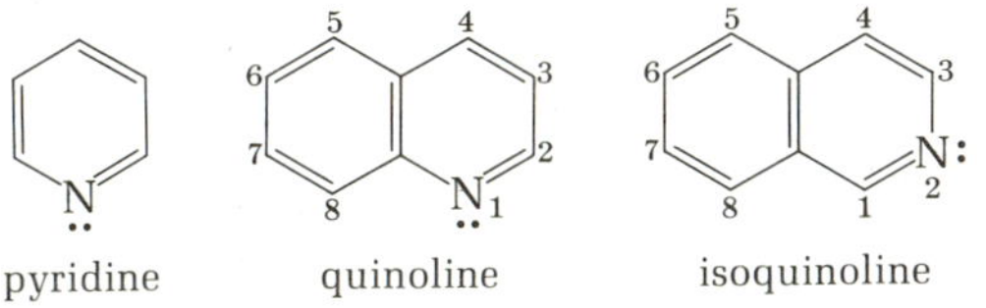

In all these compounds, nitrogen possesses a nonbonded electron pair that is orthogonal to the π-system and the molecules are therefore weak bases. They are weaker bases than trialkyl amines because the nonbonding electrons occupy an sp^2- rather than an sp^3-orbital.

The π-electron system of pyridine closely resembles that of benzene, and pyridine has a high resonance energy. The resemblance is not complete, however; since nitrogen is more electronegative than carbon, the π-electrons are less nucleophilic and, furthermore, they are not distributed equally around the ring. Those carbon atoms *ortho* and *para* to the nitrogen atom are the most depleted in π-electron density.

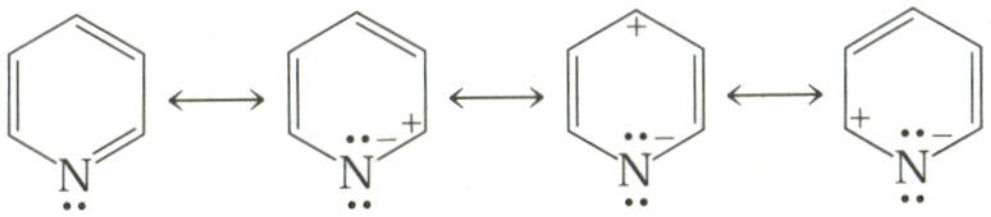

Due to its nonbonding electron pair, nitrogen may become attached to a proton or other electrophile, in the course of which it will acquire a positive charge. The π-electron system of this *pyridinium ion* will be even more polarized than before and the ring will now be quite susceptible to nucleophilic attack.

(23.48) 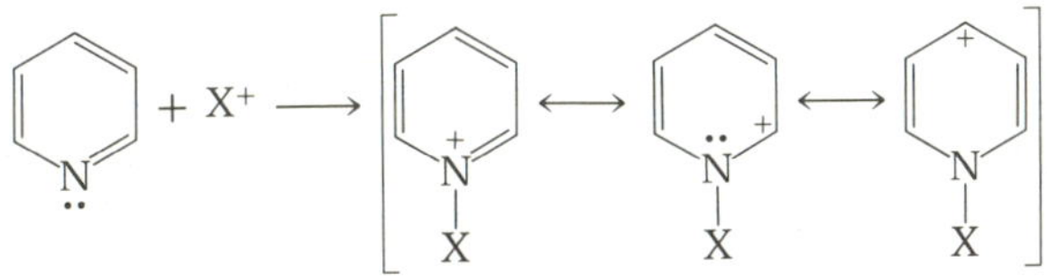

ELECTROPHILIC SUBSTITUTION OF PYRIDINE AND QUINOLINE

Pyridine is about as deactivated as nitrobenzene toward electrophilic attack. Under the conditions necessary for electrophilic substitution of a deactivated aromatic nucleus (high concentration of strong proton or Lewis acid catalyst), the ring is completely converted to a pyridinium ion, one of the most deactivated aromatic systems known. Conditions for nitration, sulfonation, and halogenation are accordingly very severe; Friedel-Crafts reactions cannot be carried out on pyridine. Substitution takes place almost exclusively at C-3 (*meta* to the nitrogen atom).

(23.49) 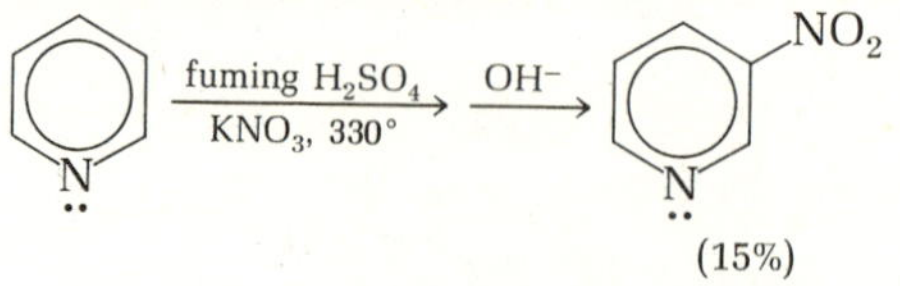

(23.50) 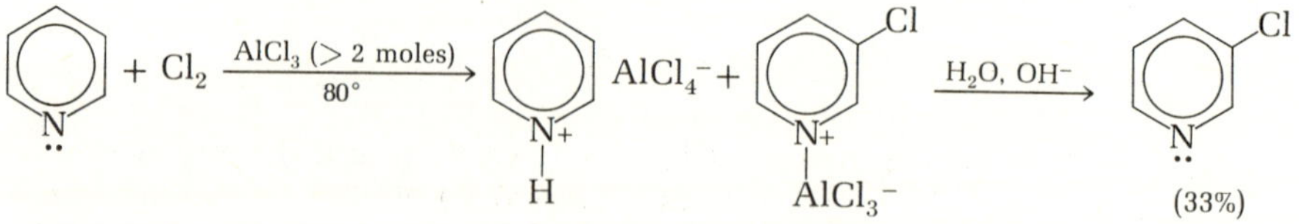

The fact that the reactivity of the heterocyclic ring can be profoundly affected by protonation of the nitrogen atom can sometimes be put to use. Bromination of quinoline in sulfuric acid, with the aid of silver sulfate, results in bromination of the benzenoid ring rather than the pyridinium ring. If the bromination is carried out in carbon tetrachloride in the presence of pyridine, which reacts with the hydrogen bromide released, then substitution takes place in the pyridinoid ring.

(23.51) 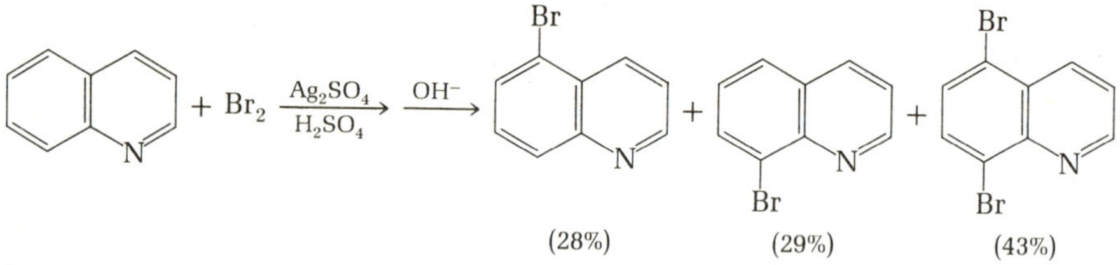

Equation 23.52 shows that quinoline is more reactive than pyridine (compare benzene and naphthalene, Chapter 15) and that the pyridinoid ring is less stable than is the benzenoid ring.

(23.52) 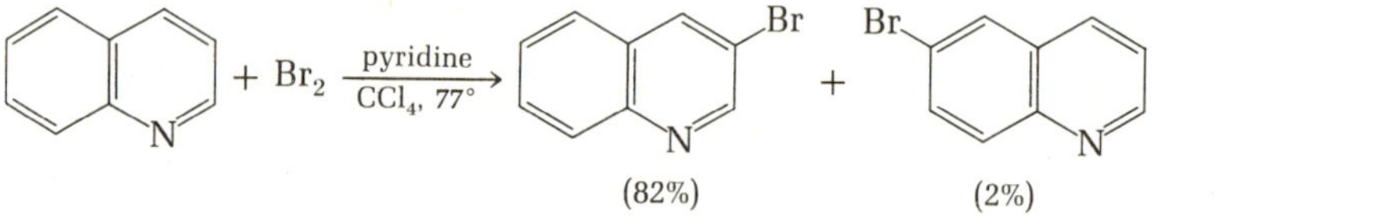

The nitrogen atom of pyridine determines the orientation of incoming substituents unless there is a very strongly activating group in the ring, in which case that group controls the substitution process. (In equation 23.53 the chlorinating agent is probably HOCl.)

(23.53) 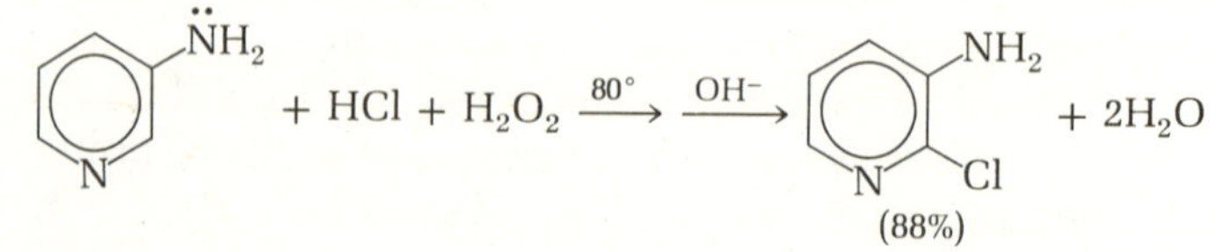

Like other tertiary amines, pyridine may be oxidized with peracids to give an N-oxide.

(23.54)

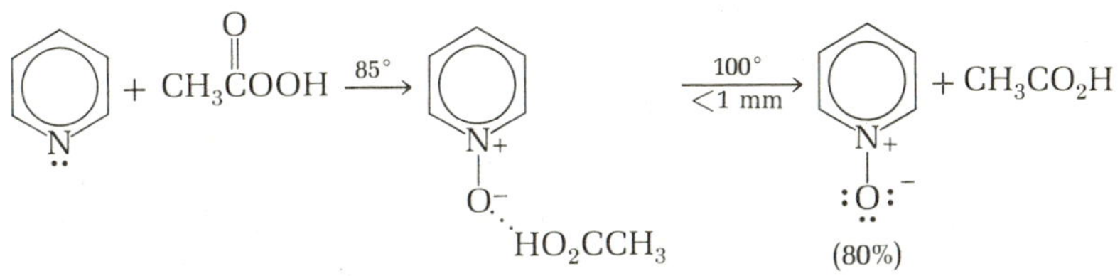

There is a formal negative charge on the oxygen atom so the oxide forms strong solvates with water and acids, and it is also a weak base ($pK_b = 13.2$). The unshared electrons on oxygen are conjugated with the ring and they have a profound effect on the pattern of electrophilic substitution. In particular, intermediates arising from electrophilic attack at C-2 or C-4 are stabilized by participation of the electrons on oxygen.

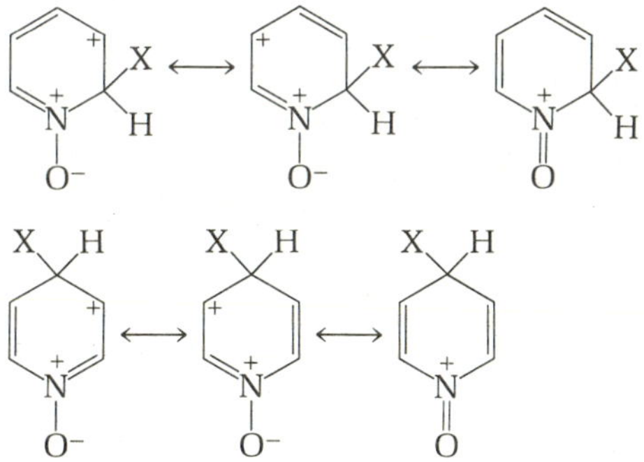

Nitration of pyridine N-oxide occurs by way of attack on the small proportion of free N-oxide present at equilibrium and the product is therefore the 4-nitro derivative. (The methylpyridines are known as picolines, which accounts for the nomenclature of the product in equation 23.56.)

(23.55)

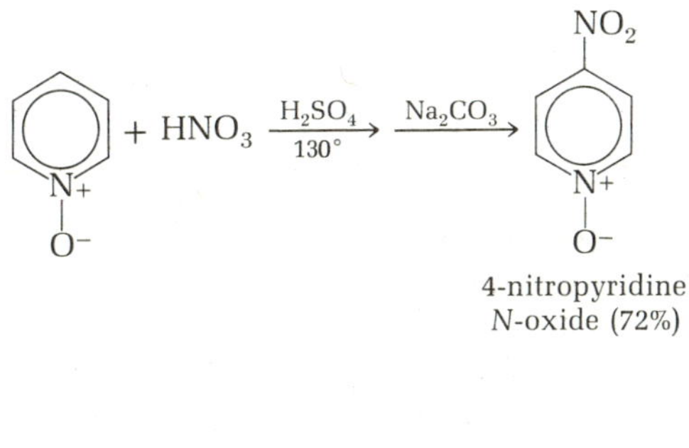

(23.56) 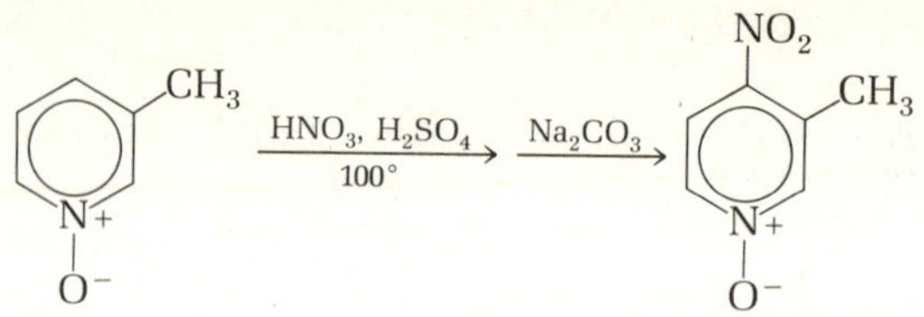

4-nitro-3-picoline *N*-oxide (70%)

The nitration of pyridine *N*-oxides, aside from its inherent theoretical interest, presents the major gateway to 4-substituted pyridines. As will shortly be apparent, the nitro group may be displaced by a variety of nucleophilic species. Reduction of the nitro group produces an amino group at C-4 that may be diazotized and replaced by the customary reagents.

(23.57) 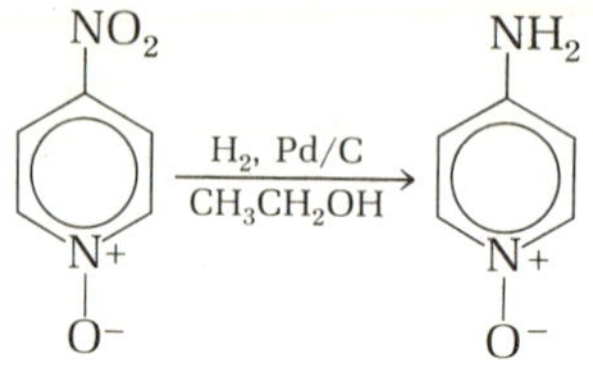

The *N*-oxide moiety may also be removed with or without concomittant reduction of the nitro group, making available a wide variety of useful synthetic intermediates.

(23.58) 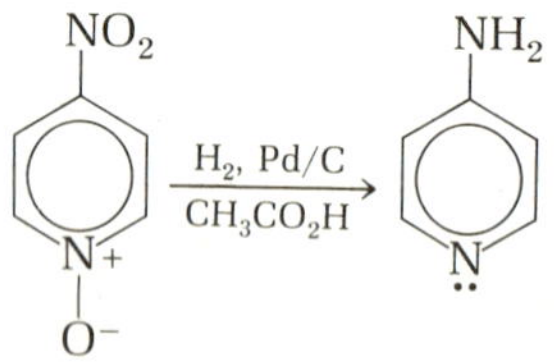

(23.59) 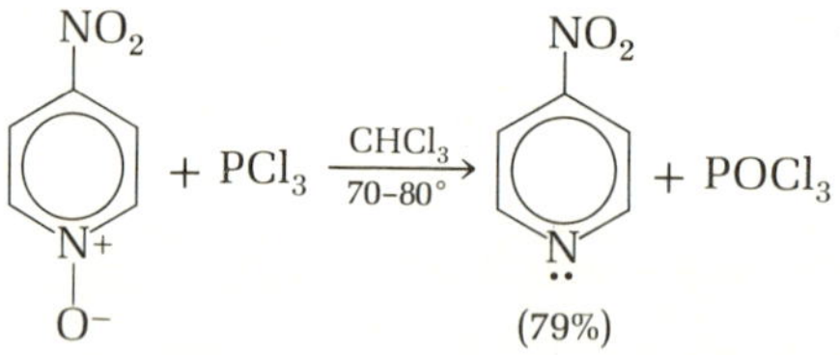

PROBLEM

3 Pyridine *N*-oxide has found some employment as an oxidizing agent. Benzylic halides may be converted to benzaldehydes by allowing them to react with the oxide and then with base (equation 2).

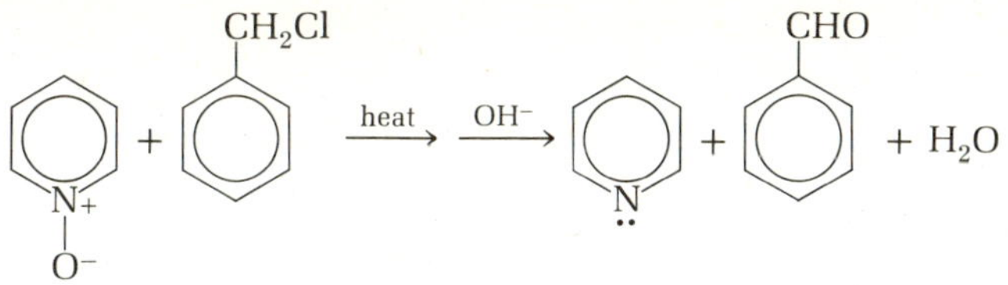

(2)

The oxide can also bring about the oxidative decarboxylation of α-halocarboxylic acids (I). Write plausible mechanisms for reactions 2 and 3.

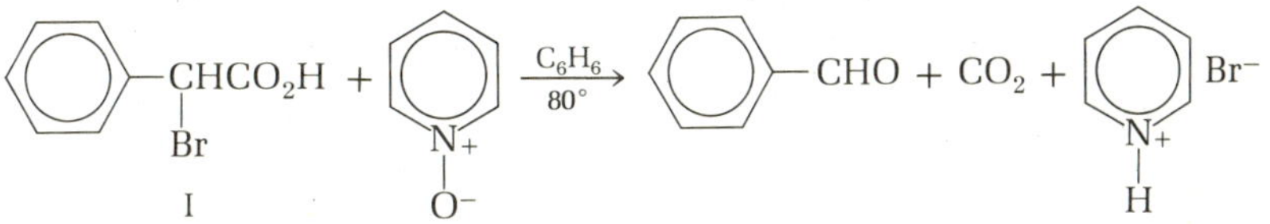

(3)

NUCLEOPHILIC ATTACK ON PYRIDINES

Because of the polarized π-electron distribution of the heterocyclic ring and the much greater ability of nitrogen to carry a negative charge compared with carbon, pyridine rings are much more hospitable to nucleophiles than unsubstituted benzene rings are. Even so, only very powerful nucleophiles can induce the formation of stable products.

A prime example of this type of substitution involves the reaction between pyridines and alkali metal amides. The mechanism is believed to be of the addition-elimination type.

(23.60)

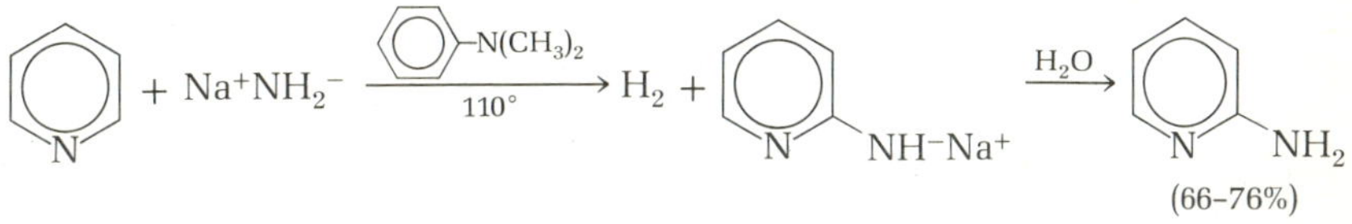

(23.61)

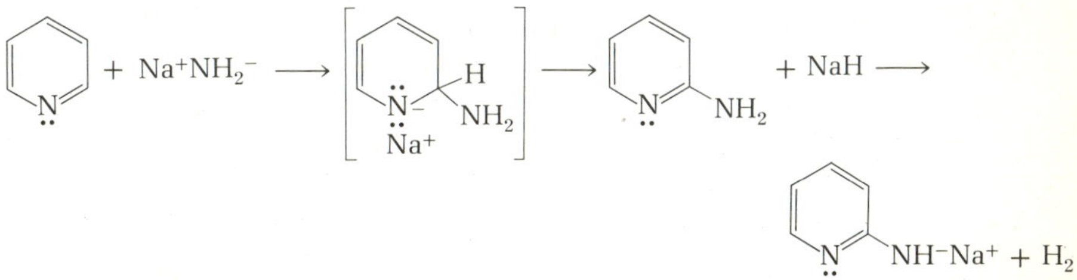

Substitution at the 2 position is strongly preferred but will take place at the 4 position if C-2 is blocked.

(23.62) 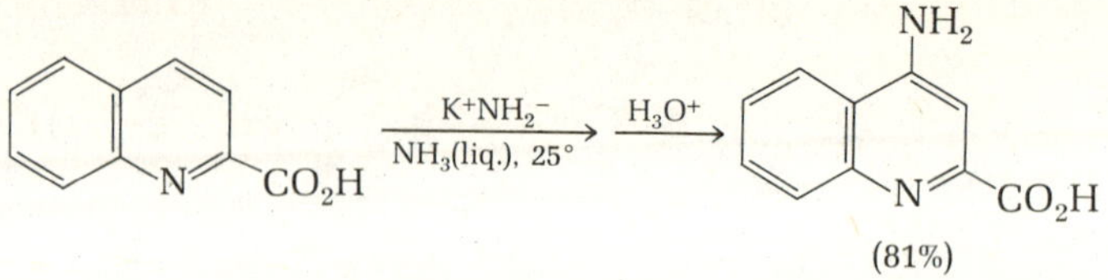

Since the 2-amino derivatives give access to 2-substituted and 3-substituted (by electrophilic attack) pyridines, this reaction is of great value. Organolithium reagents also qualify as being powerful enough to react with pyridine.

(23.63) 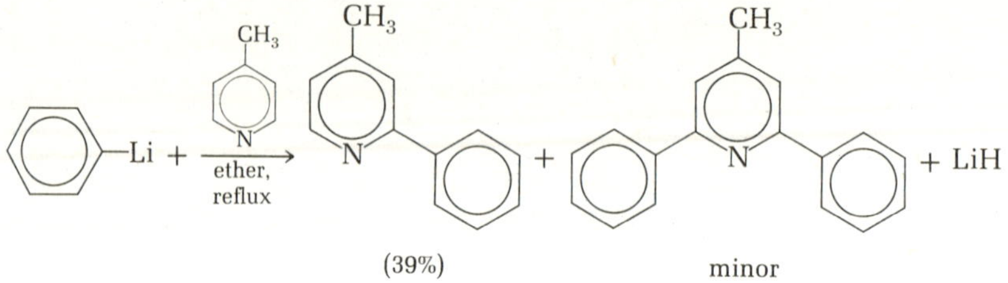

NUCLEOPHILIC ATTACK ON PYRIDINIUM IONS

π-Electron polarization and ring electrophilicity will clearly be rather high in pyridinium ions, and we can expect them to be reactive toward species with a more "normal" nucleophilicity. Such is indeed the case. Many of these studies were made with compounds that contain additional electron-withdrawing groups since the presence of these groups facilitates nucleophilic attack. Derivatives of the pyridinecarboxylic acids are popular substrates and their widely used nonsystematic names are shown below.

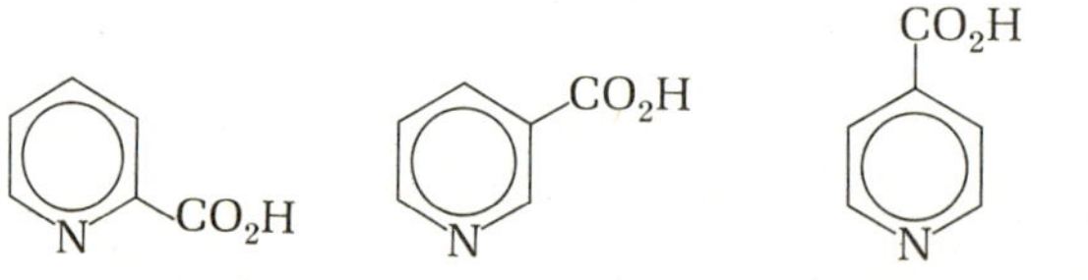

picolinic acid   nicotinic acid   isonicotinic acid

(23.64) 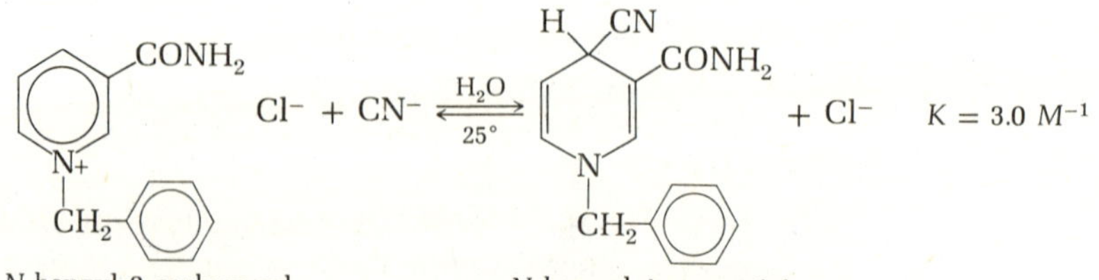

N-benzyl-3-carbamoyl pyridinium chloride

N-benzyl-4-cyano-1,4-dihydronicotinamide

The product in equation 23.64 is a derivative of 1,4-dihydropyridine. It is stable and may be isolated. If the product is allowed to stand in aqueous solution, however, it rearranges to a mixture of 1,2- and 1,6-dihydropyridines. The three isomers may be differentiated by means of their characteristic uv spectra.

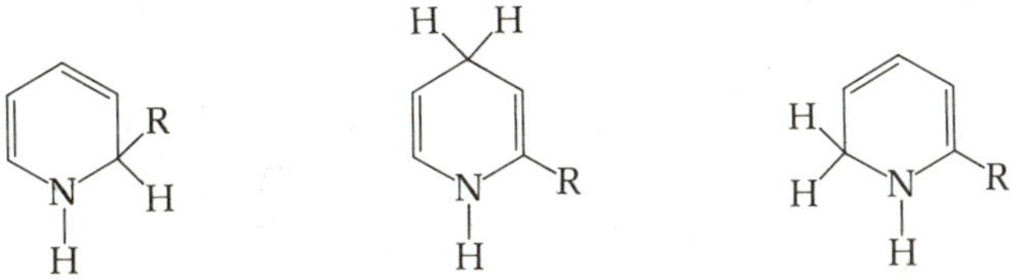

a 1,2-dihydropyridine   a 1,4-dihydropyridine   a 1,6-dihydropyridine

Grignard reagents and hydroxide ion are among the nucleophiles that have frequently been allowed to react with pyridinium ions. The dihydropyridine products are sensitive toward oxidizing agents, and one may achieve a *net* displacement of hydride ion by another nucleophile.

(23.65) 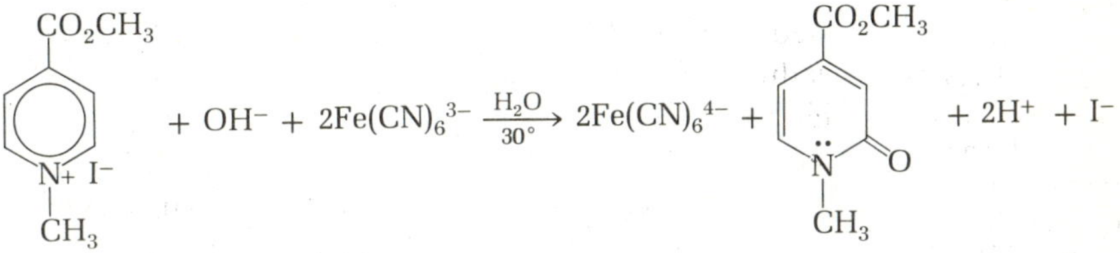

N-methyl-4-carbomethoxy-2-pyridone (96%)

One of the more intensely researched addition reactions with pyridinium ions is that involving hydride ion (or its equivalent), for the information bears on the action of the ubiquitous biological oxidation-reduction coenzymes, nicotinamide adenine dinucleotide and nicotinamide adenine dinucleotide phosphate (Figure 23-5). *N*-Alkylpyridinium ions, including NAD^+, are reduced to 1,4-dihydropyridines by sodium dithionite in basic solution. The reaction is very specific in forming the 1,4-isomer, and the NADH produced displays the same biological activity as that obtained by enzymatic reduction.

(23.66) 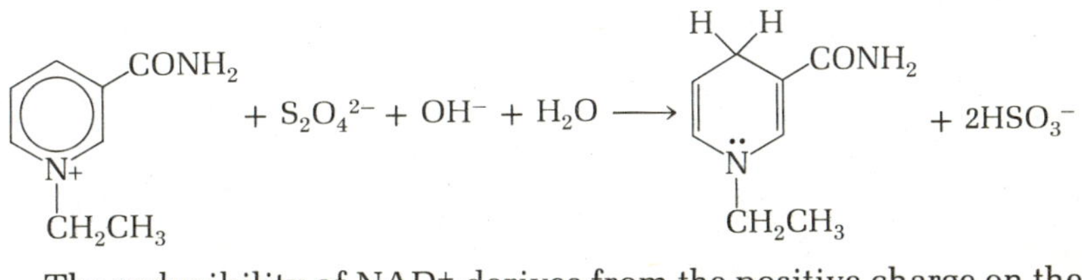

The reducibility of NAD^+ derives from the positive charge on the nitrogen atom and the electron-withdrawing effect of the amide group. On the other hand, the oxidation of NADH is aided by the aromatization of the ring. These

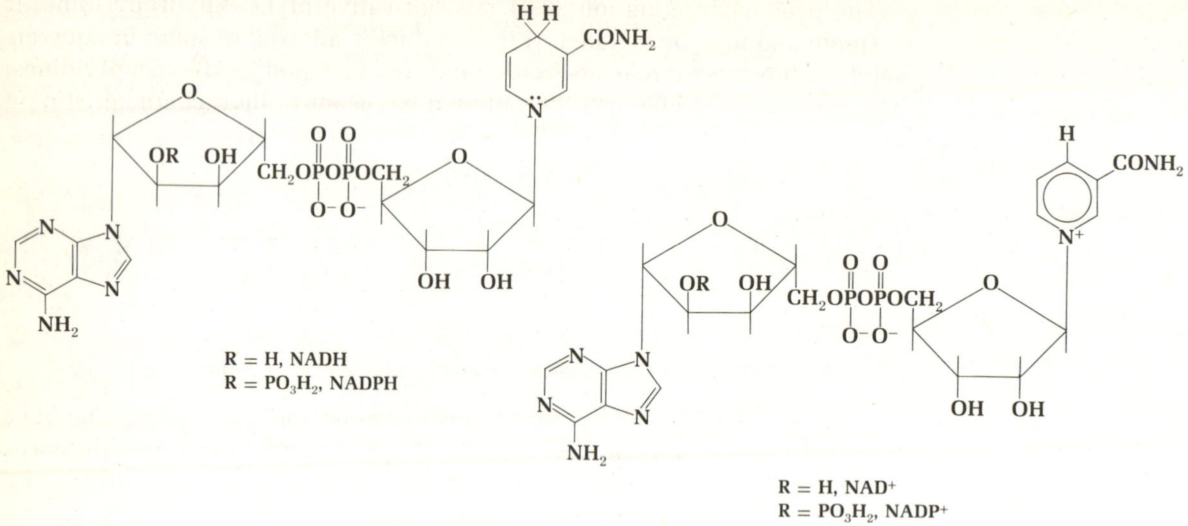

FIGURE 23-5

Oxidized and Reduced Forms of Nicotinamide Adenine Dinucleotide (NAD^+, NADH) and Nicotinamide Adenine Dinucleotide Phosphate ($NADP^+$, NADPH).

two driving forces are nearly equal in magnitude, and the oxidation-reduction equilibrium may be driven in either direction by relatively small chemical changes (Chapter 11).

Sodium borohydride displays a different selectivity toward pyridinium ions than does sodium dithionite, attacking primarily at the carbon atoms proximate to the nitrogen. Since the 1,2- and/or 1,6-dihydropyridines are dienamines, they are easily protonated even by the solvent, and the *imminium ions* so produced are further reduced to tetrahydropyridines (equation 23.67).

(23.67)

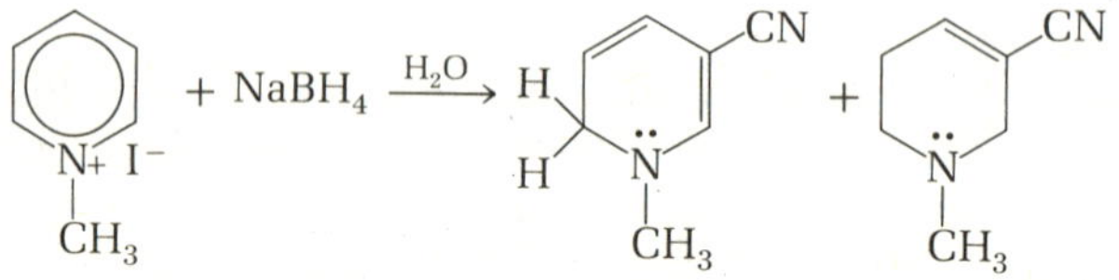

The balance between dihydropyridine and tetrahydropyridine may usually be altered by adjusting the pH, low pH's favoring protonation.

(23.68)

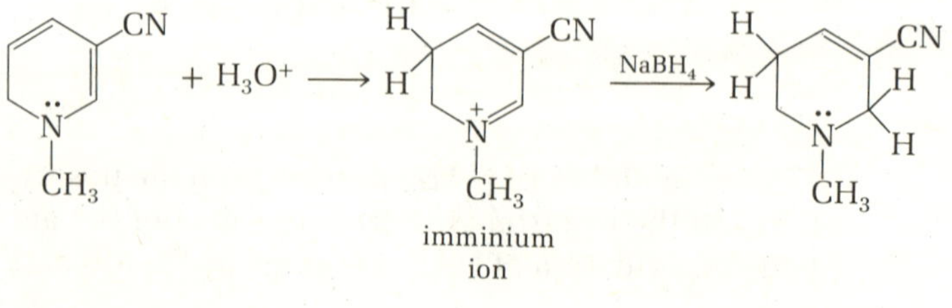

NUCLEOPHILIC SUBSTITUTION OF *N*-OXIDES

N-Oxides contain a cationic nitrogen atom and so display many of the characteristics of pyridinium ions. They are prone to undergo nucleophilic attack.

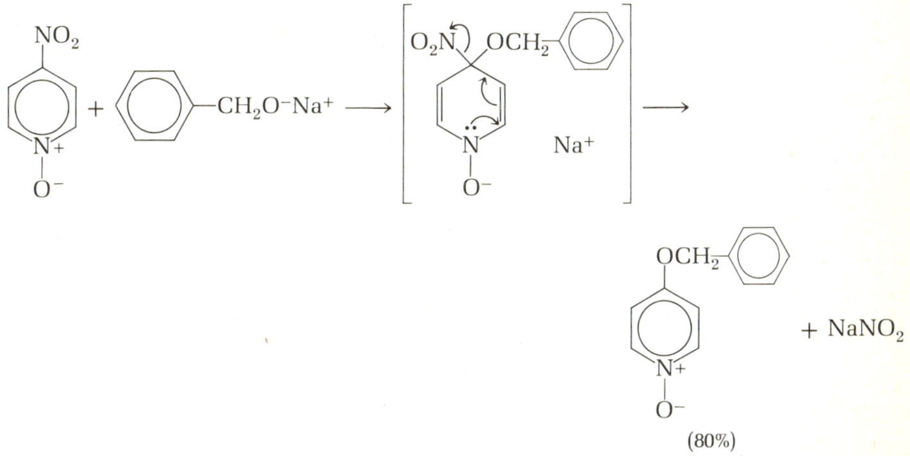

(23.69)

If the oxygen atom becomes bonded to another group, then it is lost in the course of the substitution (equation 23.70).

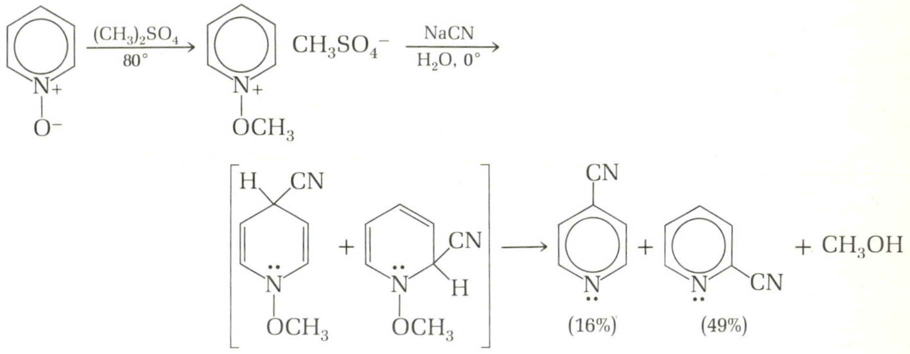

(23.70)

PYRIDINYL RADICALS

The reduction of pyridinium ions and the oxidation of dihydropyridines are processes that involve the overall addition or removal of two electrons. We have depicted these reactions as occurring with simultaneous transfer of both electrons. Electron transfer could occur in two steps, however, with *pyridinyl radicals* as intermediates. It is not really known to what extent the two different reaction modes are important in pyridinium ion chemistry, including the reactions of substances such as NAD^+ and NADH. A critical factor must be the magnitude of the one-electron and two-electron oxidation and reduction potentials of the other reactants.

That pyridinyl radicals are plausible intermediates in these reactions is confirmed by the preparation of several stable (in the absence of oxygen) pyridinyl radicals such as *N*-ethyl-4-carbomethoxypyridinyl **(16).** The stability of **16** is due in part to the extensive delocalization of the unpaired electron and in part because the ester group provides a steric bar to dimerization of the radical.

(23.71) 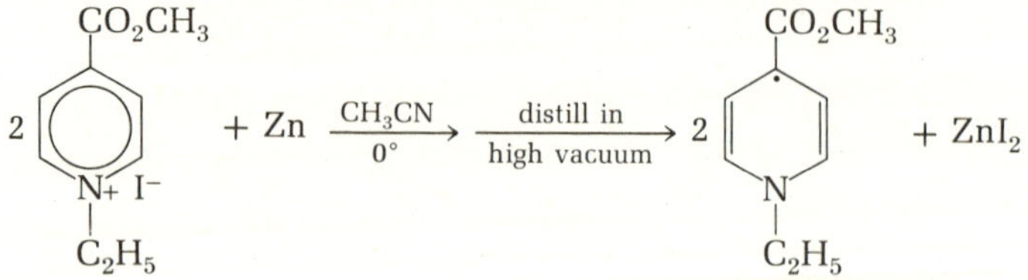

16, emerald green

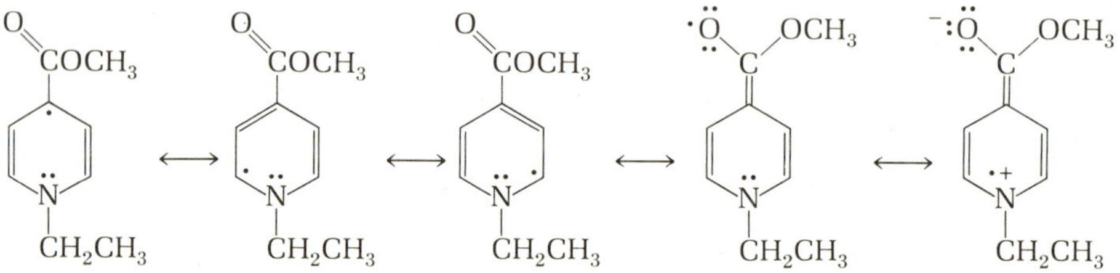

PYRIDOXAL PHOSPHATE COENZYME

A final example that illustrates the biological verstaility of the pyridine ring as an *electron source* (dihydropyridine) and as an *electron sink* (pyridinium ion) is provided by pyridoxal phosphate **(17),** a coenzyme derived from pyridoxine (vitamin B_6, **18**).

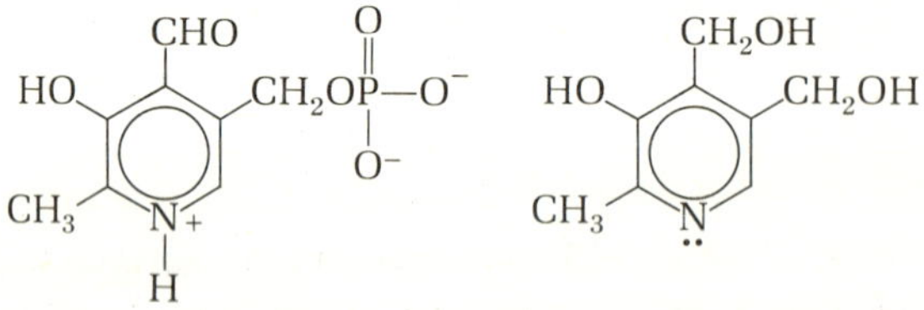

17 **18**

Pyridoxal phosphate serves as a coenzyme for a variety of reactions, among them the decarboxylation of α-amino acids (equation 23.72) and the transamination reaction between an α-amino acid and an α-keto acid (equation 23.73).

(23.72) $$\underset{\displaystyle NH_2}{\underset{|}{RCHCO_2H}} \xrightarrow{\text{enzyme}} RCH_2NH_2 \quad \text{and/or} \quad RCHO$$

(23.73) $$\underset{\displaystyle NH_2}{\underset{|}{RCHCO_2H}} + R'\overset{\displaystyle O}{\overset{\|}{C}}CO_2H \xrightarrow{\text{enzyme}} R\overset{\displaystyle O}{\overset{\|}{C}}CO_2H + \underset{\displaystyle NH_2}{\underset{|}{R'CHCO_2H}}$$

The initial step in all these cases seems to be Schiff base formation between the amino acid and the coenzyme. The internal hydrogen bonding apparently exerts an important effect on the stability and reactivity of the Schiff base, which may lose a proton from the α-position of the amino acid portion.

(23.74)

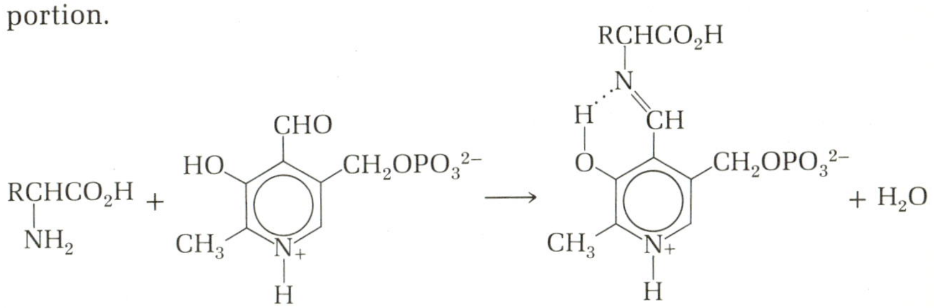

(23.75)

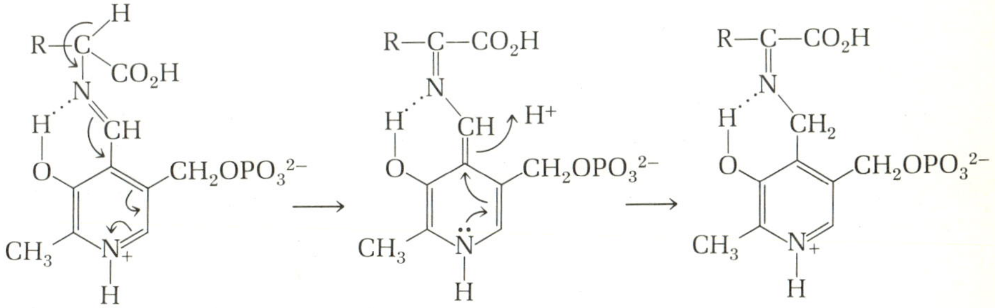

Since the product is a dihydropyridine, it is readily reprotonated on the exocyclic carbon atom, yielding a new Schiff base (equation 23.75).

Hydrolysis of this new Schiff base shows that the amino acid has been oxidized to a keto acid, while pyridoxal phosphate has been reduced to pyridoxamine phosphate **(19)**.

(23.76)

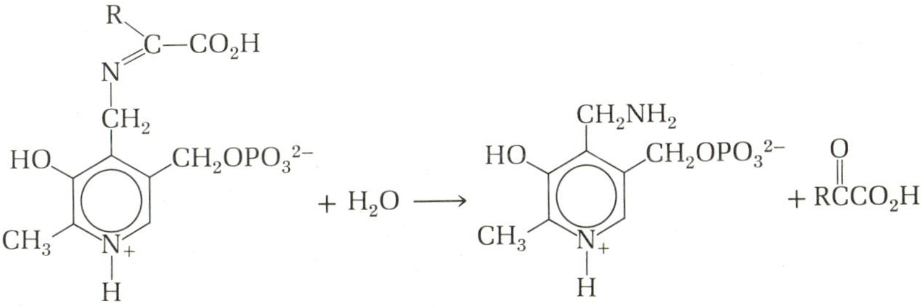

19

If the pyridoxamine phosphate reacts with a different α-keto acid ($R'\overset{\overset{O}{\|}}{C}CO_2H$ in equation 23.73), then by a reversal of equations 23.76 to 23.74 the acid will be converted to an α-amino acid and pyridoxal phosphate will be regenerated.

The Schiff base initially formed in the amino acid-pyridoxal phosphate condensation (equation 23.74) need not lose a proton; it can also decarboxylate (equation 23.77).

(23.77) 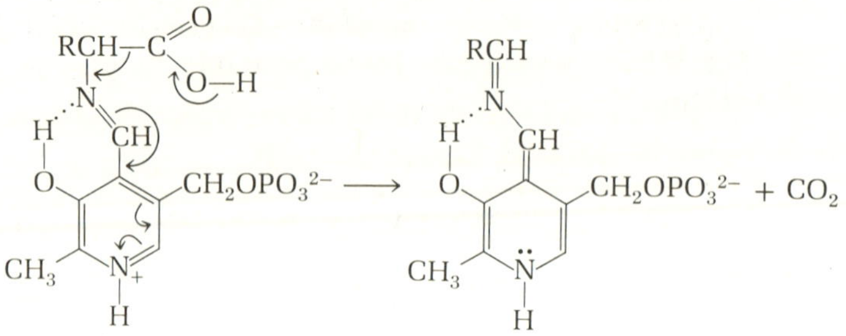

Appropriate tautomeric shifts in the dihydropyridine intermediate lead eventually to the appearance of an amine or an aldehyde.

(23.78) 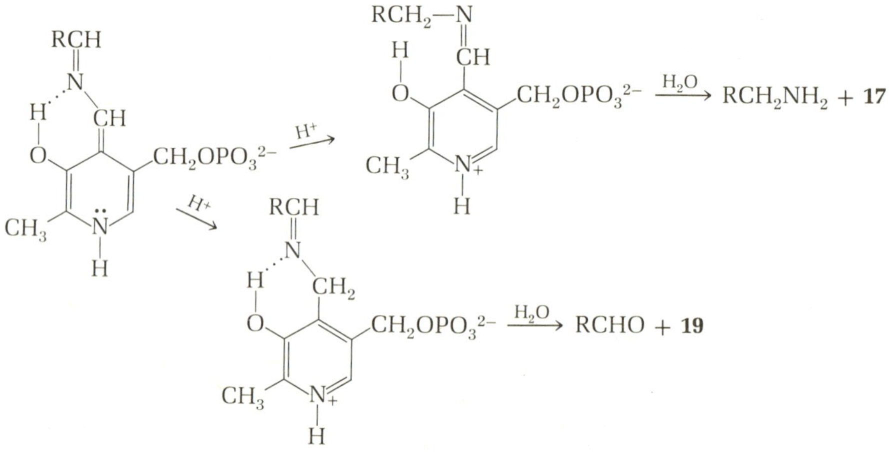

OTHER HETEROAROMATIC SYSTEMS

There are a great many heterocyclic systems, many of them aromatic, that have been passed over here. We close this chapter by mentioning several other ring systems that are particularly interesting. Replacing carbon atoms by other elements that flank them in the periodic table is a pursuit that enlarges our understanding of aromatic compounds and aromaticity. A boron-nitrogen atom pair is isoelectronic with a pair of carbon atoms, and a number of compounds are known in which the heteroatom pair is incorporated in a ring system.

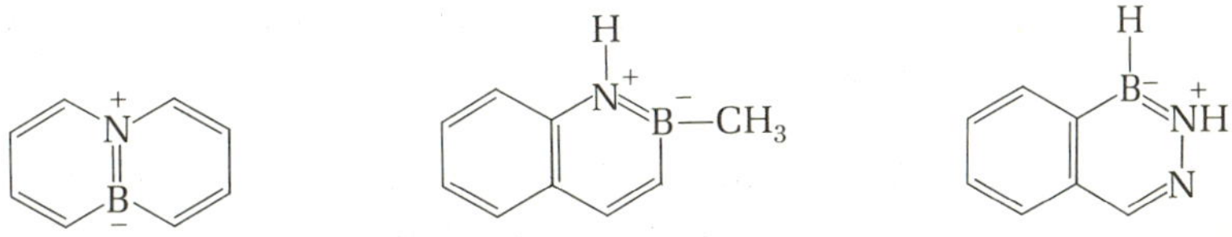

10,9-borazaronaphthalene 2-methyl-2,1-borazaronaphthalene 4,3-borazaroisoquinoline

Many of these compounds show typical aromatic spectral properties and may be nitrated, halogenated, etc.

Of the polyazaheterocycles, two of the most important are pyrimidine and purine (the latter consists of a pyrimidine ring and an imidazole ring fused together).

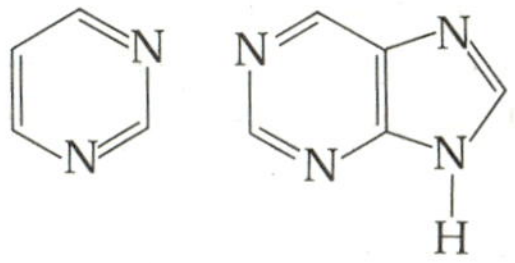

pyrimidine purine

There are two series of bases derived from these compounds that are the vital components of the *nucleic acids,* building blocks of the cellular genetic material and regulators of protein synthesis within the cell (see Chapter 24).

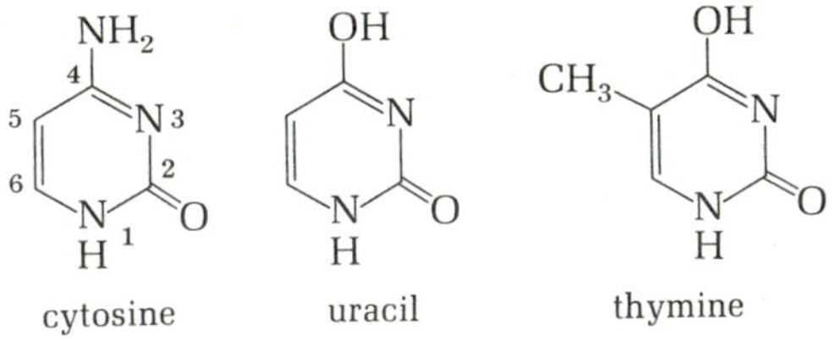

cytosine uracil thymine

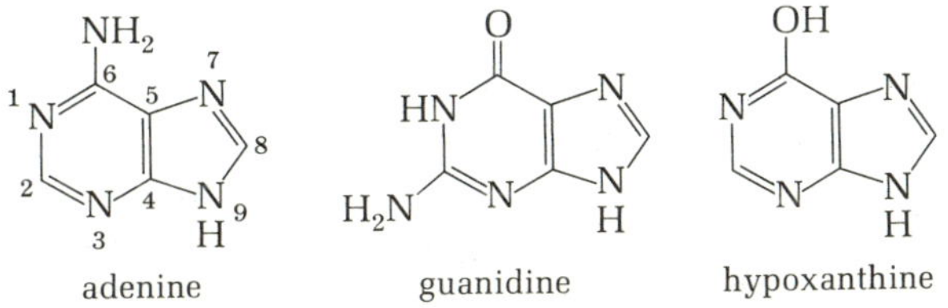

adenine guanidine hypoxanthine

Note that several of the oxygen-substituted bases exist preferentially in the carbonyl (amide) form rather than in the hydroxy form.

SUMMARY

1 When assessing the conformational behavior of saturated six-membered heterocycles, one must take account of the requirements of the lone pair(s) on the heteroatom. With nitrogen compounds the nonbonding electron pair seems to offer comparable resistance to occupying the axial position when measured against a proton.

2 Five-membered fully unsaturated heterocycles exhibit many aromatic properties; the heteroatom contributes one pair of electrons to the cyclic π-electron sextet. There are indications that the sulfur atom in thiophene can act as an electron acceptor as well as an electron donor. These heterocycles can be subjected to a number of electrophilic substitution reactions characteristic of aromatic compounds (proton exchange, nitration, halogenation, etc.). The α-positions are the preferred sites for electrophilic attack, except that a powerfully electron-withdrawing substituent at one of the α-positions may discourage attack at the remaining one.

3 Furan is the least aromatic of the monoheteroatomic five-membered heterocycles. The cation that arises from electrophilic attack on this ring will often combine with a nucleophile more rapidly than it will lose a proton. Even when furan undergoes an electrophilic substitution, the mechanism may well involve an initial addition followed by an elimination.

4 Indole combines many of the properties of benzene and pyrrole but there are important differences. The most reactive position toward electrophilic attack in the indole molecule is C-3 since attack here allows charge delocalization within the five-membered ring. Aryl hydrazones provide the starting materials for the synthesis of a large number of indoles substituted in both rings. Proton and Lewis acids serve as catalysts for this transformation, which involves loss of one nitrogen atom as ammonia.

5 Pyrrole rings are found in Nature arranged in cyclic units containing four pyrroles connected by one-carbon bridges. These units contain extensively conjugated π-systems and serve as ligands for various types of metal ions in a number of fundamentally important biological molecules.

6 Thiazolium ions readily lose their protons at C-2 to weak bases, giving rise to resonance-stabilized ylids. The ylid is reasonably nucleophilic and will attack carbonyl groups. The coenzyme thiamine pyrophosphate contains a thiazolium ring and it catalyzes the decarboxylation of α-keto acids. Nucleophilic attack of the ylid on the keto group is followed by decarboxylation, the cationic nitrogen atom acting as the electron acceptor. The intermediate may be protonated to give acetaldehyde eventually or be oxidized by lipoic acid and then ultimately converted to acetyl coenzyme A.

7 Pyridine is highly deactivated toward electrophilic attack. The positions *ortho* and *para* to the nitrogen atom are especially unreactive because attachment of an electrophile to one of these positions would result in the accumulation of positive charge on the nitrogen atom. Pyridinium ions are even more strongly deactivated than neutral pyridine rings.

8 Pyridine *N*-oxide is more readily substituted than pyridine itself and C-4 is the exclusive site of reaction. The nonbonded electron pairs on oxygen, which are conjugated with the ring, help disperse the charge in the transition state arising from electrophilic attack at C-4. Removal of the *N*-oxide function after substitution has taken place is possible; other functional groups may simultaneously be reduced under the right conditions.

9 Due to the electronegativity of nitrogen, pyridine is subject to addition by strong nucleophiles such as amide ions and organolithium reagents. The ring can usually rearomatize by expelling an alkali metal hydride. Pyridinium ions are much more electrophilic than are pyridines and the former will react with nucleophiles such as cyanide and hydroxide ions. The dihydropyridine products are of great interest in connection with studies of certain redox coenzymes. Sodium borohydride in protic solvents may reduce pyridinium ions to tetrahydropyridines.

10 Pyridinium ions can be induced to undergo one-electron reduction to pyridinyl radicals. These radicals may, in some cases, be intermediates in the interconversion of pyridinium ions and dihydropyridines.

11 Pyridoxal phosphate forms Schiff bases with α-amino acids. Thanks to the electron-accepting and electron-donating abilities of the nitrogen atom, these compounds may be tautomerized to the Schiff bases of α-keto acids; the acids are released upon hydrolysis of the condensation product. α-Amino acids may also be decarboxylated by pyridoxal phosphate, the nitrogen atom acting as an electron sink.

PROBLEMS

Note: Problems 1 through 3 appear in text.

4 The conjugate acid of tetrahydropyran can exist in two conformationally isomeric chair forms. Draw their structures. How would the conformational equilibrium in this compound compare with the related equilibrium in piperidine?

5 There are two tautomeric isomers of pyrrole, 2-pyrrolenine (II) and 3-pyrrolenine (III).

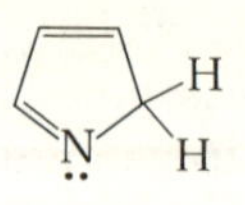

II

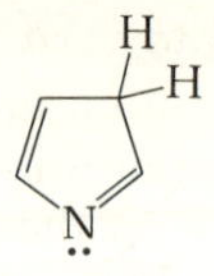

III

Attempts to prepare II or III always result in the isolation of pyrrole. Explain.

6 As noted, pyrrole is a very weak base ($pK_b = 13.6$). When subjected to mineral acids, it may be protonated; nmr studies reveal that protonation takes place at C-2 or C-3 rather than at nitrogen. Why should this be so?

7 a. When pyrrole is heated to reflux in THF over sodium hydride, gas is evolved. Addition of allyl bromide to the solution, followed by a standard work-up, leads to the isolation of two major products (and several minor ones). One of these products is found to react with a fresh batch of sodium hydride, while the second product will not. The products are isomeric and have the molecular formula C_7H_9N. Propose structures for both products, and write equations for all the chemistry involved.

b. Treatment of pyrrole with ethylmagnesium bromide in ether results in vigorous gas evolution. Addition of isopropyl bromide to the solution yields, after work-up, two isomeric products in the approximate ratio 2:1. These products both have molecular formulas $C_7H_{11}N$ and both will react with sodium hydride. Write chemical equations that account for the results.

c. What factor(s) might account for the difference in product distributions in experiments (a) and (b)?

8 When pyrrole is iodinated with potassium triodide in aqueous ethanol, the product obtained in high yield is tetraiodopyrrole. Similarly, 2,4- and 2,5-dimethylpyrroles give diiodinated products and attempts to synthesize monoiodo dimethylpyrroles were fruitless. 1,2,4-Trimethylpyrrole could, however, be monoiodinated. Account for the behavior of these compounds; note particularly the difference between the *N*-substituted and *N*-unsubstituted species.

9 a. The bromination of thiophene in acetic acid proceeds 1.7×10^9 times as fast as the bromination of benzene under the same circumstances. Reaction takes place only at the α-positions of the thiophene. Calculate the partial rate factor for α-bromination of thiophene relative to benzene.

b. Bromination of methyl thiophene-2-carboxylate in acetic acid yields only the 5-bromo derivative. The same is true of methyl furan-2-carboxylate, which reacts 120 times faster than the thiophene ester. Methyl pyrrole-2-carboxylate brominates overall 2.4×10^9 times as rapidly as the thiophene ester does. In this case, 77% of the product is the 4-bromo ester and 23% is 5-bromo ester. Calculate partial rate factors for the bromination of furan and pyrrole in the α-position.

c. Rationalize the reactivity order thiophene < furan < pyrrole, which is also found for other electrophilic substitutions. (Note that in these other reactions *N*-methyl pyrroles are not very different in reactivity from *N*-unsubstituted pyrroles.)

10 Predict the major product(s) from mononitration of the following compounds.

a. 2-Methylfuran
b. 3-Chlorothiophene
c. 5-Methoxy-2-methylpyrrole
d. 3-Chloro-4-nitrothiophene
e. Methyl 2-methylpyrrole-4-carboxylate
f. 3-Methylindole
g. 3-Nitroindole
h. Dibenzothiophene

11 Diethyl furan-3,4-dicarboxylate may be prepared from furan by means of a Diels-Alder reaction. Suggest a method for the conversion of one to the other. (Remember that acetylenes may be employed as dienophiles.)

12 The reaction of indole with ethyl nitrate ($CH_3CH_2ONO_2$) in the presence of sodium ethoxide yields 3-nitroindole. Write a plausible mechanism for this reaction.

13 Treatment of 2,3-dimethylindole with ethylmagnesium bromide results in gas evolution. After addition of methyl iodide to the solution and work-up, two isomeric products are obtained, **A** and **B**. **A** reacts with butyl bromide to give a salt, while **B** will not. Write equations that cover the reactions put forth, and account for the difference in behavior of **A** and **B** toward butyl bromide.

14 Draw in the missing structures and reagents in the following synthetic sequence.

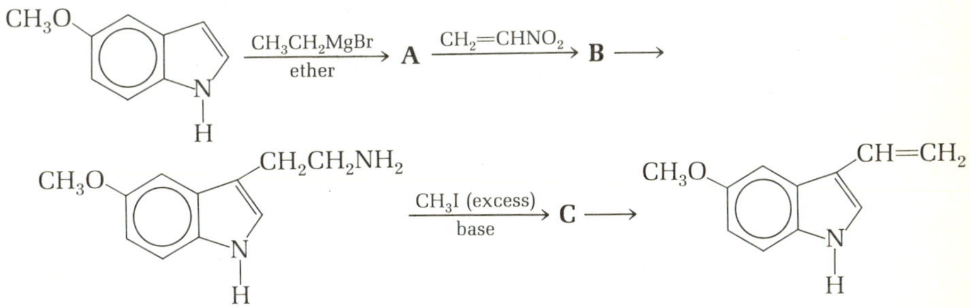

15 Studies of the base-catalyzed proton isotope exchange in 3-methylthiazolium ion show that in addition to the rapid exchange at C-2 a much slower exchange is occurring at C-5. Under these conditions (25°, pH 8 to 10) exchange at C-4 is not detected. What do these observations indicate about the relative importance of the various factors promoting exchange?

16 a. Pyridine reacts rapidly with butyl bromide in dioxane solution at 100° to give the pyridinium bromide. 3-Picoline and 4-picoline react somewhat more rapidly than pyridine but 2-picoline reacts rather more slowly. 2,6-Lutidine (2,6-dimethylpyridine) does not react at all under these conditions. Rationalize these findings.

b. The heat of reaction (neutralization) between methanesulfonic acid (CH_3SO_3H) and pyridine bases increases in the series pyridine, 2-picoline, 2,6-lutidine. The heat of reaction between boron trifluoride and these same bases *decreases* as we traverse the same series. Explain.

c. When pyridine or 2,6-lutidine are treated with sulfur trioxide in liquid sulfur dioxide (−10°), ring sulfonation does not take place; the bases may be recovered as their sulfur trioxide complexes. On the other hand, 2,6-di-*t*-butylpyridine is converted to its 3-sulfonic acid under the same conditions. Explain these data—what are the species present in solution in each case? Suggest an efficient synthesis of 2,6-di-*t*-butylpyridine.

d. Pyridine *N*-oxide and 2,6-lutidine *N*-oxide require severe conditions for sulfonation: 20% fuming sulfuric acid, mercuric sulfate catalyst, and a temperature of 220° for long periods. The products are the 3-sulfonic acids. What are the actual species undergoing sulfonation?

17 Propose efficient syntheses of the following compounds, starting with pyridine, one of the picolines, or quinoline.

a. Pyridine-3-carboxylic acid

b. Pyridine-2-aldehyde

c. 4-Methyl-3-pyridinesulfonic acid

d. 4-Nitroquinoline

e. 4-Cyanopyridine

f. 6-Bromo-2-methylpyridine

g. 6-*t*-Butyl-3-methylpyridine

h. 4-Bromo-1-propyl-1,2-dihydroquinoline

i. 4-Cyano-*N*,2-dimethylpyridinium iodide

18 In an attempted synthesis of 6-*t*-butyl-2-methylpyridine, 2-*t*-butylpyridine was allowed to react with methyllithium. None of the desired product was obtained, and a sizable quantity of starting material was recovered. The reaction mixture also yielded a modest quantity of a higher molecular weight material that proved to be 6,6′-di-*t*-butyl-2,2′-dipyridyl (IV).

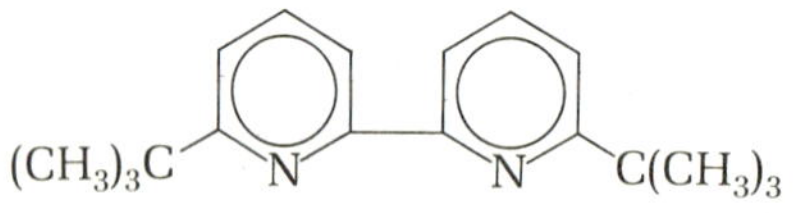

IV

Explain how IV was formed.

19 4*H*-Pyran-4-one (V) and 4*H*-thiopyran-4-one (VI) are readily accessible.

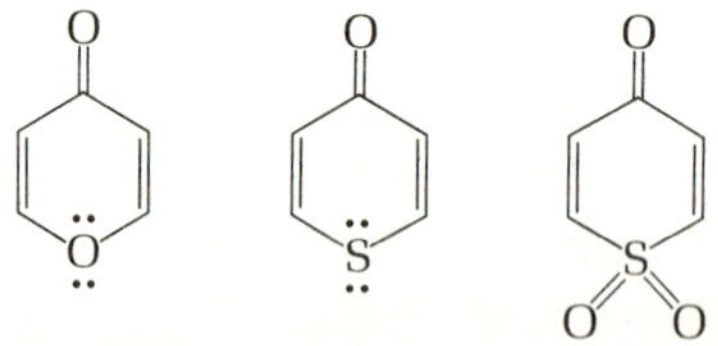

V VI VII

The carbonyl groups exhibit low ir frequencies (1658 cm^{-1}, 1609 cm^{-1}, respectively) and they react sluggishly with normal carbonyl reagents; their oximes cannot be prepared, for instance. 4*H*-Thiopyran-4-one 1,1-dioxide (VII), by contrast, forms normal carbonyl derivatives. Compound VI forms a stable hydrochloride in which the proton is attached to the carbonyl oxygen atom. Rationalize these data.

20 Add the missing intermediates and reagents in the following synthetic sequences.

a.

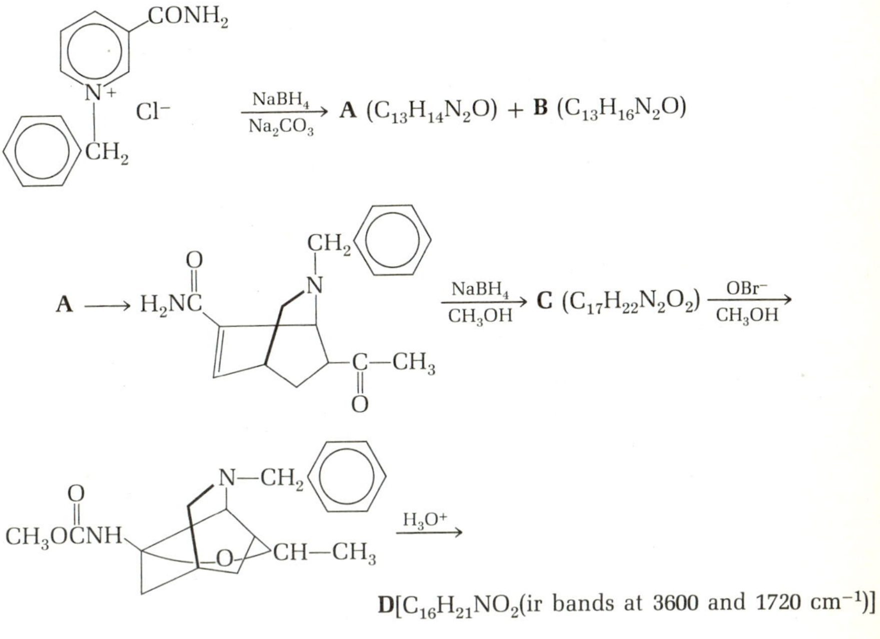

b.

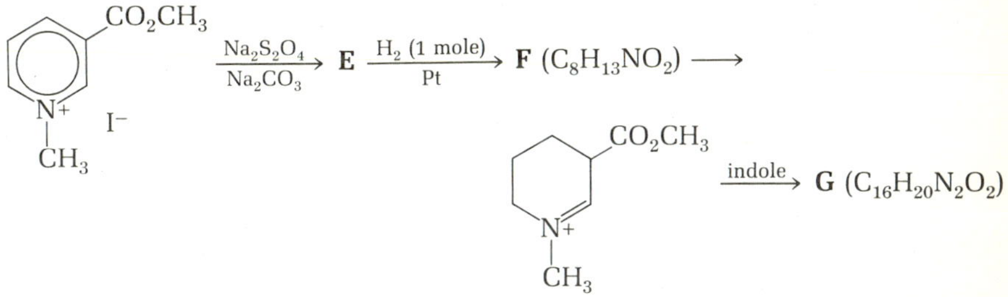

c.

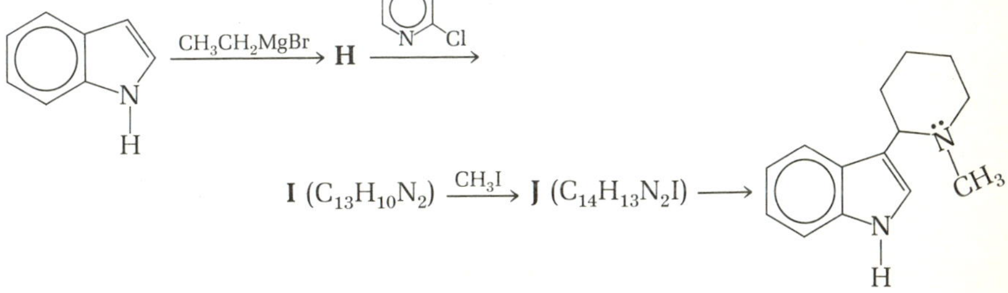

These syntheses are all concerned with intermediates used to construct various indole alkaloid skeletons.

SUPPLEMENTARY READINGS

A. R. Katritzky and J. M. Lagowski, *The Principles of Heterocyclic Chemistry*. New York: Academic Press, 1967.

L. A. Paquette, *Principles of Modern Heterocyclic Chemistry*. New York: W. A. Benjamin, Inc., 1968.

F. Freeman, "Linear Free Energy Relationships in the Furan and Thiophene Ring Systems," *J. Chem. Ed.*, **47**, 140 (1970).

E. M. Kosower, *Molecular Biochemistry*. New York: McGraw-Hill Book Co., 1962.

CHAPTER 24

SUGARS — CONFORMATION AND STEREOCHEMISTRY OF POLYFUNCTIONAL MOLECULES

Sugars are the primary units of that class of compounds known as *carbohydrates* — substances that conform to the empirical formula $C_n(H_2O)_n$. Sugar molecules are among the most important structural building blocks utilized by Nature, and their enzymatic degradation provides the life-sustaining energy for a large number of organisms. Furthermore, the synthesis of these compounds from water and carbon dioxide by green plants provides an efficient method of storing solar energy as chemical energy, as well as releasing into the atmosphere the oxygen upon which all higher forms of life are dependent.

Substances that have the empirical formula $C_nH_{2n}O_n$ may, among others, possess the structures of polyhydroxy aldehydes and ketones. Since the sugars exhibit many of the chemical properties expected of compounds with these structures, their study will be approached from that point of view. This viewpoint will be subsequently modified in accord with the results of modern investigations.

TWO- AND THREE-CARBON COMPOUNDS

Formaldehyde is the simplest compound that fits the empirical formula, and the carbohydrates as a class may be considered polymers of formaldehyde. Such a proposition is not so arbitrary as it may at first appear because formaldehyde can indeed be polymerized to yield a mixture of sugars, under the appropriate conditions.

The next members of the series are glycolaldehyde, glyceraldehyde, and dihydroxyacetone. The last two have already been shown to be intermediates in the enzymatic synthesis and degradation of fructose (Chapter 20).

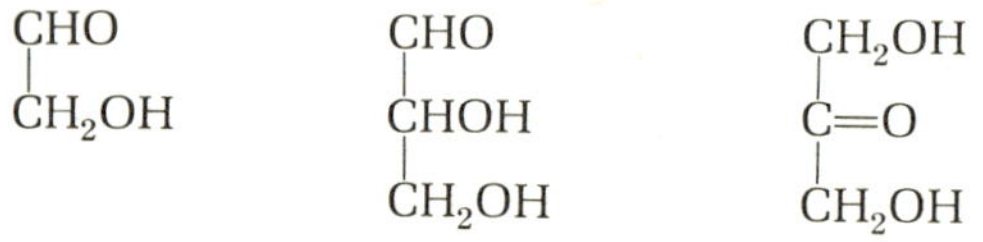

glycolaldehyde glyceraldehyde dihydroxyacetone

These low molecular weight species show certain types of chemical behavior that are highly characteristic of their higher homologues.

As an example, we may cite the fact that all the compounds mentioned above exist preferably as their dimers in the solid state. These dimers are formed by intermolecular hemiacetal formation, as shown for those derived from glycolaldehyde **(1)** and dihydroxyacetone **(2)**.

(24.1)

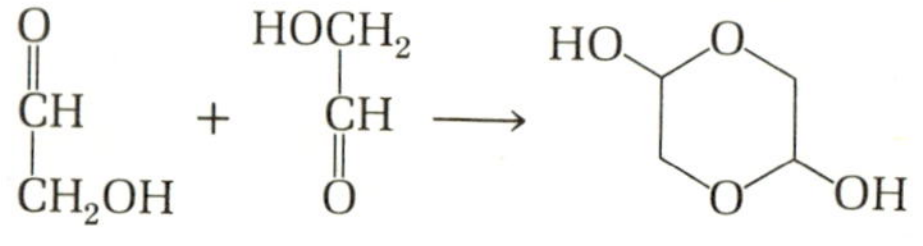

1

(24.2)

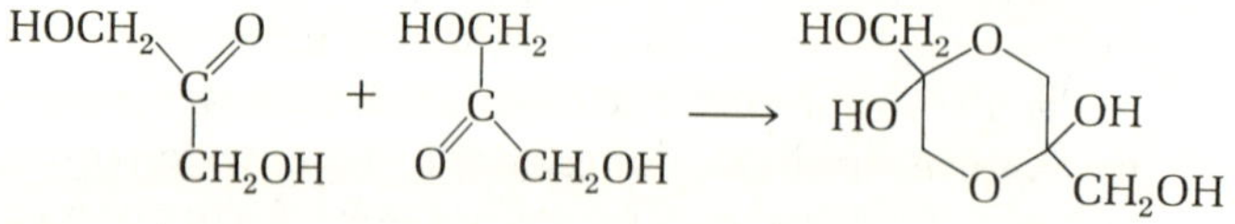

2

Sugars do not exist primarily in the carbonyl form; the carbonyl groups are, rather, masked as hemiacetals. Glycolaldehyde, glyceraldehyde, and dihydroxyacetone revert to their monomers when dissolved in water, but even in solution they probably exist as hydrates rather than as free carbonyl compounds.

There are two other chemical properties of sugars that may be exemplified by reference to these molecules. The first characteristic is their cleavage by periodic acid. Note that periodic acid will cause carbon-carbon bond scission between two hydroxyl groups, a hydroxyl and a carbonyl group, and two carbonyl groups; it will not cause bond cleavage adjacent to a carboxyl group under normal conditions (25° or less) (equation 24.4).

(24.3)
$$\begin{array}{l} \mathrm{CHO} \\ | \\ \mathrm{CHOH} \\ | \\ \mathrm{CH_2OH} \end{array} \xrightarrow{\mathrm{NaIO_4,\ H^+}} \mathrm{CH_2O} + \left[\begin{array}{l} \mathrm{CHO} \\ | \\ \mathrm{CHO} \end{array}\right] \xrightarrow{\mathrm{NaIO_4,\ H^+}} \mathrm{2HCO_2H}$$

(24.4)
$$\begin{array}{l} \mathrm{CH_2OH} \\ | \\ \mathrm{C{=}O} \\ | \\ \mathrm{CH_2OH} \end{array} \xrightarrow{\mathrm{NaIO_4,\ H^+}} \begin{array}{l} \mathrm{CO_2H} \\ | \\ \mathrm{CH_2OH} \end{array} + \mathrm{CH_2O}$$

The other feature of these compounds particularly worthy of mention is their sensitivity to base-catalyzed isomerization. It has already been noted that glyceraldehyde and dihydroxyacetone are easily interconvertible in the presence of an enzyme catalyst (Chapter 20); the same equilibration can be achieved with the aid of basic catalysts such as pyridine or dilute alkali. An enediol is presumably the critical intermediate.

(24.5)
$$\begin{array}{l} \mathrm{CHO} \\ | \\ \mathrm{CHOH} \\ | \\ \mathrm{CH_2OH} \end{array} \underset{}{\overset{\mathrm{OH^-}}{\rightleftharpoons}} \begin{array}{l} \mathrm{CHOH} \\ \| \\ \mathrm{COH} \\ | \\ \mathrm{CH_2OH} \end{array} \rightleftharpoons \begin{array}{l} \mathrm{CH_2OH} \\ | \\ \mathrm{C{=}O} \\ | \\ \mathrm{CH_2OH} \end{array}$$

ABSOLUTE CONFIGURATIONS BASED ON GLYCERALDEHYDE

Among five- and six-carbon sugars there are a large number of possible stereoisomers, due to the substantial number of asymmetric atoms in each molecule. It became clear quite early in the history of carbohydrate chemistry that it would be of the utmost importance to be able to correlate configurations between different sugars. Since there were no methods available for determining absolute configurations, it was necessary to choose a standard against which comparisons could conveniently be made.

The problem was solved by Emil Fischer, the greatest single pioneer in carbohydrate chemistry and one of the giants of organic chemistry (he received the second Nobel Prize in chemistry in 1902). Fischer took as his standard the aldohexose (+)-glucose (see below) and decreed that the highest numbered asymmetric carbon atom had a specified absolute configuration.

Because glucose contains several asymmetric atoms it is not a totally suitable standard, and in 1906 Rosanoff suggested that (+)-glyceraldehyde be adopted as the standard because it contained only one asymmetric atom and was capable of transformation into a number of other asymmetric molecules of interest. Since the absolute configurations of (+)-glyceraldehyde and C-5 of (+)-glucose can be correlated chemically, Fischer's original configurational assignments may be retained; the absolute configuration of (+)-glyceraldehyde was already implied in Fischer's original proposal.

It was necessary for Fischer to make one assumption in his method. Without in any way being available to verify it, he *assumed* that (+)-glyceraldehyde had the absolute configuration shown, and he denoted this absolute configuration by the symbol D. It has recently been shown that Fischer's guess was fortuitously correct and that D-glyceraldehyde has indeed the *R* configuration.

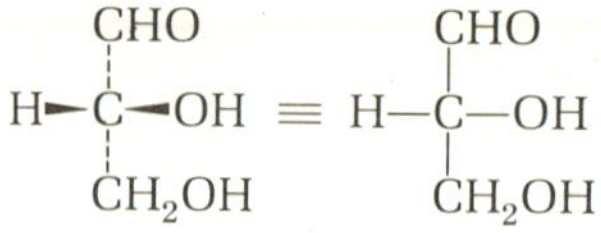

D-(+)-glyceraldehyde

One of the earliest known optically active compounds was (+)-tartaric acid (ordinary tartaric acid), obtained as a by-product from the fermentation of wine. Racemic tartaric acid was also discovered as a by-product in the manufacture of ordinary tartaric acid (Latin *racemus* — a bunch of grapes). Pasteur showed that when the sodium ammonium salt of the racemic acid was crystallized from water at a temperature below 28°, two types of crystals separated that proved to be the mirror images of one another. He painstakingly separated them by hand and from the separated crystals was able to regenerate pure (+)- and (−)-tartaric acid. This experiment constituted the first resolution of a racemic mixture into pure enantiomers.

The relationship of (+)-glyceraldehyde and (−)-tartaric acid was established by chemical means, as shown in equation 24.6.

(24.6)

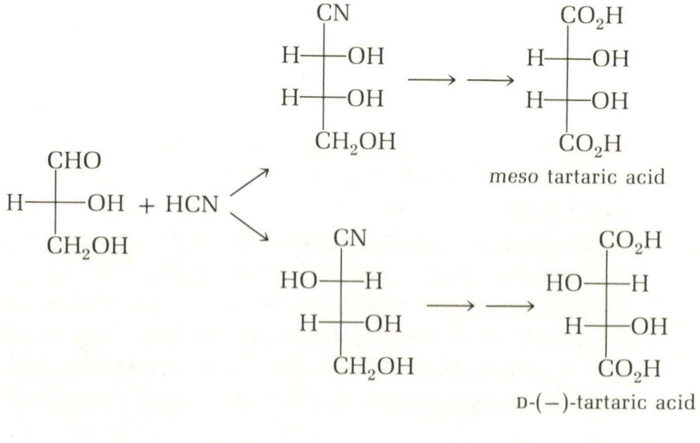

It then followed that ordinary tartaric acid must have had the relative L configuration (in this system, stereochemistry is assigned on the basis of the configuration of the *lower* asymmetric center when the structure of the molecule is written in the ordinary Fischer manner).

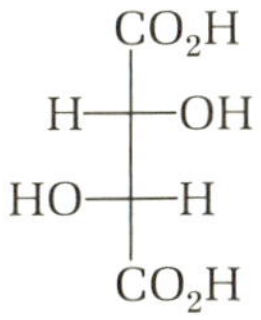

L-(+)-tartaric acid

The final link in the chain of configurational correlations was established in 1951 when it was shown that sodium rubidium (+)-tartrate had the *R,S* absolute configuration, thus confirming Fischer's original assumption about D-glyceraldehyde.

THE TETROSES

Excluding optical isomers, there are three possible four-carbon sugars — two aldehydes (*aldotetroses*) and one ketone (*ketotetrose*). Their structures are conventionally written with the carbonyl group as near the top of the page as possible; the D or L designation is then conferred on the entire molecule on the basis of the configuration of the bottommost asymmetric carbon atom.

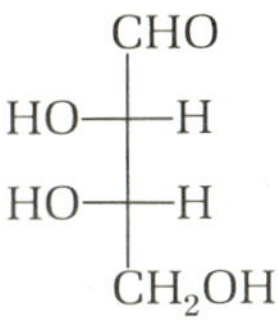

L-erythrose

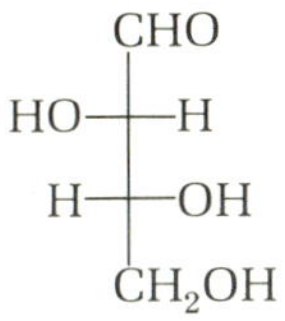

D-threose

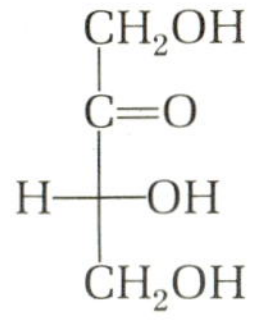

D-erythrulose

One of the most important techniques for differentiating among isomeric straight-chain aldehyde sugars involves converting *both* terminal carbon atoms to the same functionality. By symmetrizing the ends of the chain, one may then draw some conclusions about the relationships among the inner hydroxyl groups. For instance, heating aldoses with nitric acid results in the selective oxidation of the aldehyde and primary alcohol functions to carboxyl groups. Application of this reaction to D-erythrose leads to the formation of optically inactive *meso* tartaric acid (equation 24.7), while D-threose gives rise to optically active D-tartaric acid under the same conditions (equation 24.8).

(24.7) 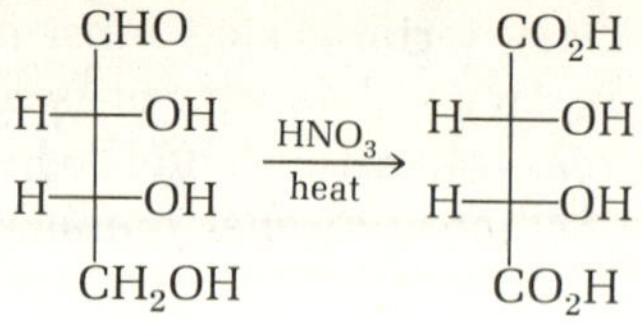

(24.8) 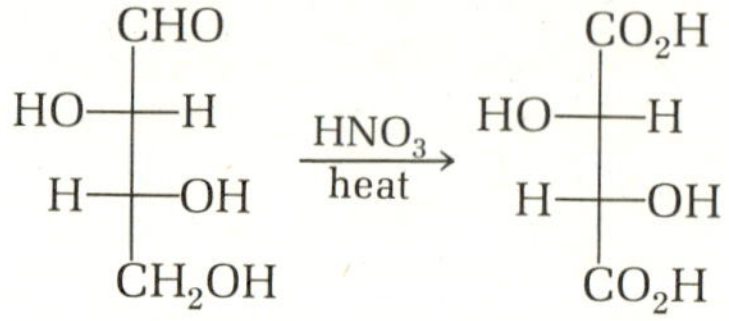

These results constitute a direct proof that the structures assigned erythrose and threose are correct.

Another approach that rests on the same logic is that in which the aldehyde group is reduced to a primary alcohol, the molecule being thereby transformed into a polyol (*alditol*). The classical reducing agent for this purpose was sodium amalgam in aqueous acid; it has been largely supplanted by sodium borohydride.

(24.9) 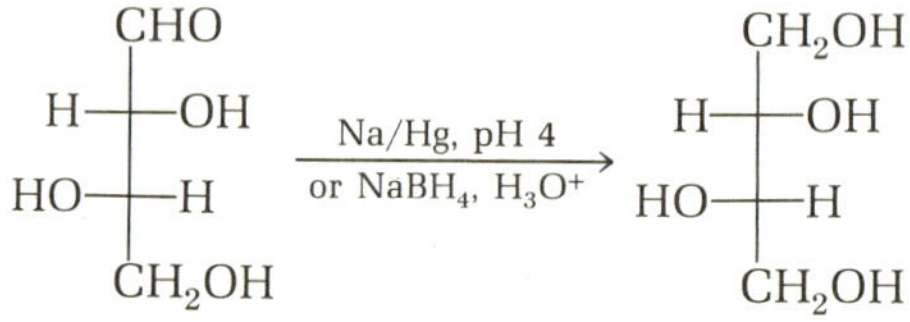

L-threitol

PROBLEMS

1 Erythritol is oxidized by sorbose bacteria to an optically active erythrulose. When the erythrulose is reduced by means of sodium amalgam, two polyols are obtained, one of which proves to be L-threitol.

a. Write the reactions described above, showing the structures of all reactants and products, including the second polyol.

b. What was the absolute configuration of the erythrulose?

c. What conclusions can be drawn about the stereospecificity of the bacterial oxidation?

2 Write the structures of the unknowns in the following equations, including their configurations, and name them. (Models may aid in solving this problem.)

a. *cis*-2-Butene-1,4-diol $\xrightarrow[CHCl_3]{Br_2}$ **A** $\xrightarrow[heat]{H_2O}$ **B** ($C_4H_{10}O_4$)

b. *trans*-1,4-Dichloro-2-butene $\xrightarrow[OH^-,\ 25°]{dil.\ KMnO_4}$ **C** $\xrightarrow{KOH}$ **D** ($C_4H_6O_2$) $\xrightarrow[heat]{H_2O}$ **E** ($C_4H_{10}O_4$)

c. *trans*-1,4-Dichloro-2-butene $\xrightarrow[H_2O]{HCO_3H}$ **F** $\xrightarrow{KOH}$ **G** ($C_4H_6O_2$) $\xrightarrow[heat]{H_2O}$ **H** ($C_4H_{10}O_4$)

All three tetroses belong to the class of *reducing sugars* — compounds that will rapidly reduce cupric, silver, or ferricyanide ions in basic medium. The classification of sugars as reducing or nonreducing is an important and useful one, as we shall subsequently see. The aldoses, erythrose and threose, may be differentiated from erythrulose, however, because of their facile oxidation to *aldonic acids* by halogens in a basic medium. Under these conditions only the aldehyde group is affected.

(24.10) 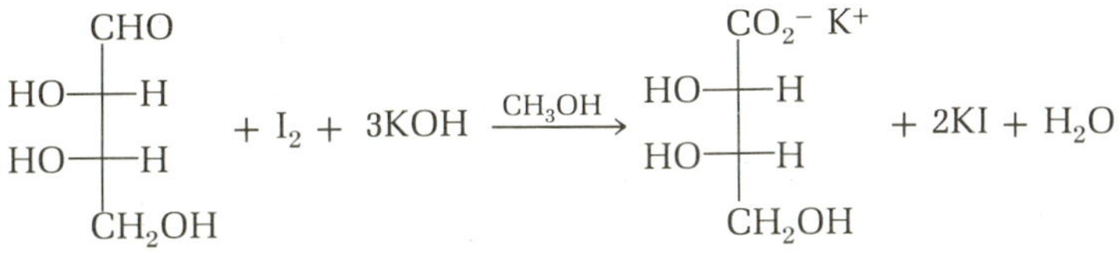

potassium L-erythronate

THE KILIANI-FISCHER SYNTHESIS

Optically active (−)-erythrose and (−)-threose may be isolated from natural sources. They may be differentiated by means of the techniques discussed above. None of the reactions so far described, however, give any information about the configurational relationship between (−)-erythrose and (−)-threose, nor on the *absolute* configuration of either isomer. Such data may be obtained by application of the *Kiliani-Fischer* synthesis.

If D-glyceraldehyde is allowed to react with an aqueous solution of sodium cyanide, a mixture of diastereomeric cyanohydrins is produced since the addition reaction involves the creation of a new asymmetric center. Hydrolysis of the mixture under basic conditions gives the salts of the diastereomeric carboxylic acids, which upon acidification are converted to the respective lactones **3** and **4**.

(24.11) 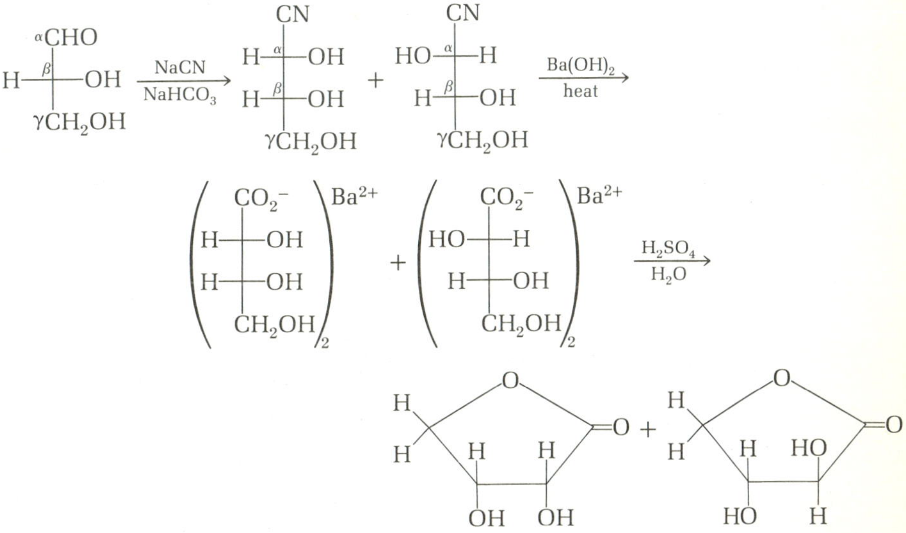

3 **4**

Either the aldonic acid salts or the aldonolactones are separable by standard techniques (fractional crystallization, chromatography, etc.). The lactones may be reduced to the corresponding aldoses by sodium amalgam in mildly acidic solution or by sodium borohydride in aqueous acid. Lactone **3** is converted to (−)-erythrose, while lactone **4** is transformed into (−)-threose.

(24.12)

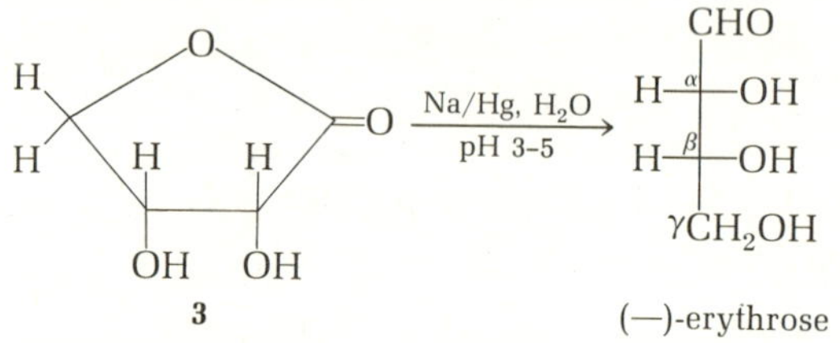

(24.13)

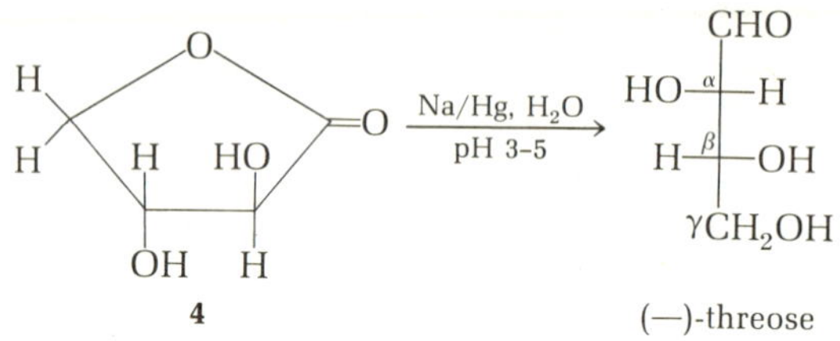

If we assume that none of the reactions in this sequence altered the configuration at what was originally the asymmetric carbon atom of D-glyceraldehyde, then we may conclude that both (−)-erythrose and (−)-threose have the same configuration at C-3 and that this configuration is D.

The experiments described above fully justify our conclusion that (−)-erythrose and (−)-threose are stereoisomers that differ only in absolute configuration at C-2. The behavior of these sugars toward phenylhydrazine, to be described below, lends further support to that conclusion.

OSAZONE FORMATION

Since sugars are carbonyl compounds, it is reasonable to expect that they would react with typical carbonyl group reagents. Indeed, Emil Fischer, the discoverer of phenylhydrazine, found that sugars would react with it to form phenylhydrazones.

(24.14)

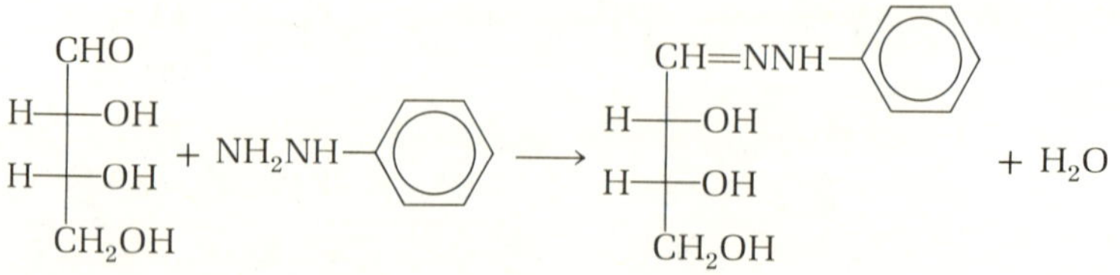

These compounds are not very useful as derivatives, however, most of them being very soluble in water. Furthermore, in the presence of excess phenylhydrazine they react further to yield **osazones** (equation 24.15).

(24.15)

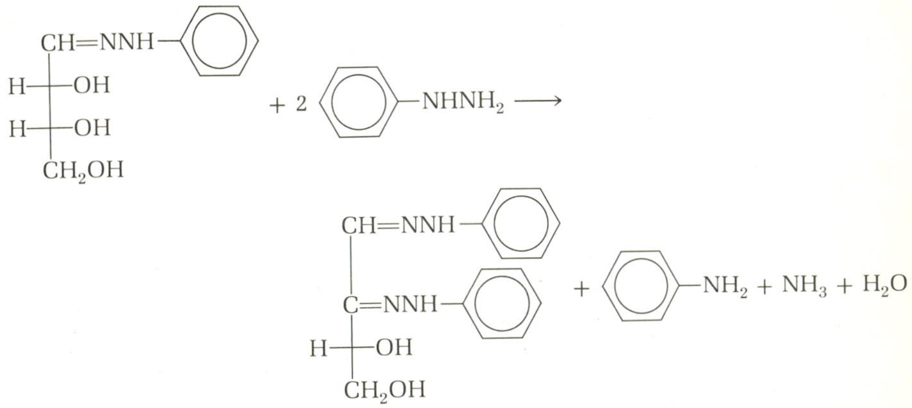

erythrosazone

A likely mechanism for osazone formation is depicted in equation 24.16; it depends a great deal on tautomeric shifts, >C=N— behaving very much like >C=O.

(24.16)

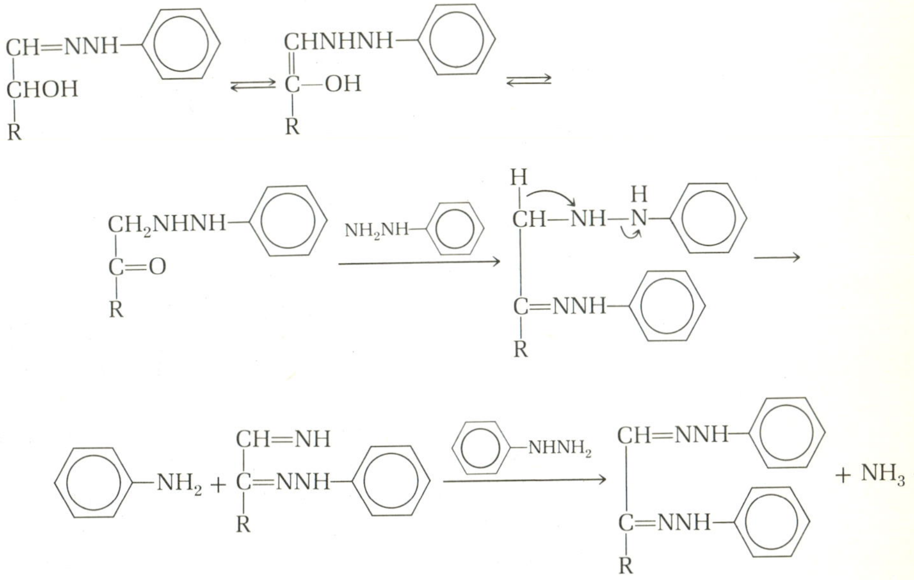

Osazone synthesis results in the loss of asymmetry at C-2 of aldoses. (−)-Erythrose and (−)-threose give rise to the identical osazone, confirming their configurational relationship.

(24.17)

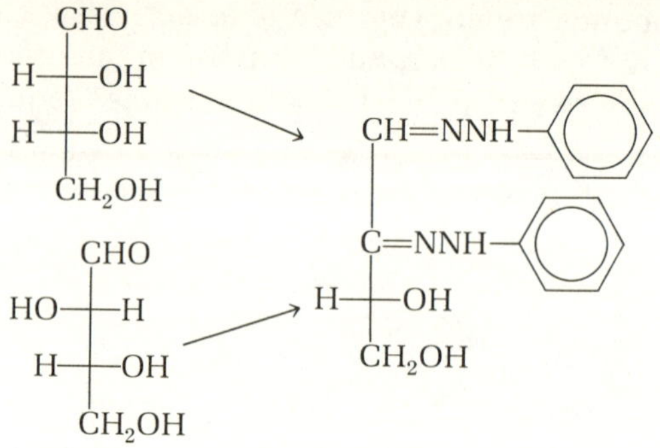

Osazones are often difficult to purify and are subject to oxidation as well. When making derivatives of sugars for identification purposes, it is often desirable to convert them to stable, crystalline *osatriazoles*, which is accomplished by heating the osazones with aqueous cupric sulfate.

(24.18)

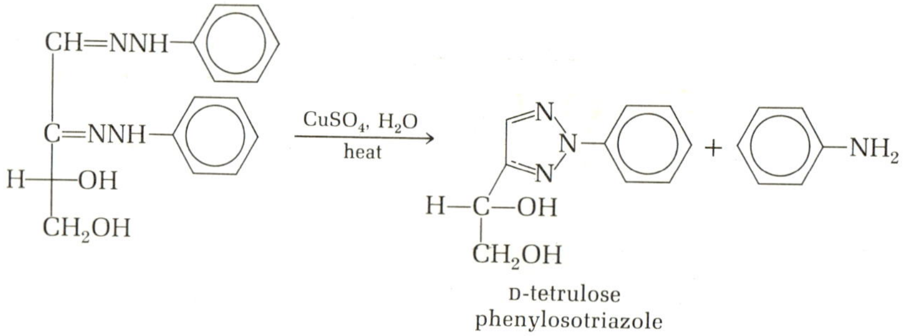

CYCLIC ACETALS AND HEMIACETALS

When it was recognized that many sugars are in fact aldehydes, they were subjected to all the varieties of reaction characteristic of aldehydes. One of these is acetal formation, and the sugars were indeed found to react with methanol and ethanol in the presence of a catalytic amount of a strong acid.

The products did not, however, conform to expectations. Analysis showed that only one molecule of alcohol had been incorporated into the product; furthermore, in most cases two or more isomeric products seemed to be produced, a finding incompatible with straightforward acetal formation. It was concluded that the products of these reactions, known generically as **glycosides**, were mixed acetals, one hydroxyl group being supplied by the external alcohol and the other by the sugar molecule itself.

Equation 24.19 depicts the acid-catalyzed reaction of D-erythrose with methanol.

(24.19)

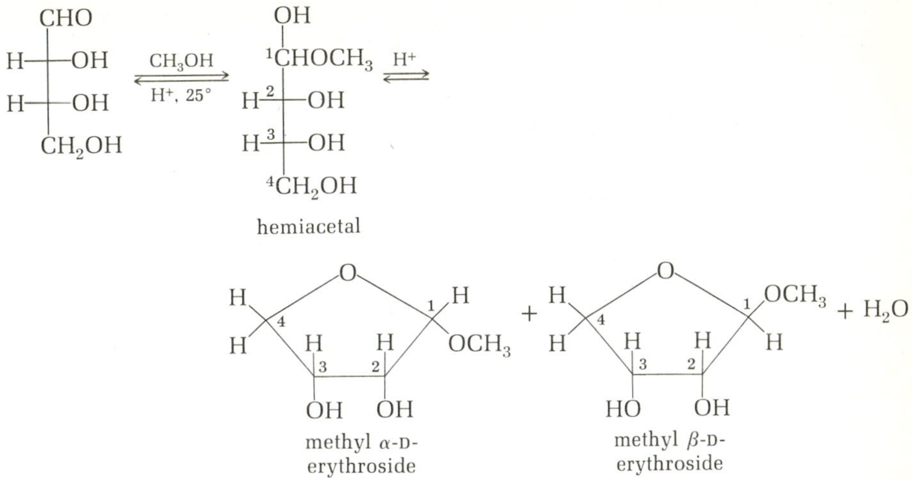

This scheme not only explains the incorporation of only one molecule of methanol, but it also accounts for the presence of two isomeric glycosides. Formation of the cyclic acetal results in the conversion of C-1 into an asymmetric carbon atom so that two configurations of the attached substituent are possible. C-1 is known as the *anomeric* carbon atom, and the α- and β-glycosides are called *anomers* (Greek *ano* — above). (With this convention, when a D sugar is depicted with C-1 to the right, as in equation 24.19, the α-isomer is the one in which the substituent at C-1 is down.)

PROBLEM 3

a. Which of the two anomers of methyl D-erythroside would you expect to be the most stable? Explain your answer.

b. Draw the anomers of methyl L-erythroside.

c. Draw the anomers of methyl D-threoside. Which of these should be the most stable, and why?

Since attempts to prepare sugar acetals or ketals almost always led to mixed acetals or ketals in which one hydroxyl group of the sugar participated, it seemed reasonable to suggest that the sugars themselves exist at least in part as internal hemiacetals. According to this theory, L-threose would exist as a mixture of the anomeric **furanoses 5** and **6.** (Cyclic sugars or their derivatives containing five-membered rings are named as furanoses since they can be viewed as substituted tetrahydrofurans.)

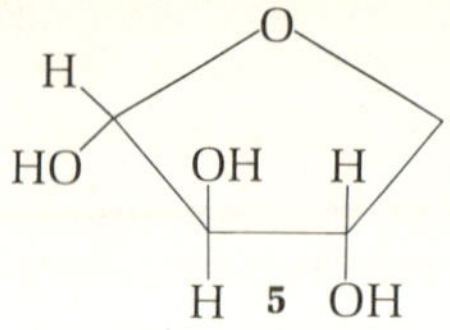

α-L-threofuranose

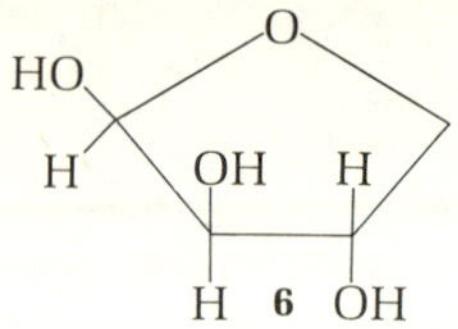

β-L-threofuranose

The evidence supporting the validity of this assumption will be developed in subsequent sections; suffice it to say at this point that most sugars appear to exist in solution and in the crystal as cyclic hemiacetals or hemiketals. The absence of a strong band in the carbonyl region in the infrared spectra of these sugars is certainly in accord with the hemiacetal theory. Due to their plenitude of hydroxyl groups, sugars are very soluble in water and correspondingly less so in most organic solvents. To obtain their infrared spectra, it is necessary to use aqueous solutions, with special cells, or to disperse the compounds in potassium bromide pellets.

PENTOSES

There are eight aldopentoses, counting all the enantiomeric possibilities. At least one representative of each enantiomeric pair may be isolated from natural sources. The D isomers of each pair are shown in their open-chain forms in structures **7** to **10.**

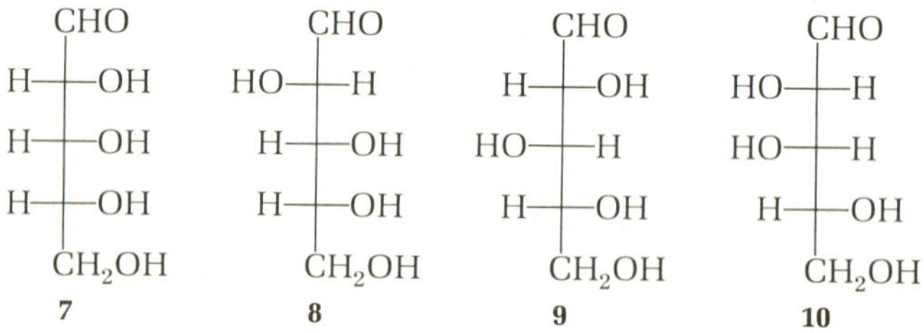

It would be instructive to consider how one might go about identifying a five-carbon aldose that had been isolated from natural sources. If, as is very likely, the isolated sugar was optically active, this fact could be exploited to effect a preliminary classification. Reduction of an aldopentose with structures **7** or **9** (or the mirror-image ones) would lead to optically inactive alditols; the other compounds would yield optically active alcohols. Furthermore, if more than one aldopentose were available, it would be possible to determine their configurational relationship at C-3 and C-4. Identical phenylosazones would be expected from sugars **7** and **8,** and a different one would be obtained from either **9** or **10.**

At this point no additional definitive information can be obtained without either lengthening or shortening the carbon chain. Chain lengthening can be achieved by way of the Kiliani-Fischer synthesis (equation 24.20). Let us assume that our unknown sugar gave rise to an optically active pentitol and must therefore have structure **8** or **10** (or their mirror images). Addition of HCN to the sugar, followed by acid hydrolysis, yields a mixture of diastereomeric hexonic acids that should be separable. Oxidation of each aldonic acid separately with nitric acid converts it to a dicarboxylic acid (an **aldaric acid**).

(24.20)

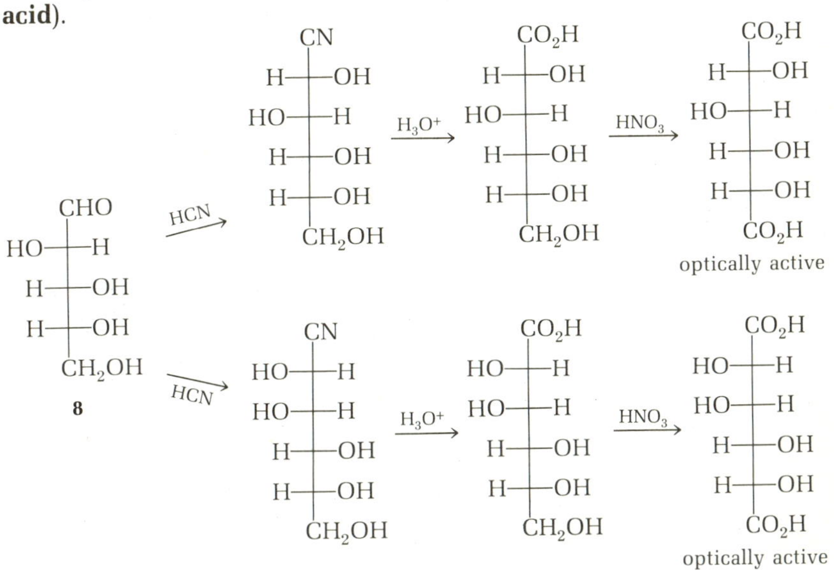

A sugar with structure **8** would serve, by this sequence, as the progenitor of two hexaric acids, *both of which would be optically active* (equation 24.20). Sugar **10,** on the other hand, would give rise to two hexaric acids, *only one of which would show optical activity* (equation 24.21).

(24.21)

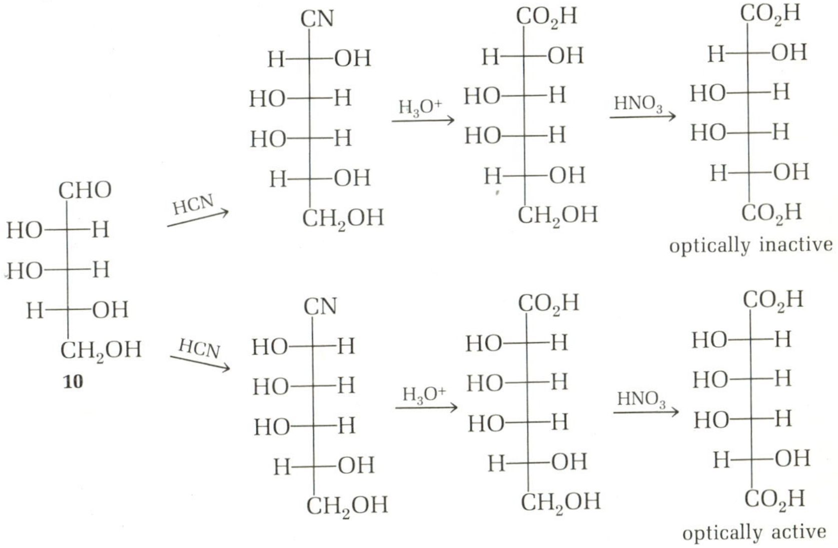

PROBLEM

4 Could the method outlined above be used to distinguish between **7** and **9**? Write the appropriate equations.

While the technique outlined above would serve to distinguish among **7**, **8**, **9**, and **10**, it could not differentiate between the mirror images corresponding to any one of these structures. To do that, one would either have to synthesize the pentose from an optically active erythrose or threose or to degrade it down to one of these compounds. Both approaches are dependent on a prior knowledge of the absolute configurations of the aldotetroses.

The simplest method for lopping one carbon atom off a sugar chain is the *Ruff degradation*. The aldose is oxidized with aqueous bromine to an aldonic acid, which is conveniently isolated as its sparingly soluble calcium salt. Further oxidation is effected by a mixture of hydrogen peroxide and ferric sulfate; the mechanism is not known but it has been shown that in the presence of ferric ions hydrogen peroxide yields hydroxyl radicals, which are the likely oxidizing agents.

(24.22)

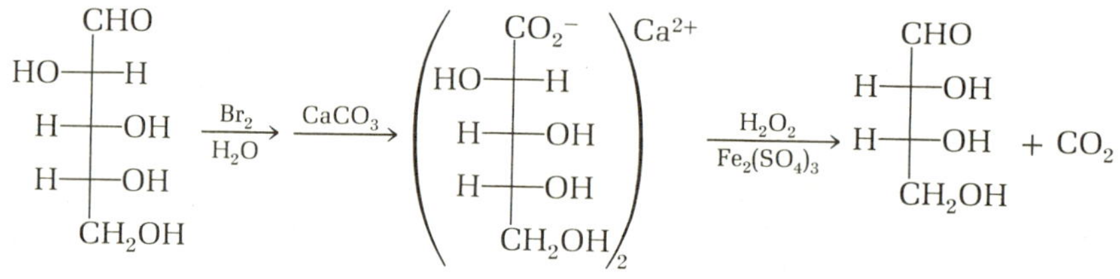

(85%)

D-erythrose (46%)

We could now say with certainty that an aldopentose that upon reduction gave an optically inactive alcohol and that could be degraded to D-erythrose must have the absolute configuration depicted by structure **7**. This sugar is the one that we call D-ribose. The structures of the other aldopentoses, arabinose, xylose, and lyxose, have been similarly confirmed; their naturally abundant mirror-image forms are shown below.

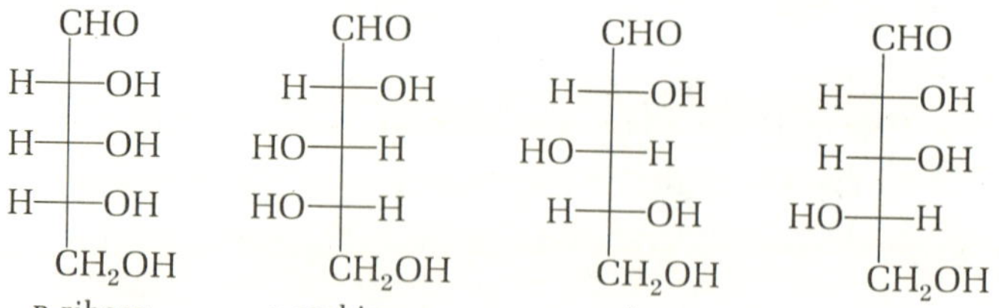

D-ribose (3) L-arabinose (12) D-xylose (111) L-lyxose (21)

As the number of carbon atoms in the sugar chain increases, so does the number of isomers, at an alarming rate. Some have found it a chore to remember the correct structure associated with each name, and several mnemonic devices have been created to help overcome these blocks. The following one, devised by L. Sattler, is based on the open-chain Fischer projection formulas for the sugars, and it is directed toward those who find it easier to remember a number than to memorize a structure.

Each sugar has associated with it a numerical code that tells the number of hydroxyl groups arrayed successively on either side of the carbon chain, in the Fischer projection. One starts from the hydroxyl group at the top of the molecule and counts only those hydroxyl groups attached to asymmetric carbon atoms. In ribose all three hydroxyl groups are on the same side of the molecule, so its code number is simply 3; in xylose, by contrast, they alternate left-right-left, so the code number is 111. For D sugars, the last hydroxyl group must be on the right, while it will be found on the left if the compound belongs to the L series. The appropriate code numbers are enclosed in parentheses beneath the structures of the aldopentoses depicted above.

CYCLIC CONFIGURATIONS OF ALDOPENTOSES

The aldopentoses exist almost exclusively as cyclic hemiacetals in solution. With a five-carbon chain there exists an additional structural possibility not encountered with the tetroses, namely, the formation of a six-membered cyclic hemiacetal, or **pyranose** (a derivative of tetrahydropyran). A study of D-ribose in heavy water showed that both furanose and pyranose forms were present at equilibrium, each isomer a mixture of anomers. The composition of the mixture was determined by means of nmr spectroscopy; the proton at C-1 had a distinct chemical shift in each of the four configurations.

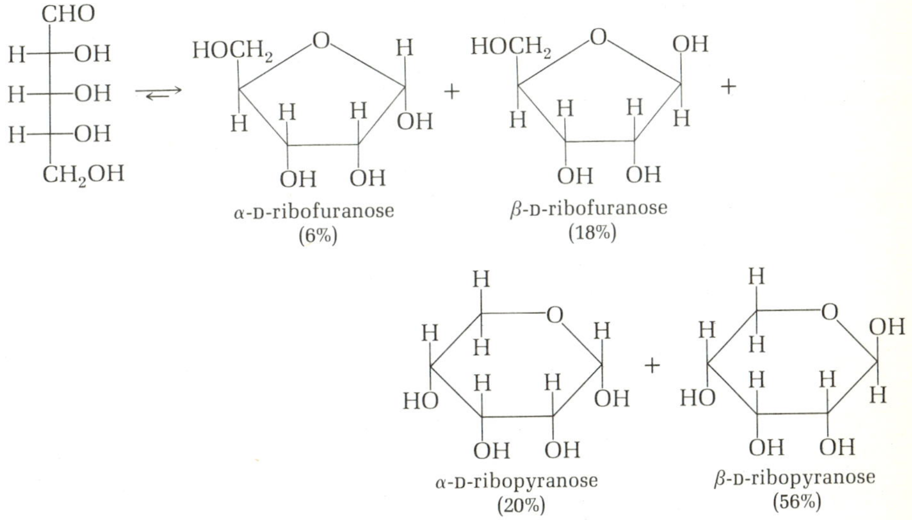

(24.23)

The furanose forms are inherently less stable than their pyranose isomers, and their total percentage does not exceed one-third in the equilibrium mixture from any common aldopentose or aldohexose in water. Ribose is an exception among the pentoses; arabinose consists of a total of only 3% furanose forms at equilibrium, and furanose isomers cannot be detected in solutions of xylose and lyxose by means of nmr spectroscopy. The furanose form is apparently destabilized by repulsions between a hydroxyl group and the hydroxymethyl group when they are *cis* to one another, as in D-arabinofuranose (**11**; attachment of a hydroxyl group by a wavy line signifies that the configuration at that carbon atom is unspecified).

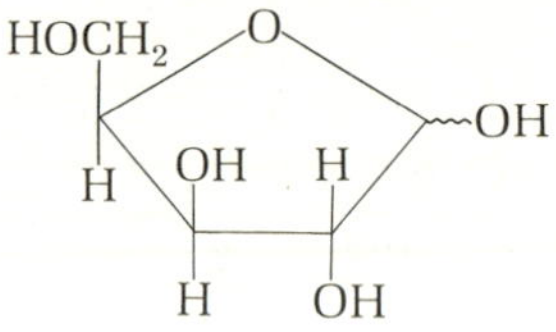

11

In equation 24.23 the pyranose rings have been depicted as being planar (*Haworth projection*). This is not likely to be their actual geometry; in fact, we have every reason to believe that they exist as chair conformers, very similar to those of substituted cyclohexanes. β-D-Ribopyranose should therefore consist of a rapidly interconverting mixture of the chair conformers **12** and **13**, in which the "normal" conformer **12** is of lower energy and should predominate. (The normal conformer is that one in which the anomeric hydroxyl group is equatorial in the β anomer and axial in the α anomer.)

(24.24)

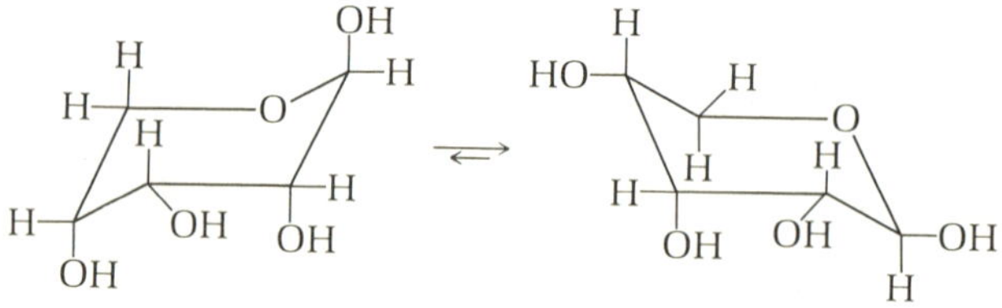

13 **12**

α-D-Ribopyranose also consists of a conformational mixture, as shown in equation 24.25. In this case the conformers should be about equal in energy.

(24.25)

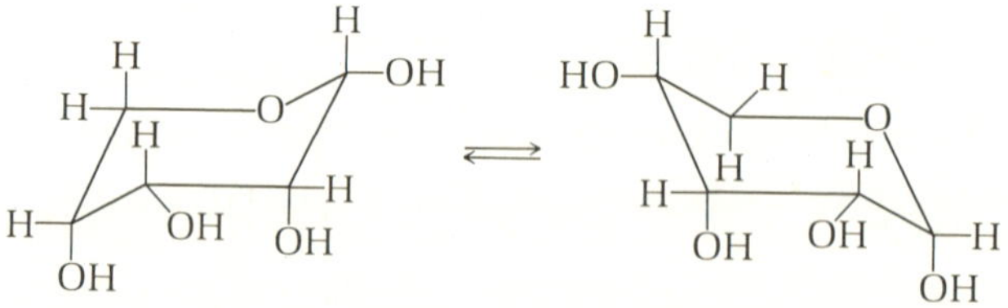

The nmr spectra are completely consistent with these interpretations, and they also confirm the lack of any appreciable concentration of the open-chain aldehyde at equilibrium.

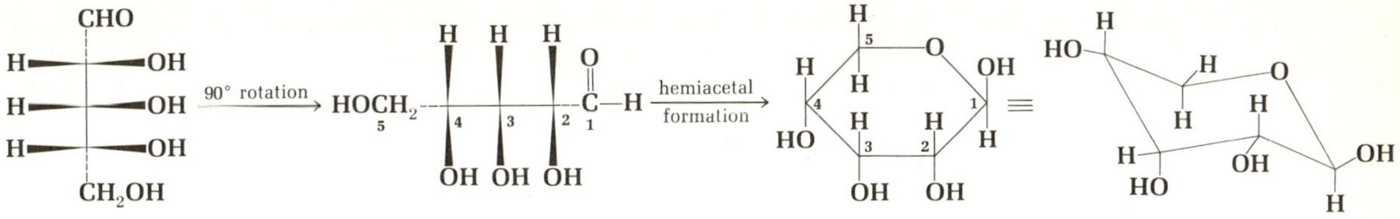

FIGURE 24-1

Derivation of the Haworth and Chair β-Pyranose Projections from the Linear Formula of D-Ribose

In order to transform the open-chair aldopentose structure into the correct Haworth pyranose structure, the following paper operations may be performed. The open-chain structure is rotated 90° clockwise, with the hydrogen and hydroxyl groups projecting above the plane of the paper, and the aldehyde and hydroxymethylene groups are then allowed to react to form the hemiacetal (Figure 24-1). To derive the furanose structure, one begins by turning the entire molecule in the same way and then performing an intramolecular rotation about the C-3–C-4 bond, which puts the hydroxyl group on C-4 behind the plane of the paper and points the hydroxymethylene group up. Hemiacetal formation once again completes the process (Figure 24-2).

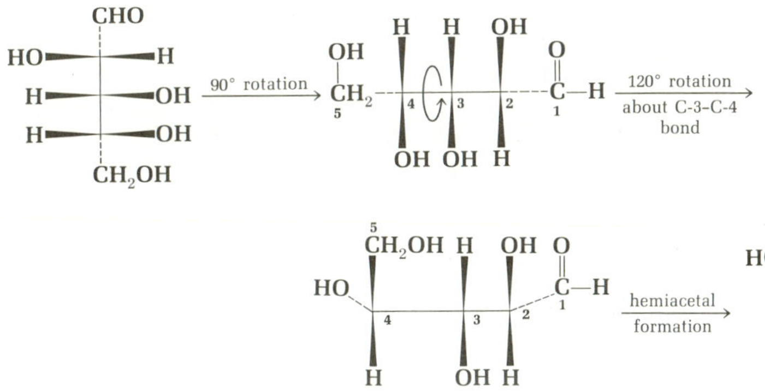

FIGURE 24-2

Derivation of the α-Furanose Form of D-Arabinose from the Linear Projection

PROBLEMS

5 Draw structures of all the cyclic isomers of D-arabinose. Which conformers of the pyranose forms should be predominant? Predict which anomer, the α or the β, will predominate for the furanose and for the pyranose forms.

6 a. What products would be expected from the reaction of D-lyxose with a 1% solution of HCl in methanol?

b. If methyl α-D-lyxofuranoside and methyl α-D-lyxopyranoside are each treated with periodic acid, what products are to be expected? Could this method be used to differentiate the two?

c. What is the relationship between the compounds obtained by periodic acid oxidation of the α- and β-furanosides? Between those obtained by oxidation of the α- and β-pyranosides? In what form are the furanoside oxidation products likely to be present in solution?

MUTAROTATION

D-Ribose may be isolated as a crystalline solid. When the solid is dissolved in water, the specific rotation of the solution is found to equal $-23.1°$. The specific rotation then begins to drop rapidly until it reaches approximately $-18°$, from which it slowly climbs back to $-23.7°$, the equilibrium value. This phenomenon, the change in specific rotation with time until an equilibrium value is obtained, is known as *mutarotation* (Latin *mutare* — to change) and is a common property of most sugar solutions.

When the solution reaches equilibrium, it consists in the main of a mixture of α- and β-furanoses and pyranoses. The crystal, however, need not contain all these forms; in many cases it consists of only one isomer. Crystalline D-ribose seems to contain solely the β-pyranose. The interpretation of the mutarotations of solutions of D-ribose is that the rapid change corresponds to equilibration of the α- and β-pyranose anomers, while the slower alteration is due to the equilibration of the furanose and pyranose ring forms.

With some sugars the change in rotation follows a simple first-order rate law for a reversible reaction (equation 24.26; $[\alpha]$ is the specific rotation of the solution while (α) and (β) are the concentrations of the two anomers).

(24.26) $$\frac{-d[\alpha]}{dt} = k_1(\alpha) - k_2(\beta)$$

This behavior seems to hold if only pyranose forms are present to any substantial extent. The equilibrium may then be depicted as the interconversion of the α- and β-pyranoses by way of the acyclic aldehyde, as shown below for D-xylose.

(24.27)

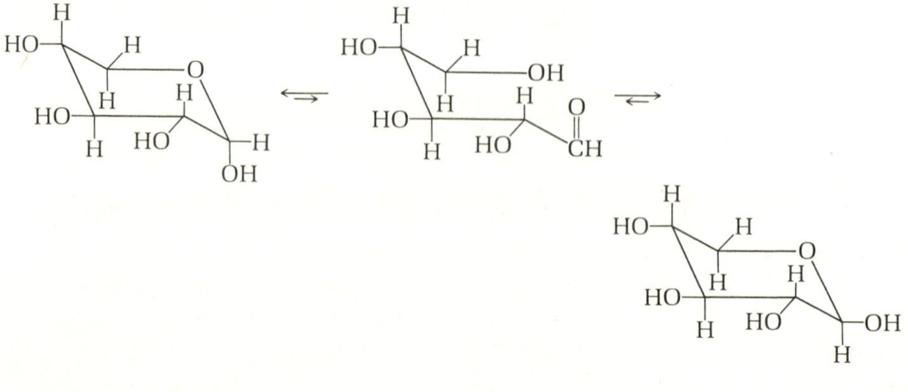

The mutarotation of lyxose also follows the rate law shown in equation 24.26. Pure, crystalline α- and β-D-lyxopyranoses may be obtained by selective crystallization of the sugar from aqueous methanol. The α anomer exhibits $[\alpha]_D^{20} = 5.6°$, while its β epimer has $[\alpha]_D^{20} = -72.6°$. When *either* anomer is dissolved in water, the rotation changes until the equilibrium value of $-13.8°$ is reached.

(24.28)

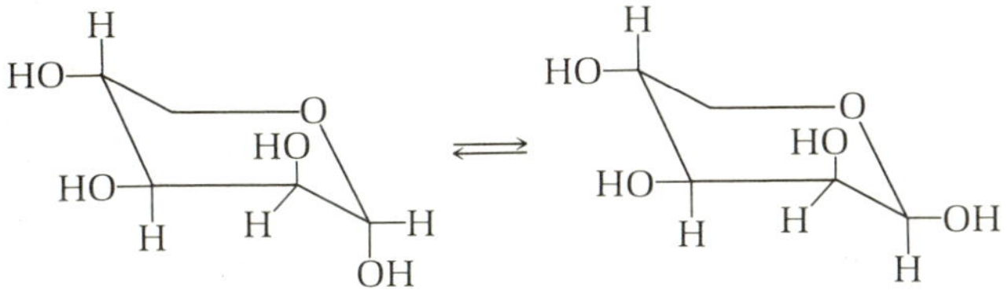

$[\alpha]_D^{20} = 5.6°$ (~70°) $[\alpha]_D^{20} = -72.6°$ (~30°)

One can calculate the percentage of α anomer present at equilibrium with the aid of these figures; it is equal to $[-72.6 - (-13.8)/(-72.6 - 5.6)] = 67\%$, at 20°. This figure agrees well with the value of 71% obtained by means of nmr spectroscopy at 40°.

THE ALDOHEXOSES

When we consider the aldohexoses, we have to deal with eight different open-chain configurations, each of which gives rise to a pair of optical isomers.

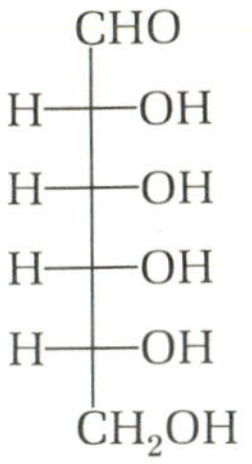

D-allose
(4)

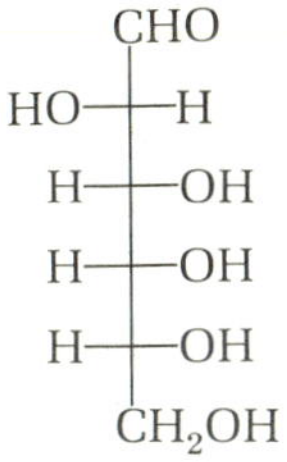

D-altrose
(13)

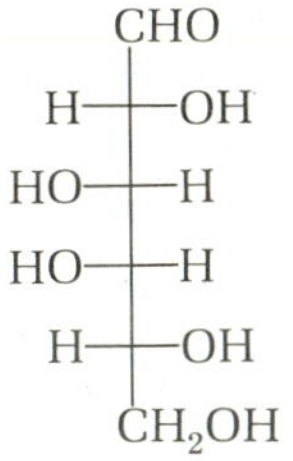

D-galactose
(121)

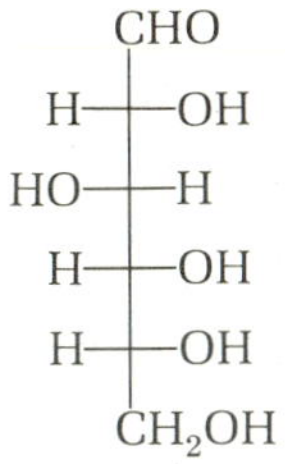

D-glucose
(112)

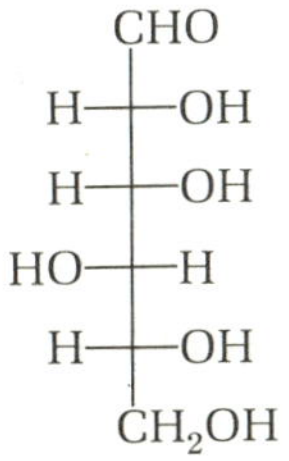

D-gulose
(211)

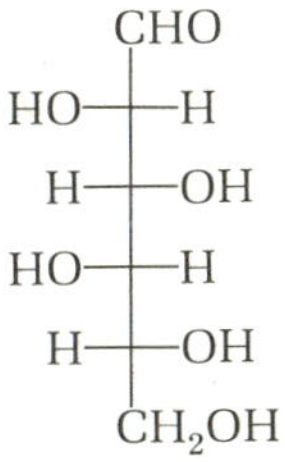

D-idose
(1111)

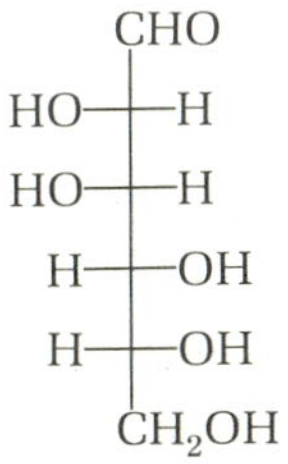

D-mannose
(22)

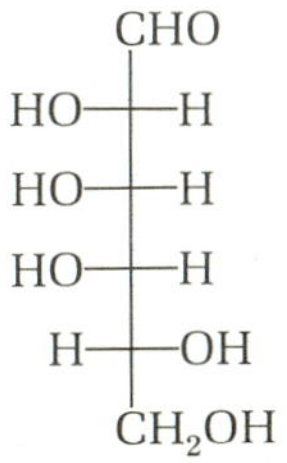

D-talose
(31)

Of all the hexoses, by far the most important is glucose. It is, in the form of its polymers, the most abundant organic compound in Nature, and one of the most heavily studied. Fischer's elucidation of its structure at the end of the nineteenth century not only opened up the entire field of carbohydrate chemistry, but more importantly it validated in a very impressive way the theory of the tetrahedral carbon atom.

PROBLEM

7 a. Group the aldohexoses according to the phenylosazones to which they give rise.

b. Certain pairs of aldohexoses will yield the same alditol upon reduction. Identify these pairs, including those arising from pairing of certain D and L sugars.

c. Indicate which aldohexoses could be obtained from which aldopentoses by way of the Kiliani-Fischer synthesis.

THE STRUCTURE OF GLUCOSE

When Fischer began his work on sugars in 1884, there already existed a small body of data about glucose and related substances. Glucose (sometimes called dextrose because of its rotatory power) had been known for some time since it is isolable from grapes, honey, and raisins. Its molecular formula was established during the nineteenth century, as was the fact that it contained five esterifiable hydroxyl groups (since it could be converted to a pentaacetate). The aldehydic nature of the carbonyl group was established when it was shown that glucose was readily oxidized and readily reduced, no carbon atoms being lost in the process.

Shortly after Fischer's entry into the field, Kiliani made two very important discoveries. Firstly, he noted that several sugars would react with hydrogen cyanide and that the resulting cyanohydrins could be hydrolyzed to polyhydroxycarboxylic acids. When this operation was performed on glucose and the resulting acid treated with hydrogen iodide and phosphorus to reductively remove the hydroxyl groups, heptanoic acid was isolated.

(24.29)
$$\begin{array}{c} CHO \\ | \\ C_5H_{11}O_5 \end{array} \xrightarrow{HCN} \xrightarrow{H_2O} \begin{array}{c} CO_2H \\ | \\ CHOH \\ | \\ C_5H_{11}O_5 \end{array} \xrightarrow{P,\ HI} \begin{array}{c} CO_2H \\ | \\ (CH_2)_5 \\ | \\ CH_3 \end{array}$$

This sequence showed that glucose possessed an unbranched carbon chain and confirmed that the carbonyl group was at the end of the chain.

Kiliani also worked on arabinose, which had been isolated from natural sources. He was able to show that the sugar was an aldopentose, and Fischer subsequently found that the alcohol derived from it by reduction was optically active.

The one other substance of vital importance to this investigation was mannose. A hexahydroxy, optically active alcohol had been isolated from the fruit of the manna ash and named mannitol. When subjected to gentle oxidation, mannitol yielded among other products a new aldohexose that Fischer was able to isolate because of the exceptional insolubility of its phenylhydrazone. He named it mannose and made the remarkable discovery that its phenylosazone was identical to that obtained from glucose. Fischer arrived at the conclusion that this result could be comfortably explained by the theory of the tetrahedral carbon atom; glucose and mannose must differ in absolute configuration at the carbon atom adjoining the aldehyde group and be otherwise identical.

This assumption also made comprehensible some other data relating to these sugars. When glucose was reduced with sodium amalgam under alkaline conditions, it gave rise to mannitol; in the absence of base, another alcohol, sorbitol (glucitol), was the product. By assuming that the base was causing a change of configuration at C-2, these results could be reconciled with the others.

(24.30)

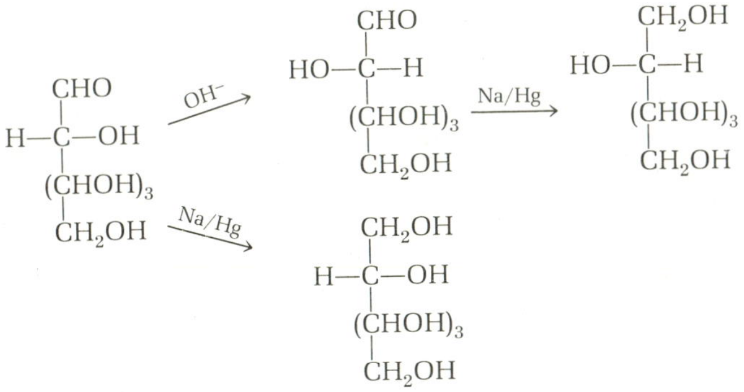

Both glucose and mannose could be oxidized to dibasic acids, called, respectively, saccharic and mannosaccharic acids (in modern nomenclature, glucaric and mannaric acids). These were, of course, nonidentical and *both* were optically active.

(24.31)

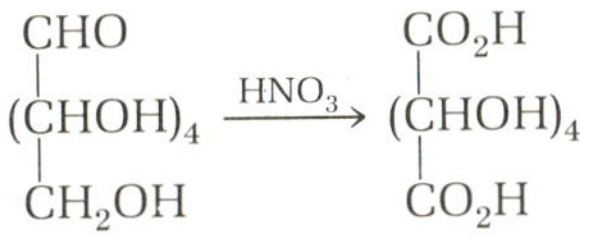

That discovery ruled out structures **14** and **15** for either glucose or mannose and, as a result, their C-2 epimers, **16** and **17**, would also be eliminated. That reduced the number of configurational possibilities to four.

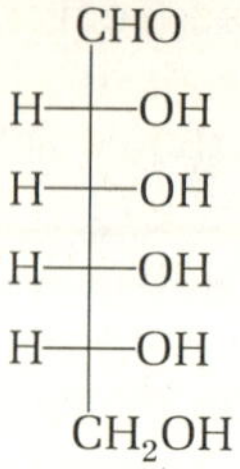

14

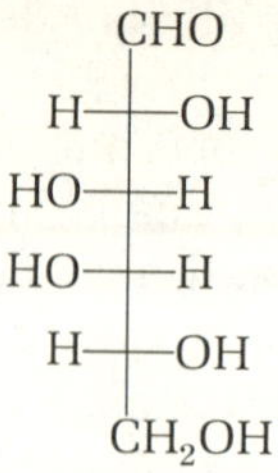

15

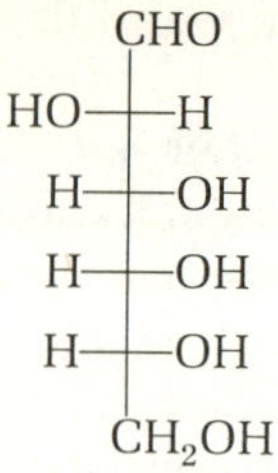

16

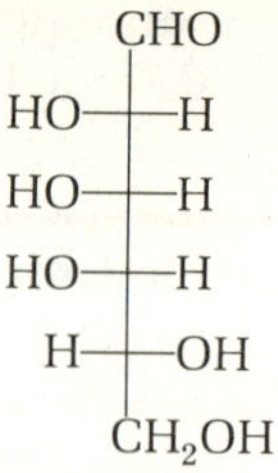

17

Further clarification came from experiments with arabinose. Since the pentose had been shown to give rise to an optically active alcohol, it followed that the hydroxyl groups at C-2 and C-4 were on opposite sides of the carbon chain, as in **18** and **19.**

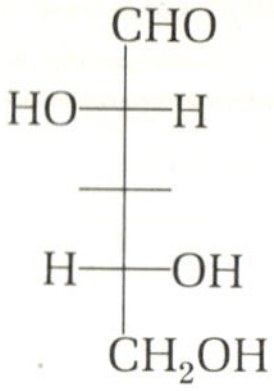

18

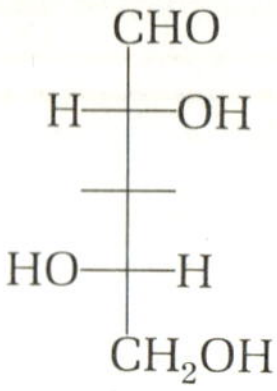

19

When Kiliani treated arabinose with hydrogen cyanide, hydrolyzed the cyanohydrin, and then oxidized the primary alcohol group with nitric acid (equation 24.34), he obtained a dibasic acid that proved to be the mirror image of mannosaccharic acid.

(24.32) $$\begin{array}{c} CHO \\ | \\ (CHOH)_3 \\ | \\ CH_2OH \end{array} \xrightarrow{HCN} \xrightarrow{H_3O^+} \begin{array}{c} CO_2H \\ | \\ (CHOH)_4 \\ | \\ CH_2OH \end{array} \xrightarrow{HNO_3} \begin{array}{c} CO_2H \\ | \\ (CHOH)_4 \\ | \\ CO_2H \end{array}$$

Fischer had arbitrarily decided that in (+)-glucose [and therefore in (+)-mannose] the hydroxyl group on the asymmetric carbon atom furthest removed from the carbonyl group lay to the right of the carbon chain. Saccharic and mannosaccharic acids could therefore be represented by formulas **20** and **21** (or vice versa), the configurations at C-3 and C-5 being fixed by the arabinose results.

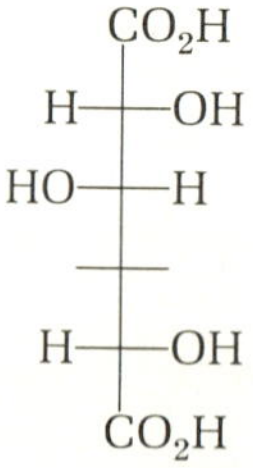

20

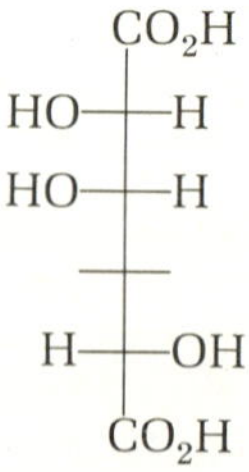

21

Since both acids were optically active, it was essential that the hydroxyl group at C-4 lie on the opposite side of the chain from the hydroxyl at C-3. The structure of glucose was thereby narrowed to a choice between **22** and **23** (the other was mannose), and the structure of the naturally occurring (+)-arabinose was shown to correspond to **24**.

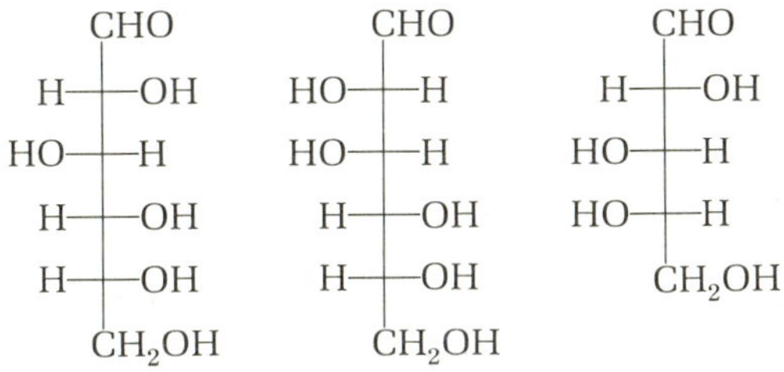

22 **23** **24**

The final link in this exquisite chain was provided by the observation that another naturally occurring aldohexose, named gulose, was converted upon nitric acid oxidation to saccharic acid, identical to that obtained from (+)-glucose. A sugar with the structure **25** could indeed give rise to acid **26**, identical to that obtainable from **22**. *There is, however, no other aldohexose that can give the same dibasic acid as that derived from* **23**. (The reader should satisfy himself or herself on this point.)

(24.33)

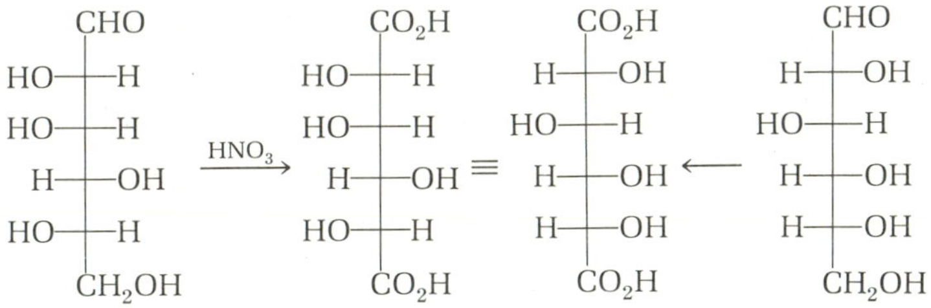

25 **26** **26** **22**

The formula of (+)-glucose was therefore established as **22** and that of (+)-mannose as **23**. With the aid of a minimum number of chemical and physical tools, some very fruitful assumptions, and a rigorous and faultless logical analysis, Fischer had brought into existence the science of carbohydrate chemistry.

CYCLIC STRUCTURE OF GLUCOSE

Many chemists had suggested, even before Fischer's time, that glucose existed partially or wholly in cyclic form. There was some dispute about the size of the ring, five being favored but a number of other sizes being proposed as well. While Fischer's elucidation of the structure of glucose would seem in itself to constitute a powerful argument for the acyclic structure, there were several characteristic properties of glucose that were difficult to reconcile with this formulation.

Fischer found that glucose reacted readily with acidic methanol to yield two isomeric substances, the molecular formulas of which indicated that 1 mole of methanol had been incorporated concomittantly with loss of 1 mole of water. These substances did not react with Tollen's or Fehling's solutions and were later shown to be the methyl α- and β-pyranosides. Had glucose behaved as a typical acyclic aldehyde, it would have reacted with 2 moles of methanol while eliminating 1 mole of water. Fischer later succeeded in isolating the anomeric glucofuranosides, which were shown to isomerize rapidly to the more stable pyranosides under acidic conditions.

Perhaps the most compelling evidence against the open-chain structure was the fact that two different forms of glucose could be obtained simply by crystallizing the sugar from water at different temperatures. These isomers had different melting points and different specific rotations, and solutions of either isomer would undergo mutarotation until the same equilibrium value was reached. It is very difficult to account for the existence of these isomers (which are the α- and β-glucopyranoses) and to explain their mutarotation without invoking cyclic hemiacetal formation.

The proof of the ring size of the glucosides made use of the unusual reactivity of the anomeric hydroxyl group. When the sugar is dissolved in acidic methanol, only this hydroxyl group is replaced by a methoxyl group because loss of the anomeric hydroxyl results in formation of a stabilized oxocarbonium ion.

(24.34)

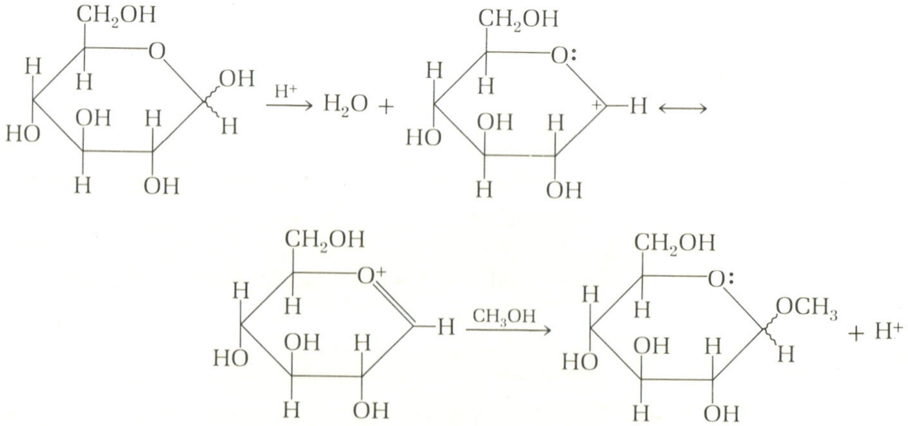

To convert the remainder of the free hydroxyl groups to methoxyl groups requires the conditions of a typical Williamson synthesis. Originally this transformation was accomplished with silver oxide and methyl iodide (equation 24.35); one might now use a mixture of methyl iodide and sodium hydride in an aprotic solvent such as THF, DMF, or DMSO.

(24.35)

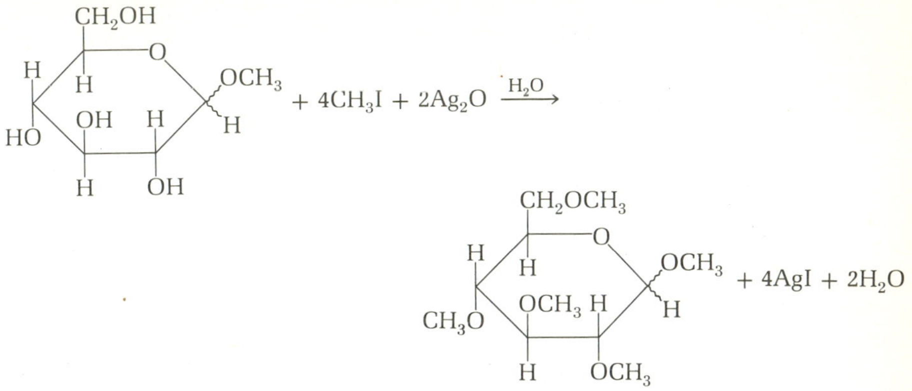

Advantage was then taken of the reactivity of the glycosidic bond to selectively hydrolyze the methoxyl group at C-1, thereby freeing the hydroxyl group that had been involved in acetal formation. Vigorous oxidation of the partially methylated glucose with nitric acid yielded, among other products, a *meso* trimethoxyglutaric acid (equation 24.36); this result is only compatible with the pyranose formulation for the original glucoside.

(24.36)

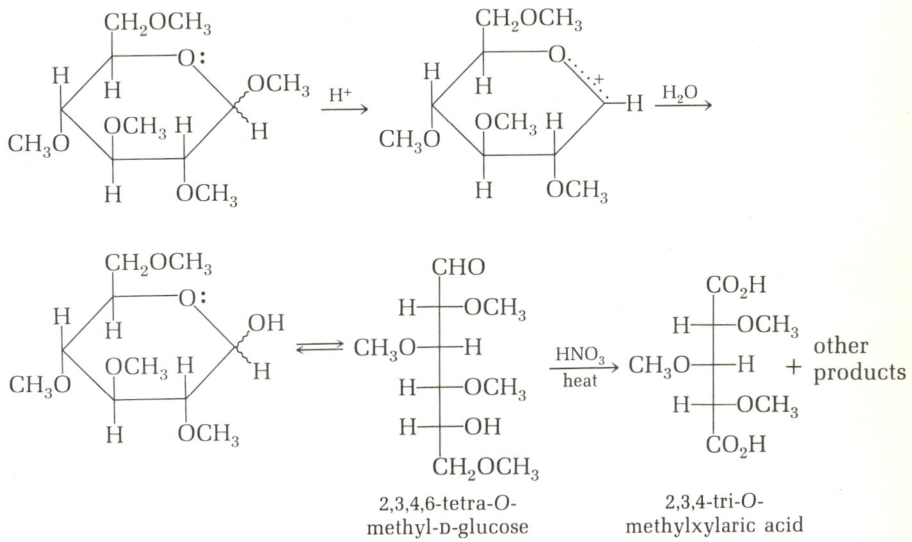

Nmr spectroscopy has confirmed the pyranose structure of glucose in solution and the spectra support our expectation that the pyranose molecule is in the chair conformation. The predominant chair conformer is that one in which the hydroxymethyl group (C-6) is equatorial since it is the largest substituent.

(24.37) 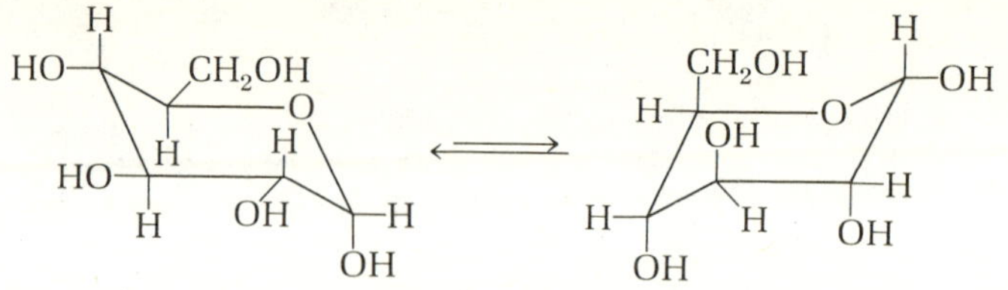

α-D-glucopyranose

All the aldohexoses favor the pyranose form and the predominant conformer of each anomer is usually the one with the equatorial hydroxymethyl group.

PROBLEMS

8 Both α- and β-D-glucopyranose have been isolated as conformationally pure crystalline compounds. The α anomer has $[\alpha]_D^{20} = 112°$, while for the β anomer $[\alpha]_D^{20} = 18.7°$. A solution of either anomer in water undergoes simple mutarotation until a value of 52.7° is reached. Calculate the proportion of each form at equilibrium, and draw the conformationally most stable form of each.

9 a. When treated with acetic anhydride under varying conditions (sodium acetate or zinc chloride catalysis), D-glucose gives rise to two pentaacetates. Draw their structures.

b. When D-galactose is acetylated with acetic anhydride in the presence of sodium acetate, two galactose pentaacetates may be isolated. Each of these may be isomerized to a different pentaacetate in the presence of zinc chloride, giving a total of four galactose pentaacetates in all. Draw their structures. What conclusions might you reach about the structure of D-galactose in solution?

c. How would you account for the difference in behavior of glucose and galactose? (*Hint:* Compare the structures of their pyranose ring forms.)

KETO SUGARS

While the aldoses are the most numerous sugars in Nature, there are several keto sugars that are of considerable importance. Perhaps the most notable of these is D-fructose (Latin *fructa*—fruit), the systematic name of which is D-*arabino*hexulose (ulose denotes a 2-keto sugar, while the prefix gives the relative configuration of the hydroxyls at the asymmetric carbon atoms).

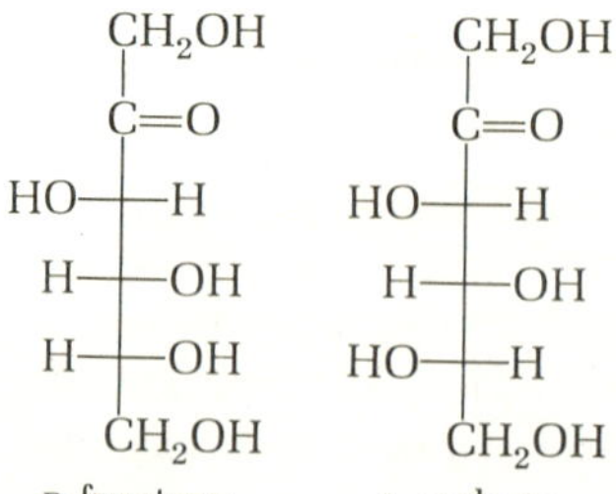

D-fructose L-sorbose

Fructose reacts with Tollen's and Fehling's reagents, and it gives the same osazone as do glucose and mannose. These results led Fischer to propose the correct structure for fructose, a proposal made more plausible by the finding that reduction of D-fructose gave a mixture of D-glucitol and D-mannitol. Interestingly, reduction of L-sorbitol also gives D-glucitol.

(24.38)

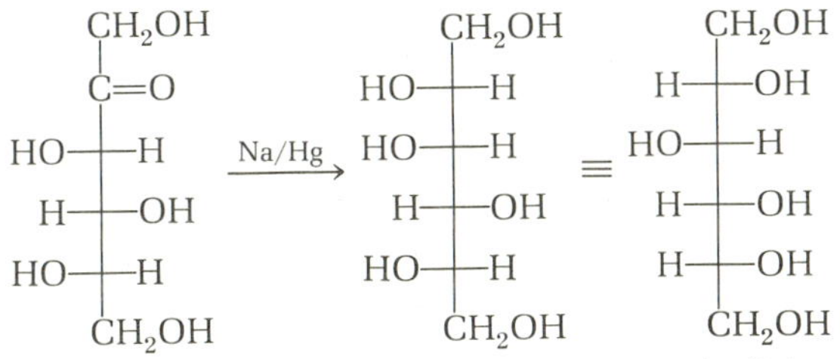

D-glucitol

Under mildly basic conditions a solution of fructose rapidly begins generating glucose and mannose, and eventually all three sugars come to equilibrium (*Lobry de Bruyn-Alberda van Eckenstein rearrangement*).

(24.39)

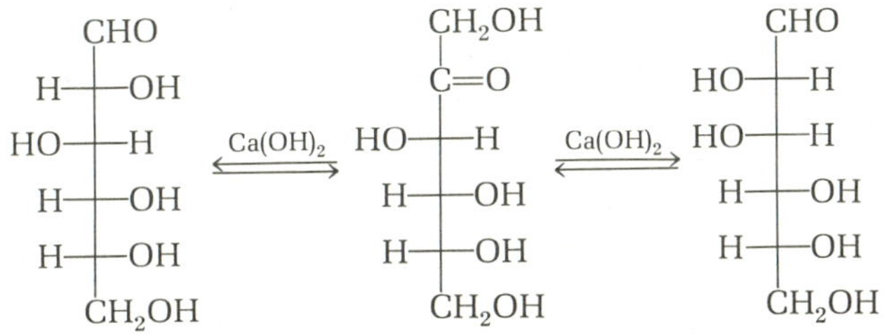

This type of equilibrium may be established among any two aldohexoses epimeric at C-2 and the corresponding 2-ketose. In all cases, the enediol **27** is almost certainly the obligatory intermediate.

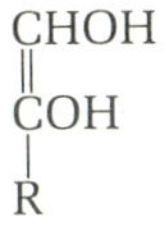

27

PROBLEM 10 Draw the structures of all the D-2-hexuloses, and indicate the D-aldohexoses from which each of them may be derived.

OLIGOSACCHARIDES AND POLYSACCHARIDES

Sugars are often found in Nature combined with one another in short chains (*oligosaccharides*) or in high molecular weight polymers (*polysaccharides*).

The most widely known oligosaccharide is *sucrose*, which is obtainable from sugar cane or sugar beets and is the ordinary table sugar in everyday use.

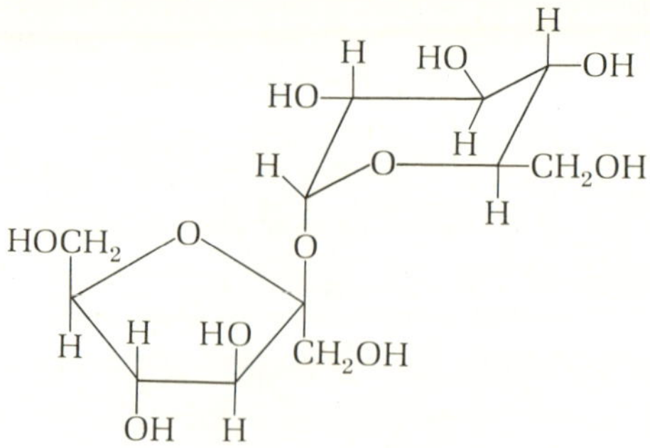

sucrose

Upon aqueous hydrolysis, sucrose yields one molecule of D-fructose and one of D-glucose. Sucrose itself does not react with either Tollen's or Fehling's reagent, indicating that neither of the component sugars is in equilibrium with its carbonyl form; they must each, therefore, be joined through their anomeric hydroxyl groups. The ring size of each monosaccharide may be determined by the usual techniques.

PROBLEM

11 Sucrose was subjected to complete methylation with methyl iodide, after which it was hydrolyzed and the hydrolysis products subjected to vigorous oxidation with nitric acid. Show how this sequence could give information about the ring size of each of the components of sucrose. (Consider the alternative possibilities.)

The determination of the stereochemistry of the glycosidic bond is perhaps the most difficult of tasks for a chemist seeking to unravel the structure of an oligosaccharide. There are no chemical methods that can elicit this information, which is often obtained by inference. The most reliable method is the time-consuming one of X-ray diffraction, which has been applied, for example, to sucrose. Carbohydrate chemists now rely heavily on enzyme-catalyzed oligosaccharide hydrolyses to provide this sort of data, taking advantage of the well-documented stereochemical specificity of most of these catalysts.

There is an enzyme present in yeast and in the human digestive tract that catalyzes the hydrolysis of sucrose to glucose and fructose. In the course of this reaction the sign of rotation of the solution undergoes an inversion from positive to negative, due to the strongly levorotatory power of D-fructose, and the enzyme has accordingly been dubbed *invertase*. In keeping with the X-ray data on sucrose, invertase has been shown to be a β-fructofuranosidase; that is, it specifically catalyzes the hydrolysis of fructofuranosides in which the anomeric oxygen is in the β configuration (equation 24.40).

(24.40) 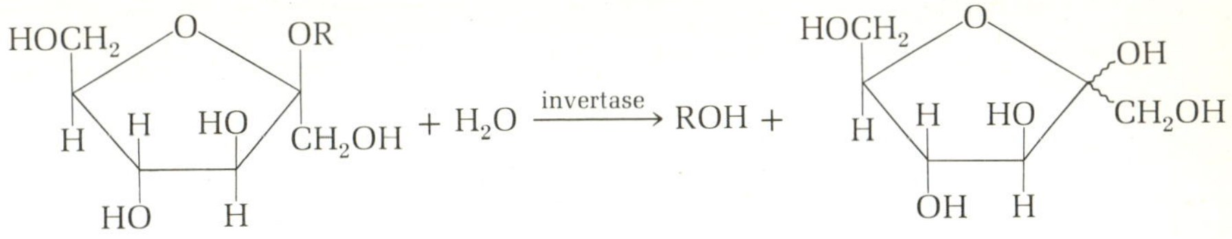

Sucrose will also undergo hydrolysis in the presence of an α-glucosidase isolated from yeast. Sucrose is therefore an α-glucopyranoside as well as a β-fructofuranoside.

Given the configuration of sucrose, its systematic name is α-D-glucopyranosyl-β-D-fructofuranoside; the -yl ending of the name of the first component indicates that it is connected to the second through its anomeric oxygen atom.

Several common reducing disaccaharides (those with free anomeric hydroxyl groups) are depicted below, along with their systematic names. The numbers connected by an arrow contained in parentheses denote the carbon atoms holding the hydroxyl groups involved in ether formation.

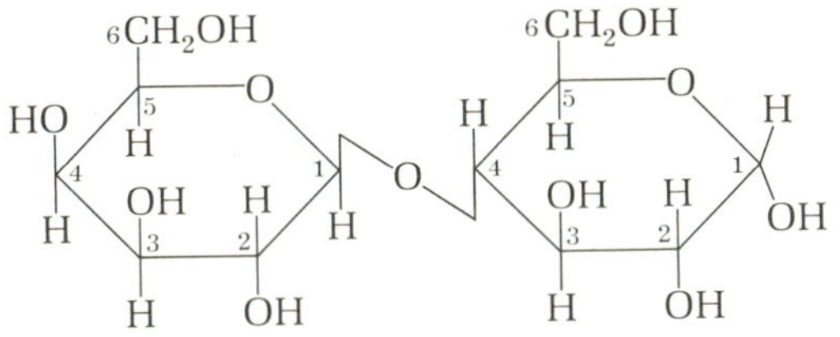

α-lactose
[O-β-D-galactopyranosyl-
(1 ⟶ 4)-α-D-glucopyranose]

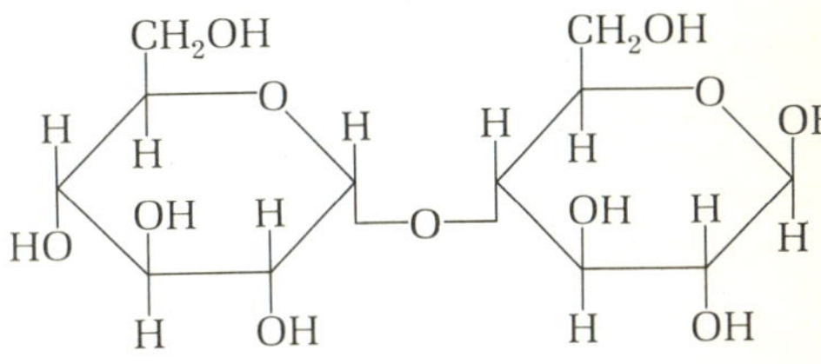

β-maltose
[O-α-D-glucopyranosyl-
(1 ⟶ 4)-β-D-glucopyranose]

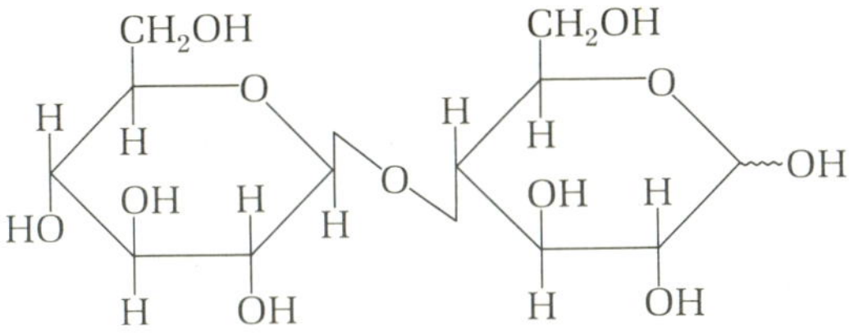

cellobiose
[O-β-D-glucopyranosyl-
(1 ⟶ 4)-D-glucopyranose]

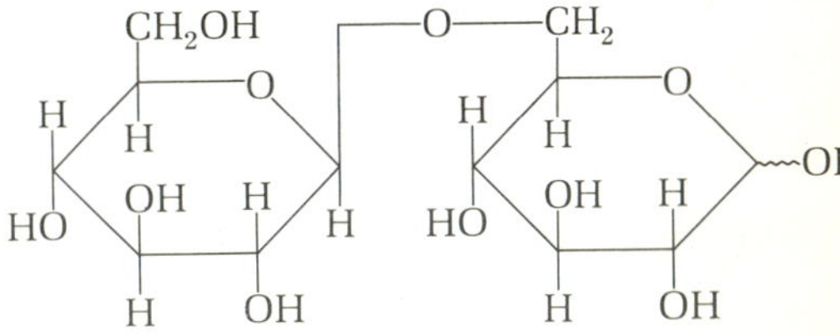

gentiobiose
[O-β-D-glucopyranosyl-
(1 ⟶ 6)-D-glucopyranose]

Maltose, cellobiose, and gentiobiose yield solely glucose upon acidic hydrolysis. Cellobiose and gentiobiose are also hydrolyzed in the presence of the enzyme emulsin, isolated from almonds, which contains a β-glucosidase. Maltose is unaffected by this enzyme but it will hydrolyze in the presence of the α-glucosidase maltase. Note that lactose is affected neither by emulsin nor by maltase since it is a *galactoside* rather than a *glucoside*. All four disaccharides are reducing since they contain free anomeric hydroxyl groups.

PROBLEM

12 a. Gentianose is a trisaccharide composed of glucose and fructose. When treated with emulsin solution, gentianose is hydrolyzed to glucose and sucrose, while invertase-catalyzed hydrolysis results in the release of gentiobiose and fructose. Write the structure of gentianose.

b. Raffinose is a nonreducing trisaccharide, which upon acid hydrolysis yields 1 mole each of fructose, galactose, and glucose. Upon treatment with invertase, raffinose gives rise to fructose and the disaccharide melibiose. After acid hydrolysis, melibiose gives glucose and galactose but it is resistant to an enzyme that hydrolyzes lactose. Suggest possible structures for melibiose and raffinose. What additional information would be needed to definitively assign structures to these compounds?

Polysaccharides are very ubiquitous in Nature; in particular, they serve as structural materials and as reserve foodstuffs. Starch and glycogen are storage forms of glucose in plants and animals, respectively. Starch contains two glucose polymers, *amylose* and *amylopectin*, each of which consists in the main of 1,4-linked α-glucosides. There are branches in these chains, however; in amylopectin, there is extensive cross-linking of different saccharide chains by way of 1,6-α-linkages (Figure 24-3). Glycogen, the principal

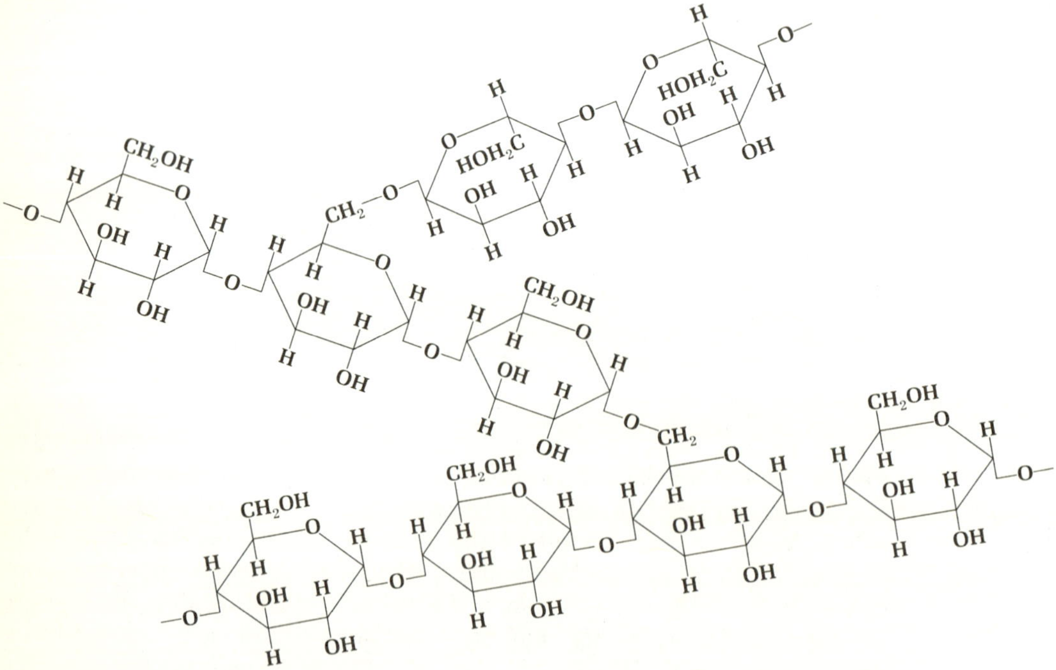

FIGURE 24-3

Section of an Amylopectose Chain Showing the Extensive Cross-Linking

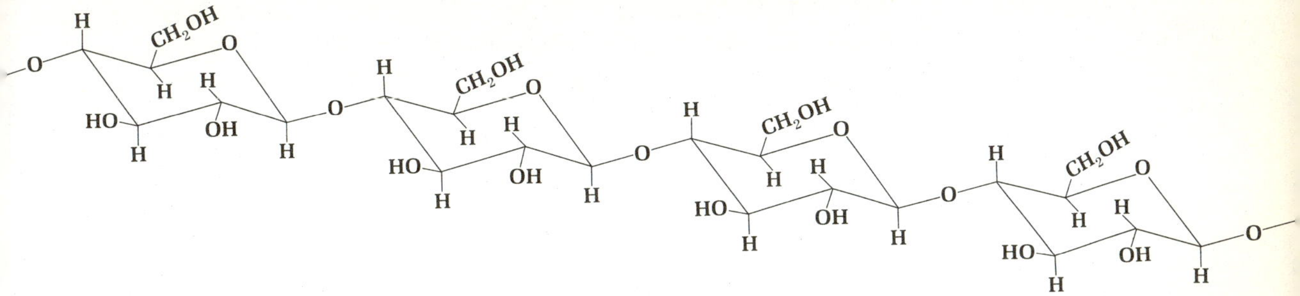

FIGURE 24-4

Section of a Cellulose Molecule

carbohydrate reserve of animals, is a heavily cross-linked poly-1,4-α-glucoside.

Cellulose is the cell wall material of higher plants. It too yields only glucose upon acid hydrolysis but, in contrast to starch and glycogen, it consists of very long 1,4-linked β-glucoside chains (Figure 24-4). The great strength and water insolubility of cellulose may be ascribed to the fact that the individual chains are held one to another by an extensive series of hydrogen bonds.

NUCLEIC ACIDS

One of the most fascinating, awesome, and potentially cataclysmic problems under investigation by chemists and biologists is the mechanism by which living cells transmit genetic information. In essence, this requires an understanding of the means by which amino acid units are linked together in the correct sequence to produce the required peptides and proteins (Chapter 25).

The substances most intimately linked with the transmission and utilization of genetic information are the *nucleic acids*. Complete hydrolysis of these high molecular weight (10^4 to 10^9) polymers reveals that they are composed of three different types of components: a purine or pyrimidine base (Chapter 23), an aldopentose, and phosphoric acid. Only two pentoses have been isolated from nucleic acids, which contain only one particular sugar in a given molecule. One of these sugars is D-ribose, which is found in **ribonucleic acids (RNA),** and the second is 2-deoxy-D-ribose **(28),** a component of the **deoxyribonucleic acids (DNA).**

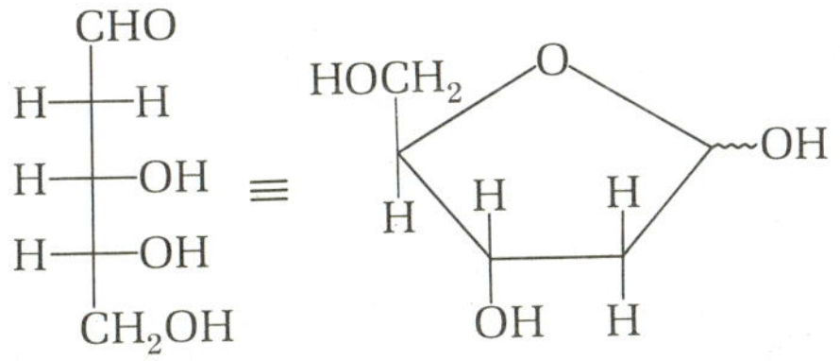

28

The fundamental repeating unit of the nucleic acids are *nucleotides* (containing D-ribose) and *deoxynucleotides* (containing 2-deoxy-D--ribose), each of which consists of a molecule of base, one of sugar, and one of phosphoric acid. The structures of some of these are depicted below. Adenine, guanidine, and cytosine are found in both RNA and DNA: uracil is found only in RNA and thymine only in DNA. (Not shown are deoxyadenylic, deoxyguanylic, and deoxycytidylic acids.)

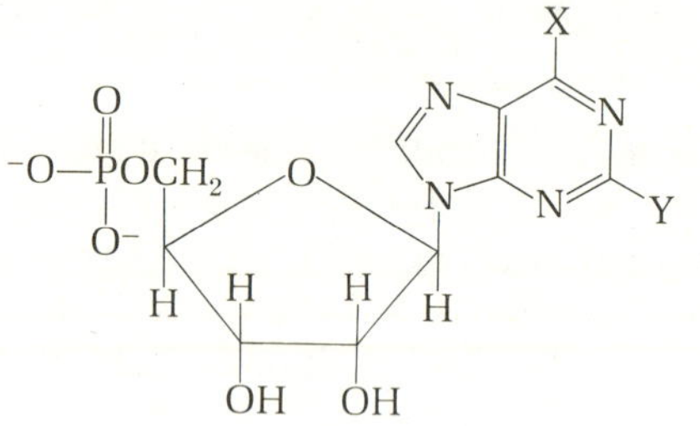

adenylic ($X = NH_2$, $Y = H$) and guanylic ($X = OH$, $Y = NH_2$) acids

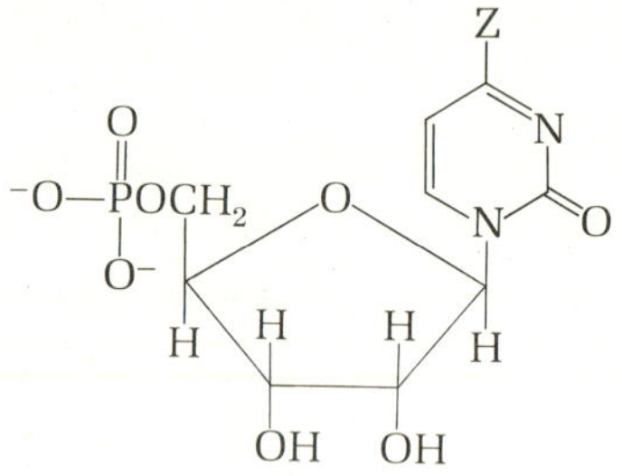

cytidylic ($Z = NH_2$) and uridylic ($Z = OH$) acids

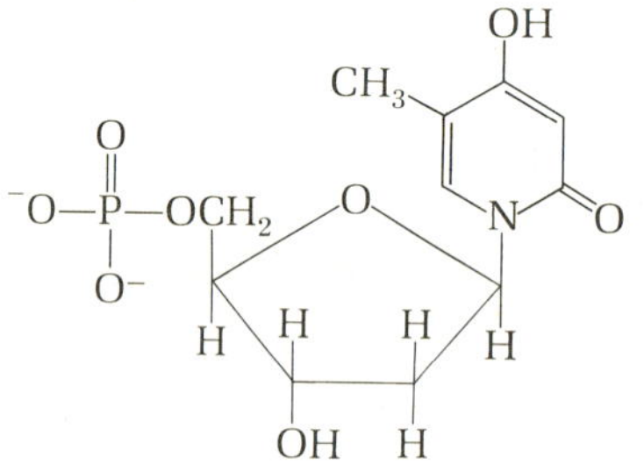

thymidylic acid

Molecules containing only a heterocyclic base and a sugar are known as *nucleosides;* the common ones are adenosine, cytidine, guanosine, thymidine, uridine, deoxyadenosine, etc. They may be isolated not only from the nucleic acids but from numerous other physiologically important molecules as well, such as adenosine triphosphate (ATP, p. 619), nicotinamide adenine dinucleotide (NAD, p. 654), and coenzyme A (p. 461).

THE GENETIC CODE

Nucleic acids are nucleotide polymers in which the phosphate group on one molecule is condensed with a hydroxyl group at C-3 of a second molecule (Figure 24-5). The genetic information is contained in the sequence of bases along the polymer chain. A set of three bases is responsible for the placing

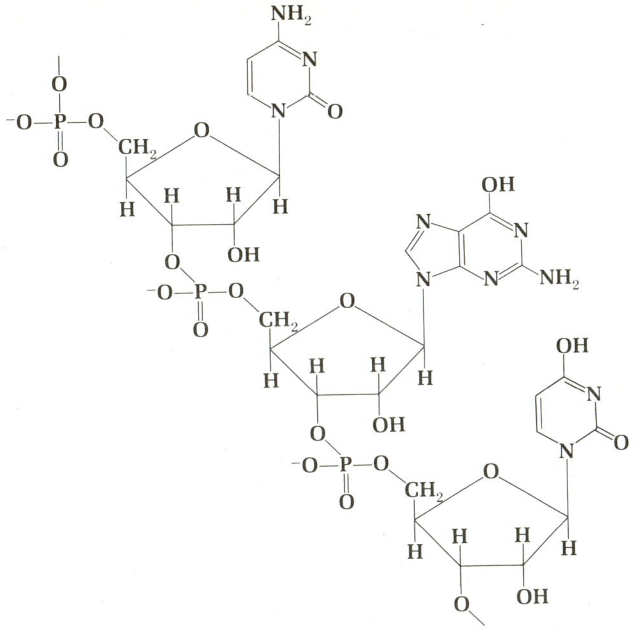

FIGURE 24-5

A Portion of a Ribonucleic Acid Molecule

of one single amino acid in a protein polymer — the correspondence of nucleic acid bases and protein amino acids gives rise to the **genetic code.**

The means by which the genetic code information is transferred from its storage site to the point of operational utilization is relatively simple but quite subtle. Data for protein synthesis are contained in the chromosomes found in the cell nucleus; each chromosome is composed of a number of genes, each of which contains the code for a single protein. Chromosomes are composed of DNA and they are the only sites within the cell in which DNA is found.

The DNA exists as a double strand twisted into the form of a helix. The two strands are held together by hydrogen bonds between complementary bases on each of the two chains. Thus, adenine bonds to thymine and guanidine with cytosine (Figure 24-6). If the strands become separated, as they do during cell division, then a new complementary strand may be synthesized, the sequence of which *is already determined by the original strand to which it is being matched.*

In the course of normal cell operation, an enzyme causes the synthesis of RNA molecules that are complementary to *one* of the strands of the DNA helix. (In RNA, uracil fulfills the role that thymidine did in DNA.) These molecules, termed *messenger RNA,* subsequently migrate from the nucleus to particles within the cytoplasm, called ribosomes. It is within the ribosomes that protein synthesis actually occurs. Special RNA molecules contained therein, known as *transfer RNA,* bring individual amino acids up to

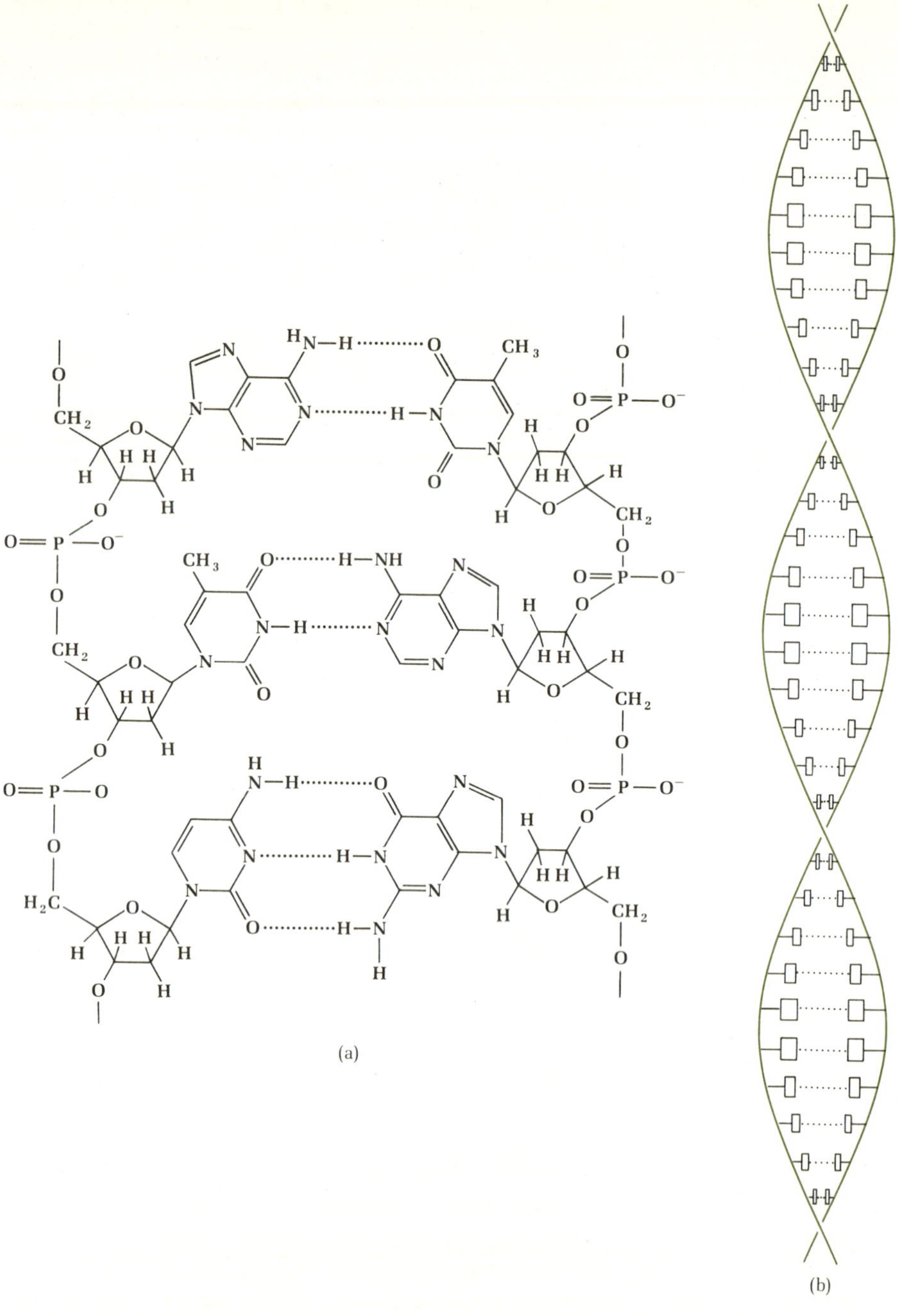

FIGURE 24-6

The Double Helix of DNA (a) A Short Section Showing the Hydrogen Bonding Between Complementary Bases in Different Chains (b) A Schematic Representation of the Two Strands Joined by Base Pair Interactions as in (a) the messenger RNA, which will cause them to be connected in the proper order. For example, if the messenger RNA molecule begins with the sequence UUGGCU,..., then the protein sequence will begin phenylalanine-glycine,.... UUG is the *codon* for phenylalanine and GCU the codon for glycine. (A stands for adenine, C for cytosine, G for guanidine, and U for uracil.) Through the operation of this sequence, either a weasel or a whale might be brought into existence.

SUMMARY

1 Simple sugars are polyhydroxy aldehydes and ketones, most of them conforming to the molecular formula $(CH_2O)_n$. They all share certain traits in common — a ready oxidizability by reagents such as Fehling's and Tollen's solutions, susceptibility to periodate oxidation, facile isomerization in base, and a tendency toward hemiacetal formation.

2 Most sugars contain several asymmetric atoms so that there are a number of geometrical isomers that correspond to the same structural formula. It was decided some time ago to relate the absolute configuration at each atom to (+)-glyceraldehyde, which was assumed to have the *R* configuration, an assumption subsequently proved to be fortuitously correct.

3 The problem of deducing the relationship among the asymmetric atoms within a molecule was classically solved by transforming the sugar into a compound with identical terminal groups and then measuring the effect of this transformation on the molecule's optical activity. With aldehyde sugars this result could be achieved either by reducing the aldehyde group to an alcohol or by oxidizing both the aldehyde and terminal hydroxyl groups to carboxyl groups. Selective oxidation of aldoses to aldonic acids is also possible by use of iodine in alkaline solution.

4 Structural relationships between sugars with different numbers of carbon atoms may be established by synthesizing a higher sugar from a lower one (Kiliani-Fischer synthesis). Addition of hydrogen cyanide to an aldose, hydrolysis of the nitrile, and reduction of the corresponding lactone result in the production of two isomeric aldoses containing one more carbon atom than the starting sugar. The products have epimeric configurations at C-2 but are otherwise identical, as confirmed by their each reacting with 3 moles of phenylhydrazine to give the same osazone. Formation of this derivative involves the loss of asymmetry at C-2 but the remainder of the asymmetric atoms are left untouched.

Another method of relating sugars depends on the selective degradation of an aldose to its next lower homologue. In the course of this procedure, C-1 is removed and C-2 is converted to a carbonyl group. A widely used method is the Ruff degradation (bromine water followed by oxidation of the aldonate salt by hydrogen peroxide).

5 When treated with an alcohol in the presence of acid, sugars form mixed acetals or ketals (glycosides). One hydroxyl group for acetal or ketal formation comes from within the molecule, while a second is supplied by the external alcohol. These derivatives may contain five-membered (furanoside) or six-membered (pyranoside) rings. Since the former carbonyl carbon atom is converted to an asymmetric atom in these derivatives, each furanoside and pyranoside isomer can exist in α and β forms (C-1 substituent down or up, respectively).

6 Most sugars appear to exist primarily as internal hemiacetals. While only one isomer is present in sugar crystals, a solution may consist of a mixture of α- and β-furanoses and pyranoses. Nmr spectra of sugar solutions are fully in accord with such an interpretation and relative proportions of each isomer in the mixture may be explained with the aid of familiar conformational concepts.

Not only the glycosides but the sugars themselves appear to exist overwhelmingly in cyclic form in both the crystal and in solution. The specific rotation of a freshly prepared sugar solution changes with time until an equilibrium value is reached. This phenomenon, known as mutarotation, served as strong evidence in favor of the cyclic hemiacetal theory. Crystalline sugars are conformationally homogenous but upon dissolution in water equilibration among the α- and β-furanoses and pyranoses sets in. Nmr spectroscopy also supports this interpretation.

7 Fischer's deduction of the structure of glucose was a classic stereochemical study. It rested in the main on the following observations. Mannose and glucose were epimeric at C-2 and otherwise identical (identical phenylosazones, etc.) and both glucaric and mannaric acids were optically active. Naturally occurring arabinose could be converted, by way of the Kiliani-Fischer synthesis, to a mannaric acid that was the mirror image of that obtained by oxidation of natural mannose. Finally, it was shown that there was another sugar, gulose, that upon oxidation gave rise to glucaric acid identical with that available from oxidation of (+)-glucose.

8 Several keto sugars are available from natural sources. The most important and widespread of these is D-fructose. Since D-fructose gives rise to the same osazone as does mannose and glucose, it must be D-*arabino*hexulose. In the presence of mild base, any one of these three hexoses may generate the other two by way of a common enediol.

9 Oligomers and polymers of simple sugars are very widely distributed in Nature. In virtually all cases when two monosaccharides are connected together, at least one of them is joined by way of its anomeric carbon atom. If all the individual sugar molecules utilize their anomeric hydroxyl groups in forming interglycosidic bonds, then the polysaccharide will be nonreducing.

The gross composition of a polysaccharide may be obtained by subjecting the compound to complete hydrolysis in aqueous acid and then analyzing the mixture of sugars so produced. If the polysaccharide is completely methylated before it is hydrolyzed, then one may deduce which hydroxyl groups had been involved in interglycosidic bonds. The determination of the stereochemistry of the interglycosidic bonds is best accomplished by making use of a range of enzymes that are specific for a given glycoside and for a given configuration at the anomeric carbon atom.

10 The genetic information necessary to the growth and reproduction of all cells is locked within the deoxyribonucleic acids (DNA) present in their nuclei. These giant molecules are polymers containing 2-deoxy-D-ribose, phosphoric acid, and the four heterocyclic bases: adenine, cytosine, guanidine, and thymine. These are arrayed in a regular manner along the polymer chain, the sequence of bases in the molecule constituting the actual genetic information. A triplet of bases, or codon, stands for a particular amino acid that will be inserted into a growing peptide chain at precisely the correct moment.

Under enzymatic control, a ribonucleic acid (RNA) is synthesized in the nucleus that is complementary to one strand of the double helical DNA. Ribonucleic acids are polymers of D-ribose, phosphoric acid, and adenine, cytosine, guanidine, and uracil. This messenger RNA migrates out of the nucleus to the ribosomes where protein synthesis is carried out. Messenger RNA serves as a template for protein synthesis, while transfer RNA's act as carriers to bring the proper amino acids over to the template.

PROBLEMS

Note: Problems 1 to 12 appear in text.

13 Write the structures of the following compounds, showing their stereochemical configuration. Is it necessary to make any assumptions in assigning these configurations? How does the D,L system of absolute configuration compare with the *R*,*S* system?

a. D-2,3-Butanediol

b. L-Malic acid (2-hydroxysuccinic acid)

c. L-2-Chloro-1-propanol

d. D-Lactic acid (α-hydroxypropionic acid)

e. D-*meso*-4-Methyl-2-hexanol

14 Show the products expected from periodic acid oxidation of the following compounds.

a. Ethylene glycol

b. Cyclopentane-1,2-diol

c. Glycerol

d. 1,2,3-Butanetriol

e. 1,2,4-Butanetriol

f. 2-Hydroxycyclohexanone

g. Pentane-2,3-dione

h. 4-Hydroxyhcptane-2,3-dione

i. Lactic acid (α-hydroxypropionic acid)

j. Tartaric acid

k. Threonic acid

15 a. What products would be expected from the oxidation of erythrose or threose with periodic acid?

b. Two sets of products are obtainable from erythrulose upon periodate oxidation, depending on the path taken. What are these two paths, and what are the products derived from each? Could the periodate method be used to distinguish isomeric aldoses and ketoses?

16 Propose structures for **A** to **E.**

$$\text{D-Erythrose} \xrightarrow{I_2,\ KOH} \xrightarrow{H^+} (+)\text{-}\mathbf{A} \xrightarrow{HNO_3} (+)\text{-}\mathbf{B}\ (C_4H_6O_6) \xrightarrow[\text{vacuum}]{\text{heat}}$$

$$\mathbf{C}\ (C_4H_4O_5) \xrightarrow{CH_3OH} \mathbf{D}\ (C_5H_8O_6) \xrightarrow[THF]{BH_3} \mathbf{E}\ (C_5H_{10}O_5) \xrightarrow{H_3O^+} (+)\text{-}\mathbf{A}$$

17 a. Propose structures for **A** and **B.** (The use of models may be beneficial in solving this problem.)

$$\text{I} \xrightarrow{NaIO_4} \mathbf{A}\ (C_{11}H_{12}O_4) \xrightarrow{H_3O^+} \mathbf{B}\ (C_4H_8O_4)$$

b. Compound I may be obtained from D-arabinose. Outline a method for bringing about that transformation.

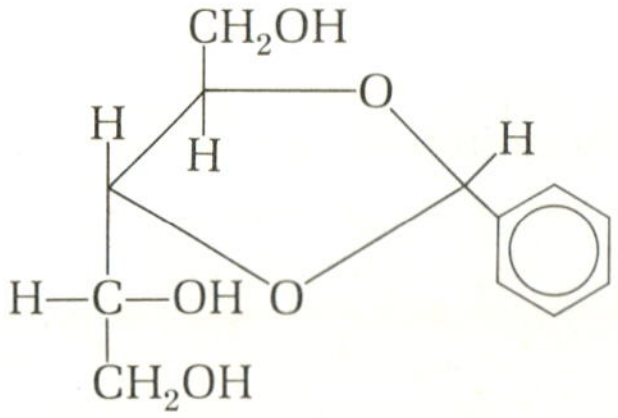

I

18 Draw likely structures for the products of the reaction shown below, keeping in mind the possibility of stereochemical isomerism.

$$\text{Glyceraldehyde} \xrightarrow[H^+]{CH_3OH} \text{several isomers of formula } C_8H_{16}O_6$$

19 Trityl chloride, $(C_6H_5)_3CCl$, is a valuable protecting group in carbohydrate chemistry. It reacts with primary hydroxyl groups in pyridine solution and converts them to trityl ethers; because of its bulk, trityl chloride usually does not react at an appreciable rate with secondary or tertiary hydroxyl groups.

$$\begin{array}{l} CH_2OH \\ | \\ CHOH \\ | \\ CH_3 \end{array} + (C_6H_5)_3CCl \xrightarrow{\text{pyridine}} \begin{array}{l} CH_2OC(C_6H_5)_3 \\ | \\ CHOH \\ | \\ CH_3 \end{array} \xrightarrow{H_3O^+} \begin{array}{l} CH_2OH \\ | \\ CHOH \\ | \\ CH_3 \end{array} + (C_6H_5)_3COH \qquad \textbf{(1)}$$

The ether is quite stable in neutral or basic media but when it is treated with dilute aqueous acid at 25°, hydrolysis rapidly takes place and the original hydroxyl group is regenerated (equation 1).

$$\text{Methyl } \beta\text{-D-glucoside} \xrightarrow[\text{pyridine}]{(C_6H_5)_3CCl} \mathbf{A}\ (C_{26}H_{28}O_6) \xrightarrow[\text{acetone}]{CrO_3} \xrightarrow[25°]{\text{dil. } H_3O^+}$$

$$\mathbf{B} + \mathbf{C} + \mathbf{D} \text{ (all isomers of } C_7H_{12}O_6)$$

$$\mathbf{B} \xrightarrow{NaBH_4} \xrightarrow[\text{heat}]{H_3O^+} \text{D-glucose and D-mannose}$$

$$\mathbf{C} \xrightarrow{NaBH_4} \xrightarrow[\text{heat}]{H_3O^+} \text{D-allose and D-glucose}$$

$$\mathbf{D} \xrightarrow{NaBH_4} \xrightarrow[\text{heat}]{H_3O^+} \text{D-galactose and D-glucose}$$

What are the structures of compounds **A** to **D**?

20 2,3,4,5-Tetra-*O*-methyl-D-glucose, when dissolved in D_2O, gives an nmr spectrum that contains C—H peaks not found in the nmr spectrum of D-glucose. What species might reasonably be present, giving rise to these peaks?

21 Upon acid hydrolysis, *lycotetraose* generates one molecule each of D-galactose and D-xylose and two molecules of D-glucose. If the acid hydrolysis is interrupted before completion, several disaccharides and trisaccharides may be isolated. Among these are *lycobiose* and *sophorose*. Lycobiose is susceptible to emulsin-catalyzed hydrolysis and yields 1 mole each of galactose and glucose. Sophorose is also hydrolyzed in the presence of emulsin, the products being two molecules of glucose. Complete methylation of lycotetraose, followed by total acid hydrolysis, yields the following products: 2,3,6-tri-*O*-methyl-D-galactose; 2,3,4,6-tetra-*O*-methyl-D-glucose; 4,6-di-*O*-methyl-D-glucose; and 2,3,4-tri-*O*-methyl-D-xylose. Write the structure of lycotetraose, specifying where possible the geometry of the glycosidic bonds.

SUPPLEMENTARY READINGS

C. S. Hudson, "Emil Fischer's Proof of the Configuration of Glucose," *J. Chem. Ed.*, **18,** 353 (1941).

E. A. Davidson, *Carbohydrate Chemistry*. New York: Holt, Rinehart and Winston, Inc., 1967.

B. Capon, "Mechanism in Carbohydrate Chemistry," *Chem. Rev.*, **69,** 407 (1969).

J. D. Watson, *The Double Helix*. New York: Athencum Pub., 1968. A personal, witty, fascinating, and iconoclastic view of the discovery of the double helical conformation of DNA, related by one of the discoverers.

CHAPTER 25

PEPTIDES, PROTEINS, AND ENZYMES

Life as we know it would not be possible in the absence of those universally important polymers of α-amino acids, the peptides and proteins. This chapter is intended to provide a selective glimpse of what is a vast and burgeoning autonomous subdiscipline of organic chemistry. It will concern itself first with properties of the peptide building blocks, the α-amino acids, and then discuss the structural analysis of peptides and proteins. Such an analysis requires not only a knowledge of the sequential order of amino acids in the protein chain, but also a picture of the three-dimensional conformation of the entire molecule as well. The biological functioning of these molecules is known to be critically dependent on their architecture. Finally, the chapter will address itself to the problem of how peptides and proteins may be synthesized in the laboratory, a task that requires a great deal of chemical sophistication.

ELECTRICAL PROPERTIES OF α-AMINO ACIDS

The amino acids that are important as protein components are all α-amino acids. There are 20 that are widely found in living systems and that may

TABLE 25-1
Fundamental Amino Acids

Name	Structure	Abbreviation
	Nonpolar Side Chains	
Alanine	$CH_3-\underset{\displaystyle NH_2}{\underset{\displaystyle \vert}{C}}HCO_2H$	Ala
Glycine	$H-\underset{\displaystyle NH_2}{\underset{\displaystyle \vert}{C}}HCO_2H$	Gly
Isoleucine	$CH_3CH_2\underset{\displaystyle CH_3}{\underset{\displaystyle \vert}{C}}H-\underset{\displaystyle NH_2}{\underset{\displaystyle \vert}{C}}HCO_2H$	Ile
Leucine	$(CH_3)_2CHCH_2-\underset{\displaystyle NH_2}{\underset{\displaystyle \vert}{C}}HCO_2H$	Leu
Methionine	$CH_3SCH_2CH_2-\underset{\displaystyle NH_2}{\underset{\displaystyle \vert}{C}}HCO_2H$	Met
Phenylalanine	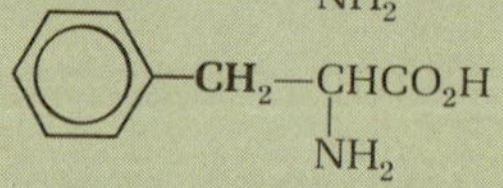	Phe
Proline	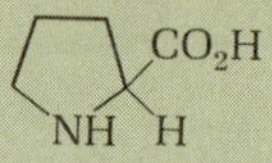	Pro
Tryptophan	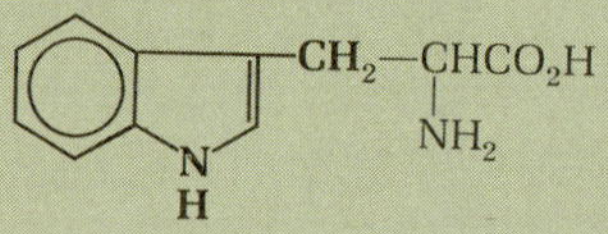	Trp
Valine	$(CH_3)_2CH-\underset{\displaystyle NH_2}{\underset{\displaystyle \vert}{C}}HCO_2H$	Val
	Polar, Nonionizing Side Chains	
Asparagine	$H_2N\overset{\displaystyle O}{\overset{\displaystyle \Vert}{C}}CH_2-\underset{\displaystyle NH_2}{\underset{\displaystyle \vert}{C}}HCO_2H$	Asn
Glutamine	$H_2N\overset{\displaystyle O}{\overset{\displaystyle \Vert}{C}}CH_2CH_2-\underset{\displaystyle NH_2}{\underset{\displaystyle \vert}{C}}HCO_2H$	Gln
Serine	$HOCH_2-\underset{\displaystyle NH_2}{\underset{\displaystyle \vert}{C}}HCO_2H$	Ser
Threonine	$CH_3\underset{\displaystyle OH}{\underset{\displaystyle \vert}{C}}H-\underset{\displaystyle NH_2}{\underset{\displaystyle \vert}{C}}HCO_2H$	Thr

Name	Structure	Abbreviation
Ionizable Side Chains		
Arginine	$HN{=}C(NH_2)NHCH_2CH_2CH_2{-}CH(NH_2)CO_2H$	Arg
Aspartic acid	$HO_2CCH_2{-}CH(NH_2)CO_2H$	Asp
Cysteine	$HSCH_2{-}CH(NH_2)CO_2H$	Cys
Glutamic acid	$HO_2CCH_2CH_2{-}CH(NH_2)CO_2H$	Glu
Histidine	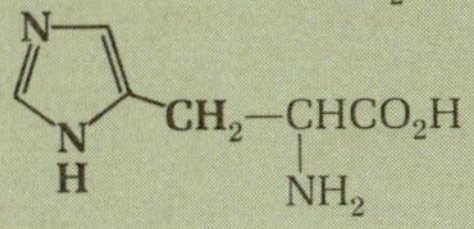	His
Lysine	$H_2NCH_2CH_2CH_2CH_2{-}CH(NH_2)CO_2H$	Lys
Tyrosine	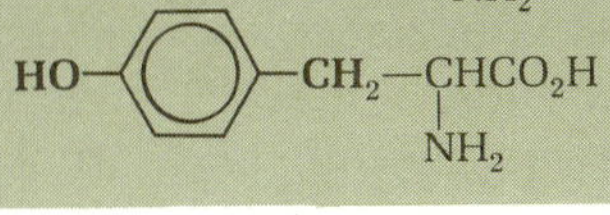	Tyr

be considered the fundamental building blocks of peptides and proteins (Table 25-1).

In Table 25-1 the amino acids have been arranged in groups based on the properties of their side chains **R** (in **R**—$C(NH_2)HCO_2H$). These side chains not only determine the properties of the individual amino acids to a great degree, they also exert a decisive influence on the reactivity and conformation of the protein chain.

Some of the amino acids are simple derivatives of other acids. Glutamine (Gln) and asparagine (Asn) are monoamides of glutamic (Glu) and aspartic (Asp) acids, respectively. Tyrosine (Tyr) is the *p*-hydroxy derivative of phenylalanine (Phe). Cysteine (Cys, pronounced sí-stĕ-ēn) is usually found in proteins in the form of its oxidative dimer *cystine* (pronounced sí-stēn).

(25.1)
$$2HSCH_2CH(NH_2)CO_2H \rightleftharpoons \begin{array}{l} SCH_2CH(NH_2)CO_2H \\ | \\ SCH_2CH(NH_2)CO_2H \end{array} + 2H$$

Cys ⟶ Cys—Cys (linked)

The two cysteine molecules may actually be located in different parts of the protein chain and formation of the —S—S— bridge serves to partially fix its conformation. Equation 25.1 may be easily made to go in either direction.

Since amino acids contain both acidic and basic groups, they exist in solution and in the crystal as internally neutralized zwitterions.

(25.2) $$\underset{\displaystyle NH_2}{RCHCO_2H} \rightleftharpoons \underset{\displaystyle NH_3^+}{RCHCO_2^-}$$

Amino acids, which are all solids, do not melt but rather decompose at high temperatures, usually abovc 250°. Such behavior is consistent with the presence of a highly ionic crystal lattice. The infrared spectra of amino acids also supports the theory that a carboxylate ion rather than a carboxylic acid is present.

For each amino acid there is a pH at which the number of ammonium groups (positive charges) and the number of carboxylate groups (negative charges on the amino acid) is exactly equal. That pH is labeled the *isoelectric point.* For simple amino acids that contain only one acidic and one basic group, the isoelectric point is in the vicinity of pH 6. Amino acids bearing additional basic groups have much higher isoelectric points, while those containing an excess of acid groups have low isoelectric points. If a solution of an amino acid is made more acidic than the pH corresponding to its isoelectric point, then part or all of the molecules will be converted to cations as the NH_3^+/CO_2^- ratio increases (equation 25.3). Amino acids containing basic side chains may be converted to dications.

(25.3)

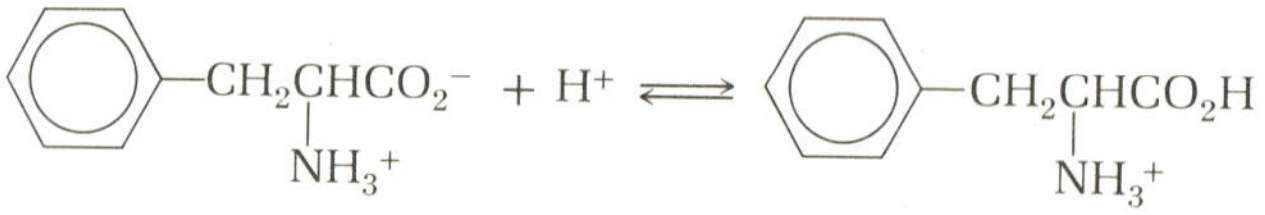

(25.4) $$H_2N(CH_2)_4\underset{\displaystyle NH_3^+}{CHCO_2^-} + 2H^+ \rightleftharpoons H_3\overset{+}{N}(CH_2)_4\underset{\displaystyle NH_3^+}{CHCO_2H}$$

PROBLEM

1 Indicate the predominant species present in solution when the following amino acids are dissolved in water at (a) pH 1; (b) pH 6; (c) pH 12.

a. Alanine
b. Isoleucine
c. Methionine
d. Threonine
e. Lysine
f. Arginine
g. Aspartic acid
h. Asparagine
i. Phenylalanine
j. Tyrosine
k. Histidine
l. Cysteine

The existence of these equilibria provides the basis for a method of separating and analyzing amino acid mixtures, known as *electrophoresis*. A sample of the amino acid mixture is placed on a piece of paper or a section of gel treated with a buffer that will maintain the desired pH. A voltage is placed across the migration medium, and those peptides that carry a net charge then begin to migrate to electrodes of opposite charge. Thus, at an acidic pH, most of the amino acids will carry a net positive charge and will therefore tend to migrate toward the cathode. The rate of migration of an ion depends on its charge and on its size and shape, and this dependency permits separation to take place.

Electrophoretic separation can also be applied to peptides and proteins since these can acquire positive and negative charges in acidic and alkaline solution, respectively, due to the basic and acidic side chains of their constituent amino acids. Peptides (Greek *peptos* — cooked) and proteins (Greek *proteios* — first, primary) are both polymers of α-amino acids in which the individual units are linked together by means of amide bonds (also called peptide bonds). The difference between peptides and proteins is only one of definition and does not involve a clear distinction.

Peptides have been defined as containing 50 or fewer amino acid *residues* (the amino acid minus a molecule of water lost when the acid is incorporated into the peptide chain); larger polymers of α-amino acids are proteins. Alternatively, a molecular weight of 10,000 has been suggested as the dividing line between peptides and proteins. In any event, low molecular weight peptides are often products of the aqueous hydrolysis of proteins, thus accounting for the derivation of their name.

ABSOLUTE CONFIGURATION OF AMINO ACIDS

Except for glycine, all the amino acids listed in Table 25.1 contain an asymmetric α-carbon atom. Several of them possess an additional asymmetric atom as well. In general, only one enantiomer of each amino acid is prevalent in Nature; furthermore, all the common amino acids have the *same* absolute configuration at the α-carbon atom. Several amino acid enantiomers with the "abnormal" configuration at the α-carbon atom have been isolated from microorganisms. Interestingly, these abnormal amino acids have been found in antibiotics of fungal and bacterial origin.

Before the advent of the *R,S* system of nomenclature, the D,L system was used. The D,L system is dependent, of course, on the adoption of one compound as a standard and is, therefore, less versatile than the *R,S* system. For α-amino acids the standard was levorotatory serine, which was *assumed* to have the configuration shown. That configuration was designated L and it was subsequently shown experimentally that all the amino acids listed in Table 25-1 occurred as the enantiomer with the L configuration at the α-position.

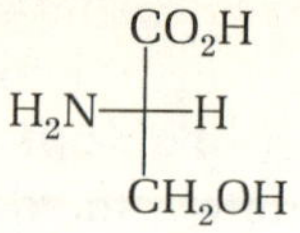

L-(−)-serine

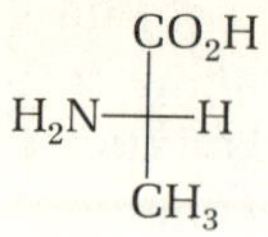

L-(+)-alanine

A straightforward example of such a configurational proof is given in equation 25.5.

(25.5)

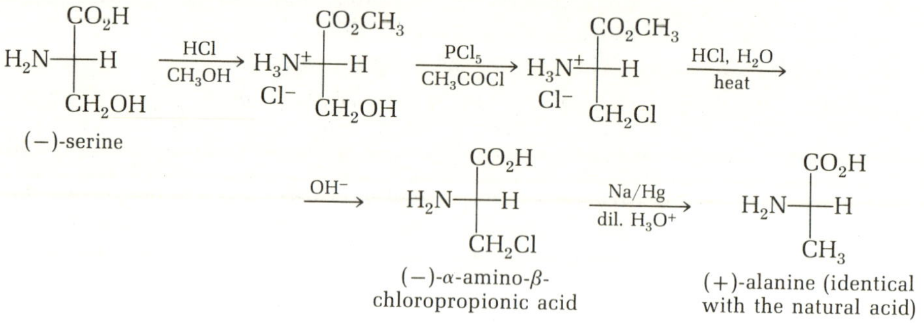

The original experiment was conducted by Emil Fischer, who was a great pioneer in protein chemistry as well as in carbohydrate chemistry.

PROBLEMS

2 The series of reactions shown in equations 1 and 2 have been carried out with the stereochemical results shown. Deduce the structures of the intermediates and the *configurational relationship* between (+)-valine and L-(+)-alanine.

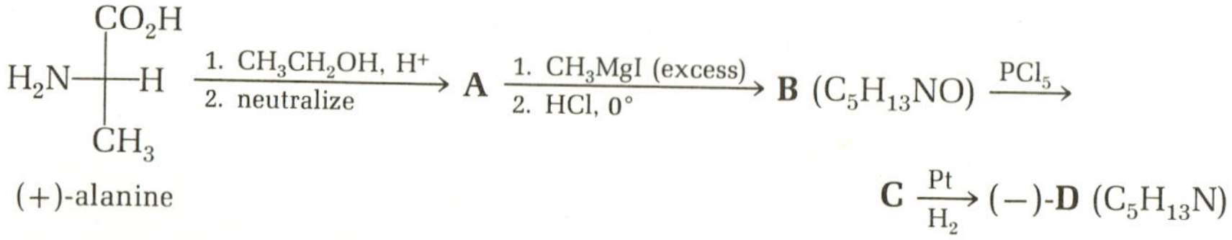

(1)

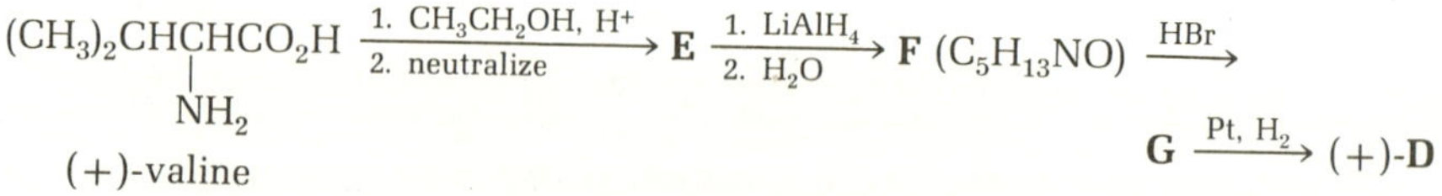

(2)

3 The absolute configuration of L-(−)-serine has been related to that of D-(+)-glyceraldehyde, in order that the configurational relationships of the amino acids and the carbohydrates be established. Based on the series of reactions listed below, indicate the configurational relationship between these two compounds, and explain your reasoning. The relationship between (+)-alanine and (−)-serine has already been established. (Working *backward* is sometimes the clearest way.)

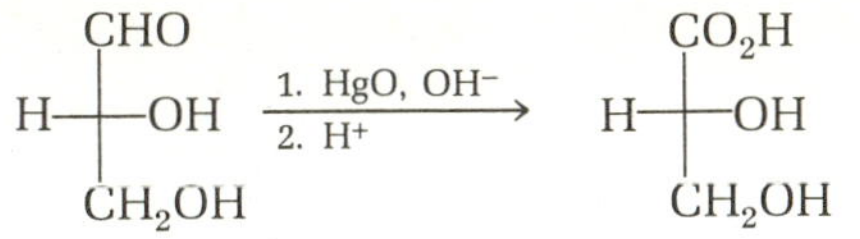 (3)

(+)-glyceraldehyde (−)-glyceric acid

(+)-β-Amino-α-hydroxypropionic acid $\xrightarrow{HNO_2}$ (−)-glyceric acid (4)
[(+)-isoserine]

(+)-Isoserine $\xrightarrow{NOBr}$ (−)-β-bromo-α-hydroxypropionic acid $\xrightarrow{Na/Hg}$ (−)-lactic acid (5)
[(−)-α-hydroxypropionic acid]

(+)-α-Bromopropionic acid $\xrightarrow{OH^-,\ H_2O}$ (+)-lactic acid (6)

(+)-α-Bromopropionic acid $\xrightarrow[H_2O]{NaN_3}$ α-azidopropionic acid $\xrightarrow{Pt,\ H_2}$ (+)-alanine (7)

TOTAL AMINO ACID ANALYSIS

The structural analysis of a protein begins with its isolation in pure form; isolation and proof of homogeneity are very often far from routine tasks. Once they have been accomplished, it is usual to ascertain the molecular weight and total amino acid content of the protein. For large molecules such as these, a knowledge of the empirical formula is virtually useless.

The familiar methods of molecular weight determination cannot be applied to the high molecular weight, nonvolatile proteins. Instead, use is made of the relationship between the molecular weight of the compound and the osmotic pressure of its solutions or the light scattering ability of those solutions as a function of the molecular weight of the scattering particles. In the *ultracentrifuge* the molecular weight is estimated from measurements of the rate of sedimentation of the protein in a very strong gravitational field (100,000 × g and greater).

In order to assay the amino acid content, it is necessary that all the peptide bonds in the protein molecule be hydrolytically cleaved. This is usually accomplished by refluxing a solution of the peptide in 6*M* hydrochloric acid for a 24-hr period. The individual amino acids may then be separated by *elution chromatography* on an ion exchange resin column.

An ion exchange resin consists of an organic polymer that contains ionic groups on its side chains. The most frequently used resin in amino acid analyses is a polymer of sodium *p*-styrenesulfonate. If an amino acid mixture at an acidic pH (so that a substantial fraction of the molecules are present as cations) is passed through such a resin, the cationic acids will be attracted to the negatively charged sulfonate groups, thereby slowing their passage through the column. The larger the *average* charge on a given amino acid

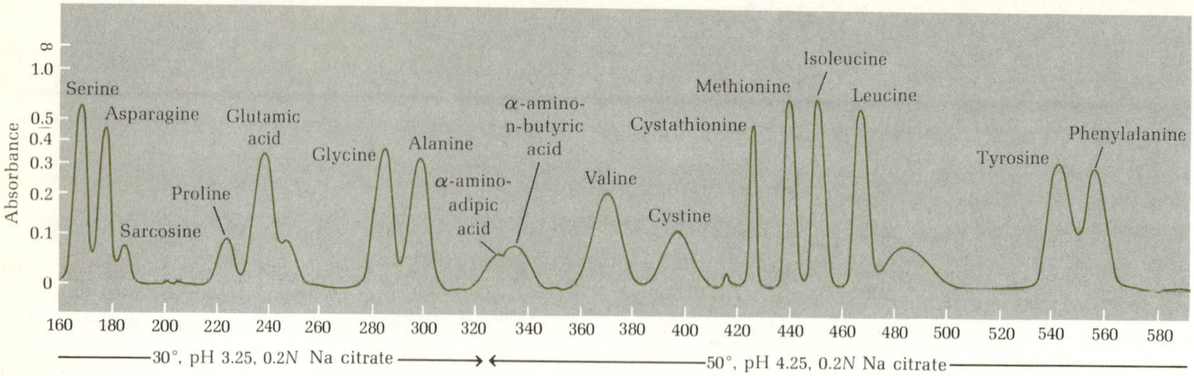

FIGURE 25-1

Elution Chromatogram of an Amino Acid Mixture from an Ion Exchange Column. The quantity actually measured is the light absorption of the ninhydrin condensation product.

at a given pH, the more sluggishly will it move through the column and the greater the volume of solvent that will be required to elute it (Figure 25-1).

The concentration of amino acid in each fraction of the eluate is measured by means of the ninhydrin reaction (see below). Other methods of amino acid separation are available, such as electrophoresis or paper chromatography, but elution chromatography remains the most widely used, particularly since it has been automated. A complete amino acid determination can be done with as little as 10^{-7} mole of protein over a 24-hr period.

THE NINHYDRIN REACTION

Once the amino acids have been separated by one or another method, it is of course necessary to locate them and to obtain a quantitative measure of their concentration. The classical method, which is still in very wide use, depends on the reaction between α-amino acids and *ninhydrin* (the triketone hydrate **1**) at pH 5 to eventually yield a colored product. The decarboxylation step (equation 25.6) is very reminiscent of one of the reactions catalyzed by pyridoxal phosphate (p. 656).

(25.6)

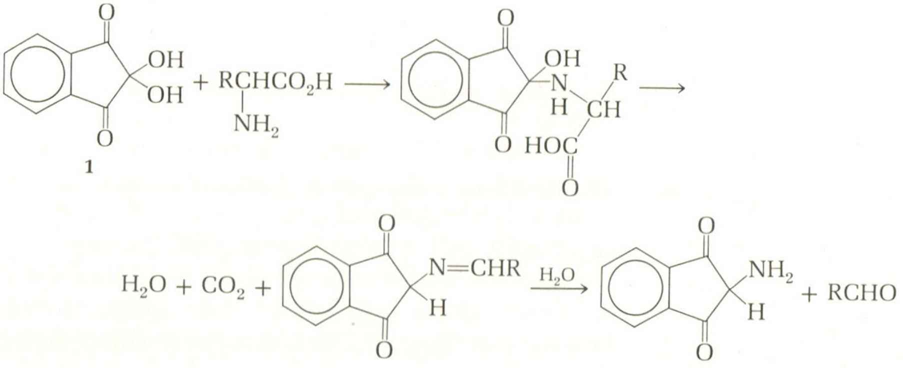

(25.7)

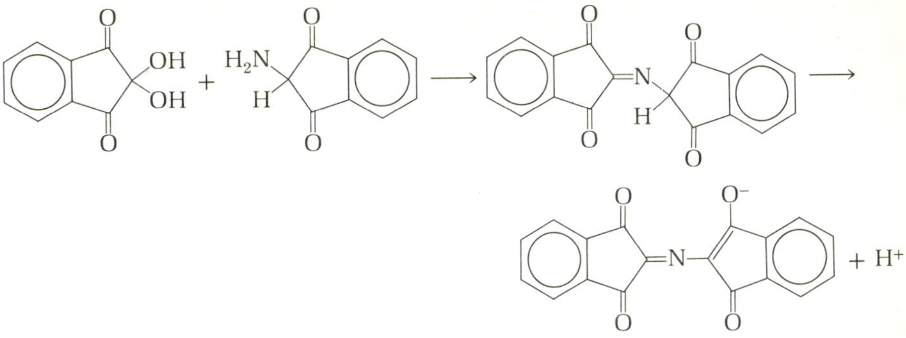

Ruhemann's purple
(λ_{max} at 570 mμ)

Quantitative analysis of the amino acid composition of proteins has now been automated. The protein hydrolyzate is placed on an ion exchange column and then continuously and automatically eluted with buffer. A stream of ninhydrin solution is continuously fed into the eluate that is then heated to 100° to effect reaction between the amino acids and the reagent. This solution is in turn pumped through a spectrophotometer that continuously scans and records the absorption at 570 mμ. In this manner a complete amino acid analysis on a complex protein may be accomplished in 24 hr with an accuracy of $\pm$3%.

Other analytical methods have been developed that are based on the ninhydrin reaction. One of them, which is quite accurate but not well suited to automated analyses, involves measuring the volume of carbon dioxide produced by reaction 25.6.

AMINO ACID SEQUENCE DETERMINATION

Once the total amino acid composition of a peptide has been established, the structural problem has just begun because the infinitely more daunting task of determining their sequence in the chain lies ahead. A measly tetrapeptide, for instance, which contains four different amino acids, can exist in no fewer than 24 different sequential arrangements.

In any noncyclic peptide, one of the terminal amino acids will have a free α-amino group not bound in an amide bond; this residue is known as the *N-terminal* amino acid. Similarly, the amino acid at the other end of the chain has a free carboxyl group and is known as the *C-terminal* amino acid. The peptide is named as a derivative of the C-terminal amino acid, the names of the other amino acids being given in sequence starting with the N-terminal acid. It is customary to write the formula of the peptide with the N-terminal acid (numbered one) on the left and the C-terminal acid on the right.

As an example, we shall use alanylglycyltyrosinylglutamine. In this tetrapeptide, alanine is the N-terminal acid and glutamine, the C-terminal residue.

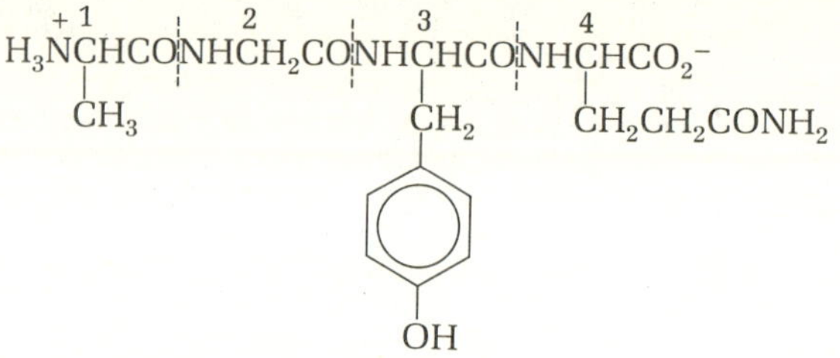

H—Ala—Gly—Tyr—Gln—OH

1 2 3 4

Since it can become exceedingly cumbersome to write the structural formula of even a relatively simple peptide, it is quite common to use the name abbreviations when indicating the sequence in a chain. With this notation, H— marks the position of the free α-amino group and —OH locates the free α-carboxyl group.

Chemically, the easiest amino acids to identify are the terminal ones, particularly the *N*-terminal residues. One oft-used method for identifying the *N*-terminal amino acid depends on the reaction between the peptide or protein and 2,4-dinitrofluorobenzene (*Sanger's method*) in sodium bicarbonate solution. The reaction is simply a nucleophilic aromatic substitution (Chapter 22) and all nucleophilic groups present (such as the ϵ-amino group of lysine) are dinitrophenylated.

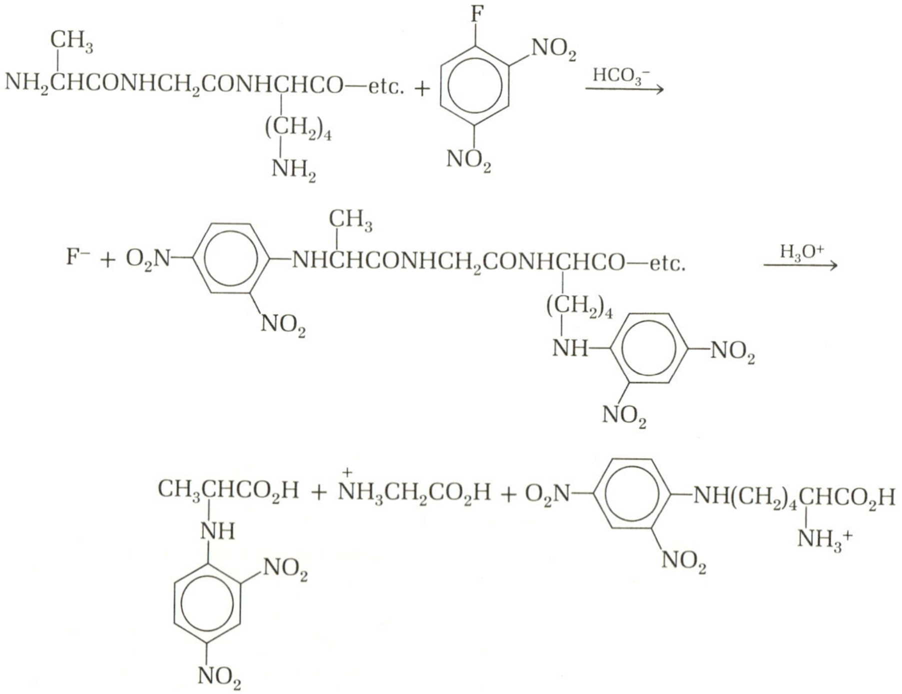

(25.8)

The terminal amino acid is the *only* one, however, that has the dinitrophenyl group attached to its α-amino group. Upon complete acid hydrolysis of the resulting protein, all the other amino acids are converted to water-soluble cations. Due to the electron-withdrawing power of the dinitrophenyl group, the (formerly) *N*-terminal amino acid is not protonated and it may be extracted into an organic solvent and identified.

PROBLEM

4 Show the structure of an arginine molecule that has been obtained by hydrolysis of a peptide molecule after a Sanger determination:

a. When arginine was the *N*-terminal residue

b. When arginine was in the center of the molecule.

Another method of *N*-terminal analysis is the *Edman degradation,* in which phenyl isothiocyanate is used to attack the *N*-terminal amino acid. Due to the nucleophilicity of sulfur, the acylated *N*-terminal amino acid breaks off as a *thiohydantoin* when the peptide is subsequently treated with HCl in an anhydrous medium.

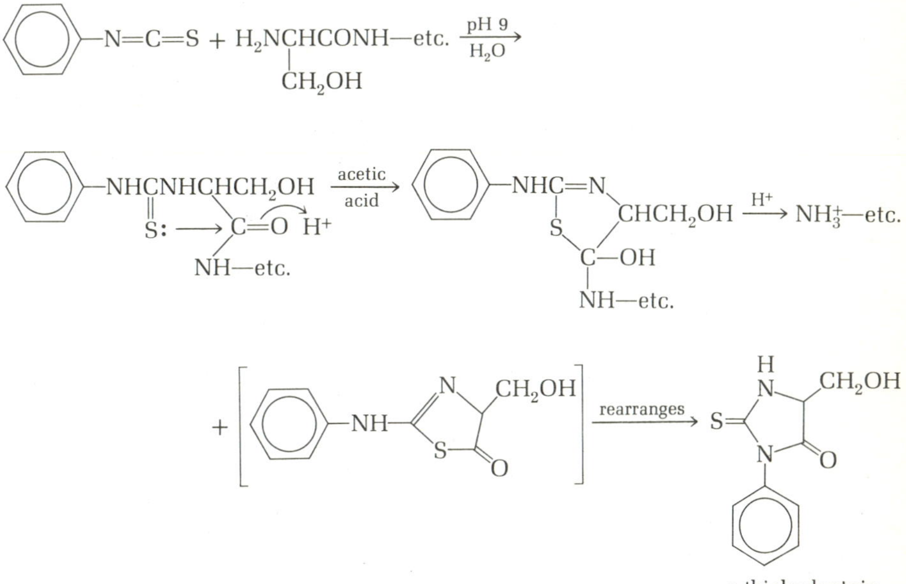

(25.9)

The advantage of the Edman method over that of Sanger is that it leaves the remainder of the peptide chain intact. This residual fragment may then be subjected to further Edman degradation. The method is useful for elucidating the entire amino acid sequence of moderately large peptide fragments since it has been partially automated. Peptides containing 50 to 60 residues have been subjected to a complete sequential analysis by this method.

PARTIAL HYDROLYSIS OF PROTEINS

When one is dealing with a peptide of any appreciable size, the key to a successful sequence analysis usually lies in its partial hydrolysis to small fragments that can themselves be completely decoded. It is also vital that the fragments overlap in amino acid content; otherwise there would be no clue as to how the fragment peptides fit together to form the complete chain. Classically the chemist used standard aqueous acid; but by decreasing the acid strength, the temperature, or the time, only partial hydrolysis of the protein chain was achieved.

Acid-catalyzed hydrolysis of peptides is rather nonspecific and produces a large number of fragments, which makes it difficult to obtain sufficient quantities of enough fragments to carry out a complete analysis. A much more subtle and economical technique is one in which various enzymes (*proteolytic enzymes*), whose specificities and limitations are known, are used to split the undecoded peptide into tractable fragments.

For investigating the *C*-terminal amino acid residues there are *carboxypeptidases*, enzymes secreted by the pancreas. They will only attack the amide bonds of terminal amino acids containing free α-carboxyl groups. A carboxypeptidase will relentlessly eat its way down a peptide chain, continually releasing the *C*-terminal amino acid. For this reason it is necessary to study the concentration of released amino acids as a function of time, in order to get information about the sequence. While it is usually not possible to unravel an entire peptide chain sequence using carboxypeptidase alone, a time study of the enzymatic hydrolysis will often suffice to completely unravel the sequence in a pentapeptide or hexapeptide fragment.

Trypsin and *chymotrypsin* also originate in the pancreas; they normally exert their digestive action in the small intestine. The action of these digestive enzymes enables the animal to utilize the amino acids from ingested protein for its own purposes. Trypsin shows a preference for hydrolyzing amides in which the carbonyl group is part of an arginine or lysine residue. Chymotrypsin tends to catalyze hydrolysis at the carbonyl groups of aromatic amino acid residues (phenylalanine, tryptophan, tyrosine), although it will also attack amide bonds from a few other amino acids.

Subtilisin and *papain* are enzymes of very low specificity and are useful for reducing the peptide chain to small fragments.

The combined deployment of a number of these techniques may be illustrated with reference to glucagon, the structure of which was solved some years ago.

Glucagon is a peptide hormone that originates in the pancreas and that sets in motion a process whereby glycogen is depolymerized and glucose released into the blood stream. A pure sample was subjected to a total amino acid analysis, and the following empirical amino acid formula was discovered:

$Ala_1Arg_2Asp_4Glu_3Gly_1His_1Leu_2Lys_1Met_1Phe_2Ser_4Thr_3Trp_1Tyr_2Val_1$.

It should be kept in mind that the conditions necessary for the total hydrolysis of the peptide will also bring about the hydrolysis of the amide groups of asparagine and glutamine. The hydrolysis of glucagon resulted in the release of four equivalents of ammonia, from which it can be deduced that four of the seven aspartic and glutamic acid residues were present as asparagine and glutamine in the peptide.

With the aid of Sanger's method it was shown that histidine was the *N*-terminal amino acid; the action of carboxypeptidase established the *C*-terminal amino acid as threonine. The peptide was then exposed to chymotrypsin, which broke the chain into six fragments. Each of these fragments was isolated in pure form and then subjected to total amino acid analysis and *N*-terminal residue determination. These fragments are shown in Figure 25-2, labeled C1 to C6; the amino acid to the left of the parenthesis is the *N*-terminal acid of that fragment, while the sequence of those contained within the parentheses is unknown.

Several deductions are immediately available to us. Since C1 is the only histidine-containing fragment, it must come at the beginning of the chain and contain amino acids 1 to 6. Analogously, fragment C6 is the only one in which threonine could come at the end of the chain, as required by the *C*-terminal residue analysis. Finally, we make use of the fact that chymotrypsin prefers to attack at the carboxyl group of aromatic amino acids. These acids, when present in a fragment (phenylalanine in C1 and C2, tyrosine in C3 and C5, tryptophan in C4), should be the *C*-terminal acids of those fragments. Incorporating these deductions, the structures of C1 to C6 may be rewritten as shown in Figure 25-3.

FIGURE 25-2

Fragments from the Chymotryptic Hydrolysis of Glucagon

His(Glu,Gly,Phe,Ser,Thr)
C1

Leu(Ala,Arg_2,Asp_2,Glu,Phe,Ser)
C2

Thr(Asp,Ser,Tyr)
C3

Val(Glu,Trp)
C4

Ser(Lys,Tyr)
C5

Leu(Asp,Met,Thr)
C6

FIGURE 25-3

Revised Structures for Chymotryptic Fragments from Glucagon

His(Glu,Gly,Ser,Thr)Phe C1	Leu(Ala,Arg_2,Asp_2,Glu,Ser)Phe C2		
Thr(Asp,Ser)Tyr C3	Val-Glu-Trp C4	Ser-Lys-Tyr C5	Leu(Asp,Met)Thr C6

Let us see where we are. By making use of the experimental data and adding a reasonable assumption based on known enzyme specificity, the complete sequences have been deduced for fragments C4 and C5 and partial sequences for the others. Fragments C1 and C6 have been identified as the *N*-terminal and *C*-terminal fragments, respectively. There still remains a good deal of sequence work within four of the fragments, particularly C1 and C6, and what is perhaps a tougher job, finding the relative order of fragments C2 to C5. In order to generate new fragments that would overlap those already known, glucagon was degraded with the aid of trypsin.

Seven fragments, shown in Figure 25-4 as *T*1 to *T*7, were isolated after tryptic hydrolysis of glucagon; their individual *N*-terminal residues were determined in the usual way. Since histidine and glycine each occur only once in the peptide chain, it is immediately apparent that *T*1 and *T*5 are fragments of C1, which may now be rewritten as shown in Figure 25-5, part (a). Similarly, the uniqueness of alanine establishes the fact that C2 and *T*2 are overlapping fragments and may be rewritten as shown in Figure 25-5, part (a). The critical points to be grasped when reconstructing fragment C2-*T*2 are (1) phenylalanine lies to the right of alanine in the chain; (2) alanine and phenylalanine define the boundaries of that section of the peptide chain that is common to fragments C2 and *T*2; (3) the remaining amino acids of C2 lie to the left of the common section, while the remaining acids of *T*2 lie to the right of it.

It seems highly probable that C3 is a fragment of *T*3, which may be rewritten to complete part (a) of Figure 25-5. Further examination of the remaining fragments shows immediately that C4 represents the final section of C2-*T*2, while C5 overlaps the tail end of *T*3. Finally, *T*4 must overlap the initial portion of C2-*T*2; if we incorporate *T*6 in here as well, which we must, then we arrive at the long fragments C2-*T*2-C4-*T*4 and *T*3-C5 of Figure 25-5, part (b).

It has already been established that C1 is the initial section of the peptide and C6-*T*7 its final section. The large fragments shown in part (b) of Figure 25-5 then represent the entire center section of the peptide. One of these fragments ends at a tyrosine residue, while the other begins with a tyrosine. Since the molecule only possesses two tyrosines, and one of them is smack

FIGURE 25-4

Fragments from the Trypsin-Catalyzed Hydrolysis of Glucagon

His(Glu,Ser) *T*1	Ala(Asp,Glu_2,Phe,Trp,Val) *T*2	Thr(Asp,Lys,Ser_2,Tyr) *T*3	
Tyr(Arg,Asp,Leu,Ser) *T*4	Gly(Phe,Thr) *T*5	Arg *T*6	Leu(Asp,Met,Thr) *T*7

A.
His-(Glu,Ser)-Gly-Thr-Phe
C1

Leu-(Arg_2,Asp,Ser)-Ala-(Asp,Glu)-Phe-(Glu,Trp,Val)
C2-T2

Thr-(Asp,Ser)-Tyr-(Lys,Ser)
T3

B.
Tyr-Leu-(Arg,Asp,Ser)-Arg-Ala-(Asp,Glu)-Phe-Val-Glu-Trp
C2-T2-C4-T4

Thr-(Asp,Ser)-Tyr-Ser-Lys-Tyr
T3-C5

C. H-His-(Glu,Ser)-Gly-Thr-Phe-Thr-(Asp,Ser)-Tyr-Ser-Lys-Tyr-Leu-(Arg,Asp,Ser)-Arg-Ala-(Asp,Glu)-Phe-Val-Glu-Trp-
C1 T3-C5 C2-T2-C4-T4

Leu-(Asp,Met)-Thr-OH
C6-T7

FIGURE 25-5
Step-by-step Reconstruction of the Peptide Chain of Glucagon

in the center of fragment *T*3-*C*5, the two terminal tyrosines must be one and the same residue. This finding provides the vital link for snapping together the central portion of the molecule and leads to the full-chain partial structure shown in Figure 25-5, part (c).

The proposed sequence was confirmed and the remainder of the structure elucidated by means of a subtilin hydrolysis of the peptide, as well as sequence determinations on fragments *C*1, *T*3, *T*4, and *C*4. Acid- and carboxypeptidase-catalyzed hydrolyses were employed. An analysis of individual fragments for ammonia release upon hydrolysis indicated which aspartic acid and glutamic acid residues were present as asparagine and glutamine in the original peptide (residues 3, 20, 24, and 28). The complete structure is shown below.

H—His—Ser—Gln—Gly—Thr—Phe—Thr—Ser—Asp—Tyr—Ser—Lys—Tyr—Leu—
1 2 3 4 5 6 7 8 9 10 11 12 13 14

Asp—Ser—Arg—Arg—Ala—Gln—Asp—Phe—Val—Gln—Trp—Leu—Met—Asn—
15 16 17 18 19 20 21 22 23 24 25 26 27 28

Thr—OH
29

It is apparent that sequence determination in peptides, even small ones, is a highly demanding and sophisticated form of problem solving. Despite the advent of automation in this field, there is still a great deal of skillful and painstaking labor involved. Not the least of the obstacles encountered in peptide sequence determination is the minute quantity of material normally available to the investigator. The purification of peptides and proteins is not easy; and, as is common when compounds of biological origin are being investigated, the substance in question is usually present in low concentration in the organism.

SOURCES OF AMINO ACIDS

There exists a strong need for enantiomerically pure amino acids for those chemists engaged in protein synthesis. One might imagine that naturally available proteins themselves would constitute a cornucopian source of these compounds, but the need is not so simply met. Most proteins contain a wide variety of amino acids, and the separation and purification of each of the components on a large scale is often too laborious to be worthwhile.

Some proteins are known, however, that are extremely rich in only a few amino acids, and they serve as very useful sources of those compounds. The protein fiber of silk, for instance, consists almost entirely of glycine, L-alanine, L-serine, and L-tyrosine, and it serves as a source for the last three. Tyrosine is easily separated because of its low solubility in water, and serine and alanine are selectively precipitated as their aromatic sulfonic acid salts.

Similarly, L-asparagine is widely distributed in plants and is conveniently obtained from certain seedlings, L-cystine is available from hair, and so on.

The latest method for obtaining amino acids, which promises to supplant all the others, is fermentation. Certain microorganisms have been cultivated that, when fed the proper diet, produce pure L-amino acids as metabolic by-products.

SYNTHESIS OF AMINO ACIDS

Despite the ready availability of a number of amino acids from natural sources, much effort has been expended on amino acid synthesis. There are a number of instances in which the use of synthetic amino acids may be necessary or desirable: (1) The natural acid is not readily available by protein hydrolysis; (2) An isotopically labeled amino acid is required; (3) A totally new amino acid is desired, etc.

The conceptually simplest route to α-amino acids involves the reaction of an α-halo acid with ammonia.

$$\underset{\displaystyle X}{\underset{|}{RCHCO_2H}} + 2NH_3 \longrightarrow \underset{\displaystyle NH_2}{\underset{|}{RCHCO_2H}} + NH_4^+ + X^- \qquad (25.10)$$

Glycine is made commercially from chloroacetic acid and ammonia, and virtually all the aliphatic amino acids have been synthesized in their racemic form from the appropriate α-bromoacid and ammonia. Other nitrogen nucleophiles, such as azide ion or potassium phthalimide, may be employed. Optically active α-bromo acids have occasionally been employed; under most conditions the reaction proceeds with inversion of configuration.

Several synthetic methods are known that involve extension of the carbon chain. One of these is the *Strecker synthesis*, in which an amino nitrile, formed analogously to a cyanohydrin, is a vital intermediate. An aldehyde or ketone is allowed to react with ammonia and hydrogen cyanide (available *in situ* from a mixture of ammonium chloride and potassium cyanide), and the resulting amino nitrile is hydrolyzed in acid.

(25.11)

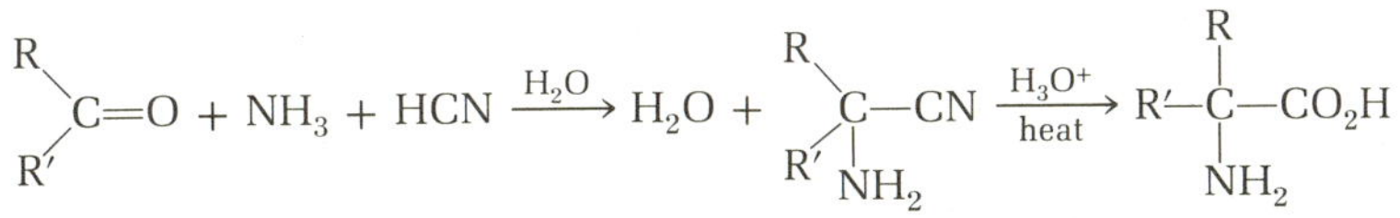

A variation on this synthesis is the *Bücherer* method, in which sodium cyanide and ammonium carbonate are the inorganic reactants and a *hydantoin* is the intermediate.

(25.12)

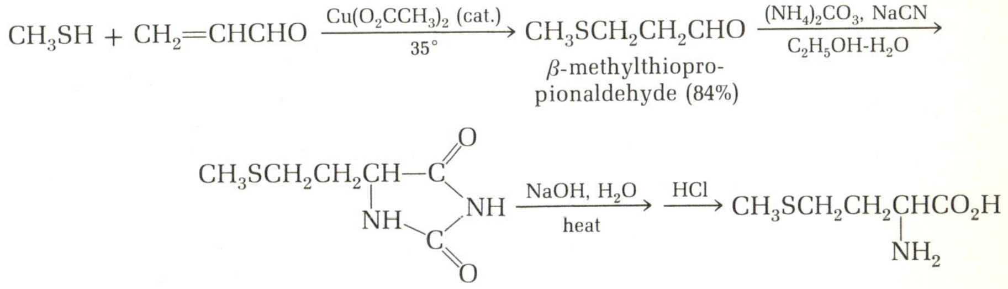

5-(2-methylthioethyl)-hydantoin (79%)

DL-methionine (85%)

A very versatile and effective amino acid synthesis involves the alkylation of an appropriate derivative of aminomalonic acid, $H_2NCH(CO_2H)_2$. Such a derivative is the readily available diethyl acetamidomalonate **(2)**.

(25.13)

$$CH_2(CO_2CH_2CH_3)_2 \xrightarrow[CH_3CO_2H]{NaNO_2} [ON{-}CH(CO_2CH_2CH_3)_2] \rightleftarrows HON{=}C(CO_2CH_2CH_3)_2$$

$$\xrightarrow[(CH_3CO)_2O]{Zn,\ CH_3CO_2H} CH_3\overset{O}{\overset{\|}{C}}NHCH(CO_2CH_2CH_3)_2$$

2 (78%)

This ester-amide is converted to its enolate by sodium ethoxide, and the enolate subsequently alkylated. Hydrolysis of the product results in saponification, deacylation, and decarboxylation.

(25.14)

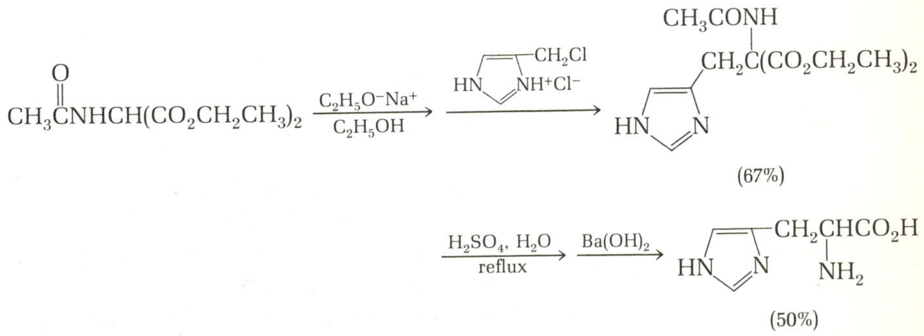

$$(25.15)\quad 2 + CH_2{=}CHCO_2CH_3 \xrightarrow[C_2H_5OH]{C_2H_5O^-Na^+} \xrightarrow[heat]{HCl,\ H_2O} \xrightarrow{base} \begin{array}{l} CO_2H \\ | \\ CHNH_2 \\ | \\ CH_2CH_2CO_2H \end{array}$$

(56%)

PROBLEMS

5 Propose syntheses of the following amino acids based on readily available starting materials.

a. Alanine
b. Valine
c. Phenylalanine
d. Serine
e. Aspartic acid
f. Lysine
g. Tyrosine
h. Proline

6 a. Draw structures for compounds **A** to **E.**

$$\text{Diethyl acetamidomalonate} \xrightarrow[2.\ CH_2{=}CHCN]{1.\ NaOC_2H_5} \mathbf{A}\ (C_{12}H_{18}N_2O_5) \xrightarrow{H_2,\ Ni}$$

$$\mathbf{B}\ (C_{12}H_{22}N_2O_5) \xrightarrow{\text{spontaneously}} \mathbf{C}\ (C_{10}H_{16}N_2O_4) \xrightarrow[heat]{HCl}$$

$$\mathbf{D}\ (C_5H_{14}N_2O_2Cl_2,\ \textit{ornithine dihydrochloride}) \xrightarrow[OH^-]{H_2NCN} \text{arginine}$$

$$\mathbf{A} \xrightarrow[heat]{HCl} \xrightarrow{NH_3} \mathbf{E}\ (C_5H_9NO_4)$$

b. When ornithine is treated with excess benzoyl chloride, it is converted to a dibenzoyl derivative. Upon hydrolysis with a limited quantity of base, one of the benzoyl groups is selectively removed. Draw the reactions. What factor(s) permit(s) the selective hydrolysis?

RESOLUTION OF SYNTHETIC α-AMINO ACIDS

Racemic amino acids are of no use for the synthesis of biologically active compounds and they must therefore be resolved. A number of synthetic α-amino acids have been resolved in the traditional way — the amino group was acylated and the *N*-acylamino acid then allowed to react with an optically active base such as brucine or strychnine. The diastereomeric salts were separated and the free acid then regenerated.

A clever method of resolution depends on the preference shown by enzymes for catalyzing reactions of L-amino acid derivatives only. In this approach the racemic amino acid is acylated with acetic anhydride and then exposed to a *deacylase*, an enzyme that catalyzes the hydrolysis of *N*-acylated amino acids. Only the L-enantiomer is affected; a mixture of L-amino acid and D-acylamino acid is obtained, the two components of which are easily separated.

(25.16)
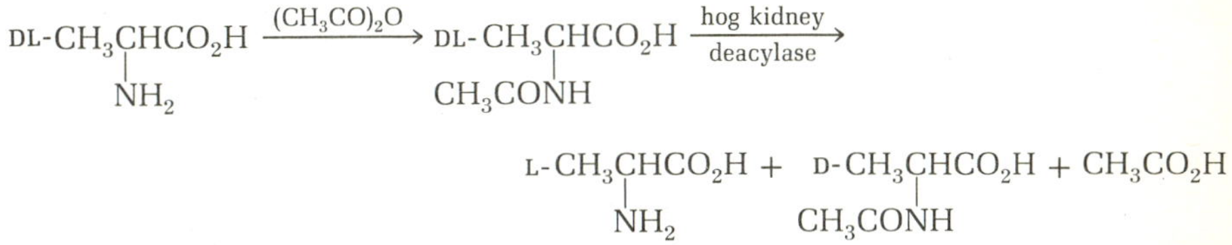

PEPTIDE SYNTHESIS: PROTECTING GROUPS

In peptide synthesis, the "trick" is to cause a number of amino acids to condense together under mild conditions *and in the correct sequence*. Thus, reaction 25.17 is to be promoted while, at the same time, reaction 25.18 is to be suppressed.

(25.17)
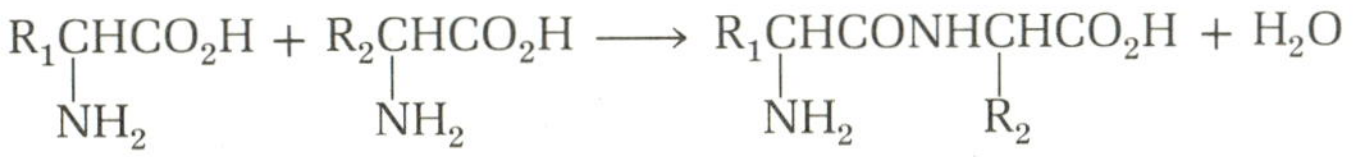

(25.18)
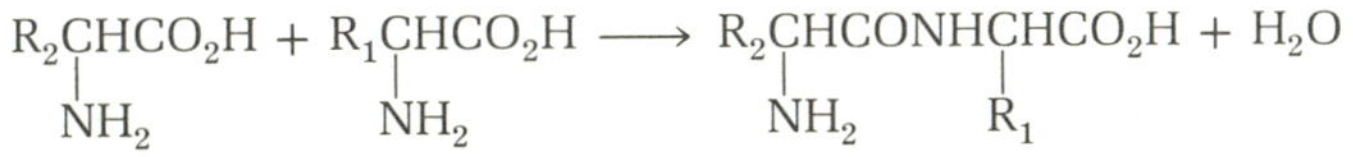

This selectivity is accomplished by the judicious use of protecting groups that deactivate nucleophilic functional groups which might interfere with the desired reaction (such as the α-amino group of $R_1C(NH_2)HCO_2H$). Selectivity is further promoted by the use of *activating groups*, which increase the electrophilicity of the carboxyl group that is to become part of the peptide bond.

The most convenient way to remove the nucleophilicity of an amine is to acylate it. As an example, the synthesis of the dipeptide alanylglycine might be undertaken as shown in equation 25.19.

(25.19)
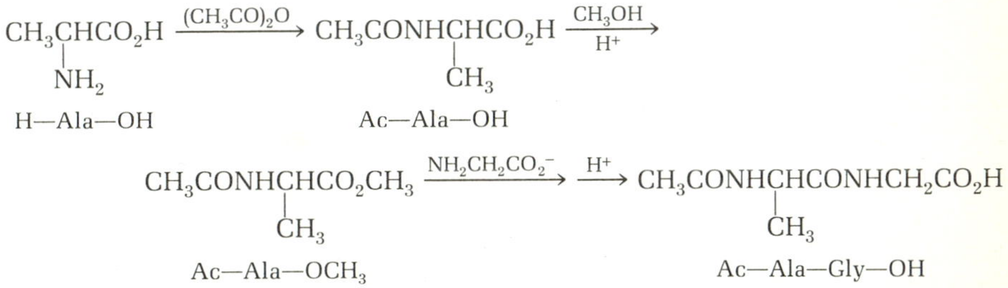

In order to obtain the desired dipeptide, the product of this reaction, *N*-acetylalanylglycine, would have to be deacetylated. Since the conditions required for this operation would also lead to cleavage of the alanine-glycine peptide bond, acetyl (or any other usual acyl group) is not a satisfactory protecting group.

It is clear that a prime requisite for a good protecting group is that it be easily and quantitatively removable, under conditions that leave the remainder of the molecule intact. Careful analysis led to the conclusion that the benzyl group would be very valuable in this capacity. It can be cleaved from a number of leaving groups under mild conditions; in particular, reductive cleavage with palladium and hydrogen or with sodium in liquid ammonia is easily accomplished. Hydrogen bromide in acetic acid is also effective for promoting benzyl cleavage.

(25.20)
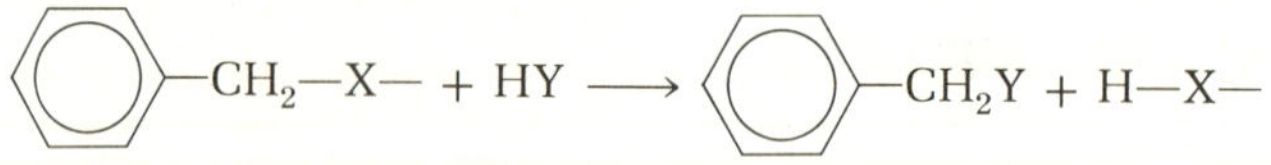

X = N, O, S
Y = H, Br

The protective group chosen for amines is the *benzyloxycarbonyl* group (also called the carbobenzoxy group and abbreviated Z- in the chemical literature), introduced by way of the acid chloride **3**. *N*-Acylation of the amino acid proceeds easily and in good yield in alkaline solution, and subsequent removal of the acyl group is equally facile.

(25.21)
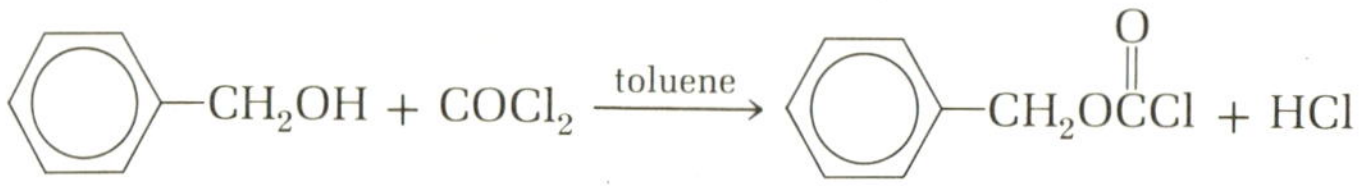

3 (90–95%)

(25.22)
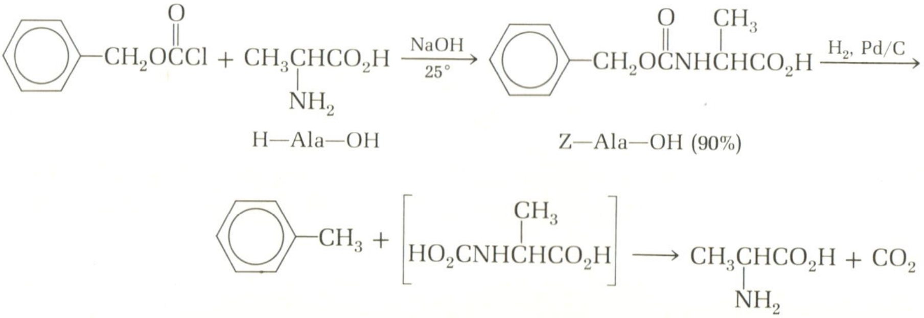

Another virtue possessed by the benzyloxycarbonyl group is the reduced reactivity toward the nucleophiles of its carbonyl group, attributable to the high resonance stabilization of *carbamates* (derivatives of carbamic acid, H_2NCO_2H).

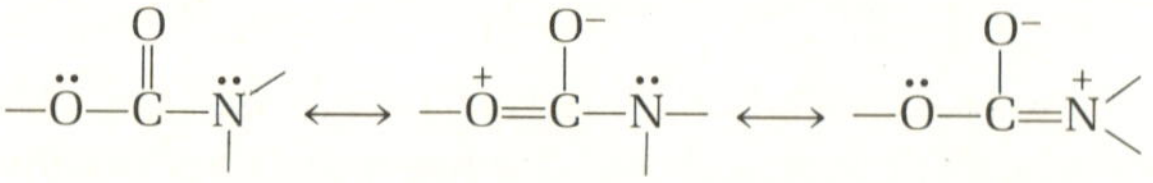

Amino groups are not the only nucleophilic centers requiring protection in amino acids. The thiol group of cysteine, the heterocyclic nitrogen atom of histidine, and the phenolic hydroxyl group of tyrosine are examples of other functionalities that must be protected. Benzyl sulfides, amines, and ethers are usually the protected forms of these functional groups.

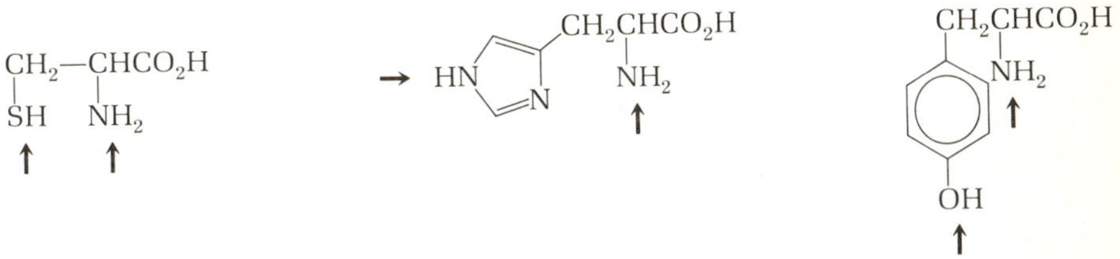

A protective group related to benzyloxycarbonyl is *t*-butoxycarbonyl (Boc), introduced by way of the azide (the chloride is quite unstable). It is somewhat complementary to benzyloxycarbonyl — it cannot be removed by hydrogenation or sodium reduction, but it is very acid-sensitive and can be detached in weak acid solution, while the Z-group remains undisturbed.

(25.23)

$$(CH_3)_3COC(=O)N_3 + H_2NCH(CH_3CHOH)CO_2H + (CH_3CH_2)_3N \longrightarrow (CH_3)_3COC(=O)NHCH(CH_3CHOH)CO_2H + (CH_3CH_2)_3NH^+N_3^-$$

Boc—Thr—OH

(25.24)

$$\text{Boc—Phe—Lys(Z)—OH} \xrightarrow[\text{acid (1}M\text{), }25^\circ]{\text{HCl in acetic}} \text{H—Phe—Lys(Z)—OH} + (CH_3)_2C{=}CH_2 + CO_2$$

The facile formation of the *t*-butyl cation makes Boc a useful protective group.

ACTIVATING GROUPS

In order to achieve amide formation between a carboxyl group and an unprotected amine, it is necessary to transform the carboxyl group to a more reactive functionality. In the synthesis of glutathione acid chlorides were used, but they are generally not very satisfactory. For one thing, the α-hydrogen atom of acid chlorides is relatively acidic and easily removed by base. This process leads to racemization of the acid chloride (equation 25.25) and a consequent diminution in the biological activity of the derived peptide.

(25.25) 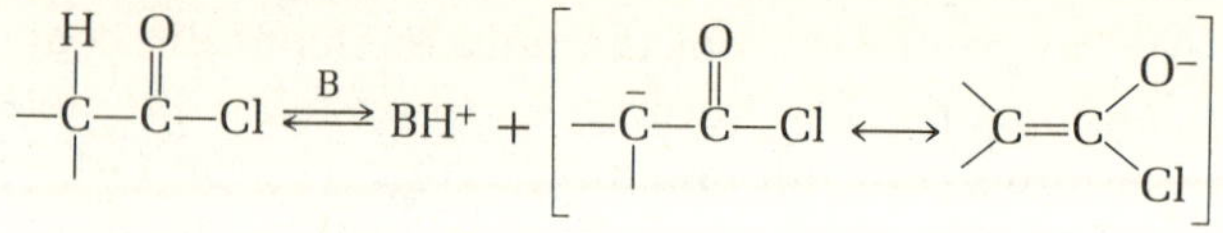

Acylaminoacid chlorides are also very prone to undergo intramolecular cyclization to *azlactones* (5-oxazolones). Azlactones will react with various nucleophiles to regenerate acyl amino acid derivatives; unfortunately, azlactones are extremely prone to racemization because of the extensive stabilization of the enolate anion.

(25.26) 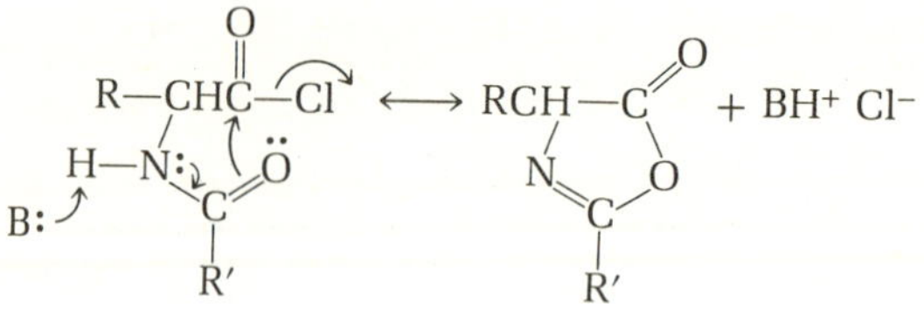

an azlactone

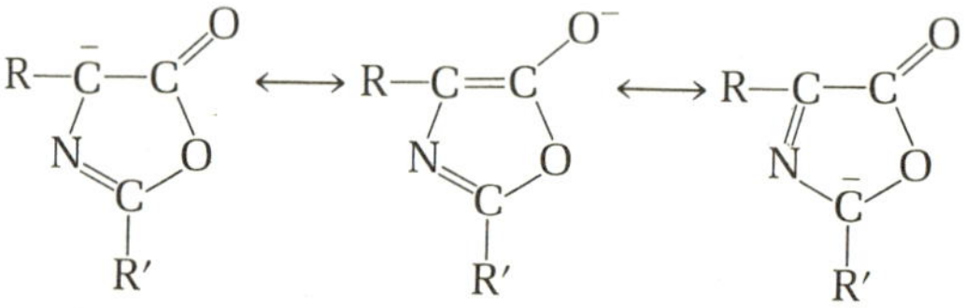

Esters are useful carboxyl derivatives for coupling purposes and, in order to maximize the rate and yield of amide formation, esters containing good leaving groups are used. Examples are *p*-nitrophenyl (—ONp) and pentachlorophenyl (—OPcp) esters.

(25.27) 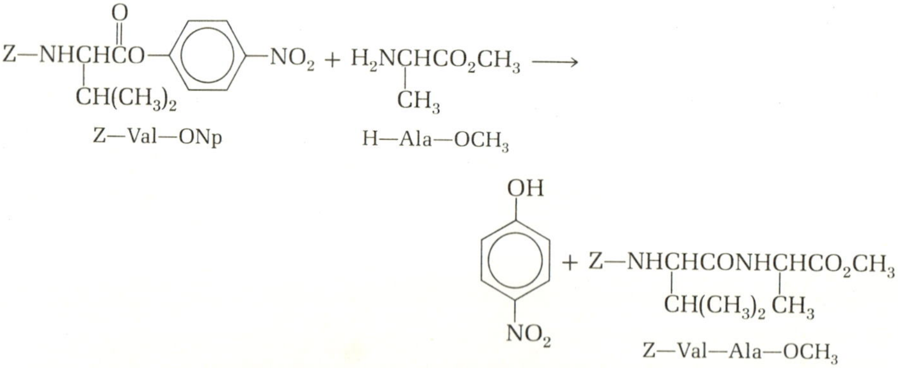

One of the most elegant methods of coupling amino acids, due to its simplicity and directness, is the *carbodiimide method*. The reactants comprise the free acid, the amine, and a carbodiimide (RN=C=NR′) and the reaction may be conducted in a variety of solvents, including water. *N,N′*-Dicyclohexylcarbodiimide (**4**, DCC) has seen the widest use in peptide synthesis.

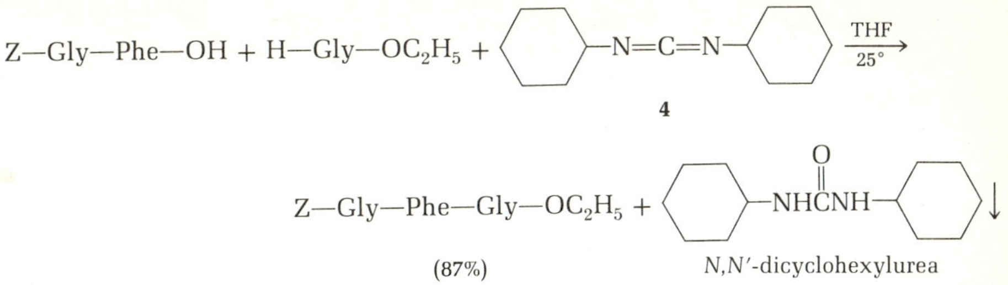

(25.28)

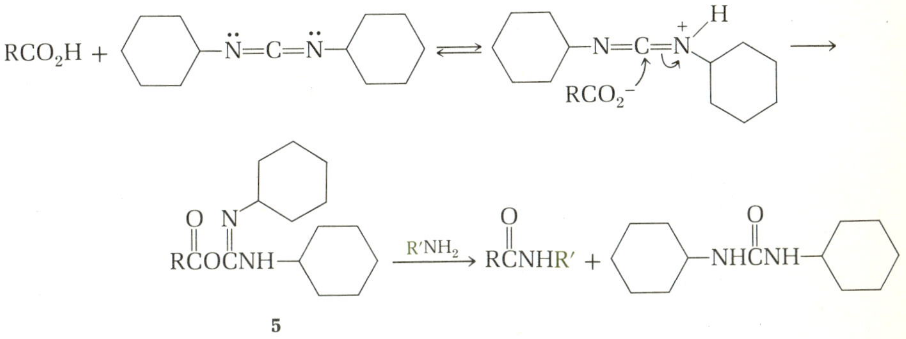

(25.29)

The initial reaction takes place between the carboxylic acid and the carbodiimide to yield an intermediate **(5).** The amine then attacks at the carbonyl group of **5,** generating the coupled product.

SOLID PHASE PEPTIDE SYNTHESIS

One of the great problems in peptide synthesis is the necessity of isolating and purifying the growing chain after each addition, in order to free it from coupling agents, remnants of protecting groups, etc. The result is a great outlay of labor and time, and usually a low overall yield.

In 1962, R. B. Merrifield proposed an ingenious technique for circumventing these obstacles. The growing peptide chain was bound to a polymeric resin that was insoluble in all solvents used in the course of the synthesis. Thus, after each coupling step the unreacted starting material, reaction by-products, etc., could be simply washed away from the peptide-containing polymer. One principal requirement was that each coupling step take place in very high yield; otherwise one would be left at the end with a mixture of peptides of varying lengths and variable composition.

Merrifield first synthesized the nonapeptide bradykinin, H-Arg-Pro-Pro-Gly-Phe-Ser-Pro-Phe-Arg-OH, by the solid phase method. Polystyrene beads were chloromethylated with chloromethyl methyl ether (a Friedel-Crafts alkylation), and the chloromethyl polymer then allowed to react with the triethylammonium salt of the *N*-protected *C*-terminal amino acid. *t*-Butoxycarbonyl-acylated amino acids were used throughout.

(25.30)

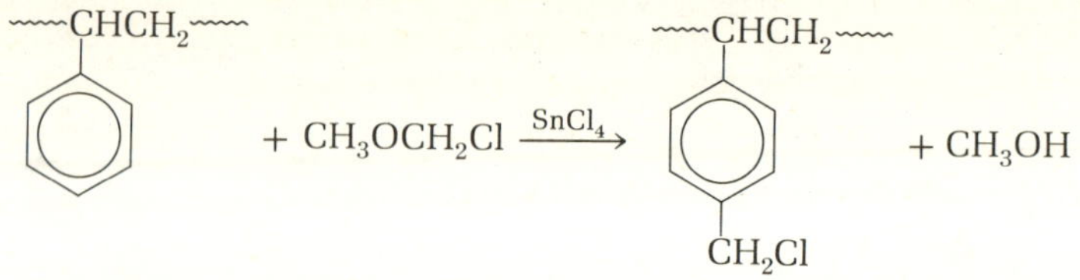

For the synthesis of bradykinin the first acid attached to the polymer was arginine (the guanadino functional group, NH=C(NH_2)—NH—, had been protected by nitration), which was converted to a polymeric benzyl ester.

(25.31)

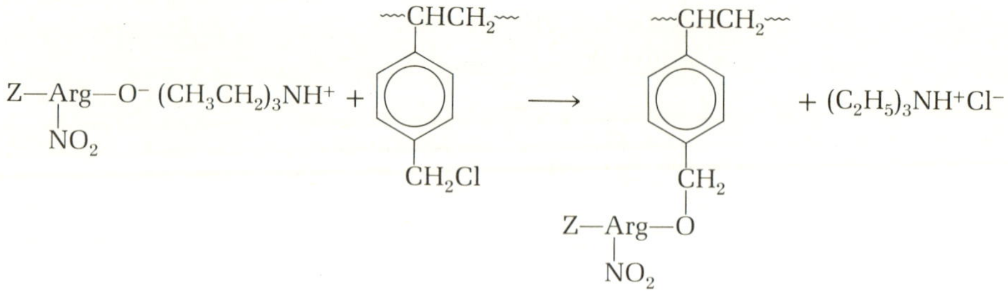

The *N*-protecting group was removed and the next amino acid in the sequence, in its *N*-acyl form, was coupled to arginine with the aid of dicyclohexylcarbodiimide.

(25.32)

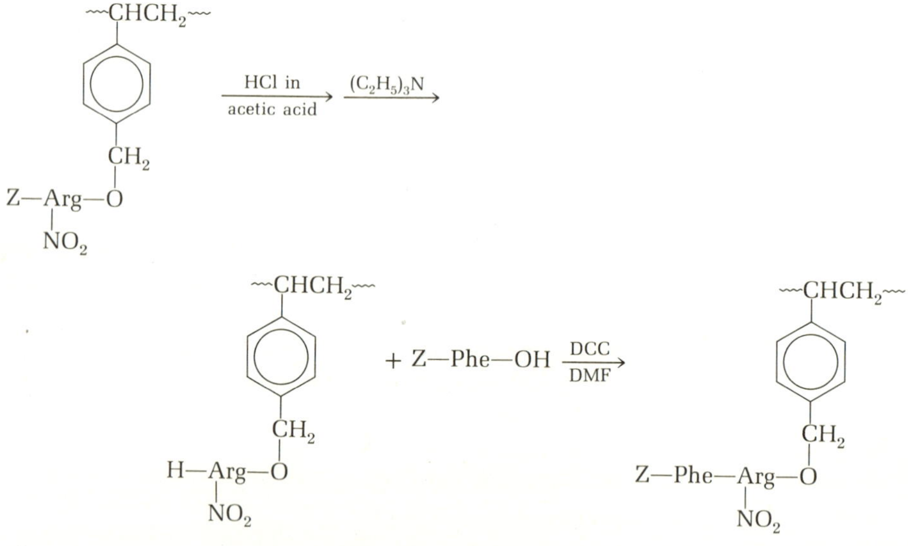

The sequence of deprotecting and DCC coupling was repeated until all the residues had been added. Other protecting groups were then removed (hydrogenation, etc.) and the completed peptide was finally cleaved from the polymer by HBr in trifluoroacetic acid.

One of the outstanding merits of the Merrifield method is that it lends itself to automation. The synthesis of bradykinin by manual manipulation required 6 days; another 2 days were spent in purification and the yield of purified material was 68%. By the automated method, bradykinin was obtained after 32 hr in comparable yield. Thanks to automation, one can think seriously of synthesizing proteins.

SECONDARY AND TERTIARY PROTEIN STRUCTURE

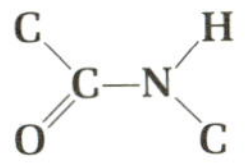

FIGURE 25-6
Preferred Stereochemistry About the Peptide Bond

Peptide chains are somewhat more rigid than analogous hydrocarbon chains because of the steric requirements for maximal overlap in the peptide bond between the nitrogen atom lone pair and the carbonyl group π-bond. All the atoms immediately attached to the peptide linkage prefer coplanarity, with the nitrogen-bound and carbon-bound carbon atoms *trans* to one another (Figure 25-6).

In small peptides, the principal conformational restraint on the molecule is that due to the hindered rotation about the CO—N bond. As the length of the peptide chain increases, there are a variety of intramolecular interactions that begin to exert a substantial influence on the conformational behavior. Chief among these is hydrogen bonding. The NH groups of amides are good hydrogen-bond donors, while the carbonyl oxygen atoms are very effective hydrogen-bond acceptors. Hydrogen bonding between *adjacent* NH and CO groups is not favored, however, because the three atoms involved in a hydrogen bond should ideally be colinear.

The criterion for colinearity can readily be satisfied if an NH group in a given peptide linkage interacts with a carbonyl group separated from it by several amino acid residues. A careful examination of molecular models has shown that an NH group can form a strong hydrogen bond to a carbonyl group three peptide linkages distant. Furthermore, this type of bonding can be established along the entire length of the chain if it assumes a spiral, or *helical*, conformation (Figure 25-7). Much experimental evidence has accumulated from measurements of optical rotation, ultraviolet and infrared absorption, viscosity changes, etc., that indicates that a large number of peptides and proteins possess at least some sections that are in the form of α-helices.

An α-helix is in itself dissymmetric, independently of the dissymmetry of its constituent units, and the presence of helical sections confers additional optical rotatory power on the molecule. This phenomenon provides a useful handle for studying helical conformations in protein chains. The helix may be right- or left-handed; when the constituents are L-amino acids, the right-handed helix is favored sterically.

The adoption of the helical conformation in proteins is usually energetically favorable because it promotes intramolecular hydrogen bonding. On the other hand, the helical molecule is more restricted than a random coil is and entropy effects work against helix formation. For a number of mole-

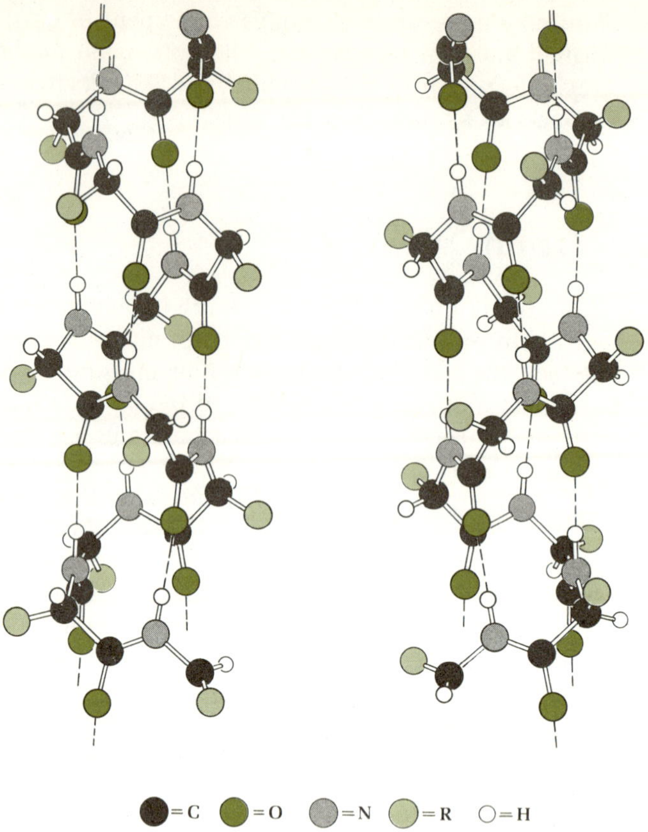

FIGURE 25-7
Models of Left- and Right-Handed Peptide α-Helices. (Pauling, Corey and Branson, *Proc. Nat. Acad. Sci. US*, **37**, 205 (1951))

cules these two tendencies are fairly well in balance and the helix rapidly forms and breaks in solution.

In any given protein not all sections of the molecule will be equally amenable to α-helix formation. The presence of a proline residue interferes with helix formation because of the rigidity of its pyrrolidine ring, while valine tends to inhibit adoption of that conformation because of the steric effect of its isopropyl group. Protein sections that do not coil into helices may still engage in intramolecular hydrogen bonding. The chain may double back on itself and the two strands clip together by way of NH·····O═C bonds to give a structural feature known as a *pleated sheet* (Figure 25-8).

The helical and pleated sheet sections provide the *secondary structure* of the protein (the actual amino acid sequence provides the primary structure). If secondary structure is disrupted by heating or by introducing a solvent that interferes with intramolecular hydrogen bonding, then the protein generally loses all its biological potency. This process is known as *denaturation* and a ubiquitous example is the irreversible change accompanying the cooking of egg white.

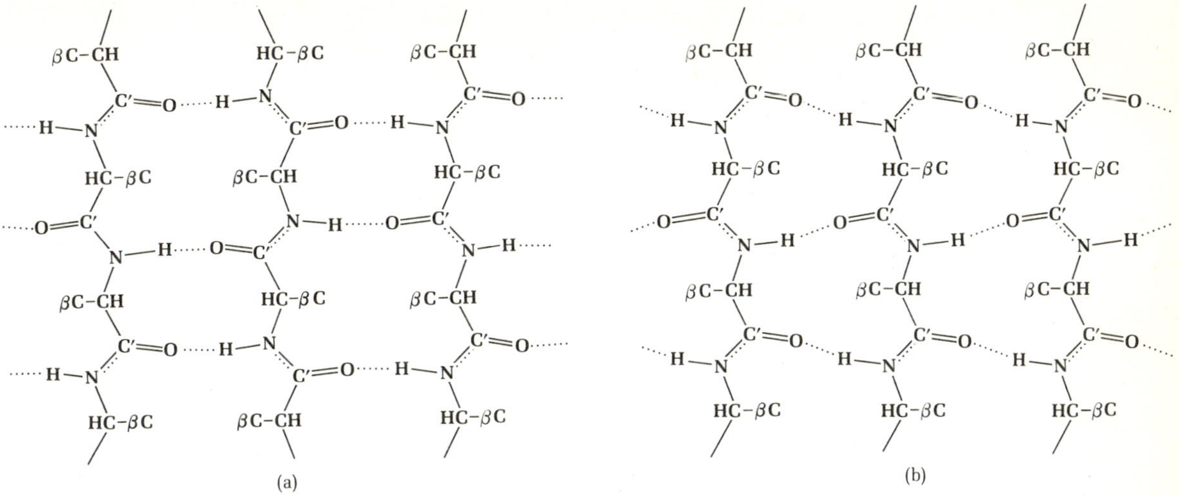

FIGURE 25-8

Parallel (a) and Anti-parallel (b) Pleated Sheet Structures in Peptide Chains. (Pauling and Corey, *Proc. Nat. Acad. Sci. US*, **37**, 729 (1951); **39**, 253 (1953))

The entire protein molecule will assume a certain overall three-dimensional shape that is also intimately connected with its biological function. This overall shape is the *tertiary structure* and is illustrated for the enzyme lysozyme, isolated from egg white (Figure 25-9). One of the important determinants of tertiary structure is the tendency of hydrophilic functional groups, such as amino and carboxyl groups, to be on the outside of the

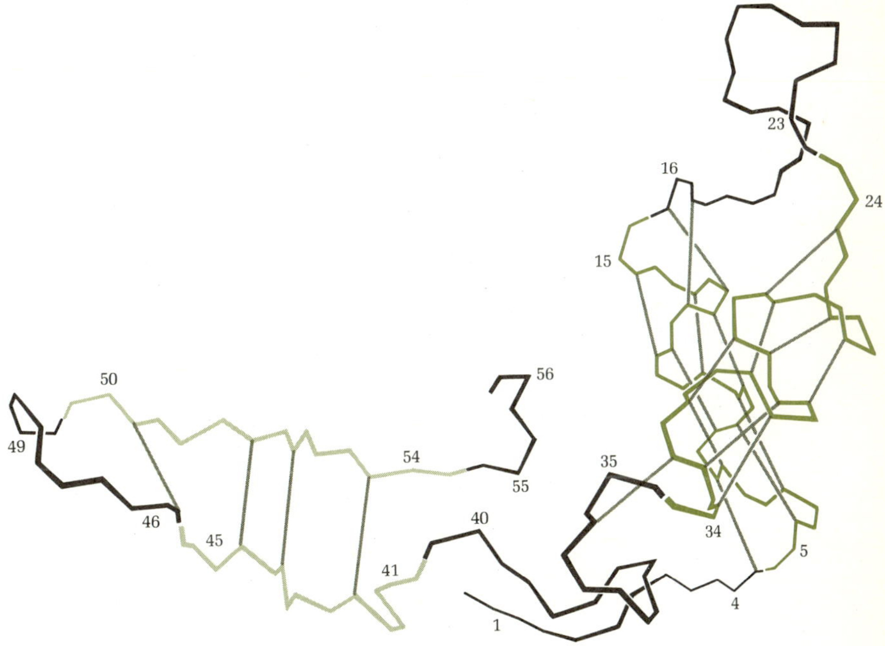

FIGURE 25-9

Conformational Diagram of a Portion of the Lysozyme Molecule. Residues 5 through 15 and 24 through 34 are wound in helical conformations. Residues 41 through 45 are bonded to residues 50 through 54 to form a pleated sheet. Light grey lines denote hydrogen bonds.

molecule and in contact with water. *Hydrophobic* side chains, such as alkyl and phenyl, prefer to huddle together in the interior of the molecule.

The only method available for seeing the secondary and tertiary structure of proteins is X-ray crystallography. It is even possible to delineate the primary structure of proteins with X rays but as the requisite resolution is difficult to obtain in practice, X-ray crystallography is used complementarily with classical methods rather than as a replacement for them.

ENZYMES: INTRODUCTION

Of all the proteins available in nature, none have engendered more interest, speculation, and investigation than the enzymes. And with good reason — life as we know it would be inconceivable without the intervention of these remarkable catalysts. At the present moment the full picture of enzyme action is just beginning to unfold, thanks to our ability to combine the data from quite sophisticated chemical and biochemical studies with the topological information provided by crystallographic observations.

Enzymes may bring about enormous increases in reaction rate compared to the rate of the uncatalyzed reaction — factors of 10^{10} to 10^{12} are not uncommon. Enzymes also possess an enormous advantage over the common laboratory catalysts (acids, bases, metals, etc.) that are the stock-in-trade of most organic chemists. Such rate accelerations are of course vitally necessary because the reactions involved must proceed rapidly at 37° at a pH near 7, with the reactants present in moderate concentration in aqueous solution. In physiological systems one cannot employ the extremes of temperature, pressure, pH, and concentration that are so easily available in the laboratory.

There are a number of factors that account, at least in part, for the efficacy of enzymatic catalysis. Some of these are:

1 The enzyme forms a complex with the substrate (one or more coenzymes may also be included) that contains all the functionalities necessary for reaction to occur; this process is more efficient than waiting for the reactants to experience a favorable collision in solution.

2 The enzyme holds the reacting groups in the proper orientation for reaction and may even align them so that the most efficient overlap of reacting orbitals is obtained.

3 The enzyme exerts a strong influence on the solvation of various groups; for instance, a reacting functionality located in the hydrophobic portion of the enzyme may be many times more reactive than if it were surrounded by water molecules.

4 The enzyme provides extremely rapid addition and removal of protons by precisely oriented acidic and basic groups.

Experiments have shown that only a few amino acid residues are intimately involved in the catalytic activity of a given enzyme. These residues are usually clustered together around the *active site*, that topographical region where the enzymatic transformation actually takes place.

THE ACTIVE SITE

When the proteases trypsin and chymotrypsin (p. 718) are treated with diisopropylfluorophosphate, they are completely deprived of catalytic activity. It has been shown that in each case the phosphate reacts irreversibly with a single serine residue in each molecule. These serine residues may be presumed to be at the active site and to play a unique role in the hydrolysis of peptides and related substances.

(25.33)

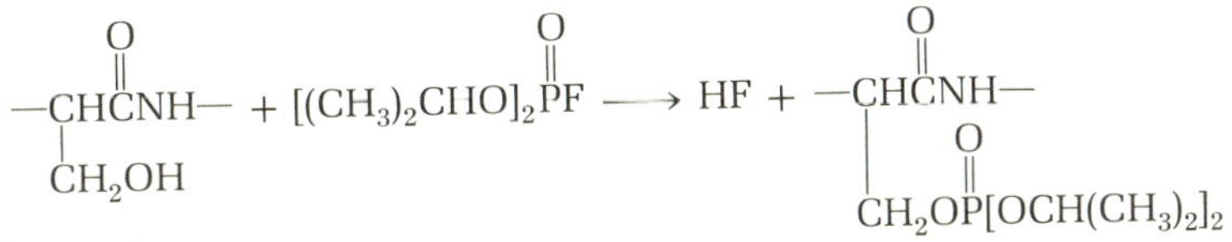

A study of the catalytic activity of chymotrypsin and trypsin as a function of pH revealed that a group with a pK_a of approximately 7 was an indispensable part of the catalytic apparatus. This behavior suggested an imidazole ring of a histidine residue. The supposition was subsequently confirmed chemically; destruction of imidazole by photooxidation, or depression of its basicity by dinitrophenylation, resulted in the disappearance of the enzymatic power of the protein.

Figure 25-10 shows the conformation of chymotrypsin; the enzyme consists of three peptide chains held together by cystine sulfur-sulfur bonds. The active site is apparently the relatively open region at the right central part of the diagram, and the catalytically active histidine and serine (as its O-tosyl derivative) residues are shown. Chymotrypsin is derived from an inactive precursor called chymotrypsinogen, which consists of a single peptide chain containing four more amino acid residues than chymotrypsin does. When the precursor (called technically a *zymogen*) is exposed to trypsin, two dipeptide units are hydrolytically removed from the molecule, resulting in the formation of the A, B, and C chains. The activation step results in the appearance of a free terminal amino group (B chain) in the vicinity of the active site; the protonated form of this amino group (part of an isoleucine residue) has been implicated in the catalytic activity.

The shape, conformation, and properties of the active sites of a number of enzymes have been probed by chemical methods. Complete X-ray pictures of enzyme conformations are available for only a few molecules and their obtention is fraught with difficulties so that the use of chemical methods

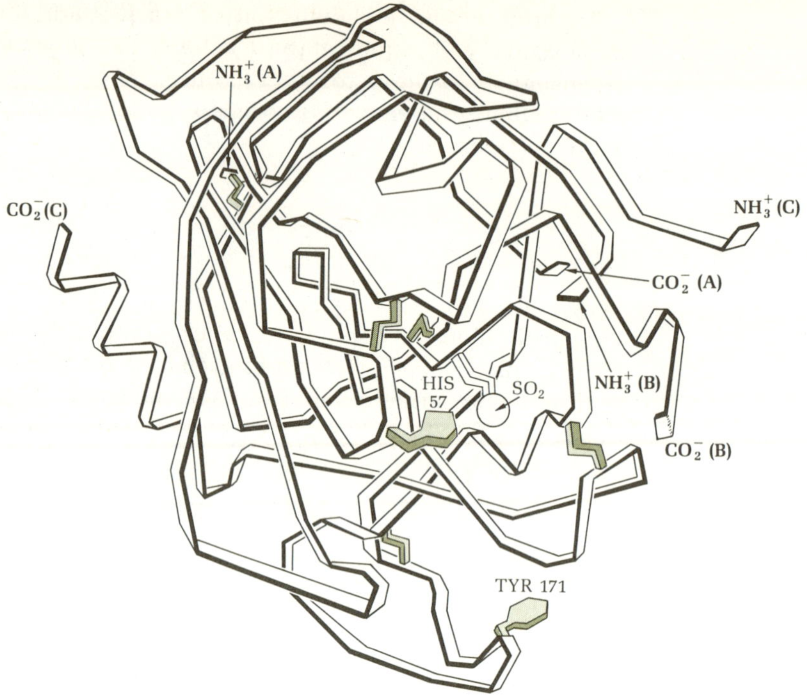

FIGURE 25-10
Conformational Model of Chymotrypsin, Showing the Three Separate Peptide Chains. SO_2 denotes the site of a catalytically-active serine residue (Ser 195) which had been converted to its tosylate. The active histidine residue (His 57) is also shown. Shaded sections depict cystinyl disulfide bridges. (Matthew *et al; Nature*, 214, 652 (1967))

is still mandatory. One very informative technique is the use of *inhibitors*, compounds that resemble the substrate in certain of its vital characteristics (dimensions, polarity, configuration, etc.) and are therefore accepted by the enzyme; the inhibitors cannot be chemically transformed, however, and they therefore "jam" the system.

THE DOUBLE DISPLACEMENT PROCESS

Consider the very generalized equation 25.34, which depicts the reaction of substrate A with a second molecule to yield product C.

(25.34) $A + B \longrightarrow C$

Under enzymatic conditions such reactions often, although not always, follow a two-step path (equations 25.35 and 25.36), in which the first step involves covalent bond formation between substrate and enzyme.

(25.35) $A + Enz \longrightarrow I$

(24.36) $I + B \longrightarrow C + Enz$

It would appear that the rates of each step of the two-step mechanism must be relatively quite fast in order for their product to exceed the rate of the one-step process. Such situations are already known in nonenzymatic catalysis. Pyridine accelerates the rate of hydrolysis of acetic anhydride because (1) it is a much more effective nucleophile toward acetic anhydride than water is, and (2) the intermediate formed is much more reactive than acetic anhydride itself (equation 25.38).

(25.37) 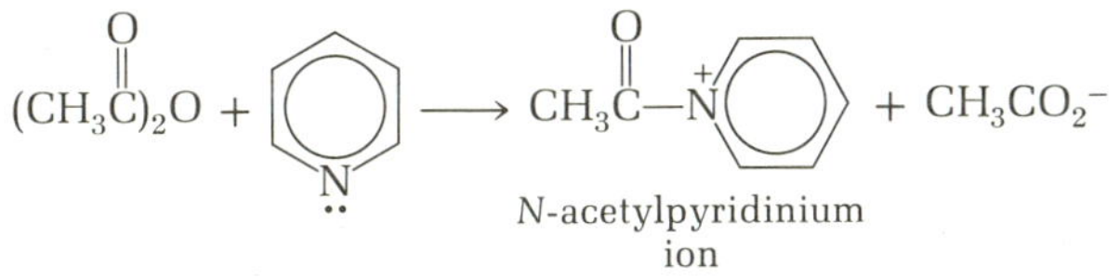

(25.38) 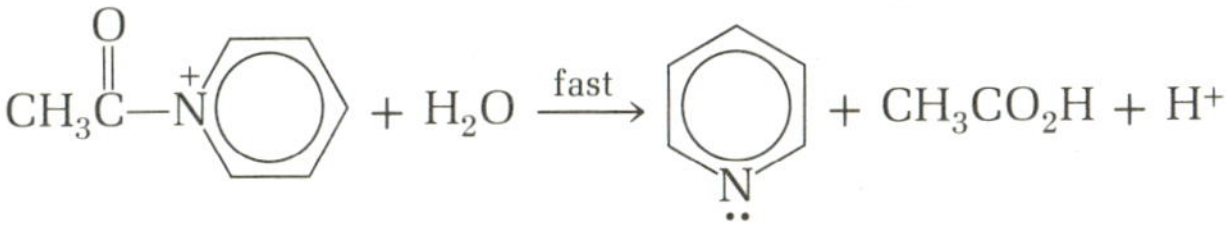

An analogous set of reactions seems to be occurring in hydrolyses catalyzed by proteases containing serine at the active site (such as trypsin and chymotrypsin). The mechanism is not known with certainty, although a number of postulated steps appear highly probable. The ester, amide, or peptide substrate is first bound to the enzyme by complexation (hydrogen bonding, etc), and an acyl group then transferred from the substrate to an enzyme serine hydroxyl group, freeing the substrate leaving group and creating an acyl enzyme. It has been suggested that this process is aided by general acid and base catalysis by the imidazole ring of the active histidine residue (equation 25.39).

(25.39) 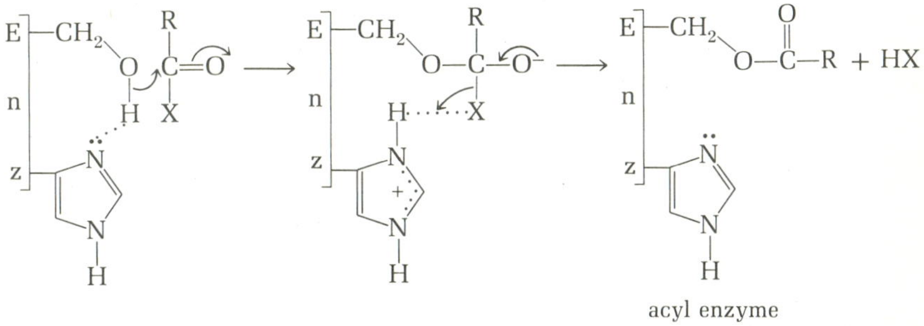

The acyl enzyme then reacts with water to regenerate the free enzyme and liberate the other molecule of product. It is presumed that the reaction is again accelerated by intramolecular catalysis by the imidazole ring. Furthermore, there is some evidence that the reactivity of the intermediate ester (acyl enzyme) is enhanced by torsional strain imposed by conformational changes within the enzyme.

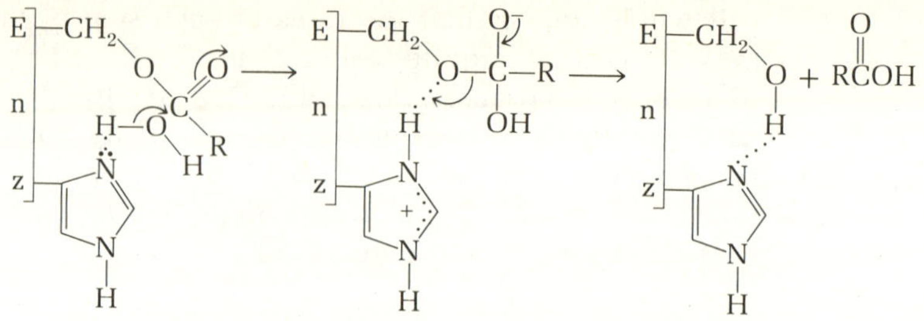

(25.40)

PROBLEM

7 Both compounds I and II undergo base-assisted hydrolysis. In each case the rate is first order in ester and first order in base.

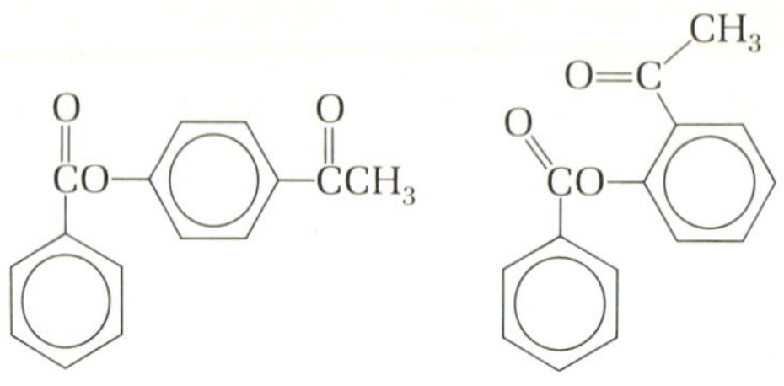

I II

Under standard conditions compound II reacts 355 times faster than compound I.

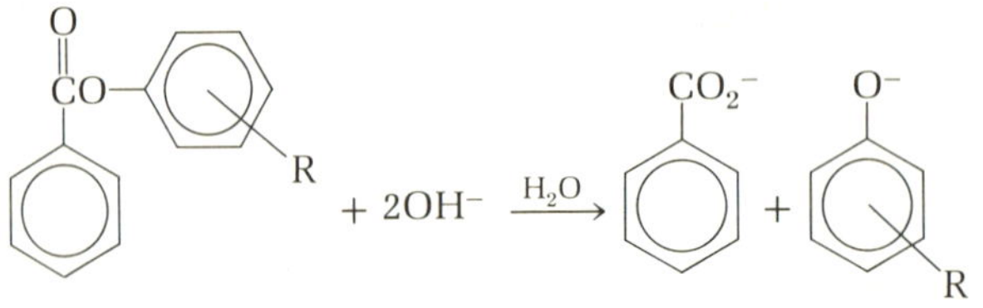

(8)

Write mechanisms for the hydrolysis of each compound. How do you account for the rate difference between I and II?

SUMMARY

1 Since α-amino acids contain both acidic and basic groups, they exist in solution and in the crystal primarily as zwitterions in which internal neutralization has taken place. The isoelectric point is that pH at which the positive and negative charges within a molecule are exactly equal; some α-amino acids have additional acidic or basic groups in their side chains, and their isoelectric points are markedly different from those of the simple

acids. Since amino acids can be converted to charged particles, they can be made to migrate in an electric field, the basis of the electrophoretic method of separation.

2 Except for glycine, all α-amino acids contain an asymmetric α-carbon atom. When isolated from proteins, these acids are each enantiomerically pure and they almost always have the *S* absolute configuration. This configuration has been related to L-glyceraldehyde and the amino acids have therefore also been given the L designation.

3 The total amino acid content of proteins is obtained by hydrolyzing a sample in aqueous hydrochloric acid and then separating the components by elution chromatography. The entire procedure has been automated and can be carried out with tiny quantities of sample. Estimation of the amount of each amino acid depends on its decarboxylative condensation with 2 moles of ninhydrin to yield Ruhemann's purple, the concentration of which may be accurately measured by spectrophotometry.

4 *N*-Terminal amino acid determination on a peptide or protein is relatively easily accomplished by allowing the molecule to react with dinitrofluorobenzene. The chain is then hydrolyzed completely and the basic acids are separated from the nonbasic, dinitrophenylated *N*-terminal acid. This approach has been somewhat superseded by the Edman method in which the *N*-terminal residue is removed without disrupting the remainder of the chain. The process may then be continued for a number of cycles; its value has increased because it has lent itself to automation.

In order to achieve a complete sequence analysis of a protein or peptide, it is usually necessary to partially hydrolyze the molecule to a set of overlapping fragments. Determination of the complete sequence of each of these fragments then allows one to deduce the complete sequence of the original molecule. The best method for effecting partial hydrolysis is to use proteolytic enzymes that bring about peptide bond scission under mild conditions. Many of these enzymes work preferentially on only a limited number of amino acid residues, thus avoiding the problem of having the chain broken down into a large number of small, random fragments.

5 Amino acids are usually obtained from natural sources because they are available in an optically pure state. The hydrolysis of appropriate proteins used to be the method of choice for obtaining most α-amino acids; it is rapidly being supplanted by fermentation.

Many laboratory syntheses of α-amino acids have been devised. The Strecker and Bücherer reactions use carbonyl compounds as starting materials and extend the carbon chain by one atom by way of reactions that are related to cyanohydrin formation. Perhaps the most versatile starting materials for amino acid synthesis are *N*-acylated derivatives of aminomalonic acid. All of these methods give racemic products. Resolution can be effected in the usual way with the aid of optically active acids and bases. Enzymes can also be used to bring about an optical resolution since most enzymes will only accept one member of an enantiomeric pair.

6 In peptide synthesis one is basically combining two amino acids to form an amide. A fundamental problem is to coax the correct amino and carboxyl groups into reacting with each other while, simultaneously, preventing undesired amide formation between the wrong amino and carboxyl groups. Prevention of the latter depends largely on the use of protecting groups that severely depress the nucleophilicity of an amino group. Foremost among these is the benzyloxycarbonyl group, which not only fulfills its protective function but may also be removed, reductively or hydrolytically, under conditions such that the amide bonds of the peptide are not disturbed. The *t*-butyloxycarbonyl group is also of value and may be easily removed by mild acid hydrolysis.

7 Free carboxyl groups will not readily condense to form amides so it is necessary to activate them. The activating group must promote rapid reaction with the nucleophile without also promoting racemization at the α-carbon atom. A very useful variant involves condensing the free acid and the amine in the presence of dicyclohexylcarbodiimide. The first step is a reaction between the carboxylic acid and the carbodiimide to yield an intermediate which subsequently reacts rapidly with a free amino group.

8 One of the most significant of recent advances in peptide synthesis is the solid state method. The *C*-terminal acid, as its *N*-Boc derivative, it attached to a chloromethylated polystyrene resin. After removal of the protecting group, another *N*-Boc acid is condensed on with the aid of dicyclohexylcarbodiimide. The insoluble product is thoroughly washed and the cycle is then repeated; isolation and purification of the growing polypeptide chain is not necessary. One of the greatest virtues of this technique is that it lends itself to automation.

9 Intramolecular hydrogen bonding is the principal structural feature that determines secondary structure — the conformation of the amino acid chain — in peptides and proteins. Two types of hydrogen-bonded structures are particularly prevalent, the α-helix and the pleated sheet. The hydrogen-bond donors are usually NH groups and the acceptors are amide CO groups. Helical folding and unfolding is often a dynamic process with a low activation barrier. Loss of secondary structure, as in denaturation, leads to loss of biological function.

The overall shape of the molecule, or tertiary structure, is dependent upon molecule-solvent interactions and upon intramolecular hydrophobic clustering.

10 The enormous rate accelerations brought about by enzymes may be understood at least in part as arising from their ability to collect together all the species necessary for the reaction, including appropriate acids and bases, and then orienting these units in the most favorable way. Most enzymes possess an active site within which the actual bond makings and breakings take place; certain specific amino acid residues seem to be critical to the functioning of that active site.

The structure of the active site may be probed by chemical methods as well as by crystallographic studies. In particular, the chemistry of enzyme inhibition sheds much light on the normal functioning of the catalyst.

11 Enzymatically-catalyzed bimolecular reactions often follow a path in which one of the reactants first becomes covalently bound to the enzyme. This step results in the production of an intermediate more reactive than the original substrate, which subsequently undergoes rapid reaction with the second component. Such catalysis is known even in nonenzymatic systems, although the rate enhancements are not nearly so dramatic.

PROBLEMS

Note: Problems 1 to 7 appear in text.

8 All the amino acids listed in Table 25-1 react with ninhydrin to eventually yield Ruhemann's purple, *with one exception.* This latter acid gives a yellow color with ninhydrin that has $\lambda_{max} = 440$ mμ. Which amino acid is the exception, and why? Suggest a structure for the product of its reaction with ninhydrin. (It is noteworthy that transient yellow colors often appear in the course of the reactions between the other amino acids and ninhydrin.)

9 Arginine contains a *guanidino* group, —NHC(=NH)NH$_2$. This group is very basic and must be protected when a peptide synthesis is undertaken in which arginine is used. The method chosen is nitration (equation 9); upon treatment of the *N*-nitroarginine with hydrogen and a catalyst, the protecting group is removed.

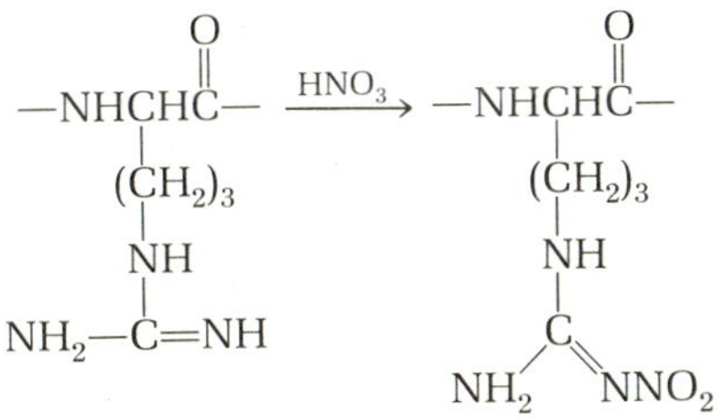

(9)

What properties of the guanidino group enable it to undergo a powerful electrophilic substitution reaction such as nitration? Use appropriate structural drawings in answering the question.

10 Chymotrypsin catalyzes the hydrolysis of a number of simple amides and esters, although not as efficiently as it catalyzes the hydrolysis of peptides. One ester that has been used a great deal in the study of chymotrypsin action is *p*-nitrophenyl acetate. When a large excess of this ester is added to a solution of the enzyme, there is initially a very rapid release of *p*-nitrophenol. Once a molar quantity of phenol equal to the number of moles of enzyme has been formed, this initial "burst" levels off and the hydrolysis then continues at a steady, markedly slower rate. How does this observation fit the proposed mechanism for chymotrypsin catalysis? What does it suggest about the relative rates of the steps involved?

SUPPLEMENTARY READINGS

K. D. Kopple, *Peptides and Amino Acids*. New York: W. A. Benjamin, Inc., 1966.

S. Bernhard. *The Structure and Function of Enzymes*. New York: W. A. Benjamin, Inc., 1968. (P)

N. F. Albertson, "Synthesis of Peptides with Mixed Anhydrides," *Org. Reactions*, **12**, 157 (1962).

W. P. Jencks, *Catalysis in Chemistry and Enzymology*. New York: McGraw-Hill Book Co., 1969.

R. E. Dickerson, and I. Geis, *The Structure and Action of Proteins*. New York: Harper and Row, 1969. (P)

"The Chemistry of Protein Synthesis," Harold Mantell, Inc., New York. An excellent color film, narrated by R. B. Merrifield, illustrating the synthesis of the enzyme ribonuclease by the automated solid phase technique.

Although it has not proved possible to discuss the subject of how enzymes come into being during the evolutionary process, this is an extremely fascinating subject that the reader may begin to explore on his own.

S. W. Fox et al., "Chemical Origins of Cells," *Chem. Engr. News*, **48**, 80 (June 22, 1970).

H. Neurath and R. A. Bradshaw, "Evolution of Proteolytic Function," *Accounts Chem. Research*, **3**, 249 (1970).

CHAPTER 26

IN QUEST OF REACTION MECHANISMS

Organic chemists of the nineteenth century made occasional observations concerning the relationship between molecular structure and chemical reactivity, but a consistent and vigorous inquiry into these connections did not begin until the second or third decade of the twentieth century. Since that time the field has flowered abundantly and the elucidation of reaction mechanisms has become a distinct subdiscipline of organic chemistry. The insights that have grown out of these labors have had a major influence on other activities of organic chemists, such as synthesis, structure elucidation, and mass spectrometry. The introduction of mechanistic concepts into biochemistry has profoundly affected the nature of the research effort in that field in modern times.

DELVING INTO THE LITERATURE

Very few inquiries into a reaction mechanism begin in the laboratory, even when one's curiosity about a given reaction has been sparked by an experimental observation. Once chemistry became an "established" science, data on various compounds and reactions were quick to accumulate and the rate of production of chemical information in our time can only be described as astronomical. Only the most foolhardy, or self-confident, investigator would neglect the opportunity to avail himself of the results and analyses generated by other workers.

The great bulk of this information is to be found in the so-called "primary" journals, which carry the detailed accounts of various chemical investigations. These journals are usually, though not always, published by various national scientific societies and a selective list of them is contained in Table 26-1.

Many topics in organic chemistry are each already the subject of a sizable body of literature. In a number of cases these topics have been surveyed in review articles that give an overview of the material and also provide a substantial number of references to the original articles. There are several journals devoted wholly or in part to this reviewing function; among the most prominent are Chemical Reviews (*Chem. Rev.*), Quarterly Reviews (*Quart. Rev., Chem. Soc.*), Angewandte Chemie (*Angew. Chem.*; published in English as the International Edition), Uspekhi Khimii (available in English

TABLE 26-1
Some Primary Journals for Organic Chemical Data

Journal Title	Chemical Abstracts Abbreviation	Publication Language(s)[a]
Journal of the American Chemical Society	*J. Amer. Chem. Soc.*	E
Journal of Organic Chemistry	*J. Org. Chem.*	E
Journal of the Chemical Society (London), Parts B and C	*J. Chem. Soc., B and C*	E
Chemical Communications (Journal of the Chemical Society, Part D)	*Chem. Commun.*	E
Bulletin de la Société Chimique de France	*Bull. Soc. Chim. Fr.*	F
Chemische Berichte	*Chem. Ber.*	G
Helvetica Chimica Acta	*Helv. Chim. Acta*	E, F, G
Recueil des Travaux Chimiques des Pays-Bas	*Rec. Trav. Chim.*	E, F
Acta Chemica Scandanavica	*Acta Chem. Scand.*	E, F, G
Zhurnal Organicheskoi Khimii	*Zh. Org. Khim.*	R[b]
Bulletin of the Chemical Society of Japan	*Bull. Chem. Soc. Jap.*	E
Tetrahedron	*Tetrahedron*	E, F, G
Tetrahedron Letters	*Tetrahedron Lett.*	E, F, G
Journal of Organometallic Chemistry	*J. Organometal. Chem.*	E, G
Journal of Heterocyclic Chemistry	*J. Heterocycl. Chem.*	E, F, G

[a]E—English; F—French; G—German; R—Russian.
[b]Available in English translation.

as *Russian Chemical Reviews*), and Accounts of Chemical Research (*Accounts Chem. Res.*).

There are a number of reviews that also appear in book form. The most distinguished series published in English is "Organic Reactions," which now comprises 18 volumes. The material chosen for inclusion is usually important in preparative chemistry and the chapters are organized around either a given type of product or a given type of reaction. A very valuable feature of these reviews is their thorough coverage of the literature and their tabular presentation of the results found in the literature. There are also a number of review series, the titles of which begin with "Advances in" or "Progress in," and among the subjects covered are organic chemistry, physical organic chemistry, heterocyclic chemistry, carbohydrate chemistry, enzymology, free radical chemistry, and stereochemistry.

There are several works available that are compilations of data concerning individual compounds and that may contain melting and boiling points, spectroscopic information, references on preparations and transformations, etc. Important examples of these compilations are the "Dictionary of Organic Compounds," "Elsevier's Encyclopedia of Organic Chemistry," and "Rodd's Chemistry of Carbon Compounds." The dean of this class of publications is Beilstein's "Handbuch der organischen Chemie." It comprises an original collection (Hauptwerk) and three supplementary editions (Ergänzungswerke), the last of which is still in the process of being published. Beilstein is reknowned for its thorough literature coverage and attention to detail. These same properties serve to ensure, however, that Beilstein will lag considerably behind the current literature. The third supplement, which is due for completion in 1980, provides literature coverage only as far as 1949!

Guides are available to initiate one into the mysteries of Beilstein, which incorporates its own system of classifying organic compounds. Once the system is grasped, the set is a wealth of readily accessible information.

If one were seeking information about a compound, one would first consult one of the compendia. Similarly, a study of a particular reaction would be well initiated by a search through the review literature as well as relevant monographs. The reviews and compendia usually contain copious references to the original literature, which must eventually be read. One cannot, however, rely totally on reviews as guides to original papers; their coverage may be only selective, and there is a cutoff date for literature references anyway, so one must eventually go directly to the primary journals.

This task is rendered feasible by the existence of several abstracting journals, of which Chemical Abstracts (*Chem. Abstr.*) is the only important one for organic chemists published in English. This publication consists of brief abstracts of articles from an extensive list of journals, reviews, and related publications, along with a set of heavily cross-referenced indices designed to permit the reader to obtain the maximum information from the abstracts. Thus, one may look up not only individual compounds but also compound classes, compound properties, individual reactions, reaction types, processes, reactants, products, solvents, concepts, mechanisms, etc. For an introduction to the use of Chemical Abstracts, see one of the articles by Hancock cited at the end of this chapter. That article also discusses the so-called alerting services, which permit up-to-the-minute coverage of the current literature.

The remainder of the chapter is devoted to somewhat detailed case histories, derived from the original literature, of two different mechanistic investigations. These expositions are aimed particularly at demonstrating the wide array of techniques and experiments that chemists can employ to illuminate the detailed course of reaction. Despite the ingenious and sometimes brilliant experiments to be described in the following pages, none of the derived mechanisms can be said to have been established with complete certainty. The inherent possibility of future revision of apparently irrefutable theories constitutes both the challenge and frustration of physical organic chemistry.

THE NITRATION OF BENZENE: REACTIVE SPECIES

When considering the mechanism of nitration of benzene and other aromatic compounds, it is convenient to divide the problem into two parts. We wish to know the identity of the active species that actually attack(s) the aromatic ring, and we are also interested in deducing the detailed course of the reaction between these entities. Perhaps the most important single study of the mechanism of this reaction, which included investigations of both aspects, was conducted by C. K. Ingold and his co-workers at University College, London, during the decade immediately following World War II.

Starting at the end of the nineteenth century, a number of chemists had speculated on the nature of the active species in nitration. Circumstantial evidence suggested that this substance was strongly electrophilic. Since sulfuric acid was very effective in increasing the rate of nitration, it was generally concluded that the active species was not nitric acid but rather some cation or highly dipolar molecule derived from it. The leading candidates were the nitronium ion **(1)**, the nitric acidium ions **2** and **3**, dinitrogen pentoxide **(4)**, and nitric sulfuric anhydride **(5)**.

NO_2^+	$H_2NO_3^+$	$H_3NO_3^{2+}$	N_2O_5	O_2NOSO_3H
1	**2**	**3**	**4**	**5**

The first important observations bearing on the identity of the active species were made by Hantzsch in 1908. He was using sulfuric acid as the solvent for *cryoscopic studies* — measurements of the effects of various solutes on the freezing point of the solvent. Pure sulfuric acid melts at 10.36°; if 1 mole of an "ideal" solute, such as sulfuryl chloride (SO_2Cl_2), is added to 1000 g of the solvent, the freezing point will be lowered by about 6°. (An ideal solute is one that does not interact with the solvent.) One mole of sodium bisulfate produces a depression *twice* as large, however, because it gives rise to 2 moles of independent particles upon solution, the ions Na^+ and HSO_4^-. We symbolize these circumstances by setting the ν factor equal to 2 for sodium bisulfate and equal to 1 for sulfuryl chloride.

A number of nonionic substances exhibit ν factors greater than 1 in sulfuric acid because they react with the solvent. Thus, for acetophenone $\nu = 2$ because the ketone is a base in sulfuric acid. Triphenylmethanol has $\nu = 4$ and it is virtually certain that the reaction shown in equation 26.2 is taking place.

(26.1)

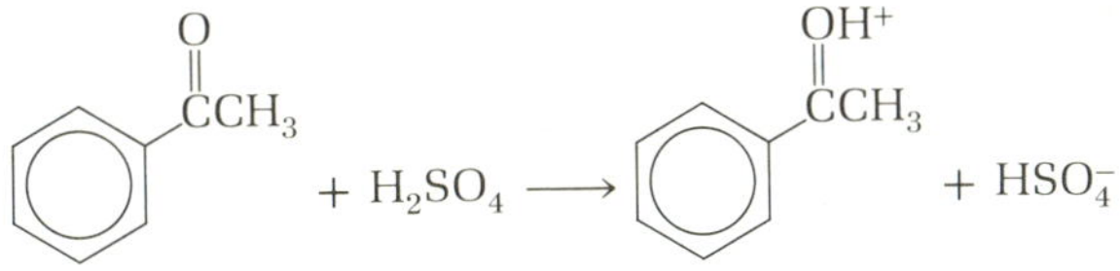

(26.2)

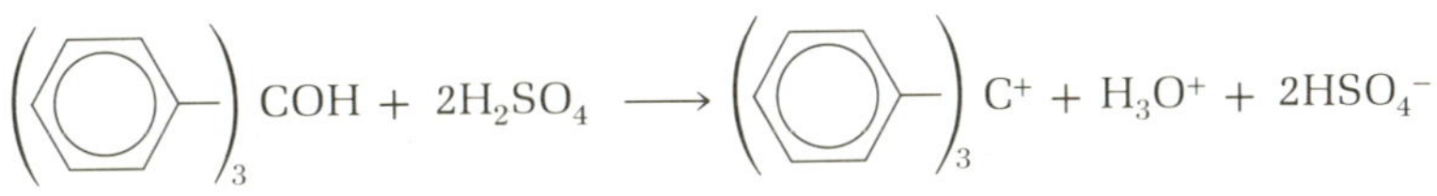

When Hantzsch dissolved nitric acid in sulfuric acid, there was a substantial freezing point depression. Considerable variation occurred in the values found for the ν factor, but the number 3 was adopted. Hantzsch believed that the important processes taking place were those depicted in equations 26.3 and 26.4.

(26.3) $HNO_3 + H_2SO_4 \longrightarrow H_2NO_3^+ + HSO_4^-$

(26.4) $HNO_3 + 2H_2SO_4 \longrightarrow H_3NO_3^{2+} + 2HSO_4^-$

Ingold and his group undertook a careful reinvestigation of this reaction, and they were to show that $\nu = 3.8$, with a high degree of reproducibility. They interpreted this result as strongly supporting reaction 26.5.

(26.5) $HNO_3 + 2H_2SO_4 \longrightarrow NO_2^+ + H_3O^+ + 2HSO_4^-$

That the ν factor fell a little short of the theoretical value of 4.0 was later shown to be due to the fact that the equilibrium depicted in equation 26.6 does not lie completely to the right.

(26.6) $H_2O + H_2SO_4 \rightleftharpoons H_3O^+ + HSO_4^-$

Additional, albeit circumstantial, evidence for the nitronium ion hypothesis was provided by Kharasch and Westheimer, who studied the rate of nitration of aromatic substrates in mixtures of water, sulfuric acid, and nitric acid. They found that the rate was inversely proportional to the water content of the medium. Furthermore, they observed that the dependence of the rate of nitration on water content closely paralleled the dependence of the ionization of triphenylmethanol in aqueous sulfuric acid (equation 26.2).

Since triphenylmethanol is a solute for which $\nu = 4$, this finding suggested that for nitric acid $\nu = 4$ as well.

Much more substantial evidence resulted from a series of spectroscopic measurements on solutions of nitric acid in sulfuric and other acids. Médard and Chédin in France, during the period 1934 to 1936, showed that solutions of nitric acid in sulfuric acid exhibited strong spectral bands at 1400 and 1050 cm^{-1}. These bands were weak in pure nitric acid but their intensity increased sharply when sulfuric acid was added. Chédin found that both bands were very intense in the spectra of solid N_2O_5 and in solutions of N_2O_5 in nitric and sulfuric acids. He thus concluded that nitric acid is converted to dinitrogen pentoxide in sulfuric acid, lending support to the hypothesis that dinitrogen pentoxide was the active nitrating agent.

There was one difficulty in Chédin's conclusion. The spectrum of N_2O_5 in solvents such as carbon tetrachloride or chloroform was bereft of the lines at 1400 and 1050 cm^{-1}. Chédin was forced to conclude that the oxide existed in different forms in nonpolar solvents, on the one hand, and in strong acids and in the solid state, on the other. It was at this point that the London group took up the problem, after a 10-year interval.

Thanks to their cryoscopic investigations they strongly suspected the nitronium ion as one source of the characteristic frequencies. Furthermore, it had already been shown that *both* the nitrate and bisulfate ions gave rise to peaks near 1050 cm^{-1}, which suggested that the 1400 cm^{-1} band was associated with the nitronium ion. This hypothesis was nicely confirmed by the spectrum of nitric acid in perchloric acid, which exhibited a strong peak at 1400 cm^{-1} and *no* absorption at 1050 cm^{-1}.

(26.7) $$HNO_3 + 2HClO_4 \longrightarrow NO_2^+ + H_3O^+ + 2ClO_4^-$$

The final link was established when it was discovered, with the aid of both vibrational spectroscopy and X-ray crystallography, that crystalline dinitrogen pentoxide was nitronium nitrate, $NO_2^+\ NO_3^-$.

The vibrational frequencies quoted above were not obtained by infrared spectroscopy but rather by *Raman spectroscopy*. In this method a monochromatic beam of light, usually one line from a mercury lamp or from a laser, is passed through the sample. The frequency of the light is such that it is too great to be absorbed but some of the photons collide with molecules of the sample and are scattered. In the course of these collisions many of the photons gain or lose small increments of energy, and these increments are exactly equal to the energies of the various vibrational modes of the molecule. Thus, some of the photons will have different frequencies before and after having been scattered, and these frequency *differences* correspond to the fundamental vibrational frequencies of the scattering molecule.

One of the virtues of Raman spectroscopy is that it is complementary to the more common infrared absorption spectroscopy. Both spectra are similar in appearance but one often finds that bands that are weak or absent in the ir spectrum are strong in the Raman spectrum and vice versa. Thus, although

Cl_2 has one fundamental vibration, at about 560 cm^{-1}, this compound does not give rise to an ir spectrum because the vibration involves no change in dipole moment (see Chapter 3). However, chlorine does produce a Raman band at this frequency. Similarly, the completely symmetrical stretching motion of the two oxygen atoms in O=C=O does not generate a band in the ir spectrum of carbon dioxide but the Raman spectrum does contain a band at 1340 cm^{-1} attributable to this vibration. The 1400 cm^{-1} band in the Raman spectrum of $O{=}\overset{+}{N}{=}O$ is almost certainly generated by the same type of motion, and the similarity of the frequencies found in the isoelectronic species CO_2 and NO_2^+ supports the general interpretation of the spectra.

With the existence of the nitronium ion firmly established spectroscopically and cryoscopically, it was natural that attempts should be made to isolate it in the form of its salts. This the Ingold group was able to do; by allowing dinitrogen pentoxide to react with perchloric acid in nitromethane, for example, they were able to isolate nitronium perchlorate. The various nitronium salts exhibited all the properties expected of them and the structure of the perchlorate was confirmed by means of X-ray crystallography. There was no question that nitric acid, upon dissolution in strong acids, was converted into nitronium ion. It remained to be shown that the intermediacy of this ion during aromatic nitration was compatible with the kinetic data.

NITRATION OF BENZENE: KINETICS

The earliest kinetic study of nitration that yielded a simple rate equation was carried out by Martinsen in 1904. Nitrobenzene was the reactant and sulfuric acid, the solvent; Martinsen found the reaction to be first order in aromatic compound and first order in nitric acid (equation 26.8).

(26.8) $$\text{Rate }(H_2SO_4) = k(C_6H_5NO_2)(HNO_3)$$

The choice of nitrobenzene as the substrate had two purposes: It nitrated at a convenient rate in this powerful nitrating system and it was much more soluble than a hydrocarbon, due to its partial protonation by sulfuric acid. Similar rate equations were found by later investigators to apply in this solvent for other deactivated molecules such as benzoic acid and anthraquinone.

The London group studied nitrations of deactivated benzenes in pure nitric acid. An overall first-order rate law was found (equation 26.9); this is to be expected from equation 26.8 since nitric acid is in enormous excess and its concentration remains effectively constant throughout the reaction.

(26.9) $$\text{Rate }(HNO_3) = k(\text{ArH})$$

Equations 26.8 and 26.9 are compatible with the intervention of a nitronium ion intermediate, but they certainly do not require it.

More significant results were obtained when the nitrations were carried out in organic solvents such as nitromethane and acetic acid. These reactions were conducted with a large excess of nitric acid so that all were pseudo-zeroth-order in nitric acid. Moderately deactivated aromatics (di- and tri-chlorobenzenes) displayed first-order kinetics, in accord with equation 26.9. Under the same conditions some more activated compounds (benzene, toluene, ethylbenzene) were nitrated in *zeroth-order reactions;* i.e., the rate was constant throughout the reaction and insensitive to the concentration and identity of the substrate (equation 26.10).

(26.10) $\text{Rate } (CH_3NO_2, \text{ active substrates}) = k$

Compounds of intermediate reactivity (monohalobenzenes) seemed to show a kinetic order intermediate between zero and one.

The discovery of the zeroth-order process was very illuminating, for it showed conclusively that there was a step preceding the attack on the aromatic ring, involving the nitric acid alone, that produced a reactive intermediate. This intermediate then reacted rapidly with activated aromatic compounds.

On chemical grounds the most plausible intermediates are the nitric acidium ion ($H_2NO_3^+$) or the nitronium ion. If the first-named ion were the reactive intermediate, then for the zeroth-order reaction the slow step would be a proton transfer (equation 26.11).

(26.11) $2HNO_3 \xrightarrow{\text{slow}} H_2NO_3^+ + NO_3^-$

Such a situation would be a highly improbable one since proton transfers between oxygen atoms are almost always very rapid.

The zeroth-order reactions may be markedly retarded by small quantities of potassium nitrate, although the addition of this salt does not lead to a change in the rate law. Furthermore, addition of comparable quantities of water has a much smaller inhibiting effect on the reaction. These data virtually require the kinetic scheme shown in equations 26.12 to 26.14.

(26.12) $2HNO_3 \overset{\text{fast}}{\rightleftharpoons} H_2ONO_2^+ + NO_3^-$

(26.13) $H_2ONO_2^+ \xrightarrow{\text{slow}} H_2O + NO_2^+$

(26.14) $NO_2^+ + ArH \xrightarrow{\text{fast}} ArNO_2 + H^+$

The first-order reaction is also subject to retardation by added nitrate salts, again without any change in kinetic order taking place. The same reactions apply, except that the formation of the nitronium ion has become fast and reversible compared to its rate of attack on the aromatic substrate. If sufficient water is added to a zeroth-order reaction, it can be converted to a first-order reaction.

The course of the nitration in different solvents can now be summarized as follows. In pure nitric acid only a few percent of the molecules are present as nitronium ions at equilibrium. Nonetheless, the equilibrium is rapidly established so that these ions are always present in constant concentration, and the rate-determining step is the reaction between the nitronium ion and the substrate (first-order kinetics). When the nitric acid is diluted with nitromethane or acetic acid, the equilibrium concentration of nitronium ion decreases and the rate of its formation also drops. If an unreactive substrate is used, then the attack of nitronium ion on this molecule will be slow and first-order kinetics will still be observed. If the aromatic compound is reactive enough to consume nitronium ion as fast as it is formed, however, then formation of the ion becomes the rate-determining step (zeroth-order kinetics).

When nitric acid is dissolved in sulfuric acid, it is converted quantitatively to nitronium ion, which then reacts with the substrate during the slow step.

(26.5) $$HNO_3 + 2H_2SO_4 \longrightarrow NO_2^+ + H_3O^+ + 2HSO_4^-$$

NITRATION OF BENZENE: ISOTOPE EFFECTS

There remains only the question of the exact mode of attack of the nitronium ion on the aromatic compound. Only two mechanisms have ever been seriously considered: a one-step, termolecular reaction (equation 26.15) and a two-step sequence comprised of two bimolecular reactions (equations 26.16 and 26.17). [Since bare protons are not encountered in solution, a base (B^-) has been explicitly included in these equations. The base could be nitrate or bisulfate ion, or even water.]

(26.15) $$ArH + NO_2^+ + B^- \longrightarrow ArNO_2 + BH$$

(26.16) $$ArH + NO_2^+ \longrightarrow \left[Ar\begin{matrix} \diagup H \\ \diagdown NO_2 \end{matrix} \right]^+$$

(26.17) $$\left[Ar\begin{matrix} \diagup H \\ \diagdown NO_2 \end{matrix} \right]^+ + B^- \longrightarrow ArNO_2 + BH$$

Although the two-step mechanism has been supported by a number of pieces of circumstantial evidence, the most convincing demonstration of its correctness comes from a study of *kinetic isotope effects.*

The potential energy functions for two isotopes of the same element are virtually identical. But molecular vibrations involving the isotopic atoms do not occur at the same frequencies because the isotopes differ in mass. An analogous mechanical situation would be one in which we had two springs of identical force constant, with a ball of different mass attached to each one. The fundamental vibrational frequency would be greater for the system with the lighter ball. (For a review of these concepts, see the first part of Chapter 3.)

A dramatic example is provided by a comparison of molecular hydrogen and deuterium. The vibrational frequency for H_2 is 4160 cm^{-1}, while that for D_2 is 2993 cm^{-1} (Raman spectroscopy). To dissociate H_2 into hydrogen atoms requires 104.2 kcal/mole, while the conversion of D_2 to deuterium atoms requires 106.0 kcal/mole. *Since H_2 in its ground state has more vibrational energy than D_2 does in its ground state, it takes less energy to cause the dissociation of H_2* (Figure 26-1).

To put the matter on a more quantitative footing, we may note that the vibrational frequency and the vibrational energy are related by equation 26.18 (for the lowest vibrational level).

(26.18) $$E_{vib} = \frac{1}{2}h\nu$$

The relationship between the vibrational frequency and the mass is given by equation 26.19; μ, the reduced mass, equals ½ for H_2 and 1 for D_2.

(26.19) $$\nu = \frac{1}{2\pi}\sqrt{\frac{f}{\mu}}$$

The force constant, f, is the *same* for both molecules since it is a function only of the potential energy. The difference in ground state vibrational energy of the two molecules is known as the *zero-point energy difference*.

We can now transfer these ideas in a qualitative way to other molecules. Consider the thermal decomposition of fluoroform, CHF_3, and fluoroform-*d*, CDF_3(equation 26.20).

(26.20) $$CF_3X \longrightarrow CF_3\cdot + X\cdot$$
(X = H or D)

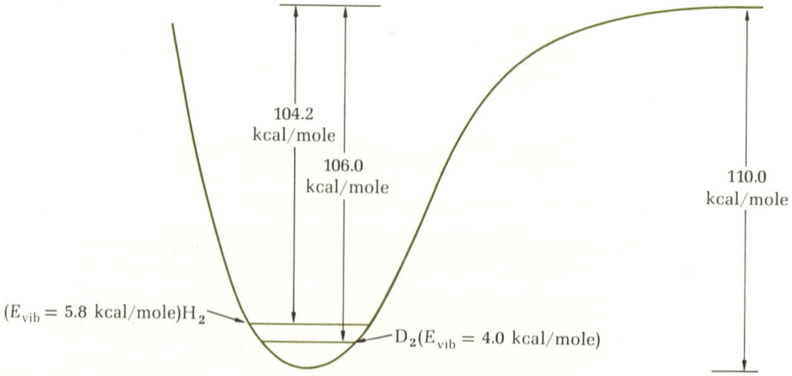

FIGURE 26-1
Ground State Vibrational Energies and Dissociation Energies of Molecular Protium and Deuterium

The hydrogen atom has three modes of motion in this molecule, and they will each be lower in energy when the deuterium isotope is present. Thus, the total vibrational energy of the C—D bond is less than that of the C—H bond and it therefore requires *more* thermal energy to break the C—D bond.

This is an example of an *isotope effect* — a change of reaction rate due to isotopic substitution of one or more atoms within the reacting molecule. In the fluoroform case, we are studying a *primary isotope effect* since the atom that has been substituted is undergoing bond making or bond breaking in the course of the reaction. This reaction would exemplify a *normal* isotope effect, in which reaction is faster when the lighter isotope is present. For the reasons discussed above, primary isotope effects are virtually always normal (expressed as $k_H/k_D > 1$). The size of the isotope effect varies from reaction to reaction and is dependent on the extent to which the bond has been broken at the transition state. Numerically, k_H/k_D might lie in the range 3 to 15 at 25°.

Melander undertook to settle the question of the mechanism of aromatic nitration. A termolecular mechanism (equation 26.15) would undoubtedly lead to the observation of a substantial isotope effect. The two-step mechanism could also result in the observation of an isotope effect, *providing* that the second step (equation 26.17) was rate determining. If the first step (equation 26.16) were rate determining, then no substantial isotope effect would be anticipated.

The radioactive hydrogen isotope, tritium, was used by Melander in his investigations. Since tritium has an atomic weight three times that of protium, very substantial isotope effects could be observed. Melander prepared benzene in which some of the molecules contained tritium at one position and subjected this material to nitration. He found that the tritium loss was statistical; i.e., of those molecules that contained tritium only one-sixth lost the heavy isotope upon nitration. This finding demonstrated that a carbon-hydrogen bond was not being broken in the slow step and that the mechanism was a two-step one in which reaction 26.16 was rate determining.

Melander studied a number of other aromatic molecules. He found that nitration at the *ortho* position of toluene and of bromobenzene, at the *meta* position of nitrobenzene, and at C-1 of naphthalene entailed no isotope effect. These results, confirmed by others working with deuterium instead of tritium, imply that in all nitrations the attack of the nitronium ion on the aromatic ring is the rate-determining step.

There is wide general agreement that the intermediate produced in this step is a cyclohexadienylium (benzenium) ion. The nitro group and the hydrogen atom that it will displace are both attached to the same sp^3-hybridized carbon atom. Olah succeeded in isolating such an ion **(6)** by nitrating benzotrifluoride with nitronium fluoborate and maintaining a temperature lower than −50°. Above that temperature the ion decomposes to *m*-benzotrifluoride.

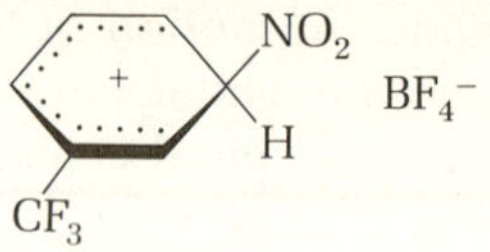

6

Some reasonable conclusions may be drawn about the transition states leading to the benzenium ion intermediates. Nitration is not a very discriminatory reaction. When nitration is carried out with nitric acid in organic solvents, it is found that toluene is only 20 to 25 times more reactive than benzene. When preformed nitronium salts are used as electrophiles, toluene proves to be less than twice as reactive as benzene. Furthermore, very substantial quantities of *o*-nitrotoluene are formed from toluene and even a few percent of *m*-nitrotoluene.

Since the nitronium ion is highly electrophilic, its reaction with benzene to form a charge-delocalized benzenium ion may well be exothermic. In that case the energy of activation would be small, and the transition state would come early before much charge had been transferred from the nitronium ion to the ring (Figure 26-2). It has been suggested that the transition state resembles a π-complex **(7)** more closely than it resembles a σ-complex (benzenium ion, **8**).

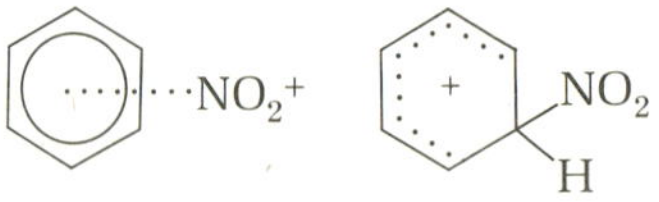

7 **8**

Such a hypothesis accords with the low selectivity noted above.

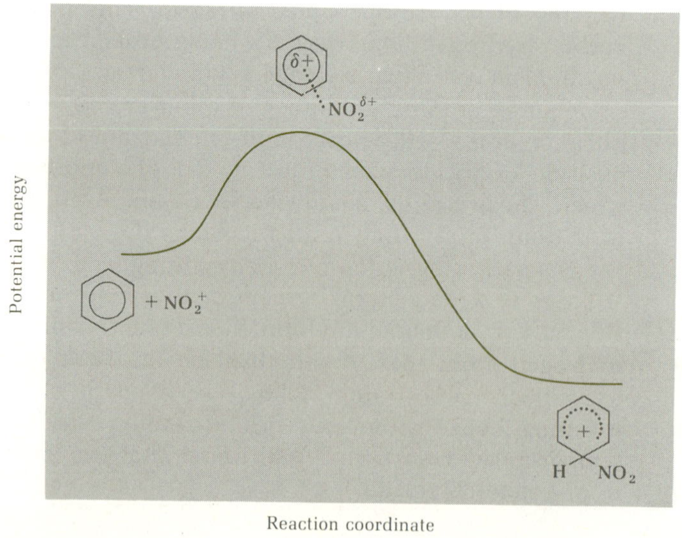

FIGURE 26-2
Energy Profile for the Reaction of Benzene with a Nitronium Ion

In 1925 Otto Diels and his co-workers at thc University of Kiel were studying the reactions of diethyl azodicarboxylate **(9).** They found that this compound reacted readily with cyclopentadiene to form a 1:1 adduct, and they were further able to show that the adduct had structure **10.**

(26.21) 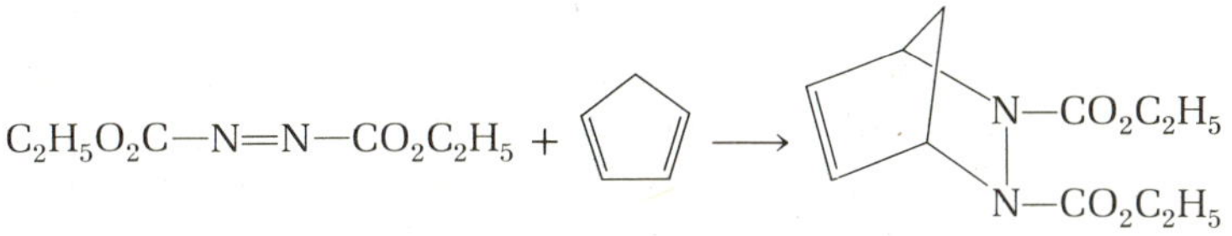

9 **10**

Isolated examples of reactions between alkenes and conjugated dienes had been known back into the nineteenth century but in many cases the structures proposed for the products were incorrect and no general conclusions were drawn.

Diels' group not only correctly deduced the course of reaction 26.21, but they provided a rationale for the process and suggested that it was merely one example of a more general addition reaction. Anschutz had shown that cyclopentadiene and *p*-benzoquinone could react to form 1:1 or 2:1 adducts. He suggested structure **11** for the 1:1 adduct, while Staudinger later proposed structure **12.** The Kiel workers advanced structure **13,** reasoning by analogy from their diethyl azodicarboxylate result.

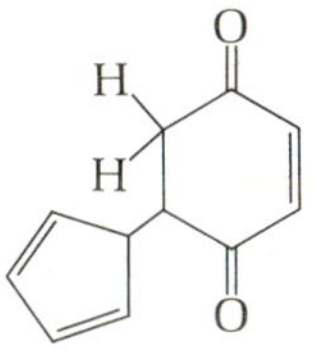

11

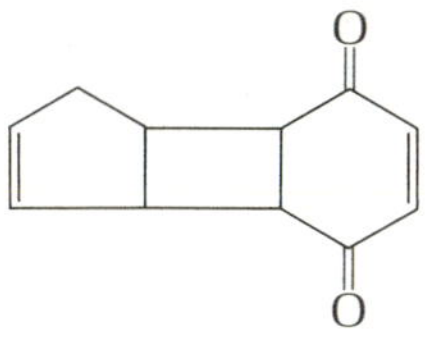

12

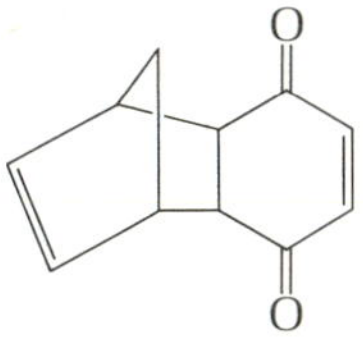

13

In 1928 Diels published a paper with his student Kurt Alder that fathered literally thousands of other publications. This paper contained two very significant results:

1 The structure of the 1:1 cyclopentadiene-benzoquinone adduct was indeed **13,** as had been earlier proposed.
2 1,3-Butadiene was also reactive, and the dienes would react with a number of alkenes, among them maleic acid and its anhydride, acrolein, and citraconic **(14)** and itaconic **(15)** anhydrides.

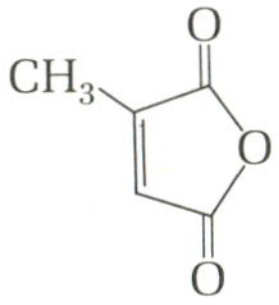

14

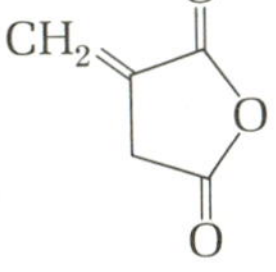

15

After the publication of this classic paper, Diels, who already had a substantial reputation for his work on the carbon skeleton of natural products, turned his attention to other areas. For Alder, however, this was the prologue to a lifetime's labor, and he played a major role in extending the scope of the reaction and in deepening our understanding of it. The discoverers referred to the reaction as the "diene synthesis," but most chemists know it as the Diels-Alder reaction. Ultimate recognition of its importance in organic chemistry came in 1950 when its originators were jointly granted the Nobel Prize for Chemistry.

Alder addressed himself, among other things, to elucidating the stereochemical course of the reaction. He discovered that it was remarkably stereospecific, a fact that accounts in good measure for its synthetic utility. Stereochemical relationships that are present in the alkene (*dienophile*) are retained in the adduct (equations 26.22 and 26.23).

(26.22)

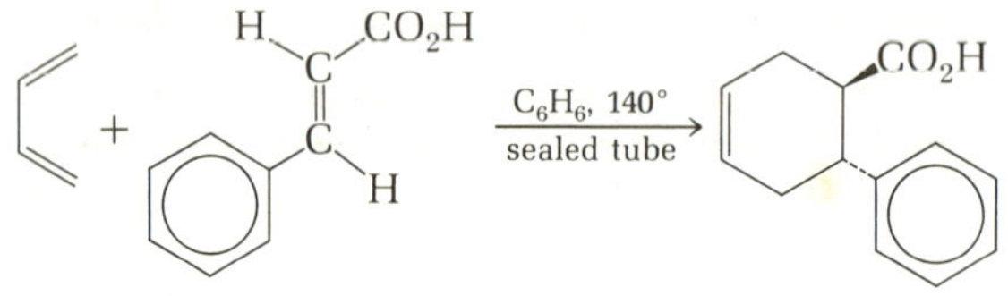

(26.23)

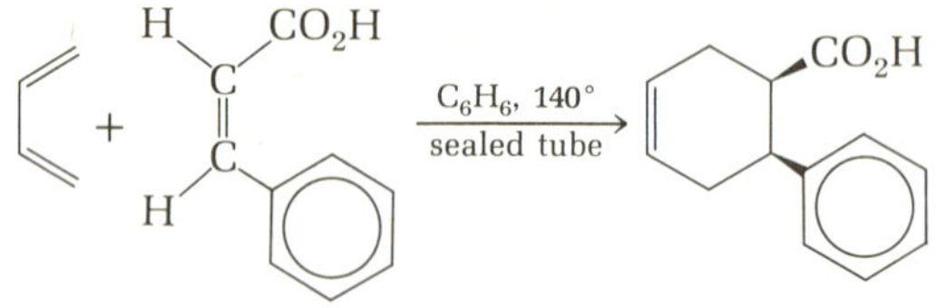

There is a similar retention of stereochemical relationships in the diene component; *trans, trans*-1,4-disubstituted butadienes yield *cis*-3,6-disubstituted cyclohexenes (equation 26.24), while *trans, cis*-1,4-disubstituted butadienes yield only *trans*-3,6-disubstituted cyclohexenes. Numerous other examples have confirmed that the Diels-Alder reaction is a highly stereospecific *cis* addition.

(26.24)

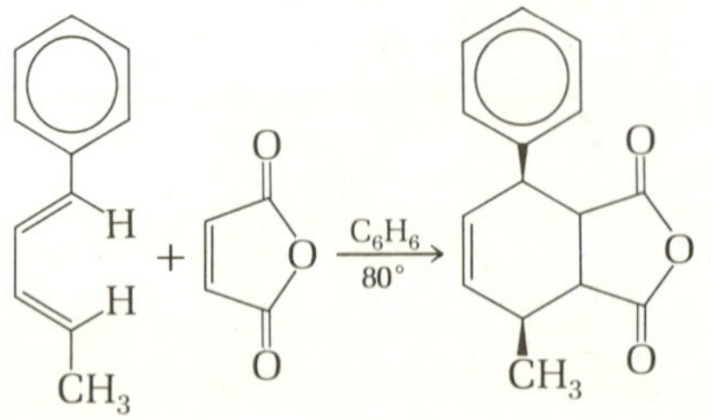

(26.25)

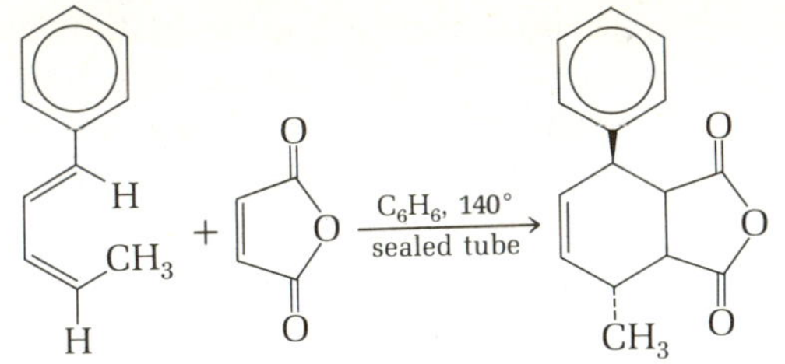

While Alder was pursuing his stereochemical studies, Albert Wassermann, in London, had undertaken an analysis of the kinetics of the reaction. He quickly established that it was second order overall, first order in each component. The retro Diels-Alder reaction, in which the adduct dissociates back into diene and dienophile, was shown to obey first-order kinetics. The rates of the forward reaction were found to be rather insensitive to the solvent polarity and they did not even differ much when the reactants were transferred from the liquid to the gas phase. It seemed highly improbable, therefore, that ionic intermediates were produced in these condensations.

Wassermann also noted that Diels-Alder reactions exhibited very small Arrhenius *A* factors; i.e., they had large *negative* entropies of activation. Furthermore, in many cases the entropy difference between reactants and transition state was nearly equal to the entropy difference between reactants and products. These results were of the utmost mechanistic significance, for they strongly suggested the creation of a highly ordered transition state.

The kinetic and stereochemical results, taken together, are very compatible with a one-step mechanism in which both new σ-bonds are partially formed in the transition state (equation 26.26).

(26.26)

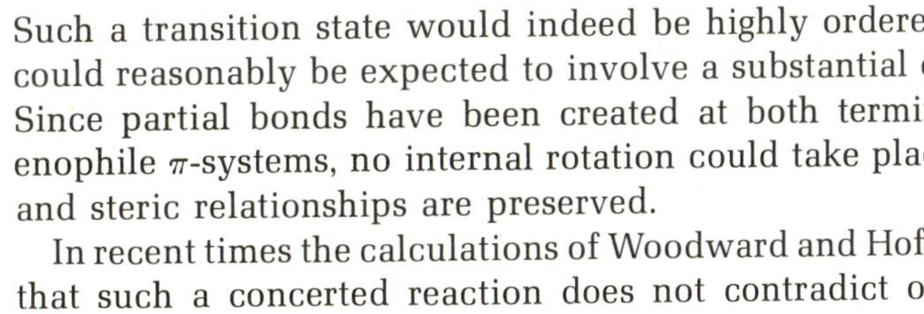

Such a transition state would indeed be highly ordered and its formation could reasonably be expected to involve a substantial decrease in entropy. Since partial bonds have been created at both termini of diene and dienophile π-systems, no internal rotation could take place in either reactant and steric relationships are preserved.

In recent times the calculations of Woodward and Hoffmann have certified that such a concerted reaction does not contradict orbital symmetry requirements. The fact that a concerted mechanism is allowed does not, of course, guarantee that it will be followed.

The principal opposition to the concerted mechanism was provided by Kistiakowsky (Harvard), who made a careful study of a number of gas-phase

Diels-Alder reactions, including the dimerization of butadiene to 4-vinylcyclohexene. He calculated the activation entropies for biradical and concerted transition states based on a number of assumed molecular frequencies and concluded that the experimental data matched the biradical case much more closely (equation 26.27).

(26.27) 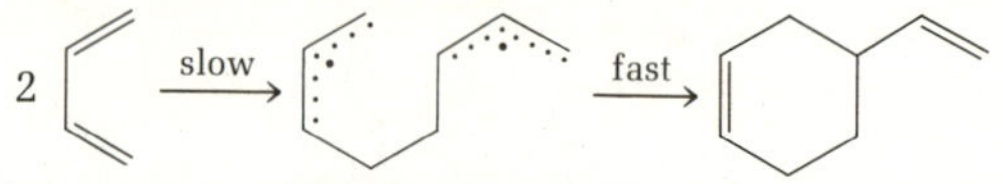

Since considerable polymerization accompanied the dimerization, Kistiakowsky concluded that both processes proceed by way of the same intermediate and the Diels-Alder reactions generally followed two-step mechanisms.

Kistiakowsky's conclusions were not widely accepted. For one thing, they failed to account for the remarkable stereospecificity of the Diels-Alder reaction; it would be anticipated that rapid rotation would take place around the single bonds in an unsymmetrical diradical, leading to a mixture of products. Furthermore, Wassermann had shown that a number of additives that normally interact with radicals (oxygen, nitric oxide, iodine, certain metal salts) did not affect the rates of Diels-Alder reactions.

By and large organic chemists in the 1930's, 40's, and 50's accepted the one-step, concerted mechanism for the Diels-Alder reaction. In 1959, however, Woodward and Katz (Harvard) published a paper that was destined to rekindle the fires of mechanistic controversy. These authors were studying the chemistry of the hydroxydiene **16,** obtained by oxidation of dicyclopentadiene. When **16** was heated to about 140°, it underwent a stereospecific rearrangement to **17.**

(26.28) 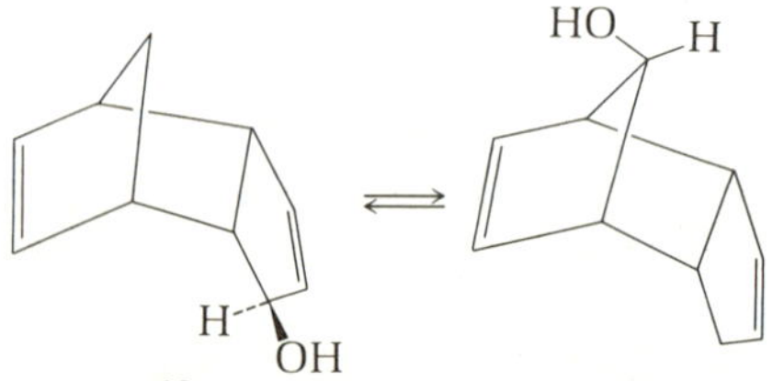

16 **17**

The system eventually reached equilibrium, at which time no other components could be detected in the mixture.

The stereospecificity of the rearrangement made it highly unlikely that the diene was undergoing a retro Diels-Alder reaction followed by recombination of the fragments (equation 26.29).

(26.29) 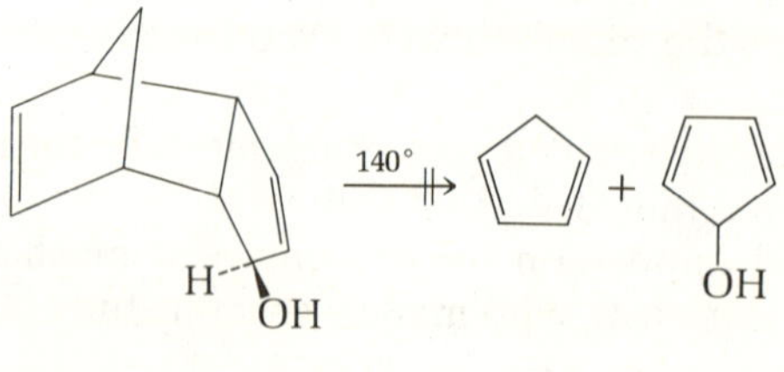

Woodward therefore proposed that bond *a* was initially broken to yield a diallylic diradical **(18).** Rotation around the remaining inter-ring bond (*b*) followed by bond formation at the termini of the π-systems (*c*) would then complete the rearrangement.

(26.30)

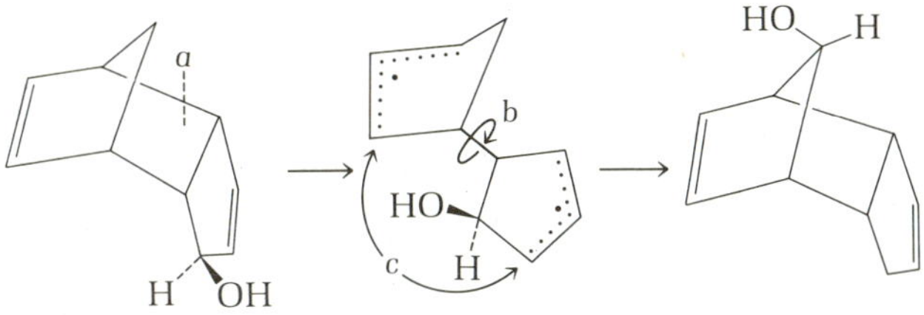

16 **18**

Woodward and Katz also knew that if the starting material were heated 20 to 30° above its rearrangement temperature, then it would undergo complete retro Diels-Alder cleavage. They argued that the same diradical **(18)** was formed initially but that at the higher temperature sufficient energy was available to allow bond *b* to break as well. Since the retro Diels-Alder reaction was therefore a two-step process, proceeding by way of diradical **18,** the principle of microscopic reversibility demanded that the forward reaction proceed by way of the same intermediate.

Shortly before the appearance of the Woodward-Katz paper, Walling (Columbia) published the first of two papers on the effect of pressure changes on the rate of some Diels-Alder reactions. It had been known for some time that numerous reactions were either accelerated or retarded by an increase in pressure. In general, if the volume of the transition state is smaller than the total volume of the reactants, then the rate of the reaction will *increase* with increasing pressure. This finding is expressed by equation 26.31, in which $\Delta V^{\ddagger}$ is the volume of activation (the difference in partial molal volume between the transition state and the reactants).

(26.31) $$\left(\frac{\delta \ln k}{\delta P}\right)_T = \frac{-\Delta V^{\ddagger}}{RT}$$

Walling and co-workers measured the variation of rate with pressure for several Diels-Alder reactions. As expected, the volume of activation was found to be negative and the reaction was accelerated by increased pressure. According to Walling's calculations, however, the volume of activation amounted to only 50 to 70% of the *overall* shrinkage in volume that occurred in the course of the reaction (ΔV). This result seemed to him incompatible with the formation of the compact transition state required by the one-step mechanism. He favored a more open transition state, easily reconciled with the biradical model of Woodward.

Thus, in the early 1960's, following the publication of the Woodward and Walling papers, it seemed as if the two-step mechanism was making a strong comeback. Both authors encountered immediate opposition, however, and the situation at once became more ambiguous.

Woodward's conclusions were questioned on two grounds. First of all, it was pointed out that even if his rearrangement mechanism were correct, there was no necessary connection between it and that of the retro Diels-Alder reaction. The fact that both processes took place at similar temperatures could well have been fortuitous. Secondly, the proposed mechanism for the rearrangement of the dicyclopentadiene derivative, with its diradical intermediate, was viewed with skepticism.

Cope had shown that derivatives of 1,5-hexadiene underwent a simple, stereospecific rearrangement (equation 26.32).

(26.32)

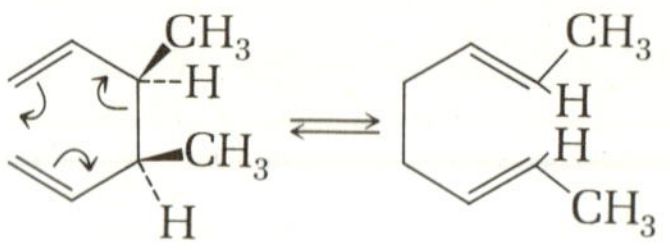

The evidence strongly suggested that the reaction was intramolecular and proceeded in one step without the intervention of intermediates. Woodward had already recognized that his dicyclopentadiene was nothing more than a bridged 1,5-hexadiene, and other workers suggested that it was undergoing a normal Cope rearrangement without any radical intermediates whatsoever. In fact, the geometry of the compound was felt to be very favorable to the reaction, the double bonds being fixed in advantageous positions.

(26.33)

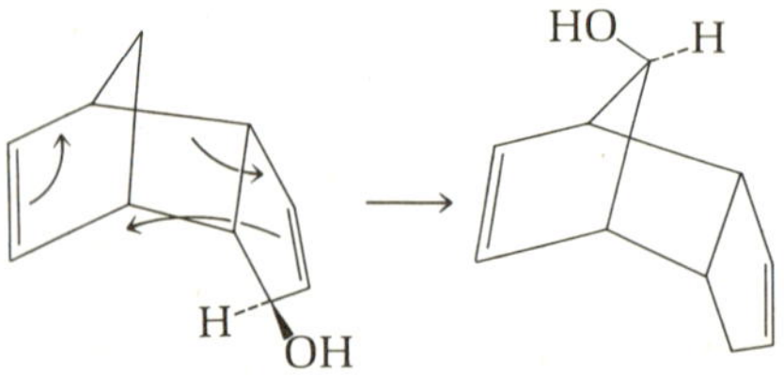

Walling's conclusions were also the subject of much controversy. The accuracy of his data was questioned and his treatment of it particularly disputed. Quite recently (1970) a set of very accurate measurements of reaction rate as a function of pressure has been obtained for several Diels-Alder reactions. In three of the four cases investigated the transition state was found to be *smaller* than the final adduct; i.e., $\Delta V^{\ddagger}$ was more negative than ΔV. These results argue strongly for a very compact transition state and, by extension, for the one-step mechanism.

At the time of writing the one-step mechanism is currently in favor, having received support from several different types of experimental approach. The concerted reaction is also favored by most theoretical calculations, although one ought to treat these conclusions with a certain degree of caution since it is not unknown for theoreticians to produce calculations that justify mutually exclusive mechanistic theories.

The two-step mechanism cannot yet be permanently removed from consideration, however; a provocative experimental result could easily breathe life into it. Furthermore, even those chemists who adhere to the one-step mechanism have had their viewpoint enlarged by the controversy, and many would concede that even though bond making is concerted in the transition state, it need not have advanced to the same extent at both bonds. In thus allowing the possibility of marked asymmetry in the degree of bond making, they have allowed a spectrum of transition states, one end of which approaches closely to the biradical intermediate.

It is precisely this lack of complete certainty in mechanistic investigations that draws so many chemists to the field. Many are the experimentalists who await that startling result that will permit them to raise an old mechanistic hypothesis, Lazarus-like, from the graveyard of rejected theories, to the discomfiture of many of their colleagues and above all to the distress of textbook authors.

SUPPLEMENTARY READINGS

General References

L. P. Hammett, *Physical Organic Chemistry*, 2nd ed. New York: McGraw-Hill Book Co., 1970.

C. K. Ingold, *Structure and Mechanism in Organic Chemistry*, 2nd ed. Ithaca, N.Y.: Cornell University Press, 1969. The first editions of this book and Hammett's were classics in the field.

E. S. Gould, *Mechanism and Structure in Organic Chemistry*. New York: Holt, Rinehart and Winston, Inc., 1959.

J. Hine, *Physical Organic Chemistry*, 2nd ed. New York: McGraw-Hill Book Co., 1962.

Specific References

J. E. H. Hancock, "An Introduction to the Literature of Organic Chemistry." *J. Chem. Ed.*, **45**, 193, 260, 336 (1968).

R. Huisgen, "Kinetic Evidence for Reactive Intermediates," *Angew. Chem. Intern. Ed.*, **9**, 751 (1970).

P. B. D. de la Mare and J. H. Ridd, *Aromatic Substitution: Nitration and Halogenation*. London: Butterworth and Co., Ltd., 1959.

A. Wassermann, *Diels-Alder Reactions*. New York: American Elsevier Publishing Co., 1965.

J. Sauer, "Diels-Alder Reactions," *Angew. Chem. Intern. Ed.*, **5**, 211 (1966); **6**, 16 (1967).

APPENDIX

TABLE A-1
Average Equilibrium Bond Lengths (Å)

C—H	1.08	C⋯C (aromatic)	1.39
C(sp^3)—C(sp^3)	1.54	C=C	1.34
C(sp^3)—C(sp^2)	1.53	C=N	1.30
C(sp^2)—C(sp^2)	1.50	C=O	1.23
C—N	1.47	C≡C	1.20
C—O	1.43	C≡N	1.16
C—F	1.38		
C—Cl	1.76		

TABLE A-2

International Atomic Weights

Name	Symbol	At. No.	At. Wt.	Name	Symbol	At. No.	At. Wt.
Actinium	Ac	89	227*	Mercury	Hg	80	200.59
Aluminum	Al	13	26.9815	Molybdenum	Mo	42	95.94
Americium	Am	95	243*	Neodymium	Nd	60	144.24
Antimony	Sb	51	121.75	Neon	Ne	10	20.183
Argon	Ar	18	39.948	Neptunium	Np	93	237*
Arsenic	As	33	74.9216	Nickel	Ni	28	58.71
Astatine	At	85	210*	Niobium	Nb	41	92.906
Barium	Ba	56	137.34	Nitrogen	N	7	14.0067
Berkelium	Bk	97	245*	Nobelium	No	102	253*(?)
Beryllium	Be	4	9.0122	Osmium	Os	76	190.2
Bismuth	Bi	83	208.980	Oxygen	O	8	15.9994
Boron	B	5	10.811	Palladium	Pd	46	105.4
Bromine	Br	35	79.909	Phosphorus	P	15	30.9738
Cadmium	Cd	48	112.40	Platinum	Pt	78	195.09
Calcium	Ca	20	40.08	Plutonium	Pu	94	242*
Californium	Cf	98	248*	Polonium	Po	84	210*
Carbon	C	6	12.01115	Potassium	K	19	39.102
Cerium	Ce	58	140.12	Praseodymium	Pr	59	140.907
Cesium	Cs	55	132.905	Promethium	Pm	61	145*
Chlorine	Cl	17	35.453	Protactinium	Pa	91	231*
Chromium	Cr	24	51.996	Radium	Ra	88	226*
Cobalt	Co	27	58.9332	Radon	Rn	86	222*
Copper	Cu	29	63.54	Rhenium	Re	75	186.2
Curium	Cm	96	245*	Rhodium	Rh	45	102.905
Dysprosium	Dy	66	162.50	Rubidium	Rb	37	85.47
Einsteinium	Es	99	247*	Ruthenium	Ru	44	101.07
Erbium	Er	68	167.26	Samarium	Sm	62	150.35
Europium	Eu	63	151.96	Scandium	Sc	21	44.956
Fermium	Fm	100	254*	Selenium	Se	34	78.96
Fluorine	F	9	18.9984	Silicon	Si	14	28.086
Francium	Fr	87	223*	Silver	Ag	47	107.870
Gadolinium	Gd	64	157.25	Sodium	Na	11	22.9898
Gallium	Ga	31	69.72	Strontium	Sr	38	87.62
Germanium	Ge	32	72.59	Sulfur	S	16	32.064
Gold	Au	79	196.967	Tantalum	Ta	73	180.948
Hafnium	Hf	72	178.49	Technetium	Tc	43	99*
Helium	He	2	4.0026	Tellurium	Te	52	127.60
Holmium	Ho	67	164.930	Terbium	Tb	65	158.924
Hydrogen	H	1	1.00797	Thallium	Tl	81	204.37
Indium	In	49	114.82	Thorium	Th	90	232.038
Iodine	I	53	126.9044	Thulium	Tm	69	168.934
Iridium	Ir	77	192.2	Tin	Sn	50	118.69
Iron	Fe	26	55.847	Titanium	Ti	22	47.90
Krypton	Kr	36	83.80	Tungsten	W	74	183.85
Lanthanum	La	57	138.91	Uranium	U	92	238.03
Lawrencium	Lw	103	259*(?)	Vanadium	V	23	50.942
Lead	Pb	82	207.19	Xenon	Xe	54	131.30
Lithium	Li	3	6.939	Ytterbium	Yb	70	173.04
Lutetium	Lu	71	174.97	Yttrium	Y	39	88.905
Magnesium	Mg	12	24.312	Zinc	Zn	30	65.37
Manganese	Mn	25	54.9381	Zirconium	Zr	40	91.22
Mendelevium	Md	101	256*(?)				

*Mass number of isotope of longest known half-life.

TABLE A-3
Some Average Bond Energies (kcal/mole)

$C(sp^3)$—H	97	C—N	78
$C(sp^2)$—H	103	C—O	86
$C(sp^3)$—$C(sp^3)$	82	C—F	106
$C(sp^3)$—$C(sp^2)$	89	C—Cl	81
$C(sp^2)$—$C(sp^2)$	100	C—Br	68
C═C	146	C—I	53
C═N	142	N—N	50–70
C═O	178	O—O	45

INDEX

B

C

D

E

I

K

L

M

N